GENETICS

From Genes to Genomes

GENETICS

From Genes to Genomes

LELAND HARTWELL
Fred Hutchinson Cancer Research Center

LEROY HOOD
University of Washington

MICHAEL L. GOLDBERG
Cornell University

ANN E. REYNOLDS
University of Washington

LEE M. SILVER
Princeton University

RUTH C. VERES

Boston   Burr Ridge, IL   Dubuque, IA   Madison, WI   New York   San Francisco   St. Louis
Bangkok   Bogotá   Caracas   Lisbon   London   Madrid
Mexico City   Milan   New Delhi   Seoul   Singapore   Sydney   Taipei   Toronto

McGraw-Hill Higher Education

A Division of The **McGraw-Hill** *Companies*

GENETICS: FROM GENES TO GENOMES

 This book is printed on recycled paper containing 10% postconsumer waste.

1 2 3 4 5 6 7 8 9 0 VNH/VNH 0 9 8 7 6 5 4 3 2 1 0

ISBN 0–07–540923–2

Vice president and editorial director: *Kevin T. Kane*
Publisher: *James M. Smith*
Developmental editor: *Jean Sims Fornango*
Marketing manager: *Martin J. Lange*
Project manager: *Cathy Ford Smith*
Production supervisor: *Enboge Chong*
Design manager: *Stuart D. Paterson*
Senior photo research coordinator: *Carrie K. Burger*
Supplement coordinator: *Brenda A. Ernzen*
Compositor: *Carlisle Communications, Ltd.*
Typeface: *10/12 Times Roman*
Printer: *Von Hoffman Press, Inc.*

Cover/interior design: *Chris Reese*
Photo research: *Jill Birschbach/Feldman & Associates, Inc.*

The credits section for this book begins on page C-1 and is considered an extension of the copyright page.

Library of Congress Cataloging-in-Publication Data

Genetics : from genes to genomes / Leland Hartwell . . . [et. al.].—
    1st ed.
        p.   cm.
    Includes index.
    ISBN 0–07–540923–2
    1. Genetics.   I. Hartwell, Leland.
   QH430.G458   2000
   576.5—dc21   99–15119
                                                      CIP

www.mhhe.com

Dr. Leland Hartwell received his Ph.D. from the Massachusetts Institute of Technology. Dr. Hartwell held assistant and associate professorships at the University of California before joining the faculty of the University of Washington, where he continues as a full professor. In 1996, Dr. Hartwell joined the Fred Hutchinson Cancer Research Center as a full member and senior advisor for scientific affairs, and was named president and director of the Center in July 1997.

Combining mutants and time-lapse photomicroscopy, Dr. Hartwell identified 32 genes in yeast that regulate the cell cycle with specific defects in spindle pole body duplication and segregation, DNA replication, mitosis, cytokinesis, and budding. He discovered a control point in the cell cycle, Start, where yeast cells exit the cell cycle to mate, arrest after nutritional starvation, and integrate growth with division. He used genetics to define many of the steps in the signal transduction pathway that feed into Start, including the cell-surface receptor for mating pheromone. The gene controlling Start, *CDC28,* was cloned in his lab and was the first CDK identified. He investigated the fidelity of chromosome transmission in the cell cycle, discovering that limitation or overexpression of many essential cell-cycle components lead to errors in chromosome transmission. Studies on how cells integrate the repair of DNA damage and cell division led to the discovery of cell-cycle checkpoints and the identification of six genes that control the DNA damage checkpoint.

Dr. Hartwell has received numerous awards and honors in the course of his career. Among them he received the Brandeis University Rosenteil Award in 1993 and the Sloan-Kettering Cancer Center Katherine Berkan Judd Award as well as the Genetics Society of America Medal in 1994. In 1995 he was awarded the MGH Warren Triennial Prize, and in 1996 he was awarded the Columbia University Horwitz Award and the Passano Award. Dr. Hartwell received the Albert Lasker Award for medical research in 1998.

Dr. Lee Hood received an M.D. from the Johns Hopkins Medical School and a Ph.D. in biochemistry from the California Institute of Technology. His research interests include immunology, development, and the development of biological instrumentation (e.g., the protein sequenator and the automated fluorescent DNA sequencer). His research played a key role in unraveling the mysteries of antibody diversity. Dr. Hood has taught molecular evolution, immunology, molecular biology, and biochemistry. He is currently the chairman (and founder) of the cross-disciplinary Department of Molecular Biotechnology at the University of Washington. Dr. Hood has received a variety of awards including the Albert Lasker Award for Medical Research and the Dickson Prize in 1987, the Cefas Award for Biochemistry in 1989, and the Distinguished Service Award from the National Association of Teachers in 1998. He is deeply involved in K–12 science education. His hobbies include running, mountain climbing, and reading.

Dr. Michael L. Goldberg is a professor at Cornell University, where he teaches introductory genetics. He was an undergraduate at Yale University and received his Ph.D. in biochemistry from Stanford University. Dr. Goldberg performed postdoctoral research at the Biozentrum of the University of Basel in Switzerland and at Harvard University. He received an NIH Fogarty Senior International Fellowship for study at Imperial College in England and at the University of Rome, Italy. His current research utilizes the tools of *Drosophila* genetics to investigate the mechanisms that ensure proper chromosome segregation during mitosis and meiosis.

Dr. Ann Reynolds is an educator and author who has been teaching genetics and biology since 1990. An affiliate faculty member of the Genetics Department at the University of Washington, her research has included studies of gene regulation in *E. coli,* chromosome structure and DNA replication in yeast, and chloroplast gene expression in marine algae. She is a graduate of Mount Holyoke College and received her Ph.D. from Tufts University. Dr. Reynolds was a postdoctoral research fellow with the Harvard University Department of Molecular Biology. Dr. Reynolds was also an author and producer of the laser disc and CD ROM *Genetics: Fundamentals to Frontiers.*

Dr. Lee M. Silver is a professor at Princeton University in the Departments of Molecular Biology, Ecology, and Evolutionary Biology and in the Program in Neuroscience. Dr. Silver graduated from the University of Pennsylvania with B.A. and M.S. degrees in physics and from Harvard University with a Ph.D. in biophysics. He was a research fellow at the Sloan-Kettering Institute for Cancer Research and a senior scientist at Cold Spring Harbor Laboratory before coming to Princeton. He is the author of

Remaking Eden: Cloning and Beyond in a Brave New World. He is also the coeditor in chief of a new international journal entitled *Cloning: Science and Policy,* and coeditor in chief of *Mammalian Genome,* the official journal of the International Mammalian Genome Society. In 1993 Dr. Silver was elected a fellow of the American Association for the Advancement of Science (AAAS).

Dr. Silver's own research has made intensive use of the mouse as a model organism to study the genetics of reproduction, development, and evolution. His current research focuses on the genetic components of behavior. At Princeton, he has taught courses in genetics, mammalian genetics, biotechnology and society, and developmental biology in the Department of Molecular Biology and human genetics, reproduction, and public policy in Princeton's Woodrow Wilson School of Public and International Affairs.

Ruth C. Veres is a science writer and editor with 25 years of experience in textbook publishing. She obtained her B.A. from Swarthmore College and M.A. degrees from Columbia University in New York and Tufts University. In addition to developing and editing more than 30 texts in the fields of political science, economics, psychology, nutrition, chemistry, and biology, she has coauthored a book on the immune system and an introductory biology text. She has also taught writing and languages at the University of California at Berkeley. She lives in San Francisco with her husband.

CONTRIBUTORS

Genetics research tends to proceed down highly specialized paths. A number of experts in specific areas generously provided information in their areas of expertise. We thank them for their contributions to this text.

Eric E. Alani, *Cornell University*
Charles F. Aquadro, *Cornell University*
Anthony B. Bleecker, *University of Wisconsin*
Deborah Brosnan, *University of Oregon*
Ronald A. Butow, *University of Texas, Southwestern Medical Center, Dallas*
Rita A. Calvo, *Cornell University*
Michael Culbertson, *University of Wisconsin*
Ian Duncan, *Washington University, St. Louis*
Sarah Elgin, *Washington University, St. Louis*
Thomas D. Fox, *Cornell University*
Leonard P. Guarente, *Massachusetts Institute of Technology*
Kenneth J. Kemphues, *Cornell University*
Joel G. Kingsolver, *University of Washington*

John T. Lis, *Cornell University*
Ross J. MacIntyre, *Cornell University*
Patrick H. Masson, *University of Wisconsin*
Jeffry B. Mitton, *University of Colorado*
Martha A. Mutschelr, *Cornell University*
June B. Nasrallah, *Cornell University*
Debra Nero, *Cornell University*
Richard D. Palmiter, *University of Washington*
Philip S. Perleman, *University of Texas, Southwestern Medical Center, Dallas*
Fabio Piano, *Cornell University*
Harry T. Stinson, Jr. *Cornell University*
William T. Sullivan, *University of California, Santa Cruz*
Volker M. Vogt, *Cornell University*
Douglas Wallace, *Emory University*
Jonathan Widom, *Northwestern University*
Mariana F. Woolfner, *Cornell University*
William B. Wood, *University of Colorado*
Andrew Wright, *Tufts University*
Stanley A. Zahler, *Cornell University*

BRIEF CONTENTS

CONTENTS

PART III

USING GENETIC ENGINEERING TO UNRAVEL THE INFORMATION IN GENOMES 260

The twentieth century witnessed the emergence of genetics as a central discipline in biology. In 1900 Gregor Mendel's laws of heredity were rediscovered; in the 1950s, James Watson and Francis Crick found that DNA, the molecule of heredity, is a double helix; and in the 1990s, the Human Genome Project progressed beyond expectations. For much of the century, the study of genetics focused on the identification of individual genes and their function. In the last decade of the century, however, another idea gained currency—the concept that no gene acts alone, instead it is through complex molecular interactions within and among vast networks of genes and proteins that organisms ultimately live and die.

Genetics: From Genes to Genomes reflects this new perspective. This book represents a new approach to an undergraduate course in genetics. It represents the way we, the authors, currently view the molecular basis of life. We integrate formal genetics—the rules by which genes are transmitted; molecular genetics—the structure of DNA and how it directs the structure of proteins; genomics and information science—the new technologies that enable gene isolation and a comprehensive analysis of the entire gene set in an organism; human genetics—how genes control health and disease; the unity of life forms—synthesis of information from many different organisms into one coherent whole; and molecular evolution—how species have evolved and diverged. The strength of this integrated approach is that students who have completed the text will have a strong command of genetics as it is practiced today by university and corporate researchers who are rapidly changing our understanding of living organisms, including ourselves; increasing our ability to prevent, treat, and diagnose disease and to engineer new life forms for food and medical uses; and, ultimately, creating the ability to replace or correct detrimental genes.

To encourage a genetic way of thinking, we begin the book with a presentation of Mendelian principles and the chromosomal basis of inheritance. From the outset, however, the integration of Mendelian genetics with fundamental molecular mechanisms is central to our approach. The Prologue presents the foundation of this integration. In Chapter 1, we tie Mendel's studies of pea-shape inheritance to the action of an enzyme that determines whether a pea is round or wrinkled. In the same chapter, we point to the relatedness of patterns of heredity in all organisms by using Mendelian principles to look at heredity in humans. Starting in Chapter 5, we focus on the physical dimensions of DNA; the implications and uses of mutations; and how the double helix of DNA encodes, copies, and transmits biological information. Beginning in Chapter 8 we also look at modern genetic techniques, including such biotechnology tools as gene cloning, hybridization, and PCR, exploring how researchers have used them to reveal the modular construction and genetic relatedness of genomes. We

then show how the modular construction of genomes has contributed to the relatively rapid evolution of life and helped generate the enormous diversity of life forms we see around us. A detailed discussion of model organisms clarifies that their use in the study of human biology is possible only because of the genetic relatedness of all organisms. Throughout our text, we present the scientific reasoning of some of the ingenious researchers who have carried out genetic analysis, from Mendel to Watson and Crick to the collaborators on the Human Genome Project.

ORGANIZATION

The Prologue outlines the central themes of *Genetics: From Genes to Genomes.* We hope students will read this section carefully because it establishes the foundation for our integrated presentation of Mendelian and molecular genetics.

Part I (Chapters 1, 2, 3, and 4) on the *Basic Principles: How Traits Are Transmitted* presents a thorough discussion of Mendelian genetics; the chromosome theory of inheritance; and linkage, recombination, and mapping.

Part II (Chapters 5, 6, and 7) covers *What Genes Are and What They Do,* including the structure and function of DNA, the role of mutation in defining genes, and the details of gene expression.

Part III (Chapters 8, 9, and 10) describes the *Use of Genetic Engineering to Unravel the Information in Genomes* and includes topics on mapping and analysis of genomes, detection of genotype, and the use of cloning, PCR, and hybridization in genetic analysis.

Part IV (Chapters 11, 12, 13, and 14) on *How Genes Travel* presents the molecular mechanisms underlying the chromosomal transmission of genetic information in eukaryotes and prokaryotes.

Part V (Chapters 15, 16, and 17) on *How Genes Are Regulated* discusses prokaryotic and eukaryotic gene regulation as well as the regulation of the cell cycle.

Part VI (Chapters 18–22) presents *Gene Regulation and Development: Portraits of Model Eukaryotic Organisms.* This **Genetic Portraits** unit contains five chapters, each one profiling a different model organism whose study has greatly contributed to genetic research. Included are

Saccharomyces cerevisiae: Genetic Portrait of Yeast
Arabidopsis thaliana: Genetic Portrait of a Model Plant
Caenorhabditis elegans: Genetic Portrait of a Simple Metazoan
Drosophila melanogaster: Genetic Portrait of a Fruit Fly
Mus musculus: Genetic Portrait of a House Mouse.

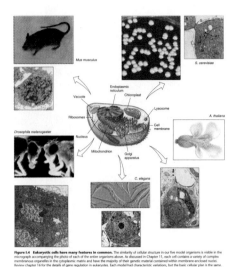

Figure I.4 Eukaryotic cells have many features in common. The similarity of cellular structure in our five model organisms is visible in the micrograph accompanying the photo of each of the entire organisms above. As discussed in Chapter 11, each cell contains a variety of complex membranous organelles in the cytoplasmic matrix and have the majority of their genetic material contained within membrane-enclosed nuclei. Review chapter 16 for the details of gene regulation in eukaryotes. Each model had characteristic variations, but the basic cellular plan is the same.

We anticipate that instructors will choose to cover one or two portrait chapters during the semester. Students may then use the specifics of the selected model organism to build an understanding of the principles and applications discussed in the text. The unique genetic manipulations and properties of each model make them important for addressing different biological questions using genetic analysis. In the portraits, we explain how biologists learned that the evolutionary relatedness of all organisms enables the extrapolation from a model to the analysis of other living forms. The portraits should thus help students understand how insights from one model organism can suggest general principles applicable to other organisms, including humans.

Part VII (Chapters 23 and 24) on *How Genes Change* explains the evolution of genes and genomes in populations and at the molecular level.

The **Epilogue** discusses *Human Genetics and the Future of Biology.* The focus of this closing essay is on the changing role of genetics research as a way to decipher biological networks and systems. Biology is now a science based on three levels of molecular information: information encoded in DNA, and information in proteins, and information encompassed in interactions among cells and tissues. The potential impact on the field of preventive medicine intensifies the need to confront many social and ethical issues.

CHAPTER FEATURES

Introduction   Each chapter begins with an engaging story related to the key ideas and principles of the chapter. This opening story is followed by a description of one or more overarching themes that unify the discussion, and then, in turn, by an advance organizer—a short, bulleted list of the chapter's topics in the order in which they appear in the text. The intent of the introduction is to create a narrative and conceptual framework that will help students organize and remember the vast amount of vocabulary and experimental data they encounter.

Feature Figures   These special two-page spreads integrate line art and text to summarize important genetic processes in detail. For example, in Chapter 5 on *DNA: How the Molecule of Heredity Carries, Replicates, and Recombines Information,* the Feature Figure details a "Model of Recombination at the Molecular Level," walking students through the basic steps of the process. In Chapter 17,

Cell-Cycle Regulation and the Genetics of Cancer, the Feature Figure details "Phenotypic Changes That Distinguish Tumor Cells from Normal Cells" outlines changes that produce uncontrolled cell growth, genomic and karyotypic instability, a potential for cellular immortality, and disruptions of local tissues that enable a tumor to invade distant tissues.

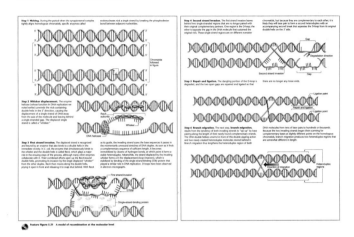

Feature Figure 5.21 A model of recombination at the molecular level.

Comprehensive Examples   These sections of the text are extensive case histories or research synopses that summarize the main points in the preceding section or chapter and show how they relate to each other. Very often these developed examples expand on the chapter's introductory story. In Chapter 6, *Anatomy and Function of a Gene: Dissection through Mutation,* for example, the opening story locates the rhodopsin gene on human chromosome 3 and explains that different mutations in the gene lead to night blindness or total blindness. The Comprehensive Example at the end of the chapter describes in detail "How Gene Mutations Affect Light-Receiving Proteins and Vision," covering such topics as the cellular and molecular basis of vision; the evolution of the rhodopsin gene family; and many of the mutations, amino-acid substitutions, and unequal crossing over events that affect both black and white and color vision.

Fast Forward Essays   This feature prefigures detailed discussions of concepts and principles in later chapters, serving as a tool to integrate Mendelian and molecular genetics. Chapter 1, *Mendel's Breakthrough: Patterns, Particles, and Principles of Heredity,* contains two Fast Forward essays, one on the fact that "Genes Encode Proteins," the other on techniques for "The Direct Analysis of Human Genotype." These essays help students understand that Mendel's laws

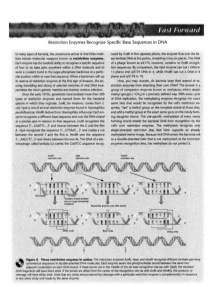

Fast Forward

Restriction Enzymes Recognize Specific Base Sequences in DNA

Figure A Three restriction enzymes in action.

have a molecular basis. In Chapter 5, *DNA: How the Molecule of Heredity Carries, Replicates, and Recombines Information,* where we present in detail the structure of the DNA molecule, the Fast Forward essay explains how "Restriction Enzymes Recognize Specific Base Sequences in DNA." This simple introduction of restriction enzymes foreshadows a discussion in Chapter 8, *DNA at High Resolution: The Use of DNA Cloning, PCR, and Hybridization as Tools of Genetic Analysis,* about the use of restriction enzymes in DNA cloning. It thus relates the basic concept of DNA structure to the tools of biotechnology that depend on a knowledge of that structure.

Genetics and Society Essays These essays explore the social and ethical issues created by the multiple applications of modern genetic research. They cover a wide variety of topics from the right to privacy to the question of who has the right to make reproductive decisions. In Chapter 9, the Genetics and Society essay asks "Does DNA Fingerprinting Serve the Interests of Justice?" In Chapter 10, the essay examines "The Patentability of DNA." In Chapter 13, it looks at "How Bacteria Can Cause Disease," presenting the mechanisms of bacterial pathogenesis step by step and describing the defense mechanisms that fight infection.

Connections and Essential Concepts Each chapter closes with a **Connections** section that serves as a bridge between the topics in the just-completed chapter and those in the up-coming chapter or chapters. The Connec-

tions section is followed by an **Essential Concepts** section that helps students focus on the most critical information—the chapter's "take-home" messages. The end-of-chapter exercises include solved problems, **Social and Ethical Issues** discussion questions, and a diverse set of problems and questions for the student to solve.

Outstanding Art Work The quality of the art is critical to the success of this text, and you will find that the photos, electron micrographs, and line art have been carefully selected and rendered to give students the best presentation possible. Color consistency has been used in rendering the line art to aid student comprehension. The following key is a guide to the use of color in our illustrations.

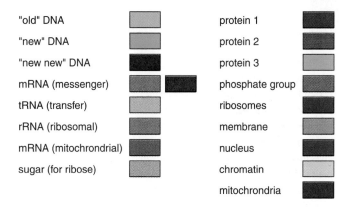

"old" DNA

"new" DNA

"new new" DNA

mRNA (messenger)

tRNA (transfer)

rRNA (ribosomal)

mRNA (mitochrondrial)

sugar (for ribose)

protein 1

protein 2

protein 3

phosphate group

ribosomes

membrane

nucleus

chromatin

mitochrondria

REFERENCE SECTION

In the back of the text, we provide a **Genetic Nomenclature Appendix.** Since the study of genetics is a relatively new science, a completely consistent nomenclature, similar to those found in more established sciences, does not exist. Instead, the details of gene notation differ from model organism to model organism. To assist students in understanding the use of gene symbols throughout the book, and particularly in the section on model organisms, this concise appendix details the minor differences in notation by organism.

Mastering the vocabulary of genetics is critical to understanding the science. To aid in that mastery, we provide a detailed **Glossary.**

The **Answer Appendix** contains answers to selected end-of-chapter problems. Students can build their problem-solving skills by working through the solved problems within each chapter and then, for the unsolved problems, checking the solutions they arrive at on their own against the answers in the Answer Appendix.

ACKNOWLEDGMENTS

The creation of a project of this scope is never solely the work of the authors. We are grateful to our colleagues around the world who took the time to review this manuscript and make suggestions for its improvement. Their willingness to share their experiences and expertise was a tremendous help to us.

Ken Belanger, *University of North Carolina, Chapel Hill*
John Belote, *Syracuse University*
Anna Berkovitz, *Purdue University*
John Botsford, *New Mexico State University*
Michael Breindl, *San Diego State University*
Bruce Chase, *University of Nebraska, Omaha*
Lee Chatfield, *University of Central Lancashire*
Alan Christensen, *University of Nebraska*
Bruce Cochrane, *University of South Florida*
James Curran, *Wake Forest University*
Rowland Davis, *University of California, Irvine*
Paul Demchick, *Barton College*
Stephen D'Surney, *University of Mississippi*
Rick Duhrkopf, *Baylor University*
Susan Dutcher, *University of Colorado*
DuWayne Englert, *Southern Illinois University*
Bentley Fane, *University of Arkansas*
Victoria Finnerty, *Emory University*
David Foltz, *Louisiana State University*
David Futch, *San Diego State University*
Ann Gerber, *University of North Dakota*
Richard Gethmann, *University of Maryland, Baltimore County*
Mike Goldman, *San Francisco State University*
Elliott Goldstein, *Arizona State University*
Nels Granholm, *South Dakota State University*
Charles Green, *Rowan College of New Jersey*
Poonam Gulati, *University of Houston*
Stephen Hedman, *University of Minnesota*
Ralph Hillman, *Temple University*
Christine Holler-Dinsmore, *Fort Peck Community College*
Martin Hollingsworth, *Tallahassee Community College*
Nancy Hollingsworth, *State University of New York, Stony Brook*
Andrew Hoyt, *Johns Hopkins University*
Lynne Hunter, *University of Pittsburgh*
Robert Ivarie, *University of Georgia*
R. C. Jackson, *Texas Technological University*
Duane Johnson, *Colorado State University*
Chris Kaiser, *Massachusetts Institute of Technology*
Kenneth J. Kemphues, *Cornell University*
Susan Kracher, *Purdue University*
Alan Koetz, *Illinois State University*
Andrew Lambertsson, *University of Oslo*
Don Lee, *University of Nebraska*
John Locke, *University of Alberta*
Larry Loeb, *University of Washington*
Robertson McClung, *Dartmouth College*
Peter Meacock, *University of Leicester*
John Merrriam, *University of California, Los Angeles*
Beth Montelone, *Kansas State University*
Patricia Moore, *Transylvania University*
Gail Patt, *Boston University*
Michael Perlin, *University of Louisville*
Richard Richardson, *University of Texas, Austin*
Mary Rykowski, *University of Arizona*
Mark Sanders, *University of California, Davis*
Randall Scholl, *Ohio State University*

David Sheppard, *University of Delaware*
Anthea Stavroulakis, *Kingsborough Community College*
John Sternick, *Mansfield University*
David Sullivan, *Syracuse University*
William Thwaites, *San Diego State University*
Akfi Uzman, *University of Houston*
Peter Webster, *University of Massachusetts*
Dean Whited, *North Dakota State University*
John Williamson, *Davidson College*
John Zamora, *Middle Tennessee State University*
Stephan Zweifel, *Carleton College*

Over the years a number of highly skilled publishing professionals helped us develop this book. We'd like to express our appreciation to Eirik Borve for his vision in launching the project, Laurel Smith for her refined and intelligent approach to art development, Marjorie Anderson for her editorial acumen, Kathi Prancan for her able and enthusiastic management (and dining-out extravaganzas), Kathy Naylor for her unusual combination of skills in art and text development; Jean Fornango for her top-notch, no-nonsense organizational skills in readying the manuscript for production; Richard Morel for his insightful preparation of the art manuscript; Ron Worthington for his scientific understanding and firm backing in a time of transition; and Jim Smith for his strong support throughout the final years of development and production. All have made a significant contribution to the final shape of this project.

SUPPLEMENTS

For the Student

■ The **Solutions Manual/ Study Guide** was written by text author Ann Reynolds, of the University of Washington. The solutions to the end-of-chapter problems and questions will aid the students in developing their problem-solving skills by providing the step-by-step logic of each solution.

■ **Genetics: From Genes to Genomes CD ROM,** developed with the content of the text, covers the most challenging concepts in the introductory genetics course. The CD attempts to make concepts more understandable by using animations of basic genetic processes and interactive exercises and simulations involving fundamental principles. Icons in the text indicate that there are related topics on the CD. A correlation guide linking text topics marked by icons to the related CD material is included in the *Instructor's Manual,* on our web site, and on the CD ROM itself. Additional quizzing options allow students to self-test and identify those areas needing additional study. Glossary definitions can be reached via hot links. The CD also has links that connect to the book's own web site.

For the Instructor

■ The **Instructor's Manual/ Test Bank** contains the CD ROM correlation guide, a list of transparencies, plus a test bank containing approximately 2000 questions. The test bank is also available in computerized form compatible with either Windows or Macintosh machines.

■ **Transparencies:** One hundred and fifty four-color illustrations from the text will be available to adopters.

■ **Visual Resource Library**: A CD ROM product containing 200 key illustrations will be available in four-color digital files. The presentation software enables you to create custom slide shows and multimedia presentations. Images

can also be exported for use in word-processing programs. Additional features enable the images to be sorted by name, type, locations, and user-defined keywords. Multiple images can be viewed at one time by using the Small Gallery View function. Jpeg files for all remaining line art is included, plus lecture outlines.

■ **Web Site:** This text-specific web site can be reached at the URL **www.mhhe.com/hartwell** and provides additional materials for both students and instructors.

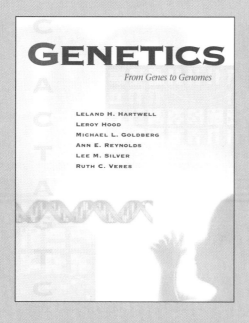

GENETICS

From Genes to Genomes

LELAND H. HARTWELL
LEROY HOOD
MICHAEL L. GOLDBERG
ANN E. REYNOLDS
LEE M. SILVER
RUTH C. VERES

MCGRAW-HILL IS PROUD TO OFFER AN EXCITING NEW SUITE OF MULTIMEDIA PRODUCTS AND SERVICES CALLED COURSE SOLUTIONS.

Designed specifically to help you with your individual course needs, **Course Solutions** will assist you in integrating your syllabus with our premier titles and state-of-the-art new media tools that support them.

AT THE HEART OF COURSE SOLUTIONS YOU'LL FIND:

- Fully integrated multimedia
- A full-scale Online Learning Center
- A Course Integration Guide

AS WELL AS THESE UNPARALLELED SERVICES:

- McGraw-Hill Learning Architecture
- McGraw-Hill Course Consultant Service
- Visual Resource Library (VRL) Image Licensing
- McGraw-Hill Student Tutorial Service
- McGraw-Hill Instructor Syllabus Service
- PageOut Lite
- PageOut: The Course Web Site Development Center
- Other Delivery Options

COURSE SOLUTIONS truly has the solutions to your every teaching need. Read on to learn how we can specifically help you with your classroom challenges.

SPECIAL ATTENTION
to your specific needs.

These "perks" are all part of the extra service delivered through *McGraw-Hill's* **Course Solutions:**

MCGRAW-HILL LEARNING ARCHITECTURE

Each McGraw-Hill *Online Learning Center* is ready to be ported into our *McGraw-Hill Learning Architecture*—a full course management software system for Local Area Networks and Distance Learning Classes. Developed in conjunction with Top Class software, *McGraw-Hill Learning Architecture* is a powerful course management system available upon special request.

McGRAW-HILL COURSE CONSULTANT SERVICE

In addition to the *Course Integration Guide*, instructors using **Course Solutions** textbooks can access a special curriculum-based *Course Consultant Service* via a web-based threaded discussion list within each *Online Learning Center*. A **McGraw-Hill Course Solutions Consultant** will personally help you—as a text adopter—integrate this text and media into your course to fit your specific needs. This content-based service is offered in addition to our usual software support services.

VISUAL RESOURCE LIBRARY (VRL) IMAGE LICENSING

Most of our **Course Solutions** titles are accompanied by a *Visual Resource Library (VRL) CD-ROM,* which features text figures in electronic format. Previously, use of these images was restricted to in-class presentation only. Now, McGraw-Hill will license adopters the right to use appropriate VRL image files—FREE OF CHARGE—for placement on their local Web site! Some restrictions apply. Consult your McGraw-Hill sales representative for more details.

McGRAW-HILL INSTRUCTOR SYLLABUS SERVICE

For *new* adopters of **Course Solutions** textbooks, McGraw-Hill will help correlate all text, supplement, and appropriate materials and services to your course syllabus. Simply call your McGraw-Hill sales representative for assistance.

PAGEOUT LITE

Free to **Course Solutions** textbook adopters, *PageOut Lite* is perfect for instructors who want to create their own Web site. In just a few minutes, even novices can turn their syllabus into a Web site using *PageOut Lite.*

PAGEOUT: THE COURSE WEB SITE DEVELOPMENT CENTER

For those that want the benefits of *PageOut Lite's* no-hassle approach to site development, but with even more features, we offer *PageOut: The Course Web Site Development Center.*

PageOut shares many of *PageOut Lite's* features, but also enables you to create links that will take your students to your original material, other Web site addresses, and to *McGraw-Hill Online Learning Center* content. This means you can assign *Online Learning Center* content within your syllabus-based Web site. *PageOut's* gradebook function will tell you when each student has taken a quiz or worked through an exercise, automatically recording their scores for you. *PageOut* also features a discussion board list where you and your students can exchange questions and post announcements, as well as an area for students to build personal Web pages.

OTHER DELIVERY OPTIONS

Online Learning Centers are also compatible with a number of full-service online course delivery systems or outside educational service providers. For a current list of compatible delivery systems, contact your McGraw-Hill sales representative.

And for your students...
McGRAW-HILL STUDENT TUTORIAL SERVICE

Within each *Online Learning Center* resides a FREE *Student Tutorial Service*. This web-based "homework hotline"—available via a threaded discussion list—features guaranteed, 24-hour response time on weekdays.

www.mhhe.com/hartwell

GENETICS:
THE STUDY OF BIOLOGICAL INFORMATION

Genetics, the science of heredity, is at its core the study of biological information. All living organisms—from single-celled bacteria and protozoa to multicellular plants and animals—must store, replicate, transmit to the next generation, and use vast quantities of information to develop, grow, reproduce, and survive in their environments. Geneticists examine how organisms pass biological information on to their progeny and how they use it during their lifetime.

This book introduces the field of genetics as it exists at the turn of the twenty-first century. Six overarching themes recur throughout the book:

■ Biological information is encoded in the DNA molecule.

■ Biological function emerges primarily from protein molecules.

■ All living forms are closely related.

■ The modular construction of genomes has allowed the relatively rapid evolution of biological complexity.

■ Genetic techniques permit the dissection of biological complexity.

■ Our focus is on human genetics.

In the remainder of the prologue, we introduce these themes. It will help to keep them in mind as you delve into the details of genetics.

BIOLOGICAL INFORMATION IS ENCODED IN THE DNA MOLECULE

The process of evolution has taken close to 4 billion years to generate the amazingly efficient mechanisms for storing, replicating, expressing, and diversifying biological information seen in organisms now inhabiting the earth. The linear DNA molecule stores biological information digitally in units known as nucleotides. Within each DNA molecule, the sequence of the four letters of the DNA alphabet—G, C, A, and T—specify which proteins an organism will make as well as when and where protein synthesis will occur. The letters refer to the bases that are components of the nucleotide building blocks of DNA. The DNA molecule itself is a double strand of nucleotides carrying complementary A–T or G–C base pairs (Fig. P.1). The DNA regions that encode proteins are called *genes*.

Within the cells of an organism, DNA molecules are assembled into *chromosomes:* organelles that package and manage the storage, duplication, expression, and evolution of DNA. The entire collection of chromosomes in each cell of an organism is its *genome*. Human cells, for example, contain 23 distinct kinds of chromosomes carrying approximately 3×10^9 base pairs and roughly 100,000 genes. The amount of information that can be encoded in this size genome is equivalent to 6 million pages of text containing 250 words per page, with each letter corresponding to one *base pair,* or pair of nucleotides.

BIOLOGICAL FUNCTION EMERGES FROM PROTEIN MOLECULES

Although there is no single characteristic that distinguishes living organisms from inanimate matter, you would have little trouble deciding which entities in a group of 20 objects are living organisms. One characteristic of nearly all living forms is an elaborate and complicated structure. Consider a fly. Not only is it complex at the resolution of what the unaided human eye can see, but the closer you look, using powerful microscopes, the more intricate its structure becomes. Another characteristic of life is the ability to move. Animals swim, fly, walk, or run, while plants grow toward or away from light. Whenever you see something move, you suspect that it is alive. Yet another characteristic of living organisms is the capacity to adapt selectively to the environment, whether it be a robin choosing materials to build a nest or a vine weaving its way up a fence. Finally, a key characteristic of living organisms is the ability to use sources of energy and matter to grow, that is, the ability to convert foreign material into their own body parts. The chemical and physical reactions that carry out these conversions are known as *metabolism*. A captured bird, for example, if unfed and thus unable to support its metabolism, will die within a few days.

Most properties of living organisms ultimately arise from a particular class of molecules: the *proteins*. Proteins are large

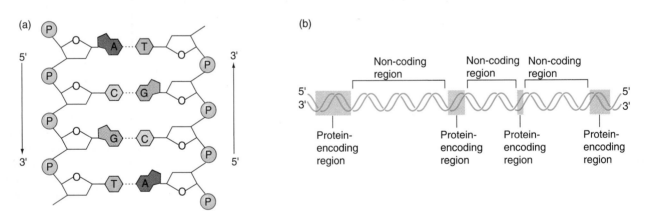

Figure P.1 Complementary base pairs are a key feature of the DNA molecule. (a) A single strand of DNA is composed of nucleotide subunits each consisting of a deoxyribose sugar (depicted here as a white pentagon), a phosphate (depicted as a yellow circle), and one of four nitrogenous bases—adenine, thymine, cytosine, or guanine (designated as lavender or green As, Ts, Cs, or Gs). The chemical structure of the bases enables A to associate tightly with T and C to associate tightly with G. As a result, A and T form one kind of complementary base pair, while C and G form another kind of complementary base pair. The association through base pairing of two complementary DNA strands produces a DNA double helix. The arrows labeled 5' to 3' show that the two strands of the double helix have opposite orientations relative to chemically distinct 5' and 3' ends. (b) Along the DNA double helix, protein-encoding regions are separated by noncoding regions. In bacteria, close to 90% of the DNA encodes proteins; in humans, the proportion of protein-coding DNA is less than 5%.

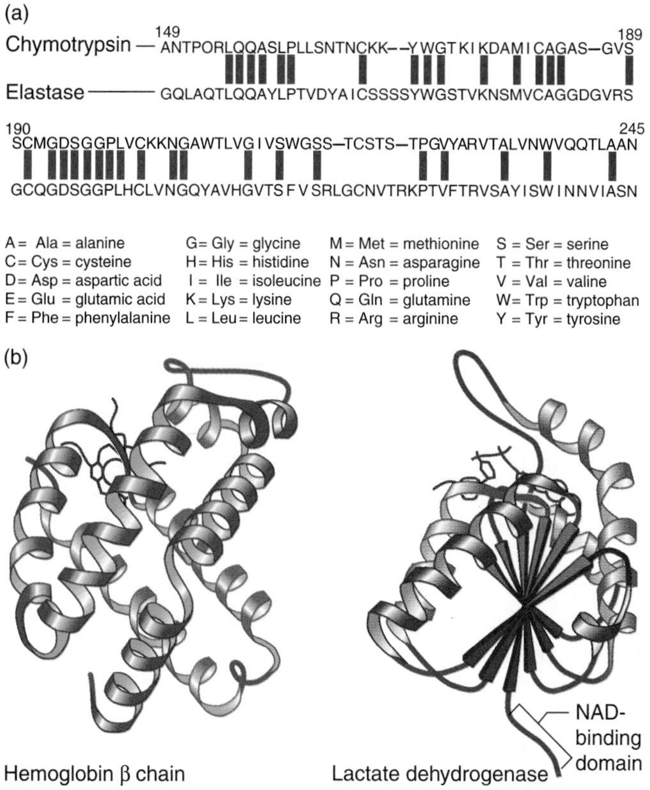
(a)
149
189
Chymotrypsin — ANTPORLQQASLPLLSNTNCKK– –YWGTKIKDAMICAGAS–GVS
Elastase — GQLAQTLQQAYLPTVDYAICSSSSYWGSTVKNSMVCAGGDGVRS
190
245
SCMGDSGGPLVCKKNGAWTLVGIVSWGSS–TCSTS–TPGVYARVTALVNWVQQTLAAN
GCQGDSGGPLHCLVNGQYAVHGVTSFVSRLGCNVTRKPTVFTRVSAYISWINNVIASN
A = Ala = alanine
C = Cys = cysteine
D = Asp = aspartic acid
E = Glu = glutamic acid
F = Phe = phenylalanine
G = Gly = glycine
H = His = histidine
I = Ile = isoleucine
K = Lys = lysine
L = Leu = leucine
M = Met = methionine
N = Asn = asparagine
P = Pro = proline
Q = Gln = glutamine
R = Arg = arginine
S = Ser = serine
T = Thr = threonine
V = Val = valine
W = Trp = tryptophan
Y = Tyr = tyrosine
(b)
NAD-binding domain
Hemoglobin β chain
Lactate dehydrogenase

Figure P.2 Proteins are polymers of amino acids that fold in three dimensions. The specific sequence of amino acids in a chain determines the precise three-dimensional shape of the protein. (a) A comparison of equivalent segments in the chains of two digestive proteins, chymotrypsin and elastase. The red lines connect sites in the two sequences that carry identical amino acids; the two chains differ at all the other sites shown. Thus, even though these two proteins are evolutionarily related to each other, they differ at enough amino acids that their structures (and functions) are not identical. (b) Schematic drawings of the hemoglobin β chain (green) and the lactate dehydrogenase NAD-binding domain (purple) show the different three-dimensional shapes determined by different amino-acid sequences. The distinct three-dimensional shapes enable the two molecules to perform distinct tasks. The β chain is part of the complex hemoglobin molecule, which binds and delivers oxygen to body tissues. The NAD-binding domain is part of the lactate dehydrogenase molecule, an enzyme that catalyzes energy conversions in microorganisms such as yeast and in the muscle cells of strenuously exercising animals.

polymers composed of hundreds to thousands of amino-acid subunits strung together in long chains; each chain folds into a specific three-dimensional conformation dictated by the sequence of its amino acids (Fig. P.2). There are 20 different amino acids. Thus, you can think of proteins as constructed from a set of 20 different kinds of snap beads distinguished by color and shape; if you were to arrange the beads in any order, make strings of a thousand beads each, and then fold or twist the chains into shapes dictated by the order of their beads, you would be able to make a nearly infinite number of different three-dimensional shapes. The astonishing diversity of three-dimensional protein structure generates the extraordinary diversity of protein function that is the basis of each organism's complex and adaptive behavior. The structure and shape of the hemoglobin protein, for example, allow it to transport oxygen in the bloodstream and release it to the tissues. The proteins myosin and actin slide together to allow muscle contraction. Chymotrypsin and elastase are enzymes that help break down other proteins. Most of the properties associated with life emerge from the constellation of protein molecules that an organism expresses (that is, synthesizes) according to instructions contained in its DNA.

ALL LIVING THINGS ARE CLOSELY RELATED

The evolution of biological information is a fascinating story spanning the 4.5 billion years of earth's history. Many biologists think that RNA was the first information-processing molecule to appear. RNA molecules are very similar to DNA and are also composed of four subunits. Like DNA, RNA has the capacity to store, replicate, mutate, and express information; like proteins, RNA can fold in three dimensions to produce molecules capable of catalyzing the chemistry of life. RNA molecules, however, are intrinsically unstable. Thus, it is probable that the more stable DNA took over the linear information storage and replication functions of RNA, while proteins, with their far greater capacity for diversity, preempted the functions derived from RNA's three-dimensional folding. The information contained in the sequence of DNA nucleotides specifies the sequence of amino acids in the proteins. With this division of labor, RNA became an intermediary in converting the information in DNA into the sequence of amino acids in protein. The separation that placed information storage in DNA and biological function in proteins was so successful that all organisms alive today descend from the first organisms that happened upon this molecular specialization. The evidence for the common origin of all living forms is present in their DNA sequences. All living organisms use essentially the same arbitrary genetic code in which various groupings of the 4 letters of the DNA and RNA alphabets encode the 20 letters of the amino-acid alphabet. Via the code, the order of bases in any organism's DNA specifies the amino-acid sequence of its proteins.

The relatedness of all living organisms is also evident from comparisons of genes with similar functions in very different organisms. For example, there is striking similarity

Alignment of cytochrome C protein sequence from yeast, *Arabidopsis*, *C. elegans*, *Drosophila*, mouse, and human.

```
S. cerevisiae      ---PGSAKKGATLFKTRCQQCHTIEEGGPNKV
A. thaliana        ----GDAKKGANLFKTRCAQCHTLKAGEGNKI
C. elegans         ---AGDYEKGKKVYKQRCLQCHVVDS-TATKT
D. melanogaster    ---AGDVEKGKKLFVQRCAQCHTVEAGGKHKV
M. musculus        ---MGDVEKGKKIFVQKCAQCHTVEKGGKHKT
H. sapiens         ---MGDVEKGKKIFIMKCSQCHTVEKGGKHKT
                        *   * *   *   * * *  *    *
```

```
S. cerevisiae      GPNLHGIFGRHSGQVKGYSYTDANINKNVKW
A. thaliana        GPELHGLFGRKTGSVAGYSYTDANKQKGIEW
C. elegans         GPTLHGVIGRTSGTVSGFDYSAANKNKGVVW
D. melanogaster    GPNLHGLIGRKTGQAAGFAYTDANKAKGITW
M. musculus        GPNLHGLFGRKTGQAAGFSYTDANKNKGITW
H. sapiens         GPNLHGLFGRKTGQAPGYSYTAANKNKGIIW
                   ** *** . ** . **     *. * ** .  *
```

```
S. cerevisiae      DEDSMSEYLTNPKKYIPGTKMAFAGLKKEKDR
A. thaliana        KDDTLFEYLENPKKYIPGTKMAFGGLKKPKDR
C. elegans         TKETLFEYLLNPKKYIPGTKMVFAGLKKADER
D. melanogaster    NEDTLFEYLENPKKYIPGTKMIFAGLKKPNER
M. musculus        GEDTLMEYLENPKKYIPGTKMIFAGIKKKGER
H. sapiens         GEDTLMEYLENPKKYIPGTKMIFVGIKKKEER
                     . .    *** * **********  * .*  *   .
```

```
S. cerevisiae      NDLITYMTKAAK---
A. thaliana        NDLITFLEEETK---
C. elegans         ADLIKYIEVESA---
D. melanogaster    GDLIAYLKSATK---
M. musculus        ADLIAYLKKATN---
H. sapiens         ADLIAYLKKATN---
                    * **   . ..
```

`*` Indicates identical and `.` indicates similar

Figure P.3 Comparisons of gene products in different species provide evidence for the relatedness of living organisms. This chart shows the amino-acid sequence for equivalent portions of the cytochrome C protein in six species: *Saccharomyces cerevisiae* (yeast), *Arabidopsis thaliana* (a weedlike flowering plant), *Caenorhabditis elegans* (a nematode), *Drosophila melanogaster* (the fruit fly), *Mus musculus* (the house mouse), and *Homo sapiens* (humans). Consult Fig. P.2 for the key to amino acid names. The cytochromes are a family of heme-containing proteins that function as electron donors and acceptors during cellular respiration and photosynthesis. Cytochrome C is the most abundant and most stable of the cytochromes; its form and function have been conserved throughout evolution. As the chart shows, there are many sequence similarities among the six types of organisms. Not included is a bacterial cytochrome C, whose sequence would show significant but less pronounced similarity with the cytochrome Cs in the chart.

between the genes for many proteins in bacteria, yeast, plants, worms, flies, mice, and humans (Fig. P.3). Moreover, it is often possible to place a gene from one organism into the genome of a very different organism and see it function normally in the new environment. For example, human genes that help regulate cell division can replace related genes in yeast and enable the yeast cells to function normally.

The close relatedness of all living organisms at the molecular level has great significance for an understanding of biology. It makes it possible to combine bits and pieces learned from different organisms into a global understanding of molecular and cellular biology that is valid for all organisms. Thus, even though controlled experimentation with humans is usually impossible, the relatedness of all organisms allows us to learn about human biology from mice, flies, worms, peas, yeast, and other organisms that are accessible to experimentation.

THE MODULAR CONSTRUCTION OF GENOMES HAS ALLOWED THE RELATIVELY RAPID EVOLUTION OF COMPLEXITY

Roughly 100,000 genes direct human growth and development. How did such complexity arise? Recent technical advances permit the complete structural analysis of the entire genome of several organisms. The information so far obtained reveals that families of genes have arisen by duplication of a primordial gene; after duplication, mutations and rearrangements may cause the two copies to diverge from each other (Fig. P.4a). In both mice and humans, for example, five different hemoglobin genes produce five different hemoglobin molecules at successive stages of development, with each protein functioning in a slightly different way to fulfill different needs for oxygen transport. The set of five hemoglobin genes arose from a single primordial gene by several duplications followed by slight divergences in structure.

Duplication followed by divergence underlies the formation of new genes with new functions. In addition, the genomes of multicellular organisms have evolved an organization that facilitates the construction of new genes from the modular components of existing genes. Many genes in multicellular organisms are assemblages of protein-encoding parts, known as *exons*, separated from each other by noncoding parts, known as *introns*. The exons often correspond to a protein's independent functional domains. When genes with this kind of modular construction undergo duplication, one of the resulting copies could ensure production of the original protein, while a shuffling of exons in the other copy could generate new genes that produce new proteins. Thus, duplications followed by rearrangements within the genome may permit old genes to continue to function and new combinations of exons to generate new genes (Fig. P4b).

The tremendous advantage of the duplication and divergence of existing pieces of genetic information is evident in

(a)

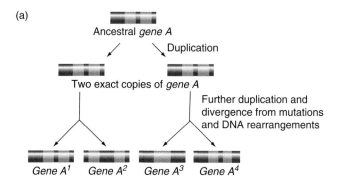

Ancestral *gene A*

Duplication

Two exact copies of *gene A*

Further duplication and divergence from mutations and DNA rearrangements

Gene A¹   *Gene A²*   *Gene A³*   *Gene A⁴*

(b)

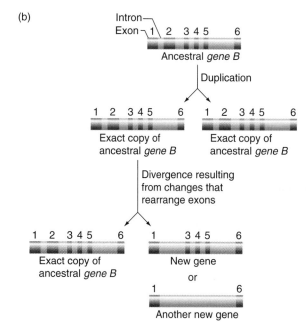

Intron
Exon 1 2 3 4 5 6
Ancestral *gene B*

Duplication

1 2 3 4 5 6   1 2 3 4 5 6

Exact copy of ancestral *gene B*

Exact copy of ancestral *gene B*

Divergence resulting from changes that rearrange exons

1 2 3 4 5 6   1 3 4 5 6

Exact copy of ancestral *gene B*

New gene

or

1   6

Another new gene

Figure P.4 How genes arise by duplication and divergence. (a) Duplications of ancestral *Gene A* followed by mutations and DNA rearrangements generate a family of related genes. (The blue and red bands indicate different regions of the genes.) (b) A duplication, followed by chromosomal changes that rearrange exons, generates new genes from an exact copy of ancestral *gene B*.

the history of life's evolution (Table P.1). *Prokaryotic* cells such as bacteria, which do not have a membrane-bounded nucleus, evolved about 3.7 billion years ago; *eukaryotic* cells such as yeast, which have a membrane-bounded nucleus, emerged around 2 billion years ago; and multicellular eukaryotic organisms appeared 600–700 million years ago. Then, at about 570 million years ago, within the relatively short evolutionary time of roughly 10 million years known as the Cambrian explosion, the multicellular life-forms diverged into a bewildering array of organisms, including primitive vertebrates. A fascinating question is: Since it took eukaryotic cells almost 2 billion years to evolve from prokaryotic cells and multicellular organisms three-quarters of a billion years to

evolve from single-celled eukaryotes, how could the multicellular forms achieve such enormous diversity in only 10 million years? The answer lies, in part, in the hierarchic organization of the information encoded in chromosomes. Exons are arranged into genes; genes duplicate and diverge to generate multigene families; and multigene families sometimes rapidly expand to supergene families containing hundreds of related genes. In both mouse and human adults, for example, the immune system is encoded by a supergene family composed of hundreds of closely related but slightly divergent genes. With the emergence of each successively larger informational unit, evolution gains the ability to duplicate increasingly complex informational cassettes through single genetic events.

TABLE P.1 Fossil Evidence for Some Major Stages in the Evolution of Life*

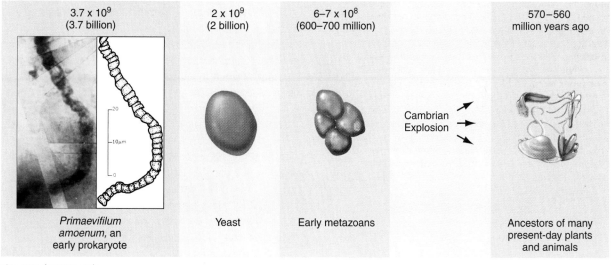

3.7×10^9 (3.7 billion)	2×10^9 (2 billion)	$6-7 \times 10^8$ (600–700 million)	570–560 million years ago
Primaevifilum amoenum, an early prokaryote	Yeast	Early metazoans	Cambrian Explosion	Ancestors of many present-day plants and animals

*Organisms not drawn to scale.

GENETIC TECHNIQUES PERMIT THE DISSECTION OF COMPLEXITY

The complexity of living systems has developed over 4.5 billion years from the continuous amplification and refinement of genetic information. The simplest bacterial cells contain about 1000 genes that interact in complex networks. Yeast cells, the simplest eukaryotic cells, contain about 6000 genes. Nematodes and fruit flies, the simplest metazoan organisms, contain roughly 15,000–20,000 genes; humans may have as many as 100,000 genes. With genetic techniques, researchers can dissect this complexity piece by piece, although the task is daunting. The logic used in genetic dissection is quite simple: inactivate a gene and observe the consequences. For example, loss of a gene for visual pigment produces fruit flies with white eyes instead of eyes of the normal red color. Over the years, geneticists have performed this type of dissection on several species, including the bacterium *Escherichia coli,* the yeast *Saccharomyces cerevisiae,* the nematode *Caenorhabditis elegans,* the fruit fly *Drosophila melanogaster,* the plant *Arabidopsis thaliana,* and the mouse *Mus musculus.* From their study of these organisms, they are amassing a detailed picture of the beauty and complexity of living systems.

However, even though the power of genetic techniques is astonishing, the complexity of biological systems is difficult to comprehend. The human nervous system, for example, is a network of 10^{11} (100,000,000,000) neurons with perhaps 10^{18} (1,000,000,000,000,000,000) connections. Knowing everything there is to know about an individual neuron provides little insight into learning, memory, emotional stability, and personality—traits that emerge from interactions among intricate networks of neurons and the human environment. Thus, after defining the single units (whether genes or neurons), it is necessary to look at how multiple units interact to generate the characteristics of the complex system. The shift from the analysis of single units to the analysis of complex systems is a major challenge in modern biology. Genetics provides unique tools for meeting this challenge, making it possible to sort out the multiplicity of genes dictating the diversity of proteins that generate complex systems.

OUR FOCUS IS ON HUMAN GENETICS

In the mid-1990s, a majority of scientists who responded to a survey conducted by *Science* magazine rated genetics as the most important field of science for the next decade. One reason is that the powerful tools of genetics open up the possibility of understanding biology, including human biology, from the molecular level up to the level of the whole organism. In combination with an appreciation of the relatedness of all living organisms, the potential of genetic analysis heralds an era that promises to help reveal who we are. The focus on human genetics in this book looks forward into that era. As we gain increasingly sophisticated knowledge about the human genetic makeup, it will become possible to cure human diseases that now resist therapy, alter the genetic makeup of our children, predict human potential, and ultimately have an impact on our own evolution. These new possibilities raise serious moral and ethical issues that will demand wisdom and humility. It is in the hope of educating young people for the moral and ethical challenges awaiting the next generation that we write this book.

BASIC PRINCIPLES: HOW TRAITS ARE TRANSMITTED

MENDEL'S BREAKTHROUGH:
PATTERNS, PARTICLES, AND PRINCIPLES
OF HEREDITY

Although Mendel's laws can predict the probability that an individual will have a
particular genetic makeup, the chance of particular male and female gametes
determine an individual's actual genetic fate.

A quick glance at an extended family portrait is likely to reveal children who resemble one parent or the other or who look like a combination of the two, with perhaps wavy hair from the father, a broad nose from the mother, and a skin color in between the two parents' (Fig. 1.1). Some children, however, look unlike any of the assembled relatives and more like a throwback to a great, great grandparent. What causes the similarities of appearance and the skipping of generations?

The answers lie in our **genes,** basic units of biological information, and in **heredity,** the way genes transmit biochemical, physical, and behavioral traits from parents to offspring. Each of us starts out as a single fertilized egg cell that develops, by division and differentiation, into a mature adult made up of 10^{14} (a hundred trillion) specialized cells, including muscle cells capable of contraction, brain cells structured for rapid communication, red blood cells tailored for transporting oxygen, and hair cells that carry pigment for black, brown, blond, or flaming red hair. By rough estimates, only 100,000 genes control this amazing developmental process. Passed from parents to offspring through egg and sperm, these genes underlie the formation of each and every heritable trait. Such traits are as diverse as the shape of your hairline, the tendency to bald as you age, the timbre of your voice, the way you clasp your hands, even your susceptibility to heart disease and certain cancers. And they all run in families in predictable patterns that impose some possibilities and exclude others.

Genetics, the science of heredity, seeks a precise explanation of the biological structures and mechanisms that determine what is inherited and how. Geneticists seek to identify genes, to learn how they determine particular traits, and to understand how genes work together to create a person, a plant, or a protozoan. In some instances, the relationship between gene and trait is remarkably simple. A change in a single gene, for example, results in sickle-cell anemia by causing construction of a defective hemoglobin molecule, the oxygen-carrying protein in red blood cells; when oxygen is in short supply, red blood cells carrying the abnormal hemoglobin become sickle shaped and clog small vessels. In other in-

Figure 1.1   A family portrait. The extended family shown here includes members of four generations.

Figure 1.2   Gregor Mendel. Photographed around 1862 holding a fuchsia, one of the many types of plants gardeners in the Brno area hybridized to create unusual varieties. At the time of this photo, Mendel was in the midst of his experiments with peas. In 1865, he introduced his study on plant hybridization before the Natural Science Society of Brno. His work formed the basis for our understanding and continued exploration of the science of genetics.

stances, the correlations between genes and traits are bewilderingly complex. An example is the genetic basis of facial features in which many genes determine a large number of molecules that interact in a variety of ways to generate the combination we recognize as a friend's face.

Gregor Mendel (1822–1884; Fig. 1.2), a stocky, bespectacled, highly respected Augustinian monk and expert plant breeder, trained in both physics and plant physiology, discovered the basic principles of genetics in the mid–nineteenth century and published his findings in 1866, just seven years after Darwin's *On the Origin of Species* appeared in print. Mendel lived and worked in Brunn, Austria (now Brno in the Czech republic), a nineteenth-century center of learning in the sciences and humanities, situated in the rich agricultural valley of the province of Moravia. Here he examined the inheritance of such clear-cut alternative traits in pea plants as purple versus white flowers or yellow versus green seeds. In so doing, he discovered why some of these traits disappeared in one generation, then reappeared in another. By rigorously analyzing the patterns of transmission through generations, he inferred genetic laws that allowed him to make verifiable predictions about which traits would appear, disappear, then reappear in which generations. Simple and straightforward, Mendel's laws are based on the hypothesis that observable traits such as seed color are determined by independent units of inheritance not visible to the naked eye. We now call these units *genes*. The concept of the gene continues to change as research deepens and refines our understanding of genetic phenomena. Today, a gene is recognized as a region of DNA that encodes a specific protein or a particular type of RNA. In the beginning, however, it was an abstraction—an imagined particle with no physical features, whose function was to control a visible trait—that Mendel proposed to explain what he saw.

Because Mendel's laws are basic to an understanding of genetics, we begin our study of genetics with a detailed look at what these laws are and how they were discovered. In subsequent chapters, we discuss logical extensions to Mendel's laws and describe how his successors grounded the abstract concept of hereditary units (genes) in an actual biological molecule (DNA). With knowledge of that molecule, geneticists devised techniques to isolate

and identify the units of heredity. Today, they integrate the analytical tools of Mendel with modern molecular techniques as they continue to probe the nature of the hereditary material and examine exactly how it is passed from parent to offspring, how it acts in organisms to produce the visible and invisible traits that define individuals, and how it evolves over time.

Four general themes emerge from our detailed discussion of Mendel's work. The first is that variation, as expressed in alternative forms of a trait (a high-pitched voice or a low one, a green pea or a yellow one), is widespread in nature. It arises from a genetic diversity that provides the raw material for the continuously evolving variety of life we see around us. Second, observable variation is essential for following genes. If all the traits of all offspring resembled their parents', Mendel would have had no basis for discerning and analyzing patterns of transmission. Third, variation is not distributed solely by chance; rather, it is inherited according to genetic laws that explain why like begets both like and unlike.

Figure 1.3   Like begets like and unlike. A Labrador retriever with her litter of pups.

Dogs beget other dogs, pea plants beget pea plants, and people, people. But there are hundreds of breeds of dogs and even within a breed—Labrador retrievers, for instance—two black dogs could have a litter of black, brown, and tan puppies (Fig. 1.3). Mendel's insights help explain why this is true. Fourth, the laws Mendel discovered about heredity apply equally well to all sexually reproducing organisms, from protozoans to peas to people.

Our presentation of Mendelian genetics examines:

■ The background: The historical puzzle of inheritance and how Mendel's new experimental approach helped resolve it.

■ The work itself: Genetic analysis according to Mendel, including a discussion of Mendel's seminal experiments and related analytical tools of genetics.

■ Repercussions: A comprehensive example of Mendelian inheritance in humans.

BACKGROUND: THE HISTORICAL PUZZLE OF INHERITANCE

There are several steps to understanding genetic phenomena: the careful observation over time of groups of organisms, such as human families, herds of cattle, or fields of corn or tomatoes; the rigorous analysis of systematically recorded information gleaned from these observations; and the development of a theoretical framework, a way of thinking about the phenomena, that can explain their origin and relation. In the mid-nineteenth century, Gregor Mendel became the first person to combine the three approaches and reveal the true basis of heredity. For many thousands of years before that, observation was the main road traveled, and random selective breeding of domesticated plants and animals, with no guarantee of what a particular mating would produce, was the only genetic practice.

Artificial Selection Was the First Applied Genetic Practice

A rudimentary use of genetics was the driving force behind a key transition in human civilization, allowing hunters and gatherers to settle in villages and survive as shepherds and farmers. Even before recorded history, people practiced applied genetics as they domesticated plants and animals for their own uses. From a large litter of semi-tamed wolves, for example, they sent the savage and the misbehaving to the stew pot while sparing the alert sentries and friendly companions for longer life and eventual mating. As a result of this **artificial selection**—purposeful control over mating by choice of parents for the next generation—the domestic dog (*Canis domesticus*) slowly arose from ancestral wolves (*Canis lupus*). The oldest bones identified indisputably as dog (and not wolf) are a skull excavated from a 20,000-year-old Alaskan settlement. Many millennia of evolution guided by artificial selection have produced massive Great Danes and minuscule Chihuahuas as well as hundreds of other modern breeds of dog. The amazing range of size, shape, and behavior bears witness to the enormous amount of genetic variation in ancient canines and the degree of differentiation that artificial selection can produce. Beginning around 10,000 years ago, people used this same kind of genetic manipulation to develop economically valuable herds of reindeer, sheep, goats, pigs, and cattle that produced life-sustaining meat, hides, and wools. By the time civilization flourished in ancient Greece, such selective animal breeding had acknowledged social value. According to Homer and other Greek poets, the wealth and power of ancient Troy arose in part from the culture's horse-breeding skills.

Similarly, farmers carried out artificial selection of plants, storing seed from the healthiest individuals with the tastiest and most succulent leaves, roots, or stems for the next planting, and eventually producing strains that grew better, produced more, and were easier to cultivate and harvest. In this way, scrawny grasses and other weedlike plants gradually, with human guidance, turned into rice, wheat, barley, lentils, and dates in Asia; corn, squash, tomatoes, potatoes, and peppers in North and South America; yams, peanuts, and gourds in Africa. Later, plant breeders recognized male and female organs in plants and carried out artificial pollination. An Assyrian frieze carved in the ninth century B.C., pictured in Fig. 1.4, is the earliest visual record of this kind of genetic experiment. It depicts a priest brushing the flowers of a female date palm with selected male pollen. By this method of artificial selection, early practical geneticists produced several hundred varieties of dates, each differing in specific observable qualities, such as the fruit's size, color, or taste. A 1929 botanical survey of three oases in Egypt turned up 400 varieties of date-bearing palms, twentieth-century evidence of the degree of natural and artificially generated variation among the early date palms.

The Puzzle of Passing on Desirable Traits

In 1822, the year of Mendel's birth, what people in Moravia understood about the basic principles of heredity was not much different from what the people of ancient Egypt and Greece had understood, even though some advances in artificial selection had occurred. By the nineteenth century, precise techniques for controlling matings in plants had given rise to the systematic creation of strains in which offspring often carried a prized parental trait. Using such strains, breeders could produce plants with desired characteristics for food and fiber, but they could not always predict why a valued trait would sometimes disappear and then reappear in only some offspring. The same problem plagued animal breeders. For example, selective breeding practices had resulted in valuable flocks of merino sheep producing large quantities of soft, fine wool, but at the 1837 annual meeting of the Moravian Sheep

Breeders Society, one breeder's dilemma epitomized the state of the art. He possessed an outstanding ram that would be priceless "if its advantages are inherited by its offspring," but "if they are not inherited, then it is worth no more than the cost of wool, meat, and skin." Which would it be? According to the meeting's recorded minutes, current breeding practices offered no assurances, no definite answers. In the concluding remarks at this same sheep-breeders meeting, the Abbot Cyril Napp pointed to a possible way out. He proposed that breeders could improve their ability to predict what traits would appear in the offspring by finding the answers to three basic questions: What is inherited? How is it inherited? What is the role of chance in heredity?

This is where matters stood in 1843 when 21-year-old Johann Mendel entered the monastery in Brno, presided over by the same Abbot Napp. The young Johann's name was changed to Gregor at this time, and thereafter that was the name he used. Although Gregor Mendel was a monk trained in theology, he was not a rank amateur in science, discovering new and exciting things as if from nowhere. Moravia was a center of learning and scientific activity, providing access to the latest techniques and literature; Mendel, for instance, was able to acquire a copy of Darwin's *On the Origin of Species* shortly after it was translated into German in 1863. As a youth, Mendel had been a student of such outstanding caliber that even when his father was disabled in an accident, Mendel, the only male child of a relatively poor peasant couple, was allowed to enter the monastery to pursue his studies, while a

(a)

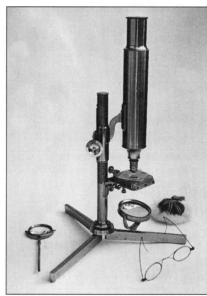
(b)

Figure 1.5   Mendel's garden and microscope. (a) Gregor Mendel's garden was part of his monastery's property in Brno. (b) Mendel used the microscope to examine plant reproductive organs and to pursue his interests in natural history.

Figure 1.4   The earliest known record of applied genetics. In this 2800-year-old Assyrian relief from the Northwest Palace of Assurnasirpal II (883–859 B.C.), priests wearing bird masks artificially pollinate flowers of female date palms.

brother-in-law took over management of the family farm. Abbot Napp, recognizing Mendel's intellectual abilities, sent him to the University of Vienna—all expenses paid—where he prescribed his own course of study based on his particular interests. His choices were an unusual mix: physics (with the renowned Christian Doppler, discoverer of the "Doppler effect"), mathematics, chemistry, botany, paleontology, and plant physiology. The cross-pollination of ideas from several disciplines would play a significant role in Mendel's discoveries. One year after he returned to Brno in 1853, he began his series of seminal genetic experiments. Figure 1.5 shows where Mendel worked and the microscope he used.

A New Experimental Approach

Before Mendel, many misconceptions clouded people's thinking about heredity. Two of the prevailing myths were particularly misleading. The first was that one parent contributes most to an offspring's inherited features. Aristotle contended it was the male, by way of a fully formed homunculus inside the sperm (Fig. 1.6). Second was the concept of *blended inheritance*, the idea that parental traits become mixed and forever changed in the offspring, as when blue and yellow pigment merge to green on a painter's palette. The theory of blending may have grown out of a natural tendency for parents to see a combination of their own traits in their offspring. While blending could account for children who look like a combination of their parents, it could not explain obvious differences between biological brothers and sisters nor the persistence of variation within extended families.

The experiments Mendel devised would lay these myths to rest by providing precise, verifiable answers to the three questions Abbot Napp had raised almost 15 years earlier: What is inherited? How is it inherited? What is the role of chance? A key component of Mendel's breakthrough was the way he set up his experiments.

What did Mendel do differently from those who preceded him? First, he chose the garden pea (*Pisum sativum*) as his experimental organism (Fig. 1.7a and b). Peas grew well in Brno and, with male and female organs on the same plant, they were normally self-fertilizing. In **self-fertilization** (or selfing), both egg and pollen come from the same plant. The particular anatomy of pea flowers, however, makes it easy to prevent self-fertilization and instead to **cross-fertilize** (synonyms: to cross or to outbreed) by brushing pollen from one plant onto a female organ of another plant, as illustrated in Fig. 1.7c. Peas offered yet another advantage. For each successive generation, Mendel could obtain large numbers of individuals within a relatively short growing season. By comparison, if he had worked with sheep, each mating would have generated only a few offspring and the time between generations would have been several years.

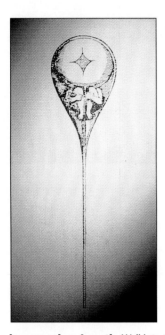

Figure 1.6 The homunculus: A myth. Well into the nineteenth century, many prominent microscopists believed they saw a fully formed, miniature fetus crouched within the head of a sperm.

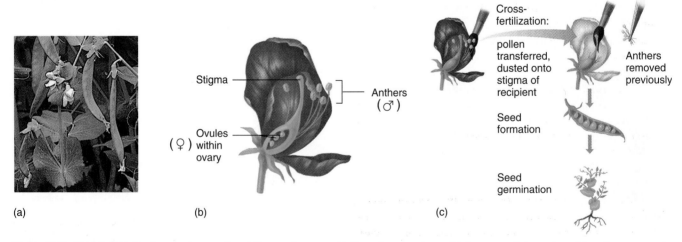

(a)　　　　(b)　　　　(c)

Figure 1.7 Mendel's experimental organism: The garden pea. (a) Flowering pea plants. (b) Anatomy of the male and female reproductive parts of a pea flower. Pollen is produced within the anthers. Mature pollen lands on the stigma, which is connected to the ovary (which becomes the pea pod). After landing, the pollen grows a tube that extends through the stigma to one of the ovules (the immature seeds), allowing fertilization to take place. (c) Method for cross-pollinating pea plants: To prevent self-fertilization, breeders remove the anthers from the female parents (here, the white flower) before the plant produces mature pollen. They then transfer pollen with a paintbrush from the anthers of the male parent (here, the purple flower) to the stigma of the female parent. Each fertilized ovule becomes an individual pea (a mature seed) that can grow into a new pea plant. All of the peas produced from one flower are encased in the same pea pod, but these peas can differ from one another as they form from different pollen grains and ovules.

Second, Mendel examined the inheritance of clear-cut alternative forms of particular traits—purple versus white flowers, yellow versus green peas. Using such "either-or" traits, he could distinguish and trace unambiguously the transmission of one or the other observed characteristic, because there were no intermediate forms. (The opposite of these so-called *discrete traits* are *continuous traits* such as height and skin color in humans. Continuous traits show many intermediate forms.)

Third, Mendel isolated and perpetuated lines of peas that bred true. Matings within such **pure-breeding lines** produce offspring carrying specific parental traits that remain constant from generation to generation. Here is how Mendel himself defined a pure-breeding line: "Permit me to state that, as an empirical worker, I must define constancy of type as the retention of a character during the period of observation." Mendel observed his pure-breeding lines for up to eight generations. Plants with white flowers always produced offspring with white flowers; plants with purple flowers produced only offspring with purple flowers. Mendel called constant but mutually exclusive, alternative traits, such as purple versus white flowers or yellow versus green seeds, "antagonistic pairs" and settled on seven such pairs for his study (Fig. 1.8). In his experiments, he not only perpetuated pure-breeding stocks for each member of a pair, but also cross-fertilized pairs of plants to produce **hybrids,** offspring of genetically dissimilar parents, for each pair of antagonistic traits. Figure 1.8 shows the appearance of the hybrids he studied.

Fourth, Mendel, the expert plant breeder, carefully controlled his matings, going to great lengths to ensure that the progeny he observed really resulted from the specific fertilizations he intended. Thus he painstakingly prevented the intrusion of any foreign pollen and assured self- or cross-pollination as the experiment demanded. Not only did this allow him to carry out controlled breedings of selected traits, he could also make **reciprocal crosses,** in which by reversing the traits of male and female parents, he controlled whether a particular trait was transmitted via the egg cell within the ovule or via the pollen. For example, he could use pollen from a purple flower to fertilize the eggs of a white flower and also use pollen from a white flower to fertilize the eggs of a purple flower. Because the progeny of these reciprocal crosses were similar, he disproved the idea that one parent contributes more to the next generation, demonstrating instead that they contribute equally to inheritance. "It is immaterial to the form of the hybrid," he wrote, "which of the parental types was the seed or pollen plant."

Fifth, Mendel worked with large numbers of plants, counted and subjected his findings to statistical analysis, and then compared his results with predictions based on his models. He was the first person to study inheritance in this manner, and no doubt his background in physics and mathematics contributed to this quantitative approach. Just as cutting and polishing a dull gray stone can reveal astonishing patterns that reflect its chemical composition, Mendel's careful quantitative analysis revealed patterns of transmission that reflected basic laws of heredity.

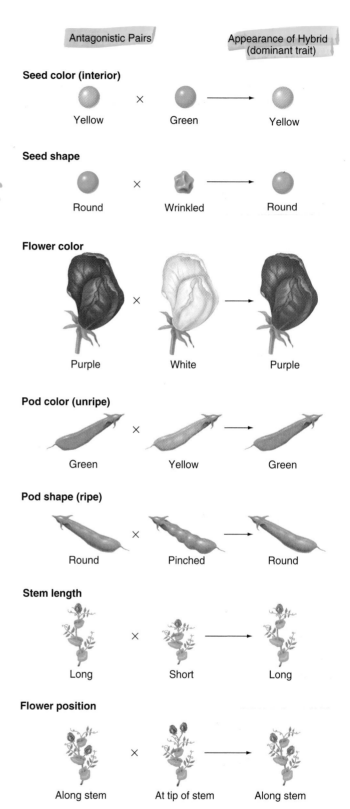

Figure 1.8 The mating of parents with antagonistic traits produces hybrids. Note that the hybrids for the seven antagonistic traits studied by Mendel resemble only one of the parents. The parental trait that shows up in the hybrid is known as the "dominant" trait.

Finally, Mendel was a brilliant practical experimentalist. When comparing tall and short plants, for example, he made sure that the short ones were out of the shade of the tall ones so their growth would not be stunted. Eventually he focused on certain traits of the pea seeds themselves, such as their color or shape, rather than on traits of the plants arising from the seeds. In this way, he could cull and observe many more individuals from the limited space of the monastery garden, and he could evaluate his results six months earlier.

In short, Mendel purposely set up a simplified "black and white" experimental system and then figured out how it worked. He did not look at the vast number of variables that determine the development of a prize ram nor at the origin of differences between species. Rather, he looked at discrete traits that came in two mutually exclusive forms and asked questions that could be answered by observation and computation. Mendel was the only botanist in the nineteenth century to conceive of and carry out these kinds of experiments.

GENETIC ANALYSIS ACCORDING TO MENDEL

In early 1865 at the age of 43, Gregor Mendel presented a paper entitled "Experiments on Plant Hybrids" before the Natural Science Society of Brno. Despite its modest heading, it was a scientific paper of uncommon clarity and simplicity that summarized a decade of original observations and experiments. In it Mendel describes in detail the transmission of visible characteristics in pea plants, defines unseen but logically deduced units that determine why these visible traits appear, and analyzes the function of these discrete determinants in simple mathematical terms to reveal precise patterns and heretofore unsuspected principles of heredity. He also tries to integrate his ideas with the newly emerging cell theory of life and the hotly debated theory of evolution.

Published the following year, the paper would become the cornerstone of modern genetics. Its stated purpose was to see whether there is a "generally applicable law governing the formation and development of hybrids." Let us examine its insights.

Monohybrid Crosses Reveal Units of Inheritance and the Law of Segregation

Once he had isolated pure-breeding lines for several sets of characteristics, Mendel carried out a series of **monohybrid crosses:** matings between individuals that differ in only one trait, such as seed color or stem length. In each monohybrid cross, one parent carries one form of the trait, and the other parent carries an alternative form of the same trait. Figure 1.9 illustrates one such monohybrid cross. Early in the spring of 1854, for example, Mendel planted pure-breeding green peas and pure-breeding yellow peas and allowed them to grow into the **parental (P) generation.** Later that spring when the plants had flowered, he dusted the female stigma of "green-pea" plant

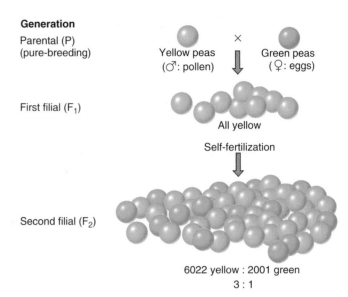

Generation

Parental (P)
(pure-breeding)

Yellow peas
(♂: pollen)   ×   Green peas
(♀: eggs)

First filial (F₁)

All yellow

Self-fertilization

Second filial (F₂)

6022 yellow : 2001 green
3 : 1

Figure 1.9   A monohybrid cross and two generations of progeny. Cross-pollination of pure-breeding parental plants produces F_1 hybrids. All the F_1 progeny resemble one of the parents. Self-pollination of F_1 plants gives rise to the F_2 generation. F_2 individuals resembling both of the original parental stocks appear in a ratio of 3:1. For simplicity, we do not show the plants that alternately produce the peas or that grow from the planted peas.

flowers with pollen from "yellow-pea" plants. He also performed the reciprocal cross, dusting "yellow-pea" plant stigma with "green-pea" pollen. In the fall, when he collected and separately analyzed the progeny peas of these reciprocal crosses, he found that in both cases, the peas were all yellow. These yellow peas, progeny of the P generation, were the beginning of what we now call the **first filial (F_1)** generation. To learn whether the green trait had disappeared entirely or remained intact but hidden in these F_1 yellow peas, Mendel planted them and allowed the F_1 plants that grew from them to self-fertilize. He then harvested and counted the peas of the resulting **second filial (F_2)** generation, progeny of the F_1 generation. Among the progeny of one series of F_1 self-fertilizations, there were 6022 yellow and 2001 green, an almost perfect ratio of 3 yellow to 1 green. F_1 plants derived from the original reciprocal cross produced a similar ratio of yellow-to-green F_2 progeny.

Reappearance of the Recessive Trait Disproves Blending
These results were irrefutable evidence that blending had not occurred. If it had, the information necessary to make green peas would have been irretrievably lost in the F_1 hybrids. Instead, the information remained intact and was able to direct the formation of some 2000 green peas actually harvested from the second filial generation. These green peas were indistinguishable from their grandparents of the P generation. Mendel concluded that there must be two types of yellow peas, those that breed true like the yellow peas of the P generation and those that can yield some green offspring like the yellow

F₁ hybrids. This second type somehow contains latent information for green peas. He called the trait that appeared in all the F₁ hybrids—in this case, yellow seeds—**dominant** (see Fig. 1.8) and the "antagonistic" green-pea trait that remained hidden in the F₁ hybrids but reappeared in the F₂ generation **recessive.** But how did he explain the 3:1 ratio of yellow to green F₂ peas?

The Discrete Units of Inheritance Are Alleles of Genes

To account for his observations, Mendel proposed that for each trait, every plant carries two copies of a unit of inheritance, receiving one from its maternal parent and the other from the paternal parent. Today, we call these units of inheritance *genes.* Each unit determines the appearance of a specific characteristic. The pea plants in Mendel's collection had two copies of a gene for seed color, two copies of another for seed shape, two copies of a third for stem length, and so forth. Mendel further proposed that each gene comes in alternative forms, and it is these alternative forms that determined the contrasting characteristics he was studying. Today we call the alternative forms of a single gene **alleles.** The gene for pea color, for example, has yellow and green alleles; the gene for pea shape has round and wrinkled alleles. (The Fast Forward box "Genes Encode Proteins," p. 20, describes the biochemical and molecular mechanisms by which different alleles determine different traits.) In Mendel's monohybrid crosses, one allele of each gene was dominant, the other recessive. In the P generation, one parent carried two dominant alleles for the trait under consideration, the other parent two recessive alleles. The hybrids of the F₁ generation carried one dominant and one recessive allele for the trait. Individuals having two different alleles for a single trait are **monohybrids.**

The Law of Segregation Explains How Genes Are Transmitted

If a plant has two copies of every gene, how does it pass only one copy of each to its progeny? And how then do the offspring end up with two copies of these same genes, one from each parent? Mendel drew on his background in plant physiology and answered these questions in terms of the two biological mechanisms behind reproduction: gamete formation—the production of eggs and pollen—and the random union of gametes at fertilization. He imagined that during the formation of pollen and eggs, the two copies of each gene in the parent separate (or *segregate*) so that each gamete receives only one allele for each trait (Fig. 1.10a). Thus, each egg and each pollen grain receives only one allele for pea color (either yellow or green). At fertilization, pollen with one or the other allele unites at random with an egg carrying one or the other allele, restoring the two copies of the gene for each trait in the fertilized egg, or **zygote** (Fig. 1.10b). If the pollen carries yellow and the egg green, the result will be a hybrid yellow pea like the F₁ monohybrids that resulted when pure-breeding parents of opposite types mated. If the yellow pollen unites with

(a) The two alleles for each trait separate during gamete formation.

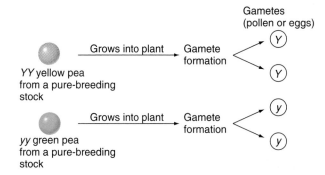

(b) Two gametes, one from each parent, unite at random at fertilization.

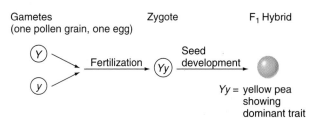

Y = yellow-determining allele of pea color gene
y = green-determining allele of pea color gene

Figure 1.10   The law of segregation. (a) The two identical alleles of pure-breeding adult plants separate (segregate) during gamete formation. As a result, each pollen grain or egg carries only one of each pair of parental alleles. (b) Cross-pollination and fertilization between pure-breeding parents with antagonistic traits result in F₁ hybrid zygotes with two different alleles, one from each type of parent, for each type of trait. For the seed color gene, a Yy hybrid zygote will develop into a yellow pea.

a yellow egg, the result will be a yellow pea that grows into a pure-breeding plant like those of the P generation that produced only yellow peas. And finally, if pollen carrying the allele for green peas fertilizes a green-carrying egg, the progeny will be a pure-breeding green pea.

Mendel's **law of segregation** encapsulates this general principle of heredity: *The two alleles for each trait separate (segregate) during gamete formation, then unite at random, one from each parent, at fertilization.* Throughout this book, the term **segregation** refers to such *equal segregation* in which one allele of each gene goes to each gamete.

Figure 1.11 shows a simple way of visualizing the events and results of the segregation and random union of alleles during gamete formation and fertilization. Mendel invented a system of symbols that allowed him to analyze all his crosses in the same way. He designated dominant alleles with a capital *A, B,* or *C* and recessive ones with a lowercase *a, b,* or *c.* Modern geneticists have adopted this convention for naming genes in peas and many other organisms, but often add some reference to the trait in question—a *Y* for yellow or an *R* for round. Throughout this book, we present gene symbols in italics. In our diagram, we denote the dominant yellow allele

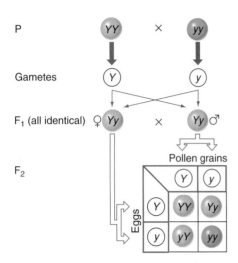

Figure 1.11   The Punnett square: Visual summary of a cross.
This Punnett square illustrates the combinations that can arise when an F₁ hybrid undergoes gamete formation and self-fertilization. The analysis shows that the F₂ generation should have a 3:1 ratio of yellow to green peas.

by a capital *Y* and the recessive green allele by a lower case *y*. The pure-breeding plants of the parental generation are either *YY* (yellow peas) or *yy* (green peas). The *YY* parent can produce only *Y* gametes, the *yy* parent only *y* gametes. You can see from the diagram why every cross between *YY* and *yy* produces exactly the same result—a *Yy* hybrid—no matter which parent (male or female) donates which particular allele.

Next, to visualize what happens when the *Yy* hybrids self-fertilize, we set up a Punnett square (named after the British mathematician who introduced it; see Fig. 1.11). The square provides a simple and convenient method for tracking the kinds of gametes produced as well as all the possible combinations that might occur at fertilization. As the Punnett square shows, each hybrid produces two kinds of gametes, *Y* and *y*, in a ratio of 1:1. Thus half the pollen and half the eggs carry *Y*, the other half *y*. At fertilization, 1/4 the progeny will be *YY*, 1/4 *Yy*, 1/4 *yY*, and 1/4 *yy*. Since the gametic source of an allele (egg or pollen) for the traits Mendel studied had no influence on the allele's effect, *Yy* and *yY* are equivalent. This means that 1/2 the progeny are yellow *Yy* hybrids, 1/4 *YY* true-breeding yellows, and 1/4 true-breeding *yy* greens. The diagram illustrates how the segregation of alleles during gamete formation and the random union of egg and pollen at fertilization can produce the 3:1 ratio of yellow to green that Mendel observed in the F₂ generation.

Mendel's Results Reflect Basic Rules of Probability

Though you may not have realized it, the Punnett square illustrates two simple laws of probability that are central to the analysis of genetic crosses. These laws of probability predict the likelihood that an event will occur. The **law of the product** states that the probability of two or more *independent*

events occurring together is the *product* of the probabilities that each event will occur by itself.

With independent events:

Probability of event 1 and event 2 =
Probability of event 1 × probability of event 2.

Consecutive coin tosses are obviously independent events; a heads in one toss neither increases nor decreases the probability of a heads in the next toss. If you toss two coins at the same time, the results are also independent events. A heads for one coin neither increases nor decreases the probability of a heads for the other coin. Thus, the probability of a given combination is the product of their independent probabilities. For example, the probability that both coins will turn up heads is

$$1/2 \times 1/2 = 1/4.$$

Similarly, the formation of egg and pollen are independent events; in a hybrid plant, the probability is 1/2 that a given gamete will carry *Y* and 1/2 that it will carry *y*. Since fertilization happens at random, the probability that a particular combination of maternal and paternal alleles will occur simultaneously in the same zygote is the product of the independent probabilities of these alleles being packaged in egg and sperm. Thus, to find the chance of a *Y* egg (formed as the result of one event) uniting with a *Y* sperm (the result of an independent event), you simply multiply 1/2 × 1/2 to get 1/4. This is the same fraction of *YY* progeny seen in the Punnett square of Fig. 1.11, which demonstrates that the Punnett square is simply another way of depicting the law of the product.

While we can describe the moment of random fertilization as the simultaneous occurrence of two independent events, we can also say that two different fertilization events are mutually exclusive. For instance, if *Y* combines with *Y*, it cannot also combine with *y* in the same zygote. A second law of probability, the **law of the sum,** states that the probability of either of two such *mutually exclusive events* occurring is the *sum* of their individual probabilities.

With mutually exclusive events:

Probability of event 1 *or* event 2 =
Probability of event 1 + probability of event 2.

To find the likelihood that an offspring of a *Yy* hybrid self-fertilization will be a hybrid like the parents, you add 1/4 (the probability of maternal *Y* uniting with paternal *y*) and 1/4 (the probability of the mutually exclusive event where paternal *Y* unites with maternal *y*) to get 1/2, again the same result as in the Punnett square. In a slightly different use of the law of the sum, you could predict the ratio of yellow-to-green F₂ progeny. The fraction of F₂s that will be yellow is the sum of 1/4 (the event producing *YY*) plus 1/4 (the mutually exclusive event generating *Yy*) plus 1/4 (the third mutually exclusive event producing *yY*) to get 3/4. The remaining 1/4 of the F₂ progeny will be green. So the yellow-to-green ratio is 3/4 to 1/4, or more simply, 3:1.

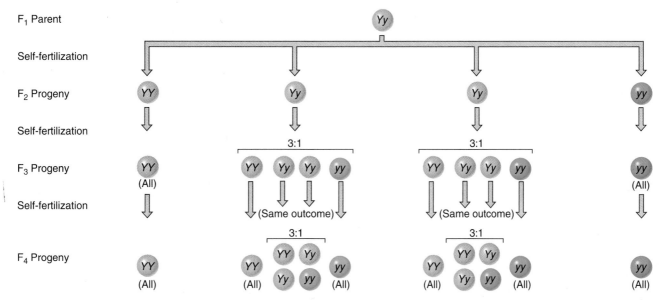

Figure 1.12 Yellow F₂ peas are of two types: Pure breeding and hybrid. An analysis of the distribution of a pair of contrasting alleles (Y and y) after four generations of self-fertilization. The homozygous individuals of each generation breed true, whereas the hybrids do not.

Further Crosses Confirm Ratios Predicted by the Law of Segregation

Although Mendel's law of segregation explains the data from his pea crosses, he performed additional experiments to rule out other possibilities. In the rigorous check of his hypotheses illustrated in Fig. 1.12, he allowed self-fertilization of all the plants in the F_2 generation, counted the types of F_3 progeny, and then allowed self-fertilization of the F_3 generation, counted the F_4 progeny, and so on. What kind of plants would you anticipate from a planting of green peas of the F_2 generation, if Mendel's laws do, in fact, apply? Mendel found that, as predicted, the plants that developed from F_2 green peas all produced only F_3 green peas, and when the resulting F_3 plants self-fertilized, the next generation also produced green peas. This is what we (and Mendel) would expect of pure-breeding lines carrying two copies of the recessive allele. The yellow peas were a different story. When Mendel allowed 518 F_2 plants that developed from yellow peas to self-fertilize, he observed that 166, roughly 1/3, were pure-breeding yellow through several generations, but the other 352 (2/3) gave rise to yellow and green F_3 peas in a ratio of 3:1. All succeeding generations where the plants from hybrid yellow peas were allowed to self-fertilize also produced a 3:1 ratio of yellow to green progeny. Of the yellow progeny, 1/3 were true-breeding yellow, while the remaining 2/3 were yellow hybrids.

It took Mendel years to conduct such rigorous experiments on seven pairs of pea traits, but in the end, he was able to conclude that the segregation of dominant and recessive alleles during gamete formation and their random union at fertilization could indeed explain the 3:1 ratios he observed whenever he allowed hybrids to self-fertilize. His results, however, raised yet another question, one of some importance to future plant and animal breeders. Plants showing a dominant trait, such as yellow peas, can be either pure-breeding

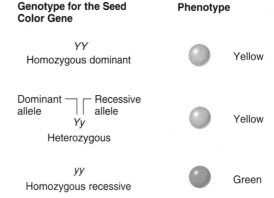

Figure 1.13 Genotype versus Phenotype in homozygotes and heterozygotes. The relationship between genotype and phenotype with a pair of contrasting alleles where one allele (Y) shows complete dominance over the other (y).

(YY) or hybrid (Yy). How can you distinguish one from the other? For self-fertilizing plants, the answer is simply to observe the appearance of the next generation. But how would you distinguish pure-breeding from hybrid individuals showing the dominant trait in species that do not self-fertilize?

Test Crosses Establish Genotype

Before describing Mendel's answer, we need to define a few more terms. An observable characteristic, such as yellow or green pea seeds, is a **phenotype,** while the actual pair of alleles present in an individual is its **genotype.** A YY or a yy genotype is called **homozygous,** because the two copies of the gene that determine the particular trait in question are the same. In contrast, a genotype with two different alleles for a trait is **heterozygous;** in other words, it is a hybrid for that trait (Fig. 1.13).

Fast Forward

Genes Encode Proteins

Genes determine traits as disparate as pea shape and the inherited human disease of cystic fibrosis by encoding the proteins that cells produce and depend on for structure and function. As early as 1940, investigators had uncovered evidence suggesting that some genes determine the formation of enzymes, proteins that catalyze specific chemical reactions. But it was not until 1991, 135 years after Mendel published his analysis of seven pairs of observable traits in peas, that a team of British geneticists identified the gene for pea shape and pinpointed how it prescribes a seed's round or wrinkled contour through the enzyme it determines. About the same time, medical researchers in the United States identified the cystic fibrosis gene and discovered how a mutant allele causes unusually sticky mucous

secretions and a susceptibility to respiratory infections and digestive malfunction, once again, through the protein the gene determines.

The pea shape gene encodes an enzyme known as SBE1 (for starch-branching enzyme 1), which catalyzes the conversion of amylose, an unbranched linear molecule of starch, to amylopectin, a starch molecule composed of several branching chains (Fig. A). The dominant *R* allele of the pea-shape gene causes the formation of active SBE1 enzyme that functions normally. As a result, *RR* homozygotes produce a high proportion of branched starch molecules, which allow the peas to maintain a rounded shape. In contrast, enzyme determined by the recessive *r* allele is abnormal and does not function effectively. In homozygous recessive *rr* peas,

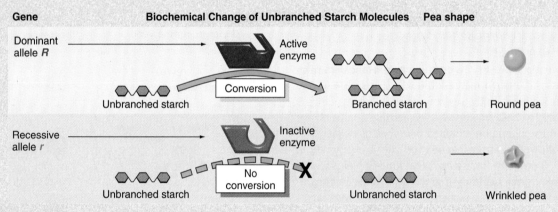

Figure A   Round and wrinkled peas: How one gene determines an enzyme that affects pea shape. The *R* allele of the pea shape gene directs the synthesis of an enzyme that converts unbranched starch to branched starch, indirectly leading to round pea shape. The *r* allele of this gene determines an inactive form of the enzyme, leading to a buildup of linear starch that ultimately causes seed wrinkling.

An individual with a homozygous genotype is a **homozygote;** one with a heterozygous genotype is a **heterozygote.** If you know the genotype and the dominance relation of the alleles, you can accurately predict the phenotype. The reverse is not true, however, because some phenotypes can derive from more than one genotype. For example, the phenotype of yellow peas can result from either the *YY* or the *Yy* genotype.

With these distinctions in mind, we can look at the method Mendel devised for deciphering the unknown genotype, we'll call it *Y—*, responsible for a dominant phenotype; the dash represents the unknown second allele, either *Y* or *y*. Called the **test cross,** it is a mating in which an individual showing the dominant phenotype, for instance, a *Y—* plant grown from a yellow pea, is crossed with an individual expressing the recessive phenotype, in this case a *yy* plant grown

from a green pea. As the Punnett squares in Fig. 1.14 illustrate, if the dominant phenotype in question derives from a homozygous *YY* genotype, all the offspring of the test cross will show the dominant phenotype and thus be yellow peas. But if the dominant parent of unknown genotype is a heterozygous hybrid (*Yy*), 1/2 the progeny will be yellow peas, the other half green. In this way, the test cross establishes the genotype behind a dominant phenotype, resolving any uncertainty.

As we mentioned earlier, Mendel deliberately simplified the problem of heredity, focusing on traits that come in only two forms. He was able to replicate his basic monohybrid findings with corn, beans, and four o'clocks (plants with tubular, white or bright red flowers). As it turns out, his concept of the gene and his law of segregation can be generalized to all sexually reproducing organisms.

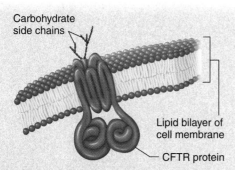

Carbohydrate side chains

Lipid bilayer of cell membrane

CFTR protein

Figure B The cystic fibrosis gene encodes a cell membrane protein. A model of the normal CFTR protein that regulates the passage of chloride ions through the cell membrane. A small change in the gene that codes for CFTR results in an altered protein that prevents proper flow of chloride ions, leading to the varied symptoms of cystic fibrosis.

where there is less starch conversion and more of the linear, unbranched starch, sucrose builds up. The excess sucrose modifies osmotic pressure, causing water to enter the young seeds. As the seeds mature, they lose water, shrink, and wrinkle. The single dominant allele in *Rr* heterozygotes apparently produces enough of the normal enzyme to prevent wrinkling. Although the biochemistry may seem complex, the genetic process is straightforward. A specific gene determines a specific enzyme whose activity affects pea shape. Knowledge of how SBE1 works to determine pea shape provides a physical reality for Mendel's concept of hereditary units.

The human disease of cystic fibrosis was first described in 1938 (see pages 27–29), but doctors and scientists did not understand the biochemical mechanism that produced the serious respiratory and digestive malfunctions associated with the disease. As a result, treatments could do little more than relieve some of the symptoms, and most CF sufferers died before the age of 30. In 1989, molecular geneticists identified and cloned the cystic fibrosis gene (see Chapters 8 and 9 for specific techniques and procedures). Once they had lo-

cated the gene, they could find the protein it encodes and clarify how a defect in just one protein can cause a cascade of biochemically determined symptoms. They could also work toward a cure.

The cystic fibrosis gene determines a protein that forges a channel through the cell membrane (Fig. B). Known as the cystic fibrosis transmembrane conductance regulator (CFTR), the protein regulates the flow of chloride ions into and out of the cell. The normal allele of this gene produces a CFTR protein that correctly regulates the back-and-forth exchange of ions, which in turn determines the cell's osmotic pressure and the flow of water through the cell membrane. In people with cystic fibrosis, however, the cells carry two recessive alleles that produce only an abnormal form of the CFTR protein. The abnormal protein creates nonfunctional chloride channels that remain shut. Since chloride ions do not flow in and out of the cells as required, the control of osmotic pressure breaks down; the cells retain water; and a thick, dehydrated mucus builds up outside the cells. In cells lining the airways and the ducts of secretory organs such as the pancreas, this single biochemical defect in an ion channel produces clogging and blockages that result in respiratory and digestive malfunction.

Cloning of the cystic fibrosis gene brought not only a protein-based explanation of disease symptoms, but also the promise of a cure. In the early 1990s, medical researchers successfully used the cloned cystic fibrosis gene to relieve disease symptoms in mice. They did this by placing the normal allele of the gene into respiratory tissue, which could then produce a functional CFTR protein. Such encouraging results in these small mammals suggested that in the not-too-distant future gene therapy might bestow relatively normal health on people suffering from this once life-threatening genetic disorder. In fact, human trials have already begun.

Although Mendel defined genes as independent units of heredity that reside in cells, he did not have the biochemical information necessary to even speculate on precisely what they are and how they work. Chapters 5–7 describe the chemical composition of genes and clarify how they carry information for making proteins.

Dihybrid Crosses Reveal the Law of Independent Assortment

Having determined from monohybrid crosses that genes are inherited according to the law of segregation, Mendel turned his attention to the simultaneous inheritance of two or more apparently unrelated traits in peas. He asked how two pairs of alleles would segregate in a **dihybrid** individual, that is, in a plant that is heterozygous for two genes at the same time. To construct such a dihybrid, he mated true-breeding plants grown from yellow round peas (*YYRR*) with true-breeding plants grown from green wrinkled peas (*yy rr*). From this cross he obtained a dihybrid F_1 generation (*Yy Rr*) showing only the two dominant phenotypes, yellow and round (Fig. 1.15). He then allowed these F_1 dihybrids to self-fertilize to produce the F_2 generation. Mendel could not predict the outcome of this mat-

ing. Would all the F_2 progeny show one or the other of the parental combinations of traits in a 3:1 ratio of yellow round to green wrinkled? Or would some new combinations of phenotypes occur that were not seen in the parental lines, such as yellow wrinkled or green round peas? New phenotypic combinations like these are called **recombinant types.** When Mendel counted the F_2 generation of one experiment, he found 315 yellow round peas, 101 yellow wrinkled, 108 round green, and 32 wrinkled green. There were, in fact, yellow wrinkled and green round recombinant phenotypes, providing evidence that some shuffling of alleles had taken place.

From the observed ratios, Mendel inferred the biological mechanism of that shuffling—the **independent assortment** of gene pairs during gamete formation. Because the genes for pea color and for pea shape assort independently, *Y* can be with

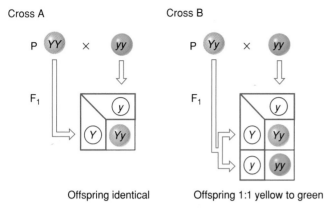

Figure 1.14   How a test cross reveals genotype. An individual of unknown genotype, but dominant phenotype, is crossed with a homozygous recessive. If the unknown genotype is homozygous, all progeny will exhibit the phenotype associated with the dominant allele, as shown in cross A. If the unknown genotype is heterozygous, half the progeny will exhibit the dominant trait, half the recessive trait, as shown in cross B.

R or *r* in any gamete with equal probability. Thus, the presence of a particular allele of one gene, say, the dominant *Y* for pea color, provides no information whatsoever about the allele of the second gene. That is, the allele for pea shape in a *Y* carrying gamete could with equal likelihood be either *R* or *r*. Each dihybrid of the F$_1$ generation can therefore make four kinds of gametes: *YR, Yr, yR,* and *yr*. In a large number of gametes, the four kinds will appear in an almost perfect ratio of 1:1:1:1, or put another way, roughly 1/4 of the eggs and 1/4 of the pollen will contain each of the four possible combinations of alleles. That "the different kinds of germinal cells [eggs or pollen] of a hybrid are produced on the average in equal numbers" was yet another one of Mendel's incisive insights.

At fertilization then, in a mating of dihybrids, 4 different kinds of eggs can combine with any one of 4 different kinds of pollen, producing a total of 16 possible zygotes. Once again, a Punnett square is a convenient way to visualize the process. If you look at the square in Fig. 1.15, you will see that some of the 16 potential allelic combinations are identical. In fact, there are only nine different genotypes—*YY RR, YY Rr, Yy RR, Yy Rr, yy RR, yy Rr, YY rr, Yy rr,* and *yy rr*—because the source of the alleles (egg or pollen) does not make any difference. If you look at the combinations of traits determined by the nine genotypes, you will see only four phenotypes—yellow round, yellow wrinkled, green round, and green wrinkled—observed in a ratio of 9:3:3:1. If, however, you look at just pea color or just pea shape, you can see that each trait is inherited in the 3:1 ratio predicted by Mendel's law of segregation. In the Punnett square, there are 12 yellows for every 4 greens and 12 rounds for every 4 wrinkleds. In other words, the ratio of each dominant trait (yellow or round) to its antagonistic recessive (green or wrinkled) is 12:4 or 3:1. This means that the inheritance of the gene for pea color is unaffected by the inheritance of the gene for pea shape and vice versa.

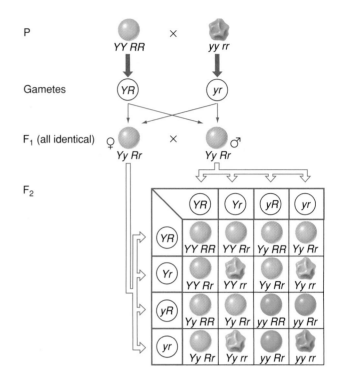

Type	Genotype	Phenotype		Number	Phenotypic ratio
Parental	*Y—R—*		yellow, round	315	9/16
Recombinant	*yy R—*		green, round	108	3/16
Recombinant	*Y—rr*		yellow, wrinkled	101	3/16
Parental	*yy rr*		green, wrinkled	32	1/16

Ratio of yellow (dominant) to green (recessive)	=	12:4 or 3:1
Ratio of round (dominant) to wrinkled (recessive)	=	12:4 or 3:1

Figure 1.15   A dihybrid cross produces parental types and recombinant types. In this analysis of a dihybrid cross showing the genotypes of the P, F$_1$, and F$_2$ generations, pure-breeding parents produce a genetically uniform generation of F$_1$ dihybrids. Self-pollination or cross-pollination of the F$_1$ plants yields the characteristic F$_2$ phenotypic ratio of 9:3:3:1.

The preceding analysis became the basis of Mendel's second general genetic principle, the **law of independent assortment:** During gamete formation, different pairs of alleles segregate independently of each other (Fig. 1.16). The independence of their segregation and the subsequent random union of gametes at fertilization determine the phenotypes observed. Using the law of the product for assessing the probability of independent events, you can see mathematically how the 9:3:3:1 phenotypic ratio observed in a dihybrid cross derives from two separate 3:1 phenotypic ratios. If the two sets of alleles assort independently, the yellow-to-green ratio

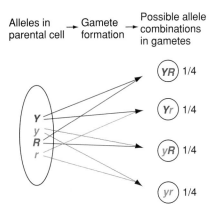

Figure 1.16 The law of independent assortment. In a dihybrid or multihybrid cross, each pair of alleles segregates independently so that in the gametes, one member of each pair is equally likely to appear with either of the two alleles of the other pair or pairs.

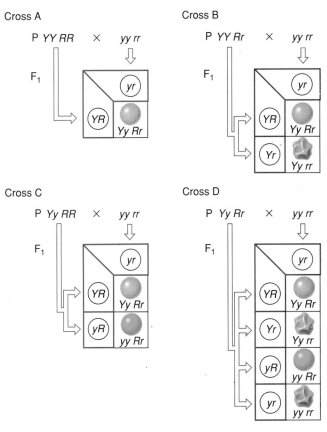

Figure 1.17 Test crosses on dihybrids. In an analysis of test crosses involving two pairs of independently assorting alleles, if all progeny show the dominant phenotype for both traits, as in cross A, the tested individual (of unknown genotype) is a homozygote for the dominant allele of both genes. If some progeny exhibit one of the recessive phenotypes, as in crosses B and C, the tested individual was heterozygous for the corresponding gene. If the progeny include double recessives, the tested individual was a dihybrid, as in cross D.

in the F_2 generation will be 3/4:1/4, and likewise, the round-to-wrinkled ratio will be 3/4:1/4. To find the probability that the two independent events of yellow and round will occur simultaneously in the same plant, you multiply as follows:

Probability of yellow round = 3/4 × 3/4 = 9/16,

Probability of yellow wrinkled = 3/4 × 1/4 = 3/16,

Probability of green round = 1/4 × 3/4 = 3/16,

Probability of green wrinkled = 1/4 × 1/4 = 1/16.

Thus, in a population of F_2 plants, there will be a 9:3:3:1 phenotypic ratio of yellow round to yellow wrinkled to green round to green wrinkled.

Suppose you work for a wholesale nursery that has only F_1 dihybrid plants, and your assignment is to grow pure-breeding plants guaranteed to produce yellow round peas. How would you proceed? One answer would be to allow the plants to self-fertilize and produce F_2 peas. You would then plant the F_2 peas with the desired yellow round phenotype. Only 1 out of 9 of the resultant F_2 plants—those grown from peas with a *YY RR* genotype—will be appropriate for your uses. To find these plants, you could subject each F_2 plant to a test cross for genotype with a green wrinkled (*yy rr*) plant, as illustrated in Fig. 1.17. If the test cross yields all yellow round offspring (test cross A), you can sell your test plant or its progeny from self-fertilization, because you know it is homozygous for both pea color and pea shape. If your test cross yields 1/2 yellow round and 1/2 yellow wrinkled (test cross B), or 1/2 yellow round and 1/2 green round (test cross C), you know that the dihybrid in question is genetically homozygous for one trait and heterozygous for the other and must therefore be discarded. Finally if the test cross yields 1/4 yellow round, 1/4 yellow wrinkled, 1/4 green round, and 1/4 green wrinkled (test cross D), you know the plant is a heterozygote for both the pea-color and the pea-shape genes.

Probabilities and Predictions

Mendel performed several sets of dihybrid crosses and also carried out **multihybrid crosses,** matings between parents that differed in three or more traits. In all of these experiments, he observed numbers and ratios very close to what he expected on the basis of his two general principles: the alleles of a gene segregate during the formation of egg or pollen, and the alleles of different genes segregate independently of each other. But his results were not perfect. They did not always conform to his predicted ratios. To understand why, let us examine both the power and the limitations of his mathematical tools.

First the power: Using simple Mendelian analysis, it is possible to make accurate predictions about the offspring of extremely complex crosses. Suppose you want to predict the occurrence of one specific genotype in a cross involving several independently assorting genes. For example, if hybrids that are heterozygous for four traits are allowed to self-fertilize—*Aa Bb Cc Dd* × *Aa Bb Cc Dd*—what proportion of their progeny will have the genotype *AA bb Cc Dd*? You could set up a Punnett square. Since for each trait there are two different alleles, the number of different eggs or sperm is found by raising 2 to the power of the number of differing traits (2^n, where n is the number of traits). By this calculation, each hybrid parent in this cross would make 16 different kinds of gametes: 4 different traits implies $2^4 = 16$ kinds of gametes. The Punnett square depicting such a cross would thus contain 256 boxes (16×16). This may be fine if you live cloistered in a monastery with plenty of time on your hands, but not if you're taking a 1-hour exam. It would be much simpler to analyze the problem by breaking down the multihybrid cross into four independently assorting monohybrid crosses. Remember that the genotypic ratios of each monohybrid cross are 1 homozygote for the dominant allele, to 2 heterozygotes, to 1 homozygote for the recessive allele $= \underline{1/4}{:}2/4{:}\underline{1/4}$. Thus you can find the probability of $AA\,bb\,Cc\,Dd$ by multiplying the probability of each independent event: AA(1/4 of the progeny produced by $Aa \times Aa$); bb (1/4); Cc (2/4); Dd (2/4):

$$1/4 \times 1/4 \times 2/4 \times 2/4 = 4/256 = 1/64.$$

The Punnett square approach would provide the exact same answer, but it would require much more time.

If instead of a specific genotype, you want to predict the probability of a certain phenotype, you can again use the product rule as long as you know the phenotypic ratios produced by each pair of alleles in the cross. For example, if in the multihybrid cross of $Aa\,Bb\,Cc\,Dd \times Aa\,Bb\,Cc\,Dd$, you want to know how many offspring will show the dominant A trait (genotype AA or $Aa = 1/4 + 2/4$, or 3/4), the recessive b trait (genotype $bb = 1/4$), the dominant C trait (genotype CC or $Cc = 3/4$), and the dominant D trait (genotype DD or $Dd = 3/4$), you simply multiply

$$3/4 \times 1/4 \times 3/4 \times 3/4 = 27/256.$$

In this way, the rules of probability make it possible to predict the outcome of very complex crosses.

Now the limitations: Like Mendel, if you were to breed pea plants, or corn or any other organism, you would most likely observe some deviation from the ratios you expected in each generation. What can account for such variation? One element is chance, as witnessed in the common coin toss experiment. With each throw, the probability of the coin coming up heads is equal to the likelihood it will come up tails. But if you toss a coin 10 times, you may get 30% (3) heads and 70% (7) tails or vice versa. If you toss it 100 times, you are more likely to get a result closer to the expected 50% heads and 50% tails. The larger the number of trials, the lower the probability that chance significantly skews the data. This is one reason Mendel worked with large numbers of pea plants. Mendel's laws, in fact, have great predictive power for populations of organisms, but they do not tell us what will happen in any one individual. With a garden full of self-fertilizing monohybrid pea plants, for example, you can expect that 3/4 of the F_2 progeny will show the dominant phenotype and 1/4 the recessive, but you cannot predict the phenotype of any particular F_2 plant. This is similar to physicists' predictions about an element's radioac-

tivity; they can predict the half-life of a radioactive element and expect to see half of the atoms decay within the specified time, but they cannot say which specific atoms will decay and which will not. (In Chapter 2, we discuss mathematical methods for assessing whether the chance variation observed in a sample of individuals within a population is compatible with a genetic hypothesis.)

Why Mendel's Work Was Unappreciated before 1900

Mendel's insights into the workings of heredity were a breakthrough of monumental proportions. By counting and analyzing data from thousands of pea plant crosses, he inferred the existence of genes—independent units that determine the observable patterns of inheritance for particular traits. His work explained the reappearance of "hidden" traits, disproved the idea of blended inheritance, and showed that mother and father make an equal contribution to the next generation. The model of heredity that he formulated was so specific that he could test predictions based on it by observation and experiment. In the process of developing his model, Mendel deduced the two basic principles of gene transmission: the segregation of each gene's alleles during gamete formation followed by a random union at fertilization, and the independent assortment of the alleles for two or more different genes. Finally, he tied the transmission of traits to the function of particular cells—the gametes—and also recognized that knowledge of the general laws of heredity were a prerequisite for understanding "the evolutionary history of organic forms [organisms]."

With the exception of Abbot Napp, none of Mendel's contemporaries appreciated the importance of his research, and Napp died less than two years after the publication of "Experiments on Plant Hybrids." Mendel, it seems, was far ahead of his time. Sadly, despite written requests from Mendel that others try to replicate his studies, no one repeated his experiments. Several citations of his paper between 1866 and 1900 referred to his expertise as a plant breeder but made no mention of his laws. Moreover, at the time Mendel presented his work, no one had yet seen the structures within cells, the *chromosomes,* that actually segregate and assort independently. That would happen only in the next few decades (as described in Chapter 3).

In an ironic twist of history, Mendel revealed a heritable source of variation—the occurrence, at fertilization, of new combinations of independently segregating alleles, which may give rise to recombinant phenotypes—but Charles Darwin (1809–1882), who was unfamiliar with Mendel's work, was plagued in his later years by critics who found his explanations for the persistence of variation in organisms insufficient. Darwin considered such variation a cornerstone of his theory of evolution, maintaining that natural selection would favor particular variants in a given population in a given environment. If the selected combinations of variant traits were passed on to subsequent generations, this transmission of variation would propel evolution. He could not, however, say how that trans-

(a)　　　　　　　　　(b)　　　　　　　　　(c)　　　　　　　　　(d)

Figure 1.18　The science of genetics begins with the rediscovery of Mendel. Working independently near the beginning of the twentieth century, Correns (b), deVries (c), and von Tschermak (d) each came to the same conclusions as those Mendel (a) summarized in his laws.

mission might occur. Had he been aware of Mendel's laws, he might not have been backed into such an uncomfortable corner. Nearly a century after Mendel published his findings, historians found an uncut copy of Mendel's paper in Darwin's study; Darwin had received but never read it.

For 34 years, Mendel's laws lay dormant—untested, unconfirmed, and unapplied. Then in 1900, 16 years after Mendel's death, Carl Correns, Hugo de Vries, and Erich von Tschermak independently rediscovered and acknowledged his work (Fig. 1.18). The scientific community had finally caught up with Mendel. Within the decade, investigators had coined many of the modern terms we have been using—phenotype, genotype, homozygote, heterozygote, gene, and genetics, the label given to the twentieth-century science of heredity. Mendel's paper provided the new discipline's foundation. His principles and analytic techniques endure today, guiding geneticists and evolutionary biologists in their studies of genetic variation. It has taken less than a century for genetics to become a central discipline of biology, its concepts and tools influencing fields as disparate as immunology and ecology, medicine and agriculture.

MENDELIAN INHERITANCE IN HUMANS: A COMPREHENSIVE EXAMPLE

Although many human traits clearly run in families, most do not show a simple Mendelian pattern of inheritance. Suppose, for example, that you have brown eyes, but both your parents' eyes are blue. Since blue is normally considered recessive to brown, does this mean that you are adopted or that your father isn't really your father? Not necessarily, because eye color is influenced by more than one gene.

Like eye color, most common and obvious human phenotypes arise from the interaction of many genes. In contrast, most of the confirmed single-gene traits in people are relatively rare and involve an abnormality that is disabling or life-threatening. Examples are the progressive mental retardation and other neurological damage of Huntington disease and the clogged lungs and potential respiratory failure of cystic fibrosis. A defective allele of a single gene gives rise to Huntington disease; defective alleles of a different gene are responsible for cystic fibrosis. There were roughly 9000 such single-gene traits known in humans in 1997, and the number continues to grow as new studies confirm the genetic basis of more traits. Table 1.1 lists some of the most common single-gene traits in humans.

Determining a genetic defect's pattern of transmission is not always an easy task since people make slippery genetic subjects. They do not base their choice of mates on purely genetic considerations. Their generation time is long, and the families they produce are relatively small, which makes statistical analysis difficult. There are no pure-breeding lines and no controlled matings. And there is rarely a true F_2 generation (like the one in which Mendel observed the 3:1 ratios from which he derived his rules) because brothers and sisters almost never mate. Geneticists circumvent these difficulties by working with a large number of families, or with several generations of a very large family. This allows them to study the large numbers of genetically related individuals needed to establish the inheritance patterns of specific traits. A family history, known as a **pedigree,** is an orderly diagram of a family's relevant genetic features, extending back to at least both sets of grandparents and preferably through as many more generations as possible. From systematic pedigree analysis in the light of Mendel's laws, geneticists can tell if a trait is determined by alternative alleles of a single gene and whether a single-gene trait is dominant or recessive. We now look at two pedigrees, one for the dominant trait of Huntington disease, the other for the recessive condition of cystic fibrosis. Because Mendel's principles are so simple and straightforward, a little logic can go a long way in explaining how traits are inherited in humans. Clear vertical or horizontal patterns in a trait's appearance in the pedigree lead to hypotheses of great predictive power.

TABLE 1.1 Some of the Most Common Single-Gene Traits in Humans

Disease	Effect	Incidence of Disease
Caused by Recessive Allele		
Thalassemia (chromosome 16 or 11)	Reduced amounts of hemoglobin; anemia, bone and spleen enlargement	1/10 in parts of Italy
Sickle-cell disease (chromosome 11)	Abnormal hemoglobin; sickle-shaped red cells, anemia, blocked circulation; increased resistance to malaria	1/625 African Americans
Cystic fibrosis (chromosome 7)	Defective cell membrane protein; excessive mucus production, digestive and respiratory failure	1/2000 Caucasians
Tay-Sachs disease (chromosome 15)	Missing enzyme; buildup of fatty deposit in brain; buildup destroys mental development	1/3000 Eastern European Jews
Phenylketonuria (PKU) (chromosome 12)	Missing enzyme; mental deficiency	1/10,000 Caucasians
Albinism (chromosome 11)	Missing enzyme; unpigmented skin, hair, and eyes	1/10,000 in Northern Ireland
Caused by Dominant Allele		
Hypercholesterolemia (chromosome 19)	Missing protein that removes cholesterol from the blood; heart attack by age 50	1/122 French Canadians
Huntington disease (chromosome 4)	Progressive mental and neurological damage; neurologic disorders by ages 40–70	1/25,000 Caucasians

A Vertical Pattern of Inheritance Indicates a Rare Dominant Trait

Huntington disease is named for George Huntington, the New York physician who first described its course. This illness usually shows up in middle age and slowly destroys its victims both mentally and physically. Symptoms include intellectual deterioration; severe depression; and jerky, irregular movements, all caused by the progressive death of nerve cells. If one parent develops the symptoms, his or her children have a 50% probability of suffering from the disease, if they too live to adulthood. Because symptoms are not present at birth and manifest themselves only later in life, Huntington disease is known as a **late-onset** genetic condition. The pedigree in Fig. 1.19 illustrates the transmission of Huntington disease through five generations of a family living in the village of San Luis on Lake Maracaibo in Venezuela. This village was founded by a small number of immigrants, including one with the Huntington trait. Isolated by their geography, the villagers are like one large extended family. For generations, most descendants have remained concentrated in one place, and the gene for Huntington disease has remained unusually prevalent among them.

Before analyzing the transmission pattern of Huntington disease, you need to know how to interpret a family pedigree (see Fig. 1.19). Squares (□) represent males, circles (○) are females, diamonds (◇) indicate that the sex is unspecified; family members affected by the trait in question are indicated by a filled-in symbol (for example, ■). A single horizontal line connecting a male and a female (□—○) represents a mating, a double connecting line (□=○) designates a **consanguineous mating,** that is, one between relatives, and a horizontal line above a series of symbols (○ □ ○) indicates the children of the same parents (a *sibship*) arranged and numbered from left to right in order of their birth. Roman numerals to the left or right of the diagram indicate the generations.

How would you proceed in assigning genotypes to the individuals in the pedigree depicted in Fig.1.19? First, you would need to find out if the disease-producing allele is dominant or recessive. Several clues suggest that Huntington disease is transmitted by a dominant allele of a single gene. Everyone who develops the disease has at least one parent who shows the trait, and in several generations, approximately half of the offspring are affected. The pattern of affected individuals is thus vertical: If you trace back through the ancestors of any affected individual, you would see at least one affected person in each generation, giving a continuous line of family members with the disease. When a disease is rare in the population as a whole, a vertical pattern is strong evidence that a dominant allele causes the trait. (A recessive trait that is extremely common might also show up in every generation.) In tracking a dominant allele through a pedigree, you can view

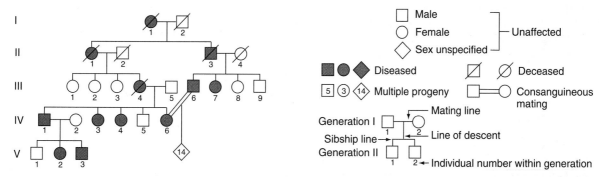

Figure 1.19 Huntington disease: A rare dominant trait. Studies of a Venezuelan family on Lake Maracaibo helped reveal the molecular basis of the disease. A pedigree of the family is shown at left; symbols used in the pedigree are defined at right. The vertical pattern of inheritance, in which some affected individuals appear in each generation, is typical of Huntington disease and other conditions caused by a rare dominant allele. All individuals represented by filled-in symbols are heterozygotes (with the possible exception of individual I-1, who could have been homozygous for the disease allele); all individuals represented by open symbols are homozygotes for the normal allele of the *HD* gene. Among the 14 children of the consanguineous mating between affected individuals, DNA testing shows that some are homozygotes for the disease allele, some are heterozygotes, and some are homozygotes for the normal allele. Human geneticists often employ the diamond designation to mask information about sex or genetic status in order to preserve patient confidentiality or indicate lack of knowledge.

every mating between an affected and an unaffected partner as analogous to a test cross. If some of the offspring do not have Huntington's, you know the parent showing the trait is a heterozygote. You can check your genotype assignments against the answers in the caption to Fig. 1.19.

There is as yet no effective treatment for Huntington disease, and because of its late onset, there was until recently no way for children of a Huntington's parent to know before middle age—usually until well after their own childbearing years—whether they carried the Huntington allele. Then, in the mid-1980s, with new knowledge of the gene, molecular geneticists developed a test that determines whether an individual carries the Huntington disease (*HD*) allele (see the Fast Forward box "The Direct Analysis of Human Genotype" on p. 28). Now a concerned couple can take the test before deciding whether to have children. Some young adults whose parents died of Huntington's, however, prefer not to be tested and thereby learn their own fate. Genetic counselors can advise these people of the probability that they have the trait. If only one of their parents was affected, the probability is 50%, and the probability that a particular child of that person will inherit the defective allele is 25%. With this information, a couple might chose to conceive a child, obtain a prenatal diagnosis of the fetus, and then, depending on their beliefs, elect an abortion if the fetus is affected. The Genetics and Society box "Developing Guidelines for Genetic Screening" on pp. 30–31 discusses some significant social and ethical issues raised by information obtained from family pedigrees and molecular tests.

A Horizontal Pattern of Inheritance Indicates a Rare Recessive Trait

Unlike Huntington disease, most confirmed single-gene traits in humans are recessive. This is because, with the exception of late-onset traits, deleterious dominant traits are unlikely to be transmitted to the next generation. For example, if people affected with Huntington disease died by the age of 10, the

trait would disappear from the population. In contrast, individuals can carry one allele for a recessive trait without knowing they have it. Figure 1.20 shows three pedigrees for cystic fibrosis (CF), the most commonly inherited recessive disease among Caucasian children in the United States. A

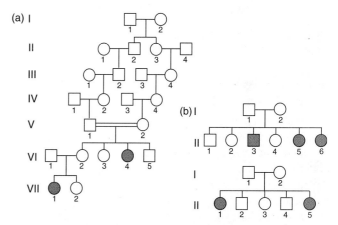

Figure 1.20 Cystic fibrosis: A recessive condition. (a) and (b) illustrate the pedigrees of families with cystic fibrosis. In (a) a consanguineous mating in generation V between unaffected third cousins (V-1 and V-2) who are carriers (heterozygotes for the dominant normal allele *CF⁺* and the recessive disease allele *CF*) produces an affected *CF CF* daughter (VI-4). Similarly, VI-1 and VI-2 are carriers and their *CF CF* homozygous daughter (VII-1) has the disease. It is probable that individuals II-2, II-3, III-2, III-4, IV-2, and IV-4 are all carriers of the same *CF* allele inherited from a common ancestor in the first generation. We cannot determine which of these founders (I-1 or I-2) was a carrier, so we must designate their genotypes as *CF⁺*—. Because the *CF* allele is relatively rare in the population, it is likely that most unaffected individuals unrelated by blood (II-1, II-4, III-1, III-3, IV-1, and IV-3) are *CF⁺ CF⁺* homozygotes. The genotype of the remaining unaffected people (VI-3, VI-5, and VII-2) is uncertain (*CF⁺*—). (b) Two families in which carrier parents produce multiple children with cystic fibrosis demonstrate horizontal patterns of inheritance. Without further information, the unaffected children in each pedigree must be regarded as having a *CF⁺*— genotype.

Fast Forward

The Direct Analysis of Human Genotype

Part of Mendel's genius was the ability to infer a *hidden* genotype from the phenotype expressed by not just one individual but by that individual's relatives as well. Human pedigrees follow this tradition, revealing genotypes on the basis of phenotypic information observed over several generations of a family. But despite the information they reveal, pedigrees do not always provide enough data for prediction. Phenotypically normal prospective parents, for example, may not know, even from an extensive family history, whether they both carry an allele for a deleterious recessive trait; they are thus unable to assess their chances of producing an afflicted child. Today, however, geneticists can go beyond deduction and inference and analyze genotype directly. With techniques developed in the mid-1970s, it is possible to tell exactly what a particular gene "looks like," pick out specific gene pieces, and assess comparable pieces for their differences or similarities. Such comparisons may reveal the presence or absence of disease-causing alleles, such as the dominant allele underlying Huntington disease or the recessive alleles giving rise to cystic fibrosis and sickle-cell anemia. For the prospective parents alluded to above, if just one of them does *not* carry a disease-causing recessive allele, the couple does not have to worry about conceiving an afflicted child.

The power of the new technology arises from its resolution and sensitivity. The resolution, or ability to detect differences between two similar substances, is as high as it can get. A combination of procedures allows detection of differences at the level of single nucleotides, the elementary building blocks of genes (see Chapters 5, 8, and 9). The sensitivity of the new technology is also as great as it can be. Today researchers can detect and analyze the one copy of each gene present in a single sperm cell.

In one application of the new technology, identification of the beta-globin (β-globin) genotype provides the basis for diagnosing sickle-cell anemia, a recessive genetic disease that afflicts roughly 1 in 600 African-Americans. Figure A shows the potential results of a test for the normal and sickle-cell alleles of the β-globin gene. The protein determined by this gene is one component of the oxygen-carrying hemoglobin molecule. In the generation of disease, the normal β-globin allele (*HbβA*, abbreviated *A*) is dominant, and the abnormal allele (*HbβS*, or simply *S*) is recessive. Because the normal *A* allele leads to the production of fully functional hemoglobin, people possessing at least one copy of this allele are healthy under most conditions. By contrast, because the abnormal *S* allele leads to the formation of defective hemoglobin molecules, people with two copies of the *S* allele suffer from sickle-cell anemia. With no normal hemoglobin, these *SS* homozygotes have a decrease in oxygen sup-

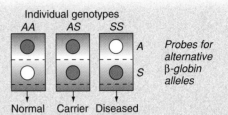

Individual genotypes

Figure A   A direct look at the β-globin genotype. Two probes, one specific for the normal *A* allele, the other specific for the abnormal sickle-cell *S* allele, are necessary to determine the genotype of a sample of DNA amplified by the polymerase chain reaction. The presence of a particular allele is indicated by reaction of the corresponding probe with the DNA sample to produce a darkened signal visible on X-ray film. In this diagram of a β-globin genotype analysis, the green circles show that a probe reacted with a specific allele, the white circles indicate no reaction. Note that only the carrier with both *A* and *S* alleles has two green circles.

ply, tire easily, and often develop heart failure from stress on the circulatory system. The test whose possible results are depicted in Fig. A was based on the polymerase chain reaction (PCR) which can replicate a single gene, or parts thereof, many times over (see Chapter 8 for details). Geneticists can find out what alleles are present in the replicated material through probes that show whether or not a particular allele is present. For example, with two allele-specific probes for the β-globin gene—one for *A*, the other for *S*—they can distinguish a normal *AA* homozygote from a healthy heterozygous *AS* carrier, and both of these from a homozygous *SS* individual afflicted with sickle-cell anemia.

Like the mutant alleles for the β-globin gene, the mutant alleles that determine Huntington disease (HD) and cystic fibrosis (CF) are distinguishable in powerful molecular tests. With HD, both heterozygotes and homozygotes for the mutant allele will eventually show symptoms of the disease. With CF, heterozygotes will be carriers that do not show symptoms, while homozygotes for the mutant allele will have the disease.

The ability to analyze genotype directly has profound social implications. This is particularly true because geneticists can use PCR and other modern techniques on fetal cells obtained from a pregnant woman, and thereby diagnose the genotype of the fetus even before it is born. The Genetics and Society box on pp. 30–31 takes a look at some of the issues related to potential uses of the new technology.

double dose of the recessive *CF* allele causes a fatal disorder in which the lungs, pancreas, and other organs become clogged with a thick, viscous mucus that can interfere with breathing and digestion. One in every 2000 white Americans is born with cystic fibrosis, and only 10% of them survive into their 30s.

There are two salient features of the CF pedigrees. First, the family pattern of people showing the trait is often horizontal: The parents, grandparents, and great-grandparents of children born with CF do not themselves manifest the disease, while several brothers and sisters in a single generation may. A horizontal pedigree pattern is a strong indication that the trait is recessive. The unaffected parents are heterozygous **carriers:** they bear a dominant normal allele that masks the effects of the recessive abnormal one. An estimated 12 million Americans are carriers of the recessive *CF* allele. The second salient feature of the CF pedigrees is that many of the couples who produce afflicted children are blood relatives, that is, their mating is consanguineous (as indicated by the double line). In Fig.1.20a, the consanguineous mating in generation V is between third cousins. Of course, children with cystic fibrosis can

also have unrelated carrier parents, but because relatives share genes, their offspring have a much greater than average chance of receiving two copies of a rare allele. Whether or not they are related, carrier parents are both heterozygotes. Thus among their offspring, the proportion of unaffected to affected children will be 3:1. To look at it another way, the chances are that one out of four children of two heterozygous carriers will be homozygous CF sufferers. You can gauge your understanding of this inheritance pattern by assigning a genotype to each phenotype in Fig. 1.20a and b and then checking your answers against the caption. Note that for several individuals, such as the generation I individuals in part b of the figure, it is impossible to assign a full genotype. We know that one of these people must be the carrier who supplied the original *CF* allele, but we do not know if it was the male or the female. As with an ambiguous dominant phenotype in peas, the unknown second allele is indicated by a dash.

Genetic researchers identified the cystic fibrosis gene in 1989, but are still in the process of developing a gene therapy that would cure the disease's debilitating symptoms (see the Fast Forward box "Genes Encode Proteins" on p.20).

Genetics and Society

Developing Guidelines for Genetic Screening

In the early 1970s, the United States launched a national screening program for carriers of the sickle-cell trait based on a simple test of hemoglobin mobility; normal and "sickling" hemoglobins move at different rates in a gel. People who participated in the screening program could use the test results to make informed reproductive decisions. A healthy man, for example, who learned he was a carrier, would not have to worry about having affected children if his mate were a noncarrier. If, however, they were both carriers, they could choose either not to conceive or to conceive in spite of the 25% risk of bearing an afflicted child. In the 1980s, the possibility of direct prenatal diagnosis of the fetal genotype, as described in the Fast Forward box "The Direct Analysis of Human Genotype" on p. 28, provided additional options. Depending on their beliefs, a couple could decide to continue a pregnancy only if the fetus were not a homozygote for the *S* allele, or knowing that their child would have sickle-cell syndrome, they could learn how to deal with the symptoms of the condition.

The original sickle-cell screening program, based on detection of the abnormal hemoglobin protein, was unfortunately not an unqualified success, largely because of insufficient educational follow-through. Many who learned they were carriers mistakenly thought they had the disease, and because employers and insurance companies obtained access to the information without receiving sufficient instruction as to its meaning, some *AS* heterozygotes were denied jobs or health insurance for no acceptable reason. Problems of public relations and education thus made a reliable screening test into a source of dissent and alienation.

Today, with the ability to look directly at the genotype of individuals born or unborn, it is becoming feasible to screen families at risk not only for sickle-cell anemia but for a growing number of other genetic disorders as well. The need to establish guidelines for genetic screening thus becomes more and more pressing. Several related questions reveal the complexity of the issue.

1. *Why carry out genetic screening at all?* There are two basic reasons. The first is to obtain information that will benefit

individuals. For example, if you learn at an early age that you have a genetic predisposition to heart disease, you can change your lifestyle to include more exercise and a low-fat diet, thereby improving your chances of staying healthy. Or, you can use the results from genetic screening to make informed reproductive decisions that reduce the probability of having children affected by a genetic disease. In Brooklyn, New York, for example, among a community of Hasidic Jews of Eastern European descent, there used to be a high incidence of a fatal neurodegenerative syndrome known as Tay-Sachs disease. In this traditional, Old World community, marriages are arranged by rabbis or matchmakers who, by encouraging testing for the abnormal allele, helped eradicate the disease. With confidential access to test results, a rabbi could counsel against marriages between two carriers. The second reason for genetic screening, which often conflicts with the first, is to benefit groups within society. Insurance companies and employers, for example, would like to be able to find out who is at risk for various genetic conditions.

2. *When is a test accurate and comprehensive enough to be used as the basis for screening?* The accuracy of standard genetic tests for cystic fibrosis is more than 90%. Because it is not 100%, a few people who test negative may actually be carriers. In contrast, the tests for Huntington disease and the sickle-cell trait pick up close to 100% of those who carry the abnormal allele. In addition to the problem of false negatives, all genetic tests occasionally produce a false positive. What all this means is that some people might decide not to have children on the basis of inaccurate information. While the accuracy of the tests continues to improve, the question remains: Do the benefits outweigh both the costs and the risks of inaccuracies?

3. *Once an accurate test becomes available at reasonable cost, should screening be required or optional?* This is partly a societal

---

CONNECTIONS

Mendel answered the three basic questions about heredity as follows: To "What is inherited?" he replied "alleles of genes." To "How are they inherited?" he responded "according to the principles of segregation and independent assortment." And to "What is the role of chance?" he said "for each individual, inheritance is determined by chance, but within a population, this chance operates in a context of strictly defined probabilities." Knowing all this, we can understand why a trait may disappear in the hybrid generation and reappear in the F_2 generation or

why with individuals that are pure-breeding, like begets like, while with hybrids, like can beget both like and unlike.

Within a decade of the 1900 rediscovery of Mendel's work, numerous breeding studies had shown that Mendel's laws hold true not only for seven pairs of antagonistic characteristics in peas, but for an enormous diversity of traits in a wide variety of sexually reproducing plant and animal species, including four o'clock flowers, beans, corn, wheat, fruits flies, chickens, mice, horses, and humans. Some of these same breeding studies, however, raised a challenge to the new ge-

decision because the public treasury bears a large part of the cost of caring for the sufferers of genetic diseases. But it is also a personal decision. For most inherited diseases there is as yet no cure. Since the psychological burden of anticipating a fatal disease for which there is no treatment can be devastating, some people might decide not to be tested. Other reasons people may not want to be tested include religious beliefs and concerns about confidentiality. On the other hand, timely information about the presence of an abnormal gene that causes a condition for which there is a therapy can save lives; an example is the genetic test for hemochromatosis, an inherited disorder affecting 1 of every 200 people in the United States. The disease deposits iron in the heart, liver, and other organs of the body, and by the fifth or sixth decade of life, those organs break down. The simple, ancient practice of bloodletting, if begun early in life, prevents the complications of hemochromatosis. Timely information may also affect childbearing decisions and thereby reduce the incidence of a disease in the population.

4. *If a screening program is established, who should be tested?* The answer depends on what the test is trying to accomplish as well as on the fact that genetic screening is expensive, a consideration that must be weighed against the usefulness of the data it provides. For the 1 in 200 people inheriting the hemochromatosis mutation, widespread testing could reduce health-care costs of affected individuals later in life. By contrast, to reduce the risk of a child being born with a rare inherited disease, a program might try to target groups with the highest incidence. In the United States, only one tenth as many African-Americans as Caucasians are affected by cystic fibrosis, and Asians almost never have the disease. Should all racial groups be tested or only Caucasians? Because of the expense, this kind of analysis has not yet been applied to large populations; instead, such direct testing has been reserved for couples or individuals whose family history puts them at risk for a severely debilitating disease.

5. *Should private employers and insurance companies be allowed to test their clients and employees?* Some employers advocate genetic screening to reduce the incidence of occupational disease, arguing that they can use data from genetic tests to make sure employees are not assigned to environments that might cause them harm. People with sickle-cell syndrome, for example, may be at increased risk for a life-threatening episode of severe sickling if exposed to carbon monoxide or trace amounts of cyanide. Critics of this position say that screening violates workers' rights, including the right to privacy, and increases racial and ethnic discrimination in the workplace. Many critics also oppose informing insurance companies of the results of genetic screening, as these companies may deny coverage to people with inherited medical problems or just the possibility of developing such problems. According to one medical ethicist, discrimination of this sort will grow unless countries pass laws, similar to one enacted in France, that ensure genetic information is confidential, to be given out only at the discretion of the tested individual. In 1998, the Clinton administration asked Congress to draft legislation that would prohibit the requirement of a genetic test as a condition of employment, and would bar employers from obtaining or disclosing genetic information about employees under most circumstances.

6. *Finally, how should people be educated about the meaning of test results?* In one small-community screening program, people identified as carriers of the recessive, life-threatening blood disorder known as β-thalassemia were ostracized and as a result, ended up marrying one another. This only made medical matters worse as it greatly increased the chances that their children would be born with two copies of the defective allele and thus the disease. By contrast, in Ferrara, Italy, where there used to be 30 new cases of β-thalassemia every year, extensive screening was so successfully combined with intensive education that the 1980s passed with no more than a few new cases of the disease.

Given all of these considerations, what kind of guidelines would you like to see established to ensure that genetic screening reaches the right people at the right time and that information gained from such screening is used for the right purposes?

netics. For certain traits in certain species, the studies uncovered unanticipated phenotypic ratios, or found F_1 and F_2 progeny with novel phenotypes that resembled neither pure-breeding parent's. Mendel himself observed such "puzzling phenomena" when he replicated his pea experiments with beans. A mating between bean plants with white flowers and those with crimson flowers produced a whole range of colors in the offspring, from purple to pale violet to white.

These phenomena could not be explained by Mendel's hypothesis that for each gene, two alternative alleles, one com- pletely dominant, the other recessive, determine a single trait. Mendel, in fact, correctly explained the unexpected outcomes he observed by proposing that two genes (instead of one) determine flower color in beans. In humans, we now know that most common traits, including skin color, eye color, and height, are determined by interactions between two or more genes (see Fig. 1.1). We also know that within a given population, there may be more than two alleles for some of those genes. Chapter 2 shows how the genetic analysis of such complex traits, that is, traits produced by complex interactions between genes and between genes and the environment, extended rather than contradicted Mendel's laws of inheritance. Each trait is still determined by the alleles of genes that behave during gamete formation and fertilization exactly as Mendel envisioned. What is new is the relationship between genotype and phenotype. It is not always as straightforward as in the analysis of pea color or of human genetic diseases such as cystic fibrosis.

ESSENTIAL CONCEPTS

1. Discrete units called *genes* control the appearance of inherited traits.

2. Genes come in alternative forms called *alleles* that are responsible for the expression of different forms of a trait.

3. Body cells of sexually reproducing organisms carry two copies of each gene. When the two copies of a gene are the same allele, the individual is *homozygous* for that gene. When the two copies of a gene are different alleles, the individual is *heterozygous* for that gene.

4. The *genotype* is a description of the allelic combination of the two copies of a gene present in an individual. The *phenotype* is the observable form of the trait that the individual expresses.

5. A cross between two parental lines (P) that are pure bred for alternative alleles of a gene will produce a *first filial* (F_1) *generation* of hybrids that are heterozygous. The phenotype expressed by these hybrids is determined by the *dominant* allele of the pair, and is the same as that expressed by individuals homozygous for the dominant allele. The phenotype associated with the *recessive* allele will reappear only in the F_2 generation in individuals homozygous for this allele. In crosses between F_1 heterozygotes, the dominant and recessive phenotypes will appear in the F_2 generation in a ratio of 3:1.

6. The two copies of each gene segregate during the formation of gametes. As a result, each egg and each sperm or pollen grain contains only one copy, and thus, only one allele, of each gene. Male and female gametes unite at random at fertilization. Mendel described this process as the *law of segregation.*

7. The segregation of alleles of any one gene is independent of the segregation of the alleles of other genes. Mendel described this process as the *law of independent assortment.* According to this law, crosses between $Aa\,Bb$ F_1 dihybrids will generate F_2 progeny with a phenotypic ratio of $9(A—B—):3(A—bb):3(aa\,B—):1(aa\,bb)$.

SOCIAL AND ETHICAL ISSUES

1. Judy and Tom were unable to have children because Judy had fibroids (benign tumors) in her uterus, making implantation of an embryo impossible. Judy was, however, able to ovulate healthy eggs. Judy and Tom decided they wanted to have a child that was genetically related to both of them. They contacted a surrogate mother, Sara, to carry an embryo conceived from their gametes by *in vitro* fertilization. A month before the child was born, Judy and Tom were killed in a plane crash. Sara decided she wanted to raise the child herself. She thought this wouldn't pose any legal problem since the genetic parents were no longer alive. But Judy's sister Ellen filed a petition with the court to gain custody of the child. Should the courts award the child to the surrogate mother or to the family of the genetic parents?

2. Louise, a single mother with four children, has a job in an aeronautics factory where she is exposed to the element beryllium. Some individuals are susceptible to the damaging effects of this element, developing a respiratory illness called chronic beryllium disease. Workers in the factory were recently tested for susceptibility and Louise tested positive. She needs the job and wants to continue working at the factory until she finds another job. The company wants her to quit immediately. Should the individual or the company have the right to decide whether a susceptible person continues to work in a potentially damaging environment?

3. Angela and Andre had difficulties conceiving and decided to have a child using *in vitro* fertilization methods. They have since gotten divorced and Angela, now a single mother of one, would like to have a second child by implantation of an embryo that was frozen when she and Andre had the first procedure. When Andre heard through a mutual friend of Angela's plan, he was upset and opposed to the idea, stating that the embryos are half his. He feels Angela needs his permission to use the embryos and he will not grant that permission. Who has rights to these embryos? Should the embryos be destroyed? Should there be a limit to how long embryos are saved?

S O L V E D   P R O B L E M S

Solving Genetics Problems

The best way to evaluate and increase your understanding of the material in the chapter is to apply your knowledge in solving genetics problems. Genetics word problems are like puzzles. Take them in slowly—don't be overwhelmed by the whole problem. Identify useful facts given in the problem, and use the facts to deduce additional information. Use genetic principles and logic to work toward the solutions. The more problems you do, the easier they become. In doing problems you will not only solidify your understanding of genetic concepts but you will also develop basic analytical skills that are applicable in many disciplines.

Solving genetics problems requires more than simply plugging numbers into formulas. Each problem is unique and requires thoughtful evaluation of the information given and the question being asked. The following are general guidelines you can follow in approaching these word problems:

a. Read through the problem once to get some sense of the concepts involved.

b. Go back through the problem, noting all the information supplied to you. For example, genotypes or phenotypes of offspring or parents may be given to you or implied in the problem. Represent the known information in a symbolic format—assign symbols for alleles; use these symbols to indicate genotypes; make a diagram of the crosses including genotypes and phenotypes given or implied.

c. Now, reassess the question and work toward the solution using the information given. Make sure you answer the question being asked!

For each chapter, the logic involved in solving two or three types of problems is described in detail.

I. In cats, white patches are caused by the dominant allele *P,* while *pp* individuals are solid-colored. Short hair is caused by a dominant allele *S,* while *ss* cats have long hair. A long-haired cat with patches whose mother was solid-colored and short-haired mates with a short-haired solid-colored cat whose mother was long-haired and solid-colored. What kinds of kittens can arise from this mating, and in what proportions?

Answer

The solution to this problem requires an understanding of dominance/recessiveness, gamete formation, and the independent assortment of alleles of two genes in a cross.

First make a representation of the known information:

Mothers:	solid, short-haired	solid, long-haired
Cross:	cat 1	cat 2
	patches, long-haired $\times$ solid, short-haired	

What genotypes can you assign? Any cat showing a recessive phenotype must be homozygous for the recessive allele. Therefore the long-haired cats are *ss*; solid cats are *pp*. Cat 1 is long-haired, so it must be homozygous for the recessive allele (*ss*). This cat has the dominant phenotype of patches and could be either *PP* or *Pp,* but since the mother was *pp,* and could only contribute a *p* allele in her gametes, the cat must be *Pp.* Cat 1's full genotype is *Pp ss.* Similarly, cat 2 is solid-colored, so it must be homozygous for the recessive allele (*pp*). Because this cat is short-haired, it could have either the *SS* or *Ss* genotype. Its mother was long-haired (*ss*) and could only contribute an *s* allele in her gamete, so cat 2 must be heterozygous *Ss.* The full genotype is *pp Ss.*

The cross is therefore between a *Pp ss* (cat 1) and a *pp Ss* (cat 2). To determine the types of kittens, first establish the types of gametes that can be produced by each cat and then set up a Punnett square to determine the genotypes of the offspring. Cat 1 (*Pp ss*) produces *Ps* and *ps* gametes in equal proportions. Cat 2 (*pp Ss*) produces *pS* and *ps* gametes in equal proportions. *Four types of kittens can result from this mating with equal probability: Pp Ss (patches, short-haired), Pp ss (patches, long-haired), pp Ss (solid, short-haired), and pp ss (solid, long-haired).*

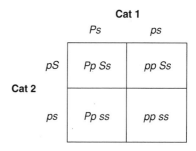

You could also work through this problem using the product rule of probability instead of a Punnett square. The principles are the same: gametes produced in equal amounts by either parent are combined at random.

Cat 1 gamete		Cat 2 gamete	Progeny	
1/2 *Ps*	$\times$	1/2 *pS*	$\longrightarrow$ 1/4 *Pp Ss*	patches, short-haired
1/2 *Ps*	$\times$	1/2 *ps*	$\longrightarrow$ 1/4 *Pp ss*	patches, long-haired
1/2 *ps*	$\times$	1/2 *pS*	$\longrightarrow$ 1/4 *pp Ss*	solid-colored, short-haired
1/2 *ps*	$\times$	1/2 *ps*	$\longrightarrow$ 1/4 *pp ss*	solid-colored, long-haired

II. In tomatoes, red fruit is dominant to yellow fruit and purple stems are dominant to green stems. The progeny from one mating consisted of 305 red fruit, purple stem plants; 328 red fruit, green stem plants; 110 yellow fruit, purple stem plants; and 97 yellow fruit, green stem plants. What was the genotype of the parents in this cross?

Answer

This problem requires an understanding of independent assortment in a dihybrid cross as well as the ratios predicted from monohybrid crosses.

Designate the alleles: R = red, r = yellow
P = purple stems, p = green stems

In genetics problems, the ratios of offspring can indicate the genotype of parents. You will usually need to total the number of progeny and approximate the ratio of offspring in each of the different classes. For this problem, in which the inheritance of two traits is given, consider each trait independently. For red fruit there are 305 + 328 = 633 red-fruited plants out of a total of 840 plants. This value (633/840) is close to 3/4. About 1/4 of the plants have yellow fruit (110 + 97 = 207/840). From Mendel's work, you know that a 3:1 phenotypic ratio results from crosses between plants that are hybrid (heterozygous) for one gene. Therefore the genotype for fruit color of each parent must have been Rr.

For stem color, 305 + 110 or 415/840 plants had purple stems. About half had purple stems and the other half (328 + 97) had green stems. A 1:1 phenotypic ratio occurs when a heterozygote is mated to a homozygous recessive (as in a test cross). The parents' genotypes must have been Pp and pp for stem color.

The complete genotype of the parent plants in this cross was $Rr\,Pp \times Rr\,pp$.

III. Tay-Sachs is a recessive lethal disease in which there is neurological deterioration early in life. This disease is rare in the population overall but is found at relatively high frequency in Ashkenazi Jews from Central Europe. A woman whose maternal uncle had the disease is trying to determine the probability that she and her husband could have an affected child. Her father does not come from a high-risk population. Her husband's sister died of the disease at an early age.
   a. Draw the pedigree of the individuals described. Include the genotypes where possible.
   b. Determine the probability that the couple's first child will be affected.

Answer

This problem requires an understanding of dominance/ recessiveness and probability. Designate the alleles: T = normal allele; t = Tay-Sachs allele.

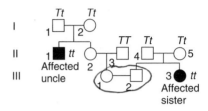

The genotypes of the two affected individuals, the woman's uncle (II-1) and the husband's sister (III-3) are tt. Because the uncle was affected, his parents must have been heterozygous. There was a 1/4 chance that these parents had a homozygous recessive (affected) child, a 2/4 chance that they had a heterozygous child (carrier), a 1/4 chance they had a homozygous dominant (unaffected) child. However, you have been told that the woman's mother (II-2) is unaffected so she could only have had a heterozygous or a homozygous dominant genotype. Consider the probability that these two genotypes will occur. If you were looking at a Punnett square, there would be only three combinations of alleles that are possible for the normal mother. Two of these are heterozygous combinations and one is homozygous dominant. There is a 2/3 chance (2 out of the 3 possible cases) that the mother was a carrier. The father was not from a high-risk population so we can assume that he is homozygous dominant. There is a 2/3 chance that the wife's mother was heterozygous and if so, a 1/2 chance that the wife inherited a recessive allele from her mother. Because both conditions are necessary for inheritance of a recessive allele, the individual probabilities are multiplied and the probability that the wife is heterozygous is 2/3 × 1/2 .

The husband (III-2) has a sister who died from the disease, therefore his parents must have been heterozygous. The probability that he is a carrier is 2/3 (using the same rationale as for II-2). The probability that the man and wife are both carriers is 2/3 × 1/2 × 2/3. Since there is a 1/4 probability that a particular child of two carriers will be affected, *the overall probability that the first child of this couple (III-1 and III-2) will be affected is 2/3 × 1/2 × 2/3 × 1/4 = 4/72, or 1/18.*

P R O B L E M S

1-1 For each of the terms in the left column, choose the best matching phrase in the right column.

a. phenotype
b. alleles
c. independent assortment
d. gametes
e. gene
f. segregation
g. heterozygote
h. dominant
i. F_1
j. test cross
k. genotype
l. recessive
m. dihybrid cross
n. homozygote

1. having two identical alleles of a given gene
2. the allele expressed in the phenotype of the heterozygote
3. alternate forms of a gene
4. observable characteristic
5. a cross between individuals both homozygous for two genes
6. alleles of one gene separate into gametes randomly with respect to alleles of other genes
7. reproductive cells containing only one copy of each gene
8. the allele that does not contribute to the phenotype of the heterozygote
9. the cross of an individual of ambiguous genotype with a homozygous recessive individual
10. an individual with two different alleles of a gene
11. the heritable entity that determines a characteristic
12. the alleles an individual has
13. the separation of the two alleles of a gene into different gametes
14. offspring of the P generation

1-2 Describe the characteristics of the garden pea that made it a good organism for Mendel's analysis of the basic principles of inheritance. Evaluate how easy or difficult it would be to make a similar study of inheritance in humans by considering the same attributes you described for the pea.

1-3 In a particular population of mice, certain individuals display a phenotype called "short tail," which is inherited as a dominant trait. Some individuals display a recessive trait called "dilute," which affects coat color. Which of these traits would be easier to eliminate from the population by selective breeding? Why?

1-4 How many genetically different eggs could be formed by women with the following genotypes?
a. *Aa bb CC DD*
b. *AA Bb Cc dd*
c. *Aa Bb cc Dd*
d. *Aa Bb Cc Dd*

1-5 A mouse sperm of genotype *a B C D E* fertilizes an egg of genotype *a b c D e*. What are all the possibilities for the genotypes of (a) the zygote and (b) a sperm or egg of the baby mouse that develops from this fertilization?

1-6 In humans, a dimple in the chin is an autosomal dominant characteristic.

a. A man who does not have a chin dimple marries a woman with a chin dimple whose mother lacked the dimple. What proportion of their children would be expected to have a chin dimple?
b. A man with a chin dimple marries a woman who lacks the dimple. Their first child lacks a dimple. What is the man's genotype?
c. A man with a chin dimple and a nondimpled woman produce eight children, all having the chin dimple. Can you be certain of the man's genotype? Why or why not? What genotype is more likely, and why?

1-7 Two short-haired cats mate and produce six short-haired and two long-haired kittens. What does this information suggest about how hair length is inherited?

1-8 Piebald spotting is a condition found in humans in which there are patches of skin that lack pigmentation. The condition results from the inability of pigment-producing cells to migrate properly during development. Two adults with piebald spotting have one child who has this trait and a second child with normal skin pigmentation.
a. Is the piebald spotting trait dominant or recessive? What information led you to this answer?
b. What are the genotypes of the parents?

1-9 Among native Americans, two types of earwax (cerumen) are seen, dry and sticky. A geneticist studied the inheritance of this trait by observing the types of offspring produced by different kinds of matings. He observed the following numbers.

Parents	Number of mating pairs	Offspring Sticky	Dry
Sticky × sticky	10	32	6
Sticky × dry	8	21	9
Dry × dry	12	0	42

a. How is earwax type inherited?
b. Why are there no 3:1 or 1:1 ratios in the data shown in the chart?

1-10 If you roll a die (singular of dice), what is the probability you will roll (a) a 6? (b) an even number? (c) a number divisible by 3? (d) If you roll a pair of dice, what is the probability that you will roll two 6's? (e) an even number on one and an odd number on the other? (f) matching numbers? (g) two numbers both over 4?

1-11 Galactosemia is a recessive human disease that is treatable by restricting lactose and glucose in the diet. Susan Smithers and her husband are both heterozygous for the galactosemia gene.
a. Susan is pregnant with twins. If she has fraternal (nonidentical) twins, what is the probability both of the twins will be girls who have galactosemia?

b. If the twins are identical, what is the probability that both will be girls and have galactosemia?

1-12 What is the probability of producing a child that will phenotypically resemble either one of the two parents?
   a. *Aa Bb Cc Dd* × *aa bb cc dd*
   b. *aa bb cc dd* × *AA BB CC DD*
   c. *Aa Bb Cc Dd* × *Aa Bb Cc Dd*
   d. *aa bb cc dd* × *aa bb cc dd*

1-13 As a *Drosophila* research geneticist, you keep stocks of flies of specific genotypes. You have a fly that has normal wings (dominant phenotype). Flies with short wings are homozygous for a recessive allele of the wing length gene. You need to know if this fly with normal wings is pure-breeding or heterozygous for the wing length trait. What cross would you do to determine the genotype and what results would you expect for each possible genotype?

1-14 A mutant cucumber plant has flowers that fail to open when mature. Crosses can be done with this plant by manually opening and pollinating the flowers with pollen from another plant. When closed × open crosses were done, all the F_1 progeny were open. The F_2 plants were 145 open and 59 closed. A cross of closed × F_1 gave 81 open and 77 closed. How is the closed trait inherited?

1-15 Imagine you have just purchased a black stallion of unknown genotype. You mate him to a red mare, and she delivers twin foals, one red and one black. Can you tell from these results how color is inherited? What crosses could you do to work this out?

1-16 A pea plant from a pure-breeding strain that is tall, has green pods, and has purple flowers that are terminal is crossed to a plant from a pure-breeding strain that is dwarf, has yellow pods, and has white flowers that are axial. The F_1 plants are all tall and have purple, axial flowers as well as green pods.
   a. What phenotypes do you expect to see in the F_2?
   b. What phenotypes and ratios would you predict in the progeny from crossing an F_1 plant to the dwarf parent?

1-17 The following chart shows the results of different matings between jimsonweed plants that had either purple or white flowers and spiny or smooth pods. Determine the dominant allele for the two traits and indicate the genotypes of the parents for each of the crosses.

Parents	Offspring phenotypes			
	purple spiny	white spiny	purple smooth	white smooth
a. purple spiny × purple spiny	94	32	28	11
b. purple spiny × purple smooth	40	0	38	0
c. purple spiny × white spiny	34	30	0	0
d. purple spiny × white spiny	89	92	31	27
e. purple smooth × purple smooth	0	0	36	11
f. white spiny × white spiny	0	45	0	16

1-18 A cross between two pea plants, both of which grew from yellow, round seeds, gave the following numbers of seeds: 156 yellow, round and 54 yellow, wrinkled. What are the genotypes of the parent plants? (Yellow and round are dominant traits.)

1-19 A third-grader decided to breed guinea pigs for her school science project. She went to a pet store and bought a male with smooth, black fur and a female with rough, white fur. She wanted to study the inheritance of those features, and was sorry to see that the first litter of eight contained only rough, black animals. To her disappointment, the second litter from those same parents contained seven rough, black animals. Soon the first litter had begun to produce F_2 offspring, and they showed a variety of coat types. Before long the child had 125 F_2 guinea pigs. Eight of them had smooth, white coats, 25 had smooth black coats, 23 were rough and white, and 69 were rough and black.
   a. How are the coat color and texture characteristics inherited?
   b. What phenotypes and proportions of offspring should the girl expect if she mates one of the smooth white females to an F_1 male?

1-20 A pea plant heterozygous for plant height, pod shape, and flower color was selfed. The progeny consisted of 272 tall, inflated pods, purple flowers; 92 tall, inflated, white flowers; 88 tall, flat pods, purple; 93 dwarf, inflated, purple; 35 tall, flat, white; 31 dwarf, inflated, white; 29 dwarf, flat, purple; 11 dwarf, flat, white. Which alleles are dominant in this cross?

1-21 The achoo syndrome (sneezing in response to bright light) and trembling chin (triggered by anxiety) are both dominant traits in humans.
   a. What is the probability that the first child of parents who are heterozygous for both the achoo gene and trembling chin will have achoo syndrome but lack the trembling chin?
   b. What is the probability that the first child will not have achoo syndrome or trembling chin?

1-22   The self-fertilization of an F_1 pea plant produced from a parent plant homozygous for yellow and wrinkled seeds and a parent homozygous for green and round seeds resulted in a pod containing seven F_2 peas. (Yellow and round are dominant.) What is the probability that all seven peas (the progeny) were yellow and round?

1-23   For each of the human pedigrees shown below, indicate whether the inheritance pattern is recessive or dominant. What feature(s) of the pedigree did you use to determine the inheritance? Give the genotypes of affected individuals and individuals who carry the disease allele.

(a)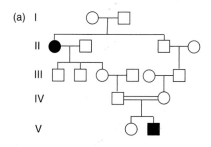

(b)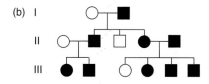

(c)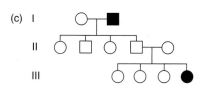

1-24   Albinism is a condition in which pigmentation is lacking. In humans, the result is white hair, nonpigmented skin, and pink eyes. The trait in humans is caused by a recessive allele. Two normal parents have an albino child. What are the parents' genotypes? What is the probability that the next child will be albino?

1-25   A young married couple went to see a genetic counselor because each had a sibling affected with cystic fibrosis. (Cystic fibrosis is a recessive disease and neither member of the couple nor any of their four parents is affected.)
   a. What is the probability that the female of this couple is a carrier?
   b. What are the chances that their child will be affected with cystic fibrosis?
   c. What is the probability that their child will be a carrier of the cystic fibrosis mutation?

1-26   Huntington disease is a rare fatal, degenerative neurological disease in which individuals start to show symptoms, on average, in their 40s. It is caused by a dominant allele. Joe, a man in his 20s, just learned that his father has Huntington disease.
   a. What is the probability that Joe will also develop the disease?
   b. Joe and his new wife have been anxious to start a family. What is the probability that their first child will develop the disease?

1-27   Is the disease shown dominant or recessive? Based on the limited information given in this pedigree, do you think the disease allele is rare or common in the population? Why?

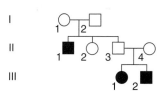

EXTENSIONS TO MENDEL:
COMPLEXITIES IN RELATING GENOTYPE
TO PHENOTYPE

Coat color in deer is one of many traits in natural populations governed by the
interactions of many genes and of these genes with the environment.

Unlike the pea traits that Mendel examined, most human characteristics—height, hair color, skin color, voice quality, artistic or athletic ability, to name a few—do not fall neatly into just two opposing phenotypic categories. And because people cannot be simply divided into two contrasting camps for each trait (brown hair versus blond, high voice versus low), they seem to defy Mendelian analysis. The same can be said of traits expressed by many of the world's food crops; the size, shape, succulence, and nutrient content of the plants' edible parts vary over a wide range of values. Lentils (*Lens culinaris*) provide a graphic illustration of this variation. A legume like peas and beans, lentils are grown in many parts of the world as a rich source of both protein and carbohydrate. The mature plants set fruit in the form of diminutive pods that contain two small round and flattened seeds, which can be ground into meal or used in soups, salads, and stews. These seeds come in an intriguing array of colors and patterns (Fig. 2.1), and since particular couplings of color and pattern are preferred in the cuisines of different cultures, there is considerable commercial interest in producing specific seed coat combinations. But crosses between pure-breeding lines of lentils result in some startling surprises in relation to Mendel's laws. A cross between pure-breeding tan and pure-breeding gray parents, for example, yields an all brown F_1 generation; and when these novel hybrids are left to self-pollinate, the F_2 plants produce not only tan, gray, and brown lentils, but also green.

Beginning with the first decade of the twentieth century, geneticists subjected all kinds of puzzling variation in plants and animals to controlled breeding tests, using Mendel's 3:1 phenotypic ratio as a guideline in interpreting the results. If the traits under analysis behaved as predicted by Mendel's laws, it meant they were determined by a single gene with alternative alleles that show clear-cut dominance and recessiveness. Many traits, however, did not behave in this way. For some, there was no clear-cut dominance and recessiveness, or there were more than two alternative alleles in a particular cross. Other traits turned out to be **multifactorial,** that is, determined by two or more factors, in-

Figure 2.1 Some phenotypic variation poses a challenge to Mendelian analysis. In this array of red, green, and brown lentils, some of the seeds have speckled patterns, while others are clear. As we see in this chapter, various combinations of alleles at two genes determine seed coat color, but various combinations of the multiple alleles of a third gene produce the pattern.

cluding multiple genes interacting with each other or one or more genes interacting with the environment. The seed coat color of lentils is a multifactorial trait. Because such traits arise from an intricate network of interactions, they do not necessarily generate straightforward Mendelian phenotypic ratios. Nonetheless, simple extensions of Mendel's hypotheses that clarify the relationship between genotype and phenotype have allowed geneticists to explain all of the observed deviations without challenging Mendel's basic laws.

As we look at the genetic factors that make it possible for plants and animals to generate the enormous amount of phenotypic variation seen in the natural world, we will encounter the following general theme: In their attempt to sort things out, geneticists limit the number of variables under investigation at any one time. Mendel, for example, compared and mated together pure-breeding, inbred strains of peas that differed from each other by one or a few traits. By establishing a genetic background that was uniform, or homogenized, for all other traits, he was able to visualize the actions of single genes in isolation. Similarly, twentieth-century geneticists have used inbred populations of fruit flies, mice, and other experimental organisms to study specific traits. Of course, geneticists cannot approach people in this way. Human populations are far from inbred, and if a phenotype arises from many genes that are not held constant, it may be difficult to see how a particular gene affects a particular trait. For this reason, until recently the genetic basis of much human variation remained a mystery. The advent of molecular biology in the 1970s, however, provided new tools that geneticists now use to unravel the genetics of complex human traits.

We divide our presentation of extensions to Mendelian analysis into two broad categories:

- Single-gene inheritance (a) in which pairs of alleles show deviations from complete dominance and recessiveness, (b) in which different forms of the gene are not limited to two alleles, and (c) where one gene may determine more than one trait.

- Multifactorial inheritance in which the phenotype arises from the interaction of one or more genes with the environment, chance, or each other.

EXTENSIONS TO MENDEL FOR SINGLE-GENE INHERITANCE

William Bateson, an early interpreter and defender of Mendel, who coined the terms "genetics," "allelomorph" (later shortened to "allele"), "homozygote," and "heterozygote," entreated the audience at a 1908 lecture: "Treasure your exceptions! . . . Keep them always un-covered and in sight. Exceptions are like the rough brickwork of a growing building which tells that there is more to come and shows where the next construction is to be." Consistent exceptions to simple Mendelian ratios revealed unexpected patterns of single-gene inheritance. By distilling the significance of these patterns, Bateson and other early geneticists extended the scope of Mendelian analysis and obtained a deeper understanding of the relationship between genotype and phenotype. We now look at the major extensions to Mendelian analysis elucidated over the last century.

Dominance Is Not Always Complete

A consistent working definition of dominance and recessiveness depends on the F_1 hybrids that arise from a mating between two pure-breeding lines. If a hybrid is identical to one parent for the trait under consideration, the allele carried by that parent is deemed dominant to the allele carried by the parent whose trait is not expressed in the hybrid. If, for example, a mating between a pure-bred blue line and a pure-bred yellow line produces F_1 hybrids that are blue, the blue allele of the gene for color is dominant to the yellow allele. If the F_1 hybrids are yellow, however, the yellow allele is dominant to the blue one (Fig. 2.2).

The determination of dominance and recessiveness does not imply anything about the actual functioning of the alleles. Although geneticists once believed that all dominant alleles were actually active in the organism, while recessive alleles were inactive, they now know of many cases where this is not true. We discuss some of these cases later in the chapter.

Mendel described and relied on complete dominance in sorting out his ratios and laws, but it is not the only kind of

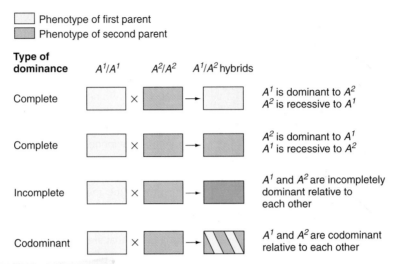

Figure 2.2 Different dominance relationships. The phenotype of the heterozygote defines the dominance relationship between pairs of alleles. Dominance is complete when the hybrid resembles one of the two pure-breeding parents. Dominance is incomplete when the hybrid resembles neither parent; its novel phenotype is usually intermediate between the parental phenotypes. Alleles are codominant when the hybrid shows the trait from both pure-breeding parents. Note that A^1 and A^2 indicate alleles of the same gene.

dominance he observed. Figure 2.2 shows two situations in which neither allele of a gene is completely dominant. Monohybrid crosses examining the effects of these alleles produce hybrids with phenotypes that differ from both parents. We now explain how these phenotypes arise.

In Incomplete Dominance, the F_1 Hybrid Resembles Neither Purebred Parent

A cross between pure late-blooming and pure early-blooming pea plants results in an F_1 generation that blooms in between the two extremes. This is just one of many examples of **incomplete dominance,** in which the hybrid does not resemble either pure-breeding parent. F_1 hybrids that differ from both parents often express a phenotype that is intermediate between those of the pure-breeding parents. Thus, with incomplete dominance, neither parental allele is dominant or recessive to the other; both contribute to the F_1 phenotype. Mendel observed plants that bloomed midway between two extremes when he cultivated various types of pure-breeding peas for his hybridization studies, but he did not pursue the implications. Blooming time was not one of the seven characteristics he chose to analyze in detail, almost certainly because in peas, the time of bloom was not as clear-cut as seed shape or flower color.

In many plant species, flower color serves as a striking example of incomplete dominance. With the tubular flowers of four o'clocks or the double-lipped floret clusters of snapdragons, for instance, a cross between pure-breeding red-flowered parents and pure-breeding white yields hybrids with pink blossoms, as if a painter had mixed red and white pigments to get pink (Fig. 2.3a). If allowed to self-pollinate, the F_1 pink-blooming plants produce F_2 progeny bearing red, pink, and white flowers in a ratio of 1:2:1 (Fig. 2.3b). This is the familiar *genotypic* ratio of an ordinary single-gene F_1 self-cross.

What is new is that because the heterozygotes look unlike either homozygote, the *phenotypic* ratios are an exact reflection of the genotypic ratios. The biochemical explanation for this type of incomplete dominance is that each allele of the gene under analysis specifies an alternative form of a protein molecule with an enzymatic role in pigment production. If the "white" allele does not give rise to functional enzyme, no pigment appears. Thus, in snapdragons and four o'clocks, two "red" alleles per cell produce a double dose of a red-producing enzyme, which generates enough pigment to make the flowers look fully red. In the heterozygote, one copy of the "red" allele per cell results in only enough pigment to make the flowers look pink. In the homozygote for the "white" allele, where there is no functional enzyme and thus no red pigment, the flowers appear white.

In Codominance, Alternative Traits Are Both Visible in the F_1 Hybrid

A cross between pure-breeding spotted lentils and pure-breeding dotted lentils produces heterozygotes that are both spotted and dotted (Fig. 2.4a). These F_1 hybrids illustrate a second significant departure from complete dominance. They look like both parents, which means that neither the "spotted" nor the "dotted" allele is dominant or recessive to the other. Since both traits show up equally in the heterozygote's phenotype, the alleles are termed **codominant.** Self-pollination of the spotted/dotted F_1 generation generates F_2 progeny in the ratio of 1 spotted:2 spotted/dotted:1 dotted. Once again, because the heterozygotes can be distinguished from both homozygotes, the phenotypic and genotypic ratios coincide.

Figure 2.2 illustrates the difference between codominance and incomplete dominance for phenotypes reflected in color differences. With codominance, the phenotypes of both pure-

(a)

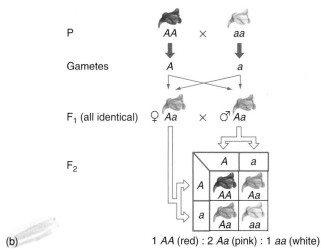

Figure 2.3 caption:

Figure 2.3   Pink flowers are the result of incomplete dominance. (a) Color differences in snapdragons reflect the activity of one pair of alleles. (b) The F₁ hybrids from a cross of pure-breeding red and white strains of snapdragons have pink blossoms, resembling neither parent. Flower colors in the F₂ appear in the ratio of 1 red: 2 pink:1 white. This ratio signifies that the alleles of a single gene determine these three colors.

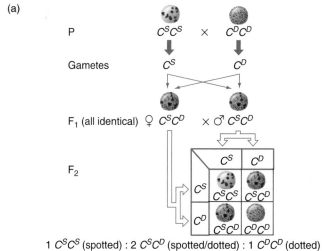

(a)

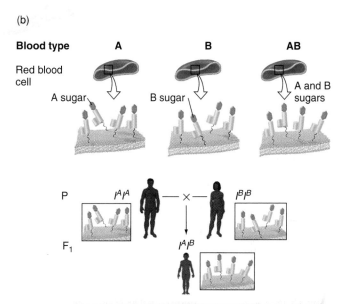

(b)

Figure 2.4   In codominance, F₁ hybrids display the traits of both parents. When alleles are codominant, each genotype has its own corresponding phenotype, so the F₂ phenotypic and genotypic ratios are identical. (a) A cross between pure-breeding spotted lentils and pure-breeding dotted lentils produces heterozygotes that are both spotted and dotted. (b) The I^A and I^B blood group alleles are codominant at both the molecular and organismal level because the red blood cells of an $I^A I^B$ heterozygote have both kinds of sugars at their surface.

breeding lines show up simultaneously in the F₁ hybrid. With incomplete dominance, neither pure-breeding trait appears in the hybrid.

In humans, some of the complex membrane-anchored molecules that distinguish different types of red blood cells exhibit codominance. For example, one gene (I) with alleles I^A and I^B controls the presence of a sugar polymer that protrudes from the red blood cell membrane. The alternative alleles each encode a slightly different form of an enzyme that causes production of a slightly different form of the complex sugar. In heterozygous individuals, the red blood cells carry both the I^A-determined and the I^B-determined sugars on their surface, whereas the cells of homozygous individuals display the products of either I^A or I^B alone (Fig. 2.4b).

It often happens that alleles are codominant at the molecular level, in terms of the protein they determine, even if that codominance is not readily apparent at the level of the whole organism. An example is sickle-cell syndrome. Homozygotes for the sickling allele have the disease; homozygotes for the normal allele do not. Heterozygotes are generally healthy under normal living conditions; thus the sickle-cell allele is recessive to the normal allele in its overall effect on well-being. But the red blood cells of a heterozygote carry both normal and sickling proteins; thus, the normal and sickling alleles are codominant at the molecular level.

This example illustrates that dominance relationships are defined in the context of the particular phenotype under analysis. When looked at from the point of view of different phenotypes, the same two alleles can display different dominance relationships. We return later to this important point.

Variations on Complete Dominance Do Not Negate Mendel's Law of Segregation

The dominance relations of a gene's alleles do not affect the alleles' transmission. Whether two alternative alleles of a single gene show complete dominance, incomplete dominance, or codominance depends in large part on the kinds of proteins determined by the alleles and the biochemical function of those proteins in the cell. These same phenotypic dominance relations, however, have no bearing on the segregation of the alleles during gamete formation. As Mendel proposed, cells

still carry two copies of each gene, and these copies—a pair of either similar or dissimilar alleles—segregate during gamete formation; fertilization then restores two alleles to each cell without reference to whether the alleles are the same or different. Variations in dominance relations thus do not detract from Mendel's laws of segregation. Rather, they reflect differences in the way gene products control the production of phenotypes, adding a level of complexity to the task of sorting out the visible results of gene transmission and inferring genotype from phenotype.

A Gene May Have More Than Two Alleles

Mendel analyzed "either-or" traits controlled by genes with two alternative alleles, but for many traits there are more than two alternatives. Human blood types provide an example. If a person with blood type A marries a person with blood type B, in some cases, it is possible for the couple to have a child that is neither A nor B, but a third blood type called O. The reason? The gene for the ABO blood types has three alleles: I^A, I^B, and i (Fig. 2.5). Allele I^A gives rise to blood type A by specifying an enzyme that adds sugar A; I^B results in blood type B by specifying an enzyme that adds sugar B; i does not produce a functional sugar-adding enzyme. Alleles I^A and I^B are both dominant to i, and blood type O is therefore a result of homozygosity for allele i.

Note in Fig. 2.5a that the A phenotype can arise from two genotypes, $I^A I^A$ or $I^A i$. The same is true for the B blood type, which can be produced by $I^B I^B$ or $I^B i$. But a combination of the two alleles $I^A I^B$ generates blood type AB.

We can draw several conclusions from these observations. First, as already stated, a given gene may have more than two, or **multiple alleles;** in our example, the series of alleles is denoted I^A, I^B, i.

Second, although the ABO blood group gene has 3 alleles, each person carries only 2 of the alternatives—$I^A I^A$, $I^B I^B$, $I^A I^B$, $I^A i$, $I^B i$, or ii. There are thus 6 possible ABO genotypes. Because each individual carries no more than two alleles for each gene, no matter how many alleles there are in a series, Mendel's law of segregation remains intact, since in any plant, person, or other sexually reproducing organism, the two alleles of a gene separate during gamete formation. Third, an allele is not inherently dominant or recessive; its dominance or recessiveness is always relative to a second allele. In other words, dominance relations are unique to a pair of alleles. In our example, I^A is completely dominant to i, but codominant with I^B. Given these dominance relations, the 6 genotypes possible with I^A, I^B, and i generate 4 different phenotypes: A, B, AB, and O. With this background, you can understand how a type A and a type B parent could produce a type O child: The parents must be $I^A i$ and $I^B i$ heterozygotes; the child receives an i allele from each parent and is thus type O.

An understanding of the genetics of the ABO system has had profound medical and legal repercussions. Matching ABO blood types is a prerequisite of successful blood transfusions, because people make antibodies to foreign blood cell molecules. A person whose cells carry only A molecules, for exam-

(a)

Genotypes	Corresponding Phenotypes: Type(s) of Molecule on Cell
$I^A I^A$ $I^A i$	A
$I^B I^B$ $I^B i$	B
$I^A I^B$	AB
ii	O

(b)

Blood Type	Antibodies in Serum
A	Antibodies against B
B	Antibodies against A
AB	No antibodies against A or B
O	Antibodies against A and B

Blood Type of Recipient	Donor Blood Type (Red Cells)			
	A	B	AB	O
A	+	−	−	+
B	−	+	−	+
AB	+	+	+	+
O	−	−	−	+

Figure 2.5   ABO blood types are determined by three alleles of one gene. (a) The 6 genotypes that produce the 4 blood group phenotypes result from all possible pairings of the three alleles of one gene: I^A, I^B, and i. (b) Blood serum contains antibodies against foreign blood cell molecules. Type AB individuals, however, make no antibodies against A or B type sugars because their own blood cell membranes carry both kinds of sugars. If a recipient's serum has antibodies against the sugars on a donor's red blood cells, the blood types of recipient and donor are incompatible and coagulation of red blood cells will occur during transfusions. In the bottom table, a plus (+) indicates compatibility and a minus (−) indicates incompatibility. In emergencies, a person with type O blood can serve as a donor to any recipient, because the cells in this type of blood will not coagulate upon contact with the blood cells of A, B, AB, or O individuals.

ple, produces anti-B antibodies; B people manufacture anti-A antibodies; AB individuals make neither type of antibody; and O individuals produce both anti-A and anti-B antibodies. These antibodies cause coagulation of cells displaying the foreign molecules (Fig. 2.5b). As a result, people with blood type O have historically been known as universal donors because their red blood cells carry no surface molecules that will stimulate an antibody attack, and people of blood type AB are considered universal recipients, because they make neither anti-A nor anti-B antibodies, which, if present, would target the surface molecules of incoming blood cells.

Information about ABO blood types can also be used as legal evidence in court, to exclude the possibility of paternity or criminal guilt. In a paternity suit, for example, if the mother is type A and her child is type B, logic dictates that the I^B allele must have come from the father, whose genotype may be $I^A I^B$, $I^B I^B$, or $I^B i$. In 1944, actress Joan Barry (phenotype A) sued Charlie Chaplin (phenotype O) for support of a child (phenotype B) whom she claimed he fathered. The scientific evidence contradicting her claim (the father carried an I^B allele and Chaplin was apparently ii) was admissible in court but did not convince the jury, and Chaplin had to pay. Today, with a public that tends to be much more scientifically literate, it is likely that members of the jury would have understood the scientific evidence, rejected Barry's claim, and found Chaplin innocent.

Lentils offer another example of multiple alleles. A gene for seed coat pattern has five alleles: spotted, dotted, clear (pattern absent), and two types of marbled. Reciprocal crosses between pairs of pure-breeding lines of all patterns (marbled-1 × marbled-2, marbled-1 × spotted, marbled-2 × spotted, and so forth) have clarified the dominance relations of all possible pairs of the alleles to reveal a **dominance series** in which alleles are listed in order from most dominant to most recessive. For example, crosses of marbled-1 with marbled-2, or of marbled-1 with spotted or dotted or clear, produce the marbled-1 phenotype in the F_1 generation and a ratio of 3 marbled-1 to 1 of any of the other phenotypes in the F_2. This indicates that the marbled-1 allele is dominant to each of the other four alleles. Analogous crosses with the remaining four phenotypes reveal the dominance series shown in Fig. 2.6. Recall that dominance relations are meaningful only when comparing two alleles; an allele, such as marbled-2, can be recessive to a second allele (marbled-1) but dominant to a third and fourth (dotted and clear). The fact that all tested pairings of lentil seed coat pattern alleles yielded a 3:1 ratio in the F_2 generation (except for spotted × dotted, which yielded the 1:2:1 phenotypic ratio reflective of codominance) indicates that these lentil seed coat patterns are determined by different alleles of the same gene.

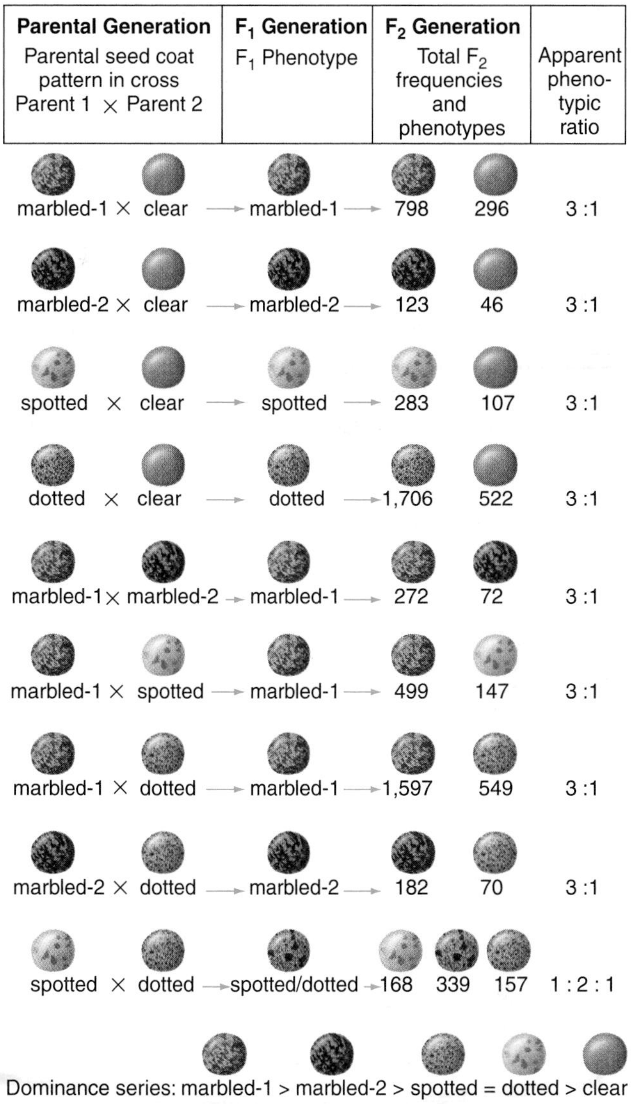

In some multiple allelic series, each allele is codominant with every other allele, and every distinct genotype therefore produces a distinct phenotype. This happens particularly with traits detectable only at the molecular level. An extreme example is the group of three major genes that encode a class of cell surface molecules in humans and other mammals known as **histocompatibility antigens.** Carried by all of the body's

Parental Generation	F₁ Generation	F₂ Generation		
Parental seed coat pattern in cross Parent 1 × Parent 2	F₁ Phenotype	Total F₂ frequencies and phenotypes		Apparent phenotypic ratio
marbled-1 × clear	marbled-1	798	296	3 :1
marbled-2 × clear	marbled-2	123	46	3 :1
spotted × clear	spotted	283	107	3 :1
dotted × clear	dotted	1,706	522	3 :1
marbled-1 × marbled-2	marbled-1	272	72	3 :1
marbled-1 × spotted	marbled-1	499	147	3 :1
marbled-1 × dotted	marbled-1	1,597	549	3 :1
marbled-2 × dotted	marbled-2	182	70	3 :1
spotted × dotted	spotted/dotted	168	339	157	1 :2 :1

Dominance series: marbled-1 > marbled-2 > spotted = dotted > clear

Figure 2.6 How to establish the dominance relations between multiple alleles. Pure-breeding lentils with different seed coat patterns (marbled-1, marbled-2, spotted, dotted, and clear) are crossed in pairs, and the F₁ progeny are self-fertilized to produce an F₂ generation. The 3:1 or 1:2:1 ratios of phenotypes among the F₂ progeny from all of these crosses indicate that different alleles of a single gene determine all the phenotypes. The phenotypes of the F₁ hybrids establish the dominance relationships between the alleles, as shown in the series at the bottom of the figure. Note that spotted and dotted alleles are codominant to each other, but each is recessive to the marbled alleles and dominant to clear.

cells, except the red blood cells and sperm, histocompatibility antigens play a critical role in stimulating a proper immune response that destroys intruders (viral or bacterial, for example) while leaving the body's own tissues intact. Because each of the three major histocompatibility genes (called *HLA-A, HLA-B,* and *HLA-C* in humans) has between 20 and 100 alleles (and these are just the ones geneticists have already discovered), the number of possible allelic combinations creates a powerful potential for the phenotypic variation of cell surface molecules. Other than identical (that is, monozygotic) twins, no two people are likely to carry the same array of cell surface molecules.

Mutations Are the Source of New Alleles

How do the multiple alleles of an allelic series arise? The answer is that chance alterations of the genetic material, known as **mutations,** arise spontaneously in nature. Once they occur in gamete-producing cells, they are faithfully inherited. Mutations that have phenotypic consequences can be counted, and such counting reveals that they occur at low frequency. The frequency of gametes carrying a mutation in a particular gene varies anywhere from 1 in 10,000 to 1 in 1,000,000. This range of rates exists because different genes have different mutation rates. Mutations make it possible to follow gene transmission. If, for example, a mutation specifies an alteration in an enzyme that normally produces yellow so that it now makes green, the new phenotype (green) will make it possible to recognize the new mutant allele. In fact, it takes at least two alleles, that is, some form of variation, to "see" the transmission of a gene. Thus, in segregation studies, geneticists can only analyze genes with variants; they have no way of following a gene that comes in only one form. If all peas were yellow, Mendel would not have been able to decipher the transmission patterns of the gene for the seed color trait. We discuss mutations in greater detail in Chapter 6.

The survival of a new allele within a population—that is, within a group of organisms of the same species that inhabit the same area—depends on many factors, particularly, its contribution to the reproductive success of the organism. Alleles that confer a survival advantage allow individuals carrying them to produce more offspring; these alleles will tend to increase in the population. Other alleles are deleterious and are lost from the population. Still other alleles are neutral; more than 99.9% of these are eventually lost from a large population because they provide no selective advantage, but occasionally, purely by chance, one may become established.

Since each organism carries 2 copies of every gene, you can calculate the number of copies of a gene in a given population by multiplying the number of individuals by 2. Each allele of the gene accounts for a percentage of the total number of gene copies, and that percentage is known as the **allele frequency.** An allele whose frequency is greater than 1% is by definition a **wild-type allele,** often designated by a superscript plus sign ($^+$). An allele with a frequency of less than 1% is considered a **mutant allele.** (A mutation is a newly arisen mutant allele.) In mice, for example, one of the main genes determining coat color is the *agouti* gene. The wild-type allele (A) produces fur with each hair having yellow and black bands that blend together from a distance to give the appearance of dark gray, or agouti. Researchers have identified in the laboratory 14 distinguishable mutant alleles for the *agouti* gene. One of these (a^t) is recessive to the wild type and gives rise to a black coat on the back and a yellow coat on the belly; another

(a) is also recessive to A and produces a pure black coat (Fig. 2.7). In nature, wild-type agoutis (AA) survive to reproduce, while very few black-backed or pure black mutants ($a^t a^t$ or aa) do so because their dark coat makes it hard for them to evade the eyes of predators. As a result, A is present at a frequency of much more than 99% and is thus the only wild-type allele in mice for the *agouti* gene. A gene with only one wild-type allele is **monomorphic.**

In contrast, some genes have more than one wild-type allele, which makes them **polymorphic.** For example, in the ABO blood type system, all three alleles—I^A, I^B, and i—have a frequency greater than 1%. They are thus all wild-type alleles and the gene is polymorphic.

A rather unusual mechanism leading to the proliferation of many different alleles occurs in the mating systems of wild species of tomatoes and petunias. Evolution of an "incompatibility" gene whose alleles determine acceptance or rejection of pollen has allowed these plants to prevent self-fertilization and promote out-breeding. In this form of incompatibility, a plant cannot accept pollen carrying an allele identical to either of its own incompatibility alleles. If, for example, pollen carrying allele S_1 of the incompatibility gene lands on the stigma of a plant that also carries S_1 as one of its incompatibility alleles, a pollen tube will not grow (Fig. 2.8). As a result, every plant is heterozygous for the incompatibility gene (since the pollen grain and female reproductive organs needed to form the plant cannot share alleles). Plants carrying rare alleles (that have arisen relatively recently by mutation and are not present in many other plants) will be able to send pollen to and receive pollen from most of the other plants in their population. In some species with this type of mating system, geneticists have detected as many as 92 alleles for the incompatibility gene, and there are probably many more alleles yet to be detected. Because this incompatibility mechanism encourages the proliferation of new mutants, there are numerous wild-type alleles as well as many mutants. This is an extreme case of multiple alleles, not seen with most genes.

One Gene May Contribute to Several Visible Characteristics

Mendel derived his laws from studies in which one gene determined one trait; but, always the careful observer, he himself noted some possible departures. In listing the traits selected for his pea experiments, he remarked that specific seed coat colors are always associated with specific flower colors.

The phenomenon of a single gene determining a number of distinct and seemingly unrelated characteristics is known as **pleiotropy.** Since geneticists now know that each gene determines a specific protein and that each protein can have a cascade of effects on an organism, we can understand how pleiotropy arises. Among the aboriginal Maiori people of New Zealand, for example, many of the men develop frequent respiratory problems and are also sterile. Researchers have found that the fault lies with the recessive allele of a single gene. The gene's normal dominant allele specifies a protein

(a)

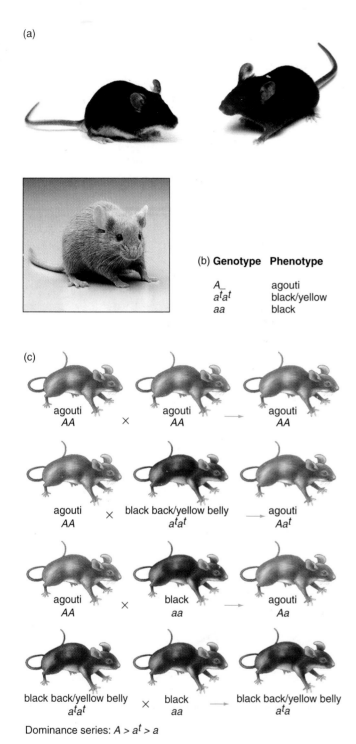

(b) **Genotype   Phenotype**

$A\_$ agouti
$a^t a^t$ black/yellow
aa black

(c)

agouti
AA ×   agouti
AA →   agouti
AA

agouti
AA ×   black back/yellow belly
$a^t a^t$ →   agouti
Aa^t

agouti
AA ×   black
aa →   agouti
Aa

black back/yellow belly
$a^t a^t$ ×   black
aa →   black back/yellow belly
$a^t a$

Dominance series: $A > a^t > a$

Figure 2.7   The mouse agouti gene: One wild-type allele, many mutant alleles. (a) Black-backed, yellow-bellied; black; and agouti mice. (b) Genotypes and corresponding phenotypes for the agouti, black/yellow, and black alleles of the *agouti* gene. (c) Four representative crosses between pure-breeding lines make it possible to arrange these three alleles in a dominance series. Interbreeding of the F₁ hybrids (not shown) yields 3:1 phenotypic ratios of F₂ progeny, indicating that *A, a^t,* and *a* are in fact alleles of the same gene.

necessary for the action of cilia and flagella, both of which are hairlike structures extending from the surfaces of some cells. In men who are homozygous for the recessive allele, however, cilia that normally clear the airways fail to work effectively and flagella that normally propel sperm fail to do their job. Thus, one gene determines a protein that indirectly affects both respiratory function and reproduction.

Some Alleles May Cause Lethality

A significant variation of pleiotropy occurs in alleles that not only produce a visible phenotype but also affect viability. Mendel assumed that all genotypes are equally viable such that representatives of each genotype have an equal rate of survival. If this were not true, and a large percentage of, say, homozygotes for a particular allele died before germination or birth, you would not be able to count them after birth, and this would alter the 1:2:1 genotypic ratios and the 3:1 phenotypic ratios predicted for the F₂ generation.

Consider the inheritance of coat color in mice. As mentioned earlier, wild-type agouti (AA) animals have black and yellow striped hairs that appear dark gray to the eye. One of the 14 mutant alleles of the *agouti* gene gives rise to mice with a much lighter, almost yellow color. When inbred AA mice are mated to yellow mice, one always observes a 1:1 ratio of the two coat colors among the offspring (Fig. 2.9a). From this result, we can draw three conclusions: (1) All yellow mice must carry the *agouti* allele even though they do not express it; (2) yellow is therefore dominant to agouti; and (3) all yellow mice are heterozygotes. (Note again that dominance and recessiveness are defined in the context of each pair of alleles. Even though agouti is dominant to the a^t and a mutations for black coat color, it can still be recessive to the yellow coat color allele.) If we designate the allele for yellow as A^y, the yellow mice in the above cross are $A^y A$ heterozygotes, and the agoutis, AA homozygotes. So far, no surprises. But a mating of yellow to yellow produces a skewed phenotypic ratio of 2 yellow mice to 1 agouti (Fig. 2.9b). Among these progeny, matings between agouti mice show that the agoutis are all pure-breeding and therefore homozygotes as expected. There are, however, no pure-breeding yellow mice among the progeny. When the yellow mice are mated to each other, they unfailingly produce 2/3 yellow and 1/3 agouti offspring, a ratio of 2:1; they must therefore be heterozygotes. In short, one can never obtain pure-breeding yellow mice.

How can we explain this phenomenon? The Punnett square in Fig. 2.9b suggests one possibility: Two copies of the A^y allele prove fatal to the animal carrying them, whereas one copy of the allele produces a yellow coat. This means that the A^y allele affects two different traits. It is dominant to A in the determination of coat color, but acts as a **recessive lethal** allele in its effect on survival.

Because the A^y allele is dominant for yellow coat color, it is easy to detect carriers of this particular recessive lethal in mice, but such is not the case for the vast majority of recessive lethal mutations that do not simultaneously show a visible

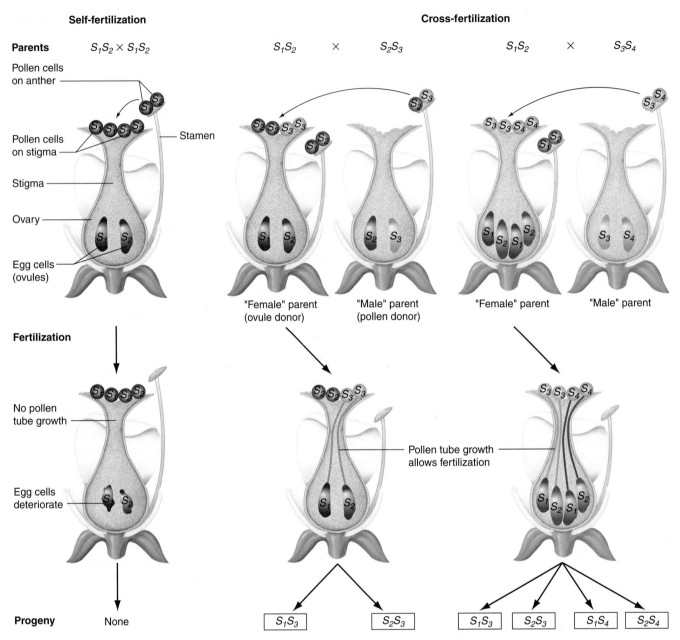

Figure 2.8 Plant incompatibility systems promote outbreeding and allele proliferation. Self-fertilization does not occur in wild petunia and wild tomato plants. A pollen grain carrying an allele of the self-incompatibility gene that is identical to either of the two alleles carried by a potential female parent is unable to grow a pollen tube; as a result, fertilization cannot take place. Because all the pollen grains produced by any one plant have one of the two alleles carried by the female reproductive parts of the same plant, self-fertilization is impossible.

dominant phenotype for some other trait. Lethal mutations can arise in many different genes, and as a result, most animals, including humans, carry some recessive lethal mutations. Such mutations usually remain "silent," except in rare cases of homozygosity, which in people are often caused by consanguineous marriages. If a mutation produces an allele that prevents production of a crucial molecule, homozygous individuals would not make any of the vital molecule and would not survive. Heterozygotes, by contrast, with only one copy of the deleterious mutation and one wild-type allele, would be able to produce 50% of the wild type amount of the normal

molecule; this is usually sufficient to sustain normal cellular processes such that life goes on.

In the preceding discussion, we have described recessive alleles that result in the death of homozygotes prenatally, *in utero*. With some mutations, however, homozygotes may survive beyond germination or birth and then die later from the deleterious consequences of the genetic defect. An example is seen in human infants with Tay-Sachs disease. The seemingly normal newborns remain healthy for five to six months, but then develop blindness, paralysis, mental retardation, and other symptoms of a deteriorating nervous sys-

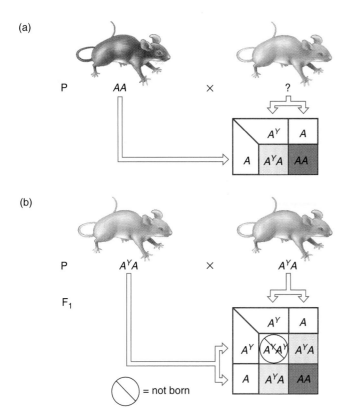

Figure 2.9 *A^Y*: **A recessive lethal allele that also produces a dominant coat color phenotype.** (a) A cross between inbred agouti mice and yellow mice yields a 1:1 ratio of yellow:agouti progeny. This indicates that the yellow mice must be A^YA heterozygotes, and that in terms of coat color, A^Y (for yellow) is dominant to A (for agouti). (b) When yellow mice mate with other yellow mice, they do not breed true. The 2:1 ratio of yellow:agouti progeny indicates that the A^Y allele is a recessive lethal: Homozygotes for this allele die *in utero* and thus do not show up among the progeny.

tem; the disease usually proves fatal by the age of six. Tay-Sachs disease results from the absence of an active lysosomal enzyme called hexosaminidase A, whose lack leads to the accumulation of a toxic waste product inside nerve cells. The approximate incidence of Tay-Sachs among live births is 1/35,000 worldwide, but it is 1/3500 among Jewish people of Eastern European descent. Reliable tests that detect carriers, in combination with genetic counseling and educational programs, have all but eliminated the disease in the United States.

Table 2.1 summarizes Mendel's basic assumptions about dominance, the number and viability of one gene's alleles, and the effects of each gene on phenotype, and compares them with the extensions contributed by his twentieth-century successors. Through carefully controlled monohybrid crosses, these later geneticists analyzed the transmission patterns of the alleles of single genes, challenging and then confirming the law of segregation.

A Comprehensive Example: Sickle-Cell Syndrome Illustrates Many Extensions to Mendel's Analysis of Single-Gene Inheritance

Sickle-cell syndrome can serve as a comprehensive example of most of the points covered in this section on single-gene traits. The disease is the result of a faulty hemoglobin molecule. Hemoglobin is composed of two types of polypeptide chains, alpha (α) globin and beta (β) globin, each specified by a different gene: $Hb\alpha$ for α globin and $Hb\beta$ for β globin. Normal red blood cells are packed full of thousands upon thousands of hemoglobin molecules, each of which picks up oxygen in the lungs and transports it to all the body's tissues.

Multiple Alleles

The β-globin gene has a normal wild-type allele ($Hb\beta^A$) that gives rise to fully functional β globin, and close to 400 mutant alleles that have been identified so far. The most common of

TABLE 2.1   For Traits Determined by One Gene: Extensions to Mendel's Analysis Explain Alterations of the 3:1 Ratio

What Mendel Described	Extension	Extension's Effect on Heterozygous Phenotype	Extension's Effect on Ratios Resulting from a $F_1 \times F_1$ Cross
Complete dominance	Incomplete dominance Codominance	Unlike either homozygote	Phenotypes coincide with genotypes in a ratio of 1:2:1
Two alleles	Multiple alleles	Multiplicity of phenotypes	A series of 3:1 ratios
All alleles are equally viable	Recessive lethal alleles	No effect	2:1 instead of 3:1
One gene determines one trait	Pleiotropy: one gene influences several traits	Several traits affected in different ways, depending on dominance relations	Different ratios, depending on dominance relations for each affected trait

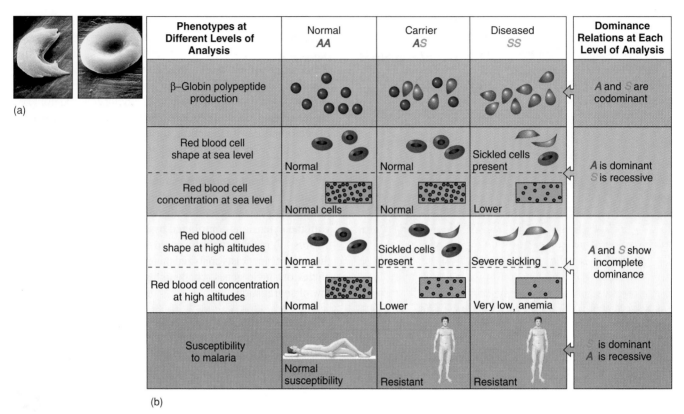

(a)

(b)

Figure 2.10   Pleiotropy of sickle-cell syndrome: Dominance relations vary with the phenotype under consideration. (a) A normal red blood cell (right) is easy to distinguish from the sickled cell in the scanning electron micrograph at the left. (b) Different levels of analysis identify various phenotypes. Dominance relationships between the *S* and *A* alleles of the *Hb*β gene vary with the phenotype under consideration. For any one phenotype, such as red blood cell shape or concentration, dominance relationships sometimes change with the environment.

these, *Hb*β^S, specifies an abnormal polypeptide that causes sickling among red blood cells (Fig. 2.10a). For simplicity, in the remainder of this discussion, we refer to the wild-type β-globin allele as *A* and the mutant sickling allele of the same *Hb*β gene as *S*.

Pleiotropy

The *S* allele of the β-globin gene affects more than one trait (Fig. 2.10b). Hemoglobin molecules in the red blood cells of homozygous *SS* individuals behave aberrantly after releasing their oxygen. Instead of remaining soluble in the cytoplasm, they aggregate to form long fibers that deform the red blood cell from a normal biconcave disk to a sickle shape (see Fig. 2.10a). The deformed cells clog the small blood vessels, reducing oxygen flow to the tissues and giving rise to muscle cramps, shortness of breath, and fatigue. The sickled cells are also very fragile and easily broken. Consumption of fragmented cells by phagocytic white blood cells leads to a low red blood cell count, a condition called anemia. On the positive side, *SS* homozygotes are resistant to malaria, because although the organism that causes the disease, *Plasmodium falciparum,* can multiply with impunity in normal red blood cells, it cannot do so in cells that sickle. Infection by *P. falciparum* causes sickle-shaped cells to break down before the organism has a chance to multiply.

Recessive Lethality

People who are homozygous for the recessive *S* allele often develop heart failure because of stress on the circulatory system. Many sickle-cell sufferers die in childhood, adolescence, or early adulthood.

Different Dominance Relations

Comparisons of heterozygous carriers of the sickle-cell allele—individuals whose cells contain one *A* and one *S* allele—with homozygous *AA* (normal) and homozygous *SS* (diseased) individuals make it possible to distinguish different dominance relations for different phenotypic aspects of sickle-cell syndrome. At the molecular level, the production of β globin, both alleles are expressed, and the *A* and *S* alleles are *codominant.* At the cellular level, in their effect on red blood cell shape, the *A* and *S* alleles show *incomplete dominance.* Although under normal oxygen conditions, the great majority of a heterozygote's red blood cells have the normal biconcave shape, when oxygen levels drop, sickling occurs in some cells. All *AS* cells, however, are resistant to malaria because like the *SS* cells described above, they break down before the malarial organism has a chance to reproduce. Thus for the trait of resistance to malaria, the *S* allele is *dominant* to the *A* allele. But luckily for the heterozygote, for the phenotype of anemia, the *S* allele is *recessive* to *A*. A corollary of this observation is that in its effect on general health under normal environmental conditions and its effect on red blood cell count, the *A* allele is *dominant* to *S*. Thus, for the β-globin gene, as for many other genes, dominance and recessiveness are not an inherent quality of alleles in isolation; rather, they are specific to each pair of alleles and to the level of physiology at which the phenotype is examined.

In the 1940s, the incomplete dominance of the *A* and *S* alleles in the expression of red blood cell shape had significant repercussions for certain soldiers who fought in World War II. Aboard transport planes flying troops across the Pacific, several heterozygous carriers suffered sickling crises similar to those usually seen in *SS* homozygotes. The explanation is as follows. The heterozygous red blood cells of a carrier produce both normal and abnormal hemoglobin molecules. At sea level, these molecules together deliver sufficient oxygen, although less than the normal amount, to the body's tissues; but with a decrease in the amount of oxygen available at the high-flying altitudes, the hemoglobin picks up less oxygen, the rate of red blood cell sickling increases, and symptoms of the disease occur.

EXTENSIONS TO MENDEL FOR MULTIFACTORIAL INHERITANCE

Although some traits are indeed determined by allelic variations of a single gene, the vast majority of common traits in all organisms are *multifactorial,* arising from the action of two or more genes, or from interactions between genes and the environment. In genetics, the term *environment* has an unusually broad meaning, which encompasses all aspects of the outside world an organism comes into contact with. These include temperature, diet, and exercise as well as the uterine environment before birth (minute chemical differences, for example, can make a difference in development) and the psychological environment afterward (the amount of stress, for example, can affect various levels of phenotype).

In this section, we examine how geneticists again used breeding experiments and the guidelines of Mendelian ratios to analyze the complex network of interactions that give rise to multifactorial traits.

Two Genes Can Interact to Determine One Trait

Two genes can interact in several ways to determine a single trait, such as the color of a flower, a seed coat, a chicken's feathers, or a dog's fur, and each type of interaction produces its own signature of phenotypic ratios. Throughout our discussion of how two genes interact to affect one trait, we use big *A* and little *a* to represent alternative alleles of the first gene and big *B* and little *b* for those of the second gene.

Novel Phenotypes Can Emerge from the Combined Action of the Alleles of Two Genes

In the chapter opening we described a mating of tan and gray lentils that produced a uniformly brown F₁ generation and

then, an F₂ generation containing lentils with brown, tan, gray, and green seed coats. An understanding of how this can happen emerges from experimental results demonstrating that the ratio of the four F₂ colors is 9 brown:3 tan:3 gray:1 green (Fig. 2.11a). Recall from Chapter 1 that this is the same ratio Mendel observed in his analysis of the F₂ generations from dihybrid crosses following two independently assorting genes. In Mendel's studies, each of the four classes consisted of plants that expressed a combination of two unrelated traits. With lentils, however, we are looking at a single trait—seed coat color. The simplest explanation for the parallel ratios is that a combination of genotypes at two independently assorting genes interacts to produce the phenotype of seed coat color in lentils.

Results obtained from self-crosses with the various types of F₂ lentil plants support this explanation. Self-crosses of F₂ green individuals show that they are pure-breeding, producing an F₃ generation that is entirely green. Tans generate either all tan or tan plus green; grays similarly produce all gray or gray plus green; and browns either breed true, or assort into two colors (either tan plus brown or gray plus brown) or into all four colors (Fig. 2.11b). The two-gene hypothesis explains why there is

- Only 1 green genotype: pure-breeding *aa bb,* but
- 2 types of tans: pure-breeding *AA bb* as well as tan- and green-producing *Aa bb;*
- 2 types of grays: pure-breeding *aa BB* and gray- and green-producing *aa Bb,* and
- 4 types of browns: true-breeding *AA BB,* brown- and tan-producing *AA Bb,* brown- and gray-producing *Aa BB,* and *Aa Bb* dihybrids that give rise to plants producing lentils of all four colors.

In short, for the two genes that determine seed coat color, both dominant alleles must be present to yield brown (*A- B-*); the dominant allele of one gene produces tan (*A- bb*); the dominant allele of the other specifies gray (*aa B-*); and the complete absence of dominant alleles (that is, the double recessive) yields green (*aa bb*). Thus, the 4 color phenotypes arise from 4 **genotypic classes,** with each class defined in terms of the presence or absence of the dominant alleles of two genes: (1) both present (*A- B-*), (2) one present (*A- bb*), (3) the other present (*aa B-*), and (4) neither present (*aa bb*). Note that the *A-* notation means that the second allele of this gene can be either *A* or *a,* while *B-* denotes a second allele of either *B* or *b.* Note also that only with a two-gene system in which the dominance and recessiveness of alleles at both genes is complete can the 9 different genotypes of the F₂ generation be categorized into the 4 genotypic classes described above. With incomplete dominance or codominance, the F₂ genotypes could not be grouped together in this simple way, as they would give rise to more than 4 phenotypes.

Further crosses between plants carrying lentils of different colors confirmed the two-gene hypothesis (review Fig. 2.11c).

(a)

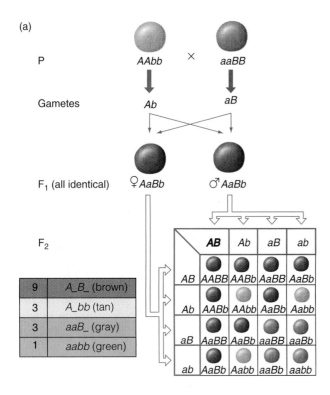

9	A_B_ (brown)
3	A_bb (tan)
3	aaB_ (gray)
1	aabb (green)

(b) Self-pollination of the F₂ to produce an F₃

Phenotypes of F₂ individual	Observed F₃ phenotypes	Expected proportion of F₂ population*
Green	Green	1/16
Tan	Tan	1/16
Tan	Tan, Green	2/16
Gray	Gray, Green	2/16
Gray	Gray	1/16
Brown	Brown	1/16
Brown	Brown, Tan	2/16
Brown	Brown, Gray	2/16
Brown	Brown, Gray, Tan, Green	4/16

*This 1:1:2:2:1:1:2:2:4 F₂ genotypic ratio corresponds to a 9 brown:3 tan:3 gray:1 green F₂ phenotypic ratio.

(c) Sorting out the dominance relations by select crosses

Seed coat color of parents	F₂ phenotypes and frequencies	Ratio
Tan × green	231 tan, 85 green	3:1
Gray × green	2586 gray, 867 green	3:1
Brown × gray	964 brown, 312 gray	3:1
Brown × tan	255 brown, 76 tan	3:1
Brown × green	57 brown, 18 gray, 13 tan, 4 green	9:3:3:1

Figure 2.11 How two genes interact to produce new colors in lentils. (a) In a cross of pure-breeding tan and gray lentils, all the F₁ hybrids are brown, but four different phenotypes appear among the F₂s. Although the color of the seed coat is perceived as a single trait, the 9:3:3:1 ratio of phenotypes in the F₂ generation suggests that seed-coat color is determined by two independently segregating genes in a dihybrid cross. (b) Expected results of selfing individual F₂s of the indicated phenotypes to produce an F₃ generation, if seed-coat color is determined by the interaction of two genes. The actual data from such an experiment (not shown) is very close to these expectations. The third column shows the expected proportion of the F₂ population that would produce the observed F₃ phenotypes. (c) Other two-generation crosses involving pure-breeding parental lines also support the two-gene hypothesis. In this chart, the F₁ hybrid generation has been omitted. Note that starting with brown and green parents yields the same F₂ results as starting with tan and gray parents (shown in part [a]). This result is expected because the genotype of the F₁s produced by both crosses would be identical, namely *Aa Bb*.

Thus, the 9:3:3:1 phenotypic ratio of brown to tan to gray to green in an F₂ descended from pure-breeding tan and pure-breeding gray lentils tells us not only that two genes assorting independently interact to produce the seed coat color, but also that each genotypic class—*A- B-*, *A- bb*, *aa B-*, and *aa bb*—determines a particular phenotype.

This is not always the case. In some two-gene interactions, the four F₂ genotypic classes defined by Mendel produce fewer than four observable phenotypes, because some of the phenotypes include two or more genotypic classes. For example, in the first decade of the twentieth century, William Bateson conducted a cross between two lines of pure-breeding white-flowered sweet peas (Fig. 2.12). Quite unexpectedly, all of the F₁ progeny were purple. Self-pollination of these novel hybrids produced a ratio of 9 purple to 7 white in the F₂ generation. The explanation? Two genes work in tandem to produce purple sweet pea flowers, and a dominant allele of both genes must be present to produce that color. A simple biochemical explanation for this type of **complementary gene action** is shown in Fig. 2.12c. Because it takes two enzymes catalyzing two separate biochemical reactions to change a col-

orless precursor into a colorful pigment, only the *A- B-* genotypic class, which produces active forms of both required enzymes, can generate colored flowers; the other three genotypic classes (*A- bb*, *aa B-*, and *aa bb*) become grouped together with respect to phenotype since they do not specify functional forms of one or the other requisite enzyme and thus give rise to no color, which is the same as white (review the Fast Forward box "Genes Encode Proteins" in Chapter 1). It is easy to see how the 7 part of the 9:7 ratio encompasses the 3:3:1 of the 9:3:3:1 ratio of two genes in action. The 9:7 ratio is the phenotypic signature of this type of complementary gene interaction in which the dominant alleles of two genes acting together (*A- B-*) produce color or some other trait, while the other three genotypic classes (*A- bb*, *aa B-*, and *aa bb*) do not.

In Epistasis, One Gene's Alleles Mask the Effects of Another Gene's Alleles

In some gene interactions, the four Mendelian genotypic classes produce fewer than four observable phenotypes because one gene masks the phenotypic effects of another. An example is seen in the sleek, short-haired coat of Labrador retrievers, which

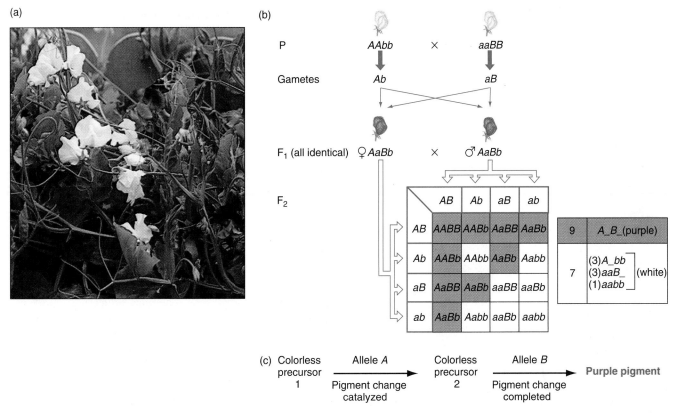

(c) Colorless precursor 1 — Allele A → Colorless precursor 2 — Allele B → **Purple pigment**

Pigment change catalyzed   Pigment change completed

Figure 2.12 Complementary gene action generates color in sweet peas. (a) White and purple sweet pea flowers. (b) The 9:7 ratio of purple to white F_2 plants indicates that at least one dominant allele for each gene is necessary for the development of purple color. (c) A possible biochemical explanation for this kind of complementary gene action: Enzymes specified by the dominant alleles of the two genes may both be necessary for completion of a biochemical pathway for pigment production.

can be black, chocolate brown, or golden yellow. Which color shows up depends on the allelic combinations of two independently assorting coat color genes (Fig. 2.13a). The dominant *B* allele of the first gene determines black, while the recessive *bb* homozygote is brown. With the second gene, the dominant *E* allele has no visible effect on black or brown coat color, but a double dose of the recessive *ee* allele hides the effect of any combination of the black or brown alleles to yield gold. A gene interaction in which the effects of an allele at one gene hide the effects of alleles at another gene is known as **epistasis;** the allele that is doing the masking (in this case, the *e* allele of the *E* gene) is **epistatic** to the gene that is being masked. In our example, where homozygosity for a recessive *e* allele of the second gene is required to hide the effects of another gene, the masking phenomenon is called **recessive epistasis** (because the allele causing the epistasis is recessive), and the recessive *ee* homozygote is considered epistatic to any allelic combination at the first gene.

Let's look at the phenomenon in greater detail. A cross between pure-breeding black retrievers (*BB EE*) and one type of pure-breeding golden retrievers (*bb ee*) creates an F_1 generation of dihybrid black retrievers (*Bb Ee*). Crosses between these F_1 dihybrids produces an F_2 generation with 9 black dogs (*B- E-*) for every 3 brown (*bb E-*) and 4 gold (*—ee*) (see Fig. 2.13a). Note that there are only three phenotypic classes because the two genotypic classes without a dominant *E*

allele—the 3 *B- ee* and the 1 *bb ee*—combine to produce golden phenotypes. The telltale ratio of recessive epistasis is thus 9:3:4, with the 4 representing a combination of 3 (*B- ee*) + 1 (*bb ee*). Because the *ee* genotype completely masks the influence of the other gene for coat color, you cannot tell by looking at a golden Labrador what its genotype is for the black or brown (*Bb*) gene.

An understanding of recessive epistasis made it possible to resolve an intriguing puzzle in human genetics. In some very rare cases, two parents who appear to have blood type O, and thus genotype *ii,* may produce a child that is either phenotype A (genotype $I^A i$) or phenotype B (genotype $I^B i$). This phenomenon occurs because an extremely rare trait, called the Bombay phenotype after its discovery in Bombay, India, superficially resembles phenotype O. As Fig. 2.13b shows, the Bombay phenotype actually arises from homozygosity for a mutant recessive allele (*hh*) of a second gene that masks the effects of any ABO alleles that might be present.

Here's how it works at the molecular level. In the construction of the red blood cell surface molecules that determine blood type, type A individuals make an enzyme that adds polysaccharide A onto a base consisting of a sugar polymer known as substance H; type B individuals make an altered form of the enzyme that adds polysaccharide B onto the base; and type O individuals make neither A-adding nor B-adding enzyme and

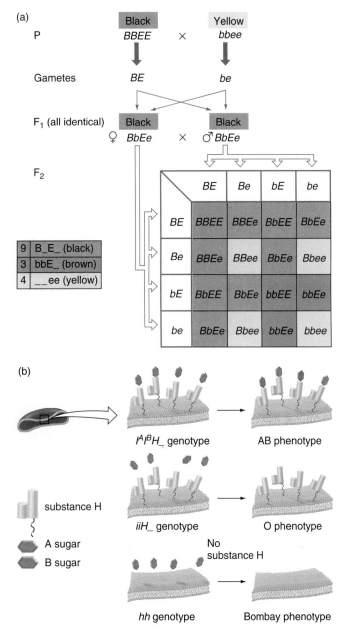

Figure 2.13 Recessive epistasis: Coat color in Labrador retrievers and a rare human blood type. (a) Golden Labrador retrievers are homozygous for the recessive *e* allele. Because this *ee* genotype masks the effects of the *B* coat color gene, golden retrievers may be of any genotype—*BB, Bb,* or *bb*—at this other gene. In *E-* dogs, a *B-* genotype produces black and a *bb* genotype produces brown. (b) Homozygosity for the *h* Bombay allele is epistatic to the *I* gene determining ABO blood types. *hh* individuals fail to produce substance H, which is needed for the addition of A or B sugars at the surface of red blood cells. The resulting Bombay phenotype resembles the type O phenotype regardless of which alleles of the *I* gene are present.

thus have an exposed substance H in the membranes of their red blood cells. All people of A, B, or O phenotype carry at least one dominant wild-type *H* allele for the second gene and thus produce some substance H. In contrast, the rare Bombay-phenotype individuals, with genotype *hh* for the second gene,

do not make substance H at all, so even if they make an enzyme that would add A or B to this polysaccharide base, they have nothing to add it onto; as a result, they appear to be type O. For this reason homozygosity for the recessive *h* allele of the H-substance gene masks the effects of the *ABO* gene, making the *hh* genotype epistatic to any combination of $I^A I^B i$ alleles. A person who carries I^A, I^B, or both I^A and I^B but is also an *hh* homozygote for the H-substance gene may appear to be type O, but he or she will be able to pass along an I^A or I^B allele in sperm or egg. The offspring receiving, let's say, an I^A allele for the ABO gene and a recessive *h* allele for the H-substance gene from its mother plus an *i* allele and a dominant *H* allele from its father would have blood type A (genotype $I^A i$, *Hh*) even though neither of its parents is phenotype A or AB.

Epistasis can also be caused by a dominant allele. In summer squash, two genes influence the color of the fruit (Fig. 2.14a). With one gene, the dominant allele (*A-*) determines yellow while homozygotes for the recessive allele (*aa*) are green. A second gene's dominant allele (*B-*) produces white, while *bb* fruit may be either yellow or green, depending on the genotype of the first gene. In the interaction between these two genes, the presence of *B* hides the effects of either *A-* or *aa,* producing white fruit, and *B-* is thus epistatic to any genotype of the *Aa* gene. The recessive *b* allele has no effect on fruit color determined by the *Aa* gene. In a self-cross of white F_1 dihybrids (*Aa Bb*), the F_2 phenotypic ratio resulting from this type of **dominant epistasis** is 12 white:3 yellow:1 green (Fig. 2.14a). The 12 includes two genotypic classes: 9 *A- B-* and 3 *aa B-*. Another way of looking at this same phenomenon is that dominant epistasis restores the 3:1 ratio for the dominant epistatic phenotype (12 white) versus all other phenotypes (4 green plus yellow). Dominant epistasis simply means that the allele causing the epistasis is dominant.

A variation of this ratio is seen in the feather color of certain chickens (Fig. 2.14b). White leghorns have a doubly dominant *AA BB* genotype for feather color; white wyandottes are homozygous recessive for both genes (*aa bb*). A cross between these two pure-breeding white strains produces an all white dihybrid (*Aa Bb*) F_1 generation, but birds with color in their feathers appear in the F_2, and the ratio of white to colored is 13:3 (see Fig. 2.14b). We can explain this ratio by assuming a kind of dominant epistasis in which *B* is epistatic to *A;* the *A* allele (in the absence of *B*) produces color; the *a, B,* and *b* alleles produce no color. The interaction is characterized by a 13:3 ratio because the 9 *A- B-*, 3 *aa B-*, and 1 *aa bb* genotypic classes combine to produce only one phenotype: white.

So far we have seen that when two independently assorting genes interact to determine a trait, the 9:3:3:1 ratio of the four Mendelian genotypic classes in the F_2 generation can produce a variety of phenotypic ratios, depending on the nature of the gene interactions. The result may be four, three, or two phenotypes, composed of different combinations of the four genotypic classes. Table 2.2 summarizes the possibilities, correlating the phenotypic ratios with the genetic phenomena they reflect.

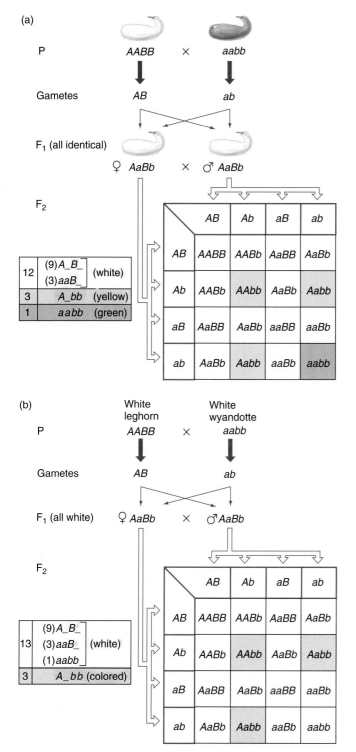

Figure 2.14 Dominant epistasis produces telltale phenotypic ratios of 12:3:1 or 13:3. (a) Summer squash comes in three colors: white, yellow, and green. One dominant *B* allele, causing white color, is sufficient to mask the effects of any combination of *A* and *a* alleles. As a result, yellow (A-) or green (*aa*) color is expressed only in *bb* individuals. (b) A cross between white leghorn and white wyandotte chickens produces an F₁ generation of white dihybrids. This is because the presence of the *B* allele results in white chickens regardless of the genotype at the other gene. In the F₂ generation resulting from a dihybrid cross (*Aa Bb × Aa Bb*), the ratio of white birds to birds with color is 13:3. This is because at least one copy of the A allele is needed to produce color.

For Some Traits, Homozygosity for a Mutant Allele at Any One of Two or More Genes Produces the Phenotype

Close to 50 different genes have mutant alleles that can cause deafness in humans. This is because it takes many genes to generate the developmental pathway that brings about hearing, and a loss of function in any part of the pathway, for instance, in one small bone of the middle ear, can result in deafness. In other words, it takes a dominant wild-type allele at each of these 50 genes (in conjunction with many others) to produce normal hearing. Thus, deafness is a **heterogeneous** trait: A mutation at any one of a number of genes can give rise to the same phenotype.

It is not always possible to determine which of many different genes has mutated in a person who expresses a heterogeneous mutant phenotype. In the case of deafness, for example, it is usually not possible to discover whether a particular man without hearing and a particular nonhearing woman carry mutations at the same gene, unless they marry and have children. If they do marry and have only children who can hear, they most likely carry mutations at two different genes, and the children carry one normal, wild-type allele at both of those genes (Fig. 2.15a). By contrast, if all of their children are deaf, it is likely that both parents are homozygous for a mutation in the same gene, and all of their children are also homozygous for this same mutation (Fig. 2.15b).

This method of discovering whether a particular phenotype arises from mutations in the same or separate genes is a naturally occurring version of an experimental genetic tool called the **complementation test.** Simply put, when what appears to be an identical recessive phenotype arises in two separate breeding lines, geneticists want to know whether mutations at the same gene are responsible for the phenotype in both lines. They answer this question by setting up a mating between animals from the two lines in which the recessive mutations arose independently. If offspring receiving the two mutations—one from each parent—express the wild-type phenotype, complementation has occurred. The observation of complementation means that the original mutations affected two different genes, and for both genes, the normal allele from one parent can provide what the mutant allele of the same gene from the other parent cannot. Figure 2.15a illustrates one example of this phenomenon in humans. By contrast, if offspring receiving two recessive mutant alleles—again, one from each parent—express the mutant phenotype, complementation does not occur because the two mutations independently alter the same gene (review Figure 2.15b). Thus, the occurrence of complementation reveals genetic heterogeneity. Note that complementation tests cannot be used if either of the mutations is dominant to the wild type. (Chapter 6 includes an in-depth discussion of complementation tests and their uses.)

In medical genetics, where researchers and doctors are always working with undefined crosses, it is very important to know that a particular mutant phenotype can result from a mutation at one of several genes. The treatment of choice often depends on which gene is involved. For example, severe mental retardation in children is a heterogeneous trait that can

TABLE 2.2 **Summary of Discussed Gene Interactions**

Gene Interaction	Example	F₂ Genotypic Ratios from an F₁ Dihybrid Cross				F2 Pheno-typic Ratio
		A- B-	A- bb	aa B-	aa bb	
None: Four distinct F₂ phenotypes	Lentil: seed coat color see p. 49	9	3	3	1	9:3:3:1
Complementary: One dominant allele of each of two genes is necessary to produce phenotype	Sweet pea: flower color see p. 50	9	3	3	1	9:7
Recessive epistasis: Homozygous recessive of one gene masks both alleles of another	Retriever: coat color see pp. 50–51	9	3	3	1	9:3:4
Dominant epistasis I: Dominant allele of one gene hides effects of both alleles of another gene	Summer squash: color see p. 52	9	3	3	1	12:3:1
Dominant epistasis II: Dominant allele of one gene hides effects of dominant allele of another gene	Chicken: feather color see p. 52	9	3	3	1	13:3

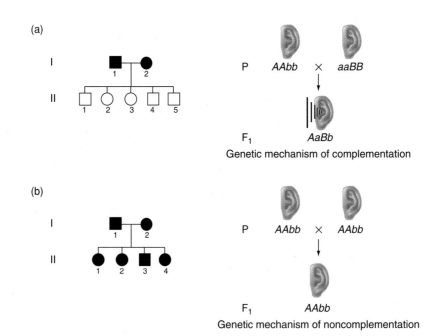

Figure 2.15 **Genetic heterogeneity in humans: Mutations in many genes can cause deafness.** Hearing requires the coordinated function of many complex structures within the ear. (a) Two deaf parents can have hearing offspring. This situation is an example of genetic complementation; it occurs if the nonhearing parents are homozygous for recessive mutations in different genes. (b) Two deaf parents may produce all deaf children. This happens when complementation does not occur because both parents carry mutations in the same gene.

be caused by many different factors, both genetic and environmental. One cause is homozygosity for a mutation that eliminates the activity of a gene encoding an enzyme known as phenylalanine hydroxylase. The enzyme normally converts the amino acid phenylalanine to tyrosine, another amino acid. Absence of the enzyme causes a buildup of phenylalanine, and after birth this buildup interferes with normal brain development, which in turn leads to mental retardation. Children born without the ability to make phenylalanine hydroxylase suffer from a genetic condition known as phenylketonuria, or PKU, which grows progressively worse in the early years of life. If a medical practitioner knows a newborn has the particular mutant genotype that gives rise to PKU, he or she can recommend the elimination of phenylalanine from the infant's diet. This dietary therapy can prevent mental retardation. Treating newborns affected by other causes of mental retardation in the same way would have no positive effect.

In sum, there are several genetic variations on the theme of multifactorial traits: (1) genes can interact to generate novel phenotypes, (2) the dominant alleles of two interacting genes can both be necessary for the production of a particular phenotype, (3) one gene's alleles can mask the effects of another's, and (4) mutant alleles at one of two or more different genes can result in the same phenotype. In examining each of these categories, for the sake of simplicity, we have looked at examples in which one allele of each gene in a pair showed complete dominance over the other. But for any type of gene interaction, the alleles of one or both genes may exhibit incomplete dominance or codominance, and these possibilities increase the potential for phenotypic diversity. For example, Figure 2.16 shows how incomplete dominance at both genes in a dihybrid cross generates phenotypic variation. Although the possibilities for variation are manifold, none of the observed departures from Mendelian phenotypic ratios contradicts Mendel's genetic laws of segregation and independent assortment. The alleles of each gene still segregate as he proposed. Interactions between the alleles of many genes simply make it harder to unravel the complex relation of genotype to phenotype.

Breeding Studies Help Decide How a Trait Is Inherited

How do you know whether a particular trait is caused by the alleles of one gene or by two genes interacting in one of a number of possible ways? Breeding tests can usually resolve the issue. Phenotypic ratios diagnostic of a particular mode of inheritance (for instance, the 9:7 or 13:3 ratios indicating that two genes are interacting) can provide the first clues and suggest hypotheses. Further breeding studies can then show which hypothesis is correct. We have seen, for example, that yellow coat color in mice is determined by a dominant allele of the agouti gene, which also acts as a recessive lethal. We now look at two other mouse genes for coat color. Since we have already designated alleles of the *agouti* gene as *Aa*, we use *Bb* and *Cc* to designate the alleles of these additional genes.

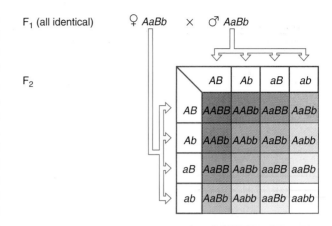

1	AABB	purple shade 9
2	AABb	purple shade 8
2	AaBB	purple shade 7
1	AAbb	purple shade 6
4	AaBb	purple shade 5
1	aaBB	purple shade 4
2	Aabb	purple shade 3
2	aaBb	purple shade 2
1	aabb	purple shade 0 (white)

Figure 2.16 With incomplete dominance, the interaction of two genes can produce 9 different phenotypes for a single trait. In this example, two genes produce different pigments that yield various shades of purple when blended in different amounts. Alleles A and a of the first gene exhibit incomplete dominance, as do alleles B and b of the second gene. Because the two alleles of each gene can combine to generate 3 different phenotypes, a double heterozygote has the potential for producing progeny of 9 (3 × 3) different colors in a ratio of 1:2:2:1:4:1:2:2:1.

A mating of one strain of pure-breeding white albino mice with pure-breeding brown results in black hybrids; and a cross between the black F₁ hybrids produces 90 black, 30 brown, and 40 albino offspring. What is the genetic constitution of these phenotypes? We could assume that we are seeing the 9:3:4 ratio of recessive epistasis and hypothesize that two genes, one epistatic to the other, interact to produce the 3 mouse phenotypes (Fig. 2.17a). But how do we know if this hypothesis is correct? We might also explain the data—160 progeny in a ratio of 90:30:40—by the activity of one gene (Fig. 2.17b). According to this one-gene hypothesis, albinos would be homozygotes for one allele (*bb*), brown mice would be homozygotes for a second allele (*BB*), and black mice would be heterozygotes (*Bb*) that have their own "intermediate" phenotype because *B* shows incomplete dominance over *b*. Under this system, a mating of black (*Bb*) to black (*Bb*) would be expected to produce 1 *BB* brown:2 *Bb* black:1 *bb* albino, or 40 brown:80 black:40 albino. Is it possible that the 30 brown, 90 black, and 40 albino mice actually counted were obtained from the inheritance of a single gene? Intuitively, the answer is yes: the ratios 40:80:40 and 30:90:40 do not seem that different. We know that

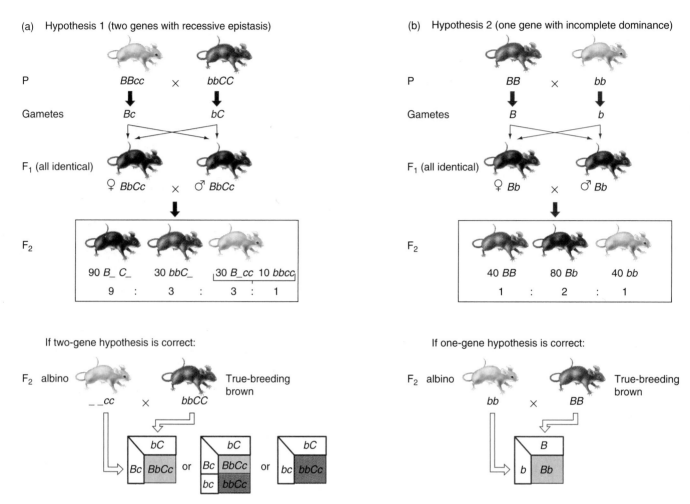

Figure 2.17 Specific breeding tests can help decide between hypotheses. Either of two hypotheses could explain the results of a cross tracking coat color in mice. (a) In one hypothesis, two genes interact with recessive epistasis to produce a 9:3:4 ratio. (b) In the other hypothesis, a single gene with incomplete dominance between the alleles generates the observed results. One way to decide between these models is to cross each of several albino F_2 mice with true-breeding brown mice. The two-gene model predicts several different outcomes depending on the —cc albino's genotype at the B gene. The one-gene model predicts that all progeny of the crosses will be black.

if we flip a coin 100 times, it doesn't always come up 50 heads:50 tails; sometimes it's 60:40 just by chance. So, how can we decide between the two-gene versus the one-gene model?

The answer is that we can use other types of crosses to verify or refute the hypotheses. For instance, if the one-gene hypothesis were correct, a mating of pure white F_2 albinos with pure-breeding brown mice similar to those of the parental generation would produce all black heterozygotes (brown [BB] × albino [bb] = all Bb) (Fig. 2.17b). But if the two-gene hypothesis is correct, with recessive mutations at an albino gene (called C) epistatic to all expression from the B gene, different matings of pure-breeding brown ($bb\ CC$) with the F_2 albinos (—cc) will give different results—all progeny are black; half are black and half brown; all are brown—depending on the albino's genotype at the Bb gene (see Fig. 2.17a). In fact, when the experiment is actually performed, the diversity of results confirms the two-gene hypothesis. The comprehensive example on pages 62 and 63 outlines some of the details of the interactions of the three mouse genes for coat color.

With Humans, Pedigree Analysis Replaces Breeding Experiments

Breeding experiments cannot be applied to humans, for obvious reasons. But a careful examination of as many family pedigrees as possible can help elucidate the genetic basis of a particular condition. In a form of albinism known as ocular-cutaneous albinism (OCA), for example, people with the inherited condition have little or no pigment in their skin, hair, and eyes (Fig. 2.18a). The three pedigrees in Fig. 2.18b trace the inheritance of OCA in three separate families. The horizontal inheritance patterns suggest that albinism is determined by the recessive allele of one gene, with albino family members being homozygotes for that allele. But a 1952 paper on albinism reported a family in which two albino parents produced three normally pigmented children (Fig. 2.18c). How would you explain this phenomenon? As Fig. 2.18d shows, the answer is that albinism is another example of heterogeneity: Mutant alleles at any one of two or more different genes can cause the condition.

Figure 2.18 Family pedigrees help unravel the genetic basis of albinism. (a) Tanzanian mother with one normally pigmented child and one albino offspring. (b) Pedigrees following the inheritance of OCA in three families indicate that the trait is recessive. The shaded symbols represent family members with OCA; the outlined symbols represent people without the inherited condition (review Fig. 1.19). (c) A family in which two albino parents have nonalbino children demonstrates that homozygosity for a recessive allele of either of two genes can cause OCA. (d) If both parents are homozygous for a recessive OCA-causing allele of the same gene, all of their children will be albinos.

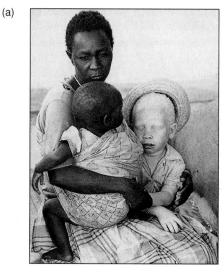

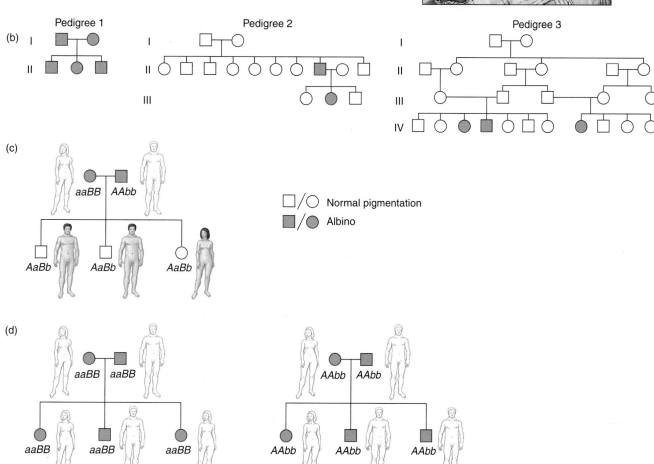

The Same Genotype Does Not Always Produce the Same Phenotype

In our discussion of gene interactions so far, we have looked at examples in which a genotype reliably fashions a particular phenotype. But this is not always what happens. Sometimes a genotype is not expressed at all, that is, even though the genotype is present, the expected phenotype does not appear; other times, the trait caused by a genotype is expressed to varying degrees or in a variety of ways in different individuals. Factors that alter the phenotypic expression of genotype include modifier genes, the environment (in the broadest sense, as defined earlier), and chance.

Phenotype Often Depends on Penetrance and Expressivity

Retinoblastoma, the most malignant form of eye cancer, arises from a dominant mutation of one gene, but only 75% of people who carry the mutant allele develop the disease. Geneticists use the term **penetrance** to describe how many members of a population with a particular genotype show the expected phenotype. Penetrance can be *complete* (100%), as in the traits that Mendel studied, or *incomplete,* as in retinoblastoma (see the Genetics and Society box "Disease Prevention versus the Right to Privacy" for another example of incomplete penetrance). For retinoblastoma, the penetrance is 75%. In some people who show the trait, only one eye is affected, while in other individuals with the phenotype, both eyes are diseased. **Expressivity** refers to the degree or intensity with which a particular genotype is expressed in a phenotype. Expressivity can be *variable,* as in retinoblastoma (one or both eyes affected), or *unvarying,* as in pea color. As we will see, the incomplete penetrance and variable expressivity of retinoblastoma are the result of chance, but in other cases, it is modifier genes or the environment that cause such variations in the appearance of phenotype.

Modifier Genes Produce Secondary Effects on Phenotype

Not all genes that influence the appearance of a trait contribute equally to the phenotype. Major genes have a large influence, while **modifier genes** have a more subtle, secondary effect. Modifier genes alter the phenotypes produced by the alleles of other genes. There is no formal distinction between major and modifier genes. Rather, there is a continuum between the two, and the cutoff is arbitrary.

Modifier genes influence the length of a mouse's tail. The mutant *T* allele of the tail-length gene causes a shortening of the normally long wild-type tail. But not all mice carrying the *T* mutation have the same length tail. A comparison of several inbred lines points to modifier genes as the cause of this variable expressivity. In one inbred line, mice carrying the *T* mu-

tation have tails that are approximately 75% as long as normal tails; in another inbred line, the tails are 50% normal length; and in a third line, the tails are only 10% as long as wild-type tails. Because all members of each inbred line grow the same length tail, no matter what the environment (for example, diet, cage temperature, or bedding), geneticists conclude it is genes and not the environment or chance that determines the length of a mutant mouse's tail. Different inbred lines most likely carry different alleles of the modifier genes that determine exactly how short the tail will be when the *T* mutation is present. Researchers have not yet identified the modifier genes or ascertained what they do, but the fact that all members of one inbred line differ in exactly the same way from all members of the second and third lines reinforces the interpretation that differences in *T*-produced tail length are caused by modifier genes.

The Environment Can Affect the Phenotypic Expression of a Genotype

Temperature is one element of the environment that has a visible effect on phenotype. For example, temperature influences the unique coat color pattern of Siamese cats (Fig. 2.19). These domestic felines are homozygous for one of the multiple alleles of a gene that encodes an enzyme catalyzing the production of the dark pigment melanin. The form of the enzyme generated by the variant "*Siamese*" allele does not function at the cat's normal body temperature. It becomes active only at the lower temperatures found in the cat's extremities, where it promotes the production of melanin, which darkens the animal's ears, nose, paws, and tail. The enzyme is thus *temperature sensitive.* Under the normal environmental conditions in temperate climates, the Siamese phenotype does not

(a)

(b)

Figure 2.19 In Siamese cats, temperature affects coat color. (a) A Siamese cat. (b) Melanin is produced only in the cooler extremities. This is because Siamese cats are homozygous for a mutation that renders an enzyme involved in melanin synthesis temperature sensitive. The mutant enzyme is active at lower temperatures, but inactive at higher temperatures.

Disease Prevention versus the Right to Privacy

In one of the most extensive human pedigrees ever assembled, a team of researchers traced a familial pattern of blindness back through five centuries of related individuals to its origin in a couple who died in a small town in northwestern France in 1495. More than 30,000 French men and women alive in 1990 descended from that one fifteenth-century couple, and within this direct lineage resided close to half of all reported French cases of a particular blindness-producing disease, a form of hereditary juvenile glaucoma. The massive genealogic tree for the trait (when posted on the office wall, it is over 100 feet long) shows that the genetic defect follows a simple Mendelian pattern of transmission determined by the dominant allele of a single gene (Fig. A). The pedigree also shows that the dominant genetic defect displays incomplete penetrance: Not all people receiving the dominant allele from one or the other of their parents become blind; these sighted carriers may unknowingly pass the blindness-causing dominant allele to their children. Thus, with hereditary juvenile glaucoma, a parent, grandparent, or great-grandparent may possess the defective dominant allele but not show up as an affected carrier in the family pedigree.

Unfortunately, people do not know they have the disease until their vision starts to deteriorate, but by that time, many of their op-

tic fibers have been irreversibly damaged and blindness is all but inevitable. Surprisingly, the existence of medical therapies that make it possible to arrest the nerve damage of hereditary juvenile glaucoma and prevent blindness have created a quandary. Because treatment, to be effective, has to begin before symptoms of impending blindness show up, information in the extensive juvenile glaucoma pedigree could help doctors pinpoint people who are at risk, even though neither of their parents is blind. However, a French law protecting personal privacy forbids public circulation of the names in the pedigree. In fact, in the late 1980s, the researchers who compiled the massive family history wanted to give physicians the names of at-risk individuals living in their area; with such information the doctors could keep a close watch on certain patients and when necessary recommend drugs or surgery to prevent blindness. But interpreters of the law maintained that if actual names were circulated, potential carriers of the disease might suffer discrimination in hiring or insurance. According to the law, individuals about whom information is collected must know how that information will be used, and at the time of the study, living participants were not told that their names would be given out to medical workers. As a result, the researchers could alert physicians to the hereditary nature and symptoms of the disease, but they could not give out names. And since only a long, arduous, expensive—and at the time infeasible—national screening program could identify affected individuals without those names, some sufferers of the disease would not receive treatment soon enough to prevent blindness.

France thus faced a serious ethical dilemma: On the one hand, giving out names could save perhaps thousands of people from blindness; on the other hand, laws designed to protect personal privacy and forestall discrimination precluded the dissemination of specific names. Until the legal problem was resolved, the only solution was a massive educational program to alert the general public to the problem so that concerned families could seek medical advice. Meanwhile, molecular geneticists were searching for the gene (or genes) that cause this particular form of juvenile glaucoma. Its (or their) identification could result in diagnostic tests based on the direct analysis of genotype (see the Fast Forward box "The Direct Analysis of Human Genotype" in Chapter 1). And these tests, if administered on a large enough scale with proper educational follow-up, could circumvent the problem of giving out names from the pedigree, by delivering genetic information directly to the parents of newborns, in plenty of time for preventive therapy.

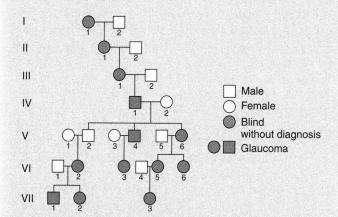

Male
Female
Blind
without diagnosis
Glaucoma

Figure A   A pedigree showing the transmission of juvenile glaucoma. A small part of the genealogic tree: The vertical transmission pattern over seven generations shows that a dominant allele of a single gene causes juvenile glaucoma. The lack of glaucoma in V-2 followed by its reappearance in VI-2 reveals that the trait is incompletely penetrant. As a result, sighted heterozygotes may unknowingly pass the condition on to their children.

vary much in expressivity from one cat to another, but one can imagine the expression of a very different phenotype—no dark extremities—in equatorial deserts where the ambient temperature is at or above normal body temperature.

Temperature can also affect survivability. In one type of experimentally bred fruit fly (*Drosophila melanogaster*), some individuals develop and multiply normally at temperatures between 18 and 29°C; but if the thermometer climbs beyond that cutoff for a short time, they become reversibly paralyzed, and if the temperature remains high for more than a few hours, they die. These insects carry a temperature sensitive allele of the *shibire* gene, which encodes a protein essential for nerve cell transmission, among other functions. This type of allele is known as a **conditional lethal** because it is lethal only under certain conditions. The range of temperatures under which the insects remain viable is the **permissive** range; the lethal temperatures above that are **restrictive.** Thus, at one temperature the allele gives rise to a phenotype that is indistinguishable from the wild type, while at another temperature, this same allele generates a mutant phenotype. Flies with the wild-type *shibire* allele are viable even at the higher temperatures. The fact that some mutations are lethal only under certain conditions clearly illustrates that the environment can affect the penetrance of a phenotype. Since flies with temperature-sensitive *shibire* mutations can survive only under carefully controlled environmental conditions, they do not survive in nature where the mutation is more likely to be penetrant at some point in the life cycle.

Even in genetically normal individuals, exposure to chemicals or other environmental agents can have phenotypic consequences that are similar to those caused by mutant alleles of specific genes. A change in phenotype arising from such environmental agents that mimics the effects of a mutation in a gene is known as a **phenocopy.** Phenocopies are not heritable because they do not arise from a change in a gene. In humans, ingestion of the sedative thalidomide by pregnant women in the early 1960s produced a phenocopy of a rare dominant trait called *phocomelia.* By disrupting limb development in otherwise normal fetuses, the drug mimicked the effect of the phocomelia-causing mutation. When this became evident, thalidomide was withdrawn from the market.

Some types of environmental change may have a positive effect on an organism's survivability, as in the following example, where a straightforward application of medical science artificially reduces the penetrance of a mutant phenotype. As discussed earlier, children born with the genetically determined disease known as phenylketonuria, or PKU, will develop a range of neurological problems, including convulsive seizures and mental retardation, unless they are put on a special diet. Today, there is a reliable blood test for detecting the condition in newborns. Once a baby with PKU is identified, the doctor or other health-care practitioner prescribes a protective diet that excludes phenylalanine; the diet must also provide enough calories to prevent the infant's body from breaking down its own proteins and thereby releasing the damaging amino acid from within. Such dietary therapy—a simple change in the environment—although somewhat unpleasant because it is highly restrictive, now enables many PKU infants to develop into healthy adults.

Finally, two of the top killer diseases in the United States—cardiovascular disease and lung cancer—also illustrate how the environment can alter phenotype by influencing both expressivity and penetrance. People may inherit a propensity to heart disease, but the environmental factors of diet and exercise contribute to the occurrence (penetrance) and seriousness (expressivity) of their condition. Similarly, some people are born genetically prone to lung cancer, but whether or not they get it (penetrance) is strongly determined by whether or not they choose to smoke.

Thus, various aspects of an organism's environment, including temperature, diet, and exercise interact with its genotype to generate the functional phenotype, the ultimate combination of traits that determines what a plant or animal looks like and how it behaves.

Chance Can Affect Penetrance and Expressivity

In the retinoblastoma example described earlier, people are born with the disease mutation in one allele of the retinoblastoma gene in every cell. But whether or not they show the phenotype and whether they show it in one or both eyes depends on additional genetic events that occur randomly to alter the second allele of the gene in specific body cells. Examples of random events that can trigger the onset of the disease include exposure to cosmic rays that alter the genetic material in retinal cells or mistakes made by retinal cell machinery during cell division. Events of this type provide the second "hit"—a mutation in the second copy of the retinoblastoma gene—necessary to turn a normal retinal cell into a cancerous one. The phenotype of retinoblastoma thus results from a specific heritable mutation in a specific gene, but the incomplete penetrance and variable expressivity of the disease depend on secondary genetic events that occur by chance in certain cells.

By contributing to incomplete penetrance and variable expressivity, modifier genes, the environment, and chance give rise to phenotypic variation. Unlike dominant epistasis or recessive lethality, however, the probability of penetrance and the level of expressivity cannot be derived from the original Mendelian principles of segregation and independent assortment; they are determined empirically by observation and counting.

Even Continuous Variation Can Be Explained by Extensions to Mendelian Analysis

In Mendel's experiments, height in pea plants was determined by two segregating alleles of one gene (in the wild, it is determined by many genes, but in Mendel's inbred populations, the alleles of all these other genes were invariant); the phenotypes that resulted were clear-cut, either short or tall, and pea plant height was therefore known as a **discontinuous trait.** In contrast, because people do not produce inbred populations, height in humans is determined by segregating alleles of many different genes whose interaction with each other and the environment produces phenotypes showing continuous variation; height in humans is thus an example of a **continuous trait.** Within most outbred human populations, individual heights vary over a range of values that when charted on a graph produce a bell curve (Fig. 2.20). In fact, many human traits, including height, weight, and skin color, show continuous variation, rather than the clear-cut alternatives analyzed by Mendel. They also appear to blend and "unblend." Think for a moment of skin color. Children of marriages between people of African and Northern European descent, for example, often seem to be a blend of their parents' skin color. Matings of these F_1 individuals produce offspring with a wide range of skin pigmentation; a few may be as light as the original Northern European parent, a few as dark as the original African parent, but most will fall in a range between the two (Fig. 2.20b). For these reasons, early human geneticists were slow to accept Mendelian analysis. Because they were working with outbred

populations, they found very few examples of "either-or" Mendelian traits in normal, healthy people.

By 1930, however, studies of corn and tobacco conclusively demonstrated that it is possible to provide a Mendelian explanation of continuous variation by simply increasing the number of genes contributing to a phenotype. The more genes, the more phenotypic categories, and the more categories, the more the variation appears continuous. As a hypothetical example, consider a series of genes all affecting the height of pole beans. For each gene, there are two alleles, a "0" allele that contributes nothing to height and a "1" allele that increases the height of a plant by one unit. All alleles exhibit incomplete dominance relative to alternative alleles at the same gene. The phenotypes determined by all these genes are additive. What would be the result of a two-generation cross between pure-breeding plants carrying only 0 alleles at each height gene and pure-breeding plants carrying only 1 alleles at each height gene? If only one gene were responsible for height, and if environmental effects could be discounted, the F_2 population would be distributed among three classes: homozygous 0,0 plants with 0 height (they lie prostrate on the ground); heterozygous 0,1 plants with a height of 1; and homozygous 1,1 plants with a height of 2 (Fig. 2.21a). This distribution of heights over 3 phenotypic categories does not make a continuous curve. But for 2 genes, there will be 5 phenotypic categories in the F_2 generation; for 3 genes, 7 categories (Fig. 2.21b–c); and for 4 genes, 9. The distributions produced by 3 and 4 genes begin to approach continuous variation, and if we add a small contribution from environmental variation, a smooth curve will appear. After all, we would expect bean plants to grow better in good soil, with ample sunlight and water. The environmental component effectively converts the stepped bar graph to a continuous curve by producing some variation in expressivity within each genotypic class. Moreover, additional variation might arise from more than two alleles at some genes (see Fig. 2.21b), unequal contribution to the phenotype by the various genes involved (review Fig. 2.16), interactions with modifier genes, and chance. Thus, from what we now know about the relation between genotype and phenotype, it is possible to see how just a handful of genes that behave according to known Mendelian principles can easily generate continuous variation.

Continuous traits (also called **quantitative traits**) vary quantitatively over a range of values and can usually be measured: the length of a tobacco flower in millimeters, the amount of milk produced by a cow per day in liters, or the height of a person in meters. They are usually **polygenic**—controlled by multiple genes—and show the additive effects of a large number of alleles, which creates an enormous potential for variation within a population. Polygenic traits are, by definition, multifactorial, but not all multifactorial traits are polygenic (some, for instance, are determined by the way the alleles of one gene interact with each other and the environment). We discuss the analysis and distribution of multifactorial traits in Chapter 23 on population genetics.

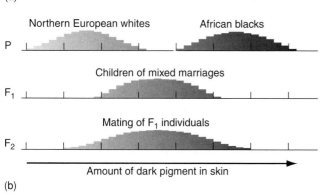

(a)

(b)

Figure 2.20 Continuous traits in humans. (a) Women runners at the start of a 5th Avenue mile race in New York City demonstrate that height is a trait showing continuous variation. (b) The skin color of most F_1 offspring is usually between the parental extremes, while the next (F_2) generation exhibits a broader distribution of continuous variation.

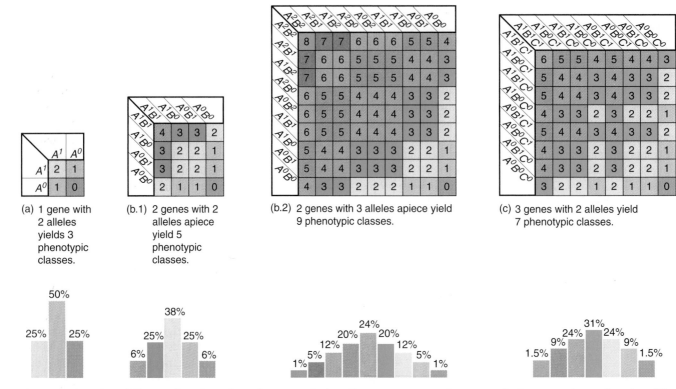

Figure 2.21 A Mendelian explanation of continuous variation. The more genes contribute to a trait, the greater the number of possible phenotypic classes, and the greater the similarity to continuous variation. In these examples, several pairs of incompletely dominant alleles have additive effects. (a) 1 allele pair yields 3 discontinuous phenotypic classes. (b) 2 allele pairs yield 5 phenotypic classes. 2 allele trios yield 9 phenotypic classes. (c) 3 allele pairs yield 7 phenotypic classes. Percentages denote fractions of the total population.

The Mouse's Coat and Tail: A Comprehensive Example of Multiple Alleles and Multifactorial Traits

Most field mice are a dark gray (agouti) and sport a long tail, but mice bred for specific mutations in the laboratory can be gray, tan, yellow, brown, black, or various combinations thereof, and their tail can be long, various degrees of short, or nonexistent. Here is a rundown of some of the genes and alleles that make such variation possible.

Gene 1—Agouti or Other Color Patterns

The agouti gene determines the distribution of color on each hair and has multiple alleles. The wild-type allele A specifies the bands of yellow and black that give the agouti appearance; A^y gets rid of the black and thus produces solid yellow; a gets rid of the yellow and thus produces solid black; and a^t specifies black on the back and yellow on the stomach. The dominance series for this set of agouti gene alleles is: $A^y > A > a^t > a$. However, although A^y is dominant to all other alleles for coat color, it is recessive to all the others for lethality: $A^y A^y$ homozygotes die before birth, while $A^Y A$, $A^Y a^t$, or $A^Y a$ heterozygotes survive.

Gene 2—Black or Brown

A second gene specifies whether the dark color of each hair is black or brown. This gene has two alleles: B is dominant and

designates black; b is recessive and generates brown. Since the A^y allele at the agouti gene completely eliminates the dark band of each hair, it acts in a dominant epistatic manner to the B gene. With all other agouti alleles, however, it is possible to distinguish the effects of the two different B alleles on phenotype. The A-B- genotype gives rise to the wild-type agouti having black with yellow hairs. The A-bb genotype generates a color referred to as cinnamon (with hairs having stripes of brown and yellow); $aa\,bb$ is all brown; and $a^t a^t bb$ is brown on the back and yellow on the stomach. A cross between two F_1 hybrid animals of genotype $A^y a, Bb$ would yield an F_2 generation with yellow (A^y-,—), black (aa, B-), and brown (aa, bb) animals in a ratio of 8:3:1. This ratio reflects the dominant epistasis of A^y and the loss of a class of 4 ($A^y A^y$,—) due to prenatal lethality.

Gene 3—Albino or Pigmented

Like other mammals, mice have a third gene influencing coat color. A recessive allele (c) abolishes the function of the enzyme that leads to the formation of dark pigment melanin, making the gene a recessive epistatic to all other coat color genes. As a result, cc homozygotes are pure white, while C-mice are agouti, black, brown, yellow, or yellow and black (or other colors and patterns), depending on what alleles they carry at the A and B genes, as well as at some 50 other genes known to play a role in determining the coat color of mice.

Adding to the complex color potential are other alleles that geneticists have uncovered for the albino gene; these cause only a partial inactivation of the melanin-producing enzyme and thus have a partial epistatic effect on phenotype.

Gene 4—Short or Long Tail

Wild-type mice (++) have a long tapering tail; but an X-ray induced mutation (*T*) of the tail-length gene makes it noticeably shorter. Fetuses homozygous for this mutation die about halfway through gestation at day 10, which makes the mutation a recessive lethal; heterozygous *T*+ mice have short tails.

As mentioned earlier, the tail-length phenotype shows variable expressivity, depending on which alleles of certain modifier genes are present. Consideration of all the known modifier genes in conjunction with *T*, makes it possible to view tail length as a quantitative trait, with values ranging continuously from zero length (taillessness) to a length of 12 cm.

This comprehensive example of coat color and tail length in mice gives some idea of the potential for variation from just a few genes, some with multiple alleles. In fact, both mice and humans carry roughly 100,000 genes, many with multiple wild-type alleles. The number of interactions that connect these genes in the expression of phenotype is in the millions, if not the billions. Thus, the potential for variation and diversity among individuals is staggering indeed.

CONNECTIONS

Part of Mendel's genius was to look at the genetic basis of variation through a very narrow window, focusing his first glimpse of the mechanisms of inheritance on such fundamental phenomena that the import of his insights remains undiminished to this day. Mendel worked on just a handful of traits in inbred populations of one species, and for each trait, one gene with one completely dominant and one recessive allele determined two distinguishable, or discontinuous, phenotypes. Both the dominant and recessive alleles showed complete penetrance and negligible differences of expressivity. He knew from other examples that phenotype does not always reflect genotype, but he probably did not foresee how many complexities could cloud the correlation between the two.

In the first few decades of the twentieth century, many questioned the general applicability of Mendelian analysis, for it seemed to shed little light on the complex inheritance patterns of most plant and animal traits or on the mechanisms producing continuous variation. Simple embellishments, however, clarified the genetic basis of continuous variation and provided explanations for other apparent exceptions to Mendelian analysis. These embellishments included the ideas that dominance need not be complete, that one gene can have multiple alleles, that one gene can determine more than one trait, that several genes can contribute to the same trait, and that the expression of genes can be affected in a variety of ways by other genes, the environment, and chance. Each embellishment extends the range of Mendelian analysis and deepens our understanding of the genetic basis of variation. And no matter how broad the view, Mendel's basic conclusions—that genes are the discrete units of heredity and that the two similar or dissimilar copies of a gene segregate during gamete formation and randomly unite at fertilization—remain valid.

But what about the law of independent assortment? As it turns out, its application is not as universal as that of the law of segregation. Many genes do assort independently, but some do not; rather, they appear to be linked and transmitted together from generation to generation. An understanding of this fact emerged from studies that located Mendel's hereditary units, the genes, in specific cellular organelles, the chromosomes. In describing how researchers deduced that genes travel with chromosomes, Chapter 3 establishes the physical basis of inheritance, including segregation, and clarifies why some genes do, while others do not, assort independently.

ESSENTIAL CONCEPTS

1. The F_1 phenotype generated by each pair of alleles defines the dominance relationship between these alleles. One allele is not always completely dominant or completely recessive to another. With *incomplete dominance,* the F_1 hybrid phenotype resembles neither parent. With *codominance,* the F_1 hybrid phenotype contains observable components from both parents. Many allele pairs are codominant at the level of protein production.

2. One gene can contribute to multiple traits; for such a gene, the dominance relation between any two alleles can vary according to the particular phenotype under consideration.

3. A single gene may have any number of alleles, each of which can cause the appearance of different phenotypes alone or through interaction with other alleles. New alleles arise by mutation. Alleles whose frequency is greater than 1% in a population are considered *wild types;* those whose frequency is below 1% are considered *mutants.* When two or more wild-type alleles exist for a gene, the gene is considered *polymorphic;* a gene with only one wild-type allele is *monomorphic.*

4. Two or more genes may interact in several ways to affect the production of a single trait. It is often possible to arrive at an understanding of these interactions by observing characteristic deviations from traditional Mendelian phenotypic ratios.

5. In *epistasis,* the action of an allele at one gene can hide traits normally caused by the expression of alleles at another gene. In *complementary gene action,* dominant alleles of two or more genes are required to generate a particular trait. In *heterogeneity,* mutant alleles at any one of two or more genes are sufficient to elicit a phenotype. The *complementation test* can reveal whether a particular phenotype arises from mutations in the same or separate genes.

6. In many cases, the route from genotype to phenotype can be modified by the environment, chance, or other genes. A phenotype shows *incomplete penetrance* when it is expressed in fewer than 100% of individuals with the same genotype. A phenotype shows *variable expressivity* when it is expressed at a quantitatively different level in different individuals with the same genotype.

7. A *continuous trait* can have any value of expression between two extremes. Most traits of this type are *polygenic,* that is, determined by the interactions of multiple genes.

SOCIAL AND ETHICAL ISSUES

1. John and Nancy's daughter is entering kindergarten, and as required by law, has to have a physical and blood test performed by their family doctor. The doctor knows that both John and Nancy have the AB blood type. But when results come back on their daughter's blood, the doctor discovers that her blood type is O. Neither parent is heterozygous for the Bombay allele. The only conclusion the doctor can reach from this result is that babies must have been switched at the hospital. The doctor is now in an awkward position. John and Nancy and their healthy daughter make a very happy family. If he explains to them the implication of his findings, their lives will be turned upside down. If not, he will be withholding personal information that one could argue rightly belongs to them. What would you do if you were the doctor?

2. A man who has committed several violent rapes is on trial and his lawyer argues that the man cannot help himself because his behavior is the result of his genetic makeup. The man's father and brother also have a history of committing violent crimes. In defense of this position, the lawyer cites studies reported in 1993 on a small Dutch family indicating that some people inherit a propensity toward violent behavior. Affected individuals in this family have a mutation in the monoamine oxidase gene involved in the breakdown of neurotransmitters. Do you think the sentencing of the man on trial should depend in part on this type of scientific study? Our judicial system is based on the assumption that people act with free will and are therefore responsible for their actions. Are people with known genetic alterations responsible for their actions? Should we make exceptions in the case of genetic predispositions? What role does society have in helping people whose behaviors may be influenced by the alleles they carry?

3. A teenage girl was diagnosed with chronic myelogenous leukemia, for which the treatment is a bone marrow transplant. A matching donor could not be found and her parents decided to have another child, hoping that the younger sibling could serve as a donor of bone marrow for the older child. Is it ethical to conceive a child for the purpose of tissue or organ donation to a sibling? Consider situations in which the donor's life will be compromised by the donation as well as cases in which the donor's life will not be compromised.

SOLVED PROBLEMS

I. Imagine you purchased an albino mouse (genotype *cc*) in a pet store. The *c* allele is epistatic to other coat color genes. How would you go about determining the genotype of this mouse at the brown locus? (In pigmented mice, *BB* and *Bb* are black, *bb* is brown.)

Answer

This problem requires an understanding of gene interactions, specifically epistasis. You have been placed in the role of experimenter and need to design crosses that will answer the question. To determine the alleles of the B gene present you need to eliminate the blocking action of the *cc* genotype. Because this is recessive epistasis (*c* allele is epistatic), when a *C* allele is present, no epistasis will occur. To introduce a *C* allele during the mating, the test mouse you mate to your albino can have the genotype *CC* or *Cc*. (If the mouse has the *Cc* genotype, half of the progeny will be albino and will not contribute useful information for you. The nonalbinos from this cross would be informative.) What alleles of the *B* gene should the test mouse carry? To make this decision, work through the expected results using each of the possible genotypes.

Test mouse genotype		Albino mouse	Expected progeny
BB	×	*BB*	All black
	×	*Bb*	All black
	×	*bb*	All black
Bb	×	*BB*	All black
	×	*Bb*	3/4 black, 1/4 brown
	×	*bb*	1/2 black, 1/2 brown
bb	×	*BB*	All black
	×	*Bb*	1/2 black, 1/2 brown
	×	*bb*	All brown

From these hypothetical crosses, you can see that using a test mouse with either the *Bb* or *bb* genotype would result in distinct outcomes depending on which of the three possible genotypes the albino mouse had. However, using the *bb* mouse is more useful and less ambiguous. First, it is easier to identify a mouse with the *bb* genotype because a brown mouse must have this homozygous recessive genotype. Second, the results are completely different for each of the three possible genotypes when you use the *bb* test mouse. (Compare to the *Bb* × *Bb* or *Bb* × *bb* crosses, which both give the same two classes of animals, with the ratio being the distinguishing feature). *To determine the full genotype of the albino mouse, you should cross it to a brown mouse (which could have the* CC *bb or* Cc *bb genotype).*

II. In a particular kind of wildflower, the wild-type flower color is deep purple and the plants are true breeding. In one true-breeding mutant stock, the flowers have a reduced pigmentation, resulting in a lavender color. In a different true-breeding mutant stock, the flowers have no pigmentation and are thus white. A lavender-flowered plant from the first mutant stock was crossed to a white-flowered plant from the second mutant stock, and the 100 F_1 plants all had purple flowers. The F_1 plants were then allowed to self-fertilize to produce an F_2 generation. The 277 F_2 plants were 157 purple:71 white:49 lavender. When the lavender F_2 plants were allowed to self-fertilize, both lavender and white-flowered plants resulted. Explain how flower color is inherited. Is this trait controlled by the alleles of a single gene?

Answer

Are there any modes of single-gene inheritance compatible with the data? The observations that the F_1 plants look different from either of their parents and the F_2 generation is composed of plants with three different phenotypes exclude complete dominance. The ratio of the three phenotypes in the F_2 plants has some resemblance to the 1:2:1 ratio expected from codominance or incomplete dominance, but the appearance of the wild-type phenotype in the F_1 generation makes it less likely that one of these modes of inheritance is operating here. The result of the self-fertilization of the F_2 lavender plants is the clincher. If this were one-gene inheritance, the lavender plants must have been true-breeding (homozygous) and all their progeny would have lavender flowers.

Consider now the possibility that two genes are involved. From a cross between plants heterozygous for two genes (*W* and *P*), the F_2 generation would contain a 9:3:3:1 ratio of the genotypes *W- P-, W- pp, ww P-,* and *ww pp* (where the dash indicates that the allele could be either a dominant or a recessive form). Are there any combinations of the 9:3:3:1 ratio that would be close to that seen in the F_2 generation in this example? The numbers seem close to a 9:4:3 ratio. What hypothesis would support combining two of the classes (3 + 1)? If *w* is epistatic to the *P* gene, then the *ww P-* and *ww pp* genotypic classes would have the same phenotype. The results of self-fertilization of the lavender F_2 plants are consistent with this interpretation. This is because the F_2 lavender plants would be a mix of genotypes *WW pp* and *Ww pp,* and plants of the *Ww pp* genotype would produce some white-flowered (*ww pp*) progeny.

III. Huntington disease (HD) is a rare dominant condition in humans that results in a slow but inexorable deterioration of the nervous system. HD shows what might be called "age-dependent penetrance," which is to say that the probability that a person with the HD genotype will express the phenotype varies with age. Assume that 50% of those inheriting the *HD* allele will express the symptoms by age 40. Susan is a 35-year-old woman whose father has HD. She currently shows no symptoms. What is the probability that Susan will show symptoms in five years?

Answer

This problem involves probability and penetrance. Two conditions are necessary for Susan to show symptoms of the disease. There is a 1/2 (50%) chance that she inherited the mutant allele from her father and a 1/2 (50%) chance that she will express the phenotype by age 40. Since these are independent events, *the probability is the product of the individual probabilities, or 1/4.*

P R O B L E M S

2-1 For each of the terms in the left column, choose the best matching phrase in the right column.

a.	epistasis	1.	one gene affecting more than one phenotype
b.	modifier gene	2.	the alleles of one gene mask the effects of alleles of another gene
c.	conditional lethal	3.	both parental phenotypes are expressed in the F_1 hybrids
d.	permissive condition	4.	a heritable change in a gene
e.	reduced penetrance	5.	cell surface molecules that are involved in the immune system and are highly variable
f.	multifactorial trait	6.	genes whose alleles subtly alter phenotypes produced by the action of other genes
g.	incomplete dominance	7.	less than 100% of the individuals possessing a particular genotype express it in their phenotype
h.	codominance	8.	environmental conditions that allow conditional lethals to live
i.	histocompatibility antigens	9.	a trait produced by the interaction of alleles of at least two genes or from interactions between gene and environment
j.	mutation	10.	the heterozygote resembles neither homozygote
k.	pleiotropy	11.	a genotype that is lethal in some situations (e.g., high temperature) but viable in others

2-2 In four o'clocks, the allele for red flowers is incompletely dominant over the allele for white flowers, so that heterozygotes have pink flowers. What ratios of flower colors would you expect among the offspring of the following crosses: (a) pink × pink, (b) white × pink, (c) red × red, (d) red × pink, (e) white × white, and (f) red × white?

2-3 A cross between two plants that both have yellow flowers produces 80 offspring plants, of which 38 have yellow flowers, 22 have red flowers, and 20 have white flowers. If one assumes that this variation in color is due to variation at a single locus, what is the genotype associated with each flower color, and how can you describe the inheritance of flower color?

2-4 In radishes, color and shape are each controlled by a single locus with two incompletely dominant alleles. Color may be red (RR), purple (Rr), or white (rr) and shape can be long (LL), oval (Ll), or round (ll). What phenotypic classes and proportions would you expect among the offspring of a cross between two plants heterozygous at both loci?

2-5 A wild legume with white flowers and long pods is crossed to one with purple flowers and short pods. The F_1 offspring are allowed to self-fertilize, and the F_2 generation has 301 long purple, 99 short purple, 612 long pink, 195 short pink, 295 long white, and 98 short white. How are these traits being inherited?

2-6

a. What is the genotype of a person who expresses the sickle-cell anemia phenotype?

b. What is the genotype of a person with a normal phenotype that has a child with sickle-cell anemia?

c. What is the *total* number of different alleles of the β-globin gene that could be carried by five children with the same mother and father?

2-7 Assuming no involvement of the Bombay phenotype,

a. If a girl has blood type O, what could be the genotypes (and corresponding phenotypes) of her parents?

b. If a girl has blood type B and her mother has blood type A, what genotype(s) and corresponding phenotype(s) could the other parent have?

2-8 There are several genes in humans in addition to the *ABO* gene that give rise to recognizable antigens on the surface of red blood cells in humans. The *MN* and *Rh* genes are two examples. The *Rh* locus can contain either a positive or negative allele with positive being dominant to negative. *M* and *N* are codominant alleles of the *MN* gene. The chart below shows several mothers and their children. For each mother–child pair, choose the father of the child from among the males in the right column, assuming one child per male.

	Mother	Child	Males
a.	O M Rh pos	B MN Rh neg	O M Rh neg
b.	B MN Rh neg	O N Rh neg	A M Rh pos
c.	O M Rh pos	A M Rh neg	O MN Rh pos
d.	AB N Rh neg	B MN Rh neg	B MN Rh pos

2-9 Alleles of the gene that determines seed coat patterns in lentils can be organized in a dominance series: marbled > spotted or dotted (codominant alleles) > clear. A lentil homozygous for the marbled seed coat pattern allele was crossed to a lentil homozygous for the spotted pattern allele. In another cross, a homozygous dotted lentil was crossed to one homozygous for clear. An F_1 plant from the first cross was then mated to an F_1 plant from the second cross.

a. What phenotypes in what proportions are expected from this mating between the two F_1 types?

b. What are the expected phenotypes of the F_1 plants from the two original parental crosses?

2-10 Some plant species have an incompatibility system different from that shown in Fig. 2.8. In this alternate kind of incompatibility, a mating cannot produce viable seeds if the male parent shares an incompatibility allele with the female parent. (Just as with the kind of incompatibility system shown in Fig. 2.8, this system also ensures that all plants are heterozygous for the incompatibility gene.) Five plants were isolated from a wild population of a species with this alternate type of incompatiblity. The results of the matings between each pair of the plants is given below (− means no seeds were produced; + means seeds were produced). How many different alleles of the incompatibility gene are present in this group of five plants? What are the genotypes of the plants?

	1	2	3	4	5
1	−	−	−	+	−
2		−	+	+	+
3			−	+	−
4				−	−
5					−

2-11 In mice, the A^y allele of the agouti gene is a recessive lethal allele but dominant for yellow coat color. What phenotypes and ratios of offspring would you expect from the cross of a mouse heterozygous at the agouti locus (genotype $A^y A$) and also at the albino locus (Cc), to an albino mouse (cc) heterozygous at the agouti locus ($A^y A$)?

2-12 In a species of tropical fish, a colorful orange and black variety called montezuma occurs. When two montezumas are crossed, 2/3 of the progeny are montezuma and 1/3 are the wild-type, dark grayish green color. Montezuma is a single gene trait and montezuma fish are never true-breeding.
a. Explain the inheritance pattern seen here and show how your explanation accounts for the phenotypic ratios given above.
b. In this same species, the morphology of the dorsal fin is altered from normal to ruffled by homozygosity for a recessive allele designated *f*. What progeny would you expect to obtain, and in what proportions, from the cross of a montezuma fish homozygous for normal fins to a green, ruffled fish?
c. What phenotypic ratios of progeny would be expected from the crossing of two of the montezuma progeny from part b?

2-13 You picked up two mice (one female and one male) that had clearly escaped from experimental cages in the animal facility. One mouse is yellow in color and the other is brown agouti. You know that this mouse colony has animals with different alleles at only three coat color genes: the agouti or nonagouti or yellow alleles of the *A* gene, the black or brown allele of the *B* gene, and the albino or nonalbino alleles of the *C* gene. However, you don't know which alleles of these genes are actually present in each of the animals that you've captured. To determine the genotypes, you breed them together. The first litter has only three pups. One is albino, one is brown (nonagouti), and the third is black agouti.
a. What alleles of the *A, B,* and *C* genes are present in the two mice you caught?
b. After raising several litters from these two parents, you have many offspring. How many different coat color phenotypes (in total) do you expect to see expressed in the population of offspring? What are the phenotypes?

2-14 A rooster with a particular comb morphology known as walnut was crossed to a hen with comb morphology known as single. The F_1 progeny all had walnut combs. When F_1 males and females were crossed to each other, 93 walnut and 11 single combs were seen among the F_2 progeny but there were also 29 birds with a new kind of comb called rose and 32 birds with a new comb called pea.
a. Explain how comb morphology is inherited.
b. What progeny would result from crossing a homozygous rose combed hen with a homozygous pea combed rooster? What phenotypes and ratios would be seen in the F_2 progeny?

2-15 A black mare was crossed to a chestnut stallion, and produced a bay son and a bay daughter. The two offspring were mated to each other several times, and produced offspring of four different coat colors: black, bay, chestnut, and liver. Crossing a liver grandson back to the black mare gave a black foal, and crossing a liver granddaughter back to the chestnut stallion gave a chestnut foal. Explain how coat color is being inherited.

2-16 Filled-in symbols in the pedigree designate individuals suffering from deafness.
a. Study the pedigree, and explain how deafness is being inherited.
b. What is the genotype of the individuals in generation V? Why are they not affected?

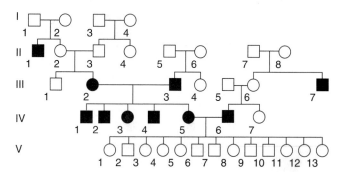

2-17 You do a cross between two true-breeding strains of zucchini. One has green fruit and the other has yellow fruit. The F_1 plants are all green, but when these are crossed, the F_2 plants consist of 9 green to 7 yellow. Indicate the phenotype, with frequencies, of the progeny of a test cross of the F_1 plants.

2-18 A student whose hobby was fishing pulled a carp out of Cayuga Lake that was very unusual: It had no scales on its body. She decided to investigate whether this strange nude phenotype had a genetic basis. She therefore obtained some inbred carp that were pure-breeding for the wild-type scale phenotype (body covered with scales in a regular pattern) and crossed them with her nude fish. To her surprise, the F_1 progeny consisted of wild-type fish and fish with a single linear row of scales on each side in a 1:1 ratio.

a. Can a single gene with two alleles account for this result? Why or why not?

b. To follow up on the first cross, the student allowed the linear fish from the F_1 generation to mate with each other. The progeny of this cross consisted of fish with four phenotypes: linear, wild type, nude, and scattered (the latter had a few scales scattered irregularly on the body). The ratio of these phenotypes was 6:3:2:1, respectively. How many genes appear to be involved in determining these phenotypes?

c. In parallel, the student allowed the phenotypically wild-type fish from the F_1 generation to mate with each other and observed, among their progeny, wild-type and scattered carp in a ratio of 3:1. How many genes with how many alleles appear to determine the difference between wild-type and scattered carp?

d. The student confirmed the conclusions of part c by crossing scattered carp with her pure-breeding wild-type stock. Diagram the genotypes and phenotypes of the parental, F_1 and F_2 generations for this cross and indicate the ratios observed.

e. The student attempted to generate a true-breeding nude stock of fish by inbreeding. However, she found that this was impossible. Every time she crossed two nude fish, she found nude and scattered fish in the progeny, in a 2:1 ratio. (The scattered fish from these crosses bred true.) Diagram the phenotypes and genotypes of this gene in a nude × nude cross and explain the altered Mendelian ratio.

f. The student now felt she could explain all of her results. Diagram the genotypes in the linear × linear cross performed by the student (in part b above). Show the genotypes of the four phenotypes observed among the progeny and explain the 6:3:2:1 ratio.

2-19 Explain the difference between epistasis and dominance. How many loci are involved in each case?

2-20 Two true-breeding white strains of the plant *Illegitamati noncarborundum* were mated and the F_1 progeny were all white. When the F_1 plants were allowed to self-fertilize, 126 white-flowered and 33 purple-flowered plants grew. How could you describe inheritance of flower color? Describe how specific alleles influence each other and therefore affect phenotype.

2-21 "Secretors" (genotypes *SS* and *Ss*) secrete their A and B blood group antigens into their saliva and other body fluids, while "nonsecretors" (*ss*) do not. What would be the apparent phenotypic blood group proportions among the offspring of an $I^A I^B Ss$ woman and an $I^A I^A Ss$ man, if typing was done using saliva?

2-22 As you will learn in later chapters, duplication of genes is an important evolutionary mechanism. As a result, many cases are known in which a species has two nearly identical genes. Suppose there are two genes, *A* and *B*, that specify production of the same enzyme. An abnormal phenotype results only if an individual does not make any of that enzyme. What ratio of normal versus abnormal progeny would result from a mating between two parents of genotype *Aa Bb*, where *A* and *B* represent alleles that specify production of the enzyme, while *a* and *b* are alleles that do not?

2-23 The table below shows the responses of blood samples from the individuals in the pedigree to anti-A and anti-B sera. A "+" in the anti-A row indicates that the red blood cells (RBCs) of that individual were clumped by anti-A serum and therefore the individual made A antigens, and a "−" indicates no clumping. The same notation is used to describe the test for the anti-B antigens.

a. Deduce the blood type of each individual from the data in the table.

b. Assign genotypes for the blood groups as accurately as you can from this data, explain the pattern of inheritance shown in the pedigree. Assume that all genetic relationships are as presented in the pedigree (no cases of false paternity).

	I-1	I-2	I-3	I-4	II-1	II-2	II-3	III-1	III-2
anti-A	+	+	−	+	−	−	+	+	−
anti-B	+	−	+	+	−	−	+	−	−

2-24 Normally wild violets have yellow petals with dark brown markings and erect stems. Imagine you discover a plant with white petals, no markings, and prostrate stems. What experiment could you perform to determine whether the non-wild-type phenotypes are due to several different mutant genes or to the pleiotropic effects of alleles at a single locus? Explain how your experiment would settle the question.

2-25 What phenotypic ratios would you predict from crossing triply heterozygous flies if three genes affect a trait and one dominant allele of *each* gene is necessary to get a wild-type phenotype?

2-26 Three different pure-breeding strains of corn that produce ears with white kernels were crossed to each other. In each case the F_1 plants were all red, while both red and white kernels were observed in the F_2 generation in a 9:7 ratio. These results are tabulated below.

	F_1	F_2
White-1 × White-2	red	9 red:7 white
White-1 × White-3	red	9 red:7 white
White-2 × White-3	red	9 red:7 white

a. How many genes are involved in determining kernel color in these three strains?
b. Define your symbols and show the genotypes for these pure-breeding strains, White-1, White-2, and White-3.
c. Diagram the cross between White-1 and White-2, showing the genotypes and phenotypes of the F_1 and F_2 progeny. Explain the observed 9:7 ratio.

2-27 You have discovered a married man and woman, both of whom are deaf, who are carrying some mutant alleles (all recessive to wild type) in three different "hearing genes." The genes are called *D1, D2,* and *D3.* Homozygosity for a mutant allele (*m*) at any one of these three genes will cause deafness through the elimination of a critical component of the developing ear. In addition, there is another interaction among these three genes in that homozygosity for any two of the three genes together in the same genome will cause prenatal lethality (and spontaneous abortion) with a penetrance of 25%. Furthermore, homozygosity for the mutant alleles of all three genes will cause prenatal lethality with a penetrance of 75%. If the genotypes of the mother and father are as indicated below, what is the likelihood that a live-born child will be deaf?

Mother:	$D1^+/D1^m, D2^+/D2^m, D3^m/D3^m$
Father:	$D1^m/D1^m, D2^+/D2^m, D3^+/D3^m$

2-28 You have come into contact with two unrelated patients who express what you think to be a rare phenotype—a dark spot on the bottom of the foot. According to a medical source, this phenotype is seen in one in every 100,000 people in the population. The two patients give their family histories to you and you generate the following pedigrees.

The Smiths

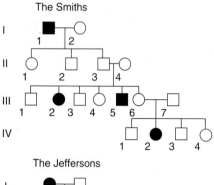

The Jeffersons

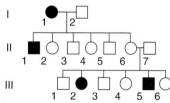

a. Given that this trait is rare, do you think the inheritance is dominant or recessive? Are there any special conditions that appear to apply to the inheritance?
b. Which nonexpressing members of these families must carry the mutant allele?
c. If this trait is instead quite common in the population, what alternative explanation would you propose for the inheritance?
d. Based on this new explanation (part c), which nonexpressing members of these families must carry the mutant allele?

THE CHROMOSOME THEORY OF INHERITANCE

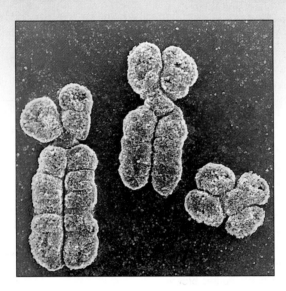

Each of these three human chromosomes carries
thousands of genes.

Down syndrome was the first human genetic disorder attributable not to a gene mutation, but to an abnormal number of chromosomes. In the spherical, membrane-bounded nuclei of plant and animal cells prepared for viewing under the microscope, *chromosomes* appear as brightly staining, threadlike, colored bodies. The nuclei of normal, healthy human cells carry 23 pairs of chromosomes for a total of 46. There are noticeable differences in size and shape among the 23 pairs, but within each pair, the two chromosomes appear to match exactly at this level of resolution. (The only exceptions are the male's sex chromosomes, designated X and Y, which constitute an unmatched pair.) Children born with Down syndrome have 47 chromosomes in each cell nucleus because they carry three, instead of the normal pair, of a very small chromosome referred to as number 21. The aberrant genotype, known as trisomy 21, gives rise to an abnormal phenotype, including a wide skull that is flatter than normal at the back; an unusually large tongue; learning disabilities caused by the abnormal development of the hippocampus and other parts of

the brain; and a propensity to respiratory infections as well as heart disorders, rapid aging, and leukemia (Fig.3.1).

How can one extra copy of a chromosome that is itself of normal size and shape cause such wide-ranging phenotypic effects? The answer has two parts. First and foremost, chromosomes are the cellular organelles responsible for transmitting genetic information. That is, each chromosome carries a number of genes that interact with each other and those of other chromosomes to determine the development, behavior, and appearance of an individual. (Note that we use the term "organelle" in its most general sense to designate a large, complex structure of characteristic shape that performs essential functions within the cell, rather than in its more specific sense of an intracellular, membrane-bound entity.) In this chapter, we describe how geneticists interpreted various bits of evidence to arrive at the idea that chromosomes are the carriers of genes, a proposal that became known as the **chromosome theory of inheritance.** The second part of the answer is that proper development depends not just on what type of genetic

Figure 3.1 Down syndrome: One extra chromosome has widespread phenotypic consequences. An extra copy of chromosome 21 (trisomy 21) usually causes changes in the physical characteristics of the head and body, as well as in the potential for learning. Despite these differences, many children with Down syndrome participate fully in regular classroom activities. The 5th grader at the photo's center in the multicolored, striped shirt with his hand raised to answer a question is one such student.

material is present but also on how much of it there is. Thus the mechanisms governing gene transmission during cell division must vigilantly maintain each cell's chromosome number.

Proof that genes are located on and travel with chromosomes comes from both breeding experiments and the microscopic examination of cells. As we analyze the evidence, you will see that the behavior of chromosomes during one type of nuclear division called **meiosis** accounts for the segregation and independent assortment of genes proposed by Mendel. Meiosis figures prominently in the process by which a sexually reproducing organism generates the gametes—eggs or sperm—that at fertilization unite to form the first cell of the next generation. This first cell is the fertilized egg, or *zygote*. Once the union of egg and sperm creates the zygote, a second kind of nuclear division, known as **mitosis,** occurs during the millions of cell divisions that propel development from a single cell to a complex multicellular organism. Mitosis is a conservative process that provides each of the many cells in an individual with the same number and types of chromosomes. Both meiosis and mitosis occur only in the eukaryotic cells of yeast

and higher organisms; these cells with a nucleus are the focus of our discussion in this chapter.

The precise chromosome-parceling mechanisms of meiosis and mitosis are so crucial to the normal functioning of an organism that when the machinery does not function properly, errors in chromosome distribution can have dire repercussions on the individual's health and survival. Down syndrome, for example, is the result of a failure of chromosome segregation during meiosis. The meiotic error gives rise to an egg or sperm carrying an extra chromosome 21, which if incorporated in the zygote at fertilization, is passed on via mitosis to every cell of the developing embryo. Trisomy—three copies of a chromosome instead of two—can occur with other chromosomes, but in nearly all such cases, the condition is prenatally lethal and results in a miscarriage.

Two themes emerge in our discussion of meiosis and mitosis. First, it was direct microscopic observations of pairs of chromosomes segregating during gamete formation that led early twentieth-century investigators to recognize that *chromosome movements parallel the behavior of Mendel's genes and are likely to carry the genetic material.* This chromosome theory of inheritance was proposed in 1902 and confirmed in the following 15 years through elegant experiments performed mainly on the fruit fly *Drosophila melanogaster.* Second, by locating the genetic material in a cell organelle specialized for the storage and transmission of genetic material, the chromosome theory transformed the concept of a gene from an abstract particle to a physical reality—part of a chromosome that could be seen and manipulated.

Our discussion of the connection between Mendel's laws of heredity and the behavior of chromosomes during cell division examines

■ Observations and experiments that placed the hereditary material in the cell nucleus, specifically on the chromosomes.

■ Mitosis, which distributes chromosomes equally to each of two daughter cells, ensuring that every cell in an organism carries the same set of chromosomes.

■ Meiosis, which distributes one member of each chromosome pair to the germ cells, generating gametes.

■ Gametogenesis, a process including both meiotic and mitotic divisions, by which specialized cells differentiate into gametes.

■ Validation of the chromosome theory of inheritance.

CHROMOSOMES CONTAIN THE GENETIC MATERIAL

One of the first questions asked at the birth of an infant—Is it a boy or a girl?—acknowledges that male and female are mutually exclusive characteristics like the yellow versus green of Mendel's peas. What's more, among humans and many other sexually reproducing species, there is a roughly 1:1 ratio of the two genders. In any randomly chosen human population, for example, approximately 50% are women and 50% men. Both males and females produce cells specialized for reproduction—sperm or eggs—that serve as a physical link to the next generation. In bridging the gap between generations, these gametes must each contribute half of the genetic material for making a normal, healthy son or daughter. Whatever part of the cell carries this material, its structure and function must be able to account for the either-or aspect of sex determination as well as the generally observed 1:1 ratio of men to women. As we will see, these two features of sex determination were among the earliest clues to the cellular basis of heredity.

Evidence That Genes Reside in the Nucleus

The nature of the link between sex and reproduction remained a mystery until Anton van Leeuwenhoek, one of the earliest and most astute of microscopists, discovered in 1667 that semen contains spermatozoa (literally "sperm animals"). He imagined that these microscopic creatures might enter the egg and somehow achieve fertilization, but it was not possible to confirm this hypothesis for another 200 years. Then, during a 20-year period starting in 1854 (about the same time Gregor Mendel was beginning his pea experiments), microscopists studying fertilization in frogs and sea urchins observed the union of male and female gametes and recorded the details of the process in a series of time-lapse drawings. These drawings, as well as later micrographs (photographs shot through a microscope), clearly show that egg and sperm nuclei are the only elements contributed equally by maternal and paternal gametes. This observation implies that something in the nucleus contains the hereditary material. In humans, the nucleus of a sperm is less than 1 millionth of a meter across at its widest point, and the nucleus of an egg, a little more than twice that size. It is indeed remarkable that the genetic link between generations is packaged within such an exceedingly small space.

Evidence That Genes Reside in Chromosomes

Further investigations, some dependent on technical innovations in microscopy, suggested that yet smaller, discrete structures within the nucleus are the repository of genetic information. In the 1880s, for example, a newly discovered combination of organic and inorganic dyes that stained the nucleus of a cell blue, the cytoplasm pink, and discrete nuclear structures brighter blue than their surroundings, revealed the existence of the long, threadlike bodies within the nucleus that we call **chromosomes** (literally "colored bodies"). It was now possible to follow the movement of chromosomes during different kinds of cell division.

In embryonic cells, the brightly staining chromosomal threads split lengthwise in two just before cell division, and each of the two newly forming daughter cells received one half of every split thread. The kind of nuclear division followed by cell division that results in two daughter cells containing the same number and type of chromosomes as the original parent cell is called **mitosis** (from the Greek *mitos* meaning "thread" and *-osis* meaning "formation" or "increase"). In contrast to mitosis, in the cells that give rise to male and female gametes, a key event is the segregation of the chromosomes composing each pair such that the resulting gametes receive only one chromosome from each chromosome pair. The gametes thus have half the number of chromosomes as other cells within the same organisms. The kind of nuclear division that generates egg or sperm cells containing half the number of chromosomes found in other cells within the same organism is called **meiosis** (from the Greek word for "diminution").

Figure 3.2 The great lubber grasshopper. In this mating pair, the smaller male is astride the female.

One Chromosome Pair Determines an Individual's Sex

In the first decade of the twentieth century, cytologists—scientists who use the microscope to study cell structure—realized that particular chromosomes carry the information for determining sex. In one study, Walter S. Sutton, a young American graduate student at Columbia University, obtained cells from the testes of the great lubber grasshopper (*Brachystola magna*) (Fig. 3.2) and followed them through the meiotic divisions that produce sperm. After Sutton's original studies had been corrected for the existence of the very small Y chromosome (which he had missed), the following picture emerged. Prior to meiosis, precursor cells within the testes of a great lubber grasshopper contain 22 chromosomes in addition to an X chromosome and a Y chromosome—a total of 24 chromosomes. After meiosis, the sperm produced within these testes are of two equally prevalent types: one half have a set of 11 chromosomes plus an X chromosome, while the other half have a set of 11 chromosomes plus a Y. By comparison, all of the eggs produced by females of the species carry an 11-plus-X set of chromosomes like the set found in the first class of sperm. When a sperm with an X chromosome fertilizes an egg, an XX female grasshopper results; when a Y-containing sperm fuses with an egg, an XY male develops. Sutton concluded that the X and the Y chromosomes determine sex.

Several researchers studying other organisms observed and argued the same point, and it soon became apparent that in most sexually reproducing multicellular species, two distinct chromosomes—known as the **sex chromosomes**—provide the basis of sex determination. One sex carries two copies of the same chromosome (a matching pair), while the other sex has one of each type of sex chromosome (an unmatched pair). The cells of normal human females, for example, contain 23 pairs of chromosomes. The two chromosomes of each pair, including the sex-determining X chromosomes, appear to be identical in size and shape. In males, however, there is one unmatched pair of chromosomes; the larger of these is the X; the smaller, the Y (Fig. 3.3a). Apart from this differ-

(a)

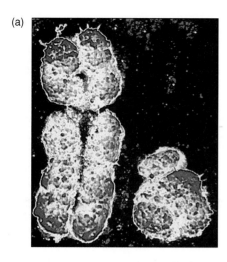

(b)

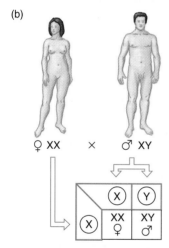

Figure 3.3 **How the X and Y chromosomes determine sex in humans.** (a) This colorized micrograph shows the differences in appearance between the human X (on the left) and Y (on the right) chromosomes. (b) Children can receive only an X chromosome from their mother, but can inherit either an X or a Y from their father. Separation (segregation) of the X and Y chromosomes during meiosis in the father ensures that roughly half of humankind will be females and the other half males.

ence in sex chromosomes, the two sexes are not distinguishable at any other pair of chromosomes. Thus, geneticists can designate women as XX and men as XY and represent sexual reproduction as a simple cross between the two: XX × XY. Now, if sex is an inherited trait subject to Mendel's rule of segregation—that is, if it is determined by a pair of sex chromosomes that separate to different cells during gamete formation—such a cross could account for both the mutual exclusion of genders and the near 1:1 ratio of males to females that are among the most striking features of sex determination (Fig. 3.3b). And if chromosomes carry information defining the two contrasting sex phenotypes, it is not extravagant to infer that chromosomes also carry genetic information specifying other characteristics as well. If so, segregation of the chromosomes of every pair—sex and nonsex alike—during gamete formation would produce eggs and sperm each contain-

ing one full set of chromosomes (one chromosome from every pair). As a result, at fertilization, each gamete would contribute one full set of chromosomes to the zygote. The fertilized egg, or zygote would thus have two sets of chromosomes.

At Fertilization, Haploid Gametes Produce Diploid Zygotes

Sutton's studies confirmed this idea by showing that the chromosomes in a fertilized egg actually consist of two matching sets, one contributed by the maternal gamete, the other by the paternal gamete. The corresponding maternal and paternal chromosomes appear alike in size and shape (with the exception of the sex chromosomes in most species). While the zygote carries two complete chromosome sets, each gamete contains one-half the number of chromosomes in the zygote, or one complete set. Gametes and other cells that carry only a single set of chromosomes are called **haploid** (from the Greek word for "single"). Zygotes and other cells carrying two matching sets are **diploid** (from the Greek word for "double"). The number of chromosomes in a normal haploid cell is designated by the shorthand symbol $n;$ the number of chromosomes in a normal diploid cell is then **2n**. Figure 3.4 shows diploid cells as well as the haploid gametes that arise from

Drosophila melanogaster

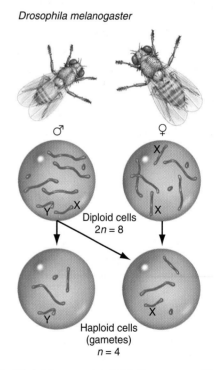

Figure 3.4 **Diploid versus haploid: 2n versus n.** Most of the cells in the body are diploid: They carry a maternal and paternal copy of each chromosome. In *Drosophila*, diploid cells have 8 chromosomes ($2n = 8$). The process of meiosis generates gametes—haploid cells containing only one copy of each chromosome. In *Drosophila*, gametes have 4 chromosomes ($n = 4$) apiece. The fusion of gametes at fertilization yields a diploid zygote (not shown). Note that the chromosomes in this diagram are pictured before their replication.

them in *Drosophila* where $2n = 8$ and $n = 4$. In humans, $2n = 46$; $n = 23$. You can see how the halving of chromosome number during meiosis and gamete formation, followed by the union of two gametes' chromosomes at fertilization, allows a constant $2n$ number of chromosomes to be maintained from generation to generation in all individuals of a species (except, of course, for those carrying rare abnormalities such as trisomy 21). The process of mitosis then ensures that all the cells of the developing individual have identical diploid chromosome sets.

The Number and Shape of Chromosomes Vary from Species to Species

Scientists analyze the chromosomal makeup of a cell by observing the nucleus when the chromosomes are most visible—at a specific moment in the cell cycle of growth and division, just before the nucleus divides. At this point, known as *metaphase* (described in detail later), individual chromosomes have duplicated and condensed from thin threads into compact rodlike structures, each one consisting of two identical halves known as **sister chromatids** attached to each other at a specific location called the **centromere** (Fig. 3.5). In **metacentric** chromosomes, the centromere is more or less in the middle; in **acrocentric** chromosomes, the centromere is very close to one end. When viewed under the microscope, some chromosomes appear to have their centromere right at one end, and geneticists originally coined the term "telocentric" to describe these chromosomes. Recent evidence from high-resolution cytolog-

ical studies, however, indicates that all apparently telocentric chromosomes have at least a small region extending beyond the centromere. As a result, even though it is not always possible to see this region under the microscope, cytogeneticists refer to all chromosomes whose centromere appears near or at one end as acrocentric.

Cells in metaphase can be fixed and stained with one of several dyes that highlight the chromosomes and accentuate the centromeres. In each kind of chromosome, under the right conditions, the dyes also produce characteristic banding patterns of lighter and darker regions, making it possible to distinguish matching chromosome pairs. Chromosomes that match in size, shape, and banding are called **homologous chromosomes,** or **homologs.** The two homologs of each pair normally contain the same set of genes, although for some of those genes, they may carry different alleles. (The differences between alleles are generally too small to show up in the microscope.)

Figure 3.5 introduces a system of notation employed throughout the remainder of the book, using color to indicate degrees of relatedness between chromosomes. Thus, sister chromatids—the two daughter molecules of a duplicated chromosome that are joined by a single centromere—which are identical at the completion of replication, appear in the same shade of the same color. Homologous chromosomes, which carry the same genes but may vary in the identity of particular alleles, are pictured in different shades (light or dark) of the same color. *Nonhomologous chromosomes,* which carry completely unrelated sets of genetic information, appear in different colors.

To study the chromosomes of a single organism, geneticists cut images of the stained chromosomes from a micrograph and, by convention, arrange them in homologous pairs of decreasing size to produce a **karyotype.** Figure 3.6 shows the karyotype of a human male, with 46 chromosomes arranged in 22 matching pairs of nonsex chromosomes and one nonmatching pair of sex chromosomes. The 44 nonsex

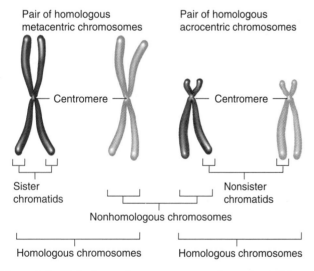

Figure 3.5   Metaphase chromosomes may be metacentric, acrocentric, or somewhere in between. By metaphase, the chromosomes have condensed into compact, discrete structures. Earlier, each chromosome has replicated into two sister chromatids connected at a centromere. The centromere can be near the middle of the chromosome (making it a metacentric chromosome), very near the end (forming an acrocentric chromosome), or anywhere in between. In a diploid cell, the chromosomes come in homologous pairs, with one chromosome in the pair from the mother and the other from the father.

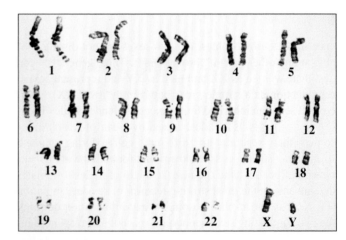

Figure 3.6   Karyotype of a human male. Computer-enhanced photos of metaphase human chromosomes are paired and arranged in order of decreasing size. In a normal human male karyotype, there are 22 pairs of autosomes, as well as an X and a Y ($2n = 46$).

chromosomes are known as **autosomes.** As you can see, autosome number 21 is one of the smallest chromosomes. Interestingly, some dyes may reveal very slight differences in banding patterns between the maternally and paternally derived chromosomes of a homologous pair, and medical researchers can use this information to trace the origin of an extra chromosome, such as the extra 21 that causes Down syndrome. Data collected in studies of this type indicate that in 80% of cases, the third chromosome 21 comes from the egg; in 20%, from the sperm.

Through thousands of karyotypes, cytologists have verified that the cells of each species carry a distinctive diploid number of chromosomes. Among three species of fruit flies, for example, *Drosophila melanogaster* carries 8 (4 pairs of) chromosomes, *Drosophila obscura* carries 10 (5 pairs), and *Drosophila virilis,* 12 (6 pairs). Differences in the size, shape, and number of chromosomes reflect differences in the assembled genetic material that determines what each species looks like and how it functions. Mendel's peas contain 14 (7 pairs of) chromosomes in each diploid cell, macaroni wheat has 28 (14 pairs), giant sequoia trees 22 (11 pairs); goldfish 94 (47 pairs), dogs 78 (39 pairs), and people 46 (23 pairs). As these figures show, the number of chromosomes does not always correlate with the size or complexity of the organism.

There Is Variation between Species in How Chromosomes Determine an Individual's Sex

As we have already seen, humans and other mammals have a pair of sex chromosomes that are identical in the XX female but different in the XY male. Several studies have shown that in humans, it is the presence or absence of the Y that actually makes the difference; that is, any person carrying a Y chromosome will look like a male, whereas XO individuals carrying an X and no second sex chromosome are females that display certain abnormalities. Other species show variations on this XX = female/XY = male chromosomal strategy of sex determination. In fruit flies, for example, even though normal females are XX and normal males XY, it is ultimately the ratio of X chromosomes to autosomes (and not the presence or absence of the Y) that determines sex. In female *Drosophila,* the ratio is 1:1 (there are 2 X chromosomes and 2 copies of each autosome); in males, the ratio is 1:2 (there is 1 X chromosome but 2 copies of each autosome). Curiously, a rarely observed abnormal intermediate ratio of 2:3 produces intersex flies that display both male and female characteristics. Although the Y chromosome in *Drosophila* does not determine whether a fly looks like a male, it is necessary for male fertility; XO flies are sterile males. Table 3.1 compares how humans and *Drosophila* respond to unusual complements of sex chromosomes. Differences between the two species arise in part because the genes they carry on their sex chromosomes are not identical and in part because the strategies they use to deal with the presence of additional sex chromosomes are not the same.

The XX = female/XY = male strategy of sex determination is by no means universal. In some species of moths, for example, the females are XX but the males are XO. In birds and butterflies, it is the males that have the matching sex chromosomes, while the females have an unmatched set; in such species, geneticists represent the sex chromosomes as ZZ in the male and WZ in the female. Yet other variations include the complicated sex-determination mechanisms of bees and wasps, in which females are diploid and males haploid, and the systems of certain fish, in which sex is determined by changes in the environment, such as fluctuations in temperature.

Despite these differences, it became clear that chromosomes carry the genetic information specifying sexual identity and probably many other characteristics as well. Sutton and other early adherents of the chromosome theory realized that because of this, the perpetuation of life itself depends on the proper distribution of chromosomes during cell division. In the next sections, as we examine step by step the microscopic maneuvers by which cells maintain their crucial chromosomal constancy, you will see that the behavior of chromosomes during mitosis and meiosis is exactly that expected of organelles carrying genes.

TABLE 3.1   Sex Determination in Fruit Flies and Humans

	Complement of Sex Chromosomes						
	XXX	XX	XXY	XO	XY	XYY	OY
Drosophila	Dies	Normal female	Normal female	Sterile male	Normal male	Normal male	Dies
Humans	Nearly normal female	Normal female	Kleinfelter male (sterile); tall, thin	Turner female (sterile); webbed neck	Normal male	Normal or nearly normal	Dies

Humans can tolerate extra X chromosomes better than *Drosophila* (compare the fate of XXX individuals). Lack of the X chromosome is lethal to both fruit flies and humans because this chromosome carries essential genes not found on other chromosomes. Additional Y chromosomes seem to have little effect in either species.

MITOSIS ENSURES THAT EVERY CELL IN AN ORGANISM CARRIES THE SAME CHROMOSOMES

The fertilized human egg is a single diploid cell that preserves its genetic identity unchanged through more than 100 generations of cells as it divides again and again to produce a full-term infant, ready to be born. As the newborn infant develops into a toddler, a teenager, and an adult, yet more cell divisions fuel continued growth and maturation. Mitosis, the nuclear division that apportions chromosomes in equal fashion to two daughter cells, is the cellular mechanism that preserves genetic information through all these generations of cells. In this section, we take a close look at how the nuclear division of mitosis fits into the overall scheme of cell growth and division.

If you were to peer through a microscope and follow the history of one cell through time, you would see that for much of your observation, the chromosomes resemble a mass of extremely fine tangled string—called **chromatin**—surrounded by the **nuclear envelope.** Each convoluted thread of chromatin is composed mainly of DNA (which carries the genetic information) and protein (which serves as a scaffold for packaging and managing that information, as described in Chapter 11). You would also be able to distinguish one or two darker areas of chromatin called **nucleoli** (singular *nucleolus,* literally "small nucleus"; see Chapter 11, page 392). Although the chromatin-laden nucleus appears quiescent during this period between cell divisions, it houses a great deal of invisible activity necessary for the growth and survival of the cell. One particularly important part of this activity is the accurate duplication of all the chromosomal material.

With continued vigilance, you would observe during a very short period in the life history of the cell a dramatic change in the nuclear landscape: The chromatin condenses into discrete threads and then each chromosome compacts even further into the twin rods clamped together at the centromere that can be identified in karyotype analysis (review Fig. 3.5). Each rod in a duo is called a **chromatid;** as described above, it is an exact duplicate of the other sister chromatid to which it is connected. Continued observation would reveal the doubled chromosomes beginning to jostle around and then lining up at the midplane of the cell. At this point, the sister chromatids comprising each chromosome separate to opposite poles of the now elongating cell, where they eventually end up as two identical sets of chromosomes enclosed in two separate nuclei in two separate cells.

The repeating pattern of cell growth (an increase in size) followed by division (the splitting of one cell into two) is called the **cell cycle** (Fig. 3.7). Only a small part of it is spent in division; the period between divisions is called **interphase.**

During Interphase, Cells Grow and Replicate Their Chromosomes

Interphase consists of three parts: gap 1 (**G_1**), synthesis (**S**), and gap 2 (**G_2**) (see Fig. 3.7a and b). G_1 lasts from the birth of a new cell to the onset of chromosome replication; for the genetic ma-

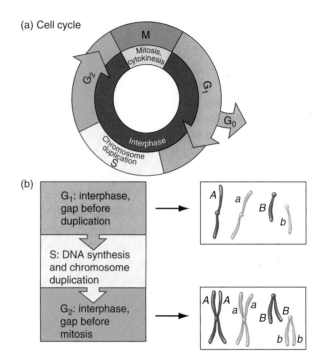

Figure 3.7   The cell cycle: An alternation between interphase and mitosis. (a) and (b) Diagram of the cell cycle. Chromosomes replicate to form sister chromatids during synthesis (S phase); the sister chromatids segregate to daughter cells during mitosis (M phase). The S and M phases are separated by gaps known as the G_1 and G_2 phases. It is during the gaps that most cell growth takes place. Interphase (b) consists of the G_1, S, and G_2 phases together. In multicellular organisms, some terminally differentiated cells may stop dividing and arrest in a so-called G_0 stage, as shown in (a).

terial, it is a period when the chromosomes are neither duplicating nor dividing. During this time, the cell makes and assembles the materials it needs to function normally and also achieves most of its growth. G_1 varies in length more than any other phase of the cell cycle. In rapidly dividing cells of the human embryo, for example, G_1 is as short as a few hours. In contrast, mature brain cells become arrested in a resting form of G_1 known as **G_0** and do not normally divide again during a person's lifetime.

Synthesis (S) is the time when the cell duplicates its genetic material by synthesizing DNA. During duplication, each chromosome doubles to produce identical sister chromatids that will become visible when the chromosomes condense at the beginning of mitosis. The two sister chromatids remain joined to each other at the centromere.

Gap 2 (G_2) is the interval between chromosome duplication and the beginning of mitosis. During this time, the cell may grow (usually less than during G_1); it also synthesizes proteins that are essential to the subsequent steps of mitosis itself.

In addition, during interphase an array of fine microtubules crucial for many interphase processes becomes visible outside the nucleus. The microtubules radiate out into the cytoplasm from a single organizing center known as the **centrosome,** usually located near the nuclear envelope. In animal cells, the discernible core of each centrosome is a pair of small, darkly staining bodies called **centrioles** (Fig. 3.8a); the

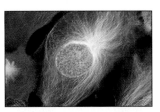
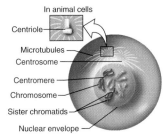

(a) Prophase: (1) Chromosomes condense and become visible; (2) centrosomes move apart toward opposite poles and generate new microtubules.

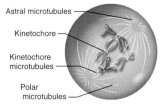

(b) Prometaphase: (1) Nuclear envelope and nucleoli disappear; (2) microtubules from the centrosomes invade the nucleus; (3) sister chromatids attach to microtubules from opposite centrosomes.

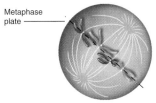

(c) Metaphase: Chromosomes align on the metaphase plate with sister chromatids facing opposite poles.

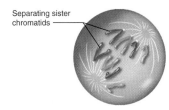

(d) Anaphase: (1) Centromeres divide; and (2) the now separated sister chromatids move to opposite poles.

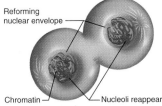

(e) Telophase: (1) Nuclear membrane and nucleoli reform; (2) spindle fibers disappear; and (3) chromosomes uncoil and become a tangle of chromatin.

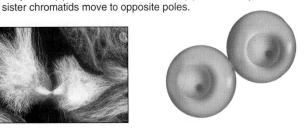

(f) Cytokinesis: The cytoplasm divides, splitting the elongated parent cell into two daughter cells with identical nuclei.

Figure 3.8   Mitosis maintains the chromosome number of the parent cell nucleus in the two daughter nuclei. In the photomicrographs (of newt lung cells), chromosomes are stained blue and microtubules appear either green or yellow.

microtubule-organizing center of plants does not contain centrioles. During the S and G$_2$ stages of interphase, the centrosomes replicate, producing two centrosomes that remain in extremely close proximity.

During Mitosis (M Phase), Sister Chromatids Separate and Are Apportioned to Different Daughter Nuclei

Although the rigorously choreographed events of nuclear and cellular division occur as a dynamic and continuous process, for the sake of clarity, we take the traditional approach of analyzing the process in separate stages marked by visible cytological events. The artist's sketches in Figure 3.8 illustrate these stages in the nematode *Ascaris,* whose diploid cells contain only four chromosomes (two homologous chromosome pairs).

Prophase: Chromosomes Condense (Fig. 3.8a)

During the whole of interphase, the cell nucleus remains intact and the chromosomes are indistinguishable aggregates of chromatin. At **prophase** (from the Greek *pro-* meaning "before"), the gradual emergence, or **condensation,** of individual chromosomes from the undifferentiated mass of chromatin marks

the beginning of mitosis. Each emerging chromosome has already been duplicated during interphase and thus consists of sister chromatids attached at the centromere. At this stage in *Ascaris* cells, there are therefore 4 chromosomes with a total of 8 chromatids. The progressive appearance of an array of individual chromosomes is a truly impressive event: Interphase molecules as long as 3–4 cm condense into discrete chromosomes whose length is measured in microns (millionths of a meter). This is equivalent to compacting a 200-m length of thin string (as long as two football fields) into a cylinder 8 mm long and 1 mm wide. With condensation, the duplicated chromosomes convert from a metabolically active state to a condition suitable for subsequent transport to daughter cells. Another visible change in chromatin also takes place during prophase: The darkly staining nucleoli begin to break down and disappear.

In addition to these nuclear events, several important processes that characterize prophase occur outside the nucleus in the cytoplasm. The centrosomes, which replicated during interphase, now move apart and become clearly distinguishable as two separate entities in the light microscope. At the same time, the interphase scaffolding of long, stable microtubules disappears and is replaced by a set of microtubules that rapidly grow from and contract back toward their centrosomal organizing centers. The centrosomes continue to move apart, migrating around the nuclear envelope toward opposite ends of the nucleus, apparently propelled by forces exerted between interdigitated microtubules extending from both centrosomes.

Prometaphase: The Spindle Forms (Fig. 3.8b)

Prometaphase ("before middle stage") begins with the breakdown of the nuclear envelope, which allows microtubules extending from the two centrosomes to invade the nucleus. Chromosomes attach to these microtubules through the **kinetochore,** a structure in the centromere region of each chromatid that is specialized for conveyance. Each kinetochore contains proteins that act as molecular motors, enabling the chromosome to slide along the microtubule. When the kinetochore of a chromosome originally contacts a microtubule at prometaphase, the kinetochore-based motor moves the chromosome toward the centrosome from which that microtubule radiates. Microtubules growing from the two centrosomes randomly capture chromosomes by the kinetochore of one of the two sister chromatids. As a result, it is sometimes possible to observe groups of chromosomes congregating in the vicinity of each centrosome. In this early part of prometaphase, for each chromosome, one chromatid's kinetochore is attached to a microtubule but the sister chromatid's kinetochore remains unattached.

During prometaphase, three different types of microtubule fibers together form the **mitotic spindle;** all of these microtubules originate from the centrosomes, which function as the two "poles" of the spindle apparatus. Microtubules that extend between a centrosome and the kinetochore of a chromatid are called **kinetochore microtubules,** or **centromeric fibers.** Microtubules from each centrosome that are directed toward the middle of the cell are **polar microtubules;** polar microtubules originating in opposite centrosomes interdigitate near the cell's equator. Finally, there are short, unstable **astral microtubules** that extend out from the centrosome toward the cell's periphery.

Near the end of prometaphase, the kinetochore of each chromosome's previously unattached sister chromatid now associates with a microtubule extending from the opposite centrosome. This event orients each chromosome such that one sister chromatid faces one pole of the cell, the other, the opposite pole. Experimental manipulation has shown that if both kinetochores become attached to microtubules from the same pole, the configuration is unstable; one of the kinetochores will repeatedly detach from the spindle until it associates with a microtubule from the other pole. Thus, attachment of sister chromatids to opposite spindle poles is the only stable arrangement.

Metaphase: Chromosomes Align at the Cell's Equator (Fig. 3.8c)

During **metaphase** ("middle stage"), the connection of sister chromatids to opposite spindle poles sets in motion a series of jostling movements that cause the chromosomes to move toward an imaginary equator halfway between the two poles. The imaginary midline is called the **metaphase plate.** When the chromosomes are aligned along it, the forces pulling and pushing them toward or away from each pole are in a balanced equilibrium. As a result, any movement away from the metaphase plate is rapidly compensated by tension that restores the chromosome to its position equidistant between the poles.

Anaphase: Sister Chromatids Move to Opposite Spindle Poles (Fig. 3.8d)

The nearly simultaneous severing of the centromeric connection between the sister chromatids of all chromosomes indicates that **anaphase** (from the Greek *ana-* meaning "up" as in "up toward the poles") is underway. The separation of sister chromatids allows each chromatid to be pulled toward the spindle pole to which it is connected by its kinetochore microtubule; as it moves toward the pole, the kinetochore microtubule shortens. Since the arms of the chromatids lag behind the kinetochores, metacentric chromatids have a characteristic V shape during anaphase. The connection of sister chromatids to microtubules emanating from opposite spindle poles means that the genetic information migrating toward one pole is exactly the same as its counterpart moving toward the opposite pole.

Telophase: Identical Sets of Chromosomes Are Enclosed in Two Nuclei (Fig. 3.8e)

The final transformation of chromosomes and the nucleus during mitosis happens at **telophase** (from the Greek *telo-* meaning "end"). Telophase is like a rewind of prophase. The spindle fibers begin to disperse; a nuclear envelope forms around the chromatids, each of which now functions as an independent chromosome; one or more nucleoli reappears; and the chromosomes decondense (uncoil) and dissolve into a tangled mass of chromatin. Mitosis, the division of one nucleus into two identical nuclei, is over.

Cytokinesis: The Cytoplasm Divides (Fig. 3.8f)

In the final stage of cell division, the daughter nuclei emerging at the end of telophase are packaged into two separate daughter cells. This final stage of division is called **cytokinesis** (literally "cell movement"). During cytokinesis, the elongated parent cell separates into two smaller independent daughter cells with identical nuclei. Cytokinesis usually begins during anaphase, but it is not completed until after telophase. The mechanism by which cells accomplish cytokinesis differs in animals and plants. In animal cells, cytoplasmic division depends on a **contractile ring** that pinches the cell into two approximately equal halves, similar to the way the pulling of a string closes the opening of a bag of marbles (Fig. 3.9a). In plants, whose cells are surrounded by a rigid cell wall, a membrane-enclosed disk, known as the **cell plate,** forms inside the cell near the equator and then grows rapidly outward, thereby dividing the cell in two (Fig. 3.9b). During cytokinesis, a large number of important organelles and other

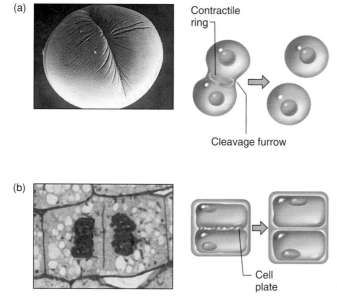

Figure 3.9 **Cytokinesis: The cytoplasm divides, producing two daughter cells.** (a) In this dividing frog's egg cell, a ring of actin filaments at the cell periphery has contracted to form a cleavage furrow visible from the exterior that will eventually pinch the cell in two. (b) In the dividing onion root cell at the center of this photo, a cell plate that began forming near the equator of the cell expands to the periphery, separating the two daughter cells.

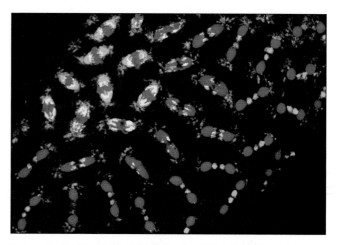

Figure 3.10 **If cytokinesis does not follow mitosis, one cell may contain many nuclei.** Immediately after fertilization of the *Drosophila* egg, 13 rounds of mitosis take place, but cytokinesis does not occur. The result is a single-celled embryo, such as the one shown in the photograph, containing several thousand nuclei. Membranes subsequently grow around these nuclei, dividing the embryo into cells.

cellular components, including ribosomes; mitochondria; membranous structures such as Golgi bodies; and in plants, chloroplasts, must be parcelled out to the emerging daughter cells. The mechanism accomplishing this task does not appear to predetermine which organelle is destined for which daughter cell. Instead, because most cells contain many copies of these cytoplasmic structures, each new cell is bound to receive at least a few representatives of each component. This original complement of structures is enough to sustain the cell until synthetic activity can repopulate the cytoplasm with organelles.

Sometimes cytoplasmic division does not immediately follow nuclear division, and the result is a cell containing more than one nucleus. An animal cell with two or more nuclei is known as a **syncytium.** The early embryos of fruit flies are multinucleated syncytia (Fig. 3.10), as are the precursors of spermatozoa in humans and many other animals. A multinucleate plant tissue is called a **coenocyte;** coconut milk is a nutrient-rich food composed of coenocytes.

In summary, through mitosis plus cytokinesis, a parental cell generates two daughter cells that have the same number and kind of chromosomes as each other and as the original parental cell.

Regulatory Checkpoints Ensure Correct Chromosome Separation During Mitosis

The cell cycle is a complex series of coordinated events that proceed in sequence with maximum precision. In higher or-

ganisms, a cell's "decision" to divide depends on both intrinsic factors, such as conditions within the cell that register a sufficient size for division, and signals from the environment, such as hormonal cues or contacts with neighboring cells that encourage or restrain division. Once a cell has initiated events leading to division, usually during the G_1 period of interphase, everything else follows like clockwork. A number of checkpoints—moments at which the cell evaluates the results of previous steps—allow the precise, sequential coordination of cell-cycle events so that, under normal circumstances, the chromosomes replicate before they condense, and the doubled chromosomes separate to opposite poles only after correct metaphase alignment of sister chromatids ensures equal distribution to the daughter nuclei (Fig. 3.11). In one illustration of the molecular basis of checkpoints, kinetochores that have not attached to spindle fibers generate a molecular signal that prevents sister chromatids from separating at the centromere. This signal makes the beginning of anaphase dependent on the prior proper alignment of chromosomes at metaphase. As a result of the multiple checkpoints operating during the cell cycle, each daughter cell reliably receives the right number of chromosomes.

Breakdown of the mitotic machinery can produce division mistakes that have crucial consequences for the cell. Improper chromosome segregation, for example, can cause serious malfunction or even the death of daughter cells. As the Fast Forward box "How Gene Mutations Cause Errors in Mitosis" explains, one source of improper segregation is gene mutations that disrupt mitotic structures, such as the spindle, kinetochores, or centrosomes. Other problems occur in cells where the normal restraints on cell division have broken down. Such cells may divide uncontrollably, leading to a tumor.

Fast Forward

How Gene Mutations Cause Errors in Mitosis

During each cell cycle, the chromosomes participate in a tightly patterned choreography that leads to duplication before condensation, midplane alignment before sister chromatid separation, and nuclear division before cytokinesis. Because of these sequential steps, synchronized in both time and space, the chromosomes convey a complete set of genes to each of two newly forming daughter cells. Not surprisingly, some of the very genes they carry encode proteins that shepherd them through the dance.

A variety of proteins, some assembled into structures such as centrosomes and microtubule fibers, make up the molecular machinery that helps coordinate the orderly progression of events in mitosis, cytokinesis, and interphase. Because a particular gene specifies each protein, we might predict that mutant alleles generating defects in particular proteins could disrupt the dance. Cells homozygous for a mutant allele might be unable to complete chromosome duplication or mitosis or cytokinesis because a required component of the molecular machinery is missing or unable to function. Experiments on organisms as disparate as yeast and fruit flies have borne out this prediction. Here we describe the effects of mutations in three *Drosophila* genes that alter chromosome segregation in different ways.

Although most mistakes in mitosis are eventually lethal to a multicellular organism, some mutant cells may manage to divide early in development, and when prepared for viewing under the microscope, these cells actually allow us to see the effects of defective mitosis. For example, Fig. A.1 shows the eight condensed metaphase chromosomes of a wild-type male fruit fly (*Drosophila melanogaster*): two pairs of large metacentric autosomes with the centromere in the center; a pair of dotlike autosomes that are so small it is not possible to see the centromere region, a telocentric X chromosome with the centromere at one end, and a metacentric Y chromosome. Because most of the Y chromosome consists of a form of chromatin known as heterochromatin, the two Y sister chromatids remain so tightly connected that they often appear as one. Figure B.1 shows the results of aberrant mitosis in an animal homozygous for a mutation that disrupted mitotic chromosome segregation during early development. The aberration, which probably occurred during anaphase, caused several chromatids to become incorporated into the wrong daughter nucleus at telophase. The result was a cell with extra autosomes and X chromosomes. Figure A.2 shows a normal anaphase separation leading to the wild-type chromosome complement seen in Fig. A.1. Figure B.2 portrays

an aberrant anaphase separation in a mutant animal that could lead to an abnormal chromosome complement similar to that seen in Fig. B.1.

A mutation in a different *Drosophila* gene active in early development blocks cytokinesis, the division of the cytoplasm that separates two newly made nuclei into two daughter cells. Failure of cytokinesis creates a cell with two nuclei and a total of 4*n*, or 16,

X chromosome

(1)

(2)

Y chromosome

Figure A   Metaphase and anaphase chromosomes in a wild-type fly. (1) Metaphase chromosomes of a wild-type male fruit fly. (2) Chromosomes at anaphase in a cell from the same fly.

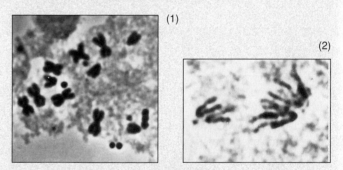

(1)

(2)

Figure B   Metaphase and anaphase chromosomes in a mutant fly. (1) Metaphase chromosomes in a cell from a *Drosophila* male hemizygous for an X-linked mutation called *zw10*. Note the presence of extra chromosomes as compared with Fig. A.1. (2) An anaphase cell in the same *zw10* mutant male as in part 1 at the left. Some chromosomes do not move properly to the poles, and can become incorporated into the wrong daughter cell.

chromosomes. Such a cell can proceed to a new round of cell division, and if cytokinesis again does not occur, an 8n-, or 32-chromosome, cell will result. Multiple rounds of mitosis not followed by cytokinesis can generate cells with hundreds of chromosomes, rather than the usual eight (Fig. C).

A completely different type of mutation affects the development of fruit fly embryos. During early embryonic development, cytokinesis does not normally follow the first 13 rounds of mitosis. The result is a giant sac, a syncytium, containing close to 6000 nuclei dispersed in a single layer covering the cytoplasm and egg yolk. At this point, an as yet undiscovered cue causes the simultaneous formation of hexagonal furrows of membrane around each nucleus and its associated cytoplasm (Fig. D). The process is analogous to a mas-

sively synchronized cytokinesis that accurately creates thousands of cells with one nucleus apiece. Several different genes encode the proteins that construct this amazing network of cell membranes at the right time and in the right place. A mutation in one of these genes, called *serendipity alpha* (*sry* α), disrupts the methodical cell-membrane construction process, causing the formation of irregularly shaped cells with as many as 25 nuclei each (Fig. E). The protein determined by the normal *sry* α gene appears in the embryonic cytoplasm immediately before cellularization and apparently regulates the positioning of the cell-dividing furrows by binding actin, a protein necessary for cytoplasmic cleavage.

The smooth unfolding of each cell cycle thus depends on a diverse array of proteins. Some form the structures that transport chromosomes or cleave cells in two; others help control when and where things happen. Particular genes specify each of the proteins active in mitosis and cytokinesis, and each protein makes a contribution to the coordinated events of the cell cycle. As a result, a mutation in any of a number of genes can disrupt the meticulously choreographed mitotic mechanism that, when functioning normally, creates generation upon generation of chromosomally identical diploid daughter cells.

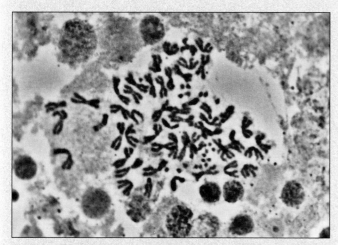

Figure C   Mitosis in the absence of cytokinesis. Mutations in a gene called *spaghetti-squash* block cytokinesis but not mitosis, resulting in cells with a geometric increase of the 2n chromosome number.

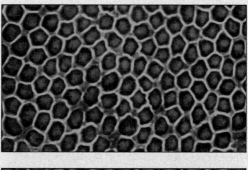

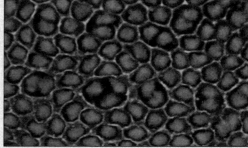

Figure E   Construction of cell membranes in a mutant *Drosophila* embryo. (*Top*) Cellularization of a wild-type embryo. Red shows the position of nuclei, while green marks the cell membranes. (*Bottom*) Aberrant cellularization of an embryo homozygous for a mutation in the *serendipity alpha* gene. Note that several of the cells contain more than one nucleus.

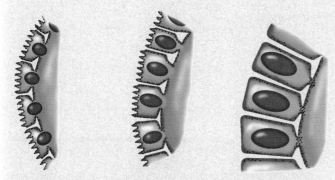

Figure D   Construction of cell membranes in a wild-type *Drosophila* embryo. Purple shows the position of nuclei; spots mark the membranes that form around the nuclei, dividing the syncytial embryo into individual cells. Each resulting cell contains one nucleus.

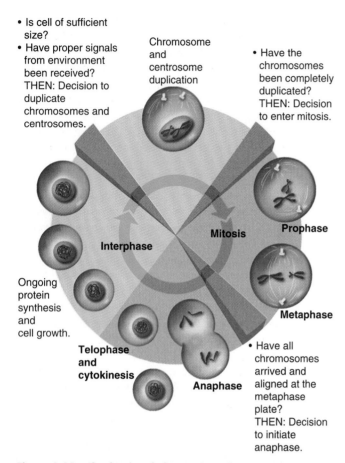

- Is cell of sufficient size?
- Have proper signals from environment been received? THEN: Decision to duplicate chromosomes and centrosomes.

Chromosome and centrosome duplication

- Have the chromosomes been completely duplicated? THEN: Decision to enter mitosis.

Prophase

Mitosis

Metaphase

Interphase

Ongoing protein synthesis and cell growth.

Telophase and cytokinesis

Anaphase

- Have all chromosomes arrived and aligned at the metaphase plate? THEN: Decision to initiate anaphase.

Figure 3.11   Checkpoints help regulate the cell cycle. Cellular checkpoints (represented by red wedges) ensure that important events in the cell cycle occur accurately and in the proper sequence. At each checkpoint, the cell evaluates one or more prior events before deciding whether to proceed to the next step of the cell cycle. (For simplicity, we show only two chromosomes per cell in this figure.)

MEIOSIS PRODUCES HAPLOID GERM CELLS, OR GAMETES

During the many rounds of cell division within an embryo, most cells either grow and divide via the mitotic cell cycle described earlier, or stop growing and become arrested in G_0. These mitotically dividing and G_0-arrested cells are the so-called **somatic cells** whose descendants continue to make up the vast majority of each organism's tissues throughout the lifetime of the individual. Early in the embryonic development of animals, however, a group of cells is set aside for a different fate. These are the **germ cells:** cells destined for a specialized role in the production of gametes. The germ cells become incorporated in the reproductive organs—ovaries and testes in animals; ovaries and anthers in flowering plants—where they ultimately undergo meiosis, the special two-part cell division that produces gametes (eggs and sperm or pollen) containing half the number of chromosomes as other body cells. Normal somatic cells have a finite lifetime; they undergo a certain number of mitotic divisions and then die. In contrast, the germ cells achieve a kind of immortality be-

cause, in giving rise to gametes, they transmit an individual's genes to the next generation.

The union of haploid gametes at fertilization yields diploid offspring that carry the combined genetic heritage of two parents. Sexual reproduction therefore requires the alternation of haploid and diploid generations. If gametes were diploid rather than haploid, the number of chromosomes would double in each successive generation such that in humans, for example, the children would have 92 chromosomes per cell, the grandchildren 184, and so on. Meiosis prevents this lethal, exponential accumulation of chromosomes. We now turn to a detailed examination of meiosis, focusing on how the first round of this two-part nuclear division gives rise to genetic variation as it halves a germ cell's chromosome number from two sets to one.

Meiosis Consists of One Round of Chromosome Replication but Two Rounds of Nuclear Division

Unlike mitosis, meiosis consists of two successive nuclear divisions, logically named **division I of meiosis** and **division II of meiosis,** or simply **meiosis I** and **meiosis II.** With each round, the cell passes through a prophase, metaphase, anaphase, and telophase. In meiosis I, the parent nucleus divides to form 2 daughter nuclei; in meiosis II, each of the 2 daughter nuclei divides, resulting in 4 nuclei (Fig. 3.12). These 4 nuclei—the final products of meiosis—then become enclosed in 4 separate daughter cells through cytokinesis. The chromosomes duplicate at the start of meiosis I, but they do not duplicate again in meiosis II, which explains why the gametes generated by meiosis contain half the number of chromosomes found in other body cells. A close look at each round of meiotic division reveals the mechanisms by which each gamete comes to receive one full haploid set of chromosomes.

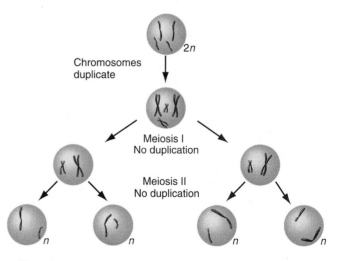

Chromosomes duplicate

$2n$

Meiosis I No duplication

Meiosis II No duplication

n n n n

 Figure 3.12   An overview of meiosis: The chromosomes replicate once, while the nuclei divide twice. In this figure, all four chromatids of each chromosome pair are shown in the same shades of the same color.

During Meiosis I, Homologous Chromosomes Pair, Exchange Parts, and Then Segregate from Each Other

The events of meiosis I are unique among nuclear divisions (Fig. 3.13, meiosis I). The process begins with the replication of chromosomes, after which each consists of two sister chromatids. A key to understanding meiosis I is the observation that the centromeres joining these chromatids remain intact throughout the entire division, rather than splitting as in mitosis. As the division proceeds, homologous chromosomes align across the cellular equator to form a coupling that ensures proper chromosome segregation to separate nuclei. Moreover, during the time homologous chromosomes face each other across the equator, the maternal and paternal chromosomes of each homologous pair exchange parts, creating a new mix of genes. Afterward, the two homologous chromosomes, each consisting of two sister chromatids connected at a single, unsplit centromere, are pulled to opposite poles of the spindle. As a result, it is homologous chromosomes (rather than sister chromatids as in mitosis) that segregate into different daughter cells at the conclusion of the first meiotic division. With this overview in mind, let us now take a closer look at the specific events of meiosis I, bearing in mind that we analyze a dynamic, flowing sequence of cellular events by breaking it down somewhat arbitrarily into the easily pictured, traditional phases.

Prophase I: During This Longest, Most Complex Phase of Meiosis, Crossing-Over Occurs

Among the critical events of **prophase** I are the condensation of chromatin, the pairing of homologous chromosomes, and the reciprocal exchange of genetic information between these paired homologs (Fig. 3.13, meiosis I). These complicated events can take many days, months, or even years to complete. For example, in the female germ cells of several species, including our own, meiosis is suspended at prophase I until ovulation. (This may sound surprising, but it will become clear when we discuss the details of egg formation in a later section.)

Leptotene (from the Greek for "thin" and "delicate") is the first definable substage of prophase I, the time when the long, thin chromosomes begin to thicken (Fig. 3.14a). Each chromosome has already duplicated prior to prophase I (as in mitosis) and thus consists of two sister chromatids affixed at a centromere. At this point, however, these sister chromatids are so tightly bound together that they are not yet visible as separate entities.

Zygotene (from the Greek for "conjugation") begins as each chromosome seeks out its homologous partner and the matching chromosomes become zipped together in a process known as **synapsis.** The "zipper" itself is an elaborate protein structure called the **synaptonemal complex** that aligns the homologs with remarkable precision, juxtaposing the corresponding genetic regions of the chromosome pair (Fig. 3.14b).

Pachytene (from the Greek for "thick" or "fat") begins at the completion of synapsis when homologous chromosomes are united along their length. Each synapsed chromosome pair

is known as a **bivalent** (because it encompasses 2 chromosomes), or a **tetrad** (because it contains 4 chromatids, which, as meiosis proceeds, will be parcelled out, one to each of the 4 products of meiosis). On one side of the bivalent is a maternally derived chromosome, on the other side a paternally derived one. Because X and Y chromosomes are not identical, they do not synapse completely; there is, however, a small region of similarity (or "homology") between the X and the Y chromosomes that allows for a limited amount of pairing.

During pachytene, structures called **recombination nodules** begin to appear along the synaptonemal complex, and an exchange of parts between nonsister (that is, between maternal and paternal) chromatids occurs at these nodules (Fig. 3.14c). Such an exchange is known as **crossing-over;** it results in the **recombination** of genetic material. You can envision the outcome of the process by imagining that the two chromatids constituting a maternal chromosome are dark blue pieces of string, and the paternal chromatids are two light blue pieces of the same length. If you cut one dark blue and one light blue string at the same position and retie dark to light and light to dark, you still have two pieces of string of equal length, but they are recombined mosaics of the originals (Fig. 3.15). As a result of crossing-over, each homologous chromosome in a pair may no longer be of purely maternal or purely paternal origin; each one, however, is still a complete entity, with no gaps in its genetic information.

Diplotene (from the Greek for "twofold" or "double") is signaled by the gradual dissolution of the synaptonemal zipper complex and a slight separation of regions of the homologous chromosomes (Fig. 3.14d). The aligned homologous chromosomes of each bivalent nonetheless remain very tightly merged at intervals along their length called **chiasmata** (singular: *chiasma*), which represent the sites where crossing-over occurred.

Diakinesis (from the Greek for "double movement") is accompanied by further condensation of the chromatids. Because of this chromatid thickening and shortening, it can now clearly be seen that each tetrad consists of 4 separate chromatids, or viewed in another way, that the 2 homologous chromosomes of a bivalent are each composed of 2 sister chromatids held together at a separate centromere (Fig. 3.14e). Nonsister chromatids that have undergone crossing-over remain closely associated at chiasmata. The end of diakinesis is analogous to the prometaphase of mitosis: The nuclear envelope breaks down and the microtubules of the spindle apparatus begin to form.

When Do Chromosome Pairing and Recombination Actually Take Place During Prophase I?

We have described the events of prophase I as they appear in the light microscope. But recent genetic experiments using the unicellular yeast *Saccharomyces cerevisiae* indicate that in this organism, chromosomes somehow begin to pair (or at least loosely align) during leptotene, well before synaptonemal complexes become visible during zygotene, and that recombination starts during leptotene or zygotene, before recombination nodules become visible during pachytene. Surprisingly, these experiments also

Meiosis I: A reduction division

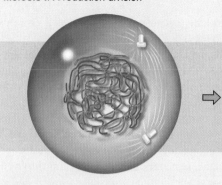

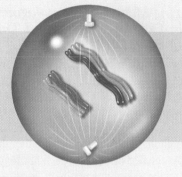

Prophase I: Leptotene
1. Chromosomes thicken and become visible but the chromatids remain invisible.
2. The nucleoli disappear.
3. Centrosomes move to each pole and produce spindle fibers.

Prophase I: Zygotene
1. The *synaptonemal complex* forms.
2. Homologous chromosomes enter *synapsis*.

Prophase I: Pachytene
1. *Crossing-over*, genetic exchange between nonsister chromatids of a homologous pair, occurs.

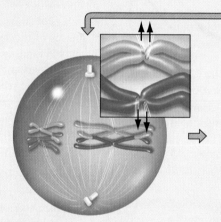

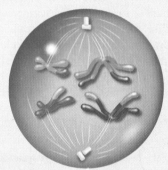

Metaphase I
1. Tetrads line up along the *metaphase plate*.
2. Each chromosome of a homologous pair attaches to fibers from opposite poles.
3. Sister chromatids attach to fibers from the same pole.

Anaphase I
1. The centromere does not divide, but the chiasmata migrate off chromatid ends and homologous chromosomes move to opposite poles.

Meiosis II: An equational division

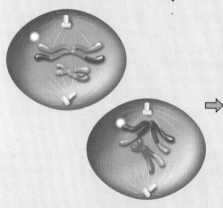

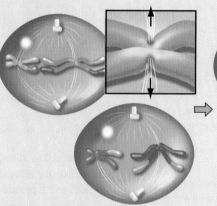

Prophase II
1. Chromosomes condense.
2. Centrioles move to the poles.
3. If they re-formed, the nucleoli and nuclear envelope disappear.

Metaphase II
1. Chromosomes align at the metaphase plate.
2. Sister chromatids attach to spindle fibers from opposite poles.

Anaphase II
1. Centromeres divide and sister chromatids move to opposite poles.

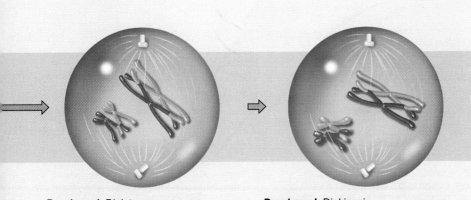

Prophase I: Diplotene
1. Synaptonemal complex dissolves.
2. A *tetrad* of four chromatids is visible.
3. Crossover points appear as *chiasmata*, which hold nonsister chromatids together.
4. Meiotic arrest occurs in species that show this phenomenon.

Prophase I: Diakinesis
1. Chromatids thicken and shorten.
2. At the end of prophase I, the nuclear membrane breaks down and the spindle begins to form.

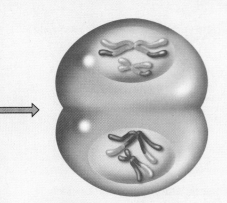

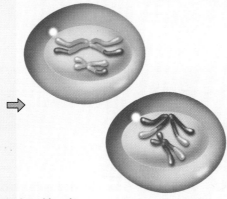

Telophase I
1. Nucleoli and nuclear envelope re-forms; does not exist in all cells.

Interkinesis
1. This is similar to interphase with one important exception: *No chromosomal duplication takes place.*

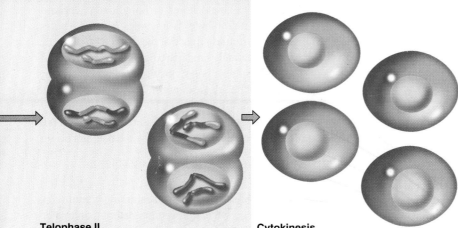

Telophase II
1. Chromosomes begin to uncoil.
2. Nuclear envelopes and nucleoli (not shown) re-form.

Cytokinesis
1. The cytoplasm divides, forming four new cells.

Feature Figure 3.13   Meiosis: One diploid nucleus produces four haploid nuclei.

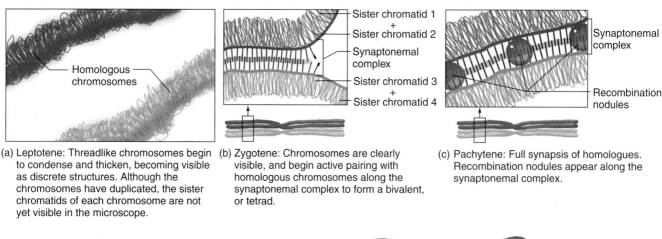

(a) Leptotene: Threadlike chromosomes begin to condense and thicken, becoming visible as discrete structures. Although the chromosomes have duplicated, the sister chromatids of each chromosome are not yet visible in the microscope.

(b) Zygotene: Chromosomes are clearly visible, and begin active pairing with homologous chromosomes along the synaptonemal complex to form a bivalent, or tetrad.

(c) Pachytene: Full synapsis of homologues. Recombination nodules appear along the synaptonemal complex.

(d) Diplotene: Bivalent appears to pull apart slightly, but remains connected at crossover sites, called chiasmata.

(e) Diakinesis: Further condensation of chromatids. Nonsister chromatids that have exchanged parts by crossing-over remain closely associated at chiasmata.

Figure 3.14 Prophase I of meiosis at very high magnification.

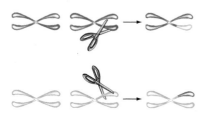

Figure 3.15 How crossing-over produces recombined chromosomes: An analogy. Crossing-over and genetic recombination occur after the chromosomes have replicated into sister chromatids (review Fig. 3.14). If we represent each pair of chromatids as a pair of strings of the same length and color, a simplified analogy for crossing-over would be the cutting of one string from each chromatid pair at the same position, and the subsequent splicing of the cut strings of different colors to produce two recombinant strings of the original size.

show that recombination can occur even if a synaptonemal complex that tightly juxtaposes the homologous chromosomes does not form. Many investigators are now trying to understand how visible structures such as the synaptonemal complex, recombination nodules, and chiasmata contribute to chromosome pairing and the initiation and completion of genetic recombination.

Metaphase I: Homologous Chromosomes Attach to Spindle Fibers Growing from Opposite Poles (Fig. 3.13, Meiosis I)

There is an essential difference between the spindle formed during meiosis I and that formed during mitosis. As we have

seen, during mitosis, each sister chromatid has a kinetochore that becomes attached to microtubules emanating from opposite spindle poles. In contrast, during meiosis I, the kinetochores of sister chromatids fuse so that each chromosome contains only a single functional kinetochore; there are thus no oppositely directed forces that could later pull the sister chromatids apart during the anaphase of meiosis I. Instead, during **metaphase I,** it is the kinetochores of homologous chromosomes that attach to microtubules from opposite spindle poles. As a result, in chromosomes aligned at the metaphase plate, the kinetochores of maternally and paternally derived chromosomes face opposite spindle poles, positioning the homologous chromosomes to which they are connected to move in opposite directions. Because the alignment and hookup of each bivalent is independent of every other bivalent's, the chromosomes facing each pole are a random mix of maternal and paternal origin.

Anaphase I: Homologous Chromosomes Move to Opposite Poles of the Spindle

At the onset of **anaphase I,** the chiasmata joining homologous chromosomes dissolve, which allows the maternal and paternal homologs to begin to move toward opposite spindle poles (see Fig. 3.13, meiosis I). Note that in the first meiotic division, the centromeres do not divide as they do in mitosis. Thus, from each homologous pair, one chromosome consisting of two sister chromatids joined at their centromere segregates to each spindle pole.

Recombination through crossing-over plays an important role in the proper segregation of homologous chromosomes during the first meiotic division. This is because chiasmata, in holding homologous chromosomes together, ensure that their kinetochores remain attached to opposite spindle poles throughout metaphase. When recombination does not occur within a bivalent, mistakes in hookup and conveyance may cause homologous chromosomes to move to the same pole instead of segregating to opposite poles. In some organisms, however, proper segregation of nonrecombinant chromosomes nonetheless occurs through other pairing processes. Investigators do not yet completely understand the nature of these processes and are currently evaluating several models to explain them.

Telophase I: Nuclear Envelopes Reform

The telophase of the first meiotic division, or **telophase I,** takes place when nuclear membranes begin to form around the chromosomes that have moved to the poles. Each of the incipient daughter nuclei contains one-half the number of chromosomes in the original parent nucleus, but each chromosome consists of two sister chromatids joined at the centromere (Fig. 3.13, meiosis I).

In most species, cytokinesis follows telophase I, with daughter nuclei becoming enclosed in separate daughter cells. A short interphase then ensues. During this time, the chromosomes usually decondense, in which case they must recondense during the prophase of the subsequent second meiotic division. In some cases, however, the chromosomes simply stay condensed. Most importantly, there is no S phase during the interphase between meiosis I and meiosis II; that is, the chromosomes do not replicate during meiotic interphase. The relatively brief interphase between meiosis I and meiosis II is known as **interkinesis.**

During Meiosis II, Sister Chromatids Separate to Produce Haploid Gametes

The second meiotic division (meiosis II) proceeds in a fashion very similar to that of mitosis, but since the number of chromosomes in each dividing nucleus has already been reduced by half, the resulting daughter cells are haploid. The same process occurs in each of the two daughter cells generated by meiosis I, producing four haploid cells at the end of this second meiotic round (Fig. 3.13, meiosis II).

Prophase II: The Chromosomes Condense as in the Prophase of Mitosis

If the chromosomes decondensed during the preceding interphase, they recondense during **prophase II.** At the end of prophase II, the nuclear envelope breaks down, and the spindle apparatus reforms.

Metaphase II: Chromosomes Align at the Metaphase Plate

The kinetochores of sister chromatids attach to microtubule fibers emanating from opposite poles of the spindle apparatus, just as in mitotic metaphase. There are nonetheless two significant features of **metaphase II** that distinguish it from mitosis:

First, the number of chromosomes is one-half that in mitotic metaphase of the same species. Second, in most chromosomes, the two sister chromatids are no longer strictly identical because of the recombination through crossing-over that occurred during meiosis I. The sister chromatids still contain the same genes, but they may carry different combinations of alleles.

Anaphase II: Sister Chromatids Separate to Opposite Spindle Poles

Just as in mitosis, severing of the centromeric connection between sister chromatids allows them to move toward opposite spindle poles during **anaphase II.**

Telophase II: Nuclear Membrane Formation Followed by Cytokinesis, Creates Four Haploid Daughter Cells

Membranes form around each of four daughter nuclei in **telophase II,** and cytokinesis places each nucleus in a separate cell. The result is four haploid gametes.

A Summary of the Significant Events of Meiosis

During meiosis, duplication of chromosomes followed by two nuclear divisions reduces the chromosome number by half (review Fig. 3.13). This reduction in chromosome number is not haphazard because in metaphase I, homologous chromosomes pair up and then separate to different daughter cells. At the end of a normal meiosis I, each daughter cell contains one complete set of chromosomes, with each chromosome consisting of two sister chromatids. During meiosis II, these two sister chromatids separate to daughter nuclei.

Segregational errors during either meiotic division can lead to aberrations, such as trisomies, in the next generation. If, for example, the homologs of a chromosome pair do not segregate during meiosis I (a mistake known as **nondisjunction**), they may travel together to the same pole and eventually become part of the same gamete. At fertilization, such an error in the segregation of homologous chromosomes may result in any one of a large variety of possible trisomies. Most autosomal trisomies, as we already mentioned, are lethal *in utero;* one exception is trisomy 21, the genetic basis of Down syndrome. Like trisomy 21, extra sex chromosomes may also be nonlethal but cause a variety of mental and physical abnormalities.

In contrast to these rare mistakes in the segregation of one pair of chromosomes, some hybrid animals carry nonhomologous chromosomes that can never pair up and segregate properly. Figure 3.16 shows the two dissimilar sets of chromosomes carried by the diploid cells of a mule; the set inherited from the donkey father contains 31 chromosomes, while the set from the horse mother has 32 chromosomes. Viable gametes cannot form in these animals and the result is sterility.

Meiosis Contributes to Genetic Diversity

The wider the assortment of different gene combinations among members of a species, the greater the chance that at least some individuals will carry combinations of alleles that allow survival in a changing environment. Two aspects of meiosis contribute to genetic diversity in a population. First,

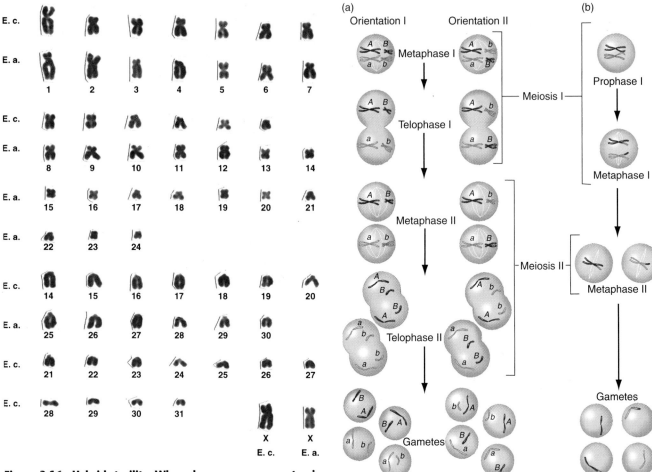

Figure 3.16 Hybrid sterility: When chromosomes cannot pair during meiosis I, they will segregate improperly. The mating of a male donkey (*E. a.*) and a female horse (*E. c.*) produces a mule with 63 chromosomes. In this mule karyotype, some of the 63 chromosomes are not sufficiently homologous to pair up during meiosis I. Where the donkey and horse chromosomes are homologous, there is one number under the pair. Starting at chromosome 14, where the parental chromosomes diverge, each set is numbered separately. Note for example that chromosome 14 from the donkey is metacentric, while chromosome 14 from the horse is larger and acrocentric.

Figure 3.17 How meiosis contributes to genetic diversity. (a) The independent assortment of different pairs of homologous chromosomes. The variation resulting from independent assortment increases with the number of chromosomes in the genome. (b) Crossing-over between homologous chromosomes ensures that each gamete produced by any individual will be unique.

since it is purely a matter of chance toward which pole the paternal and maternal homologs of each bivalent migrate during the first meiotic division, different gametes carry a different mix of maternal and paternal chromosomes. Figure 3.17a shows how two different patterns of homolog migration produce two different mixes of parental chromosomes in the gametes. The amount of potential variation generated by this random independent assortment increases with the number of chromosomes. In *Ascaris,* for example, where $n = 2$, the random assortment of homologs could produce only 2^2, or 4 types of gametes; in a person, however, where $n = 23$, this same mechanism alone could generate 2^{23}, or more than 8 million genetically different kinds of gametes. A second feature of meiosis—the reshuffling of genetic information through crossing-over during prophase I—ensures an even greater amount of genetic diversity in gametes. Because crossing-over recombines maternally and paternally derived genes, each

chromosome in each different gamete produced by a person could consist of different combinations of maternal and paternal information (Fig. 3.17b). Of course, sexual reproduction adds yet another means of generating genetic diversity. At fertilization, any of a vast number of genetically diverse sperm can fertilize an egg with its own unique genetic constitution.

Meiosis and Mitosis: A Comparison

Mitosis occurs in all types of eukaryotic cells (that is, cells with a membrane-bounded nucleus) and is a conservative mechanism that preserves the genetic status quo. Mitosis produces growth by increasing the number of cells. It also promotes the continual replacement of roots, stems, and leaves in plants and the regeneration of blood cells, intestinal tissues, and skin in animals. Meiosis, on the other hand, occurs only in sexually reproducing organisms, in just a few specialized germ cells within the reproductive organs that produce haploid gametes. It is not a conservative mechanism; rather, the extensive combinatorial

TABLE 3.2   Mitosis and Meiosis: A Comparison

Mitosis	Meiosis
Occurs in somatic cells Haploid and diploid cells can undergo mitosis One round of division	Occurs in germ cells as part of the sexual cycle Two rounds of division, meiosis I and meiosis II Only diploid cells undergo meiosis

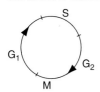 Mitosis is preceded by S phase (chromosome duplication)

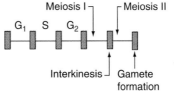 Chromosomes duplicate prior to meiosis I but not before meiosis II

 Homologous chromosomes do not pair

 During prophase of meiosis I, homologous chromosomes pair (synapse) along their length

Genetic exchange between homologous chromosomes is very rare

 Crossing-over occurs between homologous chromosomes during prophase of meiosis I

 Sister chromatids attach to spindle fibers from opposite poles during metaphase

 Homologous chromosomes (not sister chromatids) attach to spindle fibers during metaphase I

 The centromere splits at the beginning of anaphase

 The centromere does not split during meiosis I

 Sister chromatids attach to spindle fibers from opposite poles during metaphase II

 The centromere splits at the beginning of anaphase II

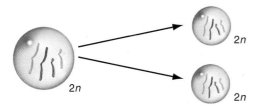

Mitosis produces two new daughter cells, identical to each other and the original cell. Mitosis is thus genetically conservative.

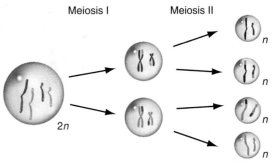

Meiosis produces four haploid cells, one (egg) or all (sperm) of which can become gametes. None of these are identical to each other or to the original cell, because meiosis results in combinatorial change.

changes arising from meiosis are one source of the genetic variation that fuels evolution. Table 3.2 illustrates the significant contrasts between the two mechanisms of cell division.

GAMETOGENESIS REQUIRES BOTH MITOTIC AND MEIOTIC DIVISIONS

In all sexually reproducing multicellular organisms, the embryonic germ cells (collectively known as the **germline**) undergo a series of mitotic divisions that yield a collection of specialized diploid cells, which subsequently divide by meiosis to produce haploid cells. As with other biological processes, there are variations on this general pattern. In some species, the haploid cells resulting from meiosis are the gametes themselves, while in other species, those cells must undergo a specific plan of differentiation to fulfill that function. Moreover, in certain organisms, the four haploid products of a single meio-

sis do not all become gametes. Thus gamete formation, or **gametogenesis,** gives rise to haploid gametes marked not only by the events of meiosis per se, but also by cellular events that precede and follow meiosis. Nevertheless, although the specifics of gametogenesis vary widely from species to species as well as from sex to sex within the same species, the general pattern of mitotic divisions followed by a meiotic halving of chromosome number is nearly universal. Here we illustrate gametogenesis with a description of egg and sperm formation in humans. The details of gamete formation in several other organisms appear throughout the text in discussions of specific experimental studies as well as in the Genetic Portraits (Chapters 18–22).

Egg Formation in Humans: Asymmetrical Meiotic Divisions Produce One Large Ovum

The end product of egg formation in humans is a large, nutrient-rich ovum whose stored resources can sustain the early embryo. The process, known as **oogenesis** (Fig. 3.18), begins

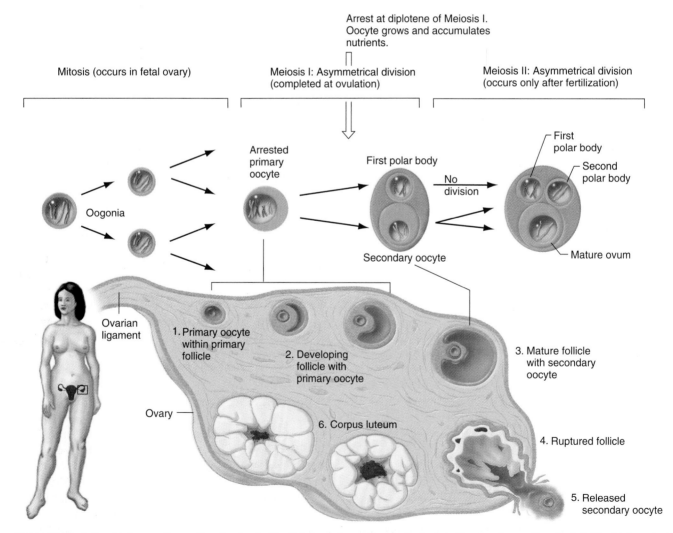

Figure 3.18 In humans, egg formation begins in the fetal ovaries and arrests during the prophase of meiosis I. The fetal ovaries contain about 500,000 primary oocytes arrested in the diplotene of meiosis I. At puberty, a young woman's ovaries begin to release one egg each month, during menstruation. If the egg is fertilized, meiosis is completed. The first polar body formed at the end of meiosis I usually arrests, and does not undergo a second meiotic division. Only one of the three cells produced by meiosis serves as the functional gamete, or ovum.

when diploid germ cells in the ovary, known as **oogonia** (singular, oogonium), multiply rapidly by mitosis and produce a large number of **primary oocytes,** which then undergo meiosis. For each primary oocyte, meiosis I results in the formation of two daughter cells that differ in size, so this division is asymmetric. The larger of these cells, the **secondary oocyte,** receives over 95% of the cytoplasm. The other, small diploid sister cell is known as the first **polar body.** During meiosis II, the secondary oocyte undergoes another asymmetrical division to produce a large haploid **ovum** and a small, haploid second polar body. The first polar body usually arrests its development and does not undergo the second meiotic division. However, the first polar body does divide in a small proportion of cases, producing two haploid polar bodies. The two (or rarely, three) small polar bodies apparently serve no function and disintegrate, leaving one large haploid ovum as the functional gamete. Thus, only one of the three (or rarely, four) products of a single meiosis serves as a female gamete. A normal human ovum carries 22 autosomes and an X sex chromosome.

Oogenesis begins in the fetus. By six months after conception, the fetal ovaries are fully formed and contain about half a million primary oocytes arrested in the diplotene of meiosis I. These cells, with their homologous chromosomes locked in synapsis, are the only oocytes the female will produce, so a girl is born with all the oocytes she will ever possess. From the onset of puberty, at about age 12, until menopause some 35 to 40 years later, most women release one primary oocyte each month (from alternate ovaries), amounting to roughly 480 oocytes released during the reproductive years. The remaining primary oocytes disintegrate during menopause. At ovulation, a released oocyte completes meiosis I and proceeds as far as the metaphase of meiosis II. If the oocyte is then fertilized, that is, penetrated by a sperm nucleus, it quickly completes meiosis II. The nuclei of the sperm and ovum then fuse to form the diploid nucleus of the zygote, and the zygote divides by mitosis to produce a functional embryo. Unfertilized oocytes exit the body during the menses stage of the menstrual cycle.

Interestingly, the long interval before completion of meiosis in oocytes released by women in their 30s, 40s, and 50s may contribute to an observed correlation between maternal age and meiotic segregational errors, including those that produce trisomies. Women in their mid-20s, for example, run a very small risk of trisomy 21; only 0.05% of children born to women of this age have this particular genetic defect. During the later childbearing years, however, the risk rapidly rises; at age 35 it is 0.9% of live births and at age 45, it is 3%. You would not expect this age-related increase in risk if meiosis were completed before birth.

Spermatogenesis in Humans: Symmetrical Meiotic Divisions Produce Four Sperm

The production of sperm, or **spermatogenesis** (Fig. 3.19), begins in the male testes in germ cells known as **spermatogonia.** As in oogenesis, mitotic divisions produce a small number of diploid **primary spermatocytes,** but unlike the diploid egg-forming cells, each spermatocyte undergoes symmetrical meiotic divisions and produces four equivalent haploid **spermatids.** These then mature by developing a characteristic whiplike tail and by concentrating all their chromosomal material in a head, thereby becoming functional **sperm**—mobile microscopic missiles capable of delivering their nuclear payload to the ovum. A human sperm, much smaller than the ovum it will fertilize, contains 22 autosomes and *either* an X *or* a Y sex chromosome.

The timing of sperm production differs radically from that of egg formation. The meiotic divisions allowing conversion of primary spermatocytes to spermatids begin only at puberty, but continue throughout a man's life. The entire process of spermatogenesis takes about 48–60 days: 16–20 for meiosis I, 16–20 for meiosis II, and 16–20 for the maturation of spermatids into fully functional sperm. Within each testis after puberty, millions of sperm are always in production, and a single ejaculate can contain up to 300 million. Over a lifetime, a man can produce billions of sperm, almost equally divided among those bearing an X and those bearing a Y chromosome.

VALIDATION OF THE CHROMOSOME THEORY

So far, we have presented two circumstantial lines of evidence in support of the chromosome theory of inheritance. First, the phenotype of sexual identity is associated with the inheritance of particular chromosomes. Second, the events of mitosis, meiosis, and gametogenesis ensure a constant number of chromosomes in the somatic cells of all members of a species over time; one would expect the genetic material to exhibit this kind of stability even in organisms with very different modes of reproduction. Final acceptance of the chromosome theory depended on researchers going beyond the circumstantial evidence to a rigorous demonstration of two key points: (1) that the inheritance of genes corresponds with the inheritance of chromosomes in every detail and (2) that the transmission of particular chromosomes coincides with the transmission of specific traits other than sex determination.

The Chromosome Theory Correlates Mendel's Laws with Chromosome Behavior during Meiosis

Walter Sutton first outlined the chromosome theory of inheritance in 1902–1903, building on the theoretical ideas and experimental results of Theodor Boveri in Germany, E. B. Wilson in New York (Sutton's mentor at Columbia and a preeminent cytologist who worked out the details of sex determination in several insect species), and others. In a 1902 paper, Sutton speculated that "the association of paternal and maternal chromosomes in pairs and their subsequent separation during the reducing division [i.e., meiosis I] . . . may constitute the physical basis of the Mendelian law of heredity." In 1903, he suggested that chromosomes carry Mendel's hereditary units because:

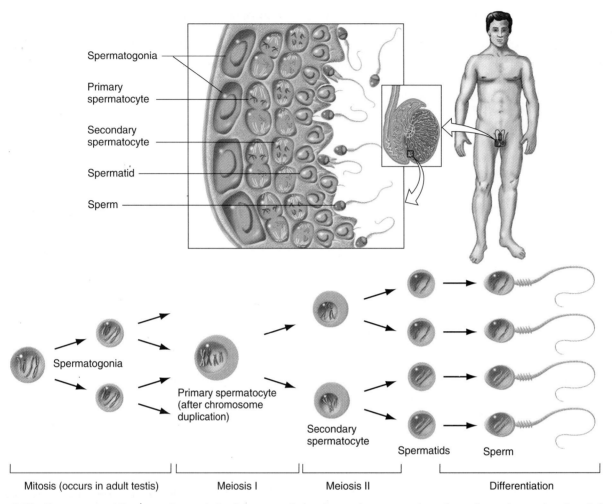

Figure 3.19 Human sperm form continuously in the testes after puberty. Spermatogonia are located near the exterior of seminiferous tubules in a human testis. Once they divide to produce the primary spermatocytes, the subsequent stages of spermatogenesis—meiotic divisions in the spermatocytes and maturation of spermatids into sperm—occur successively closer to the middle of the tubule. This allows for release of the mature sperm into the central lumen of the tubule for ejaculation. This figure illustrates the major events of spermatogenesis at the cellular and chromosomal levels.

1. Every cell contains 2 copies of each kind of chromosome and there are 2 copies of each kind of gene.

2. The chromosome complement, like Mendel's genes, appear unchanged as they are transmitted from parents to offspring through generations.

3. During meiosis, homologous chromosomes pair and then separate to different gametes just as the alternative alleles of each gene segregate to different gametes.

4. Maternal and paternal copies of each chromosome pair separate to opposite spindle poles without regard to the assortment of any other homologous chromosome pair, just as the alternative alleles of unrelated genes assort independently.

5. At fertilization, an egg's set of chromosomes unites with a randomly encountered sperm's set of chromosomes, just as alleles obtained from one parent unite at random with those from the other parent.

6. In all cells derived from the fertilized egg, one-half of the chromosomes and one-half of the genes are of maternal origin, the other half of paternal origin.

Table 3.3 illustrates the strong parallels between the behavior of genes and chromosomes, which led Sutton to propose that they are in some manner physically connected. Figure 3.17a follows different chromosomes carrying the alleles of different genes through the process of meiosis, demonstrating Sutton's theory.

From a review of Figure 3.17, you might wonder whether crossing-over abolishes the clear correspondence between Mendel's laws and the movement of chromosomes. The answer is no. Each chromatid of a homologous chromosome pair contains only one copy of a given gene, and only one chromatid from each pair of homologs is incorporated into each gamete. Because alternative alleles remain on different chromatids even after crossing-over has occurred, alternative alleles still segregate to different gametes as demanded by Mendel's first law. And because the orientation of nonhomologous chromosomes is completely random with respect to each other during both meiotic divisions, even if crossing-over occurs, the genes on different chromosomes assort independently, as demanded by Mendel's second law.

Specific Traits Are Transmitted with Specific Chromosomes

The fate of a theory depends in part on whether its predictions can be validated. Since genes determine traits, the prediction that chromosomes carry genes could be tested by breeding experiments that would show whether transmission of a specific chromosome coincides with transmission of a specific trait. Cytologists knew that one pair of chromosomes, the sex chromosomes, determines whether an individual is male or female. Would similar correlations exist for other traits?

In Drosophila, a Gene Determining Eye Color Resides on the X Chromosome

Thomas Hunt Morgan, an American experimental biologist with training in embryology, headed the research group whose findings eventually established a firm experimental base for the chromosome theory. Morgan chose to work with the fruit fly *Drosophila melanogaster* because it is extremely prolific and has a very short generation time, taking only 12 days for a fertilized egg to develop into a mature adult capable of producing hundreds of offspring. Morgan fed his flies mashed bananas and housed them in empty milk bottles capped with a wad of cotton.

In 1910 a white-eyed male appeared among a large group of flies with brick-red eyes. A mutation had apparently altered a gene determining eye color, changing it from the normal wild-type allele specifying red to a new allele that produced white. When Morgan allowed the white-eyed male to mate with its red-eyed sisters, all the flies of the F_1 generation had red eyes; the red allele was clearly dominant to the white (Fig. 3.20, cross A).

Establishing a pattern of nomenclature for *Drosophila* geneticists, Morgan named the gene identified by the abnormal white eye color, the *white* gene, for the mutation that revealed its existence. The normal wild-type allele of the *white* gene, abbreviated w^+, is for brick-red eyes, while the counterpart mutant w allele results in white eye color. The + signifies the wild type. By writing the gene name and abbreviation in lower case, Morgan symbolized that the mutant w allele is recessive to the wild-type w^+. (If a mutation results in a dominant nonwild phenotype, the first letter of the gene name or of its abbreviation is written in uppercase; thus the mutation known as *Bar* eyes is dominant to the wild-type Bar^+ allele.)

Morgan then crossed the red-eyed males of the F_1 generation with their red-eyed sisters (Fig. 3.20, cross B) and obtained an F_2 generation with the predicted 3:1 ratio of red to white eyes. But there was something askew in the pattern: Among the red-eyed offspring, there were two females for every one male, and all the white-eyed offspring were males. This result was surprisingly different from the observed transmission of Mendelian traits discussed in Chapters 1 and 2. In these fruit flies, the ratio of various phenotypes was not the same in male and female progeny. By mating F_2 red-eyed fe-

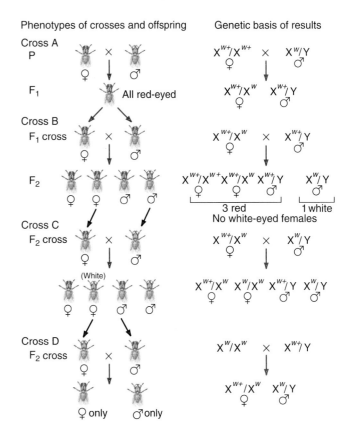

Figure 3.20   A *Drosophila* eye color gene is located on the X chromosome.   X-linkage explains the inheritance of alleles of the *white* gene.

males with their white-eyed brothers (Fig. 3.20, cross C), Morgan obtained some females with white eyes, which then allowed him to mate a white-eyed female with a red-eyed wild-type male (Fig. 3.20, cross D). The result was exclusively red-eyed daughters and white-eyed sons. The pattern seen in cross D is known as **crisscross inheritance** because the males inherit their eye color from their mothers, while the daughters inherit their eye color from their fathers. Note in Fig. 3.20 that the results of the reciprocal crosses red female × white male (cross A) and white female × red male (cross D) are not identical, again in contrast with Mendel's findings.

From the data, Morgan reasoned that the *white* gene for eye color is **X linked,** that is, carried by the X chromosome (note that while symbols for genes and alleles are italicized, symbols for chromosomes are not). The Y chromosome carries no allele of this gene for eye color. Males, therefore, have only one copy of the gene, which they inherit from their mother along with their only X chromosome (their Y chromosome must come from their father). Such males are **hemizygous** for this eye-color gene, because their diploid cells have *half* the number of alleles carried by the female on her two X chromosomes. If the single *white* gene on the X chromosome of a male is the wild-type w^+ allele, he will have red eyes and a genotype that can be written X^{w^+}/Y. (Here we designate the chromosome [X or Y] together with the allele it carries, to emphasize that certain genes are X linked. The slash [/] separates what is carried on the

TABLE 3.3 **Parallels between the Behavior of Genes and the Behavior of Chromosomes**

Mendel's Observations and Conclusions about Genes

Mendel observed phenotypes produced by experimental crosses. From his results, he inferred a mechanism of inheritance.

Mendel crossed true-breeding round-seeded plants with true-breeding wrinkle-seeded plants. All of the offspring in the F_1 generation had round seeds.

Mendel then crossed F_1 individuals with each other. The results of this cross were 3:1 round- to wrinkle-seeded plants in the F_2 generation.

Mendel explained these results by the law of segregation: In each individual, the two alleles for each trait segregate during gamete formation, then unite at random, one from each parent, at fertilization.

Mendel explained the results of his dihybrid crosses (shown below) by the law of independent assortment: During gamete formation, different pairs of alleles segregate independently of each other.

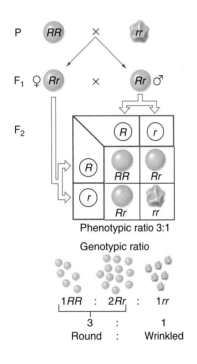

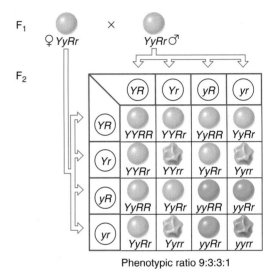

Sutton's Observations Explain Mendel's Results by the Chromosome Theory of Inheritance

Sutton observed that within each cell, the chromosomes are present in pairs. At the end of meiosis, each haploid gamete has one copy of each chromosome, or one complete set of chromosomes. At fertilization, the union of male and female

gametes restores the full diploid count of two chromosome sets per cell. One of these sets comes from the mother via the egg, the other set comes from the father via the sperm. This pattern of chromosome inheritance, combined with the

X chromosome [here, the w^+ allele for eye color] from what is carried on the Y chromosome [in this case, no allele for the gene in question]. With a pair of homologous X chromosomes, the symbol to the right of the slash [X^{w+}/X^w] would represent the second X and the allele it carries.) In contrast to an X^{w+}/Y male, a hemizygous X^w/Y male whose lone X carries the w allele would have a phenotype of white eyes. Females with two X chromosomes can be one of three genotypes: X^w/X^w (white-eyed), X^w/X^{w+} (red-eyed because w^+ is dominant to w), or X^{w+}/X^{w+} (red-eyed). As shown in Fig. 3.20, Morgan's assumption that the gene for eye color is X linked explains the results of his breeding experiments. Crisscross inheritance, for

example, results because the only X chromosome in sons of a white-eyed mother (X^w/X^w) must carry the w allele, so the sons will be white-eyed. In contrast, because daughters of a red-eyed (X^{w+}/Y) father must receive a w^+-bearing X chromosome from their father, they will have red eyes.

Analysis of Rare Mistakes in Meiosis Provided Further Support for the Chromosome Theory

Although Morgan's work strongly supported the hypothesis that the gene for eye color lies on the X chromosome, he himself continued to question the absolute validity of the chromosome theory until Calvin Bridges, one of his top students, found

TABLE 3.3 (Continued)

Sutton's Observations Explain Mendel's Results by the Chromosome Theory of Inheritance

dominance of round over wrinkled, explains why all of Mendel's F_1 plants had one phenotype.

Sutton also explained what Mendel observed in the F_2 generation by the meiotic mechanism of chromosome segregation. During meiosis, each F_1 hybrid produces two types of gametes, one carrying the round-gene chromosome, the other carrying the wrinkled-gene chromosome. At fertilization, each gamete has a 50% chance of uniting with a gamete carrying a round-gene chromosome and a 50% chance of uniting with a gamete carrying a wrinkled-gene chromosome.

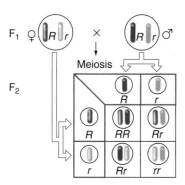

Four possible types of diploid zygotes are equally likely to result, but only one of these, in which both homologous chromosomes carry a wrinkled gene, will become a wrinkled pea; this explains Mendel's 3:1 ratio of phenotypes. Sutton thus concluded that the arrays of phenotypes observed in the F_1 and F_2 generations are the consequences both of chromosome segregation during meiosis to create haploid gametes and of the restoration of two sets of chromosomes at fertilization to create a diploid zygote.

The chromosome theory explains Mendel's law of independent assortment as follows: One pair of homologous chromosomes carries the information for seed texture. A second homologous pair carries the information for seed color. Each homologous pair aligns at random at the metaphase plate, orienting its two chromosomes toward opposite poles of the cell independently of all other homologous pairs. As a result, the assortment of traits carried by one pair of chromosomes is independent of the assortment of traits carried by all other homologous pairs.

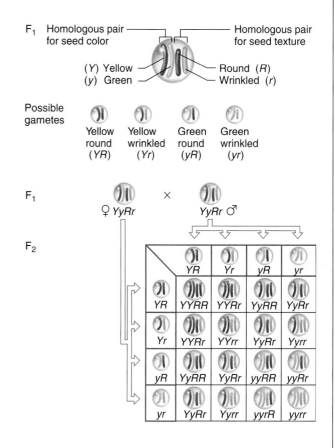

another key piece of evidence. Bridges repeated the cross Morgan had performed between white-eyed females and red-eyed males, but this time on a larger scale. As expected, the progeny of this cross consisted mostly of red-eyed females and white-eyed males. However, about 1 in every 2000 males had red eyes, and about the same small fraction of females had white eyes.

Bridges hypothesized that these exceptions arose through rare events in which the X chromosomes fail to separate during meiosis in females. He called such failures in chromosome segregation *nondisjunction*. As Fig. 3.21 shows, nondisjunction would result in some eggs with two X chromosomes and others with none. Fertilization of these chromosomally abnor-

mal eggs could produce four types of zygotes: XXY (with two X chromosomes from the egg and a Y from the sperm), XXX (with two Xs from the egg and one X from the sperm), XO (with the lone sex chromosome from the sperm and no sex chromosome from the egg), and OY (with the only sex chromosome again coming from the sperm). When Bridges examined the sex chromosomes of the rare white-eyed females produced in his large-scale cross, he found that they were indeed XXY individuals who must have received two X chromosomes and with them two w alleles from their white-eyed X^w/X^w mothers. The exceptional red-eyed males emerging from the cross were XO; their eye color showed that they must

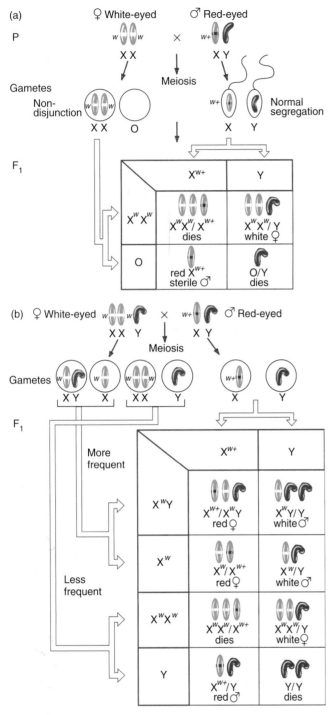

Figure 3.21 Nondisjunction: Rare mistakes in meiosis help confirm the chromosome theory. (a) Rare events of nondisjunction in an XX female produce XX and XO eggs. The results of normal disjunction in the female are not shown. XO males are sterile because the missing Y chromosome is needed for male fertility. (b) In an XXY female, the three sex chromosomes can pair and segregate in two ways. Most frequently, the two X chromosomes will pair with each other and go to opposite poles at the first meiotic division, yielding XY- and X-bearing gametes. Less frequently, the two X chromosomes will go to the same pole, creating XX- and Y-bearing eggs.

have obtained their sole sex chromosome from their X^{w+}/Y fathers. In this study, transmission of the *white* gene alleles followed the predicted behavior of X chromosomes during rare meiotic mistakes, indicating that the X chromosome carries the gene for eye color. These results also suggest that zygotes with the two other abnormal sex chromosome karyotypes (XXX and OY) expected from nondisjunction in females die during embryonic development and thus produce no progeny.

Note that the sexual identity of XO males and XXY females as well as the lack of XXX and OY adults is consistent with the sex determination information presented in Table 3.1 on page 75. OY flies do not survive because the X chromosome carries genes necessary for viability that are not found on the Y. Why XXX flies do not survive is not yet well understood. Researchers currently think that the flies die because of a lethal imbalance between proteins determined by autosomal genes and proteins determined by X-linked genes. (For further details of *Drosophila* sex determination and viability, see the Genetic Portrait in Chapter 21.)

Because the XXY white-eyed females have three sex chromosomes rather than the normal two, Bridges reasoned they would produce four kinds of eggs: XX and Y, or XY and X. You can visualize the formation of these four kinds of eggs by imagining that when the three chromosomes pair and disjoin during meiosis, two chromosomes must go to one pole and one chromosome to the other. With this kind of segregation, there are only two possible results; either the two Xs go to one pole and the Y to the other (yielding XX and Y gametes), or one X and the Y go to one pole, and the second X to the other (yielding XY and X gametes).

Bridges next predicted that fertilization of these four kinds of eggs by normal sperm would generate an array of sex chromosome karyotypes associated with specific eye color phenotypes in the progeny. Bridges verified all his predictions when he analyzed the eye color and sex chromosomes of a large number of offspring. For instance, he showed cytologically that all of the white-eyed females emerging from the cross in Fig. 3.21b had two X chromosomes and one Y chromosome, while one-half of the white-eyed males had a single X chromosome and two Y chromosomes. Bridges' painstaking observations provided compelling evidence that specific genes do in fact reside on specific chromosomes.

X- and Y-Linked Traits in Humans are Identified by Pedigree Analysis

A person unable to tell red from green would find it nearly impossible to distinguish the rose, scarlet, and magenta in the flowers of a garden bouquet from the delicately variegated greens in their foliage, or to complete a complex electrical circuit by fastening red-clad metallic wires to red ones and green to green. Such a person has most likely inherited some form of red-green colorblindness, a recessive condition that runs in families and affects mostly males. Among Western Europeans in North America and Europe, 8% of men but only 0.44% of women have this vision defect. Figure 3.22 allows you to determine the status of your color vision.

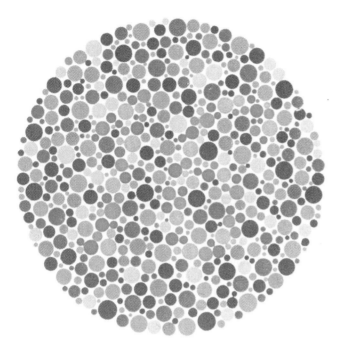

Figure 3.22   Red-green colorblindness is an X-linked recessive trait in humans. A test for colorblindness. If you see the number 16, you have normal color vision. If you do not see a number, you have red-green blindness. The above has been reproduced from *Ishihara's Tests for Colour Blindness* published by Kanehara & Co. Ltd., Tokyo, Japan, but tests for color blindness cannot be conducted with this material. For accurate testing, the original plates should be used.

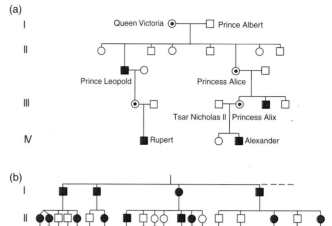

Figure 3.23   X-linked traits may be recessive or dominant. (a) Pedigree showing the inheritance of the recessive X-linked trait hemophilia in Queen Victoria's family. Dotted circles indicate carrier females. (b) Pedigree showing the inheritance of the dominant X-linked trait hypophosphatemia, commonly referred to as vitamin D–resistant rickets.

In 1911, E. B. Wilson, a contributor to the chromosome theory of inheritance, combined familiarity with studies of colorblindness and recent knowledge of sex determination by one pair of chromosomes to make the first assignment of a human gene to a particular chromosome. The gene for red-green colorblindness, he said, lies on the X because the condition usually passes from a maternal grandfather through an unaffected carrier mother to roughly 50% of the grandsons.

Several years after Wilson made his colorblindness gene assignment, pedigree analysis established that various forms of hemophilia, or "bleeders disease" (in which the blood fails to clot properly), also result from mutations on the X chromosome that give rise to a relatively rare, recessive trait. In this context, rare means "infrequent in the population." The family histories under review, including one following the descendants of Queen Victoria of England, showed that most relatively rare X-linked traits exhibit five basic properties (Fig. 3.23a).

1. The trait appears in more males than females since a female has to receive two copies of the rare defective allele to show it, whereas a hemizygous male with only one copy will display the phenotype.

2. The mutation and trait never pass from father to son because sons receive only a Y chromosome from their fathers.

3. An affected male, however, does pass the X-linked mutation to all his daughters, who serve as heterozygous carriers not manifesting the trait but passing it on to their offspring. One-half of the sons of these carrier females will inherit the defective allele and the trait.

4. Thus, the trait often skips a generation as the mutation passes from grandfather through a carrier daughter to grandson.

5. The only time the trait appears in successive generations is when a sister of an affected male is a carrier. If she is, one-half of her sons will show the trait.

Unlike colorblindness and hemophilia, some—although very few—of the known rare mutations on the X chromosome are dominant to the wild-type allele. With such dominant X-linked mutations, more females than males show the aberrant phenotype. This is because all of the daughters of an affected male but none of his sons will have the condition, while one-half of the sons and one-half of the daughters of an affected female will receive the dominant allele and therefore show the phenotype. Vitamin D–resistant rickets, or hypophosphatemia, is an example of an X-linked dominant trait. Figure 3.23b presents the pedigree of a family affected by this disease.

Theoretically, phenotypes caused by mutations on the Y chromosome should also be identifiable by pedigree analysis. Such traits would pass from an affected father to all of his sons, and from them to all future male descendants. Females would neither exhibit nor transmit a Y-linked phenotype. However, besides the determination of maleness itself as well as a contribution to sperm formation and thus male fertility, no clear-cut Y-linked visible traits have turned up. The paucity of known Y-linked traits reflects the fact that the small Y chromosome contains very few genes. Indeed, one would expect the Y chromosome to have only a limited effect on phenotype because normal XX females do perfectly well without it.

Autosomal Genes Can Also Determine Differences between the Sexes

Not all genes that have a differential effect on the two sexes reside in the X or Y chromosomes. Some autosomal genes govern traits that appear in one sex but not the other, or that are expressed differently in the two sexes. **Sex-limited traits** affect a structure or process that is found in one sex but not the other. Mutations in genes for sex-limited traits can influence only the phenotype of the sex that expresses those genes. Ovary development in females and sperm production in males are examples of sex-limited traits. **Sex-influenced traits** show up in both sexes but their expression may differ between the two sexes because of hormonal differences. Early balding, known as patterned baldness, is a sex-influenced trait in humans. Men heterozygous for the balding allele lose their hair while still in their 20s, whereas heterozygous women do not show any significant hair loss. Homozygotes in both sexes, however, become bald. This sex-influenced trait is thus dominant in men, recessive in women.

The Chromosome Theory Integrates Many Aspects of Gene Behavior

Mendel had assumed that genes are located in cells. The chromosome theory assigned the genes to a specific organelle and explained alternative alleles as physically matching parts of homologous chromosomes. In so doing, the theory provided an explanation of Mendel's laws. Because the mechanism of meiosis ensures that the matching parts of homologous chromosomes will segregate to different gametes (except in rare instances of nondisjunction), meiosis in specialized germ cells ensures that alleles segregate from each other as predicted by Mendel's first law. Because each homologous chromosome pair aligns independently of all others at meiosis I, genes carried on different chromosomes will assort independently, as predicted by Mendel's second law. The chromosome theory is also able to explain the creation of new alleles through mutation, a spontaneous change in a particular gene, that is, in a particular part of a chromosome. If a mutation occurs in the germline, it can be transmitted to subsequent generations. Finally, through mitotic cell division in the embryo and after birth, each cell in a multicellular organism receives the same chromosomes—and thus the same maternal and paternal alleles of each gene—as the zygote received from the egg and sperm at fertilization. This means that, with a few exceptions (to be discussed later), an individual's genome—the chromosomes and genes he or she carries—remains constant throughout life.

CONNECTIONS

T. H. Morgan and his students, collectively known as the *Drosophila* group, acknowledged that Mendelian genetics could exist independently of chromosomes. "Why then, we are often asked, do you drag in the chromosomes? Our answer is that since the chromosomes furnish exactly the kind of mechanism that Mendelian laws call for; and since there is an ever-increasing body of information that points clearly to the chromosomes as the bearers of the Mendelian factors, it would be folly to close one's eyes to so patent a relation. Moreover, as biologists, we are interested in heredity not primarily as a mathematical formulation, but rather as a problem concerning the cell, the egg, and the sperm."

The *Drosophila* group went on to find several X-linked mutations in addition to white eyes. One made the body yellow instead of brown, another shortened the wings, yet another made bent instead of straight body bristles. These findings raised several compelling questions. First, if the genes for all of these traits are physically linked together on the X chromosome, does this linkage affect their ability to assort independently, and if so, how? Second, does each gene have an exact chromosomal address, and if so, does this specific location in any way affect its transmission? Chapter 4 describes how the *Drosophila* group and others analyzed the transmission patterns of genes on the same chromosome in terms of known chromosome movements during meiosis, and then used the information obtained to localize genes at specific chromosomal positions.

ESSENTIAL CONCEPTS

1. *Chromosomes* are cell organelles specialized for the storage and transmission of genetic material. Genes are located on chromosomes and travel with them during cell division and gamete formation.

2. In sexually reproducing organisms, *somatic cells* carry a precise number of homologous pairs of chromosomes, which is characteristic of the species. One chromosome of each pair is of maternal origin; the other, paternal.

3. During the first division of *meiosis*, homologous chromosomes in *germ cells* segregate from each other. As a result, each gamete receives one member of each matching pair, as predicted by Mendel's first law.

4. Also during the first meiotic division, the independent alignment of each pair of homologous chromosomes at the cellular midplane results in the independent assortment of genes carried on different chromosomes, as predicted by Mendel's second law.

5. *Crossing-over* and independent alignment of homologs during the first meiotic division generate diversity.

6. The second meiotic division generates gametes with a *haploid* number of chromosomes (n).

7. Fertilization—the union of egg and sperm—restores the *diploid* number of chromosomes ($2n$) to the zygote.

8. *Mitosis* underlies the growth and development of the individual. Through mitosis, diploid cells produce identical diploid progeny cells. During mitosis, the sister chromatids of every chromosome separate to each of two daughter cells; before the next cell division, the chromosomes again duplicate to form sister chromatids.

S O C I A L   A N D   E T H I C A L   I S S U E S

1. Gregg and Jennifer are in their late 20s and just had a baby born with cri du chat syndrome. The physical and mental manifestations are microcephaly (small head), mental retardation, and a catlike cry. This dominant condition is due to a loss of a portion of one copy of chromosome 5 and is readily detectable by karyotype analysis. The policy of the health organization to which Gregg and Jennifer belong is to do routine amniocentesis on pregnant women over 35 years of age and on patients from a family with a history of a detectable genetic condition. Jennifer fit into neither category and so was not tested. Jennifer and Gregg feel their doctor was remiss in not ordering the amniocentesis and want to bring a malpractice suit against the doctor and the health organization for which he works. Is there just cause to bring this lawsuit? Should people have a say in which tests they receive? Should doctors be required to give all tests possible? What effect could this have on the cost of health insurance?

2. In 1965, a study on mentally subnormal males with violent tendencies in a Scottish institution was published, indicating that 7 of 197, or 3.5%, had the karyotype XYY. These individuals were, on average, 6 inches taller than the XY males in the institution. This percentage (3.5%) of XYY males in this institution seemed abnormally high when compared to that seen in the general population (about 0.1%). A large scale study was initiated in 1968 in Boston to screen newborns for the XYY karyotype and to follow up on the development of these boys. In 1974, a group of scientists opposed this study, feeling that it was based on preliminary observations and could be damaging to the boys involved. A major concern was that the boys would be unfairly branded as abnormal and violent. In addition, the parents had not been fully informed of the purpose of the testing. What is a fair way to approach the scientific question of the effects of an extra Y chromosome in humans? Can one separate social and cultural impacts from biological impacts of an extra Y chromosome?

3. Jack and Jessica are going to have a baby, and because Jessica is 38 years old, she had amniocentesis. The karyotype results indicate that the baby has three copies of chromosome 21 (trisomy 21 or Down syndrome). A genetic counselor and doctor have discussed with them the potential emotional and financial concerns in raising a Down syndrome child at home versus sending the child to an institution (many of which are state-supported). Jack and Jessica have decided to continue the pregnancy and raise their child at home. Given their decision to have the child knowing it will be affected, should they be eligible to receive state funds to defray the costs of special educational and therapeutic programs that they may need to raise their child at home? Should society pay for expenses accrued by children with genetic disorders if it is possible to screen for these disorders before birth?

S O L V E D   P R O B L E M S

I. In humans, chromosome 16 sometimes has a heavily staining area in the long arm near the centromere. This feature can be seen through the microscope but has no effect on the phenotype of the person carrying it. When such a "blob" exists on a particular copy of chromosome 16, it is a constant feature of that chromosome and is inherited. A couple conceived a child, but the fetus had multiple abnormalities and was miscarried. When the chromosomes of the fetus were studied, it was discovered that it was trisomic for chromosome 16, and

that two of the three chromosome 16s had large blobs. Both chromosome 16 homologs in the mother lacked blobs, but the father was heterozygous for blobs. Which parent experienced nondisjunction, and in which meiotic division did it occur?

Answer

This problem requires an understanding of nondisjunction during meiosis. When individual chromosomes contain some distinguishing feature that allows one homolog to be distinguished from another, it is possible to follow the path of that homolog through an event such as nondisjunction. In this case, since two copies had a characteristic blob that served as a marker, we can conclude that the extra copy came from the father (who has the chromosome with this same marker). In which meiotic division did the nondisjunction occur? When nondisjunction occurs during meiosis I, homologs fail to segregate to opposite poles. If this occurred in the father, the chromosome with the blob and the normal chromosome 16 would segregate into the same cell. After meiosis II, the gametes resulting from this cell would carry both types of chromosomes. After fertilization with a normal egg, there would be two copies of the normal chromosome 16 and one of the chromosome with a blob. On the other hand, if nondisjunction occurred during meiosis II in the father in a cell containing the blob 16 chromosome, sperm with two copies of the altered chromosome would be produced. After fertilization with a normal egg, the result would be a zygote of the type seen in this spontaneous abortion. *Therefore the nondisjunction occurred in meiosis II in the father.*

II. (a) What sex ratio would you expect among the offspring of a cross between a normal male mouse and a female mouse heterozygous for a recessive X-linked lethal gene? (b) What would be the expected sex ratio among the offspring of a cross between a normal hen and a rooster heterozygous for a recessive Z-linked lethal?

Answer

This problem deals with sex-linked inheritance and sex determination.

a. Mice have a sex determination system of XX = female and XY = male. A normal male mouse ($X^R Y$) × heterozygous female mouse ($X^R X^r$) would result in $X^R X^R$, $X^R X^r$, $X^R Y$, and $X^r Y$ mice. The $X^r Y$ mice would die, so there would be a 2:1 ratio of females to males.

b. The sex determination system in birds is ZZ = male and ZW = female. A normal hen ($Z^R W$) × heterozygous rooster ($Z^R Z^r$) would result in $Z^R Z^R$, $Z^R Z^r$, $Z^R W$, and $Z^r W$ chickens. Because the $Z^r W$ offspring do not live, the ratio of females to males would be 1:2.

III. A woman with normal color vision whose father was color-blind marries a man with normal color vision.

a. What do you expect to see among their offspring?

b. What would you expect if it was the normal man's father who was color-blind?

Answer

This problem involves sex-linked inheritance.

a. The father has a genotype of $X^{cb}Y$. Because the woman had to inherit an X from her father, she must have an X^{cb} chromosome, but because she has normal color vision, her other X chromosome must be X^{CB}. The man she marries has normal color vision and therefore has an $X^{CB}Y$ genotype. Their children could be $X^{CB}X^{CB}$ (normal female), $X^{CB}X^{cb}$ (carrier female), $X^{CB}Y$ (normal male), or $X^{cb}Y$ (color-blind male).

b. If the man with normal color vision had a color-blind father, the X^{cb} chromosome would not have been passed on to him, since a man does not inherit an X chromosome from his father. The man has the genotype $X^{CB}Y$ and cannot pass on the color-blind allele.

P R O B L E M S

3-1 Choose the best matching phrase in the right column for each of the terms in the left column.

a. meiosis	1. X and Y
b. gametes	2. chromosomes that do not differ between the sexes
c. karyotype	3. one of the two identical halves of a replicated chromosome
d. mitosis	4. the time during mitosis when sister chromatids separate
e. interphase	5. one diploid cell gives rise to four haploid cells
f. syncytium	6. division of the cytoplasm
g. synapsis	7. haploid cells specialized for reproduction
h. sex chromosomes	8. an animal cell containing more than one nucleus
i. cytokinesis	9. pairing of homologous chromosomes
j. anaphase	10. one diploid cell gives rise to two diploid cells
k. chromatid	11. the array of chromosomes in a given cell
l. autosomes	12. the part of the cell cycle during which the chromosomes are not visible

3-2 One oak tree cell with 14 chromosomes undergoes mitosis.

a. How many daughter cells are formed, and what is the chromosome number in each cell?

b. If the same cell had undergone meiosis, how many cells would have resulted, and what would have been the chromosome number in each?

3-3 Humans have 46 chromosomes in each somatic cell.

a. How many chromosomes does a child receive from its father?

b. How many autosomes and how many sex chromosomes are present in each somatic cell?

c. How many chromosomes are present in a human ovum?

d. How many sex chromosomes are present in a human ovum?

3-4 In the moss *Polytrichum commune,* the haploid chromosome number is 7. A haploid male gamete fuses with a haploid female gamete to form a diploid cell that divides and develops into the multicellular sporophyte. Cells of the sporophyte then undergo meiosis to produce haploid cells called spores. What is the probability that an individual spore will contain a set of chromosomes all of which came from the male gamete?

3-5 A system of sex determination known as haplodiploidy is found in honey bees. Females are diploid and males (drones) are haploid. Male offspring result from the development of unfertilized eggs. Sperm are produced by mitosis in males and fertilize eggs in the females. Ivory eye is a recessive characteristic in honey bees. Wild-type eyes are brown.
   a. What progeny would result from an ivory-eyed queen and a brown-eyed drone? (Give both genotype and phenotype.)
   b. What would result from crossing a daughter from the mating in (a) with a brown-eyed drone?

3-6 a. What are the four major stages of the cell cycle?
   b. Which stages are included in interphase?
   c. What events distinguish G_1, S and G_2?

3-7 Indicate which of the cells shown below matches each of the following stages of mitosis.
   a. anaphase
   b. prophase
   c. metaphase
   d. G_2
   e. telophase/cytokinesis

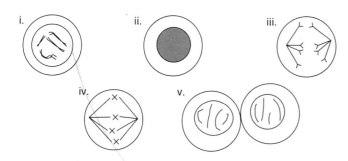

3-8 Somatic cells of chimpanzees contain 48 chromosomes. How many chromatids are present at (a) anaphase of mitosis, (b) anaphase I of meiosis, (c) anaphase II of meiosis, (d) G_1 of mitosis, (e) G_2 of mitosis, (f) G_1 just prior to meiosis I, and (g) prophase of meiosis I?

3-9 Complete the following statements using as many of the following terms as are appropriate: mitosis, meiosis I (first meiotic division), meiosis II (second meiotic division), and none (neither mitosis nor meiosis I or meiosis II).
   a. The spindle apparatus is present in cells undergoing
   _____ .

   b. Chromosome replication occurs just prior to

   _____ .
   c. The cells resulting from _____ in a haploid cell have a ploidy of *n*.
   d. The cells resulting from _____ in a diploid cell have a ploidy of *n*.
   e. Homologous chromosome pairing regularly occurs during _____ .
   f. Nonhomologous chromosome pairing regularly occurs during _____ .
   g. Physical recombination leading to the production of recombinant progeny classes occurs during

   _____ .
   h. Centromere division occurs during _____ .
   i. Nonsister chromatids are found in the same cell during _____ .

3-10 The five cells shown are all from the same individual. For each cell, indicate whether it is in mitosis, meiosis I, or meiosis II. What is *n* in this organism?

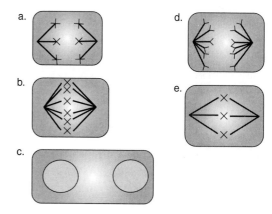

3-11 Females with Down syndrome are sometimes fertile. A 20-year-old woman has Down syndrome due to trisomy for chromosome 21. She wants to have a child. What is the probability that the child will have Down syndrome? How could this happen?

3-12 Some organisms, such as the single-celled alga *Chlamydomonas reinhardtii,* can reproduce sexually or asexually. Sexual reproduction is triggered by stressful conditions. In terms of the survival of the species, what could be an advantage obtained from sexual reproduction in these circumstances?

3-13 In humans:
   a. How many sperm develop from 100 primary spermatocytes?
   b. How many sperm develop from 100 spermatids?
   c. How many ova develop from 100 primary oocytes?
   d. How many ova develop from 100 polar bodies?

3-14 Female mammals, including women, sometimes develop benign tumors called ovarian teratomas or dermoid cysts in their ovaries. Such a tumor begins when a primary oocyte escapes from its prophase I arrest and finishes meiosis I within the ovary. (Normally meiosis I does not finish until the primary oocyte is expelled from the ovary upon ovulation.) The secondary oocyte then develops as if it were an embryo, and implants and develops within the follicle. Development is disorganized, however, and results in a tumor containing a wide variety of differentiated tissues, including teeth, hair, bone, muscle, nerve, and many others. If a dermoid cyst forms in a woman whose genotype is *Aa,* what are the possible genotypes of the cyst?

3-15 In a certain strain of turkeys, unfertilized eggs sometimes develop parthenogenetically to produce diploid offspring. (Females have WZ and males have the ZZ sex chromosomes. Assume that WW cells are inviable.) What distribution of sexes would you expect to see among the parthenogenetic offspring according to each of the following models for how parthenogenesis occurs?
a. The eggs develop without ever going through meiosis.
b. The eggs go all the way through meiosis, and then duplicate their chromosomes to become diploid.
c. The eggs go through meiosis I, and the chromatids separate to create diploidy.
d. The egg goes all the way through meiosis, and then fuses at random with one of its three polar bodies (this assumes the first polar body goes through meiosis II).

3-16 Imagine you have two pure-breeding lines of canaries, one with yellow feathers and the other with brown feathers. In crosses between these two strains, yellow female × brown male gives only brown sons and daughters, while brown female × yellow male gives only brown sons and yellow daughters. Propose a hypothesis to explain these results.

3-17 In *Drosophila,* a cross was made between a yellow-bodied male with vestigial (not fully developed) wings and a wild-type female (brown body). The F_1 generation consisted of wild-type males and wild-type females. F_1 males and females were crossed and the F_2 progeny consisted of 16 yellow-bodied males with vestigial wings, 48 yellow-bodied males with normal wings, 15 males with brown bodies and vestigial wings, 49 wild-type males, 31 brown-bodied females with vestigial wings, and 97 wild-type females. Explain the inheritance of these genes based on these results.

3-18 When Calvin Bridges observed a large number of offspring from a cross of white-eyed female *Drosophila* to red-eyed males, he observed very rare white-eyed females and red-eyed males among the offspring. He was able to show that these exceptions resulted from nondisjunction, such that the white-eyed females had received two Xs from the egg and a Y from the sperm, while the red-eyed males had received no sex chromosome from the egg, and an X from the sperm. What progeny would have arisen from these same kinds of nondisjunctional events if they had occurred in the male parent? What would their eye colors have been?

3-19 Barred feather pattern is a Z-linked dominant trait in chickens. What offspring would you expect from (a) the cross of a barred hen to a nonbarred rooster? (b) the cross of an F_1 rooster to one of his sisters?

3-20 Each of the four pedigrees shown here represents a human family within which a genetic disease is segregating. Affected individuals are indicated by filled-in symbols. One of the diseases is transmitted as an autosomal recessive condition, one as an X-linked recessive, one as an autosomal dominant, and one as an X-linked dominant. (Assume all four traits are rare in the population.)
a. Indicate which pedigree represents which mode of inheritance, and explain how you know.
b. For each pedigree, how would you advise the parents of the chance that their child (indicated by the hexagon shape) will have the condition?

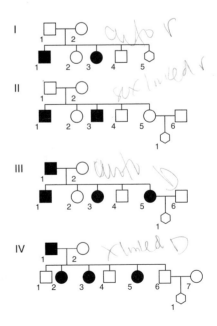

3-21 The pedigree shown here indicates the occurrence of albinism in a group of Hopi Indians, among whom the trait is unusually frequent. Assume that the trait is fully penetrant (all individuals with a genotype that could give rise to albinism will display this condition).

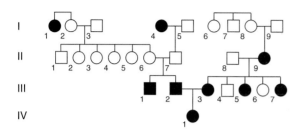

a. Is albinism in this population caused by a recessive or dominant allele?
b. Is the gene sex-linked or autosomal?

What are the genotypes of the following individuals?
c. Individual I-1.
d. Individual I-8.
e. Individual I-9.
f. Individual II-6.
g. Individual II-8.
h. Individual III-4.

3-22 Consider the following pedigrees from human families containing a male with Klinefelter syndrome (a set of abnormalities seen in XXY individuals; indicated with shaded boxes). In each, *A* and *B* refer to codominant alleles of the X-linked *G6PD* gene. The *phenotypes* of each individual (A, B, or AB) are shown on the pedigree. Indicate if nondisjunction occurred in the mother or father of the son with Klinefelter's syndrome for each of the three examples. Can you tell if the nondisjunction was in the first or second meiotic division?

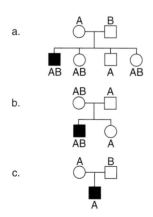

3-23 The following is a pedigree of a family in which colorblindness is found (filled-in symbols). Indicate as much as you can about the genotypes of all the individuals in the pedigree.

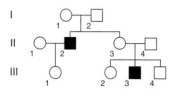

3-24 Duchenne muscular dystrophy (DMD) is caused by a relatively rare X-linked recessive allele. It causes progressive muscular wasting, and usually leads to death before age 20.
a. What is the probability that the first son of a woman whose brother is affected will be affected?
b. What is the probability that the second son of a woman whose brother is affected will be affected if her first son was affected?
c. What is the probability that a child of an unaffected man whose brother is affected will be affected?

3-25 Several different antigens can be detected in blood tests. The following four traits were tested for each individual shown below.

ABO type	$[I^A$ and I^B codominant, i recessive]
Rh type	$[Rh^+$ dominant to $Rh^-]$
MN type	[M and N codominant]
Xg(a) type	$[X^{g(a+)}$ dominant to $X^{g(a-)}]$

All of these blood type genes are autosomal, except for $X^{g(a)}$ which is X linked.

Mother	AB	Rh$^-$	MN	X$^{g(a+)}$
Daughter	A	Rh$^+$	MN	X$^{g(a-)}$
Alleged father 1	AB	Rh$^+$	M	X$^{g(a+)}$
Alleged father 2	A	Rh$^-$	N	X$^{g(a-)}$
Alleged father 3	B	Rh$^+$	N	X$^{g(a-)}$
Alleged father 4	O	Rh$^-$	MN	X$^{g(a-)}$

a. Which, if any, of the alleged fathers could be the real father?
b. Would your answer to part (a) change if the daughter had Turner syndrome (the abnormal phenotype seen in XO individuals)? If so, how?

3-26 The ancestry of a white female tiger bred in a city zoo is depicted in the following pedigree. (As you can see, there was considerable inbreeding in this lineage. For example, Mohan was mated with his daughter.) In answering the following questions, assume that "white" is determined by allelic differences at a single gene and that the trait is fully penetrant. (White tigers are indicated with unshaded symbols). Base your answers only on information in the pedigree.

a. Could white coat color be caused by a Y-linked allele?

b. Could white coat color be caused by a dominant X-linked allele?

c. Could white coat color be caused by a dominant autosomal allele?

d. Could white coat color be caused by a recessive X-linked allele?

e. Could white coat color be caused by a recessive autosomal allele?

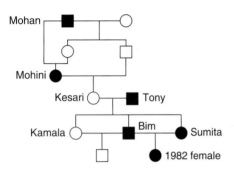

3-27 The Fast Forward box in this chapter described mutations that interfere with chromosome behavior or cytokinesis during mitosis in *Drosophila*. Researchers have also been interested in mutations that disrupt chromosome segregation during meiosis. A useful tool that can be used as a genetic marker for these studies is a special Y chromosome called $y^+ \cdot Y$, which contains a small bit of the X chromosome including the wild-type y^+ allele of the yellow (*y*) gene. The wild-type allele is dominant and produces a brown body color. The *y* allele yields yellow body color. How could you use the $y^+ \cdot Y$ chromosome to find mutants that have increased levels of nondisjunction during meiosis in *Drosophila* females?

LINKAGE, RECOMBINATION, AND THE MAPPING OF GENES ON CHROMOSOMES

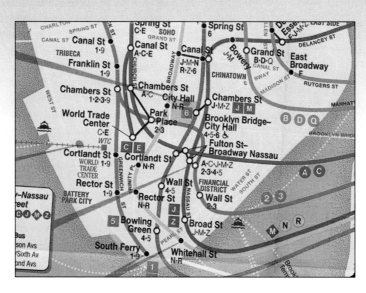

Maps illustrate the spatial relationships of objects, such as the locations of subway stations along subway lines. Genetic maps portray the positions of genes along chromosomes.

In 1928, doctors completed a four-generation pedigree tracing two known X-linked traits: red-green colorblindness and hemophilia A (the more serious X-linked form of "bleeders disease"). The maternal grandfather of the family exhibited both traits, which means that his single X chromosome carried mutant alleles of the two corresponding genes. As expected, neither colorblindness nor hemophilia showed up in his sons and daughters, but two grandsons and one great-grandson inherited both of the X-linked conditions (Fig. 4.1a). The fact that none of the descendants of this family's maternal grandfather manifested one of the traits without the other suggests that the mutant alleles did not assort independently during meiosis. Instead they traveled together in the gametes forming one generation and then into the gametes forming the next generation, producing grandsons and great-grandsons with an X chromosome specifying both colorblindness and hemophilia. Genes that travel together more often than not exhibit **genetic linkage.**

In contrast, another pedigree following colorblindness and the slightly different B form of hemophilia, which also arises from a mutation on the X chromosome, revealed a different inheritance pattern. A grandfather with hemophilia B and colorblindness had four grandsons, but only one of them exhibited both conditions. In this family, the genes for colorblindness and hemophilia appeared to assort independently, producing in the male progeny all four possible combinations of the two traits—normal vision and normal blood clotting, colorblindness and hemophilia, colorblindness and normal clotting, and normal vision and hemophilia—in approximately equal frequencies (Fig. 4.1b). Thus, even though the mutant alleles of the two genes were on the same X chromosome in the grandfather, they had to separate to give rise to grandsons III-2 and III-3. This separation of genes on the same chromosome is the result of **recombination,** the occurrence in progeny of new gene combinations not seen in previous generations. (Note that *recombinant progeny* can result in either of two

Figure 4.1 Pedigrees indicate that colorblindness and the two forms of hemophilia are X-linked traits. (a) Transmission of red-green colorblindness and hemophilia A. Symbols with a vertical bar indicate the hemophilia A phenotype; those with a horizontal bar denote colorblindness. Individuals with a cross inside the corresponding symbol have both conditions. Note that colorblindness and hemophilia A travel together through the pedigree, indicating their genetic linkage. (b) Transmission of red-green colorblindness (horizontal bars) and hemophilia B (vertical bars). Even though the genes for both conditions are on the X chromosome, the mutant alleles are inherited together in only one of four grandsons of the propositus. Together, pedigrees (a) and (b) indicate that the hemophilia A and colorblindness genes are close together, while the hemophilia B and colorblindness genes are far apart, on the human X chromosome.

ways: from the recombination of genes on the same chromosome during gamete formation, discussed in this chapter, or from the independent assortment of genes on nonhomologous chromosomes, described in Chapter 3.)

In this chapter, as we look at the tools geneticists devised to follow the transmission of genes linked on the same chromosome, you will see that recombination may separate those genes when homologous chromosomes exchange parts during meiosis. The farther apart two genes are, the greater the probability of separa-

tion through recombination. Extrapolating from this rule of thumb, you can see that the gene for hemophilia A must be very close to the gene for red-green colorblindness, because, as Fig. 4.1a shows, the two rarely separate. By comparison, the gene for hemophilia B must lie far away from the colorblindness gene, because, as Fig. 4.1b indicates, new combinations of alleles of the two genes occur quite often. Geneticists can use data about how often genes separate during transmission to map the genes' relative locations on a chromosome. Such mapping is a key to sorting out and tracking down the components of complex genetic networks; it is also crucial to geneticists' ability to isolate and characterize genes at the molecular level.

Three general themes emerge from our discussion of genetic linkage and recombination. First, maps reflecting the frequency with which linked genes travel together helped validate the chromosome theory of inheritance by confirming that each chromosome carries many genes in a particular order. Second, the twin concepts of linkage and recombination help explain the genetic underpinnings of evolution. Linkage creates a potential for the simultaneous transmission of blocks of genes that function well together, while recombination produces a potential for gene reshuffling that may enhance the chances of survival under changing conditions. Third, chromosome maps led to further refinement in the concept of a gene. For Mendel, the gene was an abstract unit that controlled a trait and segregated independently of other such units. With the chromosome theory, the gene became part of a chromosome. With mapping, the gene acquired a chromosomal address—a precise location that became the focus of more accurate predictions about inheritance.

Our presentation of the transmission patterns of genes on the same chromosome describes

- Linkage and recombination: Genes linked on the same chromosome usually assort together, instead of independently; such linked genes may nonetheless become separated through recombination.

- Mapping: The frequency with which linked genes become separated through recombination generally reflects the physical distance between them; recombination data thus make it possible to locate genes along chromosomes.

GENE LINKAGE AND RECOMBINATION

If people have roughly 100,000 genes but only 23 pairs of chromosomes, most human chromosomes must carry hundreds, if not thousands, of genes. By the late 1990s, more than 700 human genes had been assigned to the X chromosome alone, and as geneticists develop new techniques of analysis, that number continues to grow. Recognition that many genes reside on each chromosome raises an important question. If genes on *different* chromosomes assort independently because

of the independent segregation of homologs during meiosis, how do genes on the *same* chromosome assort?

Some Genes on the Same Chromosome Assort Together More Often Than Not

We begin our analysis with X-linked *Drosophila* genes because they were the first to be assigned to a specific chromosome and because it is easy to design and follow multigenerational crosses yielding large numbers of *Drosophila* progeny. As we outline various crosses, remember that females carry two alleles for each X-linked gene, while males carry only one. This, of course, is because each female has two X chromosomes, one from each parent, whereas each male has only one, the X inherited from his mother.

In Dihybrid Crosses, Departures from a 1:1:1:1 Ratio of F_1 Gametes Indicates That the Two Genes Are on the Same Chromosome

We look first at two X-linked genes that determine a fruit fly's eye color and body color. (Here and in subsequent discussion, lowercase italicized y and y^+ refer to alleles of the gene for body color, while a capital Y refers to the Y chromosome, which does not carry genes for eye or body color.) In a cross between a female with mutant white eyes and a wild-type brown body ($w w y^+ y^+$) and a male with wild-type red eyes and a mutant yellow body ($w^+ y$ plus a Y chromosome, denoted as $w^+ y / Y$), the F_1 offspring are evenly divided between brown-bodied females with normal red eyes ($w^+ w y^+ y$), and brown-bodied males with mutant white eyes ($w y^+ / Y$) (Fig. 4.2). Note that the male progeny look like their mother because their phenotype directly reflects the genotype of the single X chromosome they received from her. The same is not true for the F_1 females, who received w and y^+ on the X from their mother and $w^+ y$ on the X from their father. With two alleles for each X-linked gene, one derived from each parent, the dominance relations of each pair of alleles determine the female phenotype.

Now comes the significant cross for answering our question about the assortment of genes on the same chromosome. If these two *Drosophila* genes for eye and body color assort independently, as predicted by Mendel's second law, the F_1 females should make four kinds of gametes, with four different combinations of genes on the X chromosome—$w y^+$, $w^+ y$, $w^+ y^+$, and $w y$. These four types of gametes should occur with equal frequency, that is, in a ratio of 1:1:1:1. If it happens this way, approximately half of the gametes will be of the two **parental** types, carrying either the $w y^+$ allele combination seen in the original mother (the female of the P generation) or the $w^+ y$ allele combination seen in the original father (the male of the P generation). The remaining half of the gametes will be of the two **recombinant** types, in which reshuffling has produced either $w^+ y^+$ or $w y$ allele combinations not seen in the P generation parents of the F_1 females.

We can see if the 1:1:1:1 ratio of the four kinds of gametes actually materializes by counting the different types of male progeny in the F_2 generation, as these sons receive their only X-linked genes from their maternal gamete. The bottom part of Fig. 4.2 depicts the results of a breeding study that produced 9026 F_2 males. The relative numbers of the four X-linked gene combinations passed on by the F_1 females' gametes reflect a significant departure from the 1:1:1:1 ratio expected of independent assortment. By far the largest numbers of gametes carry the parental combinations $w y^+$ and $w^+ y$. Of the total 9026 flies counted, 8897, or almost 99%, had these genotypes. In contrast, the new combinations $w^+ y^+$ and $w y$ made up little more than 1% of the total. We can explain this outcome in one of two ways. Either the $w y^+$ and $w^+ y$ combinations are

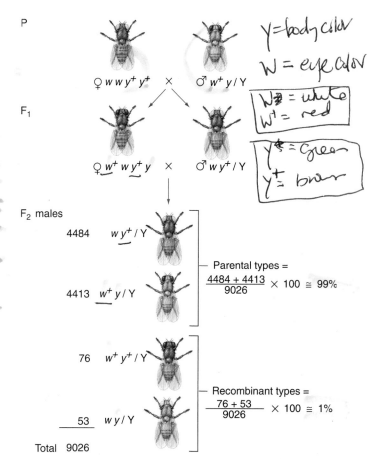

Handwritten annotations:
Y = body color
W = eye color
W = white
W^+ = red
Y^* = green
Y^+ = brown

Figure 4.2 When genes are linked, parental combinations outnumber recombinant types. The doubly heterozygous $w^+ w y^+ y$ F_1 females produce four types of sons. Sons that look like the father ($w^+ y / Y$) or mother ($w y^+ / Y$) of the F_1 females are parental types. These sons inherit the combination of alleles that an F_1 female received from one of her parents. Other sons ($w^+ y^+ / Y$ or $w y / Y$) are recombinant types. They receive an allele of one gene that came from their maternal grandmother and an allele of the other gene that came from their maternal grandfather. For these two genes that are very close together on the *Drosophila* X chromosome, there are many more parental than recombinant types among the progeny.

preferred because of some intrinsic chemical affinity between these particular alleles, or it is the parental combination of alleles the F_1 female receives from one or the other of her P generation parents that shows up most frequently.

A Preponderance of Parental Genotypes in the F_2 Generation Defines Linkage

The correct explanation of departures from a 1:1:1:1 ratio of F_1 gametes emerges from a second set of crosses (Fig. 4.3) in which the original parental generation consists of red-eyed, brown-bodied $w^+ w^+ y^+ y^+$ females and white-eyed, yellow-bodied $w y / Y$ males, producing $w^+ y^+$ and $w y$ gametes, respectively. All the F_1 females are still all $w^+ w y^+ y$ dihybrids, as in the first set of crosses. But what about the telltale F_2 males? What ratios of the four possible F_1 female gametes do they reflect?

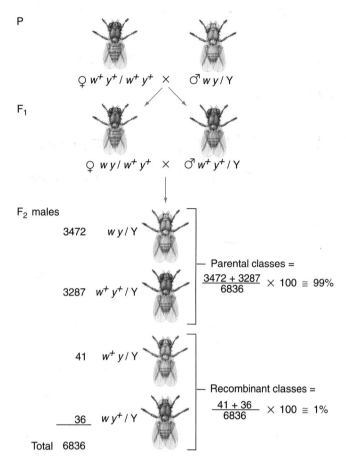

P

$♀ \, w^+ y^+ / w^+ y^+ \times ♂ \, w \, y / Y$

F_1

$♀ \, w \, y / w^+ y^+ \times ♂ \, w^+ y^+ / Y$

F_2 males

3472 $w \, y / Y$

3287 $w^+ y^+ / Y$

— Parental classes =
$$\frac{3472 + 3287}{6836} \times 100 \cong 99\%$$

41    $w^+ y / Y$

36    $w \, y^+ / Y$

— Recombinant classes =
$$\frac{41 + 36}{6836} \times 100 \cong 1\%$$

Total   6836

Figure 4.3 Designations of "parental" and "recombinant" relate to past history. The parental and recombinant classes are the opposite of those in Fig. 4.2, because the F_1 females in each experiment received different allelic combinations of the *white* and *yellow* genes from their parents. The percentages of recombinant and parental types among the total progeny are essentially the same in Figs. 4.2 and 4.3, showing that the frequency of recombination is independent of the particular arrangement of alleles.

This time, as Fig. 4.3 shows, $w^+ y$ and $w \, y^+$ are the recombinants that account for little more than 1% of the total, while $w \, y$ and $w^+ y^+$ are the parental combinations, which again add up to almost 99%. You can see that there is no preferred association of w^+ and y or of y^+ and w in this cross. Instead, comparisons of the two experiments illustrated in Figs. 4.2 and 4.3 show that the observed frequencies of the various types of progeny depend on how the arrangement of alleles in the F_1 females originated. It is the **parental classes,** the combinations originally present in the P generation, that show up most frequently in the F_2 generation. The reshuffled **recombinant classes** occur less frequently. It is important to appreciate that the designation of "parental" and "recombinant" gametes or progeny of a double heterozygous F_1 female is operational, that is, determined by the particular set of alleles she receives from each of her parents. To keep track of parental allelic combinations, we can represent the genotype of the F_1 female in Fig. 4.3 as $w \, y / w^+ y^+$, with the alleles she received from different parents shown on different sides of the slash (/).

As explained in Chapter 3, each side of the slash shows the alleles of one chromosome in a homologous pair.

When genes assort independently, the numbers of parental and recombinant F_2 progeny are equal, because a doubly heterozygous F_1 individual produces an equal number of all four types of gametes. By comparison, two genes are considered **linked** when the number of F_2 progeny with parental genotypes exceeds the number of F_2 progeny with recombinant genotypes. Instead of assorting independently, the genes behave as if they are connected to each other much of the time. The genes for eye and body color that reside on the X chromosome in *Drosophila* are an extreme illustration of the linkage concept. The two genes are so tightly coupled that the parental combinations of alleles—$w^+ y$ and $w \, y^+$, or $w^+ y^+$ and $w \, y$, depending on the cross—are reshuffled to form recombinants in only 1 out of every 100 gametes formed. In other words, the two parental allele combinations of these tightly linked genes are inherited together 99 times out of 100.

Percentages of Parental and Recombinant Classes Vary with the Gene Pair

Linkage is not always this tight. In *Drosophila*, a mutation for miniature wings (*m*) is also found on the X chromosome. A cross of red-eyed females with normal wings ($w^+ m^+ / w^+ m^+$) and white-eyed males with miniature wings ($w \, m / Y$) yields an F_1 generation containing all red-eyed, normal-winged flies. The genotype of the F_1 females is $w^+ m^+ / w \, m$ (Fig. 4.4). Of the F_2 males, 67.2% are parental types ($w^+ m^+$ and $w \, m$), while the remaining 32.8% are recombinants ($w \, m^+$ and $w^+ m$). This preponderance of parental combinations among the F_2 genotypes reveals that the two genes are linked: The parental combinations of alleles travel together more often than not. But compared to the 99% linkage between the *w* and *y* genes for eye color and body color, the linkage of *w* to *m* is not that tight. The parental combinations for color and wing size are reshuffled in, not 1, but roughly 33 out of every 100 gametes.

To determine whether two human genes on the X chromosome are linked, and if so, how tightly, geneticists analyze

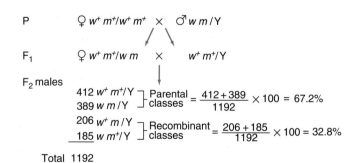

P $♀ \, w^+ m^+ / w^+ m^+ \times ♂ \, w \, m / Y$

F_1 $♀ \, w^+ m^+ / w \, m \times w^+ m^+ / Y$

F_2 males

412 $w^+ m^+ / Y$ — Parental classes = $\frac{412 + 389}{1192} \times 100 = 67.2\%$
389 $w \, m / Y$

206 $w^+ m / Y$ — Recombinant classes = $\frac{206 + 185}{1192} \times 100 = 32.8\%$
185 $w \, m^+ / Y$

Total   1192

Figure 4.4 The recombination frequency depends on the gene pair. The recombination frequency—the percentage of total progeny that are recombinant—for the *w* and *m* genes (32.8%) is very different from the recombination frequency for *w* and *y* (about 1%; see Figs. 4.2 and 4.3). This implies that the *w* and *y* genes are very close together, while the *w* and *m* genes are farther apart.

pedigrees similar to the one depicted in Fig. 4.1a. It takes many such pedigrees to develop a sample of adequate size. Such studies have shown that in families where a man of the P generation has both colorblindness and hemophilia A so that his daughter (the F_1) must be the carrier of a twice defective X chromosome, roughly 97% of her sons (the F_2) are either affected by both conditions like their grandfather or show no phenotypic abnormality. The remaining 3% have *either* hemophilia *or* colorblindness, not both. These phenotypic ratios indicate that the genes underlying colorblindness and hemophilia A in humans are closely linked on the X chromosome. The reason no recombination showed up in the pedigree depicted in Fig. 4.1a is that the sample size was too small.

Autosomal Traits Also Exhibit Linkage

In the decades following the 1900 rediscovery of Mendel's laws, certain departures from the 9:3:3:1 ratio expected of two independently assorting non-sex-linked traits did not yield to analysis in terms of the gene interactions discussed in Chapter 2. T. H. Morgan and his colleagues realized that the concept of linkage might explain some of these anomalous results. Because all progeny receive two copies of genes carried by autosomes, it is harder to follow the transmission of non-sex-linked traits than those determined by the X chromosome. But a test cross in which a homozygous parent contributes only recessive alleles makes it possible to analyze the gene combinations transmitted by the gametes of the doubly heterozygous F_1 parent.

Fruit flies, for example, carry an autosomal gene for body color (in addition to the X-linked one); the wild type is once again brown, but the recessive mutation in this gene gives rise to black (*b*). A second autosomal gene helps determine the shape of a fruit fly's wing, with the wild type having straight edges and a recessive mutation (*c*) producing curves. Figure 4.5 depicts a cross between black-bodied females with straight wings ($b\,c^+/b\,c^+$) and brown-bodied males with curved wings (b^+c/b^+c). All the F_1 progeny are double heterozygotes ($b\,c^+/b^+c$) that are phenotypically wild type. In a test cross of the F_1 females with $b\,c/b\,c$ males, all of the offspring receive the recessive *b* and *c* alleles from their father. The phenotypes of the offspring thus indicate the kinds of gametes received from the mother. For example, a black fly with normal wings would be genotype $b\,c^+/b\,c$; since we know it received the $b\,c$ combination from its father, it must have received $b\,c^+$ from its mother. As Fig. 4.5 shows, roughly 77% of the test cross progeny in one experiment received parental gene combinations (that is, allelic combinations transmitted into the F_1 females by the gametes of each of her parents), while the remaining 23% were recombinants. The parental classes thus outnumber the recombinant classes, and we can conclude that the autosomal genes for black body and curved wings are linked.

To summarize, many pairs of genes on both autosomes and sex chromosomes exhibit linkage; that is, they do not assort independently, but instead are transmitted together more than 50% of the time. (Do not confuse the general concept of linkage with the idea that some genes are X linked, which simply means that they reside on the X chromosome.) For each pair of

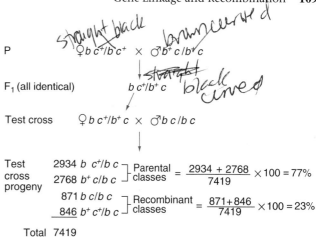

Figure 4.5 Autosomal genes can also exhibit linkage. Genes for body color (*b*) and wing shape (*c*) are both autosomal. A test cross shows that the recombination frequency for this pair of genes is 23%. Because parentals outnumber recombinants, the *b* and *c* genes are genetically linked and must be on the same *Drosophila* autosome.

genes, a particular linkage percentage indicates how often parental combinations travel together. In the preceding examples, the autosomal genes for body color and wing shape were transmitted together 77% of the time, the X-chromosome genes for eye color and wing size were transmitted together 67.2% of the time, and the X-chromosome genes for eye and body color were transmitted together 99% of the time. Linkage is never 100%. No matter how tightly two genes are linked, if you observe enough individuals, you will find some recombinants.

The Chi Square Test Pinpoints the Probability That Observed Percentages Are Evidence for Linkage

How do you know from a particular experiment whether two genes assort independently or are genetically linked? At first glance, this question should pose no problem. Discriminating between the two possibilities involves straightforward calculations based on assumptions well supported by observations. For independently assorting genes, a dihybrid F_1 female produces four types of gametes in equal numbers so that one-half of the F_2 progeny are of the parental classes and the other half of the recombinant classes. In contrast, for linked genes, the two types of parental classes by definition always outnumber the two types of recombinant classes in the F_2 generation.

The problem is that because real-world genetic transmission is based on chance events, in a particular study, even unlinked, independently assorting genes can produce deviations from the 1:1:1:1 ratio, just as in 10 tosses of a coin, you may easily get 6 heads and 4 tails (rather than the predicted 5 and 5). Thus, if a breeding experiment analyzing the transmission of two genes shows a deviation from the equal ratios expected of independent assortment, can we necessarily conclude the two genes are linked? Is it instead possible that the results represent a statistically acceptable chance fluctuation from the mean values expected of unlinked genes that assort independently? Such questions become more pressing in cases where linkage is not all that tight, so that even though the genes are linked, the percentage of recombinant classes approaches 50%.

To answer these kinds of questions, statisticians have devised a quantitative measure that indicates how often an experimentally observed deviation from the predictions of a particular hypothesis will occur solely by chance. This measure of the "goodness of fit" between observed and predicted results is a probability test known as the **chi square test.** The test is designed to account for the fact that the size of an experimental population (the "sample size") is an important component of statistical significance. To appreciate the role of sample size, it is useful to return to the proverbial coin toss before examining the details of the chi square test.

We have seen that in 10 tosses of a coin, because of chance, an outcome of 6 heads (60%) and 4 tails (40%) is not unexpected. In contrast, with 1000 tosses of the coin, a result of 600 heads (60%) and 400 tails (40%) would intuitively be highly unlikely. In the first case, a change in the results of one coin toss would alter the expected 5 heads and 5 tails to the observed 6 heads and 4 tails. In the second case, 100 tosses would have to change from tails to heads to generate the stated deviation from the predicted 500 heads and 500 tails. It is reasonable, even likely, that chance events could cause 1 deviation from the predicted number, but not 100. Two important concepts emerge from this simple presentation. First, a comparison of percentages or ratios alone will never allow you to determine whether or not *observed* data are significantly different from *predicted* values. Second, the absolute numbers obtained are important because they reflect the size of the experiment. The larger the sample size, the closer the observed percentages can be expected to match the values predicted by the experimental hypothesis, *if the hypothesis is correct.* The chi square test is therefore always calculated with numbers —actual data—and not percentages or proportions.

A critical prerequisite of the chi square test is the framing of a hypothesis that leads to clear-cut predictions. Although contemporary geneticists use the chi square test to interpret many kinds of genetic experiments, they use it most often to discover whether data obtained from breeding experiments provide evidence for or against the hypothesis that two genes are linked. But the problem with the general hypothesis that "genes *A* and *B* are linked" is that there is no precise prediction of what to expect in terms of breeding data. This is because, as we have seen, the frequency of recombination varies with each linked gene pair. In contrast, the alternative hypothesis "that genes *A* and *B* are *not* linked" gives rise to a precise prediction: that alleles at different genes will assort independently and produce 50% parental and 50% recombinant progeny. So, whenever a geneticist wants to determine whether two genes are linked, he or she actually tests whether the observed data are consistent with a **null hypothesis** of no linkage. If the chi square test shows that the observed data differ significantly from those expected with independent assortment, that is, they differ enough not to be reasonably attributable to chance alone, then the researcher can reject the null hypothesis of no linkage and accept the alternative of linkage between the two genes.

The box on page 111 presents the general protocol of the chi square test. The final result of the calculations outlined in steps 1 to 5 is the determination of the numerical probability—the *p* value—that a particular set of observed experimental results represents a chance deviation from the values predicted by a particular hypothesis. If the probability is high, it is likely that the hypothesis being tested explains the data, and the observed deviation from expected results is considered *insignificant.* If the probability is very low, the observed deviation from expected results becomes *significant.* When this happens, it is unlikely that the hypothesis under consideration explains the data, and the hypothesis can be rejected.

Applying the Chi Square Test

Figure 4.6 depicts two sets of data obtained from test-cross experiments asking whether genes *A* and *B* are linked. We first apply the chi square analysis to data accumulated in the first experiment. The total number of offspring is 50, and there are four classes of offspring, containing respectively 18, 15, 7, and 10 observed individuals. Dividing 50 by 4, you get 12.5, the number of offspring expected in each class according to the null hypothesis (which predicts a 1:1:1:1 ratio and thus equal numbers of offspring in the four classes). Now, you take one class at a time, square the observed deviation from the expected value, divide the result by the expected value, and sum all of the values obtained:

$$\chi^2 = \frac{(18 - 12.5)^2}{12.5} + \frac{(15 - 12.5)^2}{12.5} + \frac{(7 - 12.5)^2}{12.5}$$

$$+ \frac{(10 - 12.5)^2}{12.5}$$

$$= 2.42 + 0.5 + 2.42 + 0.5$$

$$= 5.84$$

Since with four classes, the number of degrees of freedom is 3, you scan the chi square table (Table 4.1) for $\chi^2 = 5.84$ and df = 3 and find by extrapolation that the corresponding *p* value is greater than 0.05. From this, you conclude that it is not possible to reject the null hypothesis on the basis of this experiment, which means that this data set is not sufficient to demonstrate linkage between *A* and *B*.

If you use the same strategy to calculate a *p* value for the data observed in the second experiment, where there are a total of 100 offspring and thus an expected number of 25 in each of the four classes, you get

$$\chi^2 = \frac{(36 - 25)^2}{25} + \frac{(30 - 25)^2}{25} + \frac{(14 - 25)^2}{25}$$

$$+ \frac{(20 - 25)^2}{25}$$

$$= 4.84 + 1.0 + 4.84 + 1.0$$

$$= 11.7$$

The Chi Square Test

The general protocol for using the chi square test and evaluating its results can be stated in a series of steps. Two preparatory steps precede the actual chi square calculation.

1. Use the data obtained from a breeding experiment to answer the following questions:
   a. What is the *total number* of offspring (events) analyzed?
   b. How many different *classes* of offspring (events) are there?
   c. In each class, what is the *number* of offspring (events) *observed*?

2. Calculate how many offspring (events) would be expected for each class if the null hypothesis (here, no linkage) were correct by multiplying the percentage predicted by the null hypothesis (here, 25%) by the total number of offspring analyzed in the experiment. You are now ready for the chi square calculation.

3. To calculate chi square, begin with one class of offspring. Subtract the expected number from the observed number (which gives the deviation from the prediction for this class), square the result, and divide this amount by the expected number. Do this for all classes and sum the individual results (the deviation^2/expected) together. The final result is the chi square (χ^2) value. This step is summarized by the equation

$$\chi^2 = \Sigma \frac{(Number\ observed\ -\ Number\ expected)^2}{Number\ expected}$$

where Σ means "for all classes."

4. Compute the **degrees of freedom (df)**. The df is a measure of the number of independently varying parameters in the experiment. For the types of experiments we are discussing, the number of degrees of freedom is one less than the number of classes. For example, if the total number of offspring in a test cross fall into four classes and you know the number present in any three, you can easily calculate the number in the fourth. Thus, with 4 (N) classes, there are only 3 ($N-1$) df.

5. Use the chi square value together with the number of degrees of freedom to determine a **p value**: the probability that a deviation from the predicted numbers at least as large as that observed in the experiment will occur by chance. Although the p value is arrived at through a numerical analysis, geneticists routinely determine the value by a quick search through a table of critical χ^2 values, such as the one in Table 4.1.

6. Evaluate the significance of the p value. You can think of the p value as the probability that the null hypothesis is true and the alternative hypothesis, in this case of linkage, is wrong. A value greater than 0.05 indicates that in more than 1 in 20 (or more than 5% of the) repetitions of an experiment of the same size, the observed data showing a deviation from the predictions of the null hypothesis could have been obtained by chance even if the null hypothesis were true; the data are therefore *not significant* for rejecting the null hypothesis and showing linkage. Statisticians have arbitrarily selected the 0.05 p value as the boundary between accepting and rejecting the null hypothesis. A p value of less than 0.05 means that you can consider the data showing deviation *significant* and reject the null hypothesis.

	Experiment 1		Experiment 2	
Genotype	Observed	Expected	Observed	Expected
A B	18	12.5	36	25
a b	15	12.5	30	25
A b	7	12.5	14	25
a B	10	12.5	20	25
Total	50	50	100	100

Figure 4.6 Applying the chi square test to see if genes *A* and *B* are linked. *A B* and *a b* are the parental types; *A b* and *a B* are the recombinant types. The null hypothesis is that the two genes are unlinked. For experiment 1, $p > 0.05$, so it is not possible to reject the null hypothesis and thus demonstrate linkage. For experiment 2, with a data set twice the size as that in experiment 1, $p < 0.05$. This is below the arbitrary boundary of significance used by most geneticists, which makes it possible to reject the null hypothesis and conclude with greater than 95% confidence that the genes are linked.

The number of degrees of freedom (df) remains 3, so you arrive at a p value greater than 0.001 but less than 0.05. In this case, you can consider the difference between the observed and expected values to be highly significant. As a result you can reject the null hypothesis of independent assortment with a high level of confidence, and conclude there is a great likelihood that genes *A* and *B* are linked.

A certain amount of subjectivity enters into any decision about the significance of a particular p value. As stated in the preceding box, statisticians have arbitrarily selected a p value of 0.05 as the boundary between significance and nonsignificance. Values lower than this indicate there would be less than 5 chances in 100 of obtaining the same results by chance if the null hypothesis were true. An extremely low p value thus suggests that the data showing deviation from values predicted by the hypothesis are significant enough to reject the null hypothesis. The lower the p value, the clearer is the case that the data cannot be regarded as chance deviations from the predictions of the hypothesis. For this reason, more conservative scientists often set the boundary of significance at $p = 0.01$. In linkage studies, such a low p value means that it is highly

TABLE 4.1 Critical Chi Square Values

[Handwritten annotations: "Not Rejected" pointing to 0.90; "Something Going On" below columns; "Due to Chance" below columns]

Degrees of Freedom	Null Hypothesis Accepted				Null Hypothesis Rejected		
	P Values						
	0.99	0.90	0.50	0.10	0.05	0.01	0.001
	χ^2 calculations						
1	—	0.02	.45	2.71	**3.84**	**6.64**	**10.83**
2	0.02	0.21	1.39	4.61	**5.99**	**9.21**	**13.82**
3	0.11	0.58	2.37	6.25	**7.81**	**11.35**	**16.27**
4	0.30	1.06	3.36	7.78	**9.49**	**13.28**	**18.47**
5	0.55	1.61	4.35	9.24	**11.07**	**15.09**	**20.52**

χ^2 values that are equal to or greater than those written in boldface allow you to reject the null hypothesis and, for recombination experiments, to postulate linkage.

unlikely (only 1 chance in 100) that the observed distribution of phenotypes arose from two unlinked genes, and we can conclude, with 99% confidence, that the genes are linked. On the other hand, higher *p* values (greater than 0.01 or 0.05, depending on the criterion used) do not necessarily mean that two genes are unlinked; it may mean only that the sample size is not large enough to provide an answer. With more data, the *p* value will normally rise if the null hypothesis of no linkage is correct, and fall if there is, in fact, linkage.

Note that in Fig. 4.6 all of the numbers in the second set of data are simply double the numbers in the first set, with the percentages remaining the same. Thus, just by doubling the sample size from 50 to 100 individuals, it was possible to go from no significant difference to a highly significant difference between the observed and the expected values. In other words, the larger the sample size, the less the likelihood that a certain percentage deviation from expected results happened simply by chance. Bearing this in mind, you can see that it is not appropriate to use the chi square test when analyzing very small samples (of less than 10), which by their nature are usually insufficient to answer questions concerning linkage. This creates a problem for human geneticists, who cannot "construct" people as they can fruit flies and other experimental organisms nor develop large populations of inbred individuals. To achieve a reasonable sample size for linkage studies in humans, they must pool data from a large number of family pedigrees.

The subjectivity involved in determining the boundary of significance means that the chi square test *does not* prove linkage or its absence. What it *does* do is provide a quantitative measure of the likelihood that the data from an experiment can be explained by a particular hypothesis. The chi square analysis is thus a general statistical test for significance; it can be used with many different experimental designs and with hypotheses other than the absence of linkage. As long as it is possible to propose a hypothesis that leads to a predicted set of values for a defined set of data classes, you can readily determine whether or not the observed data are consistent with the

hypothesis. If what you are looking for is any one of a range of outcomes that is not the outcome predicted by the hypothesis, that is, if the hypothesis serves as a straw man, set up only to be knocked down, it is considered a null hypothesis.

When experiments lead to rejection of a null hypothesis, you may need to confirm the alternative. For instance, if you are analyzing the inheritance of two opposing traits on the basis of a null hypothesis that says the two phenotypes result from the segregation of two equally viable alleles of a single gene, you would expect a test cross between an F_1 heterozygote and a recessive homozygote to produce a 1:1 ratio of the two traits in the offspring. If instead, you observe a ratio of 6:4 and the chi square test produces a *p* value of 0.009, you can reject the null hypothesis. But you are still left with the question of what the absence of a 1:1 ratio means. There are actually two alternatives: Either the reverse of the null hypothesis is true and the two alleles of the single gene are not equally viable *or* more than one gene encodes the trait. The chi square test cannot tell you which possibility is correct, and you would have to study the matter further. The problems at the end of this chapter illustrate several applications of the chi square test that are pertinent to genetics.

Recombination Results When Crossing-Over During Meiosis Separates Linked Genes

It is easy to understand how genes that are physically connected on the same chromosome can be transmitted together and thus show genetic linkage. It is not as obvious why all linked genes always show some recombination in a sample population of sufficient size. Do the chromosomes participate in a physical process that gives rise to the reshuffling of linked genes that we call recombination? The answer to this question is of more than passing interest as it provides a basis for gauging relative distances between pairs of genes on a chromosome.

In 1909 the cytologist F. Janssens described structures he had observed in the light microscope during prophase of the first meiotic division. He called these structures **chiasmata;** as described in Chapter 3, they seemed to represent regions in which nonsister chromatids of homologous chromosomes cross over each other (review Fig. 3.14b). Making inferences from a combination of genetic and cytological data, Morgan suggested that the chiasmata observed through the light microscope were sites of chromosome breakage and exchange resulting in genetic recombination.

Reciprocal Exchanges between Homologous Chromosomes Are the Physical Basis of Recombination

Morgan's idea that the physical breaking and rejoining of chromosomes during meiosis was the basis of genetic recombination seemed reasonable. But although Janssens's chiasmata could be interpreted as signs of the process, before 1930 no one had produced visible evidence that crossing-over between homologous chromosomes actually occurs. The identification of **physical markers,** or cytologically visible abnormalities that make it possible to keep track of specific chromosome parts from one generation to the next, enabled researchers to turn the logical deductions about recombination into facts derived from experimental evidence. In 1931 Harriet Creighton and Barbara McClintock, who studied corn, and Curt Stern, who worked with *Drosophila,* published the results of experiments showing that genetic recombination depends on the reciprocal exchange of parts between maternal and paternal chromosomes. Stern, for example, bred female flies with two different X chromosomes, each containing a distinct physical marker. These same females were also doubly het-

erozygous for two X-linked **genetic markers**—genes that could serve as points of reference in determining whether particular progeny were the result of recombination.

Figure 4.7 diagrams the chromosomes of these heterozygous females. One X chromosome carried mutations producing carnation eyes (a dark ruby color, abbreviated *car*) that were kidney shape (*Bar*); in addition, this chromosome was marked physically by a visible discontinuity. The other X chromosome had wild-type alleles (+) for both the *car* and the *Bar* genes, and its physical marker consisted of part of the Y chromosome that had become connected to the X chromosome centromere.

Figure 4.7 illustrates how the chromosomes in these *car Bar/car⁺ Bar⁺* females were transmitted to male progeny. In the experiment, all sons showing a phenotype determined by one or the other parental combination of genes (either *car Bar* or *car⁺ Bar⁺*) had an X chromosome that was structurally indistinguishable from one of the original X chromosomes in the mother. In recombinant sons, however, such as those that manifested carnation eye color and normal eye shape (*car Bar⁺*/Y), an identifiable exchange of the abnormal features marking the ends of homologous X chromosomes accompanied the recombination of genes. The evidence thus tied a particular instance of phenotypic recombination to the crossing-over of particular genes located in specifically marked parts of particular chromosomes. This was an elegant demonstration that genetic recombination is associated with

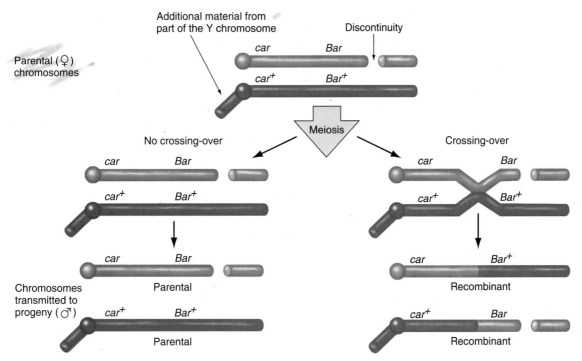

Figure 4.7   Evidence that recombination results from reciprocal exchanges between homologous chromosomes. Genetic recombination between the *car* and *Bar* genes on the *Drosophila* X chromosome is accompanied by the exchange of physical markers observable in the microscope. One marker is at one end of the maternally derived chromosome; the other marker is at the other end of the paternally derived chromosome. Note that this depiction of crossing-over is a simplification, as genetic recombination actually occurs after each chromosome has replicated into sister chromatids.

Genetics and Society

Mitotic Recombination and Cancer Formation

Although the recombination of genetic material during mitosis is a rare event, occurring no more frequently than once in a million somatic cell divisions, it can have major repercussions on health and development. Curt Stern, an early *Drosophila* geneticist, originally inferred the existence of mitotic recombination from observations of "twin spots" in a few fruit flies. **Twin spots** are adjacent islands of tissue that differ both from each other and from the tissue surrounding them. The distinctive patches arise from homozygous cells with a recessive phenotype growing amid a generally heterozygous cell population displaying the dominant phenotype. Medical geneticists later recognized that mitotic recombination can help explain the formation of certain tumors. These tumors arise from cells that become homozygous for the recessive alleles of a particular gene in a person whose other, normal body cells are heterozygous for the gene in question.

In *Drosophila,* the *yellow* (*y*) mutation changes body color from normal brown to yellow, while the *singed bristles* (*sn*) mutation causes body bristles to be short and curled rather than long and straight. Both of these genes are on the X chromosome. In 1936, Stern examined *Drosophila* females of genotype $y \; sn^+/y^+ sn$. These heterozygotes were generally wild type in appearance, but Stern noticed that some flies carried patches of yellow body color, others had small areas of singed bristles, and still others displayed twin spots: adjacent patches of yellow cells and cells with singed bristles (Fig. A.1). He assumed that mistakes in the mitotic divisions accompanying fly development could have led to these **mosaic** animals containing tissues of different genotypes. Individual yellow or singed patches could arise from chromosome loss or by mitotic nondisjunction (review Chapter 3, page 80). These errors in mitosis would yield XO cells containing only *y* (but not y^+) or *sn* (but not sn^+) alleles; such cells would show one of the recessive phenotypes.

The twin spots must have a different origin. Stern reasoned that they represented the reciprocal products of mitotic crossing-over between the *sn* gene and the centromere. The mechanism is as follows. During mitosis in a diploid cell, after chromosome duplication, homologous chromosomes occasionally—very occasionally—pair up with each other. While the chromosomes are paired, nonsister chromatids (that is, one chromatid from each of the two homologous chromosomes) can exchange parts by crossing-over. The pairing is transient, and the homologous chromosomes soon resume their positions on the mitotic metaphase plate. There, the two chromosomes can line up relative to each other in either of two ways (Fig. A.2). One of these orientations would yield two daughter cells that remain heterozygous for both genes and thus be indistinguishable from the surrounding wild-type cells. The other orientation, however, will generate two homozygous daughter cells, one $y \, sn^+/y \, sn^+$, the other $y^+ sn/y^+ sn$. Since the two daughter cells would lie next to each other, subsequent mitotic divisions would produce adjacent patches of *y* and *sn* tissue. Note that if crossing-over occurs between *sn* and *y*, single spots of yellow tissue can form, but a reciprocal singed spot cannot be generated in this fashion (Fig. A.2, lower).

The size of the spots indicates when mitotic recombination took place. If they are large, it happened early in development, giving the resulting daughter cells a long time to proliferate; if they are small, it happened later. As Chapter 21 explains, mitotic crossing-over has been of great value in the study of *Drosophila* development.

In humans, some tumors may arise as a result of mitotic recombination. Recall from Chapter 2 that retinoblastoma is the most malignant form of eye cancer. In the United States, a genetic predisposition to the cancer is seen in about 1 in 20,000 live births. The retinoblastoma gene (*RB*) resides on chromosome 13, where the normal wild-type allele (RB^+) encodes a protein that regulates retinal growth and differentiation. Cells in the eye need at least one copy of the normal wild-type allele to maintain control over cell division. A mutation in RB^+ leading to loss of normal gene function is designated RB^-. If a cell loses both copies of RB^+, it loses regulatory control, and a tumor results. The normal, wild-type RB^+ allele is thus known as a tumor-suppressing gene.

People with a genetic predisposition to retinoblastoma are born with only one functional copy of the normal RB^+ allele; their second chromosome 13 carries either a nonfunctional RB^- allele or no *RB* gene at all. If a mutagen (such as radiation) or a mistake in gene replication or segregation destroys the single remaining normal copy of the gene in a retinal cell in either eye, a retinoblastoma tumor will develop at that site. In one study of people with a genetic predisposition to retinoblastoma, cells taken from eye tumors were RB^- homozygotes, while white blood cells from the same people were RB^+/RB^- heterozygotes. As Fig. A.3 shows, mitotic recombination provides one mechanism by which a cell in an RB^+/RB^- individual could become RB^-/RB^-, leading to tumor formation.

Only 40% of retinoblastoma cases follow the preceding scenario. The other 60% occur in people who are born with two normal copies of the *RB* gene. In such people, it takes two mutational events to cause the cancer. The first of these must convert an RB^+ allele to RB^-, while the second could be a mitotic recombination producing daughter cells that become cancerous because they are homozygous for a nonfunctional allele.

Interestingly, the role of mitotic recombination in the formation of retinoblastoma helps explain the incomplete penetrance and variable expressivity of the disease. People born as RB^+/RB^- heterozygotes may or may not develop the condition (incomplete penetrance). If, as usually happens, they do, they may have it in one or both eyes (variable expressivity). It all depends on whether and in what cells of the body mitotic recombination (or some other "homozygosing" event that affects chromosome 13) occurs.

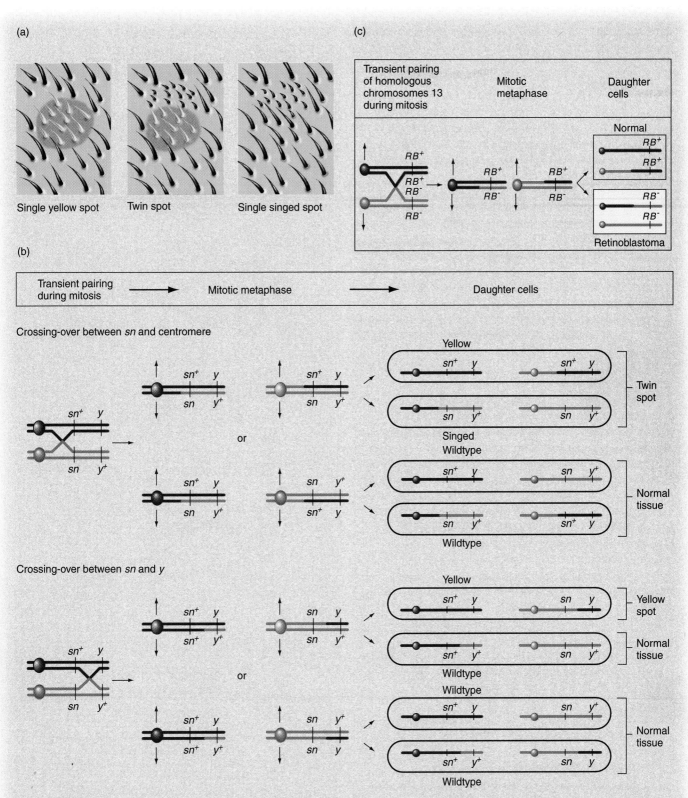

Figure A Mitotic crossing-over. (a) In a $y\,sn^+/y^+\,sn$ *Drosophila* female, most of the body is wildtype, but aberrant patches showing either yellow or singed bristles sometimes occur. In some cases, yellow and singed patches are adjacent to each other, a configuration known as *twin spots.* (b) (*top*) Mitotic crossovers in the region between the centromere and the *singed* gene can produce two daughter cells, one homozygous for *y* and the other homozygous for *sn.* As these daughter cells continue to divide, they develop into adjacent aberrant patches (twin spots). Note that this outcome depends on a particular distribution of chromatids at anaphase: If the chromatids are arranged in the equally likely opposite orientation, only phenotypically normal cells will result. (*bottom*) Crossovers between the *sn* and *y* genes can generate single yellow patches. However, a single mitotic crossover in these females cannot produce a single singed spot as long as the *singed* gene is closer to the centromere than the *yellow* gene. See if you can demonstrate this fact. (c) How mitotic crossing-over during retinal growth in an RB^-/RB^+ heterozygote can produce an RB^-/RB^- daughter cell which lacks a functional retinoblastoma gene and thus divides out of control. Only the arrangement of chromatids yielding this result is shown.

the actual reciprocal exchange of segments between homologous chromosomes during meiosis.

Through the Light Microscope: Chiasmata Mark the Sites of Recombination

Figure 4.8 outlines what is currently known about the steps of recombination as they appear in chromosomes viewed through the light microscope. (As mentioned in Chapter 3, although this low-resolution view may not represent certain details of recombination with complete accuracy, it nonetheless provides a useful frame of reference.) In Fig. 4.8a, the two homologs of each chromosome pair have already replicated so that there are now two pairs of sister chromatids or a total of four chromatids within each bivalent. In Fig. 4.8b, the synaptonemal complex zips together homologous chromosome pairs along their length. The synaptonemal zipper aligns homologous regions of all four chromatids such that allelic DNA sequences are physically near each other (see Fig. 3.14b for a detailed depiction). This proximity facilitates crossing-over between homologous sequences. In Fig. 4.8c, the synaptonemal complex begins to disassemble. Although at least some steps of the recombination process occurred while the chromatids were zipped in synapsis, it is only now that the recombination event becomes apparent. As the zipper dissolves, homologous chromosomes remain attached at chiasmata, the actual sites of crossing-over. Visible in the light microscope, chiasmata indicate where chromatid sections have switched from one molecule to another. In Fig. 4.8d, during anaphase I, as the two homologues separate, starting at their centromeres, the ends of the two recombined chromatids pull free of their respective sister chromatids and the chiasmata shift from their original positions toward a chromosome end, or telomere. This shifting of chiasmata is known as **terminalization.** In Fig. 4.8e and f, meiosis continues and eventually produces four haploid cells that contain one chromatid—now a chromosome—apiece. Non-homologous chromosomes have exchanged parts. Recombination can also take place apart from meiosis. Indeed, as The Genetics and Society box "Mitotic Recombination and Cancer Formation" explains, it sometimes, though rarely, occurs during mitosis. It also occurs with the circular chromosomes of prokaryotic organisms and cellular organelles such as mitochondria and chloroplasts, which do not undergo meiosis and do not form chiasmata (see Chapters 13 and 14).

Recombination Frequencies for Pairs of Genes Reflect the Distances between Them Along a Chromosome

T. H. Morgan's belief that chiasmata represent sites of physical crossing-over between chromosomes and that such crossing-over may result in recombination, led him to the following logical deduction. He hypothesized that different gene pairs exhibit different linkage rates because genes are arranged in a line along a chromosome, and the closer together two genes are, the less their chance of being separated by an event that cuts and recombines the line of genes. To look at it another way, if we assume for the moment that chiasmata can form

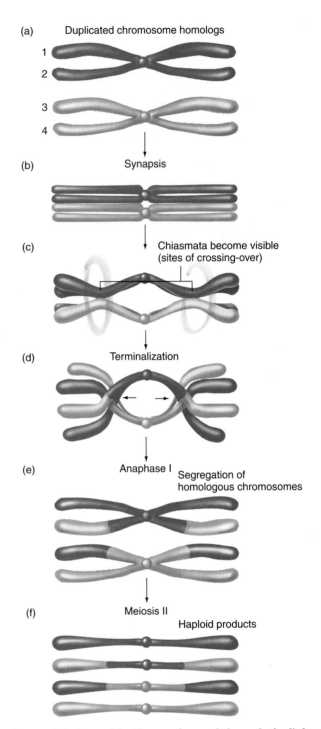

Figure 4.8 Recombination as observed through the light microscope. (a) A pair of duplicated homologous chromosomes very early in prophase of meiosis I. (b) During leptotene and zygotene of prophase I, the synaptonemal complex helps align corresponding regions of homologous chromosomes; this alignment allows recombination. (c) As the synaptonemal complex disassembles during diplotene, homologous chromosomes remain attached at chiasmata. (d) and (e) The chiasmata terminalize (move toward the chromosome ends), allowing the recombined chromosomes to separate during anaphase and telophase. (f) The result of the process is recombinant gametes.

anywhere along a chromosome with equal likelihood, then the probability of a crossover occurring between two genes increases with the distance separating them. If this is so, then the frequency of genetic recombination also must increase with the distance between genes. To illustrate the point, imagine pinning to a wall 10 inches of ribbon with a line of tiny black dots along its length and then repeatedly throwing a dart to see where you will cut the ribbon. You would find that practically every throw of the dart separates a dot at one end of the ribbon from a dot at the other end, while few if any throws separate any two particular dots positioned right next to each other.

A. H. Sturtevant, one of Morgan's students, took this idea one step further, proposing that the percentage of recombination, or **recombination frequency (RF)**, could be used as a gauge of the physical distance separating any two genes on the same chromosome. Sturtevant arbitrarily defined one RF percentage point as the unit of measure along a chromosome; later, another geneticist named the unit a **centimorgan (cM)** after T. H. Morgan. Mappers often refer to a centimorgan as a **map unit (m.u.).** Although the two terms are interchangeable, researchers prefer one or the other, depending on their experimental organism. *Drosophila* geneticists, for example, use map units while human geneticists use centimorgans. In Sturtevant's system, 1% RF = 1 cM = 1 m.u. A review of the two pairs of X-linked *Drosophila* genes we analyzed earlier shows how his proposal works. Because the X-linked genes for eye color (*w*) and body color (*y*) recombine in 1.1% of F_2 progeny, they are 1.1 m.u. apart (Fig. 4.9a). In contrast, the X-linked genes for eye color (*w*) and wing size (*m*) have a recombination frequency of 32.8 and are therefore 32.8 m.u. apart (Fig. 4.9b). As a unit of measure, the map unit is simply an index of recombination probabilities assumed to reflect distances between genes. According to this index, the *y* and *w* genes are much closer together than the *m* and *w* genes. Geneticists have used this logic to map thousands of genetic markers to the chromosomes of *Drosophila,* building recombination maps step by step with closely linked markers. And as we see next, they have learned that markers very far apart on the same chromosome may appear unlinked even though their recombination distances relative to closely linked intervening markers confirm that they are indeed on the same chromosome.

Experimental Recombination Frequencies between Two Genes Are Never Greater than 50%

If the definition of linkage is that the proportion of recombinant classes is less than that of parental classes, a recombination frequency of less than 50% indicates linkage. But what can we conclude about the relative location of genes if there are roughly equal numbers of parental and recombinant progeny? And does it ever happen that recombinants are in the majority? To answer these questions, we analyze the results of two simple experiments.

Consider first a cross following two recessive mutations in *Drosophila*—rosy eye color (*ry*) and thick wing veins (*tkv*)—that are known to lie on different autosomes. A female with rosy eyes (*ry/ry; tkv⁺/tkv⁺*) is mated to a male with thick wing veins (*ry⁺/ry⁺; tkv/tkv*); and the F_1 males and females are heterozygous for both traits (*ry/ry⁺; tkv/tkv⁺*). (Note that the semicolons separate genes on different chromosomes—in *Drosophila,* chromosomes 1 (the X), 2, 3, or 4—while the slashes separate alleles on the two homologs of a chromosome pair.) Figure 4.10a shows the distribution of progeny from a test cross between the dihybrid F_1 females (*ry/ry⁺; tkv/tkv⁺*) and males homozygous for both recessive traits (*ry/ry; tkv/tkv*). The result depicted is not surprising. Because the genes are not linked, they obey Mendel's second law of independent assortment and give rise to the four possible types of F_1 gametes with approximately equal frequency.

Compare this with a second cross following a different pair of recessive mutations known to lie at opposite ends of the same autosome: aristaless (*al*), which produces flies lacking part of the antennal structure, and blistered wings (*bs*). A test cross of doubly heterozygous F_1 females of genotype *al bs/al⁺ bs⁺* with doubly recessive *al bs/al bs* males yielded 48.2% recombinant progeny (*al bs⁺* and *al⁺ bs*) in one experiment, for a recombination frequency of 48.2% (Fig. 4.10b). Thus, with genes located very far apart on the same chromosome, the recombination frequency obtained from a single crossover between the two is also approximately 50%.

Researchers have never observed statistically significant recombination frequencies between two genes greater than 50%, which means that in any cross or series of crosses, recombinants among the F_2 progeny are never in the majority. Genes on different chromosomes yield recombination frequencies of 50% as a result of independent assortment. Genes sufficiently far apart on the same chromosome show the same percentage. Later in the chapter we examine why 50% is the upper limit on recombination frequency between two genes, even for genes on the same chromosome that are separated by very large physical distances. For now, simply note that recombination frequencies near 50% suggest either that two genes are on different chromosomes or that they lie far apart on the same chromosome. The only way to tell whether the two genes are on the same chromosome is through a series of matings showing definite linkage with other genes that lie between them. In short, even though crosses between two genes lying very far apart on a chromosome may show no linkage at all (because recombinant and

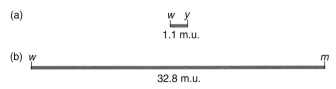

(a)   *w   y*
                              ⌐———⌐
                              1.1 m.u.

(b) *w*   *m*
    ⌐——————————————————————————————⌐
              32.8 m.u.

Figure 4.9   Recombination frequencies provide the basis of genetic mapping. (a) One percent of the gametes produced by a female doubly heterozygous for the genes *w* and *y* are recombinant. The recombination frequency is thus 1%, and the genes are approximately 1 map unit (m.u.) or 1 centimorgan (cM) apart. (b) The distance between the *w* and *m* genes is longer: 32.8 m.u. (or 32.8 cM).

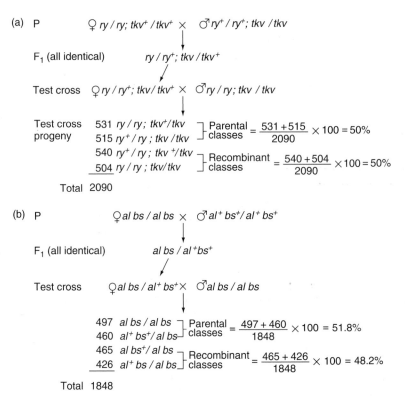

Figure 4.10 Unlinked genes show a recombination frequency of 50%. (a) The genes *ry* and *tkv* are genetically unlinked because they lie on different chromosomes. (b) The genes *al* and *bs* are genetically unlinked because they lie far apart on the same chromosome. As an exercise, use the chi square test to show that you cannot reject the null hypothesis of no linkage in either of these experiments.

parental classes are equal), you can demonstrate that they are on the same chromosome if you can tie each of the widely separated genes to one or more common intermediaries.

Linkage and Recombination: A Summary

The important conclusions from the experimental data presented thus far in this chapter can be summarized as follows. Gene pairs that are close together on the same chromosome are linked and do not follow Mendel's law of independent assortment. Instead, alleles of linked genes are more or less tightly coupled during transmission, leading to a preponderance of parental classes among the progeny of double heterozygotes. Linked genes, however, can become separated through recombination, and the frequency with which this happens is different for each pair of genes. There is a correlation between recombination frequency and the distance separating two genes.

The mechanism of recombination is crossing-over: Some genes move from the maternal to the paternal chromosome and vice versa when homologs exchange parts during meiosis. Chiasmata are the visible signs of crossing-over, and if we assume that chiasmata can occur anywhere along a chromosome, the farther apart two genes are, the greater the opportunity for chiasmata to form between them. This explains why larger recombination frequencies reflect greater distances between genes. Recombination frequencies in pairwise crosses vary from 0% to 50%. Statistically significant

values of less than 50% indicate that two genes are linked, and must therefore be on the same chromosome. Two genes that show a recombination frequency of 50% are genetically unlinked, either because they are on different chromosomes *or* because they reside far apart on the same chromosome.

Knowledge of linkage and recombination paves the way for understanding how geneticists assign genes a relative chromosomal position by comparing the recombination frequencies of many gene pairs. The chromosomal position assigned to a gene is its **locus.** Each gene's locus is the same in all individuals of a species, and the process of determining that locus is known as **mapping.** As you will see, the results of mapping experiments consistently verify the idea that genes are arranged in a line along a chromosome.

MAPPING: LOCATING GENES ALONG A CHROMOSOME

Maps are images of the relative positions of objects in space. Whether depicting the floorplan of New York's Metropolitan Museum of Art, the layout of the Roman Forum, or the location of cities served by the railways of Europe, maps turn measurements (how far apart two rooms, ruins, or railway stops are from one another) into patterns of spatial relationships that add a new level of meaning to the original data of distances. Maps that assign genes to locations on particular chromosomes are no exception. By transforming genetic data into spatial arrangements, they sharpen our ability to predict the inheritance patterns of specific traits.

We have seen that recombination frequency (RF) is a measure of the distance separating two genes along a chromosome. We now examine how data from many crosses following two and three genes at a time can be compiled and compared to generate accurate, comprehensive gene/chromosome maps.

Two-Point Crosses: Comparisons Help Establish Relative Gene Positions

In his senior undergraduate thesis, Morgan's student A. H. Sturtevant asked whether data obtained from a large number of two-point crosses (crosses tracing two genes at a time) would support the idea that genes form a definite linear series along a chromosome. He began by looking at X-linked genes in *Drosophila*. Figure 4.11a lists his recombination data for several two-point crosses. Recall that the distance between two genes that yields 1% recombinant progeny—an RF of 1%—is 1 m.u. As an example of Sturtevant's reasoning, consider the three genes *w*, *y*, and *m*. If these genes are arranged in a line (instead of a more complicated branched structure, for example), then one of them must be in the middle, flanked on either side by the other two. The greatest genetic distance should separate the two genes on the outside, and this value should roughly equal the sum of the distances separating the middle gene from each outside gene. The data Sturtevant obtained are consistent with this idea, implying that *w* lies between *y* and *m* (Fig. 4.11b). Note that the left-to-right orientation of this map was

selected at random; the map in Fig. 4.11b would be equally correct if it portrayed *y* on the left and *w* on the right.

Recombination Mapping Supports the Idea That Genes Are Arranged in a Line Along a Chromosome

By following exactly the same procedure for each set of three genes, Sturtevant established a self-consistent order for all the genes he investigated on *Drosophila*'s X chromosome (Fig. 4.11c; once again, the left-to-right arrangement is an arbitrary choice). By checking the data for every combination of three genes, you can assure yourself that this ordering makes sense. The fact that the recombination data yield a simple linear map of gene position supports the idea that genes reside in a unique linear order along a chromosome.

Two-Point Crosses Have Their Limitations

Though of great importance, the pairwise mapping of genes has several shortcomings that limit its usefulness. First, in crosses involving only two genes at a time, it may be difficult to determine gene order if some gene pairs lie very close together. For example, in mapping *y*, *w*, and *m*, 34.3 m.u. separate the outside genes *y* and *m*, while nearly as great a distance (32.8 m.u.) separates the middle *w* from the outside *m*. Before being able to conclude with any confidence that *y* and *m* are truly farther apart, that is, that the small difference between the values of 34.3 and 32.8 is not the result of sampling error, you would have to examine a very large number of flies and subject the data to a statistical test, such as the chi square test.

A second problem with Sturtevant's mapping procedure is that the actual distances in his map do not always add up, even approximately. As an example, suppose that the locus of the *y* gene at the far left of the map is regarded as position 0 (Fig. 4.11c). The *w* gene would then lie near position 1, and *m* would be located in the vicinity of 34 m.u. But what about the *r* gene, named for a mutation that produces rudimentary wings? Based solely on its distance from *y*, as inferred from the *y* ↔ *r* data in Fig. 4.11a, we would place it at position 42.9 (Fig. 4.11c). However, if we calculate its position as the sum of all intervening distances inferred from the data in Figure 4.11a, that is, as the sum of *y* ↔ *w* plus *w* ↔ *v* plus *v* ↔ *m* plus *m* ↔ *r*, the locus of *r* becomes 1.1 + 32.1 + 4.0 + 17.8 = 55.0 (Fig. 4.11d). What explains this difference, and which of these two values is closer to the truth? Three-point crosses help provide some of the answers.

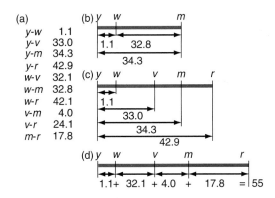

Figure 4.11 Mapping genes by comparisons of two-point crosses. (a) Sturtevant's data for the distances between pairs of genes on the *Drosophila* X chromosome. (b) Because the distance between *y* and *m* is greater than the distance between *w* and *m*, the order of genes must be *y-w-m*. Although this assignment is not completely secure because it is based on a small difference between 32.8 and 34.3 m.u., it has been verified by the results of three-point crosses (see later). (c) and (d) Maps for positions of five genes on the *Drosophila* X chromosome. The left-to-right orientation is arbitrary. Note that the numerical position of the *r* gene depends on how it is calculated. The best genetic maps are obtained by summing many small intervening distances as in (d).

Three-Point Crosses: A Faster, More Accurate Way to Map Genes

The simultaneous analysis of three markers makes it possible to obtain enough information to position the three genes in relation to each other from just one set of crosses. To describe this procedure, we look at three genes linked on one of *Drosophila*'s autosomes.

A homozygous female with mutations for vestigial wings (*vg*), black body (*b*), and purple eye color (*pr*) was mated to a wild-type male (Fig. 4.12a). All the triply heterozygous F_1 progeny, both male and female, had normal phenotypes for the

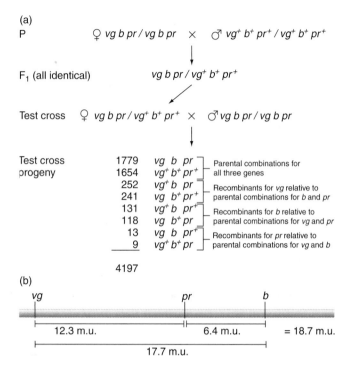

(b)

12.3 m.u.   6.4 m.u.   = 18.7 m.u.

17.7 m.u.

Figure 4.12 Analyzing the results of a three-point cross. (a) Results of a three-point cross. A test cross identifies parental or recombinant gametes produced by F_1 females triply heterozygous for *vg, b,* and *pr.* (b) A genetic map based on this data set. You can infer that *pr* is the gene in the middle from the observation that the longest distance is between the other two genes: *vg* and *b.* Because the most accurate map distances are calculated by summing shorter intervening distances, 18.7 m.u. is a more accurate estimate of the genetic distance between *vg* and *b* than 17.7 m.u.

three characteristics, indicating that the mutations are autosomal recessive. In a test cross of the F_1 females with males having vestigial wings, black body, and purple eyes, the progeny were of eight different phenotypes reflecting eight different genotypes. The order in which the genes in each phenotypic class are listed in Fig. 4.12a is completely arbitrary; instead of *vg b pr,* one could write *b vg pr* or *vg pr b* to indicate the same genotype. This is because at the outset, we do not know the gene order. Deducing it is the goal of the mapping study. Thus, the way the data are tabulated does not necessarily indicate the actual order of genes along the chromosome.

In analyzing the data, we look at two genes at a time (recall that the recombination frequency is always a function of a pair of genes). For the pair *vg* and *b,* the parental combinations are *vg b* and *vg$^+$b$^+$*; the nonparental recombinants are *vg b$^+$* and *vg$^+$b.* To determine whether a particular class of progeny is parental or recombinant for *vg* and *b,* we do not care whether the flies are *pr* or *pr$^+$*. Thus, to the nearest tenth of a map unit, the *vg* ↔ *b* distance, calculated as the percentage of recombinants in the total number of progeny, is

$$\frac{252 + 241 + 131 + 118}{4197} \times 100$$

$$= 17.7 \text{ m.u } (vg\text{–}b \text{ distance})$$

Similarly, since recombinants for the *vg–pr* gene pair are *vg pr$^+$* and *vg$^+$pr,* the interval between these two genes is

$$\frac{252 + 241 + 13 + 9}{4197} \times 100 = 12.3 \text{ m.u. } (vg\text{–}pr \text{ distance})$$

while the distance separating the *b–pr* pair is

$$\frac{131 + 118 + 13 + 9}{4197} \times 100 = 6.4 \text{ m.u. } (b\text{–}pr \text{ distance})$$

These recombination frequencies show that *vg* and *b* are separated by the largest distance (17.7 m.u., as compared to 12.3 and 6.4) and must therefore be the outside genes, flanking *pr* in the middle (Fig. 4.12b). But as with the X-linked genes analyzed by Sturtevant, the distance separating the outside *vg–b* genes (17.7) does not equal the sum of the two intervening distances (12.3 + 6.4 = 18.7). In the next section we learn that the reason for this discrepancy is the rare occurrence of double crossovers.

Three-Point Crosses Allow Correction for Double Crossovers

Figure 4.13 depicts the homologous autosomes of the F_1 females that are heterozygous for the three genes: *vg, pr,* and *b.* A close examination of the chromosomes reveals the kinds of crossovers that must have occurred to generate the classes and numbers of progeny observed. In this and subsequent figures, the chromosomes depicted are in late prophase/early metaphase of meiosis I, when there are four chromatids for each pair of homologous chromosomes. As we demonstrate later, prophase I is the stage at which recombination takes place. Note that we call the space between *vg* and *pr* "region 1" and the space between *pr* and *b* "region 2."

Recall that the progeny from a test cross between triply heterozygous F_1 females and males homozygous for the recessive allele of all three traits fall into eight groups (review Fig. 4.12). Flies in the two largest groups carry the same configurations of genes as did their grandparents of the P generation: *vg b pr* and *vg$^+$b$^+$pr$^+$*; they thus represent the parental classes (Figure 4.13a). The next two groups—*vg$^+$b pr* and *vg b$^+$pr$^+$*—are composed of recombinants that must be the reciprocal products of a crossover in region 1 between *vg* and *pr* (Fig. 4.13b).

Similarly the two groups containing *vg$^+$b pr$^+$* and *vg b$^+$pr* flies must have resulted from recombination in region 2 between *pr* and *b* (Fig. 4.13c). But what about the two smallest groups made up of rare *vg b pr$^+$* and *vg$^+$b$^+$pr* recombinants? What kinds of chromosome exchange could account for them? Most likely, they result from two different crossover events occurring simultaneously, one in region 1, the other in region 2 (Figure 4.13d). The gametes produced by such double crossing still have the parental configuration for the outside genes *vg* and *b,* even though not one but two exchanges must have occurred.

Because of the existence of double crossovers, the *vg–b* distance of 17.7 m.u. calculated in the previous section does

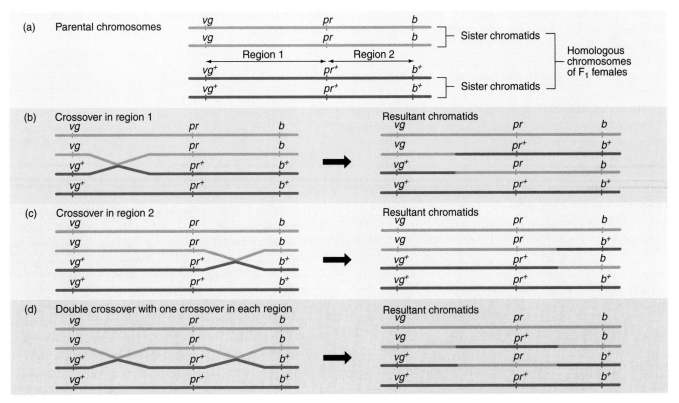

Figure 4.13 **Inferring the location of a crossover event: Different locations generate different genotypes.** This figure is based on the same set of data as Fig. 4.12. Once you establish the order of genes involved in a three-point cross, it is easy to determine where crossover events must have occurred to give rise to particular recombinant gametes. Note that double crossovers are needed to generate gametes in which the gene in the middle has recombined relative to the parental combinations for the genes at the ends.

not reflect all of the recombination events producing the gametes that gave rise to the observed progeny. To correct for this oversight, it is necessary to adjust the recombination frequency by adding the double crossovers twice, since each individual in the double crossover groups is the result of two exchanges between vg and b. The corrected distance is

$$\frac{252 + 241 + 131 + 118 + 13 + 13 + 9 + 9}{4197} \times 100$$

$$= 18.7 \text{ m.u.}$$

This value makes sense because you have accounted for all of the crossovers that occur in region 1 as well as all of the crossovers in region 2. As a result, the corrected value of 18.7 m.u. for the distance between vg and b is now exactly the same as the sum of the distances between vg and pr (region 1), and between pr and b (region 2).

As Fig. 4.11 shows, when Sturtevant originally mapped several X-linked genes in *Drosophila,* the locus of the rudimentary wings (r) gene was ambiguous. A two-point cross involving y and r gave a recombination frequency of 42.9, but the sum of all the intervening distances was 55.0 (see page 119). This discrepancy occurred because the two-point crosses ignored any double crossovers that might have occurred in the large interval between y and r. The data summing the smaller distances accounted for at least some of these double

crossovers by catching recombinations of gene pairs between y and r. Moreover, each smaller distance is less likely to encompass a double crossover than a larger distance, so each number for a smaller distance is inherently more accurate. Note that even a three-point cross like the one for $vg\,pr$ and b ignores the possibility of two recombination events taking place in, say, region 1. For greatest accuracy, it is always best to construct a map using many genes separated by relatively short distances.

Interference: The Number of Double Crossovers May Be Less than Expected

In a three-point cross following three linked genes, of the eight possible genotypic classes, the two parental classes contain the largest number of progeny, while the two double recombinant classes, resulting from double crossovers, are always the smallest (see Fig. 4.12). We can understand why double-crossover progeny are the rarest by looking at the probability of their occurrence. If an exchange in region 1 of a chromosome does not affect the probability of an exchange in region 2, the probability that both will occur simultaneously is the product of their separate probabilities (recall the law of the product, Chapter 1, page 18). For example, if progeny resulting from recombination in region 1 alone account for 10% of the total progeny (that is, if region 1 is 10 m.u.), and progeny resulting from recombination in region 2 alone account for 20%, the probability of a double crossover (one event in region 1, the second in region 2) is $0.10 \times 0.20 = 0.02$, or 2%. This makes sense because the likelihood of two rare events occurring simultaneously is even less than that of either rare event occurring alone.

As we have seen, in a three-point cross of the *b pr vg* trio of genes, the two classes of progeny arising from double crossovers contain the fewest progeny. The numerical frequencies of observed double crossovers, however, do not coincide with expectations derived from the law of the product. Let's look at the actual numbers. The probability of a single crossover between *b* and *pr* is 0.123 (corresponding to 12.3 m.u.), and the probability of a single crossover between *vg* and *pr* is 0.064 (6.4 m.u.). The product of these probabilities is

$$0.123 \times 0.064 = 0.0079 = 0.79\%$$

But the observed proportion of double crossovers (see Fig. 4.12) was

$$\frac{13 + 9}{4197} \times 100 = 0.52\%$$

The fact that the number of observed double crossovers is less than the number expected if the two exchanges are independent events suggests that the occurrence of one crossover reduces the likelihood that another crossover will occur in an adjacent part of the chromosome. This phenomenon—of crossovers not occurring independently—is called **chromosomal interference.** Interference may exist to ensure that every pair of homologous chromosomes undergoes at least one crossover event. The reasoning behind this hypothesis is straightforward: It is critical that every pair of homologous chromosomes sustain one or more crossover events because such events help the chromosomes orient properly at the metaphase plate during the first meiotic division; indeed homologous chromosome pairs without crossovers segregate improperly at high frequencies. If only a limited number of crossovers can occur during each meiosis and interference lowers the number of crossovers on large chromosomes, then interference also raises the probability of crossovers occurring on small chromosomes, increasing the likelihood of at least one crossover per homologous pair. Though the exact molecular mechanism underlying interference is not yet clear, recent experiments in yeast have shown that interference is mediated by the synaptonemal complex.

Interference is not uniform and may vary even for different regions of the same chromosome. Investigators can obtain a quantitative measure of the amount of interference in different chromosomal intervals by first calculating a **coefficient of coincidence,** defined as the ratio between the actual frequency of double crossovers observed in an experiment and the number of double crossovers expected on the basis of independent probabilities:

$$\text{Coefficient of coincidence} = \frac{\text{Frequency observed}}{\text{Frequency expected}}$$

For the three-point cross involving *b, pr,* and *vg,* the coefficient of coincidence is

$$\frac{0.52}{0.79} = 0.66$$

Interference itself is defined as 1 − the coefficient of coincidence; in this case,

$$\text{interference} = 1 - \text{coefficient of coincidence}$$
$$= 1 - 0.66 = 0.34$$

To understand the meaning of interference, it is helpful to contrast what happens when there is no interference with what happens when it is complete. If interference is 0, the frequency of observed double crossovers equals expectations, and crossovers in adjacent regions of a chromosome occur independently of each other. If interference is complete, that is, if interference = 1 because the coefficient of coincidence = 0, no double crossovers occur in the experimental progeny because one exchange effectively prevents another. As an example, in some mice, the recombination frequency for one pair of genes is 20, and for a second pair of genes, it is also 20. Without interference, the expected rate of double crossovers for these two gene pairs is

$$0.20 \times 0.20 = 0.04, \text{ or } 4\%$$

but when investigators observed the results of 1000 meiotic events, they found 0 double recombinants instead of the expected 40.

The Arrangement of Alleles in Double Recombinants Indicates the Relative Order of Three Genes

As we pointed out earlier, in three-point crosses of linked genes, the smallest classes of progeny are the two that contain double recombinants generated by double crossovers. It is possible to use the composition of alleles in these double crossover classes to determine which of three genes lies in the middle, even without calculating any recombination frequencies. Consider again the progeny of a three-point test cross looking at the *vg, pr,* and *b* genes. The F_1 females are $vg\,pr\,b / vg^+ pr^+ b^+$. As Figs. 4.12 and 4.13d show, test cross progeny resulting from double crossovers in the trihybrid females of the F_1 generation received gametes from their mothers carrying the allelic combinations $vg\,pr^+ b$ and $vg^+ pr\,b^+$. In these individuals, the alleles of the *vg* and *b* genes retain their parental associations ($vg\,b$ and $vg^+ b^+$), while the *pr* gene has recombined with respect to both the other genes ($pr\,b^+$ and $pr^+ b$; $vg\,pr^+$ and $vg^+ pr$). The same is true in all three-point crosses: In those gametes formed by double crossovers, the gene whose alleles have recombined relative to the parental configurations must be the one in the middle.

Comprehensive Example: Using a Three-Point Cross to Reanalyze Sturtevant's Map of the Drosophila X Chromosome

The technique of looking at double recombinants to discover which gene has recombined with respect to both other genes allows immediate clarification of gene order even in otherwise difficult cases. Consider the three X-linked genes y, w, and m that Sturtevant located in his original mapping experiment (*see* Fig. 4.11a and b). Since the distance between y and m (34.3 m.u.) appeared slightly larger than the distance separating w and m (32.8 m.u.), he concluded that w was the gene in the middle. But because of the small difference between the two numbers, his conclusion was subject to questions of statistical significance. If, however, we look at a three-point cross following y, w, and m, these questions disappear.

Figure 4.14 tabulates the classes and numbers of male progeny arising from females heterozygous for the y, w, and m genes. Since these male progeny receive their only X chromosome from their mothers, their phenotypes directly indicate the gametes produced by the heterozygous females. In each row of the table, the genes appear in an arbitrary order, which does not presuppose knowledge of the actual map. As you can see, the two classes of progeny listed at the top of the table outnumber the remaining six classes, which indicates that all three genes are linked to each other. Moreover, these largest groups, which are the parental classes, show that the two X chromosomes of the heterozygous females were w^+y^+m and wym^+.

Among the male progeny in Fig. 4.14, the two smallest classes, representing the double crossovers, have X chromosomes carrying w^+ym^+ and wy^+m combinations, in which the w alleles are recombined relative to those of y and m. The w gene must therefore lie between y and m, verifying Sturtevant's original assessment.

To complete a map based on the wym three-point cross, you can calculate the interval between y and w (region 1)

$$\frac{49 + 41 + 1 + 2}{6823} \times 100 = 1.3 \text{ m.u.}$$

as well as the interval between w and m (region 2)

$$\frac{1203 + 1092 + 2 + 1}{6823} \times 100 = 33.7 \text{ m.u.}$$

The genetic distance separating y and m is the sum of

$$1.3 + 33.7 = 35.0 \text{ m.u.}$$

Note that you could also calculate the distance between y and m directly by including double crossovers twice, to account for the total number of recombination events detected between these two genes:

$$\text{RF} = (1203 + 1092 + 49 + 41 + 2 + 2 + 1 + 1/6823) \\ \times 100 = 35.0 \text{ m.u.}$$

This method yields the same value as the sum of the two intervening distances (region 1 + region 2).

Further calculations show that interference is considerable in this portion of the *Drosophila* X chromosome, at least as inferred from the set of data tabulated in Fig. 4.14. The percentage of observed double recombinants was

$$3/6823 = 0.00044 \times 100 = 0.044\%$$

(rounding to the nearest tenth of a percent), while the percentage of double recombinants expected on the basis of independent probabilities by the law of the product is

$$0.013 \times 0.337 = 0.0044 \times 100 = 0.44\%$$

Thus, the coefficient of coincidence is

$$0.044/0.44 = 0.1$$

and the interference is

$$1 - 0.1 = 0.9$$

How Close Is the Correlation between a Genetic Map and Physical Reality?

In 1911, when Sturtevant first devised his system of genetic mapping based on recombination frequencies, he acknowledged there was no way to correlate a specific gene position with a physical part of the chromosome. Subsequent experiments, however, have shown that genetic maps do in fact accurately depict the order of genes along chromosomes. Thus, as we will see again and again in the many types of experiments presented throughout the remainder of this book, the *order of genes* revealed by mapping techniques corresponds to the order of those same genes along the DNA molecule of a chromosome.

In contrast, the *actual physical distances between genes*—that is, the amount of DNA separating genes—does not always correspond to genetic map distances. This is largely because the relationship between recombination frequency and physical distance along a chromosome is not simple. One complicating factor is the existence of double, triple, and even more crossovers. When genes are separated by 1 m.u. or less, double crossovers are not significant because the $0.01 \times 0.01 = 0.0001$ probability of their occurring is so small. But for genes separated by 20, 30, or 40 m.u., the probability of double

$$♀\, w^+\, w\, y^+\, y\, m^+\, m \quad \times \quad ♂\, X\, /\, Y$$

Before data analysis, you do not know the gene order or allele combination on each chromosome.

Male progeny		
2278	$w^+ y^+ m$ / Y	Parental class
2157	$w\ y\ m^+$ / Y	(noncrossover)
1203	$w\ y\ m$ / Y	Crossover in region 2
1092	$w^+ y^+ m^+$ / Y	(between w and m)
49	$w^+ y\ m$ / Y	Crossover in region 1
41	$w\ y^+ m^+$ / Y	(between y and m)
2	$w^+ y\ m^+$ / Y	Double
1	$w\ y^+ m$ / Y	crossover
6823		

After data analysis, you can conclude that the gene order and allele combinations on the X chromosomes of the F$_1$ females were $y\ w\ m^+$ / $y^+\ w^+\ m$.

Figure 4.14   How three-point crosses verify Sturtevant's map. The identity of the parental classes indicates the composition of the two X chromosomes of the F$_1$ female. The makeup of the double recombinant classes shows that w must be the gene in the middle, between y and m.

Gene Mapping Leads to a Possible Cure for Cystic Fibrosis

For forty years after the symptoms of cystic fibrosis were first described in 1938, there was no molecular clue—no visible chromosomal abnormality transmitted with the disease, no identifiable protein defect carried by affected individuals—suggesting the genetic cause of the disorder. As a result, there was no effective treatment for the 1 in 2000 Caucasian Americans born with the disease, most of whom died before they were 30. In the 1980s, however, geneticists were able to combine recently invented techniques for looking directly at DNA with maps constructed by linkage analysis to pinpoint a precise chromosomal position, or locus, for the cystic fibrosis gene. Knowledge of this locus made it possible to identify and clone the gene, discover the protein it encodes, understand the biochemical mechanisms disrupted by faulty alleles, and develop better therapies as well as a potential cure for this life-threatening condition.

The mappers of the cystic fibrosis gene faced an overwhelming task. They were searching for a gene that encoded an unknown protein, a gene that had not yet even been assigned to a chromosome. It could lie anywhere amid the 23 pairs of chromosomes in a human cell. Imagine looking for a close friend you lost track of years ago, who might now be anywhere in the world. You would first have to find ways to narrow the search to a particular continent (the equivalent of a specific chromosome in the gene mappers' search); then to a country (the long or short arm of the chromosome); next to

the state or province, county, city, or town, and street (all increasingly narrow bands of the chromosome); and finally, to a house address (the locus itself). Here, we briefly summarize how researchers applied some of these steps in mapping the cystic fibrosis gene.

■ After a review of many family pedigrees containing first-cousin marriages had confirmed that cystic fibrosis is most likely determined by a single gene (the *CF* gene), investigators collected white blood cells from 47 families with 2 or more affected children, obtaining genetic data from 106 patients, 94 parents, and 44 unaffected siblings.

■ They next tried to discover if any other trait is reliably transmitted with the condition. Analyses of the easily obtainable serum enzyme paroxonase showed that its gene (known as the *PON* gene) is indeed linked to the *CF* gene. At first, this knowledge was not that helpful, because the *PON* gene had not yet been assigned to a chromosome.

■ Then, in the early 1980s, geneticists developed a large series of DNA markers, based on new techniques that enabled them to recognize variations in the genetic material. A **DNA marker** is a piece of DNA of known size, representing a specific locus, that comes in identifiable variations. These allelic variations segregate according to Mendel's laws, which

---

crossovers skewing the data takes on greater significance. A second confounding factor is the 50% limit on the recombination frequency observable in a cross. This limit reduces the precision of RF as a measure of chromosomal distances. No matter how far apart two genes are on a long chromosome, they will never recombine more than 50% of the time.

Ever since Morgan, Sturtevant, and others began mapping, geneticists have generated mathematical equations called **mapping functions** to compensate for the inaccuracies inherent in relating recombination frequencies to physical distances. These equations generally make large corrections for RF values of widely separated genes, while barely changing the map distances separating genes that lie close together. This reflects the fact that multiple recombination events and the 50% limit on recombination do not confound the calculation of distances between closely linked genes. However, the corrections for large distances are at best imprecise, because mapping functions are based on simplifying assumptions (such as no interference) that rarely apply to the specific situation under consideration. Thus, the best way to create an accurate map is still by summing many smaller intervals, locating widely separated genes through linkage to common interme-

diaries. Because the most accurate genetic maps are drawn from a large series of genetic crosses, they are subject to continual refinement as more and more newly discovered genes are included.

Multiple Factor Crosses Help Establish Linkage Groups by Inference

Suppose you discovered a never-before-analyzed mouse trait—a bright pink tail—determined by the recessive mutation *e*. In a cross between agouti mice with pink tails (genotype *A A e e*) and black mice with wild-type tails (*a a E E*), the recombination frequency is 50%. How can you tell whether the *A* and *E* genes reside on separate chromosomes or far apart on the same chromosome? Like Sturtevant, you could conduct a series of crosses involving other genes (say, *B*, *C*, and *D*) that are known to reside on the same chromosome as the well-characterized *A* gene. If you can demonstrate that *A* is linked to *B*, *B* to *C*, *C* to *D*, and *D* to *E*, you can conclude that *A* and *E* are on the same chromosome. The Fast Forward box "Gene Mapping Leads to a Possible Cure for Cystic Fibrosis" describes how researchers used such linkage information to locate the gene for cystic fibrosis.

means it is possible to follow their transmission as you would any gene's. For example, if one allelic form of DNA marker A cosegregates with the abnormal *CF* allele, while another form of marker A assorts with the normal copy of the gene, these identifiable variations in DNA serve as molecular landmarks located near the *CF* gene. Chapters 9 and 10 explain the discovery and use of DNA markers in greater detail; for now it is only important to know that they exist and can be identified.

By 1986, linkage analyses of hundreds of DNA markers had shown that one marker, known as D7S15, is linked with both the *PON* gene and the *CF* gene. In these linkage analyses, researchers first determined whether linkage exists at all by calculating a **lod score:** the equivalent in human genetics of the chi square test's *p* value (see Chapter 10 for details). A lod score greater than 3 indicates linkage (just as a *p* value of less than 0.05 indicates linkage). In looking at linkage between the *CF* gene, the *PON* gene, and the D7S15 marker, geneticists obtained lod scores greater than 6.

Once they had established the fact of linkage, the researchers computed recombination frequencies and found that the distance from the DNA marker to the *CF* gene was 15 cM; from the DNA marker to the *PON* gene, 5 cM; and from *PON* to *CF,* 10 cM. After repeating their analyses with an ever-extending group of families, they concluded that the order of the three loci was D7S15-*PON*-*CF.* Since the *CF* gene could lie 15 cM in either of two directions from the DNA marker, the area under investigation was approximately 30 cM. And since the human genome consists of roughly 3000 cM, this step of linkage analysis narrowed the search to 1% of the human genome.

■ Next, the DNA marker D7S15 was localized to the long arm of chromosome 7, which meant that the gene for cystic fibrosis also resides in that same region of chromosome 7. Researchers had now placed the *CF* gene in a certain country on a particular genetic continent.

■ Finally, investigators discovered linkage with several other markers on the long arm of chromosome 7. These included a DNA marker known as D7S8, as well as the genes for one kind of collagen and for a key molecule of the immune system (the β chain of the T-cell receptor). Two of the markers turned out to be separated from the *CF* gene by a distance of only 1 cM. With all these markers, it became possible to place the *CF* gene in the middle third of chromosome 7's long arm, on band 31. For families with at least one child who has cystic fibrosis, geneticists using DNA analyses of these closely linked markers could now identify carriers of an abnormal copy of the *CF* gene with substantial confidence.

By 1989, researchers had used this mapping information to identify and clone the *CF* gene on the basis of its location. And by 1992, they had shown it encodes a cell membrane protein that regulates the flow of chloride ions into and out of cells (review the Fast Forward box "Genes Encode Proteins" in Chapter 1). This knowledge has become the basis of new therapies to open up ion flow (older therapies treated the results of flow blockage with only minimal success), as well as gene therapies to introduce normal copies of the *CF* gene into the cells of CF patients. Although in the early stages of development, such gene therapy holds out hope of a cure for cystic fibrosis.

Genes chained together by linkage relationships are known collectively as a **linkage group.** When enough genes have been assigned to a particular chromosome, the terms chromosome and linkage group become synonymous. And when the genetic map of a genome becomes so dense that it is possible to show that any gene on a chromosome is linked to another gene on the same chromosome, the number of linkage groups equals the number of pairs of homologous chromosomes in the species. Humans have 23 linkage groups, mice have 20, and fruit flies 4 (Fig. 4.15).

Tetrad Analysis in Fungi: A Powerful Tool for Mapping and for Understanding the Mechanisms of Recombination

With *Drosophila*, mice, peas, people, and other diploid organisms, each individual represents only one of the four potential gametes generated by each parent in a single meiotic event. Thus, until now, our presentation of linkage, recombination, and mapping has depended on inferences derived from examining the phenotypes of diploid progeny resulting from random unions of random products of meiosis. For such diploid organisms, we do not know which, if any, of the parents' other

progeny arose from gametes created in the same meiosis. Because of this limitation, the analysis of random products of meiosis in diploid organisms is based on statistical samplings of large populations.

In contrast, various species of fungi provide a unique opportunity for genetic analysis because they house all four haploid products of each meiosis in a sac called an **ascus.** These haploid cells, or **ascospores** (also known as **haplospores**), can germinate and survive as viable haploid individuals that grow and perpetuate themselves by mitosis. The phenotype of such haploid fungi is a direct representation of their genotype, without complications of dominance. Figure 4.16 illustrates the life cycles of two fungal species that preserve their meiotic products in a sac: one, the normally unicellular baker's yeast (*Saccharomyces cerevisiae*), is sold in supermarkets and contributes to the texture, shape, and flavor of bread; it generates four ascospores with each meiosis. The other, *Neurospora crassa,* is a bread mold that renders the bread on which it grows inedible; it too generates four ascospores with each meiosis, but at the completion of meiosis, each of the four haploid ascospores immediately divides once by mitosis to yield four pairs, for a total of eight haploid cells. The two cells in each pair of *Neurospora* ascospores have the same genotype, because they arose from mitosis. Haploid cells of both yeast and *Neurospora* normally reproduce vegetatively (that is, asexually) by mitosis. However, sexual reproduction is possible because the haploid cells come in two mating types, and cells of opposite mating types can fuse to form a diploid zygote. In baker's yeast, these diploid cells are stable and can reproduce through successive mitotic cycles. Stress, such as that caused by a scarcity or lack of essential nutrients, induces the diploid cells of yeast to enter meiosis. In bread mold, the diploid zygote immediately undergoes meiosis.

Mutations in haploid yeast cells affect many different traits, including the appearance of the cells and their ability to grow under particular conditions. For instance yeast cells with the *his4* mutation are unable to grow in the absence of the amino acid histidine, while yeast with the *trp1* mutation cannot grow without an external source of the amino acid tryptophan. Geneticists who specialize in the study of yeast have devised a system of representing genes that is slightly differ-

ent from the one for *Drosophila* and mice. They use capital letters (*HIS4*) to designate dominant alleles and lowercase letters (*his4*) to represent recessives. For most of the yeast genes we will discuss, the wild-type alleles are dominant, and may be represented by the alternative "+", while the symbol for the recessive alleles remains the lowercase abbreviation (*his4*).

After meiosis, the assemblage of four ascospores (or four pairs of ascospores) in a single ascus is called a **tetrad.** (Note that this is a second meaning for the term *tetrad.* In Chapter 3, a tetrad was the four homologous chromatids—two in each chromosome of a bivalent—synapsed during the prophase and metaphase of meiosis I. Here, it is the four products of a single meiosis held together in a sac. Since the four chromatids of a bivalent give rise to the four products of meiosis, the two meanings of tetrad refer to almost the same things.) In yeast, each tetrad is **unordered,** that is, the four meiotic products, known as spores, are arranged at random. In *Neurospora crassa,* each tetrad is **ordered,** with the four pairs, or eight haplospores, arranged in a line. To analyze both unordered and

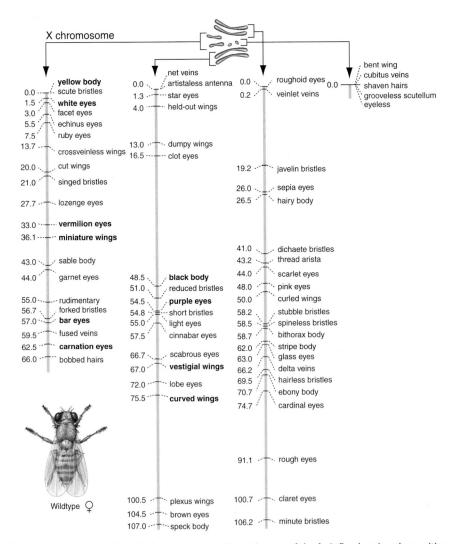

Figure 4.15 *Drosophila melanogaster* **has four linkage groups.** A genetic map of the fruit fly, showing the positions of many genes influencing visible phenotypes of body morphology, including those used as examples in this chapter (highlighted in bold). Because so many *Drosophila* genes have been mapped, each of the four chromosomes can be represented as a single linkage group.

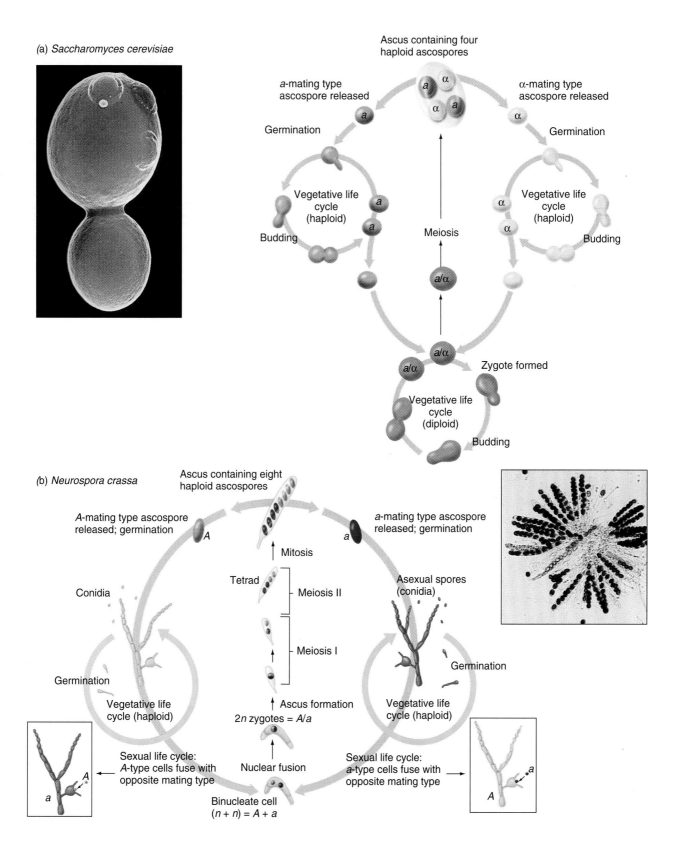

Figure 4.16   The life cycles of the yeast *Saccharomyces cerevisiae* and the bread mold *Neurospora crassa*. Both *S. cerevisiae* and *N. crassa* have two mating types that can fuse to form diploid cells that undergo meiosis to produce haploid ascospores. There are however two important differences from the genetic point of view: (a) Yeast cells can grow vegetatively either as haploids or diploids, while the diploid state in *Neurospora* exists only for a short period, and (b) ascospores are arranged randomly in the unordered yeast ascus, while the ordered arrangement of spores in the *Neurospora* ascus reflects the geometry of the meiotic spindles. The inset showing a budding (mitotically dividing) yeast cell is at much higher magnification than the inset displaying *Neurospora* asci.

ordered tetrads, researchers can release the spores of each ascus, induce the haploid cells to germinate under appropriate conditions, and then analyze the genetic makeup of the resulting haploid cultures. The data they collect in this way enable them to identify the four products of a single meiosis and compare them with the four products of many other distinct meioses. Ordered tetrads offer another possibility. With the aid of a dissecting microscope, investigators can recover the ascospores in the order in which they occur within the ascus, and thereby obtain additional information that is useful for mapping. We look first at the analysis of randomly released spores, using the unordered tetrads of yeast as an example. We then describe the additional information that can be gleaned from the microanalysis of ordered tetrads, using *Neurospora* as our model organism.

Tetrads Can Be Characterized by the Number of Parental and Recombinant Spores They Contain

What kinds of tetrads arise when diploid yeast cells heterozygous for two genes on different chromosomes are induced to undergo meiosis? Consider a mating between a haploid strain of yeast of mating type *a,* carrying the *his4* mutation and the wild-type allele of the *TRP1* gene, and a strain of the opposite mating type α that has the genotype *HIS4 trp1.* The resulting *a* / α diploid cells are *his4* / *HIS4; trp1* / *TRP1,* as shown in Fig. 4.17a. When conditions promote meiosis, the two unlinked genes will assort independently to produce equal frequencies of two different kinds of tetrads. In one kind, all of the spores are parental in that the genotype of each spore is the same as one of the parent's: *his4 TRP1* or *HIS4 trp1* (Fig. 4.17b). A tetrad that contains four parental class haploid cells is known as a **parental ditype (PD).** Note that *di-,* meaning two, indicates there are two possible parental combinations of alleles; the PD tetrad contains two of each combination. The second kind of tetrad, arising from the equally likely alternative distribution of chromosomes during meiosis, contains four recombinant spores: two *his4 trp1* and two *HIS4 TRP1* (Fig. 4.17c). This kind of tetrad is termed a **nonparental ditype (NPD),** because the two parental classes have recombined to form two reciprocal nonparental combinations of alleles.

A third kind of tetrad also appears when *his4* / *HIS4; trp1* / *TRP1* cells undergo meiosis. Called a **tetratype (T)** from the Greek word for "four," it carries four kinds of haploid cells: two different parental class spores (one *his4 TRP1* and one *HIS4 trp1*) and two different recombinants (one *his4 trp1* and one *HIS4 TRP1*). Tetratypes result from a crossover between one of the two genes and the centromere of the chromosome on which it is located (Fig. 4.17d).

Figure 4.17e displays the data from one experiment. Bear in mind that the column headings of PD, NPD, and T refer to tetrads (the group of four cells produced in meiosis), and not to individual haploid cells. Because the spores released from a yeast ascus are not arranged in any particular order, the order in which the spores are listed does not matter. The classification of a tetrad as PD, NPD, or T is based solely on the number of parental and recombinant spores found in the ascus.

When PD = NPD, Two Genes Are Unlinked

A cross following two unlinked genes must give equal numbers of individual parental and recombinant spores. This is simply another way of stating Mendel's second law of independent assortment, which predicts a 50% recombination frequency in such cases. Now, since T tetrads, regardless of their number, contain two recombinant and two nonrecombinant spores, and since all four spores in PD tetrads are parental, the only way 50% of the total progeny spores could be recombinant (as demanded by independent assortment) is if the number of NPD's (with four recombinant spores apiece) is roughly the same as the number of PDs (review Figure 4.17). For this reason, if PD = NPD, the two genes must be unlinked, either because they reside on different chromosomes or because they lie very far apart on the same chromosome.

When PDs Significantly Outnumber NPDs, It Is a Sign of Linkage

The genetic definition of linkage is the emergence of more parental types than recombinants among the progeny of a doubly heterozygous parent. In the preceding section, we saw that tetratypes always contribute an equal number of parental and recombinant spores. Thus, in terms of tetrads, linkage exists only when PD >> NPD, that is, when the number of PD tetrads (carrying only parental type spores) substantially exceeds the number of NPD tetrads (containing only recombinants). By analyzing an actual cross involving linked genes, we can see how this follows from the events occurring during meiosis.

A haploid yeast strain containing the *arg3* and *ura2* mutations was mated to a wild-type *ARG3 URA2* haploid strain (Fig. 4.18). When the resultant *a* / α diploid was induced to sporulate (that is, undergo meiosis), the 200 tetrads produced had the distribution shown in Fig. 4.18. As you can see, the 127 PD tetrads far outnumber the 3 NPD tetrads, suggesting that the two genes are linked.

Figure 4.19 shows how we can explain the particular kinds of tetrads observed in terms of the various types of crossovers that could occur between the linked genes. If no crossing-over occurs between the two genes, the resulting tetrad must be PD; since none of the four chromatids participates in an exchange, all of the products are of parental configuration (Fig. 4.19a). A single crossover between *ARG3* and *URA2* will generate a tetratype, containing four genetically different spores (Fig. 4.19b). But what about double crossovers? There are actually four different possibilities, depending on which chromatids participate, and each of the four should occur with equal frequency. A double crossover involving only two chromatids produces only parental type progeny, generating a PD tetrad (Fig. 4.19c). Three-strand double crossovers can occur in the two ways depicted in Fig. 4.19d and e; either way, a tetratype results. Finally, if all four strands take part in the two crossovers, all four progeny spores will be recombinant and the resulting tetrad is NPD (Fig. 4.19f). Thus, if two genes are linked, the only way to generate an NPD tetrad is through a four-strand double exchange.

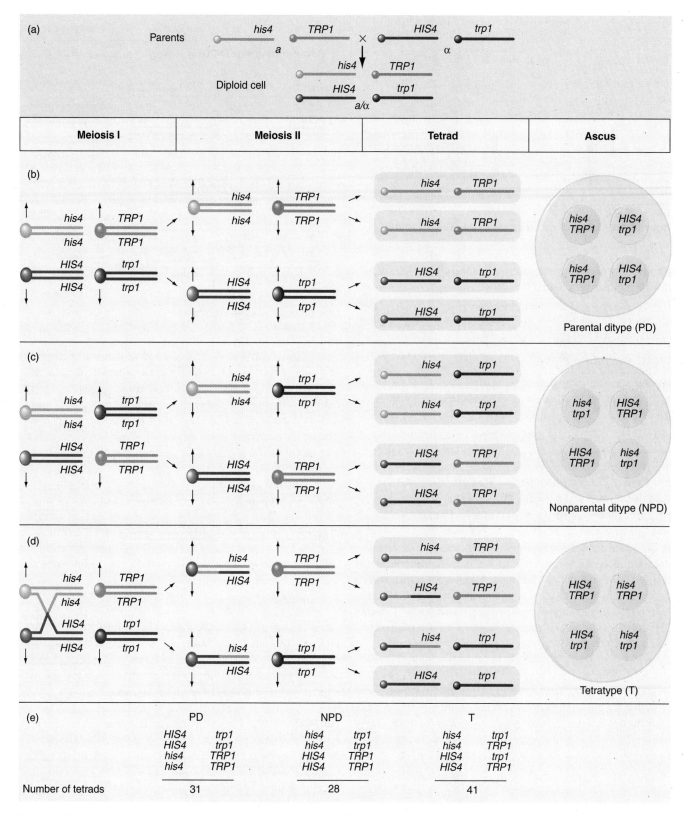

Figure 4.17   How meiosis can generate three kinds of tetrads when two genes are on different chromosomes. (a) Parental cross. (b) and (c) In the absence of recombination, the two equally likely alternative arrangements of two pairs of chromosomes yield either PD or NPD tetrads. T tetrads are made only if either gene recombines with respect to its corresponding centromere, as in (d). Numerical data in (e) show that the number of PD and NPD tetrads are approximately equal when the two genes are unlinked.

P	arg3 ura2 (a-mating type)	×	ARG3 URA2 (α-mating type)

Diploid cell   arg3 ura2 / ARG3 URA2

Meiosis

Products of meiosis	PD	NPD	T
	arg3 ura2	arg3 URA2	arg3 ura2
	arg3 ura2	arg3 URA2	arg3 URA2
	ARG3 URA2	ARG3 ura2	ARG3 ura2
	ARG3 URA2	ARG3 ura2	ARG3 URA2
Number of tetrads	127	3	70

Figure 4.18   When Genes are Linked, PDs Exceed NPDs

Meioses with crossovers generating such a specific kind of double recombination must be a lot rarer than no crossing-over or single crossovers, which produce PD and T tetrads, respectively. This explains why, if two genes are linked, PD must greatly exceed NPD.

Calculating the Recombination Frequency

Because we know that all of the spores in an NPD tetrad are recombinant, and half of the four spores in a tetratype are recombinant, we can say that

$$RF = \frac{NPD + 1/2T}{Total\ tetrads} \times 100$$

For the *ARG3 URA2* example in Fig. 4.17,

$$RF = \frac{3 + (1/2)70}{200} \times 100 = 19\ m.u.$$

It is reassuring that this formula gives the exact same result as calculating the RF as the percentage of individual recombinant spores. For example, the 200 tetrads analyzed in this experiment contain 800 (200 × 4) individual spores; each NPD ascus holds 4 recombinant ascospores, and each T tetrad contains 2 recombinants. Thus,

$$RF = \frac{(4 \times 3) + (2 \times 70)}{800} \times 100 = 19\ m.u.$$

Tetrad Analysis Confirms That Recombination Occurs at the Four-Strand Stage

Both T and NPD tetrads contain recombinant spores, and when tetrad analysis reveals linked genes, the T tetrads always outnumber the NPDs, as in the example we have been discussing. This makes sense, because all single and some double crossovers yield tetratypes, while only 1/4 of the rare double crossovers produce NPDs. The very low number of NPDs establishes that recombination occurs after the chromosomes have replicated, when there are four chromatids for each pair of homologs. If recombination took place before chromosome duplication, every single crossover event would yield four recombinant chromatids and generate an NPD

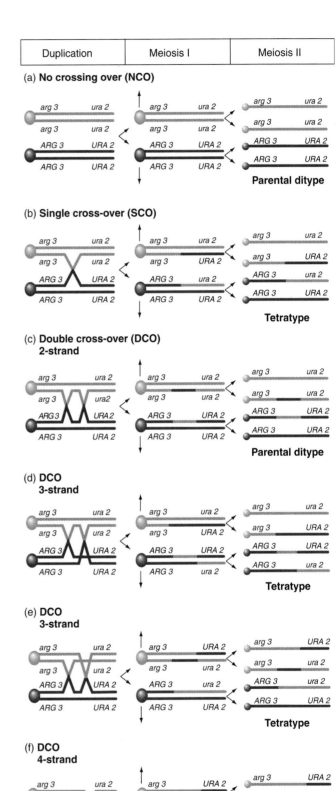

Duplication	Meiosis I	Meiosis II

(a) No crossing over (NCO)

Parental ditype

(b) Single cross-over (SCO)

Tetratype

**(c) Double cross-over (DCO)
2-strand**

Parental ditype

**(d) DCO
3-strand**

Tetratype

**(e) DCO
3-strand**

Tetratype

**(f) DCO
4-strand**

Nonparental ditype

Figure 4.19   How crossovers between linked genes generate different tetrads. (a) PDs arise when there is no crossing-over. (b) Single crossovers between the two genes yield tetratypes. (c) to (f) Double crossovers between linked genes can generate PD, T, or NPD tetrads, depending on which chromatids participate in the crossovers.

Recombination	Duplication	Meiosis I	Meiosis II

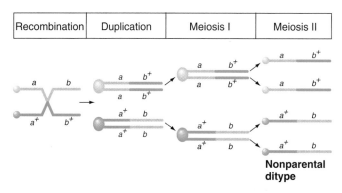

Nonparental ditype

Figure 4.20 Recombination before chromosome replication: A mistaken model. If recombination occurred before the chromosomes duplicated, you would expect, when dealing with linked genes, that most tetrads containing recombinant spores would be NPDs instead of Ts. Actual results, such as those shown in Fig. 4.18, show that the opposite is true.

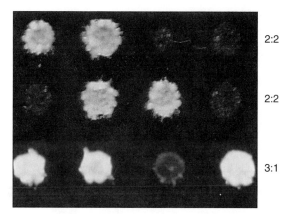

2:2

2:2

3:1

Figure 4.21 In rare tetrads, the two alleles of a gene do not segregate 2:2. In this experiment, researchers sporulated a *HIS4/his4* diploid strain of yeast, and dissected the four haploid spores from each of three different tetrads. They then plated these spores on petri plates containing medium without histidine, so that each row on the petri plate presents the results obtained with the four spores of a single tetrad. The top two rows show the normal 2:2 segregation of the two alleles of a single gene: two of the spores are *HIS4* and form colonies, whereas the other two spores are *his4* and cannot grow into colonies. The bottom row displays a rare tetrad with an unusual segregation of 3 *HIS4* :1 *his4*. This aberrant segregation pattern results from the molecular processes responsible for recombination.

tetrad (Fig. 4.20). A model assuming that recombination occurs when there are two rather than four chromatids per pair of homologous chromosomes would thus predict more NPD than T tetrads, but experimental observations show just the opposite (compare Fig. 4.20 with Fig. 4.19).

The fact that recombination takes place after the chromosomes have replicated explains the 50% limit on recombination for genes on the same chromosome. Single crossovers between two genes generate T tetrads containing two out of four spores that are recombinant. Thus, even if one crossover occurred between two such genes in every meiosis, the observed recombination frequency would be 50%. The four kinds of double crossovers depicted in Fig. 4.19 yield either PD tetrads with 0/4 recombinants; T tetrads carrying 2/4 recombinants; differently derived T tetrads, still carrying 2/4 recombinants; and NPD tetrads with 4/4 recombinants. Because these four kinds of double crossovers almost always occur with equal frequency, no more than 50% of the progeny resulting from double (or, in fact, triple or more) crossovers can be recombinant.

Tetrad Analysis also Demonstrates That Recombination Is Usually Reciprocal

Suppose you are following linked genes *A* and *B* in a cross between *A B* and *a b* strains of yeast. If the recombination that occurs during meiosis is reciprocal, every tetrad with recombinant progeny should contain equal numbers of both classes of recombinants. Observations have confirmed this prediction: every T tetrad carries one *A b* and one *a B* spore, while every NPD tetrad contains two of each type of recombinant. We can thus conclude that meiotic recombination is reciprocal, generating two homologous chromosomes that are inverted images of each other. There are, however, exceptions. Very rarely, a particular cross produces tetrads containing unequal numbers of reciprocal classes, and such tetrads cannot be classified as PD, NPD, or T. In these exceptional tetrads, the two input alleles of one of the genes, instead of segregating at

a ratio of 2*A*:2*a*, produce ratios of 1*A*:3*a* or 3*A*:1*a* or even 0*A*:4*a* or 4*A*:0*a* (Fig. 4.21). In these same tetrads, markers such as *B/b* and *C/c* that flank the *A* or *a* allele on the same chromosome, still segregate 2*B*:2*b* and 2*C*:2*c*. Moreover, careful phenotypic and genetic tests show that even when alleles do not segregate 2:2, only the original two input alleles occur in the progeny. Thus, recombination, no matter what ratios it generates, does not create new alleles. Geneticists believe that the unusual non-2:2 segregation ratios observed in rare instances result from molecular events at the site of recombination. We discuss these events at the molecular level in Chapter 5. For now, it is simply necessary to know that the unusual ratios exist but are quite rare.

Ordered Tetrads Help Locate Genes in Relation to the Centromere

Analyses of ordered tetrads, such as those produced by the bread mold *Neurospora crassa*, allow you to map the centromere of a chromosome relative to other genetic markers, information that you cannot obtain from unordered yeast tetrads. As described earlier, immediately after specialized haploid *Neurospora* cells of different mating types fuse at fertilization, the diploid zygote undergoes meiosis within the confines of a narrow ascus (review Fig. 4.16b). At the completion of meiosis, each of the four haploid meiotic products divides once by mitosis, yielding an **octad** of eight haploid ascospores. Dissection of the ascus at this point allows one to determine the phenotype of each of the eight haploid cells.

The cross-sectional diameter of the ascus is so small that cells cannot slip past each other. Moreover, during each division after fertilization, the microtubule fibers of the spindle extend outward from the centrosomes parallel to the long axis of the ascus, so that when each of the four products of meiosis divides once by mitosis, the two genetically identical cells that result lie adjacent to each other (Fig. 4.22). Because

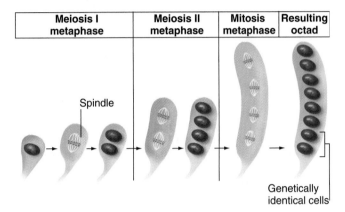

Meiosis I metaphase	Meiosis II metaphase	Mitosis metaphase	Resulting octad

Figure 4.22   How ordered tetrads form. During development of the *Neurospora* ascus, the spindles form parallel to the long axis of the ascus, such that cells cannot slide around each other. As a result, the order of ascospores reflects the geometry of the meiotic spindles. After meiosis, the haploid cells undergo a mitotic division, producing an ascus of eight cells (an octad). The octad consists of four pairs of cells; the two cells of each pair are genetically identical.

of this feature, starting from either end of the ascus, you can count the octad of ascospores as four cell pairs and analyze it as a tetrad. And from the precise positioning of the four ascospore pairs within the ascus, you can infer the arrangement of the four chromatids of each homologous chromosome pair during the two meiotic divisions.

To understand the genetic consequences of the geometry of the ascospores, it is helpful to consider what kinds of tetrads you would expect from the segregation of two alleles of a single gene. The mutant white-spore allele (*ws*) alters ascospore color from wild-type black to white. In the absence of recombination, the two alleles (ws^+/ws) separate from each other at the first meiotic division because the centromeres to which they are attached separate at that stage. The second meiotic division and subsequent mitosis create asci in which the top four ascospores are of one genotype (for instance ws^+) and the bottom four of the other (*ws*). Whether the top four are ws^+ and the bottom four *ws*, or vice versa, depends on the random metaphase I orientation of the homologs that carry the gene relative to the long axis of the developing ascus.

The segregation of two alleles of a single gene at the first meiotic division is thus indicated by an ascus in which an imaginary line drawn between the fourth and the fifth ascospores of the octad cleanly separates haploid products bearing the two alleles. Such an ascus displays a **first-division segregation** pattern (Fig. 4.23a).

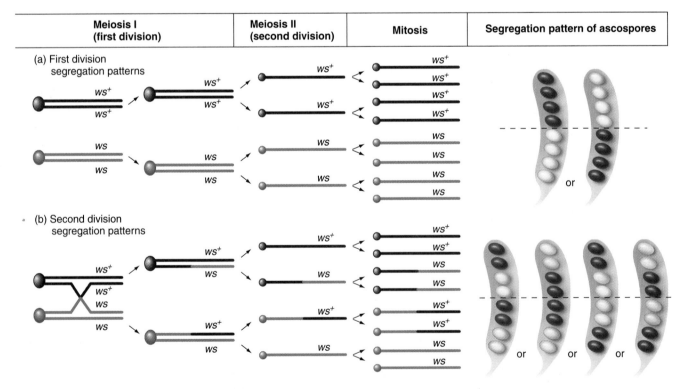

	Meiosis I (first division)	Meiosis II (second division)	Mitosis	Segregation pattern of ascospores

Figure 4.23   The absence or presence of crossovers between a gene and its centromere yield two segregation patterns. (a) In the absence of a crossover between a gene and its centromere, the two copies of a gene will separate at the first meiotic division. The result is a first-division segregation pattern in which each allele appears in spores located on only one side of an imaginary line through the middle of the ascus. (b) A crossover between a gene and its centromere produces a second-division segregation pattern in which both alleles appear in spores on the same side of the middle line.

Suppose now that during meiosis I, a crossover occurs in a heterozygote between the white-spore gene and the centromere of the chromosome on which it travels. As Fig. 4.23b illustrates, this can lead to four equally possible ascospore arrangements, each one depending on a particular orientation of the four chromatids during the two meiotic divisions. In all four cases, both ws^+ and ws spores are found on both sides of the imaginary line drawn between ascospores 4 and 5, because cells with only one kind of allele do not arise until the end of the second meiotic division. Octads carrying this configuration of spores display a **second-division segregation** pattern (review Fig. 4.23b).

Since second-division segregation patterns result from meioses in which there has been a crossover between a gene and its centromere, the relative number of asci with this pattern can be used to determine the gene–centromere distance. In an ascus showing second-division segregation, one-half of the ascospores are derived from chromatids that have exchanged parts, while the remaining half arise from chromatids that have not participated in crossovers leading to recombination. To calculate the distance between a gene and its centromere, you therefore simply divide the percentage of second-division-segregation octads by 2. Geneticists use information about the location of centromeres to make more accurate genetic maps as well as to study the structure and function of centromeres.

A Numerical Example of Ordered-Tetrad Analysis

In one experiment, a $thr^+ arg^+$ wild-type strain of *Neurospora* was crossed with a *thr arg* double mutant. The *thr* mutants cannot grow in the absence of the amino acid threonine, while *arg* mutants cannot grow without a source of the amino acid arginine; cells carrying the wild-type alleles of both genes can grow in medium that contains neither amino acid. From this cross, 105 octads, considered here as tetrads, were obtained. These tetrads were classified in seven different groups—A, B, C, D, E, F, and G—as shown in Fig. 4.24a. For each of the two genes, we can now find the distance between the gene and the centromere of the chromosome on which it is located.

To do this for the *thr* gene, we count the number of tetrads with a second-division-segregation pattern for that gene. Drawing an imaginary line through the middle of the tetrads, we see that those in groups B, D, E, and G are the result of second-division segregations, while the remainder show first-division patterns. The centromere–*thr* distance is thus

Percentage of second-division patterns =

$$\frac{16 + 2 + 2 + 1}{105} \times 100 = 20; \quad \frac{20}{2} = 10 \text{ m.u.}$$

Similarly, the second-division tetrads for the *arg* gene are in groups C, D, E, and G, so the distance between *arg* and its centromere is

$$\frac{11 + 2 + 2 + 1}{105} \times 100 = 15.2; \quad \frac{15.2}{2} = 7.6 \text{ m.u.}$$

(a)

Tetrad group	A	B	C	D	E	F	G
Segregation pattern	thr arg	thr arg	thr arg	thr arg⁺	thr arg⁺	thr arg⁺	thr arg
	thr arg	thr⁺arg	thr arg⁺	thr⁺arg	thr⁺arg	thr arg⁺	thr⁺arg⁺
	thr⁺arg⁺	thr⁺arg⁺	thr⁺arg⁺	thr⁺arg⁺	thr⁺arg	thr⁺arg	thr⁺arg⁺
	thr⁺arg⁺	thr arg⁺	thr⁺arg⁺	thr arg	thr arg⁺	thr⁺arg	thr arg
Total in group	72	16	11	2	2	1	1

(b)

Figure 4.24 Genetic mapping by ordered tetrad analysis: An example. (a) Data from a sample *Neurospora* cross. Note that for ordered tetrad analysis, tetrad classes are defined not only on the basis of the numbers of recombinant and nonrecombinant spores (i.e., whether they are PD, NPD, or T), but also according to whether they show a first- or second-division-segregation pattern. (b) Genetic map derived from the data in part (a). Ordered-tetrad analysis allows determination of the centromere's position as well as distances between genes.

To ascertain whether the *thr* and *arg* genes are linked, we need to evaluate the seven tetrad groups in a different way, looking at the combinations of alleles for the two genes to see if the tetrads in that group are PD, NPD, or T. We can then ask whether PD $>>$ NPD. Referring again to Fig. 4.24, we find that groups A and G are PD, because all the ascospores show parental combinations, while groups E and F, with four recombinant spores, are NPD. PD is thus $72 + 1 = 73$, while NPD is $1 + 2 = 3$. From these data, we can conclude that the two genes are linked.

What is the map distance between *thr* and *arg*? For this calculation, we need to find the numbers of T and NPD tetrads. Tetratypes are found in groups B, C, and D, and we already know that groups E and F carry NPDs. Using the same formula for map distances as the one previously utilized for yeast,

$$RF = \frac{NPD + 1/2T}{Total \ tetrads} \times 100$$

we get

$$RF = \frac{3 + 1/2(16 + 11 + 2)}{105} \times 100 = 16.7 \text{ m.u.}$$

Since the distance between *thr* and *arg* is larger than that separating either gene from the centromere, the centromere must lie between *thr* and *arg*, yielding the map in Fig. 4.24b. The distance between the two genes calculated by the formula (16.7 m.u.) is smaller than the sum of the two gene-centromere distances $(10.0 + 7.6 = 17.6 \text{ m.u.})$ because the formula does not account for double crossovers. As always, calculating map positions for more genes with shorter distances between them produces the most accurate picture.

CONNECTIONS

Medical geneticists have used their understanding of linkage, recombination, and mapping to make sense of the pedigrees presented at the beginning of this chapter (see Fig. 4.1). The X-linked gene for red-green colorblindness must lie very close to the gene for hemophilia A because the two are tightly coupled. In fact, an examination of many pedigrees has shown that if a doubly heterozygous woman receives defective alleles of both genes from the same parent, as few as 3% of her sons will have only one of the conditions, as a result of recombination. This means that the genetic distance between the two genes is only 3 m.u. On the other hand, even though hemophilia B is also on the X chromosome, it lies far enough away from the red-green colorblindness locus that the two genes recombine relatively freely and thus, in a small sample, may appear genetically un-linked. The recombination distance separating the genes for colorblindness and hemophilia B is about 36 m.u. Pedigrees pointing to two different forms of hemophilia, one very closely linked to colorblindness, the other almost not linked at all, provided one of several indications that hemophilia is determined by more than one gene (Fig. 4.25).

Refining the human chromosome map poses a continual challenge for medical geneticists. The newfound potential for finding and fitting more and more DNA markers into the map (review the Fast Forward box in this chapter) enormously improves the ability to assign precise locations to genes that cause disease. If several markers flank one such gene and it is possible to test a fetus for those markers, it is also possible to estimate the likelihood that the fetus has received the closely linked disease gene. DNA markers are also of great importance in the construction of a physical map of the human genome, a process we discuss in Chapter 10.

The simultaneous occurrence of genetic linkage and recombination is universal among life-forms and presumably confers some advantage to a species beyond allowing geneticists to map its genes. Although definitive conclusions are beyond our reach, speculation about the possible advantages can help us understand why genomes are arranged as they are. Linkage provides the potential for transmitting favorable combinations of genes intact to successive generations, while recombination produces great flexibility in generating new combinations of alleles. Some new combinations may help a species adapt to changing environmental conditions, whereas

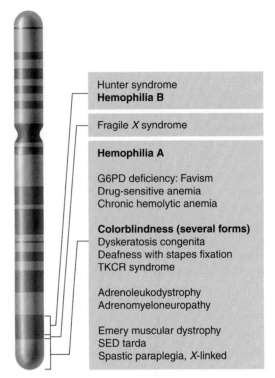

Figure 4.25   A genetic map of part of the human X chromosome.

the inheritance of successfully tested combinations can preserve what has worked in the past. Linkage and recombination thus help chromosomes function as organelles that not only transmit but also evolve genetic information.

In the first part of this book (Chapters 1–4), we have examined how genes and chromosomes are transmitted. As important and useful as this knowledge is, it tells us very little about the structure and mode of action of the genetic material. In the next section (Chapters 5–7) we carry our analysis to the level of DNA, the actual molecule of heredity. In Chapter 5, we look at DNA structure and learn how the DNA molecule carries genetic information. In Chapter 6, we describe how geneticists defined the gene as a localized region of DNA containing many nucleotides that together encode the information to make a protein. In Chapter 7, we examine how the cellular machinery interprets the genetic information in genes to produce the multitude of phenotypes that make up an organism.

ESSENTIAL CONCEPTS

1. Gene pairs that are close together on the same chromosome are genetically linked because they are transmitted together more often than not. The hallmark of linkage is that among the progeny of double heterozygotes produced from test crosses, the number of parental types is greater than the number of recombinant types.

2. The recombination frequencies of pairs of genes indicate how often two genes are transmitted together. For linked genes, the recombination frequency is less than 50%.

3. Gene pairs that assort independently exhibit a recombination frequency of 50%. This is because the number of parental types equals the number of recombinants. Genes may assort independently either because they are on different chromosomes or because they are far apart on the same chromosome.

4. Statistical analysis helps determine whether or not two genes assort independently. The probability value (p) calculated by the chi square test measures the likelihood that a particular set of data supports the null hypothesis of independent assortment, or no linkage. The lower the p value, the less likely is the null hypothesis, and the more likely the linkage. The chi square test can also be used to determine how well the outcomes of crosses fit other genetic hypotheses.

5. The greater the physical distance between linked genes, the higher the recombination frequency. However, recombination frequencies become more and more inaccurate as the distance between genes increases. Recombination occurs because chromatids of homologous chromosomes exchange parts, that is, cross over, during the prophase of meiosis I, after the chromosomes have replicated.

6. Genetic maps are a visual representation of relative recombination frequencies. The greater the density of genes on the map (and thus the smaller the distance between the genes), the more accurate and useful the map becomes in predicting inheritance.

7. Organisms that retain all the products of one meiosis within an ascus reveal the relation between genetic recombination and the segregation of chromosomes during the two meiotic divisions. Organisms like *Neurospora* that produce ordered octads make it possible to locate a chromosome's centromere on the genetic map.

SOCIAL AND ETHICAL ISSUES

1. Ophelia has been taking a genetics course and has learned about inherited predisposition to breast cancer. Realizing that many members of her family have had breast cancer, she wants to have testing done to determine if there is an inherited form of the *BRCA1* gene that is responsible for the prevalence of cancer in her family. One of the very important individuals that will need to be tested is her great-aunt who had breast cancer many years ago. Unfortunately, this aunt is showing early signs of senile dementia. Her great-aunt has periods of lucidity when she seems to understand the need for testing and agrees to give a blood sample for the DNA testing. At other times, she is confused in her thinking overall and at these times refuses to have a sample taken for these purposes. She does routinely have blood drawn for other medical tests. Since she sometimes agrees to the DNA testing, is it ethical to do the test on the blood samples even though at other times she refuses?

2. Dean Hamer and colleagues from the National Institutes of Health in 1993 reported evidence that gene(s) in a region of the X chromosome correlated with homosexuality in males. To demonstrate this association, Hamer selected for his analysis those families in which there were two gay brothers and other gay males in the maternal lineage only. Is there scientific merit in preselecting your study group in this way? Why would a human geneticist feel it is necessary to study a subset of individuals?

3. In 1996, evidence was presented for linkage of a marker to a predisposition to prostate cancer. Using 66 families in which at least three males had prostate cancer, a genetic marker was found that is inherited with high frequency in those men with prostate cancer. Somewhere in a region of about 10 million base pairs, is a gene(s) to which this marker is linked, that predisposes to prostate cancer. The estimate is that this genetic region is involved in about 3% of the *total cases* of prostate cancer. National newspapers announced these findings with the following headlines.

"Scientists Find Proof That Mutant Gene Can Increase Risk of Prostate Cancer"

"Prostate Cancer Gene Evidence Found"

"Scientists Zero in on Gene Tied to Prostate Cancer"

Do these headlines accurately represent the situation? Could they be misleading in any way? What would you use as a headline for an article on this study? What is the responsibility of the press in presenting new findings? Consider the responsibility of journalists/reporters to their employers (newspapers), their profession, and to the public.

S O L V E D   P R O B L E M S

I. The *Xg* locus on the human X chromosome has two alleles, a^+ and a. The a^+ allele causes the presence of the Xg surface antigen on red blood cells, while the recessive *a* allele does not allow antigen to appear. The *Xg* locus is 10 m.u. from the *Sts* locus. The *Sts* allele produces normal activity of the enzyme steroid sulfatase, while the recessive *sts* allele results in the lack of steroid sulfatase activity and the disease ichthyosis (scaly skin). A man with ichthyosis and no Xg antigen has a normal daughter with Xg antigen, who is expecting a child.

a. If the child is a son, what is the probability he will lack antigen and have ichthyosis?

b. What is the probability that a son would have both the antigen and ichthyosis?

c. If the child is a son with ichthyosis, what is the probability he will have Xg antigen?

Answer

a. This problem requires an understanding of how linkage affects the proportions of gametes made. First designate the genotype of the individual in which recombination during meiosis affects the transmission of alleles. In this problem, that individual is the daughter. The X chromosome she inherited from her father (who had icthyosis and no Xg antigen) must be *sts a*. (No recombination could have separated the genes during meiosis in her father since he has only one X chromosome.) Because the daughter is normal and has the Xg antigen, her other X chromosome (inherited from her mother) must contain the *Sts* and a^+ alleles. Her X chromosomes can be diagrammed as:

$$\frac{sts \quad a}{Sts \quad a^+}$$

Because the *Sts* and *Xg* loci are 10 m.u. apart on the chromosome, there is a 10% recombination frequency. Ninety percent of the gametes will be parental type gametes: *sts a* or *Sts a$^+$* (or 45% of each of these two types) and 10% will be recombinant type gametes: *sts a$^+$* or *Sts a* (5% of each type). The phenotype of a son directly reflects the genotype of the X chromosome from his mother. *Therefore, the probability that he will lack antigen and have icthyosis (genotype: a sts / Y) is 45/100.*

b. *The probability that he will have the antigen and ichthyosis (genotype: a^+ sts / Y) is 5/100.*

c. There are two classes of gametes in which the ichthyosis allele is present: *sts a* 45% and *sts a$^+$* 5%. If you think of the total number of gametes as 100, then 50 will have the *sts* allele. Of those gametes, 5 (or 10%) will have the a^+ antigen. *Therefore there is a*

1/10 probability that a child with the sts *allele will have the Xg antigen.*

II. *Drosophila* females of wild-type appearance but heterozygous for three autosomal genes are mated with males showing three autosomal recessive traits: glassy eyes, coal-colored bodies, and striped thoraxes. 1000 progeny of this cross are distributed in the following phenotypic classes:

Wild-type	27
Striped thorax	11
Coal body	484
Glassy eye, coal body	8
Glassy eye, striped thorax	441
Glassy eye, coal body, striped thorax	29

a. Draw a genetic map based on this data.

b. Show the arrangement of alleles on the two homologous chromosomes in the parent female.

c. Normal appearing males containing the same chromosomes as the parent female in the preceding cross are mated with females showing glassy eyes, coal-colored bodies, and striped thoraxes. Of 1000 progeny produced, indicate the numbers of various phenotypic classes you would expect.

Answer

A logical, methodical way to approach a three-point cross is described below.

a. Designate the alleles:

t^+ = wild-type thorax	t = striped thorax
g^+ = wild-type eye	g = glassy eye
c^+ = wild-type body	c = coal-colored body

In solving a three-point cross, designate the types of events that gave rise to each group of individuals and the genotypes of the gametes obtained from their mother. (The paternal gametes contain all the recessive alleles [*t g c*]. They do not change the phenotype and can be omitted.)

Progeny	Number	Type of event	Genotype		
1. wild type	27	single crossover	t^+	g^+	c^+
2. striped thorax	11	single crossover	t	g^+	c^+
3. coal body	484	parental	t^+	g^+	c
4. glassy eye, coal body	8	single crossover	t^+	g	c
5. glassy eye, striped thorax	441	parental	t	g	c^+
6. glassy eye, coal body, striped thorax	29	single crossover	t	g	c

Picking out the parental class is easy. If all the other classes are rare, the two most abundant categories are those gene combinations that have not undergone recombination. Then there should be two sets of two phenotypes that correspond to a single crossover event between the first and second genes, or between the second and third genes. Finally there should be two classes containing small numbers that result from double crossovers. In this example, there are no flies in the double crossover classes. Two phenotypic combinations are missing: glassy eye and coal body, striped thorax. These represent the result of double crossovers in this mating.

Look at the most abundant class to determine which alleles were on each chromosome in the female heterozygous parent. One parental class had the phenotype of coal body (484 flies), one chromosome in the female must have contained the t^+ g^+ and c alleles (notice that we cannot yet say what order these alleles are in on the chromosome). The other class was glassy eye and striped thorax and corresponds to a chromosome with the t, g, and c^+ alleles. To determine the distance between genes and the order of the genes, consider two genes at a time and total the recombination events that lead to nonparental combinations of those two genes. Between the c^+ and g^+ genes, progeny classes 1, 2, 4, and 6 all have nonparental combinations of alleles of these genes. The recombination frequency is $27 + 11 + 8 + 29 = 75/1000$ or 7.5%. Genes c^+ and g^+ are therefore 7.5 m.u. apart. Between t^+ and g^+, classes 2 and 4 have nonparental combinations of these genes. The total is $11 + 8 = 19/1000$ or 1.9%. The distance between g^+ and t^+ is thus 1.9 map units. For genes t^+ and c^+, classes 1 and 6 are nonparental. The total is $27 + 29 = 56/1000$ or 5.6% recombination, so the t^+ and c^+ loci are 5.6 m.u. apart. Because c^+ and g^+ are separated by the greatest map distance, they must be the outside genes, with t^+ between them.

The genetic map is

c^+————5.6————————t^+——1.9——g^+

b. The types of alleles on each female chromosome were already determined (c, g^+, t^+ and c^+, g, t) and now that the order of loci has also been determined, the arrangement of the alleles can be indicated:

$$\frac{c \quad\quad t^+ \quad\quad g^+}{c^+ \quad\quad t \quad\quad g}$$

c. Males of the same genotype as the starting female (ct^+g^+/c^+tg) could produce only two types of gametes: parental types ct^+g^+ and c^+tg because there is no recombination in male *Drosophila*. The progeny expected from the mating with a homozygous recessive female are thus 500 coal- and 500 glassy-eyed, striped flies.

III. The following asci were obtained in *Neurospora* when a wild-type strain (ad^+ leu^+) was crossed to a double auxotrophic mutant strain (ad^- leu^-). Only one member of each spore pair produced by the final mitosis is shown, since the two cells in a pair have the same genotype.

Spore pair	Ascus type				
1–2	ad^+ leu^+	ad^+ leu^+	ad^+ leu^+	ad^+ leu^-	ad^- leu^+
3–4	ad^+ leu^+	ad^+ leu^-	ad^+ leu^-	ad^- leu^+	ad^+ leu^+
5–6	ad^- leu^-	ad^- leu^+	ad^- leu^+	ad^- leu^-	ad^- leu^-
7–8	ad^- leu^-	ad^- leu^+	ad^- leu^-	ad^+ leu^+	ad^+ leu^-
Number of asci	30	30	40	2	18

a. What genetic event causes the alleles of two genes to segregate to different cells at the second meiotic division and when does this event occur?

b. Provide the best possible map for the two genes and their centromere(s).

Answer

This problem requires an understanding of tetrad analysis and the process (meiosis) that produces the patterns seen in ordered asci.

a. *A crossover between a gene and its centromere causes the segregation of alleles at the second meiotic division. The crossover event occurs during prophase of meiosis I.*

b. Using ordered tetrads you can determine whether two genes are linked, the distance between two genes, and the distance between a gene and its centromere. First designate the five classes of asci shown above. The first class is a parental ditype (spores contain the same combinations of alleles as their parents); the second is a nonparental ditype; the last three are tetratypes. Next determine if these genes are linked. The number of PD = number of NPD, so the genes are not linked. When genes are unlinked, the tetratype asci are generated by a crossing-over event between a gene and its centromere. Looking at the *leu* gene, there is a second-division-segregation pattern of that gene in the third and fourth asci types. Therefore, the percent of second-division segregation is

$$\frac{40 + 2}{120} \times 100 = 35\%.$$

Because only half of the chromatids in the meioses that generated these tetratype asci were involved in the crossover, the map distance between *leu* and its centromere is 35/2 or 17.5 m.u.

Asci of the fourth and fifth types above show a second-division-segregation pattern for the *ad* gene.

$$\frac{2 + 18}{120} \times 100 = 16.6\%.$$

Dividing 16.6% by 2 gives the recombination frequency and map distance of 8.3 m.u.

The map of these two genes is the following:

○—8.3————*ad*
○————————17.5————*leu*

PROBLEMS

4-1 Choose the phrase from the right column that best fits the term in the left column.

a. recombination		1.	a statistical method for testing the fit between observed and expected results
b. linkage		2.	an ascus containing spores of four different genotypes
c. chi square test		3.	one crossover along a chromosome makes a second nearby crossover less likely
d. chiasma		4.	when two loci recombine in less than 50% of gametes
e. tetratype		5.	the relative chromosomal location of a gene
f. locus		6.	the ratio of observed double crossovers to expected double crossovers
g. coefficient of coincidence		7.	fungal spores contained in a sac
h. interference		8.	formation of new genetic combinations by exchange of parts between homologs
i. parental ditype		9.	when the two alleles of a gene are segregated into different cells at the first meiotic division
j. ascospores		10.	an ascus containing only two nonrecombinant kinds of spores
k. first-division segregation		11.	structure formed at the spot where crossing-over occurs between homologs

4-2 Do the data that Mendel obtained fit his hypothesis? For example, Mendel obtained 315 round yellow, 101 wrinkled yellow, 108 round green, and 32 wrinkled green seeds from the selfing of $Yy\ Rr$ individuals (a total of 556). His hypotheses of segregation and independent assortment predict a 9:3:3:1 ratio in this case. Use the chi square test to determine whether Mendel's data are significantly different from what he predicted. (The chi square test did not exist in Mendel's day, so he was not able to test his own data for goodness of fit to his hypotheses.)

4-3 Two genes control color in corn snakes as follows: $O\text{-}B\text{-}$ snakes are brown, $O\text{-}bb$ are orange, $oo\ B\text{-}$ are black, and $oo\ bb$ are albino. An orange snake was mated to a black snake, and a large number of F_1 progeny were obtained, all of which were brown. When the F_1 snakes were mated to one another, they produced 100 brown offspring, 25 orange, 22 black and 13 albino.
- a. What are the genotypes of the F_1 snakes?
- b. What proportions of the different colors would have been expected among the F_2 snakes if the two loci assort independently?

- c. Do the observed results differ significantly from what was expected assuming independent assortment is occurring?
- d. What is the probability that differences this great between observed and expected values would happen by chance?

4-4 A mouse with normal gait was crossed to a mouse displaying an odd gait called "dancing." The F_1 animals all showed normal gait.
- a. If dancing is caused by homozygosity for the recessive allele of a single gene, what proportion of the F_2 mice should be dancers?
- b. If mice must be homozygous for recessive alleles of both of two different genes to have the dancing phenotype, what proportion of the F_2 should be dancers if the two genes are unlinked?
- c. When the F_2 mice were obtained, 42 normal and 8 dancers were seen. Use the chi square test to determine if these results better fit the one gene model or the two gene model.

4-5 The a and the b loci are 40 cM apart, and an $AA\ BB$ individual and an $aa\ bb$ individual mate:
- a. What gametes will the F_1 individuals produce, and in what proportions?
- b. If the original cross was $AA\ bb \times aa\ BB$, what gametic proportions would emerge from the F_1?

4-6 $CC\ DD$ and $cc\ dd$ individuals were crossed to each other, and the F_1 generation was backcrossed to the $cc\ dd$ parent. 903 $Cc\ Dd$, 897 $cc\ dd$, 98 $Cc\ dd$ and 102 $cc\ Dd$ offspring resulted.
- a. How far apart are the c and d loci?
- b. What progeny would you expect to result from backcrossing the F_1 generation from a $CC\ dd \times cc\ DD$ cross to $cc\ dd$?

4-7 If the a and b loci are 20 m.u. apart in humans, and an AB/ab woman marries an ab/ab man, what is the probability that their first child will be Ab/ab?

4-8 In a particular human family, John and his mother both have brachydactyly (a rare autosomal dominant causing short fingers). John's father has Huntington Disease (another rare autosomal dominant). John's wife is phenotypically normal, and is pregnant. Two-thirds of people who inherit the Huntington (HD) allele show symptoms by age 50, and John is 50 and has no symptoms. Brachydactyly is 90% penetrant.
- a. What are the genotypes of John's parents?
- b. What are the possible genotypes for John?
- c. What is the probability the child will express both brachydactyly and Huntington Disease by age 50 if the two genes are unlinked?
- d. If these two loci are 20 m.u. apart, how will it change your answer to part c?

4-9 In mice, the autosomal locus coding for the β-globin chain of hemoglobin is 1 m.u. from the albino locus. Assume for the moment that the same is true in humans. The disease sickle-cell anemia is the result of homozygosity for a particular mutation in the β-globin gene.

  a. A son is born to an albino man and a woman with sickle-cell disease. What kinds of gametes will the son form, and in what proportions?

  b. A daughter is born to a normal man and a woman who has both albinism and sickle-cell disease. What kinds of gametes will the daughter form, and in what proportions?

  c. If the son of the first marriage grows up and marries the daughter of the second marriage, what is the probability that a child of theirs will be an albino with sickle-cell anemia?

4-10 In corn, the allele *A* allows the deposition of anthocyanin (blue) pigment in the kernels (seeds), while *aa* plants have yellow kernels. At a second gene, *W-* produces smooth kernels, while *ww* kernels are wrinkled. A plant with blue, smooth kernels was crossed to a plant with yellow, wrinkled kernels. The F_1 generation consisted of 1447 blue, smooth; 169 blue, wrinkled; 186 yellow, smooth; and 1510 yellow, wrinkled.

  a. Are the *a* and *w* loci linked? If so, how far apart are they?

  b. What was the genotype of the blue, smooth parent? Include the chromosome arrangement of alleles.

  c. If a plant grown from a blue, wrinkled F_1 seed is crossed to a plant grown from a yellow, smooth F_1 seed, what kinds of kernels would be expected, and in what proportions?

4-11 Albino rabbits (lacking pigment) are homozygous for the recessive *c* allele (*C* allows pigment formation). Rabbits homozygous for the recessive *b* allele make brown pigment, while those with at least one copy of *B* make black pigment. True-breeding brown rabbits were crossed to albinos which were *BB*. F_1 rabbits, which were all black, were crossed to the double recessive (*bb cc*). The progeny obtained were 34 black, 66 brown and 100 albino.

  a. What phenotypic proportions would have been expected if the *b* and *c* loci were unlinked?

  b. How far apart are the two loci?

4-12 Cinnabar eyes (*cn*) and reduced bristles (*rd*) are autosomal recessive characters in *Drosophila*. A homozygous wild-type female was crossed to a reduced, cinnabar male, and the F_1 males were then crossed to the F_1 females to obtain the F_2. Of the 400 F_2 offspring obtained, 292 were wildtype, 9 were cinnabar, 7 were reduced, and 92 were reduced, cinnabar. Explain these results and estimate the distance between the *cn* and *rd* loci.

4-13 A DNA variant has been found linked to a rare autosomal dominant disease in humans and can be used as a marker to follow inheritance of the disease allele. In an informative family (in which one parent is heterozygous for both the disease allele and the DNA marker in a known chromosomal arrangement of alleles, and his or her mate does not have the same alleles of the DNA variant), the reliability of such a marker as a predictor of the disease in a fetus is related to the map distance between the DNA marker and the gene causing the disease. Imagine that a man affected with the disease (genotype *Dd*) is heterozygous for the A and B forms of the DNA variant, with form A on the same chromosome as the *D* allele and form B on the same chromosome as *d*. His wife is CC *dd,* where C is another allele of the DNA marker. Typing of the fetus by amniocentesis reveals that the fetus has the B and C DNA variants. How likely is it that the fetus has inherited *D* if the distance between the *D* locus and the marker locus is (a) 0 m.u., (b) 1 m.u., (c) 5 m.u., (d) 10 m.u., (e) 50 m.u.?

4-14 In the tubular flowers of foxgloves, wild-type coloration is red while a mutation called *white* produces white flowers. Another mutation, called *peloria,* causes the flowers at the apex of the stem to be huge. Yet another mutation, called *dwarf,* affects stem length. You cross a white-flowered plant (otherwise phenotypically wildtype) to a plant that is dwarf and peloria, but has wild-type red flower color. All of the F_1 plants are tall with white, normal-sized flowers. You cross an F_1 plant back to the dwarf parent, and see the progeny shown in the chart. (Only mutant traits are noted.)

dwarf, peloria	172
white	162
dwarf, peloria, white	56
wildtype	48
dwarf, white	51
peloria	43
dwarf	6
peloria, white	5

  a. Which alleles are dominant?

  b. What were the genotypes of the parents in the original cross?

  c. Draw a map showing the linkage relationships of these three loci.

  d. Is there interference?

  e. Calculate the coefficient of coincidence.

4-15 In *Drosophila,* three autosomal genes have the following map:

$$\underset{a}{\underline{\quad\quad\quad}}\underset{b}{\overset{20 \text{ m.u.}}{}}\underset{c}{\overset{10 \text{ m.u.}}{}}$$

Provide the data, in terms of the expected number of flies in the following phenotypic classes, when

$a^+b^+c^+/abc$ females are crossed to *abc/abc* males. Assume 1000 flies were counted.

a^+	b^+	c^+
a	b	c
a^+	b	c
a	b^+	c^+
a^+	b^+	c
a	b	c^+
a^+	b	c^+
a	b^+	c

4-16 A cross was performed between a haploid strain of yeast with the genotype $\alpha + +$ and another haploid strain with the genotype *afg* (a and α are mating types.) The resulting diploid was sporulated, and a random sample of 100 of the resulting haploid spores was analyzed. The following genotypic frequencies were seen: 31 $\alpha + +$; 29 *afg,* 13 $\alpha + g$; 14 *af* +; 6 *a* + *g*; 6 αf +; 1 *a* + +; 1 αfg. Map the loci involved in the cross.

4-17 Male *Drosophila* expressing the recessive mutations *sc* (*scute*), *ec* (*echinus*), *cv* (*crossveinless*), and *b* (*black*) were crossed to phenotypically wild-type females, and the 3288 progeny shown in the chart were obtained.

653	black, scute, echinus, crossveinless
670	scute, echinus, crossveinless
675	wildtype
655	black
71	black, scute
73	scute
73	black, echinus, crossveinless
74	echinus, crossveinless
87	black, scute, echinus
84	scute, echinus
86	black, crossveinless
83	crossveinless
1	black, scute, crossveinless
1	scute, crossveinless
1	black, echinus
1	echinus

a. Diagram the genotype of the female parent.
b. Map these loci.
c. Is there evidence of interference? Justify your answer with numbers.

4-18 A snapdragon with pink petals, black anthers, and long stems was allowed to self-fertilize. From the resulting seeds, 650 adult plants were obtained. The phenotypes of these offspring are shown in the chart.

78	red	long	tan
26	red	short	tan
44	red	long	black
15	red	short	black
39	pink	long	tan
13	pink	short	tan
204	pink	long	black
68	pink	short	black
5	white	long	tan
2	white	short	tan
117	white	long	black
39	white	short	black

a. Using *P* for one allele and *p* for the other, indicate how flower color is inherited.
b. What numbers of red: pink: white would have been expected among these 650 plants?
c. How are anther color and stem length inherited?
d. What was the genotype of the original plant?
e. Do the three genes show independent assortment?

4-19 The following list of four *Drosophila* mutations indicates the symbol for the mutation, the name of the gene, and the mutant phenotype:

Allele symbol	Gene name	Mutant Phenotype
dwp	dwarp	small body, warped wings
rmp	rumpled	deranged bristles
pld	pallid	pale wings
rv	raven	dark eyes and bodies

You perform the following crosses with the indicated results:

Cross #1:
dwarp, rumpled females × pallid, raven males → dwarp, rumpled males and wild-type females

Cross #2:
pallid, raven females × dwarp, rumpled males → pallid, raven males and wild-type females

F_1 females from cross #1 above were crossed to males homozygous for *dwarp rumpled pallid raven*. The 1000 progeny obtained were as follows:

pallid	3
pallid, raven	428
pallid, raven, rumpled	48
pallid, rumpled	23
dwarp, raven	22
dwarp, raven, rumpled	2
dwarp, rumpled	427
dwarp	47

Indicate the best map for these four genes, including all relevant data.

4-20 Two crosses were made in *Neurospora* involving the mating type locus and either the *ad* or *p* genes. In both cases, the mating type locus (*A* or *a*) was one of the loci whose segregation was scored. One cross was *adA* × + *a* (cross a), and the other was × + *a* (cross b). From cross a, 10 parental ditype, 9 nonparental ditype, and 1 tetratype asci were seen. From cross b, the results were 24 parental ditype, 3 nonparental ditype, and 27 tetratype. What are the linkage relationships between the mating type locus and the other two loci?

4-21 *Neurospora* of genotype *a + c* are crossed with *Neurospora* of genotype *+ b +*. The following tetrads are obtained (note that the genotype of the four spore *pairs* in an ascus are listed, rather than listing all eight spores):

$a + c$	$a\ b\ c$	$+ + c$	$+\ b\ c$	$a\ b\ +$	$a + c$
$a + c$	$a\ b\ c$	$a + c$	$a\ b\ c$	$a\ b\ +$	$a\ b\ c$
$+\ b\ +$	$+ + +$	$+ + +$	$+ + +$	$+ + c$	$+ + +$
$+\ b\ +$	$+ + +$	$a\ b\ +$	$a + +$	$+ + c$	$+\ b\ +$
137	141	26	25	2	3

a. In how many cells has meiosis occurred to yield this data?

b. Give the best genetic map to explain these results. Indicate all relevant genetic distances, both between genes and between each gene and its respective centromere.

4-22 The a, b and c loci are all on different chromosomes in yeast. When $a+$ yeast were crossed to $+b$ yeast, sporulated, and the tetrads analysed, it was found that the number of nonparental ditype tetrads was equal to the number of parental ditypes, but that there were no tetratype asci at all. On the other hand, many tetratype asci were seen in the tetrads formed after crosses of $a + x + c$ and $b + x + c$. Explain these results.

4-23 A cross was performed between a yeast strain that requires methionine and lysine for growth (met^- lys^-) and another yeast strain which is met^+ lys^+. One hundred asci were dissected and colonies grew from the four spores in each ascus. Cells from these colonies were tested for their ability to grow on petri plates containing either minimal medium (min), min + lysine (lys), min + methionine (met), or min + lys + met. The asci could be divided into two groups based on this analysis:

Group 1: In 89 asci, cells from two of the four spore colonies could grow on all four kinds of media, while the other two spore colonies could grow only on min + lys + met.

Group 2: In 11 asci, cells from one of the four spore colonies could grow on all four kinds of petri plates. Cells from a second one of the four spore colonies could grow only on min + lys plates and on min + lys + met plates. Cells from a third of the four spore colonies could only grow on min + met plates and on min + lys + met. Cells from the remaining colony could only grow on min + lys + met.

a. What are the genotypes of the spores within each of the two types of asci?

b. Are the *lys* and *met* genes linked? If so, what is the map distance between them?

c. If you could extend this analysis to many more asci, you would eventually find some asci with a different pattern. For these asci, describe the phenotypes of the four spores. List this as the ability of dissected spores to form colonies on the four kinds of petri plates.

4-24 Write the number of <u>different kinds</u> of phenotypes excluding gender you would see among a large number of progeny from an F_1 mating between individuals of identical genotype that are heterozygous for one or two genes (i.e., Aa or $Aa\ Bb$) as indicated. No gene interactions means that the phenotype determined by one gene is not influenced by the genotype of a second gene.

a. One gene; A completely dominant to a.

b. One gene; A and a codominant.

c. One gene; A *in*completely dominant to a.

d. Two unlinked genes; no gene interactions; A completely dominant to a, and B completely dominant to b.

e. Two genes, 10 m.u. apart; A completely dominant to a, and B completely dominant to b.

f. Two unlinked genes; no gene interactions; A and a codominant, and B *in*completely dominant to b.

g. Two genes, 10 m.u. apart; A completely dominant to a, and B completely dominant to b; and with recessive epistasis between the genes.

h. Two unlinked duplicated genes (i.e., A and B perform the same function); A and B completely dominant to a and b, respectively.

i. Two genes, 0 m.u. apart; A completely dominant to a, and B completely dominant to b. (There are two possible answers to part i.)

4-25 A single yeast cell placed on a solid agar will divide mitotically to produce a colony of about 10^7 cells. A haploid yeast cell that has a mutation in the $ade2$ gene will produce a red colony; an $ade2^+$ colony will be white. Some of the colonies from a diploid yeast cell with a genotype of $ade2^+/ade2^-$ will contain sectors of red within a white colony. How would you explain these sectors?

4-26 A diploid strain of yeast has a wild-type phenotype but the following genotype:

$$—b—a—\bullet—c—leth—d—e—$$
$$—b^+—a^+—\bullet—c^+—leth^+—d^+—e^+—$$

a, b, c, d, and e all represent recessive alleles that yield a visible phenotype, and *leth* represents a recessive lethal mutation. All genes are on the same chromosome, and a is very tightly linked to its centromere (indicated by dark circle). Which of the following phenotypes could be found in sectors resulting from mitotic recombination in this cell? 1) a 2) b 3) c 4) d 5) e 6) $b\ e$ 7) $c\ d$ 8) $c\ d\ e$ 9) $d\ e$ 10) $a\ b$. (Assume that double crossovers are too rare to be observed.)

WHAT GENES ARE AND WHAT THEY DO

Tobacco plant glowing in the dark because a firefly gene has been incorporated into
its DNA and is being translated into enzyme Luciferase.

CHAPTER

5

DNA: How The Molecule of Heredity Carries, Replicates, and Recombines Information

For nearly 4 billion years the double-stranded DNA molecule has served as the almost universal bearer of genetic information; 3.7 billion years ago the earliest bacterial cells incorporated it into their chromosomes, and about 2 billion years ago, when the eukaryotic precursors of plants, animals, and fungi evolved from these simple cells, their chromosomes also carried a DNA molecule. Since that time, the hardware—the structure of the molecule itself—has changed very little, although evolution has honed and expanded the software—the programs by which the molecule stores, transmits, and expresses genetic information.

Under special conditions of little or no oxygen, DNA can withstand a wide range of temperature, pressure, and humidity and remain relatively intact for hundreds, thousands, even millions of years. Molecular sleuths of the 1980s and 1990s have retrieved the evidence: 100-year-old DNA from preserved tissue of the quagga (a partially striped horselike creature that became extinct in the final years of the nineteenth century); 8000-year-old DNA from human skulls found in the swamps of Florida (Fig. 5.1); and 18-million-year-

old DNA from the compressed remnants of magnolia leaves fossilized in an ancient Idaho lake bed. Surprisingly, this ancient DNA still carries readable sequences—shards of decipherable information that become veritable time machines for the direct viewing of genes in long-vanished organisms and species. Comparisons with equivalent expanses of modern DNA make it possible to identify the precise mutations that have fueled evolution.

Francis Crick, codiscoverer with James Watson of DNA's double helical structure and a leading theoretician of molecular biology, has written that "almost all aspects of life are engineered at the molecular level, and without understanding molecules, we can only have a very sketchy understanding of life itself." In Chapters 1–4 we examined how Mendel and his successors used data from breeding experiments to deduce the presence and activity of genes, analyze the effect of alternative alleles on phenotype, and tie genes to the movements of chromosomes in time and space. With this approach, it is possible to predict the outcomes of genetic crosses in the absence of detailed knowledge of the mole-

(a)

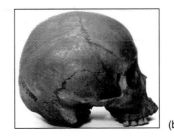

(b)

Figure 5.1 Ancient DNA still carries information. Molecular biologists have successfully extracted DNA from (a) a 100-year-old quagga, and (b) 8000-year-old human skulls. These findings attest to the chemical stability of DNA, the molecule of inheritance.

cules and biochemical reactions that underlie the events observed. But it is impossible to decipher the exact mechanisms by which genes determine phenotypes, transmit instructions between generations, and evolve new information, without understanding what they are made of. For this reason, we shift our perspective in this chapter to an examination of DNA, the molecule that composes the genetic material.

As we extend our analysis to the molecular level, two general themes emerge. First, DNA's genetic functions flow directly from its molecular structure—the way its atoms are arranged in space. Second, most of DNA's genetic functions depend on specialized proteins that interact with the molecule of heredity. One such protein, for example, binds to DNA to begin the process of replication.

In our discussion of the structure and function of DNA, we describe

- How investigators pinpointed DNA as the genetic material.
- The elegant Watson-Crick model of DNA structure.
- How DNA structure provides for the storage of genetic information.
- How DNA structure gives rise to the semiconservative mode of molecular replication.
- How DNA structure promotes the recombination of genetic information.

EXPERIMENTS DESIGNATE DNA AS THE GENETIC MATERIAL

Though accepted as fact today, it was not always obvious that DNA is the genetic material. It took a cohesive pattern of results from experiments performed over more than 50 years to convince the scientific community that DNA is the molecule of heredity. We now present key pieces of the evidence.

Chemical Characterization Localizes DNA in the Chromosomes

In 1869 Friedrich Miescher extracted a weakly acidic, phosphorus-rich material from the nuclei of human white blood cells and named it "nuclein." It was unlike any previously reported chemical compound, and its major component turned out to be DNA, although it also contained some contaminants. The full chemical name of DNA is **deoxyribonucleic acid,** reflecting three characteristics of the substance: it is acidic; one of its constituents is a sugar known as deoxyribose; and it is found mainly in cell nuclei.

After purifying DNA from the nuclein and treating it with various reagents, researchers established that DNA is composed of four different subunits linked in a long chain (Fig. 5.2). The four different subunits belong to a class of compounds known as **nucleotides,** the bonds joining one nucleotide to another are covalent **phosphodiester bonds,** and the linked chain of repeating subunits is a type of **polymer.** A procedure first reported in 1923 made it possible to discover where in the cell DNA resides. Named the Feulgen reaction after its designer, the procedure relies on a chemical called the Schiff reagent, which stains DNA red. In a preparation of stained cells, the chromosomes redden, while other areas of the cell remain relatively colorless. The reaction shows that DNA is localized almost exclusively within specific nuclear organelles—the chromosomes.

The finding that DNA is a component of chromosomes does not prove that the molecule has anything to do with genes. Eukaryotic chromosomes also contain an equal, if not greater, amount of protein by weight. Because proteins are built of 20 different amino acids whereas DNA carries just

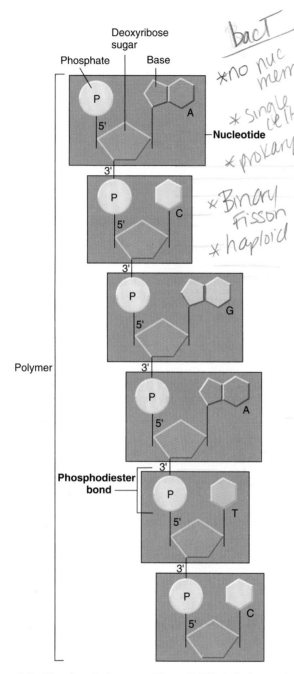

(handwritten annotations:) bacT *no nuc mem. * single celled * prokaryote * Binary fission * haploid

Figure 5.2 The chemical composition of DNA. A single strand of a DNA molecule consists of a chain of nucleotide subunits (colored boxes). Each nucleotide is made of the sugar deoxyribose connected to an inorganic phosphate group, and to one of four nitrogenous bases. The phosphodiester bonds that link the nucleotide subunits to each other attach the phosphate group of one nucleotide to the deoxyribose sugar of the preceding nucleotide.

four different subunits, many researchers thought proteins had greater potential for diversity and were better suited to serve as the genetic material. These same scientists assumed that even though DNA was an important part of chromosome structure, it was too simple to specify the complexity of genes.

Bacterial Transformation Implicates DNA as the Substance of Genes

Several studies dispelled the idea that DNA cannot be the genetic material. Some of the most important of these used single-celled bacteria as their experimental organism. Bacteria carry their genetic material in a single circular chromosome that lies in the nucleoid region of the cell without being enclosed in a nuclear membrane. The single circular chromosome (which makes them haploid) and lack of a nuclear membrane (which makes them *prokaryotes,* meaning "before a true nucleus") are two features that distinguish bacteria from the eukaryotic cells of higher organisms. Another feature is their method of cell division. With only one chromosome, bacteria do not undergo meiosis to produce germ cells, and they do not apportion their replicated chromosomes to daughter cells by mitosis; rather they divide by a process known as binary fission. Even with these acknowledged differences, at least some investigators in the first half of the twentieth century believed that the genetic material of bacteria was similar to that found in eukaryotic organisms.

As with any species, one prerequisite of genetic studies in bacteria is the detection of alternative forms of a trait within a population. In a 1923 study of *Streptococcus pneumoniae* bacteria grown in laboratory media, Frederick Griffith distinguished two bacterial forms: smooth (S) and rough (R). S is the wild type; a mutation in S gives rise to R. From observation and biochemical analysis, Griffith determined that S forms appear smooth because they synthesize a polysaccharide capsule that surrounds the cell. R forms, which arise spontaneously as mutants of S, are unable to make the capsular polysaccharide and as a result, their colonies appear to have a rough surface (Fig. 5.3a). We now know that the R form lacks an enzyme necessary for synthesis of the capsular polysaccharide. Because the polysaccharide capsule helps protect the bacteria from an animal's immune response, the S bacteria are virulent and kill most laboratory animals exposed to them (Fig. 5.3b.1); by contrast, the R forms fail to cause infection (Fig. 5.3b.2). In humans, the virulent S forms of *S. pneumoniae* can cause pneumonia.

Transformation

In 1928, Griffith published the astonishing finding that genetic information from dead bacterial cells could somehow be transmitted to live cells. He was working with two kinds of bacteria—live R forms and heat-killed S forms. Neither the heat-killed S forms nor the live R forms produced infection when injected into laboratory mice (Fig. 5.3b.2 and 3); but a mixture of the two killed the animals (Fig. 5.3b.4). Bacteria recovered from the blood of the dead animals were living S forms (Fig. 5.3b.4). The ability of a substance to change the

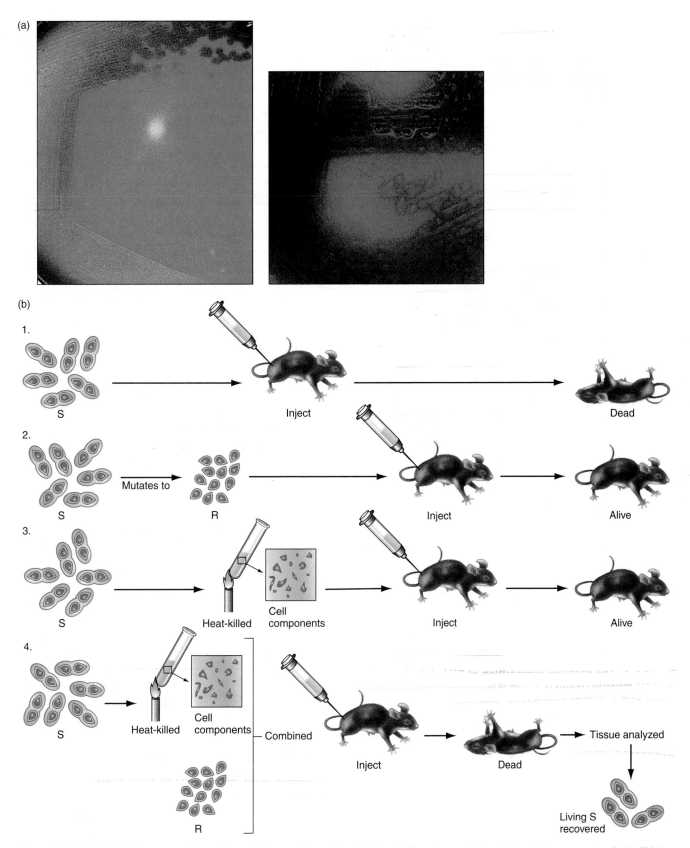

Figure 5.3   Griffith's demonstration of bacterial transformation. (a) Rough (R) and smooth (S) colonies of *S. pneumoniae*. (b) Griffith's experiment: (1) S bacteria are virulent and can cause lethal infections when injected into mice. (2) Injections of R mutants by themselves do not cause infections that kill mice. (3) Similarly, injections of heat-killed S bacteria do not cause lethal infections. (4) Lethal infection does result, however, from injections of live R bacteria mixed with heat-killed S strains; the blood of the dead host mouse contains living S-type bacteria.

genetic characteristics of an organism is known as **transformation.** Something from the heat-killed S bacteria must have transformed the living R bacteria into S. This transformation was permanent and most likely genetic, because all future generations of the bacteria grown in culture were S.

Bacterial Transformation Is Caused by DNA

By 1929, two other laboratories had repeated these results, and in 1931, investigators in Oswald T. Avery's laboratory found they could achieve transformation without using any animals at all, simply by growing R-form bacteria in medium in the presence of components from dead S forms (Fig. 5.4a). Avery then embarked on a quest that would remain the focus of his work for almost 15 years: "Try to find in that complex mixture, the active principle!" In other words, try to identify the heritable substance in the bacterial extract that induces transformation from harmless R bacteria to pathogenic S. Avery dubbed the substance he was searching for the "transforming principle" and spent many years trying to purify it sufficiently to be able to identify it unambiguously. He and his coworkers eventually prepared tangible, active transforming principle. In the final part of their procedure, a long whitish wisp material-

ized from ice cold alcohol solution and wound around the glass stirring rod to form a fibrous wad of nearly pure principle (Fig. 5.4b).

Once purified, the transforming principle had to be characterized. In 1944 Avery and two coworkers, Colin MacLeod and Maclyn McCarty, published the cumulative findings of experiments designed to determine the transforming principle's chemical composition (Fig. 5.4c). In these experiments, the purified transforming principle was active at the extraordinarily high dilution of 1 part in 600 million. Although the preparation was almost pure DNA, the investigators nevertheless exposed it to various enzymes to see if some molecule other than DNA could cause transformation. Enzymes that degraded RNA, protein, or polysaccharide had no effect on the transforming principle, but an enzyme that degrades DNA completely destroyed its activity. The tentative published conclusion was that the transforming principle appeared to be DNA. In a personal letter to his brother, Avery went one step further and confided that the transforming principle "may be a gene."

Despite the paper's abundance of concrete evidence, many within the scientific community still resisted the idea

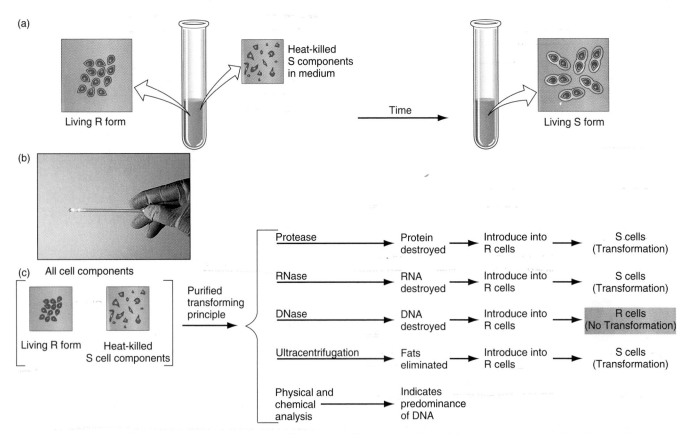

Figure 5.4 The transforming principle is DNA: experimental confirmation. (a) Bacterial transformation occurs in culture medium containing the remnants of heat-killed S bacteria. This indicates that some "transforming principle" from the heat-killed S bacteria is taken up by the live R bacteria, converting (transforming) them into virulent S strains. (b) A wad of purified DNA from heat-killed S-type *S. pneumoniae* wound around a glass stirring rod. Small amounts of this preparation are highly efficient in causing bacterial transformation. (c) Chemical fractionation of the transforming principle. Treatment of purified DNA with a DNA-degrading enzyme destroys its ability to cause bacterial transformation, while treatment with enzymes that destroy other macromolecules has no effect on the transforming principle. In contrast to the purified DNA shown in part (b), highly purified preparations of protein, lipids, or carbohydrates do not contain active transforming principle.

that DNA is the molecule of heredity. They argued that perhaps Avery's results reflected the activity of contaminants; or perhaps genetic transformation was not happening at all, and instead, the transforming principle was somehow triggering a physiological switch in the transformed bacteria. Unconvinced for the moment, these scientists remained attached to the idea that proteins are the prime candidates for the genetic material.

Convincing Evidence That Genes Are DNA: The Molecule Carries the Information Required for the Replication of Bacterial Viruses

Not everyone shared this skepticism. Alfred Hershey and Martha Chase anticipated that they could assess the relative importance of DNA and protein in gene transmission by infecting bacterial cells with viruses called **phages,** short for **bacteriophages** (literally "bacteria eaters"). Viruses are the simplest of organisms. By structure and function, they fall somewhere between living cells capable of reproducing themselves and macromolecules such as proteins. Because viruses depend on their host cell for most of the machinery required for growth and replication, they are very small and contain only a small number of genes. For many kinds of phage, each particle consists of roughly equal weights of protein and DNA (Fig. 5.5a). These phage particles can reproduce themselves only after infecting a bacterial cell. Thirty minutes after infection, the cell bursts and hundreds of newly made phages spill out (Fig. 5.5b). The question is: What substance directs the production of the new phage particles—DNA or protein?

With the invention of the electron microscope in 1939, it became possible to see individual phages, and surprisingly, electron micrographs revealed that the entire phage does not enter the bacterium it infects. Instead, a viral shell—called a *ghost*—remains attached to the outer surface of the bacterial

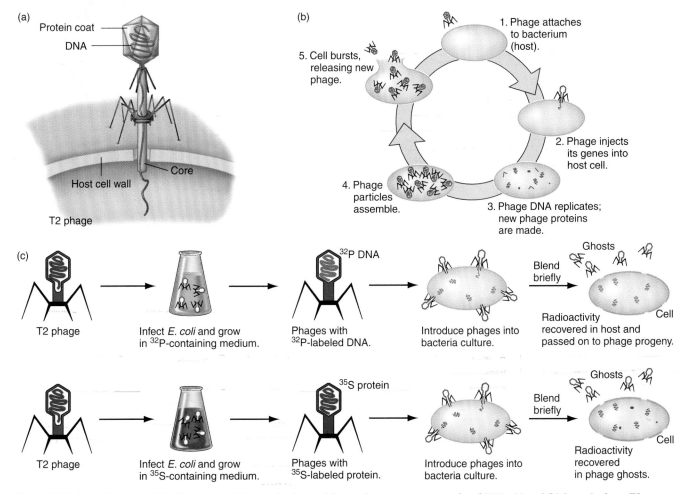

Figure 5.5   Experiments with viruses provide convincing evidence that genes are made of DNA. (a) and (b) Bacteriophage T2 structure and life cycle. The phage particle consists of DNA contained within a protein coat. The virus attaches to the bacterial host cell and injects its genes (the DNA) through the bacterial cell wall into the host cell cytoplasm. Inside the host cell, these genes direct the formation of new phage DNA and proteins, which assemble into progeny phages that are released into the environment when the cell bursts. (c) The Hershey-Chase Waring blender experiment. T2 bacteriophage particles either with ^{32}P-labeled DNA or with ^{35}S-labeled proteins were used to infect bacterial cells. After a short incubation, Hershey and Chase shook the cultures in a Waring blender and spun the samples in a centrifuge to separate the empty viral ghosts from the heavier infected cells. Most of the ^{35}S-labeled proteins remained with the ghosts, while most of the ^{32}P-labeled DNA was found in the sediment with the T2 gene-containing infected cells.

cell wall. Because the empty phage coat remains outside the bacterial cell, one investigator likened phage particles to tiny syringes that bind to the cell surface and inject transforming principle into the host cell.

In their famous Waring blender experiment of 1952, Alfred Hershey and Martha Chase tested the idea that the ghost left on the cell wall is composed of protein while the injected transforming principle consists of DNA (Fig. 5.5c). A type of phage known as T2 served as their experimental organism. They grew two separate sets of T2 in bacteria maintained in two different culture media, one infused with radioactively labeled phosphorus (^{32}P), the other with radioactively labeled sulfur (^{35}S). Since proteins incorporate sulfur but no phosphorus and DNA contains phosphorus but no sulfur, phages grown on ^{35}S would have radioactively labeled protein while particles grown on ^{32}P would have radioactive DNA. The radioactive tags would enable the investigators to determine the location of each material when the phages infected fresh cultures of bacterial cells.

After exposing one culture of bacteria to ^{32}P-labeled phage and another culture to ^{35}S-labeled phage, Hershey and Chase used a Waring blender to disrupt each culture, effectively separating the viral ghosts from the bacteria harboring the viral genes. Centrifugation of the cultures then separated the heavier infected cells, which ended up in a pellet, from the lighter phage ghosts, which remained suspended in the supernatant. Most of the radioactive ^{32}P (in DNA) went to the pellet, while most of the radioactive ^{35}S (in protein) remained in the supernatant. This confirmed that the extracellular ghosts were indeed mostly protein while the injected viral material specifying production of more phages was mostly DNA. Bacteria containing the radiolabeled phage DNA behaved just as in a normal phage infection, producing and disgorging hundreds of progeny particles. From these observations, Hershey and Chase concluded that phage genes are made of DNA.

The Hershey-Chase experiment, although less rigorous than the Avery project, had an enormous impact. In the minds of many investigators, it confirmed Avery's results and extended them to viral particles. The spotlight was now on DNA.

THE WATSON-CRICK MODEL: DNA IS A DOUBLE HELIX

Under appropriate conditions, DNA can align in fibers to produce an ordered structure. And just as a crystal chandelier scatters light to produce a distinctive pattern on the wall, DNA fibers scatter X rays to produce a characteristic diffraction pattern (Fig. 5.6). A knowledgeable X-ray crystallographer can interpret DNA's diffraction pattern to elucidate selected aspects of the molecule's three-dimensional structure. When in the spring of 1951 the 23-year-old James Watson learned that the genetic material could project a diffraction pattern, he realized that DNA "must have a regular structure that could be solved in a straightforward fashion."

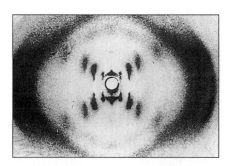

Figure 5.6   X-ray diffraction patterns reflect the helical structure of DNA. Photograph of an X-ray diffraction pattern produced by oriented DNA fibers, taken by Rosalind Franklin and Maurice Wilkins in late 1952. The crosswise pattern of X-ray reflections indicates that DNA is helical.

In this section, we analyze DNA's three-dimensional structure, looking first at significant details of the nucleotide building blocks; then at how those subunits are linked together in a polynucleotide chain; and finally, at how two chains associate to form a double helix.

Nucleotides Are the Basic Building Blocks of DNA

DNA is a long polymer composed of subunits known as *nucleotides.* Each nucleotide consists of a deoxyribose sugar, a phosphate, and a nitrogenous base. Detailed chemical knowledge of these constituents and the way they interact to form nucleotides played an important role in Watson and Crick's model building. Figure 5.7 depicts the chemical composition of deoxyribose, phosphate, and the four nitrogenous bases; how these components come together to form a nucleotide; and how sugar-phosphate bonds link the nucleotides in a chain.

As Fig. 5.7 shows, a DNA chain composed of many nucleotides has **polarity:** an overall direction. The consistent orientation of the nucleotide building blocks gives the chain this overall direction, and the two ends of a single chain are thus chemically distinct. At one end, called the 5′ end, the sugar of the terminal nucleotide has a free 5′ carbon atom, free in the sense that it is not linked to another nucleotide. Depending on how the DNA is synthesized or isolated, the 5′ carbon of the nucleotide at the 5′ end may carry either a hydroxyl or a phosphate group. At the other—3′—end of the chain, it is the 3′ carbon of the final nucleotide that is free. Along the chain between the two ends, this 5′-to-3′ polarity is conserved from nucleotide to nucleotide. By convention, a DNA chain is described in terms of its bases, written with the 5′ → 3′ direction going from left to right (unless otherwise noted). The chain depicted in Fig. 5.7c, for instance, would be TACG.

A Directional Base Sequence Can Carry Information

Information can be encoded only in a sequence of symbols whose order varies with the message to be conveyed. Without this sequence variation, there is no potential for carrying information. Because DNA's sugar-phosphate backbone is

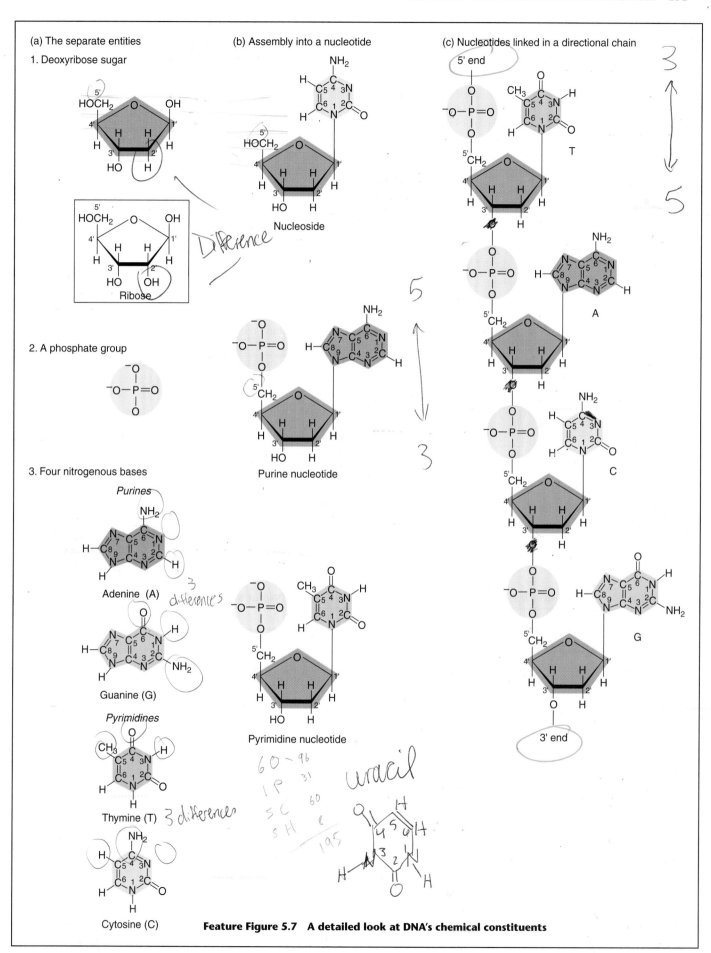

Feature Figure 5.7   A detailed look at DNA's chemical constituents

chemically identical for every nucleotide in a DNA chain, the only difference between nucleotides is in the identity of the nitrogenous base. Thus, if DNA carries genetic information, that information must consist of variations in the sequence of A, G, T, and C bases. The information constructed from the 4-letter language of DNA bases is analogous to the information built from the 26-letter alphabet of English or French or Italian. Just as you can combine the 26 letters of the alphabet in different ways to generate the words of a paper or book, so too different combinations of the four bases in very long sequences of nucleotides can create the information for constructing an organism.

The Double Helix Contains Two Antiparallel Chains That Associate by Complementary Base Pairing

Watson and Crick's discovery of the structure of the DNA molecule ranks with Darwin's theory of evolution by natural selection and Mendel's laws of inheritance in its contribution to our understanding of biological phenomena. The Watson-Crick structure, first embodied in a model that superficially resembled the Tinker Toys of preschool children, was based on an understanding and interpretation of all the chemical and physical data available at the time. Watson and Crick published their findings in the scientific journal *Nature* in April 1953.

The Physical Data: X-Ray Diffraction Patterns Indicate Molecular Parameters

The diffraction patterns of oriented DNA fibers do not, on their own, contain sufficient information to reveal structure. For instance, the number of diffraction spots, whose intensities and positions constitute the X-ray data (review Fig. 5.6),

is considerably less than the number of unknown coordinates of all the atoms in an oriented DNA molecule. Nevertheless, the photographs do reveal a wealth of structural information to the trained eye. Excellent X-ray images produced by Rosalind Franklin and Maurice Wilkins showed that the molecule is spiral-shaped, or helical; the spacing between repeating units along the axis of the helix is 3.4 Å; the helix undergoes one complete turn every 34 Å; the diameter of the molecule is ~20 Å; given this diameter, the molecule must consist of more than one polynucleotide chain.

Key Chemical Data Show That Nucleotides Exhibit Complementarity

If DNA contains more than one chain of nucleotides, what forces hold these chains together? Erwin Chargaff obtained data on the nucleotide composition of DNA from various sources that provided an important clue. Despite large variations between species in the relative amounts of the bases, in every individual, the ratio of A to T is always about 1:1, and the ratio of G to C is the same (Table 5.1). Watson grasped that the 1:1 ratios of A to T and of G to C reflect a significant aspect of the molecule's inherent structure.

To explain Chargaff's ratios in terms of chemical affinities between A and T and between G and C, Watson made cardboard cutouts of the bases in their proper physiological shapes and tried to match these up in various combinations, like pieces in a jigsaw puzzle. He knew that the substituents on purines and pyrimidines play a crucial role in molecular interactions as they can participate in the formation of **hydrogen bonds:** weak electrostatic bonds that result in a partial sharing of hydrogen atoms between reacting groups (Fig. 5.8). Some substituents on the nitrogenous bases are hydro-

TABLE 5.1   Chargaff's Data on Nucleotide Base Composition in the DNA of Various Organisms

Organism	Percentage of Base in DNA				Ratios	
	A	T	G	C	A:T	G:C
Staphylococcus afermentams	12.8	12.9	36.9	37.5	0.99	0.99
Escherichia coli	26.0	23.9	24.9	25.2	1.09	0.99
Yeast	31.3	32.9	18.7	17.1	0.95	1.09
*Caenorhabditis elegans**	31.2	29.1	19.3	20.5	1.07	0.96
*Arabadopsis thaliana**	29.1	29.7	20.5	20.7	0.98	0.99
Drosophila melanogaster	27.3	27.6	22.5	22.5	0.99	1.00
Honey bee	34.4	33.0	16.2	16.4	1.04	0.99
Mus musculus (mouse)	29.2	29.4	21.7	19.7	0.99	1.10
Human (liver)	30.7	31.2	19.3	18.8	0.98	1.03

*Data for *C. elegans* and *A. thaliana* is based on that for close relative organisms.

Note that even though the level of any one nucleotide is different in different organisms, the amount of A always approximately equals the amount of T, and the level of G is always similar to that of C. Moreover, as you can calculate for yourself, the total amount of purines (A plus G) nearly always equals the total amount of pyrimidines (C plus T).

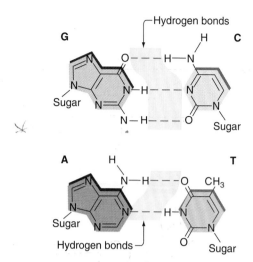

Figure 5.8 Complementary base pairing. An A on one strand can form two hydrogen bonds with a T on the other strand. A G on one strand can form three hydrogen bonds with a C on the other strand. Note that the size and shape of A–T and G–C base pairs are similar, allowing both to fill the same amount of space between the two backbones of the double helix.

A—T ᶾHydrogen bond
G-C

gen bond donors, while others are acceptors. The donors bind more tightly to the hydrogen atoms in a bond, the acceptors less tightly. Watson saw that A and T could be paired together such that two hydrogen bonds formed between them. If G and C were similarly paired, hydrogen bonds could also easily connect the nucleotides carrying these two bases. (Watson originally posited two hydrogen bonds between G and C, but there are actually three.) Remarkably, the two pairs—A–T and G–C—had essentially the same shape (Fig. 5.8). This meant that the two pairs could fit in any order between two sugar-phosphate backbones without distorting the structure. It also explained the Chargaff ratios—always equal amounts of A and T and of G and C. Note that both of these base pairs consist of one purine and one pyrimidine. Crick connected the chemical facts with the X-ray data, recognizing that because of the geometry of the base-sugar bonds in nucleotides, the orientation of the bases in Watson's pairing scheme could only arise if the bases were attached to backbones running in opposite directions. Figure 5.9 illustrates and explains the model Watson and Crick proposed in April 1953: DNA as a double helix.

The Double Helix May Assume Alternative Forms

Watson and Crick arrived at the double helix model of DNA structure by building models, not by a direct structural determination from the data alone. And even though Watson has written that "a structure this pretty just had to exist," the beauty of the structure is not necessarily evidence of its correctness. At the time of its presentation, the strongest evidence for its correctness was its physical plausibility, its chemical and spatial compatibility with all available data, and its capacity for explaining many biological phenomena.

The Biological Significance of Alternative Overall Configurations Remains Unknown

The majority of naturally occurring DNA molecules have the configuration suggested by Watson and Crick. Such molecules are known as **B-form** DNA; they spiral to the right (Fig. 5.10a). DNA is, however, more polymorphic than originally assumed. Some of the small double-stranded molecules examined by X-ray crystallography look completely different. One type, for example, contains nucleotide sequences that cause the DNA to assume a **Z form** in which the helix spirals to the left and the backbone takes on a zigzag shape (Fig. 5.10b). Researchers have observed many kinds of unusual non-B structures *in vitro* (in the test tube, literally "in glass"), and they speculate that some of these might occur at least transiently in living cells. There is some evidence, for instance, that Z DNA might exist in certain chromosomal regions *in vivo* (in the living organism). Whether or not the Z form and other unusual conformations have any biological role remains to be determined.

Some DNA Molecules Are Circular Instead of Linear

The nuclear chromosomes of all eukaryotic organisms are long, linear double helixes, but some chromosomes are circular (Fig. 5.11a and b). These include the chromosomes of prokaryotic bacteria, the chromosomes of organelles such as the mitochondria and chloroplasts that are found inside eukaryotic cells, and the chromosomes of some viruses, including the papovaviruses that can cause cancers in animals and humans. Such circular chromosomes consist of covalently closed, double-stranded circular DNA molecules. Although neither strand of these circular double helixes has an end, the two strands are still antiparallel in polarity.

Some Viruses Carry Single-Stranded DNA

In some viruses, the genetic material consists of relatively small, single-stranded DNA molecules. Once inside a cell, the single strand serves as a mold for making a second strand, and the resulting double-stranded DNA then governs the production of more virus particles. Examples of viruses carrying single-stranded DNA are bacteriophages ϕX174 and M13, and mammalian parvoviruses, which are associated with fetal death and spontaneous abortion in humans. In both ϕX174 and M13, the single DNA strand is in the form of a covalently closed circle; in the parvoviruses, it is linear (Fig. 5.11c and d).

Alternative B and Z configurations; circularization of the molecule; and single strands that are converted to double helixes before replication and expression—these are minor variations on the double-helical theme. Despite such experimentally determined departures of detail, the Watson-Crick double helix remains *the* model for thinking about DNA structure. This model describes those features of the molecule that have changed very little throughout billions of years of evolutionary history.

DNA Structure Is the Foundation of Genetic Function

Without sophisticated techniques for determining base sequence, one cannot distinguish bacterial DNA from human

Feature Figure 5.9   The structure of DNA: A double helix composed of antiparallel chains associated by complementary base pairing.

(a) In a leap of imagination, Watson and Crick took the known facts about DNA's chemical composition and physical arrangement in space and constructed a wire-frame model that not only united the evidence but also served as a basis for explaining the molecule's function.

(b) In the model, two DNA chains spiral around an axis with the sugar-phosphate backbones on the outside and pairs of bases (one from each chain) meeting in the middle. Although both chains wind around the helix axis in a right-handed sense, chemically one of them runs 5′ to 3′ upward, while the other runs in the opposite direction of 5′ to 3′ downward. In short, the *two chains are antiparallel.* The base pairs are essentially flat and perpendicular to the helix axis, and the planes of the sugars are roughly perpendicular to the base pairs. As the two chains spiral about the helix axis, they wrap around each other once every 10 base pairs, or once every 34 Å. The result is a double helix that looks like a twisted ladder with the two spiraling structural members composed of sugar-phosphate backbones and the rungs consisting of base pairs.

(c) In a space-filling representation of the model, the overall shape is that of a cylinder with a diameter of 20 Å whose axis is the axis of the double helix. The backbones spiral around the axis like threads on a screw, but because there are two backbones, there are two threads, and these two threads are vertically displaced from each other. This displacement of the backbones generates two grooves, one much wider than the other, that also spiral around the helix axis. Biochemists refer to the wider groove as the **major groove** and the narrower one as the **minor groove.**

The *two chains of the double helix are held together by hydrogen bonds between complementary base pairs, A to T and G to C* (see Fig. 5.8). Since the overall shapes of the two base pairs are quite similar, either pair can fit into the structure at each position along the DNA. Moreover, each base pair can be accommodated in the structure in two ways that are the reverse of each other: an A purine may be on strand 1 with its corresponding T pyrimidine on strand 2 or the T pyrimidine may be on strand 1 and the A purine on strand 2. The same is true of G and C base pairs. With complementary base pairing, the sugar-phosphate backbone remains in a relatively regular conformation along the entire length of both DNA chains, and the broad outlines of that conformation are independent of base sequence.

(●) Interestingly, within the double helical structure, the spatial requirements of the base pairs are satisfied if and only if each pair consists of one small pyrimidine and one large purine, and even then, only for the particular pairing of A–T and G–C. Pyrimidine–pyrimidine pairs are too small for the structure, and purine–purine pairs are too large. In addition, A–C and G–T pairs do not fit well together; that is,

(a)

they do not easily form hydrogen bonds. Complementary base pairing is thus a logical outgrowth of the molecule's steric requirements. Although any one nucleotide pair forms only two or three hydrogen bonds, the sum of these connections between successive base pairs in a long DNA molecule composed of thousands or millions of nucleotides is one basis of the molecule's great chemical stability.

(d)

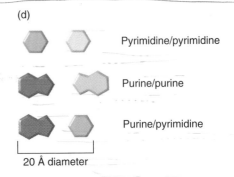

Pyrimidine/pyrimidine

Purine/purine

Purine/pyrimidine

|← 20 Å diameter →|

DNA. This is because all DNA molecules have the same general chemical properties and physical structure. Proteins, by comparison, are a much more diverse group of molecules with a much greater complexity of structure and function. In his account of the discovery of the double helix, Crick referred to this difference when he said that "DNA is, at bottom, a much less sophisticated molecule than a highly evolved protein and for this reason reveals its secrets more easily."

There are four basic DNA "secrets," embodied in four questions:

1. How does the molecule carry information?

2. How is that information copied for transmission to future generations?

3. What mechanisms allow the information to change?

4. How does the information govern the expression of phenotype?

The double helical structure of DNA provides a potential solution to each of these questions, endowing the molecule with the capacity to carry out all the critical functions required of the genetic material.

In the remainder of this chapter, we describe how DNA's structure enables it to carry genetic information, replicate that information with great fidelity, and change the information through recombination. How the information changes through mutation and how the information determines phenotype are the subjects of Chapters 6 and 7.

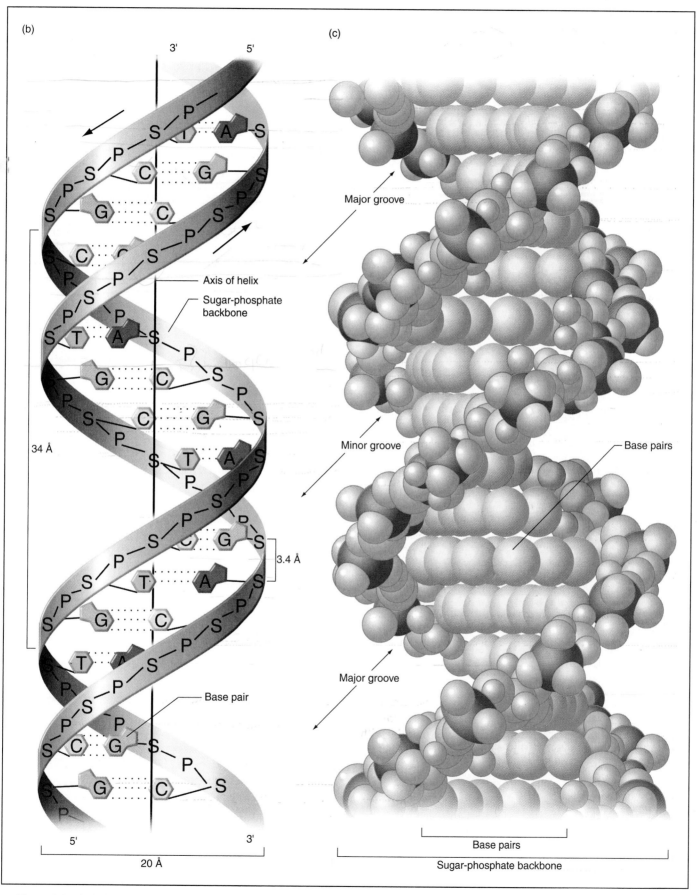

(b)

3' 5'

A—S

C ⋯⋯ G

G ⋯⋯ C

C ⋯⋯ G

Axis of helix

Sugar-phosphate backbone

T ⋯⋯ A

G ⋯⋯ C

C ⋯⋯ G

34 Å

T ⋯⋯ A

C ⋯⋯ G

T ⋯⋯ A   3.4 Å

G ⋯⋯ C

T ⋯⋯ A

Base pair

C ⋯⋯ G

G ⋯⋯ C

5' 3'

20 Å

(c)

Major groove

Minor groove

Major groove

Base pairs

Base pairs

Sugar-phosphate backbone

Figure 5.9   (Continued)

(a)

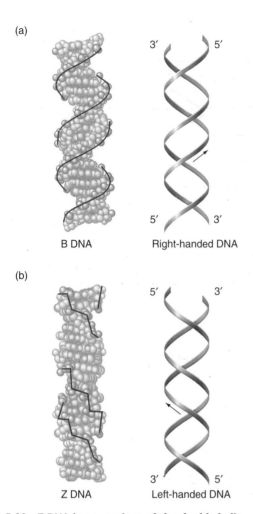

B DNA   Right-handed DNA

(b)

Z DNA   Left-handed DNA

Figure 5.10   Z DNA is one variant of the double helix.
(a) Normal Watson-Crick B-form DNA forms a right-handed helix with
a smooth backbone. (b) Z-form DNA is left-handed and has an
irregular backbone. Certain short synthetic DNA polymers (and
perhaps certain chromosomal regions) can adopt the alternative
conformation of Z DNA.

DNA STORES INFORMATION IN THE SEQUENCE OF ITS BASES

The information content of DNA resides in the sequence of its bases. The four bases in each chain are like the letters of an alphabet; they may follow each other in any order, and different sequences spell out different "words." Each "word" has its own meaning, that is, its own effect on phenotype. AGTCAT, for example, means one thing, while CTAGGT means another. Although DNA has only four different letters, or building blocks, the potential for different combinations and thus different sets of information in a long chain of nucleotides is staggering. Some human chromosomes, for example, contain chains that are 250 million nucleotides long; because the different bases may follow each other in any order, such chains contain $4^{250,000,000}$ (which translates to 1 followed by 150,515,000 zeros) potential nucleotide sequences.

Much of DNA's Sequence-Specific Information Is Accessible Only When the Double Helix Is Unwound

The unwinding of a DNA molecule exposes a single file of bases on each of two strands (Fig. 5.12a). Proteins "read" the information in DNA and carry out its instructions by binding to a specific sequence or by synthesizing a stretch of RNA or DNA complementary to a specific sequence (Fig. 5.12b).

Some Genetic Information Is Accessible Even in Intact, Double-Stranded DNA Molecules

This information emerges in part from differences between the four bases that appear in the major and minor grooves and in part from conformational irregularities in the sugar-phosphate backbone. Within the grooves, certain atoms at the periphery of the bases are exposed, and particularly in the major groove, these atoms may provide chemical information. Such information is in the form of spatial patterns of hydrogen bond donors and acceptors as well as in the distinct patterns of particular substituent shapes. Proteins can access this information to "sense" the base sequence in a stretch of DNA without disassembling the double

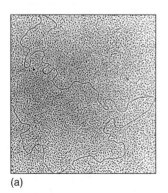

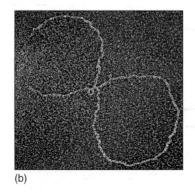

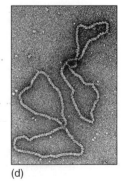

(a)   (b)   (c)   (d)

Figure 5.11   DNA molecules may be linear or circular, double-stranded or single-stranded. These electron micrographs of naturally occurring DNA molecules show (a) a fragment of a long, linear double-stranded human chromosome, (b) a circular double-stranded papovavirus chromosome, (c) a linear single-stranded parvovirus chromosome, and (d) a circular single-stranded bacteriophage M13 chromosome.

(a)

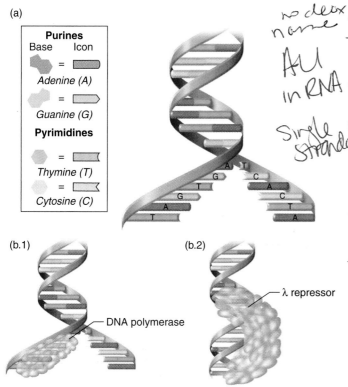

Purines

Base   Icon

= Adenine (A)

= Guanine (G)

Pyrimidines

= Thymine (T)

= Cytosine (C)

(b.1)

DNA polymerase

(b.2)

λ repressor

Figure 5.12 DNA stores information in the sequence of its bases. (a) A partially unwound DNA double helix. Note that different structural information is available in the double-stranded and unwound regions of the molecule. (b.1) Unwinding of the helix allows the enzyme DNA polymerase to copy a strand of DNA. (b.2) Some proteins can bind to specific sequences within double-helical DNA. Here, a protein called GCN4 interacts closely with features of a particular sequence of bases that are accessible to the protein in the major groove of the double helix.

helix (Fig. 5.12b.2). The proteins that help turn genes on and off make use of these subtle conformational differences. Nevertheless, the amount of information exposed in either groove or on the backbone as distinct shapes is relatively limited.

Some proteins can recognize specific base sequences in double-stranded DNA. The Fast Forward box "Restriction Enzymes Recognize Specific Base Sequences in DNA" explains how bacteria use enzymatic proteins of this type to stave off viral infection and how geneticists use these same enzymes to cut DNA at particular sites.

A Few Viruses Use RNA as the Repository of Genetic Information

In all cellular forms of life and many viruses, DNA carries the genetic information. Prokaryotes such as *Escherichia coli* bacteria carry their DNA in a double-stranded, covalently closed circular chromosome. Eukaryotic cells package their DNA in two or more double-stranded linear chromosomes. DNA viruses carry it in small molecules that are single- or double-stranded, circular, or linear.

By contrast, some viruses, including those that cause polio and AIDS, use RNA as their genetic material (Fig. 5.13).

There are three major chemical differences between RNA and DNA. First, RNA takes its name from the sugar ribose, which it incorporates instead of the deoxyribose found in DNA (Fig. 5.13a.1). Second, RNA usually contains the base uracil (U) instead of the base thymine (T); U, like T, base pairs with A (Fig. 5.13a.3). Finally, most RNA molecules are single-stranded and contain far fewer nucleotides than the very long DNA molecules found in nuclear chromosomes. Some completely double-stranded RNA molecules do nonetheless exist, although they are rare. Even within a single-stranded molecule, however, if folding brings two oppositely oriented regions that carry complementary nucleotide sequences alongside each other, they can form a double-stranded, base-paired stretch within the molecule. This means that, compared to the relatively simple, double-helical shape of a DNA molecule, many RNAs have a complicated structure of short double-stranded segments interspersed with single-stranded loops (Fig. 5.13c).

RNA has the same ability to carry information in the sequence of its bases, but is much less stable than DNA. In addition to serving as the genetic material for an array of viruses, RNA fulfills several vital functions in all cells. For example, it participates in gene expression and protein synthesis. It also plays a significant role in DNA replication, which we now describe.

DNA REPLICATION: COPYING GENETIC INFORMATION FOR TRANSMISSION TO THE NEXT GENERATION

In one of the most famous sentences of the scientific literature, Watson and Crick wrote at the end of their 1953 paper proposing the double helix model: "It has not escaped our notice that the specific pairing we have postulated immediately suggests a possible copying mechanism for the genetic material." This copying, we saw in Chapter 3, precedes the transmission of chromosomes from one generation to the next via meiosis, and is also the basis of the chromosome duplication prior to each mitosis that allows all the daughter cells of a developing organism to receive a complete copy of the genetic information.

Complementary Base Pairing Produces Semiconservative Replication: An Overview

In the process of replication postulated by Watson and Crick, the double helix unwinds to expose the bases in each strand of DNA. Each of the two separated strands then acts as a **template,** or molecular mold, for the synthesis of a new second strand (Fig. 5.14). The new strand forms as complementary bases align opposite the exposed bases on the parent strand. That is, an A at one position on the original strand signals the addition of a T at the corresponding position on the newly forming strand; a T on the original signifies addition of an A; similarly G calls for C, and C calls for G. This type of base pairing, which determines the nucleotide sequence of the new strand, is known as **complementary base pairing.** Once the appropriate base has aligned

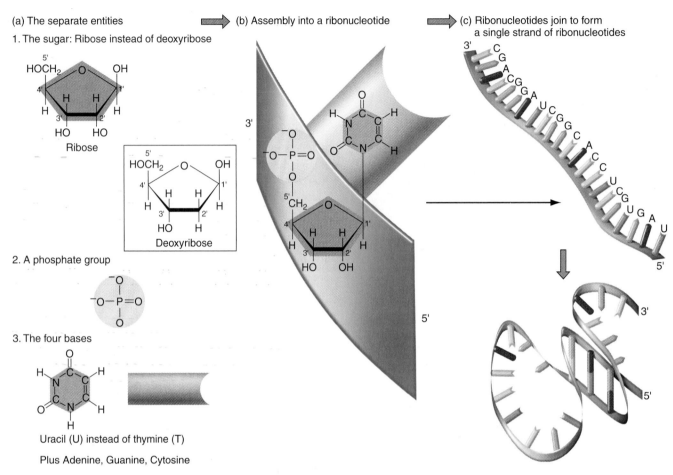

(a) The separate entities ➡ **(b) Assembly into a ribonucleotide** ➡ **(c) Ribonucleotides join to form a single strand of ribonucleotides**

1. The sugar: Ribose instead of deoxyribose

Ribose

Deoxyribose

2. A phosphate group

3. The four bases

Uracil (U) instead of thymine (T)

Plus Adenine, Guanine, Cytosine

Figure 5.13 RNA: Chemical constituents and complex folding pattern. (a) and (b) Each ribonucleotide contains the sugar ribose, an inorganic phosphate group, and a nitrogenous base. RNA contains the pyrimidine uracil (U) instead of the thymine (T) found in DNA. (c) Phosphodiester bonds join ribonucleotides into an RNA chain. Most RNA molecules are single-stranded, but are sufficiently flexible so that some regions can fold back and form base pairs with other parts of the same molecule.

opposite and formed hydrogen bonds with its complement, enzymes join the base's nucleotide to the preceding nucleotide by a phosphodiester bond, eventually linking a whole new line of nucleotides into a continuous strand. This mechanism of complementary base pairing followed by the linkage of successive nucleotides yields two "daughter" double helixes that each contain one of the original DNA strands intact (that is, "conserved") and one completely new strand (Fig. 5.15a). For this reason, such a pattern of double helix duplication is called **semiconservative replication:** a copying in which one strand is conserved from the parent molecule and the other is newly synthesized.

Watson and Crick's proposal is not the only replication mechanism imaginable. Figure 5.15b and c illustrates two possible alternatives. With *conservative* replication, one of the two "daughter" double helixes would consist entirely of original DNA strands, while the other helix would consist of two newly synthesized strands. With *dispersive* replication, both "daughter" double helixes would carry blocks of original DNA interspersed with blocks of newly synthesized material. These alternatives are less satisfactory than semiconservative replication because they do not immediately suggest a mechanism for copying the information in the sequence of bases,

and they do not explain the research data (presented later) as well.

Experimental Proof of Semiconservative Replication
In 1958 Matthew Meselson and Franklin Stahl performed an experiment that confirmed the semiconservative nature of DNA replication (Fig. 5.16). The experiment depended on being able to differentiate preexisting "parental" DNA from newly synthesized daughter DNA. To accomplish this, they controlled the isotopic composition of the nucleotides incorporated in the newly forming daughter strands as follows. They grew *E. coli* bacteria for many generations on media in which all the nitrogen was the normal isotope ^{14}N; these cultures served as a control. They grew other cultures of *E. coli* for many generations on media in which the only source of nitrogen was the heavy isotope ^{15}N. After several generations of growth on heavy-isotope medium, essentially all the nitrogen atoms in the DNA of these bacterial cells were labeled with (that is, contained) ^{15}N. The cells in some of these cultures were then transferred to new medium in which all the nitrogen was ^{14}N. Any DNA synthesized after the transfer would contain the lighter isotope.

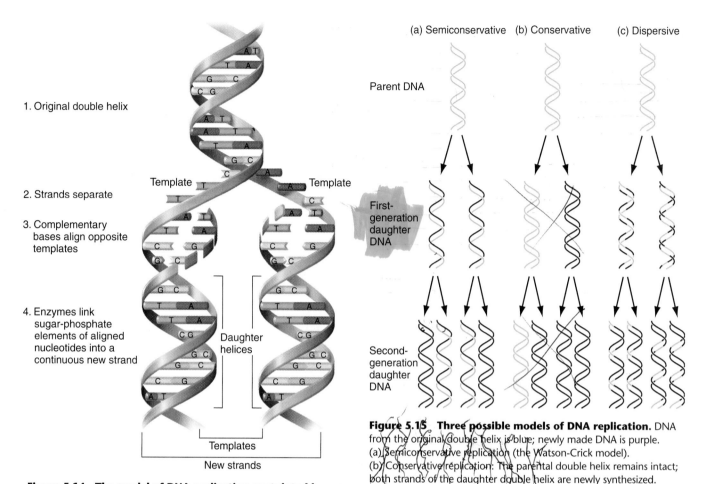

Figure 5.14 The model of DNA replication postulated by Watson and Crick. Unwinding of the double helix allows each of the two strands to serve as a template for the synthesis of two new strands by complementary base pairing. The end result: A single double helix becomes two identical daughter double helices.

Figure 5.15 Three possible models of DNA replication. DNA from the original double helix is blue; newly made DNA is purple. (a) Semiconservative replication (the Watson-Crick model). (b) Conservative replication: The parental double helix remains intact; both strands of the daughter double helix are newly synthesized. (c) Dispersive replication: At completion, both strands of both double helices contain both original and newly synthesized material.

Meselson and Stahl isolated DNA from cells grown in the different nitrogen-isotope cultures, and then subjected these DNA samples to *equilibrium density gradient centrifugation,* an analytic technique they had just developed. In a test tube, they dissolved the DNA in a solution of the dense salt cesium chloride (CsCl) and spun these solutions at very high speed (about 50,000 revolutions per minute) in an ultracentrifuge. Over a period of two to three days, the centrifugal force (roughly 250,000 times the force of gravity) causes the formation of a stable gradient of CsCl concentrations, with the highest concentration, and thus highest CsCl density, at the bottom of the tube. The DNA in the tube forms a sharply delineated band at a position where its own density equals that of the CsCl. Since DNA containing ^{15}N is denser than DNA containing ^{14}N, pure ^{15}N DNA will form a band lower, that is closer to the bottom of the tube, than pure ^{14}N DNA (see Fig. 5.16).

As Fig. 5.16 shows, when cells with pure ^{15}N DNA were transferred into ^{14}N medium and allowed to divide once, DNA from the resultant first-generation cells formed a band at a density intermediate between that of pure ^{15}N DNA and that of pure ^{14}N DNA. A logical inference is that the DNA in these cells contains equal amounts of the two isotopes. This finding invalidates the "conservative" model, which predicts the appearance of bands reflecting only pure ^{14}N and pure ^{15}N with no intermediary band. In contrast, DNA extracted from second-generation cells that had undergone a second round of division in the ^{14}N medium produced two observable bands, one at the density corresponding to equal amounts of ^{15}N and ^{14}N, the other at the density of pure ^{14}N. These observations invalidate the "dispersed" model, which predicts a single band between the two bands of the previous generation.

Meselson and Stahl's observations are consistent only with semiconservative replication: In the first generation after transfer from the ^{15}N to the ^{14}N medium, one of the two strands in every daughter DNA molecule carries the heavy isotope label; the other, newly synthesized strand carries the lighter ^{14}N isotope. The band at a density intermediate between that of ^{15}N DNA and ^{14}N DNA represents this isotopic hybrid. In the second generation after transfer, half of the DNA molecules have one ^{15}N strand and one ^{14}N strand, while the remaining half carry two ^{14}N strands. The two observable bands—one at the hybrid position, the other at the pure ^{14}N position—reflect this mix.

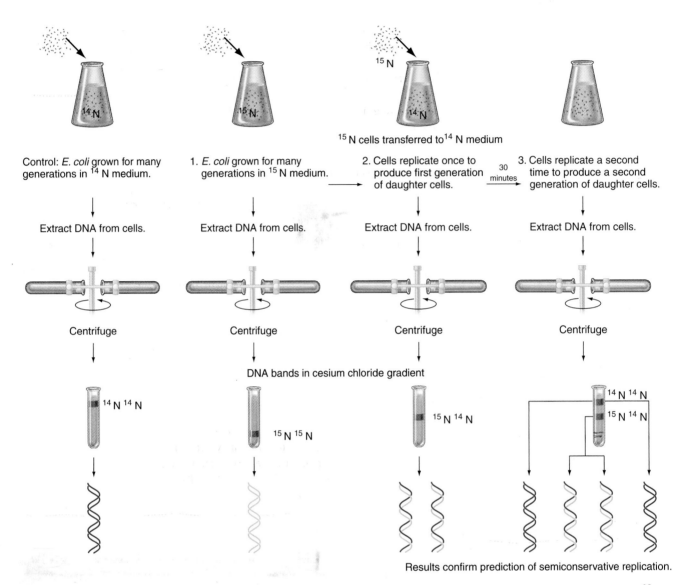

Control: *E. coli* grown for many generations in ^{14}N medium.

1. *E. coli* grown for many generations in ^{15}N medium.

^{15}N cells transferred to ^{14}N medium

2. Cells replicate once to produce first generation of daughter cells.

3. Cells replicate a second time to produce a second generation of daughter cells.

30 minutes

Extract DNA from cells.   Extract DNA from cells.   Extract DNA from cells.   Extract DNA from cells.

Centrifuge   Centrifuge   Centrifuge   Centrifuge

DNA bands in cesium chloride gradient

^{14}N ^{14}N

^{15}N ^{15}N

^{15}N ^{14}N

^{14}N ^{14}N
^{15}N ^{14}N

Results confirm prediction of semiconservative replication.

Figure 5.16   How the Meselson-Stahl experiment confirmed semiconservative replication. (1) *E. coli* cells were grown in heavy ^{15}N medium. (2) and (3) Some of these cells were transferred to ^{14}N medium and allowed to divide either once or twice. When DNA from each of these sets of cells was prepared and centrifuged in a cesium chloride gradient, the density of the extracted DNA conformed to the predictions of the semiconservative mode of replication, as shown at the bottom of the figure, where blue indicates heavy original DNA and purple depicts light, newly synthesized DNA. These results are inconsistent with the conservative and dispersive models for DNA replication (compare with Fig. 5.15b and c).

By confirming the predictions of semiconservative replication, the Meselson-Stahl experiment disproved the conservative and dispersive alternatives. We now know that the semiconservative replication of DNA is nearly universal. To understand material presented in the remainder of this book, it is important to understand precisely how semiconservative replication relates to the structure of chromosomes in eukaryotic cells during the mitotic cell cycle (see Fig. 3.7). Early in interphase, each eukaryotic chromosome contains a single continuous linear double helix of DNA. Later, during the S phase portion of interphase, the cell replicates the double helix semiconservatively; after this semiconservative replication, each chromosome is composed of two sister chromatids joined at the centromere. Each sister chromatid is a double helix of DNA, with one strand of parental DNA and one strand

of newly synthesized DNA. At the conclusion of mitosis, each of the two daughter cells receives one sister chromatid from every chromosome in the cell. This process preserves chromosome number and identity during mitotic cell division because the two sister chromatids are identical in base sequence to each other and to the original parental chromosome.

The Molecular Mechanism of Replication: Doubling the Double Helix

Although Watson and Crick's model for semiconservative replication, depicted in Fig. 5.15a, is a simple concept to grasp, the process by which it occurs is quite complicated. Replication does not happen spontaneously any time a mixture of DNA and nucleotides is present. Rather, it occurs at a precise moment in the cell cycle; depends on a network of interacting regulatory elements; requires considerable input of energy; and involves a complex array of the cell's molecular machinery, including a variety of enzymes.

Here we present highlights of replication's two basic steps: **initiation,** during which proteins open up the double helix and prepare it for complementary base pairing, and **elongation,** during which proteins connect the correct sequence of nucleotides into a continuous new strand of DNA. Figure 5.17 illustrates the molecular mechanism of replication using *E. coli* bacteria as the model organism. Although the bacterial chromosome is circular, Fig. 5.17 presents a blow-up of only a very small segment of that chromosome, which appears linear in the drawing. Biochemists, notably the laboratory of Nobel laureate Arthur Kornberg, deduced the steps by which each bacterial cell copies its single circular chromosome from biochemical studies in which they purified individual components of the replication machinery. Remarkably, they were eventually able to mix these purified molecules together in a test tube and observe DNA replication proceeding outside the cell. One notable feature of replication is that many aspects of the process rely on the requirements of DNA polymerase, particularly for a template and primer.

As Fig. 5.17 shows, DNA replication depends on the coordinated activity of many different proteins, including DNA polymerase. One unwinds the double helix; another binds to single-stranded DNA to keep the helix open; and still others make RNA primers, help copy the sequence of bases, and weld together Okazaki fragments. It took many years for biochemists and geneticists to discover how the tight collaboration of many proteins drives the intricate mechanism of DNA replication. Today they believe that this type of programmed molecular interaction underlies most of the biochemical processes that occur in cells. In these processes, a group of proteins, each performing a specialized function, like the workers on an assembly line, cooperate in the manufacture of complex macromolecules.

The Mechanics of DNA Replication at the Chromosomal Level

Not surprisingly, local unwinding of the double helix at a replication fork affects the chromosome as a whole. In *E. coli,* the unwinding of a section of a covalently closed circular chromosome distorts the entire molecule (Fig. 5.18a and b). It does this by diminishing the number of helical turns to less than the one turn every 10.5 nucleotides characteristic of B-form DNA. The chromosome accommodates the strain of distortion by twisting back upon itself. You can envision the effect by imagining a coiled telephone cord that overwinds and bunches up with use. The additional twisting of the DNA molecule is called **supercoiling.** Movement of the replication fork causes more and more supercoiling. This cumulative supercoiling, if left unchecked, would wind the chromosome up so tight that it would impede the progress of the replication fork. A group of enzymes known as **DNA topoisomerases** helps relax the supercoils by nicking one or both strands of the DNA (Fig. 5.18c). Just as a telephone cord freed at the receiver end can

unwind and restore its normal coiling pattern, the DNA strands, after nicking, can rotate relative to each other and thereby restore the normal coiling density of one helical turn per 10.5 nucleotide pairs. The activity of topoisomerases allows replication to proceed through the entire chromosome by preventing supercoils from accumulating in front of the replication fork.

Recall that the origin of replication has two forks. As a result, replication is generally **bidirectional,** with the replication forks moving in opposite directions (Fig. 5.18d). At each fork, polymerase copies both template strands, one in a continuous fashion, the other discontinuously as Okazaki fragments. As replication proceeds, one protein progressively unwinds DNA ahead of each fork, while another protein separates and stabilizes the two single strands; behind each fork, as elongation replicates chains complementary to both templates, double helixes form spontaneously. In the circular *E. coli* chromosome, there is only one origin of replication. When its two forks, moving in opposite directions, meet at a designated *termination region* about halfway around the circle from the origin of replication, replication is complete (Fig. 5.18 d–f). Replication of the original double helix sometimes produces intertwined daughter molecules whose clean separation depends on topoisomerase activity.

In the much larger, linear chromosomes of eukaryotic cells, bidirectional replication proceeds roughly as described above, but from many origins of replication, to ensure that copying reaches completion within the time allotted (that is, within the S period of the cell cycle). Because of the three rules governing DNA polymerase activity (see Fig. 5.17 a), replication of the very ends of linear chromosomes is problematic. But eukaryotic chromosomes have evolved specialized termination structures known as **telomeres,** which ensure the maintenance and accurate replication of the two ends of each linear chromosome (see Chapter 11 for details of eukaryotic chromosome replication).

Throughout replication in both circular and linear chromosomes, the disentangling of parent strands and respiraling of daughter helixes occurs continuously as enzymes, including helicases and topoisomerases, wind tight, unwind, and rewind the double helix before and after the passage of polymerase, in part by nicking and rotating the two strands in relation to each other. At completion, replication has produced two identical double-stranded daughter DNA molecules.

Cells Must Ensure the Accuracy of Their Genetic Information—Before, During, and After Replication

DNA is the sole repository of the vast amount of information required to specify the structure and function of an organism. In some species, this information may lie in storage for many years and undergo replication many times before it is called on to generate progeny. During this time, the organism must protect the integrity of the information, for even the most minor change—the simple substitution of one base for another—can have disastrous consequences, producing a severe genetic

(a) **Initiation: Preparing the double helix for complementary base pairing.** A prerequisite of DNA replication is the unwinding of a portion of the double helix. Such unwinding exposes the bases in each DNA strand, making it possible for them to pair with complementary nucleotides. Initiation begins with the unwinding of the double helix at a particular short sequence of nucleotides known as the **origin of replication.** Each circular *E. coli* chromosome has a single origin of replication. Several proteins bind to the origin,

forming a stable complex in which a small region of DNA is unwound and the two complementary strands are separated. The first of the proteins to recognize and bind to the origin of replication is called the *initiator protein.* A DNA-bound initiator attracts an enzyme called DNA helicase, which catalyzes the localized unwinding of the double helix. The opening up of a region of DNA creates two Y-shaped areas, one at either end of the unwound area, or **replication bubble.** Each Y is called a **replication fork** and consists of the two unwound DNA strands branching out into unpaired (but complementary) single strands. These single strands will serve as **templates**—molecular molds—for fashioning new strands of DNA. The molecule is now ready for replication.

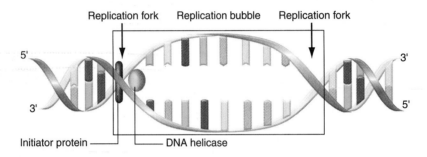

Actual formation of new DNA strands depends on the action of a complex enzyme known as **DNA polymerase,** which adds nucleotides, one after the other, to the end of a growing DNA strand. DNA polymerase operates according to three strict rules: First, it can copy only DNA that is unwound and maintained in the single-stranded state; second, it adds nucleotides only to the end of an existing chain (that is, it cannot establish the first link in the chain); and third, it functions in only one—the 5'-to-3'—direction. The requirement for a single-stranded template is satisfied by formation of the replication bubble. The requirement for an already existing chain end, however, means that something else must start the about-to-be-constructed chain. That "something else" is RNA. Construction of a very short new strand consisting of a few nucleotides of RNA provides an "end" to which DNA polymerase can add the nucleotides of DNA. Since this short stretch of RNA prepares the ground for DNA polymerase activity, it is called an RNA **primer.** An enzyme called primase synthesizes the RNA primer at the replication fork where the primer then base pairs with the single-stranded DNA template. With the double helix unwound and the primer in place, DNA replication can proceed. The third characteristic of DNA polymerase activity—one way only—determines some of the special features of subsequent steps.

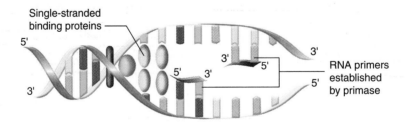

(b) **Elongation: Connecting the correct sequence of nucleotides into a continuous new strand of DNA.** Elongation—the linking together of appropriately aligned nucleotide subunits into a continuous new strand of DNA—is the heart of replication. We have seen that the lineup of bases is determined by complementary base pairing with the template strand. An A on the template strand will form hydrogen bonds only with a T on the forming strand, and a C on the template will pair only with a G. Thus, the order of bases in the template specifies the order of bases in the newly forming strand. Once complementary base pairing has determined the next nucleotide to be added to the newly forming strand, DNA polymerase catalyzes the actual joining of this properly

Feature Figure 5.17   The mechanism of DNA replication

positioned nucleotide to the preceding nucleotide. The linkage of subunits through the formation of phosphodiester bonds is known as **polymerization.** The DNA polymerase enzyme first joins the correctly paired nucleotide to the 3′ hydroxyl end of the RNA primer, and then continues to add the appropriate nucleotides to the 3′ end

of the growing chain. As a result, the DNA strand under construction grows in the 5′-to-3′ direction. The fact that the new strand is antiparallel to the template strand means that the DNA polymerase molecule actually moves along that template strand in the 3′-to-5′ direction.

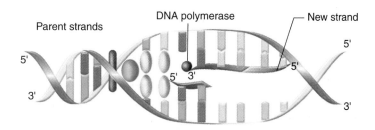

As DNA replication proceeds, helicase progressively unwinds the double helix to expose successive series of nucleotides in their unpaired, single-stranded state. DNA polymerase can then move in the same direction as the fork to synthesize one of the two new chains under construction at the fork. The enzyme encounters no problems in the polymerization of this chain—called the **leading strand**—because it can add nucleotides continuously to the growing 3′ end as soon as the unraveling fork exposes the corresponding bases on the template strand. The movement of the replication fork, however, presents problems for the synthesis of the second new DNA chain: the **lagging**

strand. The polarity of the lagging strand is opposite that of the leading strand, yet as we have seen, DNA polymerase functions only in the 5′-to-3′ direction. This means that in synthesizing the lagging strand, the polymerase must travel in a direction opposite to that of the replication fork. How can the polymerase link the nucleotides of the lagging strand if it cannot approach that strand's recently unwound template at the replication fork?

The answer is that the lagging strand is synthesized *discontinuously* as small fragments of about 1000 bases called **Okazaki fragments** (after two of their discoverers Reiji and Tuneko Okazaki).

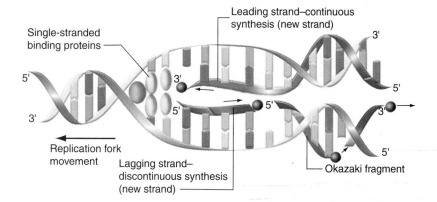

DNA polymerase still synthesizes these small fragments in the normal 5′-to-3′ direction, but because the enzyme can add nucleotides only to the 3′ end of an existing strand, each Okazaki fragment is initiated by the synthesis of a short RNA primer. The primase enzyme catalyzes formation of the RNA primer for each upcoming Okazaki fragment as soon as the replication fork has progressed a sufficient distance along the DNA. Polymerase then adds nucleotides to this new primer,

creating an Okazaki fragment that extends as far as the 5′ end of the primer of the previously synthesized fragment. Finally, DNA polymerase and other enzymes replace the RNA primer of the previously made Okazaki fragment with DNA, and an enzyme known as DNA ligase covalently joins successive Okazaki fragments into a continuous strand of DNA. With the completion of both leading and lagging strands, DNA replication is complete.

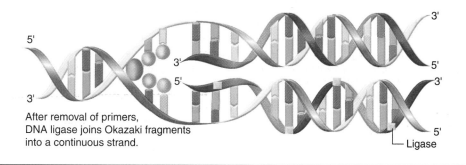

After removal of primers, DNA ligase joins Okazaki fragments into a continuous strand.

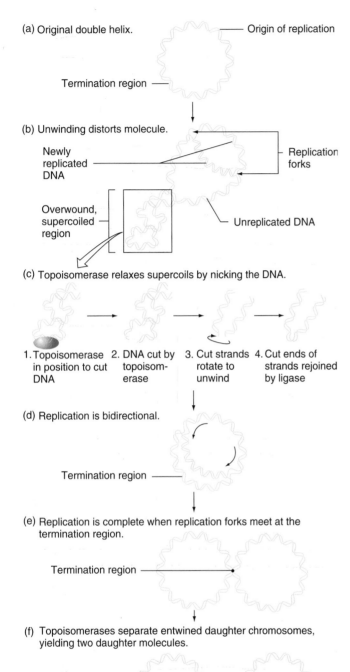

(a) Original double helix.

Origin of replication

Termination region

(b) Unwinding distorts molecule.

Newly replicated DNA

Replication forks

Overwound, supercoiled region

Unreplicated DNA

(c) Topoisomerase relaxes supercoils by nicking the DNA.

1. Topoisomerase in position to cut DNA
2. DNA cut by topoisomerase
3. Cut strands rotate to unwind
4. Cut ends of strands rejoined by ligase

(d) Replication is bidirectional.

Termination region

(e) Replication is complete when replication forks meet at the termination region.

Termination region

(f) Topoisomerases separate entwined daughter chromosomes, yielding two daughter molecules.

Figure 5.18   The bidirectional replication of a circular bacterial chromosome: An overview. (a) and (b) Replication proceeds in two directions from a single origin of replication, creating two replication forks that move in opposite directions around the circle. Local unwinding of DNA at the replication forks creates supercoiled twists in the DNA in front of the replication fork. (c) The action of topoisomerase enzymes that nick the DNA helps reduce this supercoiling. (d) and (e) When the two replication forks meet at the replication termination region, the entire chromosome has been copied. (f) Topoisomerase enzymes separate the two daughter chromosomes, which are usually entwined.

disease or even death. Each organism ensures the informational fidelity of its DNA in three important ways.

Redundancy

Since either strand of the double helix can specify the sequence of the other, each double-stranded DNA molecule is informationally redundant. Built into the structure of the molecule itself, this redundancy provides a basis for the repair of errors arising either from chemical alterations sustained during storage or from malfunctions of the replication machinery.

Enzymes That Repair Chemical Damage to DNA

The cell has an array of enzymes devoted to the repair of nearly every imaginable type of chemical damage. We describe how these enzymes carry out their corrections in Chapter 6. The evolution of sophisticated enzymatic mechanisms for detecting and correcting DNA damage underlines the importance of maintaining the fidelity of genetic information.

The Remarkable Precision of the Cellular Replication Machinery

Evolution has perfected the cellular machinery for DNA replication to the point where errors that could creep in during the copying of the information are exceedingly rare. For example, DNA polymerase has acquired the ability to remove unmatched nucleotides from a new strand of DNA; as a result, the enzyme attaches a free nucleotide to a growing leading or lagging strand only if the nucleotide's base is correctly paired with its complementary base on the parent strand. We examine the molecular mechanisms of this type of DNA repair in Chapter 6.

All of these safeguards help ensure that the information content of DNA will be transmitted intact from generation to generation of cells and organisms. However, as we see next, new combinations of existing information arise naturally as a result of recombination.

RECOMBINATION RESHUFFLES THE INFORMATION CONTENT OF DNA

Mutation, the ultimate source of all new alleles, is a relatively rare phenomenon. In sexually reproducing species, the production of individuals carrying new combinations of *already existing alleles* is actually the most important mechanism for generating genetic diversity. This diversity is essential for species survival in a changing environment, because it provides a large amount of the variation on which evolution depends.

New combinations of already existing alleles arise from two different meiotic events: crossing-over, in which two homologous chromosomes exchange parts; and independent assortment, in which each pair of homologous chromosomes segregates free from the influence of other pairs, via random spindle attachment (see Chapter 3). Independent assortment can produce gametes carrying new allelic combinations of genes on different chromosomes; but for genes linked on the same chromosome, independent assortment alone can only conserve the existing combinations of alleles. Crossing-over, however, can generate new allelic combinations of linked genes. The evolution of crossing-over thus compensated for a significant disadvantage of linkage, which without recombination would prevent the reshuffling of genes on the same chromosome.

Historically, geneticists have used the term recombination to indicate the production of new combinations of alleles by any means, including independent assortment. In the remainder of this chapter, we use **recombination** to mean the generation of new allelic combinations—new combinations of nucleotides along a single DNA molecule—through crossing-over between homologous chromosomes. In our discussion, we refer to the products of recombination as **recombinants:** chromosomes that carry a mix of alleles derived from different homologs.

As we examine recombination at the molecular level, we look first at experiments demonstrating that crossing-over occurs and then at the molecular details of a crossover event.

During Recombination, DNA Molecules Break and Rejoin

Indirect Physical Evidence of Breaking and Rejoining

When viewed through the light microscope, recombinant chromosomes bearing physical markers appear to result from two homologous chromosomes breaking and exchanging parts as they rejoin (see Figs. 4.7 and 4.8 in Chapter 4). Since the recombined chromosomes, like all other chromosomes, are composed of one long DNA molecule, a logical expectation is that they should show some physical signs of this breakage and rejoining at the molecular level. To evaluate this hypothesis, researchers selected a bacterial virus, **lambda,** as their model organism. Lambda had a distinct experimental advantage for this particular study: It is about half DNA, so the density of the whole virus reflects the density of its DNA.

The experimental approach was similar in principle to the one in which Meselson and Stahl monitored a change in DNA density to follow DNA replication, only in this case, the researchers used the change in DNA density to look at recombination (Fig. 5.19). They grew two bacterial viruses that were genetically marked to keep track of recombination, one in medium with a heavy isotope, the other in medium with a light isotope. They then infected the same bacterial cell with the two viruses under conditions that permitted little if any viral replication. With this type of coinfection, recombination could occur between "heavy" and "light" viral DNA molecules. After allowing time for recombination and the repackaging of viral DNA into virus particles, the experimenters isolated the viruses released from the lysed cells and analyzed them on a density gradient. Those viruses that had not participated in recombination formed bands in two distinct positions, one heavy and one light, as expected. Those viruses that had undergone recombination, however, banded at intermediate densities, which corresponded to the position of the recombination event. If the recombinant derived most of its alleles and hence most of its chromosome from a "heavy" DNA molecule, its

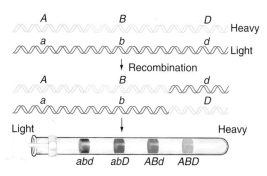

Figure 5.19 DNA molecules break and rejoin during recombination. The Experimental Evidence. Matthew Meselson and Jean Weigle infected *E. coli* cells with two different genetically marked strains of bacteriophage lambda previously grown in the presence of heavy (^{13}C and ^{15}N) or light (^{12}C and ^{14}N) isotopes of carbon and nitrogen. They then spun the progeny bacteriophages released from the cells on a CsCl density gradient. The positions of phages in the gradients reflects the relative density of their DNA. Meselson and Weigle found that genetic recombinants had densities intermediate between the heavy and light parents. The genetic constitution of the recombinants obtained at various intermediate densities reflect the fraction of the bacteriophage lambda chromosome that must have been derived from each parental chromosome.

density was skewed toward the gradient's heavy region; by comparison, if it derived most of its alleles and chromosome from a "light" DNA molecule, it had a density skewed toward the light region of the gradient. These experimental results demonstrated that recombination at the molecular level results from the breakage and rejoining of DNA molecules.

Extreme Magnification Would Reveal Heteroduplexes That Mark the Spots of Crossing-over

Recall that chiasmata, which are visible in the light microscope, indicate where chromatids from homologous chromosomes have crossed over, or exchanged parts (see Fig. 4.8). A 100,000-fold magnification of the actual site of recombination within a DNA molecule would reveal the breakage, exchange, and rejoining that constitute the molecular mechanism of crossing-over according to the lambda study. Although it is not yet possible to magnify DNA to the point of distinguishing base sequences under the microscope, a variety of molecular and genetic procedures enable geneticists to make deductions equivalent to such a 100,000-fold magnification. The data obtained provide the following two clues about the mechanism of recombination. First, the products of recombination are almost always in exact register, with not a single base pair lost or gained. Geneticists originally deduced this from observing that recombination usually does not cause mutations; today we know this to be true from analyses of DNA sequence (which we discuss in Chapter 8). Second, the two strands of a recombinant DNA molecule do not break and rejoin at the same location on the double helix. Instead, the initial break and the final rejoining occur at two different locations, offset from each other by hundreds or even thousands of base pairs. The segment of the DNA molecule located between the two break

Fast Forward

Restriction Enzymes Recognize Specific Base Sequences in DNA

In many types of bacteria, the unwelcome arrival of viral DNA mobilizes minute molecular weapons known as **restriction enzymes.** Each enzyme has the twofold ability to recognize a specific sequence of four to six base pairs anywhere within a DNA molecule and to sever a covalent bond in the sugar-phosphate backbone at a particular position within or near that sequence. When a bacterium calls up its reserve of restriction enzymes at the first sign of invasion, the ensuing shredding and dicing of selected stretches of viral DNA incapacitates the virus's genetic material and thereby restricts infection.

Since the early 1970s, geneticists have isolated more than 300 types of restriction enzymes and named them for the bacterial species in which they orginate. *Eco*RI, for instance, comes from *E. coli; Hpa*I is one of several restriction enzymes found in *Haemophilus parainfluenzae; Hind*III derives from *Haemophilus influenzae.* Each enzyme recognizes a different base sequence and cuts the DNA strand at a precise spot in relation to that sequence. *Eco*RI recognizes the sequence 5'...GAATTC...3' and cleaves between the G and the first A. *Hpa*I recognizes the sequence 5'...GTTAAC...3' and makes a cut between the second T and the first A. *Hind*III sees the sequence 5'...AAGCTT...3' and shears between the two As. The DNA of a bacteriophage called lambda (λ) carries the GAATCC sequence recog-

nized by *Eco*RI in five separate places; the enzyme thus cuts the linear lambda DNA at five points, shredding it into six pieces. The DNA of a phage known as ϕX174, however, contains no *Eco*RI recognition sequences. By comparison, the *Hpa*I enzyme can cut λ DNA in 13 places and ϕX174 DNA in 4, while *Hind*III can cut λ DNA in 6 places and ϕX174 in 10.

How, you may wonder, do bacteria keep their arsenal of restriction enzymes from attacking their own DNA? The answer is a group of companion enzymes known as *methylases,* which attach methyl groups (−CH₃) in a precisely defined way. With every cycle of DNA replication, the methylating enzymes recognize the exact same sites that would be recognized by the cell's restriction enzymes, "see" a methyl group on the template strand at those sites, and add a methyl group at the exact same spots on the newly forming daughter strand. This site-specific methylation of every newly forming strand shields the bacterial DNA from recognition by the cell's own restriction enzymes. The methylases recognize only single-stranded restriction sites that form opposite an already methylated mirror image. Because viral DNA enters the bacterial cell as a double-stranded helix that is not methylated at the restriction enzyme's recognition sites, the methylases do not protect it.

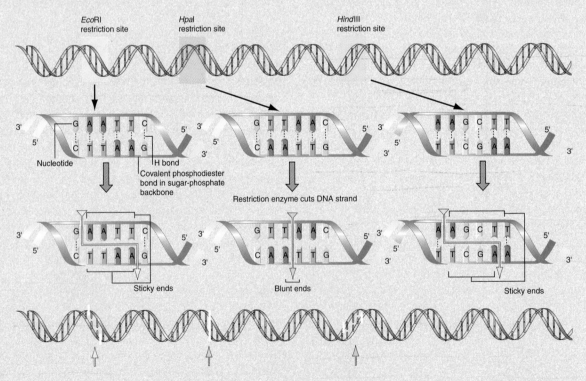

Figure A   Three restriction enzymes in action. The restriction enzymes *Eco*RI, *Hpa*I, and *Hind*III recognize different six-base-pair-long symmetrical sequences in double-stranded DNA molecules. Each enzyme severs the phosphodiester bonds between the same two adjacent nucleotides on each DNA strand. If these bonds are in the middle of the six base recognition site (as with *Hpa*I), the resultant DNA fragments will have blunt ends. If the bonds are offset from the center of the recognition site (as with *Eco*RI and *Hind*III), the products of cleavage will have sticky ends. Note that any sticky end produced by cleavage with a particular restriction enzyme is complementary in sequence to any other sticky end made by the same enzyme.

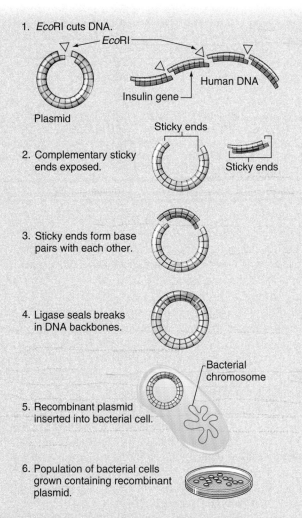

1. *Eco*RI cuts DNA.

*Eco*RI

Human DNA

Insulin gene

Plasmid

2. Complementary sticky ends exposed.

Sticky ends

Sticky ends

3. Sticky ends form base pairs with each other.

4. Ligase seals breaks in DNA backbones.

Bacterial chromosome

5. Recombinant plasmid inserted into bacterial cell.

6. Population of bacterial cells grown containing recombinant plasmid.

Figure B   One use of recombinant DNA technology: Harnessing bacteria to copy the human insulin gene.
Geneticists cleave both a small circular chromosome called a plasmid and human chromosomal DNA with the *Eco*RI restriction enzyme. This cleavage converts the plasmid to a linear piece of DNA with *Eco*RI sticky ends, while converting the human DNA into thousands of fragments with *Eco*RI sticky ends; one of these human chromosomal fragments contains the insulin gene. When these pieces of DNA are mixed together, they can adhere to each other in various combinations because of the complementarity of their sticky ends; the enzyme DNA ligase will seal these fragments together by forming the proper phosphodiester bonds. One possible outcome of this process is the creation of a recombinant DNA molecule in which the human insulin gene is spliced into the circular plasmid. Any *E. coli* cells transformed with this recombinant plasmid will become miniature factories for the synthesis of insulin.

Figure A illustrates the results of three restriction enzymes in action. Note that each recognition sequence in double-stranded DNA is symmetrical; that is, the base sequences on the two strands are identical when read in the 5′ → 3′ direction. Thus, each time an enzyme recognizes a short 5′ → 3′ sequence on one strand, it finds the exact same sequence in the 5′ → 3′ direction of the complementary antiparallel strand. In other words, viewed in the 5′ → 3′ direction, the recognition sequence is palindromic; like the phrase "TAHITI HAT" or the number 1881, it reads the same backward and forward (although the analogy is not exact because the language

of DNA is double-stranded, while English follows only a single line at a time). Note also that the cuts made by restriction enzymes leave two types of DNA ends. *Hpa*I cleaves right in the middle of the six-base sequence, making a smooth cut straight through both chains of the double helix to create what geneticists call "blunt" ends. In contrast, *Eco*RI and *Hind*III do not cleave the two strands of the double helix at the same spot in the middle of a recognition sequence, and as a result, the cuts they make are offset in opposite directions on the two strands. When the weak hydrogen bonds between the strands dissociate, these cuts leave short protruding single-stranded flaps known as **sticky,** or **cohesive, ends.** Like a tiny finger of velcro, each flap can stick to—that is, re-form hydrogen bonds with—a complementary sequence protruding from the end of another piece of DNA.

Even though restriction enzymes evolved in bacteria as a means of protection against viruses, given the appropriate base sequence, they can cut the DNA of any organism. And as Fig. B shows, any two DNA molecules with complementary protruding sequences can join together, no matter how unrelated the rest of their base sequences may be. The cutting and splicing together of DNA from two origins creates a **recombinant DNA** molecule.

In the mid-1970s, geneticists took advantage of the activity of restriction enzymes to splice together DNA from any two organisms. In their hands, the enzymes served as precision scissors that, in effect, revolutionized the study of life and gave birth to recombinant DNA technology. Although the sticky ends created by restriction enzymes enable two unrelated DNA molecules to come together by base pairing, another enzyme, known as DNA ligase, is required to stabilize the recombinant molecule. The ligase seals the breaks in the backbones of both strands.

Figure B illustrates one application of recombinant DNA technology: the splicing of the human gene for insulin into a small circle of DNA, known as a plasmid, that can replicate in bacterial cells. When the bacteria copy their own chromosome in preparation for cell division, they also make copies of any resident plasmids along with all the genes the plasmids contain. Using restriction enzymes as scissors and DNA ligase as glue, investigators can insert a copy of the human insulin gene into a large number of plasmids and then treat bacteria with these recombinant plasmids exactly as Avery did with his "transforming" DNA. The recombinant DNA molecules will enter some cells and replicate as the cultured bacterial cells grow. If the plasmids carry sequences that direct the expression of the human gene into protein, eventually a population of bacteria will grow up in which every cell not only contains a copy of the human gene but also makes the insulin encoded by that gene.

Techniques for designing and constructing recombinant DNA molecules and for harnessing bacteria to produce large quantities of a particular gene and its protein product hold great promise for medicine, agriculture, and industry. Researchers are currently using the technology to try to develop, among many other projects, a relatively inexpensive nose spray able to kill the viruses that cause the common cold, drugs to help fight cancer and AIDS, corn containing proteins with the nutritional value of beef proteins, and a process for making automobile fuel from discarded corn stalks. These and other recombinant DNA possibilities depend in large part on the original use of restriction enzymes to recognize and cut specific base sequences.

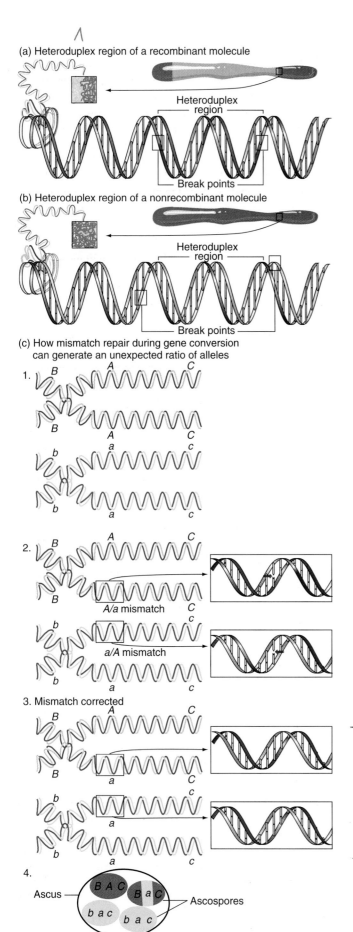

(a) Heteroduplex region of a recombinant molecule

Heteroduplex region

Break points

(b) Heteroduplex region of a nonrecombinant molecule

Heteroduplex region

Break points

(c) How mismatch repair during gene conversion can generate an unexpected ratio of alleles

1.

2.

A/a mismatch

a/A mismatch

3. Mismatch corrected

4.

Ascus — Ascospores

Figure 5.20 Mismatch repair of the heteroduplex regions that form during recombination may result in gene conversion. (a) Heteroduplex regions of recombinant molecules form at the junction between the parental chromosome sequences. (b) Heteroduplex region of a nonrecombinant molecule: Sequences from the same parental molecule are found on both sides of the heteroduplex region. The heteroduplexes depicted in (a) and (b) are thought to be two alternative products of the same molecular intermediate (see Fig. 5.21). (c.1–c.3) Mismatch repair of heteroduplex regions such as those in (a) or (b) can (c) generate gene conversion, that is, deviations from the expected 2:2 segregation of parental alleles.

points may reanneal to form a **heteroduplex region** (from the Greek *hetero* meaning "other" or "different") (Fig. 5.20). This name applies not only because one strand of the double helix in this region is of maternal origin, while the other is paternal, but also because the pairing of maternal and paternal strands may produce mismatches in which bases are not complementary. Such mismatches occur at sites in the sequence that differ between the maternal and paternal homologs involved in the recombination event. The base sequences of most homologs differ by roughly 0.1% to 1.0%. Within a heteroduplex, these mismatches prevent proper pairing at the mismatched base pairs, but they do not prevent double helix formation along the neighboring complementary nucleotides. Mismatched heteroduplex molecules do not persist for long. The same DNA repair enzymes that operate to correct mismatches during replication can move in to resolve them during recombination. The outcome of their work depends on which strand they correct. For example, a repaired G–T mismatch could become either G–C or A–T.

The heteroduplex region of a recombined DNA molecule has one break point on each strand of the double helix (Fig. 5.20a). Beyond the heteroduplex region, *both* strands of one DNA molecule have switched places with both strands of its homolog. There is, however, an alternative type of heteroduplex region in which the initiating and resolving cuts are on the same DNA strand (Fig. 5.20b). With this type of heteroduplex, only one short segment of one strand has traded places with one short segment of a homologous nonsister strand. Like the first type of heteroduplex, a short heteroduplex arising from a single-strand exchange may also contain a mismatch.

In either type of heteroduplex, mismatch repair may alter one allele to another. If for example, the original homologs carried the *A* allele in one segment of two chromatids and the *a* allele in the corresponding segment of the two homologous chromatids, the *A:a* ratio of alleles in these original homologs would be 2:2. Mismatch repair might change that *A:a* allele ratio from 2:2 to 3:1 (that is, three *A* alleles for every one *a* allele) or 1:3 (one *A* allele for every three *a* alleles; Fig. 5.20c). Any deviation from the expected 2:2 segregation of parental alleles is known as **gene conversion,** because one allele (such as *A*) has been converted to the other (in this case *a*). Since mismatch repair happens at random, it is a matter of chance which mismatch is repaired in which direction; the actual repair, however, determines the final ratio of alleles.

Although the unusual ratios resulting from gene conversion occur in many types of organisms, geneticists have studied them most intensively in yeast where tetrad analysis makes it possible to follow all four meiotic products from a single cell (Fig. 5.20c). Interestingly, observations in yeast indicate that gene conversion is associated with crossing-over about 50% of the time, but is an isolated event not associated with a crossover between flanking markers the other 50% of the time. As we see later, both outcomes derive from the same proposed molecular intermediate, which may or may not lead to a crossover.

A Molecular Model of Crossing-Over

A variety of experimental observations provide the framework for a detailed model of crossing-over. The observation that recombination occurs during meiotic prophase, after completion of DNA replication, suggests that it involves the breakage and reunion of DNA molecules from homologous nonsister chromatids. The observation that recombination occurs only between homologous regions and is highly accurate, that is, in exact register, suggests an important role for base pairing between complementary strands derived from the two homologs. The observation that recombination sites are always associated with heteroduplex regions further supports the role of base pairing in the recombination process; it also implies that the process is initiated by single-strand breaks. Finally, the observation of heteroduplex regions associated with gene conversion in the absence of recombination indicates that in some instances, recombination events are initiated but not consummated.

The model we now present derives from studies by Matthew Meselson, C. M. Radding, and others on recombination in a variety of systems. It differs in only small details from the original model formulated by Robin Holliday in 1964, and geneticists currently believe that with only minor modifications it applies to most types of molecular recombination. Figure 5.21 illustrates this consensus model of the recombination process. In the figure, we focus on the two nonsister chromatids involved in a single recombination event, showing the two noncrossover chromatids only at the beginning of the process. These two noncrossover chromatids, depicted in the outside positions in Fig. 5.21, step 1, remain unchanged throughout recombination.

As Fig. 5.21 illustrates, recombination is probably always initiated by the nicking of DNA. During meiosis in eukaryotes and the completely different sexual cycle of bacteria (described in Chapter 13), the cell programs specify enzymes to do the nicking, and in many organisms, the enzymes introduce nicks at specific sites on the DNA. In mammals, there are intermittent "hot spots," or preferred sites of recombination. In the bacterium *E. coli,* recombination begins at special eight-nucleotide-long "chi" sites where an enzyme known as recBCD both nicks and displaces a strand. In *Drosophila,* by contrast, recombination is initiated at random sites along a chromosome. Clearly, in those organisms where the breaks occur at designated sites, the distribution of different recombinations is not entirely random. However, because the completion of recombination can occur far from the site of the initial break, recombination can probably occur anywhere along the DNA.

Unlike the start of recombination during meiosis, the nicking that initiates recombination during mitosis (see the Genetics and Society box in Chapter 4) does not occur in response to an instruction from the normal cell-cycle program; instead, it follows from damage to the DNA. X rays, for example, break DNA directly and ultraviolet damage generates breaks as the cell's enzymatic machinery works to remove the damage. Since the physical and chemical insults that damage DNA occur at random along the molecule, they result in a random distribution of reparative recombination events. Moreover, since nonmeiotic recombination depends on DNA damage, the amount of such recombination is low in the absence of damage and increases in proportion to the amount of damage sustained by the cell.

The general molecular framework by which recombination occurs (depicted in Fig. 5.21) is now clear, but a few questions remain about the first step of the process. In yeast, for example, there is evidence that a certain proportion of recombination events begin with a break in both strands of one homolog, rather than with a single-strand break. In such cases, however, it still appears that only one of the two broken strands invades the second homolog, just as in the consensus model. The remaining steps in the process—repair, ligation, branch migration, and resolution—also appear to be exactly the same as those presented in Fig. 5.21. Thus, whatever the exact details of initiation turn out to be, and they may differ from one organism to another, geneticists agree that on the molecular level, recombination consists in the breaking, invasion, migration, and disengagement of complementary DNA strands originating in different homologous chromosomes. They also agree that both base pairing and the repair of DNA mismatches play a significant role in the process. The revised Holliday, or *consensus model* for recombination has become established because it explains all the data so far obtained from molecular studies as well as the five properties of recombination deduced from breeding experiments:

1. Homologs physically break, exchange parts, and rejoin.

2. Breakage and repair create reciprocal products of recombination.

3. Recombination events can occur anywhere along the DNA molecule.

4. Precision in the exchange—no gain or loss of nucleotide pairs—prevents mutations from occurring during the process.

5. Gene conversion—in which a small segment of information from one homologous chromosome transfers to the other—can give rise to an unequal yield of two different alleles. Gene conversion is associated with crossing-over, resulting in the recombination of flanking markers about 50% of the time, but is an isolated event not associated with a crossover between flanking markers the other 50% of the time.

Step 1 Nicking. During the period when the synaptonemal complex tightly aligns homologous chromatids, specific enzymes called endonucleases nick a single strand by breaking the phosphodiester bond between adjacent nucleotides.

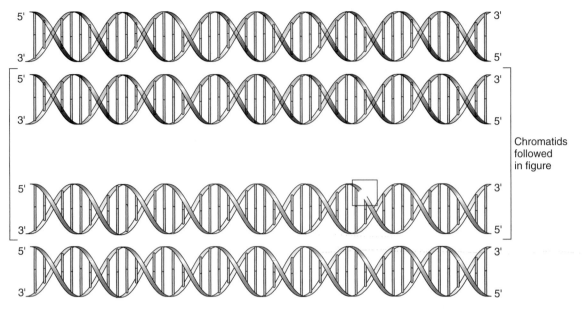

Chromatids followed in figure

Step 2 Whisker displacement. The enzyme helicase (whose function in DNA replication we noted earlier) unwinds the nick-containing double helix in the 5′ direction, causing the displacement of a single strand of DNA away from the axis of the molecule and leaving behind a single-stranded gap. The displaced single strand is called a "whisker."

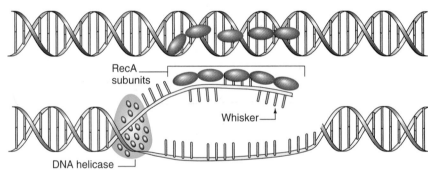

RecA subunits

Whisker

DNA helicase

Step 3 First strand invasion. The displaced strand is recognized and bound by an enzyme that also binds to a double helix in the immediate vicinity. In *E. coli,* the enzyme that simultaneously binds to the whisker and the double helix is called RecA, which plays a major role in the ensuing steps of the process, although many other enzymes collaborate with it. Their combined efforts open up the RecA-bound double helix, promoting its invasion by the single displaced "whisker" from the other duplex. RecA then moves along the double helix, prying it open in front and releasing it to snap shut behind. With RecA as its guide, the invading strand scans the base sequences it passes in the momentarily unwound stretches of DNA duplex. As soon as it finds a complementary sequence of sufficient length, it becomes immobilized by dozens of hydrogen bonds, at which point it forms a stable heteroduplex. Meanwhile, the strand displaced by the invading whisker forms a D (for displacement)-loop (maroon), which is stabilized by binding of the single-strand-binding (SSB) protein that played a similar role in DNA replication. D-loops have been observed in electron micrographs.

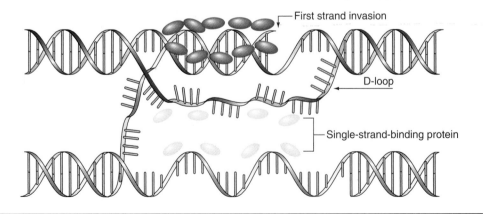

First strand invasion

D-loop

Single-strand-binding protein

Feature Figure 5.21   A model of recombination at the molecular level

Step 4 Second strand invasion. The first strand invasion leaves behind two single-stranded regions that are no longer paired with their original complementary partners. One region is the D-loop; the other is opposite the gap in the DNA molecule that sustained the original nick. These single-strand regions are on different nonsister chromatids, but because they are complementary to each other, it is likely they will base pair to form a second heteroduplex with an accompanying second break that separates the D-loop from its original double helix on the 3′ side.

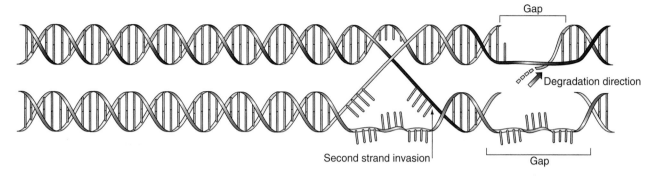

Step 5 Repair and ligation. The dangling portion of the D-loop is degraded, and the two open gaps are repaired and ligated so that there are no longer any loose ends.

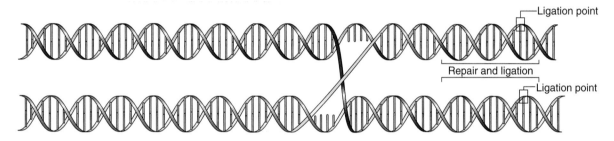

Step 6 Branch migration. The next step, **branch migration,** results from the tendency of both invading strands to "zip up" by base pairing along the length of their newly found complementary strands. The DNA double helixes unwind in front of this double zipping action and two newly created heteroduplex molecules rewind behind it. Branch migration thus lengthens the heteroduplex region of both DNA molecules from tens of base pairs to hundreds or thousands. Because the two invading strands began their scanning for complementary bases at slightly different points on the homologous chromatids, branch migration produces two heteroduplex regions that are somewhat different in length.

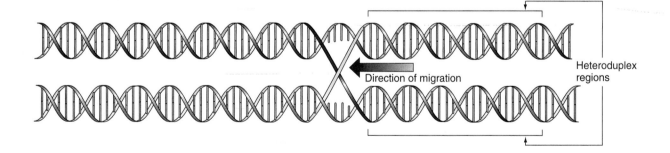

Step 7 The Holliday intermediate. For meiosis to proceed, the two interlocked nonsister chromatids must disengage, and there are two equally likely paths to such a resolution of crossing-over. To understand these alternative resolutions, it is helpful to modify the way in which we view the interlocked intermediate structure. By pushing out each of the four arms of the interlocked structure into an X pattern and then rotating one set of arms from the same original chromatid 180°, you obtain the "isomerized cross-strand exchange topology" structure pictured in step 7, commonly referred to as the "Holliday intermediate." It is important to realize that this is simply a different way of looking at the structure for pedagogical purposes. In reality, there is no preferred conformation of chromatid arms relative to each other in this small localized region. Rather, the arms are free to move about at random, constrained only by the strands that connect the two DNA molecules to each other. The view of the Holliday intermediate, however, clearly reveals that the four single-stranded regions all play an equal role in holding the structure together.

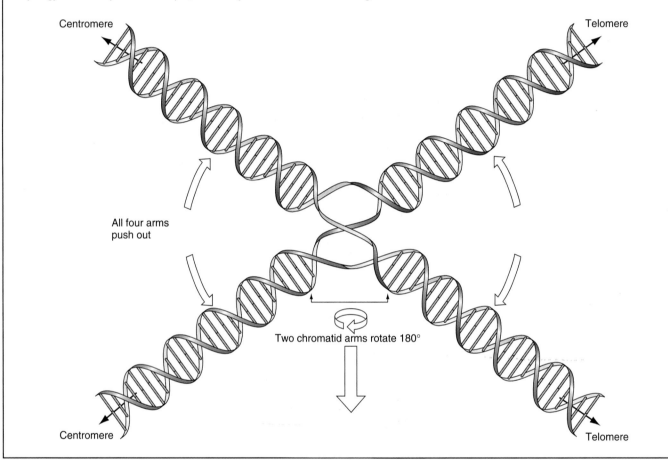

Centromere

Telomere

All four arms push out

Two chromatid arms rotate 180°

Centromere

Telomere

Figure 5.21   (Continued)

Step 8 Alternative resolutions. Now, suppose that endonucleases make a vertical cut across the noninvading strands. Such a cut, followed by repair and ligation, would produce the recombinant chromatids labeled A1 and A2 in step 8; these are similar to the recombinant molecule pictured in Fig. 5.20a. An alternative is to suppose instead that the endonucleases make a horizontal cut across both of the invading strands, severing the two single-stranded regions pictured to the left and right of the Holliday intermediate. Such a cut, followed by repair and ligation, would produce a gene conversion but no recombination, as shown in B1 and B2 of step 8; this resolution is identical to the one illustrated in Fig. 5.20b. Any set of two or more cuts other than those described and depicted in step 8 would destroy the continuity of one or both chromatids and lead to the formation of inviable gametes.

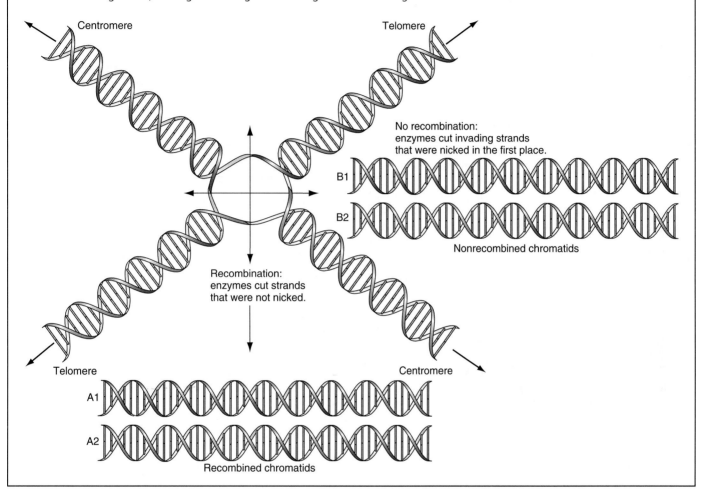

Centromere

Telomere

No recombination:
enzymes cut invading strands
that were nicked in the first place.

B1

B2

Nonrecombined chromatids

Recombination:
enzymes cut strands
that were not nicked.

Telomere

Centromere

A1

A2

Recombined chromatids

CONNECTIONS

The Watson-Crick model for the structure of DNA, the single most important biological discovery of the twentieth century, clarified how the genetic material fulfills its primary function of carrying information: Each long, linear or circular molecule carries one of a vast number of potential arrangements of the four nucleotide building blocks (A, T, G, and C). The model also suggested how base complementarity could provide a mechanism for both accurate replication and changes in base sequence that arise from recombination events.

Unlike its ability to carry information, DNA's capacities for replication and recombination are not solely properties of the DNA molecule itself. Rather they depend on the cell's complex enzymatic machinery. But even though they rely on the complicated orchestration of many different proteins acting on the DNA, they occur with extremely high fidelity—normally not a single base pair is gained or lost. Rarely, however, errors do occur, providing the genetic basis of evolution. While most of these errors are detrimental to the organism, a very small percentage produce dramatic changes in phenotype without killing the individual. For example, although most parts of the X and Y chromosomes are not similar enough to recombine, occasionally an "illegitimate" recombination does occur. Depending on the site of crossing-over, such illegitimate recombination may give rise to an XY individual who is female or an XX individual who is male (Fig. 5.22). The explanation is as follows. In the first six weeks of development, a human embryo has the potential to become either male or female, but in the critical seventh week, information from a small segment of DNA—the sex-determining region of the Y chromosome, or the *SRY* gene—determines the embryo's sex. An illegitimate recombination between the X and the Y that shifts the *SRY* gene from the Y to the X chromosome, creates a Y chromosome lacking *SRY* and an X chromosome with *SRY.* Fertilization of eggs by sperm with a Y chromosome lacking *SRY* generates XY individuals that de-

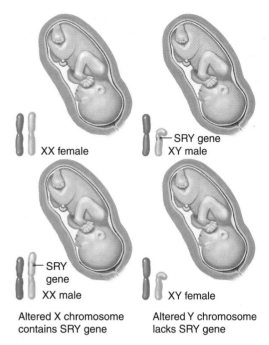

Figure 5.22   Illegitimate recombination may produce an XY female or an XX male. The *SRY* gene, normally located on the Y chromosome, dictates male development; in the absence of *SRY,* an embryo develops as a female. Rare, illegitimate recombination between the X and Y chromosome can create a Y chromosome without *SRY* and an X chromosome with *SRY.* Fertilization with gametes containing these unusual chromosomes will produce XY females or XX males.

velop as females; fertilization of eggs by sperm with an X chromosome containing *SRY* produces XX individuals that develop as males.

How do genes such as *SRY* produce their phenotypic effects? We begin to answer this question in Chapter 6 where we describe how geneticists learned that a gene is a specific sequence of nucleotides in a discrete region of DNA that encodes a particular protein.

ESSENTIAL CONCEPTS

1. DNA is the nearly universal genetic material. Experiments showing that DNA causes bacterial transformation and is the agent of virus production in phage-infected bacteria demonstrated this fact.

2. According to the Watson-Crick model, proposed in 1953 and confirmed in the succeeding decades, the DNA molecule is a double helix composed of two antiparallel strands of nucleotides; each nucleotide consists of one of four nitrogenous bases (A, T, G, or C), a deoxyribose sugar, and a phosphate. An A on one strand pairs with a T on the other, and a G pairs with a C.

3. DNA carries information in the sequence of its bases, which may follow one another in any order.

4. The DNA molecule reproduces by semiconservative replication. In this type of replication, the two DNA strands separate and the cellular machinery then synthesizes a complementary strand for each. By producing exact copies of the base sequence information in DNA, semiconservative replication allows life to reproduce itself.

5. Recombination arises from a highly accurate cellular mechanism that includes the base pairing of homologous strands of nonsister chromatids. Recombination generates new combinations of alleles.

SOCIAL AND ETHICAL ISSUES

1. In 1986, a published paper from a highly respected cancer research laboratory was retracted because the validity of some of their data had been questioned. The research was in a very specialized area of immunology and involved a collaboration between a cellular immunology laboratory and a molecular immunology laboratory. Is the head scientist of each laboratory responsible for understanding and evaluating the details of each part of a published work, including research in specialized areas outside his or her own area of expertise? Because federal grant money was used to fund the research, NIH and Congress set up a special committee to determine if there had been scientific misconduct. The scientists under investigation felt they were considered guilty until proven innocent. What is a good system for monitoring scientific research? Are scientists the best policemen of themselves or should there be a nonscientist oversight committee? Would an oversight committee understand the research and the research working community well enough to monitor activities of the scientific community? (Epilogue: The scientists were found to not have committed fraud.)

2. In 1990, researchers reported the recovery and amplification of a small fragment of DNA from a fossilized leaf that was 20 million years old. Findings such as the recovery of small bits of ancient DNA inspired the writing of the best-selling novel, *Jurassic Park* by Michael Crichton, in which dinosaurs were recreated by isolating and piecing together bits of DNA. The necessity of recovering enough intact long stretches of dinosaur DNA make this a highly improbable scenario, yet the movie portrayed this work with an aura of authenticity. Does a science fiction writer have a responsibility to indicate what is truly possible? Does this become more of an imperative when a book is made into a movie? Can the reality of what is currently possible be conveyed within the framework of science fiction?

3. Debra is a new graduate student who just joined a research laboratory. She will be following up on work done by a graduate student, Dan, who completed his Ph.D. degree and left the laboratory to do a postdoctoral fellowship at another university. When Debra begins doing the experiments needed to complete a publication that would be a joint publication by herself, Dan, and the head of the laboratory, she finds she cannot repeat some of Dan's experiments. She feels the protocols and experimental data he left behind were incomplete. When she contacts Dan to get more information, he is extremely defensive and reluctant to send her information that would enable her to repeat his work. He feels his analysis was complete and doesn't need to be repeated. Debra starts to suspect that Dan fabricated some of the data. What should she do? Should she confront Dan? Should she go to the head of the lab with the problem? Should she try to work out the protocols herself and not raise the issue with anyone?

SOLVED PROBLEMS

5′ TAAGCGTAACCCGCTAA CGTATGCGAAC GGGTCCTATTAACGTGCGTACAC 3′

3′ ATTCGCATTGGGCGATT GCATACGCTTG CCCAGGATAATTGCACGCATGTG 5′

I. Imagine that the double-stranded DNA molecule shown above was broken at the sites indicated by spaces in the sequence, and that before the breaks were repaired, the DNA fragment between the breaks was reversed. What would be the base sequence of the repaired molecule? Explain your reasoning.

Answer

To answer this question, you need to keep in mind the polarity of the DNA strands involved.

The top strand has the polarity left to right of 5′ to 3′. The reversed region must be rejoined with the same polarity. Label the polarity of the strands within the inverting region. To have a 5′-to-3′ polarity maintained on the top strand, the *fragment that is reversed must be flipped over,* so the strand that was formerly on the bottom is now on top (see below).

II. A new virus has recently been discovered that infects human lymphocytes. The virus can be grown in the laboratory using cultured lymphocytes as host cells. Design an experiment using a radioactive label that would tell you if the virus contains DNA or RNA.

5′	3′5′	3′5′	3′

TAAGCGTAACCCGCTAA**GTTCGCATACG**GGGTCCTATTAACGTGCGTACAC

ATTCGCATTGGGCGATT**CAAGCGTATGC**CCCAGGATAATTGCACGCATGTG

3′	5′3′	5′3′	5′

Answer

Use your knowledge of the differences between DNA and RNA to answer this question. RNA contains uracil instead of the thymine found in DNA. *You could set up one culture in which you add radioactive uracil to the media and a second one in which you add radioactive thymine to the culture.* After the viruses have infected cells and produced more new viruses, collect the newly synthesized virus. Determine which culture produced radioactive viruses. If the virus contains RNA, the collected virus grown in media containing radioactive uracil will be radioactive, but the virus grown in radioactive thymine will not be radioactive. If the virus contains DNA, the collected virus from the culture containing radioactive thymine will be radioactive, but the virus from the radioactive uracil culture will not. (You might also consider using radioactively labeled ribose or deoxyribose to differentiate between an RNA and DNA containing virus. Technically this does not work as well, because the radioactive sugars are processed by cells before they become incorporated into nucleic acid, thereby obscuring the results.)

III. If you expose a culture of human tissue culture cells (e.g., HeLa cells) to ^{3}H-thymidine during S phase, how would the radioactivity be distributed over a pair of homologous chromosomes at metaphase? Would the radioactivity be in (a) one chromatid of one homolog, (b) both chromatids of one homolog, (c) one chromatid each of both homologs, (d) both chromatids of both homologs, or (e) some other pattern? Choose the correct answer and explain your reasoning.

Answer

This problem requires application of your knowledge of the molecular structure and replication of DNA and how it relates to chromatids and homologs. DNA replication occurs during S phase, so the ^{3}H-thymidine would be incorporated into the new DNA strands. A chromatid is a replicated DNA molecule and each new DNA molecule contains one new strand of DNA (semiconservative replication). *The radioactivity would be in both chromatids of both homologs (d).*

PROBLEMS

5-1 For each of the terms in the left column, choose the best matching phrase in the right column.

a. transformation	1. the strand that is synthesized discontinuously during replication
b. bacteriophage	2. the sugar within the nucleotide subunits of DNA
c. pyrimidine	3. a nitrogenous base containing a double ring
d. deoxyribose	4. noncovalent bonds that hold the two strands of the double helix together
e. hydrogen bonds	5. Meselson and Stahl experiment
f. complementary bases	6. Griffith experiment
g. origin	7. structures at ends of eukaryotic chromosomes
h. Okazaki fragments	8. two nitrogenous bases that can pair via hydrogen bonds
i. purine	9. a nitrogenous base containing a single ring
j. topoisomerases	10. a short sequence of bases where unwinding of the double helix for replication begins
k. semiconservative replication	11. a virus that infects bacteria
l. lagging strand	12. short DNA fragments formed by discontinuous replication of one of the strands
m. telomeres	13. enzymes involved in DNA supercoiling

5-2 Griffith, in his 1928 experiments, demonstrated that bacterial strains could be genetically transformed. The evidence that DNA was the "transforming principle" responsible for this phenomenon came later. What was the key experiment that Avery, MacCleod, and McCarty performed to prove that DNA was responsible for the genetic change from rough cells into smooth cells?

5-3 During bacterial transformation, DNA that enters a cell is not an intact chromosome but consists of randomly generated fragments of chromosomal DNA. In a transformation where the donor DNA was from a bacterial strain that was $a^+b^+c^+$ and the recipient was $a^-b^-c^-$, 55% of the cells that became a^+ were also transformed to c^+, but only 2% of the a^+ cells were b^+. Is gene b or c closer to gene a?

5-4 Nitrogen and carbon are more abundant in proteins than sulfur. Why did Hershey and Chase use radioactive sulfur instead of nitrogen and carbon to label the protein portion of their bacteriophages in their experiments to determine whether parental protein or parental DNA is necessary for progeny phage production?

5-5 Imagine you have three test tubes containing identical solutions of purified, double-stranded human DNA. You expose the DNA in tube 1 to an agent that breaks the sugar-phosphate (phosphodiester) bonds. You expose the DNA in tube 2 to an agent that breaks the bonds that attach the bases to the sugars. You expose the DNA in tube 3 to an agent that breaks the hydrogen bonds. After treatment, how would the structures of the molecules in the three tubes differ?

5-6 What information about the structure of DNA was obtained from X-ray crystallographic data?

5-7 If 30% of the bases in human DNA are A, (a) what percentage are C? (b) what percentage are T? and (c) what percentage are G?

5-8 Which of the following statements are true about double-stranded DNA?
a. A + C = T + G
b. A + G = C + T
c. A + T = G + C
d. A/G = C/T
e. A/G = T/C
f. (A + G)/T = 1

5-9 DNA in cells is most often double-stranded but viruses contain either double-stranded or single-stranded DNA. How could you determine if the nucleic acid in a DNA virus is single-stranded or double-stranded?

5-10 When a double-stranded DNA molecule is exposed to high temperature, the two strands separate and the molecule loses its helical form. We say the DNA has been denatured. (Denaturation also occurs when DNA is exposed to acid or alkaline solutions.)
a. Regions of the DNA that contain many A–T base pairs are the first to become denatured as the temperature of a DNA solution is raised. Thinking about the chemical structure of the DNA molecule, why do you think the A–T-rich regions denature first?
b. If the temperature is lowered, the original DNA strands can reanneal or renature. In addition to the full double-stranded molecules, some molecules of the type shown below are seen when the molecules are examined under the electron microscope. How can you explain these structures?

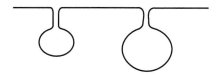

5-11 An RNA virus that infects plant cells is copied into a DNA molecule once it enters the plant cell. What would be the sequence of bases in the first strand of DNA made complementary to the section of viral RNA shown here?

5′ CCCUUGGAACUACAAAGCCGAGAUUAA 3′

5-12 The bacterial transformation and bacteriophage labeling experiments proved that DNA was the hereditary material in bacteria and in DNA containing viruses. Some viruses do not contain DNA but have RNA inside the phage particle. An example is the tobacco mosaic virus (TMV) that infects tobacco plants, causing lesions in the leaves. Two different variants of TMV exist that have different forms of a particular protein in the virus

particle that can be distinguished. It is possible to reconstitute TMV *in vitro* (in the test tube) by mixing purified proteins and RNA. The reconstituted virus can then be used to infect the host plant cells and produce a new generation of viruses. Design an experiment to show that RNA acts as the hereditary material in TMV.

5-13 The bases of one of the strands of DNA in a region where DNA replication begins are shown below. What is the sequence of the primer that is synthesized complementary to the bases in bold? (Indicate the 5′ and 3′ ends of the sequence.)

5′ AGGCCTCG**AATTCGTATA**GCTTTCAGAAA 3′

5-14 If you expose human tissue culture cells (e.g., HeLa cells) to ^{3}H-thymidine just as they enter S phase, then wash this material off the cells and let them go through a second S phase before looking at the chromosomes, how would you expect the ^{3}H to be distributed over a pair of homologous chromosomes? (Ignore the effect recombination could have on this outcome.) Would the radioactivity be in (a) one chromatid of one homolog, (b) both chromatids of one homolog, (c) one chromatid each of both homologs, (d) both chromatids of both homologs, or (e) some other pattern? Choose the correct answer and explain your reasoning. (This problem extends the analysis begun in solved problem III.)

5-15 When Meselson and Stahl grew *E. coli* in ^{15}N medium for many generations, and then transferred to ^{14}N medium for one generation, they found that the bacterial DNA banded at a density intermediate between that of pure ^{15}N DNA and pure ^{14}N DNA following equilibrium density centrifugation. When they allowed the bacteria to replicate one additional time in ^{14}N medium, they observed that half of the DNA remained at the intermediate density, while the other half banded at the density of pure ^{14}N DNA. What would they have seen after an additional generation of growth in ^{14}N medium? After two additional generations?

5-16 Replicating structures in DNA can be observed in the electron microscope. Regions being replicated appear as bubbles.

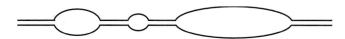

a. Assuming bidirectional replication, how may origins of replication are active in this DNA molecule?
b. How many replication forks are present?
c. Assuming that all replication forks move at the same speed, which origin of replication was activated last?

5-17 Draw a bidirectional replication fork and label the origin of replication, the leading strands, lagging strands, and the 5′ and 3′ ends of all strands shown in your diagram.

5-18 Diagram replication occurring at the end of a double-stranded linear chromosome. Show the leading and lagging strand with their primers. (Indicate the 5′ and 3′ ends of the strands.) What difficulty is encountered in producing copies of both DNA strands at the end of a chromosome?

5-19 Indicate the role of each of the following in DNA replication: (a) topoisomerase, (b) helicase, (c) primase, and (d) ligase.

5-20 Researchers have discovered that during replication of the circular DNA chromosome of the animal virus SV40, the two newly completed daughter double helices are intertwined. What would have to happen for the circles to come apart?

5-21 It has been shown in yeast that gene conversion occurs equally frequently with recombination of genetic markers flanking the region of gene conversion and without it. Why is this so?

5-22 Imagine that you have done a cross between two strains of yeast, one of which has the genotype *ABC* and the other *abc*, where the letters refer to three rather closely linked genes in the order given. You examine many tetrads resulting from this cross, and you find two that do not contain the expected two *B* and two *b* spores. In tetrad I, the spores are *ABC, ABC, aBc,* and *abc*. In tetrad II, the spores are *ABC, Abc, abC,* and *abc*. How have these unusual tetrads arisen?

5-23 What properties would you expect of a strain of *E. coli* that has a mutant allele (null or nonfunctional) of the *recA* gene? Explain.

5-24 From a cross between $e^+f^+g^+$ and $e^-f^-g^-$ strains of *Neurospora*, recombination between these linked genes resulted in a few octads containing the following ordered set of spores:

$$e^+f^+g^+$$
$$e^+f^+g^+$$
$$e^+f^-g^+$$
$$e^+f^-g^+$$
$$e^-f^-g^-$$
$$e^-f^-g^-$$
$$e^-f^-g^-$$
$$e^-f^-g^-$$

a. Where was recombination initiated?
b. Where did the resolving cut get made?
c. Why do you end up with 2:6 f^+:f^- but 4:4 e^+:e^-?

CHAPTER

6

ANATOMY AND FUNCTION OF A GENE:
DISSECTION THROUGH MUTATION

A scale played on a piano keyboard and a gene on a chromosome are both a series
of simple elements (keys or nucleotide pairs) whose use in a linear order produces
information. A wrong not or an altered nucleotide pair can call attention to the
structure and composition of the musical scale or the gene.

Human chromosome 3 consists of approximately 210 million base pairs and carries roughly 7000 genes. Somewhere on the long arm of the chromosome resides the gene for rhodopsin, a light-sensitive protein active in the rod cells of our retinas. Under the microscope, a lone chromosome 3 released from the nucleus of a human cell looks like a long, undifferentiated looping string of DNA (Fig. 6.1). At very high magnification—unfortunately higher than that obtainable with today's microscopes—this genetic twine would appear as a double helix composed of two antiparallel strands of nucleotides held together by the hydrogen bonds of complementary base pairs. But which nucleotides within the scramble make up the rhodopsin gene, and what does the gene look like? Knowledge that DNA is the molecule of heredity does not in itself answer the question "What is a gene?" From Mendel on, geneticists who delineated genes through breeding studies and pedigree analyses considered a gene to be a unit of information that affects phenotype; but the string of DNA in chromosome 3 does not resolve

into 7000 observable units. Detailed studies of mutations helped resolve this dilemma.

The rhodopsin gene determines perception of low-intensity light. People who carry the normal, wildtype allele of the gene see well in a dimly lit room and on the road at night. One simple change—a mutation—in the rhodopsin gene, however, diminishes light perception just enough to lead to night blindness; other alterations in the gene cause the destruction of rod cells, resulting in total blindness. Medical researchers have so far identified more than 30 mutations in the rhodopsin gene that affect vision. How can one gene sustain so many different mutations? And how can those mutations have such different phenotypic effects?

In this chapter, we describe how geneticists examined the relation between mutations, genes, and phenotype as they tried to understand, at the molecular level, what genes are and how they function. The researchers considered **mutations** to be heritable changes in base sequence that affect phenotype. From ingenious experiments engineered to reveal how different mutations can

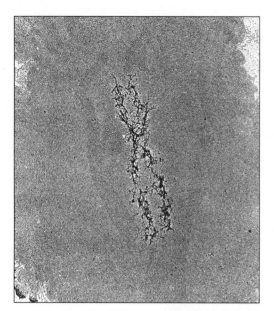

Figure 6.1 The DNA of each human chromosome contains thousands of genes. The DNA of this human chromosome has been spread out and magnified to 50,000 times its original size. Even at this level of resolution in the electron microscope, there are no topological signs that reveal where along the DNA the genes reside.

These findings provided fresh insight into gene function, clarifying that the nucleotide sequence of a gene specifies the amino acids that must be strung together to make a protein. The order of amino acids determines each protein's three-dimensional structure—how it folds in space—which in turn, determines how the protein functions in the cell.

One general theme emerges from our discussion. Knowledge of what genes are and how they work deepens our understanding of Mendelian genetics by providing a biochemical explanation for how genotype influences phenotype. One mutation in the rhodopsin gene, for example, causes the substitution of one particular amino acid for another in the construction of the rhodopsin protein, and this specific substitution changes the protein's ability to absorb photons and thus a person's ability to perceive light.

In our presentation of how genes govern the relation between mutation and phenotype, we examine

- What mutations are; how often they occur; some of the events that cause them; and how they affect the survival of an individual and the evolution of species.

- What mutations tell us about gene structure: A gene is a specific sequence of nucleotide pairs in a discrete region of DNA that acts as a unit of function, usually by encoding the instructions for making a particular protein.

- What mutations tell us about gene function: Genes encode proteins by directing the correct assembly of the amino acids that compose each polypeptide; one or more polypeptides composes a protein.

- How genotype correlates with phenotype: Phenotype depends on the structure and amount of protein produced; mutations that alter a gene's instructions for the corresponding sequence of amino acids alter protein structure and protein function, and as a result, phenotype.

occur in different parts of the same gene, and how various mutations in the same gene can have different phenotypic effects, the researchers inferred that physically, a **gene** is a specific segment of DNA composed of a distinctive set of nucleotide pairs (also referred to as base pairs) in a discrete region of a chromosome that encodes a particular protein. From these same studies, they concluded that a gene is not simply a bead on a string, changeable only as a whole and only in one way, as some had believed. Rather, genes are divisible, and each gene's subunits—the individual nucleotide pairs—can mutate independently and recombine with each other.

MUTATIONS: PRIMARY TOOLS OF GENETIC ANALYSIS

Mutations Are Heritable Changes in Base Sequences That Modify the Information Content of DNA

We saw in Chapter 2 that any allele existing at a frequency of more than 1% in a natural population under study is known as a **wildtype** allele. Many experimental geneticists, however, have a slightly different definition of a "wildtype allele," considering it to be the one allele that dictates the most commonly found phenotype in a population; with this definition, all other alleles—no matter what their frequency in the population—would be mutant alleles. Under either definition, a mutation that changes the wildtype allele of a gene to a different allele is called a **forward mutation.** The resulting novel mutant al-

lele can be either recessive or dominant to the original wildtype. Geneticists diagram forward mutations as $A^+ \rightarrow a$ when the mutation is recessive, and as $b^+ \rightarrow B$ when the mutation is dominant. Mutations can also cause a novel mutant allele to revert back to wildtype ($a \rightarrow A^+$, or $B \rightarrow b^+$) in a process known as **reverse mutation,** or **reversion.** In this chapter, we designate most wildtype alleles, whether recessive or dominant, with a "+."

Mendel originally defined genes by the visible phenotypic effects—green or yellow, round or wrinkled—of their alternative alleles. In fact, the only way he knew that genes existed at all was because alternative alleles for seven particular pea genes had arisen through forward mutations in the genetic material. Close to a century later, knowledge of DNA structure clarified that such *mutations* are heritable changes in DNA base sequence. DNA thus carries the potential for genetic change in the same place it carries genetic information—in the sequence of its bases.

One Way Geneticists Classify Mutations Is by Their Effect on the DNA Molecule

A **substitution** occurs when a base at a certain position in one strand of the DNA molecule is replaced by one of the other three bases (Fig. 6.2a); during DNA replication, a base substitution in one strand will cause a new base pair to appear in the daughter molecule generated from that strand. Substitutions

can be subdivided into *transitions,* in which one purine (A or G) replaces the other purine, or one pyrimidine (C or T) replaces the other, and *transversions,* in which a purine changes to a pyrimidine or vice versa. Other types of mutations produce more complicated rearrangements of DNA sequence. A **deletion** occurs when a block of one or more nucleotide pairs is lost from a DNA molecule; an **insertion** is just the reverse—the addition of one or more nucleotide pairs (Fig. 6.2b and c). Deletions and insertions can be as small as a single base pair or as large as megabases (that is, millions of base pairs). Researchers can see the larger changes under the microscope, in the observation of karyotypes. Even more complex mutations include **inversions,** 180° rotations of a segment of the DNA molecule (Fig. 6.2d), and **translocations,** in which parts of two nonhomologous chromosomes change places (Fig. 6.2e). Large-scale DNA rearrangements, including megabase deletions and insertions as well as inversions and translocations, cause major genetic reorganizations that change either the order of genes along a chromosome or the number of chromosomes in an organism. We discuss these **chromosomal rearrangements,** which affect many genes at a time, in Chapter 12. In this chapter, we focus on mutations that alter only one gene at a time.

Only a small fraction of the mutations in a genome actually alter the nucleotide sequences of genes in a way that affects gene function. By changing one allele to another, these mutations modify a gene's protein product, and the modified product influences phenotype. All other mutations either alter genes in a way that does not affect their function, or change the DNA between genes. We discuss mutations without observable phenotypic consequences in Chapters 7 and 8. In the remainder of this chapter, we focus on those mutations that have an impact on gene function and thereby influence phenotype.

Spontaneous Mutations Affecting Genes Occur at a Very Low Rate

Mutations that modify gene function happen so infrequently that geneticists must examine a very large number of individuals from a formerly homogeneous population to detect the new phenotypes that reflect these mutations. In one ongoing study, dedicated investigators have monitored the coat colors of millions of specially bred mice and discovered that on average, a single gene mutates to a recessive allele at the rate of 1×10^{-5} per gene per generation (Fig. 6.3). In the study, roughly 11 out of every 1 million gametes produced in a generation have a mutation in a given gene. Studies of several other organisms have yielded similar results: an average spontaneous rate of 2 to 12×10^{-6} mutations per gene per generation.

Looking at the mutation rate from a different perspective, you could ask how many mutations there might be in the genes of an individual. Since the human genome, for example, contains approximately 100,000 genes, you could extrapolate from 2 to 12×10^{-6} mutations per gene to a rate of 2 to 12 mutations per million genes. This very rough calculation would mean that, on average, 1 mutation affecting phenotype could arise in every 1 to 5 human gametes.

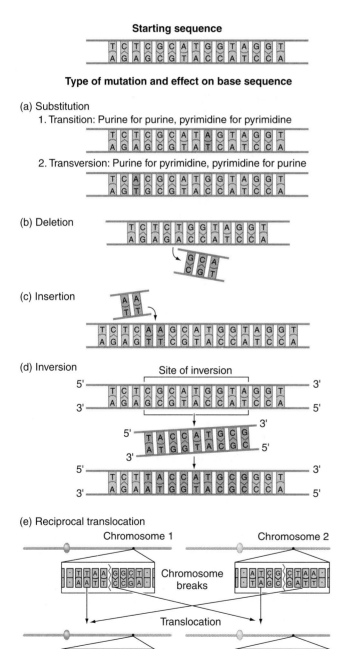

Figure 6.2   Mutations classified by their effect on DNA.

Locus[a]	Number of gametes tested	Number of mutations	Mutation rate ($\times 10^{-6}$)
a^- (albino)	67,395	3	44.5
b^- (brown)	919,699	3	3.3
c^- (nonagouti)	150,391	5	33.2
d^- (dilute)	839,447	10	11.9
ln^- (leaden)	243,444	4	16.4
	2,220,376	25	11.2 (average)

[a] Mutation is from wildtype to the recessive allele shown.

Figure 6.3   Rates of spontaneous mutation. (a) Mutant mouse coat colors: albino (*left*), brown (*right*). (b) Mutation rates from wildtype to recessive alleles of five coat-color genes. Geneticists took animals from highly inbred strains with wildtype coat colors and mated them with mice homozygous for recessive alleles of the five coat-color genes. The appearance of progeny with mutant coat colors indicated the presence of recessive mutations in gametes produced by the inbred mice.

The Mutation Rate Varies from Gene to Gene

Experiments with mice and other organisms show that mutation rates range from 10^{-8} to 10^{-4} per gene per generation. Such marked variation in rate reflects different mutation rates in different organisms, differences in gene size (larger genes are larger targets that sustain more mutations), and differences in the susceptibility to mutation of different gene functions.

Forward Mutations Occur More Often than Reverse Mutations

In the mouse coat-color study, when researchers allowed brother and sister mice homozygous for a recessive mutant allele of one of the five mutant coat-color genes to mate with each other, they could estimate the rate of reversion by examining the F₁ offspring. Any progeny expressing the dominant wildtype phenotype for a particular coat color, of necessity, carried a gene that had sustained a reverse mutation. Calculations based on observations of several million F₁ progeny revealed a reverse mutation rate ranging from 0 to 2.5×10^{-6} per gene per generation; the rate of reversion varied somewhat from gene to gene. In this study, then, the rate of reversion was significantly lower than the rate of forward mutation, most likely because there are many ways to disrupt gene function, but only a few ways to restore function once it has been dis-

rupted. Other studies have shown that the reversion rate may be more than 1000-fold lower than the rate of forward mutation, depending on the type of mutation; the reversion rate of certain kinds of insertions known as transposable elements, for example, is quite high, whereas the reversion rate of deletions is usually 0.

Although all of these estimates of mutation rates are extremely rough, they nonetheless support three general conclusions: (1) Mutations affecting phenotype occur very rarely, (2) different genes mutate at different rates, and (3) the rate of forward mutation (usually a consequence of a disruption of gene function) is higher than the rate of reversion (which requires a restoration of function to a previously disrupted gene).

Mutations Arise from Many Kinds of Random Events

Because spontaneous mutations affecting a gene occur so infrequently, it is very difficult to study the events that produce them. To overcome this problem, researchers turned to bacteria as the experimental organisms of choice. With these single-celled microbes, it is easy to grow many millions of individuals and then search rapidly through enormous populations to find the few that carry a novel mutation. In one study, investigators spread wildtype bacteria on the surface of agar containing sufficient nutrients for growth as well as a large amount of a bacteria-killing substance, such as an antibiotic or a bacteriophage. While most of the bacterial cells died, a few showed resistance to the bactericidal substance and continued to grow and divide. The descendants of a single resistant bacterium, produced by many rounds of binary fission, formed a mound of genetically identical cells called a **colony.**

The few bactericide-resistant colonies that appeared presented a puzzle. Had the cells in the colonies somehow altered their internal biochemistry to produce a life-saving response to the antibiotic or bacteriophage? Or did they carry heritable mutations conferring resistance to the bactericide? And if they did carry mutations, did those mutations arise by chance from random spontaneous events that take place continuously, even in the absence of a bactericidal substance, or did they only arise in response to environmental signals (in this case, the addition of the bactericide) that elicit a compensating change in genetic information?

Mutations Are Chance Occurrences Modifying the Genome at Random

In 1943 Salvador Luria and Max Delbrück devised an experiment to examine the origin of bacterial resistance (Fig. 6.4). According to their reasoning, if bacteriophage-resistant colonies arise in direct response to infection by bacteriophages, separate suspensions of bacteria containing equal numbers of cells will generate similar, small numbers of resistant colonies when spread in separate petri plates on nutrient agar suffused with phages. By contrast, if resistance arises from mutations that occur spontaneously even when the phages are not present, different liquid cultures, when spread on separate petri plates, will generate very different numbers

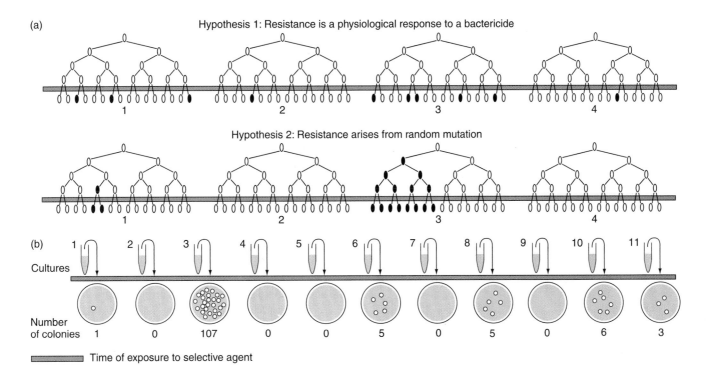

Figure 6.4 The Luria-Delbrück fluctuation experiment. (a) Predictions of two hypotheses. Hypothesis 1: If resistance arises only after exposure to a bactericide, all bacterial cultures of equal size should produce roughly the same number of resistant colonies. Hypothesis 2: If random mutations conferring resistance arise before exposure to bactericide, the number of resistant colonies in different cultures should vary (fluctuate) widely, reflecting when during the growth of the culture the random mutations occurred. There will be some cultures with a large number of resistant colonies, but they will be rare because to produce many resistant colonies, a mutation must occur early in the growth of a culture, when there are far fewer cells than later. (b) The results of a Luria-Delbrück experiment: Large fluctuations support the idea that mutations in bacteria occur as spontaneous mistakes independent of exposure to a selective agent.

of resistant colonies. This is because the mutation conferring resistance can, in theory, arise at any time during the growth of the culture; if it happens early, the cell in which it occurs will produce many mutant progeny prior to petri plating; if it happens later, there will be far fewer mutant progeny when the time for plating arrives. After plating, these numerical differences will show up as fluctuations in the numbers of resistant colonies growing in the different petri plates. The results of this **fluctuation test** were clear: Most plates supported zero to a few resistant colonies, but a few harbored hundreds of resistant colonies. From this observation of a substantial fluctuation in the number of resistant colonies in different petri plates, Luria and Delbrück concluded that bacterial resistance arises from mutations that exist before exposure to bacteriophage. After exposure, the bactericide in the petri plate becomes a selective agent that kills off nonresistant cells, allowing only the preexisting resistant ones to survive. Figure 6.5 illustrates how researchers used another technique, known as *replica plating*, to demonstrate even more directly that the mutations conferring bacterial resistance occur before the cells "see" the bactericide that selects for their resistance.

These key experiments showed that bacterial resistance to phages and other bactericides is the result of mutations. And these mutations do not arise in particular genes as a directed response to environmental change; instead they occur sponta-

neously as a result of random processes that can happen at any time and hit the genome at any place. Once such random changes occur, however, they usually remain stable. If the resistant mutants of the Luria-Delbrück experiment, for example, were grown for many generations in medium that did not contain bacteriophages, they would nevertheless remain resistant to this bactericidal virus.

We now describe some of the many kinds of random events that cause mutations, and how cells cope with the damage.

Hydrolysis, Radiation, Ultraviolet Light, and Oxidation Can Alter the Information Stored in DNA

Chemical and physical assaults on DNA are quite frequent. Geneticists estimate, for example, that the hydrolysis of a purine base, A or G, from the deoxyribose-phosphate backbone occurs 1000 times an hour in every human cell. This kind of DNA alteration is called **depurination** (Fig. 6.6a). Because the resulting *apurinic site* cannot specify a complementary base, the DNA replication process sometimes introduces a random base opposite the apurinic site, causing a mutation in the newly synthesized complementary strand three-quarters of the time. Another naturally occurring process that may modify DNA's information content is **deamination:** the removal of an amino ($-NH_2$) group. Deamination can change cytosine to uracil (U), the nitrogenous base found in RNA but not in DNA,

(a)

1. Invert master plate; pressing against velvet surface leaves an imprint of colonies. Save plate.

2. Invert second plate (replica plate); pressing against velvet surface picks up colony imprint.

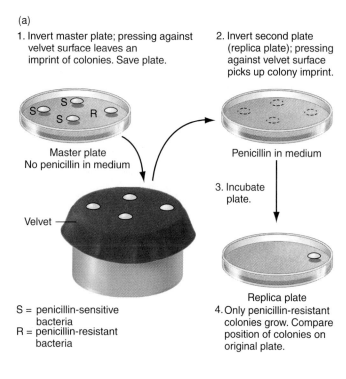

Master plate
No penicillin in medium

Penicillin in medium

3. Incubate plate.

Velvet

Replica plate

S = penicillin-sensitive bacteria
R = penicillin-resistant bacteria

4. Only penicillin-resistant colonies grow. Compare position of colonies on original plate.

(b)

10^7 colonies of penicillin-sensitive bacteria

Make three replica plates. Incubate to allow penicillin-resistant colonies to grow.

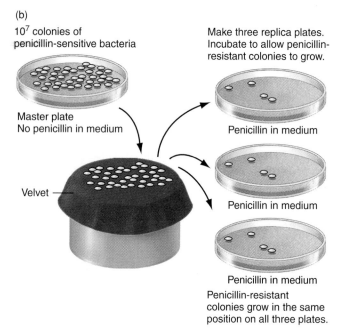

Master plate
No penicillin in medium

Penicillin in medium

Velvet

Penicillin in medium

Penicillin in medium

Penicillin-resistant colonies grow in the same position on all three plates.

Figure 6.5 Replica plating verifies that bacterial resistance is the result of preexisting mutations. (a) The replica plating technique. Pressing a *master* plate containing bacterial colonies onto a velvet surface transfers some of the cells in each colony onto the velvet. Pressing another, so-called *replica* plate onto the velvet surface then transfers some of the cells in each colony from the velvet onto the replica plate. By keeping track of the colony locations on the two plates, investigators can determine which colonies on the master plate are able to grow on the replica plate (in this example, only the penicillin-resistant colonies). (b) Using replica plating to explore the nature of bacterial resistance. A large number of colonies on a master plate without penicillin are sequentially transferred to three replica plates with penicillin. Resistant colonies grow in the same positions on all three replicas, showing that some colonies on the master plates had multiple resistant cells before exposure to antibiotic.

and because U pairs with A rather than G, deamination followed by replication may alter a C–G base pair to a T–A pair in future generations of DNA molecules (Fig. 6.6b); such a C–G-to-T–A change is a transition mutation. Other assaults include naturally occurring radiation such as cosmic rays and X rays, which breaks the sugar-phosphate backbone (Fig. 6.6c); ultraviolet light, which causes adjacent thymine residues to become chemically linked into thymine–thymine dimers (Fig. 6.6d); and oxidative damage to any of the four bases (Fig. 6.6e). All of these changes alter the information content of the DNA molecule.

Cells have evolved many mechanisms for dealing with these problems. In one, specialized proteins patrol the DNA for irregularities, detecting and repairing chemical changes that disrupt the normal Watson-Crick base pairing. When they find such a change, they cut on either side of the damage and remove the intervening bases; DNA polymerase then fills in the gap by complementary base pairing with the information in the undamaged strand (Fig. 6.7a). This repair mechanism is known as **excision repair** since it excises the damage. Excision repair utilizes the redundancy of information in the double helix as the basis for filling in the excised information. The significance of this correction device is evident from the genetic disease *xeroderma pigmentosum*. People with the disease lack the enzymatic ability to recognize and excise thymine–thymine dimers produced by ultraviolet light. Unless they avoid all exposure to sunlight, their inability to correct damaged DNA results in a whole series of mutations that eventually leads to skin cancer (Fig. 6.7b). Even in people with normal genes for excision enzymes, excision repair, though doggedly persistent, is not 100% effective. As a result, a small number of the changes caused by chemical assaults, ultraviolet light, and radiation slip through to become established mutations transmitted from one generation to the next.

Mistakes during DNA Replication Can Also Alter Genetic Information

If the cellular machinery for some reason incorporates the wrong base during replication, for instance, a C opposite an A instead of the expected T, then during the next replication cycle, one of the daughter DNAs will have the normal A–T base pair while the other will have a mutant G–C. Careful measurements of the fidelity of replication *in vivo*, in both bacteria and human cells, show that such errors are exceedingly rare, occurring less than once in every 10^9 base pairs. That is equivalent to typing this entire book 1000 times while making only one typing error. Considering the complexities of helix unwinding, base pairing, and polymerization, this level of accuracy is amazing. How do cells achieve it?

The replication machinery minimizes errors through successive stages of correction. In the test tube, DNA polymerases replicate DNA with an error rate of about 1 mistake in every 10^6 bases copied. This rate is about 1000-fold worse than that achieved by the cell. Even so, it is impressively low and is only attained because polymerase molecules provide, along with their polymerization function, a proofreading/

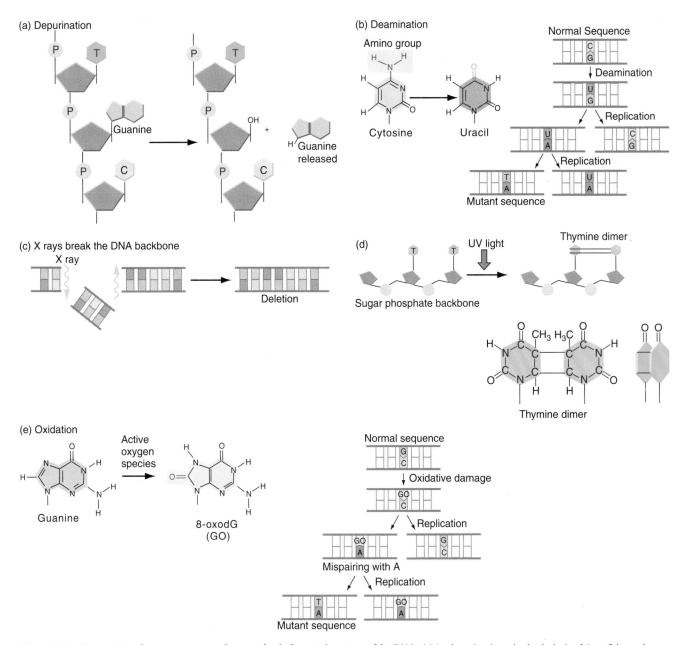

Figure 6.6 How natural processes can change the information stored in DNA. (a) In depurination, the hydrolysis of A or G bases leaves a DNA strand with a continuous sugar-phosphate backbone but an unspecified base at one nucleotide position. (b) In deamination, the removal of an amino group from cytosine initiates a process that eventually (after DNA replication) causes a transition—the change of a C–G base pair to a T–A base pair. (c) X rays have sufficient energy to break the sugar-phosphate backbone, splitting a DNA molecule into smaller pieces. If these pieces are not spliced back together properly, mutations such as the deletion shown here arise. (d) One effect of ultraviolet (UV) radiation is to cause the dimerization of adjacent thymine bases. If not repaired, a dimer can disrupt the readout of genetic information. (e) Irradiation can cause the formation of *free radicals* (such as oxygen molecules with an unpaired electron) that can affect the chemistry of individual bases. In the scenario depicted here, the pairing of the altered base GO with A creates a transversion by changing a G–C base pair into T–A.

editing function in the form of a nuclease portion that is activated whenever the polymerase makes a mistake. This nuclease portion of the polymerase molecule recognizes a mispaired base and excises it, allowing the polymerase to try again. Without its nuclease portion, DNA polymerase would have an error rate of 1 mistake in every 10^4 bases copies, so its editing function improves the fidelity of replication 100-fold. *In vivo,* DNA polymerase is part of a replication system in-

cluding many other proteins that collectively improve on the error rate another 10-fold, bringing it to within about 100-fold of the fidelity attained by the cell.

The 100-fold higher accuracy of the cell depends on a backup system that notices and corrects residual errors in the newly replicated DNA. Active only *after* polymerization, the backup system has a difficult problem to solve. Suppose that an G–C pair were copied to produce two daughter molecules,

(a)

1. Exposure to UV light

2. Dimer forms

3. Enzyme nicks strand containing dimer

4. Fragment is released from DNA

5. DNA polymerase fills in the gap

6. Ligase seals repaired strand

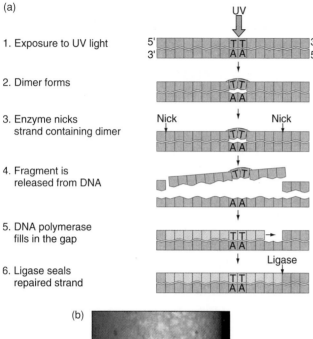

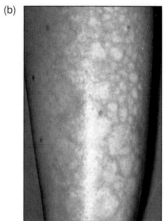

(b)

Figure 6.7 Excision repair removes damaged DNA and fills in the gap. (a) If DNA damage such as formation of a thymine dimer occurs, excision repair enzymes cut the affected strand, releasing the damaged region. DNA polymerases and DNA ligase then resynthesize the missing information and reseal the now-corrected strand. (b) Skin lesions in a patient with xeroderma pigmentosum, who lacks a critical excision repair enzyme.

(a) Parental strands are marked with methyl groups.

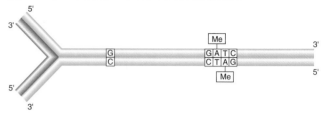

(b) Enzyme system recognizes mismatch in replicated DNA.

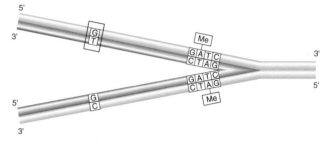

(c) DNA on unmarked new strand is excised.

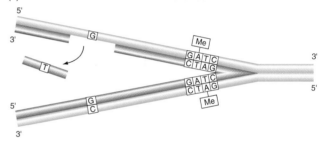

(d) Repair and methylation of newly synthesized DNA strand.

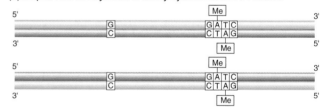

Figure 6.8 Methyl-directed mismatch repair corrects mistakes in replication. (a) The sequence GATC is always marked with a methyl group on parental strands, but the methyl groups are not immediately added to newly synthesized strands. If a substitution occurs during replication, producing a mismatched base pair (b), the enzymes active in methyl-directed mismatch repair will remove nucleotides from the mismatched region of the unmarked, newly synthesized strand (c). Just as they do in excision repair (see Fig. 6.7a), DNA polymerase and DNA ligase restore the missing information and reseal the corrected strand (d).

one of which has the correct G–C base pair, while the other has an incorrect G–T. The backup system can recognize the incorrectly matched G–T base pair because the improper base pairing distorts the double helix. But how does the system know whether to correct the pair to a G–C or an A–T? Bacteria solve this problem by placing a distinguishing mark on the parental DNA strands; everywhere the sequence GATC occurs, an enzyme system puts a methyl group on the A (Fig. 6.8a). Shortly after replication, the old template strand bears the methyl mark, while the new daughter strand is as yet unmarked. Whenever the enzymatic system detects an improper pair, it excises a few nucleotides around the mispaired nucleotides from the new, unmethylated strand and resynthesizes the information using the old, methylated strand as a template (Fig.

6.8b–d). Later, with the completion of replication and repair, enzymes mark the new strand as well, so that its parental origin will be evident in the next round of replication. This repair system is known as **methyl-directed mismatch repair.** Like excision repair, it uses the redundancy of DNA's double strands as the basis for restoring information. Unlike excision repair, which recognizes and corrects irregularities caused by chemical or physical assaults, methyl-directed mismatch repair recognizes base pairs mismatched during replication. Such mismatches show up as bulges and hollows in the newly replicated double-stranded DNA. Eukaryotic cells also have a mismatch correction system, but it is not yet known how they distinguish template from newly replicated strands. Researchers have nevertheless discovered that mismatch repair in humans has medical repercussions. Defects in mismatch repair enzymes are associated with colon and other forms of cancer.

Unequal Crossing-Over and Transposons Cause DNA Changes That Are Not Susceptible to Excision or Mismatch Repair

Some mutations arise from events other than chemical and physical assaults or replication errors. Significant among the mechanisms giving rise to such mutations is erroneous recombination. For example, in **unequal crossing-over,** which occurs during meiosis, one homologous chromosome ends up with a duplication (a kind of insertion), while the other homolog sustains a deletion. As Fig. 6.9a shows, some forms of red-green colorblindness arise from deletions and duplications in the genes that enable us to perceive red and green wavelengths of light; these reciprocal informational changes are the result of unequal crossing-over. Another notable mechanism for altering DNA sequence involves whole units of DNA

known as **transposons.** Transposons move from place to place within the genome, sometimes causing a change in DNA function when they insert themselves in a new chromosomal location (Fig. 6.9b). Chapter 12 discusses the genetic consequences of transposon behavior. Finally, researchers do not yet understand the biochemical basis of some naturally occurring DNA sequence changes, among them a type of mutation that is at the root of one of the most common forms of mental retardation in humans as well as many other disorders of the nervous system (see the Genetics and Society box "A New Class of Human Mutation: Amplified Repeats with Medical Consequences").

Mutagens Induce Mutations

Mutations make genetic analysis possible, but appear spontaneously at such a low rate that researchers have looked for controlled ways to increase their occurrence. H. J. Muller, an original member of T. H. Morgan's *Drosophila* group, first showed that exposure to a dose of X rays higher than the naturally occurring level increases the mutation rate in fruit flies (Fig. 6.10). He exposed male *Drosophila* to large doses of X rays and then mated these males with females that had one X chromosome containing an easy-to-recognize dominant mutation causing Bar eyes. This X chromosome (called a *balancer*) also carried chromosomal rearrangements known as inversions that prevented it from crossing-over with the other X chromosome (Chapter 12 explains the details of this

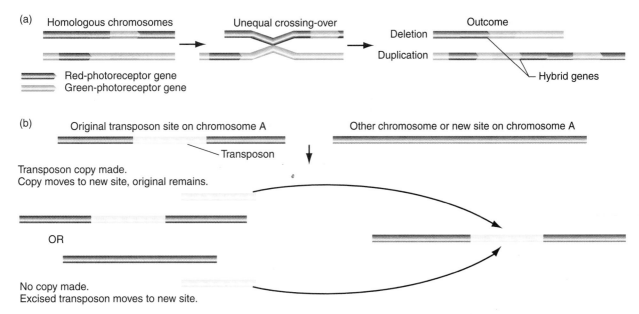

Figure 6.9 How unequal crossing-over and transposon movement change the information content of DNA. (a) If two nearby regions on the same chromosome contain a similar DNA sequence, the sequences on two homologous chromosomes can pair out of register during meiosis, to produce gametes that have either a deletion or a reciprocal duplication of the intervening region. On the human X chromosome, unequal crossing-over between the nearby and highly similar genes for red and green photoreceptors can delete one of the photoreceptor genes or make novel hybrid genes. (b) Transposable genetic elements (transposons) can move around the genome. Some kinds of transposons copy themselves before moving, while others are excised from their original positions during the transposition process. In either case, insertion of the transposon in a new position changes DNA's information content; phenotypic consequences result if the transposon becomes inserted into or nearby a gene.

Genetics and Society

A New Class of Human Mutation: Amplified Repeats with Medical Consequences

In 1992 a group of molecular geneticists discovered an unusual type of mutation: the excessive amplification of a base triplet normally repeated only a few to 50 times in succession. If, for example, a normal allele of a gene carries 5 consecutive repetitions of the base triplet CGG, an abnormal allele resulting from mutation could carry 200 repeats in a row.

Genetic alterations of this kind, the researchers showed, cause a heritable disorder known as *fragile X syndrome.* Adults affected by this syndrome manifest several physical anomalies, including an unusually large head, long face, large ears, and in men, large testicles. They also exhibit moderate to severe mental retardation. Fragile X syndrome has been found in men and women of all races and ethnic backgrounds. The fragile X mutation is, in fact, a leading genetic cause of mental retardation worldwide, second only to the trisomy 21 that results in Down syndrome.

Specially prepared karyotypes of cells from people with fragile X symptoms reveal a slightly constricted, so-called fragile site near the tip of the long arm of the X chromosome (Fig. A). In some pictures of these karyotypes, the X has actually broken at this site, releasing a small piece containing the end of the chromosome (not shown). The fragile X mutation apparently produces a localized constricted region that breaks easily. Geneticists named the fragile X disorder for this specific pinpoint of fragility more than 20 years before they identified the mutation that gives rise to it.

The gene in which the fragile X mutation occurs is called *FMR-1* (for fragile-X-associated <u>m</u>ental <u>r</u>etardation). Near one end of the gene, different people carry a different number of repeats of the sequence CGG. Normal alleles contain 5–54 of these triplet repeats, while the *FMR-1* gene in people with fragile X syndrome contains 200–4000 repeats of the exact same triplet (Fig. B.1 and 2). The rest of the gene's base sequence is the same in both normal and abnormal alleles. The precise mechanism of triplet repeat amplification has not yet been determined. One possibility is that GC-rich regions may be hard to replicate, forcing the copying machinery to slip off, then hop back on, slip off, then hop back on when it gets to this sequence. Such stopping and starting may produce a replication stutter that causes synthesis of the same triplet repeat over and over again.

In addition to its unusual variation in copy number, the triplet repeat mutation that underlies fragile X syndrome has a surprising transmission feature. Alleles with a full-blown mutation are foreshadowed by *premutation alleles* that carry an intermediate number of repeats—more than 50 but fewer than 200 (Fig. B.2). Premutation alleles do not themselves generate fragile X symptoms in most carriers, but they show significant instability and thus forecast the risk of genetic disease in a carrier's progeny. The greater the num-

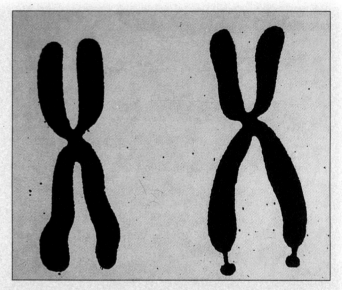

Figure A   A karyotype reveals a fragile X chromosome. The fragile X site is seen on the bottom of both chromatids of the X chromosome at the right.

ber of repeats in a premutation allele, the higher the risk of disease in that person's children. For example, if a woman carries a premutation allele with 60 CGG repeats, 17% of her offspring run the risk of exhibiting fragile X syndrome. If she carries a premutation allele with 90 repeats, close to 50% of her offspring will show symptoms. In short, the change that produces a premutation allele increases the likelihood that the *FMR-1* gene will incur more mutations; and the larger the original number of additional CGG repeats, the larger the number of subsequent additions. Interestingly, the expansion of *FMR-1* premutation alleles has some as-yet-unexplained relation to the parental origin of the repeats. Whereas most male carriers transmit their *FMR-1* allele with only a small change in the number of repeats, many women with premutation alleles bear children with 250–4000 CGG repeats in their *FMR-1* gene (Fig. B.3). One possible explanation is that whatever conditions generate fragile X mutations occur most readily during oogenesis.

This same transmission pattern does not hold for all triple repeat mutations. In 1993, geneticists found that the mutations causing Huntington disease consist of an expanded CAG trinucleotide repeat within the Huntington chorea gene; but in this case, the amplification occurs most often when the gene is transmitted by the father. The mutations resulting in Huntington disease may thus occur during spermatogenesis. Several other genetic disorders of the nervous system also arise from fluctuating numbers of triplet re-

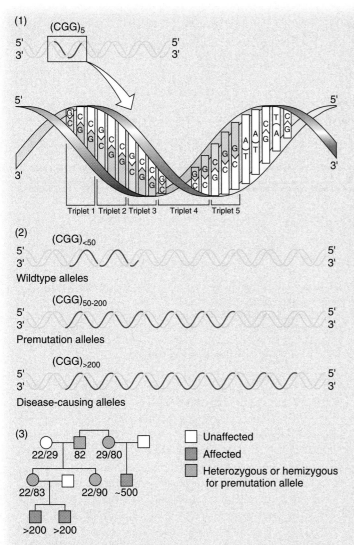

Figure B   Amplification of the triplet repeat CGG correlates with the fragile X syndrome. (1) Region of an X chromosome containing a normal *FMR-1* gene with 5 repeats of the CGG sequence on one strand. (2) *FMR-1* genes in unaffected people generally have fewer than 50 repeats; unstable premutation alleles of the gene have between 50 and 200 copies of the repeat. Full-blown disease-causing alleles have more than 200 CGG repeats; some mutant alleles have more than 4000. (3) A fragile X pedigree showing the number of CGC repeats in different individuals. Note that individuals affected by fragile X syndrome are almost always the progeny of mothers who carried premutation alleles.

peats. Why these repeats are so unstable and prone to expansion (and occasionally shrinkage) is an area of active research.

Geneticists used a mix of Mendelian and modern techniques to show that fragile X syndrome is a genetic disease and to confirm that the number of CGG repeats in the *FMR-1* gene is its ultimate cause. First, from pedigrees of families with heritable mental retardation, they learned that one type of retardation is an X-linked

disorder showing incomplete penetrance. Like other X-linked conditions, this disorder, now known to be fragile X syndrome, affects more men than women. Starting in 1969, special karyotypes—prepared from cells grown in cultures deprived of folic acid and thymidine (routinely prepared karyotypes do not detect the fragile X)—located the fragile site as a constriction near the tip of the long arm of the X chromosome. Later, mapping by linkage analysis showed that the disease gene maps to the fragile X site. Finally, molecular techniques that directly detect DNA sequence enabled researchers to find and count the exact number of triplet repeats in normal, premutation, and mutation alleles. In further studies, restriction enzymes, including *Eco*RI, cut the DNA containing the fragile X site into fragments of different lengths; those from normal people were shorter than those from people whose family had a history of retardation.

Knowledge of the genetic cause of fragile X syndrome combined with methods for quantifying the exact number of CGG repeats in an allele has interesting implications for genetic counseling. Because it is a triplet repeat mutation in the *FMR-1* gene that causes fragile X syndrome and different numbers of repeats generate different alleles, the disease phenotype is, in effect, a quantitative trait determined by a single gene with multiple alleles. As the smallest normal allele isolated to date has 5 triplet repeats and the largest abnormal allele so far isolated has roughly 4000 repeats, there are nearly 4000 potential *FMR-1* alleles. This enormous number of alleles gives rise to a slow gradation of normal-to-abnormal phenotypes, with the abnormal phenotypes progressing from mild retardation without other evidence of disease to severe retardation combined with physical anomalies. The relationship between genotype and phenotype is clear at both extremes of the triplet repeat spectrum: Up to 55 repeats produces a normal phenotype; more than 200 repeats gives rise to moderate to severe mental retardation. With an intermediate number of repeats, however, expression of the mental retardation phenotype is highly variable, depending to an unknown degree on chance, the environment, and modifier genes.

This range of variable expressivity leads to an ethical dilemma: Where should medical geneticists draw the line in their assessment of risk? Prospective parents with a history of mental retardation on one or both sides of the family may consult with a counselor to determine their options. The counselor might begin by testing one or both parents for premutation alleles to confirm that the cause of the mental retardation is indeed fragile X syndrome. But since premutation alleles do not provide an accurate basis for a forecast, it is also likely that the counselor will want to analyze the fetal cells directly. With amniocentesis, he or she can determine how many CGG repeats a fetus carries in its *FMR-1* gene. In the end, the parents' decision of whether or not to continue the pregnancy may depend in part on the counselor's overall evaluation of risk. Because of this, genetic counseling for fragile X syndrome needs to acknowledge that the middle range of triplet repeats is a gray area in which prediction can be only imperfect.

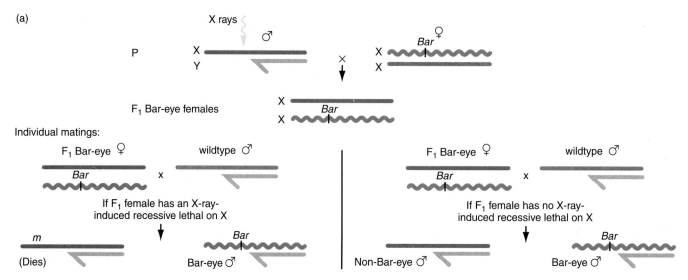

Figure 6.10 Exposure to X rays increases the mutation rate in *Drosophila*. Protocol designed by H. J. Muller. F₁ females are constructed that have an irradiated X chromosome from their father (red line), and a *Bar*-marked "balancer" X chromosome from their mother (wavy blue line). These chromosomes cannot recombine with each other during meiosis because the balancer chromosome has multiple inversions (explained in Chapter 12). If the chromosome from the father has an X-ray-induced recessive lethal mutation (*m*), the only male progeny the F₁ female can produce will be Bar-eyed. If this X chromosome has no such mutation, the F₁ female will produce both Bar-eyed and non-Bar-eyed sons. Muller found that the frequency at which recessive lethal mutations are included on the X chromosome increases linearly with X-ray dosage.

phenomenon). Some of the F₁ daughters of this mating were heterozygotes carrying a mutagenized X from their father and a *Bar*-marked X from their mother. If X rays induced a recessive lethal mutation anywhere on the paternally derived X chromosome, then these F₁ females would be unable to produce non-Bar-eyed sons. Thus simply by noting the presence or absence of non-Bar-eyed sons, Muller could establish whether a mutation had occurred in any of the more than 1000 genes on the X chromosome that are essential to *Drosophila* viability. He concluded that an increase in X-ray dose correlates linearly with an increase in the frequency of recessive lethal mutations (Fig. 6.10b).

Any physical or chemical agent that raises the frequency of mutations above the spontaneous rate is called a **mutagen.** Researchers use many different mutagens to produce mutations for study. With the Watson-Crick model of DNA structure as a guide, they can understand the action of mutagens at the molecular level. The X rays used by Muller to induce mutations on the X chromosome, for example, can break the sugar-phosphate backbones of DNA strands, sometimes at the same position on the two strands of the double helix. Multiple double-strand breaks produce DNA fragmentation, and the improper "stitching back together" of the fragments can cause inversions, deletions, or other rearrangements (see Fig. 6.6c). Another example of the molecular mechanism of mutagenesis involves mutagens known as **base analogs,** which are so similar in chemical structure to the normal nitrogenous bases that the replication machinery can incorporate them into DNA in place of the normal bases (Fig. 6.11a). Since a base analog may have pairing properties different from those of the base it replaces, it can cause base substitutions on the complementary

strand synthesized in the next round of DNA replication. Other chemical mutagens generate substitutions by directly altering a base's chemical structure and properties (Fig. 6.11b). Again, the effects of these changes become fixed in the genome when the altered base causes incorporation of an incorrect complementary base during a subsequent round of replication. Yet another class of chemical mutagens consists of compounds known as **intercalators:** flat, planar molecules that can sandwich themselves between successive base pairs and disrupt the machinery for replication, recombination, or repair (Fig. 6.11c). The disruption may eventually generate deletions or insertions of a single base pair.

Thus, even though the cells of *Drosophila,* mice, humans, and all other organisms have mechanisms that keep mutations to a minimum, allowing them to achieve a maximum of fidelity in the use and transmission of their genes, some changes in DNA information content do occur. The very low level of spontaneous mutation is the best most cells can achieve under normal conditions, and exposure to mutagens raises that rate.

Impact: Mutations Have Consequences for the Evolution of Species and the Survival of Organisms

"The capacity to blunder slightly is the real marvel of DNA. Without this special attribute, we would still be anaerobic bacteria and there would be no music." In these two sentences, the eminent medical scientist and self-appointed "biology watcher" Lewis Thomas acknowledges that changes in DNA are behind the phenotypic variations that are the raw material on which natural selection has acted for billions of years to drive evolution. The wide-ranging variation in the genetic makeup of the human population—and other populations as well—is, in fact, the result of a balance between the continuous introduction of new mutations, the loss of deleterious mutations because of the selective disadvantage they impose on the individuals that carry them, and the increase in frequency of rare mutations that provide a selective advantage to the individuals carrying them (or that spread through a population by other means; see Chapter 23 for the details of population genetics).

In sexually reproducing multicellular organisms, only germline mutations that can be passed on to the next generation play a role in the evolution of a species. Nevertheless, even mutations in somatic cells can have an impact on the well-being and survival of individuals. Somatic mutations in genes that help regulate the cell cycle may, for example, lead to cancer. The U.S. Food and Drug Administration tries to identify potential cancer-causing agents (known as carcinogens) by using the Ames test to screen for chemicals that cause mutations in bacterial cells (Fig. 6.12). This test asks whether a particular chemical can induce histidine$^+$ revertants of a special histidine$^-$ auxotrophic mutant strain of the bacterium *Salmonella typhimurium*. The advantage of the test is that only revertants can grow on petri plates containing minimal medium without histidine, so it is possible to examine large numbers of cells from an originally his$^-$ culture to find the rare his$^+$ revertants induced by the chemical in question. To increase the sensitivity of mutation detection, the his$^-$ auxotrophic strain used in the Ames test system contains a second mutation inactivating the excision repair system, which prevents the ready repair of mutations, and a third mutation causing defects in the cell wall, which allows chemicals easier access to the cell interior.

Since most agents that cause mutations in bacteria should also damage the DNA of higher eukaryotic organisms, any mutagen that increases the rate of mutation in bacteria might be expected to cause cancer in people and other mammals. Mammals, however, have complicated metabolic processes capable of inactivating some otherwise hazardous chemicals, and of creating a mutagenic substance from some otherwise nonhazardous chemicals. To simulate the action of mammalian metabolism, toxicologists often add a solution of rat liver enzymes to the chemical under analysis by the Ames test (Fig. 6.12). Because this simulation is not perfect, Food and Drug Administration agents assess the potential of bacterial mutagens identified by the Ames test to cause cancer in rodents when included in their diet.

In summary, random errors that change the information content of DNA are the ultimate source of variation within and between species. But even though chance mutations provide the raw material of evolution, most changes in genes are deleterious to an organism or its progeny. Cells have evolved several enzymatic systems that either prevent mutations from forming or eliminate those that do. These safeguards enable organisms to keep mutations to a minimum.

WHAT MUTATIONS TELL US ABOUT GENE STRUCTURE

One way to learn how DNA sequences along a chromosome constitute genes is to collect a large series of mutations in the same gene and analyze how these mutations are arranged with respect to each other. For this approach to be successful, it is necessary to establish that various mutations are, in fact, in the same gene. This is not a trivial exercise, as illustrated by the following situation.

Early *Drosophila* geneticists identified a large number of X-linked recessive mutations affecting the normally red wild-type eye color (Fig. 6.13a). The first of these to be discovered produced the famous white eyes studied by Morgan's group. Other mutations caused a whole palette of hues to appear in the eyes: darkened shades such as garnet and ruby; bright colors such as vermilion, cherry, and coral; and lighter pigmentations known as apricot, buff, and carnation. Were these mutations multiple alleles of a single gene, or did they affect more than one gene?

Complementation Testing Reveals Whether Two Mutations Are in the Same or Different Genes

Researchers sometimes define a gene as a functional unit that directs the appearance of a molecular product that in turn contributes to a particular phenotype. They can then use this definition to determine whether two mutations are in the same or different genes. If two homologous chromosomes in an individual each carries a mutation recessive to wildtype, a normal phenotype will result if the mutations are in different genes. The normal phenotype occurs because the dominant wildtype alleles on each of the two homologs make up for, or **complement,** the defect in the other chromosome, generating enough of both gene products to yield a normal phenotype (Fig. 6.13b; left). By contrast, if the recessive mutations on the two homologous chromosomes are in the same gene, no wildtype allele of that gene exists in the individual and neither mutated copy of the gene will be able to perform the normal function; as a result, there will be no complementation and no normal gene product, and a mutant phenotype will appear (Fig. 6.13b; right). Ironically, a collection of mutations that do *not* complement each other is known as a **complementation group.** Geneticists often use "complementation group" as a synonym for "gene" because the mutations in a complementation group all affect the same unit of function, thus, the same gene.

A simple test based on this idea of a gene as a unit of function can determine whether or not two mutations are alleles of the same gene. You simply examine the phenotype of a heterozygous individual in which one homolog of a particular chromosome carries one of the recessive mutations and the other homolog carries the other recessive mutation. If the phenotype is wildtype, the mutations cannot be in the same gene. This technique for determining whether two homologous chromosomes carry nonallelic mutations of different genes

How Chemical Mutagens Increase the Rate of Change

Type of action/agents	Example
(a) **Replace a base**: Base analogs have a chemical structure almost identical to that of a DNA base.	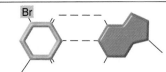 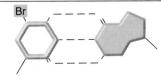 5-Bromouracil–normal state, behaves like thymine · Adenine · 5-Bromouracil–rare state, behaves like cytosine · Guanine 5-Bromouracil: almost identical to thymine. Normally pairs with A; in transient state, pairs with G.

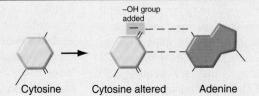

(b) **Alter base structure and properties**:

1. *Hydroxylating agents*: add a hydroxyl (–OH) group

−OH group added

Cytosine   Cytosine altered   Adenine

Hydroxylamine adds –OH to cytosine; with the –OH, hydroxylated C now pairs with A instead of G.

2. *Alkylating agents*: add ethyl (–CH₂–CH₃) or methyl (–CH₃) groups

$-CH_2-CH_3$

$-CH_3$

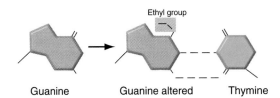

Ethyl group

Guanine   Guanine altered   Thymine

Ethylmethane sulfonate adds an ethyl group to guanine or thymine. Modified G pairs with T, and modified T pairs with G (not shown).

3. *Deaminating agents*: remove amine (–NH₂) groups

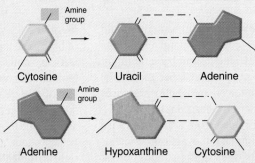

Amine group

Cytosine   Uracil   Adenine

Amine group

Adenine   Hypoxanthine   Cytosine

Nitrous acid modifies cytosine to uracil, which pairs with A instead of G; modifies adenine to hypoxanthine, a base that pairs with C instead of T.

(c) **Insert between bases**: Intercalating agents

H_3C CH_3

N   N   N

CH_3

H_3C

Acridine orange intercalates into the double helix. This disrupts DNA metabolism, eventually resulting in deletion or addition of a base pair.

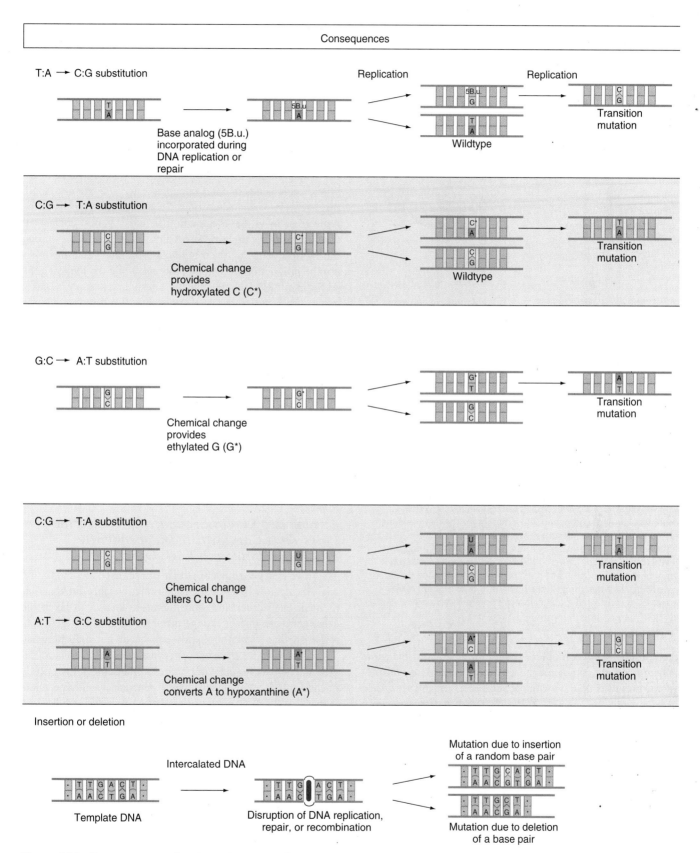

Figure 6.11   How mutagens alter DNA. (a) Base analogs incorporated into DNA may have unusual base pairing properties, which often allow the addition of incorrect nucleotides to the opposite strand during replication. (b) Some mutagens alter the structure of bases such that they pair inappropriately in the next round of replication. (c) Intercalating agents are roughly the same size and shape as a base pair of the double helix. The manner by which such intercalation eventually produces insertions or deletions of single base pairs is not precisely known.

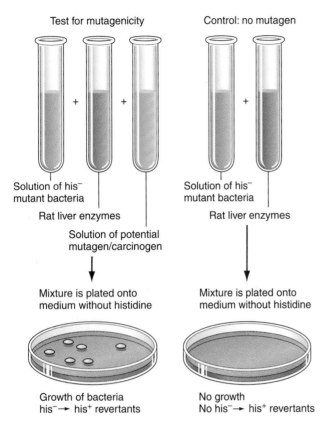

Figure 6.12 The Ames test identifies potential carcinogens through their mutagenicity. Bruce Ames devised a test for potential carcinogens based on their ability to cause mutations in *Salmonella typhimurium* bacteria. A compound to be tested is mixed with cells of a histidine⁻ auxotrophic bacterial strain, as well as with a solution of rat liver enzymes (which can sometimes convert an otherwise harmless compound into a mutagen). The mixture is then plated on a petri plate without histidine, so that only colonies that are histidine⁺ revertants are able to grow. If examination of this plate reveals more histidine⁺ revertants than a similar control plate made without the compound (right-hand part of figure), the compound, or a derivative of it that forms in the body through the action of liver enzymes, is considered mutagenic and a potential carcinogen.

that complement each other or allelic mutations of the same gene that do not complement each other is known as **complementation testing.** For example, because a female *Drosophila* heterozygous for garnet and ruby (*garnet ruby⁺/garnet⁺ ruby*) has wildtype brick-red eyes, it is possible to conclude that the mutations causing garnet and ruby colors are in different genes. Complementation testing has in fact shown that garnet, ruby, vermilion, and carnation pigmentation are governed by separate genes. But chromosomes carrying mutations yielding white, cherry, coral, apricot, and buff phenotypes fail to complement each other. These mutations therefore make up different alleles of a single gene. *Drosophila* geneticists named this gene the *white,* or *w,* gene after the first mutation observed; they designate the wildtype allele as w^+ and the various mutations as w^1 (the original white mutation discovered by T. H. Morgan, often simply designated as *w*), w^{cherry}, w^{co-}

ral, $w^{apricot}$, and w^{buff}. The eyes of a $w^1/w^{apricot}$ female are a dilute apricot color. Figure 6.13c illustrates how researchers collate data from many complementation tests in a **complementation table.** Such a table helps visualize the relationships among a large group of mutants.

In *Drosophila*, mutations in the *w* gene map in the same chromosomal region of the X chromosome, while mutations in other eye color genes lie elsewhere on the chromosome (Fig. 6.13d). This result suggests that genes are not disjointed entities with parts spread out from one end of a chromosome to another; each gene, in fact, occupies only a relatively small, discrete area of a chromosome. Studies defining genes at the molecular level have shown that most genes consist of 1000–20,000 contiguous base pairs (bp). In humans, among the shortest genes are the roughly 500-bp-long genes that govern the production of histone proteins, while the longest gene so far identified is the Duchenne muscular dystrophy (*DMD*) gene, which has a length of more than 2 million nucleotide pairs. All known human genes fall somewhere between these extremes. To put these figures in perspective, an average human chromosome is approximately 130 million base pairs in length.

Although complementation testing makes it possible to distinguish mutations in different genes from mutations in the same gene, it does not clarify how the structure of a gene can accommodate different mutations and how these different mutations can alter phenotype in different ways. Does each mutation change the whole gene at a single stroke in a particular way, or does it change only a specific part of a gene, while other mutations alter other parts?

A Gene Is a Linear Sequence of Nucleotide Pairs That Can Mutate Independently and Recombine with Each Other

The American geneticist Seymour Benzer used recombination analysis to show that two different mutations that did not complement each other and were therefore known to be in the same gene can in fact change different parts of that gene. He reasoned that if recombination can occur not only between genes, but within a gene as well, crossovers between homologous chromosomes carrying different mutations known to lie in the same gene could in theory generate a wildtype allele (Fig. 6.14). Because mutations affecting a single gene lie very close together on a chromosome, it is necessary to examine a very large number of progeny to see even one crossover event between them. The resolution of the experimental system must thus be extremely high, allowing rapid detection of rare genetic events. For his experimental organism, Benzer chose bacteriophage T4, a DNA virus that infects *Escherichia coli* cells. Because each T4 phage that infects a bacterium generates 100–1000 phage progeny in less than an hour, Benzer could easily produce enough rare recombinants for his analysis. Moreover, by exploiting a peculiarity of certain T4 mutations, he devised conditions that allowed only recombinant phages, and not parental phages, to proliferate.

(a)

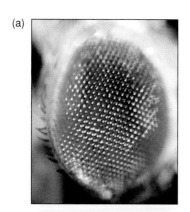

Figure 6.13 Complementation testing of *Drosophila* eye color mutations. (a) Flies carrying different X-linked eye-color mutations; a wildtype eye is at the bottom right. (b) The theoretical basis of complementation analysis: Consider a heterozygote that has one mutation (m_1) on one chromosome and a different mutation (m_2) on the homologous chromosome. If the mutations are in different genes, the heterozygote will be wildtype because the mutations complement each other (*left*). This would be the case if m_1 were in the *garnet* gene and m_2 in the *ruby* gene. If, however, both mutations affect the same gene, the phenotype will be mutant because the mutations do not complement each other (*right*). Note that the complementation testing makes sense only when both mutations are known to be recessive to the wildtype. (c) A complementation table for X-linked eye-color mutations in *Drosophila.* This table reveals the existence of five different complementation groups, that is, five different genes. Several mutations (causing buff, coral, apricot, white, and cherry phenotypes) are different alleles of the same gene (*white*). (d) A recombination map showing the positions of these eye color mutations. Mutations in different genes are generally far apart, while different mutations in the same gene are very close together.

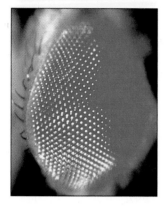

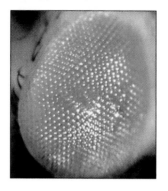

(b)

1. Complementation

Maternal chromosome — Defective gene m_1 / Functional gene, G, R

Paternal chromosome — Functional gene / Defective gene m_2, G, R

Conclusion: m_1 and m_2 are in different genes. m_1/m_2 has wildtype phenotype because one chromosome supplies gene G function while the other supplies gene R function.

2. No complementation

Defective gene m_1 / Functional gene, G, R

Defective gene m_2 / Functional gene, G, R

Conclusion: m_1 and m_2 are in the same gene. m_1/m_2 has mutant phenotype because organism has no gene G function.

(c)

Mutation	white	garnet	ruby	vermilion	cherry	coral	apricot	buff	carnation
white	−	+	+	+	−	−	−	−	+
garnet		−	+	+	+	+	+	+	+
ruby			−	+	+	+	+	+	+
vermilion				−	+	+	+	+	+
cherry					−	−	−	−	+
coral						−	−	−	+
apricot							−	−	+
buff								−	+
carnation									−

(d)

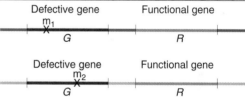

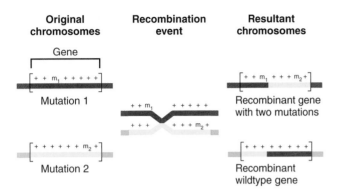

Figure 6.14   How recombination within a gene could generate a wildtype allele. Suppose a gene, indicated by the region between brackets, is composed of many sites that can mutate independently. In this gene, a mutation (m_1) occurs at one such site on one chromosome and a second mutation (m_2) occurs at another site in the same gene on the homologous chromosome. Recombination between these sites can produce a chromosome with a wildtype allele of the gene and a homologous chromosome with a reciprocal allele containing both mutations.

The Experimental System: Analyzing Multiple rII^- Mutations of Bacteriophage T4

Even though bacteriophages are too small to be seen without the aid of an electron microscope, they leave a trail of activity detectable with the unaided eye (Fig. 6.15a). To observe this trail, researchers mix a population of bacteriophage particles with a much larger number of bacteria, and then pour this mixture onto a petri plate where the cells are immobilized in a nutrient agar. If a single phage infects a single bacterial cell somewhere on this so-called **lawn** of bacteria, the cell produces and releases progeny viral particles that diffuse away to infect adjacent bacteria, which in turn produce and release yet more phage progeny. With each release of virus particles, the bacterial host cell dies. Thus, several cycles of phage infection, replication, and release produce a circular cleared area in the plate, called a **plaque,** devoid of living bacterial cells. The rest of the petri plate surface is covered by an opalescent lawn of living bacteria. Most plaques contain from 1 million to 10 million viral progeny of the single bacteriophage that originally infected a cell in that position on the petri plate. Sequential dilution of phage-containing solutions makes it possible to measure the number of phages in a particular plaque, and arrive at a countable number of viral particles (Fig. 6.15a.4).

In looking for genetic traits associated with bacteriophage T4, Benzer found mutants that when added to a lawn of *E. coli* B strain bacteria produced larger plaques with sharper, more clearly rounded edges than those produced by the wildtype bacteriophage (Fig. 6.15b.1). Because these changes in plaque morphology seemed to result from the abnormally rapid lysis of the host bacteria, Benzer named the mutations giving rise to this phenotype *r* for "rapid lysis." Many *r* mutations map to a region of the T4 chromosome known as the *rII* region; these are called rII^- mutations. An additional property of rII^- mutations

makes them ideal for the genetic *fine structure mapping* (the mapping of mutations within a gene) undertaken by Benzer. Wildtype rII^+ bacteriophages form plaques of normal shape and size on cells of both the *E. coli* B strain and a strain known as *E. coli* K(λ). These rII^- mutants, however, have an altered host range: They cannot form plaques with *E. coli* K(λ) cells, although as we have seen, they produce large, unusually distinct plaques with *E. coli* B cells (Fig. 6.15b.1 and 2). The reason that rII^- mutants are unable to infect cells of the K(λ) strain was not clear to Benzer, but their inability to do so enabled him to develop an extremely simple and effective test for rII^+ gene function that he could use to understand gene structure.

The rII Region Has Two Genes

Before he could check whether two mutations in the same gene could recombine, Benzer had to be sure he was really looking at two mutations in the same gene. To verify this, he performed customized complementation tests tailored to two significant characteristics of bacteriophage T4: They are haploid, and they can replicate only in a host bacterium. Because T4 phages are haploid, Benzer needed to ensure that two T4 chromosomes entered the same bacterial cell in order to test for complementation between the mutations. In his complementation tests, he simultaneously infected *E. coli* K(λ) cells with two types of T4 chromosomes—one carried one rII^- mutation, the other carried a different rII^- mutation—and then looked for cell lysis (Fig. 6.15c.1). To ensure that the two kinds of phages would infect almost every bacterial cell, he added many more phages of each type than there were bacteria. If the two rII^- mutations were in different genes, each of the mutant T4 chromosomes would supply one wildtype rII^+ gene function, making up for the lack of that function in the other chromosome and resulting in lysis. On the other hand, if the two rII^- mutations were in the same gene, no plaques would appear, because neither mutant chromosome would be able to supply the missing function.

Benzer had to satisfy one final experimental requirement: For the complementation test to be meaningful, he had to make sure that the two rII^- mutations under analysis were each recessive to wildtype and did not interact with each other to produce an rII^- phenotype dominant to wildtype. He checked these points by a control experiment in which he placed the two rII^- mutations on the same chromosome and then simultaneously infected *E. coli* K(λ) with these double rII^- mutants and with wildtype phages (Fig. 6.15c.2). If the mutations were recessive and did not interact with each other, the cells would lyse, in which case the complementation test would be interpretable.

The significant distinction between the actual complementation test and the control experiment is in the placement of the two rII^- mutations. In the complementation test, one rII^- mutation is on one chromosome, while the other rII^- mutation is on the other chromosome; two mutations arranged in this way are said to be in the *trans* configuration. In the control experiment, the two mutations are on the same chromosome, in the so-called *cis* configuration. Benzer thus named the complete test including complementation test and control experiment a *cis-trans* test and called any complementation group identified in this way a **cistron.** Some geneticists still use the term "cistron" as a synonym for "gene."

Tests of many different pairs of rII^- mutations showed that they fall into two complementation groups: *rIIA* and *rIIB*. With this knowledge, Benzer could look for two mutations in the same gene and then see if they ever recombine to produce wildtype progeny.

A Gene Is Composed of Mutable Subunits That Can Recombine

When Benzer infected *E. coli* B strain bacteria with a mixture of phages carrying different mutations in the same gene ($rIIA_1$ and $rIIA_2$, for example), he did observe the appearance of rII^+ progeny (Fig. 6.15d). He knew these wildtype progeny resulted from recombination and not reverse mutations because the frequencies of the rII^+ phage particles he observed were much higher than the frequencies of rII^+ revertants seen among progeny produced by infecting B strain bacteria with either mutant alone. On the basis of these observations, he drew three conclusions about gene structure: (1) A gene consists of different parts that can each mutate; (2) recombination between different mutable sites in the same gene can generate a normal, wildtype allele; and (3) a gene performs its normal function only if all of its components are wildtype.

From what we now know about the molecular structure of DNA, this all makes perfect sense. The separate mutable sites are different nucleotides within the gene, and recombination can occur between nucleotides within a gene as well as between genes.

But how are the nucleotide pairs that make up a gene arranged—in a continuous row or dispersed in precise patterns around the genome? And do the mutations that affect gene function alter many nucleotides or only a small subset within each gene?

A Gene Is a Discrete Linear Set of Nucleotides

To answer these questions about the arrangement of nucleotides in a gene, Benzer eventually obtained thousands of spontaneous and mutagen-induced rII^- mutations that he mapped with respect to each other. To map the location of a thousand mutants through comparisons of all possible two-point crosses, he would have had to set up a million ($10^3 \times 10^3$) matings. But by taking advantage of deletion mutations, he could obtain the same information with far fewer crosses.

Using Deletions to Map Mutations and Define a Gene

As we have seen, deletions are mutations that remove contiguous nucleotide pairs along a DNA molecule. In crosses between bacteriophages carrying a mutation and bacteriophages carrying deletions of the corresponding region, no wildtype recombinant progeny can arise. However, if the mutation lies outside the region deleted from the homologous chromosome, wildtype progeny can appear (Fig. 6.16a). This is true whether the mutation is a **point mutation,** that is, a mutation of one nu-

cleotide, or is itself a deletion. Crosses between any uncharacterized mutation and a known deletion thus immediately reveal whether or not the mutation resides in the region deleted from the other phage chromosome. This method of deletion mapping provides a rapid way of finding the general location of a mutation. Using a series of overlapping deletions, Benzer divided the *rII* region into a series of intervals. He could then assign any point mutation to an interval by observing whether or not it recombined to give rII^+ progeny when crossed with the series of deletions (Fig. 6.16b).

Benzer mapped 1612 spontaneous point mutations and several deletions in the *rII* locus of bacteriophage T4 through recombination analysis. He first used recombination to determine the relationship between the deletions. He next found the approximate location of individual point mutations by observing which deletions could recombine with each mutant to yield wildtype progeny. He then performed recombination tests between all point mutations known to lie in the same small region of the chromosome. These results produced a map of the "fine structure" of the region (Fig. 6.16c).

From the observation that the number of mutable sites in the *rII* region is very close to the number of nucleotides estimated to be in this region, Benzer inferred that a mutation can arise from the change of a single nucleotide, and that recombination can occur between adjacent nucleotide pairs. From the observation that mutations within the *rII* region form a self-consistent, linear recombination map, he concluded that a gene is composed of a continuous linear sequence of nucleotide pairs within the DNA. And from observations that the positions of mutations in the *rIIA* gene did not overlap those of the *rIIB* gene, he inferred that the nucleotide sequences composing those two genes are separate and distinct. A *gene* is thus a linear set of nucleotide pairs within a discrete region of a chromosome that serves as a unit of function.

Hot Spots Are More Prone to Mutation

Some sites within a gene spontaneously mutate more frequently than others and as a result are known as **hot spots** (Fig. 6.16c). The existence of hot spots suggests that certain nucleotides can be altered more readily than others. Treatment with mutagens also turns up hot spots, but because mutagens have specificities for particular nucleotides, the more mutable sites that turn up with various mutagens are often at different positions in a gene than the hot spots resulting from spontaneous mutation.

Nucleotides are chemically the same whether they lie within a gene or in the DNA between genes, and as we see here and in Chapter 8, the molecular machinery responsible for mutation and recombination does not discriminate between those nucleotides that are intragenic (within a gene) and those that are extragenic (outside of a gene). The main distinction between DNA within and DNA outside a gene is that the array of nucleotides composing a gene has evolved a function that determines phenotype. We now describe how geneticists discovered what that function is.

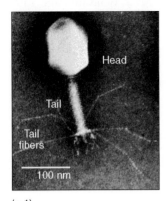

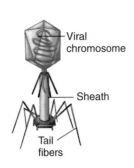

- Viral chromosome
- Sheath
- Tail fibers

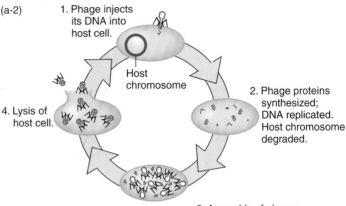

(a-2)

1. Phage injects its DNA into host cell.

Host chromosome

2. Phage proteins synthesized; DNA replicated. Host chromosome degraded.

3. Assembly of phages within host cell.

4. Lysis of host cell.

Head

Tail

Tail fibers

100 nm

(a-1)

(a-3)

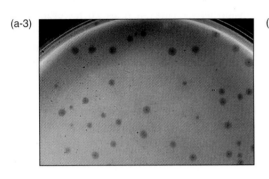

(a-4)

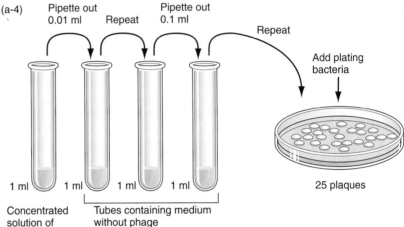

Pipette out 0.01 ml

Repeat

Pipette out 0.1 ml

Repeat

Add plating bacteria

1 ml   1 ml   1 ml   1 ml

Concentrated solution of bacteriophages

Tubes containing medium without phage

25 plaques

(a) Working with bacteriophage T4

1. Bacteriophage T4 as it appears in the electron microscope (at a magnification of approximately 100,000×) and in an artist's rendering. The viral chromosome is contained within a protein head. Other proteinaceous parts of the phage particle include the tail fibers, which help the phage attach to host cells, and the sheath, a conduit for injecting the phage chromosome into the host cell.

2. The lytic cycle of bacteriophage T4. A single phage particle infects a host cell; the phage DNA replicates and directs the synthesis of viral protein components using the machinery of the host cell; the new DNA and protein components assemble into new bacteriophage particles. Eventual lysis of the host cell releases up to 1000 progeny bacteriophages into the environment.

3. Clear plaques of bacteriophages in a lawn of bacterial cells. A mixture of bacteriophages and a large number of bacteria are poured onto the agar surface of a petri plate. Uninfected bacterial

cells grow, producing an opalescent lawn. A bacterial cell infected by even a single bacteriophage will lyse and release progeny bacteriophages, which can infect adjacent bacteria. Several cycles of infection result in a plaque: a circular cleared area containing millions of bacteriophages genetically identical to the one that originally infected the bacterial cell.

4. Counting bacteriophages by serial dilution. A small sample of a concentrated solution of bacteriophages is transferred to a test tube containing fresh medium, and a small sample of this dilution is transferred to another tube of fresh medium. Successive repeats of this process increase the degree of dilution. A sample of the final dilution, when mixed with bacteria and poured on the agar of a petri plate, yields a countable number of plaques from which it is possible to extrapolate back and calculate the number of bacteriophage particles in the starting solution. The original 1 ml of solution in this illustration contained roughly 2.5×10^7 bacteriophages.

(b) Phenotypic properties of *rII⁻* mutants of bacteriophage T4

1. *rII⁻* mutants, when plated on *E. coli* B cells, produce plaques that are larger and more distinct (with sharper edges) than plaques formed by *rII⁺* wildtype phage.

2. *rII⁻* mutants are particularly useful for looking at rare recombination events, because they have an altered host range. In contrast to *rII⁺* wildtype phages, *rII⁻* mutants cannot form plaques in lawns of *E. coli* strain K(λ) host bacteria.

(b-1)

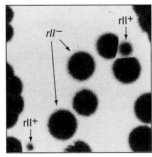

rII⁺

rII⁻

rII⁺

(b-2)

T4 strain	E. coli strain	
	B	K(λ)
rII⁻	Large, distinct	No plaques
rII⁺	Small, fuzzy	Small, fuzzy

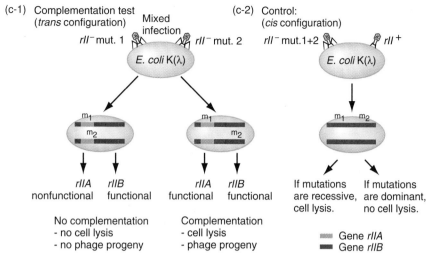

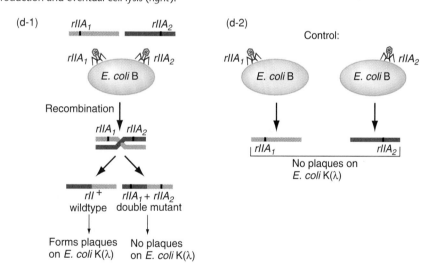

(c) A customized complementation test between *rII⁻* mutants of bacteriophage T4

1. *E. coli* K(λ) cells are simultaneously infected with an excessive amount of each of two different *rII⁻* mutants (one with mutation 1, the other with mutation 2). Inside the cell, the two *rII⁻* mutations will be in the *trans* configuration, that is, they will lie on different chromosomes. If the two mutations are in the same gene, they will affect the same function and therefore cannot complement each other (*left*). As *rII⁻* mutants cannot grow in *E. coli* strain K(λ) host bacteria, no progeny phages will arise. By contrast, if the two mutations are in different genes (*rIIA* and *rIIB*, for example), they will complement each other, leading to progeny phage production and eventual cell lysis (*right*).

2. An important control for this complementation test is the simultaneous infection of *E. coli* K(λ) bacteria with a wildtype T4 strain and a T4 strain containing both mutation 1 and mutation 2. Inside the infected bacterial cells, the two mutations will be in the *cis* configuration; that is, they will lie on the same chromosome. Cell lysis accompanied by release of phage progeny would show that both mutations are recessive to wildtype and that there is no unexpected interaction between the two mutations that prevents the cells from producing progeny phages. This control experiment is critical because complementation tests are meaningful only if the two mutations tested are both recessive to wildtype.

(d) Detecting recombination between two mutations in the same gene

1. *E. coli* B cells are simultaneously infected with a large excess of two different *rIIA* mutants (*rIIA₁* and *rIIA₂*). As the bacteriophage chromosomes replicate inside the infected cells, recombination between the two types of chromosomes may occur. If no recombination between the two *rIIA* mutations takes place, progeny phages will carry either of the original mutations and will therefore be phenotypically *rII⁻*. If recombination between the two mutations occurs, one of the products will be an *rII⁺* recombinant, while the reciprocal product of recombination will be a double mutant chromosome containing both *rIIA₁* and *rIIA₂*. When the phage progeny produced in these simultaneously

infected *E. coli* B cells subsequently infect *E. coli* K(λ) bacteria, only *rII⁺* recombinant phages will be able to form plaques.

2. As a control, *E. coli* B cells are infected with a large amount of only one kind of mutant bacteriophages (*rIIA₁* or *rIIA₂*). The only *rII⁺* phages that can result are revertants of either mutation. This control experiment shows that such revertants are extremely rare, and can be ignored among the *rII⁺* progeny of the simultaneously infected *E. coli* B bacteria in the recombination experiment. Even if the two *rIIA* mutations are in adjacent base pairs, the number of *rII⁺* recombinants obtained is more than 100 times higher than the number of *rII⁺* revertants these cells infected by a single type of mutant could produce.

Feature Figure 6.15   How Benzer Analyzed the *rII* Genes of Bacteriophage T4

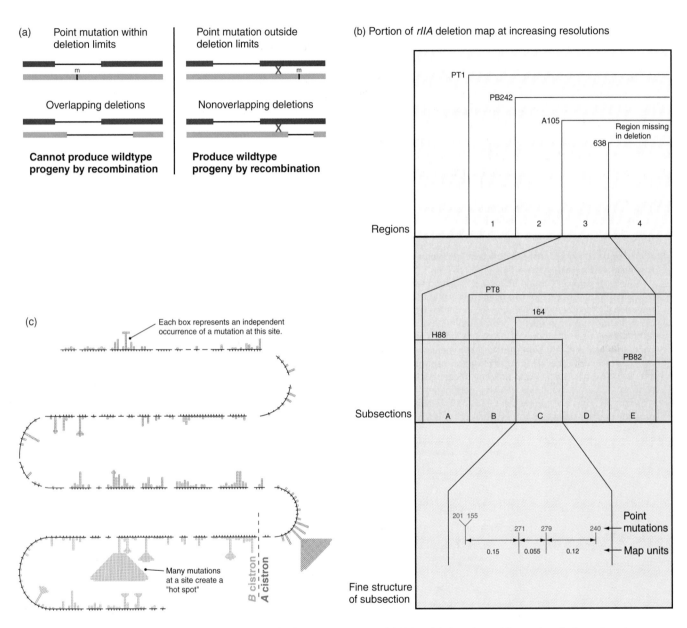

Figure 6.16 Fine structure mapping of the bacteriophage T4 *rII* genes. (a) Using deletions for rapid mapping. A phage cross between a point mutation and a deletion removing the mutated DNA sequence cannot yield wildtype recombinants because neither chromosome has wildtype information at that position. The same is true if two different deletion mutations overlap each other; neither chromosome has wildtype information in the region of the overlap. By contrast, a cross between a deletion and a mutation on the homologous chromosome that lies outside the deleted region, can produce wildtype progeny. The same is true of crosses between nonoverlapping deletions. (b) The *rII* map at increasing levels of resolution. Large deletions divide the *rII* locus into regions. Finer deletions divide each region into subsections. Researchers can map point mutations to regions and subsections. For example, point mutations, such as 201 and 271 (numbered in red at the bottom), map to region 3 if they do not recombine with deletions PT1, PB242, or A105 but do recombine with deletion 638 (top). These point mutations can be mapped to subsections of region 3 using other deletions such as PT8, 164, and H88 (middle). Recombination tests give the relative positions of point mutations in the same subregion (bottom). In rare cases, two independent point mutations (such as 201 and 155) cannot recombine to yield wildtype recombinants; this is because the two mutations affect the same nucleotide pair. (c) Benzer's gine structure map of point mutations in the *RIIA* and *rIIB* genes. Hot spots are locations with many independent mutations.

WHAT MUTATIONS TELL US ABOUT GENE FUNCTION

In the early 1900s, the British physician Dr. Archibald Garrod showed that a human genetic disorder known as *alkaptonuria* is determined by the recessive allele of an autosomal gene. Garrod analyzed family pedigrees and performed biochemical analyses on family members with and without the trait. The urine of people with alkaptonuria turns black on exposure to air. Garrod found that a substance known as homogentisic acid, which blackens upon contact with oxygen, accumulates in the urine of alkaptonuria patients. These same people excrete all of the homogentisic acid they ingest, while people not suffering from the condition do not have any homogentisic acid in their urine even after ingesting the substance. From these observations, he concluded that people with alkaptonuria are incapable of metabolizing homogentisic acid to the breakdown products generated by normal individuals (Fig. 6.17). Because many biochemical reactions within the cells of organisms are catalyzed by enzymes, Garrod hypothesized that lack of the enzyme that breaks down homogentisic acid is the cause of alkaptonuria; in the absence of this enzyme, the acid accumulates and causes the urine to turn black on contact with oxygen. He called this condition an "inborn error of metabolism."

Garrod studied several other inborn errors of metabolism and suggested that all arose from mutations that prevented a particular gene from producing an enzyme required for a specific biochemical reaction. In today's terminology, the wild-type allele of the gene would allow production of functional enzyme (in the case of alkaptonuria, the enzyme is homogentisic acid oxidase), whereas the mutant allele would not. Because the single wildtype allele in heterozygotes generates sufficient enzyme to prevent accumulation of homogentisic acid and thus alkaptonuria, the mutant allele is recessive.

The One Gene, One Enzyme Hypothesis: A Gene Contains the Information for Producing a Specific Enzyme

George Beadle and Edward Tatum carried out a series of experiments on the bread mold *Neurospora crassa* (see Chapter 4) in the 1940s that more clearly demonstrated a direct relation between genes and the enzymes that catalyze specific biochemical reactions. Their strategy was simple. They first isolated a number of mutations that disrupted synthesis of the amino acid arginine, a compound needed for *Neurospora* growth. They next hypothesized that different mutations blocked different steps in a particular **biochemical pathway:** the orderly series of reactions that allows *Neurospora* to obtain simple molecules from the environment and convert them step by step into successively more complicated molecules culminating in arginine. A study showing that mutations can block biochemical steps in the pathway resulting in arginine would constitute strong evidence that different genes control the appearance of different enzymes, each of which catalyzes a specific reaction in the biochemical pathway.

The Experimental Evidence for "One Gene, One Enzyme"

Figure 6.18a illustrates the experiments Beadle and Tatum performed to test their hypothesis. They first obtained a set of mutagen-induced mutations that prevented *Neurospora* from synthesizing arginine. Cells with any one of these mutations were unable to make arginine and could therefore grow on a minimal medium containing salt and sugar only if it had been supplemented with arginine. A mutant microorganism that can grow on minimal medium only if it has been supplemented with one or more growth factors not required by wildtype strains is known as an **auxotroph.** Recombination analyses located the arginine-blocking mutations in four distinct regions of the genome, and complementation tests showed that each of the four regions correlated with a different complementation group. On the basis of these results, Beadle and Tatum concluded that at least four genes support the biochemical pathway for arginine synthesis. They named the four genes *ARG-E, ARG-F, ARG-G,* and *ARG-H.*

They next asked whether any of the mutant *Neurospora* strains could grow in minimal medium supplemented with any of a number of compounds that are intermediates in the biochemical pathway leading to arginine, instead of with arginine itself. If a mutant grew on medium supplemented in such a way, it would indicate that *Neurospora* is able to convert the intermediate compound into arginine. After finding three such intermediates—ornithine, citrulline, and argininosuccinate—Beadle and Tatum compiled a table describing which arginine-related

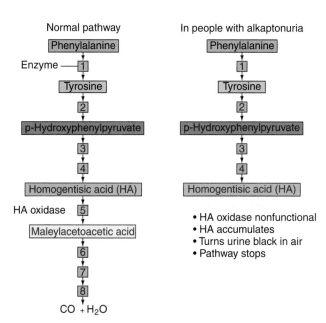

Figure 6.17 Alkaptonuria: An inborn error of metabolism. This figure depicts the biochemical pathway in humans that helps degrade phenylalanine and tyrosine via homogentisic acid (HA). In patients with alkaptonuria, the enzyme HA hydroxylase is not functional and thus does not catalyze the conversion of HA to maleylacetoacetic acid, the next compound in the pathway. As a result, HA, which oxidizes to a black compound, accumulates in the urine.

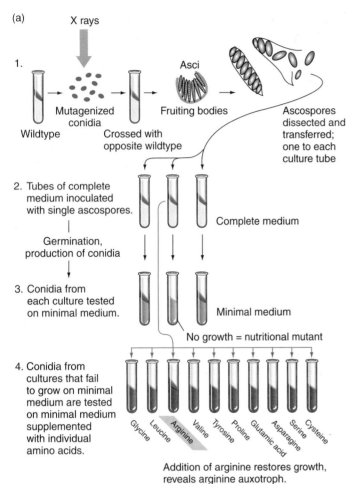

(b) **Growth response if nutrient is added to minimal medium**

Mutant strain	Nothing	Ornithine	Citrulline	Arginino-succinate	Arginine
Wildtype: *Arg⁺*	+	+	+	+	+
Arg-E⁻	−	+	+	+	+
Arg-F⁻	−	−	+	+	+
Arg-G⁻	−	−	−	+	+
Arg-H⁻	−	−	−	−	+

Figure 6.18   Beadle and Tatum's experiment supporting the "one gene, one enzyme" hypothesis. (a) Isolation of arginine auxotrophs. Beadle and Tatum mutagenized one strain of *Neurospora* with X rays, and after mating the mutagenized strain with another strain, isolated haploid ascospores that could grow on complete medium. Samples of cultures grown from individual ascospores were then tested on minimal medium; those that failed to grow were nutritional mutants. Nutritional mutants that could grow on minimal medium supplemented with arginine are arginine auxotrophs. (b) Table depicting the ability of wildtype and mutant strains to grow on minimal medium supplemented with each of the intermediates in the arginine pathway. (c) Inferred biochemical pathway for arginine biosynthesis. Each of the four genes encodes an enzyme needed to convert one intermediate to the next.

mutants were able to grow on minimal medium supplemented with each of the intermediates (Fig. 6.18b).

Interpretation of Results: Genes Encode Enzymes

On the basis of these results, Beadle and Tatum proposed a model of how *Neurospora* cells synthesize arginine (Fig. 6.18c): In the linear progression of biochemical reactions by which a cell constructs arginine from the constituents of minimal medium, each intermediate is both the product of one step and the substrate for the next. Each reaction in the precisely ordered sequence is catalyzed by a specific enzyme, and the presence of each enzyme depends on one of the four arginine genes. A mutation in one gene blocks the pathway at a particular step because the cell lacks the corresponding enzyme; this prevents the cell from growing. Supplementing the medium with any intermediate that occurs beyond the blocked reaction will restore growth, because the organism has all the remaining enzymes required to convert the intermediate to arginine. Supplementation with an intermediate that occurs before the missing enzyme will not work because the cell is unable to convert the intermediate into arginine.

By inference, because each mutation abolishes the cell's ability to make an enzyme capable of catalyzing a certain reaction, each gene controls the synthesis or activity of an enzyme,

or as stated by Beadle and Tatum: one gene, one enzyme. Of course, the gene and the enzyme are not the same thing; rather, the sequence of nucleotides in a gene contains information that somehow encodes the structure of an enzyme molecule.

While the analysis of the arginine pathway studied by Beadle and Tatum was straightforward, studies of biochemical pathways are not always so easy to interpret, because some biochemical pathways are not linear progressions of stepwise reactions. For example, a branching pathway occurs if different enzymes act on the same intermediate to convert it into two different end products. If the cell requires both of these end products for growth, a mutation in a gene encoding any of the enzymes required to synthesize the intermediate would make the cell dependent on the addition to minimal medium of both end products. A second possibility is that a cell might employ either of two independent, parallel pathways to synthesize a needed end product. In such a case, a mutation in a gene encoding an enzyme in one of the pathways would be without effect. Only a cell with mutations for genes specifying enzymes in both pathways would display an aberrant phenotype.

Even with nonlinear progressions such as these, careful genetic analysis can reveal the nature of the biochemical pathway on the basis of Beadle and Tatum's insight that genes encode proteins.

Genes Direct the Synthesis of Proteins by Specifying the Identity and Order of Amino Acids in a Polypeptide Chain

Although the one gene, one enzyme hypothesis was a critical advance in our understanding of how genes influence phenotype, it is an oversimplification. Not all genes govern the construction of enzymes active in biochemical pathways. Enzymes are one class of the molecules known as proteins, and cells contain many different kinds of protein, only some of which behave as enzymes. Among the other types are proteins that provide shape and rigidity to a cell, proteins that transport molecules in and out of cells, proteins that help fold DNA into chromosomes, and proteins that act as hormonal messengers. Genes direct the synthesis of all proteins, enzymes and nonenzymes alike. Moreover, as we see next, genes actually determine the construction of polypeptides, and because some proteins are composed of more than one type of polypeptide, more than one gene determines their construction.

Proteins Are Linear Polymers of Amino Acids Linked by Peptide Bonds

To understand how the information in a gene specifies the production of a particular protein, it is necessary to review the chemical composition of proteins. Proteins are polymers composed of building blocks known as **amino acids.** Cells use mainly 20 different amino acids to synthesize the proteins they need. All of these amino acids have certain basic features, encapsulated by the formula NH_2–CHR–COOH (Fig. 6.19a). The –COOH component, also known as *carboxylic acid,* is as the name implies, acidic; the –NH_2 component, also known as an *amino group,* is basic. The R refers to side chains that distinguish each of the 20 amino acids (Fig. 6.19b). An R group can be as simple as a hydrogen atom (in the amino acid glycine), or as complex as a benzene ring (in phenylalanine). Some side chains are relatively neutral and nonreactive; others are acidic and still others are basic.

During protein synthesis, a cell's protein-building machinery links amino acids by constructing covalent **peptide bonds** that join the -COOH group of one amino acid to the –NH_2 group of the next (Fig. 6.19c). A pair of amino acids connected in this fashion is a **dipeptide;** several amino acids linked together constitute an **oligopeptide.** The amino acid chains that make up proteins contain hundreds to thousands of amino acids joined by peptide bonds and are known as **polypeptides.** Proteins are thus linear polymers of amino acids. Like the chains of nucleotides in DNA, polypeptides have a chemical polarity. One end of a polypeptide is called the **N terminus** because it contains a free amino group that is not connected to any other amino acid. The other end of the polypeptide chain is the **C terminus,** because it contains a free carboxylic acid group.

The Primary Business of Most Genes Is to Specify the Amino Acid Sequence of the Proteins They Encode

Each protein is composed of a unique sequence of amino acids. In fact, the chemical properties that enable structural proteins to give a cell its shape, or enzymes to catalyze specific reactions, or hormones to act as messengers are a direct consequence of the identity, number, and linear order of amino acids in the protein.

If genes encode proteins, then at least some mutations could be changes in a gene that alter the proper sequence of amino acids in the protein encoded by that gene. In the mid-1950s, Vernon Ingram began to establish what kinds of changes particular mutations cause in the corresponding protein. Using recently developed techniques for determining the sequence of amino acids in a protein, he compared the amino acid sequence of the normal adult form of hemoglobin (HbA) with that of hemoglobin in the bloodstream of people homozygous for the mutation that causes sickle-cell anemia (HbS). Remarkably, he found only a single amino-acid difference between the wildtype and mutant proteins (Fig. 6.20a). The sixth amino acid from the N terminus of one of the two polypeptide chains making up hemoglobin was glutamic acid in normal individuals but valine in sickle-cell patients. Ingram thus established that a mutation substituting one amino acid for another had the power to change the structure and function of hemoglobin and thereby alter the phenotype from normal to sickle-cell disease (Fig. 6.20b). We now know that the glutamic acid-to-valine change affects the solubility of hemoglobin within the red blood cell. The hemoglobin molecule consists of two alpha (α) and two beta (β) chains complexed in a tetramer of four chains. At low concentrations of oxygen, the less soluble sickle-cell form of hemoglobin aggregates into long chains that deform the red blood cell (review Fig. 6.20a).

Because people suffering from a variety of inherited anemias also had defective hemoglobin molecules, Ingram and other geneticists were able to determine how a large number of different mutations affected the amino-acid sequence of hemoglobin (Fig. 6.20c). Most of the altered hemoglobins had a change in only one amino acid. In different patients, the alteration was generally in different amino acids, but occasionally, two independent mutations resulted in different substitutions for the same amino acid.

The Sequence of Amino Acids in a Polypeptide Determines a Protein's Three-Dimensional Shape and Thus Its Function

Despite the uniform nature of protein construction—a line of amino acids joined by peptide bonds—each polypeptide folds into a unique three-dimensional shape. The linear sequence of amino acids within a polypeptide is its **primary structure.** Each unique primary structure places restraints on how a chain can arrange itself in three-dimensional space. Because the R groups distinguishing the 20 amino acids have dissimilar chemical properties, some amino acids form hydrogen bonds or electrostatic bonds when brought into proximity with other amino acids. Nonpolar amino acids may become associated with each other by hydrophobic interactions that "hide" them from water. Two cysteine amino acids can form covalent disulfide bridges (-S-S-) through the oxidation of their -SH groups. All of these interactions (Fig. 6.21a) help stabilize the polypeptide in a specific three-dimensional conformation. The

(a) Generic amino acid structure

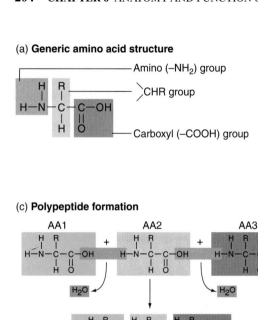

(c) Polypeptide formation

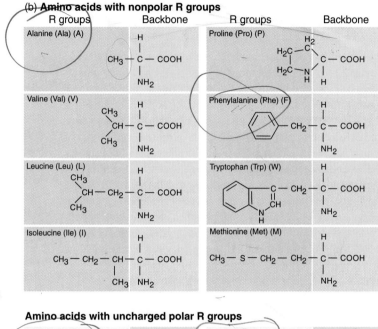

Amino acids with uncharged polar R groups

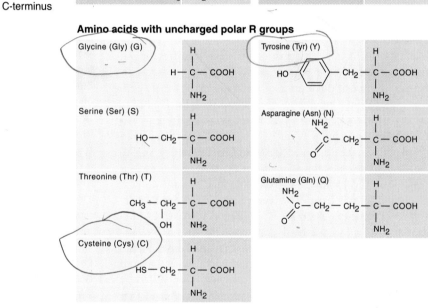

Amino acids with basic R groups

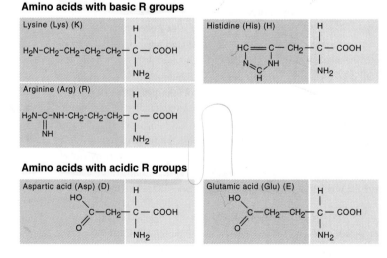

Amino acids with acidic R groups

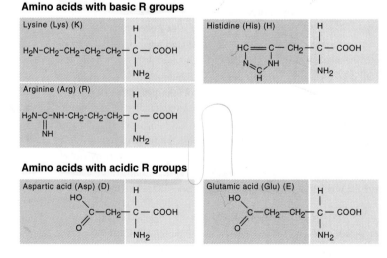

Figure 6.19 Proteins are chains of amino acids linked by peptide bonds. (a) Structure of an amino acid. Amino acids contain a basic amino group (-NH₂), an acidic carboxylic acid group (-COOH), and a >CHR moiety, where R stands for one of the 20 different side chains that distinguish the various amino acids. (b) Table of amino acids commonly found in proteins, arranged according to the properties of their R group side chains. (c) Synthesis of peptide bonds. Formation of each covalent amide linkage between the carboxylic acid of one amino acid and the amino group of another amino acid is accompanied by the loss of one molecule of water. Polypeptides such as the tripeptide shown here have polarity; they extend from an N terminus (with a free amino group) to a C terminus (with a free carboxylic acid group).

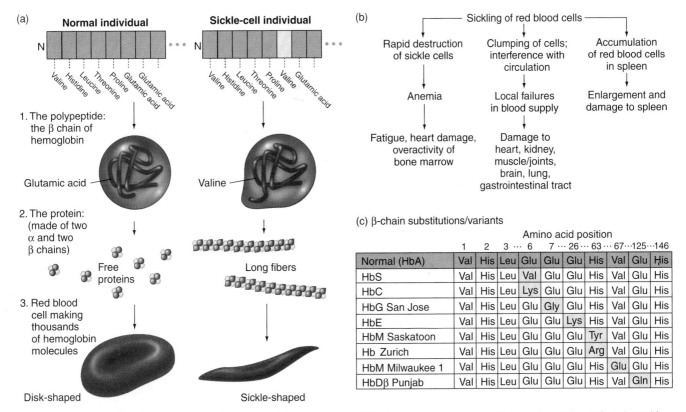

Figure 6.20 The molecular basis of sickle-cell and other anemias. (a) Substitution of glutamic acid with valine at the sixth amino acid from the N terminus affects the three-dimensional structure of the β chain of hemoglobin. Hemoglobins incorporating the mutant β chain form partially insoluble aggregates that cause red blood cells to assume the unusual sickle shape. (b) The sickling of red blood cells has many effects on the phenotype of a sickle-cell patient. (c) Other mutations in the β-chain gene can also cause anemias of various severities. These mutations can alter many different amino acids in the hemoglobin β chain; some can substitute different amino acids at the same position.

primary structure (Fig. 6.21b) determines three-dimensional shape by generating both localized regions with a characteristic geometry known as **secondary structure** (Fig. 6.21c) and the folds and twists that produce the ultimate three-dimensional, or **tertiary structure** (Fig. 6.21d), of the entire polypeptide. Normal tertiary structure—the way a long chain of amino acids naturally folds in three-dimensional space under physiological conditions—is known as a polypeptide's *native configuration*. Various forces, including hydrogen bonds, electrostatic bonds, hydrophobic interactions, and disulfide bridges, help stabilize the native configuration.

It is worth repeating that primary structure—the sequence of amino acids in a polypeptide—directly determines secondary and tertiary structure because inherent in the linear sequence of amino acids is the information required for the chain to fold into its native configuration. Several examples illustrate this fact. For instance, protein biologists know that a polypeptide chain is unable to form an α-helix in the vicinity of a proline, and on the basis of this and other structural information they have devised computer programs that examine the primary sequence of a protein for amino acids that favor or disrupt α-helix formation. They can use these programs to predict the α-helical composition of a polypeptide in its native state with considerable accuracy. In a different example, many

proteins unfold, or become **denatured,** when exposed to urea and mercaptoethanol or to increasing heat or pH because these treatments disrupt the interactions between amino acids that normally stabilize the secondary and tertiary structures. When conditions return to normal, some proteins spontaneously refold into their native configuration (Fig. 6.21e).

Some Proteins Consist of More than One Polypeptide

Certain proteins, such as the rhodopsin that promotes black and white vision, consist of one polypeptide. Many others, however, such as the lens crystallin proteiin, which provides rigidity and transparency to the lenses of our eyes, or the hemoglobin molecule that transports oxygen to and from our tissues, are composed of two or more polypeptide chains that aggregate in a specific way (Fig. 6.22a). The individual polypeptides in an aggregate are known as *subunits* and the complex of subunits is often referred to as a *multimer.* The three-dimensional configuration of subunits in a multimer is a complex protein's **quaternary structure.** The same forces that stabilize the native form of a polypeptide (that is, hydrogen bonds, electrostatic bonds, hydrophobic interactions, and disulfide bridges) also contribute to the maintenance of quaternary structure. As Fig. 6.22a shows, in some multimers, the two or more interacting subunits are identical polypeptides.

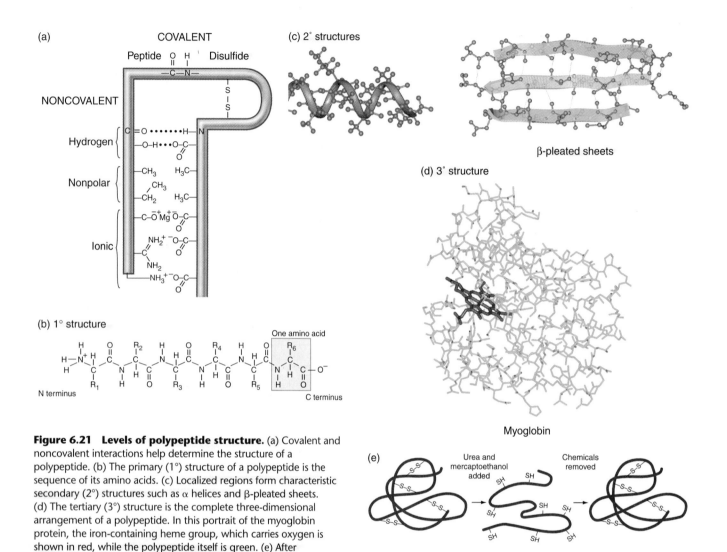

Figure 6.21 Levels of polypeptide structure. (a) Covalent and noncovalent interactions help determine the structure of a polypeptide. (b) The primary (1°) structure of a polypeptide is the sequence of its amino acids. (c) Localized regions form characteristic secondary (2°) structures such as α helices and β-pleated sheets. (d) The tertiary (3°) structure is the complete three-dimensional arrangement of a polypeptide. In this portrait of the myoglobin protein, the iron-containing heme group, which carries oxygen is shown in red, while the polypeptide itself is green. (e) After denaturation, some polypeptides spontaneously fold back (renature) into their correct configuration. Thus, the information dictating the three-dimensional structure of these polypeptides is inherent in the primary sequence of their amino acids.

These identical chains are encoded by one gene. In other multimers, by contrast, more than one kind of polypeptide makes up the protein (Fig. 6.22b). The different polypeptides in these multimers are encoded by different genes.

Alterations in just one kind of subunit, caused by a mutation in a single gene, can affect the function of a multimer. The adult hemoglobin molecule, for example, consists of two α and two β subunits, with each type of subunit determined by a different gene—one for the α chain and one for the β chain. A mutation in the β gene resulting in an amino-acid switch at position 6 in the β chain causes sickle-cell anemia. A single mutation may also affect several multimeric proteins that share a common subunit. An example is an X-linked mutation in mice and humans that incapacitates several different proteins all known as interleukin (IL) receptors. Since all of these receptors are essential to the normal function of immune-system cells that fight infection and generate immunity, this one mutation causes the life-threatening condition

known as X-linked severe combined immune deficiency (XSCID) (Fig. 6.22c).

The polypeptides of complex proteins can assemble into extremely large structures capable of changing with the needs of the cell. For example, the microtubules that make up the spindle during mitosis are gigantic assemblages of mainly two polypeptides: α- and β-tubulin (Fig. 6.22d). The cell can assemble these subunits into very long hollow tubes that grow or shrink as needed at different stages of the cell cycle.

Summation: Each Gene Directs the Synthesis of a Particular Polypeptide

Because more than one gene governs the production of some multimeric proteins, and because not all proteins are enzymes, the "one gene, one enzyme" hypothesis is not broad enough to define gene function. A more accurate statement is "one gene, one polypeptide": Each gene governs the construction of a

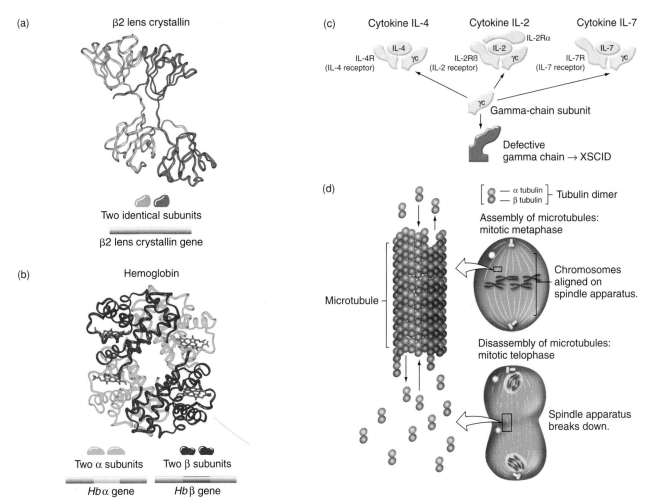

Figure 6.22 Multimeric proteins. (a) Some multimeric proteins contain multiple copies of one kind of subunit; the subunits in these proteins are all the product of a single gene. The protein β2 lens crystallin, composed of two copies of one polypeptide, is an example of this kind of multimer. The peptide backbone of the two subunits in the structure are shown in different shades of purple. (b) Other multimeric proteins are composed of different subunits, each encoded by a different gene. Hemoglobin, with its two α chains and two β chains, is an example of this type of multimer. (c) The three distinct protein receptors for immune-system molecules known as IL-4, IL-2, and IL-7 all contain a gamma chain, in combination with other peptides specific to each receptor. Production of a mutant gamma chain can thus block the function of all three receptors, leading to X-linked severe combined immune deficiency (XSCID), a complete failure of the immune system. (d) One α-tubulin polypeptide and one β-tubulin polypeptide associate to form a tubulin dimer. Many tubulin dimers are required for the formation of a single microtubule. The spindle responsible for chromosome segregation during cell division is a structured assembly of many microtubules.

particular polypeptide. As we will see in Chapter 7, even this reformulation does not encompass the function of all genes, as a few genes in all organisms do not determine the construction of proteins; instead, they encode RNAs that are not translated to polypeptides.

HOW GENOTYPE CORRELATES WITH PHENOTYPE

Cells contain a staggering variety of proteins, each of which performs a specific developmental or biochemical role (see the Fast Forward box "Using Mutagenesis to Look at Biological Processes"). Knowledge that genes encode proteins by determining the sequence of amino acids in a polypeptide enabled geneticists to begin to understand how genes control

complex phenotypes, and how gene function determines the recessiveness or dominance of an allele.

Mutations can affect phenotype by altering the amino-acid composition of a protein or by altering the amount of normal protein produced. For mutations that alter a protein's amino-acid sequence, it is possible to explain the differing phenotypic effects of mutations in the same gene by assuming that each amino acid in a polypeptide has its own specific effect on protein structure. Thus changes in different types of amino acids at different positions in a polypeptide chain will have different effects on protein function that will in turn influence phenotype in different ways.

Most proteins have an active site that carries out a particular task while other parts of the protein support the shape and position of that site. Mutations that change the identity of amino acids at the active site may have more serious consequences

Fast Forward

Using Mutagenesis to Look at Biological Processes

Geneticists can use mutations to dissect complicated biological processes into their protein components. To determine the specific, dedicated role of each protein, they introduce mutations into the genes encoding the protein. The mutations knock out, or delete, functional protein either by preventing protein production altogether or by altering it such that the resulting protein is nonfunctional. The researchers then observe what happens when the cell or organism attempts to perform the biological process without the deleted protein.

In the 1960s, Robert Edgar set out to delineate the function of the proteins determined by all the genes in the T4 bacteriophage genome. As we have seen, the T4 virus infects *Escherichia coli* bacteria. After a single viral particle infects a bacterium, the host cell stops producing bacterial proteins and becomes a factory for making only viral proteins. Thirty minutes after infection, the bacterial cell lyses, releasing 100 new viral particles. The head of each particle carries a DNA genome 200,000 base pairs in length that encodes at least 120 genes.

Edgar's experimental design was to obtain many different mutant bacteriophages, each containing a mutation that inactivates one of the genes essential for viral reproduction. By analyzing what went wrong with each type of mutant during the infective cycle, he would learn something about the function of each of the proteins produced by the T4 genome.

There was just one barrier to implementing this plan. A mutation that prevents viral reproduction by definition makes the virus unable to reproduce and therefore unavailable for experimental growth and study. The solution to this dilemma came with the discovery of *conditional lethal* mutants: microbes or other organisms (in this case, viral particles) carrying mutations that are lethal to the organism under one condition but not another. One type of conditional lethal mutant used by Edgar was temperature sensitive; that is, the mutant T4 phage could reproduce at low, but not at high, temperatures. The temperature-sensitive mutations producing these mutants changed one amino acid in a polypeptide such that the protein was stable and functional at a low temperature but became unstable and nonfunctional at a higher temperature. Temperature-sensitive mutations can occur in many,

but not all, genes. Edgar isolated thousands of conditional lethal bacteriophage T4 mutants, and using complementation studies, discovered that they fall into 65 complementation groups. These complementation groups defined 65 genes, which were then mapped onto the T4 genome.

Edgar next studied the consequences of infecting bacterial cells under *restrictive conditions*, that is, under conditions in which the mutant protein could not function. For the temperature-sensitive mutants, the restrictive condition was high temperature. He found that mutations in 17 genes prevented viral DNA replication and concluded that these 17 genes contribute to that process. Mutations in most of the other 48 genes did not impede viral DNA replication but were necessary for the construction of complete viral particles. Electron microscopy showed that mutations in these 48 genes caused the accumulation of partially constructed viral particles. Edgar assumed that these partial particles were normal intermediates in the virus construction pathway, and on the basis of this assumption, he used the incomplete particles to plot the path of viral assembly. As Fig. A illustrates, three subassembly lines—one for the tail, one for the head, and one for the tail fibers—come together during the assembly of the viral product. Once the heads are completed and filled with DNA, they attach to the tails, after which, attachment of the fibers completes particle construction. It would have been very difficult to discern this trilateral assembly pathway by any means other than mutagenesis-driven genetic dissection.

Between 1990 and 1995, molecular geneticists determined the complete DNA sequence of the T4 genome, and then using the genetic code dictionary (described in Chapter 7) translated that sequence into coding regions for proteins (see Chapters 7 and 10 for details). In addition to the 65 genes identified by Edgar, another 55 genes became evident from the sequence. Edgar did not find these genes because they are not essential to viral reproduction under the conditions used in the laboratory. The previously unidentified genes most likely play important roles in the T4 life cycle outside the laboratory, perhaps when the virus infects hosts other than the *E. coli* strain normally used in the laboratory or when the virus grows under different environmental conditions and is competing with other viruses.

than those that affect amino acids outside the active site. Some kinds of amino-acid substitution, such as replacement of an amino acid having a basic side chain with an amino acid having an acidic side chain, would be more likely to compromise protein function than substitutions that retain the chemical characteristics of the original amino acid.

Some mutations do not affect the amino-acid composition of a polypeptide but still generate an abnormal phenotype. As we see in Chapter 7, such mutations change the amount of nor-

mal polypeptide produced by generating altered DNA sequences that disrupt or abolish the genetic processes responsible for decoding a gene into a polypeptide.

Dominance Relations between Alleles Depend on the Relation between Protein Function and Phenotype

The one gene, one polypeptide hypothesis helps account for the dominance relationships between the two alleles of a

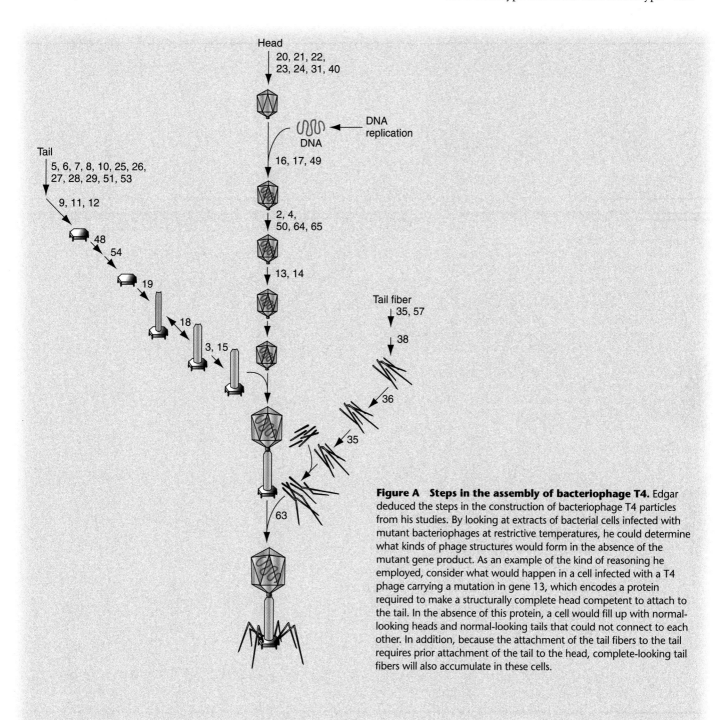

Figure A Steps in the assembly of bacteriophage T4. Edgar deduced the steps in the construction of bacteriophage T4 particles from his studies. By looking at extracts of bacterial cells infected with mutant bacteriophages at restrictive temperatures, he could determine what kinds of phage structures would form in the absence of the mutant gene product. As an example of the kind of reasoning he employed, consider what would happen in a cell infected with a T4 phage carrying a mutation in gene 13, which encodes a protein required to make a structurally complete head competent to attach to the tail. In the absence of this protein, a cell would fill up with normal-looking heads and normal-looking tails that could not connect to each other. In addition, because the attachment of the tail fibers to the tail requires prior attachment of the tail to the head, complete-looking tail fibers will also accumulate in these cells.

gene in a diploid organism. The theory is simple: Phenotype depends on protein activity, and protein activity varies with the allele; an alternate allele can change the amino acid sequence or the amount of polypeptide produced. Applications of the theory, however, take some unexpected turns, because the amount of protein required to generate a particular phenotype varies from protein to protein and thus the gene activity required to produce a normal phenotype varies from gene to gene.

Alleles That Produce Nonfunctional Proteins Are Usually Recessive

Mutations that abolish the function of a protein encoded by the wildtype allele are known as **null** mutations (Fig. 6.23). Such mutations either prevent synthesis of the protein or promote synthesis of a protein incapable of carrying out any function. For example, a deletion of a gene would by definition be a null allele. In an A^+a heterozygote, in which allele a is recessive to wildtype allele A^+, the A^+ allele would generate functional

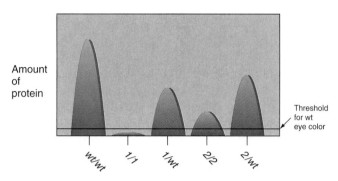

Figure 6.23 Why some mutant alleles are recessive.
Researchers subjected extracts made from flies of various genotypes to rocket immunoelectrophoresis; the size of the rocket reflects the amount of an enzyme called xanthine dehydrogenase. Flies need only 5% of the xanthine dehydrogenase produced by wildtype homozygotes (+/+) to express normal eye color. Flies homozygous for mutation 1, a null allele of the gene encoding this enzyme, produce undetectable amounts of xanthine dehydrogenase and thus have mutant eye color. Allele 1 is recessive to the wildtype allele because heterozygotes have more than the very small threshold amounts of enzyme needed for normal eye color. Mutation 2 is hypomorphic, resulting in the production of about 10% of the amount of xanthine dehydrogenase directed by the wildtype allele. Though the eye color of homozygotes or heterozygotes for mutation 2 is normal, this allele causes other recessive phenotypes that are more sensitive to reductions in the amount of the enzyme.

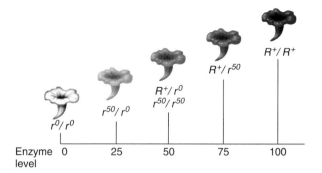

Figure 6.24 When a phenotype varies continuously with levels of protein function, incomplete dominance results.

protein, while the null a allele would not. If the amount of protein produced by the single A^+ allele (usually, though not always, half the amount produced in an $A^+ A^+$ cell) is above the threshold amount sufficient to fulfill the normal biochemical requirements of the cell, then the phenotype of the $A^+ a$ heterozygote will be wildtype. For the large number of genes that function in this way, it is clear that $A^+ A^+$ cells actually make more than twice as much of the protein needed for the normal phenotype.

A **hypomorphic** mutation produces either much less of a protein or a protein with very weak but detectable function (Fig. 6.23). In a $B^+ b$ heterozygote, where b is a hypomorphic allele recessive to wildtype allele B^+, the amount of protein activity will be somewhat greater than half the amount in a $B^+ B^+$ cell. Usually, this is enough activity to fulfill the normal biochemical requirements of the cell. Most hypomorphic mutations are detectable only in homozygotes, and only if the reduction in protein amount or function is sufficient to cause an abnormal phenotype.

Incomplete Dominance Can Arise When the Phenotype Varies in Proportion to the Amount of Functional Protein
Some combinations of alleles generate phenotypes that vary continuously with the amount of functional gene product. For example, mutations in a single pigment-producing gene can generate a red-to-white spectrum of flower colors, with the white resulting from the absence of an enzyme in a biochemical pathway (Fig. 6.24). Consider three alleles of the gene encoding this enzyme: R^+ specifies a high, wildtype amount of

the enzyme, r^{50} generates half the normal amount of the same enzyme; and r^0 is a null allele. $R^+ r^0$ heterozygotes produce pink flowers whose color is halfway between red and white because one half the $R^+ R^+$ level of enzyme activity is not enough to generate a full red. Combining R^+ or r^0 with the r^{50} allele produces pigmentation intermediate between red and pink or between pink and white.

Dominance Can Reflect Several Kinds of Phenomena
With phenotypes that are exquisitely sensitive to the amount of functional protein produced, even a relatively small change of twofold or less can cause a switch between distinct phenotypes. For example, a heterozygote for a null or hypomorphic mutation that generates only half the normal amount of functional gene product may look completely different from the wildtype. The T locus in mice has just such a mutation, with an easy-to-visualize dominant phenotype (Fig. 6.25a). Mice require the wildtype protein product of the T-locus gene during embryogenesis for the normal development of the posterior portion of the spinal cord and tail. Embryos heterozygous for a null mutation at the T locus produce only half the normal amount of the T-determined protein and mature into viable offspring that are normal in all respects except for the absence of the distal two-thirds of their tail. The notably shortened tail reflects the embryo's sensitivity to the level of T-gene product available during morphogenesis; half the normal amount of T protein is below the threshold needed for normal development. Only a minority of phenotypes are so sensitive to the amount of a particular protein. Thus, as described earlier, null and hypomorphic alleles usually produce phenotypes that are recessive to wildtype.

Another mechanism that can result in dominance is the extremely rare **hypermorphic** mutation that produces an allele generating either more protein than the wildtype allele or the same amount of a more efficient protein. If excess protein activity alters phenotype, the hypermorphic allele is dominant. A hypermorphic mutation in the rhodopsin gene produces a rhodopsin protein that is activated whether or not light is present, resulting in constant, low-level stimulation of rhodopsin in the photoreceptor cells that detect black and white. These cells, known as rod cells, function primarily at night. People

(a)

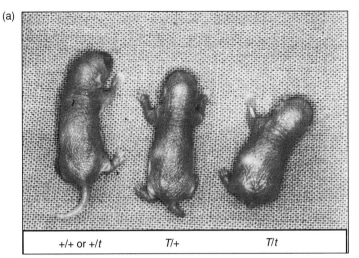

| +/+ or +/t | T/+ | T/t |

(b)

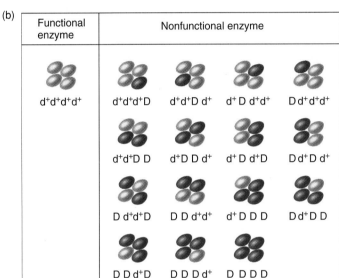

| Functional enzyme | Nonfunctional enzyme | | | |

d⁺d⁺d⁺d⁺   d⁺d⁺d⁺D   d⁺d⁺D d⁺   d⁺ D d⁺d⁺   D d⁺d⁺d⁺

d⁺d⁺D D   d⁺D D d⁺   d⁺ D d⁺D   D d⁺D d⁺

D d⁺d⁺D   D D d⁺d⁺   d⁺ D D D   D d⁺D D

D D d⁺D   D D D d⁺   D D D D

D = dominant mutant subunit
d⁺ = wildtype subunit

(c)

(d)

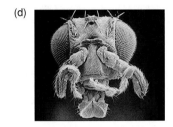

Figure 6.25 Why some mutant alleles are dominant.
(a) Mice heterozygous for a null mutation of the *T* locus have tails much shorter than normal, because the one-half normal amount of T-locus protein in these animals is insufficient to make a wildtype tail. The unusual *t* allele has no effect on its own, but enhance the phenotype caused by the dominant null *T* allele, for unknown reasons. (b) With proteins composed of four subunits encoded by the same gene, a dominant negative mutant may inactivate 15 out of every 16 multimers. (c) The *Kinky* allele in mice is a dominant negative mutation affecting a protein composed of more than one type of polypeptide; the resulting phenotype is a kink in the tail. (d) A neomorphic dominant mutation in the *Antennapedia* gene of *Drosophila* causes ectopic expression of this gene (which directs leg development) in larval "imaginal disks" that would otherwise develop into antennae.

with the mutation can still see in bright daylight, but have congenital night blindness. The blindness probably arises because the constant rhodopsin stimulation prevents adaptation of the rod cells to the very low light intensities present at night.

In yet another mechanism leading to dominance, some alleles of genes encode subunits of multimers that block the activity of the subunits produced by normal alleles; such blocking alleles are called **dominant negative.** Consider, for example, a gene encoding a polypeptide that associates with three other identical polypeptides in a four-subunit enzyme. All four subunits are products of the same gene. If a dominant mutant allele *D* directs the synthesis of a polypeptide that can still assemble into aggregates but whose presence in the multimer—even as one subunit out of four—abolishes enzyme function, the chance of a heterozygote producing a multimer composed solely of functional wildtype d^+ subunits is 1 in 16: $(1/2)^4 = 1/16 = 6.25\%$ (Fig. 6.25b). As a result, total enzyme activity in $D d^+$ heterozygotes is far less than that seen in wild-

type $d^+ d^+$ homozygotes. Dominant negative mutations can also affect subunits in multimers composed of more than one type of polypeptide. The *Kinky* allele at the *fused* locus in mice is an example of a dominant negative mutation (Fig. 6.25c).

Finally, a very rare class of dominant alleles arises from **neomorphic** mutations that generate a novel phenotype. Some neomorphic mutations produce proteins with a new function, while others cause genes to produce the normal protein but at an inappropriate time or place. A striking example of inappropriate protein production is the *Drosophila Antennapedia* (*Antp*) gene, active during embryonic and larval stages. Normally, the gene makes its protein product in tissues destined to become legs; the protein ensures that these tissues develop into legs and not, for example, head structures such as antennae. Dominant mutations of the gene cause production of the protein in the head region of the animal, where the *Antennapedia* gene is not normally active. Here, the misplaced protein causes tissues that would normally develop into antennae to develop into legs (Fig. 6.25d). Production of a protein outside of its normal place or time is called *ectopic expression.*

In summary, mutations alter phenotypes by altering the structure, amount, or place of protein produced. Some mutations cause a change in amino-acid sequence that alters protein function. Other mutations cause a change in amino-acid sequence that completely disrupts protein function. Still other mutations prevent a gene from making protein in the first place. Dominance relations between the wildtype and mutant alleles of genes in diploid organisms depend on how drastically a mutation influences protein production and how thoroughly phenotype depends on the normal wildtype amount of protein.

HOW GENE MUTATIONS AFFECT LIGHT-RECEIVING PROTEINS AND VISION: A COMPREHENSIVE EXAMPLE

Researchers first described anomalies of color perception in humans close to 200 years ago. Since that time, they have discovered a large number of mutations that modify human vision. By examining the phenotype associated with each mutation and then looking directly at the DNA alterations inherited with the mutation, they have learned a great deal about the genes influencing human visual perception and the function of the proteins they encode.

Several attributes of human subjects facilitate the experimental analysis of vision. First, people can recognize and describe variations in the way they see, from trivial differences in what the color red looks like, to not seeing any difference between red and green, to not seeing any color at all. Second, the highly developed science of psychophysics provides sensitive, noninvasive tests for accurately defining and comparing phenotypes. One diagnostic test, for example, is based on the fact that people perceive each color as a mixture of three different wavelengths of light—red, green, and blue—and can adjust ratios of red, green, and blue light of different intensities to match an arbitrarily chosen fourth wavelength such as yellow. The mixture of wavelengths does not combine to form the fourth wavelength; it just appears that way to the eye. A person with normal vision, for instance, will select a well-defined proportion of red and green lights to match a particular yellow, but a person who can't tell red from green will permit any proportion of these two color lights to make the same match. Finally, since inherited variations in the visual system rarely affect fecundity or longevity in modern human societies, mutations generating many of the new alleles that change visual perception remain in a population over time.

The Cellular and Molecular Basis of Vision
The Cells

People perceive light through neurons in the retina at the back of the eye (Fig. 6.26a). These neurons are of two types: rods and cones. The rods, which make up 95% of all light-receiving neurons, are stimulated by weak light over a range of wave-

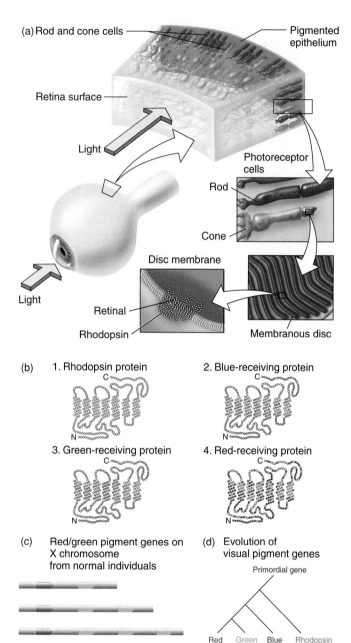

Figure 6.26 The cellular and molecular basis of vision.
(a) Rod and cone cells in the retina carry millions of membrane-bound photoreceptor proteins. (b) The photoreceptor in rod cells is rhodopsin. The blue-, green-, and red-receiving proteins in cone cells are polypeptides that are related to, but distinct from, rhodopsin. (c) Genes for the red and green photoreceptors are found in a cluster on the human X chromosome. Most people have one red gene and between one and three green genes. (d) The relatedness of rhodopsin and the three color receptors suggests that the genes for these proteins evolved from the gene for a primordial photoreceptor through three independent gene duplication events followed by divergence of the duplicated copies.

lengths. At higher light intensities, the rods become saturated and no longer send meaningful information to the brain. This is when the cones take over, processing wavelengths of bright light that enable us to see color. The cones come in three forms—one specializes in the reception of red light, a second in the reception of green, and a third in the reception of blue. For each photoreceptor cell, the act of reception consists of absorbing photons from light of a particular wavelength, transducing information about the number and energy of those photons to electrical signals, and transmitting the signals via the optic nerve to the brain.

Four Genes, Four Polypeptides

The protein that receives photons and triggers the processing of information in rod cells is rhodopsin. It consists of a single polypeptide chain containing 348 amino acids that snakes back and forth across the cell membrane (Fig. 6.26b.1). One lysine within the chain associates with retinal, a carotenoid pigment molecule that actually absorbs photons. The amino acids in the vicinity of the retinal constitute rhodopsin's active site; by positioning the retinal in a particular way, they determine its response to light. Each rod cell contains approximately 100 million molecules of rhodopsin in its specialized membrane. As we have seen, the gene governing the production of rhodopsin is on chromosome 3.

The protein that receives and initiates the processing of photons in the blue cones is a relative of rhodopsin, also consisting of a single polypeptide chain containing 348 amino acids, and also encompassing one molecule of retinal. Slightly less than half of the 348 amino acids in the blue-receiving protein are the same as those found in rhodopsin; the rest are different and account for the specialized light-receiving ability of the protein (Fig. 6.26b.2). The gene for the blue protein is on chromosome 7.

Similarly related to rhodopsin are the red- and green-receiving proteins in the red and green cones. These are also single polypeptides, 364 amino acids in length, associated with retinal and embedded in the cell membrane (Fig. 6.26b.3 and 4). Like the blue protein, the red and green proteins differ from rhodopsin in slightly more than half of their amino acids; they differ from each other in only 4 amino acids out of every hundred. Even these small differences, however, are sufficient to differentiate the light sensitivities of the two types of cones, and confer on them distinct spectral sensitivities. The genes for the red and green proteins both reside on the X chromosome in a tandem head-to-tail arrangement. Most individuals have one red gene and one to three green genes on their X chromosomes (Fig. 6.26c).

The Rhodopsin Gene Family Evolved by Duplication and Divergence

The similarity in structure and function between the four rhodopsin proteins suggests that the genes encoding these polypeptides arose by duplication of an original photoreceptor gene and then divergence through the accumulation of many mutations. Many of the mutations that promoted the ability to

see color must have provided advantages that evolutionary forces selected for over a period of millions of years. The red and green genes are the most similar, differing by less than five nucleotides out of every hundred. This suggests that they diverged from each other only in the relatively recent evolutionary past. Lower amino acid similarity of the red or green proteins with the blue protein, and even lower relatedness between rhodopsin and any color photoreceptor, reflect earlier duplication and divergence events (Fig. 6.26d).

How Mutations in the Rhodopsin Family Influence the Way We See

Many Amino-Acid Substitutions in Rhodopsin Result in Partial or Complete Blindness

At least 29 different single nucleotide substitutions in the rhodopsin gene cause an autosomal dominant vision disorder known as *retinitis pigmentosa* that begins with an early loss of rod function followed by a slow progressive degeneration of the peripheral retina. Figure 6.27a shows the location of the amino acids affected by these mutations. These amino-acid changes result in abnormal rhodopsin proteins that either do not fold properly, or once folded are unstable. While normal rhodopsin is an essential structural element of rod cell membranes, these nonfunctional mutant proteins are retained in the body of the cell where they remain unavailable for insertion into the membrane. Rod cells that cannot incorporate enough rhodopsin into their membrane eventually die. Depending on how many rod cells die, partial or complete blindness ensues.

Other mutations in the rhodopsin gene cause the far less serious condition of night blindness (Fig. 6.27a). These hypomorphic mutations change the protein's amino-acid sequence so that the threshold of stimulation required to trigger the vision cascade increases. With the change, very dim light is no longer enough to initiate vision.

Mutations in the Genes for the Cone-Cell Pigments Alter Color Vision in Predictable Ways

Vision problems caused by mutations in the cone-pigment genes are less severe than vision problems caused by similar defects in the rod cells' rhodopsin genes, most likely because the rods make up 95% of a person's light-receiving neurons, while the cones comprise only about 5%. Some mutations in the blue gene on chromosome 7 cause *tritanopia*, a defect in the ability to discriminate between colors that differ only in the amount of blue light they contain (Fig. 6.27b). Mutations in the red gene on the X chromosome can modify or abolish red protein function and as a result, the red cone cells' sensitivity to light. For example, a change at position 203 in the red-receiving protein from cysteine to arginine disrupts one of the disulfide bonds required to support the protein's tertiary structure (Fig. 6.27c). Without that bond, the protein cannot stably maintain its native configuration, and a person with the mutation has red colorblindness. In other mutations affecting the red gene, small deletions from neighboring DNA that does not specify amino acids reduces or eliminates gene function, probably by incapacitating control elements

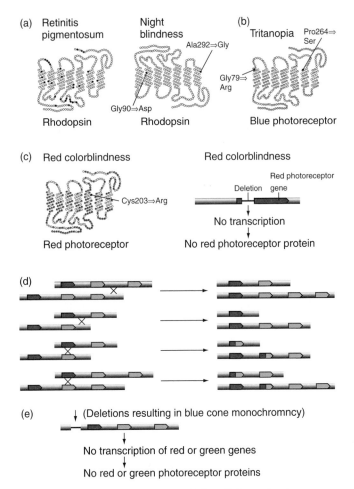

Figure 6.27 How mutations modulate light and color perception. (a) Various amino-acid substitutions (black dots) result in retinitis pigmentosum by disrupting the three-dimensional structure of rhodopsin (*left*). Two different substitutions diminish rhodopsin's sensitivity to light, causing night blindness (*right*). (b) Several different substitutions in the blue pigment produce a kind of colorblindness called tritanopia. (c) A mutation affecting the amino acid at position 203 of the red photoreceptor destabilizes this protein, leading to red colorblindness (*left*). Deletion of nucleotides in DNA regions that control the expression of the red photoreceptor can prevent synthesis of this pigment and cause red colorblindness (*right*). (d) Unequal crossing-over made possible by the high level of homology between the red and green genes can yield various arrangements of these genes on the X chromosome. (e) Deletions that remove a region controlling expression of both the red and green genes prevent the synthesis of bothphotoreceptors, resulting in blue cone monochromacy.

needed for gene activity (Fig. 6.27c). People whose cells carry one of these small deletions are partially or completely red color-blind.

Unequal Crossing-Over between the Red and Green Genes Produces Most of the Variations in Red-Green Perception

People with normal color vision have a single red gene; some of these normal individuals also have a single adjacent green gene, while others have two to five green genes. The red and green genes are 96% identical in DNA sequence; the different green genes, 99.9% identical. The proximity and high degree of homology make these genes unusually prone to unequal crossing-over, and a variety of unequal recombination events produce DNA containing no red gene, no green gene, various combinations of green genes, or a hybrid red-green gene (Fig. 6.27d). These different DNA combinations account for the large majority of the known aberrations in red-green color perception, with the remaining abnormalities stemming from point mutations, as described earlier. Because the accurate perception of red and green depends on the differing ratios of red and green light processed, people with no red or no green gene perceive red and green as the same color.

Some Mutations Knock Out the Ability to See Both Red and Green

Geneticists have detected seven different deletions that produce an X-linked trait in which green and red cones do not function. The trait is called *blue cone monochromacy*, because people with the condition see only blue-related colors in bright light. Characterization at the DNA level has shown that all seven deletions overlap in a 600-bp segment outside the coding region of the red and green genes (Fig. 6.27e). It is likely that this 600-bp region serves as a long-range control element necessary for the activity of all the genes in the red-green array.

In short, we see the way we do in part because four genes direct the production of four polypeptides in the rod and cone cells of our retina. Mutations that alter those polypeptides or their amounts change or destroy our perception of light or color.

CONNECTIONS

Careful studies of mutations showed that genes are linear arrays of mutable elements that direct the assembly of amino acids in a polypeptide. The mutable elements are the nucleotide building blocks of DNA.

Biologists call the parallel between the sequence of nucleotides in a gene and the order of amino acids in a polypeptide **colinearity.** In the next chapter, we see that colinearity arises from base pairing, a genetic code, and specific enzymes that guide the flow of information from DNA through RNA to protein. How cells accomplish the DNA-directed assembly of polypeptides is the subject of Chapter 7.

ESSENTIAL CONCEPTS

1. Mutations are alterations in the DNA molecule that occur by chance and modify the genome at random. Once they occur, they can be transmitted from generation to generation.
   a. Mutations that affect phenotype occur naturally at a very low rate, which varies from gene to gene. Forward mutations occur more often than reversions.
   b. The agents of spontaneously occurring mutations include chemical hydrolysis, radiation such as cosmic rays and ultraviolet light, and mistakes during DNA replication. Mutagens raise the frequency of mutation above the spontaneous rate. Cells have evolved enzyme systems that minimize mutations.
   c. Mutations are the raw material of evolution. The action of natural selection on heritable mutations is a major agent of evolution.

2. What mutations tell us about gene structure:
   a. Mutations within the same gene usually fail to complement each other. The concept of a complementation group thus defines the gene as a unit of function.
   b. A gene is composed of a linear sequence of nucleotide pairs in a discrete, localized region of a chromosome. Recombination can occur within a gene, even between adjacent nucleotide pairs.

3. What mutations tell us about gene function:
   a. The function of most genes is to specify the linear sequence of amino acids in a particular polypeptide.
   b. The sequence of amino acids in a polypeptide determines the polypeptide's three-dimensional structure, which in turn determines its function. Each protein consists of one, two, or more polypeptides. One gene encodes one polypeptide. Proteins composed of two or more different subunits are encoded by two or more genes.

4. How genotype correlates with phenotype:
   a. The functions of proteins produced within an organism determine phenotype.
   b. Some mutations modify phenotype by altering the sequence of amino acids and thus the function of a protein. Other mutations do not affect a polypeptide's amino-acid sequence, but instead affect the amount, time, or place of protein production.
   c. Dominance relations between alleles depend on the relation between protein function and phenotype. Recessive alleles often produce nonfunctional proteins. Incomplete dominance arises when phenotype varies with the amount of functional protein produced. Dominant alleles reflect either situations in which small perturbations in the amount of protein produced disrupt normal phenotypes or other rare events that affect protein structure or synthesis.

SOCIAL AND ETHICAL ISSUES

1. Sean was diagnosed with AIDS in 1986. The drug AZT (azidothymidine) had been shown to be effective in inhibiting replication of the HIV-1 virus in 1985 and thus was a potential therapeutic agent against AIDS. Sean read about research indicating this effect and wanted to begin taking the drug as soon as possible. However, the standard procedure for testing new drugs involves extensive clinical trials that can take years and can require frequent monitoring. One way Sean could get the drug before it was commercially available was to volunteer for trial drug testing. However, Sean lives in a small community far from the research universities where testing was done. Is it fair that those people who do not live near a major research facility are unable to get drugs that are in clinical trials? Does an individual have a right to demand that the medical community get the drug for them? Should physicians be allowed to prescribe drugs for AIDS patients before the drugs are approved, if the patients would like to try them?

2. The American Cancer Society (ACS) is a nonprofit organization that funds much cancer research, treatment, and public education. In 1996, the ACS decided to offer exclusive endorsements to *Nicoderm* nicotine patches. The ACS logo would appear on the packaging of the patches and the ACS would, in return, receive a sum of money that will be used to support its cancer-fighting programs. Is this an appropriate activity for a charitable organization such as the American Cancer Society? Could these endorsements be damaging to the reputation of the organization? What criteria could or should the ACS use in deciding whose product they will endorse?

3. Chemicals that are mutagenic are identified by tests such as the Ames test, which measures the level of mutagenesis in bacteria. The susceptibility of humans to mutagenic chemicals may vary depending on the genetic makeup of the individual. The dose that affects one person may be different from that which affects another. In some cases, there are no tests that determine level of susceptibility. How can we translate the results of the Ames test from bacteria to carcinogenicity in humans? Often reports of Ames tests on a chemical make newspaper headlines. Is this a useful and honest way to report findings that could affect human health?

S O L V E D   P R O B L E M S

I. Mutations can often be reverted to wildtype by treatment with mutagens. The type of mutagen that will reverse a mutation gives us information about the nature of the original mutation. The mutagen EMS almost exclusively causes transitions; proflavin is an intercalating agent that causes insertion or deletion of a base; ultraviolet (UV) light causes single-base substitutions. Several *E. coli* met^- mutants were treated with three mutagens separately and spread onto a plate lacking methionine to look for revertants. (In the chart − indicates no growth, and + indicates that some colonies grew.)

Mutant number	Mutagen treatment		
	EMS	Proflavin	UV light
1	+	−	+
2	−	+	−
3	−	−	−
4	−	−	+

a. Given the results, what can you say about the nature of the original mutation in each of the strains?

b. Experimental controls are designed to eliminate possible explanations for the results, thereby ensuring that data is interpretable. In the experiment described, we scored the presence or absence of colonies. How do we know if colonies that appear on plates are mutagen-induced revertants? What else could they be? What control would enable us to be confident of our revertant analysis?

Answer

To answer this question, you need to understand the concepts of mutation and reversion.

a. Mutation 1 is reverted by the mutagen that causes transitions, *so mutation 1 must have been a transition.* Consistent with this conclusion is the fact the UV light can also revert the mutation and the intercalating agent proflavin does not cause reversion. *Mutation 2 is reverted by proflavin and therefore must be either an insertion or a deletion of a base.* The other two mutagens do not revert mutation 2. Mutation 3 is not reverted by any of these mutagenic agents. It is therefore not a single-base substitution, a single-base insertion, or a single-base deletion. *Mutation 3 could be a deletion of several bases or an inversion.* Mutation 4 is reverted only by UV light so it is a single-base change, but not a transition, since EMS did not revert the mutation. Mutation 4 must be a transversion.

b. *The colonies on the plates could arise by spontaneous reversion of the mutation.* Spontaneous reversion should occur with lower frequency than mutagen-induced reversion. The important control here is to *spread each mutant culture without any mutagen treatment onto selective media to assess the level of spontaneous reversion.*

II. Imagine that 10 independently isolated recessive lethal mutations (l^1, l^2, l^3, etc.) map to chromosome 7 in mice. Complementation testing was done by mating all pairwise combinations of heterozygotes bearing these lethal mutations. Absence of complementation was scored by examining pregnant females for dead fetuses. A "+" in the chart means that the two lethals complemented, and dead embryos were not found. A "−" indicates that dead embryos were found, at the rate of about 1 in 4 conceptions. (The crosses between heterozygous mice would be expected to yield the homozygous recessive showing the lethal phenotype in 1/4 of the embryos.) The genotypes of the parental heterozygotes in each cross are listed across the top and down the left side of the chart.

	l^1/+	l^2/+	l^3/+	l^4/+	l^5/+	l^6/+	l^7/+	l^8/+	l^9/+	l^{10}/+
l^1/+	−	+	+	+	+	−	−	+	+	+
l^2/+		−	+	+	+	+	+	+	+	−
l^3/+			−	−	−	+	+	−	−	+
l^4/+				−	−	+	+	−	−	+
l^5/+					−	+	+	−	−	+
l^6/+						−	−	+	+	+
l^7/+							−	+	+	+
l^8/+								−	−	+
l^9/+									−	+
l^{10}/+										−

How many genes do the 10 lethal mutations represent? What are the complementation groups?

Answer

This problem involves the application of the complementation concept to a set of data. There are two ways to analyze these results. You can focus on the mutations that do complement each other, conclude that they are in different genes, and begin to create a list of mutations in separate genes. Alternatively, you can focus on mutations that do not complement each other and therefore are alleles of the same genes. The latter approach is more productive when several mutations are involved. For example, l^1 does not complement l^6 and l^7. These three alleles are in one complementation group. l^2 does not complement l^{10}; they are in a second complementation group. l^3 does not complement l^4, l^5, l^8, or l^9, so they form a third complementation group. *There are three complementation groups.* (Note also that for each mutant, the cross between individuals carrying the same alleles resulted in no complementation, because the homozygous recessive lethal was generated.) *The three complementation groups consist of:* (1) l^1, l^6, l^7; (2) l^2, l^{10}; (3) l^3, l^4, l^5, l^8, l^9.

III. W, X, and Y are the intermediates (in that order) in a biochemical pathway whose product is Z. Z^- mutants are found in five different complementation groups. *Z1* mutants will grow on Y or Z, but not W or X. *Z2* mutants will grow on X, Y, or Z. *Z3* mutants will only grow on Z. *Z4* mutants will grow on Y or Z. Finally, *Z5* mutants will grow on W, X, Y, or Z.

  a. Order the five complementation groups in terms of the steps they block.

  b. What does this genetic information reveal about the nature of the enzyme that carries out the conversion of X to Y?

Answer

This problem requires that you understand complementation and the connection between genes and enzymes in a biochemical pathway.

  a. A biochemical pathway represents an ordered set of reactions that must occur to produce a product. This problem gives the order of intermediates in a pathway for producing product Z. The lack of any enzyme along the way will cause the phenotype of Z⁻ but the block can occur at different places along the pathway.

If the mutant grows when given an intermediate compound, the enzymatic (and hence gene) defect must be before production of that intermediate compound. The *Z1* mutants that grow on Y or Z (but not on W or X) must have a defect in the enzyme that produces Y. *Z2* mutants have a defect prior to X; *Z3* mutants have a defect prior to Z; *Z4* mutants have a defect prior to Y; *Z5* have a defect prior to W. *The five complementation groups can be placed in order of activity within the biochemical pathway as follows:*

$$Z5 \quad\quad Z2 \quad\quad Z1, Z4 \quad\quad Z3$$
$$\longrightarrow W \longrightarrow X \longrightarrow Y \longrightarrow Z$$

  b. Mutants *Z1* and *Z4* affect the same step, but because they are in different complementation groups we know they are in different genes. *Mutations Z1 and Z4 are probably in genes that encode subunits of a multisubunit enzyme that carries out the conversion of X to Y.*

P R O B L E M S

6-1 The following is a list of mutational changes. For each of the specific mutations described below, indicate which of the following terms could apply, either as a description of the mutation or as a possible cause. More than one term can apply.

  a. transition

  b. base substitution

  c. transversion

  d. inversion

  e. translocation

  f. deletion

  g. insertion

  h. deamination

  i. X-ray irradiation

  j. intercalator

  k. unequal crossing-over

  1. an AT base pair in the wildtype gene is changed to a GC pair

  2. an AT base pair is changed to a TA pair

  3. the sequence AAGCTTATCG is changed to AAGCTATCG

  4. the sequence AAGCTTATCG is changed to AAGCTTTATCG

  5. the sequence AACGTTATCG is changed to AATGTTATCG

  6. the sequence AACGTCACACACACATCG is changed to AACGTCACATCG

  7. the gene map in a given chromosome arm is changed from *bog-rad-fox¹-fox²-try-duf* (where *fox¹* and *fox²* are highly homologous, recently diverged genes) to *bog-rad-fox¹-fox³-fox²-try-duf* (where *fox³* is a new gene with one end similar to *fox¹* and the other similar to *fox²*)

  8. the gene map in a chromosome is changed from *bog-rad-fox¹-fox²-try-duf* to *bog-rad-fox²-fox¹-try-duf*

  9. the gene map in a given chromosome is changed from *bog-rad-fox¹-fox²-try-duf* to *bog-rad-fox¹-mel-qui-txu-sqm*

6-2 The DNA sequence of a gene from three independently isolated mutants is given below. Using this information, what is the sequence of the wildtype gene in this region?

  mutant 1   ACCGTAATCGACTGGTAAACTTTGCGCG

  mutant 2   ACCGTAGTCGACCGGTAAACTTTGCGCG

  mutant 3   ACCGTAGTCGACTGGTTAACTTTGCGCG

6-3 Over a period of several years, a large hospital kept track of the number of births of babies displaying the trait achondroplasia. Achondroplasia is a very rare autosomal dominant condition resulting in dwarfism with abnormal body proportions. After 120,000 births, it was noted that there had been 27 babies born with achondroplasia. One physician was interested in determining how many of these dwarf babies result from new mutations, and whether the apparent mutation rate in his area was higher than normal. He looked up the families of the 27 dwarf births, and discovered that 4 of the dwarf babies had a dwarf parent. What is the apparent mutation rate of the achondroplasia gene in this population? Is it unusually high or low?

6-4 Among mammals, measurements of the rate of generation of autosomal recessive mutations have been made almost exclusively in mice, while many measurements of the rate of generation of dominant mutations have been made both in mice and in humans. Why do you think there has been this difference?

6-5 In a genetics lab, Kim and Maria infected a sample from an *E. coli* culture with a particular virulent bacteriophage. They noticed that most of the cells were lysed, but a few survived. The survival rate in their sample was about 1×10^{-4}. Kim was sure the bacteriophage induced the resistance in the cells, while Maria thought that resistant mutants probably already existed in the sample of cells they used. Earlier, for a different experiment, they had spread a dilute suspension of *E. coli* onto solid medium in a large petri dish, and, after seeing that about 10^5 colonies were growing up, they had replica plated that plate onto three other plates. Kim and Maria decided to use these plates to test their theories. They pipette a suspension of the bacteriophage onto each of the three replica plates. What should they see if Kim is right? What should they see if Maria is right?

6-6 Suppose you wanted to study genes controlling the structure of bacterial cell surfaces. You decide to start by isolating bacterial mutants that are resistant to infection by a bacteriophage that binds to the cell surface. The selection procedure is simple: spread cells from a culture of sensitive bacteria on a petri plate, expose them to a high concentration of phages, and pick the bacterial colonies that grow. To set up the selection you could (1) spread cells from a single liquid culture of sensitive bacteria on many different plates and pick every resistant colony *or* (2) start many different cultures, each grown from a single colony of sensitive bacteria, and spread one plate from each culture. Pick a single mutant from each plate. Which method would ensure that you are isolating many independent mutations?

6-7 A wildtype male *Drosophila* was exposed to a large dose of X rays, and was then mated to an unirradiated female, one of whose X chromosomes carried both a dominant mutation for the trait *Bar* eyes and several inversions. Many F_1 females from this mating were recovered who had the *Bar*, multiply inverted X chromosome from their mother, and an irradiated X chromosome from their fathers. (The rearrangements on the maternal X chromosome ensure that, if recombination occurs between the irradiated X chromosome and the unirradiated X chromosome in these F_1 females, recombinant X chromosomes will not be recovered in viable offspring, for reasons explained in Chapter 12.) These F_1 females were mated to normal males, and their sons were examined. Most females produced *Bar* and wildtype sons in equal proportions. There were three exceptional females, however. Female A produced as many sons as daughters but half of the sons had *Bar* eyes, and the other half had white eyes. Female B produced half as many sons as daughters, and all of the sons had *Bar* eyes. Female C produced 75% as many sons as daughters. Of these sons, 2/3 had *Bar* eyes and 1/3 had wildtype eyes. Explain the results obtained with each exceptional F_1 female.

6-8 A wildtype *Drosophila* female was mated to a wildtype male that had been exposed to X rays. One of the F_1

females was then mated with a male that had the following recessive markers on the X chromosome: *yellow* body (*y*), *crossveinless* wings (*cv*), *cut* wings (*ct*), *singed* bristles (*sn*), and *miniature* wings (*m*). These markers are known to map in the order:

$$y \text{ - } cv \text{ - } ct \text{ - } sn \text{ - } m$$

The progeny of this second mating were unusual in two respects. First, there were twice as many females as males. Second, while all of the males were wildtype in phenotype, 1/2 of the females were wildtype and the other 1/2 exhibited the *ct, sn* phenotype.
a. What did the X rays do to the irradiated male?
b. Draw the X-chromosome pair present in a progeny female fly that was phenotypically *ct* and *sn*.
c. If a *ct, sn* female fly derived from the cross described here were then crossed to a wildtype male, what phenotypic classes would you expect to find among the progeny males?

6-9 When the *his⁻* *Salmonella* strain used in the Ames test is exposed to substance X, no *his⁺* revertants are seen. If, however, rat liver supernatant is added to the cells along with substance X, revertants do occur. Is substance X a potential carcinogen for human cells? Explain.

6-10 When a particular mutagen identified by the Ames test is injected into mice, it causes the appearance of many tumors, showing that this substance is carcinogenic. When cells from these tumors are injected into other mice not exposed to the mutagen, almost all of the new mice develop tumors. However, when mice carrying mutagen-induced tumors are mated to unexposed mice, virtually all of the progeny are tumor free. Why can the tumor be transferred horizontally (by injecting cells) but not vertically (from one generation to the next)?

6-11 When 1 million cells of a culture of haploid yeast carrying a *met⁻* auxotrophic mutation were plated on petri plates lacking methionine (met), five colonies grew. You would expect cells in which the original *met⁻* mutation was reversed (by a base change back to the original sequence) would grow on the media lacking methionine, but some of these apparent reversions could be due to a mutation in a different gene that somehow suppresses the original *met⁻* mutations. How would you be able to determine if the mutations in your five colonies were due either to a precise reversion of the original *met⁻* mutation or to the generation of a suppressor mutation in a gene on another chromosome?

6-12 Imagine that you caught a female albino mouse in your kitchen, and decided to keep it for a pet. A few months later, while vacationing in Guam, you caught a male albino mouse, and decided to take it home for some interesting genetic experiments. The question you are wondering about is whether the two mice are both albino due to mutations in the same gene. What could you do to find out the answer to this question? (Assume that both mutations are recessive.)

6-13 In a haploid yeast strain, eight recessive mutations were found that resulted in a requirement for the amino acid lysine. All the mutations were found to revert at a frequency of about 1×10^{-6}, except mutations 5 and 6 which did not revert. Matings were made between a and α cells carrying these mutations. The ability of the resultant diploid strains to grow on minimal medium in the absence of lysine is shown in the chart (+ means growth and − means no growth).

a. How many complementation groups were revealed by this data? Which point mutations are found within which complementation groups?

	1	2	3	4	5	6	7	8
1	−	+	+	+	+	−	+	−
2	+	−	+	+	+	+	+	+
3	+	+	−	−	−	−	−	+
4	+	+	−	−	−	−	−	+
5	+	+	−	−	−	−	−	+
6	−	+	−	−	−	−	−	−
7	+	+	−	−	−	−	−	+
8	−	+	+	+	+	−	+	−

b. The same diploid strains are now induced to undergo sporulation. The vast majority of resultant spores cannot form colonies when plated on minimal medium minus lysine (auxotrophic spores). However, particular diploids can produce rare spores that do form colonies when plated on minimal medium minus lysine (prototrophic spores). The table shows whether (+) or not (−) any prototrophic spores are formed upon sporulation of the various diploid cells.

	1	2	3	4	5	6	7	8
1	−	+	+	+	+	+	+	+
2	+	−	+	+	+	+	+	+
3	+	+	−	+	−	+	+	+
4	+	+	+	−	−	−	+	+
5	+	+	−	−	−	−	+	+
6	−	+	+	−	−	−	+	+
7	+	+	+	+	+	+	−	+
8	+	+	+	+	+	+	+	−

When prototrophic spores occur during sporulation of the diploids discussed, what ratio of auxotrophic:prototrophic spores would you generally expect to see in any tetrad containing such a prototrophic spore? Explain the ratio you expect.

c. Utilizing the data from all parts of this question, draw the best map of the eight lysine mutations under study. Show the extent of any deletions involved, and indicate the boundaries of the various complementation groups.

6-14 You have five T4 rII^- mutants that will not grow on *E. coli* K(λ). You mixed together two mutants (indicated below), added them to *E. coli* K(λ), and scored for the ability of the mixture to grow and make plaques (indicated as a + below).

	1	2	3	4	5
1	−	+	+	−	+
2		−	−	+	−
3			−	+	−
4				−	+
5					−

a. How many genes were identified by this analysis?
b. Which mutants belong to the same complementation groups?

6-15 The following nine rII^- mutants of bacteriophage T4 were used in pairwise infections of *E. coli* K(λ) hosts. Six of the mutations in these phages are point mutations; the other three are deletions. The ability of the doubly infected cells to produce progeny phages in large numbers is scored below in the chart.

	1	2	3	4	5	6	7	8	9
1	−	−	+	+	−	−	−	+	+
2		−	+	+	−	−	−	+	+
3			−	−	+	−	+	−	−
4				−	+	−	+	−	−
5					−	−	−	+	+
6						−	−	−	−
7							−	+	+
8								−	−
9									−

The same nine mutants were then used in pairwise infections of *E. coli* B hosts. The production of progeny phage that can subsequently infect *E. coli* K(λ) hosts is now scored. In the table, 0 means the progeny do not produce any plaques on *E. coli* K(λ) cells; − means that only a very few progeny phages produce plaques; and + means that many progeny produce plaques (more than $10\times$ as many as in the − cases).

	1	2	3	4	5	6	7	8	9
1	−	+	+	+	+	−	−	+	+
2		−	+	+	+	+	−	+	+
3			0	−	+	0	+	+	−
4				−	+	−	+	+	+
5					−	+	−	+	+
6						0	0	−	+
7							0	+	+
8								−	+
9									

a. Which are the three deletions? What criteria did you use to reach your conclusion?

b. If you know that mutation 9 is in the *rIIB* gene, draw the best genetic map possible to explain the data, including the positions of all point mutations and the extent of the three deletions.

c. There should be one uncertainty remaining in your answer to part b. How could you resolve this uncertainty?

6-16 The *rosy* (*ry*) gene of *Drosophila* encodes an enzyme called xanthine dehydrogenase. Flies homozygous for *ry* mutations exhibit a rosy eye color. Heterozygous females were made that had ry^{41} *Sb* on one homolog and *Ly* ry^{564} on the other homolog, where ry^{41} and ry^{564} are two independently isolated alleles of *ry* and *Ly* (*Lyra* [narrow] wings) and *Sb* (stubble bristle) are dominant markers to the left and right of *ry*, respectively. These females are now mated to males homozygous for ry^{41}. Out of 100,000 progeny, 8 have wildtype eyes, Lyra wings, and stubble bristles, while the remainder have rosy eyes.

a. What is the genetic distance separating ry^{41} and ry^{564}?

b. What is the order of these two *ry* mutations relative to the genes *Ly* and *Sb?*

6-17 The pathway for arginine biosynthesis in *Neurospora crassa* involves several enzymes that produce a series of intermediates.

$$\overset{argE}{N\text{-acetylornithine} \rightarrow} \overset{argF}{\text{ornithine} \rightarrow} \overset{argG}{\text{citrulline} \rightarrow} \overset{argH}{\text{argininosuccinate} \rightarrow} \text{arginine}$$

a. If you did a cross between *argE*⁻ and *argH*⁻ *Neurospora* strains, what would be the distribution of Arg⁺ and Arg⁻ spores within parental ditype and nonparental ditype asci? (Give the spore types in the order in which they would appear in the ascus.)

b. For each of the spores, what nutrients could you supply in the media to get spore growth?

6-18 In corn snakes, the wildtype color is brown. One autosomal recessive mutation causes the snake to be orange, and another causes the snake to be black. An orange snake was crossed to a black one, and the F_1 offspring were all brown.

a. Indicate what phenotypes and ratios you would expect in the F_2 generation of this cross if there is one pigment pathway, with orange and black being different intermediates on the way to brown.

b. Indicate what phenotypes and ratios you would expect in the F_2 generation if orange pigment is a product of one pathway, black pigment is the product of another pathway, and brown is the effect of mixing the two pigments in the skin of the snake.

6-19 In a certain species of flowering plants with a diploid genome, four enzymes are involved in the generation of flower color. The genes encoding these four enzymes are on different chromosomes. The biochemical pathway involved is as follows:

$$\text{white} \longrightarrow \text{green} \longrightarrow \text{blue} \overset{\longrightarrow}{\underset{\longrightarrow}{}} \text{purple}$$

(Either of two different enzymes is sufficient to convert a blue pigment into a purple pigment.) A true-breeding green-flowered plant is mated with a true-breeding blue-flowered plant. All of the plants in the resultant F_1 generation have purple flowers. F_1 plants are allowed to self-fertilize, yielding an F_2 generation. Indicate the fraction of F_2 plants with the following phenotypes: white flowers, green flowers, blue flowers, and purple flowers.

6-20 The intermediates A, B, C, D, E, and F all occur in the same particular biochemical pathway. G is the product of the pathway, and mutants 1 through 7 are all G⁻, meaning that they cannot produce substance G. The table shows which intermediates will promote growth in each of the mutants. Arrange the intermediates in order of their occurrence in the pathway, and indicate the step in the pathway at which each mutant strain is blocked. A "+" in the table indicates that the strain will grow if given that substance, a "O" means lack of growth.

Mutants	Supplements						
	A	B	C	D	E	F	G
1	+	+	+	+	+	O	+
2	O	O	O	O	O	O	+
3	O	+	+	O	+	O	+
4	O	+	O	O	+	O	+
5	+	+	+	O	+	O	+
6	+	+	+	+	+	+	+
7	O	O	O	O	+	O	+

6-21 The following noncomplementing *E. coli* mutants were tested for growth on four known precursors of thymine, A–D.

	Precursor/Product				
Mutant	A	B	C	D	Thymine
9	+	−	+	−	+
10	−	−	+	−	+
14	+	+	+	−	+
18	+	+	+	+	+
21	−	−	−	−	+

a. Show a simple linear biosynthetic pathway of the four precursors and the end product, thymine. Indicate which step is blocked by each of the five mutations.

b. What precursor would accumulate in the following double mutants? 9 and 10; 10 and 14?

6-22 The pathways for the biosynthesis of the amino acids glutamine (gln) and proline (pro) involve one or more common intermediates. Auxotrophic yeast mutants are isolated that require either glutamine or proline or both amino acids for their growth, as shown in the table (+ means growth; − , no growth). These mutants are also tested for their ability to grow on the shown intermediates A–E. What is the order of these intermediates in the glutamine and proline pathways, and at which point in the pathway is each mutant blocked?

Mutant	A	B	C	D	E	gln	pro	gln+pro
1	+	−	−	−	+	−	+	+
2	−	−	−	−	−	−	+	+
3	−	−	+	−	−	−	−	+
4	−	−	−	−	−	+	−	+
5	−	−	+	+	−	−	−	+
6	+	−	−	−	−	−	+	+
7	−	+	−	−	−	+	−	+

6-23 Antibodies were made that recognize six proteins that are part of a complex inside the *Caenorhabditis elegans* one-cell embryo. The mother produces proteins that are believed to assemble stepwise into a structure in the egg, beginning at the embryo surface. The antibodies were used to detect the protein location in embryos produced by mutant mothers (homozygous recessive for the gene[s] encoding each protein). The *C. elegans* mothers are self-fertilizing hermaphrodites so no wildtype copy of a gene will be introduced during fertilization. In the table, * means the protein was present and at the embryo surface; − means that the protein was not present, and + means that the protein was present but not at the embryo surface. Assume all mutations prevent production of the corresponding protein.

Mutant in gene for protein	Protein production and location					
	A	B	C	D	E	F
A	−	+	*	+	*	+
B	*	−	*	*	*	*
C	*	+	−	+	*	+
D	*	+	*	−	*	+
E	+	+	+	+	−	+
F	*	+	*	*	*	−

Complete the figure below, showing the construction of the hypothetical protein complex, by writing the letter of the proper protein in each circle. The two proteins marked with arrowheads can assemble into the complex independently of each other, but both are needed for the addition of subsequent proteins to the complex.

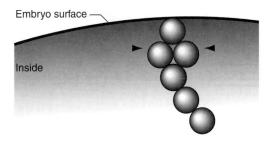

6-24 Adult hemoglobin is a multimeric protein with four polypeptides, two of which are β-globin and two of which are α-globin. How many loci are needed to define the structure of the hemoglobin protein?

6-25 In addition to the predominant adult hemoglobin, HbA, which contains two α-globin chains and two β-globin chains ($\alpha_2\beta_2$), there is a minor hemoglobin, HbA$_2$, composed of two α and two δ chains ($\alpha_2\delta_2$). The β- and δ-globin genes are arranged in tandem, and are highly homologous. Draw the chromosomes that would result from an event of unequal crossing-over between the β and δ genes.

6-26 a. What is the difference between null, hypomorphic, hypermorphic, dominant negative, and neomorphic mutations?
b. For each of these kinds of mutations, would you predict they would be dominant or recessive to a wildtype allele in producing a mutant phenotype?

6-27 The α- and β-thalassemias can arise from deletions in the α- or β-globin genes or from mutations that lead to underexpression of the α or β globins. In this problem, assume that contributions of two alleles are additive. Persons who have less than 35% of normal levels of β globin are defined as having clinical β-thalassemia. Alice is clinically affected, with only 20% of normal levels of β globin. Her father (Bill) has 70%, her mother (Carol) has 50%, her two brothers (David and Earl) have 50% and 70%, and her sister (Fran) has 100% of normal β globin.
a. What are the genotypes of Bill and Carol?
b. Define the levels of expression associated with each allele in each parent.
c. If Bill and Carol have another baby, what is the probability it will have clinical β-thalassemia?
d. What is the probability this next baby will have normal levels of β-globin?

GENE EXPRESSION: THE FLOW OF GENETIC INFORMATION FROM DNA VIA RNA TO PROTEIN

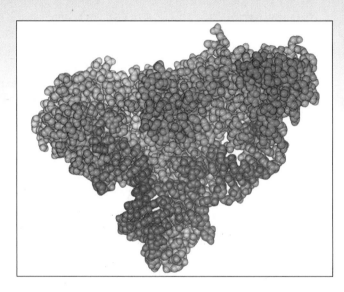

The ability of an aminoacyl-tRNA synthetase (*red*) to couple a particular tRNA (*blue*)
to its corresponding amino acid is central to the molecular machinery that converts
the language of nucleic acids into the language of proteins.

A dedicated effort to determine the complete nucleotide sequence of the haploid genome in a variety of organisms has been underway since 1990. With this sequence information in hand, geneticists can consult the universal dictionary equating nucleotide sequence with amino acid sequence to decide what parts of a genome are likely to be genes. They can also identify genes through matches with nucleotide sequences already known to encode proteins in other organisms. As a result, they can predict the total number of genes in an organism from the complete nucleotide sequence of its genome, and by extension, identify the number and amino acid sequences of all the polypeptides that determine phenotype. Knowledge of DNA sequence thus opens up powerful new possibilities for understanding an organism's growth and development at the molecular level.

Studies of the tiny nematode *Caenorhabditis elegans* illustrate the kind of insights researchers can gain from this DNA-sequence-based approach. *C. elegans* is a roundworm 1 mm in length that lives in soils throughout the world (Fig. 7.1a). Feeding on bacteria,

it grows from fertilized egg to adult—either hermaphrodite or male—in just three days. At the end of this time, each hermaphrodite produces between 250 and 1000 progeny. Because of its small size, short life cycle, and capacity for prolific reproduction, *C. elegans* is an ideal subject for genetic analysis.

The haploid genome of *C. elegans* contains 110 million base pairs distributed among six chromosomes (Fig. 7.1b). In the mid-1990s, a group of investigators reported the sequencing and preliminary analysis of 2.2 million base pairs on chromosome III. Using their knowledge of the concepts explored in this chapter, they found that this 2% of the nematode genome carries about 480 genes. Interestingly, at least 20% of the genes recognized as having a known function encode molecules that play some role in **gene expression:** the process by which cells convert DNA sequence information to RNA and then decode the RNA information to the amino acid sequence of a polypeptide (Fig. 7.2). The fact that 20% of the genes in this sequenced region encode components of gene expression suggests the importance of the process

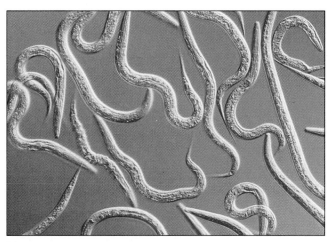

(a)

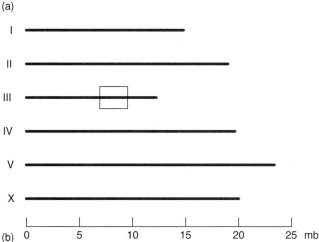

(b)

Figure 7.1 *C. elegans:* An ideal subject for genetic analysis.
(a) Micrograph of several adult worms. (b) Six chromosomes form the haploid genome of *C. elegans*. The highlighted region depicts a 2.2 million base pair portion of chromosome III that has been analyzed and found to encode about 480 genes.

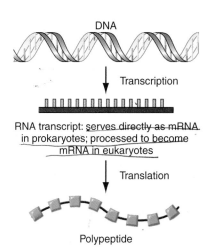

Figure 7.2 Gene expression: The flow of genetic information from DNA via RNA to protein. Transcription and translation convert the information encoded in DNA into the order of amino acids in a polypeptide. In transcription, an enzyme known as RNA polymerase catalyzes production of an RNA transcript. In translation, the cellular machinery uses instructions in mRNA to synthesize a polypeptide, following the rules of the genetic code.

DNA-encoded information to its RNA-encoded equivalent is known as **transcription.** The product of transcription is a **transcript:** a molecule of **messenger RNA (mRNA)** in prokaryotes; a molecule of RNA that undergoes processing to become an mRNA in eukaryotes. In the second stage of gene expression, the cellular machinery translates the mRNA to its polypeptide equivalent in the language of amino acids. This decoding of nucleotide information to a sequence of amino acids is known as **translation.** It takes place on molecular workbenches called **ribosomes,** which are composed of proteins and ribosomal RNAs (rRNAs); and it depends on the universal dictionary known as the **genetic code,** which defines each amino acid in terms of a specific sequence of three nucleotides. It also depends on **transfer RNAs (tRNAs),** small RNA adaptor molecules that place specific amino acids at the correct position in a growing polypeptide chain. The tRNAs can bring amino acids to the right place on the translational machinery because tRNAs and mRNAs have complementary nucleotides that can base pair with each other.

Four general themes emerge from our discussion of gene expression. First, the pairing of complementary bases figures prominently in the precise transfer of information from DNA to RNA and from RNA to polypeptide. Second, the polarities of DNA, RNA, and protein molecules help guide the mechanisms of gene expression: the 3'-to-5' transcription of a template DNA strand yields a polar mRNA that grows from its 5' to its 3' end; the 5'-to-3' translation of this mRNA yields a polar protein running from amino terminal to carboxyl terminal. Third, like DNA replication and recombination (discussed in Chapter 5), gene expression requires an input of energy and the participation of several specific proteins at different points in the process. Fourth, since the accurate one-way flow of genetic information determines protein structure, mutations

to the life of the organism. If this ratio holds for the rest of the worm's genome, about 3600 of its estimated 18,000 genes generate the machinery that enables genes to be interpreted as proteins.

In this chapter, we describe the cellular mechanisms that carry out gene expression. As intricate as some of the details may appear, the general scheme of gene expression is elegant and straightforward: *Within each cell, genetic information flows from DNA to RNA to protein.* In 1957 Francis Crick proposed that genetic information flows in only one direction, and named his concept of a one-way molecular flow the "Central Dogma" of molecular biology. As Crick explained, "once 'information' has passed into protein, it cannot get out again."

Inside most cells, as the Central Dogma suggests, genetic information flows from one class of molecule to another in two distinct stages (see Fig. 7.2). If you think of genes as instructions written in the language of nucleic acids, the cellular machinery first transcribes a set of instructions written in the DNA dialect to the same instructions written in the RNA dialect. The conversion of

that change this information or obstruct its flow can have dramatic effects on phenotype.

As we examine how cells use the sequence information contained in DNA to construct proteins, we present

■ The genetic code: How triplets of the 4 nucleotides unambiguously specify 20 amino acids, making it possible to translate information from a nucleotide chain to a sequence of amino acids.

■ Transcription: How RNA polymerase, guided by base pairing, synthesizes a single-stranded mRNA copy of a gene's DNA template.

■ Translation: How base pairing between mRNA and tRNAs directs the assembly of a polypeptide on the ribosome.

■ A comprehensive example of gene expression in *C. elegans*.

■ How mutations affect gene information and expression.

THE GENETIC CODE: HOW PRECISE GROUPINGS OF THE 4 NUCLEOTIDES SPECIFY 20 AMINO ACIDS

A code is a system of symbols that equates information in one language with information in another. A useful analogy for the genetic code is the Morse code, which uses dots and dashes or short and long sounds to transmit messages over radio or telegraph wires. Various groupings of the dot-dash/short-long symbols represent the 26 letters of the English alphabet. Because there are many more letters than the two dot or dash symbols, groups of up to four dots, four dashes, and varying combinations of the two represent some letters. And because anywhere from one to four symbols specify each letter, the Morse code requires a symbol for "pause" to signify where one letter ends and the next begins.

In the Genetic Code, a Triplet Codon Represents Each Amino Acid

The language of nucleic acids is written in four nucleotides—A, G, C, and T in the DNA dialect; A, G, C, and U in the RNA dialect—while the language of proteins is written in amino acids. To understand how the sequence of bases in DNA or RNA encodes the order of amino acids in a polypeptide chain, it is essential to know how many distinct amino acids there are. Watson and Crick produced the now accepted list of the 20 amino acids that are genetically encoded by DNA or RNA sequence over lunch one day at a local pub. They created the list by analyzing the amino acid sequence of a variety of naturally occurring polypeptides. Amino acids that are present in only a small number of proteins or in only certain tissues or organisms did not qualify as standard building blocks; Crick and Watson correctly assumed that such amino acids arise when proteins undergo modification after their synthesis. By contrast, amino acids that are present in most, though not necessarily all, proteins made the list. The question then became: How can 4 nucleotides encode 20 amino acids?

Just as the Morse code conveys information through different groupings of dots and dashes, the 4 nucleotides encode 20 amino acids through specific groupings of A, G, C, and T or A, G, C, and U. Researchers initially arrived at the number of letters per grouping by deductive reasoning, and later con-

firmed it by experiment. They reasoned that if only one nucleotide represented an amino acid, there would be information for only four amino acids: A would encode one amino acid; G, a second amino acid; C, a third; and T, a fourth. If two nucleotides represented each amino acid, there would be $4^2 = 16$ possible combinations of couplets.

Of course, if the code consisted of groups containing one *or* two nucleotides, it would have $4 + 16 = 20$ groups and could account for all the amino acids, but there would be nothing left over for the pause denoting where one group ends and the next begins. Groups of three nucleotides in a row would provide $4^3 = 64$ different triplet combinations, more than enough to code for all the amino acids. If the code consisted of doublets and triplets, a signal denoting pause would once again be necessary. But a triplets-only code would require no symbol for "pause" if the mechanism for counting to three and distinguishing among successive triplets were very reliable.

Although this kind of reasoning—explaining the unknown in terms of the known by looking for the simplest possibility—generates a theory, it does not prove it. As it turned out, however, the experiments described later did indeed demonstrate that groups of three nucleotides represent all 20 amino acids. Each nucleotide triplet is called a **codon.** Each codon, designated by the bases defining its three nucleotides, specifies one amino acid. For example, GAA is a codon for glutamic acid (Glu), and GUU is a codon for valine (Val). Because the code comes into play only during the translation part of gene expression, that is, during the decoding of messenger RNA to polypeptide, geneticists usually present the code in the RNA dialect of A, G, C, and U, as depicted in Fig. 7.3. However, when speaking of genes, they can substitute T for U to show the same code in the DNA dialect.

If you knew the sequence of nucleotides in a gene or its transcript as well as the sequence of amino acids in the corresponding polypeptide, you could deduce the genetic code without understanding how the cellular machinery uses the code to translate from nucleotides to amino acids. Although techniques for determining both nucleotide and amino-acid sequence are available today, this was not true when researchers cracked the genetic code in the 1950s and 1960s. At that time, they could establish a polypeptide's amino-acid sequence, but not the nucleotide sequence of DNA or RNA. Because of their inability to read nucleotide sequence, they used

	Second letter			
First letter	U	C	A	G	Third letter

	U	C	A	G	
U	UUU ⎫ Phe UUC ⎭ UUA ⎫ Leu UUG ⎭	UCU ⎫ UCC ⎪ Ser UCA ⎪ UCG ⎭	UAU ⎫ Tyr UAC ⎭ UAA Stop UAG Stop	UGU ⎫ Cys UGC ⎭ UGA Stop UGG Trp	U C A G
C	CUU ⎫ CUC ⎪ Leu CUA ⎪ CUG ⎭	CCU ⎫ CCC ⎪ Pro CCA ⎪ CCG ⎭	CAU ⎫ His CAC ⎭ CAA ⎫ Gln CAG ⎭	CGU ⎫ CGC ⎪ Arg CGA ⎪ CGG ⎭	U C A G
A	AUU ⎫ AUC ⎪ Ile AUA ⎭ AUG Met	ACU ⎫ ACC ⎪ Thr ACA ⎪ ACG ⎭	AAU ⎫ Asn AAC ⎭ AAA ⎫ Lys AAG ⎭	AGU ⎫ Ser AGC ⎭ AGA ⎫ Arg AGG ⎭	U C A G
G	GUU ⎫ GUC ⎪ Val GUA ⎪ GUG ⎭	GCU ⎫ GCC ⎪ Ala GCA ⎪ GCG ⎭	GAU ⎫ Asp GAC ⎭ GAA ⎫ Glu GAG ⎭	GGU ⎫ GGC ⎪ Gly GGA ⎪ GGG ⎭	U C A G

Figure 7.3   The genetic code: 61 codons represent the 20 amino acids, while 3 codons signify stop. To read the code, find the first letter in the left column, the second letter along the top, and the third letter in the right column; this reading corresponds to the 5′-to-3′ direction along the mRNA. Although most amino acids are encoded by two or more codons, the genetic code is unambiguous because each codon specifies only one amino acid.

an assortment of genetic and biochemical techniques to fathom the code. They began by examining how different mutations in a single gene affected the amino-acid sequence of the gene's polypeptide product, using the abnormal (specific mutations) to understand the normal (the general relationship between genes and polypeptides).

Mapping Studies Confirmed That a Gene's Nucleotide Sequence Is Colinear with a Polypeptide's Amino-Acid Sequence

We have seen that DNA is a linear molecule with base pairs following one another down the intertwined chains. Proteins, by contrast, have complicated three-dimensional structures. Even so, if unfolded and stretched out from amino terminus to carboxyl terminus, proteins have a one-dimensional, linear structure—a specific sequence of amino acids. If the information in a gene and its corresponding protein are colinear, the consecutive order of bases in the DNA from the beginning to the end of the gene would stipulate the consecutive order of amino acids from one end to the other of the outstretched protein. Note that this hypothesized relationship implies that both a gene and its protein product have definite polarities with an invariant relation to each other.

Charles Yanofsky, in studying the *Escherichia coli* gene for a subunit of the enzyme tryptophan synthetase, was the first to compare maps of mutations within a gene to the particular amino-acid substitutions that resulted. He began by generating a large number of *trp*⁻ auxotrophic mutants that carried mutations in the *trpA* gene for the tryptophan synthetase subunit. He next made a fine structure recombinational map of these mutations; and then he purified and determined the amino acid sequence of the mutant tryptophan synthetase subunits. As Fig. 7.4a illustrates, Yanofsky's data showed that the order of mutations mapped within the DNA of the gene by recombination was colinear with the positions of the amino-acid substitutions occurring in the resulting mutant proteins.

Genetic Analysis Revealed That Nonoverlapping Codons Are Set in a Reading Frame

By carefully examining the results of his analysis, Yanofsky, in addition to confirming the existence of colinearity, deduced key features of codons and helped establish many parameters of the genetic code relating nucleotides to amino acids.

A Codon Is Composed of More Than One Nucleotide

Yanofsky observed that different *point mutations* (changes in only one nucleotide pair) may affect the same amino acid. In one example shown in Fig. 7.4a, mutation #23 changed the glycine (Gly) at position 211 of the wildtype polypeptide chain to arginine (Arg), while mutation #46 yielded glutamic acid (Glu) at the same position. In another example, mutation #78 changed the glycine at position 234 to cysteine (Cys), while mutation #58 produced aspartic acid (Asp) at the same position. In both cases, Yanofsky also found that recombination could occasionally occur between two mutations that changed the identity of the same amino acid, and such recombination would produce a wildtype tryptophan synthetase gene (Fig. 7.4b). Because the smallest unit of recombination is the base pair, two mutations capable of recombination—in this case, in the same codon because they affect the same amino acid—must be in different (although nearby) nucleotides. Thus, a codon contains more than one nucleotide.

Each Nucleotide Is Part of Only a Single Codon

As Fig. 7.4a illustrates, each of the point mutations in the tryptophan synthetase gene characterized by Yanofsky alters the identity of only a single amino acid. This is also true of the point mutations examined in many other genes, such as the human genes for rhodopsin and hemoglobin (see Chapter 6). Since point mutations change only a single nucleotide pair and most point mutations affect only a single amino acid in a polypeptide, each nucleotide in a gene must influence the identity of only a single amino acid. If, on the contrary, a nucleotide were part of more than one codon, a mutation in that nucleotide would affect more than one amino acid.

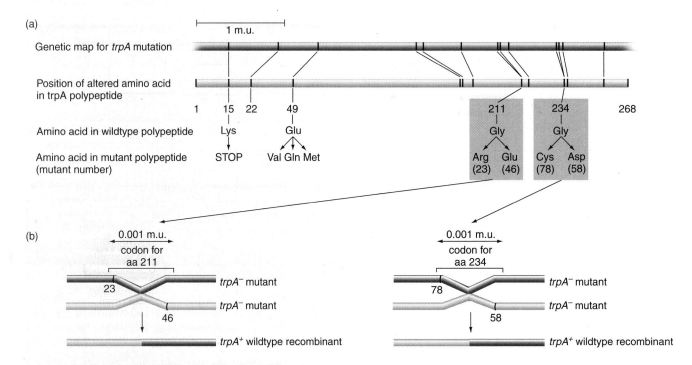

Figure 7.4   Experiments analyzing the E. coli gene for a subunit of tryptophan synthetase confirm colinearity and reveal significant features of the genetic code. (a) A genetic map of the trpA gene of E. coli, identifying the amino-acid substitutions that characterize several of Yanofsky's mutant strains. The positions of the mutations and amino-acid substitutions are colinear. These mutations change only a single amino acid, suggesting that each nucleotide is part of only a single codon. (b) Confirmation that codons must include two or more base pairs came from crosses between two strains that carried an altered amino acid at the same position. Since wildtype progeny occasionally appeared, each strain had a point mutation at a slightly different site. Crossing-over between the mutant sites could produce a wildtype allele.

A Codon Is Composed of Three Nucleotides, and the Designated Starting Point for Each Gene Establishes the Reading Frame for These Triplets

Although the most efficient code that would allow 4 nucleotides to specify 20 amino acids requires 3 nucleotides per codon, more complicated scenarios are possible. Francis Crick and Sydney Brenner obtained convincing evidence for the triplet nature of the genetic code in studies of mutations in the bacteriophage T4 rIIB gene (Fig. 7.5). They induced the mutations with proflavin, an intercalating mutagen that can insert itself between the paired bases stacked in the center of the DNA molecule (Fig. 7.5a). Their assumption was that proflavin would act like other mutagens, causing single-base substitutions. If this were true, it would be possible to generate revertants through treatment with any mutagen. Surprisingly, genes with proflavin-induced mutations did not revert to wildtype upon treatment with other mutagens known to cause nucleotide substitutions. Only further exposure to proflavin caused proflavin-induced mutations to revert to wildtype (Fig. 7.5b). Crick and Brenner had to explain this observation before they could proceed with their phage experiments. With keen insight, they correctly guessed that proflavin does not cause base substitutions; instead, it causes insertions or deletions. This hypothesis explained why base-substituting mutagens could not revert proflavin-induced mutations; it was also consistent with the structure of proflavin. By intercalating between base pairs, proflavin would distort the double helix and thus interfere with the action of enzymes that function in the repair, replication, or recombination of DNA, eventually causing the deletion or addition of one or more nucleotide pairs to the DNA molecule.

Crick and Brenner began their experiments with a particular proflavin-induced rIIB⁻ mutation. They next treated this mutant strain with more proflavin to isolate an rIIB⁺ revertant (see Fig. 7.5b), and showed that the revertant's chromosome actually contained two different rIIB⁻ mutations: One was the original mutation (FC0 in the figure); the other was newly induced (FC7). Either mutation by itself yields a mutant phenotype, but their simultaneous occurrence in the same gene yielded an rIIB⁺ phenotype. Crick and Brenner reasoned that if the first mutation was the deletion of a single base pair, represented by the symbol (−), then the counteracting mutation must be the insertion of a base pair, represented as (+). The restoration of gene function by one mutation canceling another in the same gene is known as **intragenic suppression**. On the basis of this reasoning, they went on to establish T4 strains with different numbers of (+) and (−) mutations in the same chromosome. Figure 7.5c tabulates the phenotypes associated with each combination of proflavin-induced mutations.

In analyzing the data, Crick and Brenner assumed that each codon is a trio of nucleotides, and for each gene there is a single starting point. This starting point establishes a **read-**

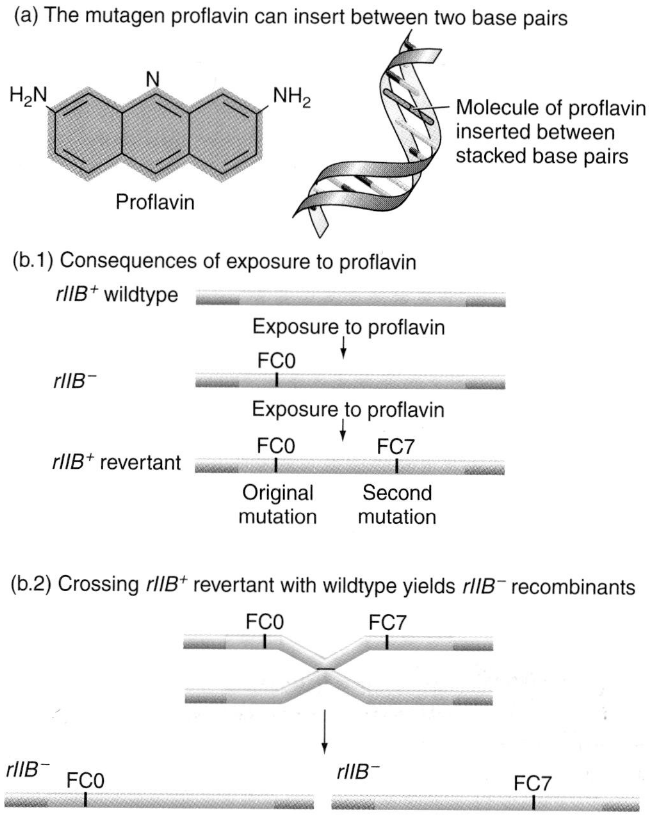

(a) The mutagen proflavin can insert between two base pairs

Proflavin

Molecule of proflavin inserted between stacked base pairs

(b.1) Consequences of exposure to proflavin

rIIB⁺ wildtype

Exposure to proflavin

FC0

rIIB⁻

Exposure to proflavin

FC0   FC7

rIIB⁺ revertant

Original mutation   Second mutation

(b.2) Crossing *rIIB*⁺ revertant with wildtype yields *rIIB*⁻ recombinants

FC0   FC7

rIIB⁻ FC0   *rIIB*⁻   FC7

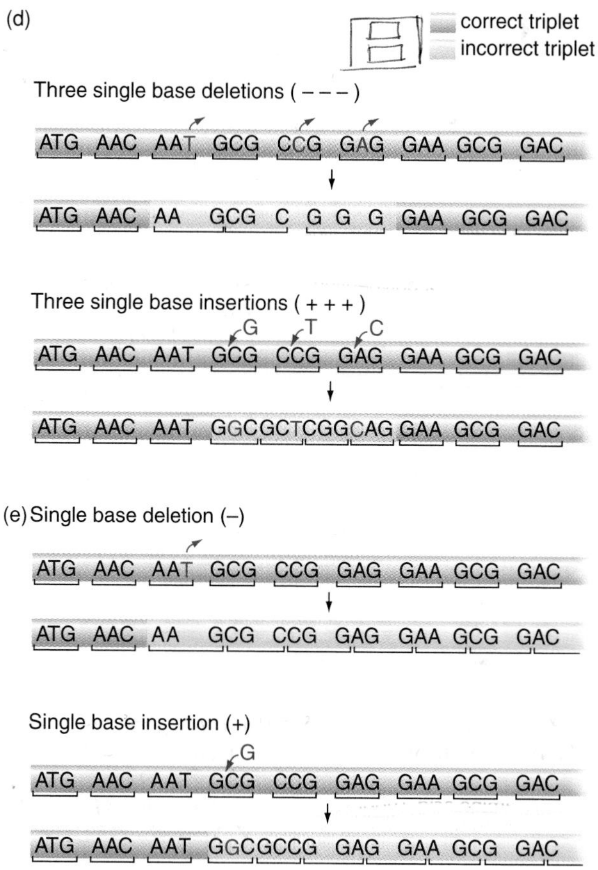

(c) Different sets of mutations generate either a mutant or a normal phenotype

Proflavin-induced mutations (+) insertion (−) deletion	Phenotype
− or +	Mutant
− − or + +	Mutant
− − − − or − − − − −	Mutant
− +	Wildtype
− − − or − − − − − or + + + or + + + + + +	Wildtype

(d)

correct triplet
incorrect triplet

Three single base deletions (– – –)

ATG AAC AAT GCG CCG GAG GAA GCG GAC

ATG AAC AA   GCG C G G G GAA GCG GAC

Three single base insertions (+ + +)

G   T   C
ATG AAC AAT GCG CCG GAG GAA GCG GAC

ATG AAC AAT GGCGCTCGGCAG GAA GCG GAC

(e) Single base deletion (−)

ATG AAC AAT GCG CCG GAG GAA GCG GAC

ATG AAC AA   GCG CCG GAG GAA GCG GAC

Single base insertion (+)

G
ATG AAC AAT GCG CCG GAG GAA GCG GAC

ATG AAC AAT GGCGCCG GAG GAA GCG GAC

Figure 7.5   Studies of frameshift mutations in the bacteriophage T4 *rIIB* gene show that codons consist of three nucleotides. (a) The mutagen proflavin can slip between adjacent base pairs, eventually resulting in a deletion or insertion. (b) Consequence of exposure to proflavin. (1) Treatment with proflavin produces a mutation at one site (FC0). Another exposure results in a second mutation (FC7) within the same gene; this second mutation suppresses the effect of the first mutation. (2) When the revertant strain is crossed with a wildtype strain, crossing-over separates the mutations such that the two *rIIB*⁻ mutations FC0 and FC7 reappear. This experiment demonstrates that the reversion to a *rIIB*⁺ phenotype is indeed a case of intragenic suppression. (c) Single or multiple proflavin-induced mutations generate either a mutant or a normal phenotype. (d) and (e) Crick and Brenner reasoned that the phenotypic effects of proflavin-induced frameshift mutations depend on whether the reading frame is restored and whether the part of the gene with an altered reading frame specifies an essential or nonessential region of the polypeptide.

ing frame: the partitioning of groups of three nucleotides such that the sequential interpretation of each succeeding triplet generates the correct order of amino acids in the resulting polypeptide chain. If codons are read in order from a fixed starting point, one mutation will counteract another if the two are equivalent mutations of opposite signs; in such a case, each insertion compensates for each deletion, and this counterbalancing restores the reading frame. The gene would only regain its wildtype activity, however, if the portion of the polypeptide encoded between the two mutations of opposite sign is not required for protein function, because in the double mutant, this region would have an improper amino-acid sequence. Similarly, if a gene sustains three or multiples of three changes of the same sign, the encoded polypeptide can still function, because the mutations do not alter the reading frame for the majority of amino acids (Fig. 7.5d). The resulting polypeptide will, however, have one extra or one fewer amino acid than normal (designated by three plus signs or three minus signs, respectively), and the region encoded by the part of the gene between the first and the last mutations will not contain the correct amino acids.

By contrast, a single nucleotide inserted into or deleted from a gene alters the reading frame and thereby affects the identity of not only one amino acid, but of all other amino acids beyond the point of alteration (Fig. 7.5e). Changes that alter the grouping of nucleotides into codons are called **frameshift mutations:** they shift the reading frame for all codons beyond the point of insertion or deletion, almost always abolishing the function of the polypeptide product.

A review of the evidence tabulated in Fig. 7.5c supports all these points. A single $(-)$ or a single $(+)$ mutation destroyed the function of the *rIIB* gene and produced an *rIIB$^-$* phage. Similarly, any gene with two base changes of the same sign $(--$ or $++)$, or with four or five insertions or deletions of the same sign (for example, $++++$) also generated a mutant phenotype. However, genes containing three or multiples of three mutations of the same sign (for example, $+++$ or $-----$) as well as genes containing a $(+ -)$ pair of mutations generated *rIIB$^+$* wildtype individuals. In these last examples, intragenic suppression allowed restitution of the reading frame and thereby restored the lost or aberrant genetic function produced by other frameshift mutations in the gene.

Most Amino Acids are Specified by More Than One Codon

As Fig. 7.5 illustrates, intragenic suppression occurs only if in the region between two frameshift mutations of opposite sign, a gene still dictates the appearance of amino acids, even if these amino acids are not the same as those appearing in the normal protein. If the frameshifted part of the gene encodes instructions to stop protein synthesis, for example, by introducing a triplet of nucleotides that does not correspond to any amino acid, then wildtype polypeptide production would not continue. This is because polypeptide synthesis would stop before the compensating mutation could reestablish the correct reading frame.

The fact that intragenic suppression occurs as often as it does suggests that the code includes more than one codon for some amino acids. Recall that there are 20 common amino acids but $4^3 = 64$ different combinations of three nucleotides. If each amino acid corresponded to only a single codon, there would be $64 - 20 = 44$ possible triplets not encoding an amino acid. These noncoding triplets would act as "stop" signals and prevent further polypeptide synthesis. If this happened, more than half of all frameshift mutations (44/64) would cause protein synthesis to stop at the first codon after the mutation, and the chances of extending the protein each amino acid farther down the chain would diminish exponentially. As a result, intragenic suppression would rarely occur. However, we have seen that many frameshift mutations of one sign can be offset by mutations of the other sign. The distances between these mutations, estimated by recombination frequencies, are in some cases large enough to code for more than 50 amino acids, which would be possible only if most of the 64 possible triplet codons specified amino acids. Thus, the data of Crick and Brenner provide strong support for the idea that the genetic code is **degenerate:** Two or more nucleotide triplets specify most of the 20 amino acids (see Fig. 7.3).

Cracking the Code: Biochemical Manipulations Revealed Which Codons Represent Which Amino Acids

Although the genetic experiments just described enabled remarkably prescient insights about the nature of the genetic code, they did not make it possible to assign particular codons to their corresponding amino acids. This awaited the discovery of messenger RNA and the development of techniques for synthesizing simple messenger RNA molecules that researchers could use to manufacture simple proteins in the test tube.

The Discovery of Messenger RNAs, Molecules for Transporting Genetic Information

In the 1950s researchers exposed eukaryotic cells to amino acids tagged with radioactivity and observed that protein synthesis incorporating the radioactive amino acids into polypeptides takes place in the cytoplasm, even though the genes for those polypeptides are sequestered in the cell nucleus. From this discovery, they deduced the existence of an intermediate molecule, made in the nucleus and capable of transporting DNA sequence information to the cytoplasm where it can direct protein synthesis. RNA was a prime candidate for this intermediary information-carrying molecule. Because of RNA's potential for base pairing with a strand of DNA, one could imagine the cellular machinery copying a strand of DNA into a complementary strand of RNA in a manner analogous to the DNA-to-DNA copying of DNA replication. Subsequent studies in eukaryotes on the incorporation of radioactive uracil (a base found only in RNA) into molecules of RNA showed that although the molecules are synthesized in the nucleus, at least some of them migrate to the cytoplasm. Among those RNA molecules that migrate to the cytoplasm are the messenger RNAs, or mRNAs, depicted in Fig. 7.2. They arise in the nucleus from the transcription of DNA sequence information through base pairing and then move, after processing, to the cytoplasm where they determine the proper order of amino acids during protein synthesis.

Synthetic mRNAs and In Vitro *Translational Systems Made It Possible to Discover Which Codons Designate Which Amino Acids*

Knowledge of mRNA served as the framework for two experimental breakthroughs that led to the deciphering of the genetic code. In the first, biochemists obtained cellular extracts that, with the addition of mRNA, synthesized polypeptides in a test tube. They called these extracts "*in vitro* translational systems." The second breakthrough was the development of techniques enabling the synthesis of artificial mRNAs containing only a few codons of known composition. When added to *in vitro* translational systems, these simple, synthetic mRNAs directed the formation of very simple polypeptides.

In 1961, Marshall Nierenberg and Heinrich Matthaei added a synthetic poly-U (5′ . . . UUUUUUUUUUUU . . . 3′) mRNA to a cell-free translational system derived from *E. coli*. With the poly-U mRNA, phenylalanine (Phe) was the only

amino acid incorporated into the resulting polypeptide (Fig. 7.6a). Because UUU is the only possible triplet in poly-U, UUU must be a codon for phenylalanine. In a similar fashion, Nierenberg and Matthaei showed that CCC encodes proline (Pro), AAA is a codon for lysine (Lys), and GGG encodes glycine (Gly) (Fig. 7.6b).

The chemist Har Gobind Khorana later made mRNAs with repeating dinucleotides, such as poly-UC (5′ . . . UCUCUCUC . . . 3′), repeating trinucleotides, such as poly-UUC, and repeating tetranucleotides, such as poly-UAUC, and used them to direct the synthesis of slightly more complex polypeptides. As Fig. 7.6b shows, his results limited the coding possibilities, but some ambiguities remained. For example, poly-UC encodes the polypeptide N . . . Ser-Leu-Ser-Leu-Ser-Leu . . . C in which serine and leucine alternate with each other. However, although the mRNA contains only two different codons (5′UCU3′ and 5′CUC3′), it is not obvious which corresponds to serine and which to leucine.

Nierenberg and Philip Leder resolved these ambiguities in 1965 with experiments in which they added short, synthetic

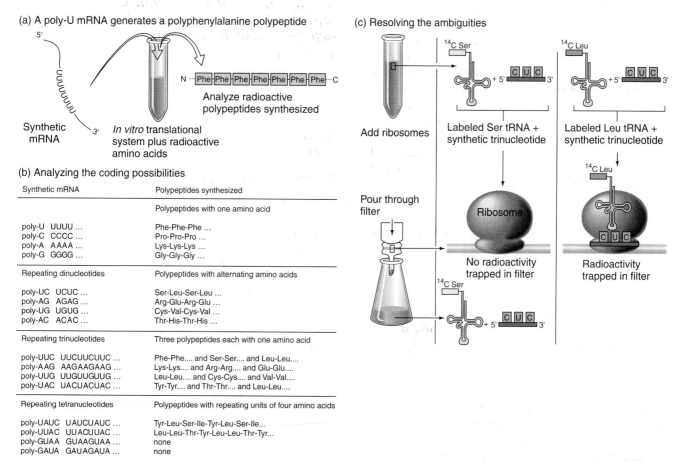

(a) A poly-U mRNA generates a polyphenylalanine polypeptide

Synthetic mRNA

N -Phe-Phe-Phe-Phe-Phe-Phe-Phe- C

Analyze radioactive polypeptides synthesized

In vitro translational system plus radioactive amino acids

(b) Analyzing the coding possibilities

Synthetic mRNA		Polypeptides synthesized
		Polypeptides with one amino acid
poly-U	UUUU ...	Phe-Phe-Phe ...
poly-C	CCCC ...	Pro-Pro-Pro ...
poly-A	AAAA ...	Lys-Lys-Lys ...
poly-G	GGGG ...	Gly-Gly-Gly ...
Repeating dinucleotides		**Polypeptides with alternating amino acids**
poly-UC	UCUC ...	Ser-Leu-Ser-Leu ...
poly-AG	AGAG ...	Arg-Glu-Arg-Glu ...
poly-UG	UGUG ...	Cys-Val-Cys-Val ...
poly-AC	ACAC ...	Thr-His-Thr-His ...
Repeating trinucleotides		**Three polypeptides each with one amino acid**
poly-UUC	UUCUUCUUC ...	Phe-Phe.... and Ser-Ser.... and Leu-Leu....
poly-AAG	AAGAAGAAG ...	Lys-Lys.... and Arg-Arg.... and Glu-Glu....
poly-UUG	UUGUUGUUG ...	Leu-Leu.... and Cys-Cys.... and Val-Val....
poly-UAC	UACUACUAC ...	Tyr-Tyr.... and Thr-Thr.... and Leu-Leu....
Repeating tetranucleotides		**Polypeptides with repeating units of four amino acids**
poly-UAUC	UAUCUAUC ...	Tyr-Leu-Ser-Ile-Tyr-Leu-Ser-Ile...
poly-UUAC	UUACUUAC ...	Leu-Leu-Thr-Tyr-Leu-Leu-Thr-Tyr...
poly-GUAA	GUAAGUAA ...	none
poly-GAUA	GAUAGAUA ...	none

(c) Resolving the ambiguities

Add ribosomes

^{14}C Ser

Labeled Ser tRNA + synthetic trinucleotide

^{14}C Leu

Labeled Leu tRNA + synthetic trinucleotide

Pour through filter

Ribosome

No radioactivity trapped in filter

^{14}C Leu

Radioactivity trapped in filter

^{14}C Ser

Figure 7.6 How geneticists used synthetic mRNAs to crack the genetic code. (a) Poly-U mRNA generates a poly-phenylalanine polypeptide. (b) Polydi-, polytri- and polytetranucleotides limit the coding possibilities. Note that in some synthetic mRNAs, such as poly-GUAA, stop codons appear in all three reading frames, so only short polypeptides can be constructed. (c) Ambiguities were resolved when researchers synthesized trinucleotides of known sequence and added them in combination with tRNAs charged with radioactive amino acid to an *in vitro* extract containing ribosomes. If the trinucleotide specified the radioactive amino acid, the amino-acid-bearing tRNA formed a complex with the ribosomes and thus became trapped when the solution was passed through a filter. The experiments shown here indicate that the codon CUC specifies leucine but not serine.

mRNAs only three nucleotides in length to an *in vitro* translational system containing 1 radioactive amino acid and 19 unlabeled amino acids, all attached to tRNA molecules. They then poured the mixture of synthetic mRNAs and translational systems containing a radioactively labeled amino acid attached to tRNA through a filter (Fig. 7.6c). tRNAs carrying an amino acid normally go right through a filter. If, however, a tRNA carrying an amino acid binds to a ribosome, it will stick in the filter, because this larger complex of ribosome, amino-acid-carrying tRNA, and small mRNA cannot pass through the filter. They could thus use this approach to see which small mRNA caused the entrapment of which radioactively labeled amino acid in the filter. For example, they knew that CUC encoded either serine or leucine. When they added the synthetic triplet CUC to an *in vitro* system where the radioactive amino acid was serine, the amino acid attached to the tRNA passed through the filter, and the filter thus emitted no radiation. But when they added the same triplet to a system where the radioactive amino acid was leucine, the filter lit up with radioactivity, indicating that the radioactively tagged leucine attached to a tRNA had bound to the ribosome-mRNA complex. Nierenberg and Leder used this technique to determine all the codon-amino acid correspondences shown in the genetic code table (Fig. 7.3).

The 5′-to-3′ Direction in mRNA Corresponds to the N-Terminal-to-C-Terminal Direction in a Polypeptide

In studies using synthetic mRNAs, when investigators added the six-nucleotide-long 5′AAAUUU3′ to an *in vitro* translational system, the product N-Lys-Phe-C emerged, but no N-Phe-Lys-C appeared. Since AAA is the codon for lysine and UUU is the codon for phenylalanine, this means that the codon closest to the 5′ end of the mRNA encoded the amino acid closest to the N terminus of the corresponding polypeptide. Similarly, the codon nearest the 3′ end of the mRNA encoded the amino acid nearest the C terminus of the resulting polypeptide.

To understand how the polarities of the macromolecules participating in gene expression relate to each other, remember that although the gene is a segment of a DAN double helix, only one of the two strands serves as a template for the mRNA. This strand is known as the **template strand.** The other strand is the **RNA-like strand,** because it has the same polarity and sequence (written in the DNA dialect) as the RNA. Figure 7.7 diagrams the respective polarities of a gene's DNA, the mRNA transcript of that DNA, and the resulting polypeptide.

HnRNA

Nonsense Codons Cause Termination of a Polypeptide Chain

Although most of the simple, repetitive RNAs synthesized by Khorana were very long and thus generated very long polypeptides, a few did not. These RNAs had signals that stopped construction of a polypeptide chain. As it turned out, three different triplets—UAA, UAG, and UGA—do not correspond to any of the amino acids. When these codons appear in frame, translation stops.

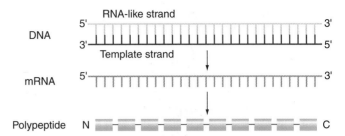

Figure 7.7   Correlation of polarities in DNA, mRNA, and polypeptide. The template strand of DNA is complementary to both the RNA-like DNA strand and the mRNA. The 5′-to-3′ direction in an mRNA corresponds to the N-terminus-to-C-terminus direction in the polypeptide.

As an example of how investigators established this fact, consider the case of poly-GUAA (review Figure 7.6b). This mRNA will not generate a long polypeptide because in all possible reading frames, it contains the **stop codon** UAA.

The three stop codons that terminate translation are also known as **nonsense codons.** For historical reasons, researchers often refer to UAA as the *ocher* codon, UAG as the *amber* codon, and UGA as the *opal* codon. The historical basis of this nomenclature is the last name of one of the early investigators—Bernstein—which means amber in German; ocher and opal derive from their correspondence with amber.

The Genetic Code: A Summary

From the experiments just described, researchers learned that the genetic code is a complete, unabridged dictionary equating the 4-letter language of the nucleic acids with the 20-letter language of the proteins. The following list summarizes the code's main features.

1. The code consists of *triplet codons,* each of which specifies an amino acid. As written in Fig. 7.3, the code shows the 5′-to-3′ sequence of the three nucleotides in each mRNA codon; that is, the first nucleotide depicted is at the 5′ end of the codon.

2. The codons are *nonoverlapping.* In the mRNA sequence 5′GAAGUUGAA3′, for example, the first three nucleotides (GAA) form one codon; nucleotides 4 through 6 (GUU) form the second; and nucleotides 7 through 9 (GAA), the third. Each nucleotide is part of only one codon.

3. The code includes three *stop, or nonsense, codons:* UAA, UAG, and UGA. These codons do not encode an amino acid and thus terminate translation.

4. The code is *degenerate,* which means that in many cases more than one codon specifies the same amino acid.

5. In reading the transcript of a gene, the cellular machinery scans a single strand of mRNA from a fixed starting point that establishes a *reading frame.* As we see later, the nucleotide triplet AUG, which specifies the amino acid methionine, serves in certain contexts as the **initiation**

codon, marking the precise spot in the nucleotide sequence of an mRNA where the code for a particular polypeptide begins. The position of the initiation codon specifies the reading frame for the rest of the sequence by determining how the nucleotides become grouped into triplets. Note that without a set reading frame established by an initiation codon, there are three possible ways to read a piece of mRNA, depending on whether you begin with the first, second, or third base in the sequence.

6. Moving from the 5′ to the 3′ end of an mRNA, each successive codon is sequentially interpreted into an amino acid, starting at the N terminus and moving toward the C terminus, of the resulting polypeptide.

7. Mutations may modify the message encoded in a sequence of nucleotides in three ways. *Frameshift mutations* are nucleotide insertions or deletions that alter the genetic instructions for polypeptide construction by changing the reading frame. *Missense mutations* change a codon for one amino acid to a codon for a different amino acid. *Nonsense mutations* change a codon for an amino acid to a stop codon.

Using Genetics to Verify the Code

The experiments that cracked the code by assigning codons to amino acids were all *in vitro* studies using cell-free extracts and synthetic mRNAs. A logical question thus arose: Do living cells construct polypeptides according to the same rules? Early evidence that they do came from studies analyzing how mutations actually affect the amino-acid composition of the polypeptides encoded by a gene. Most mutagens change a single nucleotide in a codon. As a result, most missense mutations that change the identity of a single amino acid should be single-nucleotide substitutions, and analyses of these substitutions should conform to the code. Yanofsky, for example, found two *trp*⁻ auxotrophic mutations in the *E. coli* tryptophan synthetase gene that produced two different amino acids at the same position—amino acid 211—in the polypeptide chain (Fig. 7.8a). According to the code, both of these mutations could have resulted from single-base changes in the GGA codon that normally inserts glycine at position 211. Even more telling were the *trp*⁺ revertants of these mutations subsequently isolated by Yanofsky. As Fig. 7.8a illustrates, single-base substitutions could also explain the amino-acid changes in these revertants.

Yanofsky obtained better evidence yet that cells use the genetic code *in vivo* by analyzing proflavin-induced frameshift mutations of the tryptophan synthetase gene (Fig. 7.8b). He first treated populations of *E. coli* with proflavin to produce *trp*⁻ mutants. Subsequent treatment of these mutants with more proflavin generated some revertants among the progeny. The most likely explanation for the revertants was that their tryptophan synthetase gene carried both a single-base-pair deletion and a single-base-pair insertion (− +). Upon determining the amino acid sequences of the tryptophan synthetase enzymes made by the revertant strains, Yanofsky

(a) Two *trp*⁻ mutations produce different amino acids at the same position in a tryptophan synthetase subunit

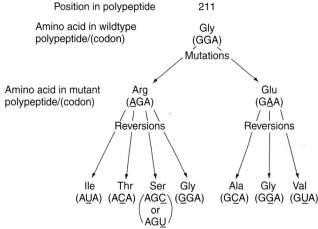

Single-base substitutions can explain the amino acid substitutions of *trp*⁺ revertants

(b) Using the genetic code to predict the amino acid alterations caused by a single-base-pair deletion and insertion

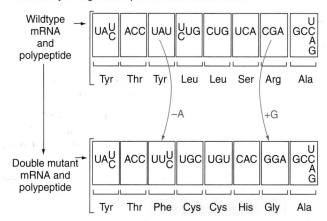

Figure 7.8 Experimental verification of the genetic code. (a) Two *trp*⁻ point mutations encode different amino acids at the same position in a polypeptide. Single-base substitutions can explain the amino-acid substitutions of these mutations and of *trp*⁺ revertants. (b) Charles Yanofsky used the genetic code to predict the amino acid alterations (yellow) that would arise from single-base-pair deletions and insertions. These predictions assumed that codons do not overlap and the code has no punctuation other than start and stop.

found that he could use the genetic code to predict the precise amino-acid alterations that had occurred by assuming the revertants had a specific single-base-pair insertion and a specific single-base-pair deletion.

Yanofsky's results helped confirm not only amino acid codon assignments, but other parameters of the code as well. His interpretations make sense only if codons do not overlap and are read from a fixed starting point with no pauses or commas separating the adjacent triplets.

In Chapter 8, we describe molecular techniques developed since the 1970s that make it possible to read the exact nucleotide sequence of a piece of DNA. In studies using these techniques, geneticists have determined the DNA sequences of genes encoding proteins of known amino-acid sequence. These nucleotide sequences verify that the codons assigned to each amino acid by *in vitro* translational systems incorporating synthetic mRNAs are in fact used for translation *in vivo*.

The Genetic Code Is Almost, But Not Quite, Universal

We now know that virtually all cells alive today—from the prokaryotic cells of *E. coli* bacteria through the range of eukaryotic cells in yeast, *C. elegans,* peas, lentils, and people—use the same genetic code. One early indication of this uniformity was that a translational system derived from one organism could use the mRNA from another organism to convert genetic information to the encoded protein. Rabbit hemoglobin mRNA, for example, when injected into frog eggs or added to cell-free extracts from wheat germ, directs the synthesis of rabbit hemoglobin proteins. More recently, comparisons of DNA and protein sequences have revealed a perfect correspondence according to the genetic code between codons and amino acids in almost all organisms examined.

The universality of the code is an indication that it evolved very early in the history of life. Once it emerged, it remained constant over billions of years, in part because evolving organisms would have little tolerance for change. A single change could disrupt the production of hundreds or thousands of proteins in a cell—from the DNA polymerase that is essential to replication to the RNA polymerase that is required for gene expression to the tubulin proteins that compose the mitotic spindle—and would therefore be lethal.

Researchers were thus quite amazed to observe a few exceptions to the universality of the code. In some species of the single-celled eukaryotic protozoans known as ciliates, the codons UAA and UAG, which are nonsense codons in most organisms, specify the amino acid glutamine; in other ciliates, UGA, the third stop codon in most organisms, specifies cysteine. These ciliates use the remaining nonsense codons as stop codons. Other systematic changes in the genetic code exist in mitochondria, the semiautonomous, self-reproducing organelles within eukaryotic cells that are the sites of ATP formation. Each mitochondrion has its own chromosomes and its own apparatus for gene expression (which we describe in detail in Chapter 14). In the mitochondria of yeast, CUA specifies threonine instead of leucine. It may be that ciliates and mitochondria tolerated these changes in the genetic code because the changes affected very few proteins. For instance, the nonsense codon UGA might have found only infrequent use in one kind of primitive ciliate, so its switch to a "sense" codon would not have made a tremendous difference in protein production. Similarly, mitochondria might have a few changes in the code because they synthesize only a handful of proteins.

The experimental evidence presented so far helped define a nearly universal genetic code. But although cracking the code made it possible to understand the broad outlines of information flow between gene and protein, it did not explain exactly how the cellular machinery accomplishes gene expression. This is our focus as we present the details of transcription and translation.

TRANSCRIPTION: RNA POLYMERASE SYNTHESIZES A SINGLE-STRANDED RNA COPY OF A GENE

Transcription is the process by which the polymerization of ribonucleotides guided by complementary base pairing produces an RNA transcript of a gene. The template for the RNA transcript is one strand of that portion of the DNA double helix that composes the gene.

Details of the Process

Fig. 7.9 depicts the basic components of transcription and illustrates key events in the process. Note in particular four main points: (1) The enzyme **RNA polymerase** catalyzes transcription. (2) DNA sequences near the beginning of genes, called **promoters,** signal RNA polymerase where to begin transcription. (3) RNA polymerase adds nucleotides to the growing RNA polymer in the 5'-to-3' direction. (4) Sequences in the RNA products, known as **terminators,** tell RNA polymerase where to stop transcription.

As you examine Fig. 7.9 to see how components of the cellular machinery work together to begin, carry out, and end transcription, bear in mind that a gene consists of two antiparallel strands of DNA, as mentioned earlier. One—the *RNA-like strand*—has the same polarity and sequence (except for T instead of U) as the emerging RNA transcript. The second—the *template strand*—has the opposite polarity and a complementary sequence that enables it to serve as the template for making the RNA transcript. When geneticists refer to the sequence of a gene, they usually mean the sequence of the RNA-like strand.

The Genetics and Society box "HIV and Reverse Transcription: An Unusual DNA Polymerase Helps Give the AIDS Virus an Evolutionary Edge" describes how the AIDS virus uses an uncommon variation of transcription, known as **reverse transcription,** to construct a double strand of DNA from an RNA template.

As Fig. 7.9d illustrates, the result of transcription is a single strand of RNA known as a **primary transcript.** In prokaryotic organisms, the RNA produced by transcription is the actual messenger RNA that guides protein synthesis. In eukaryotic organisms, by contrast, most primary transcripts undergo *processing* in the nucleus before they migrate to the cytoplasm to direct protein synthesis. This processing has played a fundamental role in the evolution of complex organisms.

In Eukaryotes, RNA Processing after Transcription Produces a Mature Messenger RNA

Some RNA processing in eukaryotes modifies only the 5' or 3' ends of the primary transcript, leaving untouched the information content of the rest of the mRNA. Other processing deletes blocks of information from the middle of the primary transcript so that the content of the mature mRNA is related, but not identical, to the complete set of instructions in the original gene.

Adding a Methylated Cap at the 5' End and a Poly-A Tail at the 3' End

The nucleotide at the 5' end of a eukaryotic mRNA is a G in reverse orientation from the rest of the molecule that is connected through a triphosphate linkage to the first nucleotide in the primary transcript. This backward G is not transcribed from the gene. Instead, a special *capping enzyme* adds it to the primary transcript after polymerization of the transcript's first few nucleotides. Enzymes known as *methyl transferases* then add methyl ($-CH_3$) groups to the backward G and to one or more of the succeeding nucleotides in the RNA, forming a so-called **methylated cap** (Fig. 7.10a). The methylated cap is critical for efficient translation of the mRNA into protein, even though it does not help specify an amino acid.

Like the 5' methylated cap, the 3' end of most eukaryotic mRNAs is not encoded directly by the gene. In a large majority of eukaryotic mRNAs, the 3' end consists of 100–200 As, referred to as a **poly-A tail** (see Fig. 7.10b). Addition of the tail is a two-step process. First, ribonuclease cleaves the primary transcript to form a new 3' end; cleavage depends on the sequence AAUAAA, which is found in poly-adenylated mRNAs 11 to 30 nucleotides upstream of the position where the tail is added. Next the enzyme *poly-A polymerase* adds As onto the 3' end exposed by cleavage. The actual function of the poly-A tail remains a matter of some controversy, but most investigators believe it has two functions: stabilizing mRNAs so that they can serve as messages for a longer time, and increasing the efficiency of the initial steps of translation.

RNA Splicing Removes Sequences Known as Introns from the Primary Transcript

Another kind of RNA processing became apparent in the late 1970s, after researchers had developed techniques that enabled them to analyze nucleotide sequence in both DNA and RNA. Using these techniques, which we describe in Chapter 8, they began to compare eukaryotic genes with the mRNAs derived from them. Their expectation was that, just as in prokaryotes, the DNA nucleotide sequence of a gene's RNA-like strand would be identical to the RNA nucleotide sequence of the messenger RNA (with the exception of U replacing T in the RNA). Surprisingly, they found that the DNA nucleotide sequences of many genes are much longer than their corresponding mRNAs, suggesting that RNA transcripts, in addition to receiving a methylated cap and a poly-A tail, undergo extensive internal processing.

An extreme example of the length difference between primary transcript and mRNA is seen in the human gene for dystrophin (Fig. 7.11). Abnormalities in the dystrophin gene underlie the genetic disorder of Duchenne muscular dystrophy (DMD). The dystrophin gene is 2.5 million nucleotides—or 2500 kilobases (kb)—long, whereas the corresponding mRNA is roughly 14,000 nucleotides, or 14 kb, in length. Obviously the gene contains DNA sequences that are not present in the mature mRNA. Those regions of the gene that do end up in the mature mRNA are scattered throughout the 2500 kb of DNA.

Sequences found in both a gene's DNA and the mature messenger RNA are called **exons** (for "expressed regions"). The DMD gene has more than 80 exons. The sequences found in the DNA of the gene but not in the mature mRNA, are known as **introns** (for "intervening regions"). Introns interrupt, or separate, the exon sequences that actually end up in the mature mRNA where they encode parts of a polypeptide or parts of the message that help control translation but are not translated. Exons vary in size from 50 to a few thousand nucleotide pairs (in the DMD gene the mean exon length is 200 bp), while introns range from 50 to over 100,000 nucleotide pairs (in the DMD gene, the mean intron length is 35 kb, but one intron is an amazing 400 kb long). The greater size variation seen in introns compared to exons reflects the fact that introns do not appear in mature mRNAs and thus do not have to encode polypeptides. Because of this, there are fewer restraints on the sequence and number of nucleotides within introns. Introns can occur anywhere within a gene: between nucleotides that eventually form part of a single codon in the mature mRNA, between nucleotides encoding adjacent codons in the message, or within regions transcribed into parts of the mRNA that help regulate translation but are not themselves translated into protein. Geneticists know very little about the rules that determine the number, size, and position of introns in different genes. In humans, although a small percentage of genes do not contain introns, the large majority of genes interrupt their coding sequences anywhere from 1 to 60 times.

How do cells make a mature mRNA from a gene whose coding sequences are interrupted by introns? The answer is that they first make a primary transcript containing all of a gene's introns and exons, and then remove the introns from the primary transcript by **RNA splicing:** a process that deletes introns and joins together adjacent exons to form a mature mRNA consisting only of exons (Fig. 7.11). Because the first and last exons of the primary transcript become the 5' and 3' ends of the mRNA, while all intervening introns are spliced out, a gene must have one more exon than introns. To construct the mature mRNA, splicing must be remarkably precise. For example, if an intron lies within a codon, splicing must remove the intron and reconstitute the codon without disrupting the reading frame of the mRNA.

Figure 7.12a, b illustrates the details of RNA splicing. Three types of short sequences within the primary transcript—**splice donors, splice acceptors,** and **branch sites**—help ensure the specificity of splicing, making it possible to sever the connections between an intron and the exons that precede and follow it, and then join the formerly separated exons. The discarded intron itself is degraded. A complicated intranuclear machine called the **spliceosome** ensures that all of the splicing reactions take place in concert (Fig. 7.12c). However, a few primary transcripts have the remarkable ability to splice themselves without the aid of a spliceosome.

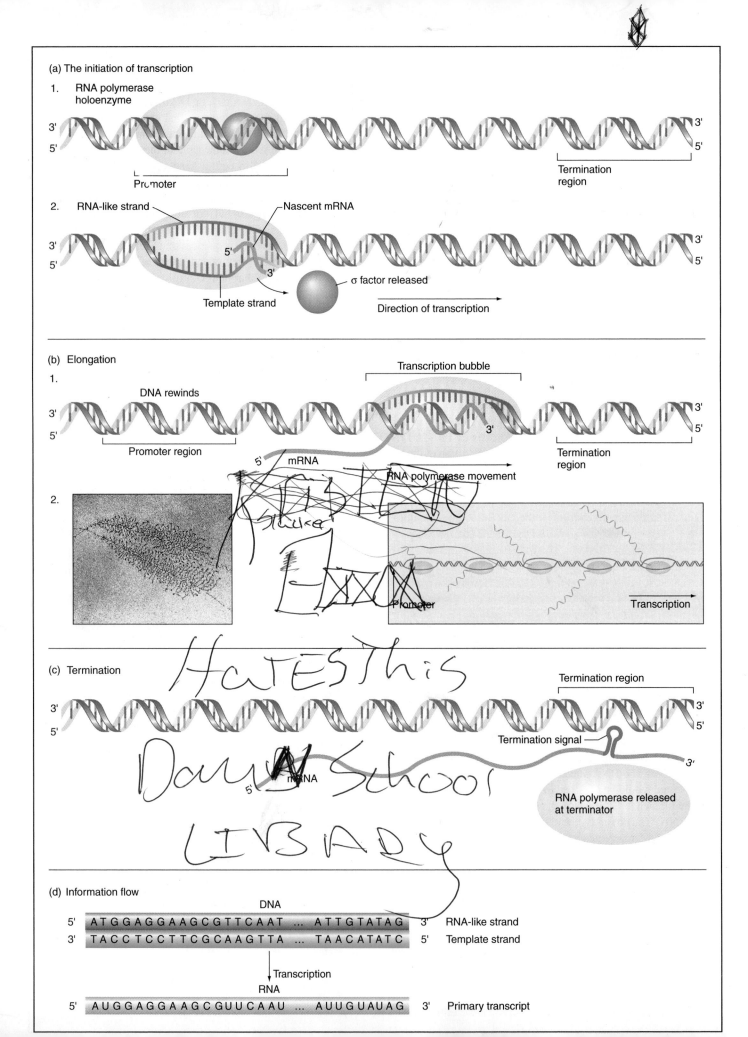

(a) The initiation of transcription

1. RNA polymerase holoenzyme

3'
5'
Promoter
Termination region

2. RNA-like strand — Nascent mRNA

3'
5'
5'
3'
Template strand
σ factor released
Direction of transcription

(b) Elongation

1. Transcription bubble

DNA rewinds

3'
5'
Promoter region
3'
Termination region
5' mRNA
RNA polymerase movement

2.

Promoter
Transcription

(c) Termination

Termination region

3'
5'
Termination signal
3'
mRNA
5'
RNA polymerase released at terminator

(d) Information flow

DNA

5' A T G G A G G A A G C G T T C A A T ... A T T G T A T A G 3' RNA-like strand
3' T A C C T C C T T C G C A A G T T A ... T A A C A T A T C 5' Template strand

↓ Transcription

RNA

5' A U G G A G G A A G C G U U C A A U ... A U U G U A U A G 3' Primary transcript

(a) The Initiation of Transcription

1. RNA polymerase binds to double-stranded DNA at the beginning of the gene to be copied. RNA polymerase recognizes and binds to *promoters,* specialized DNA sequences near the end of a gene where transcription will start. Although the promoters of different genes vary substantially in size and sequence, all promoters contain several characteristic short sequences of 6–10 nucleotide pairs that help bind RNA polymerase; the short sequence at a particular position is nearly identical in different promoters. In bacteria, the complete RNA polymerase (the *holoenzyme*) consists of a core enzyme, plus a σ subunit involved only in initiation. The σ subunit reduces RNA polymerase's general affinity for DNA, allowing the enzyme to home in on a promoter. The σ subunit also increases RNA polymerase's affinity for the promoter, making the enzyme bind much more tightly to the DNA at this location than it would in the subunit's absence. How strongly RNA polymerase binds to a particular promoter is determined by interactions between DNA, RNA polymerase, and several other molecules that help regulate gene expression.

2. After binding to the promoter, RNA polymerase unwinds part of the double helix, exposing unpaired bases on the strand that will serve as a template for copying. The enzyme identifies the template strand and chooses the two nucleotides with which to initiate copying on the basis of the geometry of its interaction with the promoter. Guided by base pairing with these two nucleotides, RNA polymerase aligns the first two ribonucleotides of the new RNA. These two ribonucleotides will be at the 5′ end of the final RNA product. The end of the gene with the promoter is sometimes called the 5′ end of the gene because it contains sequences that are transcribed into the 5′ end of the mRNA. The opposite end of the gene is the 3′ end. The mRNA grows in the 5′-to-3′ direction because RNA polymerase, like DNA polymerase, adds nucleotides to the 3′ end of the growing RNA chain. After aligning the first two ribonucleotides, RNA polymerase catalyzes the formation of a phosphodiester bond between them, resulting in a diribonucleotide. Soon thereafter, the RNA polymerase releases the σ subunit. This release marks the end of initiation.

(b) Elongation: Constructing a Complete RNA Copy of the Gene

1. When the σ subunit separates from the RNA polymerase, the enzyme loses its enhanced affinity for the promoter sequence and regains its strong generalized affinity for any DNA. These changes in affinity enable the core enzyme to leave the promoter yet remain bound to the gene being transcribed. The result is a three-part complex composed of DNA, core RNA polymerase enzyme, and growing RNA polymer. The core enzyme now moves along the chromosome, unwinding the double helix to expose the next single-stranded region of the DNA template. The enzyme extends the RNA by steadying and linking to the 3′ end of the growing chain a ribonucleotide already positioned by complementary base pairing with the template strand. As the enzyme extends the mRNA in the 5′-to-3′ direction, it moves in the antiparallel 3′-to-5′ direction along the DNA template strand. Its working speed is impressive. The *E. coli* RNA polymerase can link about 40 nucleotides a second. At this rate, it would take about 25 seconds to complete transcription of the 1000 nucleotides in an average-sized bacterial gene.

The region of DNA unwound by RNA polymerase is called the **transcription bubble.** Within the bubble, the nascent RNA chain remains base paired with the DNA template, forming a DNA-RNA hybrid. However, in those parts of the gene behind the bubble, which have already been transcribed, the DNA couble helix re-forms, displacing the RNA which hangs out of the transcription complex as a single strand with a free 5′ end.

2. Once an RNA polymerase has moved off the promoter, other RNA polymerase molecules can move in to initiate transcription. If the promoter is very strong, that is, if it can rapidly attract RNA polymerase, the gene can undergo transcription by many RNA polymerases simultaneously. Here we show an electron micrograph and an artist's interpretation of simultaneous transcription by several RNA polymerases. As you can see, there is a range of RNAs in production; those in the early steps of elongation are shorter than those in later steps of elongation. The promoter for this gene thus lies very close to where the shortest RNA is emerging from the DNA.

Geneticists often use the direction traveled by RNA polymerase as a reference when discussing various features within a gene. If for example, you started at the 5′ end of a gene at point A and moved along the gene in the same direction as RNA polymerase to point B, you would be traveling in the **downstream** direction. If by contrast, you started at point B and moved in the opposite direction from RNA polymerase to point A, you would be traveling in the **upstream** direction.

(c) Termination: The End of Transcription

RNA sequences that signal the end of transcription are known as *terminators.* There are two types of terminators: *intrinsic terminators* that cause the RNA polymerase core enzyme to terminate transcription on its own and *extrinsic terminators* that require proteins other than RNA polymerase—particularly a polypeptide known as *rho*—to bring about termination. Regardless of the precise mechanism, such as whether a terminator is intrinsic or extrinsic, all terminators in all organisms are specific sequences in the mRNA, transcribed from specific DNA regions, that signal the termination of transcription. Upon termination, RNA polymerase and a completed RNA chain are both released from the DNA.

(d) The Product of Transcription Is a Single-Stranded Primary Transcript

The RNA produced by the action of RNA polymerase on a gene is a single strand of nucleotides known as a *primary transcript.* The bases in the primary transcript are complementary to the bases between the initiation and termination sites in the template strand of the gene. The nucleotides carrying these bases include groupings for a start codon, codons specifying all the amino acids in the polypeptide to be built, and a stop codon.

Feature Figure 7.9   Transcription: How cells copy the DNA of a gene to the RNA of a transcript.

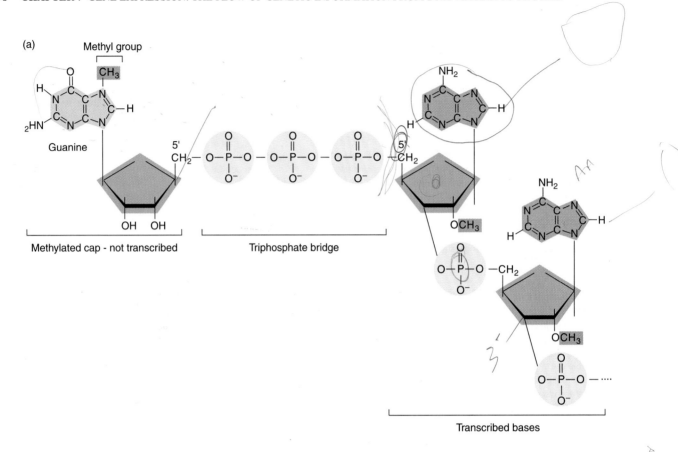

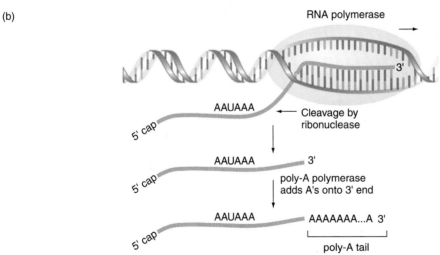

Figure 7.10 How RNA processing adds a cap and a tail to a primary transcript in eukaryotes. (a) Structure of the methylated cap added to the 5' end of a primary transcript. Capping enzyme connects a backward G to the first nucleotide of the primary transcript through a triphosphate linkage. Methyl transferase enzymes then add methyl groups to this G and to one or two of the nucleotides first transcribed from the DNA template. (b) Addition of a poly-A tail to the 3' end of a primary transcript. A ribonuclease recognizes AAUAAA in a particular context of the primary transcript, and cleaves the transcript 11–30 nucleotides downstream to create a new 3' end. The enzyme poly-A polymerase then adds 100–200 A's onto this new 3' end.

Splicing removes introns from a primary transcript

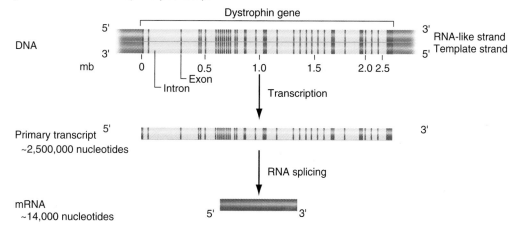

Figure 7.11   The human dystrophin gene: an extreme example of RNA splicing. Though the dystrophin gene is 2500 kb [or 2.5 mb (megabase pairs)] long, the dystrophin mRNA is only 14 kb long. More than 80 introns are removed from the 2500 kb primary transcript to produce the mature mRNA (which is not drawn to scale).

It might seem strange that eukaryotic genes incorporate DNA sequences that are spliced out of the mRNA before translation and thus do not encode amino acids. No one knows exactly why introns exist. One hypothesis proposes that they make it possible to assemble genes from various exon building blocks, which encode modules of protein function. This type of assembly would allow the shuffling of exons to make new genes, a process that appears to have played a key role in the evolution of complex organisms. The exon-as-module proposal is attractive because it is easy to understand the selective advantage of the potential for exon shuffling. Nevertheless, it remains a hypothesis without proof. There is no hard evidence for or against the hypothesis, and introns may have become established through means that scientists have yet to imagine.

Alternative Splicing Often Produces Different mRNAs from the Same Primary Transcript

Normally, RNA splicing joins together the splice donor and splice acceptor at the opposite ends of an intron, resulting in removal of the intron and fusion of two successive—and now adjacent—exons in the same gene. In some genes, however, RNA splicing during development is regulated so that some splicing signals may be ignored. For example, splicing may occur between the splice donor site of one intron and the splice acceptor site of a different intron downstream. Such **alternative splicing** produces different mRNA molecules that may encode related proteins with different—though partially overlapping—amino-acid sequences. In effect then, alternative splicing can tailor the nucleotide sequence of a primary transcript to produce more than one polypeptide.

In mammals, alternative splicing of the gene encoding the antibody heavy chain determines whether the antibody proteins become embedded in the membrane of the B lymphocyte that makes them or are instead secreted into the blood. The antibody heavy-chain gene has eight exons and

seven introns; exon number 6 has a splice site within it. To make the membrane-bound antibody, all exons except for the right-hand part of number 6 are joined to create an mRNA encoding a hydrophobic (water-hating), lipid-loving C terminus (Fig. 7.13a). For the secreted antibody, only the first six exons (including the right part of 6) are spliced together to make an mRNA encoding a heavy chain with a hydrophilic (water-loving) C terminus. The alternate forms of RNA splicing thus allow physiological compartmentalization of the antibody molecule. Membrane-bound antibody molecules serve as receptors for foreign molecules (termed *antigens*); the binding of antigen triggers proliferation and differentiation of the lymphocyte into an antibody-secreting factory. The secreted antibodies can interact directly with antigens, including portions of viruses and bacteria, targeting them for destruction.

A rare and unusual strategy of alternative splicing, seen in *C. elegans* and a few other eukaryotes, is **trans-splicing,** in which the spliceosome joins an exon of one gene with an exon of another gene (Fig. 7.13b). Special nucleotide sequences in the RNAs make trans-splicing possible.

In summary, RNA processing may add a methylated cap or a poly-A tail, delete introns (or even part of an exon) from a primary transcript, determine which exons from a primary transcript are joined together to form a mature mRNA, and even join the exons of different genes.

In *E. coli* and other prokaryotes, transcription takes place in an open intracellular space undivided by a nuclear membrane, and there is little if any RNA processing; translation occurs in the same open space, sometimes simultaneously with transcription. By contrast, in *C. elegans* and other eukaryotes, transcription and RNA processing take place in the nucleus, while the mature messenger RNA must pass through the nuclear membrane to the cytoplasm before it can participate in polypeptide synthesis (Table 7.1).

(a) Short sequences dictate the sites of splicing

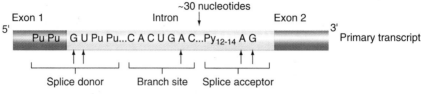

(b) Two sequential cuts remove the intron

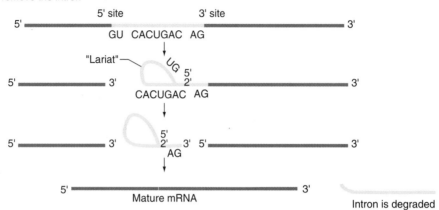

(c) Splicing is catalyzed by the spliceosome

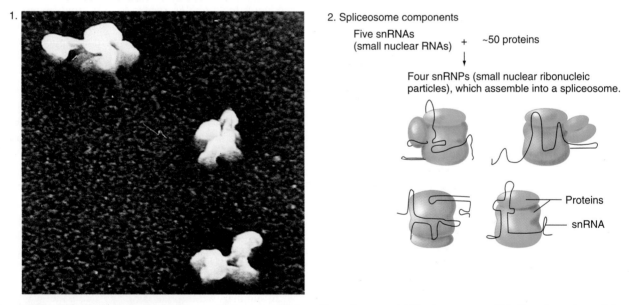

1.

2. Spliceosome components

Five snRNAs
(small nuclear RNAs)   +   ~50 proteins

Four snRNPs (small nuclear ribonucleic
particles), which assemble into a spliceosome.

Proteins

snRNA

Figure 7.12 How RNA processing splices out introns and joins adjacent exons. (a) Short sequences within the primary transcript dictate the sites of splicing. One type of short sequence that helps ensure the specificity of splicing is called the splice donor; it occurs where the 3′ end of an exon abuts the 5′ end of an intron. The splice donors of many genes in many different organisms are similar: A GU dinucleotide that begins the intron is flanked on either side by a few purines (As or Gs). A second critical region, the splice acceptor, is located at the 3′ end of the intron where it joins with the succeeding exon. The final nucleotides of the intron are always AG preceded by a run of 12–14 pyrimidines (C and U). A third required sequence for splicing is the branch site, which sits within the intron about 30 nucleotides upstream of the splice acceptor. The branch site, although quite variable, must include an A; many tend to be rich in pyrimidines. Mutations that alter the sequence of a splice donor, splice acceptor, or branch site, abolish splicing altogether or diminish its accuracy and occurrence. (b) Two sequential cuts remove the intron. The first cut is at the splice-donor site; the 5′ end on the intron becomes attached via a novel 2′–5′ phosphodiester bond to an A at the branch site, forming a so-called "lariat" structure. The second cut, at the splice acceptor site, removes the intron (which is subsequently degraded) and is accompanied by precise splicing of adjacent exons. (c) A complex of snRNPs and protein splicing factors form a spliceosome. (1) A view of three spliceosomes in the electron microscope. (2) The spliceosome consists of at least four subunits known as small nuclear ribonucleoproteins, or snRNPs (pronounced "snurps") Each snRNP contains one or two small nuclear RNAs (snRNAs) 100–300 nucleotides long, associated with proteins in a discrete particle. Some snRNAs can base pair with the splice donor and acceptor sequences in the primary transcript, and can thus bring two intron-exon junctions together.

TABLE 7.1   Differences between Prokaryotes and Eukaryotes in the Details of Gene Expression

	Prokaryotes	Eukaryotes
Overview	1. No nucleus. Transcription and translation thus take place in the same cellular compartments, and translation is often coupled to transcription.	1. Nucleus separated from the cytoplasm by a nuclear membrane. Transcription takes place in the nucleus, while translation occurs in the cytoplasm. Coupling of transcription and translation is thus not possible.
	2. Genes are not divided into exons and introns.	2. The DNA of a gene consists of exons separated by introns; the exons are defined by posttranscriptional splicing, which deletes the introns.
Transcription	1. One RNA polymerase consisting of five subunits.	1. Several kinds of RNA polymerase, each containing 10 or more subunits; different polymerases transcribe different genes.
	2. Primary transcripts are the actual mRNAs; they have a triphosphate start at the 5′ end and no tail at the 3′ end.	2. Primary transcripts undergo processing to produce mature mRNAs that have a methylated cap at the 5′ end and a poly-A tail at the 3′ end.
Translation	1. Unique initiator tRNA carries formylmethionine. 2. mRNAs have multiple ribosome binding sites and can thus direct the synthesis of several different polypeptides. 3. Small ribosomal subunit immediately binds to the mRNA's ribosome binding site.	1. Initiator tRNA carries methionine. 2. mRNAs have only one start site and can thus direct the synthesis of only one kind of polypeptide. 3. Small ribosomal subunit binds first to the methylated cap at the 5′ end of the mature mRNA and then scans the mRNA to find the ribosome binding site.

(a) Alternative splicing produces two different
    mRNAs from the same gene

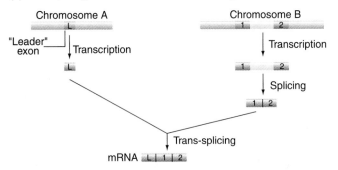

**Figure 7.13   Different mRNAs can be produced from the
same primary transcript.** (a) Alternative splicing of the gene for the
antibody heavy chain produces different mRNAs. (b) Rare trans-splicing
events combine exons from different genes into one mature mRNA.

TRANSLATION: BASE PAIRING BETWEEN mRNA AND tRNAS DIRECTS ASSEMBLY OF A POLYPEPTIDE ON THE RIBOSOME

Virtually all cells—both prokaryotic and eukaryotic—use the
same genetic code to translate the sequence of nucleotides in
a messenger RNA to the sequence of amino acids in the corresponding polypeptide. This process of translation takes place
on ribosomes that coordinate the movements of transfer RNAs
carrying specific amino acids with the genetic instructions of
an mRNA. As we examine the cell's translation machinery, we
first describe the structure and function of tRNAs and ribosomes, and then explain how these components interact during translation.

Transfer RNAs Mediate the Translation of mRNA Codons to Amino Acids

There is no obvious chemical similarity or affinity between the
nucleotide triplets of mRNA codons and the amino acids they
specify. Rather, *transfer RNAs (tRNAs)* serve as adaptor molecules that mediate the transfer of information from nucleic
acid to protein.

tRNA Structure: A Compact L Carrying an Anticodon at One End and an Amino Acid at the Other

tRNAs are short, single-stranded RNA molecules 74 to 95 nucleotides in length. Several of the nucleotides in tRNAs carry
modified bases produced by chemical alterations of the principal A, G, C, and U nucleotides (Fig. 7.14a). Each tRNA carries 1 particular amino acid, and cells must have at least one
tRNA for each of the 20 amino acids specified by the genetic
code. The name of a tRNA reflects the amino acid it carries.
For example tRNA^{Gly} carries the amino acid glycine.

The nucleotide sequence of a tRNA constitutes its primary structure (Fig. 7.14b.1). Short complementary regions
within a tRNA's single strand can form base pairs with each
other to create a characteristic cloverleaf shape (Fig. 7.14b.2);
this is the tRNA's secondary structure. Folding in three-
dimensional space creates a tertiary structure that looks like a
compact letter L (Fig. 7.14b.3).

At one end of the L, the tRNA carries an **anticodon:**
three nucleotides complementary to an mRNA codon specifying the amino acid carried by the tRNA (Fig. 7.14b.2 and
3). The anticodon never forms base pairs with other regions
of the tRNA; it is always available for base pairing with its
complementary mRNA codon. As with other complementary base sequences, during pairing at the ribosome, the
strands of anticodon and codon run antiparallel to each
other. If, for example, the anticodon is 3′CCU5′, the complementary mRNA codon is 5′GGA3′, specifying the amino
acid glycine.

At the other end of the L—where the 3′ end of the tRNA
is—the tRNA connects to the amino acid specified by the
codon. Enzymes known as **aminoacyl-tRNA synthetases**
are the catalysts of this attachment. The enzymes are extraordinarily specific, recognizing unique features of a particular tRNA—despite its general structural similarities with all
other tRNAs—while also recognizing the corresponding
amino acid. Aminoacyl-tRNA synthetases are, in fact, the
only molecules that read the languages of both nucleic acid
and protein. They are thus the actual molecular translators.
There is at least one aminoacyl-tRNA synthetase for each of
the 20 amino acids, and like tRNA, each synthetase functions with only one amino acid. Figure 7.14c shows the
two-step process that establishes the covalent bond between
an amino acid and the 3′ end of its corresponding tRNA.
A tRNA covalently coupled to its amino acid is called a
charged tRNA.

(a) Some tRNAs contain modified bases

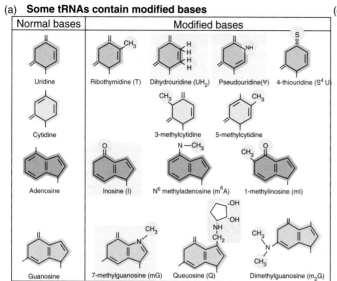

(b) Each tRNA has a primary, secondary, and tertiary structure

1. Primary structure

2. Secondary structure

Yeast tRNA^{Ala}

Amino acid attachment site

3. Tertiary structure

Amino acid attachment site

mRNA

Codon for Ala

Anticodon

(c) Aminoacyl-tRNA synthetases catalyze the attachment of a tRNA to its cognate amino acid

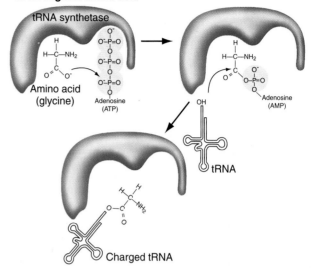

tRNA synthetase

Amino acid (glycine)

Adenosine (ATP)

Adenosine (AMP)

tRNA

Charged tRNA

Figure 7.14 tRNAs mediate the transfer of information from nucleic acid to protein. (a) Many tRNAs contain modified bases produced by chemical alterations of the principal A, G, C, and U structures. The modified bases used in the tRNA^{Ala} depicted in part (b) are highlighted. (b) Each tRNA has (1) a primary structure, (2) a secondary structure, and (3) a tertiary structure. (c) Aminoacyl-tRNA synthetases catalyze the attachment of tRNAs to their corresponding amino acids. The aminoacyl tRNA synthetase first uses the energy generated by ATP hydrolysis to activate the amino acid, forming an AMP-amino acid. The enzyme then transfers the carboxyl group of the amino acid from AMP to the hydroxyl (−OH) group of the ribose at the 3′ end of the tRNA, producing a charged tRNA.

Base Pairing between an mRNA Codon and a tRNA Anticodon Directs Incorporation of an Amino Acid into a Growing Polypeptide

While attachment of the appropriate amino acid charges a tRNA, the amino acid itself does not play a significant role in determining where it becomes incorporated in a growing polypeptide chain. It is the specific interaction between a tRNA's anticodon and an mRNA's codon that makes that decision. A simple experiment illustrates this point (Fig. 7.15). Researchers can subject charged tRNA to chemical treatments that, without altering the structure of the tRNA, change the amino acid it carries. One treatment replaces the cysteine carried by tRNA^{Cys} with alanine. When investigators then add the tRNA^{Cys} charged with alanine to a cell-free translational sys-

tem, the system incorporates *alanine* into the growing polypeptide wherever the mRNA contains a cysteine codon complementary to the anticodon of the tRNA^{Cys}.

Wobble: Some tRNAs Can Recognize More than One Codon

Although there is at least one kind of tRNA for each of the 20 amino acids, cells do not necessarily carry tRNAs with anticodons complementary to all of the 61 possible codon triplets in the degenerate genetic code. *E. coli*, for example, makes 79 different tRNAs containing 42 different anticodons. Although several of the 79 tRNAs in this collection obviously have the same anticodon, 61 − 42 = 19 of 61 potential anticodons are not represented. Thus 19 mRNA codons will not

Genetics and Society

HIV and Reverse Transcription: An Unusual DNA Polymerase Helps Give the AIDS Virus an Evolutionary Edge

The AIDS-causing human immunodeficiency virus (HIV) is the most intensively analyzed virus in history. From laboratory and clinical studies spanning more than a decade, researchers have learned that each viral particle is a rough-edged sphere consisting of an outer envelope enclosing a protein matrix, which in turn surrounds a cut-off cone-shaped core (Fig. A). Within the core lies an enzyme-studded genome: two identical single strands of RNA associated with many molecules of an unusual DNA polymerase known as **reverse transcriptase.**

During infection, the AIDS virus binds to and injects its cone-shaped core into cells of the human immune system (Fig. B). It next uses reverse transcriptase to copy its RNA genome into double-stranded DNA molecules, in the cytoplasm of the host cell. The double helixes then travel to the nucleus where another enzyme inserts them into a host chromosome. Once integrated into a host-cell chromosome, the viral genome does one of two things. It can commandeer the host cell's protein synthesis machinery to make hundreds of new viral particles that bud off from the parent cell, taking with them part of the cell membrane and sometimes killing the host cell in the process. Alternately, it lies latent inside the host chromosome, which then copies and transmits the viral genome to two new cells with each cell division.

The events of this life cycle make HIV a **retrovirus:** an RNA virus that after infecting a host cell copies its own single strands of RNA into double helixes of DNA that a viral enzyme then integrates

into a host chromosome. RNA viruses that are not retroviruses simply infect a host cell and then use the cellular machinery to make more of themselves, often killing the host cell in the process. The viruses that cause hepatitis A, the common cold, and rabies are this type of RNA virus. Unlike retroviruses, they are not transmitted by cell division to a geometrically growing number of new cells.

Reverse transcription, the foundation of the retroviral life cycle, is inconsistent with the one-way, DNA-to-RNA-to-protein flow of genetic information described in this chapter, and because of this, the possibility of the phenomenon encountered great resistance in the scientific community when first reported by Howard Temin of the University of Wisconsin and David Baltimore, then of MIT. It is now an established fact. In the normal order of molecular genetics, DNA polymerases construct DNA polymers from a DNA template, replicating a cell's double-helical genome for transmission to daughter cells during the cell cycle; and RNA polymerases construct RNA polymers from a DNA template, copying the information of genes into

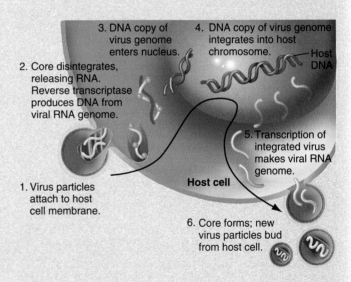

Figure B Life cycle of the AIDS virus.

HIV viral particle

Core
Protein matrix
RNA
Reverse transcriptase
Bilipid outer layer

Figure A Structure of the AIDS virus.

find a complementary anticodon in the *E. coli* collection of tRNAs. Another bacterial species, *Mycoplasma capricolum*, makes 30 different tRNAs carrying 29 different anticodons. In this species, only two tRNAs have the same anticodon, but $61 - 29 = 32$ mRNA codons will not find a complementary anticodon in this collection of tRNAs. Although it is understandable that different species assemble their repertoire of tRNAs in different ways, how can an organism construct

proper polypeptides if some of the codons in its mRNAs cannot locate tRNAs with complementary anticodons?

The answer is that some tRNAs can recognize more than one codon for the amino acid with which they are charged. That is, the anticodons of these tRNAs can interact with more than one codon for the same amino acid. Although researchers do not fully understand this "promiscuous" base pairing between codons and anticodons, Francis Crick spelled out a few

an RNA transcript that is then translated into protein. Reverse transcriptase, by comparison, is a remarkable DNA polymerase that can construct a DNA polymer from either an RNA or a DNA template.

In addition to its comprehensive copying abilities, reverse transcriptase has another feature not seen in most DNA polymerases: inaccuracy. As we saw in Chapter 5, normal DNA polymerases replicate DNA with an error rate of 1 mistake in every million nucleotides copied. Reverse transcriptase, however, introduces 1 mutation in every 5000 incorporated nucleotides.

HIV uses this capacity for mutation, in combination with its ability to integrate its genome into the chromosomes of immune-system cells, to gain a tactical advantage over the immune response of its host organism. Cells of the immune system seek to overcome an HIV invasion by multiplying in response to the proliferating viral particles. The numbers are staggering. Each day of infection in every patient, from 100 million to a billion HIV particles burst from infected cells. As long as the immune system is strong enough to withstand the assault, it may respond by producing as many as 2 billion new cells daily. Just when an immune response to a particular HIV protein wipes out those viral particles carrying the targeted protein, however, virions incorporating new forms of the protein resistant to the current immune response make their appearance. After many years of this complex chase, capture, and destruction by the immune system, the changeable virus outruns the host's immune response and gains the upper hand. Thus, the intrinsic infidelity of HIV's reverse transcriptase, by enhancing the virus's ability to compete in the evolutionary marketplace, increases its threat to human life and health.

The inherent mutability arising from reverse transcriptase's error-prone copying has undermined two therapeutic approaches that have helped control infections other than AIDS: drugs and vaccines. Some of the antiviral drugs approved in the United States for treatment of HIV infection—AZT (zidovudine), ddC (dideoxycytidine), and ddI (dideoxyinosine)—block viral replication by interfering with the action of reverse transcriptase. Each drug is similar to one of the four nucleotides, and when reverse transcriptase incorporates one of the drug molecules rather than a genuine nucleotide into a growing DNA polymer, the enzyme cannot extend the chain any further. However, the drugs are toxic at high doses and thus can be administered only at low doses that do not destroy all viral particles. Because of this limitation and the virus's high rate of mutation, mutant reverse transcriptases soon appear that work even in the presence of the drugs.

Similarly, researchers are having trouble developing safe, effective vaccines. Since HIV infects cells of the immune system, some of

the vaccines tested so far actually increase the activity of the virus; others have only a weak effect on viral replication. Moreover, if it were possible to produce a vaccine that could generate a massive immune response against one, two, or even several HIV proteins at a time, such a vaccine might be effective for only a short while—until enough mutations built up to make the virus resistant.

For these reasons, the AIDS virus will most likely not succumb entirely to drugs or vaccines that target proteins active at various stages of its life cycle, although combinations of these therapeutic tools will remain an important part of the medical arsenal for prolonging an AIDS patient's life. In 1996, for example, medical researchers found that a therapeutic "cocktail" including at least one anti-reverse-transcriptase drug and a relatively new kind of drug known as a protease inhibitor (which blocks the enzymes that cleave a long, inactive polyprotein into shorter, functional viral proteins) could reduce the viral load of some very sick AIDS patients to undetectable levels, thereby relieving their disease symptoms. One year later, however, a clinical study revealed that for slightly more than 50% of patients receiving the drug cocktail in a San Francisco hospital, the treatment lost its effectiveness after six months. For the roughly 50% of patients for whom the cocktail was still effective, the new therapy offers some hope for controlling AIDS unless and until the virus in them evolves resistance to the new treatment.

Thus, after more than a decade of cooperative research and clinical experience, scientists know a great deal about the structure and life cycle of the AIDS virus, but they have been unable to develop drugs and vaccines that stop or even significantly slow its spread for more than a short while. In the mid-1990s, epidemiologists estimated that new infections of the AIDS virus occurred once every 15 seconds. By 1995, that rate of infection added up to nearly 20 million HIV-infected people worldwide.

A self-preserving capacity for mutation, perpetuated by reverse transcriptase, is surely one of the main reasons for HIV's success. Ironically, it may also provide a basis for its subjugation. Researchers are studying what happens when the virus increases its mutational load. Their reasoning is as follows: If reverse transcriptase's error rate determines the size and integrity of the viral population in a group of host organisms, greatly accelerated mutagenesis might push the virus beyond the error threshold that allows it to function; in other words, too much mutation might destroy the virus's infectivity, virulence, or capacity to reproduce. If geneticists could figure out how to make this happen, they might be able to give the human immune system the advantage it needs to overrun the virus.

---

of the rules that govern it. The rules are consistent with the genetic code. By analyzing the genetic code, Crick concluded that the 3′ nucleotide in many codons adds nothing to the specificity of the codon. For example, 5′GGU3′, 5′GGC3′, 5′GGA3′, and 5′GGG3′ all encode glycine (review Fig. 7.3). It does not matter whether the 3′ nucleotide is U, C, A, or G as long as the first two letters are GG. The same is true for other amino acids encoded by four different codons, such as valine,

where the first two bases must be GU, but the third base can be U, C, A, or G. For amino acids specified by two different codons, the first two bases of the codon are, once again, always the same, while the third base must be either one of the two purines (A or G) or one of the two pyrimidines (U or C). Thus, 5′CAA3′ and 5′CAG3′ are both codons for glutamine; 5′CAU3′ and 5′CAC3′ are both codons for histidine. If Pu stands for either purine and Py stands for either pyrimidine,

Figure 7.15   Base pairing between an mRNA codon and a tRNA anticodon determines where an amino acid becomes incorporated in a growing polypeptide. A tRNA with an anticodon for cysteine, but carrying the amino acid alanine, adds alanine wherever the mRNA codon for cysteine appears.

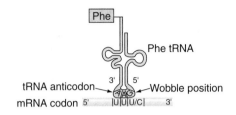

Wobble rules		
5' end of anticodon	can pair with	3' end of codon
G		U or C
C		G
A		U
U		A or G
I		U, C, or A

Figure 7.16   Wobble: Some tRNAs recognize more than one codon for the amino acid they carry. The G at the 5' end of the anticodon shown here can pair with either U or C at the 3' end of the codon. The chart shows the pairing possibilities for other nucleotides at the 5' end of the anticodon. Note that the wobble possibilities are consistent with the genetic code. For example, because methionine is specified by a single codon (5'AUG3'), methionine-specific tRNAs must interact with that codon and no others. To satisfy this requirement, methionine-specific tRNAs have a C at the 5' end of their anticodons (5'CAU3'), and this 5' C can base pair only with a G at the 3' end of the codon. By contrast, a single isoleucine-specific tRNA with inosine (I) at the 5' position of the anticodon can recognize all three codons (5'AUU3', 5'AUC3', and 5'AUA3') for isoleucine.

then CAPu represents the codons for glutamine, while CAPy represents the codons for histidine.

In fact, the 5' nucleotide of a tRNA's anticodon can often pair with more than one kind of nucleotide in the 3' position of an mRNA's codon. (Recall that after base pairing, the bases in the anticodon run antiparallel to the bases in the codon.) A tRNA charged with a particular amino acid can thus recognize several or even all of the codons for that amino acid. This flexibility in base pairing between the 3' nucleotide in the codon and the 5' nucleotide in the anticodon is known as **wobble** (Fig. 7.16).

Ribosomes Are the Sites of Polypeptide Synthesis

Ribosomes facilitate polypeptide synthesis in various ways. First, they recognize mRNA features that signal the start of translation. Second, they help ensure accurate interpretation of the genetic code by stabilizing the interaction between tRNAs and mRNA; without a ribosome, codon-anticodon recognition, mediated by only three base pairs (one of which may wobble), would be extremely weak. Third, they supply the enzymatic activity that links the amino acids in a growing polypeptide chain. Fourth, by moving 5' to 3' along an mRNA molecule, they expose the mRNA codons in sequence, ensuring the linear addition of amino acids. Finally, ribosomes help end polypeptide synthesis by dissociating from both the mRNA directing polypeptide construction and the polypeptide product itself.

Ribosomes are Complex Structures Composed of RNA and Protein

In *E. coli*, ribosomes consist of 3 different **ribosomal RNAs (rRNAs)** and 52 different ribosomal proteins (Fig. 7.17a). These components associate to form two different ribosomal subunits called the 30S subunit and the 50S subunit (with S designating a coefficient of sedimentation related to the size and shape of the subunit; the 30S subunit is smaller than the 50S subunit). Before translation begins, the two subunits exist as separate entities in the cytoplasm. Soon after the start of

translation, they come together to reconstitute a complete ribosome. Eukaryotic ribosomes have more components than their prokaryotic counterparts, but they still consist of two dissociable subunits.

Different Parts of the Ribosome Have Different Functions

The small 30S subunit, for example, is the part of the ribosome that initially binds to mRNA (Fig. 7.17a). The larger 50S subunit contributes an enzyme known as **peptidyl transferase,** which catalyzes formation of the peptide bonds joining adjacent amino acids (Fig. 7.17b). Both the small and the large subunits contribute to two distinct areas known as the **peptidyl (or P) site** and the **aminoacyl (or A) site,** which simultaneously bind to two different tRNA molecules (Fig. 7.17b). Finally, other regions of the ribosome serve as points of contact for some of the additional proteins that play a role in translation.

The Mechanism of Translation

Translation consists of an **initiation** phase that sets the stage for polypeptide synthesis; **elongation,** during which amino acids are added to a growing polypeptide; and a **termination** phase that brings polypeptide synthesis to a halt and enables the ribosome to release a completed chain of amino acids. Figure 7.18 illustrates the details of the process, focusing on translation as it occurs in bacterial cells. As you examine the figure, note the following points about the flow of information during translation: First, the first codon to be translated—the initiation codon—is an AUG set in a special context at the 5' end of

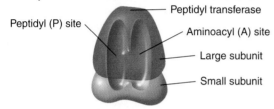

A ribosome has two subunits composed of RNA and protein

Complete ribosomes	Subunits	Nucleotides	Proteins
Prokaryotic 70S	50S	23S RNA 3000 nucleotides	31
	30S	16S RNA 1700 nucleotides / 5S RNA 120 nucleotides	21
Eukaryotic 80S	60S	28S RNA 5000 nucleotides / 5.8S RNA 160 nucleotides / 5S RNA 120 nucleotides	~ 45
	40S	18S RNA 2000 nucleotides	~ 33

Different parts of a ribosome have different functions

Peptidyl (P) site
Peptidyl transferase
Aminoacyl (A) site
Large subunit
Small subunit

Figure 7.17   The ribosome: Site of polypeptide synthesis. (a) A ribosome has two subunits, each composed of rRNA and various proteins. (b) The small subunit initially binds to mRNA. The large subunit contributes the enzyme peptidyl transferase, which catalyzes the formation of peptide bonds in the growing polypeptide chain. The two subunits together form the A and P tRNA binding sites.

the gene's reading frame (which is *not* directly at the 5′ end of the mRNA). Second, special initiating tRNAs carrying a modified form of methionine called formylmethionine (fMet) recognize the initiation codon. Thus, fMet is the first amino acid incorporated into all bacterial polypeptides, and defines their N termini. Third, the ribosome moves along the mRNA in the 5′-to-3′ direction, revealing successive codons in a stepwise fashion. Fourth, at each step of translation, the polypeptide grows by the addition of the next amino acid in the chain to its C terminus. Finally, translation terminates when the ribosome reaches a UAA, UAG, or UGA nonsense codon at the 3′ end of the gene's reading frame. These points explain the biochemical basis of colinearity: the correspondence between the 5′-to-3′ direction in the mRNA and the C-terminus-to-N-terminus direction in the resulting polypeptide.

During elongation, the translation machinery adds about 2–15 amino acids per second to the growing chain. The speed is higher in prokaryotes and lower in eukaryotes. At these rates, construction of an average size 300-amino-acid polypeptide (from an average-length mRNA that is somewhat longer than 1000 nucleotides) could take as little as 20 seconds or as long as 2.5 minutes.

There Are Some Significant Differences in Translation between Prokaryotes and Eukaryotes

One important difference is the absence or presence of a nuclear membrane (Fig. 7.18 and Table 7.1). In prokaryotic

cells, transcription and translation take place in the same undivided cellular interior and translation is often coupled to transcription. This is possible because transcription extends mRNAs in the same 5′-to-3′ direction as the ribosome moves along the mRNA; as a result, ribosomes can begin to translate a partial mRNA that the RNA polymerase is still in the process of transcribing from the DNA. This coupling of transcription and translation has significant consequences for the regulation of gene expression in prokaryotes, which we describe in Chapter 15. In eukaryotes, such coupling cannot occur because the nuclear envelope physically separates the site of transcription in the nucleus from the site of translation in the cytoplasm.

A second significant translational difference between prokaryotes and eukaryotes is in the mechanism of initiation. In prokaryotes, translation begins at a ribosome binding site on the mRNA, defined by a short, characteristic sequence of nucleotides called a Shine-Dalgarno sequence adjacent to an initiating AUG codon. There is nothing to prevent an mRNA from having more than one ribosome binding site, and in fact, many prokaryotic messages are **polycistronic:** they contain the information of several genes (sometimes referred to as *cistrons*), each of which can be translated independently starting at its own ribosome binding site. In eukaryotes, by contrast, the small ribosomal subunit first binds to the methylated cap at the 5′ end of the mature mRNA and then migrates to the initiation site. This is almost always the first AUG codon encountered by the ribosomal subunit as it moves along, or scans, the mRNA in the 5′-to-3′ direction. The mRNA region between the 5′ cap and the initiation codon is sometimes referred to as the **5′-untranslated leader.** Because of this initiation mechanism, initiation in eukaryotes takes place at only a single site on the mRNA, and each mRNA contains the information for synthesizing only a single polypeptide.

Yet another difference between prokaryotes and eukaryotes is in the composition of the initiating tRNA. In prokaryotes, it carries a modified form of methionine known as formylmethionine, while in eukaryotes it carries methionine. This difference means that immediately after translation, prokaryotic polypeptides all have fMet at their N termini while eukaryotic polypeptides have Met at that same location.

Processing After Translation Can Change a Polypeptide's Structure

Protein structure is not irrevocably fixed at the completion of translation. Several different processes may subsequently modify a polypeptide's structure. Cleavage may remove amino acids, such as the N-terminus Met or fMet, from a polypeptide or generate several smaller polypeptides from one larger product of translation (Fig. 7.19a). The addition of chemical constituents, such as phosphate groups, methyl groups, or even carbohydrates, to specific amino acids may also modify a polypeptide after translation (Fig. 7.19b). Such additions are known as **posttranslational modifications.**

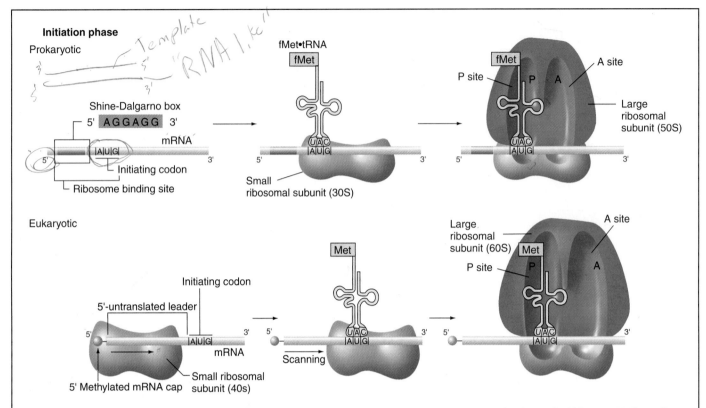

(a) Initiation: Setting the stage for polypeptide synthesis The first three nucleotides of an mRNA do not serve as the first codon to be translated into an amino acid. Instead, a special signal indicates where along the mRNA translation should begin. In prokaryotes, this signal is called the **ribosome binding site.** This mRNA site has two important elements. The first is a short sequence of six nucleotides—usually 5′ . . . AGGAGG . . . 3′— named the **Shine-Dalgarno box** after its discoverers.

The second element in an mRNA's ribosome binding site is the triplet 5′AUG3′. This AUG triplet in the ribosome binding site serves as the initiation codon. A special initiator tRNA, whose 5′CAU3′ anticodon is complementary to AUG, recognizes an AUG preceded by the Shine-Delgarno box of a ribosome binding site. The initiator tRNA carries **N-formylmethionine (fMet),** a modified methionine whose amino end is blocked by a formyl group. The specialized fMet tRNA functions only at an

initiation site. An AUG codon located within an mRNA's reading frame is recognized by a different tRNA charged with an unmodified methionine. This unmodified methionine cannot start translation.

During initiation, the 3′ end of the rRNA in the 30S ribosomal subunit binds to the mRNA's Shine-Dalgarno box, the fMet tRNA^Met binds to the mRNA's initiation codon, and the large 50S ribosomal subunit associates with a small subunit to round out the ribosome. At the end of initiation, the fMet tRNA^Met sits in the P site of the completed ribosome. Proteins known as **initiation factors** play a transient role in the initiation process.

In eukaryotes, the small ribosomal unit binds first to the methylated cap at the 5′ end of the mature mRNA. It then migrates to the initiation site—usually the first AUG it encounters as it scans the mRNA in the 5′-to-3′ direction. The initiator tRNA in eukaryotes carries unmodified methionine (Met) instead of fMet.

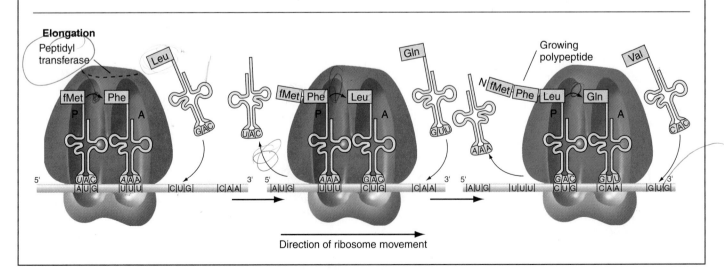

Direction of ribosome movement

Feature Figure 7.18   Translation: How cells use mRNA and charged tRNAs to assemble a polypeptide on the ribosome.

(b) Elongation: The addition of amino acids to a growing polypeptide (facing page at bottom) With the mRNA bound to the complete two-subunit ribosome and with the initiating tRNA in the P site, the elongation of the polypeptide begins. A group of proteins known as **elongation factors** usher the appropriate tRNA into the A site of the ribosome. The anticodon of this charged tRNA must recognize the next codon in the mRNA. The ribosome holds the initiating tRNA at its P site and the second tRNA at its A site so that peptidyl transferase can catalyze formation of a peptide bond between the amino acids carried by the two tRNAs. This bond-forming reaction connects the fMet at the P site to the amino acid carried by the tRNA at the A site; it also disconnects fMet from the initiating tRNA. As a result, the tRNA at the A site now carries two amino acids. The N terminus of this dipeptide is fMet; the C terminus is the second amino acid, whose carboxyl group remains covalently linked to its tRNA.

Once formation of the first peptide bond causes the initiating tRNA in the P site to release its amino acid, the ribosome moves, exposing the next mRNA codon. Like the first steps of elongation, the ribosome's movement requires the help an elongation factors and an input of energy. As the ribosome moves, the initiating tRNA in the P site, which no longer carries an amino acid, dissociates from the ribosome, and the other tRNA carrying the dipeptide shifts from the A site to the P site.

The empty A site now receives another tRNA, whose identity is determined by the next codon in the mRNA. Peptidyl transferase then catalyzes formation of a second peptide bond, generating a chain of three amino acids. This tripeptide is connected at its C terminus to the tRNA currently in the A site; the fMet at its N terminus hangs free. With each subsequent round of ribosome movement and peptide bond formation, the peptide chain grows one amino acid longer.

Because the elongation machinery adds amino acids to the C terminus of the lengthening polypeptide, polypeptide synthesis proceeds from the N terminus to the C terminus. As a result, fMet in prokaryotes (Met in eukaryotes), the first amino acid in the growing chain, will be the N-terminal amino acid of all finished polypeptides that do not undergo protein processing (see main text). Moreover, because the nucleotide triplets encoding amino acids progressively closer to a polypeptide's C terminus are found successively farther from the 5' end of the mRNA, the ribosome must move along the mRNA in the 5'-to-3' direction.

The movement of ribosomes along the mRNA has important implications. Once a ribosome has moved far enough away from the mRNA's ribosome binding site, that site becomes accessible to other ribosomes. In fact, several ribosomes can simultaneously work on the same mRNA. A complex of several ribosomes translating from the same mRNA is called a **polyribosome.** This complex allows the simultaneous synthesis of many copies of a polypeptide from a single mRNA.

Polyribosome

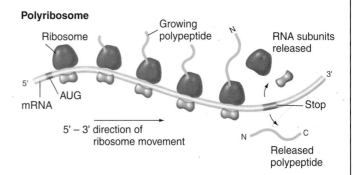

(c) Termination: The ribosome releases the completed polypeptide No normal tRNAs carry anticodons complementary to the three nonsense (stop) codons UAG, UAA, and UGA. Thus, when movement of the ribosome brings a nonsense codon into the ribosome's A site, no tRNAs can bind to that codon through complementary base pairing. Instead, proteins called **release**

factors recognize the termination codons and bring polypeptide synthesis to a halt. Researchers know little about the details of termination, except that three events occur: the tRNA specifying the C-terminal amino acid releases the completed polypeptide, the same tRNA as well as the mRNA separates from the ribosome, and the ribosome dissociates into its large and small subunits.

Termination

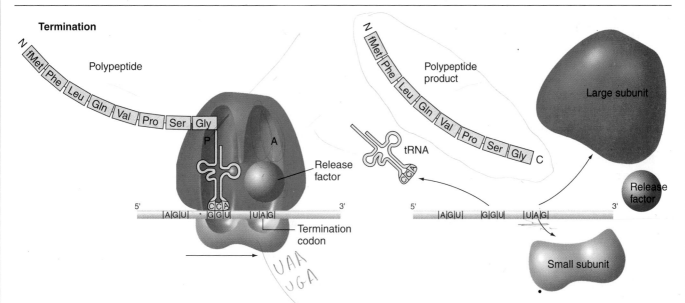

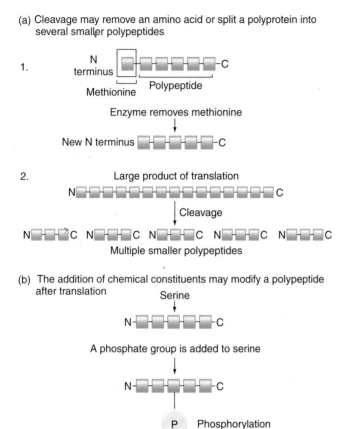

(a) Cleavage may remove an amino acid or split a polyprotein into several smaller polypeptides

1.

N terminus

Methionine   Polypeptide

Enzyme removes methionine

New N terminus   C

2.   Large product of translation

N   C

Cleavage

N   C   N   C   N   C   N   C   N   C

Multiple smaller polypeptides

(b) The addition of chemical constituents may modify a polypeptide after translation

Serine

N   C

A phosphate group is added to serine

N   C

P   Phosphorylation

Figure 7.19 Posttranslational processing can modify polypeptide structure. (a) Cleavage may remove an amino acid from the N-terminus end of a polypeptide (1) or split a larger *polyprotein* into two or more smaller functional proteins (2). (b) Chemical reactions may add a phosphate or other functional group to an amino acid in the polypeptide.

COMPREHENSIVE EXAMPLE: A COMPUTERIZED ANALYSIS OF GENE EXPRESSION IN C. ELEGANS

At the beginning of this chapter we saw that geneticists determined the precise sequence of 2.2 million nucleotides on chromosome III of the tiny nematode *C. elegans*. Using their knowledge of gene structure and gene expression, they programmed computers to locate the sequences likely to be genes. Their programs included instructions to search for possible exons by looking for strings of amino-acid-encoding nucleotide triplets uninterrupted by in-frame nonsense (stop) codons, and to ignore potential introns, identified as sequences lying between likely splice-donor and splice-acceptor sites. Once the computer had retrieved regions likely to be genes, the researchers asked it to use the genetic code to project the amino-acid sequences of the polypeptides encoded by these genes. Finally, they scanned computerized databases for similar amino-acid sequences in the polypeptides of other organisms. If they found a similar sequence in a polypeptide of known function in another organism, they could conclude that the worm's polypeptide probably has a parallel function.

In this way, they discovered 480 genes for particular *C. elegans* RNAs and proteins, many of which encode components of the worm's gene-expression machinery. One of the expression-related genes encodes a subunit of the worm's RNA polymerase, a multimer consisting of more than 10 subunits. Among the other genes located in the sequenced region of *C. elegans'* chromosome III, 4 encode parts of the ribosome workbench, 14 encode transfer tRNAs, 1 encodes aspartyl-tRNA synthetase (the enzyme that attaches aspartic acid to the corresponding tRNA), and a number encode **transcription factors:** DNA-binding proteins that regulate transcription.

Researchers can predict from the DNA sequence alone a great deal about the proteins encoded by many of these expression-related genes. They cannot, however, deduce all the details of their transcription and translation without isolating and characterizing the corresponding mRNAs. For example, although they know the locations of the protein-coding exons of the RNA polymerase subunit gene, without further experiments, they cannot discover where the mRNA starts and stops. For now, they have determined that the gene has a minimum of three exons and two introns, but it may contain additional exons and introns at its 5′ or 3′ ends. In a similar fashion, it is not possible to know whether alternative splicing produces different mRNAs, without biochemical analysis of the gene's RNA products.

By contrast, researchers have obtained both the genomic DNA and the mRNA sequences of another *C. elegans* gene on chromosome III. The gene encodes one form of the protein collagen. Figure 7.20 shows a diagram of the gene as well as the complete sequence of the gene, its primary RNA transcript, the mature mRNA, and the polypeptide product. As you can see, the gene's structural features include three exons and two introns, as well as the signals that allow transcription, RNA processing, and translation into collagen. This single-polypeptide protein is a component of the hard cuticle that surrounds and protects the worm. Related forms of collagen occur in all multicellular animals. In vertebrates, collagen is the most abundant protein, found in bones, teeth, cartilage, tendons, and other tissues.

The general structure of the collagen gene is similar to the structure of most eukaryotic genes. This is because the basic pattern of gene expression has remained substantially the same throughout evolution, even though the details, such as gene length, exon number, and the spacing or size of the untranslated 5′ and 3′ ends, vary from gene to gene and from organism to organism.

The sequencing of 2.2 million nucleotides on *C. elegans* chromosome III, in addition to defining 480 genes, helped reveal some uncommon features in the way the worm expresses its genes. Like bacteria, it transcribes some groups of adjacent genes as one long polycistronic primary transcript; it is the only eukaryotic organism in which researchers have observed this predominantly prokaryotic phenomenon. In rare instances, it uses trans-splicing to create an mRNA from the primary transcripts of two different genes (review Fig. 7.13b); before observing trans-splicing in the nematode, researchers had seen it mainly in trypanosomes, the single-celled proto-

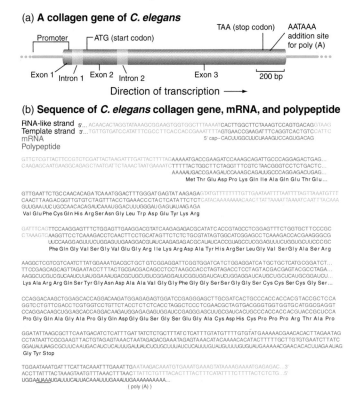

(a) A collagen gene of *C. elegans*

Promoter — ATG (start codon) — TAA (stop codon) — AATAAA addition site for poly (A)

Exon 1 — Intron 1 — Exon 2 — Intron 2 — Exon 3 — 200 bp

Direction of transcription ⟶

(b) Sequence of *C. elegans* collagen gene, mRNA, and polypeptide

RNA-like strand
5′...ACAACACTAGGTATAAAGCGGAAGTGGTGGCTTTAAAATCACTTGGCTTCTAAAGTCCAGTGACAGGTAAG
Template strand
3′...TGTTGTGATCCATATTTCGCCTTCACCACCGAAATTTAGTGAACCGAAGATTTCAGGTCACTGTCCATTC
mRNA
5′ cap–CACUUGGCUUCUAAAGUCCAGUGACAG
Polypeptide

GTTCTCGTTACTTCCGTCTCGATTACTAAGATTTGATTACTTTTAGAAAAATGACCGAAGATCCAAAGCAGATTGCCCAGGAGACTGAG...
CAAGAGCAATGAAGGCAGAGCTAATGATTCTAAACTAATGAAAATCTTTTTACTGGCTTCTAGGTTTCGTCTAACGGGTCCTCTGACTC...
AAAAAUGACCGAAGAUCCAAAGCAGAUUGCCCAGGAGACUGAG...
Met Thr Glu Asp Pro Lys Gln Ile Ala Gln Glu Thr Glu...

GTTGAATTCTGCCAACACAGATCAAATGGACTTTGGGATGAGTATAAGAGGTATGTTTTTTTTGTTGAATAATTTAATTTTAGTTAAATGTTT
CAACTTAAGACGGTTGTGTCTAGTTTACCTGAAACCCTACTCATATTCTCTCATCAAAAAAAACAACTTATTAAAATTAAAATCAATTTACAAA
GUUGAAUUCUGCCAACACAGAUCAAAUGGACUUUGGGAUGAGUAUAAG AGA
Val Glu Phe Cys Gln His Arg Ser Asn Gly Leu Trp Asp Glu Tyr Lys Arg

GATTTCAGTTCCAAGGAGTTTCTGGAGTTGAAGGACGTATCAAGAGAGACGCATATCACCGTAGCCTCGGAGTTTCTGGTGCTTCCCGC
CTAAAGTCAAGGTTCCTCAAAGACCTCAACTTCCTGCATAGTTCTCTCTGGTATAGTGGCATCGGAGCTCAAAGACCAGCAAGGGGCG
UUCCAAGGAGUUUCUGGAGUUGAAGGACGUAUCAAGAGAGACGCACAUAUCACCGUAGCCUCGGAGUUUCUGGUGCUUCCCGC
Phe Gln Val Val Ser Gly Val Val Gly Gly Arg Ile Lys Lys Arg Asp Ala Tyr His Arg Ser Leu Gly Val Ser Gly Ala Ser Arg

AAGGCTCGTCGTCAATCTTATGGAAATGACGCTGCTGTCGGAGGATTCGGTGGATCATCTGGAGGATCATGCTGCTCATGCGGATCT...
TTCCGAGCAGCAGTTAGAATACCTTTACTGGCGACGACAGCCTCCTAAGCCACCTAGTAGACCTCCTAGTACGACGAGTACGCCTAGA...
AAGGCUCGUCGUCAAUCUUAUGGAAAUGACGCUGCUGUCGGAGGAUUCGGUGGAUCAUCUGGAGGAUCAUGCUGCUCAUGCGGAUCU...
Lys Ala Arg Arg Gln Ser Tyr Gly Asn Asp Ala Ala Val Gly Gly Phe Gly Gly Ser Ser Gly Gly Ser Cys Cys Ser Cys Gly Ser...

CCAGGACAAGCTGGAGCACCAGGACAAGATGGAGAGAGTGGATCCGAGGGAGCTTGCGATCACTGCCCACCACCACGTACCGCTCCA
GGTCCTGTTCGACCTCGTGGTCCTGTTCTACCTCTCTCACCTAGGCTCCCTCGAACGCTAGTGACGGGTGGTGGTGCATGGCGAGGT
CCAGGACAAGCUGGAGCACCAGGACAAGAUGGAGAGAGUGGAUCCGAGGGAGCUUGCGAUCACUGCCCACCACCACGUACCGCUCCA
Pro Gly Gln Ala Gly Ala Pro Gly Gln Asp Gly Glu Ser Gly Ser Glu Gly Ala Cys Asp His Cys Pro Pro Pro Arg Thr Ala Pro

GGATATTAAGCGCTTCAATGACATCTCATTTGATTATCTCTGCTTTATCTCATTTGTATGTTTTGTGTATGAAAAACGAACACACTTAGAATAG
CCTATAATTCGCGAAGTTACTGTAGAGTAAACTAATAGAGACGAAATAGAGTAAACATACAAAACACATACTTTTTGCTTGTGTGAATCTTATC
GGAUAUUAAGCGCUUCAAUGACAUCUCAUUUGAUUAUCUCUGCUUUAUCUCAUUUGUAUGUUUUGUGUAUGAAAAACGAACACACUUAGAA AUAG
Gly Tyr Stop

TGGAATAAATGATTTCATTACAAATTTGAAATTGAATAAGACAAATGTGAAATGAAAGTATAAAAGAAAATGAGAGAC...3′
ACCTTATTTACTAAAGTAATGTTTAAACTTTAACTTATTCTGTTTAACATTTTCTATATTTTCTTTTACTCTCTG...5′
UGGAAUAAAUGAUUUCAUUACAAAUUUGAAAUUGAAAAAAAAAAA...
(poly (A))

Figure 7.20 Expression of a *C. elegans* gene for collagen. (a) Landmarks in a collagen gene. (b) Comparison of the complete sequence of base pairs in both strands of DNA with the sequences of nucleotides in the mature mRNA pinpoints the start of transcription, the location of exons and introns, and the position of the AAUAAA poly-A addition signd. Translation of the mRNA according to the rules of the genetic code indicates the positions of the initiating and stop codons, and confirms the colinearity of the gene with its protein product.

zoans that cause sleeping sickness. The introns that *C. elegans* deletes through RNA splicing are fewer and smaller than the introns found in other animal genes, and they are unusually rich in As and Ts. Researchers will be able to apply these insights clarifying *C. elegans'* mechanisms of gene expression to studies of the worm's growth and development (see the genetic portrait of *C. elegans* in Chapter 20).

HOW MUTATIONS AFFECT GENE EXPRESSION

We have seen that the information in DNA is the starting point of gene expression. The cell transcribes that information into mRNA and then translates the mRNA information into protein. Mutations that alter the nucleotide pairs of DNA may modify any of the steps or products of gene expression (Fig. 7.21).

Mutations in a Gene's Coding Sequence Can Alter the Gene Product

Because of the nature of the genetic code, mutations in a gene's amino-acid-encoding exons generate a range of repercussions (Fig. 7.21a).

Silent Mutations Do Not Alter the Amino Acid Specified

One consequence of the code's degeneracy is that some mutations direct the inclusion of the same amino acid as the unmutated DNA and therefore have no effect on the amino-acid composition of the encoded polypeptide. The majority of such **silent mutations** change the third nucleotide of a codon, the position at which most codons for the same amino acid differ. For example, a change from GCA to GCC in a codon would still yield alanine in the protein product.

Missense Mutations Replace One Amino Acid with Another

Missense mutations that cause substitution in the polypeptide of an amino acid with chemical properties similar to the one it replaces may have little or no effect on protein function. Such substitutions are *conservative*. By contrast, *nonconservative* missense mutations that cause substitution of an amino acid with very different properties are likely to have more noticeable consequences. The effect on phenotype is difficult to predict since it depends on how a particular amino-acid substitution changes a protein's structure and function.

Nonsense Mutations Change an Amino-Acid-Specifying Codon to a Stop Codon

Nonsense mutations therefore result in the production of proteins smaller than those encoded by wildtype alleles of the same gene. The truncated proteins lack all amino acids between the amino acid encoded by the mutant codon and the C terminus of the normal polypeptide. The mutant polypeptide will be unable to function if it requires the missing amino acids for its activity.

Frameshift Mutations Result from the Insertion or Deletion of Nucleotides within the Coding Sequence

As we have seen, if the number of extra or missing nucleotides is not divisible by 3, the insertion or deletion will skew the reading frame downstream of the mutation. As a result, **frameshift mutations** cause unrelated amino acids to appear in place of amino acids critical to protein function, destroying or diminishing polypeptide function.

Mutations in a Gene Outside the Coding Sequence Can Also Alter Gene Expression

Mutations that produce a variant phenotype are not restricted to alterations in codons. Since gene expression depends on several signals other than the actual coding sequence, changes in any of these critical signals can disrupt the process (Fig. 7.21b).

We have seen that promoters and termination signals in the DNA of a gene instruct RNA polymerase where to start and stop transcription. Changes in the sequence of a promoter that make it hard or impossible for RNA polymerase to recognize the site diminish or prevent transcription. Mutations in a termination signal can diminish the amount of mRNA produced and thus the amount of gene product.

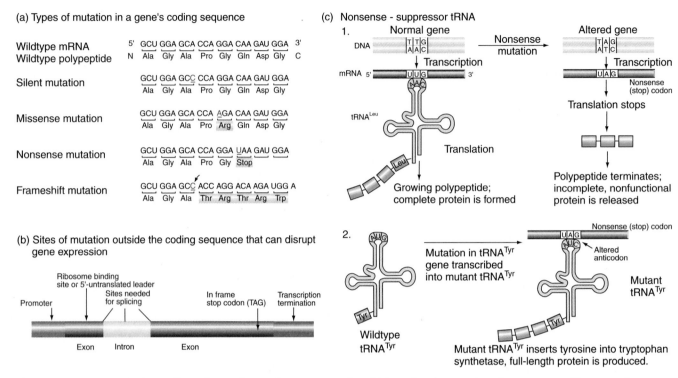

Figure 7.21 How mutations affect the products of gene expression. (a) Mutations in a gene's coding sequences. Silent mutations are not expressed because they do not alter the protein's primary structure. Missense mutations replace one amino acid with another. Nonsense mutations shorten a polypeptide by replacing a codon with a stop signal. Frameshift mutations result in a change in reading frame downstream of the addition or deletion. (b) Mutations in sites outside of the coding region can disrupt gene expression. (c) Mutations in genes encoding components of the machinery for transcription, splicing, or translation can also disrupt expression of a gene. (1) A nonsense mutation that generates a stop codon causes production of an incomplete, nonfunctional polypeptide. (2) A second nonsense-suppressing mutation in a tRNA gene causes addition of an amino acid at the stop codon, allowing production of a full-length polypeptide.

In eukaryotes, most primary transcripts have splice acceptor sites, splice donor sites, and branch sites that allow splicing to join exons together with precision in the mature mRNA. Changes in a splice acceptor or donor site can obstruct splicing, resulting in no mature mRNA and thus no normal polypeptide.

Mature mRNAs have ribosome binding sites and in-frame stop codons indicating where translation should start and stop. Mutations affecting a ribosome binding site would lower the affinity of the mRNA for the small ribosomal subunit; since this affinity helps determine the efficiency of translation, such mutations are likely to diminish the amount of polypeptide product. Mutations in a stop codon would produce longer-than-normal proteins that might be unstable or nonfunctional.

Mutations in Genes Encoding the Molecules That Implement Expression May Affect Transcription, mRNA Splicing, or Translation

Gene expression depends on an astonishing number and variety of macromolecules (Table 7.2). A separate gene encodes the subunits of each macromolecule. The genes for all the proteins are transcribed and translated the same as any other gene. The genes for all the rRNAs, tRNAs, and snRNAs are transcribed but *not* translated. Many mutations in these genes have a dramatic effect on phenotype.

Mutations Altering Genes Encoding Proteins or RNAs Involved in Gene Expression Are Usually Lethal

This is because such mutations adversely affect the synthesis of all proteins in a cell. Even a 50% reduction in the amount of some of the proteins enumerated in Table 7.2 can have severe repercussions. In *Drosophila,* for example, a null mutation in a gene encoding one of the ribosomal proteins is lethal when homozygous. This same mutation in a heterozygote causes a dominant *Minute* phenotype in which the slow growth of cells delays the fly's development.

Mutations in tRNA Genes Can Suppress Mutations in Protein-Coding Genes

If more than one gene encoded a molecule with the same role in gene expression, a mutation in one of these genes would not be lethal, and might even be useful. Bacterial geneticists have found, for example, that mutations in certain tRNA genes can suppress the effect of a nonsense mutation in other genes. The tRNA-gene mutations that have this effect give rise to **nonsense suppressor tRNAs.** Consider, for instance, an otherwise wildtype *E. coli* population with an in-frame UAG nonsense mutation in the tryptophan synthetase gene. All cells in this population make a truncated, nonfunctional form of the tryptophan synthetase enzyme and are thus tryptophan auxotrophs (*trp⁻*) unable to synthesize tryptophan (Fig. 7.21c.1).

TABLE 7.2   The Cellular Components of Gene Expression

Function	Cellular Components
Transcription*	core RNA polymerase
	sigma subunit
	rho factor
Splicing and RNA Processing	snRNAs
	protein components of spliceosomes
	additional splicing factors
	capping enzyme
	methyl transferases
	poly-A polymerase
Translation	mRNAs
	tRNAs
	aminoacyl-tRNA synthetases
	rRNAs
	protein components of ribosomes
	translation factors
Protein Processing	deformylases
	amino peptidases
	proteases
	methylases
	hydroxylases
	glycosylases
	kinases
	phosphatases

*For simplicity, we list here only proteins from prokaryotic organisms involved in transcription. The cellular components needed for transcription in eukaryotic organisms are more complex; for example, eukaryotes have three different kinds of RNA polymerase each made of numerous subunits.

Subsequent exposure of these auxotrophs to mutagens, however, generates some trp^+ cells that carry two mutations: one is the original tryptophan synthetase nonsense mutation, the second is a mutation in the gene that encodes a tRNA for the amino acid tyrosine. Evidently, the mutation in the tRNA gene suppresses the effect of the nonsense mutation, restoring the function of the original tryptophan synthetase gene. As Fig. 7.21c.2 illustrates, the basis of this nonsense suppression is that the tRNA^{Tyr} mutation changes an anticodon that recognizes the codon for tyrosine to an anticodon complementary to the UAG stop codon. The mutant tRNA can therefore insert tyrosine into tryptophan synthetase at the position of the in-frame UAG nonsense mutation, allowing the cell to make at least some full-length enzyme. Similarly, mutations in the anticodons of other tRNA genes can suppress UGA or UAA nonsense mutations.

Cells with a nonsense-suppressing mutation in a tRNA gene can only survive if two conditions coexist with the mutation. First, the cell must have other tRNAs that recognize the same codon as the suppressing tRNA recognized before mutation altered its anticodon. Without such tRNAs, the cell has no way to insert the proper amino acid in response to that codon (in our example, the codon for tyrosine). Second, the suppressing tRNA must have only a weak affinity for the translation termination signals at the ends of most mRNAs. If this were not the case, the suppressing tRNA would wreak havoc in the cell, producing a whole array of aberrant polypeptides that are longer than normal. One way cells guard against this possibility is that for many genes, termination depends on two stop codons in a row. Because a suppressing tRNA's chance of inserting an amino acid at both of these codons is very low, only a small number of extended proteins arise.

CONNECTIONS

The promoter, a particular region of the DNA double helix, signals the starting point of gene expression. Starting at this site, RNA polymerase unwinds the DNA and transcribes a simple linear sequence of nucleotides complementary to the gene's template strand into a messenger RNA containing all the information required for construction of a complex polypeptide. The cell then uses the genetic code to translate the mRNA nucleotide sequence into a colinear chain of amino acids whose sequence dictates its three-dimensional folding into a functional protein. Interestingly, the DNA within a cell is much more stable than either the RNA or proteins constructed from it. The half-life of most prokaryotic mRNAs is measured in minutes, while that of eukaryotic mRNAs is mea-

sured in hours. The half-life of many proteins is about a day. DNA is not recycled in the same way. It remains as long as the cell survives and grows.

Our knowledge of gene expression enables us to redefine the concept of a gene. A gene is not simply the DNA that is transcribed into the mRNA codons specifying the amino acids of a particular polypeptide. Rather, *a gene is all the DNA sequences needed for expression of the gene into a polypeptide product.* A gene therefore includes the promoter sequences that govern where transcription begins, and at the opposite end, signals for the termination of transcription. A gene also includes sequences dictating where translation starts and stops; such information is an important part of the mRNA. In addition to all of these features, eukaryotic genes contain introns that are spliced out of the primary transcript to make the mature mRNA. Because of introns, most eukaryotic genes are much larger than prokaryotic genes.

Even with introns, a single gene carries only a very small percentage of the nucleotide pairs in the chromosomes that make up a genome. The average gene in *C. elegans* is about 4000 nucleotide pairs in length, while the worm's haploid genome contains approximately 110 million (110,000,000) nucleotide pairs distributed among six chromosomes containing an average of 17 million nucleotide pairs apiece. In humans, where genes tend to have more introns, the average gene is 16,000 nucleotide pairs in length, and the haploid genome has roughly 3 billion (3,000,000,000) nucleotide pairs distributed among 23 chromosomes containing an average of 120 million nucleotide pairs a piece. How can geneticists sort through these enormous haystacks of genetic information to find the pinpoints of interest: the DNA of a single gene or the small changes within that gene that alter a healthy phenotype to disease?

In Chapters 8–10, we describe how researchers analyze the mass of genetic information in the chromosomes of a genome as they try to discover what parts of the DNA are genes and how those genes influence phenotype. They begin their analysis by breaking the DNA into pieces of manageable size, making many copies of those pieces to obtain enough material for study, and characterizing the pieces down to the level of nucleotide sequence. They then try to reconstruct the DNA sequence of an entire genome by determining the spatial relationship between the many pieces.

ESSENTIAL CONCEPTS

1. *Gene expression* is the process by which cells convert the DNA sequence of a gene to the RNA sequence of a transcript, and then decode the RNA sequence to the amino-acid sequence of a polypeptide.

2. The nearly universal *genetic code* consists of 64 *codons* composed of three nucleotides apiece. Of these codons, 61 specify amino acids, while 3—UAA, UAG, and UGA—are *nonsense* or *stop codons* that do not specify an amino acid.
   a. The code is *degenerate:* More than one codon specifies every amino acid except methionine and tryptophan.
   b. AUG in the context of a ribosome binding site is the initiation codon; it establishes a *reading frame* that determines the grouping of nucleotides into triplet codons.
   c. The code is *nonoverlapping.* Within a reading frame, the first three nucleotides constitute one codon; the next three, the second codon; and so forth.

3. Gene expression based on the genetic code produces the *colinearity* of a gene's nucleotide sequence and a protein's sequence of amino acids.

4. *Transcription* is the first stage of gene expression. During transcription, *RNA polymerase* synthesizes a single stranded *primary transcript* from a DNA template.
   a. RNA polymerase initiates transcription by binding to the *promoter* sequence of the DNA and unwinding the double helix to expose bases for pairing.
   b. RNA polymerase extends the mRNA in the 5'-to-3' direction by catalyzing formation of phosphodiester bonds between successively aligned nucleotides.
   c. *Terminator* sequences in the RNA cause RNA polymerase to dissociate from the DNA.
   d. In prokaryotes, the primary transcript is the mRNA that guides polypeptide synthesis.

5. In eukaryotes, *RNA processing* after transcription produces a mature mRNA that travels from the nucleus to the cytoplasm to direct polypeptide synthesis.
   a. RNA processing adds a methylated cap to the 5' end and a poly-A tail to the 3' end of the eukaryotic mRNA.
   b. The *spliceosome* removes *introns* from the primary transcript and precisely splices together the remaining *exons.* *Alternative splicing* makes it possible to produce different mRNAs from the same primary transcript.

6. *Translation* is the stage of gene expression when the cell synthesizes proteins according to instructions in the mRNA.
   a. tRNAs carry amino acids to the translation machinery. *Aminoacyl-tRNA synthetases* connect amino acids to their corresponding tRNAs. Each tRNA molecule has an *anticodon* complementary to the mRNA codon specifying the amino acid it carries. Because of *wobble,* some tRNA anticodons recognize more than one mRNA codon.
   b. Translation occurs on complex molecular machines called *ribosomes.* Ribosomes have two binding sites for tRNAs—P and A—and an enzyme known as *peptidyl transferase* that catalyzes formation of a peptide bond between amino acids carried by the tRNAs at these two sites.
   c. Initiation: To start translation, part of the ribosome binds to a *ribosome binding site* on the mRNA, which includes the AUG initiation codon. Special initiating tRNAs with codons complementary to AUG carry the amino acid fMet in prokaryotes or Met in eukaryotes to the ribosomal P site. This amino acid will become the N terminus of the growing polypeptide.
   d. Elongation: When the carboxyl group of the amino acid connected to a tRNA at the ribosome's P site becomes attached through a peptide bond to the amino

acid carried by the tRNA at the A site, the ribosome travels three nucleotides toward the 3′ end of the mRNA. This movement exposes the next codon and allows the next round of amino-acid addition. This mechanism of translation dictates that the 5′-to-3′ direction in the mRNA corresponds to the N-terminus-to-C-terminus direction in the polypeptide under construction.

e. Termination: When the ribosome encounters in-frame nonsense codons, it ends translation by releasing the mRNA and disconnecting the complete polypeptide from the tRNA.

7. *Processing after translation* may alter a polypeptide by adding or removing chemical constituents to or from particular amino acids, or by cleaving the polypeptide into smaller molecules.

8. Mutations affect gene expression in several ways.
   a. Mutations in a gene may modify the message encoded in a sequence of nucleotides. *Silent mutations* usually change the third letter of a codon and have no effect on polypeptide production. *Missense mutations* change the codon for one amino acid to the codon for another amino acid and thereby direct incorporation of a different amino acid. *Nonsense mutations* change a codon for an amino acid to a stop codon, causing synthesis of a truncated polypeptide. *Frameshift mutations* change the reading frame of a gene, altering the identity of amino acids downstream of the mutation.
   b. Mutations outside of coding sequences that alter signals required for transcription, mRNA splicing, or translation, will also disrupt gene expression.
   c. Mutations in genes encoding molecules of the gene expression machinery are often lethal. Among the exceptions to this rule are mutations in tRNA genes that suppress mutations in polypeptide-encoding genes.

SOCIAL AND ETHICAL ISSUES

1. Scientists have been exploring the possibility that chemicals similar in structure to human hormones (hormone mimics) affect physiological processes such as fertility levels, cancer rates, and intelligence. Hormone mimics are found in several synthetic compounds such as pesticides, drugs, and plastics that have become increasingly widespread in our environment. The evidence is still viewed by most scientists as preliminary. Chemical companies that produce pesticides that are hormone mimics have a major stake in these studies. What is a company's responsibility to society when this type of controversy is raised? Should it immediately stop production of a potentially harmful chemical even if this stoppage leads to a large loss of jobs? When should a chemical be banned and what criteria should be used in calling for a ban?

2. An ethnobotanist who works with native people of Peru on medicinal plants of the rainforest approached a large pharmaceutical corporation with his newly acquired knowledge. He hoped to get some protection for the native Peruvians as compensation for conveying knowledge on medicinal plants to the company. Scientists and executives in the company found several potential drugs based on information from the indigenous people. They tried expressing the genes to produce the drugs but found that their yield was low. As an alternative, chemists in the company have already begun to synthesize the compounds that they have identified in some of the medicinal plants. Because they are synthesizing the compounds, they say they owe neither the native peoples nor the country of Peru anything. Is this a responsible policy for the company? Consider the responsibility the company has to the ethnobotanist, to its stockholders, to the native people, and to Peru.

3. In 1992, the biotechnology company Calgene applied for approval of their rot-resistant tomato, the Flavr Savr® tomato. The tomato plant had been engineered by introducing a copy of the gene encoding a softening enzyme downstream of, and in its reverse orientation, upstream to a strong promoter active in the fruit. When the engineered gene is transcribed, an "antisense RNA," complementary in sequence to the softening enzyme mRNA is produced. The antisense RNA binds to the normal mRNA (sense RNA) and blocks translation of the softening enzyme so the tomatoes stay ripe longer. The FDA approved the sale of the tomato without any additional labeling because the DNA that had been introduced into the tomato was tomato DNA. (The FDA's policy on bioengineered foods is to not label food as genetically engineered except when a new substance has been introduced that could cause an allergy or there has been a change in the food's nutritional value.) Several consumer groups were upset by the FDA's action and wanted to see labeling appear on the tomatoes in the stores. The sale of these tomatoes was significantly stalled by this protest. Was this consumer group's action, which blocked an advance that many people would have wanted, warranted? Are these groups important watchdogs of everyone's well-being or are they hindering progress?

SOLVED PROBLEMS

I. A geneticist examined the amino acid sequence of a particular protein in a variety of mutants in *E. coli*. The amino acid in position 40 in the normal enzyme is glycine. The table below shows the substitutions the geneticist found at amino acid position 40 in six mutant forms of the enzyme.

mutant 1	cysteine
mutant 2	valine
mutant 3	serine
mutant 4	aspartic acid
mutant 5	arginine
mutant 6	alanine

Determine the nature of the base substitution that must have occurred in the DNA in each case. Which of these mutants would be capable of recombination with mutant 1 to form a wildtype gene?

Answer

To determine the base substitution that occurred in each case, use the genetic code table (Fig. 7.3). The original residue was glycine, which can be encoded by any one of the four codons: GGU, GGC, GGA, GGC. Mutant 1 results in a cysteine residue at position 40. Cys codons are either UGU or UGC. A change in the base pair in the DNA encoding the first position in the codon (a transversion GC to TA) must have occurred. This mutant also narrows down the particular codon for the glycine in this protein, which must have been either GGU or GGC. Valine (in mutant 2) is encoded by four different codons, GUX (with X representing any one of the four bases), but since we know the original glycine was GGU or GGC and, assuming that the mutation is a single base change, the Val codon must be either GUU or GUC. The change must have been a GC to TA transversion in the DNA corresponding to the second position of the codon. To get from glycine to serine (mutant 3) with only one base change, the GGU or GGC would be changed to AGU or AGC, respectively. There was a transition at the first position. Aspartic acid (mutant 4) is encoded by GAU or GAC, so the DNA of mutant 4 is the result of a GC to AT transition at position 2. Arginine (mutant 5) is encoded by CGX, so the DNA of mutant 5 must have undergone a GC to CG transversion at position 1. Finally, alanine (mutant 6) is encoded by GCX, so the DNA of mutant 6 must have undergone a GC to CG transversion at position 2. Mutants 2, 4, and 6 affect a base pair different from that affected by mutant 1, so they could recombine with mutant 1.

In summary, the sequence of nucleotides on the RNA-like strand of the wildtype and mutant genes at this position must be:

wildtype	5' G G T/C 3'
mutant 1	5' T G T/C 3'
mutant 2	5' G T T/C 3'
mutant 3	5' A G T/C 3'
mutant 4	5' G A T/C 3'
mutant 5	5' C G T/C 3'
mutant 6	5' G C T/C 3'

II. The double-stranded DNA molecule that forms the genome of the SV40 virus can be denatured into single-stranded DNA molecules. Because the percentage of GC bases in the two strands differs, the strands can be separated on the basis of their density into two strands designated W(atson) and C(rick). When each of the purified preparations of the single strands was mixed with mRNA from cells infected with the virus, hybrids were formed between the RNA and DNA. Closer analysis of these hybridizations showed that RNAs that hybridized with the W preparation were different from RNAs that hybridized with the C preparation. What does this tell you about the transcription templates for the different classes of RNAs?

Answer

An understanding of transcription and the polarity of DNA strands in the double helix are needed to answer this question. *Some genes use one strand of the DNA as a template, others use the opposite strand as a template.* Because of the different polarities of the DNA strands, one set of genes would be transcribed in a clockwise direction on the circular DNA and the other set would be transcribed in a counterclockwise direction.

III. Geneticists interested in human hemoglobins have found a very large number of mutant forms. Some of these mutant proteins are of normal size, with amino-acid substitutions, while others are short, due to deletions or nonsense mutations. The first extralong example was named Hb Constant Spring, in which the β globin has several extra amino acids attached at the C terminal end. Could Hb Constant Spring have arisen from failure to splice out an intron? What is a plausible explanation for its origin?

Answer

An understanding of the principles of transcription and translation are needed to answer this question. Because there is an extension on the C terminal end of the protein, *the mutation probably affected the termination (nonsense) codon rather than affecting splicing of the RNA.* This could have been a base change or a frameshift or a deletion that altered or removed the termination codon. The information beyond the normal stop codon would be transcribed and translated until another stop codon in the mRNA was reached.

P R O B L E M S

7-1 For each of the terms in the left column, choose the best matching phrase in the right column.

a. codon

1. removing base sequences corresponding to introns from the primary transcript

b. colinearity

2. UAA, UGA or UAG

c. reading frame

3. the strand of DNA that has the same base sequence as the primary transcript

d. frameshift mutation

4. a transfer RNA molecule to which the appropriate amino acid has been attached

e. degeneracy of the genetic code

5. a group of three mRNA bases signifying one amino acid

f. nonsense codon

6. most amino acids are not specified by a single codon

g. initiation codon

7. using the information in the nucleotide sequence of a strand of DNA to specify the nucleotide sequence of a strand of RNA

h. template strand

8. the grouping of mRNA bases in threes to be read as codons

i. RNA-like strand

9. AUG in a particular content

j. intron

10. the linear sequence of amino acids in the polypeptide corresponds to the linear sequence of nucleotide pairs in the gene

k. RNA splicing

11. produces different mature mRNAs from the same primary transcript

l. transcription

12. addition or deletion of a number of base pairs other than three into the coding sequence.

m. translation

13. a sequence of base pairs within a gene that is not represented by any bases in the mature mRNA

n. alternative splicing

14. the strand of DNA having the base sequence complementary to that of the primary transcript

o. charged tRNA

15. using the information encoded in the nucleotide sequence of an mRNA molecule to specify the amino-acid sequence of a polypeptide molecule.

7-2 Match the hypothesis from the left column below to the observation from the right column that gave rise to it.

a. existence of an intermediate messenger between DNA and protein

1. two mutations affecting the same amino acid can recombine to give wildtype

b. the genetic code is nonoverlapping

2. one or two base deletions (or insertions) in a gene disrupt its function; three base deletions (or insertions) are often compatible with function

c. the codon is more than one nucleotide

3. artificial messages containing certain codons produced shorter proteins than messages not containing those codons

d. the genetic code is based on triplets of bases

4. protein synthesis occurs in the cytoplasm, while DNA resides in the nucleus

e. stop codons exist and terminate translation

5. artificial messages with different base sequences gave rise to different proteins in an *in vitro* translation system

f. the amino-acid sequence of protein depends on base sequence of mRNA

6. base substitutions affect only one amino acid in the protein chain

7-3 How would the artificial mRNA GUGUGUGUGU etc. be read according to each of the following models for the genetic code?
a. two base, not overlapping
b. two base, overlapping
c. three base, not overlapping
d. three base, overlapping
e. four base, not overlapping.

7-4 An example of a portion of the T4 *rIIB* gene in which Crick and Brenner had recombined one + and one − mutation is shown below. (The RNA-like strand of the DNA is shown.)
a. Where are the + and − mutations in the mutant DNA?
b. What alterations in amino acids occurred in this double mutant, which produces wildtype plaques?
c. How can you explain the fact that amino acids are different in the double mutant compared with the wildtype sequence, yet the phage is wildtype?

wildtype   AAA AGT CCA TCA CTT AAT GCC

mutant   AAA GTC CAT CAC TTA ATG GCC

7-5 In the *HbS* allele (sickle-cell allele) of human β globin, the sixth amino acid in the β-globin chain is changed from glutamic acid to valine. In *HbC* the sixth amino acid in β globin is changed from glutamic acid to lysine. What would be the order of these two mutations within the map of the β-globin gene?

7-6 The amino-acid sequence of part of a protein has been determined:

Gly Ala Pro Arg Lys

A mutation has been induced in the gene encoding this protein using the mutagen proflavin. The resulting mutant protein can be purified and the amino-acid sequence determined. The amino-acid sequence of the mutant protein is exactly the same as the amino-acid sequence of the wildtype protein from the N terminus of the protein to the glycine in the sequence above. Starting with this glycine, the sequence of amino acids is changed to the following:

Gly His Gln Gly Lys

Using the amino-acid sequences, one can determine the sequence of 14 nucleotides from the wildtype gene encoding this protein. What is this sequence?

7-7 When using the artificial message UCUCUCUCUCUC etc. in an *in vitro* protein synthesis system, investigators found that proteins composed of alternating leucine and serine were made. What experiments were done to determine whether leucine was specified by CUC and serine by UCU, or vice versa?

7-8 The following diagram describes the mRNA sequence of part of the A protein gene and the beginning of the B protein gene of phage φX174. In this phage, there are some genes that are read in overlapping reading frames. For example, the code for the A gene is used for part of the B gene but the reading frame is displaced by one base. Shown below is the single mRNA with the codons for proteins A and B indicated.

aa	1	2	3	4	5	6	7	8	9	10	11	12
A	Ala	Lys	Glu	Trp	Asn	Asn	Ser	Leu	Lys	Thr	Lys	Leu

mRNA GCUAAAGAAUGGAACAACUCACUAAAAACCAAGCUG

B		Met	Glu	Gln	Leu	Thr	Lys	Asn	Gln	Ala
aa		1	2	3	4	5	6	7	8	9

Given the following amino-acid changes, indicate the base change that occurred (state this as the mRNA codon) and the consequences for the other protein sequence.
a. Asn at position 6 in protein A is changed to Tyr.
b. Leu at position 8 in protein A is changed to Pro.
c. Gln at position 8 in protein B is changed to Leu.
d. The occurrence of overlapping reading frames is very rare in nature. When it does occur, the extent of the overlap is not very long. Why do you think this is the case?

7-9 The coding sequence for gene *F* is read from left to right. The coding sequence for gene *G* is read from right to left. Which strand of DNA serves as the template for transcription of each gene?

7-10 If you mixed the mRNA of a human gene with the genomic DNA for the same gene and allowed the RNA and DNA to form a hybrid, what would you be likely to see in the electron microscope?

7-11 How many possible open reading frames (frames without stop codons) are there that extend through the sequence below?

5′ CTTACAGTTTATTGATACGGAGAAGG 3′

3′ GAATGTCAAATAACTATGCCTCTTCC 5′

7-12 In prokaryotes, a search for genes in a DNA sequence involves scanning the DNA sequence for long open reading frames. What problem can you see with this approach in eukaryotes?

7-13 The sequence of a segment of mRNA, beginning with the initiation codon, is given below, along with the corresponding sequences from several mutant strains.
a. Indicate the type of mutation present in each and translate the mutated portion of the sequence into an amino-acid sequence in each case.
b. Which of the mutations could be reverted by treatment with EMS? with proflavin?

Normal	AUGACACAUCGAGGGGUGGUAAACCCUAAG...
Mutant 1	AUGACACAUCCAGGGGUGGUAAACCCUAAG...
Mutant 2	AUGACACAUCGAGGGGUGGUAAACCCUAAG...
Mutant 3	AUGACGCAUCGAGGGGUGGUAAACCCUAAG...
Mutant 4	AUGACACAUCGAGGGGUUGGUAAACCCUAAG...
Mutant 5	AUGACACAUUGAGGGGUGGUAAACCCUAAG...
Mutant 6	AUGACAUUUACCACCCCUCGAUGCCCUAAG...

7-14 The sequence of a complete eukaryotic gene encoding the small protein Met Tyr Arg Gly Ala is shown at the top of the next page. All of the written sequences on the template strand are transcribed into RNA.
a. Which strand is the template strand? Which direction (right to left or left to right) does RNA polymerase move along the template as it transcribes this gene?
b. What is the sequence of the nucleotides in the processed mRNA molecule for this gene? Indicate the 5′ and 3′ directions of this gene.
c. A single base mutation in the gene results in synthesis of the peptide Met Tyr Thr. What is the sequence of nucleotides making up the mRNA produced by this mutant gene?

5′ CCCCTATGCCCCCCTGGGGGAGGATCAAAACACTTACCTGTACATGGC 3′

3′ GGGGATACGGGGGGGACCCCCTCCTAGTTTTGTGAATGGACATGTACCC 5′

7-15 Describe the steps in transcription and translation that require complementary base pairing.

7-16 The yeast gene encoding a protein found in the mitotic spindle was cloned by a laboratory studying mitosis. The gene encodes a protein of 477 amino acids.

a. What is the minimum length in nucleotides of the protein-coding part of this yeast gene?

b. A partial sequence of one DNA strand in an exon containing the middle of the coding region of the yeast gene is given below. What is the sequence of amino acids in that part of the yeast mitotic spindle protein?

c. What is the sequence of nucleotides of the mRNA in this region of the gene? Show the 5′ and 3′ directionality of your strand.

5′ GTAAGTTAACTTTCGACTAGTCCAGGGT 3′

7-17 A particular protein has the amino-acid sequence . . . Ala-Pro-His-Trp-Arg-Lys-Gly-Val-Thr . . . within its primary structure. A geneticist studying mutations affecting this protein discovered that several of the mutants produced shortened protein molecules that terminated within this region. In one of them, the His became the terminal amino acid.

a. What DNA single-base change(s) would cause the protein to terminate at the His residue?

b. What other possible termination sites do you see in this sequence?

7-18 In studying normal and mutant forms of a particular human enzyme, a geneticist came across a particularly interesting mutant form of the enzyme. The normal enzyme is 227 amino acids long, but the mutant form was 312 amino acids long, having that extra 85 amino acids as a block in the middle of the normal sequence. The inserted amino acids do not correspond in any way to the normal protein sequence. What are some possible explanations for this phenomenon? How would you distinguish among them?

7-19 a. A mutant *B. adonis* bacterium has a nonsense suppressor tRNA that inserts glutamine (Gln) to match a UAG (but not other nonsense) codons. What is the anticodon of the suppressing tRNA? (Indicate the 5′ and 3′ ends.)

b. What is the sequence of the template strand of the wildtype tRNA^{Gln}-encoding *gene* that was altered to

produce the suppressor, assuming that only a single-base-pair alteration was involved?

c. What is the *minimum* number of tRNA^{Gln} genes that could be present in a wildtype *B. adonis* cell?

7-20 You are studying mutations in a gene that codes for an enzyme whose amino-acid sequence is known. In the wildtype protein, proline is the fifth amino acid from the amino terminal end. In one of your mutants, you find a serine at position number 5. You subject this mutant to further mutagenesis, and recover three different strains. Strain A has a proline at position number 5 and acts just like wildtype. Strain B has tryptophan at position number 5 and also acts like wildtype. Strain C has no detectable enzyme function at any temperature and you can't recover any protein that resembles the enzyme. You mutagenize strain C and recover a strain (C-1) that has enzyme function. The second mutation does not map at the enzyme locus.

a. What is the nucleotide sequence in both strands of the wildtype gene at this location?

b. What is the nature of the mutation in strain C?

c. What is the second mutation that arose in C-1?

7-21 Another class of suppressor mutations, not described in the text, are mutations that suppress missense mutations.

a. Why would bacterial strains carrying such missense suppressor mutations generally grow more slowly than strains carrying nonsense suppressor mutations?

b. What kinds of mutations can you imagine in genes encoding components needed for gene expression that would suppress a missense mutation in a protein-coding gene?

7-22 An investigator was interested in studying UAG nonsense suppressor mutations in bacteria. In one species of bacteria, she was able to select two different mutants of this type, one in the tRNA^{Tyr} gene and the other in the tRNA^{Gln} gene, but in a second species she was not able to obtain any such mutations, even after very extensive effort. What could explain the difference between the two species?

7-23 Do you think each of the following types of mutations would have very severe effects, mild effects, or no effect at all?

a. nonsense mutation occurring in the sequences encoding amino acids near the N terminus of the protein

b. nonsense mutation occurring in the sequences encoding amino acids near the C terminus of the protein
c. frameshift mutation occurring in the sequences encoding amino acids near the N terminus of the protein
d. frameshift mutation occurring in the sequences encoding amino acids near the C terminus of the protein
e. silent mutation
f. conservative missense mutation
g. nonconservative missense mutation affecting the active site of the protein

h. nonconservative missense mutation not in the active site of the protein.

7-24 What would be the effect of each of the following types of mutations on RNA produced from this gene? (Assume the mutations are loss of function mutations.)
a. promoter mutations
b. splice donor or acceptor mutants
c. ribosome binding site mutation
d. Of the three types of mutations described in parts a–c, only the splice donor-acceptor mutation could result in a dominant negative allele. Why is this?

PART

III

USING GENETIC ENGINEERING
TO UNRAVEL THE
INFORMATION IN GENOMES

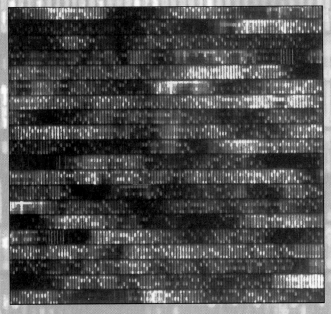

False colored image of a DNA chip designed to screen for mutation in the ATM gene. Areas of greatest hybridization signal appear in white and lighter colors while those with lower hybridization signal appear as black and the darkest colors. Each spot on the chip is 50 microns on each edge and composed of oligonucleotides of the same sequence. There are close to 45,000 different oligonucleotides on the surface of the chip.

DNA AT HIGH RESOLUTION: THE USE OF DNA CLONING, PCR, AND HYBRIDIZATION AS TOOLS OF GENETIC ANALYSIS

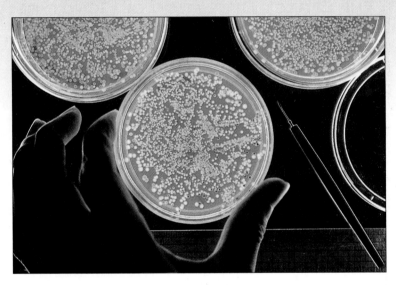

Colonies of bacterial cell clones containing recombinant DNA molecules.

The vivid red color of our blood stems from its life-sustaining ability to carry oxygen; and this ability, in turn, derives from trillions of red blood cells suspended in proteinaceous solution, each one packed with close to 280 million copies of the protein pigment known as hemoglobin. The hemoglobin picks up oxygen in the lungs and transports it to tissues throughout the body, where, after its release, the oxygen contributes to a multitude of metabolic reactions. A normal adult hemoglobin molecule consists of four polypeptide chains, two alpha (α) and two beta (β), each surrounding an iron-containing entity known as a heme (Fig. 8.1a). The iron atom within the heme sustains a reversible interaction with oxygen, binding it firmly enough to hold it on the trip from lungs to body tissue, but loosely enough to release it where needed. The intricately folded alpha and beta chains protect the iron-containing hemes from molecules in the cell's interior. Each hemoglobin molecule can carry up to four oxygen atoms, one per heme; and it is these oxygenated hemes that impart a scarlet hue to the pigment molecules and thus to the blood cells that carry them.

Surprisingly, the genetically determined molecular composition of hemoglobin changes several times during human development (Fig. 8.1b). In the first five weeks after conception, the red blood cells carry *embryonic hemoglobin,* which consists of two alpha-like zeta (ζ) chains and two beta-like epsilon (ϵ) chains. Thereafter, throughout the rest of gestation, the cells contain *fetal hemoglobin,* composed of two bona fide α chains and two beta-like gamma (γ) chains. Then, shortly before birth, production of *adult hemoglobin,* composed of two α and two β chains, begins. By three months of age, almost all of an individual's hemoglobin is of this adult type.

What structure and arrangement of hemoglobin genes facilitates the vital switches in the type of protein produced? And what can go wrong with these genes to generate hemoglobin disorders? Such disorders are the most common genetic diseases in the world and include sickle-cell anemia, which arises from an altered β chain, and thalassemia, which results from decreases in the amount of either α- or β-chain production.

Figure 8.1 Hemoglobin is composed of four polypeptide chains that change during development. (a) Scanning electron micrograph of adult human red blood cells. This cell type is unique in the human body; it has eliminated its nucleus and is loaded with one kind of protein—hemoglobin. Adult human hemoglobin consists of two α and two β polypeptide chains, each enclosing one molecule of the iron-containing pigment, heme. (b) The hemoglobin, carried by red blood cells, switches during human development from an embryonic form containing two α-like ζ chains and two β-like ϵ chains, to a fetal form containing two α chains and two β-like γ chains, and finally to the adult form containing two α and two β chains. In a small percentage of adult hemoglobin molecules, a β-like δ chain is used in place of an actual β chain.

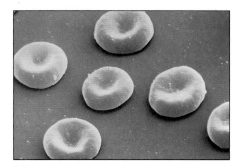

(a)

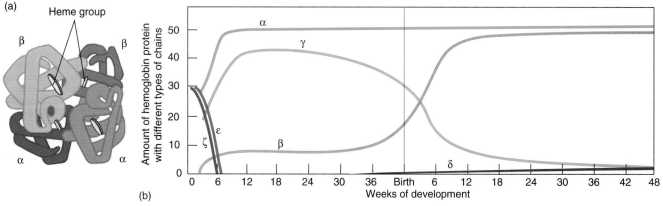

(b)

To answer these questions, researchers must have a way of looking at the hemoglobin genes of individuals with normal and abnormal phenotypes. But these genes lie buried in a diploid human genome containing 6 billion base pairs distributed among 46 different strings of DNA (the chromosomes) that range in size from 60 million to 360 million base pairs. In this chapter, we describe the powerful new tools of genetic analysis that medical researchers can now use to search through these enormously long strings of information for genes that may be only several thousand base pairs in length. Ideally suited to the analysis of complex biological systems, the new tools employ several enzymes that operate on DNA: restriction enzymes that cut the molecule, ligases that join disparate pieces, and polymerases that synthesize new parts. Researchers refer to the whole kit of new tools as **recombinant DNA technology.**

Developed during a technological revolution that began in the mid-1970s, recombinant DNA technology makes it possible to characterize DNA molecules directly, rather than indirectly through the phenotypes they produce. Geneticists use recombinant DNA techniques to gather information unobtainable in any other way, or to analyze the results of breeding and cytological studies with greater speed and accuracy than ever before.

Two general themes recur in our discussion. First, the new tools of genetic analysis emerged from a knowledge of DNA structure and function. For example, complementary base pairing between similar or identical stretches of DNA is the basis of so-called

hybridization techniques that use labeled DNA probes to identify related DNA molecules. **Hybridization** is the natural propensity of complementary single-stranded molecules to form stable double helixes. Second, the speed, sensitivity, and accuracy of the new tools make it possible to answer questions that were impossible to resolve just a short time ago. In one recent study, for instance, researchers used a protocol for making many copies of a specific DNA segment to trace the activity of the AIDS virus and found that, contrary to prior belief, the virus, after infecting a person, does not become latent inside all the cells it enters; it remains active in certain cells of the lymph nodes and adenoids.

As we illustrate the power of the new tools for analyzing a seemingly simple biological system—the hemoglobins—that turned out to be surprisingly complex, we describe how researchers use recombinant DNA technology to carry out four basic operations:

■ Cut the enormously long strings of DNA into much smaller fragments.

■ Amplify the fragments for storage and subsequent analysis.

■ Identify and isolate fragments containing a gene, a control element, or some segment of unknown import away from all other fragments.

■ Characterize fragments of interest by size, genomic location, and sequence.

CUTTING THE DNA: RESTRICTION ENZYMES SERVE AS MOLECULAR SCISSORS

Every intact human body cell, including the precursors of red blood cells, carries two nearly identical sets of 3 billion base pairs of information that when unwound extend 2 m in length and contain two copies of roughly 100,000 genes. If you could enlarge the cell nucleus to the size of a basketball, the unwound DNA would have the diameter of a fishing line and a length of 200 km. This is much too much material to study as a whole. To reduce its complexity, researchers first cut the genome into "bite size" pieces.

Restriction Enzymes Fragment the Genome at Specific Sites

Researchers use restriction enzymes to cut the DNA released from the nuclei of cells at specific sites; these well-defined cuts generate fragments suitable for manipulation and characterization. A **restriction enzyme,** as described in the Genetics and Society box "Serendipity in Science: The Discovery of Restriction Enzymes," recognizes a specific sequence of bases anywhere within the genome, then severs two covalent bonds (one in each strand) in the sugar-phosphate backbone at particular positions within or near that sequence. The fragments generated by restriction enzymes are referred to as **restriction fragments,** and the act of cutting is often called **digestion.**

Restriction enzymes originate in and can be purified from bacterial cells. In these prokaryotic cells, the enzymes protect against invasion by viruses through the digestion of viral DNA. Bacteria shield their own DNA from digestion by indigenous restriction enzymes through the selective addition of methyl groups ($-CH_3$) to the restriction recognition sites in their DNA. In the test tube, restriction enzymes from bacteria recognize target sequences of 4 to 8 base pairs in DNA isolated from any other organism, and cut the DNA at or near these sites. Table 8.1 lists the names, recognition sequences, and microbial origins of just 10 of the more than 100 commonly used restriction enzymes. For the majority of these enzymes, the recognition site contains 4–6 base pairs and exhibits a kind of palindromic symmetry in which the base sequences of each of the two DNA strands are identical when read in the $5' \rightarrow 3'$ direction. Because of this, base pairs on either side of a central line of symmetry are mirror images of each other. Each enzyme always cuts at the same place relative to its specific recognition sequence, and most enzymes make their cuts in one of two ways: either straight through both DNA strands right at the line of symmetry to produce fragments with **blunt ends,** or displaced equally in opposite directions from the line of symmetry by one or more bases to generate fragments with single-stranded ends (Fig. 8.2). Geneticists often refer to these protruding single strands as **sticky ends.** They are considered "sticky" because they are free to base pair with a complementary sequence from the DNA of *any* organism cut by the same restriction enzyme.

TABLE 8.1	Ten Commonly Used Restriction Enzymes	
Enzyme	**Sequence of Recognition Site**	**Microbial Origin**
TaqI	5' T C G A 3' / 3' A G C T 5'	*Thermus aquaticus* YTI
RsaI	5' G T A C 3' / 3' C A T G 5'	*Rhodopseudomonas sphaeroides*
Sau3AI	5' G A T C 3' / 3' C T A G 5'	*Staphylococcus aureus* 3A
EcoRI	5' G A A T T C 3' / 3' C T T A A G 5'	*Escherichia coli*
BamHI	5' G G A T C C 3' / 3' C C T A G G 5'	*Bacillus amyloliquefaciens* H.
HindIII	5' A A G C T T 3' / 3' T T C G A A 5'	*Haemophilus influenzae*
KpnI	5' G G T A C C 3' / 3' C C A T G G 5'	*Klebsiella pneumoniae* OK8
ClaI	5' A T C G A T 3' / 3' T A G C T A 5'	*Caryophanon latum*
BssHII	5' G C G C G C 3' / 3' C G C G C G 5'	*Bacillus stearothermophilus*
NotI	5' G C G G C C G C 3' / 3' C G C C G G C G 5'	*Nocardia otitidiscaviarum*

Different Restriction Enzymes Produce Fragments of Different Lengths

For each restriction enzyme, it is possible to calculate the average length of the fragments the enzyme will generate and then use that information to estimate the approximate number and distribution of recognition sites in a genome. The estimate depends on two simplifying assumptions: First, that each of the four bases occurs in equal proportions such that a genome is composed of 25% A, 25% T, 25% G, and 25% C; second, that the bases are randomly distributed in the DNA sequence. Although these assumptions are never precisely valid, they enable us to determine the average distance between recognition sites of any length by the general formula 4^n, where **n** is the number of bases in the site (Fig. 8.3a).

(a) Blunt ends (*Rsa*I)

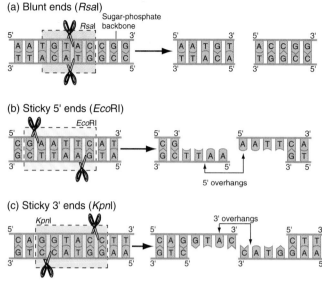

(b) Sticky 5' ends (*Eco*RI)

(c) Sticky 3' ends (*Kpn*I)

Figure 8.2   Restriction enzymes cut DNA molecules at specific locations to produce restriction fragments with either blunt or sticky ends. (a) The restriction enzyme *Rsa*I produces blunt-ended restriction fragments. (b) *Eco*RI produces sticky ends with a 5' overhang. (c) *Kpn*I produces sticky ends with a 3' overhang.

The Size of the Recognition Site is the Primary Determinant of Fragment Length

According to the formula, *Rsa*I, which recognizes the four-base-sequence GTAC, will cut on the average once every 4^4, or every 256 base pairs (bp), creating fragments averaging 256 bp in length. By comparison, the enzyme *Eco*RI, which recognizes the six-base-sequence GAATTC will cut on the average once every 4^6, or 4096 bp; since 1000 base pairs = 1 kilobase pair, researchers often round off this large number to roughly 4.1 kilobase pairs, abbreviated 4.1 kb. Similarly, an enzyme such as *Not*I, which recognizes the eight bases GCGGCCGC will cut on the average every 4^8 bp, or every 65.5 kb. Note, however, that because the actual distances between restriction sites for any enzyme vary considerably, very few of the fragments produced by the three enzymes mentioned here will be precisely 65.5 kb, 4.1 kb, or 256 bp in length (Fig. 8.3b).

The Timing of Exposure to a Restriction Enzyme Helps Determine Fragment Size

Geneticists often need to produce DNA fragments of a particular length—larger ones to study the organization of a chromosomal region, smaller ones to examine a whole gene, and ones that are smaller still to analyze particular regions within a gene. If their goal is 4-kb fragments, they have a range of six-base-cutter enzymes to choose from. By exposing the DNA to a six-base cutter for a long enough time, they give the restriction enzyme ample opportunity for digestion. The result is a **complete digest** in which the DNA has been cut at every one of the recognition sites it contains. But what if the researchers need 20-kb fragments—a length in between that produced by

(a) 1. Probability that a four-base recognition site will be found in a genome =

$$1/4 \times 1/4 \times 1/4 \times 1/4 = 1/256$$

2. Probability that a six-base recognition site will be found =

$$1/4 \times 1/4 \times 1/4 \times 1/4 \times 1/4 \times 1/4 = 1/4,096$$

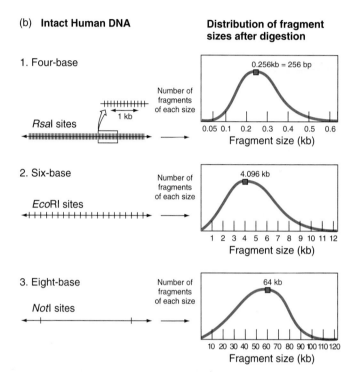

Figure 8.3   The number of base pairs in the restriction enzyme recognition site determines the average distance between restriction sites and thus the size of fragments produced. (a) You can estimate the average length of a restriction fragment from the number of base pairs in a restriction enzyme's recognition site. Since there are four different bases, the probability that any one of them will be present at any particular site is one in four, or 1/4. And since the law of the product states that two or more independent events occurring simultaneously is the product of their individual probabilities, you can use this principle to deduce the probability that a particular sequence of bases—GTAC, for example—will occur. If the probability of a G is 1/4, the probability that a G will be followed by a T is $1/4 \times 1/4 = 1/16$; the probability that a G will be followed by T, which is followed by an A, is $1/4 \times 1/4 \times 1/4 = 1/64$; finally, the probability of any four-base sequence is $(1/4)^4$ or 1/256. What this means is that, on average, you are likely to find a specific four-base sequence once in every 256 bases. If you apply the same logic to the six-base sequence recognized by the *Eco*RI restriction enzyme, you will find that this sequence occurs, on average, once every 4096 bases, or roughly once every 4.1 kb (kilobases). (b) The formula 4^n, where n = the number of bases in the recognition site, can be used to predict the average distance between restriction sites; this distance is equivalent to the average length of restriction fragments that result from complete digestion with a particular restriction enzyme. It is important to remember, however, that the result obtained with this formula is just a probability-based determination of an average. The actual size of restriction fragments obtained in any experiment will always show a broad distribution around this average.

Genetics and Society

Serendipity in Science: The Discovery of Restriction Enzymes

Most of the tools and techniques for cloning and analyzing DNA fragments emerged from studies of bacteria and the viruses that infect them. Molecular biologists had observed, for example, that viruses able to grow abundantly on one strain of bacteria grew poorly on a closely related strain of the same bacteria, and while examining the mechanisms of this discrepancy, they discovered restriction enzymes.

To follow the story, one must know that researchers compare rates of viral proliferation in terms of *plating efficiency:* the fraction of viral particles that enter and replicate inside a host bacterial cell, causing the cell to lyse and release viral progeny; the released progeny generate a visible clearing, or plaque, on the bacterial lawn. The plating efficiency of lambda virus grown on *E. coli* C is nearly 1.0. The plating efficiency of the same virus grown on *E. coli* K12 is only 1 in 10^4 or 0.0001. The bacterial capacity for limiting viral growth, in this case the growth of lambda on *E. coli* K12, is called **restriction.**

Interestingly, restriction is rarely absolute. Although lambda virus grown on *E. coli* K12 produces almost no progeny, a few viral particles do manage to proliferate. If their progeny are then tested on *E. coli* K12, the plating efficiency is nearly 1.0. The phenomenon in which growth on a restricting host modifies a virus so that succeeding generations grow more efficiently on that same host is known as **modification.**

What mechanisms account for restriction and modification? Studies following viral DNA after bacterial infection found that during restriction, the viral DNA is broken into pieces and degraded (Fig. A). Purification of the enzyme responsible for the initial breakage isolated an endonuclease, an enzyme that breaks the phosphodiester bonds in the viral DNA molecule, usually making double-strand cuts at a specific sequence in the viral chromosome. Because this breakage restricts the biological activity of the viral

DNA, researchers called the enzymes that accomplish it *restriction enzymes.* Subsequent studies showed that the small percentage of viral DNA that escapes digestion and goes on to generate new viral particles has been modified by the addition of methyl groups during its replication in the host cell. Researchers named the enzymes that add methyl groups to specific DNA sequences **modification enzymes.**

In nature, pairs of restriction-modification enzymes protect bacterial cells from foreign DNA. Unmodified DNA of any type falls prey to digestion by restriction enzymes. However, just as modification guards the bacterial DNA from destruction by its own restriction enzymes (see Fast Forward Box, p. 166), it also ensures the survival of those few viral DNA molecules that happen to evade digestion. These viral DNA molecules, modified by methylation, then use the host cell's machinery to produce new viral particles.

Biologists have identified a large number of restriction-modification systems in a variety of bacterial strains. Purification of the systems has yielded a mainstay of recombinant DNA technology: the battery of restriction enzymes used to cut DNA *in vitro* for cloning, mapping, and ligation.

This example of serendipity in science sheds some light on the debate between administrators who distribute and oversee research funding and scientists who carry out the research. Microbial investigators did not set out to find restriction enzymes; they could not have known that these enzymes would be one of their finds. Rather, they sought to understand the mechanisms by which viruses infect and proliferate in bacteria. Along the way, they discovered restriction enzymes and how they work. Their insights have had broad applications in genetic engineering and biotechnology. DNA cloning, for example, has produced, among many other medically useful products, human insulin for diabetics, which provokes far fewer allergic reactions than the pig insulin that preceded it; tissue plas-

---

a six-base cutter and an eight-base cutter? The answer is a **partial digest:** a cutting obtained by controlling the amount of enzyme or the amount of time the DNA is exposed to the restriction enzyme. With *Eco*RI, for example, exposing the DNA just long enough for the enzyme to cut on average only one of five target sites would produce fragments averaging 20 kb in length (Fig. 8.4). Researchers determine the correct timing by trial and error. Complete or partial digestion with different enzymes produces fragments of appropriate average lengths.

Different Restriction Enzymes Produce Different Numbers of Fragments from the Same Genome

We have seen that the four-base cutter *Rsa*I cuts the genome on average every 4^4 (256) bp. If you exposed the haploid human genome with its 3 billion bp to *Rsa*I for a sufficient time under appropriate conditions, you would ensure that all of the recognition sites in the genome that can be cleaved will be cleaved, and you would get

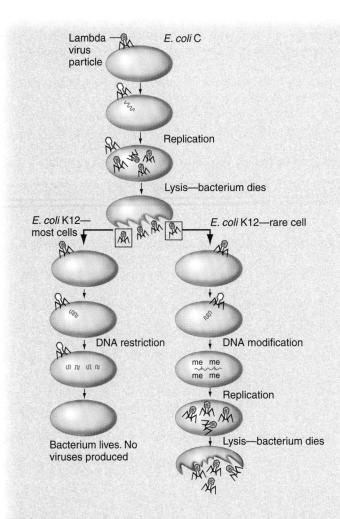

Figure A   Operation of the restriction enzyme/modification system in nature. *E. coli* C does not have a functional restriction enzyme/modification system and is susceptible to infection by the lambda phage. In contrast, *E. coli* K12 generally resists infection by the viral particles produced from an *E. coli* C phage infection. This is because *E. coli* K12 makes several restriction enzymes, including *Eco*RI, which cut the lambda DNA molecule before it can express itself. However, in rare K12 cells, the lambda DNA replicates before restriction enzymes can attack it, and the newly replicated lambda DNA is then modified by an enzyme that protects its recognition sites from attack by the host cell's restriction enzymes. This lambda DNA can now generate phage particles, which eventually destroy the bacterial cell.

minogen activator (TPA), a protein that helps dissolve blood clots; and gamma interferon, a protein produced by the human immune system that protects laboratory grown human cells from infection by herpes and hepatitis viruses. The politicians and administrators in charge of allocating funds often want to direct research spending to urgent health or agricultural problems, assuming that much of the basic research upon which such studies depend has been com-

pleted. The scientists in charge of laboratory research call for a broad distribution of funds to all projects investigating interesting biological phenomena, arguing that we still know relatively little about molecular biology and that the insights necessary for understanding disease may emerge from studies on any organism. The validity of both views suggests the need for a balanced approach to the funding of research activities.

$$\frac{3{,}000{,}000{,}000 \text{ bp}}{\sim 256 \text{ bp}} = \sim 12{,}000{,}000 \text{ fragments}$$

$$\sim 256 \text{ bp in length.}$$

By comparison, the six-base cutter *Eco*RI cuts the DNA on average every 4^6 (4096) bp, or every 4.1 kb. If you exposed the haploid human genome with its 3 billion bp, or 3 million kb, to *Eco*RI in the proper way, you would get

$$\frac{3{,}000{,}000 \text{ kb}}{\sim 4.1 \text{ kb}} = \sim 700{,}000 \text{ fragments} \sim 4.1 \text{ kb in length.}$$

And if you exposed the same haploid human genome to the eight-base cutter *Not*I, which cuts on average every 4^8 (65,536) bp, or 65.5 kb, you would obtain

$$\frac{3{,}000{,}000 \text{ kb}}{\sim 65.5 \text{ kb}} = \sim 46{,}000 \text{ fragments} \sim 65.5 \text{ kb in length.}$$

Clearly, the larger the recognition site, the smaller the number of fragments generated by enzymatic digestion.

Although restriction fragments cut DNA *in vitro* into thousands or millions of discrete restriction fragments, these fragments remain mixed together in a single test tube. To study

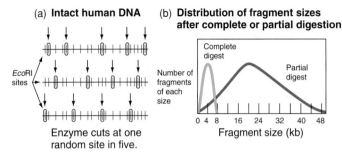

(a) Intact human DNA

*Eco*RI sites

Enzyme cuts at one random site in five.

(b) Distribution of fragment sizes after complete or partial digestion

Number of fragments of each size

Complete digest

Partial digest

Fragment size (kb)
0 4 8   16   24   32   40   48

Figure 8.4 Comparison of results from partial and complete digests. (a) By reducing the time available for the reaction to occur, you can ensure that an enzyme actually cuts only a subset of the total recognition sites within a DNA sample. In this example, the chosen reaction time allowed only 1/5 of all *Eco*RI sites to be cut. The particular 20% of sites at which the cuts occur is totally random and different even on identical DNA molecules. (b) Most of the restriction fragments produced by partial digestion are larger than those produced by complete digestion with the same restriction enzyme.

any one fragment within this complex mixture, it is necessary to purify it away from all the other fragments and amplify it, that is, make many identical copies of the single purified molecule. Researchers can then use biochemical techniques to analyze the many copies of the purified molecule. We now describe the methods used to purify and amplify isolated restriction fragments.

PURIFICATION AND AMPLIFICATION OF FRAGMENTS FOR STORAGE AND ANALYSIS

The multistep process that takes a collection of restriction fragments and uses living cells to purify and make many exact replicas of individual genomic fragments is known as **molecular cloning.** It consists of two basic steps. In the first, you take DNA fragments that fall within a specified range of sizes and insert the fragments into specialized chromosome-like carriers called *vectors* that ensure the transport, replication, and purification of individual DNA fragments. In the second step, you allow the vectors to transport individual fragments into living cells, which make many copies of the vectors along with the inserted piece of genomic DNA they carry. Since all the copies are identical, the replicated fragments are known as **DNA clones.** These DNA clones may be purified for immediate study or stored as collections of clones known as libraries for future analysis. We now describe each step.

Cloning Step 1: Ligation of Fragments to Cloning Vectors Creates Recombinant DNA Molecules

On their own, restriction fragments cannot reproduce themselves in a cell. To make replication possible, it is necessary to splice each fragment to a **vector:** a specialized DNA sequence that can enter a living cell, signal its presence to an investiga-

tor by conferring a detectable property on the host cell, and provide a means of replication for itself and the foreign DNA inserted into it. A vector must also possess distinguishing physical traits, such as size or shape, by which it can be purified away from the host cell's genome. There are several types of vectors (Table 8.2), and each one behaves as a chromosome capable of accepting foreign DNA inserts and replicating independently of the host cell's genome. The cutting and splicing together of vector and inserted fragment—DNA from two different origins—creates a **recombinant DNA molecule.**

Sticky Ends Facilitate Recombinant DNA Fabrication

To produce a vector-insert recombinant, you take advantage of two characteristics of single-stranded, or "sticky" ends: They are available for base pairing and no matter what the origin of the DNA, sticky ends produced with the same enzyme are complementary in sequence. You thus simply cut the vector with the same restriction enzyme used to generate the fragment of genomic DNA, then allow time for the base pairing of complementary sticky ends from the DNA of two different origins and for DNA ligase to stabilize the molecule by sealing the breaks in the sugar-phosphate backbones. Certain laboratory "tricks" (discussed later) help prevent two or more genomic fragments from joining with each other rather than with vectors. It is also possible to ligate restriction fragments having blunt ends onto vectors that also have blunt ends, although without the sticky ends, the ligation process is much less efficient.

There Are Several Types of Vectors

Available vectors differ from one another in origin, construction, and carrying capacity. The simplest vectors are minute circles of double-stranded DNA known as **plasmids** that can gain admission to and replicate in the cytoplasm of many kinds of bacterial cells, independently of the nuclear chromosomes (Fig. 8.5). The most useful plasmids contain several recognition sites, one for each of several different restriction enzymes; for example, one *Eco*RI site, one *Hpa*I site, and so forth. Exposure to any one of these restriction enzymes opens up the vector at the corresponding recognition site, allowing the insertion of a foreign DNA fragment, without at the same time splitting the plasmid into many pieces and thereby destroying its continuity and integrity (Fig. 8.5a); a plasmid carrying a foreign insert is known as a *recombinant plasmid.* Each plasmid vector also carries an origin of replication and a gene for antibiotic resistance. The origin of replication enables it to replicate inside a bacterium. The gene for antibiotic resistance signals its presence by conferring on the host cell the ability to survive in a medium containing antibiotics; the resistance gene thereby enables experimenters to select for propagation only those bacterial cells that contain a plasmid (Fig. 8.5b). Vector genes that make it possible to pick out cells harboring a particular DNA molecule are called **selectable markers.** Plasmids fulfill the final requirement for vectors—ease of purification—because they can be purified away from the ge-

TABLE 8.2   Various Vectors and the Size of the Inserts They Carry

Vector	Form of Vector	Host	Typical Carrying Capacity (Size of Insert Accepted)	Major Uses
Plasmid	Double-stranded circular DNA	*E. coli*	Up to 15 kb	cDNA libraries; subcloning
Bacteriophage lambda	Virus (linear DNA)	*E. coli*	Up to 25 kb	Genomic and cDNA libraries
Cosmid	Double-stranded circular DNA	*E. coli*	30–45 kb	Genomic libraries
Phagemid	Virus convertible to plasmid	*E. coli*	Up to 12 kb	cDNA and genomic libraries
Bacteriophage P1	Virus (circular DNA)	*E. coli*	70–90 kb	Genomic libraries
BAC	Bacterial artificial chromosome	*E. coli*	100–500 kb	Genomic libraries
YAC	Yeast, artificial chromosome	Yeast	250–1000 kb (1 megabase)	Genomic libraries

nomic DNA of the bacterial host by several techniques (described later).

Common plasmid vectors are 2–4 kb in length and can carry up to 15 kb of foreign DNA. Larger capacity vectors have been constructed from the genome of **phage lambda (λ)**, a naturally occurring double-stranded DNA virus that infects *E. coli*. Each phage λ chromosome is 50 kb long and can be engineered to receive DNA fragments up to 25 kb in length. Figure 8.6a illustrates the construction of a phage λ vector and the use of that vector to amplify a DNA fragment through the infection of *E. coli* bacterial cells.

Still larger capacity vectors include the **cosmids:** hybrid plasmid-phage vectors that, like phage λ, make use of a virus capsule to infect bacteria. Cosmids are constructed with plasmid-derived selectable markers and two specialized segments of phage λ DNA known as *cos* (for cohesive end) sites, which allow their packaging into viral shells. Once the virus capsule introduces the cosmid into the host cell, the DNA molecule replicates as a plasmid. Cosmids can carry up to 45 kb of insert DNA. Figure 8.6b shows the construction and use of a cosmid.

The largest capacity vectors currently available are **yeast artificial chromosomes (YACs):** hybrids built on a plasmid foundation that contain, instead of phage λ DNA, yeast DNA sequences. The yeast DNA enables YACs to segregate properly inside eukaryotic yeast cells, as if they were natural chromosomes. The big advantage of YAC vectors is that they can carry very large DNA inserts—up to 1 million base pairs, or 1 megabase (1 Mb), in length. Figure 8.7 illustrates the construction and use of a YAC.

The utility of these different cloning vectors becomes apparent when you consider the structure of the β-globin locus. Situated on chromosome 11, the entire locus, including multiple β-globin-like genes and all the regulatory information for

proper expression of these genes, spans 50 kb of DNA. Only a few vectors (P1, BAC, or YAC) can ferry this much material (review Table 8.2). By contrast, the collection of exons and introns that constitute a single β-globin gene is only 1.4 kb in length and can thus fit in a lambda cloning vector or even a plasmid. Most other genes are larger than individual β-globin genes, in part because they have greater coding capacity and many more introns; as a result, to be cloned as a single insert, they require a lambda or even a YAC vector.

Cloning Step 2: Host Cells Take Up Vector-Insert Recombinants and Amplify Them When They Copy Their Own Chromosomes

Although each type of vector functions in a slightly different way and enters a specific host, the general scheme of entering a host cell and taking advantage of the cellular environment to replicate itself is the same for all. We divide our discussion of this step of cloning into three parts: (I) getting foreign DNA into the host cell; (II) selecting cells that have received a DNA molecule; and (III) distinguishing insert-containing recombinant molecules from vectors without inserts. Figure 8.8 illustrates the three-part process with a plasmid vector containing an origin of replication, the gene for resistance to ampicillin (*amp*^r), and one end of the *E. coli lacZ* gene, which encodes the enzyme β-galactosidase. By constructing the vector with a series of common restriction sites right in the middle of the *lacZ* gene fragment, researchers can insert foreign DNA into the gene fragment at those points and then use the disruption of *lacZ* gene function to distinguish insert-containing recombinant molecules from vectors without inserts (as described in the caption and later in the text). Many of the plasmid vectors used today incorporate most if not all of the features depicted in Fig. 8.8.

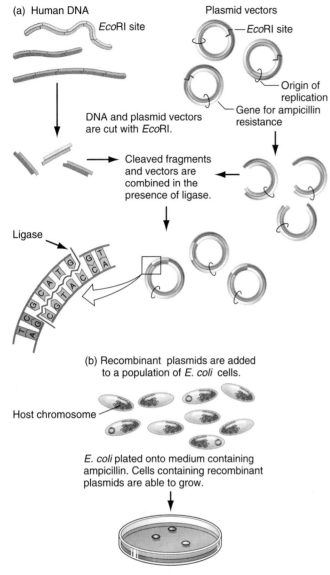

(a) Human DNA

*Eco*RI site

Plasmid vectors

*Eco*RI site

Origin of replication

Gene for ampicillin resistance

DNA and plasmid vectors are cut with *Eco*RI.

Cleaved fragments and vectors are combined in the presence of ligase.

Ligase

(b) Recombinant plasmids are added to a population of *E. coli* cells.

Host chromosome

E. coli plated onto medium containing ampicillin. Cells containing recombinant plasmids are able to grow.

Figure 8.5   How to use plasmid vectors. (a) DNA from human tissue is cut with *Eco*RI to reduce the genome into a mixture of *Eco*RI restriction fragments. A plasmid vector containing an origin of replication and the ampicillin resistance genes is also cut with *Eco*RI. The plasmid has a single *Eco*RI recognition site and digestion with the enzyme converts it into a single linear molecule. The *Eco*RI site is in a position separate from both the ampicillin resistance gene and the origin of replication. The cut genomic DNA and the cut plasmid vectors are mixed together in the presence of the enzyme ligase, which sutures the two types of molecules to each other to form a circular recombinant DNA molecule. (b) The recombinant DNA circles are added to a culture of *E. coli* cells under conditions that promote entry of the DNA circles into the cells. The *E. coli* cells are then spread on a plate containing nutrients and ampicillin. Only those cells that have picked up a plasmid will grow in the presence of ampicillin. Within 24 hours, each plasmid-containing cell will grow into a colony of identical cells that is visible to the naked eye.

Transformation: Vectors Carry Insert DNA into Cells

Transformation, we saw in Chapter 5, is the process by which a substance changes the genetic characteristics of a cell or organism. The introduction of foreign DNA molecules into a host cell thus transforms that cell.

What we now describe is similar to what Avery and colleagues did in the transformation experiments that determined DNA was the molecule of heredity (see Chapter 5), but the method outlined here is more efficient. The plasmid vector is first ligated to insert DNA from any source. The vector-insert recombinants are then mixed and shaken with a suspension of the specially prepared *E. coli*. Under conditions favoring entry, such as suspension of the bacterial cells in a cold $CaCl_2$ solution or treatment of the solution with high-voltage electric shock (a technique known as electroporation), the plasmids will enter about 1 in a 1000 cells. These protocols increase the permeability of the bacterial cell membrane, in essence, punching temporary holes through which the DNA gains entry. The rate of entry is sufficiently inefficient to ensure that each receiving bacterial cell takes up only one molecule of DNA (Fig. 8.8, part I).

Plasmids thus gain entry to cells in an inefficient random process; the same is true of YACs. By contrast, lambda and cosmid molecules, which are packaged in lambda virus particles *in vitro,* enter host cells via specific receptors for the virus and the viral injection apparatus. This much more efficient delivery system can achieve 10 to 1000 times as many transformations per unit of DNA as the nonviral systems.

How You Know Which Cells Have Been Transformed

To identify the 0.1% of cells housing a plasmid, you decant the bacteria-plasmid mixture onto a plate containing agar, nutrients, and ampicillin. Only cells transformed by a plasmid that provides resistance to ampicillin will be able to grow and multiply in the presence of the antibiotic. The plasmid's origin of replication enables it to replicate in the bacterial cell independently of the bacterial chromosome; in fact, most plasmids replicate so well that a single bacterial cell may end up with hundreds of identical copies of the same plasmid molecule. Each viable plasmid-containing bacterial cell will multiply to produce a colony of tens of millions of genetically identical cells. The colonies show up on the agar plate as spots about 1 mm in diameter, easily big enough to see, yet small enough for many distinct ones to appear on a plate (Fig. 8.8, part II).

How You Know Whether the Plasmids Inside Those Cells Contain an Insert

If prepared under proper conditions (including, for example, a controlled ratio of fragments to vectors during the ligation procedure), most plasmids will contain an insert. Some, however, will slip through without one. Figure 8.8, part III, shows how the system we are discussing distinguishes cells with just vectors from cells with vectors containing inserts. The medium on which the transformed, ampicillin-resistant bacteria grow contains, in addition to nutrients and ampicillin, a chemical compound known as X-Gal. This compound serves as a substrate

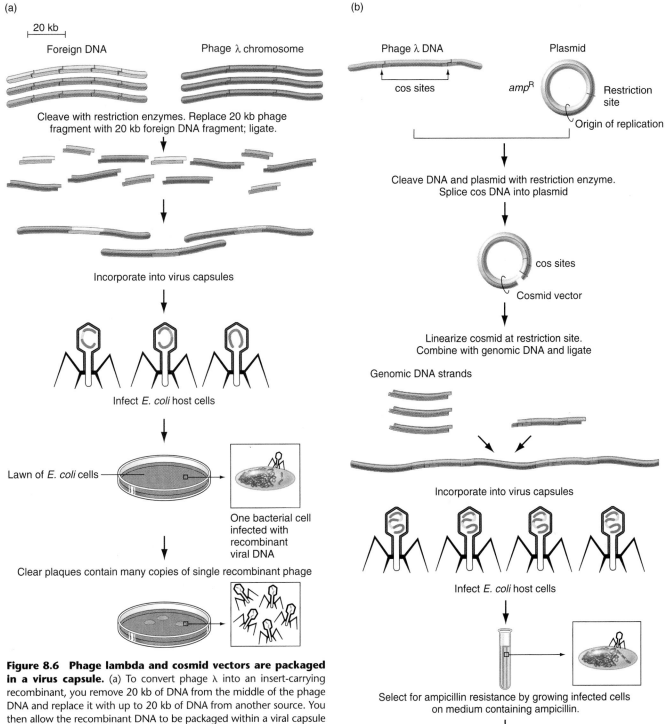

(a)

20 kb

Foreign DNA
Phage λ chromosome

Cleave with restriction enzymes. Replace 20 kb phage
fragment with 20 kb foreign DNA fragment; ligate.

Incorporate into virus capsules

Infect *E. coli* host cells

Lawn of *E. coli* cells

One bacterial cell
infected with
recombinant
viral DNA

Clear plaques contain many copies of single recombinant phage

(b)

Phage λ DNA
Plasmid

cos sites

*amp*R

Restriction
site

Origin of replication

Cleave DNA and plasmid with restriction enzyme.
Splice cos DNA into plasmid

cos sites

Cosmid vector

Linearize cosmid at restriction site.
Combine with genomic DNA and ligate

Genomic DNA strands

Incorporate into virus capsules

Infect *E. coli* host cells

Select for ampicillin resistance by growing infected cells
on medium containing ampicillin.

Surviving cells contain recombinant plasmids which may be cloned.

Figure 8.6 Phage lambda and cosmid vectors are packaged in a virus capsule. (a) To convert phage λ into an insert-carrying recombinant, you remove 20 kb of DNA from the middle of the phage DNA and replace it with up to 20 kb of DNA from another source. You then allow the recombinant DNA to be packaged within a viral capsule by combining it *in vitro* with extracts containing viral head and tail proteins. Such packaging makes it possible to take advantage of the virus' natural ability to infect living cells. When exposed to a population of *E. coli* under the proper conditions, almost all the viruses will infect bacteria; by contrast, only about 1 in 1000 naked DNAs, whether of plasmid or phage origin, enters cells. You now infect a dense culture of bacteria for ~ 20 minutes, mix this solution with molten agar, and pour the mixture onto a petri dish that already contains a hardened layer of agar containing nutrients for bacterial growth. Within a day, the top layer of bacteria (called a "lawn") is so dense that it appears opaque. The lawn, however, has some clear, transparent circles a few millimeters in diameter. Visible to the naked eye, each clearing is the result of a solitary phage that initially infected a single bacterial cell, multiplied within that cell, and burst out of the cell and repeated the cycle in neighboring cells. Repetition of this process every 20–30 minutes produces the small clearing in the lawn where live bacteria are no longer present. The transparent clearing is called a plaque. (b) Cosmid vectors are hybrids of plasmids and phages that combine the most useful cloning properties of each in a single molecule. The relatively short cos sites obtained from the phage are the only molecular components required for packaging any DNA molecule into a viral capsule. The drug-resistance gene and origin of replication obtained from the plasmid provide a means for the selection and propagation of the recombinant DNA molecule.

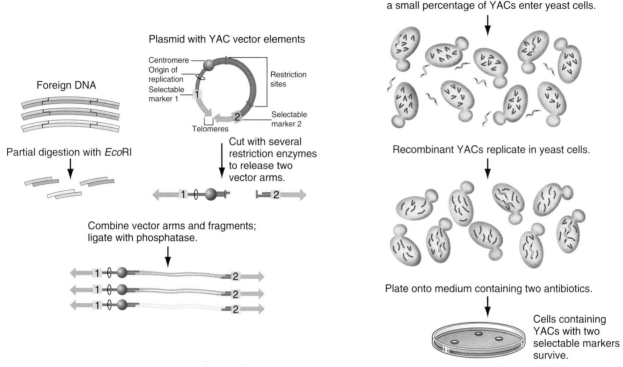

Figure 8.7 YAC vectors take advantage of DNA elements used for normal chromosome segregation within yeast cells. Two distinct arms make up each YAC vector. At the end of one arm is a telomere followed by a selectable marker, then a centromere, and finally a restriction site. The second arm lacks a centromere but has a telomere at one end, a restriction site at the other, and a second selectable marker in the middle. One of the two arms must also contain a yeast origin of replication. To make YAC-insert recombinants, you cut the two YAC arms and large foreign genomic fragments with the same restriction enzyme, mix the YAC arms with the foreign restriction fragments, and treat the mixture with phosphatase. As with bacteria exposed to plasmids, a small percentage of yeast cells exposed to YAC-insert recombinants will take up the recombinant molecules. And like bacteria that harbor plasmid vectors, yeast cells transformed by properly constructed recombinant YACs containing two selectable markers will survive and propagate in a medium infused with two antibiotics. Yeast cells with one or no marker will not. The properly constructed YAC recombinants will replicate and be transmitted along with other chromosomes inside the surviving yeast cells. Such proper YACs must meet three requirements: (1) They must contain an insert; (2) they must carry one—and only one—centromere, since those with more than one centromere will not segregate properly during mitosis; and (3) they must have a telomere at both ends. Tips without a telomere will fuse with another chromosome or decay. Since only those recombinants composed of two different arms flanking an insert will satisfy these requirements, the ability to segregate properly after replication ensures the reproduction of mostly single vector–single insert recombinants.

for the reaction catalyzed by the intact β-galactosidase enzyme (the protein encoded by the *lacZ* gene), and one product of the reaction is a new chemical with a blue pigment. Cells containing vectors without inserts turn blue because they carry the original intact β-galactosidase gene component; cells containing plasmids with inserts remain colorless, because the interrupted *lacZ* gene does not allow production of functional β-galactosidase enzyme (see Fig. 8.8, part III).

Amplification of the Recombinant DNA Molecule: The Actual Cloning

Each time a plasmid-containing bacterial cell divides, it replicates its own chromosome as well as the plasmid—an independent DNA molecule—established in the cell. The millions of cells arising from a single cell by consecutive divisions collectively form a colony, also referred to as a **cellular clone.** For clarity, we use the term "colony" throughout this discussion. The many copies of independently replicating DNA mol-

ecules in the cells of a particular colony are, as we have seen, *DNA clones.* Some of these DNA clones may be plasmids without inserts. With proper cloning procedures, however, most of the DNA clones are a combination of vector and insert, or **recombinant DNA clones.** All the recombinant DNA clones in a colony of cells contain exactly the same DNA insert. The process of producing recombinant DNA clones inside the multiplying cells of a colony derived from a single bacterial progenitor is called **cloning.** After cloning, each healthy bacterium within a colony may contain up to 50 copies of a recombinant DNA clone.

Cloned DNA Is Purified by Various Means That Separate Recombinant Plasmid from Host DNA, Then Insert from Vector

For DNA clones of interest, you can use the physical and biological characteristics of the vector to purify the vector-insert DNA from the cells that copied it. You can then take advantage

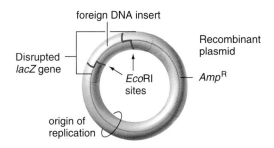

I. Transformation: foreign DNA enters the host cell

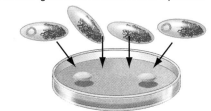

II. Selecting cells that have received a plasmid

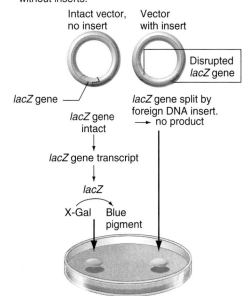

III. Distinguishing cells carrying insert-containing recombinant molecules from cells carrying vectors without inserts.

Figure 8.8   How to identify transformed bacterial cells containing plasmids with DNA inserts. Under proper conditions (including, for example, exposure to phosphatase and a controlled ratio of fragments to vectors), most plasmids will accept an insert. Some, however, will slip through without one. To identify such interlopers, plasmid vectors are constructed so that they contain the *lacZ* gene with a series of common restriction sites right in the middle of the gene. If the vector reanneals to itself without inclusion of an insert, the *lacZ* gene will remain uninterrupted; if it accepts an insert, the gene will be interrupted. (I) Transformation: When added to a culture of bacteria, plasmids enter about 1 in a 1000 cells. (II) Only cells transformed by a plasmid carrying a gene for ampicillin resistance will grow. (III) Cells containing vectors that have reannealed to themselves without the inclusion of an insert will express the uninterrupted *lacZ* gene that they carry. The polypeptide product of the gene is one component of a complete, functional β-galactosidase. Reaction of this enzyme with a substrate known as X-Gal produces a molecule that turns the cell blue. By contrast, if the opened ends of the plasmid vector reannealed to a fragment of foreign DNA, this insert will interrupt and inactivate the *lacZ* gene on the plasmid, rendering it unable to produce a functional polypeptide. As a result of this process, called insertional inactivation, any cells containing recombinant plasmids will not generate active β-galactosidase, and will therefore not turn blue. It is thus possible to visually distinguish colorless colonies that contain inserts from blue colonies that do not.

of specific restriction sites to separate insert from vector. At the end of this process, molecular cloning yields large amounts of purified fragments—billions of identical copies of the original small piece of genomic DNA that are now ready for study.

Purifying Recombinant Plasmid from Host DNA

Methods for purifying cloned DNA depend on chemical differences between the host chromosome and the recombinant plasmid. These chemical differences result from physical differences in the size and form (linear versus circular) of the various DNA molecules. In the plasmid-*E. coli* system, the bacterial chromosome is approximately 4000 kb long, whereas the recombinant plasmid can be from 4 to 20 kb in length. In one separation method (shown in Fig. 8.9a), researchers use the equilibrium density gradient centrifugation described in Chapter 6 in conjunction with ethidium bromide, an intercalating agent that inserts itself between DNA base pairs.

Purifying Inserts from Plasmids

To separate inserts from vectors, you simply cut the purified vector-insert molecules with the same restriction enzyme used to make the recombinants in the first place. You then subject the resulting mixture of insert DNA disengaged from vector DNA to any one of a number of protocols that separates DNA molecules on the basis of size (Fig. 8.9b). Electrophoresis is one such procedure; we describe it in detail later in the chapter. Separation can eventually yield a test tube containing many molecules of insert DNA purified away from the vector DNA. Although the harvest of many copies of the foreign inserts is a major goal of cloning, geneticists also preserve some cells of a cellular clone as a factory for propagating insert-carrying plasmids on future demand.

Libraries Are Collections of Cloned Fragments

Moving step by step from the DNA of any organism to a single purified DNA fragment is a long and tedious process.

(a) **Separating plasmid from bacterial chromosome**

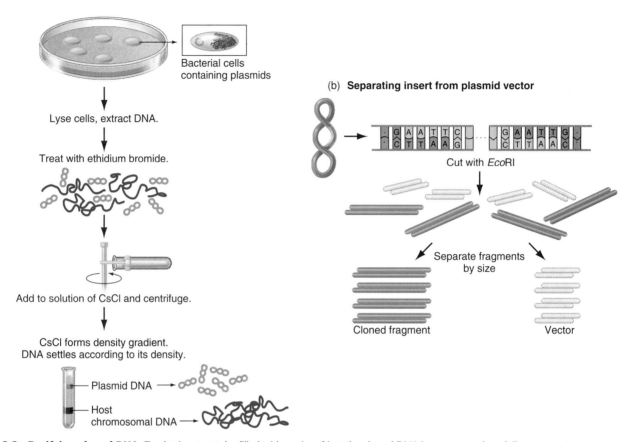

Bacterial cells
containing plasmids

Lyse cells, extract DNA.

Treat with ethidium bromide.

Add to solution of CsCl and centrifuge.

CsCl forms density gradient.
DNA settles according to its density.

Plasmid DNA →

Host
chromosomal DNA →

(b) **Separating insert from plasmid vector**

Cut with *Eco*RI

Separate fragments
by size

Cloned fragment

Vector

Figure 8.9 Purifying cloned DNA. To obtain a test tube filled with copies of just the cloned DNA insert, researchers follow a two-step process: (a) separation of the plasmid from the bacterial chromosome and (b) separation of the DNA insert from vector DNA. In the plasmid-*E. coli* system, the bacterial chromosome is about 4000 kb long, while the recombinant plasmid is from 4 to 20 kb in length. In one separation method, researchers use equilibrium density gradient centrifugation (described in Chapter 6) in conjunction with ethidium bromide, an intercalating agent that inserts itself between DNA base pairs and causes the double helix to unwind. In the first step of the procedure, shown in (a), you lyse the bacterial cells (rupture their membranes) to release the DNA. The *E. coli* chromosome, because of its large size, fragments during lysis, producing linear pieces 100–200 kb in length. The much smaller recombinant plasmids remain as undamaged, supercoiled circles. Treatment with ethidium bromide elicits a different response from the two types of DNA. The linear DNA molecules continue to bind ethidium bromide until they reach a saturation point with all bonding sites occupied. By contrast, the small closed circles of plasmid DNA compensate for intercalation-caused unwinding by negative supercoiling. After a while, the physical strain of supercoiling prevents further intercalation. Because the plasmids thus bind much less ethidium bromide than the bacterial chromosome fragments, the buoyant density of the plasmids is distinguishable from that of the chromosomes when the two are added to a cesium chloride (CsCl) gradient saturated with ethidium bromide. In the second step of the procedure, shown in (b), the purified plasmids are exposed to the specific restriction enzyme used to construct the recombinant plasmids. This enzyme will cut the recombinant plasmids into vector and insert fragments that differ in size. Various methods can then purify the insert away from the vector.

Fortunately, scientists do not have return to step 1 every time they need to purify a new genomic fragment from the same organism. Instead, they can build a **library:** a collection of DNA clones that contains multiple copies of nearly every fragment in the whole genome inserted into a suitable vector and placed into storage. Like traditional libraries, DNA libraries store large amounts of information for retrieval upon request. They make it possible to start a new project at an advanced stage of cloning, with the retrieval of a ready-made vector-insert molecule.

Genomic Libraries Are Random Collections of the DNA Fragments from the Chromosomes of a Given Species, Inserted in a Suitable Vector

If you cut an organism's genome with a restriction enzyme and ligated every fragment to a vector that then replicated the vector-insert recombinant inside the appropriate host, ideally, each of the resulting DNA clones would represent a single fragment of the original genomic DNA. A hypothetical collection of DNA clones that included one copy of every sequence in the entire genome would be a **complete genomic library.** Researchers store the DNA clones of a library in a refrigerator or freezer for future use. They may, for example, search the library a year after storage for clones containing a fragment of the α-globin locus that they wish to analyze. The form of storage for vector-insert recombinants depends on the vector. Plasmid and YAC libraries contain collections of cells, whereas phage and cosmid libraries consist of virus particles ready to infect their host. Researchers often prefer to use vectors delivered by viruses because of their greater efficiency at transformation.

How many genomic clones should a library contain to give you a reasonable chance of finding one with a particular fragment of DNA? If you started with the 3,000,000 kb of DNA from a haploid human sperm and reliably cut it into a series of 20-kb restriction fragments, you would generate 3,000,000/20 = 150,000 genomic fragments. If you could create a library by placing each and every one of these fragments into a phage lambda cloning vector, you would create a perfect library of 150,000 clones that collectively carry every locus in the genome. The number of clones in this perfect library defines a **genomic equivalent.** For any library, to find the number of clones that constitute one genomic equivalent, you simply divide the length of the genome (here, 3,000,000 kb) by the average size of the inserts carried by the library's vector (in this case, 20 kb).

In real life, it is impossible to obtain a perfect library. Each step of cloning is far from 100% efficient, and the DNA of a single cell does not supply sufficient raw material for the process. Researchers must thus harvest DNA from the hundreds of thousands of cells in a particular tissue or organism. If you make a genomic library with this DNA by collecting only one genomic equivalent (150,000 clones for a human library in bacteriophage lambda vectors), the uncertainties of random DNA sampling from the different cells in the original tissue would mean that some DNA fragments would appear more than once, while others might not be present at all. To increase the chance that a library contains the clone of a particular genomic fragment, the number of clones in the library must exceed one genomic equivalent. Including four to five genomic equivalents produces an average of four to five clones for each locus, and a 95% probability that any individual locus is present at least once.

Genomic libraries based on different cloning vectors serve as references for different levels of analysis. The first libraries, established in the late 1970s, used phage lambda vectors carrying inserts of approximately 20 kb. Although this size is suitable for many applications, there are many situations that call for clones carrying larger inserts. For example, many important mammalian and some *Drosophila* genes are larger than 20–25 kb, too large to fit in a single bacteriophage lambda clone. Researchers thus developed vectors capable of cloning larger pieces, among them the cosmids (45-kb insert capacity) and YACs (1000-kb capacity) listed in Table 8.2. One advantage of the larger inserts is that you need to examine far fewer clones to find a sequence of interest. There are only ~ 3000 1-Mb YACs in a human genomic equivalent, so

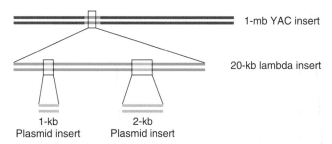

Figure 8.10 Subcloning: From YAC clones to lambda clones to plasmids. The large genomes of higher eukaryotes (such as those of mice and humans) can be readily divided into a few thousand YAC clones that each contain ~ 1 Mb of genomic DNA. To study gene-size regions from a particular YAC, however, it is convenient to subclone 20-kb fragments into a lambda vector. To analyze particular regions of a gene, it is useful to perform another round of subcloning, placing fragments that are 2 kb or smaller into a plasmid vector.

it would require only 12,000 (3000 × 4) YACs—as compared to 600,000 phages—to construct a reasonable library.

A major disadvantage of YACs and other clones carrying very large inserts is that it is hard to analyze such large pieces of DNA. To look more closely at a particular DNA segment in a YAC, researchers use restriction enzymes to dissect the YAC insert into smaller pieces that they then clone into a different kind of vector, for example, a cosmid or a lambda phage. This process of **subcloning** establishes smaller and smaller clones that are enriched for sequences of interest (Fig. 8.10).

cDNA Libraries Carry Information from the RNA Transcripts Present in a Particular Tissue

Often, only the information in a gene's coding sequence is of experimental interest, and it would be advantageous to limit analysis to the gene's exons without having to determine the structure of the introns as well. Because coding sequences account for a very small percentage of genomic DNA in higher eukaryotes, however, it is inefficient to look for them in genomic libraries. The solution is to generate **cDNA libraries,** which store sequences copied into DNA from RNA transcripts. Because they are obtained from RNA transcripts, these sequences carry only the exon information for making proteins.

To produce DNA clones from mRNA sequences, researchers rely on a series of *in vitro* reactions that mimic several stages in the life cycle of viruses known as **retroviruses.** Retroviruses, which include among their ranks the viruses that cause AIDS and certain human leukemias, transmit their genetic information in molecules of RNA and carry as part of their gene transmission kit the unusual enzyme known as **RNA-dependent DNA polymerase,** or simply **reverse transcriptase** (review the Genetics and Society box in Chapter 7). Once retroviruses infect a cell, they use reverse transcriptase to copy their single strand of RNA information into a mirror-image strand of complementary DNA; this complementary DNA is known as **cDNA.** The reverse transcriptase, functioning as a DNA polymerase, then makes a second strand of DNA complementary to this first strand. Finally, the double-stranded DNA copy of the retroviral RNA chromosome integrates into the host cell's genome. Although the designation cDNA originally meant the single strand of DNA complementary to the RNA transcript, it now refers to any DNA—single- or double-stranded—derived from RNA. In 1975, researchers began to use reverse transcriptase to convert mRNA sequences from any tissue in any organism into cDNA sequences for cloning.

Suppose you were interested in studying the structure of a mutant β-globin protein. You have already analyzed hemoglobin obtained from a patient carrying this mutation and found that the alteration affects the structure of the protein itself and not its regulation, so you now need only look at the structure of the gene's coding sequence. To establish a library enriched for the mutant gene sequence and lacking all the extraneous information from genomic introns, you would first obtain mRNA from the cytoplasm of the patient's red blood cell precursors (Fig. 8.11a). About 80% of the total mRNA in red blood cells is from the α and β hemoglobin genes, so the mRNA preparation contains a much higher proportion of the sequence corresponding to the β-globin gene than the genomic sequences found in a cell's nuclear DNA. The addition of reverse transcriptase in the presence of ample amounts of the four deoxyribose nucleotide triphosphates, DNA polymerase, and primers to initiate synthesis would generate single-stranded cDNA (Fig. 8.11b). If you used the enzyme RNase to digest the RNA, you could then make a second DNA strand by exposing the cDNA to DNA polymerase, again in the presence of the requisite nucleotide triphosphates and a primer (Fig. 8.11c).

After using restriction enzymes and ligase to insert this double-stranded cDNA in a suitable vector (Fig. 8.11d) and storing the vector-insert recombinants in an appropriate manner, you would have a library of double-stranded cDNA fragments, each one corresponding to an mRNA molecule in the red blood cells that served as your sample. In contrast to a genomic library, this cDNA library does not contain all the DNA sequences in the genome. Instead, it includes only the exons for that part of the genome that the red blood cells were actively transcribing for translation into protein. Since most fully spliced RNA transcripts are at most a few kilobases in length, the cDNA fragments constructed in their image fall within that size range as well. The percentage of clones in the library with cDNA specific for the β-globin protein will reflect the fraction of red blood cell mRNAs that are β-globin messages. For genes expressed infrequently or in very few tissues, you will have to screen many molecular clones from a cDNA library to find the gene of interest. For highly expressed genes, such as the β-globin gene, you will only have to screen very few clones.

Genomic versus cDNA Libraries

Figure 8.12 compares genomic and cDNA libraries. The main advantage of genomic libraries is that the genomic clones within them represent all regions of DNA equally and show what the intact genome looks like in the region of each clone. The chief advantage of cDNA libraries is that the cDNA clones reveal which parts of the genome encode the information used

(a) Red blood cell precursors

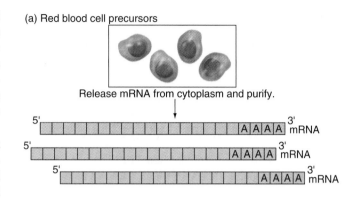

Release mRNA from cytoplasm and purify.

(b) Add oligo(dT) primer. Treat with reverse transcriptase in presence of four nucleotides.

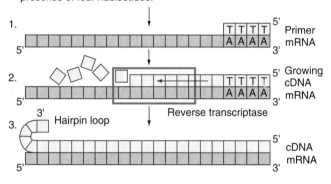

(c) Digest mRNA with RNase. cDNA directs synthesis of second cDNA strand in the presence of four nucleotides and DNA polymerase.

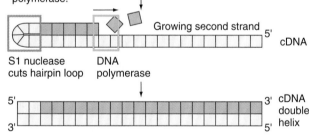

(d) Insert cDNA into vector.

Figure 8.11 Converting RNA transcripts to cDNA. (a) Obtain mRNA from red blood cell precursors (which still contain a nucleus). (b) Add oligo(dT) primers to initiate reverse transcription. Such transcription produces a strand of cDNA complementary to the mRNA. (c) Eliminate the mRNA transcript through digestion with RNase. Add DNA polymerase and the four nucleotides to produce a second cDNA strand complementary to the first. (d) Insert the newly created double-stranded DNA molecule into a vector for cloning.

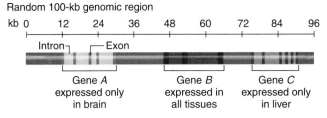

Random 100-kb genomic region

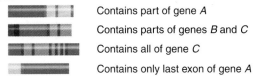

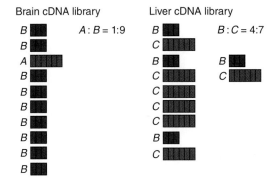

Figure 8.12 **A comparison of genomic and cDNA libraries.** Every tissue in a multicellular organism can generate the same genomic library, and the DNA fragments in that library collectively carry all the DNA of the genome. Individual genomic fragments may contain no gene, one gene, more than or less than one gene, introns that are transcribed into RNA but spliced out before translation, regulatory sequences that are not transcribed at all, or any combination of these possibilities. Moreover, on average, the clones of a genomic library represent every locus an equal number of times. By contrast, every tissue in a multicellular organism generates a different cDNA library. Clones of this library represent only the fraction of the genome that is being actively transcribed. Moreover, within a cDNA library derived from any single tissue type, the frequency with which particular fragments appear is proportional to the level of the corresponding mRNA within that tissue. Thus, the differential expression of genes within different tissues can be ascertained by the numbers of clones of each type that show up in each library.

in making protein products in specific tissues, as determined from the prevalence of the mRNAs for the genes involved. To gain as much information as possible about a gene's structure and function, researchers rely on both types of libraries.

IDENTIFYING AND ISOLATING CLONES OF INTEREST

Once you have collected the hundreds of thousands of human DNA fragments in, for example, a genomic cosmid library, how do you find the gene you wish to study, the proverbial

"needle in the haystack"? Suppose you want to find the genomic clone or clones containing the introns and exons for the hemoglobin genes—α and β. The first step is to spread the library's clones out on petri plates and place a piece of nitrocellulose paper over each plate, allowing some, but not all, of the cells from each colony to transfer to the filter replica (Fig. 8.13). You now lyse the bacteria on the filter with an alkaline solution, which both releases and denatures the DNA into single strands. Baking the filter or treating it with ultraviolet (UV) light will attach the single-stranded DNA molecules from the cells of each colony to the filter. The library's DNA is now ready for screening. Screening can be accomplished with DNA probes (as illustrated in Figs. 8.13–8.15) or with antibodies to polypeptides expressed by cloned genes (Fig. 8.16).

Screening with DNA Probes: Hybridization to Complementary Sequences Picks Out Fragments of Interest

Screening is a simple procedure. It consists of obtaining a probe complementary to the genes of interest, labeling it in some manner, and then using the labeled probe to identify one or more clones with sequences that base pair, or hybridize, with the probe.

What a DNA Probe Is

DNA probes are short single-stranded stretches of DNA of known composition, from 10 to several thousand nucleotides in length. To visualize probes, it is necessary to label them. Researchers often label them with radioactive nucleotides containing ^{32}P, or with fluorescent dyes (Fig. 8.14a). DNA probes have various uses, including the identification of clones that contain complementary DNA sequences and the diagnosis of DNA alterations that underlie genetic diseases (discussed in Chapter 9).

How to Obtain a DNA Probe

Among the methods for obtaining a DNA probe are chemical synthesis, subcloning, and the production of cDNA (Fig. 8.14b). In chemical synthesis, an automated DNA synthesizer produces probes of specific sequence that range in size from 10 to 100 nucleotides. These relatively short synthesized chains are known as **oligonucleotides.** In subcloning, once a recombinant DNA subclone has been obtained and its vector separated from its insert, exposure to heat denatures the insert's double helix, generating single-stranded probes that are hundreds to thousands of nucleotides in length. In producing cDNA probes, reverse transcriptase is used to synthesize cDNAs from RNA transcripts; the resulting cDNAs are generally a few hundred to a few thousand nucleotides long. It is also possible to obtain probes through the polymerase chain reaction, which we describe later.

To identify clones containing the hemoglobin genes, cDNA probes are the probes of choice, because it is possible to isolate mRNA from red blood cell precursors, where the hemoglobin mRNA makes up more than 80% of the total mRNA, and then use reverse transcriptase to make a radioactively labeled DNA copy of the mRNA. Nearly half of this

Master plate containing genomic
library of clones inside *E. coli* cells.

Overlay a nitrocellulose disk
to make a replica of the plate.

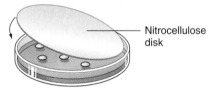

Nitrocellulose
disk

Remove disk from plate and lyse cells on
it and denature DNA with NaOH. Bake
and treat with UV light to bind DNA strand
to disk.

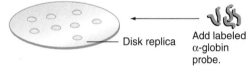

Disk replica

Add labeled
α-globin
probe.

Colonies with complementary DNA
sequences hybridize and anneal to probe.

Filter

Wash filter, expose to X-ray film.

Original plate

Filter

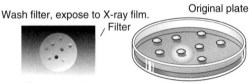

Compare with original plate to locate bacterial
clone with desired genomic clone.

Figure 8.13   Screening a library of clones by hybridization to a labeled probe. To determine which clone in a plasmid library has a DNA fragment of interest, you place a wet nitrocellulose filter onto a bacterial plate and then remove it. This procedure allows most cells within the original colonies on the plate to remain in their original positions, but transfers a small portion of each colony to the filter paper. Treating the filter paper under special conditions causes the cells to lyse and release their DNA; the DNA is then denatured with NaOH and attached to the filter through baking and exposure to UV light. Next, a labeled DNA probe in solution is applied directly to the filter paper; the solution is left on the filter for a period of time long enough to promote hybridization between the probe and any complementary strands of DNA that may be present within one or more spots that correspond to one or more of the DNA clones. Autoradiography identifies spots where hybridization has occurred, and a comparison of spots on the filter with their counterparts on the original bacterial plate identifies the clone or clones that contain DNA fragments of interest.

cDNA will be complementary to the α-globin gene, the other half to the β-globin gene. In locating the hemoglobin genes, it is also possible to use cDNA probes of hemoglobin genes from other higher eukaryotic species, since these genes have been conserved throughout vertebrate evolution.

In addition to the direct methods of obtaining probes illustrated in Fig. 8.14b, investigators can derive DNA probes from polypeptides in a process known as *reverse translation* (Fig. 8.14b). To perform a reverse translation, they translate the amino acid sequence of a protein into a DNA sequence via the genetic code dictionary. Recall, however, that the genetic code is "degenerate," with more than one codon for each amino acid. As a result, investigators must choose peptide sequences containing amino acids encoded by as few potential codons as possible, and they must then synthesize a mixture of oligonucleotides containing all possible codons for each amino acid. With this indirect method of obtaining a DNA probe, researchers can locate and clone genes even if they have only partial information about the proteins the genes encode.

Screening with DNA Probes

To find the hemoglobin genes in the genomic library you have plated out onto filter paper, you simply expose the library's DNA to a solution of α- and β-globin probes labeled with radioactivity or fluorescence. These labeled probes form base pairs, or hybridize, with only those clones containing complementary sequences (Fig. 8.14c). After incubation with radioactively labeled probe, you wash away all probe molecules that have not attached to complementary DNA strands stuck to the filter, expose the filter to X-ray film, develop the film into an autoradiograph, and look for dark spots (see Fig. 8.13); if you use fluorescence-labeled probes, you can see the hybridizing clones directly by shining ultraviolet light on the filters. The spots revealed by hybridization probes correspond to genomic clones containing sequences complementary to the α- and β-globin probes. Knowing the position on the filter that gave rise to these spots, you can return to the original petri plates and pick out the corresponding colonies containing the cloned inserts of interest. With this screening procedure, you would discover that the human genome carries one gene for β globin, but two similar α-globin genes back to back, named α1 and α2. You can harvest and purify the cloned DNA identified by the globin probes and proceed with further analysis.

Requirements and Limitations of Hybridization Probes

Hybridization, we have seen, is the natural propensity of complementary single-stranded molecules to form stable double helixes. Hybridization has a single critical requirement: The region of complementarity between two single strands must be sufficiently long to enable a large number of hydrogen bonds to provide a cohesive force; this cohesive force, in turn, will counteract the thermal forces that tend to disrupt the double helix. Individual hydrogen bonds are very weak, and the two (A:T) or three (G:C) bonds that form between complementary bases are far too weak to prevent thermal forces from pulling them apart. But, if two single strands form hydrogen bonds between hundreds or thousands of bases, the combined force is sufficient to counteract the thermal forces that operate even at room temperature. Hybridization can occur between any two single strands of nucleic acid: DNA/DNA, DNA/RNA, or RNA/RNA. In this chapter we focus on DNA/DNA hybridization.

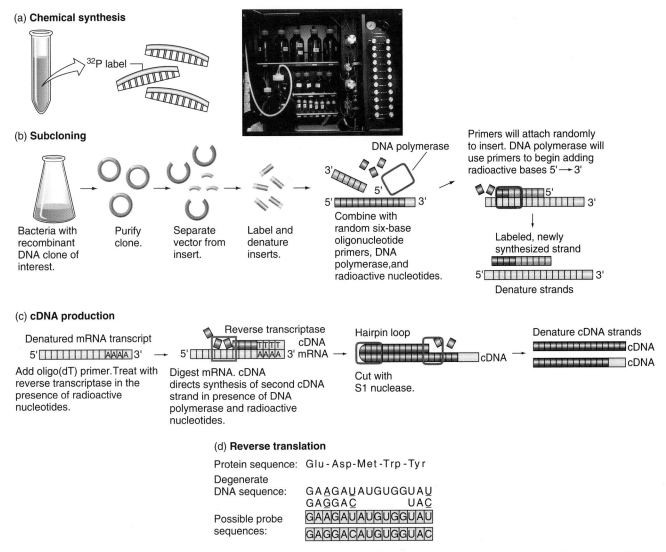

(a) Chemical synthesis

³²P label

(b) Subcloning

Bacteria with recombinant DNA clone of interest.

Purify clone.

Separate vector from insert.

Label and denature inserts.

Combine with random six-base oligonucleotide primers, DNA polymerase, and radioactive nucleotides.

DNA polymerase

Primers will attach randomly to insert. DNA polymerase will use primers to begin adding radioactive bases 5' → 3'

Labeled, newly synthesized strand

Denature strands

(c) cDNA production

Denatured mRNA transcript

Add oligo(dT) primer. Treat with reverse transcriptase in the presence of radioactive nucleotides.

Reverse transcriptase

Digest mRNA. cDNA directs synthesis of second cDNA strand in presence of DNA polymerase and radioactive nucleotides.

Hairpin loop

Cut with S1 nuclease.

Denature cDNA strands

(d) Reverse translation

Protein sequence: Glu - Asp - Met - Trp - Tyr

Degenerate DNA sequence: GAAGAUAUGUGGUAU
                            GAGGAC   UAC

Possible probe sequences: GAAGAUAUGUGGUAU
                            GAGGACAUGUGGUAC

Figure 8.14 DNA probes pick out fragments of interest by complementary base pairing. A probe is a short length of DNA with a known base sequence and a radioactive label. There are two types of methods for obtaining a probe—direct and indirect. Direct methods include (a) chemical synthesis, (b) subcloning, and (c) production of cDNA. The indirect method of reverse translation (d) in which a polypeptide is reverse translated back to DNA, is used when only the purified gene product is available. However, because the genetic code is degenerate, several possible oligonucleotides can result from reverse translation. Investigators use the automated DNA synthesizer and equimolar concentrations of the required nucleotides to produce all possibilities.

Of great importance to geneticists is the fact that hybridization can occur between single strands that are not completely complementary. In general, two single DNA strands that are longer than 50–100 bp will hybridize so long as the extent of their complementarity is more than 80%, even though mismatches may appear throughout the resulting hybrid molecule. Imperfect hybrids are less stable than perfect ones, but geneticists can exploit this difference in stability to evaluate the similarity between molecules from two different sources. Hybridization, for example, occurs between the mouse and human genes for the cystic fibrosis protein (Fig. 8.15). Researchers can thus use the human genes to identify and isolate the corresponding mouse sequences, and then use these sequences to develop a mouse containing a defective cystic

fibrosis gene. Such a mouse provides a model for cystic fibrosis in a species that, unlike humans, is amenable to experimental analysis.

Screening through Expression: Genes Cloned in Specialized Vectors Produce Proteins That Light Up with Specific Labeled Antibodies

A second approach to screening a library is based on the use of cloning vehicles known as **expression vectors,** which promote the expression of a polypeptide product from the gene inserts they carry. To make an expression vector, you insert the cDNA of a gene into a plasmid already carrying the regulatory sequences required for transcription in the host cell. Recombinant plasmids of this kind, when cloned under appropriate

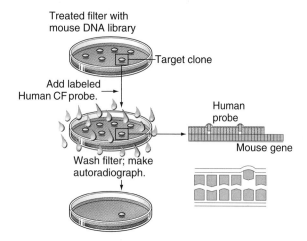

Figure 8.15 Using the human cystic fibrosis gene as a probe to screen the mouse genome. Filters containing DNA from replicas of a mouse genomic library are probed with labeled DNA fragments from the human cystic fibrosis (*CF*) gene. Clones that contain the mouse homolog of this gene will hybridize to the probes. Autoradiography can then detect these clones.

conditions, will promote the expression of the gene, under the control of the regulatory sequences. Within a cDNA library, any sequence inserted by design or by chance in an expression vector will produce the protein it encodes in the host cell. To find the gene sequence producing the protein, you simply make a replica of the host cell colonies, lyse these replicas, and expose their contents to an antibody labeled with fluorescence. The antibody, by interacting specifically with the protein, identifies the cell colony containing the sequence responsible for the protein (Fig. 8.16).

To summarize, the most effective way to isolate specific, uncharacterized segments of DNA for analysis is to cut the DNA with restriction enzymes, clone the fragments into a vector, and screen libraries of recombinant molecules for clones of interest. Once you have isolated a clone of interest, you can then analyze its DNA content to distinguish, for example, a gene's introns from its exons or a regulatory region's normal sequence from a disease-causing mutant sequence.

CHARACTERIZING CLONED FRAGMENTS BY THEIR SIZE, POSITION, AND SEQUENCE

The same restriction enzymes that enable you to construct plasmid clones, phage clones, or other types of DNA clones enable you to deconstruct them for analysis. Suppose you obtained a clone from a phage lambda library of human genomic DNA partially digested with *Eco*RI. Cutting the recombinant phage DNA with the same *Eco*RI restriction enzyme would separate vector from insert DNA. But because the enzyme's six-base GAATTC recognition sequence occurs on average every 4.1 kb, if the insert were much larger than a few kilobases in length, the enzyme might also cut the insert into several pieces. How big is each piece and

where within the clone does it lie? To answer these questions, you must first separate the pieces and determine their length.

Gel Electrophoresis Distinguishes DNA Fragments According to Size

Electrophoresis is the movement of charged molecules in an electric field. Biologists use it to separate many different types of molecules, for example, DNA of one length from DNA of other lengths, DNA from protein, or one kind of protein from another. In this discussion, we focus on its application to the separation of DNA fragments of varying length in a gel (Fig. 8.17). To carry out such a separation, you place a solution of DNA molecules into indentations called wells at one end of a porous gel-like matrix. When you place the gel in a buffered aqueous solution and set up an electric field between electrodes affixed at either end, the electric field causes all charged molecules in the wells to migrate in the direction of the electrode having an opposite charge. DNA, for example, which because of its phosphate-containing backbone carries a net negative charge in a solution near neutral pH will be pulled through the gel toward the positive electrode.

Several variables determine the rate at which DNA molecules move during electrophoresis. These variables are the strength of the electric field applied across the gel, the composition of the gel, the charge per unit volume of molecule (known as charge density), and the physical size of the molecule. For a set of linear DNA fragments migrating in a single gel, only the variable of size is different for different molecules. This is because all molecules placed in a well are subjected to the same electric field and the same matrix; and all DNA molecules have the same charge density (because the charge of all nucleotide pairs is nearly identical). As a result, only differences in size cause different linear DNA molecules to migrate at different speeds during electrophoresis.

With linear DNA molecules, differences in size are proportional to differences in length: the longer the molecule, the larger the volume it will occupy as a random coil. The larger the volume a molecule occupies, the less likely it is to find a pore in the matrix big enough to squeeze through and the more often it will bump into the matrix. And the more often the molecule bumps into the matrix, the lower its rate of migration (also referred to as its mobility). With this background, you can follow the steps of Fig. 8.17a to determine the length of the restriction fragments in the clone under consideration.

Different Techniques of Electrophoresis Separate DNA Molecules in Different Windows of Size

Double-stranded DNA molecules range in size from small fragments of less than 10 bp to whole human chromosomes that have an average length of 130,000,000 bp. No one sizing procedure has the capacity to separate molecules throughout this enormous range. To detect DNA molecules in different size ranges, researchers use a variety of protocols based

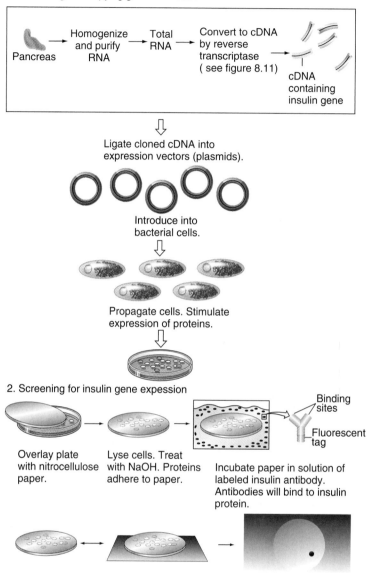

1. Obtaining and copying gene for insulin

Pancreas → Homogenize and purify RNA → Total RNA → Convert to cDNA by reverse transcriptase (see figure 8.11) → cDNA containing insulin gene

Ligate cloned cDNA into expression vectors (plasmids).

Introduce into bacterial cells.

Propagate cells. Stimulate expression of proteins.

2. Screening for insulin gene expession

Overlay plate with nitrocellulose paper.

Lyse cells. Treat with NaOH. Proteins adhere to paper.

Incubate paper in solution of labeled insulin antibody. Antibodies will bind to insulin protein.

Binding sites

Fluorescent tag

Wash filter. Make autoradiograph. Compare with original plate in order to find bacterial clone containing human insulin gene.

Figure 8.16   Using antibodies to libraries of expression vectors. If you wanted to analyze the gene for insulin, you could first copy RNA transcripts from pancreatic cells into cDNA. Next, you would insert the cDNA into an expression vector containing genetic regulatory elements that will cause host bacteria to transcribe the inserted DNA segment. When the host-cell translational machinery sees the transcripts, it will translate them into insulin protein. You now transfer the cells to a nitrocellulose filter, lyse the cells on the filter, and incubate the filter with a labeled antibody to insulin. Finally, you use autoradiography to identify colonies expressing the gene product recognized by the antibody. The recombinant DNA clones in these colonies must carry the DNA sequence that encodes the protein recognized by the antibody.

mainly on only two kinds of gels: polyacrylamide (formed by covalent bonding between acrylamide monomers) and agarose (formed by noncovalent association of agarose polymers). Figure 8.17b illustrates these differences.

Pulsed-field gel electrophoresis is a special type of electrophoretic protocol that operates on standard agarose gels but allows an extreme extension of the normal separating capacity of these gels (Fig. 8.17c). With this protocol, a DNA sample is subjected to pulses of electrical current that alternate between two directions. The range of sizes separated in this manner is a function of pulse length. Short pulses can separate DNA molecules 20 to 200 kb in length. With very long pulses, the molecules can be 9000 kb or longer. There is no theoretical limit to the size of molecules that can be separated

by this technique. The only limitation is the length of time one is willing to run a gel. Using pulse lengths of an hour or more and running times of several months, researchers have separated DNA fragments of up to 20,000 kb, as well as entire yeast chromosomes from each other. (When preparing large pieces of DNA for this type of analysis, it is possible to prevent shear forces from breaking the chromosomes into large pieces at random by manipulating the DNA inside an agarose matrix.)

Electrophoresis enables you to determine the size of restriction fragments of almost any length. Once you know the size of all the fragments in a clone, you can use that information to determine the fragments' relative positions within the clone.

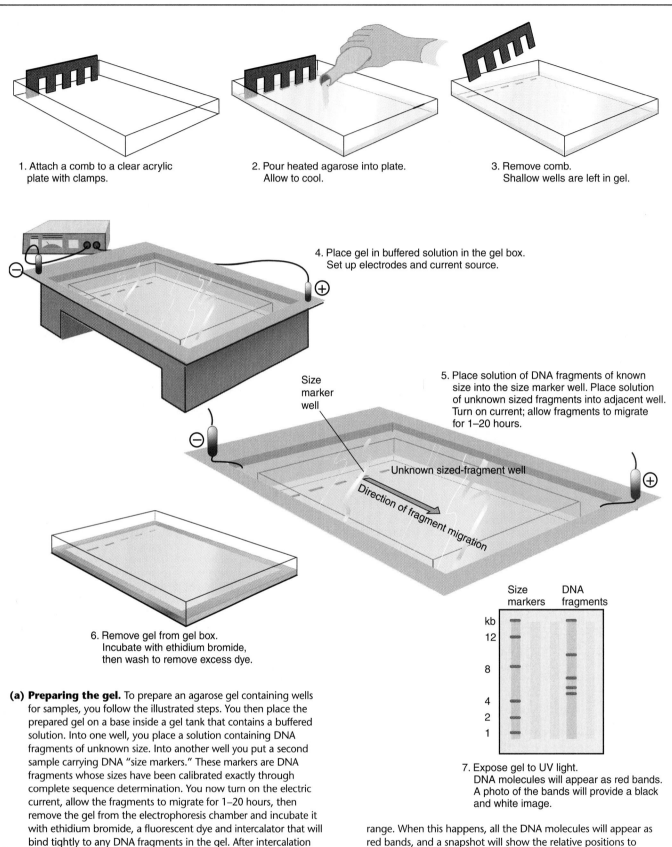

1. Attach a comb to a clear acrylic plate with clamps.

2. Pour heated agarose into plate. Allow to cool.

3. Remove comb. Shallow wells are left in gel.

4. Place gel in buffered solution in the gel box. Set up electrodes and current source.

Size marker well

5. Place solution of DNA fragments of known size into the size marker well. Place solution of unknown sized fragments into adjacent well. Turn on current; allow fragments to migrate for 1–20 hours.

Unknown sized-fragment well

Direction of fragment migration

6. Remove gel from gel box. Incubate with ethidium bromide, then wash to remove excess dye.

Size markers   DNA fragments

kb
12
8
4
2
1

7. Expose gel to UV light. DNA molecules will appear as red bands. A photo of the bands will provide a black and white image.

(a) Preparing the gel. To prepare an agarose gel containing wells for samples, you follow the illustrated steps. You then place the prepared gel on a base inside a gel tank that contains a buffered solution. Into one well, you place a solution containing DNA fragments of unknown size. Into another well you put a second sample carrying DNA "size markers." These markers are DNA fragments whose sizes have been calibrated exactly through complete sequence determination. You now turn on the electric current, allow the fragments to migrate for 1–20 hours, then remove the gel from the electrophoresis chamber and incubate it with ethidium bromide, a fluorescent dye and intercalator that will bind tightly to any DNA fragments in the gel. After intercalation has occurred, immersion of the gel in water will wash away any unbound dye molecules. Then, with exposure to ultraviolet light, the bound dye will absorb photons in the UV range of the electromagnetic spectrum and give off photons in the visible red

range. When this happens, all the DNA molecules will appear as red bands, and a snapshot will show the relative positions to which they have migrated in the gel.

To determine the length of a DNA fragment, you chart the mobility of the band composed of that fragment relative to the migration of the size marker bands in the adjacent gel lane.

Feature Figure 8.17   Gel electrophoresis separates DNA molecules according to size.

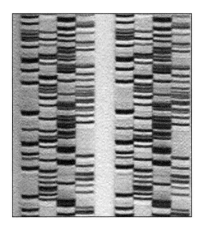

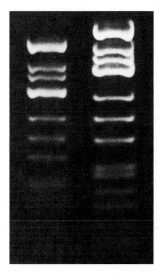

(b) Different types of gels. It is possible to separate DNA molecules in gels composed of two different types of chemical matrices: polyacrylamide and agarose. The separating capability of both types of gels varies with the concentration of matrix material used to compose the gel. The higher the concentration, the smaller the average size of the pores in the matrix, and the smaller the pores, the smaller the range of molecules that can be separated. Geneticists use high-density polyacrylamide gels to separate the smallest DNA molecules from each other. These gels have a window of resolution in the range of 1–250 bases and can distinguish molecules that differ in length by just a single base. With lower density polyacrylamide gels, the range of size resolution moves upward to as high as 1000 bp. High-density agarose gels separate DNA molecules in the 500- to 2000-bp (or 0.05- to 2-kb) range; moderate-density agarose gels accomplish separation in the 1- to 2-kb range; and low-density agarose gels, in the 4- to 20-kb range. Because electrophoretic resolution is based on a percentage of the differences in size, gels that separate larger pieces cannot distinguish between molecules that differ by only 1 bp. Rather, they have the capacity to differentiate molecules that vary by about 1% of their length. For example, a low-density gel can distinguish between molecules 18 kb (18,000 bp) and 18.5 kb (18,500 bp) long but not between molecules 18,000 and 18,001 bp in length.

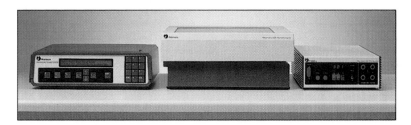

(c) Pulsed-field gel electrophoresis. A specialized protocol known as pulsed-field gel electrophoresis (PFGE) expands the separating capability of agarose gels to accommodate DNA molecules up to 9000 kb in length. In PFGE, timed pulses of current are applied along two alternating paths through the gel; the paths are separated by an angle of 90° to 120°. Each DNA molecule must change the direction in which it is moving with each change in the direction of the current. The larger the DNA molecule, the longer it takes to change direction and begin to move forward again. This difference in adaptability to current changes accounts for the separation capacity of the PFGE protocol. Researchers use longer pulse periods to separate larger DNA molecules.

Restriction Maps Provide a Rough Roadmap of a Clone

A **restriction map** shows the order of different size fragments within a clone. It consists of a linear diagram of the sites along a DNA segment at which one or another restriction enzyme cleaves the molecule.

There are numerous approaches to the derivation of a restriction map. Here, we describe one of the most commonly used procedures. In it researchers infer relative positions by cutting the same DNA sample with two or more restriction enzymes, singly and in combination. They then subject these various digested samples (referred to as *digests*) to electrophoresis in adjacent wells of a gel and compare the number and sizes of the pieces migrating in each lane. If the relative arrangements of sites for the different restriction enzymes employed does not create too many fragments, the data obtained can provide enough information to piece together a unique, self-consistent linear map showing the order and position of each DNA fragment in the clone. Figure 8.18 shows the gel pattern and inferred restriction map for a recombinant phage analyzed with the enzymes *Eco*RI and *Bam*HI. Working through the problems related to restriction mapping will help you gain a more thorough understanding of the protocol.

A restriction map provides landmarks that are useful for further analysis. You might want to know, for example, where within a DNA clone a particular gene is located. By developing a restriction map and then using the Southern blot analysis described next to determine which restriction fragment carries a segment complementary to the gene, you can position the gene relative to the rest of the genetic material in the clone.

Hybridization Can Also Serve as a Tool of Characterization

We have seen how researchers use hybridization to identify within the thousands of clones in a library, the particular clones carrying DNA sequences complementary to specific probes. Hybridization can also provide information on the location of a gene.

Suppose you had two clones—a genomic cosmid DNA clone, already mapped with restriction enzymes, that you knew contained a copy of the 1.4-kb human β-globin gene, and a cDNA clone corresponding to the smaller β-globin RNA transcript. You want to use these clones to discover the location of the gene within the cosmid as well as to learn which parts of the full gene are exons and which parts are introns. To answer both of these questions in a very straightforward fashion you could turn to a hybridization technique called the **Southern blot,** named for Edward Southern, the British scientist who developed it. With a Southern blot, you can transfer DNA fragments from a gel after electrophoresis to a wafer-thin nitrocellulose filter paper in such a way that the filter records the migration positions of all the DNA fragments. Hybridization of a probe to the blot then shows the sizes of restriction

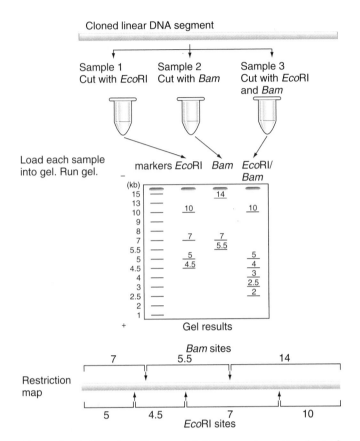

Figure 8.18 How to infer a restriction map from the sizes of restriction fragments produced by two restriction enzymes. Divide a purified preparation of cloned DNA into three aliquots; expose the first aliquot to *Eco*RI, the second to *Bam,* and the third to both enzymes. Now separate by gel electrophoresis the restriction fragments that result from each digestion and determine their sizes in relation to defined markers separated on the same gel. Finally, use a process of elimination to derive the only arrangement that can account for the results obtained with all three samples. To make sure that you understand how the restriction map at the bottom of the figure is the only arrangement that can account for all the data, explain how the pattern of restriction fragments in all three lanes of the gel is consistent with the map.

fragments containing sequences homologous to the probe. Figure 8.19 illustrates the details of the technique.

To locate the β-globin gene among the fragments of the cosmid clone, you would incubate a Southern blot of the restriction-enzyme-digested and electrophoresed cosmid fragments with purified β-globin cDNA that had been denatured and labeled with radioactivity. The cDNA probe would hybridize only to those restriction fragments in the blot that contained homologous DNA. You would now expose the Southern blot to a piece of X-ray film for several hours. After exposure, the film would show a set of dark bands representing the fragments to which the labeled cDNA probe had hybridized. By lining these bands up with their counterparts in a photograph of the original ethidium-bromide-stained gel (see Fig. 8.17a), you could discover which of the ethidium-bromide-stained fragments corresponds to the fragments that hybridized to the cDNA probe.

It is possible to interpret the information from a Southern blot hybridization in light of the restriction map of the genomic DNA insert in the cosmid. The span between the left- and rightmost fragments that hybridize to the cDNA probe define the transcriptional unit, that is, the region of genomic DNA that is expressed as the primary transcript of the β-globin gene. Fragments within this span that do not hybridize to the probe must be introns that are spliced out of the primary transcript to make the mature mRNA molecule, which served as the source of the cDNA. Fragments that hybridize to the cDNA must contain exon material, but the precise arrangement of introns and exons in these fragments is not clear. The use of additional restriction enzymes that subdivide such fragments in subsequent Southern blots could reveal more information about the arrangement of the gene's introns and exons. Ultimately, however, it is necessary to sequence both cDNA and genomic clones to define with precision the borders of introns and exons.

Southern blotting can even locate the β-globin gene among the entire, uncloned expanse of DNA obtained from a human cell. Cutting the total human DNA with *Eco*RI would produce ~ 732,000 fragments. If you ran the fragments out on a gel and photographed the result, all you would see would be a smear, because it is impossible to distinguish 732,000 fragments spread over a distance of some 10 cm. But you could still blot the smear of fragments to a nitrocellulose paper and probe the resulting Southern blot with a labeled β-globin clone, which would pick out the bands containing the β-globin gene. If the DNA is cut into large fragments, the gene might reside in a single band. Thus, the Southern blot makes it possible to start with a very complex mixture and find the one or two fragments in a million that are of interest to you. Once you find them, you can determine their size. If the DNA of an individual produces two fragments of different sizes bearing the β-globin gene, this indicates either that the restriction enzyme has cut the gene into two fragments or that the individual is a heterozygote for a sequence within or flanking the β-globin gene.

Sequencing Provides the Highest Resolution of a Cloned DNA Fragment: A Complete Description of Its Genetic Information

Determining the nucleotide sequence of a genome's cloned counterparts makes it possible to understand the ultimate structure and evolutionary history of the original DNA. In principle, there is no limit to the length of DNA that can be sequenced, although current sequencing technologies can analyze approximately 600–800 bp at a shot. Since the average unit of sequencing is thus around 700 bp, but there are, for example, 130 million nucleotides in the average human chromosome, sequencing is a bit like looking through the microscope at the encyclopedia and seeing only five letters at a time. Given this limitation, how do you determine the sequence of

nucleotides across the entire length of an insert derived from a lambda, cosmid, or YAC recombinant clone?

Sequencing: General Principles of the Procedure

DNA sequence analysis is a three-component process (Fig. 8.20). One component is the conversion of the fragment under analysis into a complete set of subfragments that differ in length by a single nucleotide from the preceding and succeeding fragments; this graduated set of fragments is known as a *nested array* (Fig. 8.20a). A second component is the labeling of each fragment with one of four tags, depending on its terminal base (Fig. 8.20b). The final component is the separation of fragments by size, followed by the reading of the tag present in each of the successively larger DNA fragments to obtain the sequence (Fig. 8.20c).

As Fig. 8.20a.1 shows, there are two basic ways to produce the fragments of a nested array: chemical cleavage, a technique developed by A. M. Maxam and W. Gilbert, and enzymatic extension, developed by F. Sanger. The chemical method uses different chemicals to cleave the DNA after a particular base or bases (either A, G, T, or C); it requires the control of conditions such that, at most, only a single cleavage occurs in each DNA molecule. The enzymatic method depends on the ability to terminate DNA synthesis with a nucleotide triphosphate lacking a 3'-hydroxyl group that is critical for chain extension; conditions are controlled so that the growing DNA chains are terminated at random at sites having the base corresponding to the chain-terminating nucleotide triphosphate.

In labeling the fragments, you can use four different fluorescent labels, one for each 3'-terminating base (as shown in Fig. 8.20b.1). With this method, you read the sequence by determining the color of the bands as they migrate past a fixed position at the bottom of the gel (the shorter fragments move faster than the longer ones) (Fig. 8.20c.1). A second approach to labeling uses just a single label (fluorescence or radioactivity) and requires the use of separate reactions to generate the fragments that end with a G, C, A, and T (Fig. 8.20b.2). You run the products of the four reactions in four adjacent gel lanes, and record the positions of the bands in the four lanes to identify the DNA sequence (Figure 8.20c.2).

Maxam and Gilbert and Sanger developed their procedures for the rapid sequencing of DNA molecules in the 1970s. Both procedures generate four populations of nested fragments, one for each of the four bases—A, T, C, or G—and both rely on electrophoresis in polyacrylamide gels as a means of determining the exact length of each of the fragments in these populations. Figure 8.21 illustrates the Sanger method, the more widely used of the two.

The accuracy of both the Sanger and the Maxam-Gilbert procedures approaches 99.9%, with only 1 misidentified base per 1000 bases sequenced. By sequencing the complementary region from both strands of a DNA segment, it is possible to achieve even greater accuracy. Various techniques of automation help increase the speed and scope of sequencing (Fig. 8.22).

To make a Southern blot, you first digest a cosmid clone with restriction enzymes whose recognition sites have been mapped; this generates a series of restriction fragments that, after gel electrophoresis, can be visualized by ethidium bromide staining.

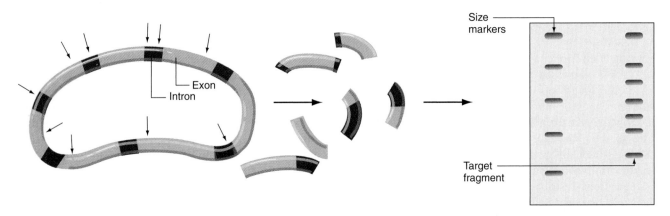

Next you place the gel in a strongly alkaline solution that causes the DNA to denature (that is, to separate into single strands), then in a neutralizing solution. You now cover the gel containing the separated DNA restriction fragments with a piece of nitrocellulose filter paper. On top of the filter paper, you place a stack of paper towels, and beneath the gel, a sponge saturated with buffer. Within this setup, the paper towels act as a blotter, pulling liquid from the buffer-saturated sponge, through the gel, and into the towels themselves. As the liquid is drawn out of the gel, so too is the DNA. But the large DNA molecules do not pass through the filter into the paper towels. Instead, they become trapped in the nitrocellulose at points directly above their locations in the gel, forming a Southern blot, which consists of the nitrocellulose filter containing DNA fragments in a pattern that is a replica of their migration pattern in the gel.

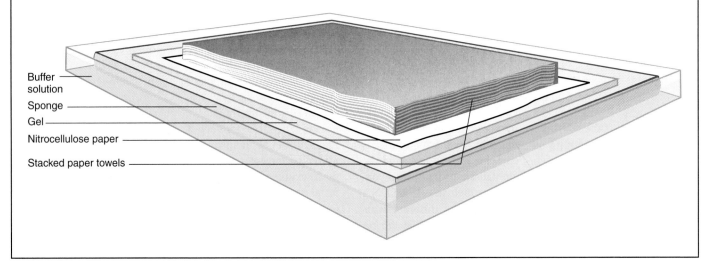

Feature Figure 8.19   Southern blot analysis.

There Are Two Basic Strategies for Sequencing a Clone: Directed and Shotgun

The **directed strategy of DNA sequencing** marches down a fragment of DNA step by step. A common approach to directed sequencing is **primer walking,** which begins with the application of two vector-based primers, flanking the two ends of an insert (Fig. 8.23a). The primers make it possible to sequence into the unknown insert for 600–800 nucleotides. You can use the information obtained at the end of each sequence to synthesize new primers, repeating the process until you have sequenced both strands of the entire insert.

The **shotgun approach to sequencing** begins with the partial digestion of the DNA with a four-base cutter or by in-

tense shearing with high-frequency, high-intensity sound waves, to break a large insert *at random* into many smaller fragments (Fig. 8.23b). The next step is to subclone a large number of these random fragments, obtain sequences for individual subclones at random, then enter these sequences into a computer programmed to look for sequence overlaps. If the number of sequenced fragments is large enough, it becomes possible to find overlaps of sequences extending across the entire original clone. The shotgun approach thus relies on redundancy—you have to gather sequence information on three to four times the actual number of base pairs in the original clone to make it work. A large sequencing project usually requires a combination of both directed and shotgun approaches.

Finally, baking the blot at 80°C or exposing the filter to ultraviolet (UV) radiation causes a crosslinking between DNA and filter that affixes the DNA fragments to the filter. The filter is then incubated with a radioactive probe that hybridizes to the sequence of interest. After immersion in water has washed away all nonhybridized probe

molecules, the hybridized probe remains attached to complementary DNA sequences that have previously crosslinked to the nitrocellulose filter. The filter is then dried and exposed to film to reveal the position or positions of the sequence of interest.

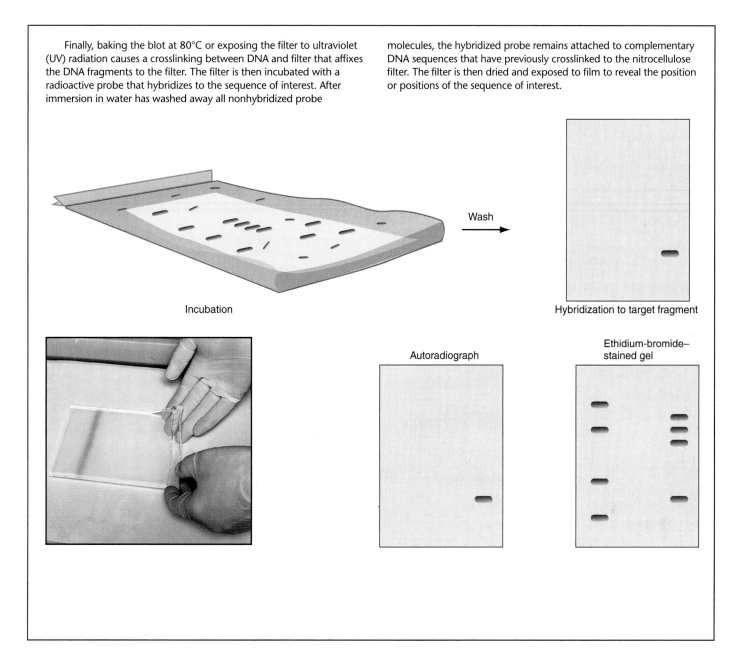

Incubation

Wash

Hybridization to target fragment

Autoradiograph

Ethidium-bromide–stained gel

Computer Analysis of DNA Sequences Can Identify Significant Genetic Motifs as well as Resemblances to Previously Determined Sequences

Data for the sequenced segments of a genome are stored in computers. In 1996, computerized databases in Europe and the United States carried more than 320 million base pairs of human genome sequences; by 1998 that number was 640 million and expected to double every two years. Teams of computer experts and geneticists have used these data to construct programs for identifying the biologically significant features of the sequences. Some of the programs already developed can distinguish exons, introns, and regulatory sequences. Others

can determine whether a new DNA sequence exhibits a statistically significant similarity to any previously identified sequence stored in a DNA database, and whether it contains a recognition site for a restriction enzyme or a binding site for a regulatory protein. Still other programs can project what type of polypeptide a particular sequence might yield.

The Polymerase Chain Reaction Can Amplify Small DNA Fragments of Partially Known Sequence for Further Analysis

Genes and sequences of interest within genes are rare targets— one among thousands or hundreds of thousands of unique

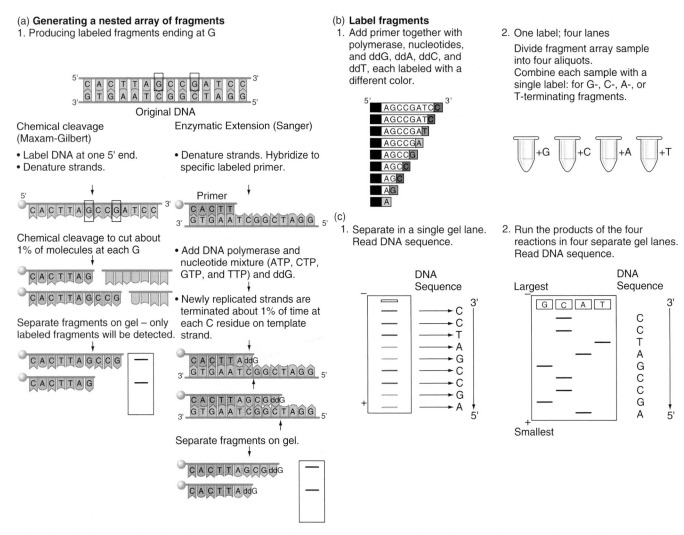

Figure 8.20   General principles of DNA sequencing. (a) All sequencing protocols rely on the production of nested arrays of fragments; the fragments all begin at the same 5′ position in a sequence but end at different 3′ positions. (1) There are two basic ways to produce the fragments: chemical cleavage, a technique developed by A. M. Maxam and W. Gilbert, and enzymatic extension, developed by F. Sanger. The chemical method uses specific chemicals to cleave the DNA after each type of base—A, G, T, or C; it requires the control of conditions such that cleavage affects only a small fraction of the bonds at a particular base. The enzymatic method depends on the ability to terminate DNA synthesis with a nucleotide base lacking a hydroxyl group that is critical for chain extension; conditions are controlled so that only a small fraction of the growing DNA chains are terminated with the addition of the chain-terminating base. (2) To generate nested arrays from the fragments produced by both of these procedures, researchers use polyacrylamide gel electrophoresis under conditions that allow the separation of DNA molecules differing in length by a single nucleotide. (b.1) It is possible to generate nested fragments ending in all four nucleotides in a single reaction, with each of the terminating nucleotides labeled with a different colored fluorescent dye. (b.2) Alternatively, four separate reactions can be performed with a different similarly labeled terminating nucleotide in each reaction. (c.1) If you generate a nested array in a single reaction, you can separate the reaction products in a single lane, with the color of each fragment indicating the associated terminating nucleotide. (c.2) Alternatively, if you use four different reactions to generate the nested array, you must use four adjacent lanes to separate the reaction products; at each position on the gel, a band will appear in only one of four lanes; its presence specifies the nucleotide at that position.

sequences—in a complex genome. Cloning overcomes this problem by reproducing large amounts of a specific DNA fragment, but it is a tedious, labor-intensive process. Once a sequence is known, or even partially known, an alternative way to achieve the same goal without cloning is the **polymerase chain reaction,** or **PCR.** First developed in 1985, PCR is faster, less expensive, and more flexible in application than cloning. It is also extremely efficient. We have seen that in creating a ge-

nomic or cDNA library, it is necessary to use a large number of cells from one or more tissues as the source of DNA or mRNA. By contrast, the single copy of a genome present in one sperm cell, or the minute amount of severely degraded DNA recovered from the wing muscle of a bee entombed in amber more than 20 million years ago provides enough material for PCR to make 100 million or more copies of a target DNA sequence in an afternoon.

How PCR Achieves the Exponential Accumulation of Target DNA

The polymerase chain reaction is a kind of reiterative loop in which an operation is repeatedly applied to the products of earlier rounds of the same operation. You can liken it to the operation of an imaginary generously paying automatic slot machine. You start the machine by inserting a quarter, at which point the handle cranks, and the machine pays out two quarters; it then reinserts those two coins, cranks, and produces four quarters; reinserts the four, cranks, and spits out eight coins, and so on. By the twenty-second round, this fantasy machine delivers more than 4 million quarters.

The PCR operation exploits the essential features of DNA replication described in Chapter 5, directing DNA polymerase to synthesize a specific region of a starting DNA molecule designated by specific primers. This initial replication is followed by subsequent rounds in which both the starting DNA and the copies synthesized in previous steps become templates for further replication, resulting in an exponential increase by doubling the number of copies of the replicated region with each step. Figure 8.24 diagrams the steps of the PCR operation, showing how you could use it to obtain many copies of a small portion of the β-globin gene for further study.

PCR Products Can Be Analyzed and Utilized Just Like Cloned Restriction Fragments

When properly executed, PCR provides all the highly enriched DNA you could want for unambiguous analyses of many types. PCR products can be digested with restriction enzymes to generate fragments that are then run on a gel and mapped. PCR products can also be labeled to produce hybridization probes, or they can function in Southern blots as the unlabeled sequence whose relatedness to other DNA molecules is checked by labeled probes. Finally, any DNA sequencing technique can be applied to a PCR product to determine the exact genetic information it contains. Because the products are obtained without cloning, it is possible to copy and learn the sequence of a specific DNA segment in a very short time. In fact, in checking for a particular hemoglobin mutation, one could start with a blood sample and determine a DNA sequence within two days.

As an analytic tool, PCR has several advantages over cloning. First, it provides the ultimate in sensitivity: The minimum input for the polymerase chain reaction to proceed is a single DNA molecule. Second, as we have seen, it is very fast, requiring no more than a few hours to generate enough amplified DNA for analysis; by comparison, the multiple steps of cloning usually require at least a week to complete. Third, in the sometimes highly competitive research world, it is an agent of democracy: Once the base sequence of the oligonucleotide primers that define a particular target sequence appears in print, anyone with funds to synthesize or buy these primers can reproduce the reaction. PCR is, nevertheless, unsuitable in certain situations. Because the protocol only copies DNA fragments up to ~ 25 kb in length, it cannot amplify larger regions of interest. And because the synthesis of PCR primers depends on sequence information for the target region, the protocol cannot serve as the starting point for the analysis of genes or genomic regions that have not yet been cloned.

PCR Has Many Uses

PCR is one of the most powerful techniques in molecular biology. Its originator, Kary Mullis, received the 1993 Nobel prize in chemistry for his 1985 invention of this tool for genetic analysis. By making it possible to amplify a specific sequence directly from complex DNAs (either genomic or cDNA), PCR has made molecular analysis an essential component of genotype detection and gene mapping; we describe its applications in these areas in Chapters 9 and 10. By making it possible to analyze traces of partially degraded DNA from ancient samples and to compare homologous sequences from a variety of sources, PCR has revolutionized evolutionary studies, enabling researchers to analyze sequences from both living and extinct organisms and to determine the relatedness between these organisms with greater accuracy than ever before. By making it possible to analyze the same sequence in hundreds or thousands of samples without having to make hundreds or thousands of libraries, PCR has changed the face of population genetics, greatly simplifying the process of monitoring the genetics of a group over time. Finally, PCR has helped bring molecular genetics to many fields outside of genetics. The following example of its use in diagnosing infectious disease provides an inkling of its potential impact on medicine.

AIDS, like other viral diseases, though not inherited, is in one sense a genetic disease, because it is caused by the activity of foreign DNA inside a subgroup of somatic cells. HIV, the virus associated with AIDS, gains entry to a person's body through the bloodstream or lymphatic system, then docks at specific membrane receptors on a few types of white blood cells, fuses with the cell membrane, and releases its RNA chromosome along with several copies of reverse transcriptase into the cell (see the Genetics and Society box in Chapter 7). Once inside the cell, the reverse transcriptase copies the RNA to cDNA. This double-stranded DNA copy of the viral genome then integrates itself into the host genome where, known as a *provirus,* it can lie latent for up to 10 years or become active at any time; when activated, it directs the cellular machinery to make more viral particles.

Standard tests for HIV detect antibodies to the virus, but it may take several months for the antibodies produced by an infected person's immune system to reach levels that are measurable in the blood. Then, in another few months, when ongoing viral activity inside many types of circulating white blood cells subsides, most of the antibodies may disappear from the circulation. This is because once the viral particles have entered the latent state, they are literally in hiding and able to avoid detection by the immune system.

With PCR, it is possible to detect small amounts of virus circulating in the blood or lymph very soon after infection,

Begin by mixing the purified, denatured DNA of a cloned insert with a labeled single-stranded primer that hybridizes (that is, base pairs) to a particular complementary site on one strand of the cloned insert. The cloned DNA has been denatured to single strands by treatment with heat.

Next divide the mixture of cloned DNA and labeled primer into four aliquots and add a special preparation of nucleotides along with DNA polymerase to catalyze the polymerization of nucleotides from the primer. The special nucleotide preparation is spiked with a chain-terminating dideoxynucleoside analog of one of the four bases. When these analog nucleosides combine with a phosphate, the result is an analog nucleotide. One aliquot, for example, receives nucleotides A, T, C, and G and an analog of A; a second aliquot receives A, T, C, and G and an analog of T; and so forth.

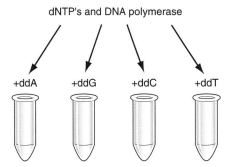

Because the analogs have no oxygen at the 3' position in the sugar, they prevent the further addition of nucleotides to the strand and thus terminate a growing chain wherever they become incorporated in place of an actual nucleotide. The aliquot that has the dideoxy form of thymidine, for example, will eventually contain a population of DNA molecules that terminate at each of the thymidines in the original template strand under analysis.

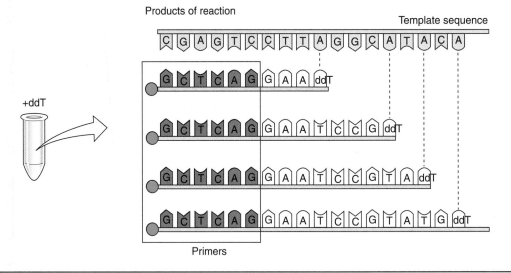

 Feature Figure 8.21   The Sanger Method of Sequencing.

Then use electrophoresis on a polyacrylamide gel to separate the fragments in each of the four aliquots and arrange them by size. The resolution of the gel is such that you can distinguish DNA molecules that differ in length by only a single base. The appearance of a DNA fragment of a particular length demonstrates the presence of a particular nucleotide at that position in the strand.

Gel analysis of fragments

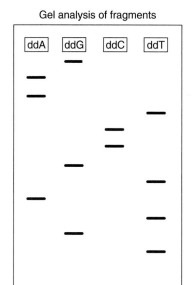

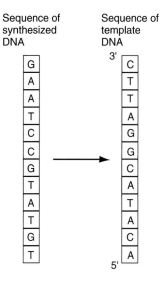

Suppose, for example, that the aliquot polymerized in the presence of dideoxythymidine shows fragments 32, 35, and 39 bases in length. These fragments indicate that thymidine is present at those positions in the strand of nucleotides. In practice, one does not independently determine the exact lengths of each fragment. Instead, one starts at the bottom of the gel, looks at which of the four lanes has a band in it, records that base, then moves up one position and determines which lane has the next band, and so on. In this way, it is possible to read several hundred bases from a single set of reactions.

(a) Automated sequencing

1. Generate nested array of fragments; each with a fluorescent label corresponding to the terminating 3' base.

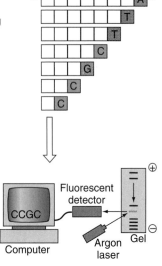

2. Fragments separated by electrophoresis in a single vertical gel lane.

3. As migrating fragments pass through the scanning laser, they fluoresce. A fluorescent detector records the color order of the passing bands. That order is translated into sequence data by a computer.

Computer · Argon laser · Gel · Fluorescent detector · CCGC

(b)

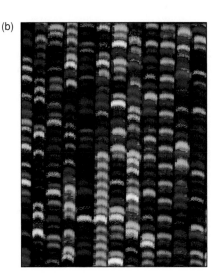

(c)

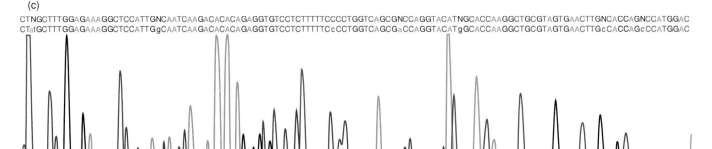

CTNGCTTTGGAGAAAGGCTCCATTGNCAATCAAGACACACAGAGGTGTCCTCTTTTTCCCCTGGTCAGCGNCCAGGTACATNGCACCAAGGCTGCGTAGTGAACTTGNCACCAGNCCATGGAC
CTatGCTTTGGAGAAAGGCTCCATTGgCAATCAAGACACACAGAGGTGTCCTCTTTTTCcCCTGGTCAGCGaCCAGGTACATgGCACCAAGGCTGCGTAGTGAACTTGcCACCAGcCCATGGAC

Figure 8.22 Automated sequencing. (a) For automated sequencing, the Sanger protocol is performed with all four individually labeled terminating nucleotides present in a single reaction. At completion of the reaction, DNA fragments terminating at every base in the sequence will be present and color coded according to the identity of the terminating nucleotide. Separation by gel electrophoresis is next. A laser and accompanying detector are positioned near the bottom of the polyacrylamide gel; as each fragment moves past the laser, the beam determines the color of the fluorescent label on the terminal base and records that color in a computer. (b) Output from automated sequencing machine. Each lane displays the sequence obtained with a seperate DNA sample and primer. (c) The computer reads the sequence of the strand complementary to the template strand from right to left, which corresponds to the 5'-to-3' direction. The machine then automatically generates the top, complementary sequence, recording any ambiguity in the base call as an "N." A technician can resolve most such ambiguities by direct examination of the fluorescence signals. If the technician concludes with high certainty that a particular N is, for example, the base G, he or she replaces that N with a "g" in the bottom sequence.

before antibody production is in full swing. PCR can also detect viral DNA incorporated in the genome of any cell, picking up as little as 1–10 copies of viral DNA per million cells. Thus, with PCR, it becomes possible to confirm and begin treating HIV infection during the critical period before antibodies reach measurable levels. It also becomes possible to follow the progress of each person's HIV infection and tailor therapies accordingly, using a large dose of certain drugs to combat a large amount of viral activity, but small doses of perhaps other drugs to prevent a small number of cells from emerging from latency.

In summary, moving from lower-resolution mapping to higher-resolution protocols that reveal the details of DNA sequence, we have seen that different techniques of characterization provide different kinds of information about the DNA of a genome. Restriction mapping helps position genomic fragments in relation to each other. Hybridization gives information about the content of a particular fragment within a DNA molecule by revealing whether or not the fragment's sequence is related to a previously characterized nucleic acid probe. Sequencing extracts the ultimate information—the order of bases in a fragment. Once the base sequence of a par-

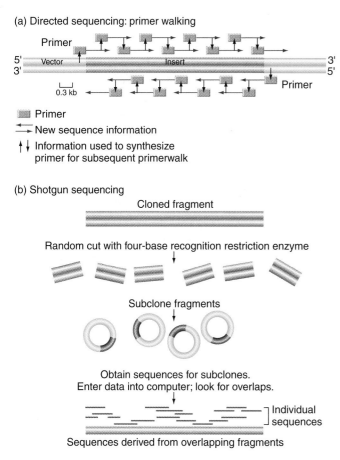

(a) Directed sequencing: primer walking

Primer

5'
3'
Vector   Insert
3'
5'

0.3 kb

Primer

■ Primer

→ New sequence information

↑↓ Information used to synthesize
primer for subsequent primerwalk

(b) Shotgun sequencing

Cloned fragment

Random cut with four-base recognition restriction enzyme

Subclone fragments

Obtain sequences for subclones.
Enter data into computer; look for overlaps.

┐ Individual
┘ sequences

Sequences derived from overlapping fragments

Figure 8.23   Two approaches to identifying the sequence of a long DNA molecule: Directed and shotgun. (a) In directed sequencing by primer walking, sequencing begins from both ends of a cloned insert with primers derived from the vector arms. After each round of sequencing, new primers are designed from the sequences obtained in the previous round. (b) In shotgun sequencing, after random fragments obtained from a cloned insert are sequenced, a computer aligns them to each other.

ticular locus has been determined, it becomes possible to develop a PCR assay capable of picking out any allelic version of that locus that is present in a sample of total genomic DNA; this is because if the primer templates find a match, PCR will amplify any sequence between the primers, even those with a slightly different order of bases. Automation of the PCR procedure makes it possible to screen thousands of samples in a short time. Researchers combine these techniques in a variety of ways to study the genes of an organism.

UNDERSTANDING THE GENES FOR HEMOGLOBIN: A COMPREHENSIVE EXAMPLE

Geneticists have used the tools of recombinant DNA technology to analyze the clusters of related genes that make up the α- and β-globin loci. In a variety of studies, they cut the chromosomal segments carrying the hemoglobin genes;

cloned, selected, and characterized the fragments; then used PCR to amplify for sequencing the genes present in many individuals, both normal and diseased. Fundamental insights from these studies have helped explain how the linear information of DNA encompasses all the instructions for the development of the hemoglobin system, including the changes in globin expression during normal development. The studies have also clarified how the globin genes evolved, and how a large number of different mutations produce the phenotypic permutations that give rise to a range of globin-related disorders.

The Genes Encoding Hemoglobin Occur in Two Clusters on Two Separate Chromosomes

An α-globin gene cluster containing three functional genes spans about 28 kb on chromosome 16 (Fig. 8.25a). Moving in the 5'-to-3' direction along the RNA-like strand, the α or α-like genes appear in the order ζ, α-2, α-1. A β-globin gene cluster covers 50 kb on chromosome 11 and contains five functional genes in the order ε, G_γ, A_γ, δ, and β (Fig. 8.25b). Geneticists refer to the chromosomal region carrying all of the α-globin-like genes as the **α-globin locus** and the region containing the β-globin-like genes as the **β-globin locus.** Note that the term **locus** signifies a location on a chromosome; that location may be as small as a single nucleotide or as large as a cluster of related genes.

The Succession of Genes in Each Cluster Correlates with the Sequence of Their Expression during Development

All of the genes in the α- and β-gene clusters point in the same direction; that is, they all use the same strand of DNA as the template for transcription. Moreover, their linear organization reflects the order in which they are expressed during development. For the α-like chains, that temporal order is ζ during the first five weeks of embryonic life, followed by the two α chains during fetal and adult life. For the β-like chains, the order is ε during the first five weeks of embryonic life; then the two γ chains during fetal life; and finally, within a few months of birth, mostly β but also some δ chains.

The fact that the order of genes on the chromosomes reflects the order of their expression during development suggests that whatever mechanism turns these genes on and off takes advantage of their relative positions. Although we still do not know what this mechanism is, the DNA structure of a mutation in the β locus provides a strong clue to one component of the control switch. Some rare adults have no β or δ chains in their red blood cells, a usually lethal situation; yet they remain healthy because their particular deletion allows expression of the fetal γ chains to continue into adulthood instead of shutting off around the time of birth. Studies have revealed that the DNA of these relatively healthy individuals has sustained large deletions that removed both the δ and the β genes; it is likely that the deleted material also carries the regulatory sequence responsible for shutting off the fetal γ genes in people without the deletions.

Suppose you are a physician and wish to understand the molecular details of the β-globin gene mutation that causes the expression of a novel form of anemia in one of your patients. Although the sequence of the normal β-globin gene as well as the sequence of many of its mutant alleles is known, you suspect that your patient may carry a new, as-yet-uncharacterized β-globin allele of unknown sequence. To characterize the potentially novel allele, you could turn to PCR. You begin by preparing a small amount of genomic DNA from skin, blood, semen, or other tissue that is easy to obtain from your patient suffering from the novel anemia. You then synthesize two specific

oligonucleotide primers, each a short single-stranded chain of 16–26 nucleotides, whose sequence is dictated by the already known sequence of the wild type β-globin allele. One of these oligonucleotides is complementary to one strand of a part of the β-globin gene to the left of the region you believe might contain the "target sequence," in this case, the mutation; the other oligonucleotide is complementary to the opposite DNA strand from a part of the β-globin gene to the right of the region of interest. As you will see, the target DNA amplified by PCR is that stretch of the genome lying between the two primers.

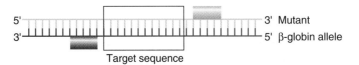

Next put the patient's genomic DNA in a test tube along with the specially prepared primers, a solution of the four deoxynucleotides, and *Taq* DNA polymerase, a specialized polymerase obtained from *Thermus aquaticus* bacteria living in hot springs. This specialized DNA polymerase remains active at the high temperatures employed during

the PCR protocol, which is not true of polymerases from organisms that live under more moderate conditions. Now place the test tube with these components in a machine called a thermal cycler, which repeatedly changes the temperature of incubation according to a preset program.

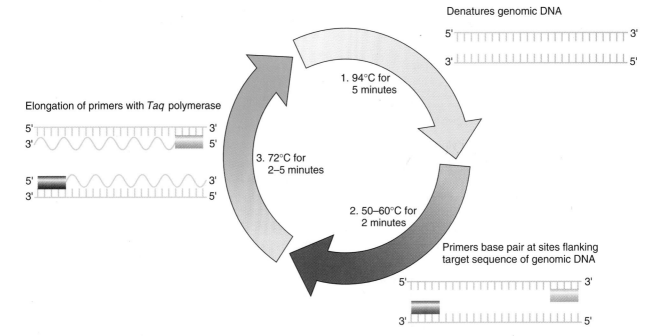

In a typical program, the cycler heats the solution to 94°C for 5 minutes. At this temperature, the target DNA separates into single strands. The temperature next lowers to 50–60°C for 2 minutes to allow the primers to base pair with complementary sequences in the single-stranded genomic DNA. Specifics of both temperature and timing within these ranges depend on the length and GC:AT ratio of

the primer sequences. The thermal cycler then raises the temperature to 72°C, the temperature at which the *Taq* polymerase functions best. Holding the temperature at 72°C for 2–5 minutes (depending on the length of the target sequence) allows DNA synthesis to proceed. At the end of this period, with the completion of DNA synthesis, the first round of PCR is over.

Feature Figure 8.24 The Polymerase Chain Reaction: How PCR works

A Variety of Mutations Accounts for the Diverse Symptoms of Globin-Related Diseases

By comparing DNA sequences from affected individuals with those from normal individuals, researchers have learned that there are two general classes of disorders arising from hemo-

globin alterations. In one, mutations change the amino-acid sequence and thus the three-dimensional structure of the α- or β-globin chain, and these structural changes result in an altered protein whose malfunction causes the destruction of red blood cells. Diseases of this type are known as hemolytic ane-

To start the next round, the cycler again raises the temperature to 94°C, but this time for only about 20 seconds, to denature the short stretches of DNA consisting of one of the original strands of genomic DNA and a newly synthesized complementary strand initiated by a primer. These short single strands become the templates for the second round of replication because the synthesized primers are able to anneal to them.

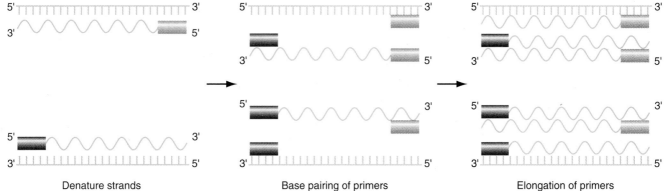

Denature strands   Base pairing of primers   Elongation of primers

The machine repeats the cycle again and again, generating an exponential increase in the amount of target sequence: 22 repetitions produce over a million copies of the target sequence; 32 repetitions over a billion. The length of the accumulating DNA strands becomes fixed at the length of the DNA between the 5' ends of the two primers, as shown. This is because, beginning with round 3, the 5' end of a majority of templates is defined by a primer that has been incorporated in one strand of a PCR product.

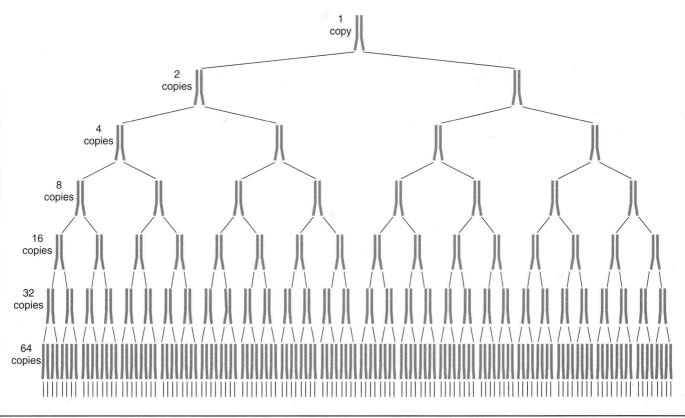

mias (Fig. 8.26a.1). An example is sickle-cell anemia, caused by an A-to-T substitution in the sixth codon of the β-globin chain (Fig. 8.26a.2 and 3). This simple change in DNA sequence alters the sixth amino acid in the chain from glutamic acid to valine, which, in turn, modifies the form and function of the affected hemoglobin molecules. Red blood cells carrying these altered molecules often have abnormal shapes that cause them to block blood vessels or be degraded.

The second major class of hemoglobin-related genetic diseases arises from DNA mutations that reduce or eliminate

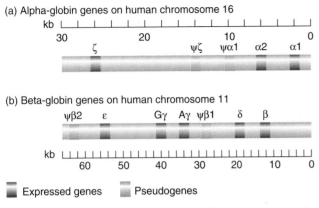

(a) Alpha-globin genes on human chromosome 16

(b) Beta-globin genes on human chromosome 11

■ Expressed genes   ■ Pseudogenes

Figure 8.25 The genes for the polypeptide components of human hemoglobin are located in two genomic clusters on two different chromosomes. (a) Schematic representation of the α-globin gene cluster on chromosome 16. The cluster contains three functional genes and two pseudogenes (designated by the symbol ψ); pseudogenes are sequences that look like but do not function as genes. (b) Schematic representation of the β-globin gene cluster on chromosome 11; this cluster has five functional genes and two pseudogenes.

the production of one of the two globin polypeptides. The disease state resulting from such mutations is known as thalassemia, from the Greek words *thalassa* meaning "sea" and *emia* meaning "blood"; the name arose from the observation that a relatively high rate of this blood disease occurs among people who live near the Mediterranean Sea. Several different types of mutation can cause thalassemia, including those that delete an entire α or β gene, those that alter the sequence in regions that are outside the gene but necessary for its regulation, or those that alter the sequence within the gene such that no protein can be produced. The consequence of these changes in DNA sequence is the total absence or a deficient amount of one or the other of the normal hemoglobin chains. Because there are two α genes that see roughly equal expression beginning a few weeks after conception, individuals carrying deletions within the α-globin locus may be missing anywhere from one to four copies of the α-globin genes (Fig. 8.26b.1). A person lacking only one would be a heterozygote for the deletion of one of two α genes; a person missing all four would be a homozygote for deletions of both α genes. The range of mutational possibilities explains the range of phenotypes seen in α-thalassemia. Individuals missing only one of four possible copies of the α genes are normal; those lacking two of the four have a mild anemia, and those without all four die before birth. The fact that the α chains are expressed early in fetal life explains why the α-thalassemias are detrimental *in utero*. By contrast, β-thalassemia major, the disease occurring in people who are homozygotes for most deletions of the single β gene, also usually results in death, but not until early childhood (Fig. 8.26b.2). Exceptions to this cause-and-effect relation between homozygous β-globin deletions and β-thalassemia major arise

when the deletions that eliminate the β-globin gene also eliminate repression in the adult of the fetal γ genes.

Comparisons of the altered DNA sequences from affected individuals with wild type sequences from normal individuals have helped illuminate the sequences necessary for normal hemoglobin expression. In some β-thalassemia patients, for example, disease symptoms arise from the alteration of a few nucleotides adjacent to the 5' end of the coding region for the β chain. Data of this type have defined sequences that are important for expression of the β-globin locus. One such segment is the TATA box, a sequence found in many eukaryotic promoters (Fig. 8.27a). In other thalassemia patients, the entire β-globin locus and adjacent regulatory segments, including the TATA box, are intact, but a mutation has altered a control region found far to the 5' side of all the β-like genes. Called the locus control region (LCR), this segment is necessary for a high level of tissue-specific expression of all β-like genes in red blood cell progenitors (Fig. 8.27b). Depending on how disruptive they are, mutations in the TATA box or the locus control region (LCR) produce β-thalassemias of varying severity. (Chapter 16 explains how these regulatory elements help control transcription.)

The α- and β-Globin Loci House Multiple Genes That Evolved from One Ancestral Gene

From sequencing the DNA of all the globin genes, researchers learned that the genes form a closely related group, or family, of genes that evolved by duplication and variation from one ancestral gene (Fig. 8.28). All gene sequences evolve continuously. The two separate products of a duplication, which start out identical, eventually diverge because their sequences evolve independently of each other. The members of a gene family may be grouped together on one chromosome (like the very closely related β-globin genes) or dispersed on different chromosomes (like the less closely related α- and β-globin genes). All the β-like genes are exactly the same length and have two introns at exactly the same positions. All the α-like genes also have two introns at exactly the same positions, but these positions are different from those of the β genes. The sequences of all the β-like genes are more similar to each other than they are to the α-like sequences, and vice versa. These comparisons suggest that an ancestral gene duplicated, and one copy of the gene moved to another chromosome. With time, one of the two gene copies gave rise to the α lineage, the other to the β lineage. Each lineage then underwent further duplications to generate the present array of three α-like and five β-like genes in humans. The duplications also produced genes that eventually lost the ability to function. Molecular geneticists made this last deduction from data showing two additional α-like sequences within the α locus and one β-like sequence within the β locus that no longer have the capacity for proper expression. Sequences that look like, but do not function as, a gene are known as **pseudogenes;** they occur in all higher eukaryotic genomes.

(a.1) Major types of structural variants causing hemolytic anemias

Name	Molecular basis of mutation	Change in polypeptide	Pathophysiological effect of mutation	Inheritance
Hb S	Single nucleotide substitution	β 6 Glu ↓ Val	Deoxygenated Hb S polymerizes → sickle cells → vascular occlusion and hemolysis	Recessive AR
Hb C	Single nucleotide substitution	β 6 Glu ↓ Lys	Oxygenated Hb C tends to crystallize → less deformable cells → mild hemolysis; the disease in Hb S:Hb C compounds is like mild sickle cell disease	Recessive AR
Hb Hammersmith	Single nucleotide substitution	β 42 Phe ↓ Ser	An unstable Hb → Hb precipitation → hemolysis; also low O_2 affinity	Dominant AD

(b.1) Clinical results of various α -thalassemia genotypes

Clinical condition	Genotype	Number of functional α-genes	α-chain production
Normal	αα/αα	4	100%
Silent carrier	αα/α–	3	75%
Heterozygous α-thalassemia— mild anemia	α–/α– or αα/– –	2	50%
HbH ($β_4$) disease— moderately severe anemia	α–/– –	1	25%
Homozygous α-thalassemia— lethal	– –/– –	0	0%

(a.2) Basis of sickle-cell anemia

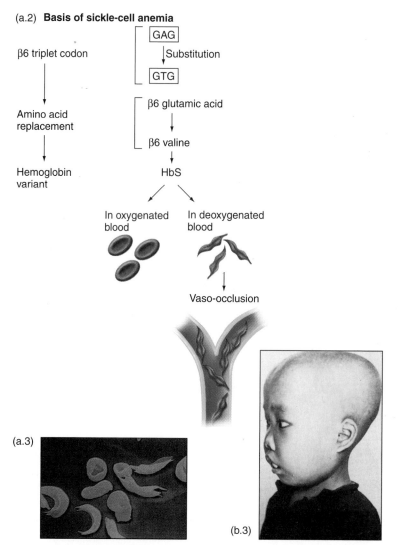

β6 triplet codon → Amino acid replacement → Hemoglobin variant

GAG ↓ Substitution GTG

β6 glutamic acid ↓ β6 valine → HbS

In oxygenated blood

In deoxygenated blood → Vaso-occlusion

(a.3)

(b.3)

(b.2) A β-thalassemia patient makes only α-globin, not β-globin.

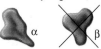

Four α subunits combine to make abnormal hemoglobin.

Abnormal hemoglobin molecules clump together, altering shape of red blood cells. Abnormal cells carry reduced amounts of oxygen.

To compensate for reduced oxygen level, medullary cavities of bones enlarge to produce more red blood cells.

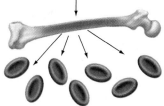

Spleen enlarges to remove excessive number of abnormal red blood cells.

Too few red blood cells in circulation result in anemia.

Figure 8.26 Mutations in the DNA for hemoglobin produce two classes of disease. (a.1) The major types of hemoglobin variants causing hemolytic anemias. (2) The basis of sickle-cell anemia. (3) Sickling red blood cells appear as crescents among more rounded nonsickling cells. (b.1) Thalassemias associated with deletions in the α-globin polypeptide. (2) The physiological basis of β-thalassemia major. (3) Child suffering from β-thalassemia major.

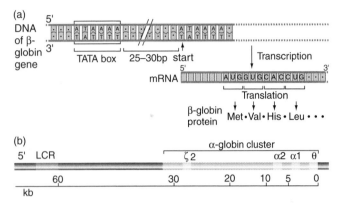

(a)

DNA of β-globin gene

TATA box   25–30bp   start

Transcription

mRNA   AUGGUGCACCUG···

Translation

β-globin protein   Met·Val·His·Leu···

(b)

α-globin cluster

5' LCR   ζ2   α2 α1 θ

60   30   20   10 5   0

kb

Figure 8.27 Mutations in globin regulatory regions can affect globin gene activity and result in thalassemia.
(a) Mutations in the TATA box associated with the β-globin gene can eliminate transcription and cause β-thalassemia. (b) A locus control region is present 25–50 kb upstream of the α-globin gene cluster. The function of the LCR is to open up the chromatin domain associated with the complete cluster of α-globin genes. In contrast, the regulatory regions next to each particular gene fine-tune gene expression during development. Mutations in the LCR can prevent expression of all the α-globin genes, resulting in severe α-thalassemia.

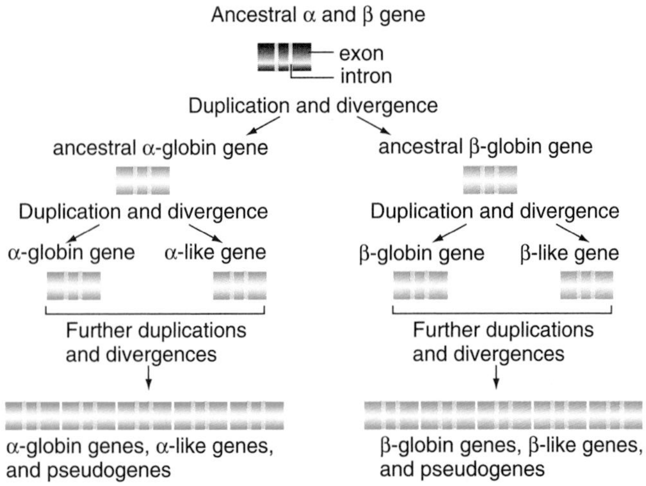

Ancestral α and β gene

exon
intron

Duplication and divergence

ancestral α-globin gene   ancestral β-globin gene

Duplication and divergence   Duplication and divergence

α-globin gene   α-like gene   β-globin gene   β-like gene

Further duplications and divergences   Further duplications and divergences

α-globin genes, α-like genes, and pseudogenes   β-globin genes, β-like genes, and pseudogenes

Figure 8.28 Evolution of the globin gene family. Duplication of an ancestral gene followed by divergence of the separate duplication products established the α- and β-globin lineages. Further rounds of duplication and divergence within the separate lineages generated the two sets of genes and pseudogenes of the globin gene family.

CONNECTIONS

The tools of recombinant DNA technology grew out of an understanding of the DNA molecule and its interaction with the enzymes that operate upon DNA in normal cells. Geneticists use the tools singly or in combination to look at DNA directly. Through cloning, restriction mapping, sequencing, hybridization, and PCR, they have been able to isolate the genes that encode, for example, the hemoglobin proteins; define their chromosomal positions; identify sequences near the genes that regulate their expression; determine the complete nucleotide sequence of each gene; and discover the changes in sequence produced by the hundreds of mutations that affect hemoglobin production. The results give a fascinating and detailed picture of how the nucleotides along a DNA molecule determine protein structure and function and how mutations in sequence produce far-ranging and varied effects on human health. The methods of classical genetics that we examined in Chapters 1–4 complement those of recombinant DNA technology to produce an integrated picture of genes and genomes at many levels.

Recombinant DNA techniques that enable us to look directly at the DNA of a chromosome make it possible to answer specific questions about the genotype of an individual, to discover, for instance, how many copies and which alleles of a gene a particular chromosome contains. These same techniques make it possible to construct maps of whole genomes and use the information in these maps to locate and isolate new genes without complex breeding experiments. The tools of recombinant DNA have thus revolutionized the discipline of genetics, and in particular, the practice of human genetics. Chapter 9 examines how geneticists use the new tools of genetic analysis to detect an individual's genotype for a particular gene or noncoding segment of DNA. Chapter 10 examines how researchers apply the new tools in conjunction with classical linkage analysis to map genes on chromosomes and to obtain the clone of a disease gene with no genetic information available other than that inferred from the observation of a disease phenotype.

ESSENTIAL CONCEPTS

1. An intact eukaryotic genome is too complex for most types of analysis. Geneticists have appropriated the enzymes that normally operate on the DNA molecule inside a bacterial cell and used them in the test tube to create the tools of recombinant DNA technology. *Restriction enzymes* cut DNA into manageable pieces, *ligase* splices the pieces together, *DNA polymerase*

makes copies, and *reverse transcriptase* copies RNA into DNA.

2. Cutting the DNA: Restriction enzymes cut DNA into restriction fragments, with each enzyme producing restriction fragments of a known average size.

3. Cloning the DNA fragments:

a. Restriction fragments can be spliced to *cloning vectors* to produce *recombinant DNA molecules.* A cloning vector is a DNA sequence that can enter a living cell, make its presence known, and provide a means of replicating and purifying both itself and any DNA to which it is spliced. Different types of vectors carry different size inserts, which can be used for different levels of analysis.

b. Once inside a living cell, vector-insert recombinants are replicated during each cell cycle (just as the cell's own chromosomes are), in a process known as cloning. The hundreds of thousands of cells arising from a single cell by consecutive divisions make up a *cellular clone.* The vector-insert recombinant molecules inside the cells of a clone, often referred to as *DNA clones,* can be purified by procedures that separate recombinant molecules from host DNA. The insert can then be cut away from the vector by restriction enzymes and purified.

4. Collecting the DNA clones:

a. *Genomic libraries* are random collections of vector-insert recombinants containing the DNA fragments of a given species. The most useful libraries carry four to five *genomic equivalents.*

b. *cDNA libraries* carry DNA copies of the RNA transcripts produced in a particular tissue. The clones in the library represent only that part of the genome transcribed and spliced into mRNA in the cells of the specific tissue.

5. Identifying and isolating desired clones: *DNA probes* can select segments of interest from a genomic or cDNA library. Antibody screening can select cDNAs encoding a particular protein from an *expression library.*

6. Characterizing the clones:

a. *Gel electrophoresis* distinguishes restriction fragments by size. Different techniques of electrophoresis separate molecules in different size ranges.

b. A *restriction map* shows the order of and distances separating restriction sites recognized by one or more enzymes.

c. *Hybridization* is the natural propensity of complementary DNA strands to form stable double helixes. Hybridization makes it possible to use labeled clones as probes. Southern blotting uses the hybridization of cloned probes to locate genes and other DNA segments of interest within complex mixtures.

d. *Sequencing* provides the ultimate description of a cloned fragment. Automation has increased the speed and scope of sequencing.

e. The *polymerase chain reaction* is a kind of reiterative loop that can make billions of copies of specific DNA segments without a requirement for DNA libraries. *PCR* products can be analyzed by all of the means used to characterize clones. Geneticists use PCR for the direct detection of genotype, mapping, medical diagnosis, and numerous other applications outside the field of genetics.

7. Geneticists use the tools of recombinant DNA technology in various combinations and orders, depending on the experimental task at hand.

SOCIAL AND ETHICAL ISSUES

1. A grocery chain in the Midwest has been receiving milk from the same dairy supplier for many years. Recently, that supplier as well as many of the dairy farmers in the Midwest have begun to give their cows the bovine somatotropin hormone (BST), a peptide that has been mass-produced using a cloned BST gene. Treatment with this growth hormone results in a 10 to 25% increase in milk production. Because milk production per cow is increased, the price of the milk is lower. The peptide is injected directly into cows and is barely detectable in the milk. Cows treated with the hormone do have an increased incidence of the relatively common disease mastitis that is treated with antibiotics. A consequence of increased mastitis is that small amounts of antibiotics may be more often found in the milk. There is concern that this exposure to antibiotics may cause an increase in antibiotic resistant bacteria in humans. The grocery chain would like to carry this milk because it has a lower price. Do they have a responsibility to customers to inform them about the conditions under which the milk was produced? How can they effectively inform the customers of the relevant scientific information?

2. Kerry Mullis, while working at Cetus Corporation, developed the polymerase chain reaction technique, one of the most powerful new tools of molecular biology in the 1980s. As an employee of the company, he had previously relinquished much of his right to royalties from the patent and received a compensation that is small relative to the value of the technique he developed. Should an employee be required to give most of the patent rights to the company? How should these rights be handled in a research university? If the idea belonged to one person, should that person alone profit? Is it possible to determine accurately the relative contributions of

individual researchers and of the institutions with which they are associated to the development of new ideas?

3. A new biotechnology company that has an interest in cloning genes involved in the immune response has approached a scientist at the university and wants to support some of his research. A stipulation of receiving the money is that the research cannot be published in journals until papers have been filed to cover any

potential patent. A postdoctoral fellow who has been doing much of the research on this project is currently looking for a job and wants to present this research in his seminars at job interviews. His boss has warned him that this may go against the boss' contract with the company. What should the postdoctoral fellow do? Is industry support of research in the university compatible with academic life in the university?

S O L V E D   P R O B L E M S

I. The following map of the plasmid cloning vector pBR322 shows the locations of the ampicillin (amp) and tetracycline (tet) resistance genes as well as two unique restriction enzyme recognition sites, one for *Eco*RI and one for *Bam*HI. You digested this plasmid vector with both *Eco*RI and *Bam*HI enzymes and purified the large *Eco*RI-*Bam*HI vector fragment. You also digested the cellular DNA that you want to insert into the vector with both *Eco*RI and *Bam*HI. After mixing the plasmid vector and the fragments together and ligating, you transformed an ampicillin-sensitive strain of *E. coli* and selected for ampicillin-resistant colonies. If you test all of your selected ampicillin-resistant transformants for tetracycline resistance, what result do you expect and why?

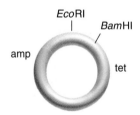

Answer

This problem requires an understanding of vectors and the process of combining DNAs using sticky ends generated by restriction enzymes.

The plasmid must be circular to replicate in *E. coli* and in this case a circular molecule will be formed only if the insert fragment joins with the cut vector DNA. The cut vector will not be able to religate without an inserted fragment because the *Bam*HI and *Eco*RI sticky ends are not complementary and cannot base pair. All ampicillin-resistant colonies therefore contain a *Bam*HI-*Eco*RI fragment ligated to the *Bam*HI-*Eco*RI sites of the vector. Fragments cloned at the *Bam*HI-*Eco*RI site interrupt and therefore inactivate the tetracycline resistance gene. *All ampicillin-resistant clones will be tetracycline sensitive.*

II. The gene for the human peptide hormone somatostatin (encoding nine amino acids) is completely contained on an *Eco*RI (5′ G^AATTC 3′) fragment, which can be cut out of the larger fragment shown below. (The ^ symbol indicates the site where the sugar-phosphate backbone is cut by the restriction enzyme.)

5′GCCGAATTCGATCCTATCAACACGAAGTGAAAGTCTTACAACCCATGAATTCGATTCG 3′

3′CGGCTTAAGCTAGGATAGTTGTGCTTCACTTTCAGAATGTTGGGTACTTAAGCTAAGC5′

a. What is the amino acid sequence of human somatostatin?

b. Indicate the direction of transcription of this gene.

c. The first step in synthesizing large amounts of human somatostatin for pharmacological treatments involves constructing a so-called "fusion gene," in which the N terminus of the protein encoded by the fusion gene consists of the N-terminal half of the *lacZ* gene (encoding β-galactosidase), while the remainder of the product of the fusion gene is human somatostatin. A family of three plasmid vectors for the construction of such a fusion gene has been created. All of these vectors have an ampicillin resistance gene and part of the *lacZ* gene encoding the first 583 amino acids of the β-galactosidase protein. The *Eco*RI fragment containing human somatostatin can be inserted into the single *Eco*RI restriction site on the vectors. The sequence of three vectors in the vicinity of the *Eco*RI site is shown below. The numbers refer to amino acids in the β-galactosidase protein with the N-terminal amino acid being number 1. The DNA sequence presented is the same as that of the *lacZ* mRNA (with Ts replacing the Us found in RNA). In which of these three vectors must the *Eco*RI fragment containing human somatostatin be inserted to generate a fusion protein with an N-terminal region from β-galactosidase and a C-terminal region from human somatostatin?

```
              582 583
              Gly Asn              EcoRI
pWR590-1  5' GGCAACCGGGCGAGCTCGAATTCG

              582 583
              Gly Asn              EcoRI
pWR590-2  5' GGCAACCCGGGCGAGCTCGAATTC

              582 583
              Gly Asn         EcoRI
pWR590-3  5' GGCAACGGGGCAGCTCGAATTCGA
```

Answer

This problem requires an understanding of the sticky ends formed by restriction enzyme digest and the requirement of appropriate reading frames for the production of proteins.

a. The only complete open reading frame (AUG start codon to a stop codon) is found on the bottom strand (underlined on the figure below). *The amino acid sequence is met-gly-cys-lys-thr-phe-thr-ser-cys.*

b. Based on the amino acid sequence determined for part (a), *the gene must be transcribed from right to left.*

c. The cut site for *Eco*RI is after the G at the 5′ end of the *Eco*RI recognition sequence on each strand. For each of the three vectors, the cut will be shifted relative to the reading frame of the *lacZ* gene by one base. The *Eco*RI fragment containing somatostatin can ligate to the vector in two possible orientations, but since we know the sequence on the bottom strand codes for the protein, a fusion protein will be produced only if the fragment is inserted with that coding sequence on the same strand as the vector coding sequences. Consider only this orientation to determine which vector will produce the fusion protein. The *Eco*RI fragment to be inserted into the vector next to the *lacZ* gene has five nucleotides that precede the first codon of the somatostatin gene (see figure below) and therefore requires one more nucleotide to match the reading frame of the vector. For pWR590-1, the cut results in an in-frame end; pWR590-2 has one base extra beyond the reading frame; pWR590-3 has a two-base extension past the reading frame. *The EcoRI fragment must be inserted in the vector pWR590-2 to get somatostatin protein produced.*

```
5′GCCGAATTCGATCCTATCAACACGAAGTGAAAGTCTTACAACCCATG^ AATTCGATTCG 3′
3′CGGCTTAAGCTAGGATAGTTGTGCTTCACTTTCAGAATGTTGGGTACTTAA ^GCTAAGC5′
```

III. Imagine you have cloned a 14.7-kb piece of DNA, which contains restriction sites as shown below

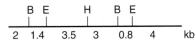

    B E   H   B E
    | | | | |
   ─┴──┴─────┴───┴──┴─────
   2 1.4 3.5 3 0.8 4   kb

B = *Bam*HI site, E = *Eco*RI site,
H = *Hin*dIII site

Numbers under the segments represent the sizes of the regions in kilobases (kb). You have labeled the left end of the molecule with ^{32}P.

a. What radioactive bands would you expect to see following electrophoresis if you did a complete digestion with *Bam*HI? *Eco*RI? *Hin*dIII?

b. What radioactive bands would you expect to see following electrophoresis if you did a partial digestion with *Bam*HI? *Eco*RI? *Hin*dIII?

Answer

This problem deals with partial and complete digests and radioactive labeling of fragments.

a. Only the leftmost fragment would be seen after complete digestion with any of the three enzymes since only the left end contains radioactivity. Radioactive bands seen after digestion with Bam*HI*: 2 kb; Eco*RI*: 3.4 kb; and Hind*III*: 6.9 kb.

b. Partial digests result in all possible combinations of adjacent fragments but again only those containing the left end of the fragment will be radioactive. For example, partial *Bam*HI digest results in the following fragments. The fragments that are radioactive have an asterisk. Bam*HI*: 2*, 7.9, 1.2, 9.9*, 12.7, and 14.7*. The radioactive fragments are Bam*HI*: 2, 9.9, 14.7 kb; Eco*RI*: 3.4, 10.7, 14.7 kb; and Hind*III*: 6.9, 14.7 kb.

P R O B L E M S

8-1 Match each of the terms in the left column to the best-fitting phrase from the right column.

a. cosmid

b. vector

c. sticky ends

d. recombinant DNA

e. YAC

f. genomic library

g. genomic equivalent

h. cDNA

i. polymerase chain reaction

j. hybridization

1. a DNA molecule used for transporting, replicating, and purifying a DNA fragment

2. a collection of the DNA fragments of a given species, inserted into a vector

3. DNA copied from RNA by reverse transcriptase

4. single-stranded DNA strands with enough base pair complementarity to form double-stranded molecules

5. efficient and rapid technique for amplifying the number of copies of a DNA fragment

6. an artificial chromosome containing centromere and telomere structural sequences from yeast

7. contains genetic material from two different organisms

8. a number of DNA fragments sufficient in aggregate length to contain the entire genome

9. short single-stranded sequences at the ends of many restriction fragments

10. a λ-plasmid hybrid vector able to contain large (45-kb) DNA fragments

8-2 Approximately how many restriction fragments would result from the complete digestion of the human genome (3×10^9 bases) with the following restriction enzymes? (The recognition sequence for each enzyme is given in parentheses.)

a. *Sau*3A (^GATC)

b. *Bam*HI (G^GATCC)

c. *Sfi*I (GGCCNNNN^NGGCC)

8-3 What purpose do selectable markers serve in vectors?

8-4 What are the functional components found in these three different vectors: plasmids, cosmids, and YACs. What size fragments would you insert into each?

8-5 A plasmid vector pBS281 is cleaved by the enzyme *Bam*HI (G^GATCC), which recognizes only one site in the DNA molecule. Human DNA is partially digested with the enzyme *Mbo*I (^GATC), which recognizes many sites in human DNA. These two digested DNAs are now ligated together. Consider only those molecules in which the pBS281 DNA has been joined with a fragment of human DNA. Answer the following questions concerning the junction between the two different kinds of DNA.

a. What proportion of the junctions between pBS281 and all possible human DNA fragments can be cleaved with *Mbo*I?

b. What proportion of the junctions between pBS281 and all possible human DNA fragments can be cleaved with *Bam*HI?

c. What proportion of the junctions between pBS281 and all possible human DNA fragments can be cleaved with *Xor*II (C^GATCG)?

d. What proportion of the junctions between pBS281 and all possible human DNA fragments can be cleaved with *Eco*RII (PuPuA^TPyPy)? (Pu and Py stand for purine and pyrimidine, respectively.)

e. What proportion of all possible junctions that can be cleaved with *Bam*HI will result from cases in which the cleavage site in human DNA was not a *Bam*HI site in the human chromosome?

8-6 Consider four different kinds of human libraries: a genomic library, a brain cDNA library, a liver cDNA library, and a unique genomic sequence (no repetitive DNA) library.

a. Assuming inserts of approximately equal size, which would contain the greatest number of different clones?

b. Would you expect any of these not to overlap the others at all in terms of the sequences it contains? Explain.

c. How do these four libraries differ in terms of the starting material for constructing the clones in the library?

8-7 As a molecular biologist and horticulturist specializing in snapdragons, you have decided you need to make a genomic library to characterize the flower color genes of snapdragons.

a. How many genome equivalents would you like to have represented in your library to be 95% confident of having a clone containing each gene in your library?

b. How do you determine the number of clones that should be isolated and screened to guarantee this number of genome equivalents?

8-8 Imagine that you are a molecular geneticist studying a particular gene in which mutations cause a serious human disease. The gene, including its flanking regulatory sequences, spans 200 kb of DNA. The distance from the first to the last coding base is 140 kb, which is divided among 10 exons and 9 introns. The exons contain a total of 38.7 kb, and the introns contain 101.3 kb of DNA. You would like to obtain the following for your work: (a) an intact clone of the whole gene, including flanking sequences; (b) a clone containing the entire coding sequences but no noncoding sequences; and (c) a clone of exon 3, which is the site of the most common disease-causing mutation in this gene. Based on the size of your inserts, what vectors would you use to obtain each of these clones and why?

8-9 Listed below are steps you would take to clone the *idiosynchratase* gene from the hoot owl. Put these into their appropriate order.

a. Incubate the probe and filters from the plates containing the phage from the hoot owl library.

b. Make the probe radioactive.

c. Package the ligated DNA into phage particles and infect *E. coli* with the mixture.

d. Extract genomic DNA from the hoot owl.

e. Digest the lambda vector with *Hin*dIII. Purify the two DNA pieces (arms of the phage DNA).

f. Plate out recombinant phage to get plaques. Make filters from each plate.

g. Partially digest the hoot owl genomic DNA with *Hin*dIII.

h. Ligate the partially digested genomic DNA to the λ arms.

8-10 Expression vectors containing sequences from the animal virus SV40 and a bacterial plasmid can be transformed and maintained in both *E. coli* and animal cells.

a. Give the general characteristics that are necessary for this type of plasmid to be used as a vector in two very different types of cells.

b. What is required to express a human protein in *E. coli* that is encoded by a gene within the insert fragment?

8-11 Why do longer DNA molecules move more slowly than shorter ones during electrophoresis?

8-12 You have a circular plasmid containing 9 kb of DNA, and you wish to map its *Eco*RI and *Bam*HI sites. When you digest the plasmid with *Eco*RI and run the resulting DNA on a gel, you observe a single band at 9 kb. You get the same result when you digest the DNA with *Bam*HI. When you digest with a mixture of both enzymes, you observe two bands, one 6 kb and the other 3 kb in size. Explain these results. Draw a map of the restriction sites.

8-13 A 49-bp *Eco*RI fragment containing the somatostatin gene was inserted into the vector pWR590 shown below. The sequence of the inserted fragment is:

5′ AATTCGATCCTATCAACACGAAGTGAAAGTCTTACAACCCATGAATTCG 3′

3′ GCTAGGATAGTTGTGCTTCACTTTCAGAATGTTGGGTACTTAAGCTTAA 5′

Distances between adjacent restriction sites are shown. What are the patterns of restriction digests with *Eco*RI (G^AATTC) or with *Mbo*I (^GATC) before and after cloning the somatostatin gene into the vector? (E = *Eco*RI, M = *Mbo*I)

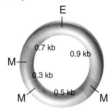

8-14 The *Notch* gene, involved in *Drosophila* development, is contained within a restriction fragment of *Drosophila* genomic DNA produced by cleavage with the enzyme *Sal*I. The restriction map of this *Drosophila* fragment for several enzymes (*Sal*I, *Pst*I, and *Xho*I) is shown below; numbers indicate the distances between adjacent restriction sites. This fragment is cloned by sticky-end ligation into the single *Sal*I site of a bacterial plasmid vector that is 5.2 kb long. The plasmid vector has no restriction sites for *Pst*I or *Xho*I enzymes.

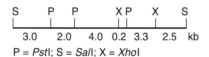

Make a sketch of the expected patterns seen after agarose gel electrophoresis and staining of a *Sal*I digest (alone), of a *Pst*I digest (alone), of a *Xho*I digest (alone), of the plasmid containing vector and *Drosophila* fragment. Indicate the fragment sizes in kilobases.

8-15 The linear bacteriophage λ genomic DNA has at each end a single strand extension of 20 bases. (These are "sticky ends" but are not, in this case, produced by restriction enzyme digestion.) These sticky ends can be ligated to form a circular piece of DNA. Both the linear and circular forms of the DNA are cut with *Eco*RI, *Bam*HI, or a mixture of the two enzymes. The results are shown below.

a. Which of the samples (A or B) represents the circular form of the DNA molecule?
b. What is the total length of the linear form of the DNA molecule?
c. What is the total length of the circular form of the DNA molecule?
d. Draw a restriction map of the linear form of the DNA molecule. Label all restriction enzyme sites as *Eco*RI or *Bam*HI.

8-16 The following fragments were found after digestion of a circular plasmid with restriction enzymes as noted. Draw a restriction map of the plasmid.

*Eco*RI: 7.0 kb
*Sal*I: 7.0 kb
*Hin*dIII: 4.0, 2.0, 1.0 kb
*Sal*I + *Hin*dIII: 2.5, 2.0, 1.5, 1.0 kb
*Eco*RI + *Hin*dIII: 4.0, 2.0, 0.6, 0.4 kb
*Eco*RI + *Sal*I: 2.9, 4.3 kb

8-17 You need to create a restriction map of a 7-kb circular plasmid you are studying. The figure on the next page shows the results of digesting the plasmid DNA with various restriction endonucleases. The figure represents a photo of a gel on which the DNA products of the digestions have been subjected to electrophoresis. The numbers across the top of the blot represent the lanes. The DNAs in each lane have been treated in the following way: lane 1: *Eco*RI; lane 2: *Hin*dIII; lane 3: *Pst*I; lane 4: *Bam*HI; lane 5: *Eco*RI + *Hin*dIII; lane 6: *Bam*HI + *Hin*dIII; lane 7: *Bam*HI + *Eco*RI; lane 8: *Pst*I + *Eco*RI; and lane 9: *Pst*I + *Bam*HI. The numbers down the left edge of the gel represent the molecular sizes of the various bands, in kilobases. Draw a map of the plasmid indicating restriction sites and their spatial relation to each other.

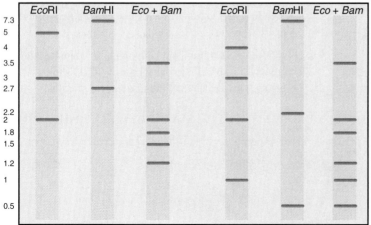

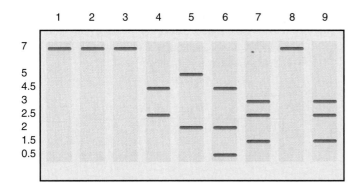

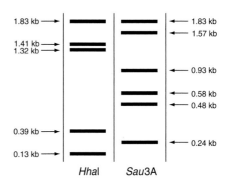

Hhal   Sau3A

8-18 The following restriction maps show the positions of *Eco*RI and *Xho*I restriction sites in the same 6-kb piece of DNA.

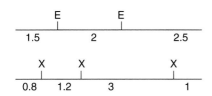

Sketch the results of electrophoresis of the following digests. Indicate which fragment is a terminal fragment.
a. partial digest with *Eco*RI
b. partial digest with *Xho*I
c. complete, double digest with *Xho*I and *Eco*RI

8-19 A fragment of human DNA with an *Eco*RI site at one end and a *Hin*dIII site at the other end is labeled with ^{32}P at the *Eco*RI end. This molecule is now partially digested with either restriction enzyme *Hha*I or *Sau*3A, and these partial digests are fractionated on an agarose gel. An autoradiograph of this gel is presented in the next column.
a. How large is the *Eco*RI to *Hin*dIII fragment in base pairs?
b. Draw a map of this fragment showing the locations of the *Hha*I and *Sau*3A recognition sites and the distances of sites from one end.
c. The same fragment, labeled with ^{32}P in the same fashion, is now digested to completion with *Hha*I, and the products are separated on an agarose gel. Sketch the expected pattern for the ethidium bromide stained gel and autoradiography of this gel.

8-20
a. Given the following restriction map of a cloned 10-kb piece of DNA, what sized fragments would you see after digesting this linear DNA fragment with each of the enzymes or combinations of enzymes listed below: (1) *Eco*RI, (2) *Bam*HI, (3) *Eco*RI + *Hin*dIII, (4) *Bam*HI + *Pst*I, and (5) *Eco*RI + *Bam*HI.
b. What fragments in the last three double digests would hybridize with a probe made from the 4.0-kb *Bam*HI fragment?

```
E      H H    E     B          P          B      E
├──────┼─┼────┼─────┼──────────┼──────────┼──────┤
  1.5  0.6 1.0  1.2      2.1        1.9       1.7  kb
```

8-21 Professor Ozone is hard at work trying to clone the *ozonase* gene. He wants to be able to produce huge amounts of this enzyme and use it to reduce smog levels in his home town of Los Angeles. He therefore needs to recover the gene from a genomic library of *E. coli* in a plasmid called pHew281. The restriction map is shown on the next page (B = *Bam*HI cloning site for the library and E = *Eco*RI).
a. A challenging aspect of cloning can be designing or obtaining a probe to use in identifying a gene. Suggest two different sources of a probe for the *ozonase* gene.
b. The professor clones the gene and finds that the *Pst*I digest of the clone produces one fragment, 12.5 kb in length. How big is the insert containing the *ozonase* gene?
c. Make a map of just the insert from the following restriction pattern obtained by digesting the plasmid clone.

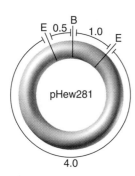

E 0.5 B 1.0 E

pHew281

4.0

B	E	H	P	BE	BH	BP	EP	EH	EHP	BHP
		12.5	12.5							
7.0										
	6.5									
5.5				5.5	5.5	5.5				5.5
					5.0					
						4.5				4.5
4.0				4.0			4.0	4.0	4.0	
							3.5	3.5		
							3.0	3.0	3.0*	
						2.5				
2.0				2.0	2.0		2.0	2.0	2.0	2.0
				1.5						
				1.0						
				0.5					0.5	0.5

8-22

a. What is the sequence of the *template strand,* given the following sequencing gel generated by the enzymatic extension method?

G   A   T   C

b. If the template for the DNA sequencing is the strand that resembles the mRNA, is this likely to be within an exon of a functional gene? Explain your answer.

8-23 You have isolated a *trp⁻ E. coli* mutant and to study the nature of the mutation, you isolate DNA from this mutant and determine the sequence of the DNA in the *trp* gene. Using enzymatic extension sequencing, the following result was obtained after electrophoresis and autoradiography.

What would be the mRNA sequence of the mutant? Where does it differ from the wild-type sequence?

G	A	T	C

8-24 The following sequence was read directly from a gel containing the products of the enzymatic extension method.

> 5′ TCTAGCCTGAACTAATGC 3′

a. Make a drawing that reproduces the gel pattern from which this sequence was read.

b. Assuming this sequence is from an exon in the middle of a gene, does this newly synthesized strand or the template strand have the same sequence as the mRNA for the gene (except that Ts are present instead of Us)? Justify your answer.

c. Using the genetic code table, give the amino-acid sequence of the hexapeptide (six amino acids) translated from the 18-base message. Indicate which is the amino terminal end of the peptide.

8-25 Which of the following set(s) of primers could be used to amplify the target DNA sequence indicated by the dots?

5′ GGCTAAGATCTGAATTTTCCGAG. . .TTGGGCAATAATGTAGCGCCTT 3′

3′ CCGATTCTAGACTTAAAAGGCTC. . .AACCCGTTATTACATCGCGGAA 5′

a. 5′ GGAAAATTCAGATCTTAG 3′; 5′ TGGGCAATAATGTAGCG 3′
b. 5′ GCTAAGATCTGAATTTₜC 3′; 3′ ACCCGTTATTACATCGC 5′
c. 3′ GATTCTAGACTTAAAGGC 5′; 3′ ACCCGTTATTACATCGC 5′
d. 5′ GCTAAGATCTGAATTTTC 3′; 5′ TGGGCAATAATGTAGCG 3′

CHAPTER 9

THE DIRECT DETECTION OF GENOTYPE

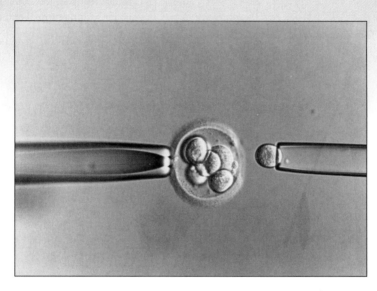

Plucking one cell from an 8-cell stage human embryo for
direct determination of genotype.

A couple whose firstborn suffers from cystic fibrosis learns from the medical diagnosis of that child's symptoms that both parents are carriers of a recessive disease-producing mutation. Together they run a 25% risk of transmitting the life-threatening hereditary condition to a second child. In 1992, one such couple did not want to take that chance, and after genetic counseling, decided to try an experimental protocol: *in vitro* fertilization combined with the direct detection *in vitro* of the embryo's genotype, before its placement in the mother's womb (Fig. 9.1).

The procedure took less than a week. At the start, a team of medical workers, including an obstetrician and a reproductive biologist, obtained close to a dozen eggs from the woman and fertilized them with sperm obtained from the man. Three days later, after the fertilized eggs had undergone several mitotic divisions to generate 6- to 10-cell embryos, a research assistant used micropipettes to remove one cell from each of six viable embryos. Earlier studies had shown that embryos containing the five to nine remaining cells could still develop normally. Next, the re-

search assistant used PCR (described in Chapter 8) to amplify from the DNA of the six isolated cells a specific segment of chromosome 7 containing the site of the most common mutation within the cystic fibrosis (*CF*) gene; previous cloning and sequencing of the gene made it possible to obtain primers bracketing this segment. Since the embryonic cells were diploid, they each supplied two alleles of the gene, one from the mother and one from the father.

To detect the genotype of the material amplified from each cell, clinicians divided each cell's PCR product into two equal portions, or aliquots, and exposed one aliquot to an allele-specific hybridization probe for the normal *CF* allele, the other to an allele-specific probe for the mutant allele. If only the first aliquot from a particular sample produced a signal, the embryo represented by the sample was a homozygote for the normal allele. If both aliquots were positive for the probe, the embryo was a heterozygote. And if only the second aliquot glowed in response to the probe, the embryo carried two copies of the disease allele.

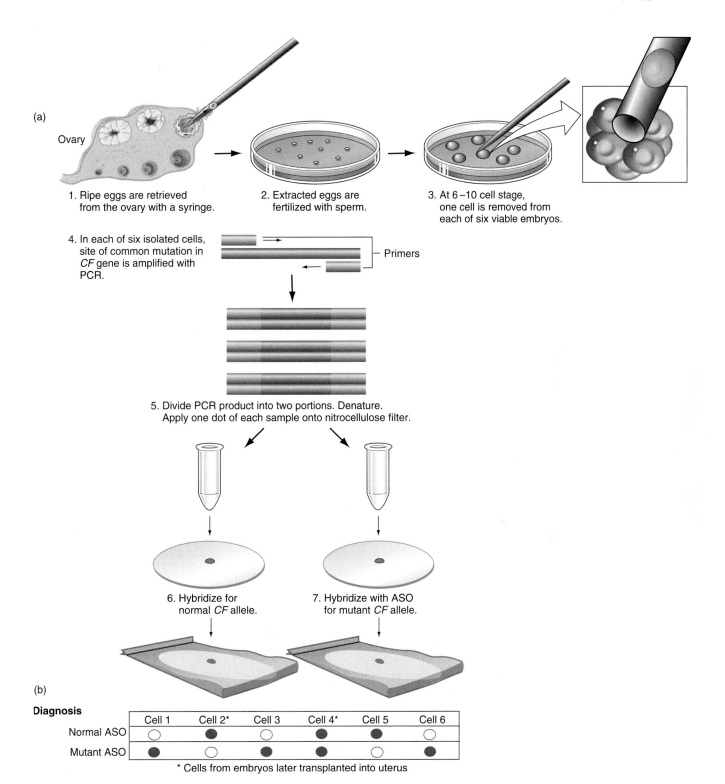

(a)

Ovary

1. Ripe eggs are retrieved from the ovary with a syringe.

2. Extracted eggs are fertilized with sperm.

3. At 6–10 cell stage, one cell is removed from each of six viable embryos.

4. In each of six isolated cells, site of common mutation in *CF* gene is amplified with PCR.

Primers

5. Divide PCR product into two portions. Denature. Apply one dot of each sample onto nitrocellulose filter.

6. Hybridize for normal *CF* allele.

7. Hybridize with ASO for mutant *CF* allele.

(b)

Diagnosis

	Cell 1	Cell 2*	Cell 3	Cell 4*	Cell 5	Cell 6
Normal ASO	○	●	○	●	●	○
Mutant ASO	●	○	●	●	○	●

* Cells from embryos later transplanted into uterus

Figure 9.1 Detecting the cystic fibrosis genotype of embryonic cells. (a) *In vitro* fertilization and preimplantation diagnosis. (b) Cell 2 is homozygous for the normal allele; cell 4 is heterozygous for the *CF* mutation.

It took about eight hours to determine the genotype of each isolated cell and, by inference, of each embryo. Thus, on the same day that the embryos were biopsied and genotyped, the doctor, in consultation with the parents, selected for placement in the mother's womb two embryos of known genotype. Of the two embryos they chose, one was a heterozygous carrier, the other a homozygote for the normal allele of the *CF* gene. The use of two to four embryos improves the chances of at least one implantation and is a part of most *in vitro* fertilization procedures; the selection of a carrier means that other embryos homozygous for the normal allele were most likely defective in some other way. As it happened, the embryo carrying two normal alleles of the *CF* gene became implanted; the carrier embryo did not.

With this approach, the woman was able to embark on a pregnancy free of genetic concern, knowing that the embryo developing within her womb carried two copies of the normal allele for the *CF* gene; as a result, this child, as well as its children, would be free of cystic fibrosis. Nine months later, the mother gave birth to a healthy 7 lb 3 oz baby girl. When evaluated by a pediatrician at her four-week checkup, the infant daughter was found to be completely normal in both physical and mental development. At the same time, testing at an independent laboratory confirmed that neither of her homologous chromosomes 7 carried the *CF* mutation. The couple in this report worked with university doctors and laboratories whose funding and operations are approved by review boards that monitor social and ethical as well as scientific issues. The Genetics and Society box "Social and Ethical Issues Surrounding Preimplantation Embryo Diagnosis" discusses several of these issues.

In this chapter, we examine how geneticists use an array of molecular tools for the direct detection of genotype within individual eukaryotic organisms, including human embryos and adults, nonhuman animals, and plants. An organism's genotype is its genetic constitution, the sequence of its DNA at one, several, and ultimately all positions in the two homologs of all chromosomes in its genome. Once cloning and sequencing have uncovered the sequence and allelic differences of a particular DNA segment, other less cumbersome protocols make it possible to establish not only which alleles an individual carries, but also what alleles exist in a population and which ones are associated with disease. The far-reaching applications of the new techniques include the diagnosis of genetic disease—*in vitro, in utero,* or after birth; and DNA fingerprinting—in the courtroom, for determining whether a criminal suspect could possibly have committed the crime for which he or she is charged—and in the laboratory, for looking at the mating behavior and genetic diversity of animals and plants.

Two general themes emerge from our discussion. First, the ability to distinguish genotypic differences of all kinds extends our concept of a locus and the alleles that define it. Mendel, Morgan, and other geneticists working before the advent of molecular technology identified a locus indirectly, inferring its existence from the observation of alternative phenotypes generated by alternative alleles of a gene; in this early era of genetic analysis, loci and genes were considered equivalents. Modern geneticists, able to look at genotype directly, can pick out genotypic differences in both genes and noncoding DNA regions on the basis of DNA variation alone. Such variation, much more common than originally believed, enables them to define a DNA locus even if it does not encode an RNA transcript or polypeptide, or help regulate gene expression. As we will see, researchers can use the DNA variations of noncoding loci as allelic markers to identify, locate, isolate, and follow the transmission of nearby genes. Thus, in modern genetic analysis, a **locus** is a designated location on a chromosome; a **gene** is a coding locus; and an **allele** is a site at which DNA—either coding or noncoding—differs among genomes. The alleles of a gene are the two or more alternative DNA sequences that affect the function of a single RNA or polypeptide product; the alleles of a noncoding sequence are simply DNA variations that occur in different genomes. The second theme is that techniques for the direct detection of genotype provide access to genetic details about human individuals never before available. This potential for gathering and storing genetic data raises questions about the information so obtained: Who should have access to it? And how should it be used?

Our presentation of the direct detection of genotype describes

■ Five classes of DNA variation that serve as the basis for the direct detection of genotype.

■ Protocols for detecting that variation, which combine techniques for amplifying isolated portions of the genome with techniques for distinguishing specific DNA variants.

■ Two applications of genotype detection: disease diagnosis and DNA fingerprinting.

DNA VARIATIONS PROVIDE THE BASIS FOR THE DIRECT DETECTION OF GENOTYPE

If a molecular geneticist were to clone the same 250-kb region of genomic DNA encompassing the cystic fibrosis gene on chromosome 7 from two healthy people and sequence the two clones in their entirety, he or she would probably find about one difference every 700 bp or 350 nucleotide differences across the length of the two sequences (Fig. 9.2). You might then wonder which of these sequences represents the true wild type, and if there are so many DNA differences between them, why doesn't one of them give rise to a disease phenotype? A geneticist would probably respond that first, both sequences are wild type because each of the nucleotide differences around and inside the *CF* gene exists in more than 1% of the population. And second, most likely none of the observed differences has any effect on the actual function of the *CF* gene, because only a small fraction of the total DNA stored in our chromosomes actually encodes or regulates the expression of proteins, and thus most nucleotide changes do not have an effect on phenotype.

In humans and many other higher eukaryotic organisms, less than 5% of the genome encodes gene products, while the short sequences that help regulate the expression of those genes make up less than 0.1%. The remaining 95% of the DNA either contributes to chromosome form and activity (as, for example, the DNA in telomeres and centromeres) or has no known function. Of the 250 kb that make up the human cystic fibrosis gene, for example, only 4.5 kb, slightly less than 2%, constitute the coding exons. Simple probability calculations thus show that the majority of mutations will occur in regions that are either noncoding or have an as yet unknown function. Because they do not affect the proteins that produce phenotype, such changes are neither deleterious nor advantageous, and as a result, they persist in a population or disappear by chance. By contrast, mutations in coding or regulatory regions can be harmful to the organism, and natural selection quickly eliminates them from the gene pool. As a consequence, the vast majority of DNA variation that remains in a population is in noncoding, nonregulatory regions.

As mentioned, geneticists consider variations at any position in the genome as alternative alleles of a particular chromosomal location, or locus. Before the development of recombinant DNA technology, the identification of a locus depended on mutations that gave rise to alternative phenotypes; investigators could follow these alternative phenotypes in breeding experiments using specific types of genetic crosses. Today, any variation in DNA sequence, once discovered, marks a locus that can be followed in a genetic cross. Geneticists consider a DNA locus that has two or more sequence variations, or alleles, each present at a frequency of 1% or more in a population, to be **polymorphic**, and they call the variations themselves **DNA polymorphisms.** (The 1% is an arbitrary cutoff point.)

When it is possible to follow the segregation of two or more alternative alleles at a particular locus with DNA probes or PCR-based analyses, researchers consider the locus a **DNA marker;** like a mile marker on a highway, it marks a point in the genome. If that position has no known function, the site is an **anonymous locus.** All anonymous loci are DNA markers, but not all DNA markers are anonymous loci; some occur within the genes themselves. Until geneticists have sequenced and analyzed the whole of a genome under investigation, they cannot be sure that an anonymous locus does not have a function.

Individual Members of the Same Species Show Enormous Sequence Variation in their Genomes

Studies comparing allelic DNA sequences throughout the genomes of numerous animal species indicate that the average frequency of polymorphism between any two homologous chromosomes is amazingly high. Such studies look only at nondomesticated organisms in nature, not at inbred laboratory animals. The data are remarkably consistent. They show that in all species evaluated, on average 1 of every 700 noncoding base pairs varies between homologous chromosomes from the same individual or from two different individuals. Using these data as a baseline, you can estimate the total number of noncoding polymorphisms that exist between homologous chromosomes in the human genome. To be on the safe side, suppose you assume that only 80% (0.8) of the 3 billion base pairs in the genome are noncoding, and on average only 1 base pair in 700 is polymorphic. With these assumptions, you can determine the frequency of polymorphism within a single individual by multiplying 3 billion by 0.8 and then multiplying that amount by 1/700:

$$(3 \times 10^9) \times 0.8 = 2.4 \times 10^9, (2.4 \times 10^9) \times \frac{1}{700} = 3.4 \text{ million.}$$

The result of 3.4 million is astonishing: It means that there are millions of differences between any two haploid sets of human chromosomes. Combined with differences in coding and regulatory sequences (which occur much less frequently), the millions of polymorphisms at anonymous loci contribute to an enormous pool of potential DNA markers.

Two cystic fibrosis genes from two healthy individuals

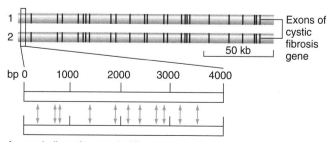

Arrows indicate base-pair differences

Figure 9.2 Base-pair differences between DNA cloned from the cystic fibrosis locus of two healthy individuals. These base-pair differences have no phenotypic effect; apparently they neither encode nor regulate expressed regions of the gene.

Genetics and Society

Social and Ethical Issues Surrounding Preimplantation Embryo Diagnosis

- In 1990, Germany passed a law that prohibits preimplantation embryo testing.

- A 1993 report from Canada's Commission on Reproductive Technologies warns against allowing market forces to determine the use of reproductive technologies. It also calls for creation of a permanent regulatory and licensing body to govern all aspects of the new reproductive practices, including sperm banks and *in vitro* fertilization.

- In 1994, France and Norway passed legislation that limits genetic testing to situations in which the results are medically therapeutic, and authorizes governmental bodies to establish the criteria for defining "therapeutic" in this context. These laws prohibit the use of genetic testing for sex selection and normal trait enhancement.

- In 1994, a U.S. National Institutes of Health advisory panel issued guidelines for federally funded research on embryos. These guidelines allow the use of preimplantation embryo testing for disease diagnosis and accept the practice of determining an embryo's gender to diagnose a sex-linked disease, such as hemophilia A. The guidelines do not accept sex selection for any other reason. An oversight committee

would monitor compliance with the guidelines to ensure the scientific qualifications of federally funded researchers as well as the likelihood that their studies will produce "significant scientific or clinical benefit" that cannot be "otherwise accomplished by using animals or unfertilized gametes."

At the same time, the United States had more than 300 privately run, unregulated *in vitro* fertilization clinics, commonly referred to as IVF Centers. Most of these centers were willing to do whatever a paying client requested, including sex selection and analysis of the genetic susceptibility for complex traits whose inheritance is not yet well understood.

This range of responses to the issues generated by the new reproductive technologies shows a diversity of approaches based on national culture and history. It also reflects international apprehension about the potential for misuse and abuse of the new technologies. Here are some of the main concerns.

When Should the Tests Be Used?

The couple in our opening story whose firstborn suffered from cystic fibrosis faced a medical problem. Preimplantation diagnosis could help them have a second child unaffected by the disease. With no cure at present for CF and no therapy that allows CF-affected

For the Purposes of Genotype Detection, Geneticists Categorize DNA Polymorphisms in Five Different Classes

As we saw in Chapter 6, DNA sequence differences arise in one of several ways—base-pair substitutions, deletions, insertions, or translocations. After a change occurs, it can be preserved by transmission from one generation to the next through DNA replication. Although the mechanism producing a mutation determines its size and shape, the same mechanism—a mistake in replication, for example—can produce different types of mutations—a substitution or a deletion; and different mechanisms—replication errors and mutagenic damage, for example—can produce the same type of mutation, such as a substitution. We now list the five classes of DNA polymorphism that geneticists use in the direct detection of genotype (Table 9.1). For each class, we describe the mechanism that can cause the variations as well as the frequency of that type of variation. All of the variations described are found distributed throughout the genomes of higher eukaryotic organisms.

Single-Base Differences

The simplest type of polymorphism results from single-base substitutions that can be triggered by mutagenic chemicals or mistakes in DNA replication. The single-base differences resulting from single-base substitutions occur throughout the genomes of all eukaryotic species at an average frequency of 1 in 700 bp. Since this type of polymorphism makes up 98% of all DNA differences, it has great importance for geneticists.

Microsatellites

DNA elements composed of 15–100 tandem repeats of one-, two-, or three-base sequences are known as **microsatellites.** Examples are AAAAAAAAAAAAAAA or CACAC-ACACACACACACACACACACACACACACACA. Also known as simple sequence repeats (SSRs), microsatellites arise spontaneously from random events that duplicate a mono-, di-, or (less often) trimeric sequence one to a few times. At some loci, these initial tandem duplications in-

people to look forward to a life of normal length, this is an example of medically therapeutic testing. Most governmental committees and bodies argue against testing for any other reason, but commercial clinics do not. Moreover, if postnatal therapies for cystic fibrosis, such as nasal sprays that introduce a normal CFTR protein into the respiratory tissues or protocols that insert normal *CF* genes in the cells of the lungs and nasal passages, become available, some medical practitioners may no longer consider preimplantation diagnosis a preferred therapy.

How Should the Tests Be Carried Out?

The couple in our opening story began by consulting a genetic counselor and then worked with medical practitioners associated with a university laboratory. Most geneticists agree that counseling before a procedure should foster an open discussion of all the issues (including the possibility that the tests might give false negatives); and that long-term follow-up should be part of the process. The preimplantation testing itself, like other forms of genetic testing, should be carried out by highly trained personnel in licensed laboratories. These accredited laboratories operate according to professional standards and have scientific and ethical review boards that monitor all work.

Who Should Have Access to the Technology?

The combination of *in vitro* fertilization and preimplantation testing cost $6000 to $10,000 in 1994. Should the government provide tests for people who cannot afford them? How should society decide this issue? (A related discussion of access to medical technology appears in the Genetics and Society box in Chapter 1.)

Should Parents Have the Right to Make Any Genetic Decision?

If, for instance, they decide to have a child affected by a genetic disease, should they bear all financial responsibility for its care, or should some form of universal health insurance provide help?

Who Should Have Access to Test Results?

Just the parents? The parents and eventually the child? The parents, the child, and certain community institutions, such as schools? Some combination of these plus commercial enterprises such as insurance companies and places of employment? (We discuss this same question of privacy in relation to other types of genetic testing in the Genetics and Society boxes in Chapters 1 and 2.)

What Constitutes a Human Individual?

Cultural and religious beliefs, rather than scientific knowledge and social customs, are the basis for answers to this question. Some people see preimplantation diagnosis as an alternative to abortion that allows a couple to make a decision before pregnancy, and thus a life, begins. Others argue that even at the eight-cell stage, a preimplantation embryo is the equivalent of a human being, and rejection of an embryo is the equivalent of killing a human being.

Although there are no simple solutions to these complex issues, geneticists around the world agree on the need for continuous open discussion and tight oversight of the development of the new reproductive technologies.

---

crease in number through errors in replication. In the mammalian genome, for example, the CA-repeat microsatellite occurs once in every 30,000 bp. Researchers have determined this frequency by probing genomic libraries and calculating the number of positive clones. Although the tandem repeats of microsatellites have no known function, they are found throughout the genomes of all vertebrates; the human genome contains roughly 100,000 microsatellite loci.

Microsatellites tend to be highly polymorphic in the number of repeats they carry, with many alleles distinguishable at each microsatellite locus. Research shows that faulty DNA replication is the major mechanism generating the many alleles (Fig. 9.3). Because the same short homologous unit (CA, for example) is repeated over and over again, DNA polymerase may develop a stutter during replication, that is, it may slip and make a second copy of the same dinucleotide, or skip over a dinucleotide.

New alleles arise at microsatellite loci at an average rate of 10^{-3} per locus per gamete (that is, one change in every thousand gametes). This empirically determined frequency is much greater than the classical gene mutation rate of 10^{-4} to 10^{-5} and results in more microsatellite variation between unrelated individuals within a population. At the same time, the rate of 1 mutation per 1000 gametes is low enough that microsatellite changes will usually not occur within a few generations of even a large family; because of this, microsatellites can be used as relatively stable markers in breeding studies and human pedigrees.

Minisatellites

DNA elements composed of longer tandemly repeating units of identical sequence lined up like the boxcars in a freight train are known as **minisatellites.** Each identical sequence unit (boxcar) is 10–40 bp in length, and each minisatellite locus (an entire train) consists of tens to thousands of the repeating

TABLE 9.1 Five Classes of DNA Polymorphism

Class	Cause	Rate of Mutation per Locus per Gamete	Frequency in Genome	Number per Human Genome (on average)
Single base	Mutagens or replication errors	10^{-8}–10^{-9}	1/700 bp	3 million
Microsatellite	Slippage during replication	10^{-3}	1/30,000 bp	100,000
Minisatellite	Unequal crossovers	10^{-3}	Unknown; discovered by chance	Fewer than 100 families known, yielding 1000 copies in all
Deletions	Mutagens; unequal crossovers	Extremely rare	Very low	0 – a few
Duplications	Mutagens; unequal crossovers	Extremely rare	Very low	0 – a few
Other insertions (excluding those resulting from micro- or minisatellite recombination)	Transposable elements	Extremely rare	Very low	0 – a few
Complex haplotype (any locus of 5 kb or more)	Any of the above	Combination of the above	Not applicable	Not applicable

units. Minisatellites are thus larger than microsatellites in every way.

There is nothing special about the sequence of repeated units in a minisatellite; many different sequences occur in such units. The defining feature of a minisatellite is simply the tandem repeats of sequences 10–40 bp in length. In fact, when minisatellites were first discovered, they were referred to as VNTRs for _variable number of tandem repeats_. Today most geneticists prefer the simpler term of minisatellites. Although the mammalian genome contains a large number of minisatellites, geneticists do not yet know whether they have a function, and if so what it is.

Like microsatellites, minisatellites arise from random events and are dispersed throughout the genomes of all vertebrates. However, since they are composed of random, repeated sequences discovered by chance, they are known only in species that have been well characterized genetically, and even then, their genomic frequency remains unknown. In the human genome, numerous minisatellites have surfaced so far; each one occurs from 1 to 50 times throughout the genome.

Minisatellite loci tend to be highly polymorphic because it is easy for homologous chromosomes with many short repeats to line up out of register along the repeated units during synapsis. Any crossing-over within the misaligned region will be unequal and produce reciprocal homologs that contain ei-

ther duplications or deletions of one or more of the repeat units. The result is two new alleles, one with fewer and the other with a greater number of units than those on the parental chromosomes (Fig. 9.4a–c).

Deletions, Duplications, and Other Insertions

The mutations in this category are the result of events that delete or insert one or more base pairs at a locus. Deletions range in size from one or a few base pairs to many megabases. The same is true of insertions. Duplications, for example, can consist of a single microsatellite unit (A or CA) or be hundreds of kilobases in length. The types of DNA change in this category of polymorphism result from the action of certain mutagens, such as ionizing radiation, or from unequal crossing-over during recombination. When the change arises from unequal crossing-over, the material deleted from one homolog is the material inserted in the other. Because of this, one product of an unequal recombination event contains a deletion, the other, a duplication. Insertions other than duplications arise most often from transposable elements that integrate at random into the genome, as described in Chapter 12. Although they occur much less frequently than single-base-pair substitutions, deletions and insertions can disrupt genes and other functional DNA segments. The insertion or deletion of a single base pair within a coding sequence, for example, causes a frameshift.

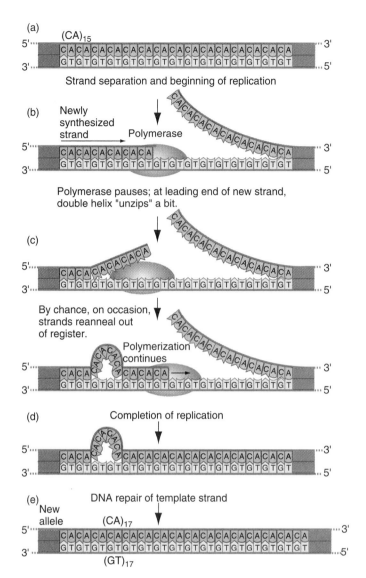

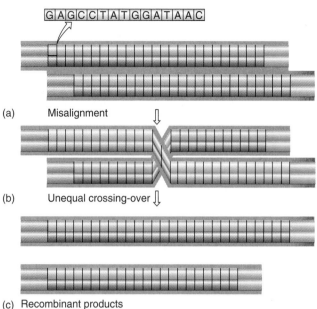

Figure 9.4 Minisatellites are highly polymorphic because of their potential for misalignment and unequal crossing-over. Minisatellites are composed of relatively long tandem repeating units of identical sequence. (a) Misalignment and (b) unequal crossing-over produce (c) recombinant products that contain different numbers of repeating units than either parental locus; each new recombinant product is a new allele.

Figure 9.3 Microsatellites are highly polymorphic because of their potential for faulty replication. (a) A microsatellite consisting of 15 tandem repeats of the CA dinucleotide sequence. (b) During replication, the strands of the DNA molecule separate and DNA polymerase moves from the 5′ to the 3′ end along each strand. (c) Throughout the replication process, pauses can occur if the required nucleotides are, by chance, not in the vicinity of the polymerase. During a pause, the newly synthesized strand may "unzip" or "breathe" apart from the template strand over a length of 5–10 bases. (d) When the required nucleotides become available, the newly synthesized strand reanneals to the template and acts as a primer for further replication. If the reannealing occurs within a microsatellite, the new strand may be out of register such that the polymerase begins by adding one or more nucleotides that it has already added. (e) The resulting DNA molecule will have one or more identical repeats in the newly synthesized strand that were not present in the template strand. DNA repair processes then adjust the template strand to make it the same length as the newly synthesized strand.

Complex Haplotypes

A contraction of the phrase "haploid genotype," the term **haplotype** refers to a specific combination of linked alleles in a cluster of related genes. Immunogeneticists often use it to describe the combination of alleles of the *major histocompatibility complex (MHC):* a large cluster of genes on human chromosome 6 that play a role in the immune response. With the resolving power to look at DNA at the level of nucleotides, "haplotype" now refers to any set of linked DNA changes along a chromosome. These changes could be in one or several genes, or in noncoding stretches. The **complex** refers to the multiple types of variation that can exist at alternative alleles, including more than one nucleotide substitution, a substitution in combination with a small deletion, duplication, or other insertion. Thus, a **complex haplotype** is a set of linked DNA variations along a chromosome, with the possibility of many differences between alternative alleles.

We present this short summary of the amount and types of polymorphisms in a genome because it is these differences in DNA sequence that provide the basis for the direct detection of genotype. Once cloning and sequencing or random searches have identified a variation or set of variations, researchers use faster, less cumbersome tools to detect which variations an individual carries and to follow the transmission of specific variants in families and populations. Which tools they select depends on what they are looking for and in what organism.

We now describe the primary techniques for the direct determination of DNA polymorphism.

PROTOCOLS OF DETECTION: IDENTIFYING SPECIFIC VARIANTS WITHIN INDIVIDUAL GENOMES

All protocols for the detection of genotype include the initial identification of the polymorphism, and the subsequent detection of alleles at the polymorphic locus in large numbers of individuals within a population. For some types of polymorphisms, microsatellites, for example, the same technique—PCR amplification of different microsatellites and gel electrophoresis to examine the size differences between the different PCR products—can serve in both stages of the analysis. For other types of polymorphisms, such as single-base substitutions, the initial detection of an allelic difference may depend on a complicated sequencing protocol that is too cumbersome and too expensive to serve as the basis of genotype detection in the larger population; instead, researchers would use discovered sequence differences to develop a simple, inexpensive, and much more rapid method for determining the genotype of large numbers of individuals. For some genotype detection protocols, DNA amplification is necessary; for others, it is not. We now look at the different protocols for allele analysis that enable rapid and efficient genotype detection for each class of polymorphism.

A Variety of Protocols Detect Single-Base Changes

The optimal techniques for detecting single-base changes vary with the circumstances. Before deciding on the most appropriate protocol, an investigator must ascertain whether the sequence of a target DNA segment is known; whether there are two, or more than two, alleles in the system under consideration; and whether there is a possibility of additional, as-yet-undiscovered alleles. We now describe five protocols for detecting single-base changes.

The Analysis of Restriction Fragment Length Polymorphisms with Southern Blots and Labeled Probes Distinguishes Allelic Differences at Restriction Sites

A small proportion of the enormous number of spontaneous mutations in a genome will eliminate or create a restriction site (Fig. 9.5a). Suppose, for example, that a local area of the genome normally contains three copies of the 6-base sequence GAATTC recognized by the *Eco*RI restriction enzyme. In most individuals, recognition site I occurs at a point arbitrarily labeled as 0; recognition site II occurs 4 kb downstream from site I; and recognition site III occurs 1 kb from site II, or 5 kb from site I. (We know this from previous cloning and restriction mapping.) If a single-base substitution changes a T to an A in site II, then the local area in some individuals will now contain only two *Eco*RI recognition sites, and they will be 5

kb apart. If, on the other hand, a GAGTTC sequence normally exists between sites I and II, and a different mutation changes the second G in the sequence to an A, the local area in some individuals will now contain a new recognition site between sites I and II. Although the vast majority of these restriction-site changes will have no effect on phenotype, they can be detected as alterations in the restriction maps that characterize the localized area where the change occurs.

Until recently, Southern blots (described in Chapter 8) were the tool of choice for viewing these localized restriction maps. The only requirement for detecting polymorphisms in the area is a DNA probe known to base pair with a region between the variant restriction site and the adjacent nonvariant restriction site. If, for example, you treat genomic DNA with *Eco*RI, then separate the restriction fragments according to size by gel electrophoresis and use Southern blotting and hybridization to a probe for viewing the region of interest, you will see the fragments depicted in Fig. 9.5b. Probe hybridization to the DNA from most individuals reveals a restriction fragment 4 kb in length. Hybridization to DNA from an individual that is homozygous for a mutation eliminating a recognition site reveals a single restriction fragment 5 kb long; hybridization to DNA from an individual homozygous for a mutation that creates a restriction site reveals a single fragment only 2 kb in length; and hybridization to DNA from individuals heterozygous for the wild-type allele and the latter mutation would reveal fragments 2 and 4 kb in length. Thus, mutations that destroy or create restriction sites generate easily detectable changes in the size of restriction fragments. These changes are visualized on Southern blots through the use of selected labeled probes. Geneticists call an allelic variation detectable in this manner a **restriction fragment length polymorphism,** or **RFLP** (pronounced "riflip"). RFLPs were the first type of DNA marker to be identified and utilized in the analysis of complex genomes.

Loci detectable through RFLPs are ideal candidates for Mendelian analysis because the DNA variations that define them act as codominant alleles through which it is possible to distinguish homozygotes from heterozygotes. Be aware, however, that the term codominant is not entirely accurate in this context, since the allelic differences under consideration are in the DNA itself and more often than not have no visible effect on phenotype. In fact, the actual variations that generate the alleles are typically 1–10 kb upstream or downstream from the region detected by the probe that is used to pick out the RFLP, so even if this probe is a clone of a gene, the existence of the RFLP does not necessarily mean the detected variation lies within the gene.

Geneticists can use RFLP analysis to follow the inheritance of specific genomic segments, because RFLPs function as markers that flag a specific portion of a chromosome through many generations of a family or among the individuals of a population. If the marker is closely associated with a disease, it may become the basis of genetic counseling predictions. To be **informative,** an RFLP must distinguish between

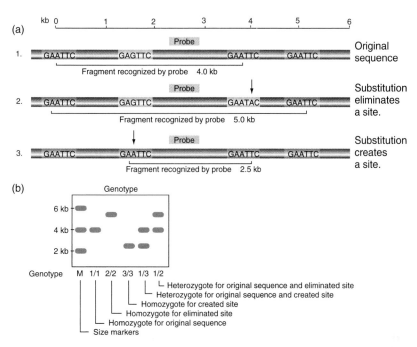

Figure 9.5 RFLPs can be generated by single-base substitutions and detected by Southern blot analysis. (a) A small proportion of the spontaneous mutations in a genome will change the original sequence (1) by either eliminating a restriction site (2) or creating a new site (3). (b) Restriction fragments are separated according to size by gel electrophoresis, blotted to filter paper, and then hybridized to a complementary probe. Autoradiography enables visualization of the RFLP genotype carried by each individual.

the two chromosomes of an individual. The most informative RFLPs are thus those for which the greatest number of individuals is heterozygous. This heterozygosity makes it possible to follow the markers' segregation and infer which offspring inherit which chromosome from which parent. The greater the number of RFLPs identified at a locus, the more informative they become as a group. Suppose, for example, that a restriction enzyme generates RFLPs in the vicinity of a disease locus visualized by a probe. The enzyme detects four different alleles— *A, B, C,* and *D*—at the locus. In one family, the father has alleles *A* and *B*, the mother has *C* and *D*, a daughter with a recessive condition has *B* and *C,* and a healthy son has *A* and *C.* From this information, it is likely that chromosomes with alleles *B* and *C* both carry the mutation responsible for expression of the trait (Fig. 9.6).

Traditional RFLP analysis has its advantages and disadvantages. Visualized by Southern blots and labeled probes, RFLPs can detect genotype in the absence of sequence information. This is their main advantage. Nonetheless, finding RFLPs within a region by trial and error is not always easy, and most of those that do surface have a limited number of polymorphisms, which means they are informative in only a minority of the families of interest. In fact, most RFLPs are diallelic, that is, come in only two alternative forms. This is because the actual site whose polymorphism underlies the RFLP is either present or absent and is flanked by two unchanging sites recognized by the same restriction enzyme. Another disadvantage of RFLP analysis is that it requires a large amount

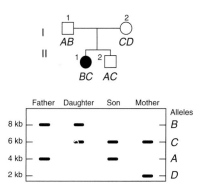

Figure 9.6 Geneticists use RFLPs as markers to follow the inheritance of linked alleles at disease loci. This pedigree shows the segregation of alleles from an RFLP locus that is closely linked to a locus containing a recessive disease-causing allele. The presence of an affected daughter and two nonaffected parents in the family shows that both parents are heterozygous carriers of the disease allele. In the daughter, the presence of the *B* and *C* alleles at the RFLP locus demonstrates an association between the *B* RFLP allele and the disease allele in the father, and between the *C* RFLP allele and the disease allele in the mother. By a process of elimination, the *A* and *D* RFLP alleles must be associated with the wild-type alleles at the locus under consideration. With this information, it is possible to diagnose the disease genotype of the son: he is a carrier with an *A*-associated wild-type allele and a *C*-associated disease allele.

of starting material as well as a great deal of time in moving from sample to result (you must purify DNA from a cell line or tissue, digest the DNA sample, run a gel, prepare a Southern blot, make a probe, hybridize probe to blot, wash blot, expose film, and so forth).

It Is Now Possible to Use Gel Electrophoresis of PCR Products to Detect Mutations in Restriction Sites

If you know the sequence of the few hundred base pairs flanking a polymorphic restriction site, you can analyze genotypic differences by amplification through PCR and subsequent analysis of the PCR products. Consider, for example, the single-base human polymorphism shown in Fig. 9.7. Allele 1 has an *Eco*RI site; in allele 2, this site has been destroyed by an A-to-G substitution. With knowledge of the nonpolymorphic sequence surrounding the site, you can select two oligonucleotides capable of acting as primers in the PCR amplification of both alleles. The PCR products from both alleles will be the same size (350 bp in Fig. 9.7a). To distinguish the two alleles, you simply digest the PCR products with *Eco*RI and then separate the digested fragments by size with gel electrophoresis. The PCR product derived from allele 2, which

does not have an *Eco*RI site, will remain in one piece and show up on the gel as a 350-bp fragment. By contrast, the PCR product from allele 1, which does have an *Eco*RI site, will be cut into two fragments 200 and 150 bp in length that are easily distinguishable on an electrophoretic gel from the 350-bp fragment of allele 2.

As Fig. 9.7b shows, to distinguish the three genotypes possible at this locus, you stain the gel with ethidium bromide and evaluate the bands that materialize. Because there is no need for blotting or hybridization, the procedure is much simpler and faster than traditional RFLP analysis. There is a one-time, up-front cost: the labor and effort to determine the sequence that surrounds the polymorphic restriction site and to synthesize the oligonucleotide primers; thus if a researcher intends to determine the genotype of only one or a few individuals, the previously described Southern blot method is more practical and cost effective. Once the sequence and primers are available, however, the much more rapid PCR-based protocol is the method of choice for detecting genotypic differences defined by restriction-site polymorphisms.

Hybridization of Allele-Specific Oligonucleotides to PCR Products Can Detect Single-Base Differences

Unfortunately for geneticists probing the genome, the great majority of single-base changes do not destroy or create known restriction sites, and thus do not reveal the variations they create through RFLP analysis. To detect simple polymorphisms that do not affect restriction sites, researchers can use PCR in conjunction with allele-specific hybridization probes. Once again, PCR serves to amplify the region encompassing the polymorphic site. The PCR products, which are greatly enriched for the target sequence, then make it possible to distinguish single-base changes through hybridization to different short oligonucleotide probes.

It is only with very short probes—oligonucleotides containing around 20 bases—that single-base changes provide a large enough difference to be readily detected. This is because for very small DNA molecules—those composed of no more than 50 bp—the length of the molecule itself plays a critical role in determining whether the double helix remains intact or falls apart. The effective length and therefore the strength of the forces holding together the double helix of a short-probe/short-target DNA hybrid is the longest stretch that does not contain any mismatches. If the two strands do not match exactly, there may not be enough weak hydrogen bonds in a row to hold them together. For example, if a 21-base probe hybridizes to a target strand that differs at a single base directly in the middle of the sequence, the effective length of the resulting double-stranded hybrids is only 10 bp. Since a 10-bp hybrid is significantly less stable than a 21-bp hybrid, it becomes easy to devise hybridization conditions (essentially by choosing the right temperatures) under which the perfect 21-bp hybrids will remain intact while the imperfect hybrids will not (Fig. 9.8a). By comparison, molecules longer than 50 bp can maintain their double helix conformation even with intermittent mismatches. In effect, the thermal stability of a 50-bp

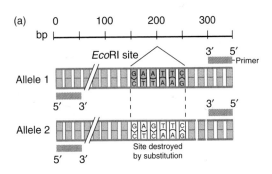

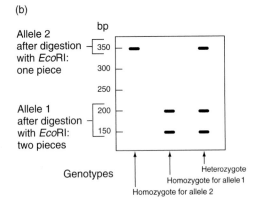

Figure 9.7 Restriction site polymorphisms can be detected most efficiently with PCR-based protocols. (a) PCR amplification of two alleles of a DNA locus with a restriction site polymorphism. Allele 1 has an *Eco*RI site that is eliminated in allele 2. The PCR products amplified from both alleles are identical in size. (b) Exposure of these PCR products to *Eco*RI causes cleavage of the allele 1 product but not the allele 2 product. Gel electrophoresis and ethidium bromide staining distinguish the three genotypes possible with the two alleles at this locus.

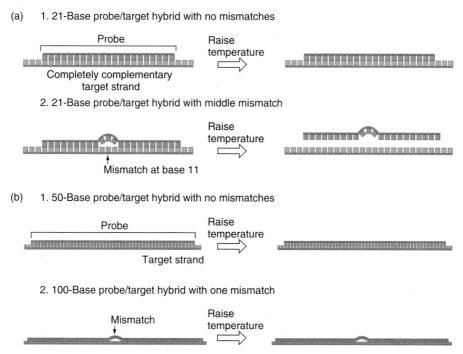

Figure 9.8   Short hybridization probes can distinguish single-base mismatches, longer probes cannot. (a) Researchers allow hybridization to occur between a short 21-base probe and two different target sequences. (1) A perfect match between probe and target extends across all 21 bases. When the temperature rises, this hybrid has enough hydrogen bonds to remain intact. (2) With a single-base mismatch in the middle of the probe, the effective length of the probe-target hybrid is only 10 bases. When the temperature rises, this hybrid does not have enough hydrogen bonds to remain intact, and it falls apart. (b) Researchers allow hybridization to occur with probes of 50 bases and 100 bases. (1) A perfect match between a 50-base probe and its target bases achieves a maximum of stability such that any extension in the length of the match would not have a significant effect on the temperature at which the hybrid falls apart. (2) Thus, it is not easy to distinguish a 100-bp hybrid with one mismatched base from a 100-bp hybrid with a perfect match.

hybrid is not sufficiently different from the thermal stability of a 100-bp hybrid (of equivalent sequence composition) for differential hybridization to distinguish between the two. This is because a critical number of hydrogen bonds is required for double helix stability, but once this number is achieved, any further increase in the number of hydrogen bonds makes no difference (Fig. 9.8b). Thus, it is best to use molecules that are 15–25 bases long in protocols relying on the differential hybridization of probe to target to pick out single-base differences on the basis of differences in effective hybrid length.

So you might think that single-base changes are easily detectable by short probes in the Southern blotting technique. However, this does not work, because only much longer probes—usually hundreds to thousands of bases in length—produce a strong enough signal to allow detection of the target sequence on a Southern blot containing samples from complex eukaryotic genomes. For example, a 1000-base (1 kb) probe will hybridize to 10 times the amount of target sequence as a 100-base probe having the same specific activity, and this will lead to a signal that is ten times stronger. But even a 1000-base target is a needle in a haystack, since it represents only a very small fraction (in the human genome, 1 part in 3 million) of the total genomic DNA. For this reason, it is not practical to perform Southern blot analysis with probes less than 100 bp in length.

PCR resolves this impasse by enormously increasing the absolute amount of target sequence and as a result, the target-to-nontarget ratio (from 1 in 3 million to 1:1 or even more). This increase allows the use of short oligonucleotides for hybridization probes, and these small probes, in turn, make it possible to detect single-base differences in a simple plus/minus assay. Short oligonucleotides that hybridize with alleles distinguished by a single base are known as **allele-specific oligonucleotides,** or **ASOs** (Fig. 9.9a). The ideal ASO is 19–21 bases in length—short enough to allow differential hybridization based on a single-base change, but long enough to provide a high probability of locus specificity, that is, a high probability that the target sequence will not be present by chance anywhere else in the genome. Figure 9.9b–e shows the results of an ASO-based test to determine individual genotypes at the *β-globin* locus.

The power of the PCR/ASO hybridization protocol is its simplicity. It does not require the running of a gel, and it can be automated if the necessary robotic resources are available. Moreover, with large amounts of PCR products serving as target sequences, probes may be labeled with nonradioactive materials that are safer and allow for their long-term storage.

Even though base-pair changes that do not create or destroy a restriction site are much more common than those that do, to identify and define those changes, it is still necessary to sequence the locus in question from multiple individuals. Moreover, since you can follow only those alleles identified by sequencing, you will not detect further alleles at the same

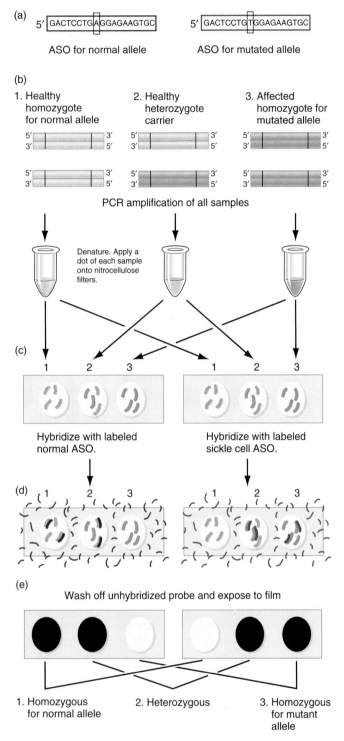

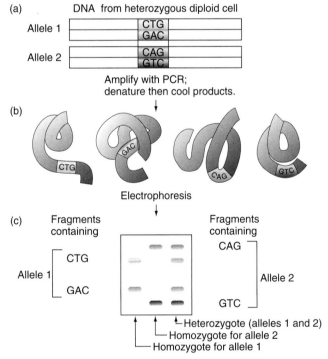

locus if these alleles were not present in the individuals initially screened to obtain the DNA variants.

Single-Strand Conformation Polymorphisms Enable Detection of Alleles That Have Not Yet Been Characterized

Once you have defined a gene of interest, it would be helpful to pick out and follow the segregation of the relatively few polymorphisms that exist for that coding locus. A perfect protocol for the detection and analysis of this limited number of polymorphisms would satisfy the following criteria. It would enable the detection of any base-pair variants within a DNA region. It would not require prior sequence information on each allele. It would not require the synthesis of ASOs. It would require no special equipment or skills beyond those found in a standard molecular biology laboratory. Finally, it would be rapid and the results would be readily reproducible.

Now a simple, relatively new protocol, illustrated in Fig. 9.10, satisfies all these criteria. It takes advantage of the fact that even single nucleotide changes can alter the three-

Figure 9.9   Using PCR with ASOs to determine genotype at the β-globin locus. (a) Before performing the genotyping protocol, it is necessary to synthesize two oligonucleotides that differ at only a single base; one of these oligonucleotides is complementary to the wild-type β-globin allele, the other is complementary to the sickle-cell allele. These two synthetic DNA molecules serve as the ASOs for the sickle-cell genotype assay. (b) Genomic DNA samples obtained from individual people are subjected to PCR amplification with primers complementary to nonpolymorphic sequences that flank the base that mutates to cause sickle-cell anemia. (c) The amplified sample from each individual is divided into two aliquots that are blotted directly to filter paper. (d) One aliquot from each sample is hybridized to the wild-type ASO; the other aliquot is hybridized to the sickle-cell ASO. (e) Autoradiography indicates the β-globin genotype of each individual.

Figure 9.10   The SSCP protocol enables detection of previously uncharacterized single-base polymorphisms. (a) The DNA region of interest is amplified by PCR using primers complementary to invariable sequences. The products amplified from each allele are identical in size. (b) The PCR products are denatured and cooled so that complementary strands do not reanneal. Single strands derived from PCR products that differ by as little as a single base (such as a T-to-A substitution) coil up in slightly different conformations. (c) Single strands with different conformations run with different mobilities during gel electrophoresis. Each double-stranded PCR product resolves into two gel bands. It is possible to distinguish different alleles by different pairs of gel bands.

dimensional conformation assumed by DNA strands when conditions allow them to remain in the single-stranded state. If, for example, you amplify a particular genomic region by PCR, then subject the PCR product to denaturation at a high temperature and immediately place it on ice, the low temperature impedes the mobility of the single strands and thereby limits their ability to find and form a double helix with their complementary strands. As a result, instead of re-forming double strands, the single strands coil up into balls, with each type of strand assuming a most-favored conformation. Hydrogen bonding between a large number of bases that happen to end up across from each other within the coils maintains the three-dimensional conformation of each single strand. The substitution of one nucleotide for another within a strand will usually alter the favored three-dimensional conformation of that strand by disrupting some hydrogen bonds and promoting others. If the new conformation is different enough from the first, it will run with an altered mobility on a gel. Different alleles of a locus that are detected in this way are known as **single-strand conformation polymorphisms, or SSCPs.**

One disadvantage of the SSCP protocol is that because of the requirement for running and analyzing gels, it is hard to automate. A second disadvantage is that it misses 10% of the variants. Advantages include its simplicity and reliability as well as its applicability to all unique sequences, both in and out of genes. In contrast to all of the other protocols described so far, which distinguish two or a handful of previously well-defined alleles, SSCP assays enable the detection of an unlimited number of alleles at a locus. Geneticists find SSCP especially useful for detecting the allelic forms of coding sequences, since observation of an SSCP implies a sequence difference within the region of DNA that has been amplified by PCR, and the PCR primers can be designed directly from exons. Thus, SSCPs can provide direct evidence of mutations within genes as well as a means of following those mutations as they segregate from parents to offspring. Suppose, for example, you wanted to look for new variants at the β-globin locus. You might screen samples from 1000 people for SSCPs and find perhaps five variants, which you would then sequence. The entire process would take no more than a few days. By comparison, without the SSCP protocol, you would have to sequence all 1000 samples to find the same five variants.

The Random Amplification of Polymorphic DNA Enables Detection of Polymorphisms in Any Uncharacterized Genome

Suppose an investigator wishes to begin genetic studies on a new species of animal or plant that is entirely uncharacterized at the molecular level (for example, one recently found in the rain forests of Central America with potential medicinal properties). She could not use any of the methods described so far because they all depend on knowledge gained from cloning and sequencing, and in this new species, no one has yet defined any loci by recombinant DNA technology. Without such knowledge, the best way to detect single-base changes at polymorphic loci throughout the genome is with a protocol known

as **random amplification of polymorphic DNA, or RAPD** (pronounced "rapid").

RAPD is based on the premise that an oligonucleotide of any random sequence that is short enough will, by chance, be complementary to random target sequences throughout the genome. If two target sequences are present on opposite strands of a genomic locus in the correct orientation and close enough to each other, PCR will amplify the DNA between them (Fig. 9.11). With randomly constructed oligonucleotides 10 bases in length serving as primers, PCR will amplify approximately 7 to 10 unrelated PCR products from a roughly billion-base-pair genome, that is, one similar in complexity to that of mammals. Different amplified fragments will represent different loci, and will by chance be of different lengths, distinguishable by gel electrophoresis and ethidium bromide staining.

Figure 9.11 shows how RAPD detects a single-base difference in the same region of two human genomes. A polymorphism at a locus detected by RAPD results from single-base substitutions that alter one of the two target sequences required for the amplification of that locus. Recall that with short DNA strands, even a single-base difference in complementarity can drastically reduce the ability of two strands to hybridize, which is the first event in each cycle of

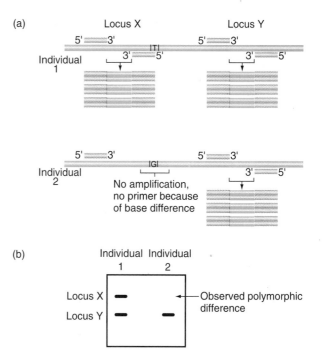

Figure 9.11 The RAPD protocol can uncover DNA polymorphisms in uncharacterized genomes. (a) Short primers are used to amplify random loci from the genomes of two individuals in a species. The genome of individual 1 has two loci—X and Y—with pairs of nearby sequences that are complementary to the primer, and both of these loci are amplified. The genome of individual 2 has a G instead of a T at one place in locus X; this substitution prevents primer binding and eliminates amplification of locus X. (b) Gel electrophoresis of the PCR products from both individuals followed by ethidium bromide staining can visualize the polymorphism in locus X.

amplification. Since there are two flanking 10-base targets to which the primers must bind to initiate the amplification of a locus, there are 20 sites at which a single-base change could prevent amplification. For each locus, RAPD will detect any 1 of these 20 potential differences as a polymorphism.

The RAPD protocol is most useful in the initial genetic analysis of an uncharacterized species. With it, you can scan an entire genome to obtain a set of polymorphic DNA markers. You begin with several randomly constructed oligonucleotides and use each one in isolation to amplify a set of random loci from a group of individuals. Each polymorphic locus detected as a gel band present in some individuals but not others can be isolated and cloned for further analysis.

Table 9.2 summarizes the basic protocols for detecting single-base changes, including RFLP analysis by Southern blots and labeled probes, or by gel electrophoresis of PCR products; PCR analysis using ASOs; distinguishing SSCPs on a gel; and evaluating the PCR products of RAPD. In analyzing genotypes, geneticists rely extensively on the detection of single-base changes because they are by far the most common form of DNA polymorphism.

For Microsatellites: PCR-Based Protocols Detect Polymorphisms

Composed most often of simple one-base or two-base units repeated 15 to 100 times in a row, microsatellites have proved to be the most important class of polymorphic markers for the development of linkage maps across the genomes of mice, humans, and other vertebrates. The reasons for their widespread use include their frequent appearance in all vertebrate genomes (where, we have seen, they are distributed at random), the ease with which they can be identified (by hybridization to simple sequence probes), their high level of polymorphism (caused by frequent length-varying mutations during replication), and the ease with which the polymorphisms can be detected (by a simple PCR and gel-based assay).

The Allelic Differences at Microsatellite Loci Reflect the Number of Units Present in the Tandem Array

It is thus the actual size of the locus, rather than a specific base substitution, that defines each allele. For this reason, the only efficient way to distinguish the alleles of a microsatellite is to measure the length of the locus.

Microsatellite Amplification Depends on Sequence Information

To distinguish the variants of a microsatellite, you must first determine the unique sequences flanking the locus on either side (Fig. 9.12a.1). You next use this information to derive oligonucleotide primers for the PCR amplification of the locus and its surrounding sequence (Fig. 9.12a.2). You then analyze the PCR products by gel electrophoresis and ethidium bromide staining, to discover the size differences related to the number of tandem repeats (Fig. 9.12b).

With the Proper Gel-Separation System, It Is Possible to Detect Size Differences as Small as 2 bp among PCR Products of Up To ~400 bp in Length

As with SSCPs, a single pair of locus-specific primers enables the detection of many different alleles, since all size differences of 2 bp or more are distinguishable from each other. Figure 9.12b shows how this protocol distinguishes the six different genotypes possible in a population with three alleles at a microsatellite locus.

TABLE 9.2 **Protocols for Detecting Single-Base Variations**

Protocol	Advantages	Disadvantages	Main Uses
RFLP analysis			
By Southern blots and labeled probes	No requirement for sequence information	Requires a large amount of DNA	Genotyping in the absence of sequence information
By restriction enzyme digestion and electrophoresis of PCR products	Rapid; no blotting and hybridization	Substitution must affect restriction site; requires sequence information from locus	Rapid analysis of restriction site polymorphisms among individuals within a population
PCR analysis using ASOs	No need to run a gel; can detect all single-base variations	Must know sequence of alternative alleles	Standard genotyping
SSCP	Useable even when allelic differences are not known at the level of sequence	Requires gel electrophoresis	Detection of new alleles at established loci; for following more than two alleles
RAPD	No requirement for sequence information	Provides information for only random loci	Analysis of genomes that are not well characterized

(a)

1. Determine sequences flanking microsatellites

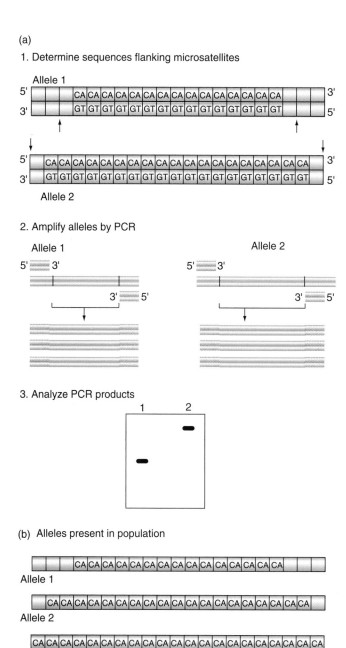

2. Amplify alleles by PCR

3. Analyze PCR products

(b) Alleles present in population

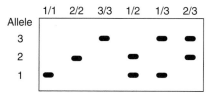

Allele 1

Allele 2

Allele 3

Diploid genotypes present in population

Figure 9.12 Detection of microsatellite polymorphisms by PCR and gel electrophoresis. (a.1) Microsatellite alleles differ from one another in length. (2) Sequence determination from both sides of a microsatellite enables the construction of primers that can be used to amplify the microsatellite by PCR. (3) Gel electrophoresis and ethidium bromide staining distinguish the alleles from each other.
(b) Microsatellites are often highly polymorphic with many different alleles present in a population. With just three alleles, there are six possible genotypes. With N (any number of) alleles, there will be $\frac{N}{2}(N + 1)$ genotypes.

For Minisatellites, Because of Their Size, Detection of Polymorphisms Depends on Southern Blotting and Hybridization Probes

Like microsatellites, minisatellites exhibit variations in size. For each minisatellite locus, different alleles contain different numbers of units. However, because minisatellites consist of units 10–40 bp in length repeated tens to thousands of times, they have an average size of 2–15 kb, and as a result, most are too long to be amplified efficiently by PCR. In addition, as mentioned, the unit sequences of most minisatellites occur at multiple locations in a genome; they thus tend to come in dispersed families of anywhere from 2 to 50 members. Gel electrophoresis and Southern blot hybridization enable the simultaneous analysis of the many individual loci.

Southern Blotting and Hybridization with a Minisatellite Probe Enable Detection of Polymorphisms at Many Different Loci Simultaneously

The protocol for detecting minisatellites begins with the digestion of genomic DNA using a restriction enzyme that does not cut within the minisatellite itself but does cut within closely flanking sites. As with any other RFLP analysis, you then run the restriction fragments on a gel, perform a Southern blot, and then probe the blot with a labeled DNA fragment derived from the minisatellite locus. The result is an autoradiograph displaying the pattern of different sized fragments representing the 2–50 loci of that minisatellite in that genome (Fig. 9.13).

Unlike traditional RFLPs, which result when point mutations create or eliminate a restriction site, minisatellite RFLPs arise from and reflect changes in the actual size of the locus itself. In this respect, they are similar to microsatellites, but on a larger scale. And since minisatellite loci come in families with individual members dispersed throughout the genome, analysis of a single Southern blot can often provide information on all members of the minisatellite family.

Deletions, Duplications, and Other Insertions: Protocols Detect Disruptions of Wild-Type Sequences

Protocols for looking at deletions, duplications, and other insertions depend on the fact that such DNA changes alter the wild type in one of two ways: they either bring together sequences that were not neighbors before or they disrupt sequences that were previously joined.

Researchers can use any protocol that exploits these changes. PCR, for example, can reveal a deletion or insertion through amplification of DNA in the interval between primers. The PCR products of both wild-type and changed DNA are run on a gel that can distinguish the differences in size (Fig. 9.14a).

It is also possible to detect deletions and insertions through RFLP analysis. If a stretch of DNA is closely flanked by restriction sites on either side, a deletion will reduce the size of the restriction fragment carrying that stretch by exactly

Digest DNA to obtain minisatellite fragments.

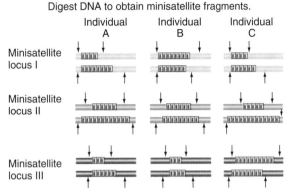

Run fragments on a gel. Perform Southern blot. Hybridize with probe containing minisatellite sequence.

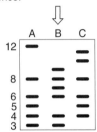

Figure 9.13 Minisatellite analysis provides a means of simultaneously detecting polymorphisms at multiple loci. A hypothetical minisatellite sequence is present at three genomic loci in three individuals. Each of the loci can be polymorphic, with different alleles defined by different numbers of repeating units. The three individuals are all heterozygous at each locus. To detect the minisatellite polymorphisms, you digest the genomic DNA from each individual with a restriction enzyme that cuts outside the minisatellite region. You then separate the digested DNA by gel electrophoresis, make a blot, and probe the blot with the minisatellite sequence. In this example, each individual produces a different pattern of six gel bands. The purpose of this protocol is not to assign a specific band to a specific locus, but rather to obtain a whole genome fingerprint based on simultaneously detected polymorphisms at multiple loci.

the size of the deletion, whereas an insertion will enlarge it by precisely the length of the inserted material (Fig. 9.14b.1). Probes that hybridize between the variant region and the closest restriction site will pick up these differences. If the DNA carries a particular restriction site within the region of deletion or insertion, digestion with the enzyme that cuts at that site will produce an altered pattern of restriction fragments that distinguish the variants (Fig. 9.14b.2 and 3).

Complex Haplotypes: For Efficient Detection, Sequencing Is the Technique of Choice

None of the protocols described so far enables researchers to detect all the differences that define each complex haplotype. Different forms of some genes, such as those encoding the major histocompatibility proteins in mammals, may differ by multiple single-base changes combined with small deletions and insertions; these differences may add up to 10% of the

DNA of certain regions. Since each allele consists of the entire complex of changes, you might imagine using a package of different assays, each one detecting one of the possible changes that could occur at the locus. In practice, however, it is more efficient and more accurate to sequence the entire polymorphic region when there is a need to know all of the changes associated with any allele at a complex locus (Fig. 9.15). The automation of sequencing makes assays of this type less formidable than they were even in the recent past.

Karyotype Analysis Detects Gross Chromosomal Rearrangements

With protocols for distinguishing single-base changes, microsatellites, minisatellites, deletions, insertions, and complex haplotypes, geneticists can analyze the vast majority of DNA variations essential to genotype detection. A very small percentage of DNA variation, however, consists of changes so big and so drastic that it is possible to detect them without recombinant DNA technology, through the microscopic examination of karyotypes. These gross aberrations include changes in chromosome number, such as trisomy 21, inversions in which segments of the same chromosome reverse themselves, and translocations in which large chunks of different chromosomes exchange places. We describe methods for detecting these and other gross chromosomal rearrangements in Chapter 12.

On the Horizon: The Simultaneous Analysis of Hundreds of Thousands of Alleles by DNA Arrays on Microchips

The detection of genotype by any PCR-based method requires a separate reaction for each locus to be analyzed from each individual in a study. Increased capacity could only be obtained by an increase in the size and complexity of the automated equipment used for analysis, but considerations of space and money would make such equipment affordable only by private companies, not basic research labs.

A new technology that combines the microprocesses of computer chip manufacturing with the chemical methods of DNA synthesis may help resolve this impasse. In 1997 scientists in both industry and academia developed tiny "chips" able to hold arrays of hundreds of thousands of different DNA sequences. With such arrays it is possible to probe a DNA sample for hundreds of thousands of alleles at a time.

There are currently several methods for manufacturing DNA arrays: one is based on the types of "masks" that computer chip makers use to etch tiny circuits, while another is based on the process by which inkjet printers deposit tiny droplets of liquid at precise locations on paper. The product of both these methods is a small DNA chip (about an inch square) with a checkerboard pattern of microscopic blocks. Each block is coated with many identical copies of just one of the two DNA strands that represent a particular segment or allele of a particular gene. Thus, each block can detect the presence or absence of a particular allele.

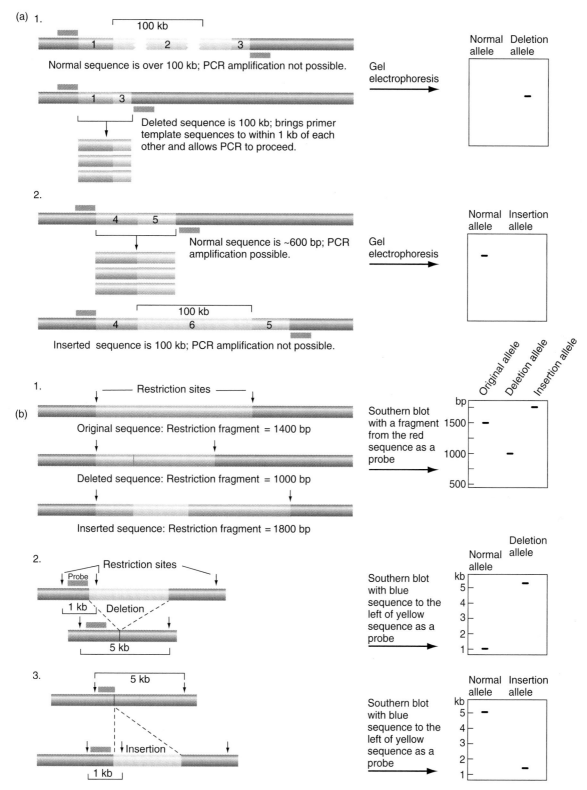

Figure 9.14 Deletions and insertions can be detected by PCR or as RFLPs. (a) The PCR protocol can identify large insertions and deletions because there is a limit on the length of DNA that can be amplified efficiently in a PCR reaction. (1) Primers that flank a large region in one allele of a locus that is deleted in a second allele will produce only a PCR product from the deleted allele. (2) Conversely, primers that flank a small region of a locus that is disrupted by a large insertion in a second allele will amplify only the nondisrupted allele at this locus. (b.1) Deletions or insertions that occur within a restriction fragment can produce RFLPs through a reduction or increase in size, which is detected by Southern blot analysis. (2) If a deleted region contains a restriction site, the deleted allele may show a larger restriction fragment than the wild-type allele (which seems counterintuitive). (3) Conversely, if an insertion contains a restriction site, the allele carrying the insertion may show a smaller restriction fragment than the wild-type allele.

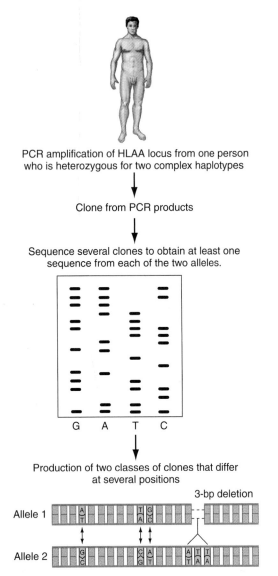

PCR amplification of HLAA locus from one person
who is heterozygous for two complex haplotypes

↓

Clone from PCR products

↓

Sequence several clones to obtain at least one
sequence from each of the two alleles.

G   A   T   C

↓

Production of two classes of clones that differ
at several positions

3-bp deletion

Allele 1

Allele 2

**Figure 9.15   The variations associated with a complex
haplotype are best defined by sequencing.** Using automated
protocols to sequence an entire polymorphic region is often the most
rapid and accurate way to detect changes associated with
polymorphic alleles at a complex locus.

To perform an analysis, one labels a DNA sample from an organism or cell with a fluorescent dye, denatures the sample by heat (which also causes the sample to fragment into many small pieces), and then places the sample on the chip for incubation at a reduced temperature. If a DNA strand in the sample is complementary to one on the chip, that sample strand will hybridize to the chip at a particular block. After a sufficient time, DNA that has not hybridized to any block is washed away, and the chip is analyzed automatically under a fluorescent microscope controlled by a computer. The computer software determines which blocks detected complementary strands (representing particular alleles or DNA segments) in the sample, and which did not.

A one-inch-square chip able to hold 400,000 independent DNA fragments has already been produced. Such a chip could screen an average of four alleles at each of the 100,000 genes in the human genome. Researchers could also use it to probe the cDNA produced from the total mRNA content of a cell or tissue; this protocol would provide a rapid readout of which genes are active, and at what level, in a particular cell at any developmental stage.

Like PCR, DNA chips will no doubt find many applications beyond those for which they were originally developed. The power of DNA chips comes from their capacity for simultaneously determining whether or not a DNA sample contains each of several hundred thousand short sequences. For example, one could use DNA chips to identify single nucleotide polymorphisms located anywhere within a particular locus; in this way it would be possible to rapidly sequence, in thousands of individuals, a particular genomic segment containing thousands of bases (across a medically important gene like the one for cystic fibrosis). With such a tool, geneticists and medical researchers will be able to perform complex genetic analyses that were not yet possible a few years ago. Technology developers are now poised to increase the capacity of DNA chips and reduce their costs by mass production. Ultimately, DNA chips will provide the necessary power for associating individual transcription units with complex traits, as we describe in Chapter 10.

TWO APPLICATIONS OF GENOTYPE DETECTION: DISEASE DIAGNOSIS AND DNA FINGERPRINTING

With reliably reproducible protocols for the rapid detection of genotype, geneticists can identify which alleles an individual carries at particular polymorphic loci. They can use such knowledge to distinguish one person from another and determine, through comparisons, how individuals and groups of individuals are related.

Disease Diagnosis: Discovering Whether an Individual Carries One, Two, or No Copies of a Particular Allele

Medical geneticists use RFLPs and PCR-based protocols to diagnose a large number of genetic conditions. Specific examples include the autosomal recessive diseases of sickle-cell anemia and cystic fibrosis, the X-linked illness of hemophilia, and the autosomal dominant Huntington disease, but there are many, many more, perhaps hundreds will be known by the time you read this book. The main prerequisite of such diagnosis is knowledge of one or more DNA polymorphisms that either cause the condition or are closely linked to the disease-causing mutation.

Direct Detection of a Disease-Causing Mutation

Sickle-cell anemia occurs, as we have seen, when a person carries two copies of a *β-globin* gene with single-base substitutions at one site. The substitution replaces an A with a T and changes the encoded amino acid from glutamic acid to valine. By chance, the mutation also destroys the recognition site of *Mst*II and several other restriction enzymes (Fig. 9.16a). Thus, a particular RFLP marks the actual mutation site and makes it possible to use restriction enzyme digestion in conjunction with Southern blotting and labeled probes to diagnose the disease. Suppose, for example, a carrier couple (genotype $Hb\beta^A$, $Hb\beta^S$ or in simplified form, *AS*) with one affected child (*SS*) want to discover the disease status of the fetus they have recently conceived (Fig. 9.16b). Because the parents are heterozygotes and the disease-related RFLPs within the *β-globin*

gene segregate according to Mendel's first law, the fetal genotype could be *AS, AA,* or *SS.* To find out which one it is, you first obtain DNA from fetal cells, which you can retrieve through amniocentesis. You then digest the DNA with *Mst*II, which recognizes a site within the normal gene that is eliminated by the disease mutation; run the resulting fragments on a gel; and use a probe for the region adjacent to the variable site to detect the genotype. Figure 9.16c shows hypothetical data for the whole family: carrier parents, affected child, and fetus. As you can see, the fetal genotype was *AA*, which means the younger sibling will neither have sickle-cell anemia nor carry the sickle-cell trait.

Sickle-cell anemia is one of the few genetic diseases where this diagnostic approach is possible. This is because the specific mutation is known; all individuals with the disease have the same mutation; and by chance, a restriction enzyme recognizes a site that disappears with the mutation.

You can also detect the sickle-cell mutation directly with PCR-based protocols, which are much faster than RFLP analysis. You could, for example, use PCR in conjunction with hybridization of allele-specific probes for the sickle-cell mutation to detect sickle-cell genotypes. You would begin by constructing two short probes for the small segment of the *β-globin* gene that carries the sickle-cell mutation. One of the probes would be an ASO complementary to the normal *β-globin* allele; the other, an ASO complementary to the mutated allele. Since the two sickle-cell alleles differ from each other in only one nucleotide, the two ASO probes would differ from each other in the same way, and you would construct them such that the difference would lie in the middle of the 19–21 base sequence. Under appropriate conditions of temperature, the ASO for the normal allele would hybridize only to the normal alleles in a sample, while the ASO reflecting the mutation would hybridize only to the disease allele.

Direct detection of the disease-causing mutation is also possible for some cystic fibrosis (*CF*) alleles. Close to 70% of CF sufferers—those affected by the most severe form of the disease—carry the same mutation: a three-base-pair deletion from chromosome 7 that eliminates the codon for the 508th amino acid—a phenylalanine—from the cystic fibrosis transmembrane regulator protein (Fig. 9.17a). Because researchers have defined the specific deletion, PCR-based protocols using ASOs are a good way to diagnose this condition. In one family with four children, for instance, one daughter had cystic fibrosis (Figure 9.17b). Although her two sisters and one brother were not affected by the disease, these three normal siblings wanted to know whether they were heterozygous carriers or homozygotes for the normal allele. To detect their genotypes, medical clinicians simply obtained a blood sample from each, used PCR to amplify a DNA region spanning the 3-bp deletion, attached the PCR products to a nitrocellulose filter, and screened them with allele-specific probes for the normal and mutated sequences. The results of this analysis appear in Fig. 9.17c. They show that one of the sisters is completely free of the disease allele, while the other sister and the brother are both carriers.

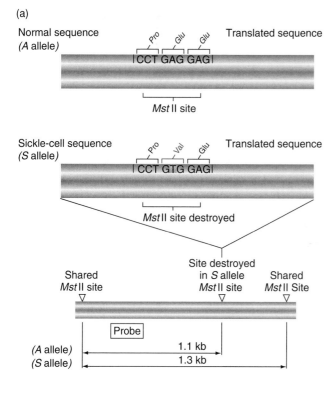

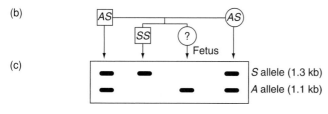

Figure 9.16 Direct detection of the sickle-cell genotype.
(a) The sickle-cell mutation destroys a recognition site for the *Mst*II restriction enzyme. Consequently, a probe that hybridizes to the left of this site recognizes a 1.1 kbp *Mst*II restriction fragment from the wild-type allele and a 1.3 kbp restriction fragment from the sickle-cell allele. (b) A family pedigree in which both parents are carriers, a first child has sickle-cell disease, and a fetus is of unknown genotype. (c) Southern blot analysis shows the RFLP genotype associated with each genotype at the sickle-cell locus. The fetus is homozygous for the wild-type allele.

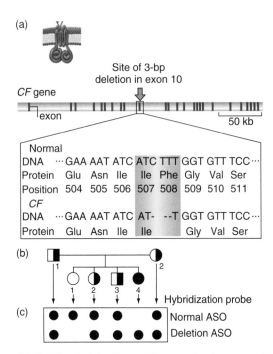

Figure 9.17 Direct detection of the most common cystic fibrosis mutation. The *CF* gene extends across 250,000 base pairs and is organized in 27 exons. It encodes a protein with 1480 amino acids. (a) The most common disease-causing mutation in the *CF* gene is a deletion of three bases in exon 10. It is possible to amplify the region containing the site of the most common mutation by PCR and divide the PCR products into two aliquots. You then blot the aliquots onto filter paper and probe with ASOs for the wild-type and mutant alleles. The ASO for the mutant allele differs by the absence of three bases from the ASO for the wild-type allele. (b) Pedigree of a family in which one daughter (child 4) has cystic fibrosis. (c) Analysis of the results of an ASO hybridization test provide direct information on the *CF* genotype of all family members.

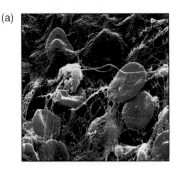

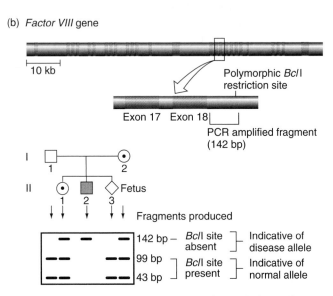

Figure 9.18 Diagnosis of hemophilia through the indirect detection of genotype at the factor VIII locus. The factor VIII protein participates in a cascade of reactions that result in formation of a blood clot. (a) A polymorphic *Bcl*I restriction site within intron 18 of the *factor VIII* gene has no effect on gene function but can provide a marker to follow the segregation of the gene from parents to children. (b) The family described by the pedigree has two healthy parents, but the mother is an obligate carrier of the disease mutation because she has passed this X-linked disease on to her son; her carrier status is signified by a circle with a dot in the middle. By comparing the RFLP pattern obtained from the mother's DNA with the pattern from her son's DNA, you can see that the disease allele is associated with the 142-bp *Bcl*I restriction fragment, and the wild-type allele in the mother's genome contains a *Bcl*I restriction site that causes this fragment to be cut into two pieces, one 43 bp and the other 99 bp in length. Using this information, you can determine that the firstborn sister is a carrier like her mother, while the male fetus will be disease free.

Only 70% of the mutant cystic fibrosis chromosomes in human populations carry a deletion of codon 508. The remaining 30% carry one of roughly 600 other mutations in the *CF* gene, each of which accounts for only a small fraction of all *CF* mutations. Because the standard test for cystic fibrosis can detect only 32 of these 600 minor variants, the diagnosis of heterozygotes or homozygotes for one of the rare *CF* alleles requires a different strategy of genotype detection. At present, the most feasible strategy is one used for the presymptomatic diagnosis of hemophilia A, described next.

Indirect Detection: Using Family History to Assess Linkage to a Polymorphic Marker That Itself Has No Effect on the Trait

In contrast to the well-defined changes that cause sickle-cell anemia and the majority of cystic fibrosis cases, many diseases arise from any one of hundreds of different mutations in the same gene, some known, some not yet defined. Hemophilia A, for example, is caused by hundreds of known mutations, none of which accounts for more than a small fraction of all cases. The disease arises from a mutation in the gene for factor VIII, a protein that forms an essential part of the blood-

clotting cascade (Fig. 9.18). In people with a defective factor VIII, clots do not form properly. The gene for factor VIII resides on the X chromosome and is 200 kb in length. There are many polymorphic nucleotides among the exons and introns of the *factor VIII* gene. These polymorphisms have no effect on phenotype. But in individuals with hemophilia, an additional change, either a missense or a nonsense mutation, affects gene expression. Even though a clone of the gene may be available, it would be difficult to sequence and synthesize probes for every potential mutation. As a result, it is not feasible to test for each and every one of the hundreds of mutations that cause the disease. And even if this were possible, there would most likely be additional mutant alleles not yet characterized that could also cause the disease. Because of these complexities, medical practitioners do not try to detect the exact mutation responsible for the disease; instead, they look for a marker, that is, for any DNA variant that is by chance closely linked to the disease allele. The marker can be an RFLP, a single-base substitution that does not affect a restriction site, or even a nearby microsatellite. As long as the marker is close enough to the gene, or better yet, inside the gene, it will cosegregate with the disease-causing mutation from one generation to the next, enabling geneticists to infer the presence of the disease mutation even when its exact nature remains unknown. For each family, trial and error is the basis for finding the most appropriate diagnostic markers.

To cull useful information concerning an X-linked disease such as hemophilia A from a two-generation pedigree, investigators must find that the mother is heterozygous for the marker under consideration. Suppose a couple with a hemophiliac son and a normal daughter wished to learn the *factor VIII* genotype and thus the disease status of the male fetus the woman was carrying. For this prenatal diagnosis, clinicians would use PCR to amplify from DNA obtained from each family member a particular 142-bp segment of the gene. In one recorded case, the amplified region contained a polymorphic *Bcl*I restriction site in intron 18 that was linked to the disease-causing mutation in this family; the nearest flanking *Bcl*I sites were not polymorphic (Fig. 9.18a). The clinicians then cut the amplified DNA with *Bcl*I and evaluated by electrophoresis in a polyacrylamide gel the restriction fragments produced in the DNA from each individual. Figure 9.18b shows the results of their analysis. DNA from the father carried the *Bcl*I restriction site; as a result the enzyme cut the 142-bp PCR products into fragments that were 99 and 43 bp long. Since hemophilia is an X-linked disease, you know that the father—a normal male— carried only the restriction site and fragments marking the normal gene. DNA from the hemophiliac son did not have the restriction site; his PCR product was not cut by *Bcl*I and thus remained 142 bp in length. From this we can conclude that the disease-producing allele is linked in this family to the *Bcl*I-negative polymorphism that generates the 142-bp fragment. DNA from the mother and daughter carried both normal and mutated sites and therefore generated three restriction fragments: 99, 43, and 142 bp in length. Finally, *Bcl*I digestion of DNA from the male fetus generated 99- and 43-bp fragments. Based on this data, the diagnosis was: *factor VIII* genotype normal; the newborn would be disease free.

This case history underscores the fact that the RFLPs associated with most genetic diseases are markers only. The RFLPs themselves do not cause the condition but are closely linked with the mutation that does. In different families, however, the disease allele may be associated with different alleles at the marker locus. (As mentioned earlier, sickle-cell anemia is the exception to this rule.) RFLP analysis is thus possible only if DNA is available from two or more generations of family members—some with, some without the condition; some homozygotes or hemizygotes (as is the case here), some heterozygotes for the particular marker sites under consideration.

In the future, techniques for DNA sequencing that use DNA chips may become so automated and inexpensive that they will replace the indirect strategy just described for detecting hemophilia A. With an automated protocol for obtaining sequence information across an entire gene, such as the *factor VIII* gene or the *CF* gene, it would be possible to diagnose directly every type of mutant allele without the need for genotype information from other family members.

Indirect Detection of Uncloned Genes May Eventually Lead to Cloning and the Possibility of Direct Detection
Because the gene for hemophilia has been cloned and characterized, the task of finding DNA markers within or very close to the gene is relatively simple and straightforward. As a result, it is possible to follow the transmission of disease alleles with nearly 100% certainty in most families. Unfortunately, for many inherited diseases, the causative gene has not been identified or cloned. The first step in understanding the genetics of such diseases is the use of linkage analysis to determine the genomic map position of the gene in question on the basis of its phenotypic expression (see Chapter 10 for details). With the chromosomal position in hand, it is possible to use the same procedure as that described for hemophilia to detect the disease in family members—with one major difference: DNA markers for an uncloned gene are inevitably farther from and therefore show some recombination with the disease gene locus. If the closest marker available, for example, lies 1 centimorgan (cM) from the disease locus, there is a 1% chance of recombination with the disease allele in every generation of transmission from parent to offspring. By definition, therefore, the accuracy of predictions based on relatively distant markers is significantly less than 100%.

The discovery of the gene for Huntington disease (HD) illustrates how the progression from mapping to cloning made it possible to move from indirect detection with some uncertainty to direct detection with close to 100% certainty. Over 30,000 Americans show one or more symptoms of Huntington disease—involuntary, jerky movements; unsteady gait; mood swings; personality changes; slurred speech; impaired judgment—and an additional 150,000 have a 50:50 chance of expressing the dominant condition as they age. Although symptoms usually show up between the ages of 30 and 50, the first signs of the disease have appeared in people as young as 2 and as old as 83. Some people with a family history of HD would like to know their genotype before deciding whether to have a family.

In 1983, researchers identified a marker close enough to the *HD* gene that it segregated with it 97% of the time. They named this marker G8 and mapped it to chromosome 4. Between 1983 and 1993, diagnosis of the probability for developing Huntington disease depended on protocols that picked out RFLPs in the G8 marker region. Then, in 1993, after 10 years of intensive research that began with the G8 marker, investigators identified

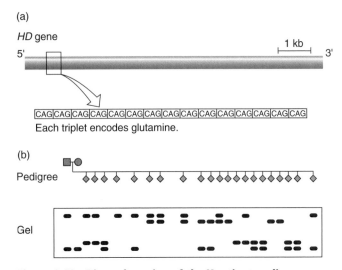

Figure 9.19 Direct detection of the Huntington disease genotype. (a) The *HD* gene contains a region with CAG repeats that each encode a glutamine in the *HD* gene product. More than 42 CAG repeats produce the *HD* phenotype. (b) Researchers can ascertain directly the number of repeats in any *HD* locus through PCR amplification and size determination by gel electrophoresis and ethidium bromide staining of the repeat region. They performed such a PCR analysis on all members of a Venezuelan family in which both parents are heterozygous for the disease alleles, analyzing two independent samples for many individuals to confirm the diagnosis. The smaller PCR fragments have fewer than 42 CAG repeats and are wild type. The larger PCR fragments have enough CAG repeats to cause the disease. One benefit of this type of analysis is its ability to predict the severity of the disease, which increases with the number of CAG repeats.

and cloned the gene (Fig. 9.19a). With the gene in hand they were able to uncover the unusual mutation—a variable tandem repeat of the CAG trinucleotide (see the Genetics and Society box in Chapter 6)—that causes the disease. With knowledge of the gene and actual mutation, it is now possible to detect *HD* alleles directly with PCR (Fig. 9.19b). The normal allele contains up to 34 repeats. Disease-causing alleles carry 42 or more, with the precise number of repeats possibly affecting expressivity. Evidence so far indicates that the greater the number of repeats, the earlier the age of disease onset. Those who inherit the disease allele invariably get the disease if they live long enough. Thus, although expressivity, which depends on the number of triplet repeats, is variable, penetrance is complete (see Chapter 2 for a discussion of expressivity and penetrance).

The cloning of the *HD* gene demonstrated that Huntington disease results from one type of mutagenic event—an expanded triplet repeat—that is directly detectable by a PCR-based assay. As it turns out, expanded triplet repeats cause a variety of neurological and other genetic disorders, including the fragile X syndrome described in the Genetics and Society box in Chapter 6. This is not true for all genetic diseases, many of which have an underlying cause that has so far evaded analysis. As data from the human genome project accumulate, however, more and more of these are likely to reveal their genetic secrets.

DNA Fingerprinting: Comparing Genotypic Patterns for Many Loci to Distinguish and Determine the Relationship of Different Individuals

For most applications of genotype detection, including the viewing of disease mutations, investigators like to have specific reagents (such as a DNA clone or a set of oligonucleotides) that in conjunction with a particular protocol (such as Southern blot analysis or PCR amplification) reveal an individual's genotype at one locus. Thus a good DNA probe is one that hybridizes only to the polymorphic restriction fragments generated from a single locus, a good pair of PCR primers are those that amplify only a single genomic site, and a desirable reagent-protocol combination is one that distinguishes between two or more alternative alleles at one site. The wherewithal to focus on the genotype at one site was the natural end point of recombinant DNA techniques aimed at reducing genomic complexity. Sometimes, however, geneticists seek the "big picture," a quick snapshot of a combination of genomic features that can identify an individual or reveal whether two individuals from the same population (be it human, other animal, or plant) are related to each other and if so to what degree. For this purpose, comparing individual genotypes at one or a few diallelic loci cannot provide sufficient information, because many unrelated individuals will by chance have the same genotype. On the other hand, as we discussed in Chapter 8, the whole genome is far too complex to picture in its entirety. The ideal solution—**DNA fingerprinting**—lies somewhere in between these extremes of analyzing one or a few loci and trying to make sense of the whole genome at a glance.

DNA Fingerprinting Is the Simultaneous Detection of Genotype at a Set of Unlinked Polymorphic Loci Dispersed at Random Throughout the Genome

How many such loci would you have to examine to discover with some certainty whether an exact match in genotype between two DNA samples reliably demonstrates that the two samples come from the same individual (or identical twins) and no one else? A simple calculation shows that the probability of two unrelated individuals having identical genotypes at a locus with two equally prevalent alleles is 37.5%—quite a high probability (Fig. 9.20a). However, the chance that the same two individuals will be identical at 10 such loci, all unlinked, is only 0.375^{10}, or 0.005%—quite a low probability (Fig. 9.20b). The result of 0.005% means there is 1 chance in 20,000 that the two will by chance have the same genotype at 10 unlinked loci. By extension, if you simultaneously detect genotype at 24 separate two-allele loci, the chance of two individuals being the same at all 24 drops to 0.375^{24}, or 1 in 17 billion. Since the total human population is less than 8 billion, the probability that two genetically unidentical individuals would have the same genotype at all 24 loci is virtually 0. In short, a relatively small number of loci is sufficient to produce a multilocus genotype pattern that, like a traditional fingerprint, will be unique for each individual (or pair of identical twins) within the population of a species.

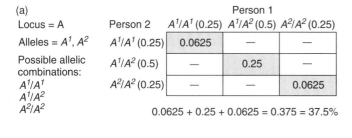

(b)

Loci	A	B	C	D	E	F	G	H	I	J

Probability of $0.375 \times 0.375 \times 0.375 \times 0.375 \times 0.375 \times 0.375 \times 0.375 \times 0.375 \times 0.375 \times 0.375$
equivalence $= 0.375^{10} = 0.005\%$

Figure 9.20 Calculating the probability of relatedness between two individuals. (a) Consider the simple case of an arbitrarily chosen locus (A) with two alleles (1 and 2) that are equally prevalent in a population of people. If you assume the parents of the population are all heterozygotes, then the population itself will be distributed among the three possible genotype classes—1/1, 1/2, and 2/2—according to the Mendelian ratio of 1:2:1 (or 1/4:1/2:1/4). You can therefore use the statistical tools described in Chapter 1 to determine the probability of two unrelated individuals having identical genotypes at the chosen locus. With two alleles, there are three possible allelic combinations—homozygous A^1A^1, heterozygous A^1A^2, homozygous A^2A^2, and they occur in the ratio 1/4:1/2:1/4. Applying the law of the product to determine the likelihood of two independent events occurring at the same time, you determine the probability that two individuals are both homozygous $A^1A^1 = 1/4 \times 1/4 = 1/16$; the probability that two individuals are both heterozygous $= 1/2 \times 1/2 = 1/4$; and the probability that both are homozygous $A^2A^2 = 1/4 \times 1/4 = 1/16$. Then, applying the law of the sum to predict the likelihood of occurrence of any one of these mutually exclusive combinations, you find: $1/16 + 1/4 + 1/4 = 3/8 = 0.375 = 37.5\%$. Hence the probability that any two unrelated offspring have identical genotypes at any one diallelic locus is quite large: 37.5%. (b) Now consider a set of 10 unlinked loci. The probability that the same two people have identical genotypes at all 10 unlinked diallelic loci is extremely remote, because the genotype at each is independent of all the others. Once again, you can use the law of the product, this time to calculate just what the probability would be for two people to have matching genotypes at 10 loci. It comes to $0.375^{10} = 0.00005$, or 0.005%. Expressed another way, 0.005% = 1 chance in 20,000.

For this hypothetical example, we have assumed that each of 24 distinct loci has only two alleles occurring in equal frequency. Such a simplifying assumption makes it possible to understand the resolving power of the numbers presented. In 1985, Alec Jeffreys and coworkers improved on the example in two ways. First, they uncovered a class of human loci with many alleles—the minisatellite loci described earlier. Second, they found that probes for many of these multiallelic loci picked out a whole group of unlinked, polymorphic loci of very similar sequence around the genome. The multilocus pattern produced by the detection of genotype at a group of unlinked, highly polymorphic loci is called a **DNA fingerprint.** Because the unit elements that define most minisatellites are present at many unlinked loci throughout the genome, and because each minisatellite locus is highly polymorphic, with many distinguishable alleles, minisatellite probes make the

perfect reagent for producing the multilocus patterns of DNA fingerprints.

Fingerprinting Protocols Use Restriction Enzymes and Southern Blotting to Detect Length Polymorphisms at Minisatellite Loci

The simplest DNA fingerprint reveals which alleles an individual carries for one family of minisatellites. To obtain such a fingerprint, you simply digest a sample of genomic DNA with a restriction enzyme that does not cut within the minisatellite itself, but does cut within closely flanking sites. As with any other RFLP analysis, you then run the restriction fragments on a gel, perform a Southern blot, and then probe the blot with a labeled DNA fragment derived from the minisatellite (Fig. 9.21a and b). The actual DNA fingerprint consists of an autoradiograph displaying the pattern of different sized fragments produced from that genome (Fig. 9.21b and c). The most useful minisatellite families are those that have 10–20 members per genome. This range of numbers is small enough to allow the resolution of all the loci as individual bands on an autoradiograph, but large enough to provide true fingerprint information. If one requires finer distinctions to resolve the relationship between two different DNA samples, one can always obtain and combine data from two, three, or even more minisatellite families.

Today, many laboratories no longer rely on the restriction analysis of minisatellites for DNA fingerprinting. Instead they use PCR to amplify the simple sequence repeats of microsatellites and then analyze the resulting PCR products by gel electrophoresis. With a sufficient source of sample material, there is no limit to the number of microsatellite loci that can be analyzed in this manner. As the number of analyzed loci increases, the probability of a coincidental match between samples from different individuals decreases in an exponential manner.

Geneticists Can Use DNA Fingerprints to Identify Individuals and Determine Parentage

DNA fingerprints are a powerful tool for forensic analysis. Prosecutors and defense attorneys alike use them to demonstrate the identity of a suspect and show the likelihood of his presence at the scene of a crime or to prove the innocence of someone falsely accused. The Genetics and Society box "Does DNA Fingerprinting Serve the Interests of Justice?" examines some of the issues raised by the forensic use of DNA fingerprinting. In one case, a man arrested on rape charges in 1981 had blood type alleles that matched those of cells in the semen found on the victim. At that time, the typing tests were too crude, that is, did not have the resolving power, to prove that the semen definitely came from the accused. But after the victim picked the accused out of a lineup, the court used the test results in conjunction with other evidence to convict the accused. Eleven years later, a defense lawyer filed an appeal. In it, he used DNA fingerprinting to establish that the semen obtained from the victim carried DNA markers the convicted man did not possess. As a result of the appeal, the court

Genetics and Society

Does DNA Fingerprinting Serve the Interests of Justice?

"Innocent until proven guilty" and "beyond a reasonable doubt"—these precepts form the keystone of the U.S. system of justice. In the twentieth century, new scientific technologies, including traditional fingerprinting; lie detection; chemical tests for the nitrite residues of explosives; blood tests; and, most recently, DNA fingerprinting, have posed a dilemma for the prosecutors, judges, defense attorneys, and jurors entrusted with seeing that justice is done. On the one hand, the technologies seem to provide "objective evidence" existing apart from the subjective perceptions of the people who witness or commit a crime. As one juror put it: "You can't argue with science." On the other hand, this presumed objectivity must be scrutinized on three counts: Is the underlying scientific theory valid? Is the procedure itself reliable? Have the results been properly interpreted?

Take, for example, the test for determining if a person has recently handled explosives. Known as the Greiss test, it can reveal the presence of nitrites on the hands or clothing of a suspect. The British government used it in its successful prosecution of six men accused of planting powerful bombs the evening of November 21, 1974. The bombs left 21 dead and 162 injured in two pubs in Birmingham, England. Largely because the scientist performing the Greiss test reported finding nitrites on the hands of two of the men, all six were convicted. They spent 16 years in jail before data from a variety of studies showed that several substances other than explosives can yield positive results. Among these substances are old playing cards, cigarette packages, lacquer, and aerosol sprays; as it happened, the men had been playing cards and smoking cigarettes just before their arrest. In 1993, the Birmingham Six were released. The Greiss test, as applied to their case, was fallible and therefore unsat-

isfactory. The theory that it would detect only the nitrites of explosives did not hold up, which of course, invalidated any forensic application of the technique.

How does DNA fingerprinting measure up in terms of theory, procedural reliability, and interpretation of results?

The Theory

There should be no better test for identifying a criminal or clearing an innocent suspect than DNA fingerprinting. No one questions the possibility of detecting DNA length polymorphisms arising from the variable number of tandem repeats in minisatellites, and the scientific reasoning from there to the establishment of DNA fingerprints is clear and logical.

Reliability of Procedures

DNA fingerprinting protocols, if carried out with care, caution, and sufficient controls, produce reliable, reproducible results. But each specific technique requires the meticulous skills of a highly trained technician, and procedural details need to be available for review. If a sample of DNA is not properly prepared, for example, its migration on a Southern blot may be aberrant, leading to incorrect conclusions about whether bands on an autoradiogram do or do not match.

The new DNA fingerprinting methodology burst on the scene so suddenly in the mid-1980s that in the first several years of its use as a forensic tool there were no standards and no regulation of public or private crime laboratories. In the early 1990s, four-fifths of forensic DNA testing took place in laboratories funded and run by police or prosecutor agencies. Many of these laboratories knew

reversed its decision and in 1993, the convicted man was set free.

That same year, the courts for the first time accepted plant DNA as evidence in a murder trial. The defendant, accused of killing a woman whose body was found abandoned in the Arizona desert, owned a pickup. When police searched the back of the truck, they found a few seed pods from a Palo Verde tree. The county sheriff's department asked a molecular geneticist if there was any way to match the pods to an individual tree at the scene of the crime. To answer that question, the geneticist had to resolve two others: Did the seed pods contain enough DNA for analysis? and Do Palo Verde plants have sufficient genetic variability to identify an individual through

DNA fingerprinting? When research showed that the trees have a high degree of variability, the geneticist turned to RAPD, the PCR-based protocol that uses short, arbitrarily chosen 10–20-bp primers to create a genetic profile from minute amounts of DNA. With multiple primers, he was able to identify the DNA of individual trees, and in blind tests he picked the correct tree from a "lineup" that included the other 11 Palo Verde trees at the crime site. Although this evidence does not necessarily place the defendant at the scene of the crime, it does strongly suggest that his truck was there.

More recently, DNA fingerprints have helped demonstrate that skeletal remains unearthed at Ekaterinburg in the Ural Mountains belonged to Czar Nicholas II and his family,

the details of a case before running the DNA tests, and some would not divulge the details of their DNA fingerprinting procedures. Currently, there is still little oversight of forensic laboratories. In fact, there is more regulation of clinical laboratories that determine, among other things, whether a sick student has mononucleosis than there is of forensic laboratories whose test results can help send a suspect to the electric chair. To correct this situation, the government, in conjunction with scientists and legal experts, began in 1992 to develop acceptable standards and criteria for evaluating DNA technology and the technicians who carry it out.

More difficult to control is the possibility of evidence tampering. A criminal might leave samples of someone else's blood or hair at the scene of a crime, or overzealous law enforcement officers might claim that they obtained a DNA sample from a location, when in fact they didn't.

Interpreting the Results of Specific Applications

This is the area of greatest controversy. The principal stumbling block is the meaning of an exact match between two DNA fingerprints obtained from correctly performed protocols. Is there a possibility that someone else in the population, other than the suspect, might have the exact same fingerprint? If a group of alleles occurs in most people, a match at these loci is meaningless. For less common alleles, at any given locus, identical twins will show a match, nonidentical siblings will have a one in four chance of a match, and the probability of finding a match in unrelated people will depend on how common the allele is in the population as a whole. To estimate this probability, most labs determine the frequency with which each of the minisatellite alleles under consideration occurs in the population at large and then use the product rule to calculate the probability of the observed pattern of multiple alleles occurring by chance. As in the calculations in Fig. 9.21, they usually arrive at a vanishingly small number.

But how accurate are these estimated probabilities and how accurate do they need to be? Since the measure of allele frequency in a population determines the reliability of the probability calculations, the question of how to determine allele frequencies becomes significant.

The FBI originally determined allele frequency in a very small, nonrandom pool of individuals—a few hundred volunteer FBI agents! Ideally, however, the reference population should be similar to the one from which the perpetrator comes, but it is often unclear what this means. Suppose, for example, that a Caucasian male born into a relatively inbred Scandinavian community, is accused of rape. At one locus, he shares an allele with the semen sample. This allele could be very rare in the general population, but much more common in the small inbred population from which the man comes. If allele frequency is determined from the general population, it will appear more likely that the suspect committed the crime than if the allele frequency is determined from the small inbred population.

To deal with such problems, geneticists are now gathering an immense amount of data on large populations as well as on dozens of isolated ethnic groups around the world. As experts analyze and correlate the information on allele frequency, the probability estimates will become increasingly accurate. Meanwhile, one solution is to use the highest possible frequency estimates for the probability of a chance match. These should be conservative enough to compensate for any ethnic substructures that might exist in the general population. Moreover, by increasing the number of loci examined, one decreases the possibility of a chance match; with enough loci, that possibility is virtually nil.

A scorecard for DNA fingerprinting would thus show that the underlying theory is valid and the technology is reliable if properly executed; but each specific application must be approached with great caution and insight. Today, the majority of geneticists and legal experts believe that this new technology, if handled with care and protected by adequate safeguards, can indeed play a role in ensuring that justice is done.

who died in 1918 during the Bolshevik revolution. Geneticists established the relationship by comparing DNA from the excavated bones with samples obtained from a number of living relatives of the Romanov family, including Prince Philip, Duke of Edinburgh. With this information, it became possible to disprove the claim of Anna Anderson that she was the Grand Duchess Anastasia; in three independent analyses, her DNA (obtained from hair and from biopsy samples removed as part of an examination for cancer years before her death in 1984) did not match that of members of the Romanov line—living or dead.

Finally, DNA fingerprinting has a number of uses in the experimental study of plants and animals other than humans,

where researchers can use it to determine the social and evolutionary relationships between individuals. In one study, they looked at the mating patterns of pilot whales whose underwater marine habitat makes their social behavior hard to observe. DNA fingerprinting of members of stranded or hunted groups showed that fetal cohorts, the equivalent of littermates, were fathered by a few unrelated or a larger number of related males, but the fathers were rarely present in the groups examined. The DNA fingerprinting evidence thus suggested that mating occurs during fleeting meetings with other groups of pilot whales. If this turns out to be true, it would mean that pilot whales display a previously unobserved mammalian system of mating.

(a) Chromosome

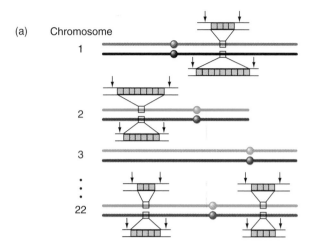

(b) Digest with restriction enzyme.
Run fragments on gel; blot, and probe.

Markers

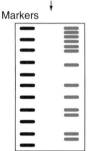

(c)

Figure 9.21 DNA fingerprinting distinguishes individuals through the simultaneous detection of polymorphisms at many loci throughout the genome. (a) Multiple members of a single minisatellite family are distributed on different chromosomes throughout the genome. (b) Digestion of genomic DNA with a restriction enzyme separates different minisatellites onto different restriction fragments. (c) The actual fingerprint.

CONNECTIONS

Armed with a growing arsenal of molecular tools that make it possible to look at DNA directly, geneticists have uncovered a large reservoir of genetic variation in the molecule of heredity. They can use their detailed knowledge of this variation to diagnose hereditary disease through the detection of genotype at single well-defined loci; and to determine the identity and degree of relatedness of individual organisms through comparisons of genotypes at multiple unlinked loci.

The new tools have broad applications in genetic research, including the mapping and sequencing of entire genomes. Before the advent of direct genotype detection, breeding studies were limited by the amount of phenotypic variation amenable to analysis. Over the years, researchers selected and preserved mutant phenotypes for peas, fruit flies, mice, and other model organisms, but it was hard to develop strains carrying and segregating alleles at more than a dozen loci. With more than 12 simultaneously segregating traits, the expression of one phenotype began to interfere with the expression of others. And a lack of predetermined matings and inbred populations made it hard to follow human traits, forcing human genetics to the periphery of the experimental loop.

Today, researchers have moved beyond a strict dependence on phenotype. In Chapter 10 we describe how they can use their newfound capacity for viewing DNA variation directly to determine the genetic configurations of both parents in a cross. With this ability, they have gone from 3-point crosses that locate three genes in relation to each other to 1000-point crosses that track all sorts of permutations from generation to generation. Geneticists use the data compiled from these many-point crosses to map and clone new genes as well as to sort out the genetic complexities of the large number of multifactorial traits that used to be experimentally out of reach. With these new approaches, the human species has become a superb system for genetic analysis.

ESSENTIAL CONCEPTS

1. *DNA polymorphisms* provide the basis for the direct detection of genotype by recombinant DNA technology.
   a. Ultimately the genotype of an organism is its entire DNA sequence, which can be divided into individual genotypes at many different loci. A *locus* is any chromosomal location. Some loci contain or occur within genes; other loci have no known function and are thus called "anonymous." Sequence

variations at a locus are DNA polymorphisms, or alleles.

b. The average frequency with which polymorphisms occur in a genome is very high, particularly in noncoding regions.

c. There are five common classes of DNA-based polymorphisms categorized according to the nature of the DNA variation and the protocols used to detect it: (1) single-base changes, (2) *microsatellites,* (3) *minisatellites,* (4) deletions and insertions, and (5) *complex haplotypes.*

2. Protocols for detecting polymorphisms combine techniques for focusing on specific DNA segments, or loci, with techniques for viewing variation at these loci.

a. For single-base changes, there are several protocols of detection:

(1) *RFLPs* are caused by any DNA mutation that creates or destroys a restriction site. Geneticists detect RFLPs by restriction enzyme digestion, gel electrophoresis, Southern blotting, hybridization to probes, and autoradiography.

(2) PCR-based protocols, in addition to detecting restriction-site polymorphisms, make it possible to follow any kind of DNA variation, including any single-base differences between two samples. Probes known as *ASOs,* are essential components of such protocols. ASOs distinguish normal from mutant alleles differing by only a single base pair.

(3) *SSCPs* do not require prior allele-specific sequence information or the synthesis of ASOs to detect single-base differences. SSCPs reflect the unique most-favored conformation assumed by each sequence of single-stranded DNA under specific conditions of denaturation followed by cooling.

(4) *RAPD* uses short oligonucleotides of random sequence that hybridize to complementary targets at random in the genome. Loci with two target sequences of correct spacing and orientation are amplified by PCR. RAPD, because it does not require any information obtained by cloning and sequencing, is most useful in the initial genetic

analysis or fingerprinting of an uncharacterized organism.

b. Microsatellites are relatively small, extremely polymorphic loci that are easily identified and typed by PCR.

c. Minisatellites are also extremely polymorphic, but because of their large size, they are detected by Southern blotting and hybridization probes.

d. Deletions and insertions are detected through protocols that reveal new juxtapositions of DNA sequence.

e. Complex haplotypes are analyzed most efficiently by sequencing.

f. Microchips of *DNA arrays* containing hundreds of thousands of independent DNA sequences will greatly increase the ability to detect and compare complex genotypes.

3. Disease diagnosis and *DNA fingerprinting* are two applications of genotype detection.

a. Medical practitioners use RFLPs and PCR-based protocols to detect directly the well-defined genotypes responsible for sickle-cell anemia and many forms of cystic fibrosis. For diseases whose molecular basis is much more complex or not yet characterized, they use indirect protocols that depend on DNA markers in conjunction with family histories that make it possible to assess the extent of linkage between the disease mutation and a marker polymorphism that may have no effect on the disease phenotype.

b. DNA fingerprinting is the simultaneous detection of genotype at a set of unlinked, highly polymorphic loci dispersed throughout the genome. The detection of polymorphisms at minisatellites via RFLP analysis was the original basis of DNA fingerprinting. Among the recently developed alternatives, RAPD uses generic PCR primers, requires much less starting material, and can be used with organisms that have not been well characterized genetically. Because DNA fingerprints can identify individuals and determine the probability of relationship, they have many applications in forensics as well as in biological research.

SOCIAL AND ETHICAL ISSUES

1. DNA fingerprinting has become accepted as evidence in many recent trials. However, jurors and judges who do not have not scientific training are being asked to evaluate the results of such tests, including how carefully a laboratory performed the test and the statistical relevance of the matches between sets of data. Scientists often differ in their assessment of such evidence. Is it fair to try someone on the basis of this kind of genetic evidence using a panel of jurors who do not have a prior

understanding of the basic principles behind the technique? Can you think of ways to improve the use of this data in criminal cases?

2. Bob and Karen have a child who was born with Tay-Sachs, a recessive genetic disorder. When Karen became pregnant again, genetic tests were done on all family members including amniocentesis on the fetus. Results showed that Karen is a carrier of a mutated allele, but Bob

is not a carrier (the fetus is unaffected). These results strongly suggest that Bob could not be the father of the first child. It is possible that a germline mutation arose in Bob's cells. Additional genetic tests could settle the paternity issue. Who should be told? The mother only? Both mother and father? Neither?

3. Mary is a 25-year-old woman who has a good job with a company where she has room to advance in her career. When she accepted her job at the age of 22, she signed a contract that stipulated that she would tell her supervisor of any change that occurred in her health status. A footnote in the contract stated explicitly that a change in health status was defined not only as a symptomatic change but also as a prognostic change. This footnote was meant to include persons who tested positive for HIV even when they did not yet exhibit the symptoms of AIDS. It seemed clear to Mary that the purpose of this stipulation was to provide the company with grounds for dismissal. Three years after the beginning of Mary's employment, Mary's 56-year-old mother Elizabeth was diagnosed with breast cancer. Because several other women in the family had breast cancer, there is a high probability that there is an inherited predisposition to breast cancer in the family. Tests indicated that her mother

does carry a mutant allele of the *BRCA1* gene, and therefore, Mary has a 50% chance of having a *BRCA1* allele that predisposes her to breast cancer. The counselor encouraged Mary to take the test since there are preventative measures that could be taken if she finds out that she is also positive. But, if she takes the test and the result shows that she carries the BRCA1 mutation, she feels that she will have to tell her boss and possibly lose her job. On the other hand, if the test shows that she doesn't have the mutation, a large load will be taken off her mind as she realizes her risk of disease is no more than that of other women without the *BRCA1* mutation. What would you do if you were she?

4. A couple who both are dwarfs want to have a child who is a dwarf. Both the parents have achondroplasia, the dominant inherited form of dwarfism. (The parents are heterozygous for the achondroplasia allele. Homozygous individuals die before age one of severe skeletal abnormalities.) They have told the genetic counselor at the testing clinic that they will abort a fetus if it *does not* carry the dwarf allele. Should the clinic perform the test for the achrondroplasia gene, knowing that the parents will choose to abort a healthy fetus simply because it will not be a dwarf?

S O L V E D   P R O B L E M S

I. When random samples of human DNAs were digested with *Bam*HI and examined for RFLPs using a cloned anonymous sequence called d17, three types of band patterns (shown below) were seen.
   a. Make a map of the region recognized by this probe showing the position of restriction sites.
   b. Which of the restriction sites on your map are varying in the population?
   c. Which patterns below represent homozygous and which represent heterozygous individuals?

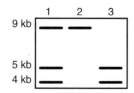

Answer

This problem requires that you understand and recognize RFLPs.

   a. There are two alleles (forms) of a polymorphism seen in these three band patterns: a 9-kb *Bam*HI fragment

and 5 + 4-kb *Bam*HI fragments. The restriction sites possible in this region are:

```
   B          B        B
   L__5.0____L__4.0__J

   B                   B
   L_____9.0_____J
```

   b. The middle *Bam*HI site of the three shown varies in the population. A chromosome containing this *Bam*HI recognition sequence will show the 5 + 4-kb *Bam*HI fragment pattern and a chromosome missing that recognition site will show the 9-kb fragment.
   c. The first pattern represents a heterozygous individual; patterns 2 and 3 represent individuals homozygous for each of the two different forms.

II. The figure shows the pedigree of a family in which an autosomal dominant disease is segregating, and a Southern blot, showing the results of digestion of the DNA of each family member with *Eco*RI and probing with a cloned anonymous sequence. Do the data from this family suggest that the locus of the RFLP detected by this anonymous probe is linked to the disease locus? Explain.

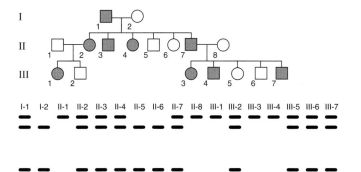

Answer

To solve this problem you will need to understand RFLPs and how they can be used to trace inheritance of a disease through a family. First examine the restriction analysis to determine what the forms of the RFLP are. Two alleles (forms) of the polymorphism represented in this pedigree: an 8-kb and a 7 + 1-kb form. To determine if the locus detected is linked to the disease, compare the restriction patterns seen in affected and unaffected individuals. The affected individuals, I-1, II-2, II-3, II-4, II-7, III-1, III-3, III-4, and III-7 all have the 8-kb allele. All of these individuals except III-7 inherited this allele from an affected parent. III-7 has the 8-kb allele but he must have inherited this allele from his unaffected mother who was homozygous for the allele. The 8-kb form is therefore not associated with the disease allele in this individual. Recombination between the polymorphic locus and the disease locus in the father would have caused the 7 + 1 form to be associated with the disease allele in III-7. *Eight out of nine affected individuals inherited the 8-kb allele*

and the disease allele, suggesting that there is linkage between these loci. Are there unaffected individuals with this allele? Yes, individuals II-1 and II-8 who marry into this lineage carry this allele. In their families, the 8-kb allele must not be associated with the disease allele, but with the normal allele instead.

III. Each of the following is a method for detecting DNA polymorphisms. Where is the polymorphism located in relation to the probe or primer DNA used in each of these techniques?
   a. ASO (allele-specific oligonucleotides)
   b. RAPD (random amplification of polymorphic DNA)
   c. RFLP (restriction fragment length polymorphism)

Answer

This problem requires an understanding of primers, probes, and three molecular detection techniques.

   a. For ASOs (allele-specific oligonucleotides), *the polymorphism is within the DNA sequence that is used as a probe.*
   b. In RAPD, *the polymorphism is usually in the sequence recognized by the PCR primers,* resulting in no amplification (depending on conditions used in PCR). This method will also detect a difference (polymorphism) in the length of DNA between the PCR primers.
   c. In RFLP analysis, *the polymorphism must be somewhere within the restriction fragment that is detected by the probe.* Because the probe is a small sequence within the restriction fragment, the polymorphism may or may not be within that portion of the restriction fragment complementary to the probe.

P R O B L E M S

9-1 Choose the phrase from the right column that best fits the term in the left column.

a. DNA polymorphism	1. DNA element composed of tandemly repeated identical short sequences
b. haplotype	2. polymorphism in the three-dimensional conformation of a single-stranded DNA fragment
c. RFLP	3. combination of linked alleles in a cluster of related genes
d. ASO	4. location on a chromosome
e. SSCP	5. one of two or more sequence variations of a DNA locus
f. DNA fingerprinting	6. a short probe specific for a given allele
g. minisatellite	7. detection of genotype at a number of unlinked polymorphic loci using one probe
h. locus	8. variation in the length of the restriction fragment detected by a particular probe

9-2 Imagine that you have discovered an RFLP closely linked to the locus causing a rare hereditary disease. You have cloned a segment adjacent to the RFLP locus from an affected person and from an unaffected person, chosen at random, and sequenced the two clones. You find that the two segments are identical for the first few hundred nucleotide pairs, and then you encounter a single-base-pair difference. You would like to think that you have discovered the locus of the disease mutation, but realize you need to consider some other possibilities.

What is another possible interpretation of the base pair difference?

9-3 Mutations recognized as polymorphisms of microsatellites occur at a frequency of 1×10^{-3}, a frequency higher than base substitutions.

a. What is the nature of these microsatellite polymorphisms?

b. By what mechanism are these polymorphisms generated?

c. How are polymorphisms of minisatellites generated?

9-4 Your thesis advisor has given you a brand new uncharacterized human cDNA clone derived from a single-copy gene and asked you to analyze it. You decide to perform a Southern blot analysis on DNA from your roommate's family and obtain the results shown below. The family consists of two parents and four children. Further analysis indicates that all six members of this family have normal karyotypes. What do these data suggest (if anything) about the gene detected by the cDNA clone?

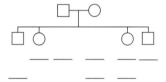

9-5 You have recently obtained a new human clone for a gene that is the subject of your senior thesis. You use this clone to probe a Southern blot of *Eco*RI-digested DNA from a family of four: two parents and two children (Sy and Tim) with the results shown in the figure.

a. How many alleles are represented in these individuals?

b. What is the genotype of each individual with respect to the RFLPs?

```
        Dad  Mom  Sy   Tim
6 kb ┌──────  ───  ───        ──┐
     │
4 kb │ ───         ───          │
3 kb │ ───   ───        ───     │
2 kb │ ───         ───          │
     └──────────────────────────┘
```

9-6 A sample of several different human DNAs was cut with the restriction enzyme *Eco*RI and subjected to gel electrophoresis. A Southern blot was made from the gel and probed with a labeled anonymous cloned sequence. Interpret the data in the blot (illustrated in the figure) in the following ways.

a. Make a drawing showing how the restriction sites are arranged along the chromosome and indicate the

region of the chromosome that must have homology with the probe.

b. How many variant restriction recognition sites are detected in these samples?

c. What is the genotype of each person whose DNA is represented on the gel?

d. Are there other genotypes possible that do not occur in this sample of the population?

e. What patterns would you expect to see in the child if the individual in lane 2 and the individual in lane 4 were to have offspring together? Explain your reasoning.

f. In fact, the person shown in lane 2 is the offspring of two of the other people whose DNA is shown. Can you identify the parents?

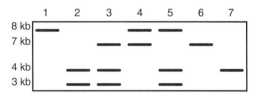

9-7 Sickle-cell anemia occurs in individuals homozygous for the autosomal recessive allele Hb^S. Analysis of Southern blots of human DNA cleaved with the enzyme *Hpa*I and probed with a radioactively labeled β-globin cDNA has revealed that *Hpa*I does not cleave within the β-globin gene. However, the positions of *Hpa*I sites surrounding the gene vary among individuals, producing at least three RFLPs in the population: the sizes of the RFLP alleles are 13.0, 7.6, and 7.0 kb. Shown on the next page are small pedigrees for families in which individuals with sickle-cell anemia are shaded in black, and the results of a Southern blot in which DNA from the individuals in these families was cut with *Hpa*I, run on a gel, blotted and probed with radioactively labeled β-globin cDNA.

a. Which RFLP allele is associated with individual II-1 in the family whose pedigree is shown on the left?

b. Which RFLP allele is associated with the Hb^S allele in individual I-1 in the family whose pedigree is shown on the right?

c. Which RFLP allele is associated with individual I-2 in the family shown on the right?

d. Use the RFLP information to determine the probability that a child of individuals II-2 from the left family and II-1 from the right family will have sickle-cell anemia.

e. Now assume you had only the pedigrees shown below as data and that it was impossible to detect Hb^S/Hb^A heterozygotes phenotypically. Using pedigree data only, what is the probability that individuals II-2 and II-1 would have a child with sickle-cell anemia?

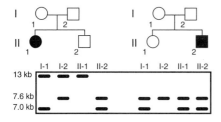

9-8 A probe has been identified that detects a polymorphism 5 m.u. away from an autosomal gene in which a dominant allele causes the disease.

   a. Which allele of the polymorphism is segregating with the disease allele in the family shown below? (The genotype below each individual in the pedigree indicates the alleles of the polymorphism present.)

   b. How can you explain the DNA genotype of individual III- 5?

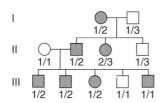

9-9 What requirements are there for an RFLP to be informative for disease diagnosis in a particular family?

9-10 A genetic counselor is examining a family in which both parents are carriers for a mutation at the cystic fibrosis (*CF*) locus. Their first child was born with the disease and the parents have come to the counselor to assess whether the new fetus inside the mother is also diseased, is a carrier, or is completely wild type at the *CF* locus. The normal *CF* gene is 250 kb in length. You have a clone from the 5′ end of the gene that you use as a probe on a Southern blot of *Bam*HI-digested DNA from the mother, father, first child, and fetus, with the following results. Does this analysis provide you with a diagnosis of the genotype at the *CF* locus for the fetus? If so, what is the genotype?

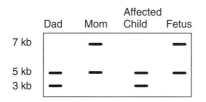

9-11 DNAs corresponding to genes affected by various hypothetical human genetic defects are available. Genomic (total) DNA was prepared from the blood cells of individuals in families, some of whom exhibit the trait, and others of whom do not. This DNA was digested with restriction enzymes, transferred to nitrocellulose filters by the Southern blot procedure, and probed with the radioactive DNAs. The resultant autoradiograms (below each pedigree) show RFLPs in the vicinity of each gene. Darker bands indicate restriction fragments that are present twice in the genome of that individual; all other fragments are present only once. Filled-in symbols indicate individuals affected by the trait. An "X" within a symbol means that person does not show the trait. Note that for this problem, blank symbols correspond to individuals who may or may not display the mutant phenotype.

What is the pattern of inheritance for each trait? Indicate the genotype of the individuals noted by a box under their name.

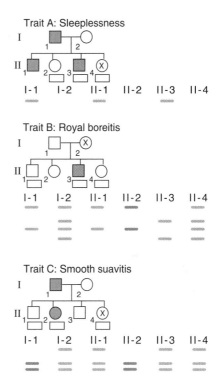

9-12 What are the advantages of PCR-based analysis of genotype compared to a Southern blot–based RFLP analysis?

9-13 Angela and George have one child, and she has sickle-cell disease. They want to have more children, but do not want any of them to have to suffer with this disease. They also do not want to be in a position of having to abort a fetus, so they elect to have *in vitro* fertilization and embryo screening. Briefly list the steps for these two procedures.

9-14 The inheritance of an autosomal recessive trait is shown on the next page for two families. An RFLP that is closely linked to the disease locus has been identified.

   a. Which allele of the RFLP cosegregates with the disease allele in the family on the left?

b. Which allele of the RFLP cosegregates with the disease allele in the family on the right?

c. Is it likely that III-1 is affected?

d. If II-4 and II-5 have a child that is a carrier, what RFLP alleles is this child likely to have?

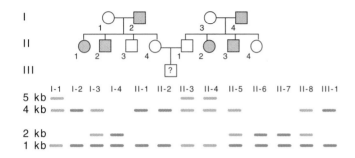

9-15 Arrange the following statements in the appropriate order for determining whether a specific allele is present in a person's DNA.

a. amplify the DNA using PCR

b. label the probe DNA

c. isolate sample DNA from an individual

d. hybridize with ASOs

e. divide the DNA sample from an individual in two

9-16 The ASO technique was used to determine the genotypes of 10 family members with regard to sickle-cell disease, as shown below. Each pair of dots represents the results of ASO analysis for the DNA from one person. The upper row represents hybridization with the normal oligo nucleotide, and the lower row represents the results of hybridization using the mutant oligonucleotide. The three replications of the assay were incubated at 100°C (upper set), 90°C (middle set), and 80°C (lower set).

a. Why do the three replications look different?

b. What are the genotypes of the individuals?

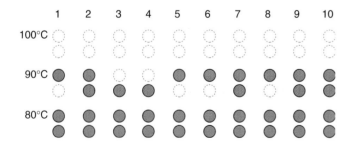

9-17 Given the sequence below and the site at which a single-base polymorphism is present in the population, what would your probe(s) be for ASO analysis? (Assume that ASO probes are usually 18 bp in length.)

allele 1
5′ GGCATTGCATGCTAACCCTATAAATGCGCTAGGCGTAGTTAGCTGGGAATAAAAAGCT 3′

allele 2
5′ GGCATTGCATGCTAACCCTATAAATG<u>G</u>GCTAGGCGTAGTTAGCTGGGAATAAAAAGCT 3′

9-18 What advantage does the SSCP technique for detecting DNA-level variation provide compared to using ASO?

9-19 The first child of the Bleebers was born with an autosomal recessive metabolic disease. They are expecting another child, and want to be certain it will not inherit the same disease. The gene in which the mutations causing this disease to occur is known and has even been cloned and sequenced, but the exact nature of the mutation is unknown. As a result, there is no direct test for a specific mutation in the gene. Suggest two approaches that might help this family.

9-20 Why would you not use PCR to detect a polymorphism in a minisatellite?

9-21 DNA fingerprinting can be used to settle cases concerning paternity. Based on the data for the two different cases below, would you conclude that it is possible that the man is the father of the child or has he been excluded as the father? Explain your reasoning.

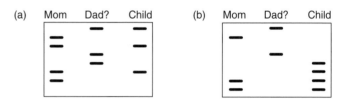

9-22 Two probes that detect polymorphisms were used in a forensic case. The incriminating alleles (based on evidence from the site of the crime) were present in the suspect. These same alleles are found at frequencies of 0.10 and 0.55 in the Caucasian population. However, among Asian-Americans, these same alleles are found at frequencies of 0.08 and 0.01.

a. If the suspect is of Caucasian ancestry, what is the likelihood that another individual in the Caucasian population will have these same two alleles?

b. If the suspect is of Asian-American ancestry, what is the likelihood that another individual in the Asian-American population will have these same two alleles?

CHAPTER
10

THE MAPPING AND ANALYSIS
OF GENOMES

The family tree of the royal Nemanjic Dynasty, Serbia (1346–1350) is an artistic
depiction of the pedigree of the family inheritance long before the familiar pedigree
chart existed. Family trees were painted, woven into tapestries, and carved to
preserve the history of family lines.

In the early era of Mendelian genetics, medical researchers learned from the analysis of pedigrees like the one shown in Fig. 10.1a that hemophilia A is an X-linked recessive trait. This chromosomal assignment and transmission pattern revealed that the hereditary condition is determined by one gene, but it did not provide a way to find the gene or its protein product. Knowledge of the disease phenotype did.

Investigators could use their observations of uncontrollable bleeding in affected individuals as the basis for an educated guess at the function of the responsible gene. **Gene function** in this context is a gene's specific contribution to phenotype. The function of the wild-type hemophilia A gene, they proposed, is production of a normal clotting factor; mutations that inactivate this factor produce hemophilia A. (This usage differs from the meaning of "gene function" in Chapters 6 and 7, where we described the *generic* function of a gene: to govern the synthesis of a polypeptide. Most genes do that. The specific function of the hemoglobin gene, however, is to

govern the synthesis of a polypeptide known as clotting factor VIII, a part of the biochemical cascade that enables blood to clot.)

With this hypothesis in mind, molecular investigators used tools developed between 1950 and 1975 for the direct analysis of proteins and nucleic acids to work out the details of the blood clotting cascade (Fig. 10.1b) and to look for clotting factors in normal individuals that were absent in hemophiliacs. In this way, they isolated a protein known as Factor VIII (Fig. 10.1c). After purifying the protein, they analyzed its amino-acid sequence and used the sequence information to develop a DNA probe that eventually picked out a clone of the gene (Fig. 10.1d). Then, using the clone itself as a probe, they verified that hemophilia A results from mutations in the gene for Factor VIII.

The search for the hemophilia A gene illustrates how researchers rise to the challenge of moving from the disease phenotype to the gene. At the same time, it is an unusual success story that cannot serve as a model for most gene searches.

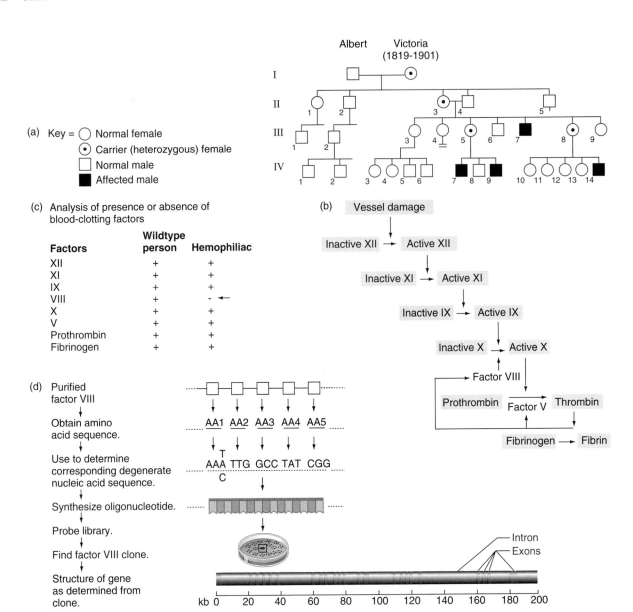

Figure 10.1 **How geneticists identified the hemophilia A gene.** (a) A pedigree of the royal family descended from Queen Victoria. This family tree uses the standard pedigree symbols. Black boxes represent males with hemophilia. (b) The blood clotting cascade. Vessel damage induces a cascade of enzymatic events that convert inactive factors to active factors. The cascade results in the transformation of fibrinogen to fibrin and the formation of a clot. (c) Many hemophiliac patients do not have an active form of Factor VIII. Blood tests can determine the presence or absence of the active form of each factor involved in the clotting cascade. The results of such analyses show that hemophiliacs, such as those found in Queen Victoria's pedigree, lack an active Factor VIII in their blood. (d) Starting with purified Factor VIII, scientists isolated DNA clones containing the *Factor VIII* gene. Researchers determined the amino-acid sequence of purified protein. Knowledge of this sequence enabled them to synthesize a degenerate oligonucleotide. They then used the oligonucleotide as a probe to screen a genomic library for clones containing all or parts of the gene. Finally, they sequenced the positive clones (that is, the clones with which the probe hybridizes) to determine the structure and coding sequence of the *Factor VIII* gene.

Unfortunately, geneticists cannot follow the hemophilia scenario to clone the genes responsible for the vast majority of inherited human diseases. The problem is that even with a clearly defined mutant phenotype, it is usually not possible to pinpoint the underlying molecular defect. For example, Duchenne muscular dystrophy (DMD) is a recessive X-linked genetic condition that strikes 1 in 3500 boys. The phenotype—muscular deterioration—results from a defect in the physiology of muscle tissue. But there are thousands of enzymes and structural proteins involved

uniquely in normal muscle function, and most of them remain unidentified. Without a way to determine which one is defective in DMD, investigators face a catch-22: They know the inheritance pattern of a disease phenotype, but cannot relate the phenotype to a specific gene function; and without an idea of gene function, they cannot work their way from gene function to protein to clone and back to a confirmation of gene function.

The dilemma for characterizing the genes associated with most human hereditary diseases thus becomes: How can you use

TABLE 10.1   A Comparison of the Developmental Complexity and Genome Size of Model Organisms

Organism			
Type	Species	Developmental Complexity	Genome Size (megabases)
Bacterium	*Escherichia coli*	One-cell prokaryote	4
Yeast	*Saccharomyces cerevisiae*	One-cell eukaryote	15
Nematode	*Caenorhabditis elegans*	~1,000 cells	100
Fly	*Drosophila melanogaster*	~50,000 cells	180
Mouse	*Mus musculus*	10^{11} cells	3,000
Human	*Homo sapiens*	10^{14} cells	3,000

a phenotype unassociated with a detectable molecular defect as a tool for discovering the gene sequence responsible for the phenotype? The answer is: through the development of high-resolution, whole-genome maps and the use of molecular techniques to analyze the genetic information provided by the maps.

In this chapter, we describe how geneticists make three types of whole-genome maps—high-density linkage maps, long-range physical maps of entire chromosomes, and complete sequence maps—and then use these maps as a starting point in their search to locate and understand the function of previously undefined genes. A concerted, international effort to develop large-scale maps of increasing resolution is underway for many plants and animals, including humans and all the organisms described in the genetic portraits in Part VI of this book. The human endeavor is referred to as the Human Genome Project; it got off to an official start in 1991 and is scheduled to reach its goal of mapping the human genome right down to the complete sequence of its 3 billion nucleotide pairs by the year 2003.

The collaborative push of whole-genome mapping projects has given rise to the new discipline of **genomics:** a branch of biology dedicated to the development and application of more effective mapping, sequencing, and computational tools. Genomicists use refined molecular techniques for linkage analysis, cloning, and the direct detection of genotype to compile a vast amount of data, which they then analyze by computer. Sophisticated programs enable them to predict the existence and, in some cases, the general functions of previously undefined genes. Their molecular biologist colleagues can then confirm the computer-generated projections through biochemical studies and genetic engineering in model organisms.

Model organisms, often used in genomic analysis, have many genetic mechanisms and cellular pathways in common with each other and with humans; but unlike humans, they are amenable to classical breeding experiments and direct manipulation of the genome. Five model organisms are under study in conjunction with the Human Genome Project (Table 10.1): *E. coli* (representing prokaryotes), *S. cerevisiae* (the bread yeast, representing single-celled eukaryotes), *Drosophila* and *C. elegans* (rep-

resenting multicellular animals of moderate complexity), and mice (representing the mammalian class of animals, which also includes humans). Researchers chose to analyze these models in conjunction with the human genome because the organisms have increasingly complex genomes that are amenable to genetic analysis and that have already been extensively mapped and studied. By comparing the DNA of model organisms to that of humans, researchers seek to uncover genes and other critical DNA elements that are conserved across evolutionary lines.

The ultimate goal of genomic mapping and genetic analysis is to view the genome of any organism as a whole, both to make a connection between every gene and its specific function and to understand how all the genes fit together on the chromosomes. With this knowledge for all 100,000 genes in the human genome, biologists may one day understand the molecular basis of every genetic disease as well as the networks of interactions that bring each person to life in a unique way.

As we describe the tools and potential applications of genomics, we encounter a compelling general theme: The social and personal repercussions of the information gained from genomic analysis are generating new areas of bioethical concern that require close attention as the new knowledge unfolds.

Our discussion of the mapping and analysis of genomes examines

- How large-scale maps—linkage, physical, and sequence—serve as guides to whole genomes.

- Positional cloning in which geneticists use large-scale maps to go from a phenotype to a clone of the responsible gene.

- How geneticists use sequence maps and molecular studies on humans and model organisms as the basis for moving from the clone of a gene to its function.

- How researchers adapt procedures for mapping and cloning single genes for the genetic dissection of complex traits.

- The scientific progress, medical promise, and potential ethical pitfalls of the Human Genome Project.

LARGE-SCALE MAPS SERVE AS GUIDES TO WHOLE GENOMES

Chromosomal maps show the location of genes, markers, centromeres, telomeres, and other points of interest along the chromosomes of an organism. In Chapters 4 and 8, we described techniques for mapping a small number of loci in a relatively small region of a genome. Here we examine how genomicists expand on these techniques to produce linkage, physical, and sequence maps for a whole genome.

High-Density Linkage Maps: Computerized Analyses of Transmission Data Position Unlimited Numbers of Markers in Relation to Each Other

We saw in Chapter 4 that a simple two-point cross can demonstrate linkage if the two loci under analysis lie close enough together on the same chromosome to show less than the 50% recombination expected with independent assortment. We also saw that the frequency of recombination between the two loci provides a direct measure of the distance separating them, as recorded in centimorgans (cM), or map units (m.u.). (Geneticists studying humans, mice, and other vertebrates use the centimorgan unit of measure, which we adopt in this chapter.) Finally, we learned that it is possible to integrate multiple pairs of linked loci into a "linkage group" by performing many different three-point crosses with overlapping sets of loci. The **linkage maps** constructed from the recombinational data of multipoint crosses depict the distances between loci as well as the order in which they occur on a chromosome.

With the traditional linkage analysis described in Chapter 4, geneticists can map only a handful of loci in any one cross because easy-to-follow phenotypes, such as those of color and gross morphology, tend to interfere with each other's expression. For example, 34 different genes cause mice to have white spots; each white-spot gene maps apart from the others, but in a two-point cross following, say, black coat color and white spots, it is impossible to tell which one of the 34 white-spot genes is contributing to the phenotype observed. This problem is particularly acute in human genetics since it is not possible to breed people with combinations of mutations appropriate for linkage analysis. As a result, with loci defined solely by variant phenotypes, the only way to construct an all-encompassing linkage map is through the tedious process of accumulating and pooling information from many two- and three-point crosses.

Geneticists have solved this problem using some of the recently developed tools for the direct detection of genotype (see Chapter 9). With these tools, they can follow the segregation of a very large number of loci in a single experimental plant or animal cross or in a human pedigree and use the transmission data to build a high-density linkage map. Such a high-density map shows the relative positions of hundreds or thousands of closely spaced (that is, densely packed) markers.

An early goal of the Human Genome Project is to publish electronically the sequences of a set of unique molecular markers spaced 100 kb apart along the entire genome; each marker has its own DNA sequence that is not duplicated anywhere else in the genome. Such one-of-a-kind markers, which tag positions along the DNA molecule, are called **sequence tagged sites,** or **STSs.**

Although an STS can be associated with any unique genomic sequence, the most useful type of STS is one that is highly polymorphic, that is, has enough distinct alleles that most individuals are likely to be heterozygous at that site. Investigators anywhere in the world interested in mapping a human disease could use STS data from electronic databases, along with machines for synthesizing oligonucleotide primers and carrying out PCR amplification, to recreate the reagents necessary for determining STS genotypes as a tool for mapping disease traits.

In the 1990s, microsatellites—those DNA elements composed of 10–100 tandem repeats of two- or three-base sequences—emerged as the best kind of STS-associated marker for generating long-range linkage maps that extend across entire chromosomes. There are several reasons for the microsatellite's mapping suitability. As mentioned in Chapter 9, most microsatellites have three, four, or even more common alleles. This large amount of polymorphism means that the probability of an individual being a heterozygote for a particular microsatellite locus is very high; in humans it is between 50% and 70%. Recall that heterozygosity is one prerequisite for inclusion of a locus in a linkage analysis. Second, there are many microsatellites dispersed throughout the genomes of all higher eukaryotes, and they are easy to identify and characterize. Third, PCR-based protocols make microsatellite genotyping, often referred to simply as "typing," very quick and efficient; to determine which alleles are present in an individual, you simply use PCR to amplify the particular microsatellite and then analyze the PCR products by gel electrophoresis and ethidium bromide staining (see Chapter 8). Different alleles are distinguishable as DNA fragments that migrate with different mobilities. Finally, because minute quantities of DNA from any tissue are sufficient to type a microsatellite locus, there is no limit to the number of different microsatellites that can be typed in an individual. Microsatellites, in fact, make it possible to type hundreds or even thousands of loci in a single cross. Unique sequences surrounding microsatellites serve as microsatellite-associated STSs.

Compiling Data from a 1000-Point Cross for Linkage Analysis

To generate and analyze mapping data from a 1000-point cross, researchers use the same methods described in Chapter 4 for a three-point cross, with two main differences: (1) They determine the genotypes of all individuals involved in the cross—locus by locus—through molecular techniques for direct marker typing and (2) they analyze the accumulated data by computer for evidence of linkage relationships.

Although this type of analysis might seem time-consuming, with today's fast computers and sophisticated programs, a computer can analyze the data from a 1000-point cross in less than a day. The computer analysis of the mapping data is thus not a limiting factor. The real limitation lies in the generation of data in the first place.

Mapping Panels Simplify the Problem of Generating Large Amounts of Data in a Reasonable Time

To type 1,000 loci in 500 individuals, you would have to perform $1,000 \times 500 = 500,000$ separate DNA tests. This means 500,000 PCR reactions for specific microsatellites, 500,000 lanes on gels, and 500,000 entries into a database. Although straightforward conceptually, this amount of work would be daunting for even the largest laboratories.

In the early 1980s, geneticists from around the world, especially those working with humans and mice, realized it would be much more efficient for researchers working with the same organism to perform all their linkage studies on samples from the same set of individuals. If, for example, 50 different laboratories each typed just 20 different loci within the same set of individuals, the combined data would be equivalent to a single linkage study of 1000 loci. The set of DNA samples used in multiple analyses of this type make up a **mapping panel.** Each laboratory participating in a particular collaboration receives aliquots of every sample in the mapping panel. After typing each sample for alleles at one or more loci, researchers send the genotype data recorded for each locus in each sample of the panel to a central database. There computers analyze the new loci for linkage with each other and with all other loci already in the database. Figure 10.2a illustrates how this division of labor would work in mapping the mouse genome.

The best characterized and most widely used human mapping panel is the CEPH reference pedigree resource. CEPH is an acronym for the Centre d'Etude du Polymorphisme Humain, a research center in Paris, France. The CEPH panel consists of DNA samples from 517 members of 40 three-generation families. CEPH researchers obtained white blood cells from each of the 517 individuals in the panel to establish permanent cell lines. Whenever a DNA sample runs out, they thaw a frozen aliquot of cells, grow up a colony in a petri dish, and extract a fresh sample of the same DNA. The panel thus provides human geneticists with unlimited material for genotypic analysis. The linkage maps depicted in Fig. 10.2b were derived from the CEPH panel as of October 1992.

A well-characterized mapping panel is a powerful resource for mapping a newly cloned locus. This is because the researcher who has cloned the new locus need only determine genotype information for this single locus and no others among the members of the mapping panel. After completing his or her single-locus analysis, the investigator can input the data over the World Wide Web to the central database. There a computer program will automatically perform a linkage analysis, assign a map position for the new locus (thus enlarging the database available to others), and transmit the results back to the investigator. With the unlimited quantities of material available from established mapping panels, the number of points in the cross can grow, one at a time, from hundreds to thousands to tens of thousands of loci. When a map shows one

gene or marker for each centimorgan of a genome it is considered a **high-density linkage map.**

One original goal of the Human Genome Project was a high-density STS-based linkage map of the human genome. As of June 1994, mappers had identified and mapped 6000 independent human microsatellite loci to serve as STS markers along the linkage groups that correspond to the human chromosomes. Since the total linkage-computed length of all 23 human chromosomes together is 4000 cM, each microsatellite locus on the map is separated from a neighboring locus on either side by only 0.675 cM on average.

The next step in the genome project is to establish 30,000 STSs at distances of 100 kb from each other along the entire genome. These 30,000 markers, instead of being derived from microsatellites, will be based on the newly developed DNA chip's ability to detect **single-nucleotide polymorphisms,** or **SNPs** (see Chapter 9 for a discussion of the DNA chip). A single DNA chip can simultaneously determine alleles at all 30,000 marker loci in a short time. A high-density linkage map provides a framework for building long-range physical maps, and for isolating the genes for polymorphic traits, including inherited diseases.

Long-Range Physical Maps: Karyotypes and Genomic Libraries Provide the Basis for Positioning Markers on Chromosomes

Physical maps, the molecular counterparts of linkage maps, are based on the direct analysis of genomic DNA. Such maps chart the actual number of base pairs (bp), kilobases (kb), or megabases (Mb) that define and separate a locus from its neighbors in a particular region of a particular chromosome. As we saw in Chapter 8, geneticists build relatively short-range physical maps by exposing DNA clones to two or more restriction enzymes and comparing the fragments produced in this way; they then use hybridization probes to ascribe a precise position to the genes and markers included in the mapped clone. The major difference between the physical maps discussed earlier and those developed by genomicists is one of scale. In Chapter 8, for example, we described the mapping of cosmid clones that spanned 30,000 bp of genomic DNA. Here we learn how molecular geneticists map a human chromosome averaging 130,000,000 bp in length (4300 times longer than a single cosmid).

The FISH Protocol Locates Cloned Loci—Genes or Markers—Directly on the Chromosomes of a Karyotype

We saw in Chapter 3 that the chromosomes of actively dividing cells in the metaphase of mitosis, when stained with a Giemsa dye and viewed in the light microscope, show an identifiable series of dark and light regions referred to as bands and interbands (Figs. 3.6 and 10.3a). The number, intensity, and width of each band and interband are highly reproducible characteristics that a skilled cytogeneticist can use to distinguish each pair of homologs from all other chromosomes in a cell. The visual description of a complete set of chromosomes in one cell of an organism is a **karyotype.** To facilitate the presentation and

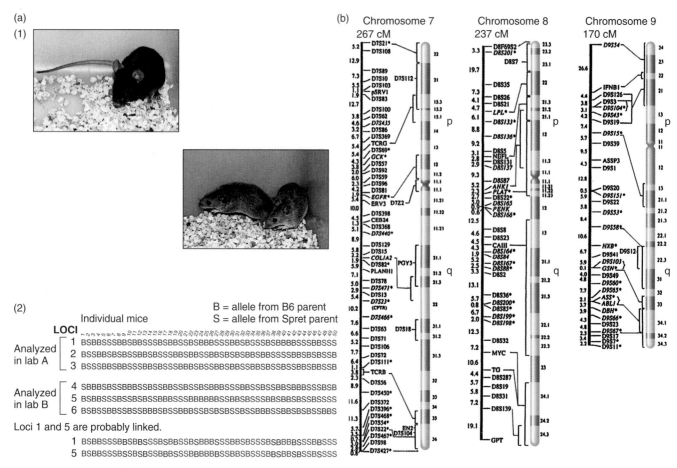

(a)

(1)

(2)

| | | B = allele from B6 parent |
| Individual mice | | S = allele from Spret parent |

LOCI

Analyzed in lab A
1 BSBBSSSBBSBBSSSBSBBSSBSBBBSSSBBSBSSSBSBBBBSSSBBSSS
2 BSBBSBSBSSSBBSBSSSBBSBBSSBBSBSBSBBSBBSSSSBBSBSBSSB
3 BSSSBSBSBSSSBBSBSSSBBBSBSSBBBSBSSSBBSBSSBSBBSBBBSSB

Analyzed in lab B
4 SBBBSBSSBBBSSSBSBSSBBSBSSSSSBBBSSSBBSSSBSBBBBSSSSSB
5 BSBBSSSBSBSBSBBSSSSBBBBBBBSSSBSSBSSSSSBBSBBSBSBSSSS
6 BSBSSBBSBSBBSBBSSBBBSBSBSBSSSSSBBBSSSBBSBSBSSSBBBS

Loci 1 and 5 are probably linked.

1 BSBBSSSBBSBBSSSBSBBSSBSBBBSSSBBSBSSSBSBBBBSSSBBSSS
5 BSBBSSSBSBSBSBBSSSSBBBBBBBSSSBSSBSSSSSBBSBBSBSBSSSS

Figure 10.2 Mapping panels provide an efficient tool for linkage analysis. (a) Using mapping panels in linkage analysis of the mouse genome. (1) Two strains of mice—B6 and Spret—served as the parents in the first generation of a two-generation cross used to create a standard mouse mapping panel. (2) Sample of mapping data generated in two different laboratories. Each mouse in the mapping panel is represented by a number from 1 to 50. Each column represents the alleles in an individual at each of the typed loci (named by a number at the left; loci 1–3 were analyzed in one lab, while loci 4–6 were analyzed in a second lab). Each row represents the data obtained in the typing of a single locus across all the individuals in the panel. A pairwise comparison of allele transmission from loci 1 and 2 shows the inheritance of two alleles from the same parental homolog (either *BB* or *SS*) in 26 instances and the inheritance of alleles from different parental homologs (either *BS* or *SB*) in 24 cases. This recombination fraction (24/50) is not significantly different from the 25/50 recombinants expected in the case of no linkage. The same type of analysis shows no evidence of linkage between 1 and 3 or 2 and 3. Similarly, there is no evidence of linkage among loci 4–6, which were typed in the second lab. But when all the loci from the first lab are analyzed for linkage to the loci typed in the second lab, the computer finds that 1 and 5 have recombined in only 12 of 50 individuals (38 showed no recombination). The data for these two loci are displayed separately in the box beneath the panel, with all instances of concordant inheritance highlighted in boldface type. You can perform a chi square analysis with one degree of freedom to test whether this result is significantly different from the one you would expect from the null hypothesis of 25:25. The result is:

$$\frac{(38-25)^2}{25} + \frac{(12-25)^2}{25} = 13.52,$$

which translates into a *p* value of less than 0.001. The analysis thus indicates a high probability that the two loci are indeed linked. It is possible to perform this type of analysis for any number of loci typed in any number of laboratories on the same set of samples, which constitute the mapping panel. (b) Linkage maps for three human chromosomes. Researchers created these linkage maps on the basis of data obtained in hundreds of laboratories that typed the CEPH panel of human pedigrees. The relationship between the linkage map and the cytogenetic map can be ascertained by direct cytogenetic localization of a subset of markers using FISH analysis. (see Fig. 10.4)

comparison of data from different karyotypes in the same and diverse species, investigators convert the light and dark bands actually observed under the microscope into black and white diagrams of the chromosomes (Fig. 10.3b and c). In these diagrams, the autosomes are numbered in descending order of length. In humans, we have seen, 1 is the longest autosome and 22 the shortest. The X and Y sex chromosomes appear apart from the autosomes. The shorter arm of each chromosome is desig-

nated "p" (for "petit") and the longer arm is "q" (for following p). Karyotypers number each band and interband, starting at the centromere and moving out along each arm toward the telomere.

How do researchers use karyotypes in making maps? They can, of course, locate the centromere and telomeres of each chromosome through microscopic observation. But what about the loci that lie between those points? A protocol known as **fluorescent *in situ* hybridization,** or **FISH,** makes it pos-

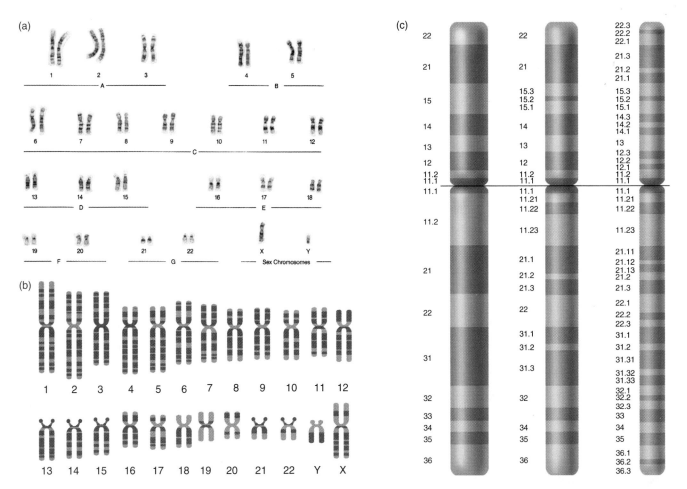

Figure 10.3   The human karyotype: Banding distinguishes the chromosomes. (a) Photograph of a complete set of human chromosomes at metaphase. Staining with Giemsa dye accentuates the bands and interbands. (b) Idiograms for the complete set of human chromosomes. An idiogram is an idealized diagram of the banding pattern associated with a stained chromosome. (c) Chromosome 7 at three different levels of banding resolution. As staining techniques improve, it becomes possible to resolve what previously appeared as a single band into a series of bands and interbands, producing more and more bands along each chromosome. Thus, at one resolution, 7q31 appears as one band. At a slightly higher resolution, 7q31 becomes two bands (7q31.1 and 7q31.3) flanking an interband (7q31.2); and at an even higher resolution, 7q31.3 itself appears as two bands (7q31.31 and 7q31.33) and an interband (7q31.32).

sible to map *any cloned locus* to a site on a metaphase chromosome. Figure 10.4 illustrates the steps of the FISH protocol. In the final step, sites of hybridization appear as spots of illumination on the chromosome under the microscope.

The FISH analysis has several advantages over linkage analysis for the mapping of a newly cloned locus to a particular chromosomal site. First, all clones can be mapped by the FISH protocol, whereas only those clones that detect polymorphisms can be mapped by linkage analysis. Second, researchers can perform a FISH analysis on any cloned locus in isolation, whereas linkage mapping requires the analysis of one locus in relation to another. Third, FISH requires only a single sample on a microscope slide, whereas linkage analysis requires genotype information from a large cohort of individuals. For this reason, human geneticists often use the FISH protocol as a rapid way to obtain the general chromosomal position of a newly cloned locus. Finally, the FISH protocol enables direct comparisons between the map positions of cloned loci and the aberrations (such as deletions or translocations)

that are visible in the chromosomes of individuals with unusual chromosomes. The main disadvantage of FISH is that its resolving power is much less than that possible with linkage analysis. The metaphase bands and interbands used as guideposts in FISH protocols contain the equivalent of 4 to 8 Mb or 5 to 10 cM of chromosomal material. In contrast, we have seen that linkage analysis can resolve map positions down to a fraction of a centimorgan.

Karyotypic analysis thus enables the generation of low-resolution physical maps that locate cloned genes and markers on particular parts of chromosomes. We now describe how geneticists use the cloning techniques described in Chapter 8 to expand these maps with a much finer resolution of loci.

Using DNA Hybridization and Restriction Mapping to Order and Analyze the Clones of a Genomic Library

It is possible to build a continuous high-resolution map across an entire chromosome by piecing together cloned fragments in a one-dimensional jigsaw puzzle. Figure 10.5 illustrates the

(a)

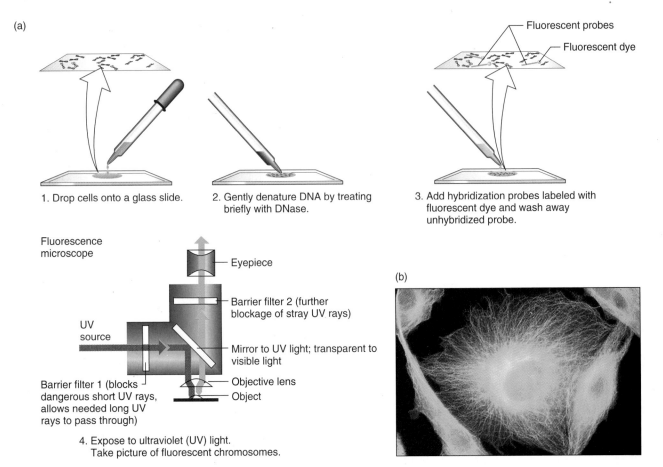

1. Drop cells onto a glass slide.

2. Gently denature DNA by treating briefly with DNase.

Fluorescent probes

Fluorescent dye

3. Add hybridization probes labeled with fluorescent dye and wash away unhybridized probe.

Fluorescence microscope

Eyepiece

Barrier filter 2 (further blockage of stray UV rays)

UV source

Mirror to UV light; transparent to visible light

Barrier filter 1 (blocks dangerous short UV rays, allows needed long UV rays to pass through)

Objective lens

Object

4. Expose to ultraviolet (UV) light. Take picture of fluorescent chromosomes.

(b)

Figure 10.4   The FISH Protocol. (a) The technique. (1) First, drop cells arrested in the metaphase stage of the cell cycle onto a microscope slide. The force of the droplet hitting the slide causes the cells to burst open with the chromosomes spread apart. (2) Next fix the chromosomes and gently denature the DNA within them such that the overall chromosomal structure is maintained even though each DNA double helix opens up at numerous points. (3) Label a DNA probe with a fluorescent dye, add it to the slide, incubate the probe with the slide long enough for hybridization to occur, and wash away unhybridized probe. (4) Now place the slide under a special microscope that focuses ultraviolet (UV) light on the chromosomes. The UV light causes the bound probe to fluoresce in the visible range of the spectrum. You can view the fluorescence through the eyepiece and photograph it. (b) A fluorescence micrograph. Photograph of a baby hamster kidney cell subjected to FISH analysis. It shows the microtubular structure.

basic steps of this kind of mapping: identifying an ordered series of genomic clones that overlap each other from one end of the chromosome to the other; analyzing each clone separately to obtain as much information as possible about its restriction sites and gene locations; aligning the overlapping ends of clones, using information from the restriction map; and combining the physical information from each clone into a physical map of the entire chromosome. Computer programs facilitate the assembly of the large amount of data.

How do researchers achieve the first step of piecing together an ordered series of overlapping DNA clones? As we saw in Chapter 8, a genomic library represents all the DNA of a particular cell incorporated into clones reproduced by cell colonies that are arranged at random on a plate. If the research goal is to identify a gene of interest, this randomness does not matter; so long as there are enough clones in the library, one or more will carry the desired gene, and hybridization probes can isolate it from the pack. But what if the goal is to identify all the overlapping clones in an ordered set, as shown in Fig.

10.5? In such a case, the significant feature of each clone is its chromosomal location, rather than its genetic function.

There are two approaches to developing an ordered set of clones: one uses genetic maps to derive physical maps, the other relies exclusively on physical mapping techniques. The two approaches are complementary. In the mapping of chromosomes, researchers use both.

The approach based on genetic mapping begins with a high-density linkage map of a selected chromosome containing evenly spaced markers no more than 1 cM apart. To generate an ordered set of clones across the chromosome, you use each marker as a hybridization probe to retrieve from the library clones carrying a complementary sequence (Fig. 10.6a.1). The clones retrieved by each marker probe are, of course, much larger than the marker itself. Consider for example, using 6000 markers evenly spaced across the human genome's 3000 Mb to screen a YAC library of the human genome. The markers will lie approximately 0.5 cM from each other; in the human genome, 0.5 cM is the equivalent of 500 kb. Since each YAC in

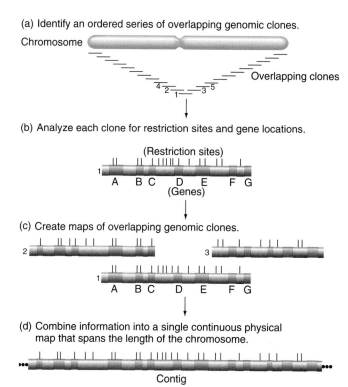

(a) Identify an ordered series of overlapping genomic clones.

Chromosome

Overlapping clones

(b) Analyze each clone for restriction sites and gene locations.

(Restriction sites)

A B C   D E   F G
(Genes)

(c) Create maps of overlapping genomic clones.

(d) Combine information into a single continuous physical map that spans the length of the chromosome.

Contig

Figure 10.5   Building a whole-chromosome physical map.
(a) To produce a whole-chromosome physical map, you first order a set of overlapping genomic clones that extend from one end of the chromosome to the other. Subsequent figures describe various methods of obtaining this ordered set of clones. (b) You next map the restriction sites of each clone in the set through restriction analysis, and analyze individual restriction fragments in other ways, such as Northern blot analysis, to identify transcription units. (c) Computers overlay the different types of maps for each clone onto the overlapping clones to obtain a continuous map. (d) The result is a single continuous map extending the length of the chromosome.

the library carries a genomic insert roughly 1 Mb in length, there is a good chance that a YAC clone picked up by screening the library with one marker (say, marker one, or M1) will overlap a YAC clone recovered by screening the same library with a neighboring marker (M2). You can use cross-hybridization between one YAC clone and the next to demonstrate an overlap between the two YAC clones (Fig. 10.6a.2). Figure 10.6b shows hypothetical data generated by screening a YAC library with seven polymorphic markers (M1–M7). The YACs recovered on screening with M1, M2, M3, and M4 contain regions of overlap, as do the YACs recovered on screening with M5, M6, and M7. A set of two or more partially overlapping cloned DNA fragments that together cover an uninterrupted stretch of the genome form a **contig** (from the word contiguous). Thus, the YACs defined by M1 through M4 make up one contig, while the YACs defined by M5 through M7 form a second contig. Each of these contigs is roughly 3 to 4 Mb in length, while the YACs of which they are composed are about 1 Mb long.

The ultimate goal of high-resolution physical mapping is the generation of one large contig for each chromosome in a genome. But because of the somewhat variable spacing between

Assigning chromosomal positions: A top-down approach

Use M1 as hybridization probe to screen YAC library.

Use M2 as hybridization probe to screen YAC library.

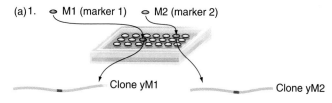

(a) 1.   M1 (marker 1)   M2 (marker 2)

Clone yM1   Clone yM2

2.   Do yM1 and yM2 overlap?

Clone M1

Isolate end fragment of yM1. Label, and use as probe of yM2.

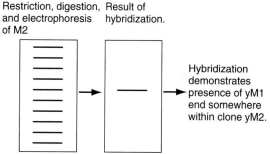

Restriction, digestion, and electrophoresis of M2

Result of hybridization.

Hybridization demonstrates presence of yM1 end somewhere within clone yM2.

(b) Data generated by screening a YAC library with seven polymorphic markers

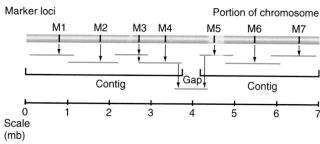

Marker loci   Portion of chromosome

M1   M2   M3   M4   M5   M6   M7

Contig   Gap   Contig

0   1   2   3   4   5   6   7
Scale
(mb)

(c) Complete contig spanning region between M1 and M7

M1   M2   M3   M4   M5   M6   M7

Figure 10.6   Using a high-density linkage map to build an overlapping set of genomic clones. (a) Marker probes retrieve and identify clones that overlap. (1) Markers that map near each other (M1 and M2) serve as probes to screen a YAC library and recover genomic clones. (2) Testing of the clones recovered by nearby markers for overlap consists in using the end fragments from each clone to probe the other. A positive hybridization signal indicates that the end fragment of one clone is present in the other clone. (b) Hypothetical data obtained from screening a YAC library with seven closely linked markers. The seven clones recovered fall into two separate overlapping sets. Considered together, a set of overlapping clones is called a "contig." The contig data show there is an uncloned genomic "gap" between markers 4 and 5. This gap can be filled by using the end fragments from each of these clones to rescreen the YAC library. (c) With the addition of the new clone, a single contig extends across the genomic region between markers M1 and M7.

linkage markers as well as variation in the length of cloned YAC fragments, "gaps" between smaller contigs arise when the distance between two markers is by chance too large to allow recovery of overlapping YACs. The gap between M4 and M5 in Fig. 10.6b is an example. The greater the density of molecular markers, the smaller the number of gaps. To fill in the gap separating two contigs, one can use small fragments from the ends of the contigs as probes to rescreen the YAC library for one or more additional clones that may bridge the gap and thereby convert two smaller contigs into one larger contig. This method of using the ends of unconnected contigs as probes to retrieve clones that extend into an unmapped region is known as **chromosome walking.** The critical feature of any contig is that through the generation of physical maps for each of the clones it contains, it is possible to piece together a physical map that spans the entire contig (Fig. 10.6c). In 1993 researchers constructed a complete physical map of the human Y chromosome using 207 molecular markers to recover 196 overlapping YAC clones that composed a contig spanning 60 Mb of genomic DNA (Fig. 10.7).

In contrast to using a high-density linkage map as the starting point for deriving a long-range physical map, it is possible to generate a whole-genome physical map by physical techniques alone. These exclusively physical approaches do not depend on linkage information derived from pedigree analysis or the initial positioning of molecular markers.

In one such approach, researchers create a cosmid library containing individual clones covering the whole genome of interest (Fig. 10.8a). They then examine each cosmid with molecular techniques that yield a unique pattern of restriction fragment sizes, or a "fingerprint," of the genomic region within the clone (Fig. 10.8b), program a computer to look for overlaps between the fingerprints (Fig. 10.8c), and use overlaps to arrange the cosmids into 24 chromosomal contigs (Fig. 10.8d). (Note that we used the term "fingerprint" the same way as we used it in Chapter 9—to mean a particular pattern of restriction fragment sizes. The only difference is that there, the DNA under analysis was the whole genome of a single individual, while here, it is the very small genomic region present in a single clone.) This bottom-up approach thus breaks the genome into a random jumble of thousands of clones and then generates sufficient fingerprinting information for each clone to allow reconstruction of the small fragments into a complete set of chromosomes.

The most common type of fingerprint is a list of fragment sizes obtained from digesting the cloned DNA with a series of restriction enzymes (review Fig. 10.8b). Each cloned region, when exposed to enough different restriction enzymes, becomes defined by a unique set of fragments of different sizes. When the computer has stored data for a sufficient number of clones (usually tens of thousands), it searches through the entire set, testing each clone against every other clone in the set, for evidence of fingerprint overlaps between pairs of clones (see Fig. 10.8c). The premise behind this strategy is that two truly overlapping genomic clones will share a subset of matching restriction fragments (from their region of overlap), whereas two unrelated clones would be unlikely to have the same number of matching fragments. With the fingerprinting of more and more clones, the gaps between contigs fill in, resulting in longer and longer contigs (Fig. 10.8d). Yeast geneticists used this bottom-up approach to generate a single contig spanning the entire length of yeast chromosome I.

Once a combination of approaches to generating whole-chromosome contigs has produced an ordered array of overlapping clones for all the chromosomes in the genome, researchers can use subcloning, restriction mapping, and hybridization (as described in Chapter 8) to study the clones of each contig in greater detail, locating and analyzing genes and markers of interest. They can then have the computer combine all the information for all the clones into a detailed physical map showing a variety of landmarks across the contigs of all the chromosomes in the genome.

Long-Range Sequence Maps: Molecular Protocols Make It Possible to Obtain a Readout of Every Nucleotide in a Chromosome

Sequence maps show the order of nucleotides in a cloned piece of DNA. The ultimate goal of the human and model-organism genome projects is the determination of the complete nucleotide sequence for every chromosome in the genome. The main hurdle is chromosome size: Current sequencing technology can reveal the order of about 700 nucleotides at a time, but the average human chromosome, for example, is 130,000,000 bp in length. The solution is to fragment a whole-chromosome contig into pieces of sequenceable size and then reintegrate the pieces into one long-running, consecutive sequence; Fig. 10.9 illustrates this process.

The starting point for developing a whole-chromosome sequence is the high-resolution physical map described previously. Since the YAC clones of a whole-chromosome contig are much too large for sequencing, researchers use the subcloning techniques described in Chapter 8 to subdivide them into smaller and smaller clones. Thus, a YAC containing a DNA insert of 1,000,000 bp is divided into an overlapping set of 50 cosmids, each containing DNA inserts of only 40,000 bp. These clones are still too large for sequencing in a single run, but researchers can sequence 600–700 bp at a time and then use the shotgun techniques described in Chapter 8 to reconstruct the complete sequence of the cosmid. Next, they integrate the sequences of the cosmid components of each YAC into a whole-YAC sequence, and finally combine the sequences of each YAC insert of a chromosomal contig into a whole-chromosome sequence.

Although the strategy for generating a whole-chromosome sequence map is easy to describe, obtaining a readout of the 3 billion base pairs in the human genome will be a daunting task. Preparation of millions of DNA samples for sequencing and the actual sequencing itself takes an inordinate amount of time, as does the compilation of 3 billion bases into the long, uninterrupted sequences of the 24 different chromosomes. Until 1998, only a few large research groups had the ability to analyze 3 million to 30 million bases per year. But in 1998, several private companies announced the development of new highly automated sequencing facilities with hundreds of robotic sequencers that could operate simultaneously, 24 hours a day, to complete the entire human genome in less than five years. By 1998, researchers had determined the complete sequence of the 16-chromosome yeast genome and finished over 90% of the readout of the 100 million nucleotide pairs in the 6-chromosome *C. elegans* genome. In 1999, *C. elegans* became the first animal to have its genome sequenced in its entirety.

How the Different Kinds of Maps Relate to Each Other

Linkage, physical, and sequence maps all locate loci along molecules of DNA. But although they map the same information, they have different resolving powers that reveal different levels of detail (Fig. 10.10). They also use different units of measure. Linkage analyses measure distances in the centimorgans determined by recombination frequencies; such analyses show the human genome to be 4000 cM long. Physical analyses measure distances in base pairs and show the human genome to have a length of 3 billion (3,000,000,000) bp, or 3000 Mb. Therefore, in maps of the human genome, 4000 cM ≈ 3000 Mb, and 1.0 cM ≈ 0.75 Mb. The megabase equivalent of centimorgan varies with a genome's overall recombination rate. The mouse, for example, has a genome that is also 3000 Mb in length, but linkage analysis allots it only 1600 cM; thus, for the mouse genome, 1 cM ≈ 2 Mb.

Now let us consider an analogy comparing the different types of maps you might use to locate a human gene with the maps you might use to find the house of a friend. The karyotypic map provides a crude localization of the gene to a 10-Mb region within the whole genome much as a general map of the United States would identify Washington as the state of residence. The linkage map positions the gene with higher resolution to a region 0.75 cM, or 500 kb, in length, much as a map of Washington would show Seattle to be the city where your friend's house is located. The physical YAC-based map provides the general details of the 500-kb locale, much as a detailed municipal map of Seattle would enable localization of the house to a specific street. Finally, the sequence map describes all genes within the 500-kb region and enables correlation of polymorphisms within a single marker or gene to the segregation of the locus of interest, just as a neighborhood street map would make it possible to locate your friend's house at a specific address.

In the next section, we examine how the different maps described in this section provide a basis for studying genes.

POSITIONAL CLONING: USING LARGE-SCALE MAPS TO MOVE FROM PHENOTYPE TO A CLONE OF THE RESPONSIBLE GENE

Throughout most of the twentieth century, the observation and analysis of variant phenotypes according to the dictates of Mendelian genetics made it possible to identify and characterize thousands of mutations in a wide variety of organisms. Mutations of agricultural import include those that make plants resistant to drought or fungi as well as those that affect the weight of chickens, the production of milk in dairy cows, and the pulp content of tomatoes. Human mutations of medical significance include those that cause the muscular degeneration of Duchenne muscular dystrophy, the neurological deterioration of Huntington disease, and the respiratory and digestive malfunctions of cystic fibrosis. For each of these diseases, researchers could clearly distinguish the phenotypes resulting from wild-type or mutant alleles, but in the vast majority of cases, the different phenotypes did not provide information about either the gene product or the DNA sequence within which the different alleles occurred.

Today, however, they can use large-scale maps to make the molecular connection between phenotype and genotype. **Positional cloning** is the process that enables them to obtain the clone of a gene without any prior knowledge of its protein product or function. In positional cloning, an investigator (a) works through a series of maps, each having a higher resolution than the preceding one, to correlate phenotypic transmission with one small area of the genome; (b) narrows the field of genes in this area to a group of likely candidates for the gene in question; and (c) determines which candidate is the sought-after locus (Fig. 10.11).

Correlating Phenotypic Transmission with One Area of the Genome (a)

This is the first step in positional cloning. (The lower case letters—here an "a"—following this and two subsequent heads refer to the steps of positional cloning just presented and shown in Fig. 10.11.)

Linkage Analysis Can Locate the General Area of the Genome where a Gene Resides

If it is possible to demonstrate linkage between a trait and one or more already mapped markers, that single linkage will place the gene for the trait on a particular chromosome. To determine such a linkage, researchers analyze a population in which some individuals express the mutant phenotype and others do not. For traits expressed in plants or animals, a specifically designed test cross can link the mutant allele to a marker. For human traits such as disease phenotypes, one can start with a small set of families that each have a large number of children or a large set of families that each have a small number of offspring, or a combination of the two. The Genetics and Society box "Using Human Pedigrees and LOD Scores to Calculate the Probability That Two Loci Are Linked" describes the mathematical approach to demonstrating linkage between a marker and an abnormal trait in humans. As mentioned, demonstration of linkage to even one previously mapped marker will locate the disease gene in one region of a particular chromosome.

There are two differences between this type of linkage analysis and the one described earlier for generating a high-resolution linkage map. First, it is not possible to use molecular methods to determine genotype at one of the loci—the disease

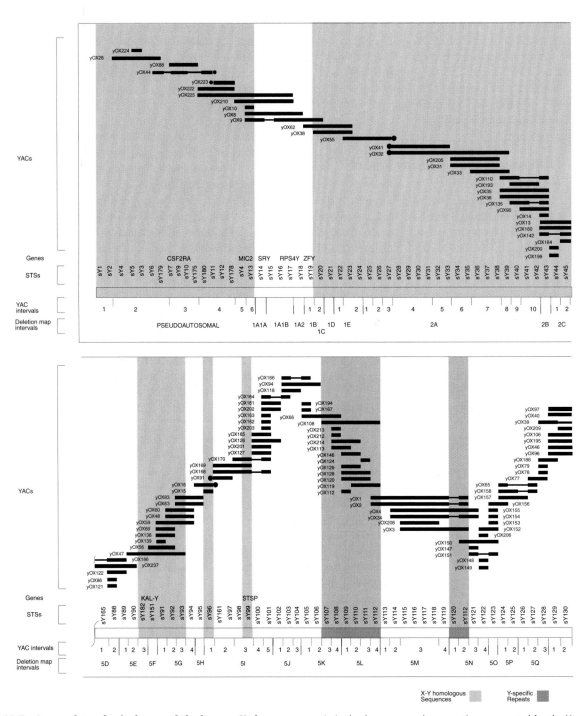

Figure 10.7 A complete physical map of the human Y chromosome. A single-chromosome-long contig was created for the Y chromosome. The contig consists of the series of overlapping YAC clones shown in the figure. The STS markers used to establish the contig are indicated by name just above the chromosome idiogram. Gene locations are indicated by name above the markers.

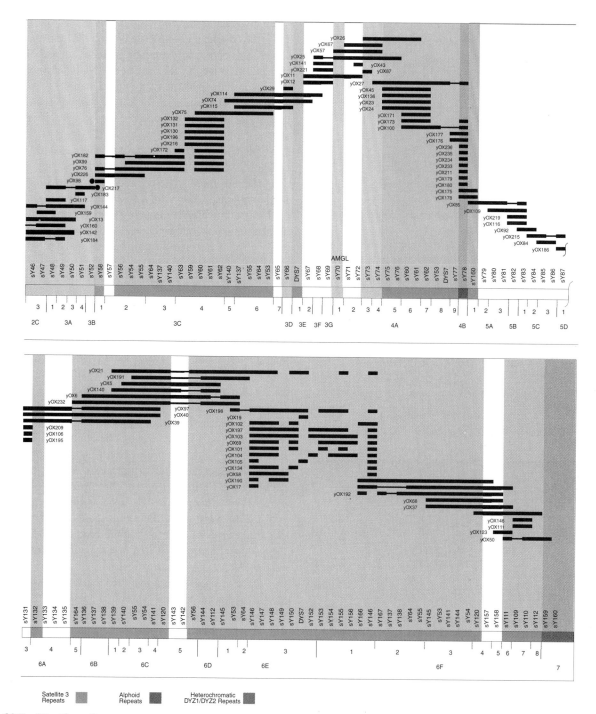

Figure 10.7   (Continued)

Bottom-up approach

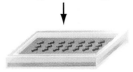

Haploid human genome

(a) Digest chromosomes with restriction enzyme. Clone fragments into a genomic cosmid library.

(b) Fingerprint each cosmid for the pattern of restriction fragment sites.

↓ – Restriction sites

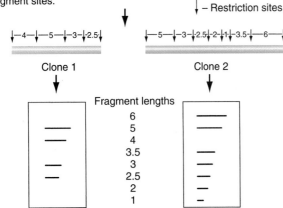

Clone 1   Clone 2

Fragment lengths
6
5
4
3.5
3
2.5
2
1

(c) Feed fingerprint data into computer programmed to look for overlaps.

(d) Arrange overlapping cosmid inserts into 24 chromosomal contigs.

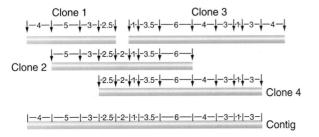

Clone 1   Clone 3

Clone 2

Clone 4

Contig

Figure 10.8 Building an overlapping set of genomic clones without linkage information. This figure illustrates an alternative method to one based on linkage mapping for building a whole-chromosome physical map. (a) The starting material is a complete library of unordered genomic clones. (b) Each clone is subjected to fingerprint analysis. The fingerprints are in the form of gel patterns of restriction fragments. (c) The fingerprint information from many clones is fed into a computer, which looks for evidence of overlaps between fingerprints from different clones. (d) With the analysis of a sufficient number of clones, the computer can generate long contigs of clones extending across each chromosome. Gaps between contigs can be filled in by the method described in Fig. 10.6c.

locus; this has no bearing on the outcome of the analysis because the computer can still use data on the transmission of observed phenotypes to determine linkage. Second, the analysis tries to determine the map position of a single locus in relation to previously mapped markers, rather than seeking to generate a long-range linkage map for a large number of markers. The disease locus therefore occupies a privileged position in the analysis; all markers are tested one at a time for linkage to it.

The mapping of a human disease locus would begin with the typing of all members of the disease-carrying families for expression of the disease phenotype as well as for a series of DNA markers already mapped to representative sites throughout the genome. Even though high-density linkage maps of the human genome contain thousands of microsatellite markers, there is no reason that this preliminary analysis has to include every one of these markers. Instead, a much smaller number of markers can provide complete coverage of the genome. **Complete coverage** means simply that the number of markers on a linkage map is sufficient for the locus controlling the phenotype in question to be linked to at least one of those markers.

If the set of markers used in an analysis provides complete genomic coverage (as shown in Fig. 10.12 for one chromosome), the disease locus will have to show linkage to at least one or two of those markers. The observation of linkage to at least one previously mapped marker will place the disease locus in a particular region of a particular chromosome.

Figure 10.13 shows a five-generation, 104-member family pedigree that demonstrates linkage between the previously mapped G8 marker and the Huntington disease locus. The location of the G8 marker on the chromosome was determined by FISH hybridization of G8 DNA to banded chromosomes. Linkage between the Huntington locus and the G8 marker places the disease gene on human chromosome 4. The first Genetics and Society box in this chapter describes the method used to demonstrate the linkage.

Once a gene has been assigned to a chromosome, the next step is to increase the resolution of the linkage map in the area of the gene. Researchers accomplish this by choosing from the complete set of already mapped markers (for instance, the 6000 available for humans), those that lie in the vicinity of the disease locus. If, for example, the first established linkage were to place a disease gene somewhere between markers M1 and M2 on the 175-cM-long chromosome shown in Fig. 10.12, you could query an electronic repository for all polymorphic STS markers mapped to the region between the 35- and 60-cM positions on the chromosome and then type each marker in all members of the pedigree used in the initial linkage analysis. With one marker every 0.5 cM, that would mean typing each of 50 markers in the more than 100 samples required in this type of analysis. The goal of this second stage of the linkage analysis is to identify a molecular marker so tightly linked to the disease locus that the two show no recombination or only a single recombination event in all the individuals tested. Suppose linkage analysis showed a map distance of 1 cM between the disease locus and a particular marker. Since the human genome contains 3000 Mb of DNA and a total-

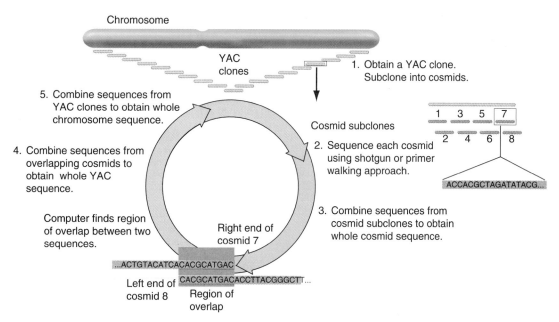

Figure 10.9 Piecing together a whole-chromosome sequence map. With a whole-chromosome contig and restriction map in place, you can generate a whole-chromosome sequence map as follows. (1) Subclone each YAC clone in the contig into smaller cosmids. (2) Sequence each cosmid by the shotgun approach or by primer walking, as described in Chapter 8. (3) Use a computer to link the sequences within each cosmid to form a continuous DNA sequence for the entire cosmid. (4) The cosmid sequences are then linked to form a whole YAC sequence. (5) Finally, the YAC sequences are linked to form a whole-chromosome sequence.

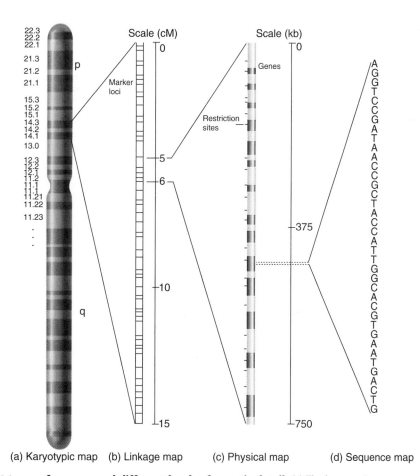

Figure 10.10 Different types of maps reveal different levels of genetic detail. (a) The karyotypic, or cytogenetic, map is at the lowest level of subchromosomal resolution. It can only distinguish loci separated by at least 5000 kb. (b) High-resolution linkage maps can resolve loci separated by a few hundred kilobases. (c) A physical YAC-based restriction map can resolve loci separated by tens of kilobases. (d) A sequence map provides the ultimate resolution down to the level of individual bases.

(a) Mapping a disease gene: a special case

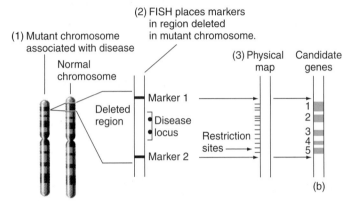

(1) Mutant chromosome associated with disease

(2) FISH places markers in region deleted in mutant chromosome.

(3) Physical map   Candidate genes

Normal chromosome

Deleted region

Marker 1

• Disease
• locus

Restriction sites

Marker 2

1
2
3
4
5

(b)

(c) Examine expression of genes 1–5 in normal and diseased individuals.

Expression

Gene	Normal	Diseased
1	+	+
2	+	+
3	+	−
4	+	+
5	+	+

Conclusion: The tested diseased individual shows a correlation between the disease phenotype and the absence of expression from the number 3 gene.

Figure 10.11 Positional cloning: From phenotype to gene.
(a) Correlating the expression of a phenotype with one small segment of the genome. (1) Some diseases, such as Duchenne muscular dystrophy, are caused by a deletion. It is sometimes possible to observe directly the absence or shortening of a band in a chromosome from an affected individual as compared to the same chromosome from a healthy individual. Even when it is not possible to observe the deletion directly, the FISH protocol can detect it. (2) A marker in the deleted region will hybridize to the chromosome from the healthy individual, but not to the same chromosome from the diseased individual. Markers associated with the disease can be used in linkage analyses of families carrying mutant disease alleles that are not deletions. (3) When linkage analysis shows that a specific chromosomal region contains the disease locus, researchers can subject the marked region to physical analysis. (b) Investigators next analyze the region between recombination sites that define the smallest area within which the disease locus can lie for the presence of candidate genes (as described later in this chapter). (c) They then compare the structure and expression of each candidate gene in many diseased and nondiseased individuals. A correlation between a mutant structure or expression for a particular candidate gene and the disease phenotype can provide evidence that a particular gene is responsible for the disease phenotype. Proof of the association, however, requires further functional studies, which we describe later in this chapter.

linkage distance of 4000 cM, these data would place the marker at approximately 750 kb from the disease gene.

Figure 10.14 shows how geneticists used this approach to map a gene with a mutant allele that increases a woman's risk of breast cancer. Unlike the link between smoking and lung cancer, no environmental factor has yet been connected with breast cancer. Rather, the much higher incidence of breast cancer in women whose mothers or aunts have the malignancy suggests that a strong genetic factor can predispose women to this condition. Initial linkage analysis placed a gene

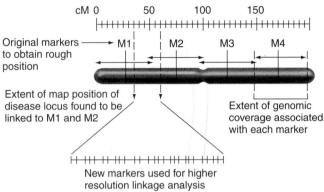

cM 0   50   100   150

Original markers to obtain rough position

M1   M2   M3   M4

Extent of map position of disease locus found to be linked to M1 and M2

Extent of genomic coverage associated with each marker

New markers used for higher resolution linkage analysis

Figure 10.12 Mapping a disease locus in the context of markers that provide complete genomic coverage.
Diagram of a human chromosome with four markers—M1, M2, M3, and M4—used in the linkage analysis of a disease phenotype. Each marker provides "coverage" of a portion of the chromosome. The region of coverage for each marker is the region within which linkage to a disease locus would be uncovered if such a location existed. The extent of coverage is related in a complex way to the number of individuals typed in a study, the relative polymorphism of the marker, and the basis of disease transmission (dominance or recessiveness, complete or incomplete penetrance). Once the parameters of a particular study are known, it becomes possible to choose markers efficiently so that a minimum number provide continuous coverage of each chromosome. In the hypothetical study illustrated here, linkage has been detected to adjacent markers M1 and M2. This suggests that the gene responsible for the disease lies between those markers. With this information, an investigator could type additional markers that lie between M1 and M2 to position the disease locus with higher resolution.

predisposing to breast cancer on chromosome 17. Further linkage analysis narrowed the region of this breast cancer locus (referred to as the *BRCA1* locus) to 1 cM, or 750 kb. From that reference point, physical mapping and ultimately sequencing enables isolation of the gene. Women who receive the mutant allele of the *BRCA1* gene have a 90% probability of developing breast cancer at some point in their lifetime. In other words, the disease trait shows 90% penetrance.

An association with karyotypic abnormalities can also locate a disease gene on a chromosome (Fig. 10.11).

Moving from a Linkage Map to a Physical Map

This step of correlating a phenotype with a narrow area of the genome is quite simple. It depends on a complete physical map of the genome in the form of overlapping series of YAC clones that span each chromosome. With such a map, all an investigator has to do is pick from a central repository the small set of YAC clones that span the genomic region defined by the markers most closely linked to the disease locus (Fig. 10.15). The Human Genome Project along with the collaborative mapping of *E. coli*, yeast, *C. elegans*, *Drosophila*, and mice will provide a complete physical map of the genomes of humans and the model organisms. Investigators analyzing variant phenotypes in species for which whole-genome maps do not yet exist will have to generate a physical map of the region of interest as described earlier in the chapter, using genetic markers closely linked to the disease locus.

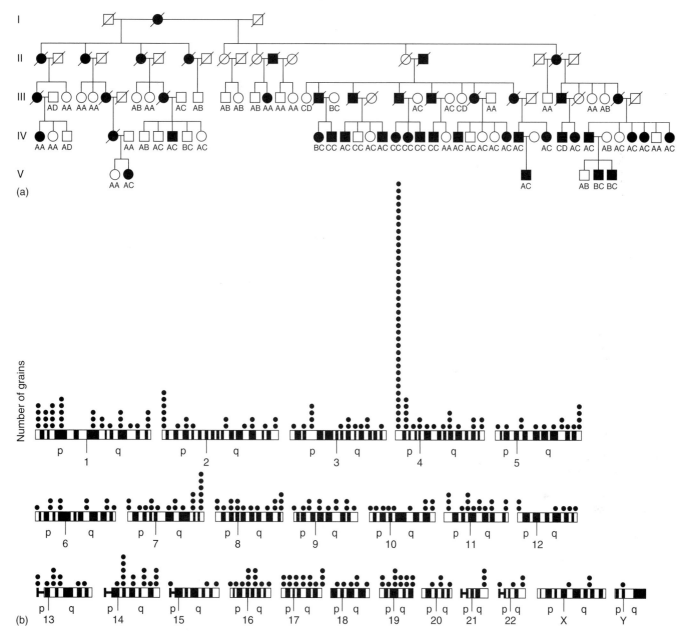

Figure 10.13   Detection of linkage between the HD locus and the marker G8 enabled investigators to map the HD locus to a region near the tip of chromosome 4. (a) Portion of a large Venezuelan pedigree affected by HD, or Huntington disease. Although many individuals in the earlier generations of the pedigree are no longer alive, it is still possible to carry out linkage analysis with molecular markers by using computer programs to predict the probabilities associated with different marker genotypes, when there is no DNA sample. For living members of the pedigree, alleles at the G8 marker locus are indicated (A, B, C, and D). Even at a glance, it is easy to see the cotransmission of marker alleles with the mutant and wild-type alleles at the HD locus. (b) Pedigree analysis shows that the HD locus is within 5 cM of the G8 marker. Thus, karyotypic localization of the G8 marker can provide a chromosomal location for the *HD* gene. Researchers carried out a karyotypic localization analysis by a precursor to the FISH technique in which radioactive probe is hybridized to metaphase chromosomes from 200 cells. Next, a liquid photographic film emulsion is poured onto the slide where it hardens. It is then placed in the dark for a few days before being developed and fixed. The radioactive probe releases beta particles that interact with the emulsion and show up as microscopic "grains" on exposure. Investigators then count the number of grains in each chromosomal region in the sets of metaphase chromosomes from all 200 cells. The results of this HD localization are shown in the figure as dots above an idiogram of the human karyotype. The tip of the small arm of chromosome 4 had many more grains than any other chromosomal region. This is the location of the G8 marker and, by inference, the *HD* locus. (All the other grains represent background noise.)

Genetics and Society

Using Human Pedigrees and LOD Scores to Calculate the Probability That Two Loci Are Linked

The demonstration of linkage between one or more mapped markers and a disease phenotype was the first step in finding and cloning the genes for cystic fibrosis, Huntington disease, and Duchenne muscular dystrophy. A geneticist cannot, however, obtain an absolute yes or no answer to the question of whether two loci are linked from experimental breeding or pedigree data alone. All he or she can do is calculate the probability with which a particular set of data is likely to occur in the absence or presence of linkage between the two loci. With experimental organisms, the statistical tool for such linkage analysis is the chi square (χ^2) test (described in Chapter 4). This test makes it possible to compare the numbers of recombinant and nonrecombinant chromosomes observed in the offspring of a planned cross, and transform the data into linkage probabilities. The χ^2 test, however, is not applicable to human pedigree analysis; because human matings are not planned experiments, it is often not possible to determine from family history alone which inherited allelic combinations are recombinant and which are not. Linkage analysis in humans thus requires specialized statistical tools. We now describe those tools.

Consider a simple two-generation human pedigree following the unusual phenotype of six fingers on each hand in a family with two parents and eight children (Fig. A). The phenotype is caused by a completely penetrant, dominant mutation at a single locus known as *SF* (for *six fingers*). The father expresses the unusual phenotype, while the mother does not. The fact that some of the children do not express the six-fingers phenotype indicates that the father is a heterozygote for the *SF* mutation with an *SF/+* genotype. The mother, with only five fingers on each hand, is a wild-type homozygote (+/+) at the *SF* locus. Like the father, the six-fingered offspring are *SF/+* heterozygotes; like the mother, the normal offspring are +/+ homozygotes.

Suppose that you want to evaluate the possibility that the *SF* locus is linked to a polymorphic marker locus (*M*) with alleles 1 and 2. A single-base-pair difference, detectable by PCR-based protocols, distinguishes the two alleles. If PCR-based analysis demonstrates that the father is heterozygous *M1/M2* while the mother is homozygous *M1/M1* (as shown in the electrophoresis results depicted below the pedigree in Fig. A), the pedigree would represent a situation analogous to the experimental test cross described in Chapter 4: one parent is heterozygous at both loci under analysis (*SF/+*;

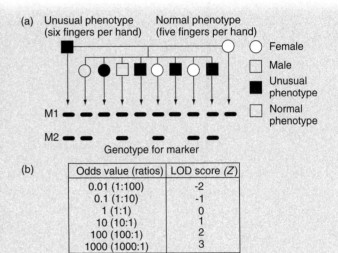

Figure A. (a) A Two-Generation Pedigree Following the Phenotype of Six Fingers on Each Hand. Transmission of the disease is shown with genotype at marker *M*. (b) Relationship between Odds Values (Ratios) and LOD Scores.

M1/M2); the other parent is homozygous at both loci (+/+; *M1/M1*); and the segregation of alleles from each paternal locus to each offspring is distinguishable with absolute confidence.

There is, however, a critical difference between the simple human pedigree and an actual test cross. In the test cross, one allele at each of the two loci is transmitted to the heterozygote parent from one of its parents, and the other set of alleles from the two loci comes from the other parent. Thus, if the two loci are linked, the pairing of alleles on each homolog is predetermined for the heterozygous parent in the test cross. In contrast, in the *SF* pedigree shown in Fig. A, if it turns out that the *SF* locus and the *M* locus are linked, it is possible that the father's mutant *SF* allele is on the same homolog as his *M2* allele, in which case his genotype would be $\frac{SF \leftrightarrow M2}{+ \leftrightarrow M1}$, but it is equally possible that his *SF* allele is on the same homolog as his *M1* allele, in which case his genotype would be $\frac{SF \leftrightarrow M1}{+ \leftrightarrow M2}$. Each of these two possible allele associations (known as *haplotype configurations*) is referred to as a **phase of linkage.**

Experimental test crosses are always set up so that the phase of linkage is known at the outset. But human pedigrees are not "set up." As a result, the phase of linkage of the first generation parents in a pedigree remains unknown, and this makes it impossible to determine which inherited allelic combinations are recombinant and which are nonrecombinant. How then is it possible to use pedigree data to calculate the probability of linkage?

Geneticists Use Odds of Linkage and LOD Scores to Discover Evidence for or against Linkage

With just a cursory look at the pedigree in Fig. A, it is clear that in seven out of eight children, the paternal *M2* allele has been inherited with the wild-type phenotype, the paternal *M1* allele with the six-finger phenotype. Intuitively, these results seem more likely if the *M* marker and the *SF* loci are linked than if they are not. But how can you convert this hunch into a numerical estimate of the probability of linkage? The answer is with a statistical approach known as **LOD score analysis.** The "LOD" of this type of analysis is an acronym for "logarithm of the <u>od</u>s"; the odds are the likelihood that linkage exists relative to the likelihood that linkage does not exist. Although gamblers speak of odds in terms of ratios, such as 3:2 or 10:1, for mathematical calculations, it is convenient to convert these ratios, through division, to real numerical values. Thus, 3:2 = 3 divided by 2 = 1.5; 10:1 = 10 divided by 1 = 10. If the calculated odds value is greater than 1.0, the likelihood of linkage is greater than the likelihood of no linkage. With an odds value of less than 1.0, the opposite is true: The likelihood of no linkage is greater than the likelihood of linkage.

The odds value obtained for the pedigree in Fig. A is 6.3, a result that confirms the hunch of linkage. But, as any gambler knows, just because the odds may favor a particular genetic model, such as linkage, does not mean that the model, in this case linkage, will always be the correct interpretation of the data. To be on the safe side, the human genetics community has set an odds value of 1000 (in other words, 1 chance in 1001 of no linkage) as the minimal cutoff for accepting odds calculations as evidence of linkage between two loci.

Unfortunately, as the *SF* example presented here shows, it is rarely possible to obtain a sufficiently high odds value with data derived from a single human pedigree. (This is an example of the problem of small numbers discussed in Chapter 4.) Large numbers, that is, segregation data from multiple *SF*-segregating families, would bolster the evidence for linkage. But how many families would it take to make the numbers large enough, and how does one combine data from different families? It is not possible simply to add up the recombinant and nonrecombinant chromosomes (since we don't know with certainty which are which) across all families, as one would for a test cross because the phase of linkage may be different in different families. Instead, it is necessary to use the law of the prod-

uct to combine probabilities derived from many independent families into a single grand probability for linkage or no linkage.

To understand why this approach is valid, it is important to view the odds value obtained for each family as a fractional number in which the probability of linkage $P(L)$ is the numerator and the probability of no linkage $P(NL)$ is the denominator. The odds value calculated for each family would then be $P(L)/P(NL)$, and the combined odds value calculated for, say, three independent families would be:

$$\left[\text{Odds} = \frac{P(L)}{P(NL)} = \frac{P1(L)}{P1(NL)} = \frac{P2(L)}{P2(NL)} = \frac{P3(L)}{P3(NL)} \right]$$

If a geneticist obtained an odds value of 6.3 with a single pedigree, she would want to figure out how many additional families of similar size she would have to analyze to have a good chance of reaching the odds = 1000 cutoff, assuming that the additional families give results similar to those she has already obtained for the first family. If she assumes that each family will provide an odds value of ~ 6.3, she can multiply $6.3 \times 6.3 \times 6.3 \ldots$ until she crosses the 1000 threshold. Although the answer is possible to find, it is not immediately obvious to most people. It is thus more convenient to work with experimental values that can be added, rather than multiplied together, to get a composite value; and there is a mathematical trick for converting multiplication to addition. The trick consists of converting odds values into LOD scores through the use of the common logarithmic function based on the following mathematical principle: $\log (A \times B \times C) = (\log A) + (\log B) + (\log C)$.

An odds value is converted into a LOD score (designated with the letter Z) by operating on it with the logarithmic function in base 10:

$$\text{LOD score} = Z = \log_{10} (\text{odds}).$$

The table in Fig. A shows the LOD scores that correspond to six key odds values. The cutoff odds value of 1000 converts to a LOD score of 3.0. The critical feature of LOD scores is that values obtained from multiple families (for the same two loci) can be summed together to produce a single composite LOD score. The odds value of 6.3, obtained from the simple *SF* pedigree, converts to a LOD score of 0.8. Thus, a total of four families with a similar LOD score would push the composite score above the 3.0 cutoff: 0.8 + 0.8 + 0.8 + 0.8 = 3.2. Three families would not: 0.8 + 0.8 + 0.8 = 2.4.

If the two loci under consideration are not linked, the assumption that additional families will have similar LOD scores would turn out to be wrong. Instead, negative LOD scores will appear that when added to the initial positive number would give a negative total, supporting a hypothesis of no linkage.

(a)

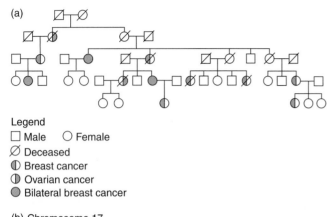

Legend
☐ Male ○ Female
⊘ Deceased
◑ Breast cancer
◐ Ovarian cancer
● Bilateral breast cancer

(b) Chromosome 17

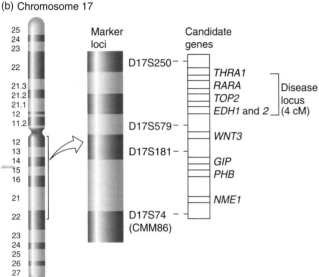

Figure 10.14 Linkage mapping of the breast cancer gene *BRCA1*. (a) Portion of a large pedigree showing transmission of the breast cancer phenotype. (b) Obtaining increasingly greater resolution in mapping the *BRCA1* locus. Researchers first mapped the *BRCA1* locus in linkage with several markers (D17S250 etc.) spread out along the long arm of human chromosome 17. Following demonstration of linkage to this region, they carried out higher resolution mapping with markers and genes located between the two original markers D17S250 and D17S579, which define the extent of the region within which *BRCA1* had to lie.

Identifying Candidate Genes (b)

Isolation of a contig of clones containing the disease locus is the straightforward part of positional cloning. Identification of the particular DNA region that corresponds to the actual gene whose alteration is responsible for the disease phenotype is much more difficult. With approximately 100,000 genes in the human genome, for example, there is roughly 1 gene every 30 kb. If the disease locus has been mapped to a 500-kb region, that region will contain approximately 15 genes. Geneticists can use several strategies to find these genes. Directed strategies identify sequences transcribed in a tissue exhibiting specific symptoms of the disease. Nondirected strategies identify as many as possible of the total number of coding units from the cloned region. Here are some examples of each approach.

One Way to Identify Candidate Genes is through Northern Blots That Pick Out Transcripts of Genes Undergoing Expression in Specific Tissues

Some phenotypes provide clues to the tissue in which a wild-type allele of the disease locus might be expressed. For example, you would expect the gene responsible for the differentiation of the immature fetal gonads into testes (rather than ovaries) to be expressed in the fetal gonad and testes, and the normal gene whose mutant alleles underlie a propensity for breast cancer to be expressed in breast tissue. The earliest and simplest protocol for determining whether a fragment of DNA is transcribed in a particular tissue is known as **Northern blot analysis.** Unlike Southern blots, which were named after Ed Southern, their inventor, Northern is a play on the common meaning of the word "southern" and not a person's name. Recall that in Southern blotting, researchers separate DNA fragments by electrophoresis; blot the separated fragments to a special filter paper; expose the filter to radioactive probes that hybridize to complementary sequences in the blot; and finally, observe a band on an autoradiograph if the probe has hybridized to a stretch of DNA on the filter. In Northern blotting, it is the RNA transcripts in the cells of a particular tissue that are separated by gel electrophoresis. The ensuing blotting, probing, and autoradiography are the same as for the Southern blotting described in Chapter 8.

There are, however, two critical differences in the way investigators interpret Northern blot data. First, since RNA samples are not enzymatically treated in any way before electrophoresis, the position of individual molecules on the blot provides a direct measure of their size; and from their size, it is possible to estimate the transcript's coding capacity and thence the size of the protein product it encodes. Second, Northern blot analysis of RNA samples from many different tissues enables one to determine the specific tissue in which a gene is expressed as well as the relative levels of that expression in all cells where transcription is occurring.

Investigators can use Northern blots to identify the various transcription units in a large cloned genomic region. They accomplish this by dividing the genomic region into many smaller fragments and labeling each fragment for use as a Northern blot probe. Genomic fragments that do not include coding regions will not light up bands in any RNA sample; fragments from genes expressed in all tissues will light up a band in all RNA samples; and fragments from genes expressed only in specific tissues will light up a band in some RNA samples but not others.

Figure 10.16a shows how Northern blot analysis supported the hypothesis that a particular genomic fragment from the Y chromosome contains the gene responsible for testes differentiation. The presence of this gene (called *TDF* for testes-determining factor) in a developing embryo causes the undifferentiated gonads to become testes; absence of the gene causes the gonads to develop into ovaries. All aspects of secondary sexual differentiation, including the formation of genitals, depend on this initial developmental event. On the basis of

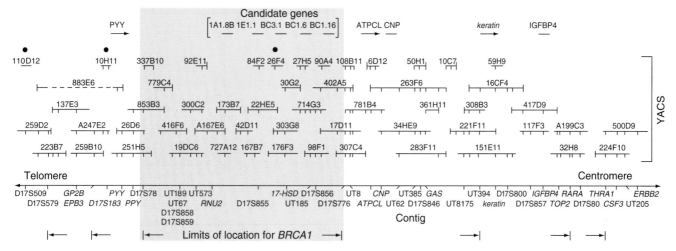

Figure 10.15 Physical Mapping of the *BRCA1* Locus. With many family pedigrees and markers in the *BRCA1* region, researchers localized the *BRCA1* locus between likely recombination sites that delineated a section less than 1 Mb in length. After cloning the entire region into a series of overlapping YAC and bacteriophage clones (41 YACs and 11 phages), they identified five candidate genes that seemed to be cotransmitted with the *BRCA1* mutation.

the foregoing information, geneticists predicted that the *TDF* gene is expressed in the testes and probably no other tissue. Observations of a Northern blot probed with a genomic fragment called pY53.3 confirmed this prediction (Fig. 10.16b). A sequence within this fragment is transcribed into a 1.1-kb RNA in the testes, but not in the ovaries, lungs, or kidneys. As a control, researchers probed the same Northern blot with a clone from the β-actin gene, known to be expressed in all tissues; observation of an actin transcript in every lane demonstrated that intact RNA had been isolated from each tissue.

It is not always possible to identify a gene on the basis of its expression. Cells may express the gene at such a low level that Northern blot hybridization cannot detect the transcripts. Newer, more sensitive techniques for analyzing transcription patterns, using RNA or cDNA as the starting material, may eliminate this barrier. However, some phenotypes provide no clue to which tissues are responsible for faulty gene function; for example, some developmental abnormalities result from mutant alleles expressed for only a fleeting moment of embryogenesis in one of many possible tissues. For these reasons, researchers have developed gene-finding strategies that do not depend on gene expression.

A Second Way to Identify Candidate Genes Is on the Basis of the Evolutionary Conservation of Coding Sequences

The human β-globin gene has the same function as the mouse β-globin gene, and the two genes encode similar sequences of amino acids. Consequently, a clone of the human β-globin gene will hybridize with the mouse β-globin gene.

Hybridization data compiled to date reveal that nearly all human genes cross-hybridize to counterparts in the mouse genome. In contrast, probes derived from the noncoding regions between genes and within introns, which make up roughly 95% of the human genome, do not hybridize to mouse

DNA. Geneticists interpret this type of information to mean that over long periods of time, the DNA of genes is under strong selective pressure not to change, while most noncoding regions can drift in sequence—as a result of mutation—without evolutionary consequences. Researchers can use this difference between coding and noncoding regions as a tool for discovering genes in cloned DNA fragments. For example, they can subdivide a YAC clone into a large number of fragments and after blocking (by prehybridization) the simple sequence microsatellite repeats, which are present in all animal species, they can use the human YAC fragments as probes to test for the presence or absence of hybridization to total mouse DNA. Fragments that hybridize are likely to contain coding (or regulatory) sequences; fragments that do not hybridize probably do not.

Figure 10.17 shows how researchers used this approach to identify the human *TDF* gene within cloned fragments of human DNA from the genomic region to which they had previously mapped the *TDF* locus. As Fig. 10.17a illustrates, one of nine tested fragments from the human *TDF* region hybridized to DNA from two other species—mouse and cow. Researchers designated this fragment as pY53.3 and tested it in both male and female samples from additional mammalian species (Fig. 10.17b). The observation that the human pY53.3 fragment hybridized to genomic DNA only in samples obtained from male animals placed the cross-hybridizing fragments on the Y chromosome in all of these species. The evolutionary conservation of the pY53.3 fragment on the Y chromosomes of various animals is of particular interest because it is this same fragment that showed testes-specific expression in the Northern blots illustrated in Fig. 10.16b. These results together provide much more support than either would alone for the hypothesis that the cloned pY53.3 fragment contains at least part of the *TDF* gene.

(a)

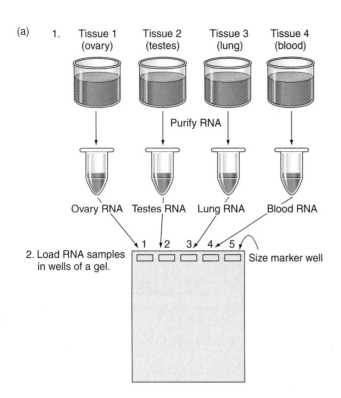

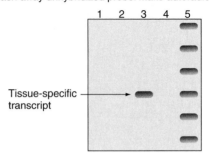
4. Wash away unhybridized probe. Make autoradiograph.

Tissue-specific transcript

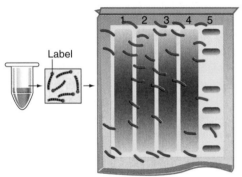

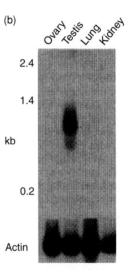

(b)

Figure 10.16 Northern blots: Snapshots of gene expression. (a) The protocol. (1) Purify RNA from each tissue to be examined for expression of the gene under investigation; here since you are looking at the *SRY* candidate for the testes-determining factor, the tissues to be examined are ovary, testes, lung, and blood. (2) Make an agarose gel and load each of the four RNA samples into a different well and load a fifth well with RNA size markers. Now subject the gel to an electric current that causes the RNA in each sample to migrate along a lane toward the bottom of the gel. The mobility of each RNA transcript in a sample depends on its size: smaller RNAs move faster, while larger RNAs migrate more slowly. When the smallest RNAs reach the bottom of the gel, turn off the current. Staining the RNAs in each lane would produce a smear reflecting the presence of so many RNAs of different sizes that they cannot be resolved from each other. (3) Blot the RNA within the gel and fix it to a filter so that each RNA molecule retains its position relative to all the other molecules. Expose the filter to labeled probe and allow the label to hybridize for several hours. (4) Wash away unhybridized probe. Place the filter on a film for autoradiography. Develop the film. You will see bands only in those lanes containing a tissue where the gene represented by the probe has been expressed. (b) Northern blot results obtained using the pY53.3 clone as a probe. This clone contains the *SRY* gene. The results show that *SRY* is expressed in the testes, but not the ovary, lung, or kidney. This result makes SRY a good candidate for the *TDF* locus. In a control experiment, researchers probed an identical blot with the same RNA samples using a clone containing the actin gene. As expected, a band of the same size appears in every lane. This control demonstrates the integrity of the RNA samples used in this study.

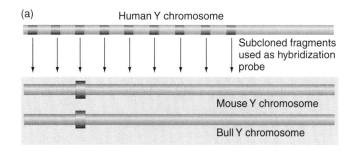

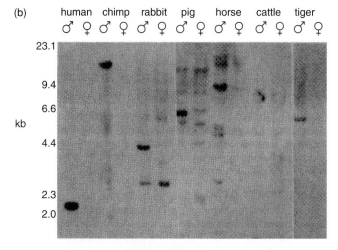

Figure 10.17 Cross-hybridization between DNAs from different species picks out genomic segments that have been conserved during evolution. (a) Researchers tested nine subcloned fragments from the region of the human Y chromosome within which *TDF* maps for hybridization to DNA from mice and cattle. Only one of the fragments—called pY53.3—showed any cross-species hybridization. (b) Researchers next used the pY53.3 fragment as a probe to test for hybridization to male and female DNA from six nonhuman species. In every species, the probe detected male-specific bands that, by definition, must represent Y-chromosome loci. In some species, additional bands were detected in both male and female DNA samples; these bands represent other members of a gene family (known as the SOX family), which contains *SRY*.

Yet Another Tool for Gene Identification Is Computer Programs That Scan a Long-Range DNA Sequence for the Exons of Coding Regions

Such programs look for long open reading frames (that is, reading frames with no stop codons) and appropriate codon usage. Even though multiple codons encode most amino acids, every species shows a bias in the codons it actually uses. The programs also look for the specific sequences that mark the splice sites at the beginning and end of each exon. Geneticists working with computer experts have created programs that at present can identify exons with 90% accuracy. Other computer programs also provide a much more sensitive means than hybridization for finding significant sequence conservation in corresponding regions of the genome from two different species.

Finding the One Gene among All the Candidates That Is Responsible for the Phenotype (c)

At the end of the second stage of positional cloning, you have a set of candidate genes in a small genomic region that you know contains the locus responsible for the phenotype under analysis (review Fig. 10.15). How do you determine which candidate is the correct gene? There is no simple answer to this question. Rather you must perform a series of tests, such as those shown in Fig. 10.18, to accumulate evidence that favors one candidate over all others.

You Can Determine if the Gene Is Expressed in the Appropriate Tissues

For candidate genes identified by nonexpression strategies, you can use sensitive PCR-based protocols that detect and quantitate very low levels of specific RNA molecules, to determine whether the wild-type allele of the gene is expressed where you would expect to see such expression (Fig. 10.18a). If a candidate gene is not expressed in the expected tissues, you can rule it out.

You can also look for variation in the expression of a candidate gene from individuals homozygous for the disease allele. For example, if a candidate gene allele fails to show expression in a diseased tissue where you observed expression in nondiseased individuals, this would provide further evidence that the particular candidate is the disease locus (see Fig. 10.18a). Or you might use sequence analysis of both DNA and polypeptide to find that the candidate disease allele has an altered nucleotide sequence, such as a nonsense or frameshift mutation, that would destroy protein function (see Fig. 10.18a).

If analysis shows that many independent disease-associated alleles of a candidate gene have changes that disrupt protein function in some way, this constitutes strong evidence that the particular candidate and the disease locus are one and the same.

Transgenic Analysis Can Confirm That You Have Picked the Right Piece of Cloned DNA as the Gene

The term **transgene** describes any piece of foreign DNA that researchers have inserted into the genome of a complex organism, such as a mouse or pea plant, through experimental manipulation of early stage embryos or germ cells. An individual carrying a transgene is known as a **transgenic** plant or animal. Although researchers had succeeded in inserting a certain type of foreign DNA in the germ line of animals during the 1970s, it was not until 1981 that workers in five independent laboratories described essentially the same method for transferring any DNA sequence into the germ line of any mammal. They accomplished the transfer by injecting a "handful" of cloned DNA molecules through very fine pipettes into the haploid egg and sperm nuclei (known as pronuclei) present in newly fertilized eggs (Fig. 10.18b). They then placed the injected eggs in the uterus of a "foster" mother and allowed them to develop. Up to 50% of the mice born to

(a) Search for hypothetical gene whose mutant allele causes pancreatic degeneration.

Test of four gene candidates: (*c1, c2, c3,* and *c4*)

1. Tissue expression

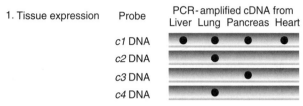

Conclusion: Rule out *c2* and *c4* – not expressed in pancreas.

2. Expression of remaining candidates (*c1* and *c3*) in normal and homozygous diseased individual pancreas.

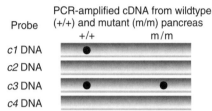

Conclusion: The tested diseased individual shows a correlation between the disease phenotype and the absence of expression from the *c1* gene.

3. Sequencing of the *c1* gene from wildtype and diseased genomes.

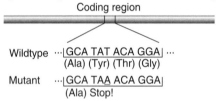

Coding region

Wildtype ⋯|GCA TAT ACA GGA| ⋯
              (Ala) (Tyr) (Thr) (Gly)

Mutant   ⋯|GCA TAA ACA GGA|
              (Ala) Stop!

(b) Transgenic analysis

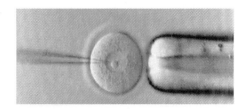

Foreign DNA — Fertilized egg
♂ Pronucleus
♀ Pronucleus

Foreign DNA injected into male pronucleus of newly fertilized egg. → Injected eggs surgically implanted into uterus of 'foster' mother and allowed to develop. → Mice are born with foreign DNA in every cell nucleus.

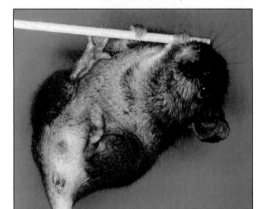

Figure 10.18   Ways to identify the candidate gene corresponding to the locus responsible for a variant phenotype. (a) In this hypothetical example, an investigator is testing four closely linked candidate genes to see which one corresponds with a disease locus that causes pancreatic degeneration. (1) An expression assay: The investigator tests PCR primers specific for transcripts from each of the four candidate genes (*c1, c2, c3,* and *c4*) for their ability to amplify a product from RNA samples obtained from liver, lung, pancreas, and heart tissues. Only two of the genes—*c1* and *c3*—are expressed in the pancreas. This result argues against the likelihood that either *c2* or *c4* is involved in the disease phenotype. (2) The researcher now examines *c1* and *c3* for expression in pancreatic RNA samples from normal and homozygous mutant individuals. She finds that *c3* is expressed in both, but *c1* is not expressed in the mutant samples. This result correlates faulty gene expression with the mutant phenotype. (3) She next sequences the *c1* gene alleles present in normal and diseased individuals to localize the genetic change responsible for the faulty gene expression. In this hypothetical example, she finds a mutation to a stop codon in the *c1* alleles of diseased individuals. This result provides strong evidence that the *c1* gene corresponds with the pancreatic degeneration locus. However, this is still only proof by correlation. Ultimately the investigator will want to confirm this finding with genetic or biochemical evidence. (b) It is sometimes possible to prove the equivalence of a candidate gene and a locus defined by phenotype alone through the use of transgenic mouse technology. To demonstrate that *SRY* was the same as *TDF*, researchers injected a clone containing the *SRY* gene into mouse embryos with two *X* chromosomes. The transgenic mice that resulted from this protocol had a penis. This observation proves that *SRY* is the equivalent of *TDF*.

these foster mothers had integrated the injected DNA into their chromosomes; they later passed the transgene on through their germ cells to their own offspring.

Since 1981 researchers have successfully inserted transgenes into the genomes of cows, pigs, goats, sheep, and other mammals. There is no limitation on the type of DNA that can serve as a transgene; it can come from any natural source, from chemical synthesis, or a combination of the two. With current protocols for the creation of transgenic mice by embryo microinjection, the site of transgene integration into the host genome is not predetermined, and for all practical purposes should be considered random.

Researchers used transgenic technology to confirm that the gene contained within the pY53.3 fragment is the same as the *TDF* locus. We have seen that the Y chromosome in all mammals carries a *TDF* locus that is required for the development of immature fetal gonads into testes. As described earlier investigators first localized the *TDF* gene to a small region of the Y chromosome in mice and then discovered a good candidate that was expressed only in the testes. To prove that this candidate was really the same as *TDF,* they created transgenic mice as follows. They first cloned a genomic fragment that overlapped with pY53.3 and contained the entire transcription unit along with its regulatory sequences. Before they knew the function of this transcription unit, they called it *SRY* for sex-determining region on the Y chromosome. They next injected two to five copies of the complete *SRY* gene into a large group of fertilized eggs from a mating between two wild-type parents. They placed the embryos in foster mothers, and then two to three weeks after the birth of the mice that developed from the embryos, they checked the animals in three ways: (1) by karyotypic analysis to determine the combination of sex chromosomes, either XX or XY; (2) by Southern blot analysis to see whether the transgene had integrated into the genome; and (3) by visual examination to check for the presence of a male sex organ. The identification of several transgenic XX animals with a penis (Fig. 10.18b) proved beyond a doubt that the *SRY* gene is the same as the *TDF* gene.

Summary: Positional Cloning of the Cystic Fibrosis Gene Leads to a Potential Disease Therapy

A description of how medical geneticists used positional cloning to identify the gene whose mutant forms are responsible for cystic fibrosis (CF) reviews the steps and benefits of positional cloning (Figure 10.19a–d). In 1985 researchers uncovered linkage between the *CF* locus and a marker called met, placing the *CF* gene in the middle of human chromosome 7. Further linkage analysis carried out on a large number of families each having two or more members affected by CF showed that the met locus is ∼ 2 cM from the *CF* gene. Next, linkage analysis using many more markers on chromosome 7 identified two additional markers—XV-2c and KM-19—that displayed no recombination with the disease gene and were thus much closer to it. Investigators used these markers as starting points for a chromosome walk extending over 280 kb of DNA to identify candidate genes. As discussed earlier, children afflicted with cystic fibrosis have a variety of symptoms arising from abnormally viscous secretions in the lungs, pancreas, sweat glands, and several other tissues. Northern blot analyses of four candidate genes in the *CF* genomic region revealed that only one, subsequently called the *CFTR* gene (for cystic fibrosis transmembrane conductance regulator), was expressed in all of these tissues and not in others. A small portion of this gene was included at the far end of the cloned 280-kb region. When medical geneticists examined the gene in many CF-affected people, they found that in every patient, small deletions or substitutions had destroyed its function by either eliminating polypeptide production or disrupting protein function (Fig. 10.19e).

Although these findings strongly suggested that the *CFTR* gene and the *CF* disease locus were the same, researchers confirmed their equivalence through experiments in mice. They engineered the mouse homolog of the *CFTR* gene to produce a mutant allele that did not generate a CFTR polypeptide, and then created mice homozygous for this mutant *CF* allele. (The genetic portrait of the mouse in Chapter 22 describes the protocol used to target mutations to a particular locus in the mouse genome.) The homozygous mice carrying two copies of the mutant *CFTR* allele showed the same basic phenotype as that observed in humans with the disease. Such engineered homozygotes continue to provide the only nonhuman model of the disease.

Medical practitioners and researchers now use PCR-based protocols to determine *CF* genotypes for families concerned about the disease: in adults (to see if they are carriers); in the fetus to establish what the disease status of the newborn will be; or in preimplantation embryos to select ones that are not afflicted for initiating a pregnancy (see in Chapter 9). They also use clones of the gene to synthesize the wild-type protein and learn how its absence or inactivation causes disease symptoms. An understanding of gene function obtained in this way suggests a potential therapy: inserting a copy of the wild-type *CFTR* gene into a small fraction of a CF patient's lung cells. Cells that expressed the inserted, wild-type gene might be able to produce enough functional protein to eliminate the buildup of fluids in the lungs, which is the leading cause of death from cystic fibrosis. Clinical trials to test the safety and efficacy of such a therapy are in progress in both mice and humans.

USING SEQUENCE MAPS AS A STARTING POINT FOR MOVING FROM THE CLONE OF A GENE TO ITS FUNCTION

Going from gene to function runs counter to the way that biomedical scientists have traditionally studied the biology of humans and mammalian models like the mouse. Scientists trying to understand the genetics of hemophilia, for example, started with an inherited phenotype, inferred the function of the responsible gene (production of a blood-clotting factor), used this hypothesis to isolate the polypeptide in the appropriate tissue (blood), purified the protein (Factor VIII), then sequenced the protein, deduced the DNA sequence that generated it, and finally, used hybridization probes to find the gene. Hemophilia is one of a small number of phenotypes for which this scenario can work.

In the late 1990s, however, with the development of the large-scale physical and sequence maps described in this chapter, biologists for the first time faced a large number of cloned genes unassociated with any function or phenotype. The human genome, for example, is likely to have roughly 100,000 genes; yet researchers have defined fewer than 1000 of these cloned genes through studies of disease phenotypes or characterization by other means, which means that more than 90% of the genes found through mapping will be of unknown function. To understand how these genes work, geneticists must move beyond genomics, using mapping data in combination with molecular techniques and recombinant DNA technology

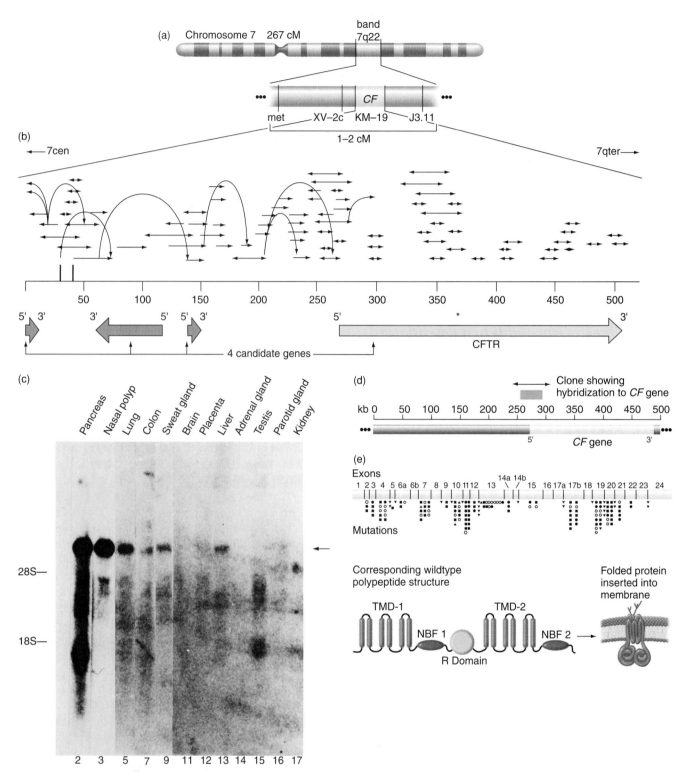

Figure 10.19 Positional cloning of the cystic fibrosis gene: A review. (a) Linkage analysis places the *CF* locus on chromosome 7. This idiogram of chromosome 7 shows the positions of markers—met, XV-2c, KM-19, and J3.11—that define the 1- to 2-cM region within which the *CF* locus must lie. (b) Walking from a linked marker to the *CF* gene. Using the closest linked marker (XV-2c) to probe a cosmid library of the genome (with average insert sizes of 30 kb), researchers pulled out clones that they characterized. They then used the ends of these clones to probe the library again. They repeated this process many times to obtain a contig of 280 kb. Within this 280-kb region, they identified four transcription units. These researchers pursued the cloning of the *CF* gene before the development of the YAC vectors that make it possible to produce large-insert genomic libraries. Today it would be easy to obtain this entire contig on a single YAC clone. (c) Northern blot analysis reveals that only one of the candidate genes is expressed in the lungs and pancreas, both affected tissues in *CF* individuals. This candidate (now known as *CFTR*) is not expressed in the brain, liver, or testes, all tissues that are not affected in diseased individuals. (d) The 5′ end of the CFTR gene was located at one end of the original 280-kb contig and found to extend over 250 kb of DNA. (e) Every CF patient has a mutated allele of the *CFTR* gene on both of her or his chromosome 7 homologs. The locations and types of mutations uncovered in different patients are shown under the diagram of the gene. The normal gene encodes the wild-type protein.

to determine the function of cloned genes. We now describe several possible strategies.

A Gene's Sequence Can Provide Insight into the Structure and Function of the Polypeptide It Encodes

The first step in the analysis of a newly cloned gene is to determine its nucleotide sequence and use this DNA sequence to establish the amino-acid sequence of the polypeptide it encodes. You can then use the amino-acid sequence to help determine the structure and general function of the polypeptide, and thus of the gene. One way to begin is by asking whether the protein sequence shows any similarity with already characterized proteins. Comparisons of the new sequence with sequences stored in electronic databases, which are readily accessible to scientists throughout the world, can help answer this question; about 50% of the polypeptides encoded by newly discovered genes display significant similarity to proteins whose function is already partially understood. Protein-sequence similarities provide a basis for inferring shared structural or functional features.

Short-Sequence Similarities Provide Clues to Protein Structure and Function

Some newly defined proteins contain short sequences that show similarity to well-defined **peptide motifs:** stretches of amino acids conserved in many otherwise unrelated polypeptides. Data compiled from thousands of proteins from many different organisms suggest that a few hundred to a thousand peptide motifs provide the fundamental building blocks for many of the polypeptides produced by eukaryotic cells. The motifs so far defined vary in length from just three amino acids to several hundred. Genetic processes such as duplication and chromosomal inversions, translocations, insertions, and deletions have mixed and matched these motif sequences, bringing them together in different combinations, like Tinker-Toy components, to build different but related proteins.

One of the simplest peptide motifs is the amino-acid set of asparagine (Asn)-X-serine (Ser) or asparagine-X-threonine (Thr), where X is any amino acid except proline; Ser and Thr are very similar residues that both contain polar hydroxyl (—OH) groups (Fig. 10.20a.1). This three-amino-acid motif is a target for the glycosylation of proteins transported to the plasma membrane through the endoplasmic reticulum; the covalent addition of sugars occurs on the asparagine and appears to protect proteins at the cell surface from degradation. Thus, the presence of this simple motif might suggest that a newly sequenced protein functions at the cell surface.

In another example, a peptide motif known as a zinc finger gives the proteins in which it occurs finger-like projections that help the proteins bind DNA (Fig. 10.20a.2). Zinc is part of the name because these peptide motifs bind zinc as a cofactor; attachment of the zinc causes the polypeptide to extrude extensions that look like fingers (see Chapter 16). The critical characteristic of all zinc fingers is a series of suitably spaced cysteine (C) and, in some cases, histidine (H) residues that capture the zinc ion at the base of the finger. Most zinc-finger proteins contain multiple zinc-finger motifs. Molecular biologists think that these proteins bind DNA by inserting the peptide fingers into the grooves of the DNA molecule. Discovery of a zinc-finger motif would suggest that a newly characterized protein participates somehow in the regulation of gene expression.

Larger Regions of Sequence Similarity May Contribute to a More Specific Definition of Function

Sometimes the translated sequence of a newly cloned gene shows similarity to a long amino-acid sequence that extends across the entire length of a polypeptide already characterized in the same or a different species. This is usually good evidence that the genes encoding the two polypeptides are **homologous,** that is, are similar because they diverged from a common ancestral gene (Fig. 10.20b).

If the observed DNA and amino acid similarities are between genes and proteins of different species, the two genes might have a common ancestor and the same function. The T locus in the mouse, for example, influences development of the posterior portion of the mouse embryo. Researchers identified the T-locus more than 50 years ago and found that embryos homozygous for a mutant null allele fail to develop their posterior region, and die midway through gestation. Much later, when they cloned the gene, they discovered expression of the wild-type product in the posterior end of the developing wild-type embryo, as predicted from the mutant phenotype. In a recent, independent study, researchers working with zebra-fish found a mutation that they named "no-tail" because homozygous mutant fish embryos fail to develop a posterior end; the no-tail mutation is lethal to homozygotes, which die before hatching. After cloning and characterizing the zebra-fish no-tail gene, the researchers checked its sequence by computer for similarity with all sequences in the worldwide electronic database known as GenBank. The computer check picked out the mouse T-locus gene and revealed the presence of a subregion in the mouse and zebra-fish polypeptides with 90% identity (Fig. 10.20b.1). This strong sequence similarity coupled with the similarity of functions (as indicated by the mutant phenotypes) show that the zebra-fish no-tail gene and the mouse T locus are homologs. As it turns out, they both produce DNA-binding proteins that regulate the same aspects of embryonic development.

By comparison, homology between two genes and proteins of the same species, coupled with some nonidentical sequences, indicates that the genes have related but not identical functions and probably arose through duplication of a whole gene. For example, the T locus in mice shows sequence similarity to a second, recently cloned and sequenced mouse gene known as *Tbx2*. The T-locus protein shows a 41% amino-acid identity with the Tbx2 protein in the same region that was 90% homologous with the zebra-fish protein (Fig. 10.20b.2). This

(a)

1.

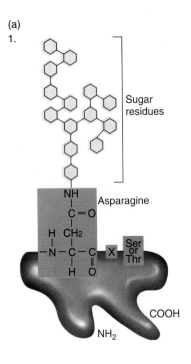

Sugar residues

Asparagine

2. Zinc fingers

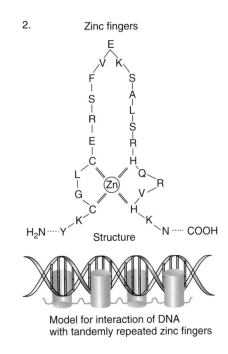

Structure

Model for interaction of DNA
with tandemly repeated zinc fingers

(b)

1. Similarity with a whole polypeptide in a different species

Amino acid sequence

|1|10|20|30|40|50|60|70|80|

Mouse T-locus YIHPDSPNFGAHWMKAPVSFSKVKLTNKLNGG.GQ...IMLNSLHKYEPRIHIVRVGGPQRM....ITSHCFPETQFIAVTAYQN
Zebra fish T-locusYIHPDSPNFGAHWMKAPVSFSKVKLSNKLNGG.GQ...IMLNSLHKYEPRIHIVKVGGIQRM....ISSQCFPETQFIAVTAYQN

2. In the same species

Amino acid sequence

Mouse *Tbx2* YIHPDSPATGEQWMAKPVAFHKLKLTNNISDKHGF...TILNSMHKYQPRFHIVRANDILKLPYSTFRTYVFPETDFIAVTAYQN
Mouse T-locus YIHPDSPNFGAHWMKAPVSFSKVKLTNKLNGG.GQ...IMLNSLHKYEPRIHIVRVGGPQRM....ITSHCFPETQFIAVTAYQN
Zebra fish T-locus YIHPDSPNFGAHWMKAPVSFSKVKLSNKLNGG.GQ...IMLNSLHKYEPRIHIVKVGGIQKM....ISSQSFPETQFIAVTAYQN

▮ Amino acids not conserved between mouse T-locus and Zebra fish T-locus

▮ Amino acids not conserved between mouse T-locus and mouse *Tbx2*

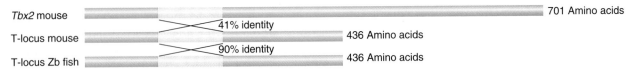

Tbx2 mouse   701 Amino acids
                          41% identity
T-locus mouse   436 Amino acids
                          90% identity
T-locus Zb fish   436 Amino acids

Figure 10.20 Polypeptide motifs and sequence similarities indicating homology are clues to protein structure and function.
(a) Short peptide motifs can indicate discrete biochemical properties. (1) The three-amino-acid sequence Asn-X-Ser/Thr is a target for the covalent addition of sugars. (2) The zinc-finger motif is defined by the presence of cysteine and histidine residues in a conformation that allows sequestration of a zinc ion and the formation of a "finger-like" projection that can interact with the bases of the DNA molecule. (b) Similarity between two whole protein sequences suggests that the genes encoding them derive from a single ancestral gene. (1) Proteins from two different species are considered homologous if they show a fundamental similarity because of descent from a common ancestor. In the T locus from the mouse and the zebra fish, more than 90% of a region that encodes a DNA-binding portion of the gene product carries the same amino acids. This strong similarity suggests the conservation of function for the T-locus gene products in these two species. Researchers have confirmed this homology by demonstrating that the T-locus protein plays the same role in the development of the early embryo in both fish and mouse.
(2) Sequence similarity between the products of two different genes in the same species suggests that a gene duplication event occurred in an ancestor of the current organisms. By comparing the degree of similarity between different sequences, it is possible to estimate the times at which particular evolutionary events occurred. For example, the mouse and zebra fish versions of the T-locus protein are 90% identical to each other, but only 40% identical to the mouse Tbx2 protein. These data indicating homology suggest that the duplication event that ultimately produced the T-locus and *Tbx2* genes currently found in the mouse genome took place before the divergence of zebra fish and mice from a common ancestor over 500 million years ago.

degree of similarity suggests that the T locus and *Tbx2* genes have related but not identical functions. Subsequent studies have shown that both proteins bind DNA to regulate gene activity during development, but they do so in different parts of the embryo. They may well have arisen through an ancient duplication event that occurred early in the evolution of animals.

The discovery of sequence similarities implying homology provides a clue to gene function, but does not establish it. Some bacteria, for example, have a gene whose function is nitrogen fixation; but while researchers have found a related gene in yeast and learned that the yeast gene is essential for cell growth, they do not yet know its precise function, because yeast cells do not fix nitrogen.

The Analysis of Expression Patterns Is Another Means of Connecting a Cloned Gene to Its Function

DNA-RNA hybridization shows where and when a gene is active. We have seen that cloned genes can serve as probes for screening RNA prepared from different tissues and from tissues at different stages of development on Northern blots. DNA probes that hybridize to RNA help reveal the expression pattern of the gene, and this expression pattern can narrow the field of possible functions. A gene expressed in all tissues at all stages of development, for example, is likely to play a cellular "housekeeping" role; the genes for the myosin and actin that make up a cell's cytoskeleton are among those that fall into this category. In contrast, a gene expressed in only one tissue is likely to have a function unique to that tissue; the DMD gene expressed only in muscle cells and the antibody genes expressed only in the immune-system cells known as B lymphocytes would be in this category. Expression occurring only at one specific stage of development is also informative. A gene expressed only during meiosis in germ cells, for example, may have a function in chromosome pairing, recombination, or segregation, but certainly not in antibody formation.

Although analyses of sequence and expression patterns may provide clues to gene function, further biochemical or genetic experiments are always necessary to demonstrate conclusively the role that a particular gene product plays in the life of a cell or organism.

Conversion of a DNA Clone to a Polypeptide Makes It Possible to Analyze the Peptide's Role in the Cell and Demonstrate Gene Function

To perform biochemical studies on a gene's product, investigators must obtain sufficient quantities of the encoded protein for analysis. They accomplish this task by cloning the coding regions of the gene into special expression vectors that they then insert into host cells, such as *E. coli* or yeast, for expression. Cells carrying these special expression vectors transcribe the gene and translate its mRNA into protein. Researchers, after purifying the protein expressed from the clone, can use it to immunize rabbits or mice, thereby inducing them to manufacture specific antibodies to the pro-

tein. These antibodies can become a tool for localizing the protein to one or more subcellular compartments (Fig. 10.21a). The antibodies can also serve as a means for purifying the native protein from the cells that produce it (Fig. 10.21b). Investigators can then analyze this native protein for specific enzymatic activity or other functions. In their search, they look for obvious characteristics, but the list can never be exhaustive. Their analyses may therefore turn out not to be informative.

In summary, sequence analyses, completed rapidly, provide clues to general gene function, suggesting, for example, that a protein is transported to the cell surface or binds to DNA. Expression analyses, completed in a matter of days, tell you where a gene is or is not expressed. Biochemical experiments, which may take a few weeks or months to complete, refine knowledge of gene function, revealing, for example, that the protein is a cell-surface receptor or a required regulator of transcription. The final confirmation of a gene's function, however, depends on genetic studies that show what happens to an organism if you "knock out" the allele under analysis, that is, prevent it from functioning.

Genetic Studies in Model Organisms Are the Ultimate Confirmation of the Function of a Cloned Gene

To understand the specific function of a gene, geneticists have traditionally generated mutations and analyzed the impact of these perturbations on the life of the organism. For ethical reasons, however, they cannot apply this approach to the study of genes in humans. The solution is to study the homologs conserved by evolution in one or more model organisms.

Let us review the models under study in conjunction with the Human Genome Project, in order of descending complexity (Table 10.1). The laboratory mouse (*M. musculus*) is the most complex of the model organisms. The only mammal among the models, it is the most closely related to humans (*H. sapiens*). Mice and humans diverged from a common ancestor that lived some 65 million years ago. Mice undergo essentially the same pathway of development and end up with the same set of tissues as humans; and as we have seen, the two species have the same size genome and a very similar set of genes, although these genes may be present in different orders or chromosomes. One great advantage of the mouse as a model for human genetics is that almost every human gene appears to have a mouse counterpart that functions in a very similar way. In fact, the majority of human genes have conserved homologs in all mammalian species. As an illustration, all mammals have α- and β-globin genes encoding polypeptides that join to form multimeric hemoglobin molecules that carry oxygen through the bloodstream. For many—though not all—genes, an understanding of gene function in the mouse will translate directly to an understanding of gene function in humans. The major exceptions to this general rule are some human genes with neurological functions. Such genes must be defined specifically in humans by clinical geneticists.

(a)

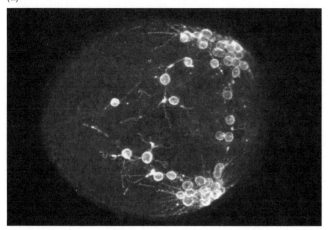

(b) Antibodies can also serve as a tool for purifying native protein from the cells that produce it.

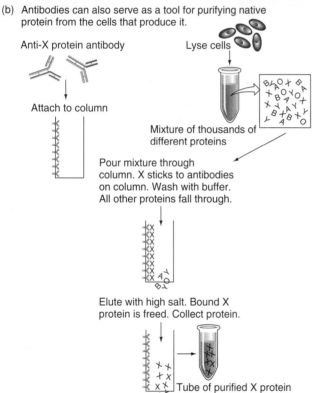

Anti-X protein antibody

Lyse cells

Attach to column

Mixture of thousands of different proteins

Pour mixture through column. X sticks to antibodies on column. Wash with buffer. All other proteins fall through.

Elute with high salt. Bound X protein is freed. Collect protein.

Tube of purified X protein

Figure 10.21   Using antibodies as reagents to determine where in the cell a protein does its work and what that work might be. (a) Labeled antibodies can serve as probes that bind to their protein counterparts within the cell. Researchers can use either light or electron microscopy to see where in the cell the proteins reside. This Confocal micrograph shows a labeled monoclonal antibody bound to the primary mesenchyme cells of the gastrular stage of sea urchin (*Lytechinus pictus*) embryonic development. In the gastrula, the antibody recognizes both the cell's plasma membrane and a centrally located organelle. The label used is a secondary antibody bound to the primary antibody which fluoresces under ultraviolet light. (b) Antibodies can also serve as a tool for purifying native protein from the cells that produce it. Here, antibodies to protein X are attached to a column through a chemical reaction; and after cells with the protein are lysed, the lysate, containing a mixture of proteins, is poured into the column. Most of the protein molecules pass right through the column without stopping, but protein X molecules stick to antibodies on the column. When the other proteins have exited the column, researchers can remove purified protein X from the column by adding a high-salt buffer, which disrupts the charge interactions that bind protein to antibody.

Moving down the evolutionary tree from mice, we come to fruit flies (*Drosophila*) and nematodes (*C. elegans*). Both these species undergo complex development, but because of their small sizes, short generation times, greater numbers of offspring, and smaller genomes, they are easier to study than the mouse. On the other hand, because they are more distantly related to humans, their genomes contain homologs to a smaller percentage of human genes. The least complex model eukaryotic organism is the unicellular yeast *Saccharomyces*. Its relatively small genome is highly amenable to experimental manipulation, but it carries homologs to fewer human genes than *C. elegans*, *Drosophila*, or *M. musculus*. Because these are the genes for the activities of single cells, however, yeast is a good model for the housekeeping genes responsible for the fundamental cell processes—progression through the cell cycle, transcription, and DNA replication—that occur in all eukaryotic cells. The simplest model organism is the prokaryote *E. coli*. It is the most amenable of all the model organisms to genetic analysis, but its genome contains a much smaller set of homologs to higher eukaryotic genes; it is useful as a model for studying basic biochemical processes such as glycolysis.

Finding the Simplest Model Organism for a Particular Study

In analyzing the function of a human gene through its homolog in a model organism, it is most efficient to begin by identifying the simplest model in which a homolog is present. Today researchers do this by probing the DNA of each model organism with a clone of the human gene. In the future, computer searches between genomic sequences in all the models under consideration will be able to reveal homologies that DNA probes cannot detect by hybridization. If the gene turns up in *E. coli* or yeast, elegant tools make it possible to mutate the homolog in specific ways to obtain a detailed profile of the gene's function. (Chapter 13 on bacterial genetics and Chapter 20 on yeast describe these tools and techniques.) Consider, for example, how geneticists used yeast to study the function of one of the first human protooncogenes to be cloned. A **protooncogene** is a gene that can mutate to an oncogene—an allele that causes a cell to become cancerous. Many protooncogenes play a central role in cell growth and division (see Chapter 17). The first cloned human protooncogene was named *RAS*. Molecular studies demonstrated the existence of a *RAS* homolog in yeast. Biochemical and genetic studies showed that the protein product of the *RAS* gene is active in the cytoplasm as a component of the intracellular pathway that relays signals regulating the cell cycle from the cell surface to the nucleus. Cells without a *RAS* gene die. Cells with an overactive *RAS* gene may become cancerous. Insertion of the human *RAS* gene into the yeast genome demonstrated that the human gene can function in place of the yeast homolog; the oncogene form of the human *RAS* gene stimulated yeast cells to divide when they should not. The interchangeability of the yeast and human RAS proteins proves that the two have the same basic function. Subsequent study of *RAS* in yeast has helped reveal the *RAS* gene's function in humans.

THE GENETIC DISSECTION OF COMPLEX TRAITS

During transmission, all genes follow Mendel's laws of segregation and independent assortment, modified only by the constraints of linkage. Nonetheless, as we saw in Chapter 2, the connection between genotype and phenotype is often more complicated than the 3:1 Mendelian ratio of alternative phenotypes produced by the segregation of dominant and recessive alleles from a single gene in both parents.

We now review the chief causes of complex inheritance and describe the problems they pose for the linkage mapping and positional cloning protocols described so far for single-gene traits (Table 10.2). We then explain how researchers adapt the procedures for analyzing single-gene traits to identify, map, and characterize the genes contributing to complex patterns of transmission.

The Causes of Complex Inheritance Patterns and How They Confound Linkage Mapping and Positional Cloning

In humans, only a small fraction of all disease traits follow the simple Mendelian pattern of single-gene inheritance. Rather, most common characteristics of human appearance, such as height; skin color; facial shape; and hair density, color, and shape (curly, wavy, or straight), and many essential measures of human physiology, such as blood pressure, cholesterol levels, basal metabolism, and susceptibility to infectious diseases, have a more complex pattern of inheritance.

Incomplete Penetrance: When a Mutant Genotype Causes the Occurrence of a Mutant Phenotype with a Probability of Less Than 100%

Sixty-six percent of all women who carry a mutant allele at the *BRCA1* locus will develop breast cancer by the age of 55. Thus, the mutant *BRCA1* allele predisposes a woman to breast cancer, although it does not guarantee that the disease phenotype will occur. By comparison, a disease such as sickle-cell anemia, in which a mutant genotype always causes a mutant phenotype, is completely penetrant.

The causes of incomplete penetrance vary from trait to trait and from individual to individual. With breast cancer, it seems that chance plays the largest role in determining which predisposed individuals get the disease—through the accumulation of secondary somatic mutations—and which do not. With heart disease, the individual's environment—especially diet and amount of exercise—plays a large role in determining whether a predisposing genotype is expressed and if so, at what age.

Incomplete penetrance hampers linkage mapping and positional cloning for one main reason: Individuals who do not express a mutant phenotype may nevertheless carry a mutant genotype. The simplest solution to this problem is to exclude all nondiseased individuals from the analysis. In the analysis of age-dependent traits like breast cancer and Huntington disease, such exclusion has meant that in disease-carrying families, the majority of children and adults under the age of 40 could not be included. As a result, many more families were required for the analyses that led to the mapping and cloning of the genes associated with both diseases.

TABLE 10.2   Complexities That Alter Traditional Mendelian Ratios

Category of Complexity	Problem in Observed Relationship between Genotype and Phenotype	Changes Necessitated in Mapping Strategy
1. Incomplete penetrance	Disease genotype can occur in an individual who does not express the disease phenotype	Eliminate nondiseased individuals from analysis
2. Phenocopy	Disease phenotype can be expressed by an individual who does not have the disease genotype	Program computer to flag unexpected cases of double recombination events that flank a putative disease locus
3. Genetic heterogeneity	In different families, different disease genotypes are responsible for the same disease phenotype	Divide complete set of disease-transmitting families into subgroups (based on various parameters such as average age of onset) and perform linkage analysis separately on each subgroup
4. Polygenic determination	Mutant alleles at more than one locus are required for expression of the disease phenotype in a single individual	Program computer to search for complex patterns of association between the disease trait and multiple unlinked loci

Variable Expressivity Occurs When Individual Expression of a Mutant Trait, Rather Than Its Expression in the Population as a Whole, Differs from Person to Person

The variation may be in age of onset, phenotypic severity, or any other measurable parameter. Variable expressivity does not normally interfere with genetic analysis, because geneticists can use any degree of mutant phenotype as evidence for the presence of a mutant allele.

Phenocopy: When the Disease Phenotype Does Not Arise from Inheritance of a Predisposing Mutation at Any Locus

The observation that 3% of women who do not carry a mutation at the *BRCA1* locus or have any family history of the disease still develop breast cancer by age 55 suggests that the disease can arise entirely from one or more somatic mutations in the breast cells themselves. This form of the disease is considered a **phenocopy** because it is indistinguishable from the inherited form of the disease yet is not caused by an inherited mutant genotype. The percentage of women who develop phenocopy breast cancer rises to 8% by age 80.

Researchers would not include women who develop breast cancer in the absence of a family history of disease in a study aimed at mapping predisposing alleles. However, even within families with a disease history whose members are included in a mapping study, there will be a small but significant fraction of women classified as carrying a mutant allele (because of disease expression) who are, in fact, wild type. Phenocopies can thus cause an artificial increase in the rate of recombination between a disease locus and a linked marker, undermining progress toward high-resolution mapping.

Genetic Heterogeneity: When Mutations at One of Two or More Loci Cause Expression of the Same Phenotype

Sometimes it is possible to use sophisticated diagnostic techniques to separate what appears to be a single disease phenotype into a set of related but clearly different diseases. For example, biochemical studies have enabled researchers to distinguish insulin-dependent from insulin-independent diabetes on the basis of their different physiological origins. However, even when the limit of disease subdivision has been reached, what appears to be a homogeneous phenotype may still arise from genetic heterogeneity. The seemingly simple disease of thalassemia is a case in point. Mutations in either the α-globin or the β-globin gene can eliminate or severely reduce the number of functional hemoglobin molecules produced in red blood cell precursors.

Genetic heterogeneity can hinder attempts to map causative loci. Although individual human families will usually segregate only a single mutation responsible for a relatively rare disease, most families do not have enough members to provide sufficient data for determining linkage (see the Genetics and Society box "Using Human Pedigrees and LOD Scores to Calculate the Probability That Two Loci Are Linked"). For this reason, linkage studies in humans almost always combine data from multiple families. But if a disease is heterogeneous, a marker linked to the disease locus in one family may assort independently from an alternative disease locus in a second family. As a result, when data from the two families are combined, the calculated probability of linkage between the marker and the disease locus will drop below that obtained with just the first family alone.

One suspects genetic heterogeneity whenever a comprehensive analysis of many families and markers fails to map a locus responsible for a disease trait that appears in a significant number of members from each affected family. When this happens, investigators can try to divide the complete set of disease-transmitting families into subsets based on any of several parameters and then combine only the families in each subset for linkage analysis. For example, when researchers combined the data from a large set of breast-cancer-prone families, there was no evidence of linkage to any marker. But when they selected a subset of families in which the average age of disease onset was less than 47 and analyzed the data from these families alone, they obtained strong evidence for a disease locus, named *BRCA1*, that mapped to chromosomal band 17q21 (Fig. 10.22a). It is now clear that mutations in *BRCA1* cause the onset of breast cancer at an earlier age than predisposing alleles at other loci. Classification of families into early-onset and late-onset groups may be helpful with any trait showing age-dependent expression. If this classification fails to produce evidence of linkage with age-dependent phenotypes or if a trait is independent of age, classification based on other variables, such as severity of the expressed phenotype, may prove helpful.

Once researchers have defined a single locus responsible for a disease, they can use closely linked markers to test whether that locus determines the disease phenotype in other families with a disease history that were not used in the original linkage analysis. A process of elimination may identify a subset of families that inherit the disease because of a predisposing allele at a different locus. This type of testing and elimination identified a group of families in which *BRCA1* could not be the locus predisposing women to breast cancer. The combined data from these families revealed a second breast cancer locus, named *BRCA2*, in the chromosomal region 13q12–13 (Fig. 10.22b). Mutations at *BRCA1* or *BRCA2* account for the majority of inherited breast cancers, but not all. As of 1998, there was still at least one more breast cancer locus yet to be discovered.

Polygenic Inheritance: When Alternative Alleles in Two or More Genes All Play a Major Role in the Expression of a Particular Phenotype

Our discussion so far has examined ways in which diseases caused by mutations in a single gene of an individual are associated with complex patterns of inheritance. But as we saw in Chapter 2, some traits arise from the interaction between two or more genes. Some polygenic traits are discrete, that is, they either occur or they don't; the occurrence of a heart attack, or a myocardial infarction, is a discrete polygenic trait. Other polygenic traits are quantitative, that is, they vary over a continuous range of measurement, from one extreme

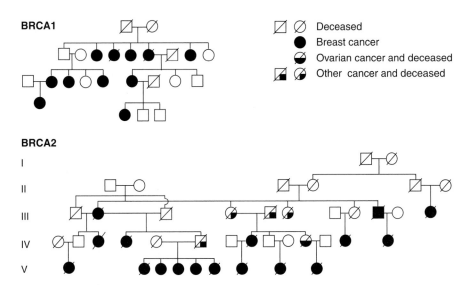

Figure 10.22 Genetic heterogeneity: Mutations at different loci can give rise to the same disease phenotype. Both pedigrees in this figure show evidence of the transmission of a dominant mutation with incomplete penetrance that causes breast cancer. Linkage analysis shows that the mutation in the first pedigree resides on chromosome 17, whereas the mutation in the second pedigree is on chromosome 13. The first family is segregating a mutant *BRCA1* allele, while the second family is segregating a mutant *BRCA2* allele.

through the normal range to the opposite extreme. Loci that control the expression of such quantitative traits are known as **quantitative trait loci, or QTLs.** In humans, such common characteristics as height, skin color, blood pressure, and cholesterol levels are quantitative traits. Although extreme values of expression of each of these traits are considered abnormal, the border between normal and abnormal is somewhat arbitrary.

There is virtually an unlimited number of transmission patterns for polygenic traits. For example, a completely penetrant discrete trait may require mutations at a complete set of loci. For other discrete polygenic traits, penetrance may increase as the number of mutant loci increases. With quantitative polygenic traits, the measured degree of expression may vary with the number of mutant loci present in the individual.

Many other factors can complicate the analysis of polygenic traits. Some members of a set of interacting polygenic loci may make a disproportionately large (or small) contribution to the penetrance or quantitative expression of the trait. Or, mutations at some loci in a set may be recessive, while mutations at other loci are dominant or semidominant. Moreover, some traits may arise from a mixture of polygenic and heterogeneous components. For example, one form of a disease may be caused by mutations at loci A, B, C, and D; a second form, from mutations at B and E; and a third form, from a single mutation at F. The more complex the inheritance pattern of a trait, the more difficult it will be to identify the loci involved.

How Geneticists Move from Complex Traits to Sets of Contributing Loci

Model mammalian organisms are powerful tools for analyzing traits that turn out to be too complex for dissection with human

pedigrees. The model organism of choice in nearly all cases has been the mouse. There are many reasons for this choice, including the existence of a large number of genetically homogeneous inbred mouse lines, many of which express a variety of traits that correspond to similar traits in humans. The analysis of complex traits in mice has many advantages over studies in humans. First, conducting the analysis on two contrasting genotypes represented by two parental inbred strains eliminates genetic heterogeneity. Second, the possibility of maintaining all animals under identical conditions eliminates environmental differences. Third, the transmission of only two alternative alleles at every marker and disease locus simplifies the analysis. In fact, an experimental cross of this type is the equivalent of a large family pedigree documenting the genotypes of several hundred children all born to a single set of parents.

Human geneticists are not the only researchers interested in the dissection of complex traits. The genetic analysis of complex traits is important in agriculture as well. For example, dairy and crop farmers are interested in milk production in cows and the taste, texture, and shape of fruits harvested for direct consumption or for processing. The contemporary method of dissecting complex traits is basically the same in all sexually reproducing experimental organisms, whether plants or animals. As an example, we examine the genetics of the pulp content of tomatoes, but you could use an analogous set of crosses and analyses to look at complex traits in mice, fruit flies, or nematodes as well.

The pulp of a tomato is the succulent, edible solid part of the fruit, everything but the water, seeds, and skin. Manufacturers of tomato-based products, such as canned tomatoes, tomato sauce, and tomato paste, look for fruit with a high pulp content because it produces a higher product yield per weight of starting material than tomatoes with a low pulp content (Fig. 10.23a). A strain of tomato plants with ideal pulp content,

Figure 10.23 **Using linkage analysis to identify loci that contribute to the expression of complex traits, such as pulp content in tomatoes.** (a) Tomatoes with an ideal amount of pulp (left) and with too little pulp (right). (b) Researchers use a two-generation cross to map loci that contribute to the amount of pulp in individual tomatoes. The first-generation cross is between parental lines that differ greatly in pulp content. The F_1 plants are all genetically identical, but their gametes contain recombinant chromosomes that assort to different offspring in the F_2 generation. The researchers correlate expression of the trait in the F_2 offspring with transmission of particular chromosomal regions. They then use statistical analysis to identify significant correlations between chromosomal regions and expression of the trait.

(a)

(b)

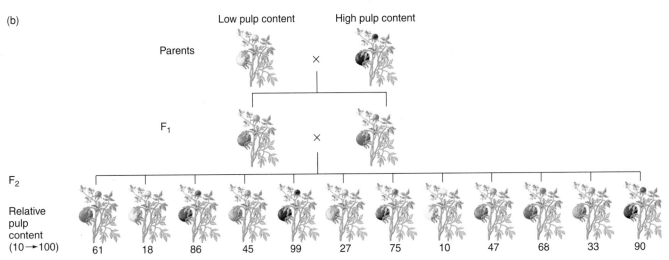

however, might not have an ideal taste or texture. If it were possible to identify molecular markers closely linked to all loci that contribute to pulp content, you could set up genetic crosses between an ideal pulp strain and other strains ideal for taste and texture, and then use marker analysis to identify F_2 generations of plants that had received only the ideal pulp alleles and were otherwise genetically the same as parents with the best taste and texture.

The first step in such an analysis is the identification of two inbred (that is completely homozygous) strains with extreme, reproducible differences in the expression of the trait of interest. In studying the pulp content of tomatoes, you would begin with two strains that showed extreme values when measured for this trait, for example, a high-pulp strain with an average pulp value arbitrarily set at 100 and a low-pulp strain with an average pulp value of 10.

The next step in the genetic analysis of pulp content is to cross plants from the high- and low-pulp strains to obtain a group of identical F_1 hybrids, which you then cross with each other to generate several hundred F_2 offspring. Because of independent assortment and recombination in the F_1 hybrids, the F_2 generation will have a range of different genotype combinations and thus exhibit a range of "pulp" phenotypes

(Fig. 10.23b). Some plants will have a pulp content similar to one parent's; others will have a pulp content similar to the other parent's. With most plants, the pulp-content value will lie somewhere between the two parental extremes. You now type each F_2 plant with molecular markers for which the parents carry different alleles. The markers, chosen from across the tomato genome at set intervals, enable detection of linkage to any intervening locus. You then analyze the data for each marker in all F_2 plants for correlations between marker genotype and pulp phenotype. A marker linked to a locus involved in pulp content would generate a significant correlation between F_2 plants that are homozygous for a parental strain marker allele and the parental strain phenotype. In other words, F_2 plants that are homozygous for the high-pulp allele at the linked marker would, as a group, have a higher pulp content than plants homozygous for the other marker allele. In contrast, markers not linked to a pulp-content locus would show no correlation between marker genotype and plant pulp phenotype. You could use this basic process to find all tomato loci with a significant effect on pulp content, and then perform further analyses to understand the interactions between the various loci. When it is possible to identify the genes contributing to every major trait of interest to tomato growers and processors, it will be possible to custom design a tomato plant possessing only the most desirable traits through the selection of individuals carrying the correct combination of marker alleles.

COMPREHENSIVE EXAMPLE: PROGRESS, PROMISE, AND PITFALLS OF THE HUMAN GENOME PROJECT

Genomics has changed the face of human genetics. If the Human Genome Project reaches its 15-year goal of completing high-resolution linkage, physical, and sequence maps for the entire genome by the year 2006, human geneticists will be able to use these maps in determining the chromosomal location and molecular composition of all 100,000 human genes and the regulatory elements that control their expression. Moving from the description of all 100,000 genes to an understanding of the complex networks that underlie most human phenotypes, however, will be very difficult and may require the development of new tools and techniques.

Progress Report
As of December 1993, geneticists had constructed contigs of overlapping YAC clones for every human chromosome. As of June 1994, they had compiled a linkage map of the whole human genome with microsatellite-based STS markers every 0.5 cM. By January 1997, the sequence map—a readout of all 3 billion base pairs in the human genome—was 2.6% complete. At the same time, the four or five laboratories with the most advanced sequencing technology could each scan 3 million to 30 million base pairs per year. Researchers are now developing technologies to increase that rate by automating many of the repetitive tasks now accomplished manually, and by integrating these tasks into a robot-intensive factory-like production line. At the same time, work on some of the model organisms has progressed faster than anticipated. Bacterial geneticists have determined the complete sequence of the genomes of more than a dozen microorganisms, including *Haemophilus influenzae* (the first free-living organism to have its genome sequenced) and *Bacillus subtilis* (the first grampositive bacterium to have its genome sequenced). A complete sequence map of the 16 chromosomes in the yeast genome was in place in 1996, and a complete readout of the 100 million nucleotide pairs on the 6 chromosomes of *C. elegans* was finished in 1999.

Promise for Genetic Research and Medical Therapies
Research
It is a curious fact that the Human Genome Project may provide the only way to identify many human genes. For example, it is possible that half the genes in the human genome are expressed only in the brain or during its development. The expression of many of these genes, however, occurs during such a narrow window of development or in so few cells that many

methods would fail to detect them or their transcripts. Accordingly, the only way to find many of these genes is by walking along each chromosome and identifying each gene in order by sequence. Computer analysis of the complete sequence of the human genome in conjunction with appropriate biological studies may also make it possible to find and analyze many of the components that regulate gene expression, that is, all the genetic components that determine when and where each gene is expressed, and how much of its product is produced at each time and place. In the future, a medical researcher may be able to type into a computer "kidney" and obtain the subset of human genes expressed in the kidney, or even more specifically, "a cell in the descending loop of Henle" and obtain the small subset of genes expressed only in that type of cell and no others. In this manner, he or she can begin to understand the network of interactions that occur among the proteins encoded by the genes.

The Human Genome Project will also provide a picture of the complete architecture of the genome and the probable pathways through which it evolved from single-cell ancestors (see Chapter 24 for details).

Finally, the complete nucleotide sequence of the human genome will make it possible to identify the DNA polymorphisms that distinguish one individual from another. Because researchers will be able to study these polymorphisms in minute detail, they will come to understand which polymorphisms are compatible with normal physiology and the implications of particular polymorphisms for pathology.

Medical Therapies
The ability to analyze human polymorphisms will eventually enable identification of all gene alleles that cause or predispose to hereditary disease. In 1994 researchers used the powerful gene-mapping techniques described in this chapter to examine the entire length of all 46 human chromosomes in the diploid cells of people with diabetes and their families; in the course of their study, they located two new genes—one on the long arm of chromosome 11, the other on the long arm of chromosome 6—that underlie type I diabetes. People with this type of diabetes require daily injections of insulin. The researchers also found evidence that yet another gene, this one on the long arm of chromosome 18, plays a role in the disease. The picture that emerges is of a complex condition caused by the interaction of many genes and environmental factors. As they uncover more and more of the contributing components, geneticists can begin to understand the complex genetic network that gives rise to normal insulin metabolism and how mutations disrupt the network to produce diabetes. Armed with this information, they may be able to fashion a different therapy that is specific for each mutant state, to prevent the condition or alleviate its symptoms.

The analysis of genomes has revealed that most common hereditary conditions—diabetes, hypertension, and allergies—show the complex inheritance patterns associated with incomplete penetrance, genetic heterogeneity, and polygenic determinants. Multiple sclerosis, another abnormal condition, may well turn out to be three or four distinct diseases

Genetics and Society

The Patentability of DNA

Some people argue that DNA, the naturally occurring raw material of life's evolution, is a common heritage that belongs to everyone. Yet since the mid-1970s, universities and biotechnology companies have sought patents on specific stretches of DNA—human, bacterial, and plasmid. To understand this apparent paradox, we take a brief look at the history and purpose of U.S. patent law and examine its application to various types of DNA fragments.

Congress passed the first patent act, penned by Thomas Jefferson, in 1790. That same year the U.S. Patent Office awarded the first patent to Samuel Hopkins of Vermont for the making of potash, a potassium-based compound used to produce fertilizer and soap. One hundred and ninety years later, in 1980, the patent office granted two patents for specific uses of DNA. The first was to a microbiologist at General Electric for a genetically engineered oil-eating bacterium that could help clean up spills; the second was to Stanford University and the University of California at San Francisco (UCSF) for the basic gene-splicing and -cloning technology invented by researchers at those institutions.

Patent examiners evaluate the patentability of a product or process by three criteria: Is it *new*? Is it *nonobvious*? Is it *useful*? In the DNA arena, the courts have made the following interpretations of patent law. Raw materials of nature, such as wild-type DNA in a living organism, are not novel and thus are not patentable; modified products, such as bacterial DNA altered by a synthetic mutation or human DNA in a mouse genome, are novel and thus eligible for patents by the novelty criterion. DNA-based processes that produce a novel material, such as clones of a gene (consisting of a DNA construct in a vector), are also patentable. Publication in a scientific journal makes an item, such as a DNA sequence, obvious, thus unpatentable per se; but a specific use of a published sequence may be nonobvious and thus patentable. A well-defined use might be a particular test for a particular genetic aberration or

a particular process for the manufacture of a particular therapeutic agent.

Since the mid-1980s, patent examiners have used these guidelines to grant DNA-related patents on a case-by-case basis. The cDNA for insulin integrated into a bacterial plasmid has received a patent; this cDNA does not occur naturally in significant quantities, but is fabricated through a human-devised protocol using reverse transcriptase. By contrast, human cDNA fragments isolated at random from brain cells did not qualify for a patent, in part because the potential use of these fragments could not be defined without identification of the complete cDNAs and the proteins they encode. In another example, researchers at the University of Utah have applied for a patent on *BRCA1* sequences to be used as a test for predisposition to breast cancer; because these researchers have mapped and sequenced the gene and incorporated that knowledge into the design of a socially useful test, their application may well pass muster. To date, the processing of DNA-related patent applications has taken 8–12 years.

The rationale for granting patents is to encourage innovation—the invention of useful contributions to society—by providing a time-limited monopoly to protect an invention from imitation. This commercial protection is given in exchange for the complete disclosure of the information related to a product or process. In fact, in pursuing a patent, a company protects its interests by making available as much information as possible about the modified product or process. Patents protect use for profit; they do not interfere with research or other noncommercial uses of the information in the patent. Institutions, companies, or individuals holding patents can profit from their inventions themselves or license the right to profit from them to another party. Stanford and UCSF, for example, originally granted many nonexclusive licenses to companies using recombinant DNA technology in the produc-

---

distinguishable by different sets of molecular markers. The specific treatment for each of these diseases will be different. Consequently, there may come a time 25–30 years in the future when 100 or more genes predisposing to cardiovascular disease, cancer, metabolic diseases, immunologic diseases, and other major human diseases will have been identified. At the age of consent, an individual could then go to a genetic counselor for an analysis of his or her genetic profile. The profile would be a readout of the *potential* "future disease history" of the individual as well as a description of preventive measures that could help circumvent some of the genetic limitations. The focus in this new era of preventive medicine will be measures that keep people healthy rather than cures for illnesses that have already occurred.

Pitfalls: Social and Ethical Problems for Which There Are No Simple Solutions

This view of medicine in the twenty-first century raises a host of pressing questions about the privacy of genetic information, limitations on the use of genetic testing, and the extent to which the human genetic engineer should seek to engineer him- or herself.

We discuss the privacy issue—the need to ensure that genetic information is confidential, to be given out only at the discretion of the tested individual—in the Genetics and Society boxes in Chapters 1 and 2. In the Genetics and Society box in Chapter 1 and in the Genetics and Society box "Social and Ethical Issues Surrounding Preimplantation Embryo Diagnosis" in Chapter 9, we outline considerations that make it necessary to establish guidelines for genetic screening; among those consid-

tion of various agents for a fee of $10,000 a year. In the first four years of its patent, Stanford received $3 million in licensing fees and patent royalties.

Different countries apply the basic tenets of patent law in different ways. In the United States, the purified form of a gene or protein is patentable because genes and proteins do not exist in nature in purified form. In England, a naturally occurring gene sequence is not patentable no matter what form it's in. And in France, the Code on intellectual property declares unpatentable "the human body, its elements and products as well as knowledge of the partial or total structure of a human gene."

There is currently healthy debate over the application of patent criteria originally developed for mechanical and chemical inventions to DNA sequences and various life-forms. The debate focuses on several areas of concern.

Openness versus Secrecy

How does the patent process affect the open exchange of ideas that is a prerequisite of basic research? A company putting a great deal of money into the research and development of what it hopes will be a patentable gene therapy, may withhold publication of its data or publish only partial results until its patent application is in the pipeline. For example, in the 1980s, before the discovery of the cystic fibrosis gene, one gene diagnostics company published an article about markers for the *CF* gene but, in an effort to protect their work, did not include the fact that those markers were on chromosome 7. As it happens, other groups subsequently found closer markers and then the gene. Still the question remains: Do commercial considerations interfere with the free exchange of ideas and results about genomes? If so, should something be done about this?

Profitability versus the Social Good

Who has control over genetic data, deciding what applications to pursue with commercial funding? What criteria do they use? Most companies consider potential profitability as the basis for pursuing research and development. Some people wonder whether they should also factor into the equation concerns about serving the poor

(who cannot pay much for drugs or therapies) and the relatively few patients suffering from uncommon diseases.

Development and Funding of Socially Useful Research Applications

Since companies must pour large sums of money into developing DNA-based drugs or therapies, testing them on large numbers of people, and bringing them to market, many researchers—at both universities and commercial enterprises—maintain that the survival of biotechnology companies, particularly small ones, depends on patent protection. Without the possibility of a patent, they argue, companies may not be able to afford to use knowledge of, for example, the *CF* gene and the transmembrane conductance regulator it encodes, to develop therapies that may ease and extend the lives of thousands of CF sufferers. Development of drugs or therapies that cost a great deal of money, they maintain, simply will not occur without patents. On the other hand, many of the patents providing the basis for development and marketing are on inventions made in university laboratories; and the investigators in some of these laboratories are nonprofit recipients of federal grants that are ultimately funded by taxpayers. Is it proper for companies to use these inventions as the basis for financial gain? Most government agencies allow such patents because they themselves do not have the funds to develop research ideas for the marketplace. Furthermore, nearly all of the major technological advances of the late 20th century in computers and electronics were based to one degree or another on research funded by federal money. We don't question the right of computer companies to profit off extensions of this research.

Thus, the availability of purified DNA, especially specific stretches of the human genome, raises several issues that the framers of patent law could not foresee. Vigorous public debate will help scientists, business people, and lawmakers distill some of the answers. One radical and perhaps impractical proposal to deal with the legitimate concerns on both sides of the public-funding-versus-private-profit issue suggests creation of an international corporation that would hold patents on all human cDNA. This corporation would license the patents for development at auction and reinvest the proceeds from licensing fees in basic research.

---

erations are who should be tested and whether testing should be carried out if no therapies or preventive measures exist to help an individual with a genetic predisposition. Most geneticists emphasize the need for educating the general public about the benefits and drawbacks of testing. Meanwhile, the U.S. Congress has passed a law ensuring the privacy of medical records (see the Genetics and Society box "The Patentability of DNA").

But what about the social and ethical concerns raised by genetic engineering, that is, the ability to alter specific genes of an organism in specific ways? Many of the preventative or remedial measures alluded to earlier constitute **somatic gene therapy** in which medical practitioners compensate for a faulty gene by inserting a replacement gene into the affected tissue where the gene is expressed. Somatic gene therapy causes biochemical and

physiological changes in the genetically modified tissue or tissues that die with the individual. A potential therapy for cystic fibrosis, for example, consists in inserting a wild-type *CFTR* gene into lung cells. This type of genetic engineering is not different in kind from drug therapies aimed at correcting a particular physiological deficiency (for example, the insulin injections aimed at treating diabetes). The alterations resulting from somatic gene therapy affect only the somatic cells of the individual undergoing the therapy and cannot be transmitted to offspring.

Most of the controversy surrounding genetic engineering stems from the potential for **germ line gene therapy:** modifications of the human germ line. Germ line gene therapy produces changes in germ cells that are passed on to progeny; such therapy can thus cause permanent changes in the gene pool.

There are two main categories of reservations related to germ line therapy: technical and ethical. Technical concerns focus on whether genetic engineers have the know-how to make the changes accurately and safely. Insertion of a gene in the wrong place in the genome could cause a worse defect than the one being treated. With somatic gene therapy, this is less of a problem because destruction of a particular gene function in a single cell does not usually affect other cells of the individual. With germ line gene therapy, however, the newly created mutant allele would be transmitted to every cell in the developing embryo and adult, and could have far-reaching consequences for a particular stage of development or for the functioning of a tissue or organ. Ethical concerns focus on what is and is not appropriate. Should, for example, a couple be able to eliminate a cancer-predisposing gene in their unborn child? Should they be allowed to alter the child's potential for obesity or longevity? Should they have the option of choosing his or her eye color? Many geneticists and bioethicists currently oppose germ line therapy. At the same time, they urge serious and open discussion of the issues before perfection of the technology overtakes our ability to control its use.

CONNECTIONS

The ultimate goal of genomics is to map right down to the level of nucleotide sequence all the DNA that makes up a genome. Mapping and sequencing projects underway in laboratories around the world promise to provide the complete nucleotide sequence of the DNA in entire genomes, and through computational analysis, to reveal each of the sequences that code for proteins. Comparisons of these sequences among organisms will reveal related families of genes, and where the function of a particular gene has been identified, may suggest the function of related family members in other organisms. Someday, it may even be possible to understand the complete network of tissue-specific regulations of gene expression, and with this knowledge, to read from genome sequence the time and place of each gene's expression in an individual.

Will this information provide a complete understanding of life? Far from it. Biologists will need much more than a knowledge of the instructions specified by nucleotide sequence to comprehend the networks of interactions that bring each individual to life. This is because chromosomes are not simply repositories of genetic information. Composed of many protein molecules in addition to DNA, they are complex organelles that package and manage the availability of DNA. In fact, the activity of large chromosomal regions is controlled not just by local DNA sequence but by the way the DNA is packaged; some regions, for example, are marked in a way that irreversibly extinguishes their activity, including gene expression. Moreover, chromosomes have a hierarchic organization that enables them to go through a cycle of condensation and decondensation with each cell cycle, and they contain specialized regions that dictate how they duplicate and segregate during nuclear division. In Chapter 11, we examine how proteins interact with DNA to generate a chromosome's functional complexity.

ESSENTIAL CONCEPTS

1. Large-scale maps are the foundation of genomics.
   a. *High-density linkage maps* developed by computer analysis of raw genotyping data show the relative positions of closely spaced DNA markers. *Mapping panels* make it possible for different laboratories to fill in information in a highly efficient manner by using the same set of pedigree samples.
   b. Long-range *physical maps* chart the features of chromosomes.
      (1) Fluorescent *in situ* hybridization, or *FISH,* locates a cloned locus to a particular band on a particular chromosome.
      (2) DNA hybridization and fingerprinting order the clones of a genomic library into *contigs* that represent the chromosomes of the species. One approach to assigning chromosomal positions uses the markers of a high-density linkage map as hybridization probes to obtain overlapping, large-insert clones, followed by chromosome walking to fill in the gaps between contigs. Another approach does not rely on linkage mapping and genetic markers. Instead, physical analysis begins with the complete set of individual clones in a whole-genome library. Researchers establish fingerprints for each of the clones, and make a computer search to determine overlaps in fingerprints that indicate overlaps in the corresponding clones. With a large enough number of clones, it is possible to form contigs that extend across every chromosome.
      (3) Researchers use restriction mapping and hybridization probes to characterize the fine details of each clone in a contig.
      (4) Computers combine the information for individual clones into detailed whole-chromosome physical maps.
   c. In the future, long-range sequence maps, compiled from the sequences of subclones will provide a readout of every nucleotide in each chromosome.

2. In positional cloning, investigators use high-resolution maps to move from a phenotype to a clone of the responsible gene.
   a. The first step is to correlate transmission of the phenotype with transmission of one area of the genome.
      (1) Linkage analysis with markers that provide complete genomic coverage places the locus responsible for a phenotype on a chromosome; further linkage analysis narrows the chromosomal area transmitted with the phenotype.
      (2) To place the locus on a physical map, researchers select from a central repository the small set of YAC clones that span the genomic region defined by the closely linked markers.
   b. With the clones in hand, they identify a group of *candidate genes* through
      (1) *Northern blots* that reveal tissue-specific transcription units.
      (2) Hybridization probes that pick out regions conserved by evolution.
      (3) Computer programs that scan a long DNA sequence for coding regions associated with the exon subunits of genes.
   c. To find the one gene actually responsible for the phenotype, researchers
      (1) Determine whether a candidate is expressed in the appropriate tissues.
      (2) Look for mutations in candidate genes that eliminate gene function and correlate with expression of the disease.
      (3) Confirm their findings with *transgenic* analysis of gene function.

3. Sequence maps provide a basis for moving from the clone of a gene to its function. To determine that function, researchers determine the nucleotide sequence of the clone and use the DNA sequence to establish the amino-acid sequence of the polypeptide it encodes. They then analyze

   a. Short-amino-acid-sequence similarities with well-defined *peptide motifs* and long-amino-acid-sequence similarities with entire polypeptides.
   b. Expression patterns revealed by Northern blot analysis.
   c. Biochemical studies that localize an encoded protein's activity in the cell and analyze the protein's function at that site.
   d. Genetic studies of homologs in model organisms.

4. Most naturally occurring genetically determined trait variation among the individuals of a species is complex.
   a. Some mutant alleles cause only a predisposition to express a mutant trait, with a probability that may increase with age. A trait determined by such an allele shows *incomplete penetrance.*
   b. Mutant traits that arise in the absence of a mutant genotype are considered *phenocopies.* Phenocopies that occur at a significant frequency can thwart attempts to map predisposing loci.
   c. Mutant traits caused by mutations at any one of two or more alternative loci show *genetic heterogeneity.*
   d. A phenotype controlled by alleles at multiple loci is a *polygenic* trait. If different combinations of alleles cause quantitative differences in trait expression, the trait is known as a *quantitative trait,* and the loci involved are known as *quantitative trait loci,* or *QTLs.*
   e. The process for the genetic dissection of complex traits is similar for all sexually reproducing organisms.

5. Participants in the Human Genome Project have already compiled a high-density linkage map of the human genome and constructed contigs of overlapping clones for all 24 human chromosomes (the 22 autosomes and the X and the Y). The speed of sequencing is poised to increase rapidly and the complete sequence map for the genome's 3 billion nucleotides is expected by the year 2006. The information flowing from the genome project will have many medical applications and may usher in an era of preventive medicine. Geneticists, as they work toward fulfilling this promise, need to be cognizant of the social and ethical problems arising from the new knowledge.

S O C I A L   A N D   E T H I C A L   I S S U E S

1. Several companies were formed in the 1990s with the purpose of sequencing portions of the human genome and using this information to devise new drugs. These companies sought to patent each bit of DNA sequence that they decoded even if they didn't know whether it encoded a gene product. Their justification was to ensure that any use of the sequence would yield payback for their research. Do you think patents should be granted for each partial gene sequence that has been determined? Do you think patents should be granted for entire gene sequences even if we don't know what the gene does? What criteria

do you think are compatible with facilitating private industry drug development, encouraging research, and protecting intellectual property?

2. Josephine read in the newspaper that an individual's propensity to engage in risk-taking behaviors is influenced by genetic makeup. She would like her children to take a more aggressive or assertive approach to new situations than she ever did. She thinks that the possibility of germ cell gene therapy (altering DNA of the egg or sperm) to modify behavioral genes such as this

would be very attractive and would like to see the technology developed to do this type of manipulation. Is it ethical to change the genetic makeup of the germ cells? Do humans have the right to change traits in their children? in themselves? What is the difference between controlling a child's genes and controlling a child's environment when the goal of each form of control is to maximize a child's chances of success? Would your answer be different if it were possible to introduce an allele that caused resistance to HIV?

3. Researchers who have similar interests and complementary expertise often collaborate on projects. The Johnston lab has recently started a collaboration with a lab in San Francisco to use a unique tissue culture cell line that they (the San Francisco lab) had established. At a scientific meeting, a graduate student from the San Francisco lab heard Marcus Johnston give a seminar talk in which he discussed future work and described this cell line, but did not mention where it came from. He also did not describe it accurately. The graduate student was very upset with this, but then thought perhaps she was being too sensitive and when the work was done and publication occurred, appropriate credit would be given to the San Francisco lab. Should the student speak up at the end of the talk to correct misinformation? Should she approach Marcus later? Should she wait and talk with her advisor first? Should she do nothing and assume the misinformation will be corrected by someone else?

SOLVED PROBLEMS

I. Four pairs of primers were used for PCR amplification of DNA from one man's somatic cells and from 21 single sperm that he provided for this research. Each of these primer pairs amplifies a different region of the human genome referred to as A, B, C, and D. The four pairs of PCR primers were used simultaneously on each sample of DNA. Each of the amplified DNAs was divided into eight aliquots (identical subsamples), and these aliquots were denatured and spotted onto eight nitrocellulose membrane strips, as shown in the figure. Each of these strips was then hybridized with a different ASO (allele-specific oligonucleotide). There are two different ASOs for each of the four regions. Oligonucleotides A1, A2, B1, B2, C1, C2, D1, and D2 reveal single-nucleotide differences for each of the regions A–D. A black circle indicates that the amplified DNA hybridized to the ASO probe.

a. Based on the results shown, which gene(s) could be X-linked?

b. Which gene(s) could be on the Y chromosome?

c. Which gene(s) must be autosomal and homozygous for a particular ASO?

d. Which two genes are linked to each other?

e. Ignoring the results from sperm number 21, what is the distance between the two linked genes?

f. Which gene(s) might be better mapped with an additional ASO?

g. What event must have occurred to give rise to sperm number 21?

ASO:	Gene A		Gene B		Gene C		Gene D	
	A1	A2	B1	B2	C1	C2	D1	D2
Somatic cell	●		●	●		●	●	●
Sperm number								
1	●		●					●
2	●			●	●		●	
3	●		●		●		●	
4	●		●		●			●
5	●			●			●	
6	●			●				●
7	●			●		●	●	
8	●			●		●	●	
9	●			●		●		●
10	●		●					●
11	●		●					●
12	●		●			●		●
13	●			●			●	
14	●			●			●	
15	●		●			●		●
16	●		●			●		●
17	●			●				●
18	●			●		●		●
19	●			●	●		●	
20	●		●			●		●
21	●	●	●	●			●	●

Answer

For this problem, you need to understand how allele-specific oligonucleotides are used and be able to follow and interpret the experimental processes of amplification and hybridization.

a. What would you expect for hybridization to an X-linked gene in somatic and sperm cell samples? An X-linked gene would be present in only about half the sperm (the other half being Y-bearing sperm). *Gene C shows this type of pattern.*

b. Similarly, a gene on the Y chromosome would be found in only half the sperm. Again gene C *is a candidate for a Y linked gene.*

c. If an individual is homozygous for a gene, all the sperm from that individual therefore have the same ASO. *Gene* A *appears to be homozygous and autosomal.*

d. The alleles of genes that are linked would segregate together more than 50% of the time and would therefore end up in the same sperm. The patterns for B1, D2 and B2, D1 are very similar. *Genes* B *and* D *are linked.*

e. In sperm 3, 9, and 18, the B1, D2 or B2, D1 pairs of alleles have not cosegregated and therefore must represent the results of recombination. Three out of twenty or 15% are recombinant. *There are therefore 15 m. u. between genes* B *and* D.

f. It is not possible to get any mapping information for *gene* A using the probes A1 and A2 because these probes are not detecting any polymorphism in that gene. For a set of probes to be useful in mapping, they should be distinguishing between the alleles on each chromosome so associations with alleles of another gene can be followed. *Gene* C *could also be better mapped with other oligonucleotides because ASO C1 does not recognize the polymorphism that prevents C2 from hybridizing.*

g. Copies of both *B* alleles and both *D* alleles are present in this sperm. The presence of both alleles suggests that two copies of the chromosome carrying *B + D* (which are on the same chromosome) are present. *Nondisjunction of the chromosome carrying genes* B *and* D *would have produced sperm having both alleles of* B *and* D.

II. A chromosomal walk was conducted in an area of a human chromosome containing three genes (and part of a fourth gene) that are transcribed within ovarian tumor tissue. The restriction map of the region is presented in next column; tick marks above the line indicate *Bam*HI restriction sites, while those below the line indicate *Xho*I sites. Sizes of DNA fragments between adjacent restriction sites are given in kilobases. Individual restriction fragments were purified, made radioactive, and used as probes for Northern blots of polyA-containing RNA derived from ovarian tumors. The resulting autoradiograms are presented below the

restriction map. Using these data, characterize the four genes within the chromosome walk in the following ways.

a. What is the length of the mRNA for each of the three complete genes?

b. What is the minimum length of the primary transcript for the largest of these RNAs?

c. What is the minimum number of exons for each gene?

d. What is the minimum number of introns for each gene?

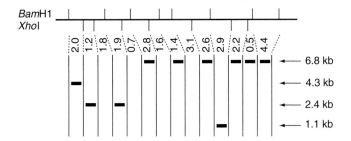

Answer

This problem requires an understanding of primary and processed transcripts and the analysis of RNAs using Northern hybridization.

a. The bands on the Northern blots that hybridize with the probes represent the mRNAs from this region. The three mRNAs corresponding to the genes in this region are *1.1, 2.4, and 6.8 kb* in length. The 4.3-kb transcript comes from a gene that is only partly contained in the DNA used as a probe since the one DNA fragment that hybridizes is only 2.0 kb in length.

b. The bands on the Northern blot represent processed transcripts (introns have been removed). Primary mRNAs are made by copying from contiguous DNA sequences, including those regions that will be removed by splicing. The minimum length of a primary transcript is based on the sizes of the restriction fragments that hybridize with the RNAs and any intervening fragments. *For the 6.8-kb largest mRNA, the minimum size is 2.8 + 1.6 + 1.4 + 3.1 + 2.6 + 2.9 + 2.2 + 0.5 + 4.4 kb or 21.5 kb.* We cannot say from these data whether transcription begins before the 2.8-kb fragment or extends beyond the 4.4-kb fragment.

c. and d. The minimum number of exons is determined by counting the number of hybridizing fragments or groups of contiguous fragments. This is a minimum number because any of the fragments could contain more than one exon. The nonhybridizing fragments that separate these must contain introns.

Transcript	Minimum Number of Exons	Minimum Number of Introns
1.1 kb	one exon	none
2.4 kb	two exons	one
6.8 kb	four exons	three

III. Reverse transcriptase, the enzyme used to synthesize cDNA starting with mRNA as a template, often falls off the template before completely copying the mRNA. When screening for a cDNA clone of a gene, it is therefore not uncommon to isolate partial cDNA clones.

What comparison could you make experimentally that would indicate if you had isolated a plasmid clone containing a partial cDNA?

Answer

For this problem, consider the two alternatives: you have the complete cDNA clone or you have a partial cDNA clone. What would be the differences that you could detect experimentally? The cDNA must be as long as its corresponding mRNA to be full length. You need to find out what the full length of the message is to do the comparison. The way to *measure the length of mRNA is to run a Northern gel and hybridize with the cDNA as a probe.* (Another alternative would be to hybridize a fragment from near the 5′ end of the gene to your clone. This would require that you have a fragment that you know is from the 5′ end of the gene.)

P R O B L E M S

10-1 For each of the terms in the left column, choose the best matching phrase in the right column.

a. sequence tagged sites (STSs)

b. chromosome walking

c. contig

d. linkage group

e. positional cloning

f. phenocopy

g. physical map

h. incomplete penetrance

i. high-density linkage map

1. set of two or more partially overlapping cloned DNA fragments

2. not all individuals with a mutant allele have a mutant phenotype

3. phenotype that has been induced by an environmental condition rather than by an inherited mutation

4. a map showing the order of cloned bits of DNA

5. set of genes in which adjacent genes are transmitted together more than 50% of the time

6. using a DNA probe from the end of one clone to isolate a clone containing adjacent chromosomal DNA

7. way to clone a gene without prior knowledge of its function or the protein product

8. a map of genetic markers that are separated by less than 1 cM

9. unique DNA sequences that serve as molecular markers

10-2 Give two different reasons for the much higher ratio of total DNA to DNA that encodes proteins in the human genome compared to bacterial genomes.

10-3 Using a microsatellite DNA probe you have isolated several clones from a library. You would like to identify unique sequences adjacent to these repeat DNAs found throughout the genome to establish some microsatellite-based STSs.
   a. How could you determine if there are unique sequences within the clones you have isolated?
   b. The most useful STSs are those associated with microsatellites that are polymorphic in the population (that is, have sequence variants). How could you determine if an STS is associated with a polymorphic microsatellite?

10-4 What are the advantages of the FISH protocol over linkage mapping for the initial characterization of the chromosomal location of a gene?

10-5 A clear limitation to gene mapping in humans is that family sizes are small, so it is very difficult to collate enough data to get accurate recombination frequencies. A technique that circumvents this problem begins with the purification of DNA from single sperm cells. (Remember that recombination occurs during meiosis. Analysis of a population of sperm provides a large data set for linkage studies.) The DNA from single sperm cells can be used for polymerase chain reaction (PCR) studies. The PCR reactions can be run such that a PCR primer will not hybridize to DNA if there is a single-

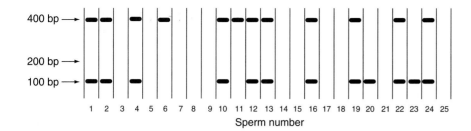

Sperm number

nucleotide difference between the sequence of the primer and the template DNA to be analyzed. A PCR product will be formed only if both primers hybridize to the template.

A series of three pairs of PCR primers were synthesized according to DNA sequences from human genomic clones. The alpha primers amplify a fragment 100 bp long; the beta primers produce a 200-bp product; the delta primers produce a 400-bp product. Twenty-five sperm obtained from one man were subjected to PCR analysis with these three pairs of primers simultaneously. The data appear above.

a. Is this man homozygous for any nucleotide polymorphism detected by this method? If so, which one?

b. Determine the genetic map information you can from this data.

10-6 The map of human chromosomes in males and females indicates that the centimorgan distances between genes is different in the two sexes.

a. What could account for this difference in distances?

b. Do you think the physical distances between genes in males and females would differ also? Why or why not?

10-7 A human genomic library was made by cutting human genomic DNA partially with the enzyme *Sau*3A (recognition site: 5′ ^GATC 3′), and then ligating the resultant fragments into the *Bam*HI site (5′ G^GATCC 3′) of the plasmid vector pCU999 shown in next column. This plasmid contains a "polylinker" in which the *Bam*HI site (B) is flanked on one side by a site for *Sal*I (S), and on the other side by a site for *Not*I (N). The distance between the *Sal*I and *Not*I sites in the vector is about 20 bp.

a. One particular clone from this library (clone 1) was isolated as part of a chromosome walk and was then cut by *Not*I (N), *Sal*I (S), and by *Not*I + *Sal*I (N + S). The fragments were separated by electrophoresis on an agarose gel and stained with ethidium bromide. A Southern blot of this DNA was made,

and the filter was probed with radioactive DNA made from the plasmid vector alone. The results are shown below. (Assume you cannot see a signal on the autoradiograph if the amount of homology with the probe is < 50 bp.) Show the positions of all *Sal*I and *Not*I restriction sites in the genomic DNA part of clone 1, and indicate the distance between adjacent sites.

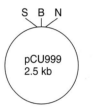

Restriction Digests of Clone 1

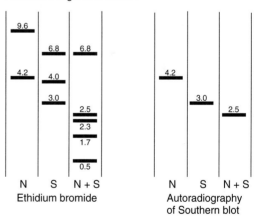

b. A clone with an overlapping fragment of human genomic DNA (clone 2) was isolated from the same genomic library. It was digested with the same enzymes as before, and electrophoresed and transferred to nitrocellulose paper as before. This time, the Southern blot was probed with radioactive clone 1 (genomic DNA + vector DNA). The results appear below. Draw a map of clone 2, showing the positions of all *Sal*I and *Not*I restriction sites in the genomic DNA part of clone 2, indicating the distance between adjacent sites.

Restriction Digests of Clone 2

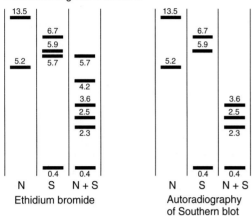

N   S   N + S
Ethidium bromide

N   S   N + S
Autoradiography of Southern blot

c. Whole genomic human DNA was now cut by *Sal*I. Identical blots of this digest are now probed with clone 1 or with clone 2. The results of this whole-genome Southern blot are shown below. The position of six *Sal*I sites in this region of the human genome can be mapped using the data above. Draw a *Sal*I map of this region.

Whole-Genome Southern Blot

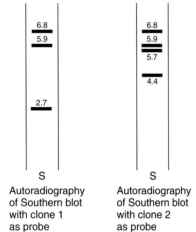

S
Autoradiography of Southern blot with clone 1 as probe

S
Autoradiography of Southern blot with clone 2 as probe

10-8 To make a set of clones more suitable for analysis of DNA sequence, a series of cosmid clones was prepared by digesting a YAC and subcloning the resulting restriction fragments. The restriction patterns of the inserted fragments are shown below. Arrange these four cosmids into a physical map, showing the order of the clones and the overlap between them.

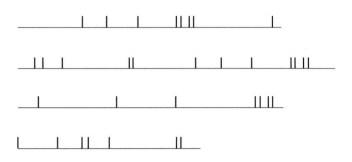

10-9 In an alternative approach to that described in problem 10-8, a cosmid library was probed with an insert from a YAC and 22 cosmids were obtained. All but two of the cosmids could be arranged as an overlapping group (a contig). This set of cosmids covered the length of the YAC insert. What explanation can you offer for the two cosmids that did not fit into this contig? Why did these hybridize with the YAC probe?

10-10 Briefly describe what is involved in using positional cloning as a way to search for a human gene. Start with the phenotype of a disease and describe the process through only to the point where you have clones that cover the genomic region within which the disease gene must lie.

10-11 List three independent techniques that can be used to identify DNA sequences encoding human genes within a cloned genomic region.

10-12 Indicate one way in which you might be able to prove that a gene you identified by positional cloning is the gene that can mutate to cause a disease phenotype in humans.

10-13 A restriction map of part of the *Drosophila* X chromosome has been determined, and is presented on the next page. Messenger RNA from *Drosophila* adults was purified, fractionated on agarose gels according to size, and then transferred to nitrocellulose filters. Radioactive recombinant DNAs corresponding to fragments A–G were hybridized to identical copies of the nitrocellulose filters containing the fractionated mRNAs. Using the map information, indicate the region of the DNA that must be transcribed to form three primary transcripts which are then processed into the three mRNAs observed. Indicate where exons and introns occur in relation to the restriction fragments.

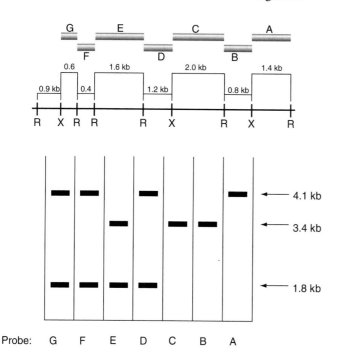

Probe: G F E D C B A

10-14 To study only the genes expressed in certain tissues, cDNA libraries are made starting with messenger RNA from a specific tissue. A mouse cDNA library was

constructed as follows: cDNA complementary to the RNAs from the pituitary gland was prepared and then digested with the restriction enzyme *Eco*RI (E). The DNA fragments generated in this manner were inserted into the single *Eco*RI site of the plasmid vector pBR319 (see the map of which is presented below). A probe of a growth factor gene found in *Drosophila* was used to isolate the recombinant plasmid called pcDNA1. A clone was also isolated from a mouse genomic library that had been prepared by digesting mouse DNA with *Eco*RI and ligating the fragments into the pBR319 plasmid vector. The plasmid that hybridized with the probe is called pGEN1.

pcDNA1 and pGEN1 were subjected to digestion with combinations of the three restriction enzymes: *Bam*HI (B), *Hin*dIII (H), and *Pst*I (P). (There are no *Hin*dIII sites in the vector.) Ethidium-bromide stained patterns of the agarose gels are presented below.

a. How large is the cDNA insert in pcDNA1?
b. How large is the genomic DNA insert in pGEN1?
c. Make a map of pcDNA1 and pGEN1 indicating the positions of all B, H, and P sites and the distances in kilobases between all adjacent sites.
d. How many introns are indicated by this data, and how long is each intron? Show the position of this (these) intron(s) on the appropriate map.
e. Could you have learned this information by DNA sequencing of the genomic clone? How would you have used the sequence information?

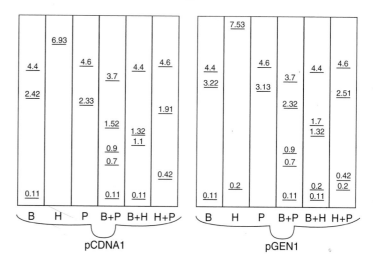

10-15 Using a cDNA library, you isolated two different cDNA clones that hybridized with your probe for a nerve growth factor. The beginning and ending sequences are the same but the middle sequence is different. How can you explain the different cDNAs?

10-16 What sequence information about a gene is lacking in a cDNA library?

10-17 An interesting phenomenon found in vertebrate DNA is the existence of pseudogenes, nonfunctional copies of a gene found elsewhere in the genome. Pseudogenes appear to be double-stranded DNA copies of mature mRNA inserted into the chromosome. What sequence information would be a clue(s) that the source of these pseudogenes is cDNA?

10-18 You constructed a mutant form of the growth hormone receptor gene that you are injecting into mice to form a transgenic mouse with a dominant negative mutant phenotype (interferes with the wild-type function). Two transgenic mice have been obtained for analysis.
   a. FISH hybridization shows positive hybridization of the growth hormone probe with chromosome 1 and chromosome 5. How would you explain the two hybridization signals?
   b. In the second mouse, there are positive hybridization signals on chromosomes 1, 2, and 20. How can you explain these hybridizations?
   c. If three tandem copies of a transgene are present at a single locus, what would you see using FISH?
   d. What would be the characteristics of a Southern blot and hybridization with the transgene if the genomic DNA from the transgenic animal with three tandem copies was digested with a restriction enzyme that cuts two times within the transgene?

10-19 With new information from the Human Genome Project, many new genes will be identified for which the function is not known.
   a. What features of the DNA sequence might help you determine the function of a newly identified gene?
   b. What are two other types of analysis that would help you learn more about a new gene?

10-20 a. If you found a zinc-finger motif in a newly identified gene, what hypothesis would you make about its function?

   b. In another gene, if you found an overall high percentage of similarity throughout the gene with a previously identified gene in the same organism, what would that suggest about the origin of the gene?

10-21 From an expression library, you isolated a clone and made an antibody to the protein produced by the cloned gene. When the antibody was used in an *in situ* hybridization experiment using a mouse embryo, the protein was found localized to the brain. Northern analysis showed that there were three mRNA transcripts present in the brain. Two of these mRNAs were also present in tissues other than the brain. Propose a hypothesis to explain the presence of RNAs in many tissues but protein only in the brain.

10-22 List an advantage of each of the following model organisms for analysis of gene function. Give an example of the type of gene you might choose to analyze in each of these model organisms.
   a. yeast (*S. cerevisiae*)
   b. nematodes (*C. elegans*)
   c. mouse (*M. musculus*)

10-23 The following factors pose some technical challenges for the design of gene therapy experiments. Suggest ways to get around these problems.
   a. large gene size in eukaryotes (up to 100 kb of DNA in the genome)
   b. the need to get genes expressed in specific tissues

10-24 Random integration is a more serious concern in considering germ line gene therapy than it is in somatic gene therapy. Why?

10-25 Many human diseases are genetically very complex. For example, mutations in two different genes or mutations in different combinations of genes can result in a disease phenotype. Suggest a way that may help identify different genetic causes of the same disease.

10-26 A wheat geneticist working for the U.S. Department of Agriculture has decided to analyze the genes involved in the quantitative trait of kernel size in wheat. What would be the genotype and phenotype of the plants he would cross to begin his analysis?

HOW GENES TRAVEL

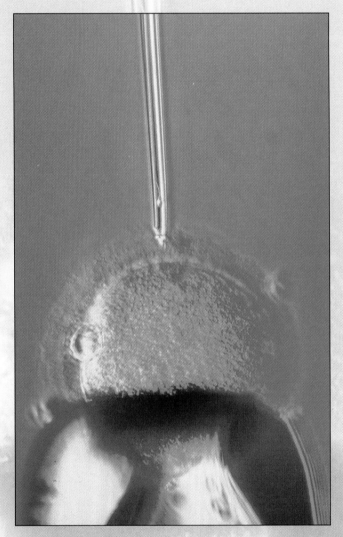

DNA being injected into a mouse egg through a small capillary pipet
(magnification ×800)

THE EUKARYOTIC CHROMOSOME:
AN ORGANELLE FOR
PACKAGING AND MANAGING DNA

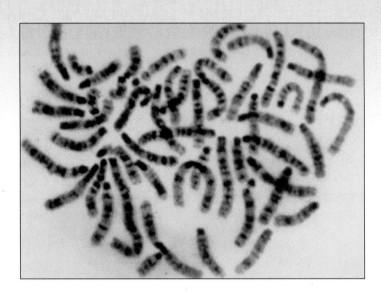

Mouse mitotic metaphase chromosomes stained for karyotype analysis.
(magnification ×600)

The linear chromosomes of eukaryotic cells terminate at both ends in protective caps called telomeres (Fig. 11.1). Composed of DNA associated with proteins, these caps contain no genes but are nevertheless crucial in preserving the structural integrity of each chromosome. Chromosomes unprotected by telomeres fuse end to end, producing entities with two centromeres. During anaphase of mitosis, if the two centromeres are pulled in opposite directions, the DNA between them will rupture, resulting in broken chromosomes that segregate poorly and eventually become lost from the cell.

Cells must thus preserve their telomeres to maintain the normal genetic complement on which their viability depends. But the replication of telomeres is problematic. We saw in Chapter 5 that DNA polymerase, a key component of the replication machinery, functions only in the 5'-to-3' direction and can add nucleotides

only to the 3' end of an existing chain. With these constraints, the enzyme cannot possibly replicate some of the nucleotides at the 5' ends of the two DNA strands; the 5' end of one of these strands resides in the telomere at one end of the chromosome, while the 5' end of the other strand is in the other telomere. Thus, if DNA polymerase were the only enzyme active in copying DNA, the 5' ends of each strand would shorten by a few nucleotides at every DNA replication during mitosis and meiosis. Many eukaryotes resolve this impasse through expression of an enzyme known as **telomerase.**

The activity of telomerase in normal yeast cells ensures the full reconstruction of each chromosome's ends with each DNA replication. In studies where researchers deleted the yeast gene for telomerase, the telomeres shortened at the rate of about 3 bp per generation, and after significant loss of telomeric length, the chro-

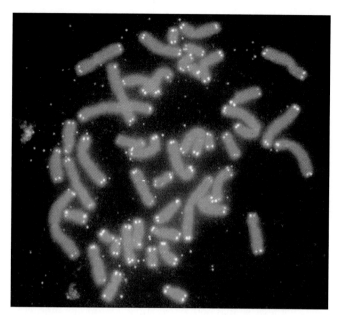

Figure 11.1   Telomeres protect the ends of eukaryotic chromosomes. Human telomeres light up in yellow upon *in situ* hybridization with fluorescent probes (FISH) that recognize the base sequence T T A G G G. The telomeres of humans and many other species contain large numbers of repeats of this six-base-pair motif.

mosomes began to break, and the yeast cells died. Telomerase activity, it seems, endows normal yeast cells with the potential for immortality; given the proper conditions and continual telomere reconstruction, the cells can reproduce forever.

In humans, the telomerase gene is part of every cell's genome. Germ line cells, which maintain their chromosomal ends through repeated rounds of DNA replication, express the gene, as do some stem cells; but many normal somatic cells, which have a finite life span, do not. In these differentiated somatic cells, the telomeres shorten slightly with each cell division, and this shortening helps determine how many times a particular cell is able to divide. In culture, most somatic cells, after dividing for 30 or 40 or 50 generations, show signs of senescence and then die. Some researchers who study aging in humans have suggested that it may be possible to forestall aging by slowing cellular senescence through the activation of telomerase expression. They reason that when cells of the immune system, for instance, lose a certain telomeric length, they divide less often, and as a result, the immune system gradually loses its capacity for protection. If it were possible to preserve the telomeres, this loss of cells and immune function might not occur.

As with many aspects of human biology, a telomerase-based boost to longevity might not be the simple solution it appears at

first glance, because it could make cells susceptible to cancerous proliferation. Many human tumor cells that become immortal exhibit telomerase activity. In such cells, the abnormal activation of the telomerase gene seems to bestow the capacity to divide indefinitely. Cells isolated from human ovarian tumors, for example, express the telomerase enzyme and maintain stable telomeres; cells from normal ovarian tissue do not. Oncologists hypothesize from these and other observations that expression of telomerase in cancerous human cells may keep those cells from losing their telomeres and eventually dying, and thereby perpetuate tumors. Some scientists even speculate that the genetic program silencing the telomerase gene in the somatic cells of differentiated human tissues may have evolved as a defense against cancer. Ironically, if this were true, the same mechanism that helps humans resist cancer blocks their biological potential for immortality.

In this chapter, we examine the structure and function of telomeres along with other significant features of the eukaryotic chromosome. Our discussion reveals that the chromosome is a dynamic organelle for the packaging, replication, segregation, and expression of the information in a single long molecule of DNA. (As noted earlier in the book, we consider chromosomes "organelles" because they are highly organized cellular structures that have important functions and are transmitted to daughter cells at cell division. This contrasts with the usage of many biologists who consider an entity to be an organelle only if it is surrounded by a membrane.) Each chromosome consists of that DNA molecule combined with a variety of proteins. Flexible DNA-protein interactions condense the chromosome for segregation during mitosis, and decondense it for replication or gene expression during interphase. Specific sequences in the chromosomal DNA dictate where spindle attachment occurs for proper segregation; others determine where replication begins.

One general theme emerges from our discussion. Chromosomes have a versatile, modular structure for packaging DNA that supports a remarkable flexibility of form and function.

Our presentation of the chromosome as an organelle for packaging and managing genetic information examines

- The components: one long DNA molecule and many kinds of proteins.

- Chromosome structure: how variable DNA-protein interactions create reversible levels of compaction.

- Specialized elements that ensure the accurate duplication and segregation of chromosomes: telomeres, origins of replication, and centromeres.

- How chromosomal packaging influences gene activity: Decondensation precedes gene expression; extreme condensation silences genes.

THE COMPONENTS OF EUKARYOTIC CHROMOSOMES: DNA, HISTONES, AND NONHISTONE PROTEINS

When viewed under the light microscope, chromosomes seem to change shape, character, and position as they pass through the cell cycle. During interphase, they look like a tangled mass of grainy string scattered throughout the nucleus. By metaphase of mitosis, they appear as a set number of paired bars facing each other across the cellular midplane. In this section, we describe what chromosomes are made of; then, in succeeding sections, we explain how these chromosomal components associate, dissociate, and reassociate to produce the observed metamorphoses of structure.

Each Chromosome Packages a Single Long Molecule of DNA

Researchers learned from chemical and physical analyses that each chromosome within a cell nucleus contains one long linear molecule of DNA. In one study, they used the base analog 5-bromodeoxyuridine to label the newly synthesized DNA in metaphase chromosomes consisting of two sister chromatids. The two chromatids stained in a fashion consistent with the idea that each chromatid contains one long double helix of DNA produced by a semiconservative mode of replication. In a second study, they placed chromosomal DNA between two cylinders, stretched the DNA by rotating one of the cylinders, and measured the DNA's rate of recoil. Shorter molecules recoil faster than longer ones. When they applied this measure to the DNA in a *Drosophila* chromosome, the length of the DNA molecule was sufficient to account for all the DNA in a single chromosome. It is also possible to use pulsed gel electrophoresis (review Fig. 8.18) to separate the DNA molecules of a relatively small genome by size; with the yeast genome, the number and sizes of the separated DNA molecules corresponds to the number and sizes expected if each chromosome contains a single DNA molecule.

The Protein Components of Eukaryotic Chromosomes: Histones and Nonhistone Proteins

By itself, DNA does not have the ability to fold up small enough to fit in the cell nucleus. For sufficient compaction, it depends on interactions with two categories of proteins: histones and nonhistone chromosomal proteins. **Chromatin** is the generic term for any complex of DNA and protein found in a cell's nucleus. Chromosomes are the separate pieces of chromatin that behave as a unit during cell division. Chromatin is the same chemical substance that Miescher extracted from the nuclei of white blood cells and named nuclein in 1869. Although chromatin is roughly 1/3 DNA, 1/3 histones, and 1/3 nonhistones by weight, it may also contain traces of RNA. Because these RNA bits result mainly from gene transcription and are probably unrelated to chromatin structure, we do not include RNA in our discussion of chromatin components.

Histone Proteins Abound in the Chromatin of all Eukaryotic Cells

Discovered in 1884, **histones** are relatively small proteins with a preponderance of the basic, positively charged amino acids lysine and arginine. The histones' strong positive charge enables them to bind to and neutralize the negatively charged DNA throughout the chromatin. Histones make up half of all chromatin protein by weight and are classified into five types of molecules: H1, H2A, H2B, H3 and H4 (Table 11.1). The last four types—H2A, H2B, H3, and H4—form the core of the most rudimentary DNA packaging unit—the **nucleosome**—and are therefore referred to as **core histones.** (We examine the role of these histones in nucleosome structure later.) All five types of histones appear throughout the chromatin of nearly all diploid eukaryotic cells. Histone synthesis occurs during S phase of the cell cycle, when chromosome replication requires more histones for the packaging of newly made DNA. Special regulatory mechanisms tightly correlate DNA and histone synthesis so that both occur at the appropriate time.

Interestingly, there is great similarity among particular histone proteins in different organisms. In the H4 proteins of pea plants and calves, for example, all but two of the sequence of 102 amino acids are identical. That histones have changed so little throughout evolution underscores the importance of their contribution to chromatin structure.

Nonhistone Proteins Are a Heterogeneous Group Named by Default

Fully half of the mass of protein in the chromatin of most eukaryotic cells is not composed of histones. Rather, it consists of hundreds or even thousands of different kinds of nonhistone proteins, depending on the organism. The chromatin of a diploid genome contains from 200 to 2,000,000 molecules of each kind of nonhistone protein. Not surprisingly, this large variety of proteins fulfills many different functions, only a few of which have been defined to date. Some nonhistone proteins play a purely structural role, helping to package DNA into structures distinct from the histone-containing nucleosomes. The proteins that form the structural backbone, or *scaffold,* of the chromosome fall in this category (Fig. 11.2a). Others, such as DNA polymerase, are active in replication. Still others are active in chromosome segregation; for example, the motor proteins of kinetochores help move chromosomes along the spindle apparatus and thus expedite the transport of chromosomes from parent to daughter cells during mitosis and meiosis (Fig. 11.2b). By far the largest class of nonhistone proteins foster or regulate transcription and RNA processing during gene expression. Mammals carry 5000 to 10,000 different proteins of this kind. By interacting with DNA, these proteins influence when, where, and at what rate genes give rise to their protein products.

Unlike the distribution of histones, which are found in similar amounts in all cells of an organism and are dispersed relatively evenly throughout the chromosomes, the distribution of nonhistone proteins is uneven. Nonhistone proteins

TABLE 11.1 Some Molecular Characteristics of the Histone Proteins Found in Calf Thymus*

Histone	Basic/Amino Acids	Molecular Weight	Total Amino Acid Residues
H1	5.4	23,000	224
H2A	1.4	13,960	129
H2B	1.7	13,774	125
H3	1.8	15,273	135
H4	2.5	11,236	102

*Note the small size of the histones; in addition, their predominantly basic amino acid constitution ensures that they will be positively charged in physiological conditions.

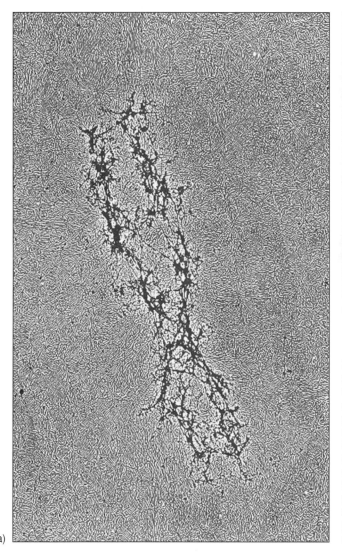

(a)

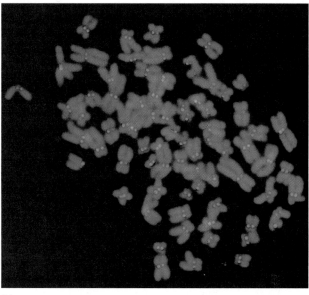

(b)

Figure 11.2 Nonhistone proteins have diverse functions.
(a) Some nonhistone proteins form the chromosome scaffold. When the human chromosome in this picture was gently treated with detergents to remove the histones and some of the nonhistone proteins, a dark scaffold composed of some of the remaining nonhistone proteins, and in the shape of the two sister chromatids, became visible. Loops of DNA freed by the detergent treatment surround the scaffold. (b) Some nonhistone proteins power chromosome movements along the spindle during cell division. In this figure, chromosomes are stained in blue and a nonhistone protein known as CENP-E is stained in red. CENP-E is located at the sister kinetochores of each duplicated chromosome and appears to play a major role in moving separated sister chromatids toward the spindle poles during anaphase.

appear in different amounts and in different proportions in different tissues within the same organism. The difference between universal and sporadic distribution no doubt reflects the functional differences between the two categories of proteins. Whereas histones have the singular task of interacting with DNA to form universal units of compaction, the nonhistone proteins fulfill diverse functions specific to different tissues (for example, tissue-specific transcription factors), to different moments in the life cycle (for example, proteins of the synaptonemal complex), or to different parts of a chromosome (for example, proteins involved in centromere function).

In summary, chromatin is roughly 1/3 DNA, 1/3 histones, and 1/3 nonhistone proteins. We now examine how the components of chromatin associate to package long strands of DNA into chromosomes that can condense or decondense as circumstances require.

CHROMOSOME STRUCTURE: VARIABLE DNA-PROTEIN INTERACTIONS CREATE REVERSIBLE LEVELS OF COMPACTION

Stretched out in a thin, straight thread, the DNA of a single human cell would be six feet in length. This is, of course, much longer than the microscopic cell itself, whose dimensions are measured in fractions of millimeters. Several levels of compaction enable the DNA to fit inside the cell (Table 11.2). First, the winding of DNA around histones forms small nucleosomes. Next, tight coiling gathers the DNA with nucleosomes together into higher-order structures. Other levels of compaction, which researchers do not yet understand, produce the metaphase chromosomes observable in the microscope.

The Nucleosome: The Fundamental Unit of Chromosomal Packaging Arises from DNA's Association with Histones

The electron micrograph of chromatin in Fig. 11.3a shows long, nub-studded fibers bursting from the nucleus of a chick red blood cell. The chromatin fibers resemble beads on a string, with the beads having a diameter of about 100 Å, and the string a diameter of about 20 Å ($1 \text{ Å} = 10^{-10} \text{ m} = 0.1 \text{ nm}$). This 20-Å string is DNA. Figure 11.3b illustrates how DNA wraps around histone cores to form the chromatin fiber's observed beads-on-a-string structure. Each bead is a nucleosome containing roughly 160 base pairs of DNA wrapped around a core composed of eight histones—two each of H2A, H2B, H3, and H4, arranged as shown in the figure. The 160 base pairs of DNA wrap twice around this core octamer. An additional 40 base pairs form **linker DNA** that connects one nucleosome with the next. Histone H1 lies outside the core, apparently associating with DNA where the DNA enters and leaves the nucleosome. When investigators use specific reagents to remove H1 from the chromatin, some DNA unwinds from each nucleosome, but the nucleosomes do not fall apart; about 140 base pairs remain wrapped around each core.

Researchers can crystallize the nucleosome cores and subject the crystals to X-ray diffraction analysis. The pictures obtained confirm the nucleosome structure just described and also indicate that the DNA does not coil smoothly around the histone core (Fig. 11.3c). Instead, it bends sharply at some positions and barely at all at others. Because the sharp bending may occur only with some DNA sequences and not others, base sequence helps dictate preferred nucleosome positions along the DNA. It is not yet known exactly what elements of base sequence dictate whether a region will or will not be wound into a nucleosome. The bending required for a nucleosome is close to the limit possible for B DNA. The stiffer forms of A DNA and Z DNA cannot bend sufficiently to form nucleosomes. Although DNA sequence and B-form structure help determine where nucleosomes occur, other as yet undetermined factors also contribute to nucleosome formation.

The spacing and structure of nucleosomes affect genetic function. The nucleosomes of each chromosome are not evenly spaced, but they do have a well-defined arrangement along the chromatin. This arrangement is transmitted with high fidelity from parent to daughter cells. The spacing of nucleosomes along the chromosome is critical because DNA in the regions between nucleosomes is readily available for the interactions with proteins that initiate expression, replication, and further compaction. The way in which DNA is wound around a nucleosome also plays a role in determining whether and how certain proteins interact with specific DNA sequences; this is because some DNA regions in the nucleosome, despite their proximity to the histone core, can still be recognized by nonhistone binding proteins.

Packaging into nucleosomes condenses naked DNA about 7-fold. With this condensation, the 2 m of DNA in a diploid human genome shortens to approximately 0.25 m (a little less than a foot) in length. This is still much too long to

TABLE 11.2 The Different Levels of Chromosome Compaction

Mechanism	Status	What It Accomplishes
Nucleosome	Confirmed by crystal structure	Condenses naked DNA 7-fold to a 100-Å fiber
Supercoiling	Hypothetical model (the 300-Å fiber predicted by the model has been seen in the electron microscope)	Causes additional 6-fold compaction, achieving a 40- to 50-fold condensation relative to naked DNA
Radial loops—scaffold	Hypothetical model (preliminary experimental support exists for this model)	Through progressive compaction of 300-Å fiber, condenses DNA to rodlike mitotic chromosome that is 10,000 times more compact than naked DNA

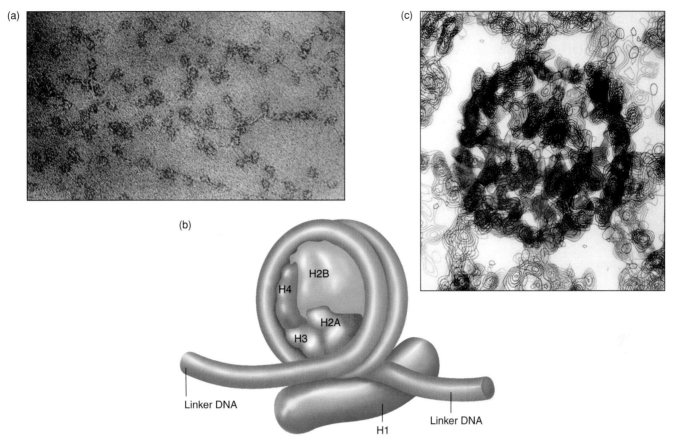

(a)

(b)

(c)

Figure 11.3   Nucleosomes: The basic building blocks of chromatin. (a) In the electron microscope, nucleosomes look like beads on a string. (b) Nucleosome structure: The DNA in each nucleosome wraps twice around a nucleosome core. An additional molecule of each histone (H2A, H2B, H3, and H4) in the nucleosome core is hidden in this view. Histone H1 associates with the DNA as it enters and leaves the nucleosome. (c) The structure of the nucleosome as determined by X-ray crystallography: In this overhead view of the core particle, you can see that the DNA (brown) actually bends sharply at several places as it wraps around the core histone octamer (blue and turquoise). The small circles in this picture represent contours of electron densities.

fit in the nucleus of even the largest body cell, and additional compaction is required.

Models of Higher-Level Packaging Seek to Explain the Extreme Compaction of Chromosomes at Mitosis

The details of chromosomal condensation beyond the nucleosome remain unknown, but researchers have proposed several models to explain the different levels of compaction (see Table 11.2).

Formation of the 300-Å Fiber through Supercoiling

One model of additional compaction beyond nucleosomal winding proposes that the 100-Å nucleosomal chromatin supercoils into a 300-Å superhelix, achieving a further 6-fold chromatin condensation. Support for part of this model comes from electron microscope images of 300-Å fibers that contain about six nucleosomes per turn (Fig. 11.4a). Whereas the 100-Å fiber is one nucleosome in width, the 300-Å fiber looks three beads wide. Removal of some H1 from a 300-Å fiber causes it to unwind to 100 Å. Adding back the H1 reinstates

the 300-Å fiber. Although electron microscopists can actually see the 300-Å fiber, they do not know exactly how it forms. Higher levels of compaction are even less well understood.

The Radial Loop-Scaffold Model Seeks to Explain Compaction of the 300-Å Fiber

This model proposes that certain nonhistone proteins, including topoisomerase II, bind to chromatin every 60–100 kilobases and tether the supercoiled, nucleosome-studded 300-Å fiber into structural loops (Fig. 11.4b). Evidence that nonhistone proteins fasten these loops comes from chemical manipulations in which the removal of histones does not cause the chromatin to unfold completely. Other nonhistone proteins may, in turn, gather the loops into daisy-like rosettes (Fig. 11.4c); and additional nonhistone proteins may then compress the rosette centers into a compact bundle. A range of nonhistones thus forms the condensation scaffold depicted in Fig. 11.2a. This proposal of looping and gathering is known as the **radial loop-scaffold model** of compaction. To visualize how it achieves condensation, imagine a long piece of string. To shorten it, you knot it at intervals to form loops separated by

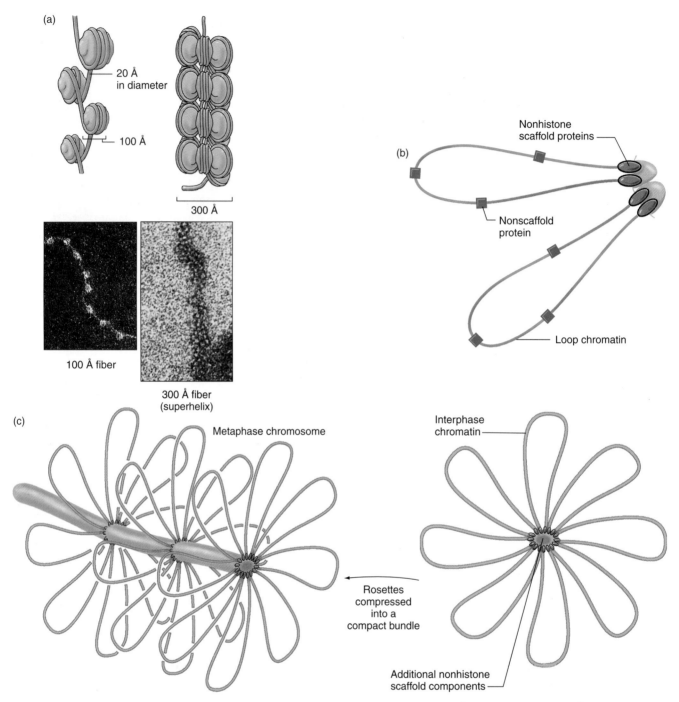

Figure 11.4 **Models of higher level packaging.** (a) Electron micrographs contrasting the 100-Å fiber (left) with the 300-Å fiber (right). The line drawings show the probable arrangement of nucleosomes (with green cores) in these structures. (b) The radial loop-scaffold model for yet higher levels of compaction. According to this model, the 300-Å fiber is first drawn into loops, each including 60–100 kb of DNA, that are tethered at their bases by nonhistone scaffold proteins (brown and orange) including topoisomerase II (brown). (c) Additional nonhistone proteins might gather several loops together into daisy-like rosettes and then compress the rosette centers into a compact bundle.

straight stretches; the knots are at the base of each loop. To shorten the string still further, you clip together sets of knots. Finally, you pin together all the clips. In this image, the knots, clips, and pins function as the condensation scaffold.

The radial loop-scaffold model of chromosome packaging offers a simple explanation of progressive chromosome compaction from interphase to metaphase chromosomes. At inter-

phase, the nucleosome-studded chromatin forms many structural loops, which are anchored together in rosettes in some areas. This initial looping and gathering compresses the genetic material sufficiently to fit into the nucleus, where it appears as a mass of tangled string. As the chromosomes enter prophase of mitosis, looping and gathering increase, and bundling through protein cross-ties begins. By metaphase, the height of looping, gathering, and bundling achieves a 250-fold compaction of the roughly 40-fold-compacted 300-Å fiber, giving rise to the highly condensed, rodlike shapes we refer to as mitotic chromosomes.

Several pieces of biochemical and micrographic evidence support the radial loop-scaffold model. For example, metaphase chromosomes from which experimenters have extracted all the histones still maintain their familiar X-like shapes (see Fig. 11.2). Moreover, electron micrographs of whole-mounted mitotic chromosomes show loops of chromatin at the periphery of the chromosomes (Fig. 11.5). And recent analyses of DNA indicate that special, irregularly spaced repetitive base sequences associate with nonhistone proteins to define the chromatin loops. These stretches of DNA are known as **scaffold-associated regions,** or **SARs.** Found at the base of the chromatin loops, SARs are most likely the sites at which the DNA is anchored to the condensation scaffold.

Despite bits and pieces of experimental evidence, studies that directly confirm or reject the radial loop-scaffold model have not yet been completed. Thus the loops and scaffold concept of higher-level chromatin packaging remains a hypothesis. The hypothetical status of this higher-level compaction model contrasts sharply with nucleosomes, which are real entities that investigators have isolated, crystallized, and analyzed in detail.

Figure 11.5 Experimental support for the radial loop-scaffold model. A close-up of the image in Fig. 11.2, this electron micrograph shows long DNA loops emanating from the protein scaffold at the bottom of the picture. The two ends of each DNA loop appear to attach to adjacent locations in the protein scaffold. Note that there are only loops—not ends—at the top of the photo.

A Closer Look at Karyotypes: Fully Compacted Metaphase Chromosomes Have Unique, Reproducible Banding Patterns That Identify Them

We have seen that different levels of packaging compact the DNA in human metaphase chromosomes 10,000-fold (see Table 11.1). With this amount of compaction, the centromere region and telomeres of each chromosome become visible. We have also seen (in Chapters 3 and 10) that various staining techniques reveal a characteristic banding pattern for each metaphase chromosome, establishing a karyotype. In G-banding, for instance, chromosomes are first gently heated and then exposed to Giemsa stain; this DNA dye preferentially darkens certain regions to produce alternating dark and light "G bands." Each G band is a very large segment of DNA from 1 to 10 mb in length, containing many loops. On analysis with low-resolution techniques, a human karyotype contains approximately 300 dark G-bands and light interbands. High-resolution G-banding techniques enable cytologists to subdivide the G bands into smaller bands and interbands that picture chromosomes at even higher resolution (Fig. 11.6a). There are nearly 2000 identifiable G-bands and interbands in the chromosomes of the standard high-resolution human karyotype. G-banding serves as the basis for most karyotyping because G-band preparations are very stable and require only a good light microscope for detection. The karyotypes described in Chapters 3 and 10 present a cell's mitotic metaphase chromosomes by size, shape, and banding patterns.

Banding Patterns Are Highly Reproducible, Although No One Knows for Sure What They Represent

Most molecular geneticists think the bands produced by staining probably do *not* embody differences in base composition over long distances, but more likely reflect an uneven packaging of loops determined in some way by the spacing and density of short repetitive DNA sequences. Although the detailed biochemical basis of banding is not yet understood, every time a chromosome replicates, whatever underlies its banding pattern is faithfully reproduced. The fact that banding patterns are so highly reproducible from one generation to the next indicates they are an intrinsic property of each chromosome, determined by the DNA sequence itself.

Researchers Can Use Bands to Locate Genes

As we saw in Chapter 10, geneticists can designate the chromosomal location of a gene by describing its position in relation to the bands on the p (short) or q (long) arm of a particular chromosome. For this purpose, the p and q arms are subdivided into regions, and within each region, the dark and light bands are numbered consecutively. The X-linked genes for color-blindness, for example, reside at q27-qter, which means they are located on the X chromosome's long (q) arm somewhere between the beginning of the seventh band in the second region and the end of the telomere (terminus, or ter; Fig. 11.6b).

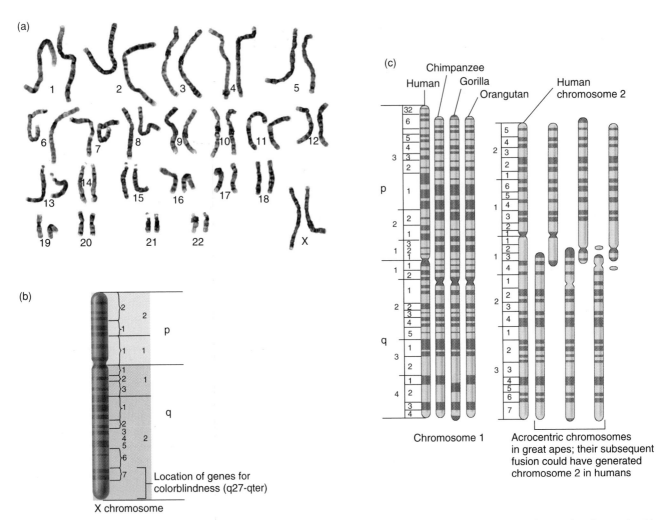

Figure 11.6   Chromosomes can be characterized by their banding patterns. (a) Karyotype of a human female examined by high-resolution G-banding techniques reveals approximately 1000 bands and interbands in the haploid complement of chromosomes. A trained eye could see the 2000 bands in this diploid set of chromosomes, although a novice may not. (b) Genes for colorblindness in humans have been located to a small region near the tip of the long arm of the X chromosome. (c) A comparison of banding patterns in the two largest human chromosomes and their counterparts in the great apes.

Geneticists Can Use Banding Patterns to Analyze Chromosomal Differences between Species

In all placental mammals, the diploid genome carries roughly 6 billion base pairs, but this amazingly similar amount of DNA is packaged into different numbers of chromosomes with different banding patterns. Deer mice have 4 chromosomes in the nucleus of each body cell; cats have 38; house mice have 40; rabbits, 44; humans 46; dogs have 48; cattle have 60; horses have 64; and hippopotamuses have 96.

High-resolution banding techniques, which divide chromosomes into so many stripes that their diagrams look a little like the universal bar codes on grocery-store items, enable geneticists to compare the banding patterns of different species. The similarities between closely related species are striking. We know, for example, that humans have 46 chromosomes. Their nearest primate relatives, the great apes (chimpanzees, gorillas, and orangutans) carry 48 chromosomes per diploid cell. As Fig. 11.6c shows, a single event—the fusion of two acrocentric chro-

mosomes in the great apes to form metacentric chromosome 2 in humans—can account for the numeric difference. Note that the banding patterns of the two great ape chromosomes that later fused are identical to those of human chromosome 2, except in the region of fusion and at the chromosome tips. Human banding patterns are almost identical to chimpanzees' for 13 chromosomes, to gorillas for 9, and to orangutans for 8. In the rest of the chromosomes, almost all the same bands are present, but they are distributed slightly differently.

Banding Pattern Differences between Individuals Can Reveal the Cause of Genetic Disease

Karyotypes are a powerful tool for research and genetic counseling. As we have seen, medical geneticists use them to diagnose abnormalities related to chromosome number. Down syndrome, for example, is the result of three chromosomes 21. Some genetic disorders result from very slight additions or deletions of genetic material somewhere within the normal number of chromosomes. With high-resolution G-banding techniques, clinicians can uncover some of these discrepancies by comparing the chromosomes of a person suffering from the genetic disorder with those of a standard karyotype.

A striking example of how genetic counselors exploit the phenomenon of exact and highly reproducible chromosomal packaging for clinical purposes is seen in the case history of a young boy identified in the medical literature as BB. With no known family history of genetic disorders, BB suffered from defects in four X-linked traits: Duchenne muscular dystrophy; chronic granulomatous disease, which impairs the infection-fighting ability of white blood cells; the rare McLeod's blood type, which results in bouts of anemia; and retinitis pigmentosa in which a deterioration of the retina eventually causes blindness. The simultaneous occurrence of four X-linked conditions is so rare that one doctor asked medical geneticists to scrutinize the boy's X chromosome for the cause of the problems. High-resolution banding revealed the answer—deletion of a small white band sandwiched between two larger dark bands, which removed the four genes in question.

In summary, the DNA-protein interactions that define chromatin structure generate several levels of packaging: tightly wound nucleosomes, some form of supercoiling, and variously gathered larger loops. The flexible but reproducible structure of chromatin confers on each chromosome a structural versatility that enables it to adapt to the widely different functions of duplication, segregation, and the regulation of gene expression. In the following sections, we examine how chromosome structure helps determine those functions.

SPECIALIZED CHROMOSOMAL ELEMENTS ENSURE ACCURATE REPLICATION AND SEGREGATION OF CHROMOSOMES

The Accurate Duplication of Chromosomal DNA Depends on Origins of Replication and Telomeres

As the chromosomes decondense for copying during replication, special DNA sequences that do not encode protein regulate the timing and accuracy of the process. Some of these sequences serve as origins of replication that signal where and when the DNA double helix opens up to form replication forks; others function as telomeres that protect the ends of individual chromosomes from progressive decay.

There Are Many Origins of Replication

During replication the enzyme DNA polymerase assembles a new string of nucleotides according to a DNA template, linking about 50 nucleotides per second in a typical human cell. At this rate and with only one origin of replication, it would take the polymerase about 800 hours, a little more than a month, to copy the 130 million base pairs in an average human chromosome. But the length of the cell cycle in actively dividing human tissues is much shorter, some 24 hours, and S

phase (the period of DNA replication) occupies only about a third of this time. Eukaryotic chromosomes meet these time constraints through multiple origins of replication that can function simultaneously (Fig. 11.7a). Most mammalian cells carry 10,000 such origins strategically positioned among the chromosomes. As we saw in Chapter 5, each origin of replication binds proteins that unwind the two strands of the double helix, separating them to produce two mirror-image replication forks. Replication then proceeds in two directions (bidirectionally), going one way at one fork and the opposite way at the other, until the forks run into adjacent forks. As replication opens up a chromosome's DNA, a replication bubble becomes visible in the electron microscope, and with many origins, many bubbles appear (see Fig. 11.7a). The DNA running both ways from one origin of replication to the endpoints where it merges with DNA from adjoining replication forks is called a **replication unit,** or **replicon.** As yet unidentified controls tie the number of active origins to the length of S phase. In *Drosophila,* for example, early embryonic cells replicate their DNA in less than 10 minutes. To complete S phase in this short a time, their chromosomes utilize many more origins of replication than are active later in development when S phase is 6 to 10 times longer. Thus, all origins of replication are not necessarily active during all the mitotic divisions that create an organism.

The 10,000 origins of replication scattered throughout the chromatin of each mammalian cell nucleus are separated from each other by 30–300 kb of DNA, which suggests that there is at least one origin of replication per loop of DNA. Origins of replication in yeast (known as autonomously replicating sequences, or ARSs) can be isolated by their ability to permit replication of plasmids in yeast cells. ARS origins are capable of binding to the enzymes that initiate replication, and consist of an AT-rich region of DNA adjacent to special flanking sequences (Fig. 11.7b). By digesting interphase chromatin with DNase I, an enzyme that fragments the chromatin only at points where the DNA is not protected inside a nucleosome, investigators have determined that origins of replication are accessible regions of DNA devoid of nucleosomes.

Telomeres Preserve the Integrity of Linear Chromosomes

We learned in Chapter 5 that DNA polymerase can travel only in the 5′-to-3′ direction, and that it can begin a new strand only by adding nucleotides to the 3′ end of the primer. This means that in duplicating an entire chromosome, DNA polymerase cannot begin the 5′ end of a new strand without a primer. Even if an RNA primer were available at the 5′ end, there is no way the enzyme could, on its own, replace the primer with DNA once the RNA is removed. In short, DNA polymerase can reconstruct the 3′ end of each newly made DNA strand in a chromosome, but not the 5′ end (Fig. 11.8). If left to its own devices, the enzyme would fail to fill in an RNA primer's length of nucleotides at the 5′ end of every new chromosomal strand with each cell cycle. As a result, the chromosomes in successive generations of cells would become shorter and shorter, losing crucial genes as their DNA diminished.

(a)

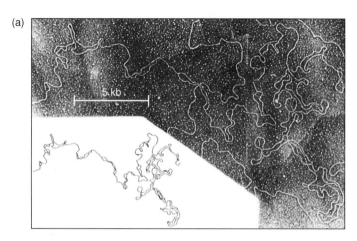

Figure 11.7 Eukaryotic chromosomes have multiple origins of replication. (a) Electron micrograph and diagrammatic interpretation, showing a region of replicating DNA from a *Drosophila* embryo. Many origins of replication are active at the same time, creating multiple replicons. (b) Structure of the yeast origin of replication ARSI. The orange-boxed sequence is the AT-rich consensus region found in all ARS elements. The blue boxes are the flanking sequences close to the ARSI consensus region that promote function.

(b)

Consensus region

5' •••CAAATTTCGTCAAAAATGCTAAGAAATAGGTTATTACTTTTATTTAAGTATTGTTTGTGCCTTTTGAAAAGCAAGCATAAAAGATCTAAACATAAAATCTGTAAAATAAC••• 3'
3' •••GTTTAAAGCAGTTTTTACGATTCTTTATCCAATAATGAAAATAAATTCATAACAAACACGGAAAACTTTTCGTTCGTATTTTCTAGATTTGTATTTTAGACATTTTATTG•••5'

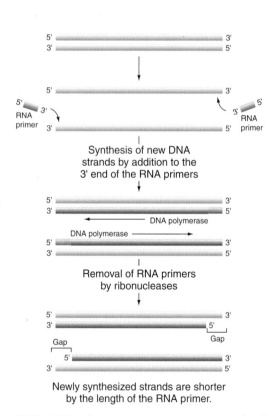

Synthesis of new DNA
strands by addition to the
3' end of the RNA primers

Removal of RNA primers
by ribonucleases

Newly synthesized strands are shorter
by the length of the RNA primer.

Figure 11.8 DNA polymerase cannot reconstruct the 5' end of a DNA strand. Even if an RNA primer at the 5' end can begin synthesis of a new strand, a gap will remain when ribonucleases eventually remove the primer. The requirement of DNA polymerase for a primer on which to continue polymerization means that the enzyme cannot fill this gap, so newly synthesized strands would always be shorter than parental strands if DNA polymerase were the only player in the production of new ends. In this figure, both parental DNA strands are shown in light blue, the two newly synthesized strands in dark blue, and the RNA primers in red.

Telomeres, as we saw at the beginning of this chapter, are a countermeasure to these problematic quirks of DNA polymerase. They consist of special repetitive DNA sequences that cap the ends of each chromosome and prevent the loss of DNA through incomplete replication (Fig. 11.9). The telomeres of human chromosomes, for example, are composed of the base sequence T T A G G G repeated 250 to 1500 times. The number of repeats varies with the cell type. Sperm have the longest telomeres. The same exact T T A G G G sequence occurs in the telomeres of all mammals as well as in birds, reptiles, amphibians, bony fish, and many plant species. Some much more distantly related organisms also have repeats in their telomeres, but with slightly different sequences. For example, the telomeric repeat in the chromosomes of the ciliate *Tetrahymena* is T T G G G G. The close conservation of these repeated sequences across phyla suggests that they perform a vital function that emerged in the earliest stages of the evolutionary line leading to eukaryotic organisms, long before dinosaurs roamed the earth.

The special repetitive sequences of telomeres contain no protein-encoding genes; in fact, telomeres prevent the transcription of genes brought into their vicinity. Instead, the T T A G G G repeat folds around itself to form a special pro-

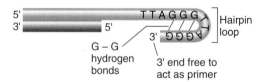

Figure 11.9 Telomeres consist of repetitive T T A G G G sequences that might protect against DNA loss by folding back on themselves. Unusual non-Watson-Crick G–G base pairing allows the multiple repeats of the telomere to fold back on themselves to form a hairpin loop. The 3' end of the hairpin loop is free to act as a primer from which DNA polymerase can synthesize DNA near the 5' end of the new strand. However, even this property of telomere repeats cannot prevent the resulting double helix from being slightly shorter than the original.

tective structure. This structure may be in the form of a hairpin loop in which some of the repeats pair with other repeats through the formation of unusual G–G hydrogen bonds (see Fig. 11.9), or it may have a more complex folding pattern.

A telomere functions in the replication of chromosome ends in two ways. First, the T T A G G G sequence, by folding back on itself, provides a 3′ end that DNA polymerase can use as a primer to complete the synthesis of the new DNA strand. If however, the hairpin is cleaved by enzymes, both the template strand and the strand newly synthesized from it will be shorter than the equivalent two strands were before chromosome replication. The cell can compensate for this loss, as well as losses resulting from normal RNA-primed synthesis in which the hairpin is not used as a primer, in subsequent rounds of DNA replication by making use of a second property of the telomere. The telomere attracts the enzyme telomerase, which extends the telomere, roughly restoring it to its original length.

Telomerase is an unusual enzyme consisting of protein in association with RNA (Fig. 11.10). Because of this mix, it is called a *ribonucleoprotein*. The RNA portion of the enzyme contains C C C T A A repeats that are complementary to the T T A G G G repeats in telomeres, and serve as a template for adding new T T A G G G repeats to the end of the telomere. In many cells, including the perpetually reproducing cells of yeast and the germ cells of humans, some kind of feedback

mechanism appears to maintain the optimal number of repeats at the telomeres. In human somatic cells where the gene for telomerase is silenced, the lack of telomerase activity similarly results in the progressive shortening of telomeres as cells divide. This telomere shortening, as we saw at the beginning of this chapter, may be one of the molecular bases of aging, but it may also provide protection against cancer.

Many studies and observations have shown that telomeres are critical to chromosome function. In addition to preventing a chromosomal shortening during replication that could dismantle vital genes nucleotide by nucleotide, the telomeres maintain the integrity of the chromosomal ends. Broken chromosomes, which lack telomeres, are recognized as defective by the cellular DNA repair machinery, which often remedies the situation by putting the broken ends back together, restoring the telomeres. Sometimes, however, the unprotected, broken ends are subject to inappropriate repair resulting in chromosome fusion, or they may attract enzymes that degrade the chromosome entirely. Both fusion and degradation disrupt chromosome number and function. Telomeric ends, however, are not recognized by the DNA repair machinery, which makes no attempt to "fix" these ends. Thus, even though they normally carry no genes, telomeres contain information essential to the duplication, segregation, and stability of chromosomes.

Proteins Are an Essential Part of Reproducing Chromatin Structure

DNA replication is only one step in chromatin duplication. The complex process also includes the synthesis and incorporation of histone and nonhistone proteins to regenerate tissue-specific chromatin structure. Researchers speculate that the process works something like this. Before DNA synthesis can take place, the chromatin fiber must unwind. Next, as DNA replication proceeds, newly formed DNA must associate with histones, either preexisting histones or recently synthesized histones that have just made their way to the nucleus. The synthesis and transport of histones must be tightly coordinated with DNA synthesis since the nascent DNA becomes incorporated into nucleosomes within minutes of its formation. Finally, the nucleosomal DNA must interact in specific ways with a variety of proteins to compact in the same pattern as before. An exception to the exact replication of compaction patterns occurs in differentiating cells. Changes in available nuclear proteins produce slightly different folding patterns that promote the expression of different genes. Studies with mammalian cells have shown that some hormones can induce changes in gene expression if and only if they are present during chromatin replication.

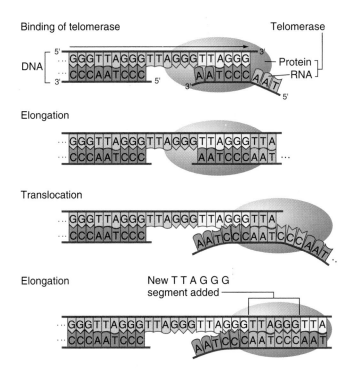

Figure 11.10 How telomerase extends telomeres. Telomerase binds to the ends of chromosomes because of complementarity between the C C C T A A repeats of telomerase RNA (orange) and the T T A G G G repeats of telomeres. Telomerase RNA C C C T A A repeats serve as templates for adding T T A G G G repeats to the ends of telomeres. After a telomere has acquired a new repeat, the telomerase enzyme moves (translocates) to the newly synthesized end, allowing additional rounds of telomere elongation.

The Segregation of Condensed Chromosomes Depends on Centromeres

When cell nuclei divide at mitosis or meiosis II, the two chromatids of each replicated chromosome must separate from one another and segregate such that each daughter cell receives one and only one chromatid from each chromosome. At meiosis I, homologous chromosomes must pair and segregate such that each daughter cell receives one and only one chromosome from each homologous pair. The centromeres of eukaryotic chromosomes ensure this precise distribution during different kinds of cell division by serving as segregation centers.

Centromeres Appear as Constrictions in the Chromosomes

These constrictions arise because centromeres are contained within blocks of repetitive, simple noncoding sequences, known as **satellite DNAs,** which have a very different chromatin structure and different higher-order packaging than other chromosomal regions (Fig. 11.11a). There are many different kinds of satellite DNA, each consisting of short sequences 5–300 bp long repeated in tandem thousands or millions of times to form large arrays. The predominant human satellite, "alphoid DNA," is a noncoding sequence 171 bp in length; it is present in a block of tandem repeats extending over a megabase of DNA in the centromeric region of each chromosome. Various human centromeres also contain several satellites unrelated to alphoid, which give their centromeric regions a complex structure. Although most satellite sequences lie in centromeric regions, some satellites are found outside the centromere on the chromosome arms.

The centromere can occur almost anywhere on a chromosome, except at the very ends (which must, instead, be telomeres). As we have seen, in a metacentric chromosome, the centromere is at or near the middle, while in an acrocentric chromosome, it is near one end.

Centromeres Have Two Functions That Ensure Proper Segregation

First, they are the sites that hold sister chromatids together—up to the beginning of anaphase in mitosis and up to anaphase II in meiosis. Researchers do not yet know how centromeres hold sister chromatids together. One model suggests that the DNA of sister chromatids somehow intertwines; another model proposes that proteins located at the centromere interconnect the two sister chromatids. *Drosophila* researchers have recently isolated a centromeric protein that offers support for the second model; found on meiotic chromosomes prior to anaphase II, the protein helps keep sister chromatids together until this time.

In addition to holding sister chromatids together, centromeres contribute to proper chromosome segregation through elaboration of a **kinetochore:** a specialized structure composed of DNA and proteins that is the site at which chromosomes attach to the spindle fibers (Fig. 11.11b). Some of the kinetochore proteins are motor proteins that help power chromosome movement during mitosis and meiosis. During mitosis, a kinetochore develops late in prophase on each sister chromatid, at the part of the centromere that faces one or the other cellular pole. By prometaphase, the kinetochores on the two sister chromatids attach to spindle fibers emanating from centrosomes at opposite poles of the cell. Although it is not yet clear what ensures this bipolar attachment, it appears that kinetochores somehow measure the tension arising when sister chromatids that are connected through their centromere are pulled in opposite directions. At the beginning of

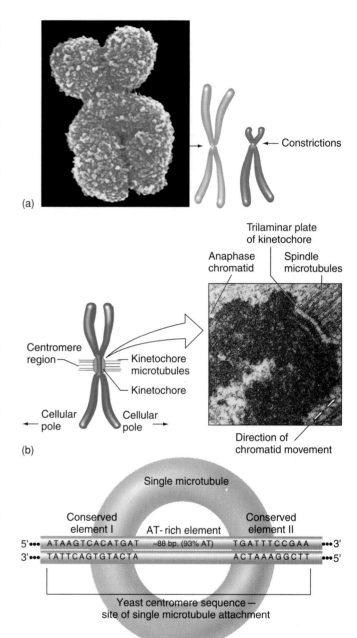

Figure 11.11   Centromere structure and function. (a) Color-enhanced scanning electron micrograph of a human metaphase chromosome and corresponding drawing show that the centromere corresponds to the major constriction of mitotic chromosomes. (b) Structure of centromeres in higher organisms. Centromeres hold sister chromatids together and contain information for the construction of a kinetochore, the structure that allows the chromosome to bind to spindle fibers. In this micrograph, an anaphase chromatid is migrating toward the spindle pole via microtubules attached to the kinetochore. The kinetochore is organized into a complicated "trilaminar plate"; note also that multiple spindle microtubules attach to a single kinetochore. (c) Structure and DNA sequence organization of yeast centromeres.

anaphase, the centromere "splits," freeing the sister chromatids to migrate toward opposite poles, with the assistance of the motor proteins in the kinetochore (review Fig. 3.8). During meiosis, controls on centromere splitting make sure that the sister chromatids of each homolog migrate together to the same pole during meiosis I, but to opposite poles during meiosis II (review Fig. 3.13).

Using Centromere Function to Analyze Centromere Structure

Investigators can exploit the centromere's role in chromosome segregation to isolate and then analyze the specific chromosomal regions that make up a centromere. If removal of a DNA sequence disrupts chromosome segregation and reinsertion of that same sequence restores stable transmission, the sequence must be the centromere. In the yeast *Saccharomyces cerevisiae*, centromeres consist of two highly conserved nucleotide sequences, each only 10–15 bp long, separated by approximately 90 bp of AT-rich DNA (Fig. 11.11c). This means that a short stretch of roughly 120 nucleotides is sufficient to specify a centromere. The centromere sequences of different yeast chromosomes are so closely related that the centromere of one chromosome can substitute for that of another. This indicates that while all centromeres play the same role in chromosome segregation, they do not help distinguish one chromosome from another.

The centromeres of higher eukaryotic organisms are much larger and more complex than those of yeast. In these multicellular organisms, the centromeres lie buried in a considerable amount of darkly staining, highly condensed chromatin, which makes it difficult to discover which specific DNA sequences are critical to centromere function. Although repetitive satellite DNA sequences in the vicinity of the centromeric constriction probably play some role in chromosome segregation, some geneticists believe that role is a secondary one and that, as in yeast, longer conserved DNA sequences determine the centromere's primary contribution to chromosome segregation. The kinetochores in higher eukaryotes attach to many spindle fibers, instead of just one, as in yeast. Researchers think that these complex kinetochores are likely to consist of repeating structural subunits, with each subunit responsible for attachment to one fiber.

Comprehensive Example: Yeast Artificial Chromosomes Confirm Which Structural Elements Are Essential for Replication and Segregation

In the mid-1980s, molecular geneticists learned how to construct artificial chromosomes, using raw materials from the yeast *S. cerevisiae*. Single-celled yeast was the organism of choice because it is easy to manipulate, its genetic machinery resembles that found in the cells of higher organisms, and it is so far the only eukaryotic organism whose origins of replication, centromeres, and telomeres have been defined as discrete, small segments of DNA. Yeast cells have 16 chromosomes that range in length from 300,000 to more than 2 million base pairs. Within each chromosome, the functional elements, such as centromeres, are only 100 to 1000 bp long. In contrast, human

chromosomes have an average length of 100 million base pairs and a centromere may span 1 million base pairs.

To build a **yeast artificial chromosome**, or **YAC**, investigators insert into bacterial plasmids a yeast centromere, a yeast origin of replication, and telomere sequences as well as suitable selectable markers, such as yeast genes conferring unusual color or the ability to grow in the absence of an essential nutrient (Fig. 11.12). They then expose yeast cells to the doctored plasmids such that individual cells become transformed by a plasmid.

Manipulation of the YAC construction process allows insights into chromosome function. Plasmids containing only origins of replication but no centromere or telomere, replicate but do not segregate properly. Plasmids with origins of replication and a centromere but no telomeres replicate and segregate fairly well only if they are circular. Small plasmids carrying all three elements replicate and segregate as linear chromosomes, but unlike natural yeast chromosomes, they do not segregate accurately; instead, they segregate at random. Studies aimed at understanding this phenomenon revealed that a minimum length of 100,000–150,000 bp is essential to normal chromosome function. YACs carrying 11,000 bp show segregation errors in 50% of cell divisions. Longer YACs containing 55,000 bp show segregation errors in 1.5% of cell divisions. When the artificial chromosomes are extended to more than 100,000 bp, however, the rate of segregation error falls to 0.3%. Although these longer artificial chromosomes function well, their frequency of segregation error is still 200 times

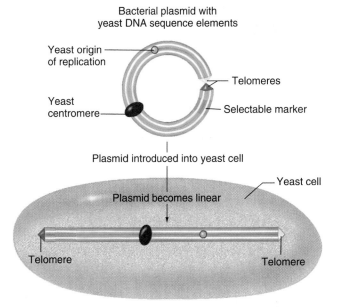

Figure 11.12 Tracking the segregation of yeast artificial chromosomes. Building a YAC. Scientists insert a selectable marker gene, a yeast origin of replication (ARS), a yeast centromere, and telomere sequences into the backbone of a bacterial plasmid. When this newly constructed circular plasmid is introduced, or transformed into a yeast cell, enzymes in the yeast cell cut the telomeric repeats, producing a small linear yeast artificial chromosome (a YAC) with telomeres at both ends.

greater than that seen with natural yeast chromosomes of normal size, indicating that some subtle aspects of chromosome structure and function remain to be discovered.

The construction and analysis of YACs confirm that three noncoding elements are essential to chromosome function: origins of replication, centromeres, and telomeres. Moreover, observations of YAC behavior show there is a minimum length requirement for accurate chromosome segregation. Geneticists can now use YACs for further studies of mitosis and meiosis. They can also use them to clone genomic regions that are too large for other vectors. As we saw in Chapters 8 and 10, cosmids can accept at most 40,000 bp of foreign DNA, while YACs can accommodate up to 1,000,000 bp.

HOW CHROMOSOMAL PACKAGING INFLUENCES GENE ACTIVITY

Origins of replication, telomeres, and centromeres are specialized chromosomal elements composed of specific noncoding DNA sequences packaged in specific ways for proper function. The chromosomal packaging of coding sequences also affects their function. We now examine how different types of chromosomal packaging influence gene expression.

Controlled Decompaction Precedes Gene Expression

Cells express their genes mainly during interphase when the chromosomes have decondensed, or decompacted. Even the relatively decompacted interphase chromatin, however, requires further unwinding to open up the higher orders of packaging and expose the DNA inside nucleosomes for transcription.

Boundary Elements Delimit Areas of Decompaction

Molecular geneticists do not yet understand the mechanism that releases the rosettes, loops, and superhelixes to expose a coding sequence for transcription, but whatever it is, it removes the proteins that gather the chromosomal loops into higher-order structures such as rosettes, and also unfastens the cross-ties that anchor the chromosome loops in the protein scaffold. Some sort of *boundary elements,* or *insulators,* prevent the decompaction from spreading beyond a certain point. These boundary elements, not yet characterized in organisms other than *Drosophila,* restrict the unwinding to a specific region and somehow insulate that region from the tighter chromatin packaging of adjacent regions.

Nucleosomes in the Decompacted Area Unwind to Allow the Initiation of Transcription

RNA polymerase, the enzyme that transcribes genes, is so large compared to a supercoil of nucleosomes in the 300-Å solenoid fiber that it cannot reach all of the sequences it needs to copy. To provide the RNA polymerase with access to the DNA, some loosening of individual nucleosomes must occur at the beginning of genes. Some of the nonhistone proteins known as transcription factors carry out this task by unwind-

ing nucleosomes and dislodging histones from DNA at the 5′ end of genes where transcription will begin. The nucleosome-free regions of the chromosome become sufficiently decompacted that they are open to interactions with RNA polymerase and other transcription factors, which can now recognize the promoter and initiate gene expression.

Researchers use DNase I, an enzyme that breaks DNA into fragments by splitting the phosphodiester bonds between nucleotides, as a tool for analyzing the patterns of chromatin compaction. DNA in highly compacted chromatin is relatively impervious to DNase I because it is shielded by the structure of the nucleosome (Fig. 11.13a). Decondensed chromatin is much more vulnerable to fragmentation. Treatment with very small amounts of DNase I reveals that while even highly compacted chromatin is punctuated with **DNase hypersensitive (DH) sites** that contain few if any nucleosomes, most (though not all) DH sites are at the 5′ ends of genes to be transcribed by the cell (Fig. 11.13b). This is evidence that the unwinding of these 5′ regions is a prerequisite of transcription.

Interestingly, it appears that once transcription begins, the transcribed part of most genes does not become entirely free of nucleosomes. However, although the body of the gene does not become DNase hypersensitive, it is still more sensitive to DNase I than is the highly condensed chromatin. Some researchers speculate that as RNA polymerase copies DNA, the histones that remain bound to the DNA somehow step around the transcribing enzyme (Fig. 11.13c).

By Influencing Gene Expression, Chromatin Packaging Affects Tissue Differentiation

Significantly, the number and distribution of DNase hypersensitive sites varies from tissue to tissue. In one type of white blood cell, for example, DH sites expose the 5′ ends of the genes for making antibodies. In young red blood cells, it is the 5′ ends of the gene for hemoglobin synthesis. In muscle cells, it is the genes for myoglobin, myosin, actin, and dystrophin; in nerve cells, the genes for neurotransmitters; and so forth. Thus, by the time an individual matures, because of slight differences in packaging, different areas of chromatin unwind for expression during interphase in different types of cells.

At present, the mechanism that selectively alters certain areas of chromatin for specific gene activity remains a black, or at least a gray, box. Molecular geneticists know what happens—chromatin unwinds; where it happens—at points of gene expression; and that it is mediated by a variety of molecular interactions (for example, the enzyme topoisomerase II can dislodge histone H1). But they are only beginning to discover how many different proteins contribute to the complex network of molecular interactions and what signals coordinate their activity. One thing is clear, however. Many DNA sequences, including a variety of repetitive AT-rich regions to which different proteins bind, collaborate to ensure the cell-to-cell transmission of particular chromosomal folding patterns. These patterns produce multiple levels of packaging that function as multiple points of control over the accessibility of genes. In each cell, this

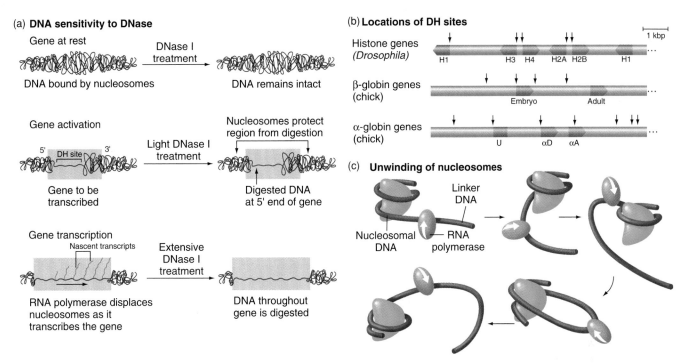

Figure 11.13   DNA compaction and transcription. (a) The sensitivity of DNA to DNase I measures levels of chromosome compaction. A gene not undergoing transcription in a particular cell is compacted into nucleosomes, which makes it relatively impervious to digestion by DNase I. Before a gene can be activated, transcription factors must displace nucleosomes from the promoter-containing region at the 5′ end of the gene. The unwound region created by this displacement is hypersensitive to digestion by DNase I and is referred to as a DH site. Transcribed regions of genes have an intermediate level of DNase I sensitivity so that only extensive exposure to the enzyme can degrade these DNA sequences. This intermediate sensitivity most likely reflects RNA polymerase's temporary modification or displacement of nucleosomes during transcription. (b) Location of DH sites in short gene-containing segments of chromosomes. Most, but not all, DH sites correspond to the 5′ end of genes (*arrows*). (c) One model of how RNA polymerase transcribes through nucleosomes: As the enzyme moves, it partially unwinds the nucleosomes. DNA sequences behind the polymerase (closer to the promoter) can now loop around and interact with the histones. Eventually, the nucleosome behind the polymerase becomes completely rewound with DNA.

transmitted chromatin structure allows the decompaction leading to expression of relatively few genes, while repressing all the others.

Extreme Condensation Silences Expression

In cells stained with certain DNA-binding chemicals, a small proportion of chromosomal regions appear much darker than others when viewed under the light microscope. Geneticists call these darker regions **heterochromatin;** they refer to the contrasting lighter regions as **euchromatin.** The distinction between euchromatin and heterochromatin also shows up in the electron microscope, where the heterochromatin appears much more condensed than the euchromatin. Microscopists first identified dark-staining heterochromatin in the decondensed chromatin of interphase cells, where it tends to localize at the periphery of the nucleus (Fig. 11.14a). Even highly compacted metaphase chromosomes show the differential staining of heterochromatin versus euchromatin (Fig. 11.14b). Most of the heterochromatin in highly condensed chromosomes is found in regions flanking the centromere, but in some animals, heterochromatin forms in other regions of the chromosomes. In *Drosophila* the entire Y chromosome, and in humans most of the Y chromosome, is heterochromatic. Chromosomal regions that remain condensed in het-

erochromatin at most times in all cells are known as **constitutive heterochromatin.**

Autoradiography reveals that cells actively expressing their housekeeping and specialty genes incorporate radioactive RNA precursors into RNA almost exclusively in regions of euchromatin (see Fig. 11.14a). This observation indicates that euchromatin contains most of the sites of transcription and thus almost all of the genes. By contrast, heterochromatin appears to be transcriptionally inactive for the most part, probably because it is so tightly packaged that the enzymes required for transcription of the few genes it contains have a hard time gaining access to the correct DNA sequences.

Two specialized phenomena—position effect variegation in *Drosophila* and Barr bodies in mammalian females—clearly illustrate the correlation between heterochromatin and a loss of gene activity.

Position Effect Variegation in Drosophila: Moving a Gene Near Heterochromatin Prevents its Expression

The *white*+ (*w*+) gene in *Drosophila* is normally located near the telomere of the X chromosome, in a region of relatively decondensed euchromatin. When a chromosomal rearrangement such as an inversion or a translocation places the gene next to or within highly compacted heterochromatin near the

(a)

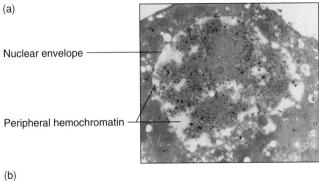

Nuclear envelope

Peripheral hemochromatin

(b)

Figure 11.14 Heterochromatin versus euchromatin. (a) In this image of decondensed interphase chromatin; areas containing heterochromatin are bleached white and tend to be packed just inside the nuclear envelope, while euchromatin is darker and granular. The cell from which this thin section was taken was pulse-labeled with [³H]uridine to mark sites of RNA synthesis (dark spots), most of which occur in euchromatin. (b) In this image, human metaphase chromosomes were stained by a special C-banding technique that darkens the constitutive heterochromatin, most of which localizes to regions surrounding the centromere.

(a) Wildtype X chromosome

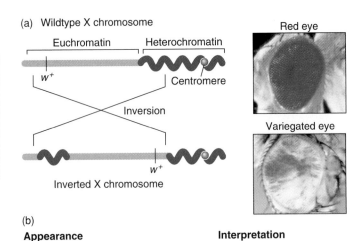

(b)

Appearance   **Interpretation**

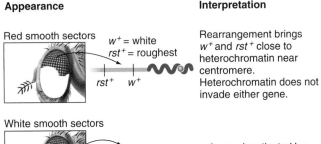

Red smooth sectors

w^+ = white
rst^+ = roughest

Rearrangement brings w^+ and rst^+ close to heterochromatin near centromere. Heterochromatin does not invade either gene.

White smooth sectors

w^+ gene inactivated by spread of heterochromatin. rst^+ gene is active.

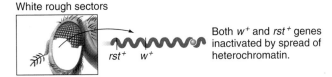

White rough sectors

Both w^+ and rst^+ genes inactivated by spread of heterochromatin.

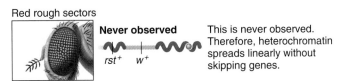

Red rough sectors

Never observed

This is never observed. Therefore, heterochromatin spreads linearly without skipping genes.

Figure 11.15 Position effect variegation in *Drosophila* is a phenotypic effect of facultative heterochromatin. (a) When the w^+ eye color gene (which specifies red color) is brought near an area of heterochromatin through a chromosomal rearrangement such as an inversion, the eyes of the fly can become variegated, with some red cells and some white cells. (b) A model for position effect variegation of the w^+ and *roughest* (rst^+) genes postulates that heterochromatin can spread from its normal location surrounding the centromere to nearby genes, causing their inactivation. Cells in which w^+ is active are red; those in which the gene is inactive are white. Cells in which rst^+ is active have a smooth surface; those in which this gene is inactive have a rough surface. Heterochromatin can spread different distances in different cells, but it cannot skip genes. This model explains the kinds of sectors seen as a result of variegation for these two genes.

centromere, the gene's expression may cease. Some rearrangements silence w^+ gene expression in some cells and not others, producing **position-effect variegation**. In flies carrying the wildtype w^+ allele, cells in the eye with an active w^+ gene are red, while cells with an inactive w^+ gene are white. Apparently, when normally euchromatic genes come into the vicinity of heterochromatin, the heterochromatin can spread into the euchromatic regions, shutting off gene expression in those cells where the heterochromatic "invasion" takes place. In such a situation, the DNA of the gene has not been altered, but the relocation has altered the gene's packaging in some cells. The phenomenon of position effect variegation thus reflects the existence of **facultative heterochromatin:** regions of chromosomes (or even whole chromosomes) that are heterochromatic in some cells and euchromatic in other cells of the same organism.

Position-effect variegation of red and white eye color in *Drosophila* produces eyes that are a mosaic of red and white patches of different sizes (Fig. 11.15a). The position and size of the patches varies from eye to eye. Such variation suggests that the "decision" determining whether heterochromatin spreads to the w^+ gene in a particular cell is the result of a random process. Because patches composed of many adjacent cells have the same

color, the decision must be made early in the development of the eye. Once made, the decision determining whether the white gene will be on or off is transmitted to all the mitotic descendants of that cell. These descendants occupy a particular region of the eye, forming a patch of white cells against a red background.

One interesting property of position effect variegation is that heterochromatin can spread over more than 1000 kb of previously euchromatic chromatin. For example, some rearrangements that bring the w^+ gene near heterochromatin also place the *roughest*$^+$ gene in the same vicinity, although a little farther away from the centromeric heterochromatin (Fig. 11.15b). The wildtype *roughest*$^+$ (rst^+) gene normally produces a smooth eye surface. In flies carrying the rearrangements, however, some large white-colored patches contain within them smaller rough-surfaced patches, indicating that the heterochromatin inactivated first the w^+ and then the rst^+ gene. Red-colored, rough-surfaced patches never form, which means that the heterochromatin does not skip over genes as it spreads along the chromosome.

Geneticists currently know very little about how heterochromatin invades euchromatin, what determines the extent of heterochromatin invasion in a particular cell, or how that cell's descendants inherit its specific heterochromatin pattern. They can, nevertheless, use the phenomenon of position effect variegation to identify the molecules involved in heterochromatin formation. In one procedure, they obtained mutations that either enhance the amount of variegation produced by genes positioned near heterochromatin, or diminish the amount of variegation. Enhancement reflects gene inactivation in more cells; diminishment reflects gene inactivation in fewer cells. The researchers later isolated by molecular cloning several of the genes that had mutated, and raised antibodies against the mutant protein products of these genes. In this way, they discovered that at least some of the genes influencing heterochromatin formation encode DNA-binding proteins that localize selectively to the heterochromatin. The researchers hypothesize that if these proteins migrate from established heterochromatin to nearby DNA sequences, position effect variegation results.

The Fast Forward box "Genomic Imprinting: Parental Origin Affects the Expression of Some Genes in Mammals" describes genomic imprinting, another way in which alterations of chromatin structure affect gene function; these chromatin alterations, however, silence genes through a very different mechanism than compaction.

Barr Bodies Are Another Example of the Correlation between Heterochromatin and a Decrease in Genetic Activity

In both fruit flies and mammals, normal males have one copy of the X chromosome, while females have two; the two sexes, however, require equal amounts of most proteins encoded by genes on the X chromosome. To compensate for the discrepancy in the dose of X-linked genes, male *Drosophila* double the rate at which they express the genes on their single X. Mammals have a different control mechanism for dosage compensation: the random inactivation of all but one X chromosome in the female's somatic cells. The inactive X chromosomes are observable in interphase cells as darkly stained heterochromatin masses. Geneticists call these densely staining X chromosomes **Barr bodies** after Murray Barr, the cytologist who discovered them (Fig. 11.16a).

The inactive X chromosome in mammalian females is another example of facultative heterochromatin. Here, a whole X chromosome becomes completely heterochromatic in some cells while remaining euchromatic in others. Most genes on this alternating X chromosome are available for transcription only in cells where the chromosome is euchromatic; very few genes are available for transcription when the X becomes heterochromatic.

An XX female has one Barr body in each somatic cell. The X chromosome that remains genetically active in these cells decondenses and stains as expected during interphase. XY male cells do not contain Barr bodies. As a result, normal male and female mammals have the same number of active X chromosomes. Females with XXX or XXXX karyotypes can survive because they have two or three Barr bodies and only a single

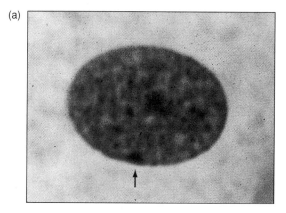

(a)

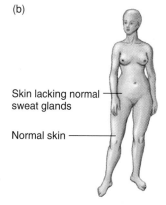

(b)

Skin lacking normal sweat glands

Normal skin

Figure 11.16 Barr bodies indicate that only one X chromosome in a mammalian cell is genetically active. (a) Micrograph showing one Barr body (arrow) in the interphase nucleus of an XX human female cell. (b) Random X-inactivation can produce skin mosaics in women. This human female is heterozygous for an X-linked mutation causing anhidrotic ectodermal displasia, which causes the absence of sweat glands. During her embryonic development, the X chromosome carrying the normal allele was inactivated in some cells, while the X chromosome carrying the mutant allele was inactivated in other cells. The decision of which X to inactivate was made randomly. The mitotic clones established from the two types of cells eventually became patches of normal or mutant skin.

Genomic Imprinting: Parental Origin Affects the Expression of Some Genes in Mammals

A major tenet of Mendelian genetics is that the parental origin of an allele—whether it comes from the mother or the father—does not affect its function in the F_1 generation. For the vast majority of genes in plants and animals, this principle still holds true today. Surprisingly, however, experiments and pedigree analyses have uncovered convincing evidence of exceptions to this general rule for some genes in mammals.

In the 1980s when embryologists learned to manipulate the haploid nuclei, or pronuclei, of eggs and sperm, they created *in vitro* fertilized eggs carrying one maternally derived and one paternally derived pronucleus; or two maternal pronuclei; or two paternal pronuclei. When placed in a mouse uterus, embryos emerging from the fusion of one maternal and one paternal pronucleus developed into viable, fertile adult mice, like embryos that arise in a similar way in nature. But embryos formed from two maternal or two paternal haploid nuclei did not develop normally. Since all of the fertilized eggs providing the pronuclei were produced by matings between inbred animals of the same strain, the only possible genetic differences between the embryos lay in the number and type of sex chromosomes. Yet, even XX embryos with two maternal or two paternal pronuclei failed to thrive. The experimenters concluded that something other than the DNA itself differentiates maternal and paternal pronuclei.

With the advent of molecular tools able to distinguish the transcription of a gene on a maternally derived chromosome from transcription of the same gene on the paternally derived homolog, geneticists observed that expression of a small number of genes—scattered around the genome, but often found in clusters—depends on whether the copy of the gene comes from the mother or the father. The phenomenon in which a gene's expression depends on the parent that transmits it is known as **genomic imprinting.** In most cases of genomic imprinting, the copy of a gene inherited from one parent is transcriptionally inactive in all or most of the tissues in which the copy from the other parent is active. The term "imprinting" signifies that whatever silences the maternal or paternal copy of an imprinted gene is not encoded in DNA sequence; rather it exercises its effect through some **epigenetic** (that is, outside-the-genes) alteration of the DNA or chromatin during gametogenesis. This epigenetic imprint remains throughout the life of the mammal, but is erased and regenerated during each passage of the gene through the germ line into the next generation (Fig. A). Some genes receive an imprint in the maternal germ line; others receive it in the paternal germ line. For each gene, imprinting occurs in either the maternal or paternal line, never in both.

Although the biochemical mechanism of genomic imprinting is not yet understood, one important component is the methylation of cytosines in C G dinucleotides within the imprinted region. The methylated C's silence the gene or genes in the region by preventing RNA polymerase and other transcription factors from gaining ac-

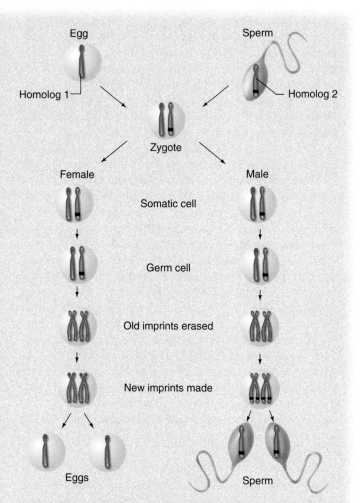

Figure A The resetting of genomic imprints during meiosis. This diagram follows the transmission of a pair of homologous chromosomes (homolog 1 from the mother and homolog 2 from the father) from gametes through fertilization, and the development of female and male progeny to meiosis and the creation of a new set of gametes. Maternally imprinted genes are shown in red, paternally imprinted genes in black. In the figure, the zygote formed by fusion of egg and sperm is a homozygote of identical wildtype homologous chromosomes. The zygotic genome contains a gene in red on homolog 1 that has a maternal imprint, and a different gene in black on homolog 2 that has a paternal imprint. The individual that develops from this zygote, whether male or female, will show no abnormal phenotype because one copy of each of the two genes under consideration remains active. During meiosis, the cellular machinery erases the old imprints and establishes new ones. In the female, all the resulting egg genomes will have inactivated the maternally imprinted gene. In the male, the resulting sperm genomes will have inactivated the paternally imprinted gene. Note that in both egg and sperm, one of the chromosomes (homolog 2 for egg, homolog 1 for the sperm) will be differently imprinted than the corresponding chromosome that entered the zygote.

cess to the DNA. A particular pattern of methylation can be transmitted during DNA replication, with the presence of a methyl group on one strand of a newly synthesized double helix signaling methylase enzymes to add a methyl group to the other strand.

Among vertebrate animals, only mammals imprint their genomes. Thus imprinting is a rather late evolutionary development. By 1995, researchers had evidence for imprinting in about 15 genes in the mouse and human genomes, and they estimate that about 1% of the mammalian genome — roughly 1000 genes—will turn out to be imprinted. Most genes imprinted in one mammalian species are also imprinted in other species of mammals, but there are exceptions.

Genomic imprinting passes unnoticed in individuals with a wildtype phenotype. This is because at each imprinted locus, an active copy of the gene on only one homolog can provide the optimal amount of gene product required. But what happens if a mutation deletes an allele of an imprinted gene? If the gene is maternally imprinted, only the paternal copy of the gene is active in a normal individual. Individuals who inherit the deleted allele from their mother will function normally because the level of gene expression is unaltered; but in individuals who inherit the deletion from their father, disaster could strike. The paternally derived gene copy that would have been active is missing and the maternally imprinted gene copy is silent; consequently there is no gene product. If the animal requires the gene product for normal development, the embryo could die or the newborn could have a severe birth defect. Thus, the same exact heterozygous genotype—*del*/+—produces either a wildtype phenotype or a disease phenotype, depending on which parent has supplied the wildtype allele. The disease-causing mutation in this example could pass unnoticed from mother to daughter for many generations. If, however, the mutation passed from mother to son, the son would have a normal phenotype (because he received an active wildtype allele from his father), but the son's children, both boys and girls, would each have a 50% chance of expressing the mutant phenotype (Fig. B).

Before the late 1980s, clinical geneticists were used to seeing sex-linked differences in inherited phenotypes related to the sex of the affected individual. With imprinting, however, it is the sex of the parent carrying a mutant allele that counts, and not the sex of the individual inheriting the mutation. After the discovery of imprinting in mice, medical geneticists reanalyzed human pedigrees and determined retroactively that what appeared to be instances of incomplete penetrance were actually manifestations of imprinting. Evidence for imprinting as a contributing factor now exists for a variety of human developmental disorders, including the related pair of syndromes known as Prader-Willi syndrome and Angelman syndrome. Children with Prader-Willi syndrome have small hands and feet, underdeveloped gonads and genitalia, a short stature, and mental retardation; they are also obese, and compulsive overeaters. Children affected by Angelman syndrome have red cheeks, a large jaw, a large mouth with a prominent tongue, and a happy disposition accompanied by excessive laughing; they also show severe mental and motor retardation. Both syndromes are often associated with small deletions in the q11–13 region of chromosome 15. When the deletions are inherited from the father, the child gets Prader-Willi

Paternal imprinting

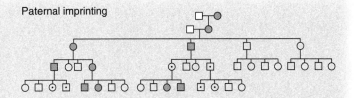

Maternal imprinting

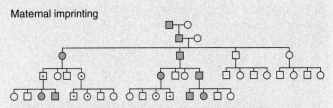

Figure B Pedigrees reflecting maternal and paternal imprinting. In both pedigrees, affected individuals (represented by filled-in, yellow circles and squares) are heterozygotes for a deletion removing a gene that has either a paternal or a maternal imprint. With paternal imprinting, affected individuals receive a deleted chromosome from their mother and an inactivated copy of the gene from their father. Therefore, they have no functional copy of the gene in question. The opposite is true with maternal imprinting. The deleted chromosome comes from the father, while the inactivated copy of the gene comes from the mother. In these pedigrees, a dotted symbol indicates individuals carrying a deleted chromosome but not displaying the mutant phenotype.

syndrome; when the deletions come from the mother, the child has Angelman syndrome. The explanation is that there are at least two genes in the region of these deletions that are differently imprinted. One gene is maternally imprinted: Children receiving a deleted chromosome from their father and a wildtype (nondeleted) chromosome with an imprinted (inactivated) copy of this gene from their mother will have Prader-Willi syndrome. The other gene is paternally imprinted: Children receiving a deleted chromosome from their mother and an imprinted (inactivated) gene from their father will have Angelman syndrome.

From an evolutionary standpoint, genomic imprinting seems to place individuals at a disadvantage for survival. Individuals with two potentially active copies of a gene have one copy to spare to random mutations. Most null mutations, for example, are recessive because with only one copy mutated, the remaining functional copy of the gene provides enough gene product for viability. By comparison, with imprinted genes, the single functional gene copy makes an individual susceptible to all mutations.

Why then does imprinting exist? Does it confer some advantage on a population of organisms that counterbalances the obvious evolutionary disadvantages? One proposed answer is that imprinting represents the endgame in a struggle between the sexes. The argument is as follows: In most mammalian species, a female in estrus can mate with multiple males, generating multiple embryos fathered by different males. If a male could cause the embryos he sires to underexpress genes that normally retard embryonic growth, those embryos would use more maternal resources and grow more rapidly *in utero* than the sibling embryos; at birth, the larger, more robust offspring would have a fitness advantage. But the over-growth of embryos and fetuses, while potentially beneficial to some offspring, would be severely draining to the mother. She would compensate by down-regulating her copies of genes that normally enhance growth. This down-regulation would give all the embryos she is carrying an equal chance of survival, and prevent a multiple pregnancy from sapping all her strength. At the evolutionary endpoint of this male-versus-female tug of war, imprinting will have silenced growth-enhancing genes in all females and growth-retarding genes in all males.

Known as the Haig hypothesis for David Haig, its originator, this line of reasoning explains why imprinting is found only in mammals: It is only in mammals that embryos can compete for maternal resources *in utero*. Interestingly, many of the imprinted genes uncovered to date are involved in embryonic growth, as predicted by the hypothesis. However, other imprinted genes are not. Only many more hours of experiment and analysis will confirm or reject the Haig hypothesis.

active X. Barr bodies are not restricted to females, but their presence does require more than one X chromosome per cell. Cells from XXY Klinefelter's males also contain a Barr body.

The "decision" determining which X chromosome in each cell becomes a Barr body occurs at random in the early stages of development and is inherited by the descendants of each cell. In humans, for example, two weeks after fertilization, when an XX female embryo consists of 500–1000 cells, one of the X chromosomes in each cell condenses to a Barr body. Each embryonic cell "decides" independently which X it will be. In some cells, it is the X inherited from the mother; in others, it is the X inherited from the father. Once the determination is made, it is clonally perpetuated so that all of the thousands or millions of cells descended by mitosis from a particular embryonic cell condense the same X chromosome to a Barr body and thereby inactivate it. Female mammals are thus a mosaic of cells containing either a maternally or a paternally derived inactivated X chromosome (Fig. 11.16b). In an individual female that is heterozygous for an X-linked gene, some cells will express one allele, while other cells will express the alternative. In females heterozygous for an X-linked mutation that would be lethal in a male, the relation between the two populations of activated X cells can make the difference between life and death. If the X chromosome carrying the wild-type allele is active in a high enough proportion of cells where expression of the gene is required, the individual will survive.

About a dozen genes on the X chromosomes of both mice and humans escape Barr body inactivation, most likely because they are in small regions that do not become completely heterochromatic. Most of these genes remain active on both X chromosomes in a female. One, however, is inactive on the euchromatic X but active on the heterochromatic X, and may contribute to the mechanism that causes X inactivation. This gene is referred to as *Xist* for X̲ i̲nactivation s̲pecific t̲ranscript. An X chromosome that does not contain *Xist* cannot be inactivated. Deletion of the *Xist* gene abolishes a chromosome's capacity for X inactivation. Cells carrying one X chromosome that lacks the *Xist* gene must inactivate the other X. The specific function of the *Xist* gene is to produce an unusually large *cis*-acting RNA transcript. Studies using fluorescent molecular probes show that this *cis*-acting *Xist* RNA, unlike most transcripts, never leaves the nucleus. One model for how the *Xist* RNA produces inactivation proposes that the unusually

large *cis*-acting RNA binds to the X chromosome that produces it, perhaps in association with other proteins; this binding makes most genes on that chromosome inaccessible to transcription factors.

Although position-effect variegation and Barr bodies are specialized phenomena, they epitomize the relationship between chromatin structure and gene function in all eukaryotic chromosomes. The compaction of chromosomes, which prepares the sister chromatids of each duplicated chromosome for accurate segregation and transport through mitosis and meiosis, makes DNA inaccessible to the proteins that initiate gene expression. Such compaction reaches its height during metaphase and anaphase, curtailing the transcription of most genes at those phases of the cell cycle. Decompaction, by contrast, precedes gene expression. The opening up and closing off of chromosomal areas depends on a dynamic network of DNA-protein interactions, which modify the density of chromatin packaging to appropriate levels for genetic function and passage through the cell cycle. Once established, the chromatin structure that determines a gene's expression is stably transmitted to daughter cells.

Unusual Chromosome Structures Clarify the Correlation between Chromosome Packaging and Gene Function

The chromosomes of some highly specialized cells present an unusual appearance, as do the regions of interphase chromosomes in normal cells that carry genes for ribosomal RNA (rRNA) synthesis. When viewed in the light microscope, for instance, chromosomes in the salivary gland cells of fruit fly larvae look like thickened, striped lengths of rope radiating from a darkened central core. Cells in interphase of the cell cycle exhibit a large, well-defined sphere within the nucleus. Detailed studies of these unusual forms have clarified how chromosome structure affects gene function.

Polytene Chromosomes Magnify Patterns of Compaction and Tie Them to Gene Expression

Within the salivary gland cells of *Drosophila* larvae, the interphase chromosomes go through 10 rounds of replication without ever entering mitosis. As a result, the sister chromatids never separate and each chromosome consists of 2^{10} (1024) double helixes. In addition, because homologous chromosomes in the somatic cells of *Drosophila* remain tightly paired throughout interphase, pairs of homologs form a cable of double thickness containing 2048 double helixes of DNA (1024 from each homolog). These giant chromosomes consisting of many identical chromatids lying in parallel register are called **polytene chromosomes** (Fig. 11.17a). When stained and viewed in the light microscope, *Drosophila* polytene chromosomes have an irregular fine grain banding pattern in which denser dark bands alternate with lighter interbands. Each chromosome has a distinctive banding pattern different from that of all other chromosomes. Microscopic examination sup-

ports the idea that the chromatin of each dark band is 10 times more condensed than the chromatin in the lighter interbands (Fig. 11.17b).

The centromeres of polytene chromosomes fuse to form a dense heterochromatic mass known as the **chromocenter.** The chromocenter locks the chromosomes into a starfishlike formation in which the undulating chromosome arms radiate from a central point (see Fig. 11.17a). The DNA in the chromocenter undergoes fewer duplications than the DNA in the remainder of the chromosome. As a result, whereas the normal somatic chromosomes in *Drosophila* have almost equal

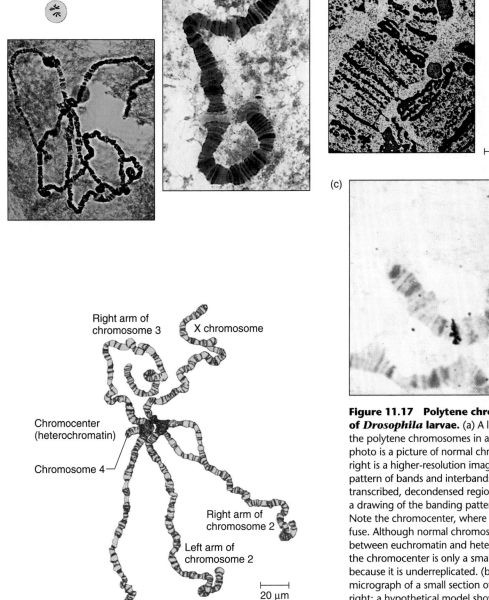

Figure 11.17 Polytene chromosomes in the salivary glands of *Drosophila* larvae. (a) A low-resolution photo (left) showing all the polytene chromosomes in a larval salivary gland cell. Above the photo is a picture of normal chromosomes at the same scale. On the right is a higher-resolution image revealing the irregular fine-grain pattern of bands and interbands and a chromosome puff: a highly transcribed, decondensed region of a polytene chromosome. Below is a drawing of the banding pattern seen in polytene chromosomes. Note the chromocenter, where the centromeres of the chromosomes fuse. Although normal chromosomes are about equally divided between euchromatin and heterochromatin, the heterochromatin at the chromocenter is only a small part of a polytene chromosome because it is underreplicated. (b) Left: a very high resolution electron micrograph of a small section of a *Drosophila* polytene chromosome; right: a hypothetical model showing how the chromatin of the bands is more condensed than the chromatin of the interbands. (c) *In situ* hybridization of a probe containing the *white* gene to a single band (3C2) near the tip of the *Drosophila* X chromosome.

amounts of heterochromatin and euchromatin, polytene chromosomes have relatively little heterochromatin by mass.

Geneticists once thought there might be a one-to-one relationship between the bands of polytene chromosomes and genes, but studies using the molecular techniques described in Chapters 8–10 show that this is not so. Some bands contain more than one gene; others do not seem to carry any. However, even though the 1:1 ratio of bands to genes has turned out to be too simplistic, most geneticists still believe that the division into bands has some functional significance. One possibility is that the bands represent units of transcriptional regulation containing genes activated at the same time.

The large size and precisely reproducible banding patterns of polytene chromosomes make them an invaluable research tool. Because the large number of bands reliably subdivides each chromosome into very small segments, geneticists can use them as a physical guide to mapping. The chromosomes of *Drosophila*, for example, collectively carry 5000 bands that range in size from 3 kb to more than 150 kb. Investigators designate these bands by numbers and letters of the alphabet. The familiar *white* gene is located in band 3C2 on the X chromosome. In mapping, the banding pattern of polytene chromosomes provides extremely high resolution for *in situ* hybridization tests (Fig. 11.17c). A particular fragment of DNA can be localized to a 30- to 100-kb region of a polytene chromosome. By contrast, *in situ* hybridization with normal metaphase chromosomes locates a fragment in a much larger 5- to 10-Mb region.

Polytene chromosomes have also provided a basis for studying the relation between chromosome structure and gene activity. For example, when a gene in a salivary gland cell undergoes a very high rate of transcription, the band containing that gene decondenses and that region of the polytene chromosome swells to form a large, diffuse structure called a **chromosomal puff** (see Fig. 11.17a). The correlation between puffs and regions of high transcriptional activity is yet another example of the general rule that chromatin must unwind to allow gene expression. The pattern of puffing indicates which genes are being expressed in the salivary gland cells at a specific time. At an early stage in the development of salivary glands, for example, one band puffs while at a later stage, this puff has disappeared and other bands are forming puffs. Researchers have recently learned that the early puffing band contains a gene encoding a protein that helps activate transcription of genes in the later puffing bands.

Decondensed Chromatin in the Nucleolus of Interphase Cells Contains rRNA Genes Actively Undergoing Transcription

The most obvious nuclear feature of nearly every interphase eukaryotic cell viewed through the light microscope is the large sphere-shaped organelle known as the **nucleolus** (Fig. 11.18a). Within the nucleolus, long loops of DNA from several

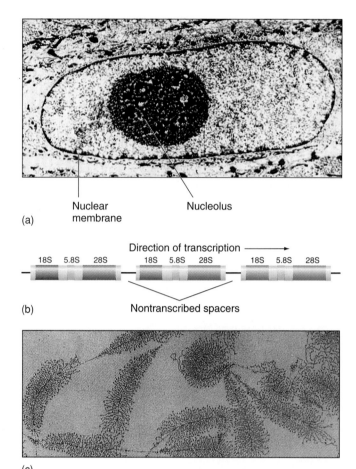

(a) Nuclear membrane   Nucleolus

(b) Direction of transcription →
18S 5.8S 28S   18S 5.8S 28S   18S 5.8S 28S
Nontranscribed spacers

(c)

Figure 11.18 The nucleolus. (a) Electron micrograph of a thin section of part of a human skin cell showing the oblong nucleus and its outstanding feature, the nucleolus (large purplish circle). (b) A nucleolus can contain hundreds of copies of rRNA genes (collectively known as the nucleolus organizer); the genes are organized as a series of transcription units separated by nontranscribed spacer regions. (c) The nucleolus is the site of rRNA gene transcription. To generate this image, a region of chromatin from the nucleolus was spread for visualization in the electron microscope. Newly transcribed rRNAs appear as short, wispy strands emanating from the DNA; the longer strands are closer to completion. Particles near the end of each RNA transcript are thought to be partially assembled ribosomal subunits. Each transcription unit of rRNA genes generates a large precursor RNA that corresponds to one of the "Christmas trees," or "feathers" in this micrograph. Processing of these large precursor RNAs produces the 18S, 5.8S, and 28S rRNAs of the ribosomal subunit.

chromosomes each carry a cluster of rRNA genes; together the genes make up the **nucleolar organizer** (Fig. 11.18b). The nucleolus also houses growing rRNA transcripts and completed transcripts from arrays of rRNA genes repeated in tandem that look like felled Christmas trees or feathers (Fig. 11.18c). Like the chromosomal puffs of polytene chromosomes, the decompacted DNA in the nucleolus is a sign of high levels of transcription.

CONNECTIONS

Eukaryotic chromosomes package and manage the genetic information in DNA through a modular chromatin design whose flexibility allows back-and-forth shifts between different levels of organization. These reversible changes in chromatin structure reliably sustain a variety of chromosome functions, producing selective unwinding for gene expression, universal unwinding for replication, and coordinated compaction for segregation and transport. Histones and nonhistone proteins provide the framework for chromatin and help regulate changes in chromosome structure and function. Noncoding sequences that specify the origins of replication, centromeres, and telomeres are essential to chromosome duplication, segregation, and integrity.

Although the faithful function, replication, and transmission of chromosomes underlies the perpetuation of life within each species, chromosomal changes do occur. We have already described two mechanisms of change: mutation of individual nucleotides (Chapter 6) and homologous recombination, which exchanges bases between homologs (Chapters 3, 4, 5, and 10). In the next chapter, we examine broader chromosomal rearrangements that produce different numbers of chromosomes, reshuffle genes between nonhomologous chromosomes, and reorganize the genes of a single chromosome. These large-scale modifications, by altering the genetic content of a genome, provide some of the important variations that fuel evolution.

ESSENTIAL CONCEPTS

1. Each chromosome consists of one long molecule of DNA compacted by *histone* and *nonhistone* proteins. The five types of histones—H1, H2A, H2B, H3, and H4—are essential to the establishment of generalized chromosome structure. There are more than 1000 kinds of nonhistone proteins: Some are structural; others are active in replication and segregation; but most help determine the time and place of gene expression.

2. DNA-protein interactions create reversible levels of compaction.
   a. The naked DNA wraps around core histones to form *nucleosomes,* which are secured by H1.
   b. Models of higher-level compaction suggest that some sort of *supercoiling* condenses the nucleosomal fiber to a wider fiber. Nonhistone proteins then anchor this fiber to form *loops* and higher levels of compaction (perhaps in the form of *rosettes* and bunches of rosettes). In metaphase chromosomes, higher levels of folding compact DNA 10,000-fold.
   c. In fully compacted metaphase chromosomes, the *centromere* and *telomeres* become visible under the microscope. Giemsa staining of metaphase chromosomes reveals highly reproducible banding patterns that researchers can use to locate genes, analyze chromosomal differences between species, and diagnose some genetic diseases.

3. Specialized elements are required for normal chromosome function.
   a. *Origins of replication* are sites accessible for the binding of proteins that initiate replication. In eukaryotic chromosomes, many origins of replication ensure timely DNA replication.
   b. Telomeres, composed of repetitive base sequences, protect the ends of chromosomes, ensuring their integrity. The enzyme *telomerase* helps reconstruct the complete telomere with each cell division.
   c. Centromeres, which appear as constrictions in metaphase chromosomes, ensure proper segregation by holding sister chromatids together and by elaborating kinetochores that properly attach sister chromatids to spindle fibers for mitosis and meiosis.

4. Chromosomal packaging influences gene activity.
   a. Conditions that decondense selected areas of chromatin precede and facilitate gene expression. Puffs in *Drosophila* polytene chromosomes and the *nucleoli* in most interphase cells contain decondensed chromatin that is highly transcribed. Boundary elements called insulators delimit these areas of decompaction. Nucleosomes in the decompacted areas unwind to allow the initiation of transcription.
   b. Extreme condensation silences gene expression; extremely condensed chromosomal areas appear as darkly staining *heterochromatin* under the microscope. Position-effect variegation in *Drosophila* and Barr bodies in mammals are examples of the correlation between heterochromatin and a loss of gene activity.

SOCIAL AND ETHICAL ISSUES

1. A city school board put forward a proposal to screen school children who have learning disabilities for the fragile X mutation (see the Genetics and Society box "A New Class of Human Mutation: Amplified Repeats with Medical Consequences" in Chapter 6). Fragile X is the most common form of inherited mental retardation and is caused by an unusual expansion of a triplet nucleotide repeat in a gene on the X chromosome. (The triplet expansion is extreme enough to be seen as a cytological abnormality.) The extent of mental impairment and other phenotypes varies with the size of the repeat. The proposal was well-intended, hoping to identify students with potential disabilities and give them extra help and consideration. However, many parents and professionals were opposed to the testing. What could the harm be to children who are selected for testing? for those identified in this way as fragile X children? What are the benefits?

2. Recent evidence has shown that the ends of chromosomes (telomeres) get shorter as cells age because of a lack of the enzyme telomerase that maintains the ends and length of chromosomes. In contrast, cancer cells that are "immortal" and rapidly dividing have telomerase activity. Whether the lack of telomeres in older cells is a cause or result of aging is still under investigation. A proposal that included experimentation on telomerase has been submitted to the National Institutes of Health in which a stated goal was exploration of ways to extend the human life span. Is this a valid way to spend federal research monies given that many companies are already exploring ways to utilize these findings to expand the human life span? What are the consequences for society if people live longer? What do you think of the goal of a longer life?

3. With the sequencing of many different genomes occurring at a rapid pace, how quickly information is made available to other researchers has become an issue. A scientist who has worked for two years on identifying a human disease gene has recently cloned the region in which the gene is located and determined the sequence of the entire region. The next steps are to use this sequence information to identify potential gene candidates and determine which of the candidates is altered by disease-causing mutations. He hesitates to release the sequence information because he thinks many other researchers will immediately use that data and design PCR primers to investigate the potential genes and mutations in affected individuals. Does he have a right to withhold the information until he has completed his analysis of mutants? Does the answer to this question depend on whether his work is supported by public (government) funding agencies or by a private biotechnology company?

SOLVED PROBLEMS

I. One can construct YACs that range in size from 15 kb to 1 Mb. Based on DNA length, what level of chromosome compaction would you predict for a YAC of 50 kb compared to a YAC of 500 kb?

Answer

To answer this question, you need to apply information about the amount of DNA needed to get different types of chromosome condensation.

The 500-kb YAC would probably be more condensed than the 50-kb YAC based on its larger size. The DNA of both YACs would be wound around histones to form the nucleosome structure (160 bp around the core histones plus 40 bp in linker region). That DNA would be further compacted into 300-Å fibers that contain six nucleosomes per turn. *The 500-kb YAC would be compacted at a higher level of order, presumably in radial loops that occur every 60–100 kb in the chromosome, but the 50-kb YAC is not large enough to be packaged in this way.*

II. To clone and express genes in yeast, researchers constructed recombinant plasmids in the 1970s. Most of these early yeast plasmids did not contain an origin of replication and therefore were not maintained as autonomous plasmids separate from the chromosomes. If one of these plasmids integrated into the chromosome by recombination, it was replicated as part of the chromosome and was stably inherited from one division to the next. Later, when yeast origins of replication (ARS elements) were characterized, researchers constructed autonomous yeast plasmids that could be maintained and transmitted independently of the chromosome. Describe how you would distinguish a plasmid integrated into a chromosome from an autonomous plasmid, using hybridization techniques.

Answer

This question requires an understanding of the use of hybridization (i.e., the source of DNA and the probe used) and the interpretation of results. Think about differences between the two plasmid states (integrated and autonomous) and how the differences might be visualized using hybridization.

If plasmid DNA is used as a probe in a hybridization with a blot of restriction enzyme-digested DNA from cells containing integrated plasmids and from cells containing autonomous plasmids, the hybridizing bands will be different (see the diagram). In the DNA sample containing the autonomous plasmid, the probe will hybridize to band(s) that together equal the total size of the original plasmid. If DNA from cells containing the integrated plasmid is digested with a restriction enzyme that cuts within the plasmid, two new fragments appear that contain both plasmid DNA and the chromosomal DNA on either side of the integration site in the chromosome. In the example in the figure, a 4-kb plasmid recombined with the chromosome at a site within a 2.5-kb chromosomal fragment. The 2.2-kb and 4.3-kb fragments are joint fragments containing both chromosomal DNA and plasmid DNA. Both bands will hybridize with the probe.

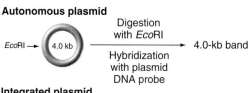

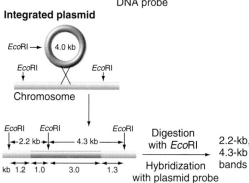

III. Mouse geneticist Mary Lyon proposed in 1961 that all but one copy of the X chromosome were inactivated in mammals.
  a. What cytological finding supports the Lyon hypothesis?
  b. What is a genetic result that supports the Lyon hypothesis?

Answer

This question requires an understanding of the experimental observations on X inactivation.

a. Microscopic examination of cells produces cytological evidence. *The number of Barr bodies seen in the cells of individuals with different numbers of X chromosomes supports the hypothesis.* For example, cells from XX females have one Barr body, whereas cells from an XXX female have two Barr bodies.

b. *Examination of the phenotype of cells from a female heterozygous for two different alleles of an X-linked gene produces genetic evidence that supports the Lyon hypothesis.* In these females, some cells have the phenotype associated with one allele, while other cells have the phenotype associated with the other allele.

PROBLEMS

11-1 For each of the terms in the left column, choose the best matching phrase in the right column.

a. telomere
b. G bands
c. polytene chromosomes
d. kinetochore
e. nucleosome
f. ARS
g. satellite DNA
h. chromatin

1. anchoring site at the base of chromatin loops that anchors loops to the scaffold
2. origin of replication in yeast
3. repetitive DNA found near the centromere in higher eukaryotes
4. specialized structure at the end of a linear chromosome
5. giant chromosomes in *Drosophila* salivary glands
6. complexes of DNA and protein in the eukaryotic nucleus
7. small basic proteins that bind to DNA and form the core of the nucleosome
8. complex of DNA and proteins where spindle fibers attach to a chromosome

i. SAR
j. histones

9. beadlike structure consisting of DNA wound around histone proteins
10. regions of a chromosome that are distinguished by staining differences

11-2 The sixteen chromosomes of the yeast *Saccharomyces cerevisiae* can be separated by size using pulse-field gel electrophoresis. You have just isolated a clone of a yeast gene and want to know on which chromosome the gene is located. How could you use pulse-field gel electrophoresis to find this out?

11-3 What difference is there in the compaction of chromosomes during metaphase and interphase?

11-4 What is the role of the core histones in compaction compared with the role of histone H1?

11-5 Many proteins other than histones are found associated with chromosomes. What roles do these nonhistone proteins play?

11-6 The mitotic cell divisions in the early embryo of *D. melanogaster* occur very rapidly (every 8 minutes).

a. If there was one bidirectional origin in the middle of each chromosome, how many nucleotides would have to be added by DNA polymerase per second to replicate all the DNA in the longest chromosome (66 mb) during the 8-minute early embryonic cell cycles? (Assume that replication occurs during the entire cell division cycle.)

b. In fact, many origins of replication are active on each chromosome during the early embryonic divisions and are spaced approximately 7 kb apart. Calculate the average rate (per second) with which DNA polymerase adds complementary nucleotides to a growing chain in the early *Drosophila* embryo.

11-7 The extreme ends of chromosomes (telomeres) contain repeated DNA sequences (e.g., 5′ TTAGGG 3′ in humans). If a blot of genomic DNA was hybridized using a probe of 3′ AATCCC 5′, all end fragments would hybridize with the probe.

a. If you wanted to examine an end of one specific chromosome by hybridization, what would you have to use as a probe?

b. When a probe specific for one chromosome end is used in Southern hybridization, a blurred band is seen (see the figure). Why is the hybridizing band not a distinct sharp band as seen with probes that hybridize with internal sequences on the chromosome?

Size standard   Genomic DNA

11-8 The enzyme telomerase consists of protein and an RNA containing a template sequence that directs the addition of appropriate end sequence for the species. Telomere sequences (TTGGGG) from the ciliated protozoan *Tetrahymena* were cloned onto the ends of a linear YAC that was then transformed into yeast. The YAC survived as a linear piece of DNA but the YAC now had TGGTGG sequences at the very ends in addition to TTGGGG. Why do you think these sequences were added?

11-9 A number of yeast derived elements were added to the bacterial plasmid pBR322. Yeast that require uracil for growth (Ura⁻ cells) were transformed with these modified plasmids and Ura⁺ colonies were selected by growth in media lacking uracil. For plasmids containing each of the elements listed, indicate whether you expect the plasmid to integrate into a chromosome or be maintained separately as a plasmid. If it is maintained autonomously, is it stably inherited by all of the daughter cells of subsequent generations when you no longer select for Ura⁺ cells (grow in media containing uracil)?

a. URA⁺ gene

b. URA⁺ gene, ARS

c. URA⁺ gene, ARS, CEN (centromere)

11-10 A protein called CBF1 was identified in yeast as a centromere binding protein. You want to know if similar proteins are present in human cells.

a. Starting with a human cDNA expression library from actively dividing cells, how could you isolate a clone containing a human gene similar to the yeast gene?

b. Imagine you obtained a clone containing a human gene similar to the yeast gene. How could you test to see if the protein encoded by the human gene is associated with a human centromere region?

c. If you had a haploid yeast strain with a temperature sensitive mutation in the CBF1 gene, what phenotype(s) can you imagine when the temperature is raised to the nonpermissive temperature?

11-11 A DNA fragment containing yeast centromere DNA was cloned into a TRP ARS plasmid, YRp7, causing the plasmid to become mitotically very stable (i.e., the plasmid was transmitted during mitotic divisions to each daughter cell). The assay for mitotic stability consists of growing a transformed cell without selection for the plasmid and determining the number of Trp⁺ colonies remaining after 20 generations of growth under conditions that are not selective for the plasmid. To identify the region of the cloned fragment that contained centromere DNA, you cut the initial fragment into smaller pieces, reclone those pieces into YRp7, and tested for mitotic stability. Based on the map that follows and results of the mitotic stability assay, where is the centromere DNA located?

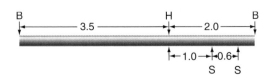

B   H   B
|——— 3.5 ———|—— 2.0 ——|
                  |—1.0—|—0.6—|
                  S   S

Results of Mitotic Stability Assay

Plasmid DNA	Percentage of Trp$^+$ colonies after 20 generations
YRp7	0.9
YRp7 + 5.5 kb *Bam*HI (B)	68
YRp7 + 3.5 kb *Bam*HI-*Hin*dIII (H)	0.5
YRp7 + 2.0 kb *Bam*HI-*Hin*dIII	80
YRp7 + 0.6 kb *Sau*3A (S)	76
YRp7 + 1.0 kb *Hin*dIII-*Sau*3A	0.7

11-12 Another vector system for cloning large fragments of DNA that investigators developed for genome analysis uses a derivative of the baculovirus that infects arthropods. One of the advantages of these BAC vectors compared to YACs is that DNA cloned into these vectors is less likely to get rearranged. Clones of a human *Sfi*I fragment were isolated from BAC and YAC based libraries using the same probe but the *Hin*dIII digests of these clones were very different from each other. How could you determine which of these two clones, if either, had the same order and size of *Hin*dIII fragments as are present in the human genome?

11-13 If a sequence that acts as a boundary for the controlled decomposition of chromatin is removed, what effect could this have on genes in adjacent regions?

11-14 Some genes are highly transcribed in cells of one tissue type but not in other tissues. Based on the hypothesis presented in this chapter, which of the patterns of DNase digestion shown below in the vicinity of gene X would you expect to see in DNA isolated from cells in which the gene X is highly expressed? (Vertical lines indicate sites where DNase I digests the DNA.)

a. Cell type I

___|_____|_____|_____

b. Cell type II

__||_|_____|||__|_____||_|_|_|_|__|_||__

11-15 Give examples of constitutive and facultative heterochromatin in
a. *Drosophila*
b. humans

11-16 *Drosophila* geneticists have isolated many mutations that modify position effect variegation. Dominant *suppressors of variegation* [*Su(var)*s] cause less frequent inactivation of genes brought near heterochromatin by chromosome rearrangements, while dominant *enhancers of variegation* [*E(var)*s] cause more frequent inactivation of such genes.
a. What effects would each of these two kinds of mutations have on position effect variegation of the *white* gene in *Drosophila*.
b. Assuming that these *Su(var)* and *E(var)* mutations are loss of function (null) alleles in the corresponding genes, what kinds of proteins do you think these genes encode?

11-17 How many Barr bodies are present in humans with the following karyotype?
a. an XX female
b. an XY male
c. an XX male (known as an exceptional male*)
d. an XXY male
e. an XXXX female
f. an XO female

_____

*In an exceptional XX male, one of the two X chromosomes has essentially all of the genes normally found on the X chromosome plus a small region from the Y chromosome that carries the testis-determining factor required for male development.

11-18 A pair of twin sisters were believed to be identical until one was diagnosed with Duchenne muscular dystrophy, an X-linked trait. (Her sister did not have the disease.) Does this finding mean they are definitely not identical twins (derived from fertilization of one egg)? Why or why not?

11-19 Females with the genotype $X^{CB}X^{cb}$ are rarely colorblind although some have only partial color vision. Speculate on why this is true.

11-20 What are two characteristics of polytene chromosomes that make them useful for cytogenetics?

11-21 If a gene is paternally imprinted and there is an affected son, from which parent did the son inherit the mutant allele?

11-22 Prader-Willi syndrome is maternally imprinted. Answer the following questions as True or False, assuming that the trait is 100% penetrant.
a. Half of the sons of affected males will show the syndrome.
b. Half of the daughters of affected males will show the syndrome.
c. Half of the sons of affected females will show the syndrome.
d. Half of the daughters of affected females will show the syndrome.

11-23 Assume that the disease illustrated with this pedigree is due to expression of a rare allele of an autosomal gene that is a paternally imprinted gene. What would you predict is the genotype of individuals (a) I-1, (b) II-1, and (c) III-2?

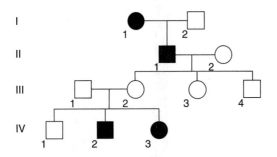

11-24 The *IGF-2R* gene is autosomal and maternally imprinted. Copies of the gene received from the mother are not expressed, but copies received from the father are expressed. You have found two alleles of this gene that encode two different forms of the IGF-2R protein distinguishable by gel electrophoresis. One allele encodes a 60K blood protein, the other allele encodes a 50K blood protein. In an analysis of blood proteins from a married couple named Bill and Joan, you find only the 60K protein in Joan's blood and only the 50K protein in Bill's blood. You then look at their children, Jill and Bill Jr. Jill is producing only the 50K protein, while Bill Jr. is producing only the 60K protein.

a. With these data alone, what can you say about the *IGF-2R* genotype of Bill Sr. and Joan?

b. Bill Jr. gets married to a woman named Sara. He has two children, Pat and Tim. Pat produces only the 60K protein and Tim produces only the 50K protein. With the accumulated data, what can you now say about the genotypes of Joan and Bill Sr.?

CHROMOSOMAL REARRANGEMENTS AND CHANGES IN CHROMOSOME NUMBER RESHAPE EUKARYOTIC GENOMES

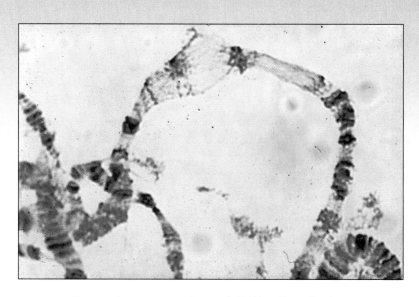

Chromosomal rearrangements can be mapped with high precision on the polytene chromosomes from *Drosophila* salivary glands.

Studies comparing the genome of the laboratory mouse (*Mus musculus*) with that of humans reveal an evolutionary paradox. At the DNA level, there is a close similarity of nucleotide sequence across hundreds of thousands of base pairs; but at the chromosomal level, normal mouse and human karyotypes bear little resemblance to each other. Researchers have, for example, sequenced regions encompassing more than 2000 kb of mouse and human DNA containing a complex of genes that encode proteins known as T-cell receptors. These receptors are key components of the mammalian immune system; after synthesis, they become embedded in the membrane of the immune system's T cells, where they recognize foreign molecules and microorganisms and orchestrate the immune responses that dispose of them. Comparisons of the corresponding mouse and human regions show that the base sequences of the T-cell receptor genes are similar (though not identical) in the two species, as are the order of the genes and the relative positions of a variety of noncoding sequences (of unknown function) along the chromosome. Comparisons of mouse

and human Giemsa-stained karyotypes, however, reveal no conservation of banding patterns between the 20 mouse and 23 human chromosomes.

Data for resolving this apparent paradox come from linkage maps, a level of genetic analysis that is intermediate between DNA sequencing and karyotyping. Because virtually all genes cloned from the mouse genome are conserved in the human genome and vice versa, it is possible to use the techniques described in Chapters 4 and 10 to construct linkage maps for the two genomes from the same set of markers. The maps become a resource for assessing the similarity of linkage groups. If the order of genes in one species were unrelated to the order of genes in the other, identical linkage relationships would occur by chance and appear only rarely. By contrast, if gene order were strictly conserved, it would be possible to align the whole-chromosome linkage maps of both species and see an exact correspondence. Actual comparisons of the mouse and human linkage maps reveal a picture of intermediate conservation. Genes closely linked in *M. musculus* tend to be

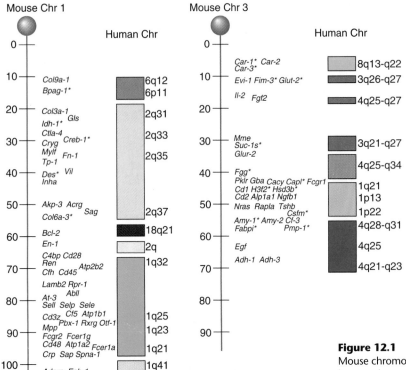

Figure 12.1   Comparing the mouse and human genomes. Mouse chromosome 1 contains large blocks of sequences found on human chromosomes 1, 2, 6, and 18; mouse chromosome 3 includes large fragments of DNA that are located on human chromosomes 1, 3, 4, and 8.

closely linked in humans, but genes that are less tightly linked in one species tend not to be linked at all in the other. From careful analysis of these linkage patterns, geneticists learned that it is possible to "reconstruct" the mouse genome by breaking the human genome into approximately 170 fragments, each an average length of 17.6 Mb, and pasting these fragments together in a different order (Fig. 12.1). Since a 17.6-Mb fragment would occupy no more than one or two bands of a stained chromosome, this level of conservation is not visible in karyotypes. It does, however, show up in the sequence of a smaller genomic region, such as that encoding the T-cell receptors.

These findings contribute to our understanding of how complex life-forms evolved. Although mice and humans diverged from a common ancestor about 65 million years ago, the DNA-derived information currently available in many regions of the two genomes is very similar. It is thus possible to hypothesize that the mouse and human genomes evolved through a series of approximately 170 reshaping events during which chromosomes broke apart and resealed end-to-end in novel ways. After each event, the newly rearranged chromosomes somehow became fixed in the genome of the emerging species. Both nucleotide sequence differences and differences in genome organization thus contribute to dissimilarities between the species.

In this chapter, we examine two types of events that reshape genomes: (1) **rearrangements,** which reorganize the DNA sequences within one or more chromosomes, and (2) losses or gains of entire chromosomes or sets of chromosomes (Table 12.1). Re-

arrangements and changes in chromosome number may affect gene activity or gene transmission by altering the position, order, or number of certain genes in a cell. Such changes often, but not always, lead to a genetic imbalance that is harmful to the organism or its progeny.

Two general themes emerge from our discussion. First, karyotypes generally remain constant within a species, not because rearrangements and changes in chromosome number occur infrequently (they are, in fact, quite common), but because the genetic instabilities and imbalances produced by such changes usually place individual cells or organisms and their progeny at a selective disadvantage. Second, despite selection against chromosomal variations, related species almost always have different karyotypes, with closely related species (such as chimpanzees and humans) diverging by only a few rearrangements and more distantly related species (such as mice and humans) diverging by a larger number of rearrangements. These observations suggest there is some correlation between karyotypic rearrangements and the evolution of new species.

In our examination of events that reshape eukaryotic genomes, we discuss

■ Rearrangements of DNA sequences within chromosomes: deletions, duplications, inversions, translocations, and movements of transposable elements

■ Changes in chromosome number: aneuploidy, including monosomy and trisomy; monoploidy; and polyploidy

TABLE 12.1 Chromosomal Rearrangements and Changes in Chromosome Number (or Ploidy).

Chromosomal Rearrangements

Before → **After**

Deletion: Removal of a segment of DNA

| 1 | 2 | 3 | 4 | 5 | 6 | 7 | 8 | → | 1 | 2 | 3 | 5 | 6 | 7 | 8 |

Duplication: Increase in the number of copies of a chromosomal region

1 2 3 4 5 6 7 8 → 1 2 3 2 3 4 5 6 7 8

Inversion: Half-circle rotation of a chromosomal region

1 2 3 4 5 6 7 8 (180° Rotation) → 1 4 3 2 5 6 7 8

Translocations:

Nonreciprocal: Unequal exchanges between nonhomologous chromosomes

1 2 3 4 5 6 7 8 / 12 13 14 15 16 17 18 → 12 13 4 5 6 7 8 / 14 15 16 17 18

Reciprocal: Parts of two nonhomologous chromosomes trade places

1 2 3 4 5 6 7 8 / 12 13 14 15 16 17 18 → 12 13 14 15 5 6 7 8 / 1 2 3 4 16 17 18

Transposition: Movement of short DNA segments from one position in the genome to another

1 2 3 4 5 6 7 8 → 1 2 4 5 6 3 7 8

Changes in Chromosome Number or Ploidy

Euploidy: Cells that contain only complete sets of chromosomes

Diploidy ($2x$): Two copies of each homolog

Monoploidy (x): One copy of each homolog

Polyploidy: More than the normal diploid number of chromosome sets

Triploidy ($3x$): Three copies of each homolog

Tetraploidy ($4x$): Four copies of each homolog

Aneuploidy: Loss or gain of one or more chromosomes producing a chromosome number that is not an exact multiple of the haploid number

Monosomy ($2n - 1$)

Trisomy ($2n + 1$)

Tetrasomy ($2n + 2$)

Note that it is more accurate to denote monoploids, triploids, and tetraploids as multiples of x, which represents the number of different chromosomes in a complete set, rather than as multiples of n, the number of chromosomes in the gametes. In this table, as throughout the chapter, nonhomologous chromosomes are drawn in different colors. Different shades of the same color highlight different regions of the same chromosome.

REARRANGEMENTS OF DNA SEQUENCES WITHIN CHROMOSOMES

All chromosomal rearrangements, no matter how small, alter DNA sequence. Some do so by removing or adding base pairs. Others relocate chromosomal regions without changing the number of base pairs they contain. The Fast Forward Box "Proper Development of the Immune System Depends on Programmed DNA Rearrangements" describes how the normal development of the human immune system depends on programmed rearrangements of this latter kind in somatic cells.

Deletions Remove Material from the Genome

Deletions, we saw in Chapter 6, remove one or more contiguous base pairs of DNA from a chromosome. They may arise from errors in replication, from faulty meiotic or mitotic recombination, and from exposure to X rays or other chromosome-damaging agents that break the DNA backbone (Fig. 12.2a). Here we use the symbol *Del* to designate a chromosome that has sustained a deletion.

Small deletions often affect only one gene, while large deletions can generate chromosomes lacking tens or even hundreds of genes. In higher organisms, geneticists usually find it difficult to distinguish small deletions affecting only one gene from point

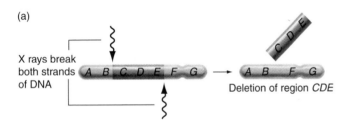

(a)

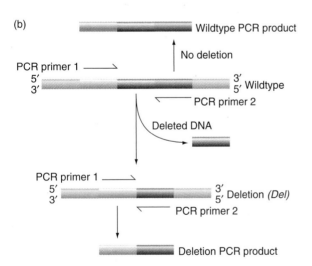

(b)

Figure 12.2   Deletions: Origin and detection. (a) If a chromosome sustains two double-stranded breaks (for example, by exposure to X rays), a deletion will result if the chromosomal fragments are not properly religated. (b) One way to detect deletions is by PCR. The two PCR primers in this figure will amplify a larger PCR product from wildtype DNA than from genomic DNA that carries a deletion.

mutations; they can resolve such distinctions only through analysis of the DNA itself. For example, deletions can result in smaller restriction fragments or PCR products, whereas most point mutations would not cause such changes (Fig. 12.2b). Larger deletions are sometimes identifiable because they affect the expression of two or more adjacent genes. Very large deletions are visible at the relatively low resolution of a karyotype, showing up as the loss of one or more bands from a chromosome.

Homozygosity for a Deletion Is Often, but Not Always, Lethal

Because many of the genes in a genome are essential to an individual's survival, homozygotes (*Del/Del*) or hemizygotes (*Del/*Y) for most deletion-bearing chromosomes do not survive. In rare cases where the deleted chromosomal region is devoid of genes essential for viability, however, a deletion hemi- or homozygote may survive. For example, *Drosophila* males hemizygous for an 80 kb deletion including the *white* (*w*) gene survive perfectly well in the laboratory; lacking the *w+* allele required for red eye pigmentation, they have white eyes.

Heterozygosity for a Deletion Is Often Detrimental

Usually, the only way an organism can survive a deletion of more than a few genes is if it carries a nondeleted wildtype homolog of the deleted chromosome. Such a *Del/+* individual is known as a *deletion heterozygote.* Nonetheless, the missing segment cannot be too large, as heterozygosity for very large deletions is almost always lethal. Even small deletions can be harmful in heterozygotes. Newborn humans heterozygous for a relatively small deletion from the short arm of chromosome 5 have *cri du chat* syndrome (from the French for "cat cry"), so named because the symptoms include an abnormal cry reminiscent of a mewing kitten. The syndrome also leads to mental retardation.

Why should heterozygosity for a deletion have harmful consequences when the *Del/+* individual has at least one wildtype copy of all of its genes? The answer is that changes in **gene dosage**—the number of times a given gene is present in the cell nucleus—can create a *genetic imbalance.* This imbalance in gene dosage alters the amount of a particular protein relative to all other proteins, and this alteration in the relative amounts of protein can have a variety of phenotypic effects, depending on the how the proteins function and how critical the maintenance of a precise ratio of proteins is to the survival of the organism. For some rare genes, the normal diploid level of gene expression is essential to individual survival; fewer than two copies of such a gene results in lethality. In *Drosophila,* a single dose of the locus known as *Triplolethal* (*Tpl*) is lethal in an otherwise diploid individual. For certain other genes, the phenotypic consequences of a decrease in gene dosage are noticeable but not catastrophic. For example, *Drosophila* containing only one copy of the wildtype *Notch* gene have visible wing abnormalities but otherwise seem to function normally (Fig. 12.3). In contrast, for most genes, diminishing gene dosage produces no obvious change in phenotype. There is a catch, however. Although a single dose of any one gene may not cause substantial harm to the individual, the

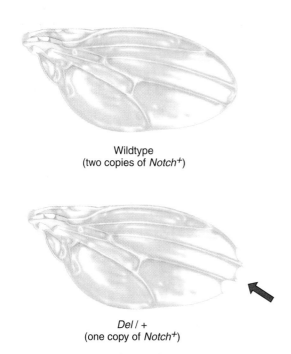

Wildtype
(two copies of *Notch⁺*)

Del / +
(one copy of *Notch⁺*)

Figure 12.3   Heterozygosity for deletions may have phenotypic consequences. Flies carrying only one copy of the *Notch⁺* gene instead of the normal two have abnormal wings.

(a)

(b)

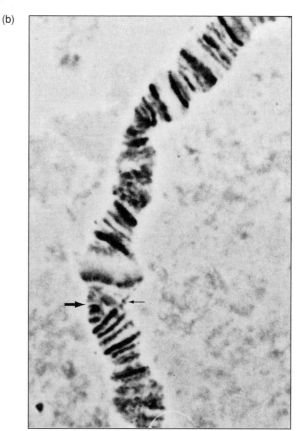

Figure 12.4   The meiotic consequences of deletions in deletion heterozygotes. (a) During prophase of the first meiotic division in a deletion heterozygote, the undeleted region of the normal chromosome has nothing with which to pair or recombine, and as a result, a deletion loop forms. Genes within the deletion loop are always transmitted to the next generation as a unit. In this simplified figure, each line represents two chromatids (since there are actually four chromatids at the first meiotic division). (b) Deletion loops also form in the polytene chromosomes of *Drosophila,* enabling researchers to determine very accurately which bands a deletion has removed. The thick arrow points to the wildtype chromosome in this *Del/+* animal; the thin arrow shows that the corresponding region is missing from the *Del* homolog.

genetic imbalance resulting from a single dose of many genes at the same time can be lethal. Humans, for example, cannot survive, even as heterozygotes, with deletions of more than about 3% of any part of their haploid genome.

There is another answer to the question of why heterozygosity for a deletion can have harmful consequences. If a somatic cell heterozygous for a deletion sustains a mutation in the remaining wildtype region of the homologous chromosome, the cell may become unable to generate the protein products of a gene in the deleted region. If this protein helps control cell division, the cell may divide out of control and generate a tumor. Thus, individuals born heterozygotes for certain deletions have a greatly increased risk of losing both copies of certain genes and developing cancer. Retinoblastoma (RB), the most malignant form of eye cancer, is a case in point (see the Genetics and Society Box in Chapter 4). Karyotypes of normal, noncancerous tissues from many people suffering from retinoblastoma reveal heterozygosity for deletions on chromosome 13. Cells from the retinal tumors of these same patients have a mutation in the remaining copy of the *RB* gene on the nondeleted chromosome 13. Chapter 17, "Cell-Cycle Regulation and the Genetics of Cancer" explains in detail how the loss of one homologous chromosomal region greatly increases the risk of cancer and how researchers have used this knowledge to locate and clone some of the genes whose mutant forms cause cancer.

Heterozygosity for Deletions Affects Mapping Distances

Because recombination between maternal and paternal homologs can occur only at regions of similarity, map distances derived from genetic recombination frequencies in deletion heterozygotes will be aberrant. For example, no recombination is

possible between genes *C, D,* and *E* in Fig. 12.4a because the DNA in this region of the normal, nondeleted chromosome has nothing with which to recombine. In fact, during the pairing of homologs in prophase of meiosis I, the "orphaned" region of the nondeleted chromosome forms a **deletion loop**—an unpaired bulge of the normal chromosome that corresponds to the area deleted from the other homolog. The progeny of a *Del/+* heterozygote will always inherit the markers in a deletion loop (*C, D,* and *E* in Fig. 12.4a) as a unit. As a result, these genes cannot be separated by recombination, and the map distances between them, as determined by the phenotypic classes in the progeny of

Fast Forward

Proper Development of the Immune System Depends on Programmed DNA Rearrangements

The human immune system is a marvel of specificity and diversity, including as it does close to a trillion lymphocytes of more than a billion different varieties. The lymphocytes, one category of white blood cell, come in two types: B cells, which make antibodies; and T cells, which coordinate and control immune responses through membrane-bound molecules, while also directly dispatching infectious agents. Both B cells and T cells respond to microorganisms by proliferating, communicating with other immune system cells, and gearing up to produce molecular missiles that subdue the invaders. To simplify our discussion, we focus on the maturation of B cells and the genetics of antibody formation, but variations of the mechanisms we describe also apply to T cells and the membrane-bound receptors they carry.

The Clonal Expansion of Specific B Cells Generates an Antibody Response Specifically Tailored to the Invader

The B cells of a normal human immune system produce antibodies of close to a billion different binding specificities. Each B cell, however, carries membrane-bound antibodies of only one specificity, which can bind to bacterial or viral proteins (called antigens in the context of immune responses) of complementary specificity. The binding of antibody to antigen causes the B cell carrying the membrane-bound antibodies to proliferate. The resulting clone of B cells matures generation by generation to a population containing two types of differentiated cells: dedicated antibody producers that every second secrete 1000 antibodies of the same binding specificity as the original membrane-bound antibodies, and memory cells that circulate in the blood and lymph ready to subdue in subsequent encounters antigens similar to those that caused the initial clonal expansion.

One intriguing question about antibody responses is: How can a genome containing only a hundred thousand (10^5) genes encode a billion (10^9) different types of antibodies? The answer, which emerged from a series of insightful studies, is that programmed gene rearrangements in conjunction with somatic mutations and the diverse pairing of polypeptides of different sizes can generate roughly a billion binding specificities from a much smaller number of genes. To understand the mechanism of this diversity, it is necessary to know how antibodies are constructed and how B cells come to express the genes determining specific antibody-binding sites.

The Genetics of Antibody Formation Produce Specificity and Diversity

Every antibody molecule consists of four polypeptides: two identical light and two identical heavy chains. Each of these four polypeptides has a constant (C) domain and a variable (V) domain. The C domain determines a chain's class (there are two classes of light chains and five classes of heavy chains), which in turn, helps determine where

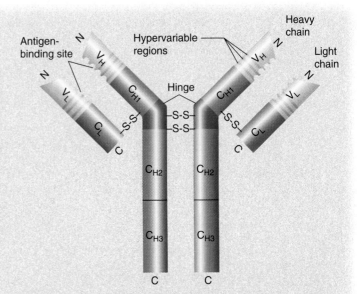

Figure A How antibody specificity emerges from molecular structure. Two heavy chains and two light chains are held together by disulfide bonds to form an antibody molecule. Both the light chain and the heavy chain have variable (V) domains near their N termini. The remainder of each chain is composed of a constant (C) domain; the C domain of the heavy chain can be subdivided into several subdomains (C_{H1}, hinge, C_{H2}, and C_{H3}). The antigen-binding site is generated by the association of the V domains of the heavy and light chains; some "hypervariable" stretches of amino acids within the V domains are particularly likely to vary between antibody molecules.

and how an antibody functions. The variable domains of light and heavy chains come together to form the antigen-binding site, which defines an antibody's specificity (Fig. A).

In each human B cell, the DNA for all domains of the heavy chain resides on chromosome 14. This heavy chain gene region consists of more than 100 V-encoding segments, each preceded by a promoter, several D (for diversity) segments, several J (for joining) segments, and nine C-encoding segments preceded by an enhancer (a short DNA segment that aids in the initiation of transcription by interacting with the promoter; see Chapter 16 for details). In embryonic B cells, these various gene segments lie far apart on the chromosome (Fig. B). During B-cell maturation, however, somatic rearrangements juxtapose random, individual V, D, and J segments together to form the particular variable region that will be transcribed, and place this newly-formed variable region next to a C segment and its enhancer. These rearrangements bring the promoter and enhancer into proximity, allowing transcription of the heavy chain gene. RNA splicing removes the introns from the primary transcript, making a mature mRNA encoding a complete heavy chain polypeptide.

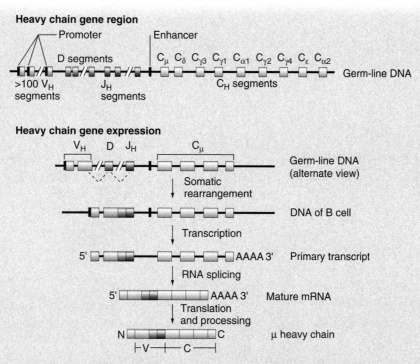

Figure B   The heavy-chain gene region on chromosome 14. The region contains more than 100 V_H segments, about 20 D segments, 6 J_H segments, and 9 C_H segments arranged in germline DNA (and in the DNA of non-antibody-producing cells) as shown at the top. Each V_H and C_H segment is composed of two or more exons, as seen in the alternate view of the same germline DNA on the next line. In antibody-producing B cells, somatic rearrangements bring together random, individual V_H, D, and J_H segments. The promoter in front of the selected V_H gene is now activated by the enhancer adjacent to C_μ, allowing transcription of the newly-constructed heavy chain gene into a primary transcript, which is subsequently spliced into a mature mRNA. Later in B cell development, other rearrangements (not shown) connect the same V-D-J variable region to other C_H segments such as C_δ, allowing the synthesis of other antibody classes.

The somatic rearrangements that shuffle the same deck of V, D, J, and C segments at random in each B cell of each individual permit expression of one, and only one, specific heavy chain. Without the rearrangements, antibody gene expression cannot occur. Random somatic rearrangements on human chromosome 2 (for one class of light chain) and human chromosome 22 (for the other light chain class) similarly generate the actual genes that will be expressed as light chains. The somatic rearrangements allowing the expression of antibodies thus generate enormous diversity of binding sites through the random selection and recombination of gene elements.

Several other mechanisms add to this diversity. First there is the imprecise joining of each gene's DNA elements, perpetrated by cutting and splicing enzymes that sometimes trim DNA from or add nucleotides to the junctions of the segments they join. Next, random somatic mutations in a rearranged gene's V region increase the variation of the antibody's V domain. Finally, in every B cell, two copies of a specific H chain that emerged from random DNA rearrangements combine with two copies of a specific L chain that emerged from random DNA rearrangements to create molecules with a specific, possibly unique, binding site. Each B cell will display antibodies of only this specificity in its membrane, and manufacture large quantities of these same antibodies when stimulated by antigen. The fact that any light chain can pair with any heavy chain exponentially increases the potential diversity of antibody types. For example, if there were 10^4 different light chains and 10^5 different heavy chains, there would be 10^9 possible combinations of the two.

Interestingly, the somatic rearrangements that lead to antibody expression may also be responsible for the phenomenon of "allelic exclusion" in which a cell expresses either the paternal or maternal allele of an antibody gene—but not both. One would expect each B cell to express several types of antibodies—some bearing a paternal H and L chain, some a maternal H and L chain, and some a mix of the two. But this rarely, if ever, happens. The "one cell, one antibody" phenomenon prevails; each B cell makes only one kind of H chain and one kind of L chain. Immunogeneticists believe that allelic exclusion is one consequence of the rearrangement process. They point out that many rearrangements are not successful because of the randomness of joining and the requirement that after joining, the downstream transcript must be in the proper reading frame. In B cells and T cells in which a successful rearrangement has occurred, the rearrangement apparatus shuts down, precluding the possibility of expression of more than one allele.

Neither somatic rearrangement nor allelic exclusion is seen in cells outside the immune system. However, both phenomena appear in the B cells and T cells of all mammals.

Mistakes by the Enzymes That Carry Out Antibody Gene Rearrangements Can Lead to Cancer

RagI and RagII are among the enzymes that interact with DNA sequences in antibody genes to catalyze the rearrangements required for gene expression. In carrying out their rearrangement activities,

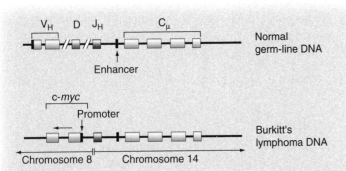

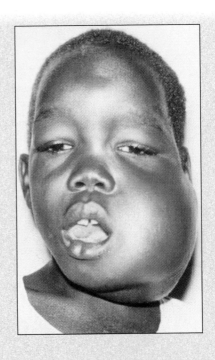

Figure C Misguided translocations can lead to Burkitt's lymphoma. In DNA from this Burkitt's lymphoma patient, translocations bring transcription of the c-*myc* gene under the control of the enhancer adjacent to C_μ. As a result, B cells produce abnormally high levels of the c-*myc* protein. The diagram suggests that the RagI and RagII enzymes have mistakenly connected a J_H segment to the c-*myc* gene region from chromosome 8, instead of to a D segment as would normally occur during B cell maturation (see Fig. B).

however, the enzymes sometimes make a mistake that results in a reciprocal translocation between human chromosomes 8 and 14. After this translocation, the enhancer of the chromosome 14 antibody gene lies in the vicinity of the unrelated c-*myc* gene from chromosome 8. Under normal circumstances, c-*myc* generates a transcription factor that turns on other genes active in cell division, at the appropriate time and rate in the cell cycle. The translocated antibody-gene enhancer accelerates expression of c-*myc*, causing B

cells containing the translocation to divide out of control. This uncontrolled B-cell division leads to a cancer known as Burkitt's lymphoma (Fig. C).

Thus, although programmed gene rearrangements contribute to the normal development of a healthy immune system, misfiring of the rearrangement mechanism can promote disease.

Chapter 24 describes the evolution of the gene families that encode antibodies and T-cell receptors.

---

a *Del*/+ individual, will be 0. In addition, the genetic distance between loci on either side of the deletion (such as between markers *B* and *F* in Fig. 12.4a) will be shorter than expected since fewer crossovers can occur between them.

Because homologous chromosomes pair in the polytene chromosomes of *Drosophila,* as they do at prophase I of meiosis in any diploid organism, deletion loops are also visible in the polytene chromosomes of a deletion heterozygote (Fig. 12.4b). The high resolution of the banding patterns in these unusual chromosomes makes it possible to pinpoint the region of the deletion in relation to missing bands. By contrast, deletion loops are not present in human karyotypes because homologous metaphase chromosomes do not pair with each other during mitosis. Geneticists can still determine the extent of a deletion in these chromosomes by observing which bands are missing from the *Del* chromosome, as described later.

Deletions in Heterozygotes Can "Uncover" Genes

A deletion heterozygote is, in effect, a hemizygote for genes on the normal, nondeleted chromosome that are missing from the deleted chromosome. If the normal chromosome carries a mutant recessive allele of one of these genes, the individual will exhibit the mutant phenotype. In *Drosophila,* for example, the *scarlet* (*st*) eye-color mutation is recessive to wildtype. However, a female heterozygous for the *st* mutation and a deletion that removes the *scarlet* gene (*st/Del*) will have bright scarlet eyes, rather than wildtype, dark red eyes (Fig. 12.5a). In these circumstances, the deletion "uncovers" (that is, reveals) the phenotype of the recessive mutation. Geneticists can look for recessive phenotypes to determine whether a deletion has removed a particular gene. If the phenotype of a recessive-allele/deletion heterozygote is mutant, the deletion has uncovered the mutated locus, which means that the gene lies inside the region of deletion. If the trait determined by the gene is wildtype, the deletion has not uncovered the recessive allele, and the gene must lie outside the deleted region.

Using Deletions to Locate Genes

Geneticists can use deletions detected through alterations in chromosomal banding patterns to map genes relative to specific regions of metaphase or polytene chromosomes. A deletion that results in the loss of one or more bands from a chromosome and also uncovers the recessive mutation of a particular gene places that gene within the missing chromosomal segment. Recessive mutations in the mouse *shaker-1* gene, for example, prevent homozygous animals from walk-

(a)

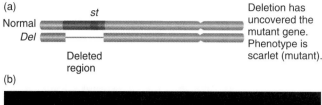

(b)

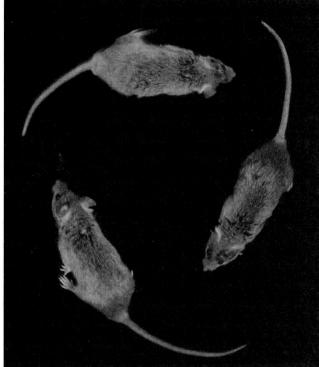

Figure 12.5   In deletion heterozygotes, pseudodominance shows that a deletion has removed a particular gene. (a) A fly of genotype *st/Del* (where *Del* is a deletion of the region in which the *scarlet [st]* gene normally lies) displays the recessive scarlet eye color. The deletion has thus "uncovered" the mutant gene. (b) Mice heterozygous for the *shaker-1* mutation and a deletion that uncovers the *shaker-1* gene display the mutant phenotype: they cannot walk in a straight line.

ing in a straight line and also cause deafness (Fig. 12.5b). Deletions that uncovered the mutant phenotypes placed the *shaker-1* gene in a 10-Mb region of chromosome 7.

The greater the number of distinguishable bands in a chromosome, the greater the accuracy of gene localization by this strategy. For this reason, the polytene chromosomes of *Drosophila,* which provide more than a 100-fold increase over metaphase chromosomes in the resolution of bands and interbands, are a prized mapping resource. If researchers determine that a small deletion removing only a few polytene chromosome bands uncovers a gene, or that several overlapping larger deletions affect the same gene, they can assign the gene to one or a small number of bands, often representing less than 100 kb of DNA. Figure 12.6a shows how *Drosophila* geneticists used this strategy to assign three genes to regions containing only one or two polytene chromosome bands on the *Drosophila* X chromosome.

Geneticists can use deletions analyzed at even higher levels of resolution to help locate genes on cloned fragments of

DNA. They must first determine whether a particular deletion uncovers a recessive allele of the gene of interest and then ascertain which DNA sequences are removed by the deletion. Several molecular techniques can show whether a particular DNA sequence is part of a deletion. Suppose you are trying to determine whether a small segment of the *Drosophila* X chromosome in the vicinity of the *white* gene has been deleted. You could use purified DNA fragments as probes for *in situ* hybridization to polytene chromosomes prepared from female flies containing various deletions in this region of both their X chromosomes. If a probe hybridizes to a *Del* chromosome, the deletion has not completely removed that particular fragment of DNA; lack of a hybridization signal, however, indicates that the fragment has been deleted (Fig. 12.6b).

Southern blot analysis of genomic DNAs from individuals with deleted chromosomes could supply similar information (Fig. 12.6c). If you probed DNA from a deletion heterozygote and compared the results with those obtained from normal DNA, you would observe a 50% decrease in band signal intensity with the deletion heterozygote if the probe hybridized to DNA within the deletion. Moreover, if one of the breakpoints of a deletion occurs within a restriction fragment homologous to the probe, a genomic Southern blot could reveal this by generating bands of unexpected sizes. This combination of genetic and molecular information can help pinpoint the DNA corresponding to genes of interest.

Duplications Add Material to the Genome

Duplications increase the number of copies of a particular chromosomal region. In **tandem** duplications, repeats of a region lie adjacent to each other, either in the same order or in reverse order (Fig. 12.7a). In **nontandem** (or *dispersed*) duplications, the two or more copies of a region are not adjacent to each other and may lie far apart on the same chromosome or on different chromosomes (Fig. 12.7a). Duplications arise by chromosomal breakage and faulty repair, unequal crossing-over, or errors in replication (Fig. 12.7b). In this book, we use *Dp* as the symbol for a chromosome carrying a duplication.

Most duplications have no obvious phenotypic consequences and can be detected only by cytological or molecular means. Sufficiently large duplications, for example, show up as repeated bands in metaphase or polytene chromosomes. During the prophase of meiosis I in heterozygotes for such duplications (*Dp/+*), the repeated bands form a **duplication loop**—a bulge in the *Dp*-bearing chromosome that has no similar region with which to pair in the unduplicated normal homologous chromosome. Duplication loops can occur in several alternative configurations (Fig. 12.7c). Such loops also form in the polytene chromosome of *Drosophila* duplication heterozygotes, where the pattern of the bands in the duplication loops is a repeat of that seen in the other copy of the same region elsewhere on the chromosome.

Duplications Can Affect Phenotype

Although duplications are much less likely to affect phenotype than are deletions of comparable size, some duplications do

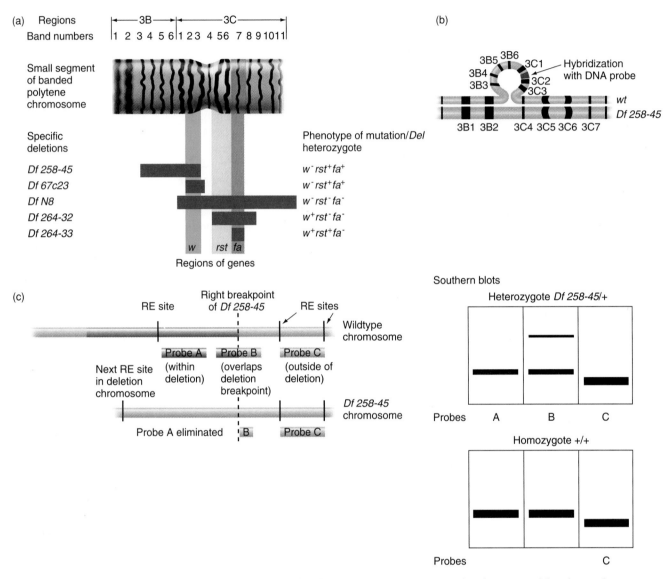

Figure 12.6 Various techniques for mapping genes with deletions. (a) Assigning genes to bands on *Drosophila* polytene chromosomes. Several deletions remove DNA from regions 3B and 3C near the tip of the X chromosome. For example, the deletion *Df258-45* is missing bands 3B3–3C3, while the deletion *Df67c23* lacks only bands 3C2 and 3C3. Complementation experiments determined whether these deletions uncovered any of three genes—*white* (*w*), *roughest* (*rst*), and *facet* (*fa*)—known to reside in this region. For example, *w/Df258-45* females have white eyes, while *rst/Df258-45* females are phenotypically wildtype: Thus, the *w* gene is removed by the deletion *Df258-45,* while the *rst* gene is not. With the analysis of many deletions, it is possible to assign the *w* gene to bands 3C2–3, the *rst* gene to bands 3C5–6, and the *fa* gene to band 3C7. (b) Using *in situ* hybridization to determine whether a specific DNA sequence is deleted. A labeled fragment of DNA from the 3C region serves as a probe of the polytene chromosomes of *Df258-45/+* heterozygotes. The probe hybridizes to the wildtype chromosome, but not to the *Df254-45* deletion chromosome, showing that the deleted chromosome lacks DNA homologous to the probe. (c) Using whole-genome Southern blots to locate a deletion breakpoint. Probes A, B, and C, prepared from genomic DNA of the 3C region, were hybridized to Southern blots of restriction enzyme (RE)-cut genomic DNA from either a *Df258-45/+* heterozygote or a wildtype (*+/+*) control. The hybridization pattern in the figure reveals that probe A sequences are deleted, while probe C sequences are not deleted. A novel band of lower intensity appears when *Df258-45/+* DNA is hybridized with probe B. Together these results show that the rightmost breakpoint of *Df258-45* is within the probe B genomic sequences.

have phenotypic consequences for visible traits or for survival, and these consequences signal the presence of the duplication. Duplications can produce a novel phenotype either by increasing the number of copies of a particular gene or set of genes, or by placing the genes bordering the duplication in a new chromosomal environment that alters their expression. These phenotypic consequences often arise even in duplication

heterozygotes (*Dp/+*). For example, *Drosophila* heterozygous for a duplication including the *Notch*[+] gene have abnormal wings that signal the three copies of *Notch*[+] (Fig. 12.8a); we have already seen that *Del/+* flies with only one copy of the *Notch*[+] gene have a different kind of wing abnormality. In another example from *Drosophila*, the locus known as *Triplolethal* (*Tpl*) is lethal when present in one or three doses in an otherwise

(a) Tandem duplication

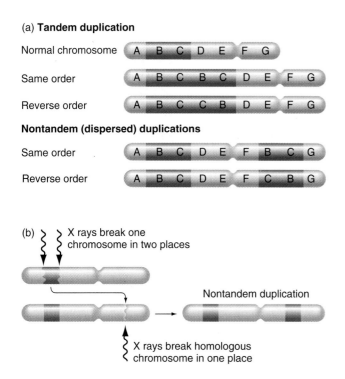

(c) Duplicated chromosome

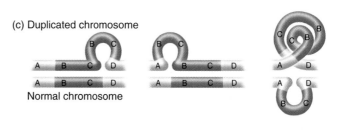

Figure 12.7 Duplications: Structure, origin, and detection. (a) The structure of duplications. For tandem duplications, the repeated regions lie adjacent to each other in the same or in reverse order. In nontandem duplications, the two copies of the same region lie far apart on the same chromosome, in the same relative order or in reverse order. (b) Duplications can arise in several ways. In one scenario, X rays break one chromosome twice and its homolog once. This allows a fragment of the first chromosome to insert into a different position on its homolog, producing a nontandem duplication. (c) Duplication loops form when chromosomes pair in duplication heterozygotes (*Dp*/+). During prophase of meiosis I, the duplication loop can assume several different configurations, each of which maximizes the pairing of related regions. (Note that a single line represents two chromatids in this simplified diagram.)

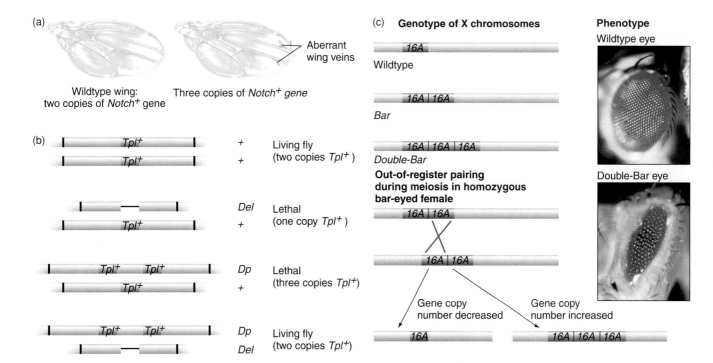

Figure 12.8 The genetic consequences of duplications. (a) Duplication heterozygotes (*Dp*/+) have three copies of genes contained in the duplication. This triple gene dose can affect visible phenotypes. An example is the aberrant wing veins seen in flies carrying three copies of the *Notch*+ gene. This phenotype differs from that caused by only one copy of *Notch*+ (see Fig. 12.3b). (b) With a very few genes, such as the *Tpl* gene in *Drosophila,* the animal must carry exactly two copies to survive; three copies or one copy of *Tpl*+ are lethal. (c) Unequal crossing-over between duplicated segments can increase or decrease copy number. Flies homozygous or hemizygous for a duplication of the X chromosome polytene region 16A have Bar eyes. Unequal pairing and crossing-over during meiosis in females homozygous for this duplication produces chromosomes that have either one copy of this region (conferring normal eye shape) or three copies of the region, causing the more abnormal double-Bar eyes.

diploid individual (Fig. 12.8b). Thus heterozygotes for a *Tpl* deletion (*Del/+*) or for a *Tpl* duplication (*Dp/+*) do not survive. Heterozygotes carrying one homolog deleted for the locus and the other homolog duplicated for the locus (*Del/Dp*) are viable because they have two copies of *Tpl.* Organisms are usually not so sensitive to additional copies of a gene, but just as for large deletions, imbalances for the many genes included in a very large duplication have additive deleterious effects that jeopardize survival. In humans, heterozygosity for duplications covering more than 5% of the haploid genome is most often lethal.

Unequal Crossing-Over between Duplications Increases or Decreases Gene Copy Number

In individuals homozygous for a tandem duplication (*Dp/Dp*), homologs carrying the duplications occasionally pair out of register during meiosis. Unequal crossing-over, that is, recombination resulting from such out-of-register pairing, generates gametes containing increases to three and reciprocal decreases to one in the number of copies of the duplicated region. In *Drosophila,* tandem duplication of several polytene bands near the X chromosome centromere produces the Bar phenotype of kidney-shaped eyes (Fig. 12.8c). *Drosophila* females homozygous for the Bar-eye duplication produce mostly Bar-eye progeny. Some progeny, however, have wildtype eyes, while other progeny have double-Bar eyes that are even smaller than Bar eyes. The genetic explanation is that flies with wildtype eyes carry X chromosomes containing only one copy of the region in question, flies with Bar eyes have X chromosomes containing two copies of the region, and flies with double-Bar eyes have X chromosomes carrying three copies. Unequal crossing-over in females homozygous for double-Bar chromosomes can yield progeny with more extreme phenotypes associated with four or five copies of the duplicated region. Duplications in homozygotes thus allow for the expansion and contraction of the number of copies of a chromosomal region from one generation to the next.

How Duplications and Deletions Affect Phenotype and Evolution: A Summary

Both duplications and deletions alter the number of genes on a chromosome and as a result may affect the phenotypes of heterozygotes. Heterozygosity for deletions or duplications produces one or three copies of a gene in an otherwise diploid organism. Such changes in gene dosage create an imbalance in gene products that can alter visible phenotypes; for a very few genes, the alterations in dosage are lethal to the organism. The deleterious effects of genetic imbalance are additive; heterozygotes for very large deletions or duplications of virtually any chromosomal region cannot survive. Both duplications and deletions can also alter phenotype by placing a gene in a new chromosomal location that modifies its expression, as discussed in Chapter 11.

Finally, deletions (by removing material from the genome) and duplications (by adding material to the genome) function as engines driving the evolution of the genome. In-

deed, the duplication of chromosomal segments followed by the expansion or contraction of the number of copies of the duplicated regions has played a significant role in the evolution of present-day genomes. Families of tandemly repeated genes, such as those for hemoglobin and color perception, arose in this way (see the comprehensive examples in Chapters 6 and 8). We discuss the molecular basis of this evolutionary mechanism in more detail in Chapter 24.

Inversions Reorganize the DNA Sequence of a Chromosome

The half-circle rotation of a chromosomal region known as an **inversion** can occur when radiation produces two double-strand breaks in a chromosome's DNA; the breaks release a middle fragment, which may turn 180° before religation to the flanking chromosomal regions, resulting in an inversion (Fig. 12.9a). Inversions may also result from rare crossovers between related DNA sequences present in two positions on the same chromosome in inverted orientation (Fig. 12.9b), or by the action of transposable genetic elements (discussed later). Inversions that include the centromere are **pericentric,** while inversions that exclude the centromere are **paracentric** (Fig. 12.9a).

Most inversions do not result in an abnormal phenotype, because even though they alter the order of genes along the chromosome, they do not add or remove DNA and therefore do not change the identity or number of genes. Geneticists can detect some inversions that do not affect phenotype, especially those that cause cytologically visible changes in banding patterns, or those that suppress recombination in heterozygotes (as described later) and thereby change the expected results of linkage analysis. In natural populations, however, many inversions that do not affect phenotype go undetected.

If one end of an inversion lies within the DNA of a gene (Fig. 12.9c), a novel phenotype can occur. Inversion following an intragenic break separates the two parts of the gene, relocating one part to a distant region of the chromosome, while leaving the other part at its original site. Such a split disrupts the gene's function. If that function is essential to viability, the inversion acts as a recessive lethal mutation, and homozygotes for the inversion will not survive.

Inversions can also produce unusual phenotypes by moving genes residing near the inversion breakpoints to chromosomal environments that alter their normal expression. For example, mutations in the *Antennapedia* gene of *Drosophila* that transform antennae into legs (review Fig. 6.23c) are inversions that place the gene in a new regulatory environment, next to sequences that cause it to be transcribed in tissues where it would normally remain unexpressed. Inversions that reposition genes normally found in a chromosome's euchromatin to a position near a region of heterochromatin can also produce an unusual phenotype; spreading of the heterochromatin may inactivate the gene in some cells, leading to position-effect variegation, as discussed in Chapter 11 (see Fig. 11.19).

(a) Pericentric inversion   Paracentric inversion

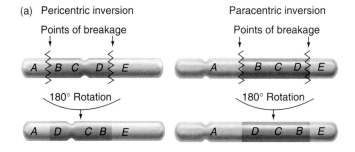

(b)

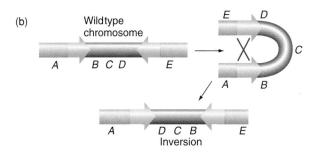

(c)

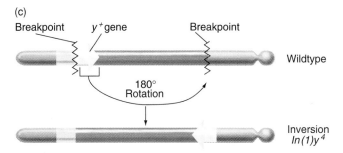

Figure 12.9 Inversions: Origins, types, and phenotypic effects. (a) Inversions can arise when agents such as X rays break a chromosome at two sites, and the DNA segment between the breaks rotates 180° before it reattaches to the remainder of the chromosome. If the rotated segment includes the centromere, the inversion is *pericentric*. If the rotated segment is completely on one arm of the chromosome and thus does not include the centromere, the inversion is *paracentric*. (b) If a chromosome has two copies of a sequence in reverse orientation with respect to each other, rare intrachromosomal recombinations between the two copies can give rise to an inversion. (c) An inversion can affect phenotype if it disrupts a gene. Here, the inversion $In(1)y^4$ has a breakpoint within the *Drosophila yellow* (*y*) gene. The inversion inactivates the *y* gene by dividing it in two, thereby preventing synthesis of a full-length transcript or protein. $In(1)y^4$ is thus a null recessive mutation of yellow, and produces light yellow rather than the usual tan body color.

Inversion Heterozygotes Reduce the Number of Recombinant Progeny

Cells heterozygous for an inversion (*In*/+) are *inversion heterozygotes.* In such cells, when a chromosome carrying an inversion pairs with its homolog at meiosis, formation of an **inversion loop** allows the tightest possible alignment of homologous regions. In an inversion loop, one chromosomal region rotates to conform to the similar region in the other

homolog (Fig. 12.10a). Crossing-over within an inversion loop produces aberrant recombinant chromatids whether the inversion is pericentric or paracentric.

If the inversion is pericentric and a crossover occurs within the inversion loop, each recombinant chromatid will have a single centromere—the normal number—but will carry a duplication of one region and a deletion of a different region (Fig. 12.10b). Gametes carrying these recombinant chromatids will have an abnormal dosage of some genes. After fertilization, zygotes created by the union of these abnormal gametes with normal gametes may die because of genetic imbalance.

If the inversion is paracentric and a crossover occurs within the inversion loop, the recombinant chromatids will be unbalanced not only in gene dosage, but also in centromere number (Fig. 12.10c). One crossover product will be an **acentric fragment** lacking a centromere; while the reciprocal crossover product will be a **dicentric chromatid** with two centromeres. Because the acentric fragment without a centromere cannot attach to the spindle apparatus during the first meiotic division, the cell cannot package it into either of the daughter nuclei; as a result, this chromosome is lost and will not be included in a gamete. By contrast, at anaphase of meiosis I, opposing spindle forces pull the dicentric chromatid toward both spindle poles at the same time with such strength that the dicentric chromatid breaks at random positions along the chromosome. These broken chromosome fragments are deleted for many of their genes. This loss of the acentric fragment and breakage of the dicentric chromatid results in genetically unbalanced gametes, which at fertilization, are lethal to the zygote's development. Consequently, no recombinant progeny resulting from a crossover in a paracentric inversion loop survive. Any surviving progeny are nonrecombinants.

In summary, whether an inversion is pericentric or paracentric, crossing-over within the inversion loop of an inversion heterozygote has the same effect: formation of recombinant gametes that after fertilization prevent the zygote from developing. Since only gametes containing chromosomes that did not recombine within the inversion loop can yield viable progeny, inversions act as **crossover suppressors.** This does not mean that crossovers do not occur within inversion loops, but simply that there are no recombinants among the viable progeny of an inversion heterozygote.

Translocations Attach Part of One Chromosome to Another Chromosome

Translocations are large-scale mutations in which part of one chromosome becomes attached to a nonhomologous chromosome, or in which parts of two different chromosomes trade places. This second type of translocation is known as a **reciprocal translocation** (Fig. 12.11a and b). It results when two breaks, one in each of two chromosomes, yield DNA fragments that do not religate to their chromosome of origin; rather, they switch places and become attached to the other chromosome. Depending on the positions of the breaks and the sizes of the exchanged fragments, the translocated chromosomes may be

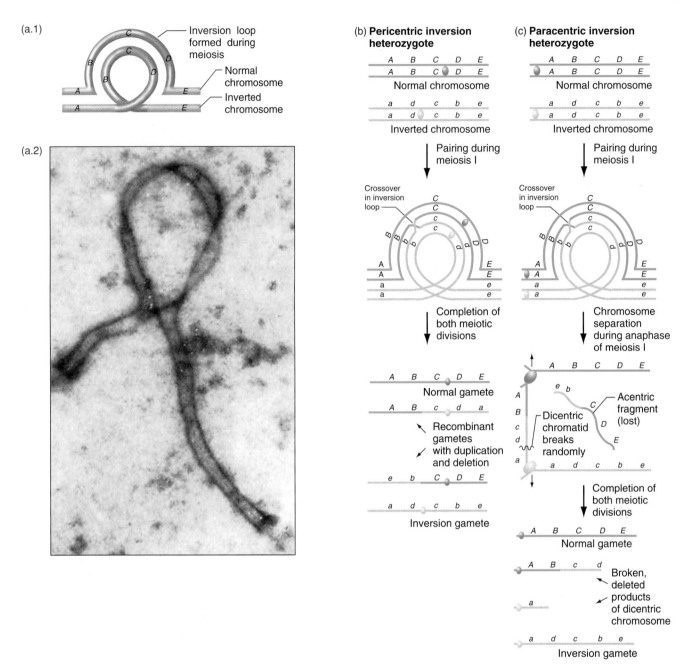

Figure 12.10   Why inversion heterozygotes produce few if any recombinant progeny. (a) To maximize pairing during prophase of meiosis I in an inversion heterozygote (*In/*+), homologous regions form an inversion loop. (1) Simplified schematic diagram in which one line represents a pair of sister chromatids. (2) Electron micrograph of an inversion loop during meiosis in an *In/*+ mouse. (b) The recombinant chromatids formed by recombination within the inversion loop of a pericentric inversion heterozygote are genetically unbalanced. (c) The recombinant chromatids formed by recombination within the inversion loop of a paracentric inversion heterozygote are not only genetically unbalanced but also contain two or no centromeres, instead of the normal one. Acentric fragments cannot connect to the spindle and are lost; dicentric chromatids break when pulled in both directions at anaphase of the first meiotic division. In parts (b) and (c), each line represents one chromatid, and different shades of green indicate the two homologous chromosomes.

so different from the original chromosomes that the translocation is visible in a cytological examination. **Robertsonian translocations,** for example, arise from breaks at or near the centromeres of two acrocentric chromosomes. The reciprocal exchange of broken parts generates one large metacentric chromosome and one very small chromosome containing few, if any, genes. This tiny chromosome may subsequently be lost

from the organism (Fig. 12.11c). Robertsonian translocations are named after W. R. B. Robertson, who in 1911 was the first to suggest that during evolution, metacentric chromosomes may arise from the fusion of two acrocentrics.

Most individuals bearing reciprocal translocations are phenotypically normal because they have neither lost nor gained genetic material. As with inversions, however, if one of

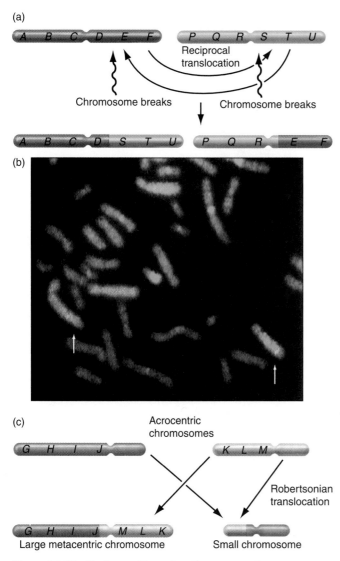

(a)

Reciprocal translocation

Chromosome breaks   Chromosome breaks

(b)

(c)

Acrocentric chromosomes

Robertsonian translocation

Large metacentric chromosome   Small chromosome

Figure 12.11 Reciprocal translocations are exchanges between nonhomologous chromosomes. (a) In a reciprocal translocation, the region gained by one chromosome is the region lost by the other chromosome. (b) Karyotype of a human genome containing a translocation. To produce the karyotype researchers used a "chromosome-painting" hybridization technique that renders particular chromosomes in different colors. Note that the two chromosomes involved in the reciprocal translocation are stained with two colors (red and green; *arrows*). Two other chromosomes are each stained entirely with one of these two colors, indicating that this person is heterozygous for the translocation. (c) In a Robertsonian translocation, reciprocal exchanges between two acrocentric chromosomes generate a large metacentric chromosome and a very small chromosome. The latter may be so small that it is invisible in karyotypes; sometimes it carries so few genes that its loss does not cause severe genetic imbalance.

the translocation breakpoints occurs within a gene, that gene's function may change or be destroyed. Or if the translocation places a gene normally found in the euchromatin of one chromosome near the heterochromatin of the other chromosome, normal expression of the gene may cease in some cells, giving rise to position-effect variegation (see Chapter 11).

Several kinds of cancer are associated with translocations in somatic cells. In normal cells, genes known as *proto-oncogenes* help control cell division. Translocations that relocate these genes can turn them into tumor-producing *oncogenes* whose protein products have an altered structure or level of expression that leads to runaway cell division. For example, in almost all patients with chronic myelogenous leukemia, a type of cancer caused by overproduction of certain white blood cells, the leukemic cells have a reciprocal translocation between chromosomes 9 and 22 (Fig. 12.12). The breakpoint in chromosome 9 occurs within an intron of a protooncogene called c-*abl*; the breakpoint in chromosome 22 occurs within an intron of the *bcr* gene. After the translocation, parts of the two genes are adjacent to one another. During transcription, the RNA-producing machinery runs these two genes together, creating a long RNA that after splicing is translated into a fused protein in which some 25 amino acids at the N terminal of the c-*abl*-determined protein are replaced by about 600 amino acids from the *bcr*-determined protein. The activity of this fused protein somehow releases the normal controls on cell division, leading to leukemia. (See the Fast Forward Box in this chapter for another example of a translocation-induced cancer.)

Medical practitioners can exploit the rearrangement of DNA sequences that accompany cancer-related translocations for diagnostic and therapeutic purposes. To confirm a diagnosis of myelogenous leukemia, for example, they obtain a blood sample from the patient, and then use a pair of PCR primers derived from opposite sides of the breakpoint—one synthesized from the appropriate part of chromosome 22, the other from chromosome 9—to carry out a polymerase chain reaction on DNA from the blood cells. The PCR will amplify the region between the primers only if the DNA sample contains the translocation (Fig. 12.12b). To monitor the effects of chemotherapy, they again obtain a blood sample and extract genomic DNA from the white blood cells. If the sample contains even a few malignant cells, a PCR test with the same two primers will amplify the DNA translocation from those cells, indicating the need for more therapy. PCR thus becomes a sensitive assay for this type of leukemic cell.

Heterozygosity for Translocations Diminishes Fertility and Results in Pseudolinkage

Translocations, like inversions, produce no significant genetic consequences in homozygotes if the breakpoints do not interfere with gene function. During meiosis in a translocation homozygote, chromosomes segregate normally according to Mendelian principles (Fig. 12.13a). Even though the genes have been rearranged, both haploid sets of chromosomes in the individual have the same rearrangement. As a result, all chromosomes will find a single partner with which to pair at meiosis, and there will be no deleterious consequences for the progeny.

In translocation heterozygotes, by contrast, certain patterns of chromosome segregation during meiosis produce genetically unbalanced gametes that at fertilization become deleterious to the zygote. In a translocation heterozygote, the two haploid sets of chromosomes do not carry the same

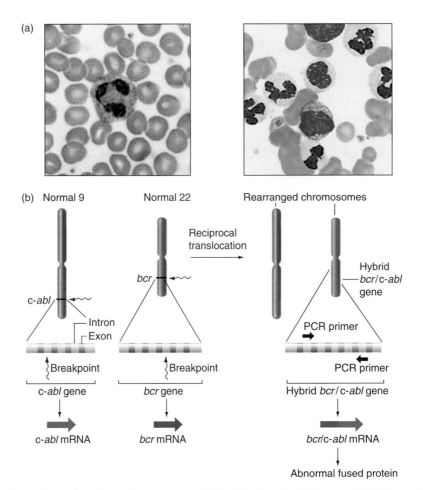

Figure 12.12 How a reciprocal translocation helps cause one kind of leukemia. (a) Uncontrolled divisions of large, dark-staining white blood cells in the blood of a leukemia patient (right) produce a higher than normal ratio of white to red blood cells than that of a normal individual (left). (b) A reciprocal translocation between chromosomes 9 and 22 contributes to chronic myelogenous leukemia. This rearrangement makes an abnormal hybrid gene composed of part of the c-*abl* gene on chromosome 9 and part of the *bcr* gene on chromosome 22. The hybrid gene produces a mRNA with sequences from both c-*abl* and *bcr,* and this hybrid mRNA is translated into an abnormal fused protein that disrupts controls on cell division. Black arrows indicate PCR primers that will generate a PCR product only in DNA containing the hybrid gene.

arrangement of genetic information. As a result, during prophase of the first meiotic division, the translocated chromosomes and their normal homologs assume a crosslike configuration in which four chromosomes, rather than the normal two, pair to achieve a maximum of synapsis between similar regions (Fig. 12.13b). To keep track of the four chromosomes participating in this crosslike structure, we denote the chromosomes carrying translocated material with a *T,* and the chromosomes with a normal order of genes with an *N.* Chromosomes *N1* and *T1* have homologous centromeres found in wildtype on chromosome 1; *N2* and *T2* have centromeres found in wildtype on chromosome 2.

During the anaphase of meiosis I, the mechanisms governing attachment to the spindle of the chromosomes in this crosslike configuration still usually ensure the disjunction of homologous centromeres, bringing homologous chromosomes to opposite spindle poles (that is, *T1* and *N1* go to opposite poles, as do *T2* and *N2*). Depending on the arrangement of the four chromosomes on the metaphase plate, this normal disjunction of homologs produces one of two patterns of seg-

regation (Fig. 12.13c). In the **alternate segregation pattern,** the two translocation chromosomes (labeled *T1* and *T2* in the figure) go to one pole, while the two normal chromosomes (*N1* and *N2*) move to the opposite pole. Both kinds of gametes resulting from this segregation (*T1,T2* and *N1,N2*) carry the correct haploid number of genes; and the zygotes formed by union of these gametes with a normal gamete will be viable. By contrast, in the **adjacent-1 segregation pattern,** homologous centromeres disjoin so that *T1* and *N2* go to one pole, while the *N1* and *T2* go to the opposite pole. As a result, each gamete contains a large duplication (of the region found in both the normal and the translocated chromosome in that gamete) and a correspondingly large deletion (of the region found in neither of the chromosomes in that gamete), which make them genetically unbalanced. Zygotes formed by union of these gametes with a normal gamete are usually not viable.

Moreover, because of the unusual crosslike pairing configuration in translocation heterozygotes, nondisjunction of homologous centromeres occurs at a measurable but low rate. This nondisjunction produces an **adjacent-2 segregation pattern** in

(a) Segregation in a translocation homozygote

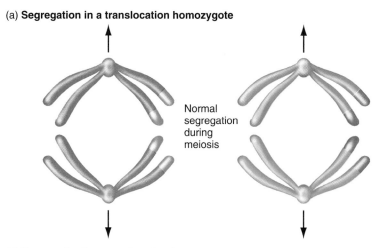

Normal segregation during meiosis

(b) Chromosome pairing in a translocation heterozygote

(c) Segregation in a translocation heterozygote

Segregation pattern	Alternate		Adjacent - 1		Adjacent - 2	
	Balanced $N1 + N2$	Balanced $T1 + T2$	Unbalanced $T1 + N2$	Unbalanced $N1 + T2$	Unbalanced $N1 + T1$	Unbalanced $N2 + T2$
Gametes	a b c d e f	p q r s t u	A B C D S T U	P Q R E F	a b c d e f	P Q R s t u	a b c d e f	A B C D S T U	p q r s t u	P Q R E F
Type of progeny when mated with normal *abcdefpqrstu* homozygote	*abcdef pqrstu*	*ABCDEF PQRSTU*	None surviving	None surviving	None surviving	None surviving

(d)

Figure 12.13 **The meiotic segregation of chromosomes that have sustained reciprocal translocations.** In all parts of this figure, each bar represents one chromatid. (a) In a translocation homozygote (T/T), chromosomes segregate normally during meiosis I because each chromosome can pair with only one other chromosome. (b) In a translocation heterozygote ($T/+$) resulting from the same translocation as in Fig. 12.11a, the four relevant chromosomes assume a cruciform configuration to maximize pairing. The alleles of genes on chromosomes in the original order ($N1$ and $N2$) are shown in lower case; the alleles of the same genes on the translocated chromosomes ($T1$ and $T2$) are in capital letters. (c) There are three possible segregation patterns in a translocation heterozygote from the cruciform configuration: alternate and adjacent-1 patterns, which occur with equally high frequency as homologous centromeres disjoin properly, and the adjacent-2 pattern, which occurs less often because it involves nondisjunction of homologous chromosomes. Only the alternate segregation pattern gives rise to balanced gametes. As a result, genes that were originally on two different chromosomes will behave as if they are genetically linked. This is because in the absence of recombination, the only gametes producing viable progeny are *abcdefpqrstu* or *ABCDEFPQRSTU*, not *abcdefPQRSTU* or *ABCDEFpqrstu*, which contrasts with expectations based on Mendel's law of independent assortment. (d) Phenotypic consequences of semisterility in corn. This ear of corn comes from a plant heterozygous for a reciprocal translocation. It has fewer kernels than normal ears because of the abortion of unbalanced ovules.

which homologous centromeres such as *N1* and *T1* go to the same spindle pole (Fig. 12.13c). The resulting genetic imbalances are lethal after fertilization to the zygotes containing them.

Thus, of all the gametes generated by translocation heterozygotes, only those arising from alternate segregation, which is slightly less than half the total, can produce viable progeny when crossed with individuals who do not carry the translocation. As a result, the fertility of most translocation heterozygotes, that is, their capacity for generating viable offspring, is diminished by at least 50%, leading to a condition known as

semisterility. Corn plants illustrate the correlation between translocation heterozygosity and semisterility: The demise of genetically unbalanced ovules produces gaps in the ear where kernels would normally appear (Fig. 12.13d); in addition, genetically unbalanced pollen grains are abnormally small.

The semisterility of translocation heterozygotes undermines their potential for independent assortment. Mendel's second law requires that gametes resulting from both possible metaphase alignments of two chromosomal pairs produce viable progeny. But as we have seen, in a translocation heterozygote, only the alternate segregation pattern yields viable progeny in outcrosses; the equally likely adjacent-1 pattern and the rare adjacent-2 pattern do not. Because of this, genes near the translocation breakpoints on the nonhomologous chromosomes participating in a reciprocal translocation behave as if they are linked (Fig. 12.13c).

Translocations Can Sometimes Help Map Important Genes
In humans, approximately 1 of every 500 individuals is heterozygous for some kind of translocation. While most such people are phenotypically normal, their fertility is diminished by a high frequency of spontaneous abortion among their progeny. Nonetheless, some people carrying a heterozygous translocation show some exceptions to the patterns of semisterility described earlier.

The genetic imbalances associated with gametes formed by adjacent-1 segregation in a translocation heterozygote are usually but not always lethal; if the duplicated or deleted regions are very small, lethality may not result. Individuals heterozygous for a reciprocal translocation involving chromosome 21, for example, are phenotypically normal but produce some gametes from the adjacent-1 segregation pattern that have two copies of a part of chromosome 21 near the tip of its long arm (Fig. 12.14). At fertilization, if a gamete with the duplication unites with a normal gamete, the resulting child will have three copies of this region of chromosome 21. A few individuals affected by Down syndrome have, in this way, inherited a third copy of only a small part of chromosome 21.

These individuals with **translocation Down syndrome** provide evidence that the entirety of chromosome 21 need not be present in three copies to generate the phenotype.

Geneticists are now mapping the chromosome 21 regions duplicated in translocation Down syndrome patients to find the one or more genes responsible for the syndrome. Although chromosome 21 is the smallest human autosome, it nevertheless contains an estimated 2000 genes, most of them in the 43 million base pairs of its long arm. The mapping of genes relative to the breakpoints of one or more translocations considerably simplifies the task of identifying those genes that in triplicate produce the symptoms of Down syndrome. One way to locate which parts of chromosome 21 are responsible for Down syndrome is to obtain cloned sequences from chromosome 21 from a genome project and then use these clones as FISH (fluorescence *in situ* hybridization, described in Chapter 10) probes for the genome of the translocation Down syndrome patient. If the probe lights up the translocation chromosome as well as the two normal copies of chromosome 21 (see Fig. 12.14), it identifies a region of the genome that is of potential importance to the syndrome.

Translocations and Inversions Have Several Effects in Common: A Summary
Reciprocal translocations, like inversions, bring together pieces of DNA that were not adjacent before the rearrangement, without altering the amount of DNA in the genome. Both reciprocal translocations and inversions may affect gene

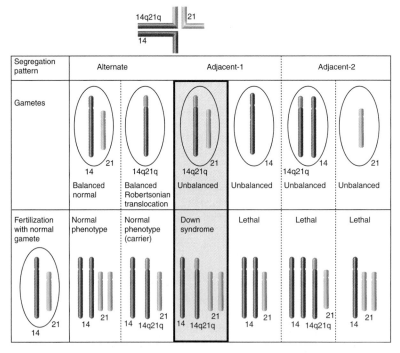

Figure 12.14   How translocation Down syndrome arises. In heterozygotes for a translocation involving chromosome 21, such as 14q21q, the Robertsonian translocation between chromosomes 21 and 14 shown here, the adjacent-1 segregation pattern can produce gametes with two copies of part of chromosome 21. If such a gamete unites at fertilization with a normal gamete, the resulting zygote will have three copies of part of chromosome 21. Depending on which region of chromosome 21 is present in three copies, this tripling may cause Down syndrome.

function and as a result phenotype if one of the breakpoints of the rearrangement is within a gene, or if the rearrangement places a gene in a chromosomal environment that modifies its expression. Homozygosity or heterozygosity for translocations and inversions that do not affect gene function is usually without obvious phenotypic effect. However, heterozygosity for both types of rearrangements can produce genetically imbalanced gametes that are deleterious to the zygotes they help create. Inversion heterozygotes produce genetically imbalanced gametes from crossing-over within an inversion loop; this results in crossover suppression. Translocation heterozygotes produce genetically imbalanced gametes from two of three possible meiotic segregation patterns; this leads to semi-sterility. Finally, translocations and inversions, by reducing production of viable progeny and heterozygotes, can be catalysts of speciation, as we explain later in the comprehensive example.

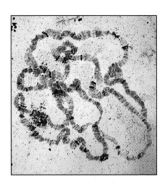
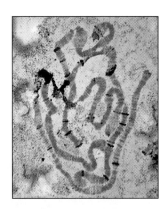

Figure 12.15   Transposable elements can move to many locations in a genome. A probe for the *copia* transposable element hybridizes to multiple sites that differ in two different fly strains.

Transposable Genetic Elements Move from Place to Place in the Genome

Large deletions and duplications, as well as inversions and translocations, are major chromosomal reorganizations visible at the relatively low resolution of a karyotype. Small deletions and duplications are lesser chromosomal reorganizations that reshape genomes without any visible effect on karyotype. Another type of cytologically invisible sequence rearrangement with a significant genomic impact is **transposition:** the movement of small segments of DNA—entities known as **transposable elements**—from one position in the genome to another.

Marcus Rhoades in the 1930s and Barbara McClintock in the 1950s inferred the existence of transposable elements from intricate genetic studies of corn. At first, the scientific community did not appreciate the importance of their work because their findings did not support the conclusion from classical recombination mapping that genes are located at fixed positions on chromosomes. Once the cloning of transposable elements made it possible to study them in detail, geneticists acknowledged their existence. In 1983 Barbara McClintock received the Nobel prize for her work on movable genetic elements in corn.

Copia is a transposable element in *Drosophila.* If you examined the polytene chromosomes from two strains of flies isolated from different geographic locations, you would find in general that the chromosomes appear identical. A probe derived from the *white* gene for eye color, for example, would hybridize to a single site near the tip of the X chromosome in both strains (review Fig. 11.17). However, a probe including the *copia* transposable element would hybridize to 30–50 sites scattered throughout the genome, and the overall positions of *in situ* hybridization would not be the same in the two strains. Some sites would be identical in the two polytene sets, but others would be different (Fig. 12.15). These observations suggest that since the time the strains were separated geo-

graphically, the *copia* sequences have moved around (transposed) in different ways in the two genomes even though most genes have remained in a fixed position.

Transposable elements are found in virtually all organisms, from bacteria to humans. Any segment of DNA that evolves the ability to move from place to place within a genome is by definition a transposable element, regardless of its origin or function. Transposable elements need not be sequences that do something for the organism; rather, many scientists regard them as "selfish" entities carrying only information that allows their self-perpetuation. Most transposable elements in nature range from 50 bp to approximately 10,000 bp (10 kb) in length. A particular transposable element can be present in a genome anywhere from one to hundreds of thousands of times. *Drosophila melanogaster,* for example, harbors some 80 different transposable elements, each an average of 5 kb in length, and each present an average of 50 times. These transposable genetic elements constitute $80 \times 50 \times 5 = 20,000$ kb, or roughly 12.5% of the 160,000 kb *Drosophila* genome. Mammals carry two major classes of transposable elements: **LINES,** or long interspersed elements; and **SINES,** or short interspersed elements. The human genome contains approximately 20,000 copies of the main human LINE—*L1*—which is up to 6.4 kb in length. The human genome also carries 300,000 copies of the main human SINE—*Alu*—which is 0.28 kb in length (Fig. 12.16a). These two transposable elements thus constitute roughly 7% of the 3,000,000-kb human genome. Because some transposable elements exist in only one or a few closely related species, it is probable that some elements arise and then disappear rather frequently over evolutionary time. (Chapter 24 describes the evolutionary origins of LINES and SINES.)

Classification of transposable genetic elements on the basis of how they move around the genome distinguishes two groups. **Retroposons** transpose via reverse transcription of an RNA intermediate. The *Drosophila copia* elements and the

(a)

0.28 kb *Alu* units: ~300,000 found
dispersed throughout human genome
at ~10-kb intervals

kb 0   10   20   30   40   50

(b)

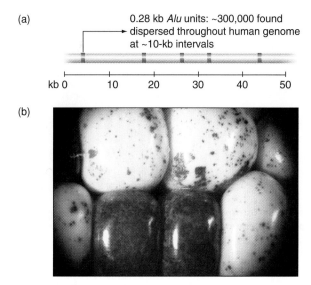

Figure 12.16   Transposable elements in human and corn genomes. (a) The human genome carries about 300,000 copies of the 0.28-kb *Alu* retroposon, the major human SINE; these are spaced, on average, about 10 kb apart. The longer transposable elements called LINEs are, on average, spaced further apart (not shown). (b) Movements of a transposon mottles corn kernels when the transposon jumps into or out of genes that influence pigmentation.

human SINEs and LINEs described earlier are retroposons. **Transposons** move their DNA directly without the requirement of an RNA intermediate. Genetic elements in corn responsible for mottling the kernels are transposons (Fig. 12.16b). Some biologists use the term "transposon" in the broader sense to refer to all transposable genetic elements. In this text, we reserve it for the direct-movement class of genetic elements, and use "transposable elements" to indicate all DNA segments that move about in the genome, regardless of mechanism.

Retroposons: Transcription Generates an RNA That Encodes a Reverse-Transcriptase-Like Enzyme

This enzyme with reverse-transcriptase activity, like the reverse transcriptase described in the Genetics and Society Box in Chapter 7 and in Chapter 8, can copy RNA into a single strand of cDNA and then use that single DNA strand for producing a double strand of cDNA. Many retroposons also encode polypeptides other than reverse transcriptase.

Some retroposons have a poly-A tail at the 3′ end of the RNA-like DNA strand, a configuration reminiscent of mRNA molecules (Fig. 12.17a). Other retroposons end in *long terminal repeats* (*LTR*s), nucleotide sequences repeated in the same orientation at both ends of the element (Fig. 12.17a). The structure of this second type of retroposon is similar to the integrated DNA copies of RNA tumor viruses (known as retroviruses), suggesting that retroviruses evolved from this kind of retroposon or vice versa. In support of this notion, re-

searchers sometimes find retroposon transcripts enclosed in virus-like particles.

The structural parallels between retroposons, mRNAs, and retroviruses and the fact that retroposons encode a reverse-transcriptase-like enzyme prompted investigators to ask whether retroposons move around the genome via an RNA intermediate. Experiments in yeast helped confirm that they do. In one study, a copy of the *Ty1* retroposon found on a yeast plasmid contained an intron in one of its genes; after transposition into the yeast chromosome, however, the intron was not there (Fig. 12.17b). Since removal of introns occurs only during mRNA processing, researchers concluded that the *Ty1* retroposon passes through an RNA intermediate during transposition.

The mechanisms by which various retroposons move around the genome resemble each other in general outline but differ in detail. Figure 12.17c outlines what is known of the process for the better understood LTR-containing retroposons. As the figure illustrates, one outcome of transposition via an RNA intermediate is that the original copy of the retroposon remains in place while the new copy inserts into another location. With this mode of transmission, the number of copies can increase rapidly with time. Human LINEs and SINEs, for example, occur in tens of thousands or even hundreds of thousands of copies within the genome. Other retroposons, however, such as the *copia* elements found in *Drosophila*, do not proliferate so profusely and exist in much more moderate copy numbers of 30–50. Currently unknown mechanisms may account for these differences by regulating the rate of retroposon transcription or by limiting the number of copies through selection at the level of the whole organism.

Transposons Encode Transposase Enzymes That Catalyze the Events of Transposition

A hallmark of transposons is that their ends are inverted repeats of each other, that is, a sequence of base pairs at one end is present in mirror image at the other end (Fig. 12.18a). The inverted repeat is usually 10–200 bp long.

DNA between the inverted repeats usually contains one gene encoding a transposase, a protein that catalyzes transposition. The exact sequence of the steps resulting in transposition is not yet known, but as Fig. 12.18a illustrates, the process includes excision of the transposon from its original genomic position and integration into a new location. The double-stranded breaks at the transposon's excision and integration sites are repaired in different ways in different cases. Figure 12.18b shows one of the possibilities so far observed. In *Drosophila*, after excision of a transposon known as a *P element*, DNA exonucleases first widen the resulting gap and then repair it using either a sister chromatid or a homologous chromosome as a template. If the template contains the *P* element and DNA replication is completely accurate, repair will restore a *P* element to the position from which it was excised;

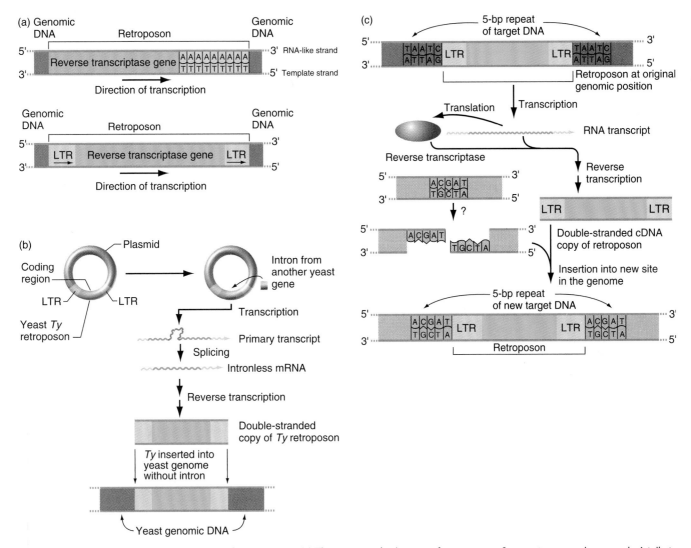

Figure 12.17   Retroposons: Structure and movement. (a) There are two basic types of retroposons. Some retroposons have a poly-A tail at the 3′ end of the RNA-like DNA strand (top). Other retroposons are flanked on both sides by long terminal repeats (LTRs; bottom). Complete retroposons of both kinds encode reverse transcriptase as well as one or more additional proteins. (b) Experiments show that retroposons move via an RNA intermediate. In one study, researchers added an intron from an unrelated yeast gene to a *Ty1* retroposon contained on a plasmid. This plasmid was then transformed into yeast cells that did not contain a *Ty1* retroposon. When new insertions of the *Ty1* element into yeast genomic DNA were isolated, they did not have the intron. This implies that the *Ty1* retroposon on the plasmid was transcribed into a primary transcript that was spliced into an intronless mRNA. Reverse transcriptase then copied the mRNA into a DNA sequence that inserted into the yeast genome. (c) Movement of the LTR-containing yeast retroposon *Ty1*. The reverse-transcriptase-like enzyme encoded by *Ty1* elements in the cell helps synthesize double-stranded *Ty1* cDNA in a series of steps. When this double-stranded cDNA inserts into a new genomic location (shown in blue), it generates a 5-bp duplication of DNA sequences at the target site. These 5 base pairs were present only once prior to the insertion, but afterward are found on both sides of the inserted retroposon. (Other kinds of retroposons cause the duplication of a different number of base pairs at the target site.) Scientists speculate that integration of the retroposon involves the introduction of a staggered cleavage of the target site, leaving "sticky ends." Polymerization to fill in the sticky ends would eventually produce two copies of the target site (not shown).

this will make it appear as if the *P* element remained at its original location during transposition. If the template does not contain a *P* element, the transposon will be lost from the original site after transposition.

Some strains of *D. melanogaster* are called "P strains" because they harbor many copies of the *P* element; "M strains" of the same species do not carry the *P* element. Virtually all commonly used laboratory flies are M strains, while many

flies isolated from natural populations since 1950 are P strains. Since Morgan and coworkers isolated the flies that have proliferated into most laboratory strains in the early part of the twentieth century, these observations suggest that *P* elements did not enter *D. melanogaster* genomes until around 1950. The prevalence of *P* elements in many contemporary natural populations attests to the rapidity with which transposable elements can spread once they are introduced into a species.

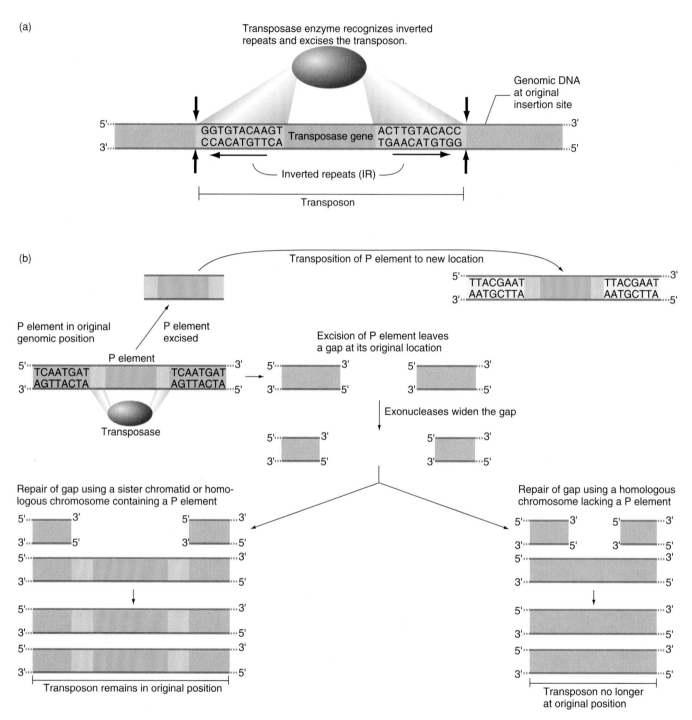

Figure 12.18 Transposons: Structure and movement. (a) Most transposons contain inverted repeats at their ends. Different kinds of transposons have different inverted repeats. Complete transposons encode a transposase enzyme that recognizes the inverted repeats at the end of the same element; some transposons also contain other genes. The transposase enzyme makes cuts at the borders between the transposon and adjacent genomic DNA sequences. These cuts excise the transposon, which then, with the help of transposase, integrates at a new site. (b) Movement of *Drosophila P* elements. Transposase-catalyzed integration of *P* elements creates a duplication of 8 bp present at the target site. This number of duplicated base pairs varies for different kinds of transposon. A gap remains when transposons are excised from their original position. After exonucleases widen the gap, cells repair the gap using related DNA sequences as templates. The molecular machinery involves strand breakage and invasion reminiscent of the mechanisms underlying recombination described in Chapter 5. Depending on whether the template for repair contains or lacks a *P* element, the transposon will appear to remain or to be excised from that location during transposition. Mistakes during the repair process can create defective transposons or cause the rearrangement of nearby chromosomal information (not shown).

Genomes Contain Defective Copies of Transposable Elements

Many copies of transposable elements sustain deletions, as a result of the transposition process itself (for example, incomplete reverse transcription of a retroposon RNA), or as a result of events following transposition (faulty repair of an excision site). If a deletion removes the promoter needed for transcription of a retroposon, that copy of the element cannot generate the RNA intermediate for future movements. If the deletion removes one of the inverted repeats at one end of a transposon, transposase will be unable to catalyze transposition of that element. Such deletions create defective genetic elements unable to transpose again. Most SINEs and LINEs are defective in this way. Other types of deletions create defective elements that are unable to move on their own, but can move if nondefective copies of the element elsewhere in the genome supply the deleted function. For example, a deletion inactivating the reverse transcriptase gene in a retroposon or the transposase gene in a transposon would "ground" that copy of the element at one genomic location if it is the only source of the essential enzyme in the genome. If reverse transcriptase or transposase were provided by other copies of the same element in the genome, however, the defective copy could move. Defective elements that require the activity of nondeleted copies of the same element for movement are called **nonautonomous elements;** the nondeleted copies that can move by themselves are **autonomous elements.**

Transposable Elements Can Generate Mutations in Adjacent Genes

Insertion of a transposable element near or within a gene can affect gene expression and alter phenotype. We now know that the wrinkled pea mutation first studied by Mendel resulted from insertion of a transposable element into the gene for starch-branching enzyme I. In *Drosophila*, a large percentage of spontaneous mutations, including the w^1 mutation discovered by T. H. Morgan, are caused by insertion of transposable elements (Fig. 12.19). Surprisingly, in light of the large numbers of LINEs and SINEs in human genomes, only a handful of mutant human phenotypes are known to result from insertion of transposable elements. Among these is a B type hemophilia caused by *Alu* insertion into a gene encoding clotting factor IX.

A transposable element's effect on a gene depends on what the element is and where within or near the gene it inserts (Fig. 12.19). If an element lands within a protein-coding exon, the additional DNA may shift the reading frame or supply an in-frame stop codon that truncates the polypeptide. If the element falls in an intron, it could diminish the efficiency of splicing. Some of these inefficient splicing events might completely remove the element from the gene's primary transcript; this would still allow some—but less than normal—synthesis of functional polypeptide. Elements that land within exons or introns may also provide a transcription stop signal that prevents transcription of gene sequences downstream of

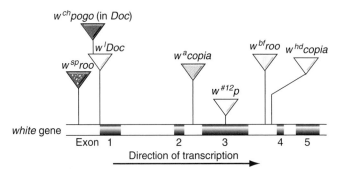

Figure 12.19 Transposable elements can cause mutations on insertion into a gene. Many spontaneous mutations in the *white* gene of *Drosophila* arise from insertions of transposable elements such as *copia, roo, pogo,* or *Doc.* The consequences for the eye color phenotype (indicated by the color in the triangles) are often hard to predict and depend on the element involved and where in the gene it inserts. Note, for example, that the mutation wch contains an insertion of a *pogo* element into the *Doc* element causing w^1, but wch has a darker eye color than w^1.

the insertion site. Finally, insertions into regions that regulate transcription, such as promoters, can influence the amount of gene product made in particular tissues at particular times during development. Some transposons insert preferentially into the upstream regulatory regions of genes and some even prefer specific types of genes, such as tRNA genes.

Transposable Genetic Elements Can also Generate Chromosomal Rearrangements

Retroposons and transposons can trigger spontaneous chromosomal rearrangements other than transpositions in several ways. Sometimes, deletion or duplication of chromosomal material adjacent to the transposon occurs as a mistake during the transposition event itself. In another mechanism, if two copies of the same transposable element occupy nearby but not identical sites in homologous chromosomes, the two copies of the transposable element in heterozygotes carrying both types of homolog may pair with each other and crossover (Fig. 12.20a). The recombination resulting from this unequal cross-over would produce one homolog deleted for the region between the two transposable elements and a reciprocal homolog with a tandem duplication of the same region. The duplication associated with the *Bar* mutation in *Drosophila* (Fig. 12.8c) apparently arose in this way.

Transposition Can Relocate Genes

When two copies of a transposon occur in nearby but not identical locations on the same chromosome, the transposon-inverted repeats are positioned such that an inverted version of the sequence at the 5' end of the copy on the left will exist at the 3' end of the copy to its right (Fig. 12.20b). If transposase acts on this pair of inverted repeats during transposition, it allows the entire region between them to move as one giant transposon, relocating any genes the region contains. Some

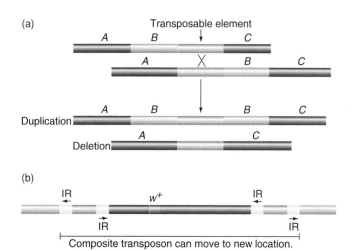

Figure 12.20 How transposable elements generate chromosomal rearrangements and relocate genes. (a) If a heterozygote has copies of a transposable element in slightly different locations on homologous chromosomes (here shown on opposite sides of segment *B*), unequal crossing-over between the elements will produce reciprocal deletions and duplications of the intervening region. (b) If two copies of the same transposon are nearby on the same chromosome, transposase can recognize the outermost inverted repeats (IRs); if it does, the entire unit between these repeats will behave as a composite transposon, allowing intervening genes such as w^+ to jump to new locations.

composite transposons carry as much as 400 kb of DNA. In prokaryotes, the capacity of two transposable elements to relocate the intervening genes helps mediate the transfer of drug resistance between different strains or species of bacteria (see Chapter 13).

Summary of Transposition

Although pieces of DNA that can move from place to place in the genome might enhance or diminish a genome's ability to respond to a changing environment, it is helpful to think of transposable genetic elements as segments of "selfish" DNA that exist for their own sake. The movement of transposable elements has three main genetic consequences.

1. The insertion of both retroposons and transposons can mutate a gene; precise removal of the element from the gene then causes a reversion to wildtype.

2. Side effects of the transposition process itself or unequal crossing-over between copies of the same transposable element can help generate chromosomal rearrangements.

3. Recognition by transposase of two nearby transposons on the same chromosome can relocate the intervening genes.

Because of these genetic consequences, transposons make a major contribution to the evolution of genomes.

Rearrangements and Evolution: A Speculative Comprehensive Example

We saw at the beginning of this chapter that roughly 170 chromosomal rearrangements could reshape the human genome to a form that resembles the mouse genome. This suggests two things. First, although most chromosomal variations, including single-base changes and chromosomal rearrangements, are deleterious to an organism or its progeny, a few are either neutral or provide an advantage for survival and manage to become fixed in a population. Second, some rearrangements contribute to the processes underlying speciation. We still do not know enough of the details to describe the precise rearrangements that separate the mouse and human genomes. It is nonetheless useful to consider here some general examples of how chromosomal rearrangements might contribute to evolution.

Deletions A small deletion that moves a coding sequence of one gene next to a promoter or other regulatory element of an adjacent gene may, rarely, allow expression of a protein at a novel time in development or in a novel tissue. If the new time or place of expression is advantageous to the organism, it might become established in the genome.

Duplications An organism cannot normally tolerate mutations in a gene essential to its survival, but duplication would provide two copies of the gene. If one copy remained intact to perform the essential function, the other would be free to evolve a new function. The genomes of most higher plants and animals, in fact, contain many **gene families**—sets of closely related genes with slightly different functions that most likely arose from a succession of gene duplication events. In vertebrates, some *multigene families* have hundreds of members.

Inversions Suppose one region of a chromosome has three mutations that together greatly enhance the reproductive fitness of the organism. In heterozygotes where one homolog carries the mutations and the other does not, recombination could undo the beneficial linkage. If, however, the three mutations are part of an inversion, crossover suppression will ensure that they remain together as they spread through the population.

Translocations Suppose that a flood or other natural disaster separates two small populations of a species. After the separation, a translocation occurs and becomes fixed in one of the populations, so that now one population is homozygous for the normal gene order and the other is homozygous for the translocated gene order. If the two populations subsequently resume contact, heterozygotes formed by matings between individuals from each population would be semisterile, providing a partial breeding barrier between the two populations. New mutations favoring matings between homozygotes (for example, a behavioral mutation that makes one population seek mates in the spring and the other seek mates in the fall) would minimize the risk of producing semisterile progeny. The rapid spread of such mutations as a result of improved fertility would increase the breeding barrier between the populations and thus contribute to their eventual separation into different species.

Transpositions   Movement of transposable genetic elements may cause novel mutations, a small proportion of which might be selected for because they are advantageous to the organism. Transposable elements can also help generate potentially useful duplications and inversions.

Because of these possibilities, many geneticists speculate that chromosomal rearrangements and transpositions are, like point mutations, important instruments of evolution.

CHANGES IN CHROMOSOME NUMBER

We have seen that in peas, *Drosophila,* and humans, normal diploid individuals carry a $2n$ complement of chromosomes, where n is the number of chromosomes in the gametes. All the chromosomes in the haploid gametes of these diploid organisms are different from one another. In this section, we examine two types of departure from chromosomal diploidy found in eukaryotes: (1) aberrations in usually diploid species that generate cells or individuals whose genomes contain one to a few chromosomes more or less than the normal $2n$, for example, $2n + 1$ or $2n - 1$, and (2) species whose genomes contain complete but nondiploid sets of chromosomes, for example, $3n$ or $4n$.

The Loss or Gain of One or More Chromosomes Results in Aneuploidy

Individuals whose chromosome number is not an exact multiple of the haploid number (n) for the species are **aneuploids** (review Table 12.1 on page 421). Individuals lacking one chromosome from the diploid number ($2n - 1$) are **monosomic,** while individuals having one chromosome in addition to the normal diploid set ($2n + 1$) are **trisomic.** Organisms with four copies of a particular chromosome ($2n + 2$) are **tetrasomic.**

Autosomal Aneuploidy Is Harmful to the Organism
Monosomy, trisomy, and other forms of aneuploidy create a genetic imbalance that is usually deleterious to the organism. In humans, monosomy for any autosome is generally lethal; medical geneticists have reported a few cases of monosomy for chromosome 21, one of the smallest human chromosomes; although born with severe multiple abnormalities, these monosomic individuals survived beyond birth. Similarly, trisomies involving a human autosome are also highly deleterious. Individuals with trisomies for larger chromosomes, such as 1 and 2, are almost always aborted spontaneously early in pregnancy. Trisomy 18 causes Edwards syndrome, and trisomy 13 causes Patau syndrome; both phenotypes include gross developmental abnormalities that result in early death.

The most frequently observed human autosomal trisomy, trisomy 21, results in Down syndrome. As one of the shortest human autosomes, chromosome 21 contains about 1.5% of the DNA in the human genome. Although there is considerable phenotypic variation among Down syndrome individuals, traits such as mental retardation and skeletal abnormalities are usually associated with the condition. Many Down syndrome babies die in their first year after birth from heart defects and increased susceptibility to infection. We saw earlier (in the discussion of translocations) that some people with Down syndrome have three copies of only part of, rather than the entire, chromosome 21. It is thus probable that genetic imbalance for only one or a few genes may be a sufficient cause of the condition.

Aneuploidy for the X Chromosome Is Often Not Lethal to the Individual Because of the Dosage Compensation Resulting from X Inactivation
Although the X chromosome is one of the longest human chromosomes and contains 5% of the DNA in the genome, individuals with X chromosome aneuploidy, such as XXY males, XO females, and XXX females, survive quite well compared with aneuploids for the larger autosomes. The explanation for this tolerance of X-chromosome aneuploidy is that X-chromosome inactivation and other dosage-compensation mechanisms equalize the expression of most X-linked genes in individuals with different numbers of X chromosomes. As we saw in Chapter 11, X-chromosome inactivation represses expression of most genes on all but one X chromosome in a cell. As a result, even if the number of X chromosomes varies, the amount of protein generated by most X-linked genes remains constant. Nonetheless, human X-chromosome aneuploidies are not without consequence. XXY men have Klinefelter syndrome, and XO women have Turner syndrome. The aneuploid individuals affected by these syndromes are usually infertile and display skeletal abnormalities, leading in the men to unusually long limbs and in the women to unusually short stature.

If X inactivation were 100% effective, we would not expect to see even the relatively minor abnormalities of Klinefelter syndrome, because the number of functional X chromosomes—one—would be the same as in normal individuals. One explanation is that during X inactivation, several genes near the telomere and centromere of the short arm of the human X chromosome escape inactivation and thus remain active. As a result, XXY males make twice the amount of protein encoded by these genes as XY males (Fig. 12.21a).

The reverse of X inactivation is X *reactivation,* which occurs in normal female germ line cells that will develop into oocytes. When the diploid germ line cells start to develop into oocytes and enter meiosis, reactivation of the previously inactivated X chromosome enables every egg cell to receive an active X. The phenomenon of X reactivation in these cells could help explain the infertility of women with Turner syndrome. With X reactivation, the developing oocytes in normal females have two functional doses of X chromosome genes; but the corresponding cells in Turner women have only one dose of the same genes.

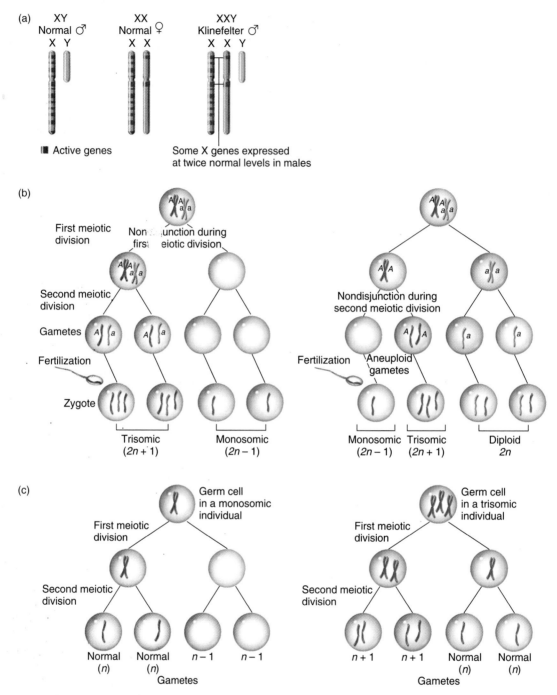

Figure 12.21 Aneuploidy: Causes and consequences. (a) Because X chromosome inactivation does not affect all genes on the X chromosome, aneuploidy for the X chromosome can have phenotypic consequences. In an XXY Klinefelter male, for example, a few X chromosome genes are expressed inappropriately at twice their level in normal males. (b) Aneuploidy can arise from nondisjunction during either meiotic division. If trisomic progeny inherit two different alleles (*A* and *a*) of a centromere-linked gene from one of their parents, the nondisjunction occurred in the first meiotic division. If the two alleles inherited from one parent are the same (*A* and *A*; or *a* and *a*), the nondisjunction occurred during the second meiotic division. (c) Because aneuploids carry chromosomes that have no homolog with which to pair, aneuploid individuals frequently produce aneuploid progeny.

Aneuploidy Results from Meiotic Nondisjunction

How does aneuploidy arise? Mistakes in chromosome segregation during meiosis produce aneuploids of different types, depending on when the mistakes occur. If homologous chromosomes do not separate, or disjoin, during the first meiotic division, two of the resulting haploid gametes will carry both homologs, and two will carry neither. Union of these gametes with normal gametes will produce aneuploid zygotes, half monosomic, half trisomic (Fig. 12.21b). By contrast, if meiotic nondisjunction occurs during meio-

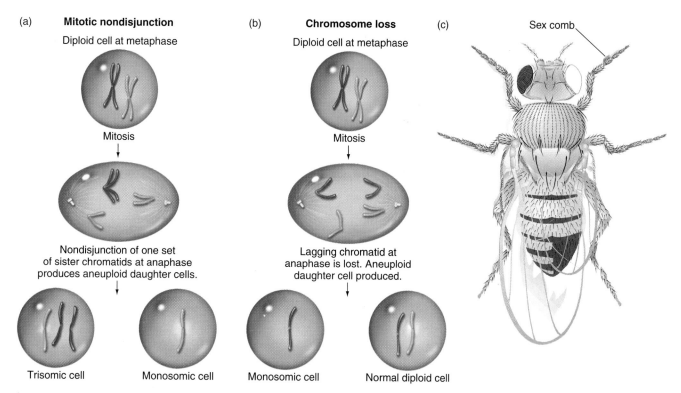

(a) **Mitotic nondisjunction**

Diploid cell at metaphase

Mitosis

Nondisjunction of one set
of sister chromatids at anaphase
produces aneuploid daughter cells.

Trisomic cell   Monosomic cell

(b) **Chromosome loss**

Diploid cell at metaphase

Mitosis

Lagging chromatid at
anaphase is lost. Aneuploid
daughter cell produced.

Monosomic cell   Normal diploid cell

(c) Sex comb

Figure 12.22   Mistakes during mitosis can generate clones of aneuploid cells. Mitotic nondisjunction (a) or chromosome loss during mitosis (b) can create monosomic or trisomic cells. Further rounds of division of the aneuploid cell generate a clone. (c) If one of the two X chromosomes is lost during the first mitotic division after fertilization in *Drosophila,* one of the daughter cells will be XX (female) while the other will be XO (male). Development of this embryo will yield a gynandromorph. In this case, the zygote was w^+m^+/wm, so the XX half of the fly has red eyes and normal wings; loss of the w^+m^+ X chromosome gives the XO half of the fly white eyes (w) and miniature wings (m).

sis II, only two of the four resulting gametes will be aneuploid (Fig. 12.21b). Abnormal $n + 1$ gametes resulting from nondisjunction in a cell that is heterozygous for alleles on the nondisjoining chromosome will be heterozygous if the nondisjunction happens in the first meiotic division, but homozygous if the nondisjunction takes place in the second meiotic division. (We assume here that no recombination has occurred between the heterozygous gene in question and the centromere, as would be the case for genes closely linked to the centromere.) It is possible to use this distinction to determine when a particular nondisjunction occurred (review Fig. 12.21b). Interestingly, the nondisjunction events in women that give rise to Down syndrome seem to occur more often during the first meiotic division than during the second. By contrast, in the small number of cases in which the nondisjunction event leading to Down syndrome takes place in men, the reverse is true.

If an aneuploid individual survives and is fertile, the incidence of aneuploidy among his or her offspring will generally be extremely high. This is because half of the gametes produced by meiosis in a monosomic individual lack the chromosome in question, while half of the gametes produced in a trisomic individual have an additional copy of the chromosome (Fig. 12.21c).

*Aneuploid Mosaics Can Result from Mitotic
Nondisjunction or Chromosome Loss*

As a zygote divides many times to become a fully formed organism, mistakes in chromosome segregation during the mitotic divisions accompanying this development may, in rare instances, augment or diminish the complement of chromosomes. In **mitotic nondisjunction,** the failure of two sister chromatids to separate during mitotic anaphase generates reciprocal trisomic and monosomic daughter cells (Fig. 12.22a). Other types of mistakes, such as a lagging chromatid not pulled to either spindle pole at mitotic anaphase, result in a **chromosome loss** that produces one monosomic daughter cell (Fig. 12.22b).

Aneuploid cells arising from either mitotic nondisjunction or chromosome loss may survive and undergo further rounds of cell division, producing clones of cells with an abnormal chromosome count. Nondisjunction or chromosome loss occurring early in development will generate larger aneuploid clones than the same events occurring later in development. The side-by-side existence of aneuploid and normal tissues results in a **mosaic** organism whose phenotype depends on what tissue bears the aneuploidy, the number of aneuploid cells, and the specific genes on the aneuploid chromosome. Many examples of mosaicism involve the sex chromosomes. If an XX *Drosophila* female loses one of the X chromosomes during the first mitotic division after fertilization, the result is a **gynandromorph** composed of equal parts male and female tissue (Fig. 12.22c).

Interestingly, in humans many Turner syndrome females are mosaics carrying some XX cells and some XO cells. These individuals began their development as XX zygotes, but with the loss of an X chromosome during the embryo's early mitotic divisions, they acquired a clone of XO cells. Similar mosaicism involving the autosomes also occurs. For example, physicians have recorded several cases of mild Down syndrome arising from mosaicism for trisomy 21. In people with Turner or Down mosaicism, the existence of some normal tissue appears to ameliorate the condition, with the individual phenotype depending on the distribution of the two types of cells.

Euploid Species Contain Complete but Nondiploid Sets of Chromosomes

In contrast to aneuploids, **euploid** cells contain only complete sets of chromosomes. Most euploid species are diploid, but some euploid species are **polyploids** that carry three or more complete sets of chromosomes (see Table 12.1 on page 421). When speaking of polyploids, geneticists use the symbol x to indicate the **basic chromosome number,** that is, the number of different chromosomes that make up a single complete set. Triploid species, which have three complete sets of chromosomes are then $3x$; tetraploid species with four complete sets of chromosomes are $4x$; and so forth. For diploid species, x is identical to n (the number of chromosomes in the gametes), because each gamete contains a single complete set of chromosomes. This identity of $x = n$ does not hold for polyploid species as the following example illustrates. Commercially grown bread wheat has a total of 42 chromosomes: 6 nearly (but not wholly) identical sets each containing 7 different chromosomes. Bread wheat is thus a hexaploid with a basic number of $x = 7$ and $6x = 42$. But each triploid gamete has one-half the total number of chromosomes, so $n = 21$. Thus, for bread wheat, x and n are not the same. Another form of euploidy, in addition to polyploidy, exists in monoploid (x) organisms, which have only one set of chromosomes.

Monoploidy and polyploidy are rarely observed in animals. Among the few examples of monoploidy are some species of ants and bees in which the males are monoploid, while the females are diploid. Males of these species develop parthenogenetically from unfertilized eggs, and these males produce gametes through a modified mitosis. Polyploidy in animals normally exists only in species with unusual reproductive cycles, such as hermaphroditic earthworms, which carry both male and female reproductive organs, and goldfish, which are parthenogenetically tetraploid species. In *Drosophila*, it is possible, under special circumstances, to produce triploid and tetraploid females, but never males. In humans, polyploidy is always lethal, usually resulting in abortion during the first trimester of pregnancy.

Monoploid Organisms Contain a Single Copy of Each Chromosome and Are Usually Infertile

Botanists can produce monoploid plants experimentally by special treatment of germ cells from diploid species that have completed meiosis and would normally develop into pollen. (Note that monoploid plants obtained in this manner can also be considered haploids because $x = n$.) The treated cells divide into a mass of tissue known as an *embryoid*. Subsequent exposure to plant hormones enables the embryoid to develop into a plant (Fig. 12.23a). Monoploid plants may also arise from rare spontaneous events in a large natural population. Most monoploid plants, no matter how they originate, are infertile. Since the chromosomes have no homologs with which to pair during meiosis I, they are distributed at random to the two spindle poles during the first division. Rarely do all chromosomes go to the same pole, and if they do not, the resulting gametes are defective as they lack one or more chromosomes. The greater the number of chromosomes in the genome, the lower the likelihood of producing a gamete containing all of them.

Despite such gamete-generating problems, monoploid plants and tissues are of great value to plant breeders. They make it possible to visualize normally recessive traits directly, without crosses to achieve homozygosity. Plant researchers can also introduce mutations into individual monoploid cells, select for desirable phenotypes, such as resistance to herbicides, and use hormone treatments to grow the selected cells into monoploid plants (Fig. 12.23b). They can then convert monoploids of their choice into homozygous diploid plants by treating tissue with *colchicine,* an alkaloid drug obtained from the autumn crocus. By binding to tubulin—the major protein component of the spindle—colchicine prevents formation of the spindle apparatus. In cells without a spindle, the sister chromatids cannot segregate after the centromere splits, so that there is often a doubling of the chromosome set following treatment with colchicine (Fig. 12.23c). The resulting diploid cells can be grown into diploid plants that will express the desired phenotype and produce fertile gametes.

Polyploidy Has Accompanied the Evolution of Many Cultivated Plant Species

Roughly one out of every three known species of flowering plants is a polyploid, and because polyploidy often increases plant size and vigor, many polyploid plants with edible parts have been selected for agricultural cultivation. Most commercially grown alfalfa, coffee, and peanuts are tetraploids ($4x$). MacIntosh apple and Bartlett pear trees that produce giant fruits are also tetraploids. Commercially grown strawberries are octaploids ($8x$) (Fig. 12.24). The evolutionary success of polyploid plant species may stem from the fact that polyploidy, like gene duplication, provides additional copies of genes; while one copy continues to perform the original function, the others become available for evolution to new functions. However, as we now show, the consequences of polyploidy depend in large part on whether the number of chromosome sets is even or odd.

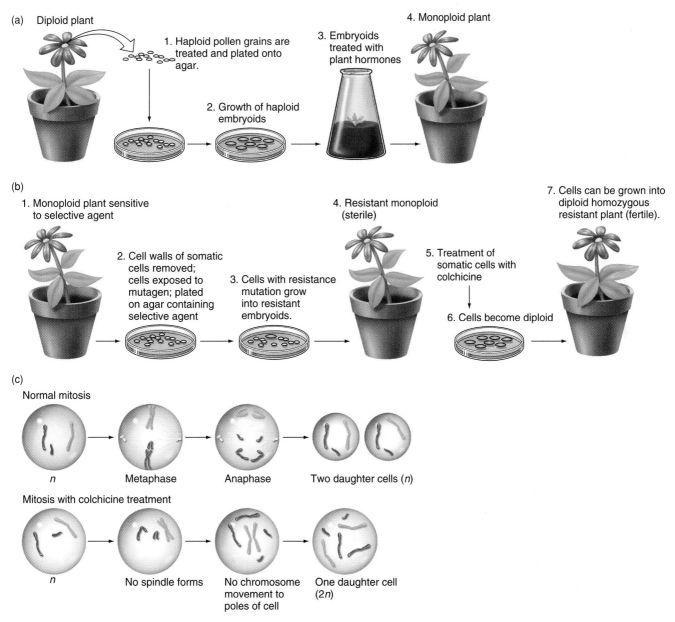

Figure 12.23   The creation and use of monoploid plants. (a) Making a monoploid plant. Under certain conditions, haploid pollen grains can grow into haploid masses of tissues called embryoids. When treated properly with plant hormones, haploid embryoids will grow into monoploid plants. (b) Somatic cells obtained from a monoploid plant are haploid, so researchers can use them to select for recessive traits expressed at the level of the cell, such as resistance to a toxin. The selected cells can then be grown into a resistant embryoid that (with hormone treatment) eventually makes a mature, resistant monoploid plant. Treating some tissue of this plant with colchicine allows a doubling of the chromosome number in a small number of cells. The resulting diploid cells can be grown in culture with hormones to make a homozygous resistant diploid plant. (c) Colchicine prevents the formation of the mitotic spindle and also blocks cytokinesis. Colchicine treatment thus doubles the number of chromosome sets in the cell. The original cell in this figure is haploid, with $n = 3$; these three chromosomes are thus nonhomologous.

Triploids Are Almost Always Sterile

Triploids ($3x$) result from the union of monoploid (x) and diploid ($2x$) gametes (Fig. 12.25a). The diploid gametes may be the products of meiosis in tetraploid ($4x$) germ cells, or the products of rare spindle or cytokinesis failures during meiosis in a diploid.

Sexual reproduction in triploid organisms is extremely inefficient because meiosis produces mostly unbalanced gametes. During the first meiotic division in a triploid germ cell,

three sets of chromosomes must segregate into two daughter cells; regardless of how the chromosomes align in pairs, there is no way to ensure that the resulting gametes obtain a complete, balanced x or $2x$ complement of chromosomes. In most cases, at the end of anaphase I, two chromosomes of any one type move to one pole, while the remaining chromosome of the same type moves to the opposite pole. The products of such a meiosis have two copies of some chromosomes and one

Figure 12.24 Many polyploid plants are larger than their diploid counterparts. A comparison of octaploid (left) and diploid (right) strawberries.

(a) **Formation of a triploid organism**

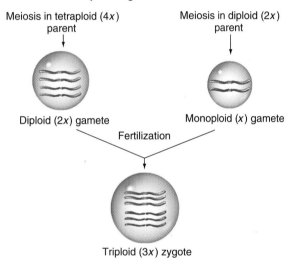

(b) **Meiosis in a triploid organism**

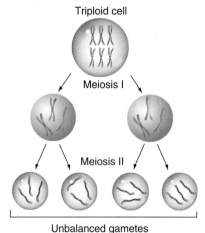

Figure 12.25 The genetics of triploidy. (a) Production of a triploid ($x = 2$) from fertilization of a monoploid gamete by a diploid gamete. Nonhomologous chromosomes are either blue or red. (b) Meiosis in a triploid produces unbalanced gametes because at the end of meiosis I, the two daughter cells cannot receive equal numbers of any one type of chromosome. If x is large, balanced gametes with equal numbers of all the chromosomes would be very rare.

copy of others (Fig. 12.25b). If the number of chromosomes in the basic set is large, the chance of obtaining any balanced gametes at all is remote. Thus, fertilization with triploid gametes does not produce many viable offspring.

It is possible to propagate some triploid species, such as bananas and watermelons, through asexual reproduction. The fruits of triploid plants are seedless because the unbalanced gametes do not function properly in fertilization or, if fertilization occurs, the resultant zygote is so genetically unbalanced that it cannot develop. Either way, no seeds form. Like triploids, all polyploids with odd numbers of chromosome sets are sterile because they cannot reliably produce balanced gametes.

Tetraploids Are Often the Source of New Species

During mitosis, if the chromosomes in a diploid tissue fail to separate after replication, the resulting daughter cells will be tetraploid (Fig. 12.26a). If such tetraploid cells arise in reproductive tissue, subsequent meioses will produce diploid gametes. Rare unions between diploid gametes produce tetraploid organisms. Self-fertilization of a newly created tetraploid organism can produce an entirely new species, because crosses between the tetraploid and the original diploid organism will produce infertile triploids. Tetraploids made in this fashion are **autopolyploids,** a kind of polyploid that derives all its chromosome sets from the same species.

Maintenance of a tetraploid species depends on the production of gametes with balanced sets of chromosomes. Most successful tetraploids have evolved mechanisms ensuring that the four copies of each group of homologs pair two by two to form two **bivalents**—pairs of synapsed homologous chromosomes (Fig. 12.26b). Because the chromosomes in each bivalent become attached to opposite spindle poles during meiosis I, meiosis regularly produces gametes carrying two complete sets of chromosomes. The mechanism requiring that each chromosome pair with only a single homolog suppresses other pairing possibilities, such as a 3-to-1, which cannot guarantee equivalent chromosome segregation.

Tetraploids, with four copies of every gene, generate unusual Mendelian ratios. For example, even if there are only two alleles of a gene (say, A and a), five different genotypes are possible: $AAAA$, $AAAa$, $AAaa$, $Aaaa$, $aaaa$. If the phenotype depends on the dosage of A, then five phenotypes, each corresponding to one of the genotypes, will appear. The segregation of alleles during meiosis in a tetraploid is similarly complex. Consider an $AAaa$ heterozygote in which the A gene is closely linked to the centromere, and the A allele is completely dominant. What are the chances of obtaining progeny with the recessive phenotype, generated by only the $aaaa$ genotype? As Fig. 12.26c illustrates, if during meiosis I, the four chromosomes carrying the gene align at random in bivalents along the metaphase plate, the expected ratio of gametes is $2AA:8Aa:2aa = 1AA:4Aa:1aa$. The chance of obtaining $aaaa$ progeny during self-fertilization is thus $1/6 \times 1/6 = 1/36$. In other words, since A is completely dominant, the ratio of dominant to recessive phenotypes, determined by the ratio of A--- to $aaaa$ genotypes is 35:1. The ratios will be different

(a) **Generation of tetraploid (4x) cells**

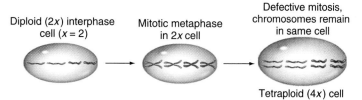

(b) **Pairing of chromosomes as bivalent**

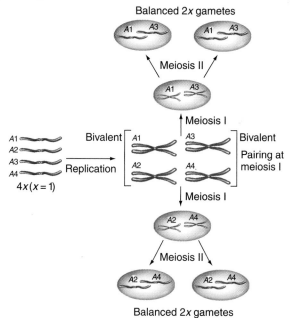

(c)

Chromosomes	Pairing	Gametes produced by random spindle attachment
1. ___A___ 2. ___A___	1 ↑___A___ 3 ↑___a___ 2 ___ 4 ___ ↓A ↓a	1 + 3 Aa 1 + 4 Aa 2 + 4 Aa or 2 + 3 Aa
3. ___a___ 4. ___a___	1 ↑___A___ 2 ↑___A___ 3 ___ 4 ___ ↓a ↓a	1 + 2 AA 1 + 4 Aa 3 + 4 aa or 2 + 3 Aa
	1 ↑___A___ 2 ↑___A___ 4 ___ 3 ___ ↓a ↓a	1 + 2 AA 1 + 3 Aa 3 + 4 aa or 2 + 4 Aa

Total:
8Aa : 2AA : 2aa = 4Aa : 1AA : 1aa

Figure 12.26 The genetics of tetraploidy. (a) Tetraploids arise from a failure of chromosomes to separate into two daughter cells during mitosis in a diploid. (b) In successful tetraploids, the pairing of chromosomes as bivalents generates genetically balanced gametes. (c) Ratios of gametes produced in a tetraploid that is heterozygous for two alleles of a gene linked to the centromere, with orderly pairing of bivalents. There are three possible ways the four chromosomes can pair to form two bivalents. For each pairing scheme, the chromosomes in the two pairs can assort with respect to each other in two different orientations just as in Mendel's law of independent assortment for meiosis in a diploid. Only one way is shown in the figure. For example, the pairing scheme at the bottom (1 pairs with 4; 2 pairs with 3) can produce either AA and aa gametes in the orientation shown, or two Aa gametes in the orientation not shown. If all pairing schemes and chromosome orientations are equally likely, the expected genotype frequency in a population of gametes will be 4Aa:1AA:1aa.

if the gene is not closely linked to the centromere or if the dominance relationship between the alleles is not so simple.

New levels of polyploidy can arise from the doubling of a polyploid genome. Such doubling occurs on rare occasions in nature; it also results from controlled treatment with colchicine or other drugs that disrupt the mitotic spindle. The doubling of a tetraploid genome yields an octaploid ($8x$). These higher level polyploids created by successive rounds of genome doubling are autopolyploids because all of their chromosomes derive from a single species.

Some Polyploids Have Agriculturally Desirable Traits Derived from Two Species

Polyploidy can also arise from crosses between members of two species with different numbers of chromosome sets. Hybrids in which the chromosome sets come from two or more distinct, though related, species are known as **allopolyploids.** In crosses between octaploids and tetraploids, for example, fertilization unites tetraploid and diploid gametes to produce hexaploid progeny. Fertile allopolyploids arise only rarely, under special conditions, because chromosomes from the two species differ in shape, size, and number, and thus cannot easily pair with each other. The resulting irregular segregation creates genetically unbalanced gametes such that the hybrid progeny will be sterile. Chromosomal doubling in germ cells, however, can restore fertility by creating a pairing partner for each chromosome. Organisms produced in this manner are termed **amphidiploids** if the two parental species were diploids; they contain two diploid genomes, each one derived from a different parent. As the following illustrations show, it is hard to predict the characteristics of an amphidiploid or other allopolyploids.

A cross between cabbages and radishes, for example, leads to the production of amphidiploids known as *Raphanobrassica*. The gametes of both parental species contain 9 chromosomes; the sterile F_1 hybrids have 18 chromosomes, none of which has a homolog. Chromosome doubling in the germ cells after treatment with colchicine, followed by union of two of the resulting gametes, produces a new species, a fertile *Raphanobrassica* amphidiploid carrying 36 chromosomes—a full complement of 18 (9 pairs) derived from cabbages and a full complement of 18 (9 pairs) derived from radishes. Unfortunately, this amphidiploid has the roots of a cabbage plant and leaves resembling those of a radish, so it is not agriculturally useful.

By contrast, crosses between tetraploid (or hexaploid) wheat and diploid rye have led to the creation of several allopolyploid hybrids with agriculturally desirable traits from both species (Fig. 12.27a and b). Some of the hybrids combine the high yields of wheat with rye's ability to adapt to unfavorable environments. Others combine wheat's high level of protein with rye's high level of lysine; wheat protein does not contain very much of this amino acid, an essential ingredient in the human diet. The various hybrids between wheat and rye form a new genus known as *Triticale*. Some *Triticale* species produce nutritious grains that already appear in breads sold in health food stores. Plant breeders are currently assessing the usefulness of various *Triticale* species for large-scale agriculture.

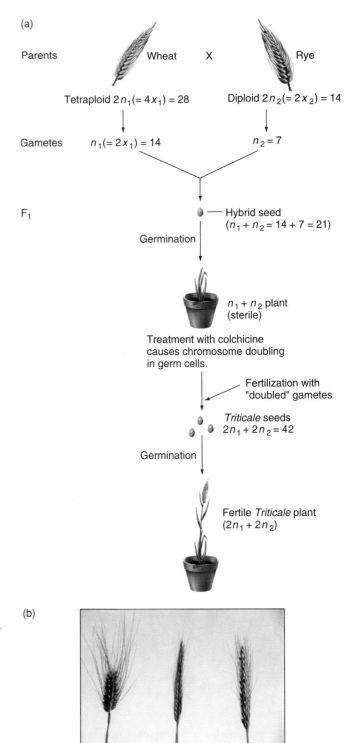

Figure 12.27 Amphidiploids in agriculture. (a) Agricultural breeders can cross wheat with rye to create an amphidiploid strain of *Triticale*. Because this strain of wheat is tetraploid, x_1 (the number of chromosomes in the basic set) is one-half n_1 (the number of chromosomes in a wheat gamete). For diploid rye, $n_2 = x_2$. Note that the F_1 hybrid between wheat and rye is sterile because the rye chromosomes have no pairing partners. Doubling of chromosome numbers by colchicine treatment of the F_1 hybrid corrects this problem, allowing regular pairing. (b) A comparison of wheat, rye, and *Triticale* grain stalks.

CONNECTIONS

The detrimental consequences of most changes in chromosome organization and number cause considerable distress in humans (Table 12.2). Approximately 4 of every 1000 individuals has an abnormal phenotype associated with aberrant chromosome organization or number. Most of these abnormalities result from either aneuploidy for the X chromosome or trisomy 21. By comparison, about 10 people per 1000 suffer from an inherited disease caused by a single-gene mutation.

The incidence of chromosomal abnormalities among humans would be much larger were it not for the fact that many fetuses with abnormal karyotypes abort spontaneously early in pregnancy. Fully 15% to 20% of recognized pregnancies end with detectable spontaneous abortions; and half of the spontaneously aborted fetuses show chromosomal abnormalities, particularly trisomy, sex chromosome monosomy, and triploidy. These figures probably underestimate the rate of spontaneous abortion caused by abnormal chromosomal variations, since fetuses carrying aberrations for larger chromosomes, such as monosomy 2 or trisomy 5, may abort so early that the pregnancy goes unrecognized.

But despite all the negative effects of chromosomal rearrangements and changes in chromosome number, a few departures from normal genome organization survive to become instruments of evolution by natural selection.

As we see in the next chapter, chromosomal rearrangements occur in bacteria as well as in eukaryotic organisms. Interestingly, in bacteria, transposable elements catalyze many of the changes in chromosomal organization; and the reshuffling of genes between different DNA molecules in the same cell catalyzes the transfer of genetic information from one bacterial cell to another.

TABLE 12.2 Data on Human Health Problems Resulting from Changes in Chromosome Organization and Number—Aneuploid Abnormalities in the Human Population

Chromosomes	Syndrome	Frequency at Birth
Autosomes		
Trisomic 21	Down	1/700
Trisomic 13	Patau	1/5,000
Trisomic 18	Edwards	1/10,000
Sex chromosomes, females		
XO, monosomic	Turner	1/5,000
XXX, trisomic		
XXXX, tetrasomic		1/700
XXXXX, pentasomic		
Sex chromosomes, males		
XYY, trisomic	Normal	1/10,000
XXYY, tetrasomic		
XXXY, tetrasomic	Klinefelter	1/500
XXXXY, pentasomic		
XXXXXY, hexasomic		

About 0.4% of all babies born have a detectable chromosomal abnormality that generates a detrimental phenotype.

ESSENTIAL CONCEPTS

1. *Rearrangements* reorganize the DNA sequences within genomes. Like point mutations, rearrangements are subject to natural selection and thus serve as instruments of evolution.

   a. *Deletions* remove DNA from a chromosome. Homozygosity for a large deletion is usually lethal, but even heterozygosity for a large deletion can create a deleterious *genetic imbalance*. Deletions may uncover recessive mutations on the homologous chromosome, and are thus useful for gene mapping at the cytological and molecular levels.

   b. *Duplications* add DNA to a chromosome. The additional copies of genes contained in a duplication are a major source of new genetic functions. Duplications sometimes produce changes in phenotype that allow their detection. Homozygosity or heterozygosity for duplications causes departures from normal gene dosage that are often harmful to the organism. *Unequal crossing-over* between

   duplicated regions expands or contracts the copy number of a gene.

   c. *Inversions* alter the order, but not the number, of genes on a chromosome. They may produce novel phenotypes by modifying the activity of genes near the rearrangement breakpoints or by disrupting genes at the breakpoints. Inversion heterozygotes act as *crossover suppressors* because progeny formed from recombinant gametes are genetically imbalanced.

   d. In *reciprocal translocations,* parts of two chromosomes trade places without the loss or gain of chromosomal material. By relocating pieces of DNA, translocations may modify the function of genes at or near the translocation breakpoints. Heterozygosity for translocations in the germ line results in *semisterility* and *pseudolinkage.*

   e. *Transposable genetic elements* are short, mobile segments of DNA that reshape genomes by generating mutations, causing chromosomal rearrangements, and relocating genes.

2. Changes in chromosome number also reshape genomes.

   a. *Aneuploidy*, the loss or gain of one or more chromosomes, creates a genetic imbalance. Aneuploidy can result from mistakes in meiosis which produce aneuploid gametes, or from mistakes in mitosis, which generate aneuploid clones of cells. Autosomal aneuploidy is usually lethal to the organism. Aneuploidy for sex chromosomes is better tolerated because of dosage compensation mechanisms; nevertheless, it can have significant consequences.

   b. *Euploid* organisms contain complete sets of chromosomes. Organisms with three or more sets of chromosomes are *polyploids*. *Autopolyploids* derive all their chromosome sets from the same species; *allopolyploids* are hybrids in which the chromosome sets come from two or more distinct, though related, species.

   c. Organisms containing odd numbers of complete chromosome sets are sterile because the chromosomes cannot pair properly during the first meiotic division.

   d. Polyploids with even numbers of chromosome sets can be fertile if proper chromosome segregation occurs sufficiently frequently. *Amphidiploids*, which are allopolyploids produced by chromosome doubling of two genomes derived from different diploid parental species, are often fertile and are sometimes useful in agriculture.

SOCIAL AND ETHICAL ISSUES

1. Until recently, female athletes were gender tested to confirm they were female using the Barr body test. In 1985, the Spanish hurdler Maria Patinez failed the Barr body test. She was encouraged by her coaches to fake an injury, quietly withdraw from the race, and retire from competition. Maria, indignant that someone suggested she was not a woman, spoke out against the testing policy. What are the problems scientifically with the Barr body test for gender identification? What do you think is a reasonable, accurate way to ascertain the sex of an individual? How should sex be defined? by karyotype? by examination of external genitalia? by internal gonadal structure?

2. Gary is a livestock farmer and has 2000 head of cattle. An outbreak of bacterial infection would be devastating to his livelihood, so he uses feed that contains antibiotics. Recently he heard a report that indicated that the use of antibiotics in livestock food was contributing to the increase in antibiotic resistance in bacteria in that infect humans. (The antibiotic resistance genes are contained on transposable elements that can be transferred from one bacteria to another.) Being a conscientious citizen, Gary is now considering eliminating the use of antibiotic feed. Will discontinuing the use of this feed make enough of a difference to justify the risk he takes? What should he do?

What are the stakes for him and for the public in this issue?

3. Jason and Sylvia received some unsettling news from amniocentesis of their baby. Karyotype analysis showed partial trisomy involving a region of chromosome 8 (a portion of one homolog was duplicated) in the baby. Genetic imbalances that can be observed cytologically are not well tolerated in humans and usually lead to medical conditions or developmental abnormalities. However, this specific partial trisomy had not been previously observed or associated with any medical conditions or developmental abnormalities. The parents decided to continue the pregnancy. Their doctor alerted other medical colleagues who now would like to follow the progress of this child after birth. In this way, more knowledge can be gained for future similar situations. Jason and Sylvia would prefer to not have their child subjected to intense medical scrutiny and are upset that their doctor gave information to other colleagues without their permission. Should the doctor have obtained permission before sharing the details with colleagues? Do Jason and Sylvia and their unborn child have a responsibility to provide information on the effects of this partial trisomy so others in the future will know what to expect with occurrence of the duplication?

S O L V E D   P R O B L E M S

I. Male *Drosophila* from a true-breeding wildtype stock were irradiated with X rays and then mated with females from a true-breeding stock carrying the following recessive mutations on the X chromosome: yellow body (*y*), crossveinless wings (*cv*), cut wings (*ct*), singed bristles (*sn*), and miniature wings (*m*). These markers are known to map in the order:

y-cv-ct-sn-m

Most of the female progeny of this cross were phenotypically wildtype, but one female exhibited a *ct, sn* phenotype. When this exceptional *ct, sn* female was mated with a male from the true-breeding wildtype stock, there were twice as many females as males among the progeny.
a. What is the nature of the X-ray-induced mutation present in the exceptional female?
b. Draw the X chromosomes present in the exceptional *ct, sn* female as they would appear during pairing in meiosis.
c. What phenotypic classes would you expect to see among the progeny produced by mating the exceptional *ct, sn* female with a normal male from a true-breeding wildtype stock? List males and females separately.

Answer

To answer this problem, you need to think about effects of different types of chromosomal mutations to deduce what the mutation is and then evaluate the consequences of the mutation on inheritance.

a. Two observations in this problem indicate that X rays induced a *deletion mutation*. The fact that two recessive mutations are now phenotypically expressed in the exceptional female suggests that a deletion was present on the X chromosome that now allowed expression of (that is, uncovered) the two mutant alleles (*ct* and *sn*) on the other copy of the X chromosome. Second, the finding that there were twice as many females as males in the cross of the exceptional female to wildtype males is consistent with the presence of a deletion. The exceptional female contains and can pass on a chromosome containing a deletion or a chromosome carrying all the mutant alleles. Because the deletion is on the X chromosome, the males (who inherit their one X chromosome from the mother) that inherit the deletion chromosome will be inviable (probably because other essential genes are located in the

region that is now deleted). When the female passes on the deletion chromosome to female progeny, those females are also receiving a normal copy of the X chromosome from their fathers, so the daughters are all viable. As a result, there are half as many male progeny as females from the cross of the exceptional female with the wildtype males.

b. During pairing, the DNA in *the normal (nondeleted) chromosome will loop out*, as shown below because there is no homologous region in the deletion chromosome. (In the simplified drawing below, each line represents both chromatids comprising each homolog.)

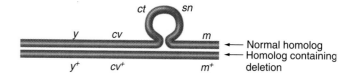

c. *Female progeny of the exceptional female crossed with the wildtype males will all be wildtype* since the wildtype male contributes wildtype copies of each of the genes. Because the males inherit their one X chromosome from their mothers, the genotypes of their X chromosome is representative of the meiotic products from the mother. This includes gametes in which no recombination occurred as well as those in which recombination leading to viable products occurred. *Males can have any of the genotypes listed below* and therefore the corresponding phenotypes. All contain the *ct sn* combination because no recombination between homologs is possible in this deleted region. Some of these genotypes require multiple crossovers during meiosis in the mother, and will therefore be relatively rare.

y	*cv*	*ct*	*sn*	*m*
+	+	*ct*	*sn*	+
+	*cv*	*ct*	*sn*	*m*
y	+	*ct*	*sn*	+
y	*cv*	*ct*	*sn*	+
+	+	*ct*	*sn*	*m*
+	*cv*	*ct*	*sn*	+
y	+	*ct*	*sn*	*m*

II. A female heterozygous for an inversion of most of the X chromosome including the X-linked genes for yellow body color (*y*), vermilion eye color (*v*), and forked bristles (*f*) was crossed to a male carrying the dominant X-linked Bar eye mutation (*B*). The cross is diagrammed below. The ● indicates the position of the centromere.

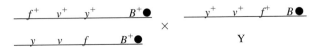

The cross produced the following male offspring:

y	*v*	*f*	*B*⁺	46
y⁺	*v*⁺	*f*⁺	*B*⁺	46
y	*v*⁺	*f*	*B*⁺	3
y⁺	*v*	*f*⁺	*B*⁺	3

a. Why are there no $y\,v\,f^+\,B^+$, $Y^+\,v^+\,f\,B^+$, $y\,v^+\,f^+\,B^+$, and $y^+\,v\,f B^+$ male offspring?

b. What kind of crossovers produced the $y^+\,v\,f^+\,B^+$ and $y\,v^+\,f B^+$ offspring? Where would the crossovers occur?

Answer

To answer this question, you need to be able to draw and interpret pairing in inversion heterozygotes. Note that this inversion is paracentric.

a. During meiosis in an inversion heterozygote, a loop of the inverted region is formed when the homologous genes align. (In the simplified drawing below, each line represents both chromatids comprising each homolog.)

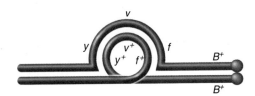

If a single crossover occurs within the inversion loop, a dicentric and an acentric chromosome are formed. Cells containing these types of chromosome are not viable. The resulting allele combinations from such single crossovers are not recovered. The four phenotypic classes of missing male offspring would be formed by single crossovers between the *y* and *v* or *v* and *f* genes in the female inversion heterozygote and therefore are not recovered.

b. *The $y^+\,v\,f^+\,B^+$ and $y\,v^+\,f\,B^+$ offspring would result from two crossover events within the inversion loop, one between the y and v genes and the other between the v and f genes.*

III. In maize trisomics, *n* + 1 pollen is not viable. If a dominant allele at the *B* locus produces purple color instead of the recessive phenotype bronze, and a *B b b*

trisomic plant is pollinated by a *B B b* plant, what proportion of the progeny produced will be trisomic and have a bronze phenotype?

Answer

To solve this problem, think about what is needed to produce a trisomic bronze progeny: three *b* chromosomes in the zygote. The female parent would have to contribute two *b* alleles, since the *n* + 1 pollen from the male is not viable. What kinds of gametes could be generated by the trisomic *B b b* purple female parent, and in what proportion? To track all the possibilities, rewrite this genotype as *B b₁ b₂*, even though b_1 and b_2 have identical effects on phenotype. In the trisomic female, there are three possible ways the chromosomes carrying these alleles could pair as bivalents during the first meiotic division so that they would segregate to opposite poles: *B* with b_1, *B* with b_2, and b_1 with b_2. In all three cases, the remaining chromosome could move to either pole. To tabulate the possibilities:

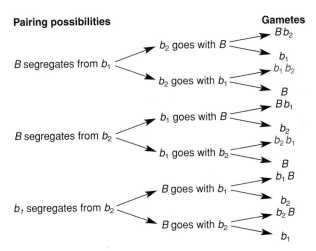

Of the twelve gamete classes produced by these different possible segregations, only the two classes written in red contain the two *b* alleles needed to generate the bronze (*b b b*) trisomic zygotes. There is thus a 2/12 = 1/6 chance of obtaining such gametes.

Although segregation in the *B B b* male parent is equally complicated, remember that males cannot produce viable *n* + 1 pollen. The only surviving gametes would thus be *B* and *b*, in a ratio (2/3 *B* and 1/3 *b*) that must reflect their relative prevalence in the male parent genome. The probability of obtaining trisomic bronze progeny from this cross is therefore the product of the individual probabilities of the appropriate *b b* gametes from the female parent (1/6) and *b* pollen from the male parent (1/3): 1/6 × 1/3 = 1/18.

PROBLEMS

12-1   For each of the terms in the left column, choose the best matching phrase in the right column.

a. reciprocal translocation	1. lacking one or more chromosomes or having one or more extra chromosomes
b. gynandromorph	2. movement of short DNA elements
c. pericentric	3. having complete extra sets of chromosomes
d. paracentric	4. exact exchange of parts of two nonhomologous chromosomes
e. euploids	5. excluding the centromere
f. polyploidy	6. including the centromere
g. transposition	7. having complete sets of chromosomes
h. aneuploids	8. mosaic combination of male and female tissue

12-2   A diploid strain of yeast was made by mating a haploid strain with a genotype w^-, x^-, y^-, and z^- with a haploid strain of opposite mating type that is wildtype for these four genes. The diploid strain was phenotypically wildtype. Four different X-ray-induced mutants with the following phenotypes were produced from this diploid yeast strain. (Assume there is a single mutation in each strain.)

Strain 1	w^-	x^+	y^-	z^+
Strain 2	w^+	x^-	y^-	z^-
Strain 3	w^-	x^+	y^-	z^-
Strain 4	w^-	x^+	y^+	z^+

When these mutant diploid strains of yeast go through meiosis, each ascus is found to contain only two viable haploid spores.
a. What kind of mutations were induced by X rays to make the listed diploid strains?
b. Why did two spores in each ascus die?
c. Are any of the genes *w, x, y,* and *z* located on the same chromosome?
d. Give the order of the genes that are found on the same chromosome.

12-3   In flies that are heterozygous for either a deletion or a duplication, there will be a looped-out region in a preparation of polytene chromosomes. How could you distinguish between a deletion or a duplication using polytene chromosome analysis?

12-4   A series of chromosomal mutations in *Drosophila* were used to map the *javelin* gene that affects bristle shape and *henna* that affects eye pigmentation. Both the *javelin* and *henna* mutations are recessive. A diagram of region 65 of the *Drosophila* polytene chromosomes is shown below.

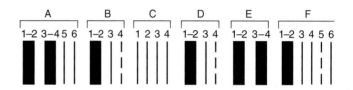

The chromosomal breakpoints for six chromosome rearrangements are indicated in the table below. (For example, deletion 1 has one breakpoint between bands A2 and A3 and the other between bands D2 and D3.)

		Breakpoints in region 65	
Deletion	A	A2–3;	D2–3
	B	C2–3;	E4–F1
	C	D2–3;	F4–5
	D	D4–E1;	F3–4
		Breakpoints	
Inversions	A	Region 65A6	Region 82A1
	B	Region 65B4	Region 98A3

Flies with a chromosome containing one of these six rearrangements (deletions or inversions) were mated to flies homozygous for both *javelin* and *henna*. The phenotypes of the heterozygous progeny of the rearrangement/*javelin, henna* flies are shown below.

		Phenotypes of F_1 flies
Deletion	A	*javelin, henna*
	B	*henna*
	C	wildtype
	D	wildtype
Inversions	A	*javelin*
	B	wildtype

Using this data, what can you conclude about the cytogenetic location for the *javelin* and *henna* genes?

12-5   Four strains of *Drosophila* were constructed in which one autosome contained recessive mutant alleles of the four genes *rolled eyes, thick legs, straw bristles,* and *apterous* wings, and the homologous autosome contained one of four different deletions (deletions 1–4). The phenotypes of the flies were as follows:

Deletion	Phenotype
1	rolled eyes, straw bristles
2	apterous wings, rolled eyes
3	thick legs, straw bristles
4	apterous wings

Whole genome DNA was prepared from the flies. The DNA was digested to completion with *Bam*HI, run on an agarose gel, and then transferred to nitrocellulose filters. The filters were then probed with a 20-kb cloned piece of wildtype genomic DNA obtained by partially digesting the plasmid clone with *Bam*HI (so the ends of the probe were *Bam*HI ends but the piece was not digested into all the possible *Bam*HI fragments). The results of this whole genome Southern blot are shown below. Dark bands indicate fragments present twice in the diploid genome; light bands indicate fragments present once in the genome.

	1	2	3	4
	6.3	6.3	6.3	6.3
	5.6	5.6	5.6	5.6
	5.1			
				4.9
	4.2	4.2	4.2	4.2
	3.8			
		3.3		
	3.0	3.0	3.0	3.0
			1.7	
	0.9	0.9	0.9	0.9

Deletion   1   2   3   4

a. Make a map of the *Bam*HI restriction sites in this 20-kb part of the wildtype *Drosophila* genome, indicating distances in kilobases between adjacent *Bam*HI sites. (Hint: The genomic DNA fragments in wildtype are 6.3, 5.6, 4.2, 3.0, and 0.9 kb long.)

b. On your map, indicate the locations of the genes.

12-6 The partially recessive, X-linked z^1 mutation of the *Drosophila* gene *zeste* (*z*) can produce a yellow (zeste) eye color only in flies that have two or more copies of the wildtype *white* (*w*) gene. Using this property, tandem duplications of the w^+ gene, called w^{+R} were identified. Males with the genotype $z^1 w^{+R}/Y$ have zeste eyes. These males were crossed to females with the genotype $y z^1 w^{+R} spl/y^+ z^1 w^{+R} spl^+$. (These four genes are closely linked on the X chromosome, in the order given in the genotype, with the centromere to the right of all these genes: *y* = yellow bodies; y^+ = tan bodies; *spl* = split bristles; spl^+ = normal bristles). Out of 81,540 male progeny of these females, the following exceptions were found.

Class A	2430 yellow bodies, zeste eyes, wildtype bristles
Class B	2394 tan bodies, zeste eyes, split bristles
Class C	23 yellow bodies, wildtype eyes, wildtype bristles
Class D	23 tan bodies, wildtype eyes, split bristles

a. What were the phenotypes of the remainder of the 81,540 males from the first cross?

b. What events gave rise to class A and B progeny?

c. What events gave rise to class C and D progeny?

12-7 Two yeast strains were mated and sporulated (allowed to carry out meiosis). One strain was a haploid with normal chromosomes and the linked genetic markers *ura3* (requires uracil for growth) and *arg9* (requires arginine for growth) surrounding their centromere. The other strain was wildtype for the two markers (*URA3* and *ARG9*) but had an inversion in this region of the chromosome (red) as shown below:

During meiosis, several different kinds of crossover events could occur. (Do not consider crossovers between sister chromatids.) For each of the events specified below, give the phenotype of the resulting four haploid spores. Assume that any chromosomal deficiencies are lethal in haploid yeast.

a. a single crossover outside the inverted region

b. a single crossover between *URA3* and the centromere

c. a double crossover involving the same two chromatids each time, where one crossover occurs between *URA3* and the centromere and the other occurs between *ARG9* and the centromere

12-8 Suppose a haploid yeast strain carrying two recessive linked markers *his4* and *leu2* were crossed with a strain that was wildtype for *HIS4* and *LEU2* but had an inversion of this region of the chromosome as shown below in blue.

Several different kinds of crossover events could occur during meiosis in the resulting diploid. For each of the events specified below state the genotype and phenotype of the resulting four haploid spores. (Do not consider crossover events between chromatids attached to the same centromere.)

a. a single crossover between the markers *HIS4* and *LEU2*

b. a double crossover involving the same chromatids each time, where both crossovers occur between the markers *HIS4* and *LEU2*

12-9 In the following group of figures, the pink lines indicate an area of a chromosome that is inverted relative to the normal (black line) order of genes. The diploid chromosome constitution of individuals 1–4 is shown. Match the individuals with the appropriate statement(s) below. (More than one diagram may correspond to the following statements and a diagram may be a correct answer for more than one question.)

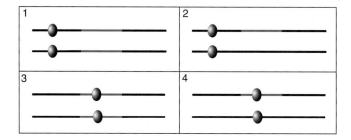

a. An inversion loop would form during meiosis I and in polytene chromosomes.
b. A single crossover involving the inverted region on one chromosome and the homologous region of the other chromosome would yield genetically imbalanced gametes.
c. A single crossover involving the inverted region on one chromosome and the homologous region of the other chromosome would yield an acentric fragment.
d. A single crossover involving the inverted region yields four viable gametes.

12-10 Three strains of *Drosophila* (Bravo, X-ray, and Zorro) are obtained that are homozygous for three variant forms of a particular chromosome. When examined in salivary gland polytene chromosome spreads, all chromosomes have the same number of bands in all three strains. When genetic mapping is performed in the Bravo strain, the following map is obtained (distances in map units).

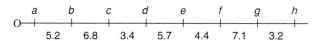

Bravo and X-ray flies are now mated to form Bravo/X-ray F_1 progeny, and Bravo flies are also mated with Zorro flies to form Bravo/Zorro F_1 progeny. In subsequent crosses, the following genetic distances were found to separate the various genes in the hybrids:

	Bravo/X-ray	Bravo/Zorro
a-b	5.2	5.2
b-c	6.8	0.7
c-d	0.2	<0.1
d-e	<0.1	<0.1
e-f	<0.1	<0.1
f-g	0.65	0.7
g-h	3.2	3.2

a. Make a map showing the relative order of genes *a* through *h* in the X-ray and Zorro strains. (Do not show distances between genes).

b. In the original X-ray homozygotes, would the physical distance between genes *c* and *d* be greater than, less than, or approximately equal to the physical distance between these same genes in the original Bravo homozygotes?
c. In the original X-ray homozygotes, would the physical distance between genes *d* and *e* be greater than, less than, or approximately equal to the physical distance between these same genes in the original Bravo homozygotes?

12-11 Semisterility in corn, as seen by unfilled ears with gaps due to abortion of approximately half the ovules, is an indication that the strain is a translocation heterozygote. The chromosomes involved in the translocation can be identified by crossing the translocation heterozygote to a strain homozygous recessive for a gene on the chromosome being tested. The ratio of phenotypic classes produced from crossing semisterile F_1 progeny back to a homozygous recessive plant indicates whether the gene is on one of the chromosomes involved in the translocation. For example, a semisterile strain could be crossed to a strain homozygous for the *yg* mutation on chromosome 9. (The mutant has yellow-green leaves instead of the wildtype green leaves.) The semisterile F_1 progeny would then be backcrossed to the homozygous *yg* mutant.
a. What types of progeny (fertile, semisterile, green, yellow-green) would you predict from the backcross of the F_1 to the homozygous *yg* mutant if the gene was not on one of the two chromosomes involved in the translocation?
b. What types of progeny (fertile, semisterile, green, yellow-green) would you predict from the backcross of the F_1 to the homozygous mutant if the *yg* gene is on one of the two chromosomes involved in the translocation?
c. If the *yg* gene is located in a region of the chromosome that was translocated, some relatively rare fertile, green and semisterile, yellow-green progeny are produced. How could these arise? What genetic distance could you determine from the frequency of these rare progeny?

12-12 During ascus formation in *Neurospora*, any ascospore with a chromosomal deletion dies and appears white in color. How many ascospores of the eight spores in the ascus would be white if the octad came from a cross of a wildtype strain with a strain of the opposite mating type carrying
a. a paracentric inversion and no crossovers occurred between normal and inverted chromosomes?
b. a pericentric inversion and a single crossover occurred in the inversion loop?
c. a paracentric inversion and a single crossover occurred outside the inversion loop?

d. a reciprocal translocation and an adjacent-1 segregation occurred with no crossovers between translocated chromosomes?

e. a reciprocal translocation and alternate segregation occurred with no crossovers between translocated chromosomes?

f. a reciprocal translocation and alternate segregation occurred with one crossover between translocated chromosomes (but not between the translocation breakpoint and the centromere of any chromosome)?

12-13 In the figure below, black and pink lines represent nonhomologous chromosomes. Which of the figures matches the description? (More than one diagram may correspond to the following statements and a diagram may be a correct answer for more than one question.)

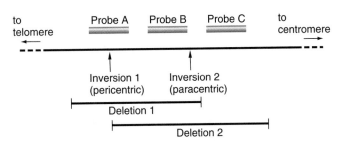

a. gametes produced by a translocation heterozygote

b. gametes that could not be produced by a translocation heterozygote

c. genetically balanced gametes produced by a translocation heterozygote

d. genetically imbalanced gametes that can be produced (at any frequency) by a translocation heterozygote

12-14 In *Drosophila*, the gene for cinnabar eye color is on chromosome 2 and the gene for scarlet eye color is on chromosome 3. A fly homozygous for recessive cinnabar and scarlet alleles (*cn/cn; st/st*) is white eyed.

a. If male flies (containing chromosomes with the normal gene order) heterozygous for *cn* and *st* alleles are crossed to white-eyed females homozygous for the *cn* and *st* alleles, what are the expected phenotypes and their frequencies in the progeny?

b. One unusual male heterozygous for *cn* and *st* alleles, when crossed to a white-eyed female, produced only wildtype and white-eyed progeny. Explain the likely chromosomal constitution of this male.

c. When the wildtype F$_1$ females from the cross with the unusual male were backcrossed to normal <CR> *cn/cn st/st* males, the following results were obtained:

wildtype	45%
cinnabar	5%
scarlet	5%
white	45%

Diagram a genetic event at metaphase I that could produce the rare cinnabar or scarlet flies among the progeny of the wildtype F$_1$ females.

12-15 Human chromosome 1 is a large, metacentric chromosome. A map of a cloned region from near the telomere of chromosome 1 is shown below. Three probe DNAs (A, B, and C) from this region were used for *in situ* hybridization to human mitotic metaphase chromosome squashes made with cells obtained from individuals with various genotypes. The breakpoints of chromosomal rearrangements in this region are indicated on the map. The black bars for Deletions (*Del*) 1 and 2 represent DNA that is deleted. The breakpoints of Inversions (*Inv*) 1 and 2 not shown in the figure are near, but not at the centromere. For each of the following genotypes, draw chromosome 1 as it would appear after *in situ* hybridization. An example is shown below for hybridization of probe A to the two copies of chromosome 1 in wildtype (+/+).

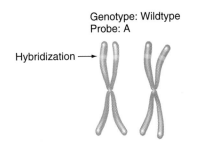

a. genotype: *Del1/Del2;* probe: B

b. genotype: *Del1/Del2;* probe: C

c. genotype: *Del1/+;* probe: A

d. genotype: *Inv1/+;* probe: A

e. genotype: *Inv2/+;* probe: B

f. genotype: *Inv2/Inv2;* probe: C

12-16 A proposed biological method for insect control involves release of insects that could interfere with the fertility of the normal resident insects. One approach is to introduce sterile males to compete with the resident fertile males for matings. A disadvantage of this strategy is that the irradiated sterile males are not very robust and can have problems competing with the fertile males. An alternate approach that is being tried is to release laboratory-reared insects that are homozygous for several translocations. Explain how this strategy will work. Be sure to mention which insects will be sterile.

12-17 A *Drosophila* male is heterozygous for a translocation between an autosome originally bearing the dominant mutation *Lyra* wings and the Y chromosome; the other copy of the same autosome is *Lyra*⁺. This male is now mated with a true-breeding, wildtype female. What kinds of progeny would be obtained, and in what proportions?

12-18 Explain how transposable elements can cause movement of genes that are not part of the transposable element.

12-19 In the 1950s, Barbara McClintock found a genetic element in corn she called *Ds* (*Dissociator*). When located at a particular location, this element could often cause chromosomal breaks at that site, but these breaks occurred only in the presence of another unlinked genetic element she called *Ac* (*Activator*). She found further that in the presence of *Ac*, *Ds* could jump to other chromosomal locations. At some of these locations (and in the presence of *Ac*) *Ds* would now cause chromosomal breakage at the new position; at other positions, it appeared that *Ds* could cause new mutations that were unstable as shown by their patchy, variegated expression in kernels. Interestingly, the position of the *Ac* element seemed to be very different in various strains of corn. Explain these results in terms of our present-day understanding of transposons.

12-20 Gerasimova and colleagues in the former U.S.S.R. characterized a mutation in the *Drosophila* cut wing (*ct*) gene called *ct*^MR2 which is associated with the insertion of a transposable element called *gypsy*. This allele is very unstable: approximately 1 in 100 of the progeny of flies bearing *ct*^MR2 show new *ct* variants. Some of these are *ct*⁺ revertants, while others appear to be more severe alleles of *ct* with stronger effects on wing shape. When the *ct*⁺ revertants themselves are mated, some of the *ct*⁺ alleles appear to be stable (no new *ct* mutants appear) while others are highly unstable (many new mutations appear). What might explain the generation of stable and unstable *ct*⁺ revertants as well as the stronger *ct* mutant alleles?

12-21 In sequencing a region of the human genome, you have come across a segment of about 200 A residues. You suspect that the sequence preceding the A residues may

have been moved here by a transposition event mediated by reverse transcriptase. If the adjacent sequence is in fact a retroposon, you might expect to find other copies in the genome. How could you determine if other copies of this DNA exist?

12-22 The *Eco*RI restriction map of the region in which a coat color gene in mice is located is presented below. The leftmost *Eco*RI site is arbitrarily labeled 0 and the other distances in kilobases are given relative to this coordinate. Genomic DNA was prepared from one wildtype mouse and 10 mice homozygous for various mutant alleles. This genomic DNA is digested with *Eco*RI, fractionated on agarose gels, and then transferred to nitrocellulose filters. The filters were probed with the radioactive DNA fragment indicated by the purple bar, extending from coordinate 2.6 kb to coordinate 14.5. The resultant autoradiogram is shown schematically below.

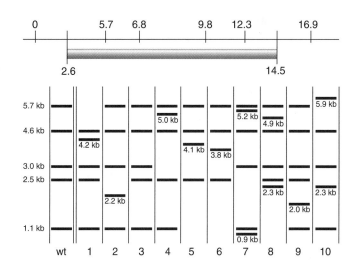

Assume that each of the mutations 1–10 is caused by one and only one of the events on the following list. Which event corresponds to which mutation?

a. a point mutation exactly at coordinate 6.8

b. a point mutation exactly at coordinate 6.9

c. a deletion between coordinates 10.1 and 10.4

d. a deletion between coordinates 6.7 and 7.0

e. insertion of a transposable element at coordinate 6.2

f. an inversion with breakpoints at coordinates 2.2 and 9.9

g. a reciprocal translocation with another chromosome with a breakpoint at coordinate 10.1

h. a reciprocal translocation with another chromosome with a breakpoint at coordinate 2.4

i. a tandem duplication of sequences between coordinates 7.2 and 9.2

j. a tandem duplication of sequences between coordinates 11.3 and 13.3

12-23 The number of chromosomes in the somatic cells of several oat varieties (*Avena* species) are given below.
   a. What is the basic chromosome number (*x*) in *Avena?*
   b. What is the ploidy for each of the different species?
   c. What is the number of chromosomes in the gametes produced by each of these oat varieties?
   d. What is the *n* number of chromosomes in each species?

   sand oats (*Avena strigosa*)—14
   slender wild oats (*Avena barata*)—28
   cultivated oats (*Avena sativa*)—42

12-24 Common red clover, *Trifolium pratense,* is a diploid with 14 chromosomes per somatic cell. What would be the somatic chromosome number of
   a. a trisomic variant of this species?
   b. a monosomic variant of this species?
   c. a triploid variant of this species?
   d. an autotetraploid variant?

12-25 Fred and Mary have a child with Down syndrome. A probe derived from chromosome 21 was used to identify RFLPs in Fred, Mary, and the child (darker bands indicate signals of twice the intensity). Explain what kind of nondisjunction events must have occurred to produce the child if the child's RFLP pattern looked like that in lanes A, B, C, or D of the figure below.

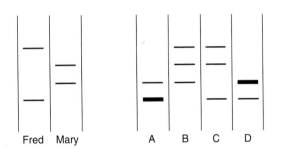

12-26 Among adults with Turner syndrome, it has been found that a very high proportion are genetic mosaics. These are of two types: in some individuals, the majority of cells are 45, XO but a minority of cells are 46, XX. In other Turner individuals, the majority of cells are 45, XO but a minority of cells are 46, XY. Explain how these somatic mosaics could arise.

12-27 You have haploid tobacco cells in culture and have made transgenic cells that are resistant to herbicide. What would you do to obtain a diploid cell line that could be used to generate a new fertile herbicide-resistant plant?

12-28 An allotetraploid species has a genome comprised of two ancestral genomes, A and B, each of which have a basic chromosome number (*x*) of 7. In this species the two copies of each chromosome of each ancestral genome pairs only with each other during meiosis. Resistance to a pathogen that attacks the foliage of the plant is controlled by a dominant allele at the *F* locus. The recessive alleles F^a and F^b confer sensitivity to the pathogen, but the dominant resistance alleles present in the two genomes have slightly different effects. Plants with at least one F^A allele are resistant to races 1 and 2 of the pathogen regardless of the genotype in the *B* genome, and plants with at least one F^B allele are resistant to races 1 and 3 of the pathogen regardless of the genotype in the *A* genome. What proportion of the self progeny of an $F^A F^a F^B F^b$ plant will be resistant to all three races of the pathogen?

12-29 Using karyotype analysis, how could you distinguish between autopolyploids and allopolyploids?

THE PROKARYOTIC CHROMOSOME: GENETIC ANALYSIS IN BACTERIA

Species of bacteria can live in environments even as seemingly hostile as hot springs (such as this beautiful pool in Yellowstone National Park, WY). Comparative genome analyses of bacteria that live in unusual environments will increase our understanding of the adaptations that allow the organisms to survive in different niches.

Gonorrhea, a sexually transmitted infection of the urogenital tract in men and women, is on the rise in many parts of the world. Caused by the bacterium *Neisseria gonorrhoeae,* gonorrhea is rarely fatal, but in men it can spread from an initial site, usually the urethra, to the prostate gland and the epididymis (two structures that play a role in sperm production), diminishing sperm count; and in women, it can move from the cervix to the uterine lining and fallopian tubes, leading to sterility. A common symptom is painful urination, although many infected people have no symptoms at all. Because infants passing through a gonorrhea-infected birth canal can contract severe eye infections, hospitals routinely treat the eyes of newborns with a few drops of silver nitrate solution or penicillin. If left untreated, gonorrhea can spread through the bloodstream to the joints, skin, bones, tendons, and heart. Until the late 1970s, a few shots of penicillin were a surefire cure for gonorrhea, but by 1995, more than 20% of *N. gonorrhoeae* bacteria isolated from patients worldwide were resistant to penicillin.

Geneticists now know that the agent of this alarming increase in antibiotic resistance was the transfer of DNA from one bacterium to another (see Chapter 5). According to epidemiologists, penicillin-resistant *N. gonorrhoeae* bacteria first appeared in Asia in the 1980s, in a patient receiving penicillin treatment for gonorrhea, who was also fighting an infection caused by another species of bacteria—*Haemophilus influenzae.* Some of the patient's *H. influenzae* bacteria apparently carried a small circular plasmid (see Chapter 8) containing a gene encoding penicillinase, an enzyme that destroys penicillin. When the doubly infected patient mounted a specific immune response to *H. influenzae* that degraded these cells, the broken bacteria released their plasmids. Some of the freed loops of DNA entered *N. gonorrhoeae* cells, transforming them to penicillin-resistant bacteria. The transformed gonorrhea bacteria then multiplied, and successive exposures to penicillin selected for the resistant bacteria. As a result, the patient transmitted penicillin-resistant *N. gonorrhoeae* to subsequent sexual partners. Thus, while penicillin treatment does not create the genes for resistance, it fast-forwards the spread of those genes; after continuous, long-term exposure to the drug, only cells resistant to penicillin survive. Today in the United States, many *N. gonorrhoeae* are simultaneously resistant to penicillin, spectinomycin, and tetracycline.

In this chapter, we survey bacterial diversity and discuss the structure of bacterial genomes. We also examine the mechanisms by which bacteria transfer genes between cells of the same species, between cells of distantly related species, and between bacterial cells and bacterial viruses. Geneticists can use their knowledge of bacterial genetics in general and of various forms of bacterial gene transfer in particular to identify, map, and characterize the genes that contribute to biological processes. Such genetic analysis has made it possible to dissect and reconstruct, for example, the biochemical pathways by which bacteria move toward food and away from toxins, pathways that resist analysis by other means.

As we describe how bacteria replicate their DNA, reproduce themselves, and transfer genes between cells, we present

- A general overview of bacteria: their numbers; their range of sizes, lifestyles, and metabolic activity; and how researchers grow them for study.

- The bacterial genome: the structure, organization, transcription, replication, and evolution of large circular bacterial chromosomes; and the structure and function of smaller circles of DNA known as plasmids.

- Gene transfer in bacteria: the mechanisms of transformation, conjugation, and transduction.

- A comprehensive example: using the tools of bacterial genetics to dissect bacterial chemotaxis.

- The future of bacterial genetics.

A GENERAL OVERVIEW OF PROKARYOTES

Bacteria such as *N. gonorrhoeae* and *H. influenzae* constitute one of the three major evolutionary lineages of living organisms (Fig. 13.1). The two other lineages are the **eukaryotes** (organisms whose cells have nuclei encased in a membrane) and the **archaea.** Bacteria and archaea are both classified as **prokaryotes:** organisms lacking a nuclear membrane, whose chromosomes are therefore not enclosed in a nucleus. Archaea and bacteria are so similar in appearance that it is often difficult to distinguish them in the light microscope. Most of the

few known groups of archaea live under extreme conditions in unusual ecological niches such as very salty surroundings or anaerobic environments that do not contain oxygen.

In 1996 researchers determined the complete DNA sequence of the genome of one organism in the archaea lineage: *Methanococcus jannaschii,* a methane-producing anaerobe. The results provided strong evidence that bacteria, archaea, and eukaryotes are of equal taxonomic rank. Some genes in archaea resemble those in bacteria, while others are similar to genes in eukaryotes. Remarkably, more than 50% of *M. jannaschii* genes are completely different from any bacterial or eukaryotic genes. On the basis of these comparisons, many evolutionary biologists propose that the earliest organisms, presumably single-celled prokaryotes, existed roughly 3.5 billion years ago. A billion years later, bacteria and archaea diverged from their common ancestor, and somewhat later eukaryotes split off from the archaea evolutionary line (Fig. 13.1).

Despite the evolutionary distance separating bacteria from archaea, living representatives of the two lineages show marked similarities in morphology, the structure of their genomes, and the mechanisms by which they transfer genes. Thus, although our discussion in the remainder of this chapter focuses on bacteria (sometimes referred to as "eubacteria"), it is possible to apply to archaea many of the lessons learned from the other prokaryotes.

Numbers, Sizes, Lifestyles, and Metabolic Diversity of Bacteria

Bacteria are crucial to the maintenance of earth's environment. Various species release oxygen into the atmosphere; recycle carbon, nitrogen, and other elements; and digest human and other animal wastes as well as pesticides and other pollutants,

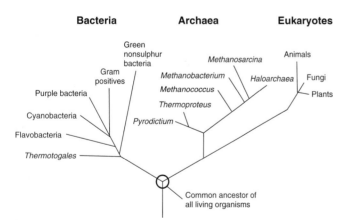

Figure 13.1   A family tree of living organisms. There are many diverse present-day forms of bacteria and archaea. The three major evolutionary lineages of living organisms are bacteria and archaea (both prokaryotes) and eukaryotes. DNA sequencing results suggest that all living forms descended from a common prokaryotic ancestor that lived 3.5 billion years ago.

which would otherwise eventually poison the air, soil, and water. Bacteria also cause hundreds of animal and plant diseases, but even so, harmful species are a small fraction of all bacteria; many species, in fact, produce vitamins and other materials essential to the health and survival of humans and other organisms. Two key features of bacteria are their astounding numbers and enormous diversity.

Bacteria Outnumber All Other Organisms on Earth by Many Orders of Magnitude

A human adult carries at least 100 grams (roughly a quarter pound) of live bacteria, mainly in the intestines. An estimated 10^{14} bacteria make up those 100 grams, a number many thousands of times as great as the number of people on the planet.

Researchers Have Identified 10,000 Species of Bacteria with an Amazing Range of Sizes and Lifestyles

The smallest bacteria are about 200 nanometers (nm = billionths of a meter) in diameter. The largest are 500 micrometers (mm = millionths of a meter) in length, which makes them 10 billion times larger in volume and mass than the smallest bacterial cells. These large bacteria are visible without the aid of the microscope. Some bacteria live independently on land, others float freely in aquatic environments, and still others live as parasites or symbionts inside other lifeforms. (Table 13.1)

Bacteria Display a Remarkable Metabolic Diversity That Enables Them to Live Almost Everywhere

Some soil bacteria obtain the energy to fuel their metabolism from the chemical ammonia, while other, photosynthesizing bacteria obtain their energy from sunlight. Because of their metabolic diversity, bacteria play essential roles in many natural processes, such as the decomposition of materials essential for nutrient cycling. The balance of microorganisms is a key to the success of these ecological processes, which help maintain the environment. In the cycling of nitrogen, for example, decomposing bacteria break down plant and animal matter rich in nitrogen, and produce ammonia. Nitrifying bacteria then use this ammonia as a source of energy and release nitrate (NO_3), which some plants can use as is; denitrifying bacteria convert the nitrate not used directly by plants to atmospheric nitrogen (N_2); and nitrogen-fixing bacteria, such as Rhizobium, that live in the roots of peas and other leguminous plants convert N_2 to ammonium, which their host plants can use.

Recently, geneticists and molecular biologists have used microbes to isolate unusual enzymes that carry out natural and industrial processes. They then clone and manipulate the genes encoding these enzymes. One use of unusual microorganisms and their enzymes is the development of bioremediation bacteria that can, for example, break down the hydrocarbons found in oil. In 1989 when the tanker Exxon

Valdez spilled millions of barrels of oil along the Alaskan coastline, cleanup crews used oil-digesting bacteria to help revive the environment.

Despite Their Diversity, Bacterial Cells Have a Number of Features in Common

Although bacteria come in a variety of shapes and sizes adapted to a range of lifestyles, all lack a defined nuclear membrane as well as membrane-bound organelles, such as the mitochondria and chloroplasts found in eukaryotic cells. Bacterial chromosomes fold to form a dense **nucleoid body** that appears to exclude ribosomes, which function in the surrounding cytoplasm. A plasma membrane encloses bacterial cells, and a membranous structure known as the *mesosome,* which apparently serves as a source of new membranes during cell division, projects from the plasma membrane into the cell's interior. In most species of bacteria, the plasma membrane is supported by a cell wall composed of carbohydrate and peptide polymers. Some bacteria have, in addition to the cell wall, a thick, mucus-like coating called a *capsule* that helps the cell resist attack by the immune system. Many bacteria have flagella that propel them toward food or light.

The Power of Bacterial Genetics Is the Potential for Studying Rare Events

Researchers grow bacteria in liquid media or on media solidified by agar (Fig. 13.2). In a liquid medium, the cells of commonly studied species, such as *E. coli,* grow to a concentration of 10^9 cells per milliliter within a day. In agar-solidified medium, a single bacterium will multiply to a visible colony containing 10^7 or 10^8 cells in less than 1 day. The ability to grow large numbers of cells is one advantage that has made bacteria, especially *E. coli,* so attractive for genetic studies.

(a)

(b)

Figure 13.2 Bacteria in the laboratory. Bacteria grow as (a) colonies on solid nutrient-agar plates or (b) as suspension in a liquid medium.

TABLE 13.1   Characteristics and Significance of Various Groups of Bacteria

Domain and Evolutionary Group	Groups Based on Structure and Physiology	Characteristics and Significance
Bacteria		
Purple bacteria and relatives	Purple photosynthetic bacteria	Some can form a purple layer in anaerobic zone of lakes; mitochondria arose from ancient purple bacteria similar to *Rhodospirillum*
	Pseudomonads	Versatile heterotrophs in the soil; useful in cleaning up industrial pollutants
	Enteric bacteria	Inhabit intestinal tracts, aiding digestion (*Escherichia coli*), or causing disease (*Salmonella*)
	Myxobacteria	Gliding bacteria whose colonies make complex multicellular spore-forming stalks
	Nitrogen fixers	Free in soil or associated with plants (*Rhizobium*); fix nitrogen in biological form
	Rickettsias	Live only inside other cells; cause typhus fever and Rocky Mountain spotted fever
Cyanobacteria	Cyanobacteria	Have photosynthetic system like a plant, and are nitrogen fixing; most are self-sufficient organisms; earliest organism to release oxygen; precursors of chloroplasts
Gram-positive bacteria	Endospore formers	Heat-resistant spores can cause botulism or tetanus (*Clostridium* and *Bacillus*)
	Lactic acid bacteria	*Lactobacillus* make yogurt; *Streptococcus* causes strep throat
	Gram-positive cocci	*Staphylococcus* causes toxic shock syndrome and other infections
	Actinomycetes	Cause leprosy and tuberculosis; produce antibiotics
	Mycoplasmas	Smallest of all cells, no cell wall; one cause of pneumonia
Chlamydiae	Chlamydia	Leading cause of blindness in humans, cause of most common sexually transmitted disease, cells live inside other cells, can't make ATP
Bacterioides and relatives	A branch of the gram-negative rods	Major anaerobic cells in the large intestine
Green sulfur bacteria	A branch of the phototropic bacteria	Live on the warm fringes of sulfer hot springs
Spirochetes	Spirochetes	Corkscrew shape; cause Lyme disease and syphilis
Deinococci and relatives	A branch of gram-positive cocci	Highly resistant to radiation; found near atomic reactors
Green nonsulfur bacteria	A branch of phototrophic bacteria	Grow in mats at the fringes of hot springs
Thermotogas	Anaerobic fermenters	Live in oceanic vents at highest temperature of any true bacteria

Genetic studies of bacteria require techniques to count these large numbers of cells and to isolate individual cells of interest. Researchers can use a solid medium to calculate the number of cells in a liquid culture. They begin with sequential dilutions of cells in the liquid medium (as illustrated in Fig. 6.15a). They then spread a small sample of the diluted solutions on agar-medium plates and count the number of colonies that form. Although it is still difficult to work with a single bacterial cell, except in very specialized studies, the cells constituting a single colony contain the genetically identical descendants of the one bacterial cell that founded the colony.

Focus on E. coli

The most studied and best understood species of bacteria is *Escherichia coli,* a common inhabitant of vertebrate intestines (Fig. 13.3). *E. coli* cells can grow in the complete absence of oxygen—the condition found in the intestines—or in air. The *E. coli* strains studied in the laboratory are not pathogenic, but other strains of the species can cause a variety of intestinal diseases, most of them mild, a few life-threatening.

Because *E. coli* encodes all the enzymes it needs for amino-acid and nucleotide biosynthesis, it can grow in **minimal** media, which contain glucose as the only source of carbon and energy, and inorganic salts as the source of the other elements that compose bacterial cells. In a minimal medium, *E. coli* cells divide every hour, redoubling their numbers 24 times a day. In a richer, more complex medium containing sugars and amino acids, *E. coli* cells divide every 20 minutes to produce 72 generations per day. Two days of growth at this rate, if unchecked by any limiting factor, would generate a mass of bacteria equal to the mass of the earth.

The rapidity of bacterial multiplication makes it possible to grow an enormous number of cells in a relatively short time, and as a result, to examine very rare genetic events. For example, wildtype *E. coli* cells are normally sensitive to the antibiotic streptomycin. By spreading a billion wildtype bacteria on an agar-medium plate containing streptomycin, you could isolate a few extremely rare mutants (that might, by chance, have arisen among the 10^9 cells) that are resistant to streptomycin. It is not as easy to find and examine such rare events

with nonmicrobial organisms; in multicellular animals, it is almost impossible.

THE BACTERIAL GENOME

The study of bacteria played a critical role in the development of the field of genetics. From the 1940s to the 1970s (considered the era of classical bacterial genetics), virtually everything researchers learned about gene organization, gene expression, and gene regulation came from analyses of bacteria and the phages that infect them. In the 1970s and 1980s, the advent of recombinant DNA technology depended on an understanding of genes and chromosomes in bacteria.

One Circular Chromosome

The essential component of a typical bacterial genome is the **bacterial chromosome:** a single circular molecule of double-helical DNA. This single closed circle of double-stranded DNA is 4–5 Mb long in commonly studied species. The circular chromosome of *E. coli,* if broken at one point and laid out in a line, would form a DNA molecule 2.4 nm wide and 1.6 mm long, almost a thousand times longer than the *E. coli* cell in which it is found (Fig. 13.4a). Inside the cell, the long, circular DNA molecule condenses by supercoiling and looping into a densely packed nucleoid body. During the bacterial cell cycle, each bacterium replicates its circular chromosome and then divides by binary fission into two identical daughter cells, each with its own chromosome, generating two organisms from one (Fig. 13.4b).

The Representation of Genes, Genotype, and Phenotype Depends on an Understanding of Mutations

Because the bacterial genome consists of one circular chromosome, it is a haploid genome, and the relation between gene mutation and phenotypic variation is relatively straightforward; that is, in the absence of a second, wildtype allele for each gene, all mutations express their phenotype.

Bacteria are so small that the only practical way to examine them is in the colonies of cells they form on a petri plate. Within this constraint, it is still possible to identify many different kinds of mutations. These include

1. Mutations affecting *colony morphology,* that is, whether a colony is large or small, shiny or dull, round or irregular.

2. Mutations conferring *resistance* to bactericidal agents such as antibiotics or bacteriophages.

3. Mutations that create *auxotrophs* unable to grow and reproduce on minimal medium; auxotrophic mutations are in genes required to synthesize relatively complex compounds such as amino acids or nucleotide components from simpler materials in the environment.

4. Mutations affecting the *ability of cells to break down and utilize complicated chemicals* in the environment; for example, the *lacZ* gene in *E. coli* encodes the enzyme β-galactosidase needed to break down the sugar lactose into

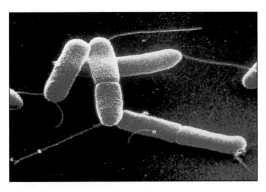

Figure 13.3   *Escherichia coli.* Scanning electron micrograph of *E. coli* (× 14,000).

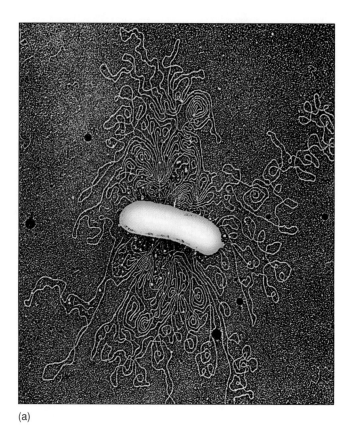

(a)

(b) Parent cell

Cell wall
Chromosome

Origin of replication — Nucleoid

Replication fork

Replication fork

Daughter cells

(c)

Diagrammatic representations
of bacterial chromosome

Figure 13.4. The bacterial chromosome: A circle of double-stranded DNA. (a) An electron micrograph of an *E. coli* cell that has been lysed open, allowing its chromosome to escape.
(b) Chromosome replication and distribution during cell division. In this figure, both strands of DNA are shown, with newly synthesized DNA shown in red. Replication of the *E. coli* chromosome starts from a single origin, from which replication forks move in both directions. After replication is complete and the cell divides, the two daughter chromosomes are parceled out to the two daughter cells. (c) The chromosomal DNA can be represented as the helix (sharing both backbone strands) or as a ring.

glucose and galactose. Wildtype cells can grow if lactose, rather than the normal glucose, is the sole source of carbon in the medium, but *lacZ⁻* mutants cannot.

5. Mutations in *essential* genes whose protein products are required under all conditions of growth; because a null mutation in an essential gene would prevent a colony from growing in any environment, bacteriologists must work with conditional lethal mutations such as

temperature-sensitive (ts) mutations that allow growth at one temperature but not at another.

Bacteriologists use different techniques to isolate rare mutations of the various classes described previously. With mutations conferring resistance to a particular agent, they can do a straightforward **selection,** that is, establish conditions in which only the desired mutant will grow. For example, if wild-type bacteria are streaked on a petri plate containing the antibiotic streptomycin, the only colonies to appear will be streptomycin resistant (Str^r). It is also possible to select for prototrophic revertants of strains carrying auxotrophic mutations by simply plating cells on minimal medium agar, which does not contain the compounds auxotrophs require for growth.

Because the key characteristic of most of the other types of mutants just described is their inability to grow under particular conditions, it is not possible to select for them directly. Instead, researchers must identify these mutations by a **genetic screen:** an examination of each colony in a population for its phenotype. They can, for example, use a toothpick to transfer cells from a colony growing on minimal medium supplemented with methionine (Met) to a petri plate containing minimal medium without methionine. Failure of those cells to grow on the unsupplemented medium would indicate that the corresponding colony on the original plate is auxotrophic for methionine.

Spontaneous mutations in specific bacterial genes occur very rarely, in 1 in 10^6 to 1 in 10^8 cells, depending on the gene. It would be virtually impossible to identify such rare mutations if you had to check the phenotype of a million to a hundred million colonies through the individual transfer of each one with a toothpick. A number of techniques simplify the process. We describe four:

1. *Replica plating* (see Chapter 5) allows the simultaneous transfer of thousands of colonies from one plate to another.

2. *Treatments with mutagens* (see Chapter 6) increases the frequency with which a mutation in a gene appears in the population.

3. *Enrichment* increases the proportion of mutant cells in a population by exposing the population to agents that kill wildtype cells. Penicillin is one agent of enrichment; it acts by disrupting the formation of the cell wall in growing cells, and thus kills cells that can grow but not

cells that cannot grow. Figure 13.5 shows how researchers use this property of penicillin to enrich the proportion of auxotrophic cells in a mixed population of auxotrophic and prototrophic cells.

4. *Testing for visible phenotypes* in cases where wildtype and mutant cells have different phenotypes on the same petri plate. In an important example, *E. coli* producing functional β-galactosidase (the product of the *lacZ* gene) cleave the colorless artificial compound X-Gal, yielding a blue product. Thus *lacZ$^+$* colonies turn blue on medium containing X-Gal, while *lacZ$^-$* colonies remain white (the usual color of the colonies; review Fig. 8.8).

Researchers designate the genes of bacteria by three lowercase, italicized letters that signify something about the function of the gene. For example, genes in which mutations result in the inability to synthesize the amino acid leucine are *leu* genes. In *E. coli*, there are four *leu* genes—*leuA, leuB, leuC,* and *leuD*—that correspond to the three enzymes (one constructed from two different polypeptides) needed for the synthesis of leucine from other compounds. A mutation in any one of the *leu* genes changes a bacterium into an auxotroph for leucine, that is, into a cell unable to synthesize the amino acid; such a cell can grow only in media supplemented with leucine. Mutations in genes required for the breakdown of a sugar (for example, the *lacZ* gene) produce cells unable to grow in medium containing only that sugar (lactose) as a source of carbon. Other types of mutations give rise to antibiotic resistance; *str^r* is a mutation producing streptomycin resistance. To designate the alleles of genes present in wildtype bacteria, researchers use a superscript "+": *leu$^+$, str$^+$, lacZ$^+$*; to designate mutant alleles they use a superscript "−", as in *leuA$^-$* and *lacZ$^-$*, or a superscript description, as in *str^r*.

The phenotype of a bacterium that is wildtype or mutant for a particular gene is indicated by the three letters that designate the gene, written, however, with an initial capital letter, no

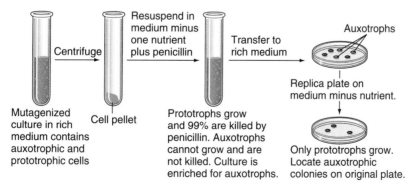

Figure 13.5   Penicillin enrichment for auxotrophic mutants. Penicillin disrupts synthesis of the bacterial cell wall, but does not affect cell walls that were previously made. Thus, penicillin selectively kills growing cells that are making new cell walls, but not bacteria whose growth is arrested. To enrich for auxotrophic mutants, a mutagenized culture of bacteria is transferred, for a period of several normal cell generations, to a medium lacking a nutrient but containing penicillin. Prototrophs can grow, but will be killed by the antibiotic, while bacteria auxotrophic for the nutrient lacking in the media will be unable to grow. Thus, auxotrophs will not be killed by the penicillin. Once enrichment has taken place, cells remaining alive in the culture must be screened by replica plating to identify auxotrophs, because the penicillin treatment does not kill 100% of the prototrophs.

italics, and a superscript of minus, plus, or a one-letter abbreviation: Leu⁻ (requires leucine for growth); Lac⁺ (grows on lactose); Str^r (is resistant to streptomycin). A Leu⁻ *E. coli* strain cannot multiply unless it grows in a medium containing leucine; a Lac⁺ strain can grow if lactose replaces the usual glucose in the medium; a Str^r strain can grow in the presence of streptomycin.

Structure and Organization of the E. coli Chromosome

The circular *E. coli* chromosome consists of about 4.6 million base pairs. In 1997, molecular geneticists completed a project to determine the sequence of bases in the entire *E. coli* K12 genome. From previous genetic work, they knew many of the genes within this sequence; and they could identify others because the polypeptides they encode have amino-acid sequences similar to those of sequenced proteins found in other bacteria as well as in other species. Some of the sequences identified as genes, however, encode proteins with functions that have not yet been assigned. Such putative genes are known as **open reading frames,** or **ORFs;** they consist of long stretches of codons in the same reading frame uninterrupted by stop codons.

Close to 90% of the DNA encodes proteins. This contrasts sharply with the human genome in which there is roughly one gene every 30 kilobases, and less than 5% of the DNA encodes proteins. One reason for this discrepancy is that *E. coli* genes have no introns. In addition, there is very little repeated DNA in bacteria, and intergenic regions tend to be very small.

The complete sequence of the *E. coli* genome identified 4288 genes, and surprisingly, the function of 40% of the genes remains a mystery at this time. Given the small genome size and the tools developed over the years for genetic analysis in *E. coli,* however, researchers can easily mutate the genes and examine the resulting cells for phenotypic effects. So far, they have grouped genes whose function is known or has been deduced on the basis of sequence into broad functional classes. The 427 genes that are postulated to have a transport function make up the largest class; and of the 182 genes involved in translation, 6 were identified by the sequence analysis. Other classes include the genes for amino-acid biosynthesis and the genes for DNA replication and recombination.

Another interesting feature of the *E. coli* genome is the existence in eight different locations of remnants of bacteriophage genomes. The presence of these sequences suggests an evolutionary history of bacteria that includes invasion by viruses on several occasions.

Insertion Sequences Dot the E. coli Chromosome

DNA sequence analysis of bacterial genomes also revealed the position of several small transposable elements called **insertion sequences (ISs).** These elements, which dot the chromosomes of many types of bacteria, are transposons that do not contain selectable markers (such as genes conferring antibiotic resistance). Researchers have identified several distinct elements ranging in length from 700 to 5000 bp; they named the elements IS1, IS2, IS3, and so forth, with the numbers designating the or-

der of discovery. Like the ends of transposons in eukaryotic cells (see Chapter 12), the ends of IS elements are inverted repeats of each other; and each IS includes a gene encoding a transposase that initiates transposition by recognizing these mirror-image ends (Fig. 13.6a). Because insertion sequences can move to other sites on the bacterial chromosome when they transpose, their distribution differs in different strains of the same bacterial species. For example, one strain of *E. coli* may have 15 insertion sequences of five different kinds, while a second strain isolated from a different population may have 25 insertion sequences, lack one of the types found in the first strain, and have a different distribution of ISs around the chromosome. Some bacteria, such as *Bacillus subtilis,* carry no insertion sequences.

Insertion sequences were first identified in the 1970s as elements that caused inactivation of genes required for galactose metabolism (Gal⁻ mutants). When an IS transposes and lands within the coding region of a gene, it disrupts the coding region and inactivates the gene (Fig. 13.6b). We now know that many of the spontaneous mutations isolated in *E. coli* are the result of IS transposition into a gene. Researchers have exploited this ability to cause mutation, using a more complex type of trans-

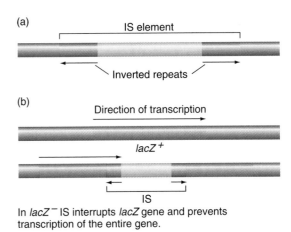

(a)

IS element

Inverted repeats

(b)

Direction of transcription

lacZ⁺

IS

In *lacZ*⁻ IS interrupts *lacZ* gene and prevents transcription of the entire gene.

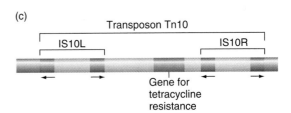

(c)

Transposon Tn10

IS10L   IS10R

Gene for tetracycline resistance

Figure 13.6 Transposable elements in bacteria. (a) An IS element. The ends of the IS element are inverted repeats of each other. Transposition depends on the ability of a transposase enzyme encoded by the IS element to recognize these inverted repeats. (b) Insertion of an IS element into a gene can be mutagenic. In this example, insertion of an IS element into the *E. coli lacZ* gene inactivates the gene because the IS element contains a transcription termination signal. (c) The composite transposon Tn10. In this transposon, two slightly different IS10 elements (IS10L and IS10R) flank almost 7 kb of DNA including a gene for tetracycline resistance. Because the ends of IS10L and IS10R are inverted with respect to each other, Tn10 can be mobilized by the IS10 transposase.

posable elements in bacteria known as Tn elements. In addition to a gene for transposase, these transposons carry genes conferring resistance to antibiotics or toxic metals such as mercury. One Tn element known as Tn10 consists of two IS10 elements flanking a gene encoding resistance to tetracycline (Fig. 13.6c). After the introduction of Tn10 into a cell, its transposition into, for example, the *lacZ* gene, produces a *lacZ⁻* mutant that is phenotypically both Lac⁻ and Tet^r (resistant to tetracycline). Because of these effects, the Tn element in the gene is an easily scored genetic marker for mapping experiments (as described later). In addition, because researchers know the sequence of Tn10, they can make a primer corresponding to the end of the Tn10 element to determine the base sequence of the adjacent DNA, and thereby identify the gene that was mutated by the Tn10 insertion. This is only one of the many ingenious uses of transposons researchers have devised.

Transcription in E. coli

During the transcription of the *E. coli* chromosome, different strands of DNA serve as the template for different genes. The transcription machinery moves clockwise on one DNA strand for a few kilobases, transcribing several contiguous and often related genes; then it switches to the other strand and moves counterclockwise for a few kilobases; then it switches back to the first strand. The only rule seems to be that the most highly transcribed genes tend to be oriented in the direction that the replication fork moves.

DNA Replication in E. coli

DNA replication begins at a site called *oriC*, for origin of chromosomal replication, and proceeds simultaneously in both directions (clockwise and counterclockwise) from that point to a termination region called *terC* (Fig. 13.7a). Genes that must undergo a particularly high rate of expression, such as the genes for ribosomal RNA and ribosomal proteins, are concentrated near the origin of replication. The reason is as follows. During rapid growth, new rounds of replication are initiated before the previous one is completed. This means that there are several replication forks progressing along the chromosome from the single origin of replication at any one time (Fig. 13.7b). As a result, genes near the origin are present in four to eight copies, while genes near the terminator are present in only one copy. The increase in copy number of genes near the origin allows the cell to produce more of the corresponding proteins.

Chromosomal Evolution: Comparing Different Strains of E. coli *and* E. coli *with* Salmonella typhimurium

Chromosomes of *E. coli* strains retrieved from different sources have minor differences in the sequences of individual genes and somewhat greater differences in the regions between genes. Most *E. coli* chromosomes carry a total of one to two dozen copies of four or five different insertion sequences. As we have seen, the IS positions and numbers vary from strain to strain.

Studies comparing the chromosome of a typical *E. coli* strain with the chromosome of a typical strain of *S. typhimurium*,

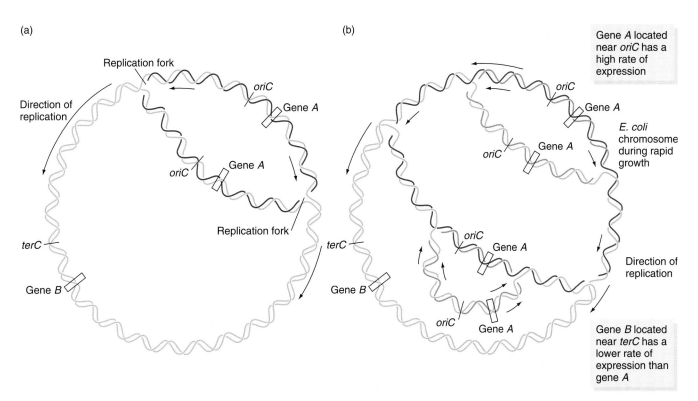

Figure 13.7 Replication of the ***E. coli*** **chromosome.** Replication begins at the origin (*oriC*) and proceeds to the terminator (*terC*). In the diagram, the arrow indicates the direction in which the replication fork moves. (a) Under conditions of slow growth, the origin of replication "fires" once in each cell division cycle. (b) Under conditions of rapid growth, replication initiates more times than the cell divides. As a result, there will be more copies of genes near the origin than genes near the terminator at any given time.

a related species that causes food poisoning in humans and more severe infections in some other mammals, reveal more sequence differences between related genes in the two species than between genes in two strains of *E. coli*. The insertion sequences found in *S. typhimurium* are also different from those found in *E. coli*. Larger chromosomal differences that distinguish the two species include several inversions of extensive chromosomal regions, the deletion from the *S. typhimurium* chromosome of whole groups of genes that are present in the *E. coli* chromosome; and *S. typhimurium*'s inclusion of several genes that *E. coli* lacks, notably genes involved in pathogenesis. Several factors help maintain the boundaries between even closely related bacterial species. For example, differences in DNA sequence make recombination between two species less efficient than recombination within a species. Moreover, unlike different strains of the same species, different species of bacteria encode restriction enzymes that can destroy the DNA of other species (see the Fast Forward Box in Chapter 5 and Chapter 8 for more about restriction enzymes).

Plasmids: Smaller Circles of DNA That Do Not Carry Essential Genes

Bacteria carry their essential genes—those necessary for growth and reproduction—in their large circular chromosome. In addition, some bacteria carry genes not needed for growth and reproduction under normal conditions in smaller circles of double-stranded DNA known as **plasmids** (Fig. 13.8a). Plasmids come in a range of sizes. The smallest are 1000 bp long; the largest are several megabase pairs in length. Bacteria usually harbor no more than one extremely large plasmid, but they can house several or even hundreds of copies of smaller DNA circles.

Although plasmids carry genes not normally needed by their bacterial hosts for growth and reproduction, these same genes may benefit the host cell under certain conditions. For example, the plasmids in many bacterial species carry genes that protect their hosts against toxic metals such as mercury. The plasmids of various soil-inhabiting *Pseudomonas* species encode proteins allowing the bacteria to metabolize chemicals such as toluene, naphthalene, or petroleum products. Since the 1980s, natural and genetically engineered plasmids of this type have become part of the tool kit for cleaning up oil spills and other contaminated sites. Plasmids thus help expand the capabilities of bacteria in nature, and also provide a rich source of unusual and useful proteins for commercial purposes.

Many of the genes that contribute to pathogenicity (that is, the capacity for causing disease in plants or animals) reside in plasmids; for example, the toxins produced by *Shigella dysenteriae*, the causative agent of dysentery, are encoded by plasmids. Genes encoding resistance to antibiotics are also often located on plasmids. The plasmid-determined resistance to multiple drugs was first discovered in *Shigella* in the 1970s. Multiple antibiotic resistance is often due to composite IS/Tn elements on a plasmid (Fig. 13.8b). As we see later, plasmids can be transferred from one bacterium to another, sometimes

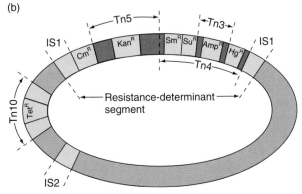

Figure 13.8   Plasmids are small circles of double-stranded DNA. (a) Electron micrograph showing plasmids and many circular DNA. (b) Some plasmids contain multiple antibiotic resistance genes. Transposons (IS and Tn elements) have played an important role in the evolution of these frightening plasmids by facilitating the movement of the antibiotic resistance genes onto the plasmid. The antibiotic resistance genes are in yellow (*Cm^R* for chloramphenicol, *Kan^R* for kanamycin, *Sm^R* for streptomycin, *Su^R* for sulfonamide, *Amp^R* for ampicillin, *Hg^R* for mercury, and *Tet^R* for tetracycline); IS elements are pale peach. Note that many antibiotic resistance genes are located between two IS1 elements, allowing them to transpose subsequently as a unit (a "resistance-determining segment").

even across species. Plasmids thus have terrifying implications for medicine. If transferred to new strains of pathogenic bacteria, the new hosts acquire resistance to many antibiotics in a single step.

One important group of plasmids allows the bacterial cells that carry them to make contact with another bacterium and transfer genes—both plasmid and bacterial—to the second cell. We describe this cell-to-cell mating, known as conjugation, in the next section.

GENE TRANSFER IN BACTERIA

Gene transfer plays an important role in the evolution of new variants in nature. Many cases of such horizontal transfer—so called because the traits involved were not transferred by inheritance from one generation to the next—have come to light through recent molecular and DNA sequencing analyses. Researchers make use of the various methods of gene transfer to map genes and to construct bacterial strains with which to test the function and regulation of specific genes.

Bacteria can transfer genes from one strain to another by three different mechanisms: transformation, conjugation, and transduction (Fig. 13.9). In all three mechanisms, one cell—the **donor**—provides the genetic material for transfer, while a second cell—the **recipient**—receives the material. In **transformation,** DNA from a donor is added to the bacterial growth medium and is then taken up from that medium by the recipient. In **conjugation,** the donor carries a special type of plasmid that allows it to come in contact with the recipient and transfer DNA directly. In **transduction,** the donor DNA is packaged within the protein coat of a bacteriophage and transferred to the recipient when the phage particle infects it. The recipients of a gene transfer are known as **transformants, exconjugants,** or **transductants,** depending on the mechanism of DNA transfer that created them.

All bacterial gene transfer is asymmetrical. Most recipients receive 3% or less of a donor's DNA; only some exconjugants contain a greater percentage of donor material. Thus, the amount of donor DNA entering the recipient is small relative to the size of the recipient's chromosome, and the recipient retains most of its own DNA. We now examine each type of gene transfer.

Transformation: Fragments of Donor DNA in the Immediate Environment Enter the Recipient and Alter Its Genotype

A few species of bacteria spontaneously take up DNA fragments from their surroundings in a process known as **natural transformation.** By contrast, the large majority of bacterial species can take up DNA from the surrounding medium only after laboratory procedures make their cell walls and membranes permeable to DNA in a process known as **artificial transformation.**

In Natural Transformation, the Recipient Cell Has the Enzymatic Machinery for DNA Import

Researchers have studied several species of bacteria that undergo natural transformation, including *S. pneumoniae,* the pathogen in which transformation was discovered by Frederick Griffith (see Chapter 5) and which causes pneumonia in humans; *B. subtilis,* a harmless soil bacterium; *H. influenzae,* a pathogen causing various diseases in humans; and *N. gonorrhoeae,* the microbial agent of gonorrhea.

In one study of natural transformation, investigators isolated *B. subtilis* bacteria with two mutations—*trpC2* and *hisB2*—that made them Trp$^-$, His$^-$ double auxotrophs. These double auxotrophs served as the recipients in the study; wildtype cells (*trp*$^+$, *his*$^+$) were the donors. The experimenters extracted and purified donor DNA and grew the *trpC2 hisH2* recipients in a suitable medium until the cells became **competent,** that is, able to take up DNA from the medium. Different bacterial species require different regimens to achieve competence. For

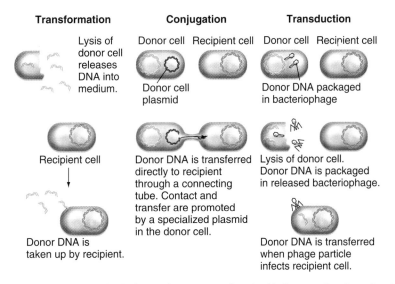

Transformation

Lysis of donor cell releases DNA into medium.

Recipient cell

↓

Donor DNA is taken up by recipient.

Conjugation

Donor cell   Recipient cell

Donor cell plasmid

Donor DNA is transferred directly to recipient through a connecting tube. Contact and transfer are promoted by a specialized plasmid in the donor cell.

Transduction

Donor cell   Recipient cell

Donor DNA packaged in bacteriophage

Lysis of donor cell. Donor DNA is packaged in released bacteriophage.

Donor DNA is transferred when phage particle infects recipient cell.

Figure 13.9   Three mechanisms of gene transfer in bacteria: An overview. In this figure, as in other related figures in this chapter, the chromosome of the donor is shown in blue, and the chromosome of the recipient is shown in orange. In transformation, fragments of donor DNA released into the medium from a previously lysed cell enters the recipient cell. In conjugation, a specialized plasmid (shown in red) in the donor cell promotes contact with another cell and initiates the transfer of DNA from the donor to the recipient. In transduction, DNA from the donor cell is packaged into bacteriophage particles (bacteriophage DNA shown in blue); when the viral particle infects a recipient cell, the donor DNA is transferred into the recipient.

(a)

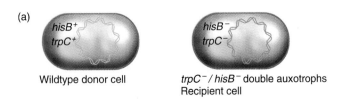

Wildtype donor cell

$trpC^-$ / $hisB^-$ double auxotrophs
Recipient cell

(b)

Competent cell recipient

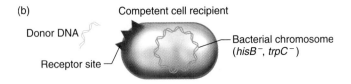

Donor DNA

Bacterial chromosome
($hisB^-$, $trpC^-$)

Receptor site

1. Donor DNA binds to recipient cell at receptor site.

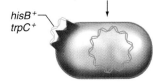

$hisB^+$
$trpC^+$

2. One donor strand is degraded. Admitted donor strand pairs with homologous region of bacterial chromosome. Replaced strand is degraded.

One strand degraded

3. Donor strand is integrated into bacterial chromosome.

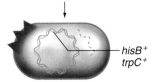

$hisB^+$
$trpC^+$

4. After cell replication, one cell is identical to original recipient; the other carries the mutant gene.

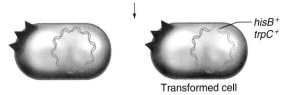

$hisB^+$
$trpC^+$

Transformed cell

Figure 13.10 Natural transformation in *B. subtilis*.
(a) Diagram of the genomes of a wildtype donor and a *hisB⁻ trpC⁻* double auxotroph. Selection for His⁺ and/or Trp⁺ phenotypes allows the identification of transformants. (b) Mechanism of natural transformation in *B. subtilis*. One strand of a fragment of donor DNA enters the recipient, while the other strand is degraded. The entering strand recombines with the recipient chromosome, producing a transformant when the recipient cell divides.

B. subtilis (Fig. 13.10a), competence occurs only in nearly starving cells at very specific times in the growth of the culture; investigators can starve the cells by growing them in a glucose-salts medium containing a limited amount of tryptophan and histidine. As growth of the culture slows toward the

end of what is known as the stage of logarithmic growth, a fraction of the bacteria—1% to 5% in *B. subtilis*—become competent and will take up DNA added to the medium at this time.

The mechanism of DNA uptake in *B. subtilis* is more complicated than was originally anticipated (Fig. 13.10b). The donor DNA fed to the competent cells usually consists of linear double-stranded molecules with an average length of 20 kb. The competent cells have enzyme complexes at their surfaces that take up one strand of a DNA fragment and simultaneously break down the other strand to its constituent nucleotides. The wildtype DNA strand admitted to the doubly auxotrophic cell quickly finds a homologous region of the bacterial chromosome and recombines with it, replacing the almost identical portion of one of the two strands of the recipient DNA (which is subsequently degraded). After the chromosome replicates and the cell divides, one of the two daughter cells becomes a transformant, carrying wildtype $trpC^+$ and/or $hisH^+$ genes from the donor.

To observe and count Trp⁺ transformants, researchers decant the liquid containing newly transformed recipient cells on a simple glucose-salts solid medium containing histidine. Recipient cells that did not take up donor DNA are unable to grow on this medium because it lacks tryptophan, but the Trp⁺ transformants can grow and be counted. To select for His⁺ transformants, researchers pour the transformation mixture on glucose-salts solid medium containing tryptophan, instead of histidine. In this study, the numbers of Trp⁺ and His⁺ transformants were equal. In conditions where *B. subtilis* bacteria become highly competent, 109 cells will produce approximately 105 Trp⁺ transformants and 105 His⁺ transformants.

To discover if any of the Trp⁺ transformants were also His⁺, the researchers used sterile toothpicks to transfer colonies of Trp⁺ transformants to a glucose-salts solid medium containing neither tryptophan nor histidine. Forty of every 100 Trp⁺ transferred colonies grew on this minimal medium, indicating that they were also His⁺. Similarly, tests of the His⁺ transformants showed that roughly 40% are also Trp⁺. Thus, in 40% of the analyzed colonies, the $trpC^+$ and $hisH^+$ genes had been cotransformed. **Cotransformation** is the simultaneous transformation of two or more genes.

Since during transformation, donor DNA replaces only a small percentage of the recipient's chromosome, it might seem surprising that the two *B. subtilis* genes are cotransformed with such high frequency. The explanation is that the *trpC* and *hisH* genes lie close together on the chromosome and are thus genetically linked. The entire *B. subtilis* chromosome is approximately 4700 kb long; only genes in the same chromosomal vicinity can be cotransformed. The closer together the genes lie, the more frequently they will be cotransformed. Thus, although the donor chromosome is fragmented into small pieces during its extraction for the transformation process, the wildtype $trpC^+$ and $hisH^+$ alleles are so close that they often appear together in the same donor DNA molecule. Sequence analysis shows that the *trpC* and *hisH* genes are only about 7 kb apart. By contrast, genes sufficiently far apart that they cannot appear together on a fragment of donor DNA will almost never be cotransformed, because transformation is so inefficient that recipient cells usually take up only a single DNA fragment.

Transformation usually incorporates a single strand of a linear donor DNA fragment into the bacterial chromosome of the recipient through recombination. However, if the donor DNA includes plasmids, recipient cells may take up an entire plasmid and acquire the characteristics conferred by the plasmid genes. Bacteriologists suspect that penicillin-resistant *N. gonorrhoeae* described in the introduction to this chapter originated through transformation by plasmids. The donors of the plasmids were *H. influenzae* cells disrupted by the immune defenses of a doubly infected patient; the plasmids carried the gene for penicillinase; and the recipient *N. gonorrhoeae* bacteria, transformed by the plasmids, acquired resistance to penicillin.

In Artificial Transformation, Damage to Recipient Cell Walls Allows Donor DNA to Enter the Cells

Researchers have devised many methods to transform bacteria that do not undergo natural transformation. The existence of these techniques was critical for the development of the gene-cloning technology described in Chapter 8. All the methods include treatments that damage the cell walls and membranes of recipient bacteria so that donor DNA can diffuse into the cells. With *E. coli*, the most common treatment consists in suspending the cells in a high concentration of calcium at cold temperature. Under these conditions, the cells become permeable to single- and double-stranded DNA. Because *E. coli* cells have efficient DNA exonucleases that degrade free ends, circular DNA is much more efficient than linear DNA at transforming *E. coli*.

Another technique of artificial transformation is *electroporation* in which researchers mix a suspension of recipient bacteria with donor DNA and then subject the mixture to a very brief high-voltage shock; the shock most likely causes

holes to form in the cell membranes. With the proper shocking conditions, recipient cells take up the donor DNA very efficiently. Transformation by electroporation works with most bacteria. The only exception is the group of helical bacteria known as spirochetes, which include the causative agents of human syphilis and Lyme disease. No one has yet developed a technique that transforms spirochetes.

Conjugation: Donor Cells Carrying Specialized Plasmids Establish Contact with and Transfer DNA to Recipients

In the late 1940s, Joshua Lederberg and Edward Tatum analyzed two *E. coli* strains that were each multiple auxotrophs and made the striking discovery that genes seemed to transfer from one *E. coli* cell to the next (Fig. 13.11a). Neither strain grew on a minimal glucose-salts medium. Strain A required supplementation with methionine and biotin; strain B required supplementation with threonine, leucine, and thiamine (vitamin B$_1$). Lederberg and Tatum grew the two strains together on supplemented medium. When they then transferred a mixture of the two strains to minimal medium, 1 in every 10^7 transferred cells proliferated to a visible colony. What were these colonies and how they did they arise?

To answer these questions, they performed the necessary control for this experiment. If they plated either strain alone on a minimal medium, nothing grew. Because they observed colonies only when they plated a mixture of the two strains on a simple medium, Lederberg and Tatum concluded that some sort of genetic exchange—rather than a highly unlikely set of simultaneous reversions—had taken place. *E. coli*, however, do not undergo natural transformation. Further experiments showed that instead, direct cell-to-cell contact was necessary for the exchange. In one experiment, Bernard Davis showed that the genetic exchange did not occur when cells of one strain were

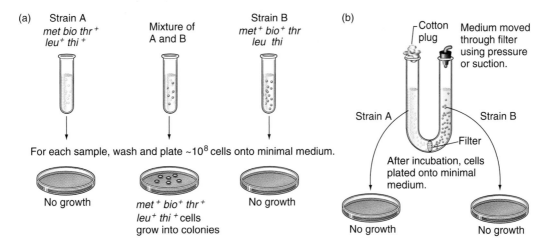

Figure 13.11 Conjugation: A type of gene transfer requiring cell-to-cell contact. (a) Lederberg and Tatum found that neither of two multiply auxotrophic strains could grow and form colonies on minimal medium. However, when they mixed cells of the two strains, gene transfer produced some prototrophic cells that could form colonies on minimal medium. (b) The U-tube experiment demonstrating that direct contact between cells is needed for conjugation. Here, the two strains used by Lederberg and Tatum were separated by a filter that allowed the passage of all molecules in the medium including DNA fragments, but prevented contact between the cells. The filter abolished gene transfer between the two strains.

mixed with a filtrate containing DNA of the other strain, or when a porous filter allowing the passage of liquid medium but not bacteria cells separated strain A from strain B in a U-shaped tube (Fig. 13.11b). More than a decade of further experiments confirmed that Lederberg and Tatum had observed what became known as bacterial conjugation: a one-way DNA transfer from donor to recipient initiated by **conjugative plasmids** in donor strains. Many different plasmids can initiate conjugation because they carry genes that allow them to transfer themselves (and sometimes some of the donor's chromosome) to the recipient.

The F Plasmid and Conjugation

Figure 13.12 illustrates the type of bacterial conjugation initiated by the first conjugative plasmid to be discovered—the F plasmid of *E. coli*. Cells carrying an F plasmid are called F^+ cells; cells without the plasmid are F^-. The F plasmid carries many genes required for the transfer of DNA, including genes for formation of an appendage, known as the pilus, by which a donor cell contacts a recipient cell, and a gene encoding an endonuclease that nicks the F plasmid's DNA at a specific site

(the origin of transfer). Once a donor has contacted a recipient cell (lacking the F plasmid) via the pilus, retraction of the pilus pulls the donor and recipient close together. The F plasmid DNA is then nicked and a single strand moves across a bridge between the two cells. When the donor DNA enters the recipient cell, it re-forms a circle and the recipient synthesizes the complementary DNA strand. In this $F^+ \times F^-$ mating, the recipient becomes F^+ and the donor remains F^+. By initiating and carrying out conjugation, the F plasmid acts in bacterial populations the way an agent of sexually transmitted disease acts in human populations. When introduced via a few donor bacteria into a large culture of cells that do not carry the plasmid, the F plasmid soon spreads throughout the entire culture.

The F Plasmid Occasionally Integrates into the E. coli Chromosome Where It Promotes the Conjugational Transfer of Essential Chromosomal Genes

The F plasmid contains three different IS elements: one copy of IS2, two copies of IS3, and one copy of the particularly long IS1000. These IS sequences on the F plasmid are iden-

---

a. The F plasmid contains genes for synthesizing connections between donor and recipient cells. The F plasmid is a 100-kb-long circle of double-stranded DNA. Host cells that carry it generally have one copy of the plasmid. By analogy with sexual reproduction, researchers think of F^+ cells as *male bacteria* because the cells can transfer genes to other bacteria. About 35% of F plasmid DNA consists of genes that control the transfer of the plasmids. Most of these genes encode polypeptides involved in the construction of a structure called the F **pilus** (plural, **pili**): a stiff thin strand of protein that protrudes from the bacterial cell. Other regions of the plasmid carry ISs and genes for proteins involved in DNA replication.

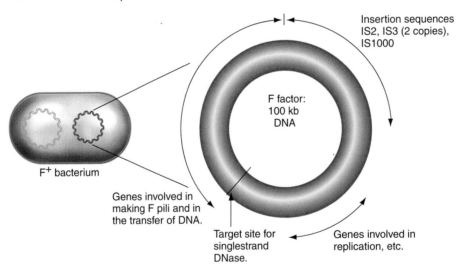

b. The process of conjugation.
1. **The pilus.** An average pilus is 1 μm in length, which is almost as long as the average *E. coli* cell. The distal tip of the pilus consists of a protein that binds specifically to the cell walls of F^- *E. coli* not carrying the F factor.
2. **Attachment to F^- cells (female bacteria).** Because they lack F factors, F^- cells cannot make pili. The pilus of an F^+ cell, on contact with an F^- cell, retracts into the F^+ cell, drawing the F^-

cell closer. A narrow passageway forms through the now adjacent F^+ and F^- cell membranes.
3. **Gene transfer: A single strand of DNA travels from the male to the female cell.** Completion of the cell-to-cell corridor signals an endonuclease encoded by the F plasmid to cut one strand of the F plasmid DNA at a specific site (*the origin of transfer*). The F^+ cell extrudes the cut strand through the passageway into the F^- cell at a rate of about 47,000 bases per minute. As it receives the single

Feature Figure 13.12   The F plasmid and conjugation

tical to copies of the same IS elements found at various positions along the bacterial chromosome. In roughly 1 of every 100,000 F$^+$ cells, homologous recombination between an IS on the plasmid and the same IS on the chromosome integrates the entire F plasmid into the *E. coli* chromosome (Fig. 13.13a). This insertion by recombination enlarges the 4700-kb chromosome by 100 kb of DNA. Cells whose chromosomes carry an integrated plasmid are called **Hfr** bacteria, because they produce a <u>h</u>igh <u>f</u>requency of <u>r</u>ecombinants for chromosomal genes in mating experiments with F$^-$ strains. Since the recombination event that results in the F plasmid's insertion into the bacterial chromosome can occur between any of the IS elements on the F plasmid and any of the corresponding IS elements in the bacterial chromosome, geneticists can isolate 20–30 different strains of Hfr cells (Fig. 13.13b). These different Hfr strains have different names; one, for example, is called HfrH after its discoverer William Hayes. Plasmids like the F plasmid that can integrate into the host chromosome are sometimes called **episomes.**

During bacterial reproduction, the integrated plasmid of an Hfr cell replicates with the rest of the bacterial chromosome. As a result, the chromosomes in daughter cells produced by cell division contain an intact F plasmid at exactly the same location that the plasmid originally integrated into the chromosome of the parental cell. All progeny of an Hfr cell are thus identical to the parent into whose chromosome the F plasmid originally inserted.

The integrated F plasmid still has the capacity to initiate DNA transfer via conjugation, but now that it is part of a bacterial chromosome, it can promote the transfer of some or all of that chromosome as well (Fig. 13.14). The transfer of DNA from an Hfr cell mated to an F$^-$ cell starts with a cut in the middle of the integrated F plasmid at the origin of transfer. Beginning at this site, half of the F plasmid moves into the recipient, followed by donor chromosomal DNA carrying bacterial genes. However, since in conjugation, the cell-to-cell connection is labile and easily disrupted, the cells often separate before transfer of the whole *E. coli* chromosome has been completed. As a result, the exconjugant is a partially diploid

strand of F plasmid DNA, the F$^-$ cell synthesizes a complementary strand so that after 2–3 minutes of contact, the formerly F$^-$ cell contains a double-stranded F plasmid and is now an F$^+$ cell.

4. **Meanwhile, back in the original F$^+$ cell, newly synthesized DNA replaces the single strand transferred**

to the previously F$^-$ cell. When the two bacteria separate at the completion of DNA transfer and synthesis, they are both F$^+$.

To summarize: Conjugation passes the F$^+$ plasmid and therefore maleness—the ability to transfer DNA to other cells—from F$^+$ to F$^-$ bacteria.

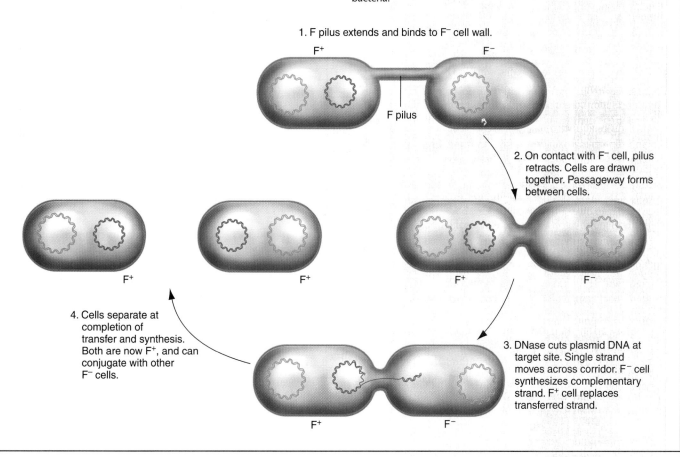

(a)

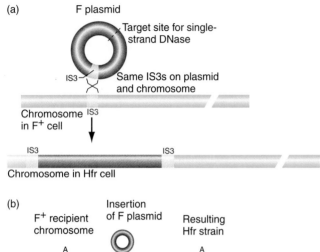

Chromosome in Hfr cell

(b)

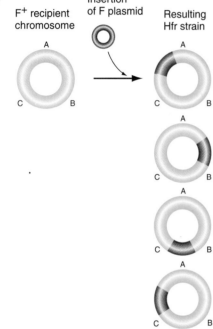

Figure 13.13   Integration of the F plasmid into the bacterial chromosome forms an Hfr bacterium. In this figure, the filled bar represents both strands of DNA. (a) Recombination between an IS element on the F plasmid and the same kind of IS element on the bacterial chromosome creates an Hfr chromosome. (b) Recombination can occur between any of the IS elements on the F plasmid and any corresponding IS element on the bacterial chromosome. This flexibility allows the creation of many different Hfr strains, with the F plasmid inserted at different sites and in different orientations (clockwise or counterclockwise) relative to the bacterial chromosome.

bacterium that carries half of the F plasmid attached to a variable amount of DNA from the Hfr cell's chromosome, as well as the F⁻ cell's entire genome. Because the transfer is usually interrupted in midstream, the second half of the F plasmid is not transferred and the recipient thus remains F⁻.

The single-stranded DNA fragment transferred to the recipient cell is copied into a double-stranded linear fragment by DNA polymerases in the new host cell. But because bacterial cells cannot properly replicate and transmit to subsequent generations linear fragments of DNA, DNA from the donor will be lost unless it is incorporated into the recipient's chromosome. Incorporation can occur through a double recombi-

nation event (catalyzed by bacterial recombination enzymes) between homologous regions on the linear DNA fragment from the donor and the circular chromosome of the recipient.

Only about 1 in 10,000 Hfr cells transfers an entire strand of its chromosome to a recipient F⁻ cell. And only those rare F⁻ cells that receive the entire Hfr chromosome, including the second half of the F plasmid, have the full 100 kb of F plasmid DNA that make them F⁺. Therefore, most of the progeny produced by an Hfr × F⁻ conjugation remain F⁻. This contrasts sharply with the result of conjugation initiated by an F⁺ cell, where almost all the exconjugants become F⁺. Thus, Hfr cells are poor donors of the entire F plasmid, but efficient donors of chromosomal genes. The situation is reversed for F⁺ donors, which can transfer bacterial genes only in those rare cells in which the F plasmid integrates to form an Hfr.

Analyses of Hfr Transfers Help Map Genes

Because genes in an Hfr chromosome are transferred into the recipient in a consistent order, researchers realized that they could use Hfr crosses to map genes. To gain useful information from a conjugational mating between an Hfr male and an F⁻ female, they analyze strains that differ from each other in the alleles of several genes. In early studies of the *E. coli* genome, for example, Elie Wollman and François Jacob used the following genotypes:

HfrH	×	F⁻
str^s (sensitive to streptomycin)		*str^r* (resistant to streptomycin)
thr^+ (able to synthesize the amino acid threonine)		*thr^-* (threonine auxotroph)
azi^r (resistant to sodium azide)		*azi^s* (sensitive to sodium azide)
ton^r (resistant to bacteriophage T1)		*ton^s* (sensitive to phage T1)
lac^+ (able to grow with lactose as sole source of carbon)		*lac^-* (unable to grow on lactose)
gal^+ (able to grow with galactose as sole source of carbon)		*gal^-* (unable to grow on galactose)

To isolate and analyze the exconjugants in which transfer and recombination had occurred, the researchers had to be able to select for only those particular cells. In one experiment using this donor-recipient combination just outlined, Wollman and Jacob mixed the two strains in a rich nonselective liquid medium that allowed the growth of both strains; they also established conditions that favored the rapid formation of cell-to-cell connections. Next, at 1-minute intervals, they violently stirred samples of the mating mixture in a kitchen blender, reasoning that the extreme agitation would separate pairs of cells before all genes from the donor had passed to the recipient, thereby interrupting the mating (Fig. 13.15a). (The experiment derives its name of *interrupted-mating experiment* from this part of the procedure.) They then poured samples of the terminated matings onto petri plates containing streptomycin, which killed the original donor cells. The plates also lacked threonine, to select against F⁻ cells that had not mated. Finally, after the exconjugants grew

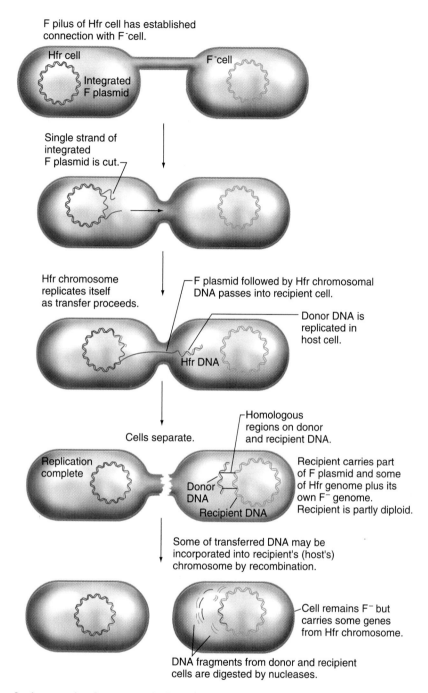

F pilus of Hfr cell has established connection with F⁻cell.

Hfr cell

Integrated F plasmid

F⁻cell

Single strand of integrated F plasmid is cut.

Hfr chromosome replicates itself as transfer proceeds.

F plasmid followed by Hfr chromosomal DNA passes into recipient cell.

Donor DNA is replicated in host cell.

Hfr DNA

Cells separate.

Homologous regions on donor and recipient DNA.

Replication complete

Donor DNA

Recipient DNA

Recipient carries part of F plasmid and some of Hfr genome plus its own F⁻ genome. Recipient is partly diploid.

Some of transferred DNA may be incorporated into recipient's (host's) chromosome by recombination.

Cell remains F⁻ but carries some genes from Hfr chromosome.

DNA fragments from donor and recipient cells are digested by nucleases.

Figure 13.14   Gene transfer in a mating between Hfr donors and F⁻ recipients. In an Hfr × F⁻ mating, single-stranded DNA is transferred into the recipient, starting with the origin of transfer on the integrated F plasmid. Thus, the first material to be transferred is part of the F plasmid, followed by adjacent genes from the Hfr donor's chromosome. Within the recipient cell, this single-stranded DNA is copied into double-stranded DNA. If mating is interrupted before the complete transfer of the Hfr chromosome, the recipient cell will contain a double-stranded linear fragment of DNA in addition to its own chromosome. Genes from the donor are retained in the exconjugant only if they recombine into the recipient's chromosome.

on these streptomycin plates lacking threonine, they used replica plating to copy each exconjugant colony onto media that would test for the four other markers present in the donor and recipient strains. In this way, they could determine which donor genes had transferred into the F⁻ recipients before the interruption of mating.

Figure 13.15b shows the frequency of recombinant colonies containing various alleles from the donor strain with each increase in the length of conjugation before violent agitation. After mating has proceeded for 3 minutes, a small fraction of the recombinants are *azi*ʳ, but not one carries the other donor alleles. At about 4 minutes, some of the recombinants

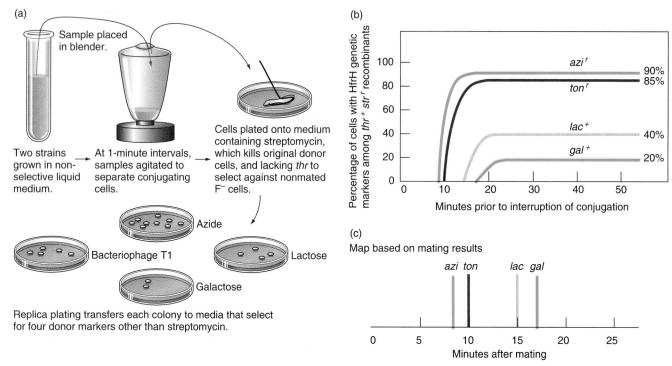

Figure 13.15 Mapping genes by interrupted mating experiments. (a) Wollman and Jacob mixed Hfr and F⁻ cells to allow mating, then agitated samples at 1-minute intervals in a kitchen blender to disrupt gene transfer. They plated the cells onto a medium that contained streptomycin (to kill the Hfr donor cells) and that lacked threonine (to prevent growth of F⁻ cells that had not mated; this was possible because earlier experiments had established that *thr*⁺ gets transferred into the recipient almost immediately after mating is initiated). Wollman and Jacob then tested the phenotypes of the exconjugants for other markers by replica plating. (b) Results of the Wollman-Jacob interrupted mating experiment. Donor genes enter the recipient at specific times after mating is initiated. The percentage of exconjugants containing a particular donor gene reaches a characteristic plateau that decreases for genes that enter the recipient later. (c) Gene order established from the data in part (b). Gene positions are determined by the number of minutes that elapse from the beginning of mating before a donor gene is first transferred into the recipient.

also have the donor's *ton*ʳ allele. By 10 minutes, the first *lac*⁺ alleles appear in the recombinant colonies; and by 17 minutes, *gal*⁺ arrives. The percentages of exconjugant colonies containing a particular gene from the Hfr donor increase with time until they reach a plateau characteristic of the gene: 90% for *azi*ʳ, 85% for *ton*ʳ, 40% for *lac*⁺, and 20% for *gal*⁺.

The presence of an integrated F plasmid in Hfr strains explains the two attributes of each gene revealed by the interrupted mating experiment: the postmating initiation time of gene transfer into the recipient cell and the plateau percentages of exconjugants that carry the donor gene. The time of gene transfer depends on the fact that all HfrH donor cells carry the F plasmid DNA at the same site on their chromosomes, and they all start to inject their DNA into the F⁻ cells at about the same time and in the same orientation. Thus, the time a gene first enters the recipient cell reflects the distance of that gene from the origin of transfer located within the integrated F plasmid. The first few HfrH donors to establish a mating connection with an F⁻ partner inject their *azi*ʳ allele through the intercellular connection by 3 minutes; more donors make the connection and transfer *azi*ʳ thereafter. The first *ton*ʳ alleles pass into F⁻ cells at about 4 minutes, the first *lac*⁺ alleles at about 8 minutes, and the first *gal*⁺ alleles at about 17 minutes after the mixing together of Hfr and F⁻

cells. This interpretation not only predicts the order of genes on the *E. coli* chromosome, it also makes it possible to map the distances between the genes. The units of distance are defined as *minutes of chromosome transfer;* each minute translates to 47,000 bases transferred (Fig. 13.15c).

The plateaus for the percentage of colonies carrying each transferred gene derive from the fragility of cell-to-cell connections. Transfer between Hfr donors and F⁻ recipients can spontaneously abort if the connected cells separate or if the transferring chromosome breaks. As a result, not all exconjugants receive even early arriving genes such as *azi*ʳ. In addition, the more time required to transfer a marker, the greater the chance that mating pairs will separate before the transfer occurs. As a result, the percentage of exconjugants receiving later arriving genes becomes successively smaller.

Comparisons of Interrupted Mating Studies Using Different Hfr Strains Confirm That the Bacterial Chromosome Is a Circle

The HfrH strain transfers genes in the order *azi-ton-lac-gal*. A different Hfr strain (strain 1 in Fig. 13.16a) transfers these genes in the opposite order: *gal-lac-ton-azi*. The maps obtained using these and other Hfr strains as donors can be related on the basis of the fact that in various Hfr strains, the F

(a)

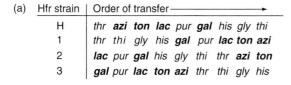

Hfr strain	Order of transfer ⟶
H	*thr* **azi ton** *lac* **pur gal** *his gly thi*
1	*thr thi gly his* **gal** *pur* **lac ton azi**
2	**lac** *pur* **gal** *his gly thi thr* **azi ton**
3	**gal** *pur* **lac ton azi** *thr thi gly his*

(b)

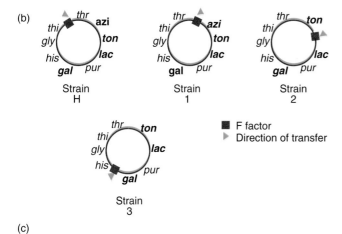

(c)

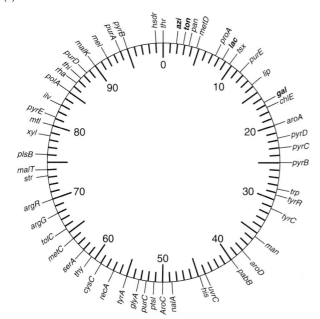

Figure 13.16 Order of transfer of genes in different Hfr strains. (a) Comparison of the order in which genes are transferred in crosses between F⁻ recipients and various Hfr strains. The genes studied by Wollman and Jacob are indicated in bold; genes mapped later by other investigators are in normal font. (b) Interpretation of the data shown in part (a). In different Hfr strains, the F plasmid integrates in different locations and in different orientations into the circular donor chromosome. (c) A partial genetic map of the *E. coli* chromosome.

plasmid integrates into the bacterial chromosome at different locations and in different orientations (Fig. 13.16b). The location of the origin within each type of F plasmid determines where transfer begins and thus the start of that particular map. The orientation of the plasmid with respect to the bacterial

chromosome determines whether transfer occurs clockwise or counterclockwise.

Comparisons of maps drawn from many Hfr × F⁻ crosses provided the first clues that the bacterial chromosome is circular (see Fig. 13.16b). The maps from different Hfr matings are all permutations of the same order of genes that would occur if the genes were located on a circular chromosome, but with transfer starting at a different position and moving in either the clockwise or counterclockwise direction. Many physical studies examining the DNA of the bacterial genome later confirmed the circularity of the bacterial chromosome.

Researchers now use well-studied collections of Hfr strains with different sites and orientations of F plasmid insertion to determine the approximate position on the *E. coli* genetic map of any gene they can follow on the basis of its phenotype (Fig. 13.16c). They designate the gene's position by the minute at which its transfer begins. The circumference of the entire circular map is 100 minutes, the time required for transfer of the complete *E. coli* chromosome. The top of the circle, where 0 minutes meets 100 minutes, is arbitrarily set at the *thr* gene, the site of F plasmid insertion in HfrH. They have so far located about 1400 of the estimated 4000–5000 *E. coli* genes in this way. A selection of these mapped genes appears in Fig. 13.16c.

Recombination Analyses of Hfr Crosses Improve Mapping Accuracy

Interrupted mating experiments cannot distinguish the relative positions of genes within about 2 minutes of each other and thus give only a crude idea of gene location. However, analyses of conjugation experiments can provide better resolution if they take into account what happens within an F⁻ exconjugant after it has received genes from the Hfr donor. Because bacteria cannot completely replicate and transmit the linear DNA from the Hfr donor remaining inside the F⁻ cell at the interruption of mating, the only way the progeny of the recipient F⁻ cell can inherit donor genes is if those genes recombine into the F⁻ cell's chromosome. Note that a single recombination event, or any odd number of recombination events, between the circular bacterial chromosome and the linear DNA fragment would be lethal: By making the chromosome linear, it would prevent multiplication of the cell. As a result, conjugation can produce recombinant cells only if an even number (2, 4, 6, and so forth) of crossovers occurs.

To map genes on the basis of the frequencies of recombinant classes emerging from a conjugation experiment, you must consider only exconjugants that received from the donor all of the genes to be mapped. Otherwise, the resulting progeny would reflect not only recombination events, but also the likelihood that a particular gene from the Hfr donor was transferred to recipient cells. To ensure that all the recipient cells you are examining received all pertinent genes from the donor, you select exconjugants that contain the last of the genes transferred from the Hfr donor, as these cells must also have received the earlier arriving genes.

With this understanding of how to select suitable exconjugants and what type of recombination occurs inside those F⁻ cells, it is possible to map closely linked genes by a type of three-point cross (Fig. 13.17). Recall that in three-point crosses in eukaryotes, you can infer that the gene in the middle is the gene that recombined with respect to the other two in the double-crossover class. You can do the same in prokaryotes, although here, the equivalent of the double-crossover class requires four crossovers (Fig. 13.17a–c). Comparisons of recombination frequencies between genes in a particular bacterial cross provide a measure of relative map distances (Fig. 13.17d).

Bacterial Geneticists Can Use a Special Type of F Plasmid Known As F′ for Genetic Complementation Studies

We saw earlier in this chapter that insertion of the F plasmid into the bacterial chromosome occurs in about 1 F⁺ cell per 100,000 to produce an Hfr cell. In approximately that same proportion of Hfr cells, an excision event causes the Hfr cell to revert to an F⁺ cell; and in a small fraction of these excision events, an error in recombination generates a plasmid containing most of the F plasmid genes plus a small region of the bacterial chromosome that had been adjacent to the integrated F plasmid. The newly formed plasmid carrying most of the genes of the F plasmid plus some bacterial DNA is known as an **F′ plasmid** (Fig. 13.18a). F′ plasmids replicate as discrete circles of DNA inside *E. coli* cells. They are transferred to recipient (F⁻) cells in the same way that F plasmids are transferred. The difference is that a few chromosomal genes will always be transferred as part of the F′ plasmid. This ability to transfer chromosomal genes on a molecule that can replicate independently of the bacterial chromosome makes the F′ plasmid a perfect vehicle for creating partial diploids for complementation analysis.

Recall from Chapter 6 that complementation tests depend on cells that are diploid for the genes under analysis. Although bacteria are haploid, F′ plasmids carrying bacterial genes can create specific regions of partial diploidy. In fact, it is possible to find F′ plasmids carrying *E. coli* genes that lie close to any of the chromosomal positions where F plasmids integrate. For example, some F′ plasmids carry the *trp* genes that control the biosynthesis of tryptophan. An F′ plasmid carrying these genes is called an F′ *trp* plasmid.

To create partial diploids in bacterial cells using F′ plasmids, researchers must transfer the F′ plasmid into a strain whose chromosome is not deleted for the genes carried by the F′ plasmid (the chromosomes of the cells in which the aberrant plasmid excision occurred are now deleted for those genes). Because the F′ plasmids can still promote conjugation, experimenters easily accomplish this by mating F′-carrying cells with F⁻ cells (Fig. 13.18b). The exconjugants from these matings are partial diploids containing two copies of certain bacterial genes—one on the F′ plasmid, the other on the bacterial chromosome. Partial diploids in which the two gene copies are identical are known as **merozygotes;**

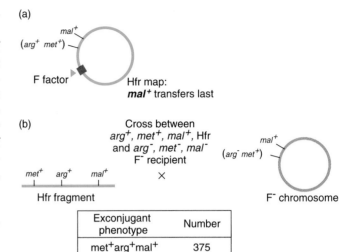

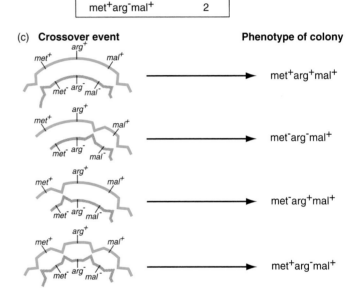

Figure 13.17   Mapping genes using a three-point cross.
(a) The order of closely linked genes *met* and *art* can be determined using a three-point cross in which a nearby marker that is known to be transferred later is selected in the cross (in this case, *mal*⁺). (b) The percentage of *mal*⁺ exconjugants having all other combinations of the *arg* and *met* alleles are counted. (c) Crossovers that occur to produce the phenotypes in the proportions they appear are shown. The *met*⁺*arg*⁻*mal*⁺ combination in the exconjugant requires quadruple crossovers so is the least frequently observed class of exconjugants.

partial diploids carrying different alleles of the same gene are **heterogenotes.**

As an example of complementation studies using F′ plasmids, consider an analysis of mutations affecting tryptophan biosynthesis (Fig. 13.18c). You first construct a partial diploid by introducing an F′ plasmid carrying a *trp*⁻ mutation into a bacterial strain that carries a different *trp*⁻ mutation in the chromosome. Growth of a particular partial

(a)

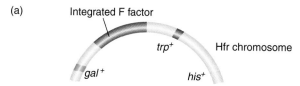

Integrated F factor

trp⁺   Hfr chromosome

gal⁺   *his⁺*

A rare recombination event between regions of limited sequence homology permits out-looping of F factor including *trp⁺* locus.

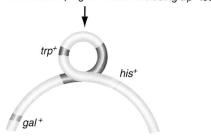

trp⁺

his⁺

gal⁺

Separation of F creates F' *trp⁺* plasmid and a chromosome deleted for the *trp* genes.

△*trp*

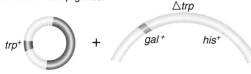

trp⁺   +   *gal⁺*   *his⁺*

(b)

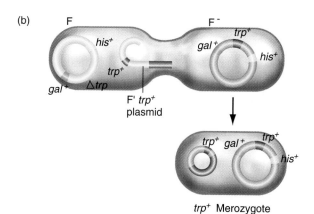

F   F⁻

his⁺   *trp⁺*   *gal⁺*   *his⁺*   *trp⁺*

gal⁺ △*trp*   F' *trp⁺* plasmid

trp⁺ gal⁺ trp⁺ his⁺

trp⁺ Merozygote

(c)

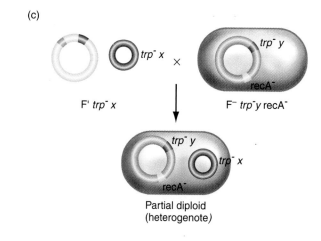

F' *trp⁻ x*   ×   *trp⁻ y*   recA⁻   F⁻ *trp⁻ y* recA⁻

trp⁻ y   *trp⁻ x*   recA⁻

Partial diploid (heterogenote)

Figure 13.18   F' plasmids. (a) Formation of an F' plasmid. Improper excision of an integrated F plasmid from an Hfr chromosome can create an F' plasmid containing adjacent bacterial genes. (b) Transfer of an F' *trp* plasmid into a *trp⁺* F⁻ recipient creates a merozygote with two copies of *trp⁺*. (c) Using F' plasmids for complementation analysis. Different *trp⁻* mutations are induced on the F' *trp* plasmid in one strain (*trp⁻x*), and in the chromosome of an F⁻ strain (*trp⁻y*). The recipient strain is defective in recombination (*recA⁻*) so no crossing-over occurs between F' and chromosome. Matings between these strains create partial diploids with both mutations; the phenotype of the partial diploid establishes whether the two mutations complement each other or not.

diploid on minimal medium without tryptophan would indicate that the mutations complement each other and are thus in the different genes. If the cells do not grow, the mutations must be in the same gene. Complementation studies using F' *trp* heterogenotes have shown that there are five different *trp* genes: *A, B, C, D,* and *E,* each one corresponding to one of the five polypeptides required for the biosynthesis of tryptophan.

Summary of Conjugation in E. coli

The F plasmid that initiates conjugation in *E. coli* functions in one of three ways. In F⁺ cells, it remains a small, independent, self-replicating circle of DNA that during conjugation transfers only its own genes. In Hfr cells, it integrates via homologous recombination into the bacterial chromosome so that during conjugation, it transfers some or all of the chromoso-

mal DNA. In F' cells, the F plasmid carries a small amount of chromosomal DNA, and during conjugation, only the chromosomal DNA on the plasmid is transferred, along with the plasmid itself. Table 13.2 compares the conjugational cycles of the three F-plasmid incarnations.

Transduction: Gene Transfer Via Bacteriophages

The bacteriophages, or phages, that infect, multiply in, and kill various species of bacteria are widely distributed in nature. Most bacteria are susceptible to one or more such viruses. During infection, a virus particle may incorporate a piece of the bacterial chromosome and introduce this piece of bacterial DNA into other host cells during subsequent rounds of infection. The process by which viral particles transfer bacterial DNA from one host cell to another is known as transduction.

TABLE 13.2 Comparing the Conjugational Cycles of F⁺, Hfr, and F′ Cells

	In an F⁺ Cell	In an Hfr Cell	In an F′ Cell
Donor and Recipient before Conjugation	F plasmid F⁺ F⁻	Hfr F⁻	F′ plasmid F′ F⁻
Events of Conjugation			
Donor and Recipient after Conjugation	F⁺ F⁺	Hfr F⁻ partial diploid (transient)	F′ F′ merozygote
Comments	Efficient transfer of F factor; poor transfer of bacterial genes.	Transfer of *entire* F factor only inefficiently. Efficient transfer of bacterial genes, but requires an even number of crossovers in recipient.	Efficient transfer of F′ element, which includes a few bacterial genes, generating stable partial diploids. Poor transfer of other chromosomal genes.

Bacteriophage Particles Are Produced by the Lytic Cycle

When a bacteriophage injects its DNA into a bacterial cell, the phage DNA takes over the cell's protein synthesis and DNA replication machinery, forcing it to express the phage genes, produce phage proteins, and replicate the phage DNA (see Fig. 6.15). The newly produced phage proteins and DNA assemble into phage particles, after which the infected cell bursts, or lyses, releasing 100–200 new viral particles ready to infect other cells. The cycle resulting in cell lysis and release of progeny phage is called the **lytic cycle** of phage multiplication. The population of phage particles released from the host bacteria at the end of the lytic cycle is known as a **lysate.**

Generalized Transduction

Many kinds of bacteriophages encode enzymes that cut up and then break down the chromosomes of the cells in which they are multiplying. Digestion of the bacterial chromosome by these enzymes sometimes generates fragments of bacterial DNA about the same length as the phage genome, and these phage-length bacterial DNA fragments occasionally get incorporated into phage particles in place of the phage DNA (Fig. 13.19). After lysis of the host cell, the phage particles can attach to and inject the DNA they carry into other bacterial cells, thereby transferring genes from the first bacterial strain (the donor) to a second strain (the recipient); recombination between the injected DNA and the chromosome of the new host completes the transfer. This process, which can result in the transfer of any bacterial gene between related strains of bacteria, is known as **generalized transduction.**

Mapping Genes by Generalized Transduction

As with cotransformation, two genes close together on the bacterial chromosome may be cotransduced. The frequency with which two bacterial genes are cotransduced depends directly on the distance between them: the closer two genes are, the more likely they are to appear on the same short DNA fragment and be packaged into the same transducing phage. Two genes that are farther apart than the amount of DNA that can be packaged into a single phage particle can never be cotransduced. For bacteriophage P1, a phage often used for generalized transduction experiments with *E. coli,* the maximum separation allowing cotransduction is about 90 kb of DNA, which corresponds to about 2% (or 2 minutes) of the bacterial chromosome. You will recall that researchers cannot use interrupted conjugation to define the relative location of genes injected less than 2 minutes apart. Thus, the two methods of genetic mapping can be used in concert to cover the entire spectrum of distances. In mapping a new *E. coli* mutation, investigators first find the approximate position of the mutation on the *E. coli* chromosome map by mating the mutant strain to a small collection of different Hfr strains. They then use P1 phage transduction to map the new mutation relative to other known genes in the region.

Consider, for example, the three genes—*thyA, lysA,* and *cysC*—that all map by interrupted mating experiments to the region between 53 and 56 minutes of the *E. coli* chromosome.

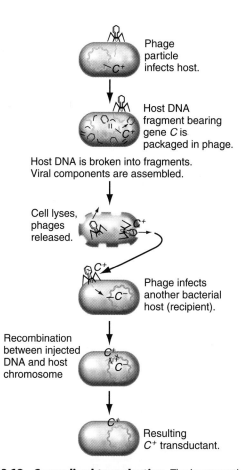

Figure 13.19 Generalized transduction. The incorporation of random fragments of bacterial DNA from a first host (the donor) into bacteriophage particles yields generalized transducing phages. When these phage particles infect a second host (the recipient), donor DNA is injected into the recipient cell. Recombination of donor DNA fragments with the recipient cell chromosome yields transductants.

(a) Donor: $thyA^+$ $lysA^+$ $cysC^+$

make P1 lysate infect recipient

Recipient: $thyA^-$ $lysA^-$ $cysC^-$

Selected marker	Unselected marker
thy^+	47% lys^+; 2% cys^+
lys^+	50% thy^+; 0% cys^+

(b) $lysA$ $thyA$ $cysC$

Figure 13.20 Mapping genes by cotransduction frequencies. (a) A $thyA^+$ $lysA^+$ $cysC^+$ donor strain of *E. coli* is infected with bacteriophage P1. The resulting lysate is used to infect a $thyA^-$ $lysA^-$ $cysC^-$ recipient strain. Either thy^+ or lys^+ cells are selected, then tested for lys and cys genotype. (b) Genetic map based on the data in part (a). The $lysA$ and $cysC$ genes were not contransduced but $thyA$ and $cysC$ were cotransduced at a low frequency, so they must be closer together than $lysA$ and $cysC$.

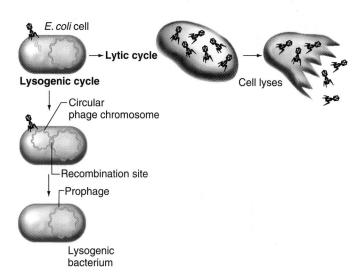

Figure 13.21 Temperate phages can choose between lytic and lysogenic modes of reproduction. Cells infected with temperate bacteriophages (whose chromosomes are shown in green) enter either the lytic or lysogenic cycles. In the lytic cycle, phages reproduce by forming new bacteriophage particles that lyse the host cell and can infect new hosts. In the lysogenic cycle, the phage chromosome becomes a prophage incorporated into the host chromosome.

Where do they lie in relation to one another? You can find out by using a P1 generalized transducing lysate from a wildtype strain to infect a $thyA^-$, $lysA^-$, $cysC^-$ strain and then selecting the transductants for either Thy$^+$ or Lys$^+$ phenotypes. After replica plating, you test each type of selected transductant for alleles of the two nonselected genes. As the phenotypic data in Fig. 13.20a indicate, $thyA$ and $lysA$ are close to each other but far from $cysC$; $lysA$ and $cysC$ are so far apart that they never appear in the same transducing phage particle; $thyA$ and $cysC$ are only rarely cotransduced. Thus, the order of the three genes must be $lysA$, $thyA$, $cysC$ (Fig. 13.20b).

Temperate Phages Can Integrate into the Bacterial Chromosome

The types of bacteriophages we have discussed so far are **virulent:** After infecting a host, they always enter the lytic cycle, multiplying rapidly and killing the cell. Other types of bacteriophages are **temperate:** Although they can enter the lytic cycle, they can also enter an alternative **lysogenic cycle,** during which

their DNA integrates into the host genome and multiplies along with it, doing little harm to the host (Fig. 13.21). The choice of lifestyle—lytic or lysogenic—occurs when a temperate phage injects its DNA into a bacterial cell, and depends on many factors including environmental conditions. Normally when temperate phages inject their DNA into host cells, some of the cells undergo a lytic cycle, while others undergo a lysogenic cycle. The integration of the viral DNA into the host chromosome is known as **lysogeny:** a stable state that is harmless and sometimes, depending on the genes carried by the phage DNA, even helpful to the host cell. The integrated phage genome is called a **prophage.** A bacterial cell that carries a prophage is a lysogen, or **lysogenic bacterium.** Many strains of bacteria in nature are lysogenic for at least one phage. The DNA of most temperate phages forms a temporary circular molecule, which subsequently recombines with the bacterial chromosome at a particular site, similar to the way in which an F plasmid inserts into the chromosome during the formation of an Hfr cell. Each type of temperate phage has a specific site of integration into the host bacterium's genome.

Phage Lambda (λ): A Model Temperate Phage

Phage λ is a temperate phage containing a linear molecule of double-stranded DNA 48 kb in length (Fig. 13.22a). Researchers have completely sequenced those 48-kb pairs and learned more about the genetics and lifestyle of phage λ than about any other temperate phage. After injection into a susceptible bacterium, the sticky ends of the phage DNA spontaneously reanneal to form a 48-kb double-stranded circle. This circle can enter the lytic cycle, multiplying rapidly and lysing the cell; or it can em-

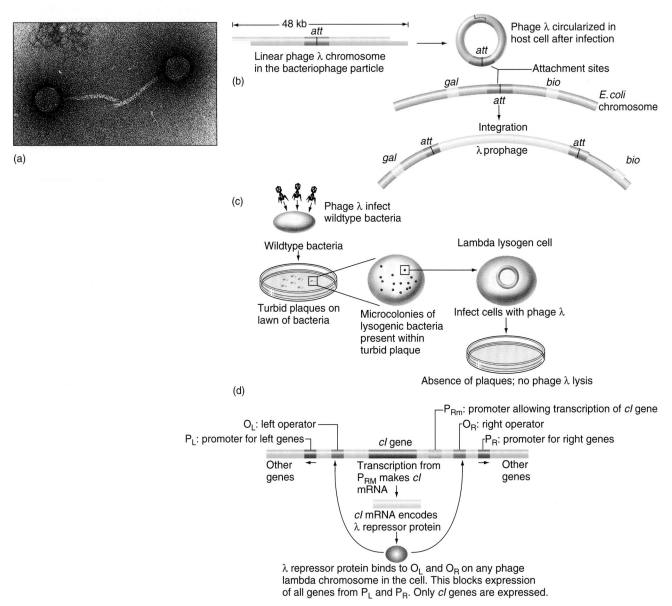

Figure 13.22 Bacteriophage λ and lysogeny. (a) Electron micrograph of phage λ particles. (b) Integration of the bacteriophage λ chromosome to initiate the lysogenic cycle. The 48-kb λ chromosome circularizes by virtue of its sticky ends. Recombination between similar *att* sites on the λ chromosome and the *E. coli* chromosome allows integration of the λ prophage. (c) λ lysogens are immune to superinfection (infection with additional λ phage). Phage λ forms turbid plaques, because even though many *E. coli* cells are killed by phage λ growing in the lytic cycle, some bacteria become lysogenized by λ prophage and grow as microcolonies within the plaque. New λ phage cannot lyse these immune lysogenic cells. (d) The λ repressor protein is responsible for maintaining the lysogenic state and for immunity to superinfection. The repressor protein, the product of the *cI* gene, binds to DNA preventing transcription of other λ genes, either on the prophage or on the genome of incoming λ particles.

bark on the lysogenic cycle, recombining through its own special λ attachment site (*att*) with a nearly identical copy of the *att* site on the bacterial chromosome located between the *gal* genes (whose products degrade galactose) and the *bio* genes (whose products synthesize biotin), at about 17 minutes on the *E. coli* map (Fig. 13.22b). Bacteria lysogenic for phage λ thus carry a chromosome that is 48 kb (about 1%) longer than the chromosome carried by cells of the same strain that lack the λ prophage. The decision to enter the lytic or the lysogenic cycle depends on many factors, but in general, the infection of one cell by a single λ particle leads to the lytic cycle, while simultaneous infection of a cell by two or more λ particles favors lysogeny.

Lysogens Are Immune to Further Infection by the Same Phage

One of the most significant differences between lysogenic and nonlysogenic bacteria is the lysogen's acquisition of immunity to further infection by the same kind of temperate phage. Phage λ, for example, can attach to the lysogen's cell wall and inject its DNA, but inside the lysogenic host, the phage DNA remains inert and incapable of entering the lytic cycle because the λ **repressor protein** is present in the lysogen (Fig. 13.22c). The λ repressor protein, which is responsible for this phenomenon, functions as follows. When the λ prophage replicates as part of the bacterial chromosome, most of the 50 or so genes it contains remain inactive. The most significant exception to this inactivity is expression of the *cI* gene, which encodes the λ repressor protein. This protein shuts down expression of the majority of λ genes by binding to prophage DNA at specific sites, thereby blocking transcription of most phage genes (Fig. 13.22d). When a λ phage particle infects a lysogen, it encounters repressor proteins produced by the λ prophage, which are floating free in the cell's cytoplasm. These repressor proteins bind to the infecting phage DNA, preventing the new phage from transcribing its genes and thus making the host cell immune to infection by incoming λ particles.

Events That Damage DNA May Induce Prophages to Enter the Lytic Cycle

In 1 in 100,000 lysogens, the repressor protein is released from the prophage DNA, and the prophage leaves the bacterial chromosome, forms a DNA circle, and enters the lytic cycle, causing the host cell to lyse and release 100–200 new phage particles (Fig. 13.23). This process of phage DNA excision from the bacterial chromosome and entry into the lytic cycle is called **phage induction.** Chemicals and physical treatments that damage DNA indirectly cause repressor inactivation and increase the frequency of phage induction from 1 per 100,000 to nearly 100%. For example, small doses of ultraviolet light or of the antibiotic mitomycin C, both of which damage DNA, can cause lysis of an entire bacterial culture and the release of 100–200 viral particles from each cell in the culture. Another efficient means of induction is to use λ phages carrying a temperature-sensitive mutation in the repressor-encoding *cI* gene; with such phages, an increase in temperature causes the

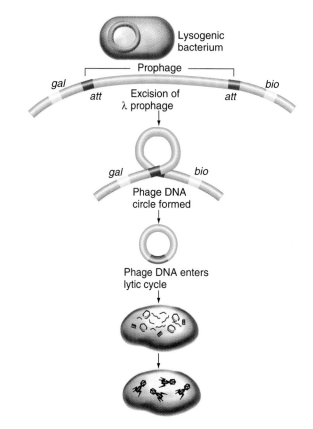

Figure 13.23 Induction of a λ lysogen. Treatments that inactivate the λ repressor protein cause a lysogenic cell to enter the lytic cycle. Reversing the steps of integration, the prophage chromosomes excises from the host chromosome to make a circular λ chromosome. Replication of this circular chromosome, and transcription and translation of the phage genes on it, produce new phage particles and the eventual lysis of the host bacterium.

repressor to become inactivated, the culture to be induced, and the cells to enter the lytic cycle. A liter of culture containing 10^9 bacteria per milliliter could produce as many as 10^{14} phage particles after induction by any of these methods.

During Induction, Phages May Acquire the Capacity for Specialized Transduction (Fig. 13.24)

The excision of integrated phage DNA by recombination between the two copies of the *att* site at the ends of the prophages is not 100% accurate. Sometimes the recombination event that would normally excise the prophages instead occurs between chromosomal DNA adjacent to the phages and DNA within the prophages. As a result, some chromosomal DNA becomes part of the phage genome incorporated into a phage particle; this is analogous to the faulty excision of an F plasmid that creates the F′ plasmid. Since phage λ integrates between the *gal* and the *bio* genes, on excision, it may pick up DNA carrying either the *gal* genes or the *bio* genes. Thus, in a lysate of phage λ induced from a donor Gal⁺ lysogenic strain of *E. coli,* you might find that roughly one phage per million carries a small region of the donor bacterial chromosome, including the wildtype

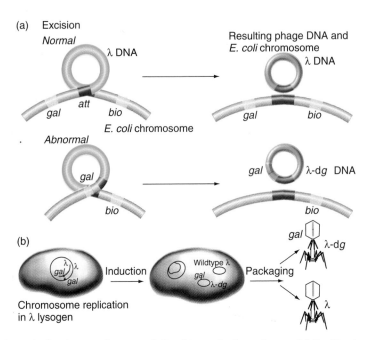

Figure 13.24 **Errors in prophage induction produce specialized transducing phages.** (a) Illegitimate recombination between DNA sequences in the λ prophage and the bacterial chromosome forms a λ-*dg* circle, which contains most but not all of the λ chromosome, along with bacterial genes including the *gal* operon. (b) Chromosome replication can generate more than one copy of the λ prophage in the same cell. Simultaneous induction of these prophages can excise both wildtype λ chromosomes and λ-*dg* chromosomes. The wildtype λ chromosome encodes all the proteins needed for packaging of phage particles. Some phage particles will be wildtype, while others are specialized transducing phages that contain λ-*dg*.

gal$^+$ gene, that it picked up from the bacterial chromosome during its transition from the lysogenic to the lytic cycle. Bacteriophages carrying mainly phage DNA but also one or a few of the bacterial genes that lie near the site of prophage insertion are known as **specialized transducing phages;** they can transfer genes from one bacterium to another in the process known as **specialized transduction.**

Specialized transducing phages include most but not all of the phage λ DNA plus a piece of the *E. coli* chromosome. The excised DNA might, for example, lack a fragment of λ DNA and include a fragment of the bacterial chromosome carrying the *gal* operon. If some essential phage genes are lost in this process, the transducing phage becomes unable to reproduce by itself and is called *defective.* Geneticists designate the excised circles that are *gal*$^+$ but deleted for some λ genes as λ-*dg* (lambda, defective, *gal*$^+$).

If the λ-*dg* phages are defective, how can they reproduce? The answer is that defective phages require help from wildtype phages, which must also be present in the cell undergoing induction. During the normal growth of *E. coli* and many other bacteria, most cells contain at any moment two or more identical copies of most chromosomal regions (see Fig. 13.7). In a growing culture of lysogenic bacteria that is induced to lyse, all of the prophages leave the chromosomes simultaneously. Thus induction can result in the excision of two kinds of circles in the same cell: λ-*dg* and wildtype λ. Both types of circles can replicate, and the wildtype λ chromosome can encode all needed λ proteins, including those that are absent from the

λ-*dg* circle. These proteins will separately package both wildtype and defective chromosomes into individual phage particles (see Fig. 13.24b).

We have seen that specialized transduction allows the transfer from donor to recipient of genes close to the site of prophage insertion. Bacteriophage λ is thus most useful for transducing the closely linked *gal* and *bio* genes adjacent to this phage's *E. coli* attachment site. However, with techniques for finding rare λ and other prophage insertions at a variety of sites on the *E. coli* chromosome, researchers can achieve specialized transduction for a variety of bacterial genes.

Summary: Comparing Generalized and Specialized Transduction

Phage particles that act as agents of generalized transduction differ in three critical ways from particles that carry out specialized transduction:

1. Generalized transducing phages pick up donor bacterial DNA during the lytic cycle, at the point when DNA is packaged inside a phage protein coat; specialized transducing phages pick up the donor bacterial DNA during the transition from the lysogenic to the lytic cycle.

2. Generalized transducing phages can transfer any bacterial gene or set of genes contained in the right size DNA fragment into the bacterial chromosome; specialized transducing phages can transfer just those genes near the site of prophage insertion.

3. Generalized transducing phages do not contain any phage DNA and therefore cannot reproduce themselves; specialized transducing phages can reproduce if helper wildtype phages are also present in the cell.

Many Naturally Occurring Mechanisms of Gene Transfer Have Been Adapted for Genetic Analysis

Expanding knowledge of gene-transfer mechanisms and the availability of data from genomic sequencing have helped simplify genetic analysis. Here we describe some of the techniques that geneticists have developed based on their growing understanding of the naturally occurring phenomena already described.

Mutant Isolation

Transposons have played the largest role in simplifying genetic analysis. The insertion of a transposon into a gene, resulting in the gene's inactivation, is the basis of many mutant screens. Transposons are useful as mutagenic agents because they contain genes for easily selectable antibiotic resistance. To carry out transposon mutagenesis, geneticists introduce a transposon into a cell as part of a DNA molecule that cannot replicate on its own inside the cell. For the transposon to be passed on during cell division, it must transpose from the incoming DNA molecule to the bacterial chromosome. By growing cells on a medium containing antibiotics, it is possible to select for those cells in which transposition has occurred. One can then screen the resulting population of cells, which contain transposons at different locations around the chromosome, for the mutant phenotype.

Analysis of the sequence of bacterial genomes has led to the identification of genes whose functions are not yet known. One approach to determining the function of such genes is to make a knockout mutation in the chromosomal gene, using recombinant DNA techniques and the homologous recombination machinery of the cell (Fig. 13.25). The gene to be mutated is first cloned on a plasmid. Next, a gene for antibiotic resistance is inserted into the middle of the cloned gene through use of a restriction enzyme that cuts only within the gene. The resulting circular molecule is then cut with a restriction enzyme that recognizes only a vector sequence. The resulting linear DNA fragment is used to transform a population of cells, after which selection for antibiotic resistance identifies those cells in which recombination between the mutated gene on the linear DNA fragment and the wildtype gene on the bacterial chromosome resulted in replacement of the wildtype gene with the mutated copy. The phenotype of the cell containing the knockout mutation provides clues about the function of the gene.

Recall, however, that knockout (complete loss-of-function) mutations are not always the most desirable class of mutation. For genes encoding essential proteins, knockout mutations cannot be isolated, and for nonessential genes, mutations that alter, rather than eliminate, the function of the gene may be more informative.

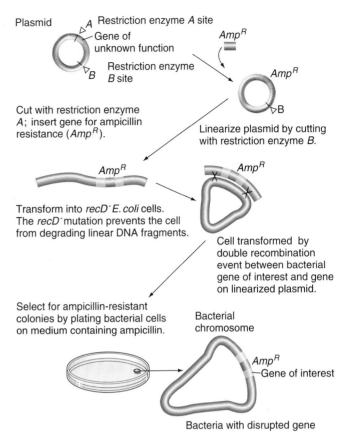

Figure 13.25   One way to characterize the roles of uncharted bacterial genes. This figure shows one protocol to disrupt and inactivate a bacterial gene of unknown function. First, an ampicillin resistance gene (*Amp^R*) is inserted into the gene of interest, located on a plasmid. The linearized plasmid is transformed into a *recD⁻ E. coli* strain where it recombines to leave a disrupted gene. It is then possible to study the phenotypic effects caused by inactivation of the gene.

Mapping

Researchers can now map mutations originally identified by transposon insertion through DNA sequence analysis. Using a primer that recognizes a DNA sequence near the end of the transposable element, they sequence into the adjacent DNA (that is, the DNA of the gene that was interrupted by the transposition). They then match this sequence with the genome sequence database to determine which gene was interrupted by the transposon. This technique not only indicates map position, but also identifies the gene involved.

The standard Hfr mapping procedures described earlier are still used for mapping a mutation that is not caused by transposon insertion. To facilitate such Hfr mapping, researchers have created a set of Hfr strains with different origins of transfer around the chromosome and a Tn10 transposon in a location in each strain that ensures transfer of the transposon-based tetracycline resistance marker early in a mating. Each Hfr strain is mated for a short period of time to an F⁻ cell containing the mutation under analysis, and the resulting cells are screened for antibiotic resistance. For one of the Hfr strains, the mating will result in loss of the mutant phenotype in most of the Tet^r exconjugants. The map location of the mutation is therefore near the origin of transfer (the site at which the F plasmid integrated into the bacterial chromosome) in this particular Hfr strain. This quick method for mapping a mutation eliminates the need for a series of interrupted matings and is thus much less labor intensive.

COMPREHENSIVE EXAMPLE: GENETIC DISSECTION HELPS EXPLAIN HOW BACTERIA MOVE

Bacteria can move up gradients of chemical attractants, such as food particles, and down gradients of chemical repellents, such as toxins. This sophisticated behavior of seeking out or avoiding chemicals diffusing through their growth medium is known as **bacterial chemotaxis.** Researchers have applied the powerful procedures for mutating, mapping, and cloning bacterial genes to identify the genes and proteins that function in bacterial chemotaxis and to discover how the proteins work together to accomplish their task.

How Bacteria Move to Achieve Chemotaxis

With specially constructed microscopes that make it possible to record the motion of individual bacteria, scientists have observed that cells swim rapidly in one direction for about 1 second, then abruptly reorient and swim in another direction (Fig. 13.26). The 1-second, rapid, one-way movement is called a "straight run"; the abrupt reorienting movement is a "tumble." The change in direction is random; thus bacteria move through space in a **random walk.** With the addition of an attractant or repellent, the behavior of individual bacteria changes. They still move in a random walk, but the time spent in a straight run is longer immediately after the addition of an attractant and shorter immediately after the addition of a repellent. This reactive behavior is known as a **biased random walk:** Each bacterium still executes a random walk with frequent changes in direction, but its movement over time is biased toward or away from a chemical gradient. The biased behavior of individual bacteria determines the net movement of a population toward an attractant and away from a repellent.

If the concentration of attractant or repellent does not change, within a short time, the bacteria resume the unbiased random walk they were doing before the addition of either type of chemical. The ability to stop responding when the stimulus is present in an unchanging concentration allows a bacterium to "sense" when it is moving up or down a concentration gradient; this aspect of chemotaxis is known as **adaptation.** It is analogous to the way you adjust to the bright light of day when you emerge from a dark movie theater—after a while in the new environment, your eyes adapt and you no longer blink.

Rotating Flagella Are the Mechanical Basis of the Random Walk

Bacteria have many long, rotating flagella distributed over their surface. When the flagella all rotate counterclockwise, they bundle together, propelling the cell in a forward run. When they rotate clockwise, the helical structure of each flagellum prevents bundling, and the cell undergoes a chaotic motion producing a tumble (Fig. 13.27).

You can observe the direction of rotation of the flagella on individual bacteria by attaching the cells to the surface of a petri dish with an antibody against the flagella, and then look-

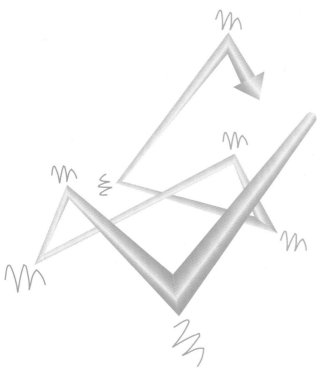

Figure 13.26   The run and tumble of a bacterium's random walk. The lack of a consistent direction is seen in the run and tumble style of movement. Squiggles indicate tumbles.

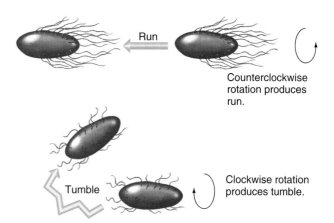

Figure 13.27   Counterclockwise rotations of bacterial flagella produce runs; clockwise rotations cause tumbles. During a run, each flagellum turns counterclockwise, allowing all the flagella to bundle together and rotate counterclockwise as a unit. When flagella turn clockwise, they cannot bundle and thus cannot go in a straight line, producing a tumble.

ing down through a microscope at the rotating cells. As with runs and tumbles, the cells spend about 1 second rotating counterclockwise, then switch and rotate clockwise for a brief period. Immediately after the addition of an attractant, cells spend longer periods rotating counterclockwise; immediately after the addition of a repellent, they spend shorter periods rotating counterclockwise. From these observations, it is possible to conclude that gradients of attractants and repellents influence the direction of rotation of the flagella.

Many Bacterial Mutants Cannot Carry Out Chemotaxis

Bacteria inoculated into the center of a nutrient plate that is wet enough for them to swim in, will multiply and use up the nutrients in the immediate vicinity. As they do so, the population of cells swarms out in concentric circles, with the leading edge moving toward the unused nutrient supply (Fig. 13.28). This behavior is the basis of a quick test for mutants that cannot move toward food; it also provides a way to isolate such mutants. After the motile cells have swarmed, it is possible to lift from the center of the plate the cells that did not move. Given the known details of how bacteria move in chemotaxis, one would anticipate a variety of mutants incapable of chemotaxis. Some might be defective in making flagella; some in turning the flagella; others in coordinating the behavior of flagella; and still others in detecting gradients of chemicals. Investigators have isolated all of these types of mutants. Genetic and biochemical analyses of the mutants' properties have produced a detailed understanding of how bacteria accomplish the sophisticated task of moving toward and away from chemicals.

Flagellum Mutants

More than 20 *fla* genes are required to generate the complex structure of a flagellum (Fig. 13.29). Most of each flagellum's extracellular portion is composed of a single protein, but electron microscopy of the intracellular structure reveals a series of rings, the only known example of a biological wheel. Mutations in the *fla* genes usually prevent production of functional flagella.

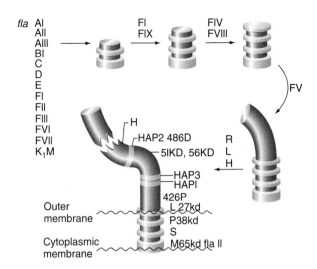

Figure 13.29 More than 20 genes are needed to generate a bacterial flagellum. Proteins are required for anchoring the flagella in the membranes and for the switch in directions, as well as the main structural protein of the flagella outside the cell.

Motor Mutants

The *mot* genes are required to turn the flagellum. Mutant *mot⁻* cells have normal-looking flagella, but they are paralyzed.

Signal Transduction Mutants

Mutants in the *che* (*chemotaxis*) genes have flagella that rotate, but only in one direction—either clockwise or counterclockwise. The *che* genes encode components of a **signal transduction system,** which relays messages from the cell surface, where nutrients are detected, to the motor, where the messages control the frequency with which the direction of rotation changes. The signal transduction system must decide whether the concentration of a chemical is increasing or decreasing, send the appropriate stimulus, and then undergo adaptation to bring the system back to its unstimulated state. The proteins encoded by the *che* genes send signals by phosphorylating one another. Phosphorylation is the signal that tells the next protein in the cascade whether to be active or inactive.

Receptor Mutants

All the mutants described so far have a defective response to all attractants and repellents. But some mutants respond in a defective way to a particular attractant or repellent, while responding normally to others. A simple "capillary test" reveals why. If a capillary pipette containing a nutrient is placed in a solution of bacteria, the cells swim into the tube (Fig. 13.30). Researchers leave the capillary tube in the solution for a standard amount of time, then remove it and determine the number of cells it contains by diluting and plating the bacteria for colony count. With this assay, they can discover the response of bacteria to specific nutrients and determine whether particular bacteria can respond to one nutrient in the capillary in the presence of a high concentration of another nutrient (both in

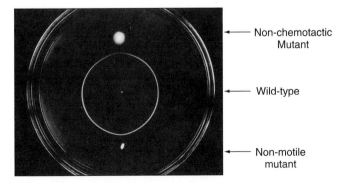

Figure 13.28 Isolating bacterial mutants that cannot move toward food. The wildtype bacteria move away from the center of the plate toward more food as nutrients are used up but the non-chemotactic and non-motile mutants are unable to sense or move toward food respectively.

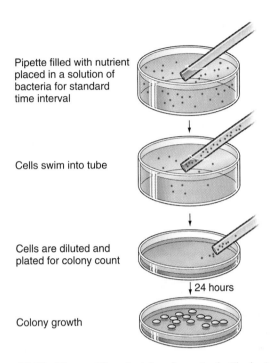

Pipette filled with nutrient placed in a solution of bacteria for standard time interval

Cells swim into tube

Cells are diluted and plated for colony count

↓ 24 hours

Colony growth

Figure 13.30 The capillary test for chemotaxis. The basis of this test is to see whether bacteria swim into a capillary tube containing attractants and/or repellents. The number of bacteria retained in the tube after a standard amount of time allows for quantitative measurements of chemotactic behavior.

the bacterial solution and in the capillary). Their observations confirm that bacteria normally respond to many different attractants and repellents independently. This suggests that the bacterial systems for detecting these chemicals are composed, in part, of different molecules.

Other studies have shown that mutants defective in specific responses have normal flagella that rotate normally and

that switch from clockwise to counterclockwise with the expected frequency for an unstimulated cell. These mutants do not, however, respond when a particular stimulus is added to their growth medium, because the membrane proteins that act as receptors for the attractant or repellent in question are defective. These receptor proteins have an extracellular domain that interacts with one or a few attractants or repellents, a transmembrane domain, and an intracellular domain that interacts with the Che proteins to stimulate signal transduction (Fig. 13.31).

Not surprisingly, the cell surface receptors that bind particular chemicals are the molecules that are modified during adaptation. The protein products of the *cheB* gene reversibly methylate positions on the intracellular domains of the receptors; the protein products of the *cheR* gene reversibly demethylate those same sites. The methylation status of the intracellular domain, in turn, determines the receptors' ability to respond to the binding of attractant or repellent by the extracellular receptor domain.

Figure 13.32 diagrams our current understanding of how the products of the *fla, mot, che,* and membrane receptor genes interact with one another to generate bacterial chemotaxis.

THE FUTURE OF BACTERIAL GENETICS

E. coli was the first organism in which the detailed genetic analysis of cellular processes was practical on a large scale. Studies of *E. coli* using the classical genetic techniques described in this chapter helped establish basic principles of cell physiology as well as protocols for genetic analysis. With the advent of recombinant DNA technology, bacteria (notably again *E. coli*) became the vehicle for cloning genes; mutants

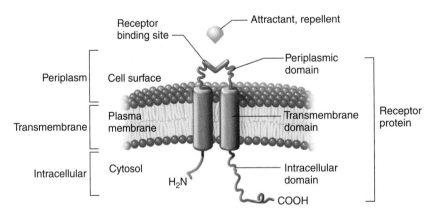

Figure 13.31 Bacteria have cell surface receptor proteins that recognize particular attractants or repellents. The binding of an attractant or repellent molecule to the periplasmic domain of a receptor specific for the compound causes changes in the conformation of the receptor's intracellular domain. The intracellular domain then interacts with the general components of the signaling pathway.

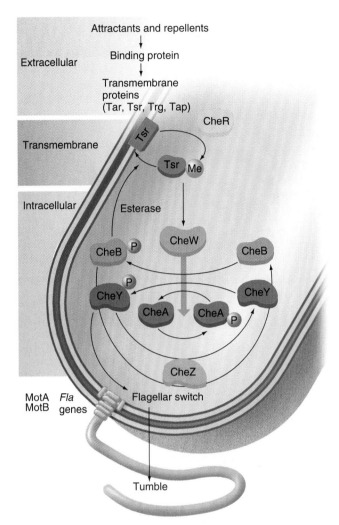

Figure 13.32   The genetic and molecular basis of bacterial chemotaxis. The *fla* genes are necessary for the structure of the flagella; *mot* genes are involved in the flagellar motor and determine direction of rotation; *che* genes process the signals from receptors and communicate with the Mot proteins to cause the appropriate response.

nomic information to resolve basic biological questions and practical social problems.

The DNA sequence of an organism's genome provides a rich, new source of data and opens up many avenues of exploration. The fruitfulness of the approach is evident in the analysis of *Haemophilus influenzae* whose DNA sequence, published in 1996, was the first full cellular genome to be completely sequenced. Knowledge of the sequence revealed several interesting features of the genome. For example, it contains many copies of the single-strand uptake sequence that is specifically recognized by the *H. influenzae* transformation system. The frequency of this sequence suggests that strand uptake is important for *H. influenzae* survival. Such uptake may be a mechanism for rapid evolutionary changes that allow the bacterium to avoid the host (human) defense systems. In another example, earlier experiments had identified particular repeated sequences preceding some of the genes required for pathogenesis. With the complete genomic sequence available, bacterial geneticists noted that these repeated sequences were associated with other genes whose function was unknown. When they mutated these genes of unknown function, they found that the genes also contribute to pathogenesis and were able to assign them a function. This knowledge increased their understanding of the way in which *H. influenzae* cause disease.

Now and in the future, human needs will undoubtedly influence the choice of which organisms to sequence. The bacteria chosen for study may be major pathogens or organisms with unusual and perhaps commercially useful gene products. With the pathogens, one hope is that knowledge of the genome will lead to the identification of vaccine candidates, an ever more pressing concern in this era of the diminishing efficacy of antibiotics due to increased bacterial resistance to such drugs.

The challenge for bacterial geneticists in the postgenomic era will be to use classical and recombinant DNA methodology in analyzing the function of genes. With the fundamental knowledge of what each gene does, they will then seek to understand how a bacterial cell integrates various gene products into its physiology. With greater knowledge of the physiology, they can progress to a better understanding of the complex interactive world of microbial ecology. Thus, it is likely that microorganisms, with their small genomes in which it is relatively easy to isolate mutants, will remain the organisms of choice for examining many fundamental questions about cell function.

that had been isolated earlier, for example, defective recombination mutants, served to maintain a collection of clones. Bacterial genetics is now in the midst of a genomic phase and on the threshold of a postgenomic phase. Researchers have already cloned the complete genomes of several bacterial species and are beginning to see how they can use this ge-

Genetics and Society

How Bacteria Cause Disease

Bacteria that cause disease in other organisms are known as *pathogens*. Their ability to initiate illness is their *virulence*. Mild pathogens cause relatively mild symptoms; virulent pathogens inflict serious damage that may threaten life. To be successful, a pathogen must outwit the varied defenses of a host organism, multiply inside the host, and be transmitted to other hosts. How do bacterial pathogens accomplish all this in humans?

The Steps of Bacterial Pathogenesis

Entry into the host. The human skin is a protective barrier that excludes most bacteria, but there are several ways to breach its defenses. Some species of bacteria gain entry to the body via wounds that damage the skin; *Clostridium tetani,* the agent of tetanus, is one such species. Other bacteria pass through openings established by insect bites; *Yersinia pestis,* the agent of bubonic plague, for example, travels from rodent hosts to humans via fleas and their bites. A third avenue of access to the human host is transmission to mucosal surfaces. *Mycobacterium tuberculosis,* the agent of tuberculosis, infects the respiratory tract via aerosols produced by sneezing or coughing; *N. gonorrhoeae,* the agent of gonorrhea, infects the urogenital tract via sexual transmission; *Vibrio cholerae,* the agent of cholera, infects the digestive tract via ingestion.

Once inside the host, some bacteria adhere to the surface of eukaryotic cells in the region of entry. Attachment depends on interactions between protein structures on the bacterial cell surface—similar to the pili that help initiate conjugation—and specific receptors, usually composed of carbohydrate, on the surface of the host's eukaryotic cells (Fig. A). Adherence is essential to pathogenesis. One illustration of this fact is the observation that the few pathogenic strains of *E. coli* are able to synthesize attachment proteins, while the large majority of nonpathogenic *E. coli* strains make no attachment proteins and are unable to adhere to cells of the intestinal mucosa. After adhesion, pathogenic bacteria begin to replicate.

Some bacteria penetrate to deeper layers of host cells. *Helicobacter pylori,* the bacteria associated with many cases of ulcers, twists through the layer of mucus protecting the stomach and adheres to cells on the surface of the stomach wall. After penetrating the stomach's mucosal lining, *H. pylori* secretes urease, an enzyme that generates ammonia, which in turn, neutralizes the stomach acid in the area. Other penetrating bacteria forge a path into tissues deep within the body by secreting enzymes such as collagenase, which break down the molecules that help hold tissues together. These penetrating bacteria then adhere to cells and begin to replicate.

Some bacteria invade host cells. Multiplication inside those cells leads to cell lysis (Fig. B). *Shigella dysenteriae,* the agent of bacterial dysentery, invades the epithelial cells of the intestines. *Legionella*

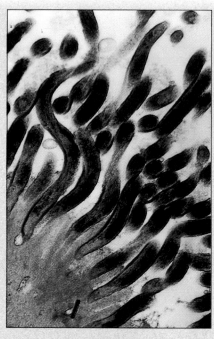

Figure A   Bacteria can adhere to the surface of eukaryotic cells. Shown here are spirochetes attached to bowel cells.

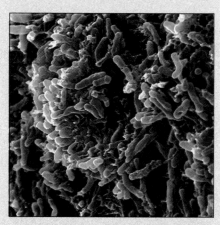

Figure B   Some bacteria, such as *Mycobacterium tuberculosis* (green cells), can invade and multiply within host cells.

pneumophila (the agent of Legionnarie's disease) and *M. tuberculosis* live and replicate within the large white blood cells known as *phagocytes,* which normally destroy bacteria.

Bacterial replication can damage the host in several ways. First, the presence of a large number of bacteria at one site can block the normal flow of materials. In the respiratory system, for example, a mass of bacteria can block the flow of oxygen through the smaller air passages; in the heart, large numbers of bacteria can clog a valve, hindering the flow of blood. Second, bacterial multiplication inside invaded cells destroys those cells, causing inflammation and disarray that undermine the function of essential tissues. Finally, some bacteria produce toxins that act on site or travel to other parts of the body. Tetanus bacteria, for example, produce a neurotoxin that interferes with the transmission of nerve impulses and causes muscle spasms in the neck and jaw. Dysentery bacteria produce a toxin that poisons intestinal cells and causes diarrhea.

In summary, adherence to the surface of host cells near the point of entry, penetration and adherence to tissues deep within the body, and invasion of host cells allow bacteria to replicate, using nutrients provided by the host. Sufficient replication produces large colonies that cause disease.

The Human Host Puts Up a Fight Based on Natural and Synthetic Defenses

Several natural components of the human immune system as well as human-made drugs act to deter and cure bacterial infections.

The skin and mucosal membranes. We have already seen that bacteria must breach the skin or mucosal linings of the mouth, nose, or digestive and urogenital tracts to gain entry into the host. The skin is a tough, nearly impervious barrier. The mucosal membranes are very acidic, providing an environment inhospitable to most bacteria. Mucus also contains sterilizing factors such as the enzyme lysozyme, which destroys bacterial cell walls.

Antibodies. The immune system responds to bacterial infection with the production of immense numbers of antibodies that bind specifically to various proteins on the invading bacteria. Some antibodies block the surface attachment molecules the bacteria use to adhere to eukaryotic cells; others may agglutinate the toxin molecules produced by the bacteria, rendering them nontoxic. The binding of antibodies to bacteria also activates the complement cascade—a series of reactions between a set of eleven proteins that ultimately form a complex on the bacteria and punch holes in their cell membranes.

Phagocytes and other cells of the immune system. Phagocytes are relatively large white blood cells that engulf and destroy bacteria. Highly differentiated phagocytes known as macrophages become activated by other cells of the immune system to pursue and dispose of specific invaders.

Drugs. In the last hundred years, pharmaceutical research has produced a large number of bactericidal drugs, including the sulfa drugs, penicillin, and other antibiotics. If administered soon enough after infection, these drugs can kill susceptible bacteria before they replicate sufficiently to cause serious symptoms.

Bacteria Fight Back Via Evasion of Immune Responses and the Development of Drug Resistance

The success of a pathogen depends on effective host entry strategies, efficient multiplication within the host, and the possibility of transmission to a new host. By these measures, *N. gonorrhoeae* is one of the most successful bacterial pathogens of humans in the world today. It has evolved impressive stratagems for bypassing the skin barrier and overcoming human immune defenses; it has also become resistant to several antibiotics.

For *N. gonorrhoeae,* transmission via sexual intercourse circumvents the skin barrier, while the unprotected sexual practices of many human cultures ensures entry into the mucosal linings of the urogenital tract of a large number of hosts worldwide. Once inside the host, the bacteria secrete an enzyme that destroys immunoglobulin A, the type of antibody that is active in the bodily secretions it encounters in the urogenital tract. The bacteria then buy time for replication by varying their cell surface molecules, the very molecules the human immune system recognizes and reacts against; because an antibody response against one molecule has little effect against a different molecule, the variation tactic helps *N. gonorrhoeae* evade the immune system. The bacteria also avoid immune destruction through the special construction of their cell membrane; although components of the complement cascade can still bind to the bacterial surface, the spatial configuration of the bound complement is aberrant and, as a result, the proteins are unable to punch holes in the bacterial membrane. The gonorrheal bacterium escapes destruction by phagocytes by invading and replicating inside other types of host cells. Finally, as we have seen, many populations of *N. gonorrhoeae* have developed resistance to three widely used antibiotics—penicillin, spectinomycin, and tetracycline—through genes acquired from other bacterial species. Thus, for the moment, *N. gonorrhoeae* seems to be winning the battle between the bacteria and their human hosts. New drugs, however, could temporarily tip the balance in the other direction.

Knowledge of Bacterial Pathogenesis Has Led to an Understanding of the Genetic Underpinnings of Virulence

Virulence, the ability of a pathogen to cause disease, depends on a bacterium's production of proteins that promote adherence to host cells, penetration of host tissues, the invasion of host cells, the production of bacterial toxins, and the avoidance of host defenses. Bacterial geneticists have begun to determine which genes contribute to the pathogenicity of particular bacterial species by using recombinant DNA techniques to transfer genes from pathogenic strains into nonvirulent strains. If the nonvirulent strains become somewhat pathogenic, the likely conclusion is that the transferred genes conferred virulence. A significant insight from this work is that the genes conferring virulence are often present as clusters on plasmids or transposons. Their tandem arrangement on small pieces of DNA that move from strain to strain underlies the one-step conversion of nonpathogenic bacteria to virulent pathogens. The medical implications of this fact merit close attention and emphasize the need to learn more about gene transfer.

C O N N E C T I O N S

The study of bacterial genetics underscores the unity of genetic phenomena in all types of living organisms. Double-stranded DNA serves as the genetic material in bacteria as it does in eukaryotes. The general mechanisms of gene expression, DNA replication, and recombination are also similar in prokaryotes and eukaryotes. The study of bacterial genetics shows, however, that within the unity of basic biological phenomena, there is a remarkable diversity of detail. Unlike most eukaryotic cells, bacteria do not produce gametes through meiosis or new generations of cells by the union of gametes. Nevertheless, bacteria can exchange genes between different strains through transformation, conjugation, and transduction. Together, these three modes of gene transfer increase the potential for evolution of the genetic material.

Knowledge of bacterial genetics raises the possibility of combatting bacteria harmful to humans, animals, and crops and utilizing beneficial bacteria for positive purposes, such as de-grading oil slicks and protecting crops from freezing. Knowledge of bacterial genetics also increases our understanding of the organelles of eukaryotic cells. Biologists believe that mitochondria, the cell organelles that produce energy for metabolic processes, and chloroplasts, the photosynthetic organelles of plant cells, are descendants of aerobic bacteria that fused with the earliest nucleated cells. Mitochondria are similar in size and shape to today's aerobic bacteria and have their own DNA, which replicates independently of the cell's nuclear genetic material. Chloroplasts are similar in shape and size to certain cyanobacteria and have self-replicating DNA carrying bacteria-like genes. Based in part on these observations, the **endosymbiont theory** proposes that chloroplasts and mitochondria originated when free-living bacteria were engulfed by primitive nucleated cells. Host and guest formed cellular communities in which each member adapted to the group arrangement and derived benefit from it. Chapter 14 examines the structure and function of the chromosomes of mitochondria and chloroplasts.

E S S E N T I A L   C O N C E P T S

1. Bacteria are prokaryotic cells with no membrane-enclosed nucleus or other cell organelles. The bacterial genome consists of a single circular chromosome in which the genes are tightly packed, with about one gene per kilobase pair.

2. During transcription in *E. coli,* different strands of DNA serve as the template for different genes, with the most frequently transcribed genes oriented in the direction of replication fork movement. Replication begins at the origin of chromosomal replication (*oriC*) and proceeds bidirectionally around the circular genome to the termination region (*terC*).

3. In addition to their chromosome, some bacteria carry plasmids: small circles of double-stranded DNA. Plasmids may include genes that benefit the bacterial host under certain conditions. One important group of plasmids promotes conjugative gene transfer between two bacteria.

4. Bacterial genomes contain IS and Tn elements, transposons that can move between sites on any DNA molecule in the cell.

5. *Transformation* is a form of gene transfer in which donor DNA that is floating free in the growth medium enters a recipient cell. Some bacteria have cellular machinery that supports efficient natural transformation. Species that do not undergo natural transformation can be induced to take up DNA by treatments that disrupt their cell walls in a process known as artificial transformation.

6. *Conjugation* is a second form of gene transfer. It depends on direct cell-to-cell contact between a donor carrying a conjugative plasmid (the F plasmid is one example) and a recipient lacking such a plasmid. In crosses between F^+ and F^- strains, only the plasmid is transferred. In crosses between Hfr and F^- strains or between F' and F^- strains, the plasmid also transfers chromosomal genes.

7. *Transduction* is a third form of gene transfer in bacteria. It depends on the packaging of bacterial donor DNA in the protein coat of a bacteriophage. In generalized transduction, phages can package any part of the donor genome. Specialized transduction is a property of lysogenic bacteriophages, which can package only host genes adjacent to the integrated prophage genome.

8. It is possible to map bacterial genes using any of the three forms of gene transfer. Interrupted mating experiments can map the approximate positions of genes. Measures of frequencies of cotransformation or cotransduction provide better resolution: The closer two genes are, the more likely they are to appear on the same short DNA fragment. Sequences of complete bacterial genomes are becoming increasingly important as mapping resources.

S O C I A L   A N D   E T H I C A L   I S S U E S

1. Bacteria have now been engineered that will break down components of oil, potentially providing a biological way to clean up an oil spill. Many environmentalists have been upset by and are opposed to this action. Why would they be disturbed by this action? How widespread do you think effects of seeding a spill site with oil-eating bacteria might be? Who would have to approve this type of action in cleaning an oil spill? What characteristics might you want to engineer into these bacteria to make them safer for the environment?

2. Sarah has been using Southern blot analysis to determine whether specific homologous genes are present in a rare thermophilic (heat-loving) bacterium. She has very little of the DNA so could only make one filter and had to reprobe the filter several times. After many months of work, she is getting her answer on what genes are present. The final blot that she does gives a large smudge in an important lane, but she feels she can still tell where there are bands and where there are not bands. The blot is not of high enough quality for reproduction in a publication. Sarah knows that reviewers and editors would probably reject it, but she thinks this piece of evidence could be referred to in her paper without showing a photo of the

blot (listed as data not shown in the paper). It would take several months to again grow enough of the bacteria to repeat the experiment, and she knows there are competitors with similar experiments to report. Should she go ahead and submit the paper for publication even though most people would say the data are messy?

3. *E. coli* strain O157 is a unique strain of this species of bacterium that has caused serious outbreaks of human illness, in some cases, leading to death of small children. Yet *E. coli* is part of the normal flora of the gut of all individuals, and researchers work on harmless laboratory strains all the time. In a college microbiology course in a large university, one student got upset at the prospect of working with *E. coli* in the lab. Even with an explanation of the difference between the toxic and nontoxic bacteria, he was not convinced that his safety was ensured. He says he will not do the lab work himself but will do analysis of the lab data obtained by his partner. The laboratory is a required part of the course and accounts for 25% of the final grade. What should and/or could the professor do with this student? Should the student have to forfeit the laboratory points in the course? How far should the professor or university go to accommodate his needs?

S O L V E D   P R O B L E M S

I. You have cloned the gene encoding the major protein in the flagella of a new bacterial strain. In screening for mutant bacteria that have a defective flagella protein, you found mutants at an exceptionally high frequency (1 in 10^3 bacterial cells). You suspect these may have been caused by insertion of a transposable element into the gene. How could you determine if this had occurred?

Answer

One way to determine if the high-frequency mutants result from insertion into the gene is to perform a Southern hybridization. *The cloned gene would be used as a probe to hybridize with DNA from the wildtype and the mutant cells.* If the mutant arises from insertion of a transposable element, the size of fragments containing the interrupted gene will be different from fragments containing the copy of the normal gene.

II. You have an F^- strain of *E. coli* that is resistant to streptomycin but requires the following amino acids for growth on minimal medium: arginine, cysteine, methionine, phenylalanine, and proline. You do a series of interrupted mating experiments with two different Hfr strains (Hfr1 and Hfr2), both of which are wildtype for all of the amino acid markers, but are Lac$^-$ and Str^s.

a. The following data are the times in minutes at which you first observed recombinant progeny containing the indicated amino-acid marker. Draw a chromosome map showing all the genes, the position of the origin of each Hfr, the direction of transfer of each Hfr, and the distances between each marker. Use a directional arrowhead to indicate each origin of transfer.

	$F^- \times$ Hfr1	$F^- \times$ Hfr2
Arginine	5	40
Cysteine	65	70
Methionine	75	60
Phenylalanine	35	10
Proline	10	35

b. You do a new mating experiment with the same strains; you interrupt the mating after 60 minutes and plate on a medium that selects for progeny that are recombinant for the phenylalanine gene (phe^+). What medium would you use to select for these progeny?

Answer

a. The order of transfer of the five genes is indicated by the time at which cells showing the marker first appear. The origin of these two Hfr strains is different;

that is, the location of the inserted F factor is different and the direction of transfer is different in the two Hfr strains.

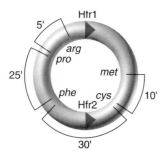

b. When the experiment is repeated selecting for progeny recombinant for phenylalanine, *the cells are plated on medium that contains arginine, cysteine, methionine, proline, and streptomycin.* The streptomycin selects against the Hfr strain and only phenylalanine is left out of the media to select for F⁻ cells that have received genes from the donor.

III. Using bacteriophage P22 you performed a three-factor cross in *Salmonella typhimurium.* The cross was between an Arg⁻ Leu⁻ His⁻ recipient bacterium and bacteriophage P22, which was grown on an Arg⁺, Leu⁺,

and His⁺ strain. You selected for 1000 Arg⁺ transductants and tested them on several selective media by replica plating. You obtained the following results:

Arg⁺ Leu⁻ His⁻	585
Arg⁺ Leu⁻ His⁺	300
Arg⁺ Leu⁺ His⁺	114
Arg⁺ Leu⁺ His⁻	1

a. What is the order of the three markers?
b. What are the cotransduction frequencies?

Answer

a. The order can be determined by looking at the relative frequencies of each phenotypic class. Arg⁺ Leu⁻ His⁻ is the largest class, with only the *arg* gene transferred. The next largest class is Arg⁺ and His⁺. Therefore *arg* and *his* are closer to each other than *arg* and *leu. The order of the genes is* arg his leu.

b. Cotransduction frequency is the percentage of cells that received two markers. For *arg* and *his,* this includes the Arg⁺ Leu⁻ His⁺ cells (300) and Arg⁺ Leu⁺ His⁺ cells (114). *The cotransduction frequency of* arg *and* his *is 414/1000 = 41.4%. The cotransduction frequency of* arg *and* leu *is 114+1 or 115/1000 = 11.5%.*

P R O B L E M S

13-1 Choose the phrase from the right column that best fits the term in the left column.

a. transformation   1. requires supplements in medium for growth

b. conjugation   2. transfer of DNA between bacteria via virus particles

c. transduction   3. small circular DNA molecule that can integrate into the chromosome

d. lytic cycle   4. transfer of naked DNA

e. lysogeny   5. transfer of DNA requiring direct physical contact

f. episome   6. integration of phage DNA into the chromosome

g. auxotroph   7. infection by phages in which lysis of cells releases new virus particles

13-2 A liquid culture of *E. coli* at a concentration of 2×10^8 cells/ml was diluted serially as shown in the following diagram and 0.1 ml of cells from the last two test tubes were spread on agar plates containing rich media. How many colonies do you expect will grow on each of the two plates?

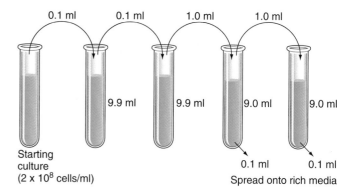

13-3 Pick out the media onto which you would spread cells from a Lac⁻ Met⁻ *E. coli* culture to
a. select for Lac⁺ cells
b. screen for Lac⁺ cells
c. select for Met⁺ cells
    i. minimal media + glucose + methionine
    ii. minimal media + glucose (no methionine)
    iii. rich media + X-Gal
    iv. minimal media + lactose + methionine

13-4 Now that the sequence of the entire *E. coli* genome (about 5 Mb) has been determined, you can determine exactly where a cloned fragment of DNA came from in the genome by sequencing a few bases and matching that data with genomic information.

a. How many nucleotides of sequence information would you need to determine exactly where a fragment is from?

b. If you had purified a protein from *E. coli* cells, roughly how many amino acids of that protein would you need to know to establish which gene encoded that protein?

13-5 The numbers of IS1 elements in different laboratory strains of *E. coli* vary. There are no recognition sites for the enzyme *Eco*RI in IS1. How could you determine the number of IS1 elements in the two strains *E. coli B* and in *E. coli K?*

13-6 There is usually one copy of the F plasmid per cell in an *E. coli* strain. You suspect you have isolated a cell in which a mutation increases the copy number of F to three to four per cell. (The copy number is determined by hybridization experiments.) How could you distinguish between the possibility that the copy number change was due to a mutation in the F plasmid versus a mutation in a chromosomal gene?

13-7 Genes encoding toxins are often located on plasmids. There has been a recent outbreak in which a bacterium that is usually nonpathogenic is producing a toxin. Plasmid DNA can be isolated from this newly pathogenic bacterial strain and separated from the chromosomal DNA. To determine if the plasmid DNA contains a gene encoding the toxin, you could determine the sequence of the entire plasmid and search for a sequence that looks like other toxin genes previously identified. There is an easier way to determine whether the plasmid DNA carries the gene(s) for the toxin that does not involve DNA sequence analysis. Describe an experiment using this easier method.

13-8 DNA sequencing of the entire *H. influenzae* genome was completed in 1995. When DNA from the nonpathogenic serotype *H. influenzae Rd* was compared to that of the pathogenic *b* strain, eight genes of the fimbrial gene cluster (located between the *purE* and *pepN* genes) involved in adhesion of bacteria to host cells were completely missing from the nonpathogenic strain. What effect would this have on cotransformation of *purE* and *pepN* genes using DNA isolated from the nonpathogenic versus the pathogenic strain?

13-9 In an Hfr × F⁻ cross, the *pyrE* gene enters the recipient in 5 minutes, but at this time point there are no exconjugants that are Met⁺, Xyl⁺, Tyr⁺, Arg⁺, His⁺, or Mal⁺. The mating is now allowed to proceed for 30 minutes and Pyr⁺ exconjugants are selected. Of the Pyr⁺ cells, 32% are Met⁺, 94% are Xyl⁺, 7% are Tyr⁺, 59% are Arg⁺, 0% are His⁺, 71% are Mal⁺. What can you conclude about the order of the genes?

13-10 You have strains with the following genotypes:
Hfr *cys⁻ str^s*
F⁻ *trp⁻ man⁻ his⁻ tyr⁻ thr⁻ str^r*
These mutations have the following phenotype:

Cys⁻	cannot synthesize cysteine
Str^s	sensitive to streptomycin
Trp⁻	cannot synthesize tryptophan
Man⁻	cannot utilize the sugar mannose
His⁻	cannot synthesize histidine
Tyr⁻	cannot synthesize tyrosine
Thr⁻	cannot synthesize threonine
Str^r	resistant to streptomycin

a. What is the simplest medium you can use to grow the Hfr strain? the F⁻ strain?

b. You mixed the two strains and did two experiments. In experiment 1, you took samples, diluted and plated on minimal medium supplemented with histidine, tyrosine, threonine, and streptomycin and glucose as a carbon source. In experiment 2, you did essentially the same experiment but before you diluted you subjected the mating mixture to treatment in the food blender. In experiment 2, you never got any colonies until after 8 minutes. In experiment 1 you got some colonies at an earlier time. Why was there a difference in the results of these two experiments?

c. What medium would you use to select the exconjugants from experiment 2 that are Man⁺?

d. The following table presents some of the data from experiment 2 in part (b).

Marker	Time when first recombinants were found in minutes
man⁺	15.5
his⁺	24
thr⁺	16
tyr⁺	16

Make the best map of the Hfr origin and genes.

e. To resolve ambiguities, you did the following crosses:

Cross 3   Hfr *man⁺thr⁻ tyr⁺* × F⁻ *man⁻ thr⁺ tyr⁻*
Cross 4   Hfr *man⁺thr⁺ tyr⁻* × F⁻ *man⁻ thr⁻ tyr⁺*

You note how many wildtype recombinants are produced and find many more in cross 3 than in cross 4. Show the order of these three genes.

13-11 You have recovered three independent mutations in one of the arginine biosynthesis genes. You now wish to map the site of the three mutations (*arg101, arg102,* and *arg103*) in the *arg* gene via the following crosses in which Lac⁺ cells were selected, then screened for Arg phenotype.

Cross	Hfr (lac^+)	F^- (lac^-)	% of Lac^+ that are Arg^+
A	$arg101^-, arg102^+$	$arg101^+, arg102^-$	0.50
B	$arg101^+, arg102^-$	$arg101^-, arg102^+$	0.02
C	$arg101^+, arg103^-$	$arg101^-, arg103^+$	0.50
D	$arg101^- arg103^+$	$arg101^+, arg103^-$	0.04
E	$arg102^+, arg103^-$	$arg102^-, arg103^+$	0.50
F	$arg102^-, arg103^+$	$arg102^+, arg103^-$	0.06

Draw a genetic map of the three *arg* mutations. Show the order of the three sites relative to the *lac* gene.

13-12 In a cross between an Hfr that has the genotype ilv^+ bgl^+ mtl^+ and an F^- that is ilv^- bgl^- mtl^-, *ilv* is known to be transferred later than *bgl* and *mtl*. To determine the order of *bgl* and *mtl* with respect to *ilv*, Ilv^+ exconjugants were selected and these colonies were screened for Bgl and Mtl phenotypes. Using the following data, what is the order of the three genes?

			Number of exconjugants
Ilv^+	Mtl^+	Bgl^+	220
Ilv^+	Mtl^-	Bgl^-	60
Ilv^+	Mtl^+	Bgl^-	0
Ilv^+	Mtl^-	Bgl^+	18

13-13 An F' element containing the maltose (mal^+) gene is discovered in *E. coli*.

a. Starting with an F^+ strain of *E. coli*, diagram the chromosomal events that must have occurred to produce this element.

b. The F' mal^+ element was introduced into a F^- mal^- strain. Most of the resulting cells transferred the element to other F^- cells. However, occasional cells appeared that transferred the entire *E. coli* chromosome, beginning with the maltose gene. Draw the chromosomal events that must have occurred to produce such a strain.

13-14 Starting with an F^+ strain that was prototrophic (i.e. had no auxotrophic mutations) and Str^s, several independent Hfr strains were isolated. These Hfr strains were mated to an F^- strain that was Str^r, Arg^- Cys^- His^- Ilv^- Lys^- Met^- Nic^- Pab^- Pyr^- Trp^-. Interrupted mating experiments showed that the Hfr strains transferred the wildtype alleles in the order indicated below as a function of time. The time of entry for the markers within parentheses could not be distinguished from one another.

Hfr strain	Order of transfer——————————→
HfrA	*pab ilv met arg nic (trp pyr cys) his lys*
HfrB	*(trp pyr cys) nic arg met ilv pab lys his*
HfrC	*his lys pab ilv met arg nic (trp pyr cys)*
HfrD	*arg met ilv pab lys his (trp pyr cys) nic*
HfrE	*his (trp pyr cys) nic arg met ilv pab lys*

a. From these data derive a map of the relative position of these markers. Indicate with labeled arrows the position and orientation of the integrated F plasmid for each Hfr strain.

b. To determine the relative order of the *trp*, *pyr*, and *cys* markers and the distances between them, HfrB was mated with the F^- strain long enough to allow transfer of the *nic* marker, after which Trp^+ recombinants were selected. The unselected markers *pyr* and *cys* were then scored in the Trp^+ recombinants, yielding the following results:

Number of recombinants	Trp	Pyr	Cys
790	+	+	+
145	+	+	−
60	+	−	+
5	+	−	−

Draw a map of the *trp*, *pyr*, and *cys* markers relative to each other. (Note that you cannot determine the order relative to the *nic* gene using this data.) Express map distances between adjacent genes as the frequency of crossing-over between them.

c. Detailed study of the genes in the *trp*, *pyr*, *cys* region would require an F' plasmid carrying the three genes. Which of the preceding strain(s) would you use to isolate such an F' plasmid?

13-15 In *E. coli*, four streptomycin-sensitive, prototrophic Hfr strains were mated with a streptomycin-resistant F^- strain auxotrophic for a number of nutritional requirements. Matings were interrupted at various intervals, and the results were plotted (shown in following diagram).

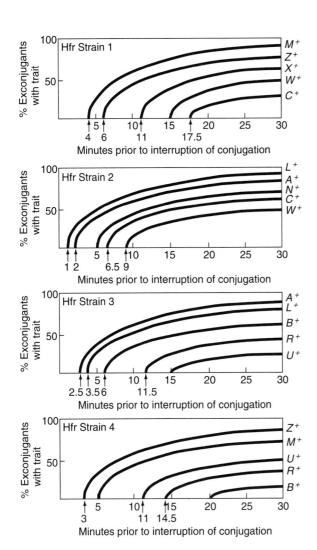

a. Draw the best map for the *E. coli* genome consistent with the data. Include distances between the genes, and label arrowheads to show the origin of transfer for the four Hfr strains.

b. Suppose you wanted to generate an F′ (*N*⁺) plasmid for use in sexduction experiments. Describe how you would obtain such a plasmid, indicating which of the four Hfr strains above you would use, and listing briefly the steps involved in isolating the F′ (*N*⁺) plasmid. (Hint: Think about how the time of transfer of gene *N* could differ in an Hfr and an F′(*N*⁺) strain.)

13-16 You can carry out matings between an Hfr and F⁻ strain by mixing the two cell types in a small patch on a plate and then replica plating to selective media. This methodology was used to screen hundreds of different cells for a recombination deficient *recA*⁻ mutant. Why is this an assay for RecA function? Would you be screening for a *recA* mutation in the F⁻ or Hfr strain using this protocol?

13-17 In two isolates (one is resistant to ampicillin and the other is sensitive to ampicillin) of a new bacterium, you found that genes encoding ampicillin resistance are being transferred into the sensitive strain. To determine if the gene transfer is transduction or transformation, you treat the mixed culture of cells with DNase. Why would this distinguish between these two modes of gene transfer? Describe the results predicted if the gene transfer is transformation versus transduction.

13-18 Suppose you have two Hfr strains of *E. coli* (HfrA and HfrB), derived from a fully prototrophic streptomycin-sensitive (wildtype) F⁺ strain. In separate experiments you allow these two Hfr strains to conjugate with an F⁻ recipient strain (Rcp) that is streptomycin resistant and auxotrophic for glycine (*gly*⁻), lysine (*lys*⁻), nicotinic acid (*nic*⁻), phenylalanine (*phe*⁻), tyrosine (*tyr*⁻), and uracil (*ura*⁻). By using an interrupted mating protocol you determined the earliest time after mating at which each of the markers can be detected in the streptomycin resistant recipient strain, as shown below.

Earliest appearance of marker from Hfr

	gly⁺	*lys*⁺	*nic*⁺	*phe*⁺	*tyr*⁺	*ura*⁺
HfrA × Rcp	3	*	8	3	3	3
HfrB × Rcp	8	3	13	8	8	8

(The * indicates that no Lys⁺ cells were recovered in the 60 minutes of the experiment.)

a. Draw the best map you can from these data, showing the relative locations of the markers and the origins of transfer in strains HfrA and HfrB. Show distances where possible.

b. To resolve ambiguities in the preceding map, you studied cotransduction of the markers by the generalized transducing phage P1. You grew phage P1 on strain HfrB and then used the lysate to infect strain Rcp. You selected 1000 Phe⁺ clones for the presence of unselected markers, with the following results:

Number of transductants	Phenotype					
	Gly	Lys	Nic	Phe	Tyr	Ura
600	–	–	–	+	–	–
300	–	–	–	+	–	+
100	–	–	–	+	+	+

Draw the order of the genes as best you can based on the preceding cotransduction data.

c. Suppose you wanted to use generalized transduction to map the *gly* gene relative to at least some of the other markers. How would you modify the cotransduction experiment just described to increase your chances of success? Describe the composition of the medium you would use.

13-19 Generalized and specialized transduction both involve bacteriophages. What are the differences between these two types of transduction?

13-20 Assume in *E. coli* there are three adjacent galactose utilization genes, *galA, galB,* and *galC*. A mutation in any one of these genes gives rise to a Gal⁻ phenotype. Recall that the *gal* cluster of genes is adjacent to the phage lambda attachment site and defective phages carrying the linked *gal* genes are found in the phage lysate at a frequency of one in 10^5 phages following induction of the lysogens with UV light.

a. If a Gal⁺ lysogen is induced and 10^7 phage are recovered and used to infect a Gal⁻ strain of *E. coli*, how many Gal⁺ colonies will be obtained?

b. You have isolated four recessive Gal⁻ mutants (strains 1–4) and wish to determine for each whether it is mutant in the *galA, galB,* or *galC* genes. Starting with nonlysogenic strains of each of the four Gal⁻ mutants and the nonlysogenic tester strains $galA^-B^+C^+$, $galA^+B^-C^+$, and $galA^+B^+C^-$, as well as a lysate of wildtype lambda phage, describe how you would accomplish this in four or five experimental steps.

c. You obtain the following numbers of Gal⁺ colonies when each of the four mutants is crossed to each of the three tester strains via specialized transduction.

	Tester strain		
Gal⁻ mutant	*galA⁻*	*galB⁻*	*galC⁻*
1	100	100	1
2	100	1	100
3	0	0	0
4	100	1	100

Which gene is mutant in strains 1, 2, and 4?

d. What is the likely basis of the mutation in strain 3?

e. Diagram the chromosome of the Gal⁺ colonies from the *gal1⁻* × *galA⁻* cross.

f. What are two possible explanations for the single Gal⁺ colonies in the

$$gal1^- \times galC^-$$
$$gal2^- \times galB^-$$
$$gal4^- \times galB^-$$

13-21 You have isolated a new temperature-sensitive mutation that affects the ability of *E. coli* to insert proteins into the membrane. At the restrictive temperature, the cells are unable to grow and thus do not form colonies.

a. Using a set of Hfr strains with origins every 5′ around the chromosome and a Tn10 insertion as a marker that is transferred in the first 5′, describe how you would rapidly narrow down the location of the new mutation. How many matings would you have to do?

b. If the Hfr set was not available, you could locate the mutation by transforming with a set of plasmids (average size: 20 kb). About how many plasmids would you have to transform into the mutant to just cover the genome? (Assume that your plasmids each overlap by 1 kb.)

13-22 a. Using the following pieces of technical information in the order given, explain how you would be able to identify the genes encoding proteins in *E.coli* cells that could bind directly to β-galactosidase: (1) β-galactosidase protein binds very tightly to a resin called APTG-agarose; (2) the 20 amino acids found in proteins vary widely in molecular weight; (3) the enzyme trypsin can cleave proteins into smaller peptides that are in the range of 3–40 amino acids long; the enzyme cleaves in a very predictable and reproducible way (after lysine and arginine amino acids); (4) modern techniques of mass spectrometry can measure the molecular weight of peptides to an accuracy of 0.01%; these machines measure the molecular weights of a large number of peptides in a complex mixture at the same time; (5) the entire *E. coli* genome has been sequenced.

b. Generalize the technique you described in part (a) to identify the genes encoding proteins that bind to any other particular protein in *E. coli*. (Hint: Use the fact that β-galactosidase binds to the APTG-agarose in your scheme.)

CHAPTER
14

THE CHROMOSOMES OF ORGANELLES OUTSIDE THE NUCLEUS EXHIBIT NON-MENDELIAN PATTERNS OF INHERITANCE

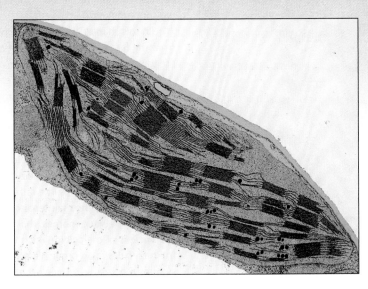

False color electron micrograph of an isolated chloroplast in a leaf cell of timothy grass (*Phleum pratense*) (×11,000).

Just nine years after the rediscovery of Mendel's laws, plant geneticists reported a perplexing phenomenon that challenged one of Mendel's basic assumptions. In a 1909 paper, they described the results of reciprocal crosses analyzing the transmission of green versus variegated leaves in flowering plants known as four-o'clocks (Fig. 14.1). Fertilization of eggs from a plant with variegated leaves by pollen from a green-leafed plant produced uniformly variegated offspring. Surprisingly, the reciprocal cross—in which the leaves of the mother plant were green and those of the father variegated—did not lead to the same outcome; instead all of the progeny from this cross displayed green foliage. From these results, it appeared that offspring inherit their form of the variegation trait from the mother only. This type of transmission, known as maternal inheritance, challenged Mendel's assumption that maternal and paternal gametes contribute equally to inheritance. Geneticists thus said that the trait in question exhibited **non-Mendelian inheritance.**

Another example of a non-Mendelian trait emerged forty years later. In 1949 French researchers published studies on the size of yeast colonies in laboratory strains of the single-celled organism *Saccharomyces cerevisiae*. Mitotically dividing cultures of these cells, when grown on plates containing the fermentable sugar glucose as the source of carbon, produced colonies of two distinctly different sizes. Ninety-five percent of the colonies were large (in French, *grande*); the remaining 5% were small (*petite*). Cells from *grande* colonies, when separated and grown on fresh plates containing glucose, yielded some *petite* colonies, but cells from *petite* colonies never generated *grande* colonies. From these observations, the researchers deduced that the founder cells of *petite* colonies arose from frequent—1 in 20 cells—mutations in cells of the *grande* colonies.

The French researchers pursued their study of the genetic basis of this difference in colony size by analyzing various matings using haploid cells from *grande* and *petite* colonies. As described in

Figure 14.1 Non-Mendelian inheritance in four-o'clocks. The first example of non-Mendelian inheritance uncovered by geneticists was seen in the flowering plant known as a four-o-clock.

Chapter 4, the diploid cells formed by mating haploid cells of opposite mating types may, under stressful conditions (not enough nutrients, for example), enter meiosis. When the French researchers mated *grande* cells of one mating type with *grande* cells of the opposite mating type, the resulting diploids were *grande*; and when these *grande* diploids sporulated via meiosis, each one yielded 4 *grande* spores (that is, spores that after germination produced *grande* colonies) and 0 *petite* spores. A cross of two cells from *petite* colonies produced only *petite* diploids, which, however, could not sporulate because of their deficiency in respiration. Matings of *grande* with *petite* generated only *grande* diploids, and each sporulation of those diploids yielded 4 *grande* spores and 0 *petites.* This 4:0 ratio consistently replaced the 2:2 ratio predicted by Mendelian genetics. From these observations, the researchers concluded that a genetic factor necessary for respiratory growth is present in *grande* cells but absent from *petite* cells. They named the factor "rho" (symbolized by "ρ"); and designated grande cells "ρ$^+$," petite cells "ρ$^-$." They also noted that because of the non-Mendelian inheritance pattern, the rho factor did not segregate at meiosis.

Is there a connection between the maternal inheritance of leaf variegation in four-o'clocks and the unusual 4:0 inheritance pat-

tern of *grande* and *petite* colony sizes in yeast? The answer is yes. Decades of experiments have shown that both traits are determined by genes that do not reside in the nucleus, but instead lie in the genomes of nonnuclear organelles. Mutations resulting in leaf variegation occur in chloroplast DNA (cpDNA), in genes that encode proteins active in photosynthesis. Mutations that diminish yeast colony size occur in mitochondrial DNA (mtDNA), in regions of the genome that influence the efficiency of a cell's energy use. While chloroplasts are the sites of photosynthesis in plant cells, mitochondria are the sites of cellular respiration—the aerobic conversion of energy stored in nutrient molecules to energy stored in the high-energy bonds of ATP—in all eukaryotic cells. The *petite* mutants form smaller colonies because they are unable to carry out cellular respiration and must obtain the energy they need for survival from the less efficient, anaerobic energy conversion pathway of fermentation.

In this chapter, we see that non-Mendelian inheritance results from the fact that the genomes of chloroplasts and mitochondria are transmitted from generation to generation separately from the nuclear genome and in a very different fashion.

Three themes surface during our detailed discussion of the genes and genomes of mitochondria and chloroplasts. First, unlike the rules governing the transmission of nuclear genes, the formal rules for the transmission of organelle genomes can vary from organism to organism. Most diploid organisms, for example, inherit non-Mendelian traits from their mother, but in some species, the traits are transmitted by the father, and in other species, by both parents. Second, the maintenance of organelles requires cooperation between two genomes: that of the organelle itself and that in the nucleus of the cell in which the organelle functions. Some of the molecules in an organelle are encoded by organelle DNA and synthesized in the organelle; other molecules active in the organelle are encoded by nuclear genes and imported into the organelle. Finally, the genomes and biochemical processes of organelles are more similar in many ways to those of bacteria than to those in other parts of the eukaryotic cell. These observations formed the basis of the now accepted endosymbiont theory of the origin of extranuclear organelles. The theory proposes that organelles are the evolutionary remnants of bacteria that became symbionts in the ancient precursors of the earliest eukaryotes.

As we examine the structure, function, and transmission of the DNA carried by mitochondria and chloroplasts, we present

- The structure and function of mitochondrial and chloroplast genomes, including a description of the size, shape, replication, and expression of those genomes and a brief discussion of the evolutionary origin of mitochondria and chloroplasts.

- How genetic studies of transmission revealed and explained non-Mendelian patterns of inheritance, including descriptions of uniparental and biparental inheritance and mitotic segregation.

- A comprehensive example of how mutations in mitochondrial DNA affect human health.

THE STRUCTURE AND FUNCTION OF MITOCHONDRIAL AND CHLOROPLAST GENOMES

Although geneticists pieced together the molecular foundation of non-Mendelian inheritance—that is, the structure and function of organelle genomes—beginning only in the 1960s, we present a summary of that molecular context up front to provide a clear framework for understanding the earlier genetic evidence. We begin with a brief description of the structure and function of mitochondria and chloroplasts, and then describe the genomes that encode the RNA and protein machinery the organelles use to sustain their activities.

Mitochondria and Chloroplasts Are Organelles of Energy Conversion That Carry Their Own DNA

Every time a bird takes flight, a person speaks, a worm turns, or a flower unfurls, the organism's cells use energy that a plant captured from sunlight and stored in nutrient molecules. The chloroplasts of plants and other photosynthesizers capture solar energy and store it in the chemical bonds of carbohydrates. The mitochondria found in most eukaryotic cells release energy from those nutrients and convert it to ATP.

Mitochondria Are Sites of the Krebs Cycle and an Electron Transport Chain That Carries Out the Oxidative Phosphorylation of ADP to ATP

Each eukaryotic cell houses many mitochondria, with the exact number depending on the energy requirements of the cell as well as the chance distribution of mitochondria during cell division. In humans, nerve, muscle, and liver cells carry more than a thousand mitochondria.

Figure 14.2 reveals the organelle's detailed structure. Separating the mitochondrion from the cell's cytoplasm is a smooth *outer membrane* perforated by protein-lined pores; molecules up to the size of small proteins can readily pass through these pores to an *intermembrane space*—a gap between the outer membrane, and a second, *inner membrane*. The inner membrane has many infoldings called *cristae* and is studded with enzymes of the electron transport chain and ATP synthesis. It is unusually rich in proteins (it is 75% protein by weight), and is impermeable to most substances, which can traverse it only by active transport. Tightly sequestered inside the inner membrane is a core compartment known as the *matrix,* which contains many enzyme systems, including one that carries out the tricarboxylic acid (TCA) cycle, commonly referred to as the Krebs cycle after its discoverer. The matrix also carries one or more circles of mitochondrial DNA.

Biologists call mitochondria the powerhouses of the cell because these organelles produce most of the cell's usable energy in the form of ATP molecules. There are two stages by which mitochondria help convert food into energy. In the first stage, mitochondria employ the Krebs cycle to metabolize pyruvate and fatty acids (the breakdown products of carbohydrates and fats, respectively) and produce the high-energy

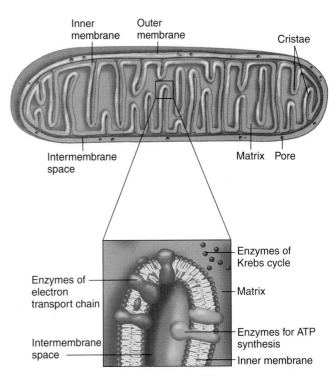

Figure 14.2 Anatomy of a mitochondrian. The organization and structure of a single mitochondrian is shown. Core regions that are entirely enclosed within an inner membrane are known as the matrix (shown in blue). The matrix contains the mitochondrial DNA and enzymes of the Krebs cycle. Individual inner membrane foldings are called cristae. A single crista is magnified to show how enzymes of the electron transport chain carry out oxidative phosphorylation between the matrix and the intermembrane space.

electron carriers NADH and $FADH_2$. In the second stage, four multisubunit enzyme complexes (referred to as I, II, III, and IV), which are embedded in the inner mitochondrial membrane, harness the energy in the high-energy electrons carried by the NADH and $FADH_2$. The process by which they harness this energy is **oxidative phosphorylation:** a set of reactions requiring oxygen that creates portable packets of energy in the form of ATP. In carrying out oxidative phosphorylation, molecular complexes I, II, III, and IV form a chain that transports the high-energy electrons from NADH and $FADH_2$ to their final electron acceptor, oxygen. A fifth protein complex (V) attached to the inner mitochondrial membrane uses the energy released by the electron transport chain to produce ATP.

To summarize: Starting with O_2 and the breakdown products of fats and carbohydrates, mitochondria produce packets of energy in the form of ATP.

Chloroplasts Are the Sites of Photosynthesis: The Capture, Conversion, and Storage of Solar Energy in the Bonds of Carbohydrates

In corn, one of the many crop plants adept at carrying out photosynthesis, each leaf cell contains 40–50 chloroplasts, and each square millimeter of leaf surface carries more than

Chloroplast

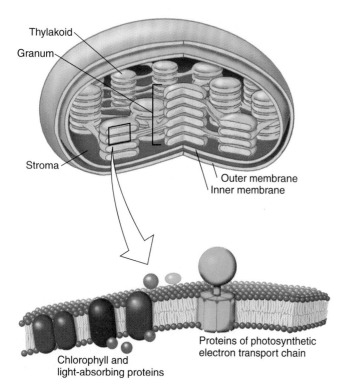

Figure 14.3   Anatomy of a chloroplast. Like a mitochondrion, a chloroplast also has an outer membrane and an inner membrane, but the inner membrane is not folded. The space found within the inner membrane—containing the chloroplast DNA and photosynthetic enzymes—is called a stroma. The stroma contains thylakoid membranes surrounding the thylakoid lumens. The magnified portion of a thylakoid membrane indicates the presence of photosynthetic enzymes.

500,000 of the organelles. Figure 14.3 illustrates the structure of a chloroplast. Like mitochondria, chloroplasts have an *outer membrane,* an *intermembrane space,* and an *inner membrane* enclosing a core compartment. In chloroplasts, this core compartment is called the *stroma.* It contains the chloroplast's DNA and various enzyme systems, including the photosynthetic enzymes that fix carbon dioxide into carbohydrates. Within the stroma lies a third set of membranes, the *thylakoid membranes,* which surround a space called the *thylakoid lumen.* In places, portions of the thylakoid membranes fuse into stacks of disk-like sacs known as *grana* (singular, *granum*). Embedded in the thylakoid membranes are chlorophyll and other light-absorbing proteins as well as proteins of the photosynthetic electron transport system. The light-absorbing proteins capture light energy from the sun and transfer this trapped solar energy to a photosynthetic electron transport chain.

Photosynthesis takes place in two parts: a light-trapping phase and a sugar-building phase. During the light-trapping phase, solar energy becomes trapped as it boosts electrons in chlorophyll or another pigment molecule to higher energy levels. The energized electrons are then conveyed to an electron

transport system that uses the energy to convert water to oxygen and H^+. Photosynthetic electron transport forms NADPH and drives the synthesis of ATP via an ATP synthase similar to the one in mitochondria. During the sugar-building phase of photosynthesis, enzymes of the Calvin cycle, located in the chloroplast stroma, use the ATP and NADPH generated by the conversion of solar to chemical energy to fix atmospheric carbon dioxide into carbohydrates. The energy stored in the bonds of these nutrient molecules fuels the activities of both the plants that make the carbohydrates and the animals that ingest them.

Mitochondria and Chloroplasts Carry Their Own DNA

Cells stained with DNA-specific dyes, when viewed under the light microscope, reveal DNA molecules in the matrix of mitochondria and the stroma of chloroplasts, as well as in the nucleus. Using methods for purifying mitochondria and chloroplasts, researchers have extracted DNA directly from these organelles and shown by analyses of base composition and buoyant density that an organism's organelle DNA differs from its nuclear DNA.

With this background in mind, we examine the structure and function of mitochondrial and chloroplast genomes. As we do, we see that although both organelles replicate and express all the genes in their own DNA, their genomes encode only some of the proteins they require for their activities.

The Genomes of Mitochondria

Mitochondrial DNA lies within the matrix of the organelle, where it appears in highly condensed structures called nucleoids. Each haploid yeast cell, for example, has about 20 mtDNA molecules in a variable number of mitochondria and nucleoids. In cells with just 1 to a few mitochondria, each organelle contains a number of nucleoids. By contrast, in cells grown on lactate medium, which carry up to 20 mitochondria, each organelle contains only 1 to a few nucleoids. Thus, the mtDNA of most cells does not reside in a single location.

Variations in the number of mitochondria, nucleoids, and mtDNA molecules are regulated by complicated means that researchers do not yet understand. Mitochondria can fuse with each other as well as divide. In general, however, mitochondria double in size and then divide in half in each cell generation. The replication of mtDNA molecules, as well as the division of the mitochondria can occur throughout the cell cycle independent of the replication of genomic nuclear DNA (which occurs only during S phase) and of the cell division at the end of mitosis. Interestingly, which mtDNA molecules undergo replication seems to be determined at random; as a result, some molecules replicate many times in each cell cycle, while others do not replicate at all. This is one cause of the mitotic segregation of mitochondrial genomes discussed later in the chapter.

The size and gene content of mitochondrial DNA varies from organism to organism (Tables 14.1 and 14.2). The mtDNAs in the malaria parasite *Plasmodium falciparum* are only 6 kb in length; those in the free-living nematode *Ascaris*

TABLE 14.1 Mitochondrial DNA Sizes

Organism	Size (kb)
Plasmodium	6
Yeast	75
Drosophila	18
Pea	110
Human	16.5

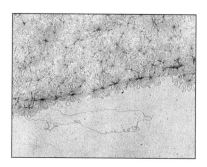

Figure 14.4 Kinetoplast DNA network. In certain protozoan parasites, there is a single mitochondrian, or kinetoplast, that contains a large interlocking network of DNA molecules present in mini- and maxicircles.

suum are 14.3 kb; those in the muskmelon *Cucumis melo* are a giant 2400 kb long. These mtDNA size differences do not necessarily reflect comparable differences in gene content. Although the large mtDNAs of higher plants do contain more genes than the smaller mtDNAs of other organisms, the 75-kb mtDNA of baker's yeast encodes fewer proteins of the respiratory chain than does the 16.5-kb mtDNA of humans. Table 14.1 summarizes the size and gene content of mtDNAs from organisms representative of the plant, animal, and fungus kingdoms.

Like size and gene content, the shape of mtDNAs varies. Biochemical analyses and mapping studies have shown that the mtDNAs of most species are circular; but the mtDNAs of the ciliated protozoans *Tetrahymena* and *Paramecium,* the alga *Chlamydomonas,* and the yeast *Hansenula* are linear. The difficulty in isolating unbroken mtDNA molecules from some organisms makes it hard to tell the shape of their mtDNA *in vivo.*

To complete this snapshot of the mitochondrial genome's variation in size, gene content, and shape, it is necessary to mention the unusually organized mtDNAs of protozoan parasites of the genera *Trypanosoma, Leishmania,* and *Crithidia.* These single-celled eukaryotic organisms carry a single mitochondrion known as a kinetoplast. Within this structure, the mtDNA exists in one place (contrary to the mtDNA of most other cells) as a large network of 10 to 25,000 "minicircles" 0.5–2.5 kb in length interlocked with 50 to 100 maxicircles 21–31 kb long (Fig. 14.4). The maxicircles contain most of the genes usually found on mtDNA, while the minicircles play a role in RNA editing, as described later.

A comparison of mitochondrial genome organization and function in humans, yeast, and the liverwort (a species of moss) illustrates some of the details of mtDNA diversity. Molecular biologists have sequenced the mtDNAs of all three organisms.

Human mtDNA Carries Closely Packed Genes

The 16.5-kb human mitochondrial genome, which accounts for 0.3% of a cell's DNA, is a circular DNA molecule (like most mtDNAs) that carries 37 genes. Thirteen of these genes encode polypeptide subunits of the protein complexes, comprising the oxidative phosphorylation apparatus embedded in the mitochondrial inner membrane; these same genes are present in the mtDNAs of all metazoans analyzed to date. In addition to the 13 polypeptide-encoding genes, there are 22 tRNA genes. The remaining 2 genes are for the large and small rRNAs found in mitochondrial ribosomes (Fig. 14.5a).

A significant feature of the human mitochondrial genome is the compactness of its gene arrangement. As Fig. 14.5a shows, adjacent genes either abut each other or slightly overlap. With virtually no nucleotides between them and no introns within them, the genes are packaged very tightly. The reason for this compact arrangement is not yet known.

The Larger Yeast Mitochondrial Genome Contains Spacers and Introns

The mitochondrial genome of the yeast *S. cerevisiae* is more than four times longer than human and other animal mtDNA (Fig. 14.5b). Two DNA elements account for the larger size of the yeast mitochondrial genome in comparison with the mitochondrial genome of mammals: long intergenic sequences and introns. Long A-T-rich sequences called "spacers" separate the genes in yeast mtDNA and account for more than half of the additional DNA. Most of the spacer DNA is transcribed, and some of it is retained in the mRNAs as long untranslated 5′ and 3′ extensions. Introns, the second DNA-lengthening element,

TABLE 14.2 Comparison of Some Functions Encoded in mtDNA

Organism	Oxidative Phosphorylation Genes	tRNAs	Genome Size (kb)
Yeast	7	25	75
Marchantia (liverwort)	14	29	186.0
Human	13	22	16.5

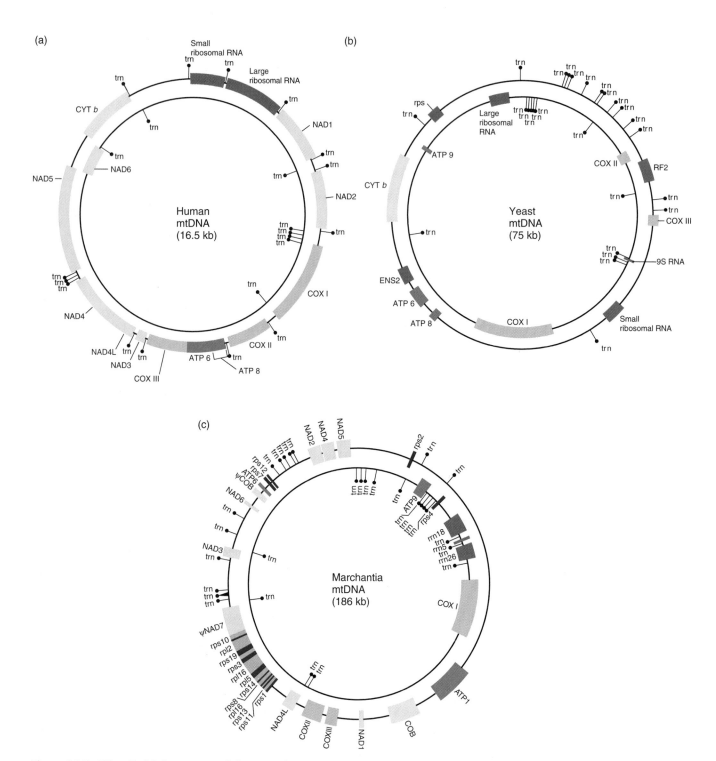

Figure 14.5   Mitochodrial genomes of three species. Comparison of mitochondrial genomes of humans, baker's yeast, and liverwort. Human mtDNA is 16.5 kb long, while that of baker's yeast is 75 kb long, and liverwort is 121 kb long. Important differences illustrated in the diagrams are the presence of long intergenic sequences in yeast mtDNA and their absence from human mtDNA and differences in the numbers of genes in the mt genomes from these three organisms. The color scheme allows you to quickly see differences in numbers of genes for various functions in these three species. The following color scheme is used: green, genes coding for cytochrome oxidase proteins; red, genes coding for ATPase subunit proteins; yellow, genes coding for NADH complex proteins; tan, genes coding for cytochrome complex proteins; purple, genes coding for ribosomal proteins or ribosomal RNAs. tRNA genes are indicated by black ball and stick. Genes shown on the outer and inner circles are transcribed in opposite directions.

form about 25% of the yeast mitochondrial genome and account for most of the remaining size difference between yeast and human mtDNA.

The 186-kb mtDNA of the Liverwort Marchantia polymorpha Carries Many More Genes than Its Animal and Fungal Counterparts

The mtDNA of *M. polymorpha* was the first plant mtDNA to be entirely sequenced. Although it is one of the smallest plant mitochondrial genomes, it is far larger and has many more genes than nonplant mtDNAs (Fig. 14.5c). The genes it carries encode 12 of the 13 electron-transport-chain proteins found in human mtDNA (lacking only the gene for subunit 8 of ATP synthase), as well as subunit 9 of ATP synthase (which is absent from humans) and subunit A of ATP synthase (which is in the nuclear genomes of all animals and fungi). In addition, the moss mitochondrial genome contains 16 genes for ribosomal proteins and 29 genes encoding proteins of unknown function.

Thus, although mitochondria in different eukaryotic organisms play similar roles in the conversion of food to energy, evolution has produced mtDNAs with an astonishing diversity in the content and organization of their genes. As we see next, mitochondrial evolution has also led to some remarkable variations on the basic mechanisms of gene expression.

Mitochondrial Transcripts Undergo RNA Editing, an Unusual Variation on the Basic Theme of Gene Expression

Researchers discovered the unexpected phenomenon of RNA editing in the mitochondria of trypanosomes. As already noted, these protozoan parasites have a single large mitochondrion—the kinetoplast—which contains much more DNA than the mitochondria of other organisms; this kDNA exists as a series of interlocking maxi- and minicircles. DNA sequencing shows that the minicircles carry no protein-encoding genes. The detection of transcripts from maxicircle DNA, however, confirms that these larger circles do carry and express genes.

Surprisingly, the sequencing of maxicircle DNA reveals only short recognizable gene fragments, instead of whole mitochondrial genes. Furthermore, the sequencing of RNA molecules in the kinetoplast revealed both RNAs that looked like the strange fragments of kinetoplast genes and related RNAs that could encode recognizable mitochondrial proteins. From these observations, investigators concluded that kDNA encodes a precursor (the strange fragment observed) for each mRNA. After transcription, the cellular machinery turns these precursors into functional mRNAs through the insertion or deletion of nucleotides.

The process that converts pre-mRNAs to mature mRNAs is **RNA editing.** It is essential for the expression of these mitochondrial genes because without RNA editing, the pre-mRNAs do not encode polypeptides. Some pre-mRNAs lack a first codon suitable for translation initiation; others lack a stop codon for the termination of transcription. RNA editing creates both types of sites, as well as many new codons within the genes.

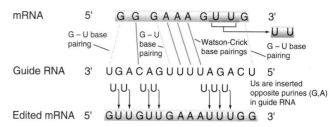

Figure 14.6 RNA editing in trypanosomes. Example of a portion of a pre-mRNA sequence is shown at the top. This pre-mRNA forms a double-stranded hybrid with a guide RNA through both standard Watson-Crick A–U and G–C base pairing, as well as atypical G–U base pairing. Unpaired G and C bases within the guide RNA initiate the insertion of U's within the pre-mRNA sequence, bringing about the final edited mRNA.

In addition to the mitochondria of trypanosomes, the mitochondria of some plants and fungi carry out RNA editing. The extent of RNA editing varies from mRNA to mRNA and from organism to organism. In trypanosomes, the RNA editing machinery adds or deletes uracils. In plants, the editing adds or deletes cytosines. At present, researchers understand the general mechanism of uracil editing, but not that of cytosine editing. As Fig. 14.6 shows, uracil editing occurs in stages in which enzymes use an RNA template as a guide for correcting the pre-mRNA. The guide RNAs are encoded by short stretches of kDNA on both maxi- and minicircles, and a structure known as an "editosome" is the workbench where the RNA editing takes place.

Translation in Mitochondria Shows That the Genetic Code is Not Universal

As mtDNA carrying its own rRNA and tRNA genes would suggest, mitochondria have their own distinct translational apparatus. Mitochondrial translation is quite unlike the cytoplasmic translation of mRNAs transcribed from nuclear genes in eukaryotes. Instead, many aspects of the mitochondrial translational system resemble details of translation in prokaryotes. For example, as in bacteria, N-formyl methionine and tRNA[fMet] initiate translation in mitochondria. Moreover, inhibitors of bacterial translation, such as chloramphenicol and erythromycin, which have no effect on eukaryotic cytoplasmic protein synthesis, are potent inhibitors of mitochondrial protein synthesis.

We saw in Chapter 7 that the genetic code is almost, but not quite, universal. The mtDNA sequences of tRNAs and protein-encoding genes in several species cannot explain the sequences of the resulting proteins in terms of the "universal" code. For example, in human mtDNA, the codon UGA specifies tryptophan rather than stop (as in the standard genetic code); AGG and AGA specify stop instead of arginine; and AUA specifies methionine rather than isoleucine (Table 14.3). No single mitochondrial genetic code functions in all organisms, and the mitochondria of higher plants use the universal code. Moreover, while an f-Met-tRNA usually initiates translation in mitochondria by reading AUG or AUA, other triplets, which do not specify methionine, often mark the site of initiation. The genetic codes of mitochondria probably

TABLE 14.3 Variations in the Genetic Code of Mitochondria

Characteristic	Universal Code	mtDNA Code
Number of tRNAs	32	22
U G G	Trp	Trp
U G A	Stop	Trp
A G G	Arg	Stop
A G A	Arg	Stop
A U G	Met	Met
A U A	Ile	Met

Altered genetic code. The mtDNA genetic code is simplified such that a modified U in the tRNA "wobble" position can read all four codons in a codon family (i.e., U U U, U U C, U U A, U U G). An unmodified U can read both purines and G can read both pyrimidines. Tryptophan tRNA has a U in the wobble position so that it will read both the traditional U G G codon and the associated U G A stop codon as tryptophan. Similarly, the methionine codon reads both A U G and the associated A U A as methionine. Finally, one of the two arginine tRNAs has been deleted such that two of the six arginine codons (A G G and A G A) now function as stop codons.

TABLE 14.4 Chloroplast DNA Sizes

Organism	Size (kb)
Chlamydomonas reinhardtii	196
Marchantia (liverwort)	121
Nicotiana tabacum (tobacco)	156
Oryza sativa (rice)	135

diverged from the universal code by a series of mutations occurring some time after the organelles became established components of eukaryotic cells.

As we see next, chloroplast DNA, although similar in many ways to mtDNA, has some remarkable features of its own.

The Genomes of Chloroplasts

Chloroplasts occur in plants and algae. The genomes they carry are much more uniform in size than are the genomes of mitochondria; although cpDNAs range in size from 120 to 217 kb, most are between 120 and 160 kb long (Table 14.4). Chloroplast DNA contains many more genes than mtDNA. Like the genes of bacteria and human mtDNA, these genes are closely packed, with relatively few nucleotides between adjacent coding sequences. Like the genes of yeast mtDNA, they contain introns. The shape of most cpDNAs is a circle.

A close look at the chloroplast genome of the liverwort *M. polymorpha*, the first cpDNA to be sequenced completely, illustrates details of cpDNA structure and function (Fig. 14.7). Liverwort cpDNA contains 92 protein-encoding genes and 36 genes for tRNAs and rRNAs, which is more genes than are carried by the mitochondrial genomes of much larger plants. The rRNA genes of the chloroplast encode four ribosomal

RNAs—16S, 23S, 4.5S, and 5S; these same four rRNAs are found in eubacteria. Another similarity with the bacterial genome is that some functionally related cpDNA genes are organized in clusters that resemble bacterial operons.

The 92 polypeptides encoded by *M. polymorpha* cpDNA account for only a small fraction of the proteins active in chloroplasts. However, cpDNA encodes proteins for a wider range of processes than does mtDNA. The cpDNA-encoded proteins include many of the molecules that carry out photosynthetic electron transport and other aspects of photosynthesis, as well as RNA polymerase, translation factors, ribosomal proteins, and other molecules active in chloroplast gene expression. The RNA polymerase of chloroplasts is similar to the multisubunit bacterial RNA polymerases. Inhibitors of bacterial translation, such as chloramphenicol and streptomycin, inhibit translation in chloroplasts, as they do in mitochondria.

The function of some chloroplast genes is not yet known. Because most of these Unidentified Reading Frames (URFs) are well conserved in many cpDNAs and have counterparts in bacteria, it is likely that they are key elements in chloroplast gene expression and organelle function.

Mitochondria and Chloroplast Functions Require Cooperation between the Organellar and Nuclear Genomes

The maintenance and assembly of functional mitochondria and chloroplasts depends on gene products from both the organelles themselves and from the nuclear genome (Fig. 14.8). In most organisms, for example, cytochrome *c* oxidase, the terminal protein of the mitochondrial electron transport chain, is composed of seven subunits, three of which are encoded by mitochondrial genes, whose mRNAs are translated on mitochondrial ribosomes. The remaining four are encoded by nuclear genes whose messages are translated on ribosomes in the cytoplasm. In all organisms, nuclear genes encode the majority of the proteins active in mitochondria and chloroplasts. For example, although mitochondrial genomes carry the rRNA genes, nuclear genomes carry the genes for most (in yeast and plants) or all (in animals) of the proteins in the mitochondrial ribosome.

Because mitochondria and chloroplasts do not carry all the genes for the proteins (and in some organisms, the tRNAs) they need to function and reproduce, they are **semiautonomous,** requiring the constant provision of proteins (and tRNAs) encoded by nuclear genes.

Origin and Evolution of Organelle Genomes: The Molecular Evidence

Mitochondria are remarkably similar in size and shape to today's aerobic bacteria, while chloroplasts in many ways resemble living photosynthetic bacteria. In addition, the genomes of both organelles bear many striking resemblances to the genomes of bacteria. Because of these morphological and molecular likenesses, it is probable that mitochondria and chloroplasts started out as free-living bacteria that merged with the ancestors of

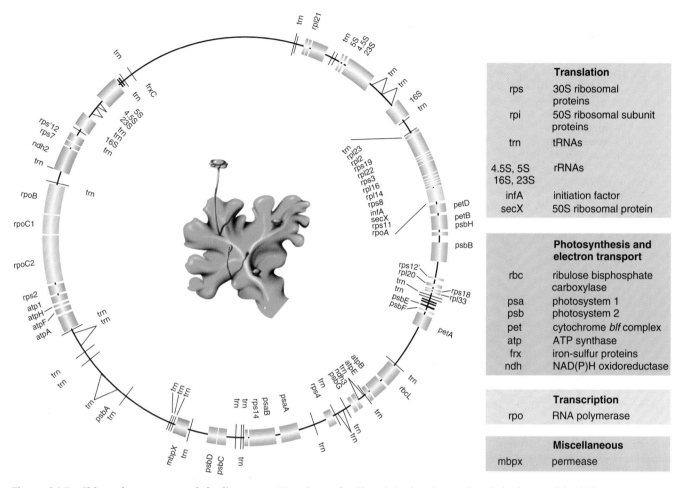

Figure 14.7   Chloroplast genome of the liverwort, *M. polymorpha*. The relative locations and symbols of some of the 128 genes are indicated. Genes are color-coded according to general function.

	Translation
rps	30S ribosomal proteins
rpl	50S ribosomal subunit proteins
trn	tRNAs
4.5S, 5S 16S, 23S	rRNAs
infA	initiation factor
secX	50S ribosomal protein

	Photosynthesis and electron transport
rbc	ribulose bisphosphate carboxylase
psa	photosystem 1
psb	photosystem 2
pet	cytochrome *b/f* complex
atp	ATP synthase
frx	iron-sulfur proteins
ndh	NAD(P)H oxidoreductase

	Transcription
rpo	RNA polymerase

	Miscellaneous
mbpx	permease

modern eukaryotic cells to form a cellular community in which host and guest benefitted from the group arrangement.

The "Endosymbiont Theory"

In the 1970s, Lynn Margulis was one of the first biologists to propose that mitochondria and chloroplasts originated more than a billion years ago when ancient precursors of eukaryotic cells engulfed and established a symbiotic relationship with some bacteria. The primitive cells carrying a mitochondrion-like or chloroplast-like bacterial cell would have gained an edge in the fierce competition for energy production and eventually evolved into complex eukaryotes. There is now so much evidence supporting this hypothesis that it is now accepted as the **endosymbiont theory.** The molecular evidence for the theory includes the fact that both mitochondria and chloroplasts have their own DNA which replicates independently of the nuclear genome. Moreover, like the DNA of bacteria, mtDNA and cpDNA are not organized into nucleosomes by histones. In addition, as we have seen, mitochondrial genomes use *N*-formyl methionine and tRNA^{fMet} in translation, and inhibitors of bacterial translation, such as chloramphenicol and erythromycin, block mitochondrial translation but have no effect on eukaryotic cytoplasmic protein synthesis. Compar-

isons of DNA sequences from organelle and bacterial rRNA genes suggest that many mitochondrial genomes derive from a common ancestor of present-day gram-negative nonsulfur purple bacteria, while chloroplast genomes derive from cyanobacteria (formerly referred to as blue-green algae) and prochlorophyte bacteria.

It is likely that the precursors of eukaryotes and the ancient bacteria that evolved into organelles interacted many times and in many different ways en route to the present-day symbiotic relations. The first interactions may have been transient; indeed, many present-day eukaryotes containing mitochondria and chloroplasts also have bacteria living inside them. Once a stable symbiotic state formed in a given lineage, however, if the engulfed cell lost some genes critical for life on the outside, it would have also lost the ability to function and compete on its own. The loss of genes could have occurred through transfer to the nucleus or deletion of redundant sequences. We have already seen that some of the genes required for oxidative phosphorylation and photosynthesis reside in the nuclear genome; they may have been transferred there from the organelle genome. The deletion of redundant sequences might have occurred when organelle genes encoding a biosynthetic pathway critical for autonomous growth were discarded

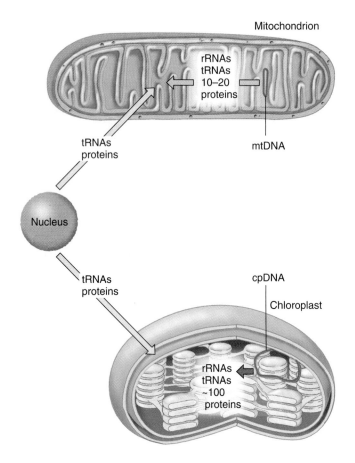

Number and genomic location of oxidative phosphorylation genes

	Number of polypeptides					
	Electron-transport chain				ATP synthase	
Genomic location	I	II	III	IV	V	Total
Mitochondrion	7	0	1	3	2	13
Nucleus	≥33	4	10	10	10	≥67
Total	≥40	4	11	13	12	≥80

Figure 14.8   Mitochondria and chloroplasts depend on gene products from the nucleus. Although some organelles in some species have many more genes than others, all are dependent on RNA and protein products encoded by nucleus genes. The location of oxidative phosphorylation (OXPHOS) genes is shown.

or lost from the organelle genome because the nuclear genome of the host encoded its own version of the pathway. In the evolution of the eukaryote's symbiotic community, the engulfed cell may have not only lost genes but also contributed coding sequences to the host cell; these sequences could have encoded enzymes that added to or replaced proteins encoded by the genes of the nuclear chromosomes.

Gene Transfer Occurs through an RNA Intermediate or by Movement of Pieces of DNA

Researchers have some understanding of the mechanisms by which genes transfer between an organelle and the nucleus. In many plants, the mitochondrial genome encodes the *COXII* gene

of the mitochondrial electron-transport chain; in other plants, the nuclear DNA encodes that same gene; and in several plant species where the nuclear *COXII* gene is functional, the mtDNA still contains a recognizable, but nonfunctional copy of the gene (that is, a *COXII* pseudogene). Remarkably, the mtDNA gene contains an intron, while the nuclear gene does not. Geneticists have interpreted this finding to mean that the *COXII* gene transferred from mtDNA to nDNA via an RNA intermediate; the RNA would have lacked the intron, and when the mRNA was copied into DNA and integrated into a chromosome in the nucleus, the resulting nuclear gene also had no intron.

There is also good evidence for the transfer of many genes at the DNA level. The fact that some plant mtDNAs carry large fragments of cpDNA shows that pieces of cpDNA can move from one organelle to another. Similarly, nonfunctional, intact or partial copies of organelle genes litter the nuclear genomes of eukaryotes. DNA sequencing reveals strong similarities between the organelle and nuclear DNAs, which means that the nuclear copies are relatively young. This, in turn, suggests that the organelle-to-nucleus transfer of DNA is still going on.

In this evolutionary perspective, the properties of mitochondrial and chloroplast genomes that vary among the organelles of present-day species are probably relatively new. These recently established features include long stretches of cpDNA incorporated in the mtDNAs of many plants, as well as many of the introns in organelle genomes. Some of these introns may have originated in the earliest bacterial symbionts; or they may have been incorporated into the organelle genomes after horizontal transfers between organelle DNAs long after the organelles were established.

Mitochondrial DNA Has a High Rate of Mutation

In the 1980s, surveys of DNA sequence variations among individuals of a given species and between closely related species showed that the mtDNA of vertebrates evolves almost 10 times more rapidly than the nuclear DNA of those same organisms. The higher rate of DNA mutation in mitochondria probably reflects more errors in replication and less efficient repair mechanisms.

Because of mtDNA's high mutation rate, the variation among mitochondrial genomes provides a valuable tool for studying the evolutionary relationships of organisms whose nuclear DNAs are very similar. Conversely, mtDNA variation, because it accumulates so rapidly, is of little value in evaluating the relationships of distant evolutionary relations, but here sequence variation data for nuclear genomes are useful (see Chapter 24). Sequence analyses of mtDNA have shown that the maternal lineage of all present-day humans, no matter what ethnic group they belong to, traces back to a few women who lived in Africa some 200,000 years ago. The Fast Forward box "Mitochondrial DNA Sequences Shed Light on Human Evolution," describes the evidence for this hypothesis of a "mitochondrial Eve." It also explains how the early representatives of modern humans may have interacted with archaic humans and other hominid species in the genus *Homo* during the evolution of modern human populations.

Cloning and PCR of mtDNA, which enable the amplification of small biological samples for analysis, have helped identify the closest mammalian relatives of the extinct *Equus quagga,* a horse-like African animal that died out in 1883 (see Chapter 5). Analyses of quagga mtDNA from preserved museum samples show that the quagga's closest living relatives are African plains zebras, and that it is less closely related to the mountain zebra and the horse. The Genetics and Society box "Mitochondrial DNA Tests Replace HLA Typing as Evidence of Kinship in Argentine Courts" describes how a human rights organization in Argentina eventually used mtDNA sequences as the legal basis for reuniting kidnapped children with their biological families. Such sequences, although they evolve very rapidly, show few if any changes in the 20–60 years it takes to produce one or two family generations.

GENETIC STUDIES OF ORGANELLE GENOMES CLARIFY KEY ELEMENTS OF NON-MENDELIAN INHERITANCE

Mutations in organelle genes produce readily detectable whole-organism phenotypes because the altered proteins and RNAs they encode disrupt the production of cellular energy. The mtDNA mutations that cause petite colonies in yeast, for example, disrupt the oxidative-phosphorylation system; the cpDNA mutations that cause variegation in four-o'clocks incapacitate proteins essential for photosynthesis. Although most mutations in the genes for these energy-producing systems are lethal, in both plants and animals, some organelle gene mutations yield detectable, nonlethal phenotypes that researchers can study genetically. Data from such genetic studies show that modes of organelle gene transmission vary among organisms. We describe the main modes of transmission from one generation to the next, which determine what is inherited. We also examine how parental contributions to the zygote and mitotic segregation influence what is available for inheritance by determining which organelles are passed on to progeny cells during cell divisions.

In Most Species, Progeny Inherit Organelle DNA from Only One—Usually the Maternal—Parent

Genetic and biochemical analyses of thousands of crosses have shown that most species transmit mtDNA and cpDNA via one parent, usually the mother; that is, they utilize a uniparental, mainly maternal, mode of transmission. Researchers have analyzed **uniparental inheritance** in several ways.

Maternal Inheritance of Preexisting Differences in Wildtype mtDNAs

In a classic experiment documenting maternal inheritance in vertebrates, investigators purified mtDNA from frog eggs, which contain a large number of mitochondria, and used hybridization tests to distinguish the mtDNA of one frog species, *Xenopus laevis,* from the mtDNA of the closely related *X. borealis.* In these tests, probes from *X. laevis* hy-

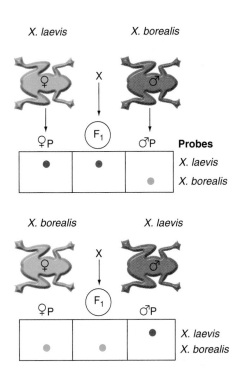

Figure 14.9 Maternal inheritance of *Xenopus* mtDNA. *X. laevis* and *X. borealis* mtDNA can be distinguished by strong hybridization only to probe made from the same species. Reciprocal crosses between two species produce F₁ hybrids. Each F₁ hybrid retains mtDNA only from its mother.

bridized more efficiently with *X. laevis* mtDNA than with *X. borealis* DNA and vice versa. Because interspecific crosses between the two species yield viable progeny, the analysis of F₁ mtDNA was one way to trace the inheritance of that DNA in frogs. Figure 14.9 diagrams the reciprocal crosses and mtDNA typing that formed the basis of the study. The first-generation progeny of both crosses carried mtDNA like that of the maternal parent. Although the analysis might have missed small contributions from the paternal genome, these *Xenopus* crosses confirmed the predominantly maternal inheritance of mtDNA in these species. They also showed that it is possible to follow preexisting differences in functionally wildtype mtDNAs in a cross. Since the 1980s, sequence differences uncovered by RFLP analyses have confirmed maternal mtDNA inheritance among horses, donkeys, and many other vertebrates.

Maternal Inheritance of Specific Genes in cpDNA

Interspecific crosses tracing biochemically detectable, species-specific differences in several chloroplast proteins provided further evidence of maternal inheritance. In the mtDNA inheritance studies cited here, the identity of the organelle gene containing the markers was not known and did not matter. By contrast, in cpDNA studies researchers identified specific organelle genes through the analysis of proteins. They began by isolating from tobacco plants (*Nicotiana* species) proteins in which interspecies differences could be distinguished by gel electrophoresis. To determine each

Fast Forward

Mitochondrial DNA Sequences Shed Light on Human Evolution

The mitochondrial DNA of all humans alive today traces back through maternal lineages to the mtDNA of a human population living in Africa some 200,000 years ago. Such is the startling conclusion of two papers published in 1987 and 1991 by Allan C. Wilson and colleagues. The carrier of this ancestral mtDNA, dubbed "mitochondrial Eve," probably lived in a population of 10,000 to 50,000 people. (In this context, a *population* is a group of interbreeding individuals of the same species that inhabit the same space.)

How mtDNA Variations Suggest Region Where Modern Humans Emerged

In their studies supporting an African origin for modern humans, Wilson and coworkers first looked at restriction fragment length polymorphisms (RFLPs) in the mtDNA of 143 subjects, including Americans of African, Asian, European, and Middle Eastern origin as well as Aboriginal women in New Guinea and Australia. Four years later, the group followed up their original 1987 study with sequence analyses of a rapidly evolving, highly polymorphic noncoding segment of mtDNA from 189 individuals, including 121 native Africans from various parts of the continent, Papua New Guineans, Europeans, Asians, African-Americans, and a native Australian. In both studies, the researchers found greater sequence differences among Africans, particularly sub-Saharan Africans, than among Asians or Europeans. Because mutations accumulate over time, they concluded that the African population has had the longest time to evolve variation and thus modern humans originated in Africa.

Statistical Calculations of the Number and Rate of Mutations Suggests When Modern Humans Appeared

Having proposed that modern humans first appeared in Africa, the researchers calculated the probable date of their origin by extrapolating the unknown from the known. They had observed the greatest human mtDNA variation in a sub-Saharan African population; about 2.8% of the base pairs in the mitochondrial genome varied among the individuals they studied from that population. They also knew that chimpanzees and humans diverged approximately 5 million years ago and that human mtDNA differs from that of chimpanzees in about 15% of the genome. Adjusting these data to account for multiple substitutions at the same base pair, they estimated that the mtDNA of humans and chimpanzees has been diverging at an average rate of about 13.8% per million years. To determine approximately how long ago the human population containing mitochondrial Eve lived, they divided the percentage of maximal human variation by the rate of chimp-human divergence:

$$2.8/13.8 = 0.20 \text{ million} = 200,000 \text{ years ago.}$$

Although there has been some controversy over the statistical methods and assumptions that formed the basis of this analysis, most geneticists now agree with the conclusion that the women carrying our ancestral mtDNA lived roughly 200,000 years ago in sub-Saharan Africa. Recent studies of parts of the Y chromosome and other pieces of nuclear DNA support this conclusion. In one 1997 study, evolutionary geneticists examined two nuclear DNA segments on chromosome 12 in the genomes of 1600 individuals from 42 populations around the world. In populations outside of Africa, one combination of the two elements showed up almost exclusively; by contrast, in sub-Saharan Africa, almost all the possible variants appeared. Recently, analyses of Y chromosome polymorphisms uncovered two ancient DNA markers shared by nonhuman primates and a small group of sub-Saharan African men. These markers sustained mutations about 100,000 to 200,000 years ago. Today all men outside of Africa as well as most African men carry the mutated markers.

Comparisons of Sequence Variations Point to How *Homo sapiens* Evolved

In addition to providing evidence for where and when modern humans first appeared, mtDNA may help evolutionary biologists determine what happened to the human populations that migrated out of Africa to other areas of the globe. At present, there are two models. One, the replacement theory, posits that as early modern humans (*H. sapiens*) migrated to Europe, Asia, and elsewhere, they eventually replaced the archaic hominid groups living in those areas. In this view, although early modern *H. sapiens* spread to parts of the world already populated by archaic species in the genus *Homo* (such as *H. erectus*), the early moderns did not intermingle with these other hominid groups. Rather, they outcompeted and eventually replaced them such that anatomically modern *H. sapiens* became the only extant human (*Homo*) species. An opposing view, the regional continuity theory, proposes that humans evolved continuously in many parts of the world, with early modern humans interbreeding with archaic peoples in Europe and Asia. These archaic peoples would have included the Neanderthals who coexisted with modern humans in the Middle East and parts of Europe and Asia for close to 70,000 years.

The first Neanderthal skeleton to be dug up was unearthed in Germany's Neander Valley (in German, Neander Tal) in 1856 (Fig. A); 141 years later (in 1997), a group of German and American biologists used state-of-the-art tools of DNA analysis to amplify and examine a minute amount of mtDNA from that skeleton. Even though mitochondrial DNA has a much greater chance of surviving in a long-dead specimen because of its abundance (500–1000 copies per cell rather than the usual 2 for nuclear DNA), it was a technical triumph to recover enough untainted mtDNA—379 bp in all—for PCR amplification and subsequent analysis. When the scientists compared

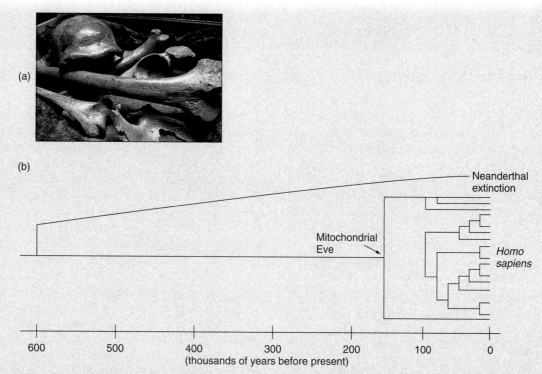

Figure A   Analysis of Neanderthal mitochondrial DNA provided proof that our closely related extinct sibling species did not contribute to the gene pool of *H. sapiens*. (a) The first Neanderthal skeleton uncovered in 1856. (b) Evolutionary relationship of Neanderthal to *H. sapiens* as established by mtDNA analysis.

the Neanderthal sequence with more than 900 distinct mtDNA sequences from living humans, they found, on average, three times as many differences between Neanderthal and modern human DNAs than between mtDNAs from pairs of modern humans. (However, since only one Neanderthal skeleton has been analyzed so far, the extent of variation among Neanderthal genomes is not yet known.) The mtDNA pairs from modern humans differed at an average of 8 positions, while the human–Neanderthal pairs differed at 25.6 positions. Furthermore, the types of nucleotide substitution and their locations were different in the two groups. According to the researchers, these data "make it highly unlikely that Neanderthals contributed to the human mtDNA pool." Using the same type of calculations that the Wilson group used, including the time when humans and chimps diverged and the rate at which the particular segments of mtDNA under analysis accumulate mutations, the German and American researchers estimated that the mtDNA sequence ancestral to that of contemporary humans and Neanderthals began to diverge 550,000 to 690,000 years ago, compared to only 120,000 to 150,000 years ago for divergence of the ancestral sequence of all modern humans. In other words, the last common ancestor of Neanderthals and modern humans is roughly four times older than the common ancestor of all modern humans.

These calculations, although again based on assumptions that are open to question, nevertheless suggest that humans and Neanderthals diverged before the first known Neanderthal appeared some 300,000 years ago and long before the first modern humans

evolved around 200,000 years ago. If further evidence and analyses support these inferences, the logical conclusion would be that there was no mixing of humans and Neanderthals, at least not in the mitochondrial genes. This, in turn, would support the replacement theory of modern human evolution.

Opponents of the replacement theory maintain that the close similarity of modern human mtDNAs from all parts of the world and its difference from that of Neanderthals could have arisen in other ways—for example, from a population bottleneck, such as a flood or drought, that left few survivors carrying Neanderthal genes, or from the disappearance by random genetic loss or negative selection of Neanderthal genes that may have been part of the early modern human gene pool. (Chapter 23 explains the details of these possibilities.)

The information needed to put this controversy to rest is not yet in hand. One step in that direction would be the unearthing and analysis of mtDNA from a North African Neanderthal, who would represent an early stage in the migration out of Africa. Another step would be the analysis of mtDNA from the already discovered 30,000-year-old Cro-Magnon, judged to be a representative of the first undisputed early modern humans. Evidence gathered in such studies could narrow the possibilities. If it supports the replacement theory, the icon of human evolution would be a bush with many twigs, of which all but modern *H. sapiens* have dried up. If it doesn't support the theory, the image would be much more complex.

Genetics and Society

Mitochondrial DNA Tests Replace HLA Typing as Evidence of Kinship in Argentine Courts

Between 1976 and 1983 the military dictatorship of Argentina kidnapped, incarcerated, and killed more than 10,000 university students, teachers, social workers, union members, and others who did not support the regime. Many very young children disappeared along with the young adults, and close to 120 babies were born to women in detention centers. In 1977, the grandmothers of some of these infants and toddlers began to hold vigils in the main square of Buenos Aires to bear witness to and inform others about the disappearance of their children and grandchildren. They soon formed a human rights group—the "Grandmothers of the Plaza de Mayo."

The grandmothers' goal was to locate the more than 200 grandchildren they suspected were still alive and reunite them with their biological families. To this end, they gathered information from eye witnesses, such as midwives and former jailers, and set up a network to monitor the papers of children entering kindergarten. Questionable birth certificates could reveal a possible missing grandchild. They also publicized their work inside Argentina, and contacted organizations outside the country, including the United Nations Human Rights Commission and the American Association for the Advancement of Science (AAAS).

What the grandmothers asked of AAAS was help with genetic analyses that would stand up in court. By the time a democracy had replaced the military regime and the grandmothers could argue their legal cases before a relatively impartial court, children abducted at age 2 or 3 or born in 1976 were 7–10 years old. Although

the grandmothers had compiled an enormous amount of circumstantial evidence, the Argentine courts did not accept such evidence as proof of a young person's identity and biological relatedness. The courts did acknowledge, however, that although the size and other external features of the children had changed, their genes—relating them unequivocally to their biological families—had not. The grandmothers, who had educated themselves about the potential of genetic tests, sought help with the details of obtaining and analyzing such tests. Starting in 1983, the courts agreed to accept their test results as proof of kinship.

In 1983, the best way to confirm or exclude the relatedness of two or more individuals (for instance, a child and her grandparents) was to compare the gene-encoded proteins called human lymphocyte antigens (HLAs). People carry a unique set of HLA markers on their white blood cells, or lymphocytes, and these markers are diverse enough to form a kind of molecular fingerprint. HLA analyses can be carried out even if a child's parents are no longer alive, because for each HLA marker, a child inherits one allele from the maternal grandparents and one from the paternal grandparents. Statistical analyses can establish the probability that a child shares genes with a set of grandparents because he or she is their grandchild compared to the probability that the child shares the genes by chance.

The AAAS put the grandmothers in touch with Mary Claire King, then at the University of California. In the 1980s, King taught

protein's mode of inheritance, they evaluated the allele expressed in the progeny of a controlled cross, carefully noting the maternal (ovum) and paternal (pollen) contributions. In one set of experiments, ribulose bisphosphate carboxylase (Rubisco, for short), the first enzyme of photosynthetic carbon fixation in plants and the most abundant protein in tobacco leaves, was the protein under observation. Rubisco has a 55-kDa <u>l</u>arge <u>su</u>bunit (called LSU) and a 12-kDa <u>s</u>mall <u>su</u>bunit (called SSU). The researchers purified Rubisco from many species of tobacco plants, digested the purified proteins with trypsin, and analyzed the digests for usable differences. When they followed the inheritance of these differences, they found that LSUs manifested patterns of maternal inheritance, while SSUs showed biparental inheritance. From these results, they concluded that a chloroplast gene encodes the LSU polypeptide, while a nuclear gene encodes the SSU.

These studies of the two Rubisco subunits reveal that organelle and nuclear genomes cooperate in specifying even a relatively simple enzyme with only two different subunits. Both genomes contribute essential information for most pho-

tosynthetic activities, including those whose elements are much more complex.

The inheritance studies just described followed preexisting differences in functionally wildtype organelle genomes. To verify and understand the details of uniparental inheritance, organelle geneticists followed the inheritance of mutations affecting phenotype at both the biochemical and the oganismal levels.

A Mutation in a Human Mitochondrial Gene Generates a Maternally Inherited Neurodegenerative Disease

<u>L</u>eber's <u>h</u>ereditary <u>o</u>ptic <u>n</u>europathy, or LHON, is a disease in which flaws in the mitochondria's electron transport chain lead to optic nerve degeneration and blindness (Fig. 14.10). Family pedigrees show that LHON passes only from mother to offspring. In the late 1980s and early 1990s, a series of molecular studies showed that a G-to-A substitution at nucleotide 11,778 in the human mitochondrial genome is a main cause of the condition. The substitution alters an arginine-specifying codon in the NADH dehydrogenase subunit 4 gene to a histi-

Argentine medical workers to analyze the diverse HLA markers on white blood cells. The grandmothers then obtained the HLA types of as many living members as possible of the missing children's families and stored that information in an HLA bank. When a child whom they believed to be one of the missing turned up, they analyzed his or her HLA type and tried to find a match among their data. Depending on the number of different alleles carried and the rarity of the variations, the probability that a tested child belonged to the family claiming him or her on the basis of eye-witness accounts of a birth or abduction varied from 75% to 99%.

By the late 1980s, the combination of circumstantial and genetic evidence had helped reunite 49 children with their biological families. In one case, a grandmother who had been kidnapped September 5, 1976, along with her pregnant daughter and then released from prison, contacted the Grandmothers of the Plaza de Mayo for help. Ten and a half years later, in April 1987, she was reunited with her granddaughter. The child had been born on November 5, 1976, and registered in the name of a detention center police officer. Her place of birth remains unknown. But, her grandmother recounted in an April 1987 interview, "It was very easy to prove she was my granddaughter because of the blood tests. They analyzed my blood, the blood of my husband and the other grandparents and they [the results] proved more than 99% positive."

As time passed and the "easier" cases had been settled, the limitations of the HLA approach became apparent. By the mid-1980s, for example, there were too few living relatives in some families to establish a reliable match through HLA typing. But the advent of new tools such as PCR and DNA sequencing made it possible to look at DNA directly.

King and two colleagues—C. Orrego and A. C. Wilson—used the new techniques to develop an mtDNA test based on the PCR amplification and direct sequencing of a highly variable noncoding region of the mitochondrial genome. Mitochondrial DNA has several characteristics that make it more powerful than HLA typing for confirming or excluding family relatedness: maternal inheritance, a lack of recombination, a highly variable noncoding region of 331 bp, and a large number of mtDNAs per cell. The maternal inheritance and lack of recombination mean that as long as a single maternal relative is available for matching, the approach can resolve cases of disputed relatedness. The extremely polymorphic noncoding region makes it possible to identify grandchildren through a direct match with the mtDNA of only one person—their maternal grandmother, or mother's sister or brother—rather than through statistical calculations assessing data from four people; the probability that any two unrelated individuals would be identical for all 331 bp by chance is almost nil. Finally, the large number of mtDNAs per cell increases the chances of isolating mtDNA from and identifying the bodies of presumed parents that are found after many years.

To validate their approach for matching maternal grandmothers to their grandchildren, King and colleagues amplified sequences from three children and their three maternal grandmothers without knowing who was related to whom. The mtDNA test unambiguously matched the children with their grandmothers. Thus, after 1989, the grandmothers included mtDNA data in their archives.

Today, the grandchildren—the children of "the disappeared"—have reached adulthood and attained legal independence. Although most of their grandmothers have died, the grandchildren may still discover their biological identity and what happened to their families through the HLA and mtDNA data the grandmothers left behind.

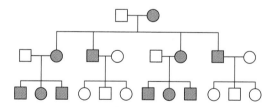

Figure 14.10 Hypothetical example of LHON pedigree. Characteristic pedigree of mitochondrial disease. All offspring of diseased mother can express the disease phenotype, while none of the offspring of the diseased fathers express the disease phenotype.

dine codon. The resulting protein product diminishes the efficiency of electron flow down the respiratory transport chain, reducing the cell's production of ATP sufficiently to cause a gradual decline in cell function and ultimately cell death. Since optic nerve cells have a relatively high requirement for energy, the genetic defect affects vision before it affects other physiological systems.

In large families, not all offspring show signs of the disease, and not all siblings manifesting the condition have symptoms of the same severity. The random allotment to daughter cells of a large number of mitochondria during mitosis (described later) helps explain these observations.

Cells Can Contain One Type or a Mixture of Organelle Genomes

A diploid cell contains dozens to thousands of organelle DNAs. It is therefore not possible to use the terms "homozygous" and "heterozygous" to describe a cell's complement of mtDNA or cpDNA. Instead, geneticists use the terms "heteroplasmic" and "homoplasmic" to describe the genomic makeup of a cell's organelles. **Heteroplasmic** cells contain a mixture of organelle genomes. **Homoplasmic** cells carry only one type of organelle DNA. Except for the rare appearance of a new mutation, the mitotic progeny of homoplasmic cells carry a single type of organelle DNA. By contrast, the mitotic progeny of heteroplasmic cells may be heteroplasmic, homoplasmic wildtype, or homoplasmic mutant. In most people affected by LHON, for example, the optic nerve cells are homoplasmic for the disease mutation; but in some LHON patients, these optic nerve cells are heteroplasmic. Homoplasmy causes earlier appearance of the disease as well as more severe symptoms.

Mitotic Segregation Produces an Uneven Distribution of Organelle Genes in Heteroplasmic Cells

The random partitioning of organelles during cell divisions is the basis of the mitotic segregation of organelle genomes. This pattern of random segregation is in sharp contrast to the mitotic transmission pattern of nuclear genes: In diploid cells that are heterozygous for a nuclear gene, the alleles of the gene almost never segregate during mitotic divisions; the rare exception in which a heterozygous diploid produces homozygous wildtype and homozygous mutant progeny is the result of mitotic crossing-over (described in the Genetics and Society box in Chapter 4).

The mitotic segregation of organelle genomes has distinct phenotypic consequences. In a woman whose cells are heteroplasmic for the LHON mutation, some ova may carry a few mitochondria with the LHON mutation and a large number of mitochondria with the wildtype gene for subunit 4 of NADH dehydrogenase; other ova may carry mainly mitochondria with LHON mutations; still others may carry only wildtype organelles. The precise combination depends on the random partitioning of mitochondria during the mitotic divisions that gave rise to the germ line. After fertilization, as a result of the mitotic divisions of embryonic development, the random segregation of mutation-carrying mitochondria from heteroplasmic cells can pro'uce tissues with completely normal ATP production and t   es of low energy production. If cells homoplasmic for lo   nergy production happen to end up in the optic nerve, LHO   will result.

In plants where cpDNA mutations leading to a defect in photosynthesis would be lethal, heteroplasmy for chloroplast genomes is prevalent. In fact, mitotic segregation of the chloroplasts of heteroplasmic cells explains the transmission of variegation in four-o'clocks, described at the beginning of this chapter (see Fig. 14.1). Most female gametes from a variegated plant are heteroplasmic for mutant and wildtype cpDNAs. Zygotes resulting from fertilization with pollen from a wildtype green plant will develop into variegated progeny. Segregation of wildtype and mutant chloroplasts during F_1 plant development may, however, generate some female gametes with only mutant cpDNA. Fertilization of these homoplasmic mutant gametes with wildtype pollen produces zygotes with only mutant cpDNA; the seedlings that develop from these zygotes cannot carry out photosynthesis and eventually die.

Experiments with Mutants of cpDNA in Chlamydomonas reinhardtii *Reveal Uniparental Inheritance of Chloroplasts*

Chlamydomonas, a unicellular alga, is easy to handle, grows abundantly in defined media, and has a tractable sexual reproduction cycle. It can live in different environmental conditions, functioning photosynthetically in the light on media where atmospheric CO_2 is the sole source of carbon, or without photosynthesis in the dark on media containing acetate as a source of carbon. These features make it a useful experimental organism. In the laboratory, *Chlamydomonas* usually grows as a haploid cell, and each haploid strain comes in one of two mating types: mt^+ or mt^-. Researchers can induce cultures of mitotically dividing haploid cells to develop into haploid gametes by a special physiological regime (Fig. 14.11a). When they then mix gametes of opposite mating types, an elaborate mating process results in the fusion of gametes of opposite types to form diploid zygotes. On suitable medium, these zygotes undergo meiosis and form unordered tetrads. For simplicity, we describe all experiments with this organism as leading to tetrads with four spores, even though in reality sometimes a mitotic division after meiosis generates eight spores.

Pioneering research on the genetics of chloroplasts in *Chlamydomonas* led to the isolation of mutants with various defects in photosynthetic growth. The first such mutant analyzed genetically was resistant to the antibiotic streptomycin, that is, it carried the *sm-r* allele, which encodes an inhibitor of prokaryotic (and also chloroplast) protein synthesis. These mutants could grow photosynthetically on medium containing streptomycin, while the drug-sensitive *sm-s* parental strain could not. The mutant strain also grew well in the absence of the drug. Researchers could thus carry out crosses in media lacking the drug and then grow the progeny on drug-containing media to determine their phenotype.

Consider a cross between mt^+ *sm-r* gametes and mt^- *sm-s* gametes in which researchers scored the resulting tetrads for the transmission of drug sensitivity (Fig. 14.11b). The investigators included a known nuclear marker affecting colony color (*yellow 1,* with alleles $y1^+$ and $y1^-$) in the parental strains and monitored the marker in the progeny. In this experiment, alleles of the mating-type gene and the yellow-marker gene segregated 2:2 in the spores, but the ratio of *sm-r* to *sm-s* was 4:0. This lack of segregation was a striking departure from Mendelian principles. Subsequent reciprocal crosses showed that the *sm-r* allele did not itself determine the nonsegregating outcome. In a reciprocal cross of mt^+ $y1^+$ *sm-s* gametes with mt^- $y1^-$ *sm-r* gametes, the tetrads were 4 *sm-s*:0 *sm-r*. Because *Chlamydomonas* lacks distinct male and female gametes, it is an *isogamous* species in which the "sex" cells are similar in size and shape but come in two mating types. In speaking of isogamous species, geneticists say that a 4:0 ratio of spore genotypes reflects *uniparental,* rather than maternal, inheritance. Researchers have isolated many other *Chlamydomonas* mutants resistant to various antibiotics or deficient in photosynthesis, and all show the uniparental mode of transmission.

Many Mechanisms Contribute to Uniparental Inheritance of Mitochondrial and Chloroplast Genes

Differences in gamete size help explain maternal inheritance in some species. In most higher eukaryotes, the male gamete is much smaller than the female gamete. As a result, the zygote receives a very large number of maternal organelles and at most, a very small number of paternal organelles. In some organisms, cells degrade the organelles or the organelle DNA of

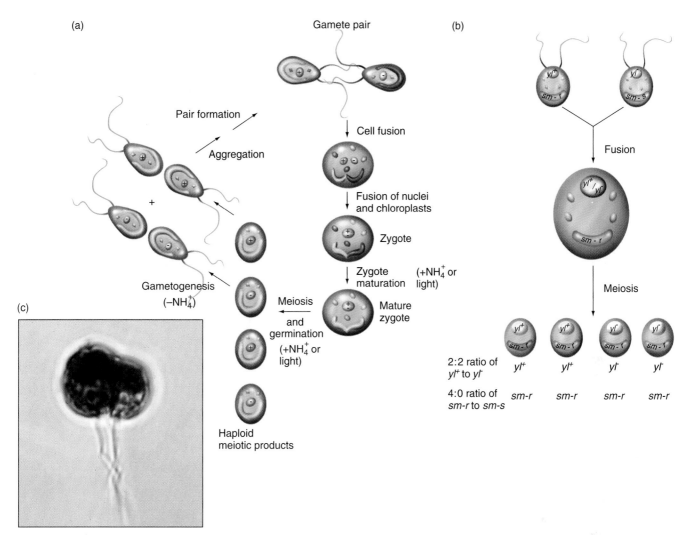

Figure 14.11 Genes in the cpDNA of *C. reinhardtii* do not segregate at meiosis. a) A cross of *C. reinhardtii* gametes is illustrated. Vegetative cells of this organism are usually haploid. Mating of those cells is controlled by the mating type locus that exists in two alleles (mt^+ and mt^-). The nucleus of each haploid cell shown is color coded (mt^+ = blue, mt^- = yellow.) Each haploid cell has one large chloroplast (light or dark green for different alleles of cpDNA) and several mitochondria (light or dark red for different alleles of mtDNA). Mating occurs by pairwise fusion of gametes of opposite mating type. Shortly after fusion, the nuclei and chloroplasts fuse so that each zygote has one diploid nucleus and one large chloroplast. The zygote matures, during which time chloroplast markers from the mt^- parent and mitochondrial markers from the mt^+ parent are inactivated. Mature zygotes then undergo meiosis, forming four haploid meiotic products each of which can germinate. If light or ammonium ions are present the haploid cells will grow vegetatively. If ammonium ions are absent the cells will undergo gametogenesis to prepare for another round of mating. b) The nuclear gene *yl* segregates 2:2 in the gametes but the chloroplast encoded *sm* gene segregates 4:0. c) When haploid cultures of gametes of opposite mating type are mixed, cells aggregate and form pairs, leading to another round of life cycles.

male gametes. In some plants, the early divisions of the zygote distribute most or all of the paternal organelle genomes to cells that are not destined to become part of the embryo. In some animals, details of fertilization prevent a paternal cell from contributing its organelles to the zygote. With tunicates, for example, events of fertilization allow only the sperm nucleus to enter the egg, physically excluding the paternal mitochondria. In some organisms where the complete gametes fuse, the zygote destroys the paternal organelles after fertilization.

These explanations do not apply to *Chlamydomonas,* in which the two gametes are of equal size and fuse completely. In this organism, the single chloroplast of one gamete fuses with the chloroplast of the other gamete shortly after cell fu-

sion. However, the zygote, able to distinguish between cpDNAs from mt^+ and mt^- cells, degrades most of the cpDNAs from the mt^- parent. Interestingly, *Chlamydomonas* mtDNA is also inherited from one parent, but here, it is the mt^- cell type whose chloroplast genome remains.

Summary of Uniparental Inheritance

The transmission of organelles and their genomes via one parent, mainly the mother, in most plants and animals results in two significant departures from Mendelian patterns of inheritance. First, organelle genes show no meiotic segregation: all F_1 individuals carry the same organelle genotype—that of the transmitting parent. Second, when the parent transmitting the organelle genome is heteroplasmic and therefore passes along organelles of more than one genotype, mitotic segregation of those genotypes occurs in the offspring. This segregation of genotypes during mitosis is a consequence of the random partitioning of organelles during cell division.

Some Organisms Exhibit Biparental Inheritance

Although uniparental inheritance is the norm among most metazoans and plants, single-celled yeast and some plants inherit their organelle genomes from both parents in a **biparental** fashion. The earliest report of biparental inheritance is a 1909 description of reciprocal crosses between green and variegated geraniums (*Pelargonium zonale*). Unlike what happens in four-o'clocks (as described earlier), both reciprocal crosses yielded green, white, and variegated seedlings in varying proportions (Fig. 14.12). Thus, variegation in geraniums is a chloroplast trait inherited from both parents. Many other plants as well as yeast similarly inherit their organelle genomes from both parents. Geneticists have devised experimental systems for analyzing biparental organelle inheritance in several organisms. We now look at some of the studies in yeast.

A Genetic System for Studying Mitochondrial Genes in Yeast

We have seen that translation in mitochondria resembles translation in bacteria, and that inhibitors of bacterial translation, such as chloramphenicol and erythromycin, are potent inhibitors of mitochondrial protein synthesis. In the early 1960s, researchers discovered that chloramphenicol inhibits the growth of wildtype yeast on media containing a nonfermentable source of carbon (either glycerol or ethanol); but the same or higher levels of the drug does not inhibit yeast growth on medium containing glucose, a fermentable carbon source. They concluded that the inhibitor acts on the then recently discovered mitochondrial translation system, and began to develop a system for studying the inheritance of yeast mtDNA. The initial step was to isolate mutants that could grow on medium containing glycerol in the presence of a drug that inhibits mitochondrial protein synthesis. The first useful mutants were resistant to chloramphenicol or erythromycin (C^r or E^r); they derived from wildtype cells that were sensitive to one drug or the other (C^s or E^s).

Like *Chlamydomonas,* yeast is isogamous. Each haploid strain has one of two mating types: MATa or MATα. The mixing of cells of opposite mating type results in the fusion of dissimilar haploid cells to form zygotes (Fig. 14.13a). Unlike *Chlamydomonas,* however, yeast cells can mate without first differentiating into gametes, and yeast crosses generate cultures of diploid cells because the zygote produced by mating does not usually enter meiosis; rather, it continues to multiply by mitosis until some sort of stress induces its diploid descendants to enter the meiotic cycle.

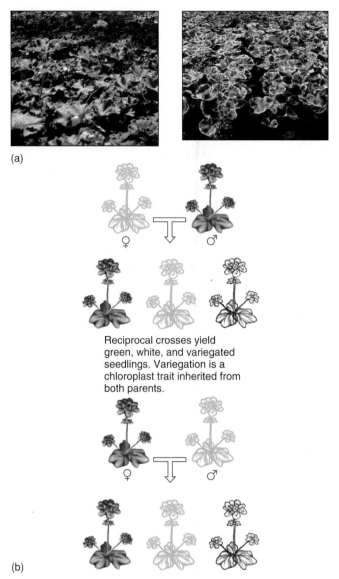

(a)

Reciprocal crosses yield green, white, and variegated seedlings. Variegation is a chloroplast trait inherited from both parents.

(b)

Figure 14.12   Biparental inheritance of variegation in geraniums. (a) Examples of green and variegated *P. zonale* plants. (b) Reciprocal crosses yield varying numbers of green, variegated and white offspring.

In Yeast, mtDNA-Encoded Traits Show a Biparental Mode of Inheritance and Mitotic Segregation

To analyze the inheritance pattern of mtDNA in yeast, researchers mixed C^r and C^s parental cells; allowed them to grow for many generations; and isolated diploids, taking advantage of nuclear auxotrophic markers in the parental strains for the selection of diploids (Fig. 14.13b). They then scored cells from the diploid progeny for the C^r or C^s phenotype by replica plating the cells to petri dishes containing glycerol medium with or without an inhibitory level of chloramphenicol.

In most progeny resulting from the crosses, about half of the diploid cells were C^s and half were C^r. This result shows that both parents transmitted organelles to the progeny, that is, inheritance was biparental. It also shows that the organelles

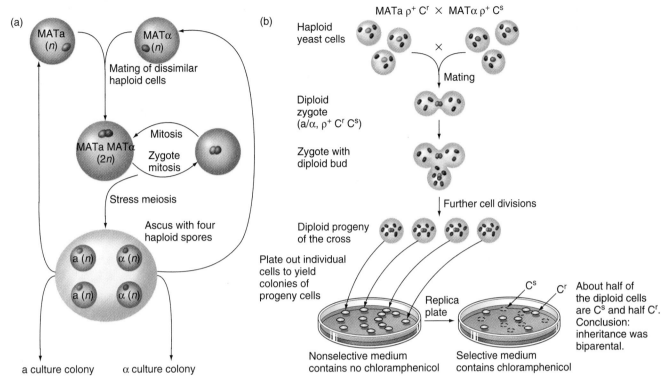

Figure 14.13   Mitochondrial transmission from the yeast *S. cerevisiae*. (a) General life cycle of *S. cerevisiae*. (b) One-factor crosses in *S. cerevisiae* show biparental inheritance of mtDNA.

from parental cells of different mating types segregated during the mitotic divisions of vegetative growth so that after several generations colonies were either C^s or C^r, but not both. Importantly, all of the progeny scored, whether C^s or C^r, had the same nuclear-determined phenotype (that is, they were prototrophic for two nutrients). This meant that the diploids carried nuclear genes from both parents in the cross, and those nuclear genes did not segregate during this period of vegetative growth.

Researchers Can Now Use Recombinant DNA Techniques to Study the Genetics of Organelles

In the early days of recombinant DNA technology, organelle researchers were frustrated by an inability to transfer cloned genes and mutated DNA fragments into organelle genomes. Researchers could deliver plasmid DNAs to yeast cell nuclei very efficiently, but they could not get it to enter the mtDNA of yeast mitochondria or the cpDNA of *Chlamydomonas* chloroplasts. Because recombination occurs readily in both organelle genomes, the problem seemed to be in getting the foreign DNA into the organelles themselves.

In the late 1980s, development of the *gene gun* (Fig. 14.14) and a gene delivery method known as "biolistic transformation" solved the problem. The basic idea is to coat small (~1 micrometer) metal particles with DNA and then shoot these DNA-carrying "bullets" at cells. First attempted in plant cells, biolistic transformation occurs when a particle lands within a cell, such as a plant protoplast, without killing it, and releases the DNA from the metal; rarely, the released DNA en-

ters the nucleus. If the DNA shot into the cell contains a strong selective marker, plant geneticists can isolate the rare transformed plant cells in which the released DNA has entered the nucleus.

Shortly after the use of the gene gun to transfer genes into plants, two groups of researchers, one working with yeast mitochondria, the other with *Chlamydomonas* chloroplasts, successfully transformed organelle DNA. In one study, researchers constructed a special, mitochondrion-translatable form of a gene-encoding green fluorescent protein (GFP); the gene product fluoresces brightly when stimulated by light of the right wavelength. The codons of the specially constructed gene were based on the mitochondrial genetic code, instead of on the "universal" genetic code. To this specially constructed gene, the experimenters fused that part of the mitochondrial *COX3* gene extending from the promoter through sequences encoding the first three amino acids of the COX3 protein. When they then bombarded mitochondria with this plasmid, they obtained transformants that expressed the *GFP* gene via the mitochondrial gene expression system. Fluorescence of the mitochondria showed that the transformed cells retained the GFP protein in their mitochondria (Fig. 14.14).

Current researchers can use yeast with this GFP construct to select for mutations that affect mitochondrial gene expression. In fact, the transformation of organelle genomes is now a reliable way to manipulate and study organelle genes in yeast and *Chlamydomonas*. Recently, researchers have achieved chloroplast transformation in higher plants and think they may

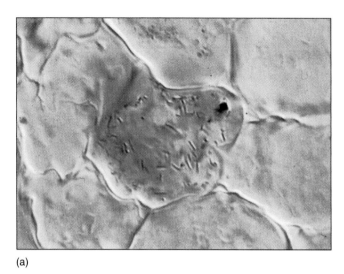

(a)

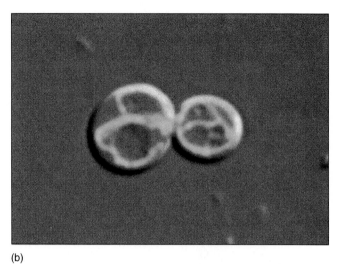

(b)

Figure 14.14 Genetic transformation of organelle genomes. (a) Foreign DNA can be delivered into organelle with the use of a gene gun that deposits tiny biolistic pellets having micrometer diameters. (b) Fluorescent green mitochondria demonstrate transformation of the mitochondrial genomes with a fusion DNA construct containing a modified form of the GFP coding region based on the mitochondrial genetic code, attached to a mitochondrial specific promotor.

some day be able to use such transformation in engineering crop plants resistant to certain herbicides.

Summary of the Genetic Principles of Non-Mendelian Inheritance

Three features distinguish the non-Mendelian traits encoded by organelle genomes from the Mendelian traits encoded by nuclear genomes.

1. In the inheritance of organelle genomes from one generation to the next, there is a 4:0 segregation of parental alleles, instead of the 2:2 pattern seen for the alleles of nuclear genes.

2. In most organisms, transmission of organelle-encoded traits is uniparental, mainly maternal, although in a few organisms, notably yeast, transmission is biparental.

3. With both uniparental and biparental inheritance, when the parents transmit organelles of more than one genotype, mitotic segregation of those genotypes occurs in the offspring. This segregation of genotypes during mitosis is a consequence of the random partitioning of organelles during cell division.

COMPREHENSIVE EXAMPLE: HOW MUTATIONS IN mtDNA AFFECT HUMAN HEALTH

Some debilitating diseases of the human nervous system pass from mother to daughters and sons, from affected daughters to granddaughters and grandsons, and so on down through the maternal line. Unexpectedly, the symptoms of these diseases vary enormously among family members, even among very close relatives. Pedigrees tracing general transmission pat-

terns, in combination with molecular studies, help clarify what is happening.

Individuals with Certain Rare Diseases of the Nervous System Are Heteroplasmic for Wildtype and Mutant mtDNAs

People with a rare inherited condition known as myoclonic epilepsy and ragged red fiber disease (MERRF) have a range of symptoms: uncontrolled jerking (the myoclonic epilepsy part of the condition); muscle weakness; deafness; heart problems, kidney problems, and progressive dementia. Affected individuals often have an unusual "ragged" staining pattern in regions of their skeletal muscles, which explains the ragged red fiber part of the condition's name (Fig. 14.15a). As the pedigree in Fig. 14.15b shows, family members inherit MERRF from their mothers; in the pedigree, none of the offspring of the affected male sibling (II4) has symptoms of the disease. The family history also reveals individual variations in the number and severity of symptoms. From these two features of transmission, clinical researchers suspected that MERRF results from mutations in the mitochondrial genome.

Molecular analyses confirm this hypothesis. The mtDNA from patients affected by MERRF carries a mutation in the gene for tRNA^{Lys} (review Fig. 14.5) or one of the other mitochondrial tRNAs. These tRNA mutations disrupt the synthesis of proteins in multiplexes I and IV of the mitochondrial electron transport chain, thereby decreasing the production of ATP.

The Proportion of Mutant mtDNAs and the Tissue in Which They Reside Influence Phenotype

In a second large family in which clinicians looked at the mitochondria of muscle cells, an individual carrying 73% mutant and 27% normal mtDNAs showed no symptoms of MERRF; a relative with 85% mutant mtDNA showed no ex-

(a)

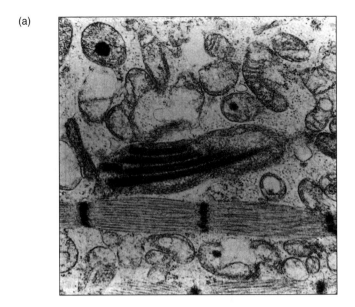

(b)

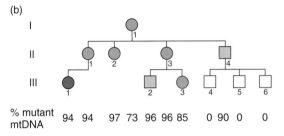

Figure 14.15   Maternal inheritance of the mitochondrial disease MERRF. (a) Transmission electron micrograph of muscle. Mitochondria from patients expressing MERRF. Mutant mitochondria are highly abnormal, showing paracrystalline arrays and crista degeneration. (b) Pedigree of family showing inheritance of MERRF. Pedigree shows typical pattern of maternal transmission observed with mitochondrial mutations. Mothers transmit the disease to all children, both boys and girls. Diseased fathers never transmit the disease to any children.

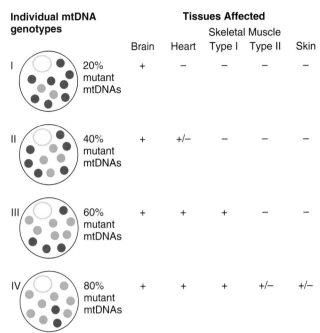

Figure 14.16   Disease phenotypes reflect the ratio of mutant-to-wildtype mtDNAs and the reliance of cell type on mitochondrial function. The proportion of mutant mitochondria determines the severity of the MERRF phenotype and the tissues that are affected. Tissues with higher energy requirements (e.g., brain) are least tolerant of mutant mitochondria. Tissues with low energy requirements (e.g., skin) are only affected when the proportion of wildtype mitochondria is greatly reduced.

ternal signs of the disease but lab tests revealed some muscle tissue abnormalities; and two family members with 98% mutant mitochondria showed serious symptoms of MERRF. This suggests that a relatively small percentage of normal mitochondria can have a strong protective effect. The random partitioning of mitochondria during the mitotic cell divisions of development, both *in utero* and after birth, explains the variation in the percentage of mutant mitochondria in individuals of the second, third, and fourth generations of family pedigrees.

It is likely that many tissues in individuals affected by MERRF are heteroplasmic for the tRNA mutations causing the disease. Within each person, the ratio of mutant to wildtype mtDNA varies considerably from tissue to tissue, and because each tissue has its own energy requirements, even the same ratio can affect different tissues differently (Fig. 14.16). Muscle and nerve cells have the highest energy needs of all types of cells and are therefore the most dependent on oxidative phosphorylation. This explains why mitochondrial mutations that

by chance segregate to these tissues generate the defining features of MERRF.

The neurodegenerative disease mentioned earlier in the chapter—Leber's hereditary optic neuropathy, or LHON—also shows the maternal inheritance and variation in symptoms (severity as well as time of onset) characteristic of conditions caused by mtDNA mutations. While MERRF is the consequence of mutations in tRNA genes, LHON, we have seen, results from mutations in the protein-encoding *ND4* gene whose product is one component of the NADH dehydrogenase enzyme active in the mitochondrial electron transport chain (see Fig. 14.16). Many people who inherit a LHON mutation are normal throughout childhood and then lose their sight and develop other neurological as well as cardiac defects in young adulthood. The random partitioning of mitochondria that leads to mitotic segregation helps explain this pattern of pathology. By the time a person carrying some mutant mitochondria reaches young adulthood, many cells in the optic nerve may have become homoplasmic for the LHON-causing mutation, or if they remain heteroplasmic, they may have such a small proportion of normal mtDNA that it is not sufficient to sustain the energy needs of the cells. Although the replication and segregation of LHON mutations has been going on for years in these people, it is only when enough cells die from lack of sufficient ATP that the optic nerve or cardiac tissue suddenly crosses the threshold from function to nonfunction.

Mitochondrial Inheritance in Identical Twins

In some pairs of identical twins, one twin manifests symptoms of neurodegenerative disease, while the other twin does not. This is because even though their nuclear genomes are identical, their mitochondrial genomes are not.

Consider the following. A fertilized human egg cell carries up to 2000 mtDNAs, 99.9% of which come from the mother. If 10% of those mtDNAs carry a MERRF mutation, at the first mitotic division, the mutant mtDNAs could by chance be partitioned in any one of a number of ways: The cell that will develop into one twin may get all the mutant mitochondria (this would be a very rare occurrence); the mtDNAs with mutations may be equally divided between the two twin-producing cells; the mutated mtDNAs may be unevenly distributed to the two twin-generating cells. Even with an even distribution at the first cell division, during the subsequent divisions of development, the mutant mitochondria in one twin may end up in cells that eventually die during normal development, while the mutant mitochondria in the other twin may end up in cells that differentiate into heart and optic nerve tissue.

Mitochondrial Mutations and Aging

We have seen that like the nuclear DNA mutations that cause cancer, some mutations in mtDNA are inherited through the germ line, while others arise sporadically in somatic cells as a result of random events, such as radiation or chemical insults. We have also seen that the rate of somatic mutations is much higher in mitochondrial DNA than in nuclear DNA. This is in part because the oxidative phosphorylation system in mitochondria generates a relatively high concentration of DNA-damaging free radicals in a membrane-confined space. In one study, mtDNA accumulated 16 times more oxidative damage than nDNA.

Some researchers focusing on the genetics of aging think that the accumulation of mutations (both base substitutions and deletions) in mtDNA over a lifetime, and the progressive enrichment of deleted mtDNAs through their tendency to replicate more rapidly than longer, nondeleted mtDNAs, result in an age-related decline in oxidative phosphorylation. This decline, in turn, accounts for some of the symptoms of aging, such as decreases in heart and brain function.

Proponents of this hypothesis suggest that individuals born with deleterious mtDNA mutations start life with a diminished capacity for ATP production, and as a result, several of their tissues may cross the threshold from function to nonfunction early or in the middle of life; by comparison, people born with a normal mitochondrial genome start life with a high capacity for ATP production, and may die before a large number of tissues dip below the required energy threshold.

Evidence in support of an association between mtDNA mutations and aging comes from a variety of studies. In one study, researchers looked at 140 hearts obtained from autopsies and found significant decreases in cytochrome *c* oxidase, a respiratory enzyme largely encoded by mtDNA. In another study, researchers analyzed a 7.4-kb and a 5-kb deletion in heart and brain mtDNAs in people of different ages. The percentage of hearts that had the 7.4-kb deletion increased with age, and the number of 5-kb deletions increased in normal heart tissue after age 40. Moreover, the 5-kb deletion was absent from the brain tissue of children but present in the brain tissue of adults. Finally, although biomedical researchers had known for decades that the brain cells of people showing symptoms of Alzheimer's disease (AD) have an abnormally low energy metabolism, they recently discovered that 20% to 35% of the mitochondria in the brain cells of most AD patients carry mutations in two of their three cytochrome *c* oxidase genes; the inefficiency of the resulting mutant cytochrome *c* oxidase enzyme impairs the brain's energy metabolism. To confirm an association between this enzymatic abnormality and AD, the researchers transferred mitochondria from AD patients into normal cultured cells from which they had removed the native mitochondria, and found that the engineered cells had defective energy production.

These data suggest that if it were possible to assess all forms of mtDNA damage, it might turn out that a significant proportion of mtDNAs are defective in elderly people. On the basis of this hypothesis, clinicians have proposed that the restoration of enzymes encoded by wildtype mtDNA but absent from mitochondria with deleted and otherwise mutated genomes might ease some of the symptoms of aging.

CONNECTIONS

Mitochondria, the sites of oxidative phosphorylation, and chloroplasts, the sites of photosynthesis, exhibit non-Mendelian patterns of inheritance that are remarkably similar across a wide range of species.

A variety of regulatory mechanisms help correlate the day-to-day operations of a cell's organelles with the needs and circumstances of different tissues. One unusual example of the fine tuning of these controls is seen in the voodoo lily (*Sauromatum guttatum*), a native of Southeast Asia that belongs to the same plant family as the skunk cabbage. The day the flower opens, a burst of mitochondrial activity in some of its cells increases the temperature inside the bloom by 22°C (72°F); the heat so generated causes odorous molecules to evaporate and attract insect pollinators, increasing the plant's chances for reproduction. Although researchers currently know very little about the regulation of mitochondrial genes, they believe that, given the endosymbiont theory of organelle evolution, the regulation of genes in prokaryotes may serve as a model for what happens in mitochondria.

In Chapter 15 we examine the molecular mechanisms that regulate gene expression in prokaryotes. Then, in Chapter 16, we explore the regulation of nuclear genes in eukaryotes. In both these chapters, we see that gene regulation controls the metabolic activities of all cells as well as the differentiation of cells in multicellular organisms.

E S S E N T I A L   C O N C E P T S

1. Mitochondria and chloroplasts are semiautonomous organelles of energy conversion. They carry their own double-stranded DNA in circular or linear chromosomes whose size and gene content varies from species to species.

2. Translation in the mitochondria of many species depends on an alternative genetic code.

3. Chloroplast genomes are relatively uniform in size and carry many more genes than mitochondrial genomes. In some chloroplasts, the genes are organized in clusters that resemble the operons of bacteria.

4. According to the *endosymbiont theory,* mitochondria and chloroplasts evolved from bacteria engulfed by the precursors of eukaryotic cells. The genomes of these organelles have probably lost more than two-thirds of their original bacterial genes in the course of evolution.

5. In most species, organelle genomes show *uniparental inheritance,* mainly through the maternal line. Cells containing a mixture of organelle genomes are *heteroplasmic.* Cells carrying only one type of organelle DNA are *homoplasmic.* The genomes of heteroplasmic cells are not evenly partitioned at mitosis.

6. The organelles of yeast and several other species exhibit biparental inheritance. The genomes of organelles inherited in a biparental fashion show mitotic segregation.

S O C I A L   A N D   E T H I C A L   I S S U E S

1. A researcher has promising results that could lead to a treatment for a neurodegenerative mitochondrial disease. When the researcher applies for a grant from a federal agency, the reviewers respond that the research proposed is more applied than basic in nature and therefore should be funded by a commercial pharmaceutical company. When she approaches drug companies with her proposal, she finds they are not interested in developing a drug that would have a rather limited market. Should mechanisms be created by the government to cover the expenses of drug development and clinical trials for drugs that will potentially not produce a profit because of relatively low use but will be of great benefit to affected people?

2. During repressive political regimes in many countries, children often are separated from their parents and/or become orphaned. Human rights groups have been in the forefront of assisting the identification and reunion of relatives using genetic markers such as mitochondrial DNA. Should the government of the country in which the atrocities were committed be obligated to pay some or all of the costs of these searches, even if the current regime is far removed in political ideology and action from the repressive regime?

3. It has been suggested that a databank consisting of DNA fingerprints of previously convicted felons be established. Both nuclear and mitochondrial fingerprints would be included to accommodate situations in which the nuclear DNA was too degraded to be used. Who should have access to such a database? Law enforcers? Psychologists? Sociologists? Is this an invasion of privacy or a protection for the public?

S O L V E D   P R O B L E M S

I. Differential hybridization of a probe to mitochondrial DNA from two *Xenopus* species was the methodology employed to demonstrate maternal inheritance in vertebrates. However, as noted in the text of this chapter, this hybridization technique was not sensitive enough to detect small amounts of paternal DNA. What technique that is more sensitive to small amounts of DNA could be used today? How could you use this technique to determine if paternal mitochondrial DNA was present in the progeny of the interspecies cross?

Answer

Polymerase chain reaction (PCR) is a sensitive technique that detects very small amounts of DNA. Oligonucleotide primers that are specific for each of the mitochondrial DNAs in each of the two different species could be used to determine if paternal DNA is present in the offspring from the interspecies cross.

II. A newly isolated *petite* (respiration deficient) mutant of yeast, when mated to a wildtype strain and sporulated, produced two *grande* and two *petite* spores. What does this meiotic segregation pattern tell you about the *petite* mutation?

Answer

The 2:2 segregation of mutant and wildtype phenotypes from meiosis indicates that the *mutation is in a nuclear gene*. The finding that a *petite* phenotype is caused by a nuclear mutation is not surprising, since functioning of the mitochondria requires both nuclear and mitochondrial genes. If the mutation were in mitochondrial DNA, a 4:0 segregation pattern would be seen instead of the 2:2 pattern.

III. a. Does the following pedigree suggest mitochondrial inheritance? Why or why not?
    b. Is there another mode of inheritance that is consistent with this data?

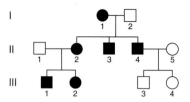

Answer

a. The data presented in this pedigree *is consistent with mitochondrial inheritance since the trait is transmitted by females; the affected males in this family did not transmit the trait; and all of the females' progeny have the trait.*

b. This inheritance pattern is *also consistent with transmission of an autosomal dominant trait.* According to this hypothesis, individuals I-1 and II-1 passed on the dominant allele to all children, but II-4 did not pass on the dominant allele to either child.

PROBLEMS

14-1 Choose the phrase from the right column that best fits the term in the left column.

a. cristae — 1. results in production of ATP

b. grana — 2. yeast cells that can respire

c. *petite* — 3. results in production of sugar

d. *grande* — 4. cell containing organelles that are genetically identical

e. oxidative phosphorylation — 5. contains multiply linked mini- and maxicircles of DNA

f. Calvin cycle — 6. folded inner membrane in mitochondria

g. kinetoplast — 7. yeast cells that cannot respire

h. heteroplasmic — 8. stacked thylakoid membranes in chloroplasts

i. homoplasmic — 9. cell containing two genetically different organelles

14-2 Does each of the following processes occur in the chloroplast, mitochondria, both, or neither?
    a. electron transport
    b. pyruvate and fatty acid breakdown
    c. sugar synthesis
    d. O_2 generation
    e. O_2 consumption

14-3 Is each of the statements below true of chloroplast or mitochondrial genomes, both, or neither?
    a. contain tRNA genes
    b. exist as condensed structures called nucleoids
    c. all genes necessary for function of the organelle are present
    d. vary in size from organism to organism

14-4 Some genes required for chloroplast function are encoded in the nuclear genome; others are encoded in the chloroplast genome. Nuclear and chloroplast DNA have different buoyant densities and can therefore be separated from each other by centrifugation based on these differences. There is a small amount of cross-contamination in the separation of nuclear and chloroplast DNAs using this technique. You have just found that a probe for a photosynthetic gene that is present in the chloroplast genome of plants hybridizes to nuclear DNA of a red alga. What experimental control(s) would you do to be more confident that your result means that the gene is encoded in the nucleus of this organism?

14-5 "Reverse translation" is a term given to the process of deducing the DNA sequence that could encode a particular protein. If you had the amino acid sequence Trp His Ile Met,
    a. What mammalian nuclear DNA sequence could have encoded these amino acids? (Include all possible variations.)
    b. What mammalian mitochondrial DNA sequence could have encoded these amino acids? (Include all possible variations.)

14-6 a. Results from hybridization using a probe for the small subunit gene of the Rubisco protein and a probe for the large subunit gene of the Rubisco protein to chloroplast and nuclear DNAs from a green, a red, and a brown alga are on following page. What conclusions would you reach about the location of the small and large subunit genes in each of the three types of algae?

(a)

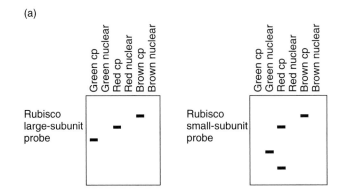

b. When RNA was extracted from the same three algal species and hybridized with a large subunit Rubisco probe and also with a small subunit probe, the following results were obtained. What conclusion would you reach about large and small subunit gene transcription in red and golden brown algae? Is this consistent with your answer in part a?

(b)

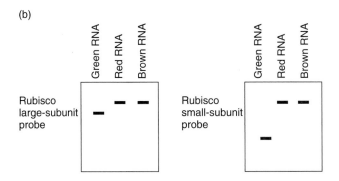

14-7 Which of the characteristics of chloroplasts and/or mitochondria listed below make them seem more similar to bacterial cells than to eukaryotic cells?
   a. Translation is sensitive to chloramphenicol and erythromycin.
   b. Alternate codons are used in mitochondria genes.
   c. Chloroplasts and mitochondria have inner and outer membranes.
   d. Introns are present in organelle genes.
   e. DNA in organelles is not arranged in nucleosomes.

14-8 A chloroplast translation elongation factor gene (*tufA*) hybridized to a fragment of nuclear DNA in the green alga *Coleochaete*. How could you determine if the entire coding sequence of the gene was intact in the nuclear fragment that hybridized with the probe?

14-9 There is a form of male sterility in corn that is inherited maternally. Marcus Rhoades first described this cytoplasmic male sterility by crossing female gametes from a male sterile plant with pollen from a male fertile plant. The resulting progeny plant was male sterile. Female gametes from this male sterile progeny

were backcrossed again with pollen from the same male fertile plant of the first cross. The process was repeated many times. What was the purpose of the many backcrosses? (Hint: Think about what is happening to the nuclear genome with repeated backcrosses.)

14-10 Which of the two methods listed would you choose to determine if organelles in an organism are heteroplasmic or homoplasmic and why?
   a. hybridize probes to cells immobilized on a slide
   b. PCR amplify DNA isolated from a population of cells

14-11 Studies distinguishing between uniparental and biparental inheritance of organelles employed a variety of detection methods. Match the system studied with the method used.

   a. *Xenopus laevis* and *X. borealis* mitochondrial DNA   1. protein analysis

   b. *Nicotiana* large and small subunits of Rubisco   2. pedigrees

   c. LHON phenotype in humans   3. differential hybridization

14-12 Describe two ways in which the contribution of mitochondrial genomes from male parents is prevented in different species.

14-13 If a human trait is determined by a factor in the cytoplasm, would an offspring more resemble its mother or its father? Why?

14-14 Why are very severe mitochondrial or chloroplast mutations usually found in heteroplasmic cells instead of homoplasmic cells?

14-15 You have isolated a new erythromycin-resistant mutant of *Chlamydomonas* in an mt^+ strain.
   a. If the mutation was in a chloroplast gene, what predictions would you make about genotype and phenotype of the diploid after mating the mutant with a mt^- ery^s strain?
   b. What would be the genotype and phenotype of the diploid if the mutation was in a nuclear gene?
   c. What additional step would enable you to determine if the mutation was in chloroplast or nuclear DNA?

14-16 You have isolated a new C^r mutant in yeast, starting with an MATa strain. How could you test the new C^r mutant to determine if the mutation is in nuclear DNA or mitochondrial DNA?

14-17 An example of a cloning vector used for biolistic transformation of chloroplasts is shown in the following diagram. The vector DNA can be prepared in large quantities in *Escherichia coli*. Once "shot"

into a chloroplast, the vector DNA integrates into the genome. Match the component of the vector with its function.

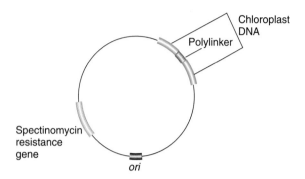

a. spectinomycin-resistance gene
b. chloroplast DNA
c. polylinker (multiple restriction sites)
d. ori

1. homologous DNA that mediates integration
2. gene used to select chloroplast transformants
3. sequence for replication in *E. coli*
4. site at which DNA can be inserted

14-18 What characteristics in a human pedigree suggest a mitochondrial location for a mutation affecting the trait?

14-19 The first person in the family represented by the pedigree below who exhibited symptoms of the mitochondrial disease MERFF was II-2.

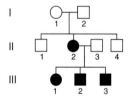

a. What are two possible explanations of why the mother I-1 was unaffected but daughter II-2 was affected?

b. How could you differentiate between the two possible explanations?

14-20 Which of the pedigrees shown could be consistent with each of the following mutation scenarios? (1) Mutation occurred in a somatic cell in an embryo. (2) Mutation occurred during development in I-1 in a precursor to all germ cells. (3) Mutation occurred during development in an egg cell.

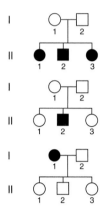

14-21 Kearns-Sayre is a disease in which mitochrondrial DNA carries deletions of up to 7.6 kb of the mitochrondrial genome. Although Kearns-Sayre is due to a mitochondrial DNA defect, it does not show maternal inheritance but arises as a new mutation in an individual. The severity of symptoms range from mild to severe and affected people can have defects in some tissues but not in others. How can you explain the variation in tissues affected and severity of symptoms? (Assume that the size of the deletion does not contribute to phenotypic differences.)

14-22 Deletions of various sizes in the mitochondrial genome have been found to increase with age in humans. Which technique would you choose to use to analyze human mtDNA samples for deletions—PCR or gel electrophoresis? Why?

HOW GENES ARE REGULATED

Biochemical signals control the flow of genetic traffic. Genes are turned
on, turned off, and compromised by the signals they receive. When
accidents occur mutations result.

GENE REGULATION IN PROKARYOTES

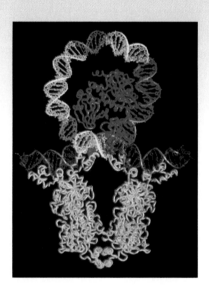

Lac repressor protein (violet) binds to specific sites in the DNA to turn off expression
of the *lac* operon in *E. coli.* Lac repressor is a tetranier with two subunits binding to
each of two operator sites 93 bares apart, causing a loop (blue and green) to form in
the DNA. This model also shows where the CAP protein (dark blue) binds to *lac* DNA.

Among the many types of bacteria that thrive in sewage water is the species *Vibrio cholerae,* the cause of the life-threatening diarrheal disease cholera. The last worldwide cholera pandemic occurred in 1961. Today, cholera is nearly absent from areas of the globe where secure sanitation systems are in place, but epidemics of the disease still devastate human populations in regions where sewage treatment and water purification programs are inadequate or nonexistent.

When a person drinks water contaminated by *V. cholerae,* the bacteria enter the digestive tract. Soon after, the bacteria encounter the "perilous" environment of the stomach whose acidity kills the majority of them. The bacteria conserve energy by curtailing production of several proteins they will use later but do not need for passage through the stomach. Only a large initial *V. cholerae* population ensures that at least a small group of cells will survive to exit the stomach and enter the small intestine.

Upon their arrival in the small intestine, these survivors come face to face with a thick mucus that coats their ultimate target—the intestinal epithelial cells. To penetrate this protective mucous layer, the *V. cholerae* cells navigate by chemotaxis (see the comprehensive example in Chapter 13) and using flagella (Fig. 15.1). They also make and secrete proteases that ease their passage by degrading the protein component of mucus.

When the bacteria at last reach their destination, they attach to the epithelial cells of the small intestine, stop fabricating flagellin (the chief protein component of flagella), and begin production of several virulence proteins, including a potent toxin that is the actual agent of cholera. Mutant bacteria that produce no toxin do not generate symptoms of disease.

The toxin secreted by cholera bacteria causes chloride ions (Cl^-) to leak from the intestinal cells. To reestablish the osmotic balance, these same cells secrete water. Symptoms of this ionic disruption and fluid flow are watery diarrhea and severe dehydration, which can lead to death within a few hours. The most effective life-saving therapy is oral rehydration: administration of an electrolyte solution consisting of glucose, table salt (NaCl), sodium carbonate

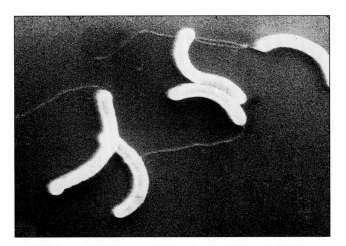

Figure 15.1 *V. cholerae* **bacteria.** *V. cholerae* invade cells in the intestine.

(NaHCO₃), and potassium chloride (KCl) dissolved in purified water. Once toxin production is in full swing, antibiotics without oral rehydration are of little benefit.

The story of *V. cholerae* infection illustrates two key aspects of the life of a unicellular prokaryote: direct contact with the external environment and the ability to respond to changes in that environment by changes in gene expression. *V. cholerae* bacteria that have entered a human host respond to rapid changes in external conditions in part by diminishing or abolishing the production of proteins not required for survival (thereby conserving energy and nutrients) and in part by initiating or increasing synthesis of proteins required in new environments (such as proteases when they contact mucus and toxin and other virulence proteins in the vicinity of intestinal epithelial cells). Such coordinated control of gene expression in a bacterial cell is an example of **prokaryotic gene regulation,** the subject of this chapter. Prokaryotes regulate gene expression by activating, increasing, diminishing, and preventing the transcription and translation of specific genes or groups of genes. Most metabolic pathways in bacteria require several gene products, and bacterial cells have evolved coordinate controls that enable them to turn the genes encoding those products on and off in concert. Bacteria attune their controls to gene function. With the catabolic pathways that break down complex substances to usable units—for example, lactose to glucose and galactose—the appearance of the substrate (lactose) induces, or turns on, coordi-

nate expression of genes whose protein products will accomplish the substrate's breakdown. With anabolic (biosynthetic) pathways that build needed compounds—for example, the amino acid tryptophan—the control mechanisms coordinately repress expression of the pathway's multiple genes when the product of the pathway (tryptophan) is present. This means that bacterial cells do not waste energy making unneeded proteins, such as the enzymes in a catabolic pathway whose substrate is absent or the enzymes in an anabolic pathway whose end product is already present in the environment. In fact, many aspects of prokaryotic gene regulation enable bacterial cells to conserve energy by distinguishing housekeeping proteins that are synthesized continuously, irrespective of environmental conditions, from proteins required only in situations signaled by specific cues.

One overarching theme emerges from our discussion. In unicellular organisms like bacteria, the regulatory mechanisms that turn genes on and off in response to environmental conditions enable the organisms to adapt and survive in a constantly changing world.

As we examine the regulation of gene expression in bacteria we describe

- An overview of prokaryotic gene expression: There are many steps in gene expression and regulation can occur at any one of them.

- The regulation of transcription: Genetic and molecular studies, particularly of the genes for lactose utilization, show that most regulation affects the initiation of RNA transcripts; "negative regulation" blocks transcription, while "positive regulation" increases it; DNA-binding proteins able to inhibit or enhance the activity of RNA polymerase at the promoter are the main agents of both types of regulation.

- The attenuation of gene expression: Controls regulating the genes of the tryptophan pathway include mechanisms that fine tune gene expression by prematurely terminating transcription.

- Global regulatory mechanisms: *Escherichia coli*'s response to heat shock is an example of the bacterial ability to coordinate the expression of different sets of genes dispersed around the chromosome.

- A comprehensive example of how *V. cholera* regulate their virulence genes.

AN OVERVIEW OF PROKARYOTIC GENE REGULATION

We saw in Chapter 7 that *gene expression* is the production of proteins according to instructions encoded in DNA. During gene expression, the information in DNA is transcribed into RNA, and the RNA message is translated into a string of amino acids.

RNA Polymerase Is the Key Enzyme for Transcription

To begin the process of gene expression in prokaryotes, RNA polymerase transcribes a gene's DNA into RNA. RNA polymerase participates in all three phases of transcription: initiation, elongation, and termination. Initiation requires a special subunit of RNA polymerase—the sigma (σ) subunit—in addition to the two alpha (α), one beta (β), and one β′ subunits that make up the core enzyme (Fig. 15.2). When bound to the core enzyme, the σ subunit recognizes and binds specific DNA sequences at the promoter; in its free form, σ does not bind DNA because its DNA binding site is obscured by its own carboxyl terminal tail. The full RNA polymerase—core enzyme plus σ—when bound to the promoter, functions as a complex that both initiates transcription by unwinding the DNA and begins polymerization of bases complementary to the DNA template strand.

The switch from initiation to elongation requires the movement of RNA polymerase away from the promoter and the release of σ. Interaction between the short nascent RNA chain and a site on the polymerase surface probably triggers this release. During elongation, the core RNA polymerase catalyzes the polymerization of ribonucleotides onto the 3′ end of the growing chain, eventually forming a complete mRNA. Elongation continues until the RNA polymerase encounters a signal in the RNA sequence that triggers termination (see Fig. 15.2). Two types of termination signals are found in prokaryotes: rho dependent and rho independent. In rho-dependent termination, a protein factor called *rho* recognizes a sequence in the newly transcribed mRNA and terminates transcription by binding to the RNA and pulling it away from the RNA polymerase enzyme. In rho-independent ter-

mination, a sequence of about 20 bases in the RNA, with a run of 6 or more Us at the end, forms a secondary structure, known as a *stem loop,* that serves as a signal for the release of RNA polymerase.

Translation in Prokaryotes Starts before Transcription Ends

Because there is no membrane enclosing the bacterial chromosome, translation of the RNA message into a polypeptide can begin while mRNA is still being transcribed. Ribosomes bind to special initiation sites at the 5′ end of the reading RNA frame while transcription of downstream regions of the RNA is still in progress. Signals for the initiation and termination of translation are distinct from signals for the initiation and termination of transcription. Because prokaryotic mRNAs are often polycistronic, that is, contain the information of several genes, ribosomes can initiate translation at several positions along a single mRNA molecule. See Fig. 7.21 for a review of how ribosomes, tRNAs, and translation factors mediate the initiation, elongation, and termination phases of mRNA translation to produce a polypeptide that grows from its N terminus to its C terminus according to instructions embodied in the sequence of mRNA codons.

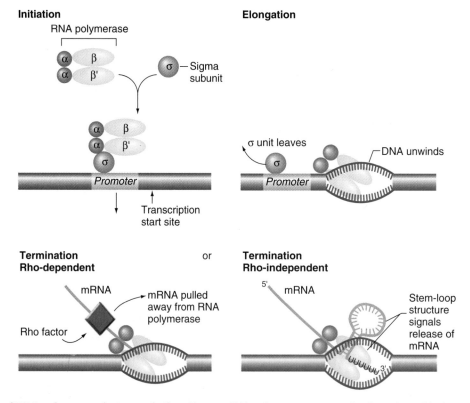

Figure 15.2 Role of RNA polymerase in transcription. The core RNA polymerase enzyme plus sigma factor bind to a promoter sequence to initiate transcription. Once productive copying of the DNA into RNA begins, RNA polymerase moves along the DNA to elongate the transcript, leaving sigma factor behind. Transcription is terminated when rho factor recognizes a sequence on the mRNA and pulls the message away from the enzyme or a stem loop (rho-independent signal) forms in the mRNA.

The Regulation of Gene Expression Can Occur at Many Steps

Many levels of control determine the amount of a particular polypeptide in a bacterial cell at any one time. Some controls affect the binding of RNA polymerase to the promoter, the shift from transcriptional initiation to elongation, and the release of the mRNA at the termination of transcription. Other controls determine the stability of the mRNA after its synthesis, the efficiency with which ribosomes recognize the various translational initiation sites along the mRNA, and the stability of the polypeptide product.

As we see next, the critical step in the regulation of most bacterial genes is the binding of RNA polymerase to DNA at the promoter. The other potential points of control, while sometimes important in the expression of certain genes, serve more often to fine-tune the amount of protein produced.

THE REGULATION OF GENE TRANSCRIPTION

Researchers delineated the principles of gene regulation in prokaryotes through studies of various metabolic pathways in *E. coli*. In this section, we focus our attention on regulation of the lactose utilization genes in *E. coli* because genetic and molecular experimentation in this system established a fundamental principle of gene regulation: The binding of regulatory proteins to DNA targets controls transcription. The DNA binding of these regulatory proteins either inhibits or enhances the effectiveness of RNA polymerase in initiating transcription. In our discussion, we consider the inhibition of RNA polymerase activity as "negative regulation" and the enhancement of RNA polymerase activity as "positive regulation."

The Utilization of Lactose by E. coli: A Model System for Studying Gene Regulation

Proliferating *E. coli* bacteria can use any one of several sugars as a source of carbon and energy. If given a choice, however, they prefer glucose. *E. coli* grown in medium containing both glucose and lactose, for example, will deplete the glucose before gearing up to utilize lactose.

Lactose is a complex sugar composed of two monosaccharides: glucose and galactose. A membrane protein, lac permease, transports lactose in the medium into the *E. coli* cell. There, the enzyme β-galactosidase splits the lactose into galactose and glucose (Fig. 15.3).

The Presence of Lactose Induces Expression of the Genes Required for Lactose Utilization

The two proteins, lac permease and β-galactosidase, both required for lactose utilization, are present at very low levels in cells grown without lactose. The addition of lactose to the bacterial medium induces a 1000-fold increase in the production of these proteins. The process by which a specific molecule stimulates synthesis of a given protein is known as **induction.**

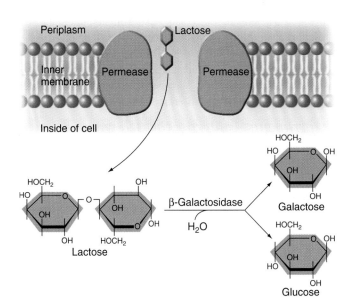

Figure 15.3   Lactose utilization in an *E. coli* cell. Lactose passes through the membranes of the cell via an opening formed by the lactose permease protein. Inside the cell, β-galactosidase splits lactose into galactose and glucose.

The molecule responsible for stimulating production of the protein is called the **inducer.** In the regulatory system under consideration, lactose modified to a derivative known as allolactose is the inducer of the genes for lactose utilization.

How lactose in the medium induces the simultaneous expression of the proteins required for its utilization was the subject of a major research effort in the 1950s and 1960s—a period some refer to as the golden era of bacterial genetics.

Analysis of the Lactose Induction System Was a Wise Choice for the Study of Gene Regulation

Like Mendel's decision to study seven easily identifiable characteristics in peas, the decision to focus on the induction of genes for lactose utilization in *E. coli* was well considered. The possibility of culturing large numbers of the bacteria made it easy to isolate rare mutants. Once isolated, the mutations responsible for the altered phenotypes could be located by Hfr mapping techniques developed in the 1950s (see Fig. 13.15).

The ability to measure levels of expression was critical for many of these experiments. To this end, chemists synthesized compounds other than lactose, such as *o*-nitrophenylgalactoside (ONPG), that could be split by β-galactosidase into products that were easy to assay. One product of ONPG splitting has a yellow color, whose intensity is proportional to the amount of product made, and thus reflects the level of activity of the β-galactosidase enzyme. A spectrophotometer can easily measure the amount of cleaved yellow product in a sample. Other substrates of the β-galactosidase enzyme that produce a color change upon cleavage include X-Gal whose cleavage produces a blue substance; as we have seen, X-Gal is often used to indicate whether a piece of DNA has been cloned into plasmid vectors containing parts of the β-galactosidase gene (see Fig. 8.8).

We now look at some of the experiments that suggested how the presence of lactose induces expression of the genes required for its own utilization. The techniques designed for these experiments and the ideas that emerged from the results were extremely influential because of their broad application to the study of gene regulation in other systems.

Experiments Analyzing the Behavior of Lactose-Utilization Mutants Reveal the Coordinate Repression and Induction of Three Genes

Jacques Monod, a man of diverse interests, was a catalyst of research on the regulation of lactose utilization. A political activist and a chief of French Resistance operations during World War II, he was also a fine musician and esteemed writer on the philosophy of science. Monod led a research effort centered at the Pasteur Institute in Paris, where scientists from around the world came to study enzyme induction. Results from many genetic studies led Monod and his close collaborator François Jacob to propose a model of gene regulation known as the *operon theory.* We describe key experiments that influenced their thinking as well as data that supported components of the theory; and then summarize the theory itself.

Jacques Monod. A key scientist in discovering principles of gene regulation, Jacques Monod was also a talented musician, philosopher, and political activist.

Complementation Analysis of Dozens of Mutants Identifies a Set of Lactose Utilization Genes

Monod and his collaborators isolated many different Lac⁻ mutants, that is, bacterial cells unable to utilize lactose. Then, using complementation analysis, they showed that the cells' inability to   break down lactose resulted from mutations in two genes: *lacZ,* which encodes β-galactosidase, and *lacY,* which encodes lac permease. They also discovered a third lac gene, *lacA,* which encodes a transacetylase enzyme that adds an acetyl (CH_3CO) group to lactose and other β-galactosides. Genetic mapping showed that the three genes appear on the bacterial chromosome in a tightly linked cluster, in the order *lacZ, lacY, lacA* (Fig. 15.4). Because the *lacA* gene product is not required for the breakdown of lactose, most studies of lactose utilization have focused on *lacZ* and *lacY.*

Figure 15.4 Lactose utilization genes in *E. coli.* Three genes, *lacZ, lacY,* and *lacA* are involved in lactose metabolism in *E. coli.*

Experimental Evidence for a Repressor Protein

Mutations in another gene named *lacI* produce constitutive mutants that synthesize β-galactosidase and lac permease even in the absence of lactose. (Recall that constitutive mutants synthesize certain enzymes irrespective of environmental conditions.) The existence of these constitutive mutants suggested that *lacI* encodes a negative regulator, or **repressor.** Cells would need such a repressor to prevent expression of *lacY* and *lacZ* in the absence of inducer. In constitutive mutants, however, a mutation in the *lacI* gene generates a defect in the repressor protein that prevents it from carrying out this negative regulatory function.

The historic PaJaMo experiment—named after Arthur Pardee (a third collaborator), Jacob, and Monod—provided further evidence that *lacI* indeed encodes this hypothetical negative regulator of the *lac* genes. Hfr matings in which the chromosomal DNA of a donor cell is transferred into a recipient cell served as the basis of the PaJaMo study. The researchers transferred the *lacI⁺* and *lacZ⁺* alleles into a cytoplasm devoid of LacI and LacZ proteins by crossing a *lacI⁺ lacZ⁺* Hfr strain with a *lacI⁻ lacZ⁻* mutant F⁻ strain, in a medium containing no lactose (Fig. 15.5). Shortly after the transfer of the *lacI⁺ lacZ⁺* genes, the researchers detected synthesis of β-galactosidase. Within about an hour, this synthesis stopped.

Pardee, Jacob, and Monod interpreted these results as follows. When the donor DNA is first transferred to the recipient, there is no repressor (LacI protein) in the cytoplasm because the recipient cell's chromosome is *lacI⁻*. In the absence of repressor, the *lacY* and *lacZ* genes are expressed. Over time, however, the host cell begins to make the LacI repressor protein from the *lacI⁺* gene introduced by the Hfr cross.

On the basis of experiments described in the following, Monod and company proposed that the repressor protein prevents further transcription of *lacY* and *lacZ* by binding to a hypothetical site: a DNA sequence near the promoter of the lactose utilization genes. They suggested that the binding of repressor to this site, termed the **operator site,** blocks the promoter, and occurs only when lactose is not present in the medium.

How the Inducer Releases Repression to Trigger Enzyme Synthesis

In the final step of the PaJaMo experiment, the researchers added the lactose inducer to the culture medium. With this addition, the synthesis of β-galactosidase resumed.

Their interpretation of this result was that the inducer binds to the repressor. This binding changes the shape of the repressor so that it can no longer bind to DNA. When the inducer is removed from the environment, the repressor, free of

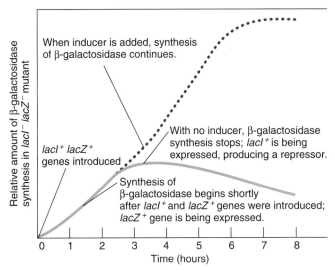

Figure 15.5   The PaJaMo experiment. When DNA carrying *lacI*⁺ *lacZ*⁺ genes was introduced into a *lacI*⁻ *lacZ*⁻ cell, β-galactosidase was synthesized from the introduced *lacZ*⁺ gene initially, but as repressor (made from the introduced *lacI*⁺ copy of the gene) accumulates, the synthesis of β-galactosidase stops (and only residual β-galactosidase is seen). If inducer is added (dotted line), the synthesis of β-galactosidase resumes.

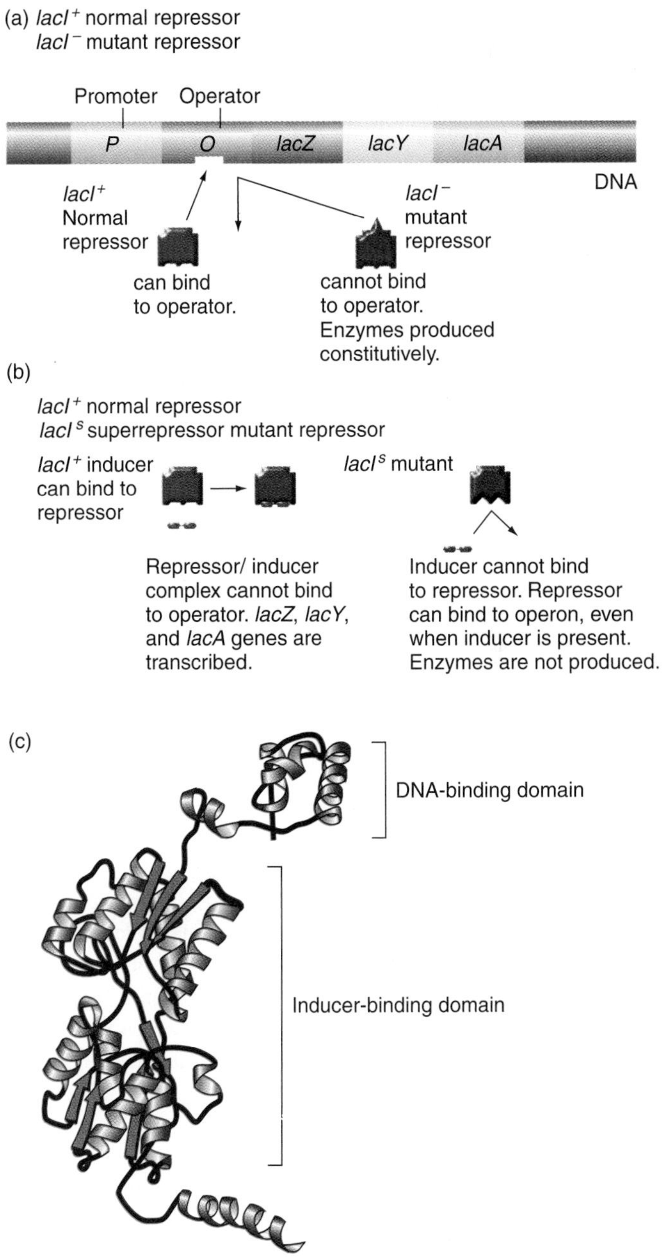

Labels in graph:

Relative amount of β-galactosidase synthesis in *lacI*⁻ *lacZ*⁻ mutant

When inducer is added, synthesis of β-galactosidase continues.

lacI⁺ *lacZ*⁺ genes introduced

With no inducer, β-galactosidase synthesis stops; *lacI*⁺ is being expressed, producing a repressor.

Synthesis of β-galactosidase begins shortly after *lacI*⁺ and *lacZ*⁺ genes were introduced; *lacZ*⁺ gene is being expressed.

Time (hours)

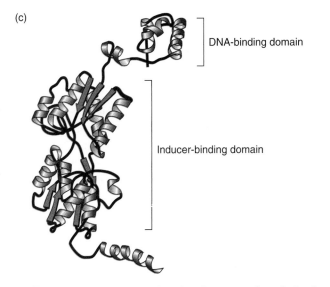

Figure 15.6   Mutational and structural analysis of repressor. (a) The *lacI*⁻ mutants. In the *lacI*⁻ mutants, the repressor cannot bind to the operator site and therefore cannot repress the operon. (b) The *lacI*ˢ mutants. Superrepressor mutants *lacI*ˢ bind to operator but cannot bind inducer, so the repressor cannot be removed from the operator and genes are continually repressed. (c) Structural analysis. X-ray crystallographic data enables the construction of a model of repressor structure that shows a region to which operator DNA binds and another region to which inducer binds.

Labels in Figure 15.6:

(a) *lacI*⁺ normal repressor
lacI⁻ mutant repressor

Promoter   Operator

P   O   *lacZ*   *lacY*   *lacA*   DNA

lacI⁺ Normal repressor   can bind to operator.

lacI⁻ mutant repressor   cannot bind to operator. Enzymes produced constitutively.

(b) *lacI*⁺ normal repressor
*lacI*ˢ superrepressor mutant repressor

lacI⁺ inducer can bind to repressor

*lacI*ˢ mutant

Repressor/ inducer complex cannot bind to operator. *lacZ*, *lacY*, and *lacA* genes are transcribed.

Inducer cannot bind to repressor. Repressor can bind to operon, even when inducer is present. Enzymes are not produced.

(c) DNA-binding domain

Inducer-binding domain

---

inducer, reverts to its DNA-bindable shape. Proteins that undergo reversible changes in conformation when bound to another molecule are called **allosteric** proteins. The binding of inducer to repressor causes an allosteric effect that abolishes the repressor's ability to bind the operator. In this sequence of events, the inducer is an effector molecule that releases repression without itself binding to the DNA.

In summary, repressor bound to inducer cannot bind to DNA; repressor not bound to inducer can bind to the DNA of the operator. The binding of repressor to the operator keeps RNA polymerase from recognizing the promoter.

The Repressor Has Distinct Binding Domains— One for the Operator and the Other for the Inducer

If the repressor protein interacts with both the operator and the inducer, what outcome would you predict for mutations that disrupt one of these interactions without affecting the other? Genetic studies showed that the constitutive *lacI*⁻ mutations we discussed earlier produce defects in the repressor's ability to bind DNA (Fig. 15.6a). By contrast, mutations isolated in the 1960s that cause a defect in the repressor's ability to bind inducer erase the possibility of induction; researchers designate the superrepressor mutations that give rise to this noninducible phenotype as *lacI*ˢ (Fig. 15.6b). The *lacI*ˢ mutants, although they cannot bind inducer, can still bind to DNA and repress the operon. In the 1980s, the mapping of large numbers of these two types of mutations by DNA sequencing showed that *lacI*⁻ missense mutations, which generate proteins incapable of binding DNA, are clustered in the codons that determine the amino terminus of the repressor, while the *lacI*ˢ mutations, which generate proteins incapable of binding inducer, cause

amino-acid alterations throughout much of the rest of the repressor. Subsequent structural analyses of the repressor protein confirmed what these mutational analyses suggest: that the repressor protein has at least two separate domains, one that binds to DNA, another that binds inducer (Fig. 15.6c).

To summarize, the repressor has separate domains correlating with two functions defined by mutations. Defects in either domain as well as the presence or absence of the inducer can affect the function of the repressor.

Changes in the Operator DNA to Which the Repressor Binds Can Affect Repressor Activity

While mutations in the DNA-binding domain of the repressor can erase repressor activity, mutations that alter the specific nucleotide sequence of the operator recognized by the repressor can have the same effect (Fig. 15.7). When mutations change the nucleotide sequence of the operator, the repressor is unable to recognize and bind to the site; the resulting phenotype is the constitutive synthesis of the lactose-utilization proteins. Researchers have, in fact, isolated constitutive mutants whose genetic defects map to the *lac* operator site, which is adjacent to the *lacZ* and *lacY* genes. They call the constitutive operator DNA alterations o^c mutations.

How can one distinguish the constitutive operator (o^c) mutants from the previously described constitutive *lacI⁻* mutants, considering that both prevent repression?

Proteins Act in Trans, but DNA Sites Act Only in Cis

Elements that act **in trans** can diffuse through the cytoplasm and act at target DNA sites on any DNA molecule in the cell. Elements that act **in cis** can influence the expression of only adjacent genes on the same DNA molecule. Studies of partial diploids constructed using F′ plasmids (see Fig. 13.18) helped distinguish mutations in the operator site (o^c), which act in *cis*, from mutations in *lacI*, which encodes a protein that acts in *trans*.

Recall from Chapter 13 that F′ factors are plasmids that carry a few chromosomal bacterial genes. When F′ *lac* plasmids are present in a bacterium, the cell has two copies of the lactose-utilization genes—one on the plasmid and one on the bacterial chromosome. Using F′ *lac* plasmids, Monod's group could create bacterial strains with diverse combinations of regulatory and structural-gene (*lacZ* and *lacY*) mutations. As described later, the phenotype of these partially diploid cells allowed Monod and his collaborators to determine whether particular constitutive mutations were in the genes that produce diffusible, *trans*-acting proteins or at *cis*-acting DNA sites that affect only genes on the same molecule.

In one experiment, Monod and colleagues used a *lacI⁻ Z⁺* bacterial strain that was constitutive for β-galactosidase

production because it could not synthesize repressor (Fig. 15.8a). The introduction of an F′ *lacI⁺ Z⁻* plasmid into this strain created a partial diploid that was phenotypically wildtype: both repressible in the absence of lactose and inducible in its presence. Its capacity for repression and subsequent induction indicated that I^+ is dominant to I^-, and that the LacI protein produced from the *lacI⁺* gene on the plasmid can bind to the operator on the bacterial chromosome. Thus, the product of the *lacI* gene is a *trans*-acting protein able to diffuse inside the cell and bind to any operator site it encounters, regardless of the chromosomal location of the operator.

In a second experiment, introduction of a *lacI^s* plasmid into a *lacI⁺* strain of bacteria that was both repressible and inducible created bacteria that remained repressible but were no longer inducible (Fig. 15.8b). This was because the mutant LacI^s repressor, while still able to bind to the operator, could no longer bind inducer. The noninducible repressor was dominant to the wildtype repressor because after a while, the mutant repressor, unable to bind inducer, occupied all the operator sites and blocked all *lac* gene transcription in the cell.

In a third set of experiments, the researchers used *lacI⁺ o^c lacZ⁺* bacteria that were constitutive for lactose digestion because the wildtype repressor they produced could not bind to the altered operator (Fig. 15.8c). Introduction of an F′ *lacI⁺ o⁺ lacZ⁻* plasmid did not change this state of affairs—the cells remained constitutive for β-galactosidase production. The explanation is that the o^+ operator on the plasmid had no effect on the *lacZ⁺* gene on the chromosomal DNA because the operator DNA acts only in *cis*. Since it was able to influence gene expression only of the *lacZ⁻* gene on its own DNA molecule, the wildtype operator on the plasmid could not override the mutated chromosomal operator to allow repression of genes on the bacterial chromosome.

A general rule derived from these experiments is that if a gene encodes a diffusible element—usually a protein—that can bind to target sites on any DNA molecule in the cell, whichever allele of the gene is dominant will override any other allele of that gene in the cell (and therefore act in *trans*). If a mutation is *cis*-acting, it affects only the expression of adjacent genes on the same DNA molecule; it does this by altering a DNA site, such as a protein-binding site, rather than by altering a protein-encoding gene.

The lac Genes Are Transcribed Together

Many of the experiments that led to an understanding of the *lac* genes focused on the expression of *lacZ* because the level of β-galactosidase is easy to measure. But Monod and coworkers also showed that the repression and induction of *lacY* and *lacA* occur in tandem with the repression and induction of *lacZ*.

Observation of the coordinate expression of the genes for lactose utilization led to the proposal that the three genes are transcribed as part of the same polycistronic mRNA. Although researchers in the 1960s hypothesized that RNA was the intermediate between DNA and protein, they had not yet demonstrated the existence of such an intermediary for any gene. In

Mutant operator (o^c)

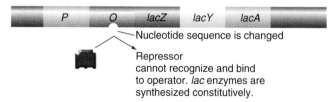

Figure 15.7 Operator mutants. Repressor cannot recognize the altered DNA sequence in the o^c mutant site, so cannot bind and repress the operon. Lactose genes are constitutively expressed.

(a) F' *lacI⁺ o⁺ Z⁻* plasmid in *lacI⁻ o⁺ Z⁺* bacteria

lacI⁺ gene encodes a diffusible element that acts in *trans*.
Inducible synthesis

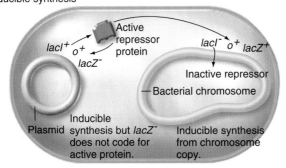

Inducible synthesis but *lacZ⁻* does not code for active protein.

Inducible synthesis from chromosome copy.

(b) F' *lacIˢ Z⁻* plasmid in *lacI⁻ Z⁺* bacteria

Noninducible—all *o⁺* sites eventually occupied by superrepressor.

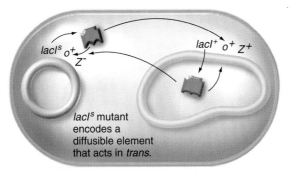

lacIˢ mutant encodes a diffusible element that acts in *trans*.

(c) F' *lacI⁺ o⁺ lacZ⁻* plasmid in *lacI⁺ oᶜ lacZ⁺* bacteria

Constitutive—presence of *o⁺* in plasmid has no effect on expression of *lacZ⁺* gene in bacterial chromosome.

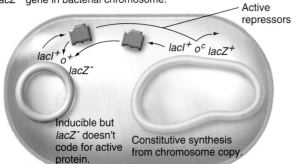

Inducible but *lacZ⁻* doesn't code for active protein.

Constitutive synthesis from chromosome copy.

Figure 15.8   *Trans*-acting proteins and *cis*-acting sites.
(a) LacI⁺ protein acts in *trans*. Repressor protein, made from the *lacI⁺* gene on the plasmid, can diffuse in the cytoplasm and bind to the operator on the chromosome as well as the operator on the plasmid. (b) LacIˢ protein acts in *trans*. The superrepressor encoded by *lacIˢ* on the plasmid diffuses and binds to operators on both the plasmid and chromosome to repress the *lac* operon. (c) *oᶜ* acts in *cis*. The *oᶜ* constitutive mutation affects only the operon of which it is a part. In this cell, only the chromosomal copy will be expressed constitutively. The copy on the plasmid will be inducible.

the 1970s, however, biochemical studies showed that RNA polymerase initiates transcription of the tightly linked *lac* gene cluster from a single promoter. During transcription, the polymerase produces a single polycistronic mRNA containing the *lac* gene information in the order 5'-*lacZ-lacY-lacA*-3'. As a result, mutations in the promoter affect the transcription of all three genes.

The Operon Theory

Based on the genetic evidence for three lactose-utilization genes that function as an ensemble, as well as the genetic evidence for a repressor that binds both an operator site in the DNA and an inducer, Monod and Jacob proposed the **operon theory** of gene regulation. Figure 15.9 shows the players in the theory and how they interact to achieve the coordinate regulation of the genes for lactose utilization. As shown, the *lacZ, lacY,* and *lacA* genes together with the promoter (*p*) and the operator (*o*) make up the ***lac* operon**: a single DNA unit enabling the simultaneous regulation of the three structural genes in response to environmental changes. Elements that interact with the operon include the repressor, which binds to the operon's operator, and the inducer, which when present, binds to the repressor and prevents it from binding to the operator.

The 1961 Paper Describing the Operon Theory Was a Landmark

In proposing the operon theory, Jacob and Monod took enzyme induction, which most of their contemporaries considered a biochemical problem, and used genetic analysis to develop a molecular model explaining how environmental changes could provoke changes in gene activity. Their theory is remarkable because the authors were working with a very abstract sense of the molecules in the bacterial cell: The Watson-Crick model of DNA structure was only eight years old, mRNA had only recently been identified, and the details of transcription had not yet been described. In fact, some of the research described earlier in the chapter that confirmed the operon theory followed the 1961 Jacob and Monod publication by years and decades. For example, although Monod was a biochemist with a special interest in allostery and its effects, the repressor itself was a purely conceptual construct; at the time of publication, it had not yet been isolated and it was unknown whether it was RNA or protein. Jacob and Monod thus made a major leap in understanding to propose the theory. In their endeavor, they applied the philosophic rule known as Occam's razor: The simplest, most economical explanation is preferable to a more complex one. As is often true in nature, there is a beauty, or elegance, in the simplicity of the solution.

We now know that a key concept of the theory—that proteins bind to DNA to regulate gene expression—holds true for the positive as well as the negative regulation of the *lac* operon. It also applies to many prokaryotic genes outside the *lac* operon and to eukaryotic genes as well. Our next section examines positive regulation of the *lac* operon.

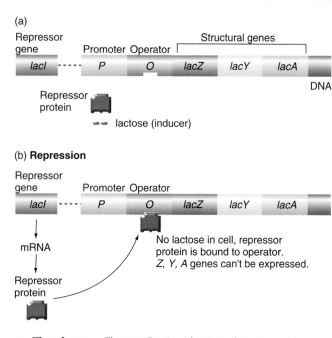

(a)

Repressor gene | Structural genes

Repressor protein

lactose (inducer)

(b) **Repression**

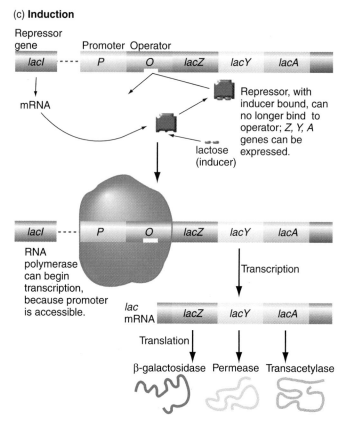

(c) **Induction**

a. The players. The coordination of various elements enables bacteria to utilize lactose in an energy-efficient way. These elements include

1. A closely linked cluster of three genes—*lacZ, lacY,* and *lacA*—that encode the enzymes active in splitting lactose into glucose and galactose.
2. A promoter site, from which RNA polymerase initiates transcription of a polycistronic mRNA. The promoter acts in *cis,* affecting the expression of only downstream structural *lac* genes on the same DNA molecule.
3. A *cis*-acting DNA operator site lying very near the *lac* operon promoter.
   The three structural genes together with the promoter and the operator constitute the *lac operon:* a regulatory unit that allows the *lacZ, lacY,* and *lacA* genes to respond to changes in the environment in the same way.
4. A *trans*-acting repressor that can bind to the operator; the repressor is encoded by the *lacI* gene, which is separate from the operon.
5. An inducer that prevents the repressor's binding to the operator. Although early experimenters thought lactose was the inducer, we now know that the inducer is actually allolactose, a molecule derived from and thus related to lactose.

b. How the players interact to regulate the lactose-utilization genes.

1. In the absence of lactose, the repressor binds to the DNA of the operator, and this binding prevents transcription. The repressor thus serves as a negative regulatory element.
2. When lactose is present, allolactose, an inducer derived from the sugar, binds to the repressor. This binding changes the shape of the repressor, making it unable to bind to the operator.
3. With the release of the repressor from the operator, RNA polymerase gains access to the *lac* operon promoter and initiates transcription of the three lactose utilization genes into a single polycistronic mRNA.

Feature Figure 15.9   The lactose operon in *E. coli.*

A Positive Control Increases Transcription of lacZ, lacY, *and* lacA

In focusing on the repression and subsequent induction of the genes for lactose utilization, the Jacob and Monod model did not address a key question: Why do *E. coli* grown in a medium containing both glucose and lactose not initiate *lac* gene transcription? If the lactose is present, why doesn't it act as an inducer? Answers to these questions emerged from molecular studies carried out long after publication of the theory. These studies showed that transcriptional initiation at the *lac* operon is a complex event. In addition to the release of repression, ini-

tiation depends on a positive regulator protein that assists RNA polymerase in the start-up of transcription. Without this assist, the polymerase does not open up the double helix very efficiently. As we see next, the presence of glucose indirectly blocks the function of this positive regulator.

The CAP Protein Becomes a Positive Regulator of the lac *and Other Operons After it Forms a Complex with cAMP*

Inside bacterial cells, the small nucleotide known as cAMP (cyclic adenosine monophosphate) binds to a protein called cAMP activator protein, or CAP. The binding of cAMP to

CAP enables CAP to bind to DNA in the regulatory region of the *lac* operon, and this DNA binding of CAP increases the ability of RNA polymerase to transcribe the *lac* genes (Fig. 15.10). Thus, CAP functions as a positive regulator that enhances the transcriptional activity of RNA polymerase at the *lac* promoter, while cAMP is an effector whose binding to CAP enables CAP to bind to DNA near the promoter and carry out its regulatory function.

Glucose indirectly controls the amount of cAMP in the cell by decreasing the availability of adenyl cyclase, the enzyme that converts ATP into cAMP. Thus, when glucose is present, the level of cAMP remains low; when glucose is absent, cAMP synthesis increases. As a result, when glucose is present in the culture medium, there is little cAMP available to bind to CAP and there is therefore little induction of the *lac* operon, even if lactose is present in the culture medium. The overall effect of glucose in preventing *lac* gene transcription is known as *catabolite repression.* Interestingly, glucose not only achieves catabolite repression of the *lac* operon by reducing the level of cAMP, it also inhibits the cellular uptake of lactose.

In addition to functioning as a positive regulator of the *lac* operon, the CAP–cAMP complex increases transcription in several other catabolic gene systems, including the *gal* operon (whose protein products help break down the sugar galactose) and the *ara* operon (contributing to the breakdown of the sugar arabinose). As you would expect, these other catabolic operons are also sensitive to the presence of glucose, exhibiting a low level of expression when glucose is present and cAMP is in short supply. Mutations in the gene encoding CAP that alter the DNA-binding domain of the protein reduce transcription of the *lac* and other catabolic operons.

While CAP is a Positive Regulator for Many Catabolic Operons, Some Positive Regulators Increase Transcription of the Genes in Only One Pathway

AraC, for example, is a positive regulatory protein specific for all the arabinose genes responsible for the breakdown of the sugar arabinose. Three arabinose genes, *araB*, *araA*, and *araD*, appearing on the chromosome in that order, constitute an operon (*araBAD*) that is regulated as a whole; additional arabinose genes at other chromosomal locations are coregulated with *araBAD*. Like the genes for lactose utilization, the arabinose genes are induced when their substrate (arabinose) is present. Evidence that the AraC protein is a positive regulator of the *araBAD* operon came from studies in which *araC⁻* mutants did not express high levels of these three arabinose genes in either the presence or absence of arabinose (Fig. 15.11). (We describe later the mechanisms by which AraC acts.) The mutations were recessive, loss-of-function mutations. When the loss of function of a regulatory protein results in no expression of the regulated genes, the protein must be a positive regulator. (By contrast, loss of function of a negative regulator causes constitutive production of the operon's gene products.)

Summary: How DNA-Binding Proteins Control the Initiation of Transcription at the Lactose and Other Operons

In bacteria, the initiation of transcription by RNA polymerase is under the control of regulatory genes whose products bind to specific DNA sequences in the vicinity of the promoter. The binding of negative regulatory proteins prevents the initiation of transcription; the binding of positive regulators assists the initiation of transcription. Regulation of the *lac* operon depends on at least two proteins: the repressor (a negative regulator) and CAP (a positive regulator). Maximum induction of the *lac* operon occurs in media containing lactose but lacking glucose. Under these conditions, the repressor binds inducer

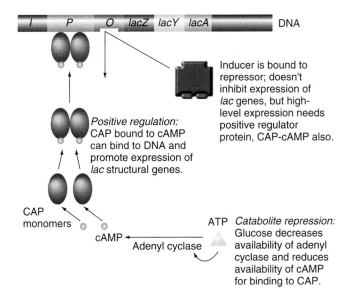

Figure 15.10 Positive regulation of the *lac* operon by CAP–cAMP. High-level expression of the *lac* operon requires that a positive regulator, the CAP–cAMP complex, is bound to the promoter region. When glucose is low in the cell, adenyl cyclase activity is high and causes synthesis of cAMP, which binds to CAP to make up the positive regulator.

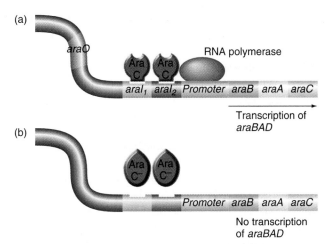

Figure 15.11 Loss of function mutations show that AraC is a positive regulator. Expression of the arabinose genes in *E. coli* requires the AraC protein bound next to the promoter. In an *araC⁻* mutant, the defective protein cannot bind and RNA polymerase will not transcribe the genes.

and becomes unable to bind to the operator, while CAP complexed with cAMP binds to a site near the promoter to assist RNA polymerase in the initiation of transcription.

Operons that function in the breakdown of other sugars are also under the control of negative and positive regulators. Transcription of the arabinose operon, for example, which is induced in the presence of arabinose, receives a boost from two positive regulators: the CAP–cAMP complex and AraC.

Thus, proteins that bind to DNA affect RNA polymerase's ability to transcribe a gene. The activity of multiple regulators that respond to different cues increases the range of gene regulation.

Molecular Studies Help Fill in the Details about Control Mechanisms

In 1966, Walter Gilbert and coworkers purified the *lac* repressor protein and determined that it is a tetramer of four *lacI*-encoded subunits, with each subunit containing an inducer-binding domain as well as a domain that recognizes and binds to DNA. (Note that we use the term "domains" for the functional parts of proteins, but the term "sites" for the DNA sequences with which a protein's DNA-binding domain interacts.) Gilbert and colleagues then used radioactively labeled repressor protein and a bacterial virus DNA that contained the *lac* operon to show that the repressor binds to operator DNA. When they combined the labeled protein and viral DNA and centrifuged the mixture in a glycerol gradient, the radioactive protein cosedimented with the DNA (Fig. 15.12). If the viral DNA contained a *lac* operon that had an o^c mutation, the protein did not cosediment with the DNA, because it could not bind to it. Subsequent sequence analysis of the isolated DNA revealed that the *lac* operator is about 26 bp in length, and includes the first nucleotides used as a template for the mRNA.

With the development of cloning, DNA sequencing, and techniques for analyzing protein-DNA interactions in the 1970s, researchers increased their ability to isolate specific macromolecules, determine the structure of each molecule, and analyze the interactions between molecules. This new experimental potential revolutionized the study of genetics even though researchers still used mutants isolated in earlier genetic analyses as the bases of many of their studies. The new generation of gene-regulation analyses confirmed the basic principles of the operon theory and elucidated compelling details of their components. We now present some of these details.

Many DNA-Binding Proteins Contain a Helix-Turn-Helix Motif

It is now possible to predict a protein's secondary structures—such as α helices and β sheets—by comparing the amino-acid sequence of a newly isolated protein with sequences of proteins whose secondary structures have already been determined. Several of the polypeptides that make up repressor proteins, including the subunits of the *lac* repressor, have the identifiable feature of two α-helical regions separated by a turn in the protein structure. This helix-turn-helix (HTH) motif in the protein fits well into the major groove of the DNA.

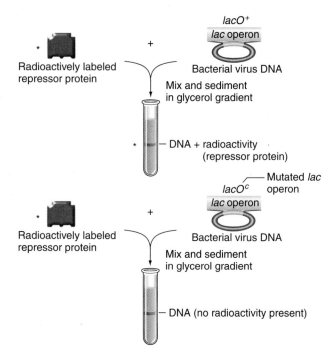

Figure 15.12 The *lac* repressor binds to operator DNA. A radioactive tag is attached to the *lac* repressor protein so it can be followed in the experiment. (a) When repressor protein from *lacI⁺* cells was purified and mixed with DNA containing the *lac* operator (on bacterial virus DNA), the protein cosedimented with the DNA. (b) When wildtype repressor was mixed with DNA containing a mutant operator site, no radioactivity sedimented with the DNA.

One of the α helices in an HTH carries amino acids that recognize and interact with a specific DNA sequence of nucleotides; thus each HTH has a specificity for DNA binding based on its sequence of amino acids (Fig. 15.13a).

Mark Ptashne and colleagues used cloned DNA to construct made-to-order mutations in the gene encoding the repressor of a bacterial virus known as 434. (This bacterial virus is similar to bacteriophage λ, which infects *E. coli*, and P22, which infects *Salmonella typhimurium*.) The 434 repressor binds to DNA of the prophage (that is, the viral DNA that has integrated into the bacterial genome) and prevents transcription and production of viral particles. After predicting that a region of the α helix of the 434 repressor recognizes the DNA of its specific operator site, Ptashne and coworkers altered the DNA sequence of the gene region encoding this α helix so that it was the same as the sequence found in the corresponding part of the repressor for another bacterial virus known as P22 (Fig. 15.13b). The resulting hybrid 434-P22 repressor protein, encoded by the altered gene, contained a P22 α helix that recognized the P22 operator *in vivo*. Binding of the hybrid repressor to the P22 prophage DNA operator region in its lysogenic bacterial host shut down transcription of most P22 viral proteins and prevented subsequent infection by the P22 virus.

The HTH motif is found in hundreds of DNA-binding proteins. Surprisingly, more than 20 different DNA-binding proteins in bacteria are very similar to the LacI repressor, not

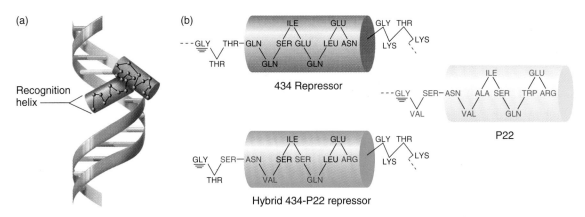

Figure 15.13   DNA recognition sequences in a helix-turn-helix section of a protein. (a) A protein motif that has the shape of a helix-turn-helix (helices shown here inside a barrel shape) fits into the major groove of the DNA helix. Specific amino acids within the helical region of the protein recognize a particular base sequence in the DNA. (b) The amino acids inside the recognition helix for phages 433 and P22 and for the hybrid 434-P22 repressor made in the Ptashne experiment are shown.

only in the HTH DNA-binding domain, but throughout much of the protein. This group of repressors is known as the LacI repressor family of proteins. Their structural similarity suggests that they evolved from a common ancestral gene whose duplication and divergence produced a family of transcriptional repressor proteins with similar overall structures but unique recognition regions. The uniqueness of their DNA-recognition regions means that they interact with different operators to regulate different groups of genes.

Most Regulatory Proteins Are Oligomeric and Contain More Than One Binding Domain

In the 1970s, geneticists studying regulation developed new *in vitro* techniques to determine where regulatory proteins bind to the DNA. Purified proteins that bind to fragments of DNA protect the region of the double helix to which they bind from digestion by enzymes such as DNase I that break the phosphodiester bonds between nucleotides. If a sample of DNA, labeled at one end of one strand and bound by a purified protein, is partially digested with DNase I, the enzyme will cleave phosphodiester bonds in at least some DNA molecules in the sample, except for those phosphodiester bonds that are in regions protected by the binding protein. After the DNase I and binding proteins are removed and the DNA is denatured into single strands, gel electrophoresis of the DNA followed by autoradiography of the gel will reveal bands at positions corresponding to the cleavage at each phosphodiester bond, except in the region where bound protein protected the DNA. Portions of the gel without bands are thus "footprints," indicating the nucleotides of the DNA fragment that were protected by the DNA-binding protein (Fig. 15.14a).

When analyzed in this way, many of the DNA sequences to which a negative or positive regulator protein binds exhibit rotational symmetry; that is, their DNA strands have a similarity of sequence when read in the 5′-to-3′ direction (although these sequences are usually not perfect palindromes). An example of such symmetry is in the *lac* operon's CAP-binding site whose sequence is

5′ GTGAGT TAGCTAC 3′
3′ CACTC AATCGATG 5′

Most regulatory proteins that bind to DNA exist as oligomers composed of two to four polypeptide subunits. The regulatory proteins, which are present in very low numbers in the cell, gain an advantage from this multimeric form: Since each polypeptide subunit of an oligomer has a DNA-binding domain, an assembled oligomer has multiple DNA-binding domains. By increasing the stability of protein-DNA interactions, these multiple binding domains collectively produce the strength of binding necessary to maintain repression or activate transcription.

The CAP protein binds to its DNA site (for example, the site in the *lac* operon just shown) as a dimer at a sequence with rotational symmetry, with one monomer of CAP binding to each side of the sequence (Fig. 15.14b). Thus, the DNA site that binds CAP actually consists of two sequences, each able to bind one subunit of the CAP dimer.

The *lac* repressor exists as a tetramer with each of its four subunits containing a DNA-binding HTH motif. This tetramer binds to two operators located far apart on the DNA, with each operator containing two recognition sequences. The binding of the tetrameric repressor to the two operators causes a loop of DNA to form between the two operator sites (Fig. 15.14c); formation of the loop, in turn, facilitates the two-position binding. There are actually three operator sites in the *lac* operon to which the repressor can bind: O_1 (the site originally identified by o^c mutations), O_2, and O_3 (see Fig. 15.14c). Site O_1 has the strongest binding affinity for the repressor and two subunits of the tetramer always bind at this site. The other two subunits bind at either O_2 or O_3. The distance between operator sites—multiples of 10 bases—allows repressor binding to the same side of the helix and thus formation of the loop. Mutations in *either* O_2 *or* O_3 have very little effect on repression. By contrast, mutations in *both* O_2 *and* O_3 make repression 50 times less effective. The conclusion is that for maximal repression, all four of the repressor's subunits must bind DNA simultaneously. Binding at four recognition sequences (in two operator sites) increases the

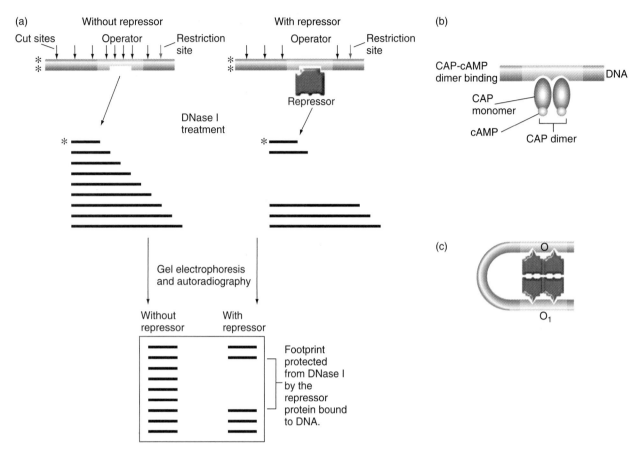

Figure 15.14 Proteins binding to DNA. (a) DNase footprint establishes the region to which a protein binds. A partial digestion with DNase I produces a series of fragments. If a protein is bound to DNA, DNase cannot digest at sites covered by the protein. Gel electrophoresis of digested products shows which products were not generated and indicates where the protein binds. (b) CAP–cAMP binds as a dimer to a regulatory region. (c) The Lac repressor is a tetrameric protein. For simplicity we previously showed a single repressor object binding to one operator site, but in reality there are two identical LacI subunits that bind to each operator site. Two of the subunits bind to the sequence in one operator site (O_1) and the other two subunits bind to a second operator (either O_2 or O_3). The model on page 530 shows binding of *lac* repressor to O_1 and O_3.

stability of the protein-DNA interactions. In fact, the DNA binding of the *lac* repressor is so efficient that only 10 repressor tetramers per cell are sufficient to maintain repression.

The Looping of DNA Is a Common Feature of Regulatory Systems

We have seen that the DNA binding of the *lac* repressor tetramer to two distant sites on the DNA results in formation of a loop. This ability to form a loop shows that DNA is not a static, stiff molecule. It has a flexibility that enables it to bend, and this potential for bending enables all four DNA-binding domains of the lac repressor tetramer to work in tandem for maximal repression.

Looping first came to light in work on AraC, the regulatory protein that helps control the arabinose operon described above. AraC functions as a dimer. As we have seen, in the presence of the inducer arabinose, AraC is a positive regulator that helps initiate transcription of the *araBAD* operon. Unexpectedly, in the absence of arabinose, AraC acts as a repressor. In this capacity, the AraC dimer binds to two sites—*araO* and *araI*—that are 194 nucleotide pairs apart (Fig. 15.15a). In one set of experiments analyzing the binding to two sites with concomitant looping of DNA, researchers altered the distance between *araO*

and *araI* by inserting several base pairs. The introduction of 11 or 31 bp—alterations that are close to integral changes in the number of turns of the double helix (a full turn of the helix = 10.5 bp)—had little effect on repression. The introduction of 5, 15, or 24 bp, however, noticeably reduced repression. These results suggest that the orientation of the binding sites is a significant variable; only when two sites have an orientation that puts them on the same side of the helix can a dimer bind simultaneously to both and cause formation of a DNA loop.

What enables the AraC dimer to function as both an activator and a repressor? The answer most likely involves allostery and AraC's different binding affinities for recognition sequences at *araI* and *araO*. The *araI* site contains two recognition sequences (*araI₁* and *araI₂*) to which AraC can bind; the *araO* site contains one recognition sequence. The AraC dimer is an allosteric protein whose structure changes with the binding of inducer (arabinose). One hypothesis of how this allosteric change alters function is that in the absence of arabinose, the size and shape of the protein unbound to inducer allow the AraC dimer to bind to two recognition sequences—*araO* and one of the two sites within *araI*—at the same time; this double binding prevents AraC from binding in a way that would enable it to as-

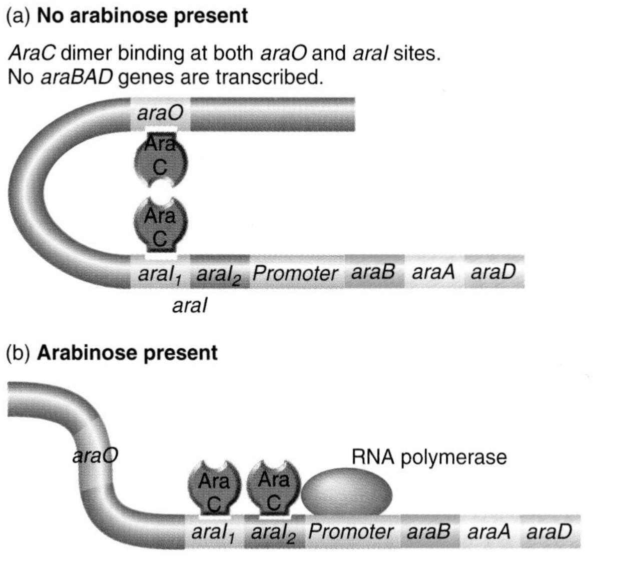

(a) **No arabinose present**

AraC dimer binding at both *araO* and *araI* sites.
No *araBAD* genes are transcribed.

(b) **Arabinose present**

Figure 15.15   How AraC acts as both a repressor and an activator. The AraC protein can bind to sites *araI₁*, *araI₂*, and *araO*. When no arabinose is present, binding of *araC* to *araO* and the *araI₁* sites causes looping of the DNA and prevents RNA polymerase from transcribing the genes. When arabinose (inducer) is present, it binds to *araI₁* and *araI₂* but not to *araO*. RNA polymerase interacts with *araC* at the *araI* sites and transcribes the genes.

sist RNA polymerase in the initiation of transcription. When arabinose is present, binding to the inducer changes the shape of the AraC dimer. In this inducer-bound conformation, the regulatory molecule does not bind to *araI* and *araO* at the same time; instead, the inducer–dimer complex binds exclusively to recognition sequences in *araI* (Fig. 15.15b). When bound to DNA at only this site, AraC's positive regulatory domain is free to interact with RNA polymerase and increase transcription.

How Regulatory Proteins Interact with RNA Polymerase

Negative regulators, such as the *lac* repressor prevent initiation by physically blocking the DNA binding site of RNA polymerase. For example, the O₁ operator site to which the *lac* repressor tightly binds consists of 27 nucleotides centered 11 bp downstream from the transcriptional start site. The operator thus includes part of the region where RNA polymerase has to bind to initiate transcription (Fig. 15. 16a). When repressor is bound to the operator, its presence on the DNA prevents the binding of RNA polymerase.

Positive regulators, by contrast, establish a physical contact with RNA polymerase that enhances the enzyme's ability to initiate transcription (Fig. 15.16b). For several positive regulators, researchers have identified points of contact between the regulator and the α, β, or σ subunits of RNA polymerase. Although RNA polymerase will bind to a promoter in the absence of a positive regulator, it is less likely to unwind DNA and begin polymerization than when it receives assistance from a positive regulator.

In Studying Gene Regulation, Researchers Use the lacZ *Gene as a Reporter of Gene Expression*

The extensive molecular knowledge of *lacZ* has enabled its use as a "reporter" gene in the study of a large variety of regulatory regions in both prokaryotes and eukaryotes. A **reporter gene** is a protein-encoding gene whose expression in the cell is quantifiable by sensitive and reliable techniques of protein detection. Fusion of the coding region of the reporter gene to *cis*-acting regulatory regions (including promoters and operators) of other genes creates a DNA construct that enables researchers to assess the activity of the regulatory elements by monitoring the amount of reporter gene product appearing in the cell. For example, with the fusion of gene *X*'s regulatory

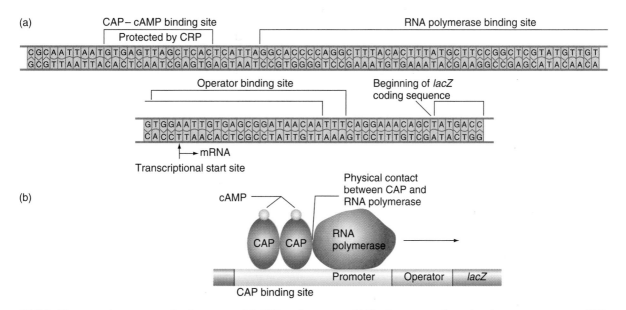

Figure 15.16   How regulatory proteins interact with RNA polymerase. (a) The *lac* repressor bound to the operator prevents RNA polymerase from binding. The binding sites for RNA polymerase and repressor (determined by DNase digestion experiments) show that there is overlap between the two sites. (b) The CAP–cAMP complex contacts RNA polymerase directly to help in transcription initiation.

region to the *lacZ* gene, one can assess the activity of the regulatory elements of gene *X* by monitoring the level of β-galactosidase expression. With this construct, conditions that induce expression of gene *X* will generate β-galactosidase (Fig. 15.17a). The use of reporter genes makes it possible to identify the DNA sites necessary for regulation as well as the genes and signals involved in that regulation. For example, you could mutate a gene's control region *in vitro,* then transform the construct back into bacterial cells and look for mutations that disrupt a particular aspect of control (as measured by levels of *lacZ*); this protocol would identify *cis*-acting sites important for the regulation. Or, you could mutate the bacteria themselves, then introduce a reporter construct into the mutated cells and look for changes in level of *lacZ* expression as measured by the blue or white color; this protocol would identify *trans*-acting genes.

Reporter constructs not only provide the basis for analyzing the regulation of one specific gene; they also make it possible to identify many genes regulated by the same stimulus. To accomplish this, researchers can use transposition to insert the *lacZ* gene without its regulatory region at various sites around the bacterial chromosome. For example, they can engineer the bacteriophage Mu to contain *lacZ* inside its genome and then infect a population of *E. coli* cells with the specially engineered phage. The phage will insert at sites around the bacterial chromosome at random, generating a collection of *E. coli* cells, some of which contain *lacZ* fused to genes (and their regulatory regions). On exposure of this collection to a stimulus such as UV light, the cells containing *lacZ* fused to the regulatory regions of genes induced by UV light will produce β-galactosidase (Fig. 15.17b). Researchers identified a set of genes activated by exposure to DNA-damaging agents by this method.

In addition to using *lacZ* as a reporter gene, geneticists studying gene regulation can use their extensive knowledge of the *lac* operon regulatory region to construct recombinant molecules carrying genes whose expression can be controlled. For example, by fusing the *lac* operon control DNA to a human gene expressed in *E. coli,* they could cause overproduction of the human protein in response to an induction cue. Controlling the expression of these foreign genes is important as protein production is not usually turned on until cells have proliferated to a high density. The overproduction in *E. coli* of human proteins, such as human growth hormone, human insulin, and other pharmacologically useful proteins is based on this strategy (Fig. 15.17c).

THE ATTENUATION OF GENE EXPRESSION: FINE-TUNING OF THE TRP OPERON THROUGH THE TERMINATION OF TRANSCRIPTION

In bacteria, the multiple genes of both catabolic and anabolic pathways are clustered and coregulated in operons. We have seen that regulators of the catabolic *lac* operon respond to the presence of lactose by inducing gene expression. By contrast, regulators of anabolic operons respond to the presence of the pathway's end product by shutting down the genes of the operon.

There are many anabolic bacterial operons involved in the synthesis of amino acids. A well-studied example is the *E. coli* tryptophan (*trp*) operon, a group of five genes—*trpE, trpD, trpC, trpB,* and *trpA*—required for construction of the amino acid tryptophan. Maximal expression of the *trp* genes occurs when tryptophan is absent from the growth medium.

The Presence of Tryptophan Activates a Repressor of the trp Operon

The *trp* operon is regulated by a protein repressor that is the product of the *trpR* gene. In contrast to the *lac* operon, where lactose functions as an inducer that prevents the repressor from binding to the operator, tryptophan functions as a **corepressor:** an effector molecule whose binding to the actual TrpR repressor protein allows the negative regulator to bind to DNA and inhibit transcription of the genes in the operon. The binding of tryptophan to the TrpR repressor causes an allosteric alteration in the repressor's shape, and only with this alteration can the TrpR protein bind to the operator site (Fig. 15.18a). Mutations in the *trpR* gene that change either the protein's tryptophan-binding domain or its DNA-binding domain destroy the TrpR repressor's ability to bind DNA, and result in the expression of the *trp* genes even when tryptophan is present in the growth medium. The TrpR-mediated repression of the *trp* operon is only one of the regulatory components controlling expression of the *trp* genes in *E. coli.*

The Termination of Transcription Fine-Tunes Regulation of the trp Operon

One would expect *trpR⁻* mutants to show constitutive expression of their *trp* genes—with or without tryptophan in the medium, with no repressor to bind at the operator, RNA polymerase would have uninterrupted access to the *trp* promoter. Surprisingly, studies show that the *trp* genes of *trp⁻* mutants are not completely derepressed (that is, turned on) when tryptophan is present in the growth medium. As Table 15.1 shows, the removal of tryptophan from the medium in which *trpR⁻* mutants are growing causes expression of the *trp* genes to increase threefold. What control mechanism is responsible for this repressor-independent change in *trp* operon expression?

Two Alternative Transcripts Lead to Different Transcriptional Outcomes

In a series of elegant experiments analyzing transcription of the *trp* operon, Charles Yanofsky and coworkers found that initiation at the *trp* promoter can produce two alternative transcripts (Fig. 15.18b). Sometimes initiation at the promoter leads to transcription of a truncated mRNA about 140 bases long from a short DNA region immediately preceding the first *trp* gene; this pregene DNA region is called a **leader sequence** and the RNA transcribed from it is the RNA leader. At other times, transcription continues beyond the end of the leader sequence to produce a full operon-length transcript. In analyzing why some mRNAs terminate before they can transcribe the structural *trp* genes, while others do not, the researchers

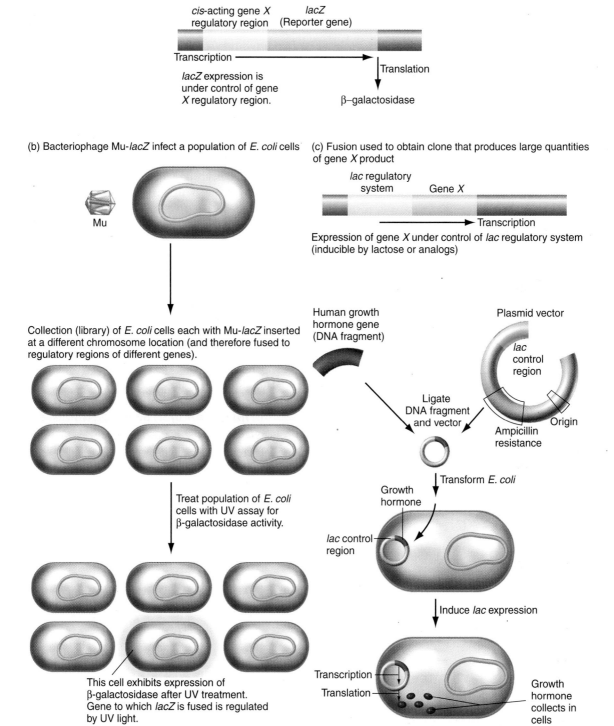

Figure 15.17   Use of the *lac* gene fusions to study gene regulation. (a) The *lacZ* structural gene can be fused to a regulatory region of gene *X*. Expression of β-galactosidase will be dependent on signals in the regulatory region to which *lacZ* is fused. (b) Creating a collection of *lacZ* insertions in the chromosome. Bacteriophage Mu containing the *lacZ* gene without its promoter integrates randomly around the chromosome. If the Mu-*lacZ* integrates within a gene in the orientation of transcription, *lacZ* expression will be controlled by that gene's regulatory region. The library of clones created can be screened to identify insertions in genes regulated by a common signal. (c) Use of fusions to overproduce a gene product. The *lac* regulatory region can be fused to gene *X* to control expression of genes. The gene encoding human growth hormone is cloned next to the *lac* control region and transformed into *E. coli*. Conditions that induce *lac* expression will cause expression of growth hormone that can be purified from the cells.

(a) Tryptophan present

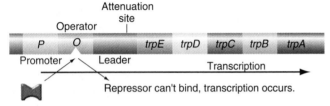

Attenuation
site

Operator

| P | O | | trpE | trpD | trpC | trpB | trpA |

Promoter   Leader

Transcription
repressed

Binding of corepressor
alters repressor's shape
allowing it to bind to operator.

TrpR
(repressor)   Tryptophan
(corepressor)

(b) Tryptophan not present

Attenuation
site

Operator

| P | O | | trpE | trpD | trpC | trpB | trpA |

Promoter   Leader

Transcription

Repressor can't bind, transcription occurs.

Figure 15.18 Tryptophan acts as a corepressor.
(a) When tryptophan is available, it binds to the *trp* repressor
causing the molecule to change shape so that repressor can
bind to the operator of the *trp* operon and repress transcription.
(b) When tryptophan is not available, repressor cannot bind to the
operator and the tryptophan biosynthetic genes are expressed.

**TABLE 15.1 Expression of *trp* Operon in *trpR*⁺
and *trpR*⁻ Strains**

	With Tryptophan* (%)	Without Tryptophan* (%)
trpR⁺	8	100
trpR⁻	33	100

*In the growth medium.

discovered **attenuation:** control of gene expression by premature termination of transcription. Whether or not transcription terminates prematurely depends on how the translation
machinery reads the secondary structure of the RNA leader.

That RNA leader can fold into two different stable conformations, each one based on the complementarity of bases in the
RNA. The first structure contains two stem-loop structures with
regions 1 and 2 associated by base pairing and regions 3 and 4
similarly associated by base pairing (Fig. 15.19a). When the transcriptional machinery "sees" the 3–4 stem-loop configuration,
which has seven Us at the end, it stops transcription, producing a
short, "attenuated" RNA. The alternative RNA structure forms

by base pairing between regions 2 and 3. In this conformation, the
leader RNA does not display the 3–4 stem-loop termination signal, and as a result, the transcription machinery reads right
through it to produce a full-length transcript.

The Simultaneous Translation of a Small Peptide in the Leader Determines Which Outcome Occurs

The early translation of a short portion of the RNA leader
(while transcription of the rest of the leader is still taking
place) determines which of the two alternative RNA structures
forms. That key portion of the RNA leader includes a short
open reading frame containing 14 codons, 2 of which are *trp*
codons (Fig. 15.19b). When tryptophan is present, the ribosome moves quickly past the *trp* codons in the RNA leader and
proceeds to the end of the leader's codons, allowing formation
of the stem-loop 3–4 structure and preventing formation of the
alternative RNA structure. As we have seen, this 3–4 RNA
structure causes the termination of transcription and hence the
attenuation of *trp* gene expression. In the absence of tryptophan, the ribosome stalls at the two *trp* codons in the RNA
leader because of the lack of charged tRNA^Trp in the cell. The
2–3 stem-loop structure depicted in Fig. 15.19c is then able to
form, and its formation prevents formation of the 3–4 stem-
loop RNA structure recognized by the transcriptional terminator. As a result, transcription proceeds through the leader into
the structural genes.

How do we know these secondary RNA structures exist *in
vivo* and that the translation of the leader RNA plays a significant role in their formation? Several experiments support this
model of attenuation. First, deletion of the complete leader sequence results in the loss of control by attenuation; in *trpR*⁻
mutants that also do not contain the leader sequence, there is
no difference in *trp* expression with or without tryptophan in
the medium. Second, a mutation that weakens the stems of the
RNA stem-loop secondary structures alters regulation but can
be compensated for by a second-site mutation that restores
base pairing. Third, mutations that change the RNA terminator structure increase the readthrough of transcription and thus
enhance gene expression. Finally, mutations of the translation-
initiating AUG codon that prevent translation of the leader sequence produce an increase in the amount of short transcript,
as predicted from the model.

Why has such a complex system evolved in the regulation
of *trp* operon and other biosynthetic pathways? While the
TrpR repressor shuts off transcription in the presence of tryptophan and allows it in the amino acid's absence, the attenuation mechanism provides a way to fine-tune this off-on switch.
It allows the cell to sense the level of tryptophan by the level
of charged tRNA^Trp, and to adjust the level of *trp* mRNA accordingly. In *E. coli,* systems for regulation by attenuation
similar to that observed for tryptophan exist for several other
amino acid biosynthetic operons, including histidine, phenylalanine, threonine, and leucine.

The attenuation mechanism is unique to prokaryotes because only in cells without a membrane-enclosed nucleus can

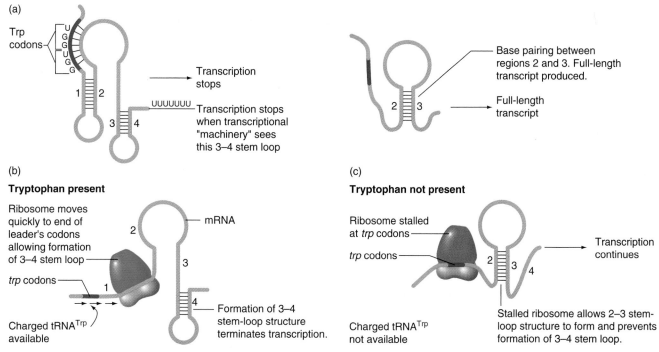

Figure 15.19 Attenuation in the tryptophan operon of *E. coli.* (a) Stem loops form by complementary base pairing in the *trp* leader RNA. Two different secondary structures are possible in the mRNA from the *trp* operon. Stem loops using complementary base pairs between regions 1 and 2 will enable the formation also of the stem loop 3–4, which is a termination signal for RNA polymerase. Base pairing between regions 2 and 3 leads to a stem loop that prevents formation of stem loop 3–4. In the early portion of the transcript there are two codons for tryptophan. (b) When tryptophan is present, the ribosome follows quickly along the transcript, preventing stem loop 2–3 from forming. Stem loop 3–4 can form and transcription is terminated. (c) If tryptophan is absent, the ribosome stalls at the *trp* codons, allowing formation of stem loop 2–3, preventing stem loop 3–4 formation. Transcription continues.

the expression machinery couple transcription and translation. The possibility of some aspect of the translational apparatus directly affecting the outcome of transcription does not exist in eukaryotes.

GLOBAL REGULATORY MECHANISMS COORDINATE THE EXPRESSION OF MANY SETS OF GENES

Dramatic shifts in environmental conditions can trigger the expression of sets of genes or operons dispersed around the chromosome. *E. coli*'s response to heat shock—exposure to extremely high temperatures (up to 45°C)—is an example of such global regulation.

An Alternative Sigma (σ) Factor Mediates E. coli's Global Response to Heat Shock

At high temperatures, most proteins denature or aggregate or both. In *E. coli,* exposure to high temperature induces the expression of several proteins that alleviate heat-shock-related damage. The induced proteins include those that recognize and degrade aberrant proteins as well as so-called *chaperone proteins* that assist in the refolding of other proteins and also prevent their aggregation.

E. coli's induction of the proteins that deal with heat shock is a highly conserved stress response. Organisms as different as bacteria, flies, and plants induce similar proteins, notably the chaperones, in response to high temperatures.

Studies of Mutants Unable to Generate a Heat-Shock Response Helped Reveal a Mechanism of Global Regulation

Conditional lethal *E. coli* mutants in which high temperatures do not induce transcription of the heat shock genes provided critical evidence for the regulatory mechanism. These conditional lethal mutants have a defect in the *rpoH* gene that encodes an alternate RNA polymerase sigma factor known as σ^{32}. The normal housekeeping sigma factor, σ^{70}, is active in the cell under normal physiological conditions. By contrast, the alternative σ^{32} can function at high temperatures; it also recognizes different promoter sequences than those recognized by σ^{70}. Genes induced by heat shock contain nucleotide sequences in their promoters that are recognized by σ^{32} (Fig. 15.20a). σ^{32} mediates the heat-shock response by binding to the core RNA polymerase and thereby allowing the polymerase to initiate transcription of the genes encoding the heat-shock proteins. The RNA polymerase σ^{32} holoenzyme is relatively resistant to heat inactivation compared to the heat-sensitive σ^{70}-dependent RNA polymerase. As a result, when temperatures rise, genes

Nitrogen Fixation Depends on Many Levels of Gene Regulation

Nitrogen, an essential component of amino acids, chlorophylls, and nucleic acids, is a growth-limiting plant nutrient—the more nitrogen available, the faster most plants grow. However, although gaseous nitrogen (N_2) makes up 78% of earth's atmosphere, plants cannot use nitrogen in this form. They can use only nitrogen that has been *fixed,* that is, converted to ammonia (NH_3) or another nitrogen-containing compound.

Several types of bacteria are agents of nitrogen fixation. These bacteria may be free-living cells (such as those in the genus *Klebsiella*) or plant symbionts (such as those in the genus *Rhizobium*). Of the many symbiotic rhizobial species, each one is able to form a working relationship with only one or a few plants, mainly legumes like peas, beans, and alfalfa. For hundreds of centuries, without understanding the microbiological and molecular bases of nitrogen fixation, farmers made use of the nitrogen-fixing abilities of rhizobial bacteria via the rotation of crops, observing that if they planted alfalfa one year, the next year's wheat crop, for example, would be much more abundant. Here we describe a few aspects of the multilayered regulatory mechanisms controlling the nitrogen-fixation activities of *Rhizobium meliloti,* a small, heterotrophic, rhizobial bacterium that lives in the soil and acquires a source of energy and amino acids through the induction of root nodules in alfalfa.

Bacteria-Plant Interactions Lead to Nitrogen Fixation

In the symbiotic relationship that develops between *R. meliloti* and alfalfa, bacterial genes produce the enzymatic machinery for nitrogen fixation while the plants provide a low-oxygen environment that allows the nitrogen-fixation enzymes to function. Achievement of the symbiotic state depends on a series of communications be-

tween plant and bacteria that lead to dramatic changes in both the anatomy of the plant and the structure of the bacteria.

Alfalfa's secretion of flavenoids triggers the events leading to nitrogen fixation. *R. meliloti* responds to this environmental signal by expressing *nodulation* (*nod*) genes whose protein products are enzymes active in the synthesis of liposaccharides known as Nod factors. Different Nod factors determine the specificity of interaction between different bacterial species and the plants with which they establish symbiosis. Release of the Nod factors from the rhizobial cells elicits a curling of root hairs and cell division in the meristem that lead to the formation of root nodules in the alfalfa plant (Fig. A).

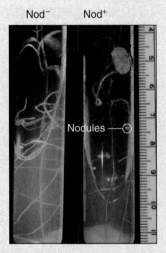

Figure A   *R. meliloti*'s release of Nod factors induces the formation of root nodules in alfalfa. Nod⁻ mutants cannot form nodules.

with a σ^{32} promoter undergo transcription, while genes with a σ^{70} promoter do not.

Levels of the σ^{32} protein and the σ^{32} RNA polymerase holoenzyme increase immediately after heat shock. Several factors cause this increase in σ^{32} activity, including

- An increase in the transcription of the *rpoH* gene.

- An increase in the translation of σ^{32} mRNA stemming from greater stability of the *rpoH* mRNA.

- An increase in the stability and activity of the σ^{32} protein. Heat-shock proteins dnaJ/K (chaperones) bind

to and inhibit σ^{32} under normal physiological conditions. When the temperature rises, these proteins bind to the large number of cellular proteins that become denatured, leaving σ^{32} free to associate with RNA polymerase.

- The inactivity of σ^{70} at high temperatures. Because of this inactivity, σ^{70} does not compete with σ^{32} in forming the RNA polymerase holoenzyme.

Given that high temperatures render σ^{70} inactive, what enables the transcription of σ^{32} during heat shock? The *rpoH* gene, which encodes σ^{32}, has a promoter sequence that is rec-

The bacteria now navigate by chemotaxis to the alfalfa host nodules and penetrate the host's root cortex with the help of a gelatinous filament secreted by the plant itself. Once inside the plant, the *R. meliloti* enter root cells where they divide and differentiate into *bacteroids,* cells that produce nitrogenase, an enzyme complex that catalyzes the conversion of N_2 to NH_3. The host plant monitors the concentration of oxygen in the area of the nodule where the bacteroids thrive to ensure that it is much lower than in the surrounding plant cells or soil. A nearly anaerobic environment in the nodule is crucial to the survival and function of nitrogenase, which is extremely sensitive to oxygen and becomes irreversibly inactivated in air. The alfalfa plant uses the nitrogen fixed by *R. meliloti* as its source of nitrogen, and in return, provides the bacteria with photosynthetic products and amino acids.

The Genetic Components and Mechanisms That Mediate Nitrogen Fixation in Rhizobial Bacteria

Three sets of genes, a special promoter, and a specific activator. The steps of nodule formation and nitrogen fixation just outlined require the coordinated expression of at least three types of *R. meliloti* genes: *nod* genes that elicit the early steps of nodule formation, *fix* genes that contribute to the development and metabolism of bacteroids and are essential to nitrogen fixation (a mutation in a *fix* gene destroys the ability to fix nitrogen), and *nif* genes that encode the polypeptide subunits of the nitrogenase complex. The *nif* genes carry a special type of promoter that RNA polymerase recognizes only when the polymerase is associated with the σ^{54} factor. Initiation of transcription at the σ^{54}-dependent promoters of the *nif* genes depends on NifA, an activator protein responsive (via various intermediaries) to specific environmental signals (Fig. B.1).

Cascades of regulatory events make nitrogen fixation as energy efficient as possible. The process of nitrogen fixation has high energy demands, and the *R. meliloti* bacteroids can only proceed using energy and materials supplied by the host plant. Under conditions of low oxygen concentration and ample nitrogen, the FixL protein in the membrane senses the O_2 concentration and activates FixJ,

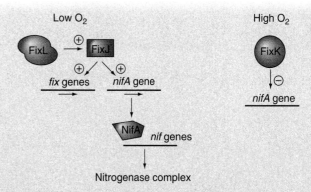

Figure B How environmental signals influence the expression of *nif* genes in *R. meliloti*. Low oxygen activates the *nif* genes via RNA polymerase associated with a σ^{54} factor. A regulatory cascade promotes expression of the *nifA* gene only under appropriate conditions.

which then turns on expression of other *fix* genes as well as *nifA*. The NifA protein product interacts with RNA polymerase associated with σ^{54} to activate the *nif* genes encoding nitrogenase. If the oxygen level rises too high, the FixK gene product turns off *nifA* expression. No NifA means no synthesis of the *nif*-gene-directed nitrogenase complex; this lack of nitrogenase production saves energy because the enzyme complex would be incapacitated by the excess oxygen. The transcriptional activator NifA is thus turned on and off in response to environmental cues reflecting the concentration of oxygen (Fig. B.2).

This brief description of the process of nitrogen fixation by *R. meliloti* gives some idea of the complex layering of gene regulation that ensures nitrogen fixation moves forward under favorable conditions but comes to a halt under conditions of too much oxygen. Other symbiotic, nitrogen-fixing bacteria also have intricate controls over the expression of their *fix* and *nif* genes. The coordinate expression of these bacterial genes contributes to the bacteria's ability to respond to environmental signals and become nitrogen-generating symbionts in their host plants.

ognized by σ^{70} and used for transcription at lower temperatures; however, at high temperatures, another sigma factor, σ^{24}, recognizes a different promoter sequence at *rpoH* and transcribes the *rpoH* gene from that promoter (see Fig. 15.20a). Although σ^{24} is always present in the cell, its own transcription (mediated by the σ^{24} holoenzyme) increases with heat shock and the appearance of denatured proteins.

In summary, *E. coli*'s heat-shock response depends on the regulation of transcription by alternative sigma factors. These alternative factors, which recognize different promoter sequences, complex with the core RNA polymerase as the temperature rises, leading to the transcription of heat-shock proteins.

The Induction of Alternative σ Factors That Recognize Different Promoter Sequences Serves as a Global Regulatory Mechanism in Many Bacteria

By coordinating the expression of different sets of bacterial genes in response to specific cues from the environment, this mechanism contributes to the control of such complex processes as sporulation, the synthesis of flagella, and nitrogen fixation (see the Genetics and Society box "Nitrogen Fixation Depends on Many Levels of Gene Regulation"). For example, under the adverse conditions of nutrient deprivation (for example, nitrogen or carbon starvation), the bacterium

(a)

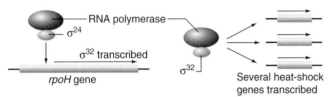

σ^{70} recognizes this promoter sequence

| T T G A C A | 16–18 bp | T A T A A T |

σ^{32} recognizes this promoter sequence

| C T T G A A | 13–15 bp | C C C C A T N T |

(b) At high temperatures σ^{24} recognizes different promoter sequence on *rpoH* gene and σ^{32} is transcribed

RNA polymerase

σ^{24}

σ^{32} transcribed

rpoH gene

σ^{32}

Several heat-shock genes transcribed

Figure 15.20 **Alternate sigma factors.** (a) Base sequences recognized by σ^{32} and σ^{70}. (The *N* indicates that any base can be found at this position.) (b) At high temperature, the *rpoH* gene (encoding σ^{32}) is transcribed. The σ^{32} interacts with RNA polymerase and transcribes the heat-shock genes.

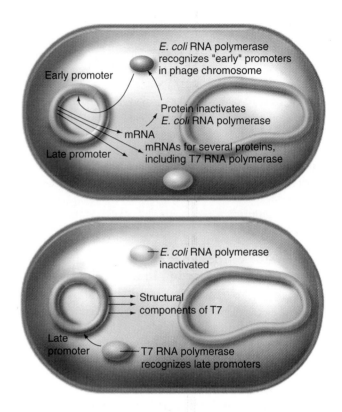

E. coli RNA polymerase recognizes "early" promoters in phage chromosome

Early promoter

Protein inactivates *E. coli* RNA polymerase

mRNA

Late promoter

mRNAs for several proteins, including T7 RNA polymerase

E. coli RNA polymerase inactivated

Structural components of T7

Late promoter

T7 RNA polymerase recognizes late promoters

Figure 15.21 **Transcription of early and late genes of T7 using different RNA polymerases.** After infection, the early genes of bacteriophage T7 are transcribed by the host *E. coli* RNA polymerase. Once T7 RNA polymerase is made and *E. coli* polymerase has been inactivated, late genes of T7 are transcribed by the T7 RNA polymerase.

Bacillus subtilis uses a cascade of sigma factors, induced in a temporal order, to turn on successive sets of genes needed to form spores. With the proper expression of these genes, the bacterial cell becomes a metabolically inert spore able to withstand heat, aridity, extreme cold, toxic chemicals, and radiation.

Bacteriophage T7 Encodes its Own RNA Polymerase, Which Allows the Sequential Expression of Different Sets of Genes during Infection

The *E. coli* DNA bacteriophage T7 provides a vivid illustration of the importance of expressing genes in a specific sequence. The genes required for phage replication must be expressed early in the process of infection, but the genes for the structural components of progeny phages must be expressed later. In particular, the phage must not synthesize the enzyme that causes lysis of the host bacterial cell until after the progeny phages have been assembled.

Bacteriophage T7 achieves this gene expression program in large part because it encodes its own RNA polymerase, which recognizes promoters different from those recognized by the *E. coli* host RNA polymerase (Fig. 15.21). Early in infection, the *E. coli* RNA polymerase recognizes the "early" promoter in the phage chromosome, resulting in an mRNA for several proteins, including the T7 RNA polymerase, and a protein that inactivates the *E. coli* RNA polymerase. The newly synthesized T7 polymerase then recognizes the "late" promoters in the phage chromosome, allowing the delayed transcription of the genes for phage structural components.

A COMPREHENSIVE EXAMPLE: THE REGULATION OF VIRULENCE GENES IN V. CHOLERAE

The principles and mechanisms of gene regulation in *E. coli* apply to gene regulation in other prokaryotic species as well, including the *V. cholerae* bacteria we described at the beginning of this chapter. As we saw, these bacterial agents of cholera are able to sense changes in their environment and transmit signals about those changes to regulators that initiate, enhance, diminish, or repress the expression of various genes. Of particular interest to epidemiologists and medical practitioners seeking to prevent or treat the symptoms of cholera are the genes bestowing virulence.

Three Regulatory Proteins—ToxR, ToxS, and ToxT—Turn on the Genes for Virulence

To understand the regulation of the genes for virulence, researchers first cloned the two genes that encode the polypeptide subunits of cholera toxin: *ctxA* and *ctxB*, which are transcribed and regulated together as an operon. They next made a *ctxA-lacZ* fusion construct that could detect changes in regulation of the operon. They then cut the *V. cholerae* DNA into pieces and cloned these into a vector that would replicate in *E. coli*. With the construction of these tools, they were able to perform experiments in *E. coli* cells, which are more amenable than *V. cholerae* to some types of genetic manipulation. To isolate a gene that regulates expression of the *ctx* operon, they transformed *E. coli* cells already containing the *ctx-lacZ* construct with the clones containing *V. cholerae* DNA. A positive cholera regulator should turn on expression of the *lacZ* part of the fusion hybrid in *E. coli*. Clones that turned on expression contained the regulatory *toxR* gene, which encodes a membrane protein (ToxR) with an amino-terminal end in the cytoplasm and a carboxy-terminal end in the periplasm (the space between the inner and outer membranes). This is a surprising location for a regulatory molecule, since most gene regulators must diffuse through the cytoplasm until they bind to DNA. However, in bacteria, unlike in eukaryotes, the chromosome not enclosed in a nuclear membrane is free to make contact with the inner cell membrane. In *V. cholerae*, *toxR⁻* mutants do not induce virulence, and the *toxR⁻* mutation is recessive as you would predict of a positive regulator.

The *V. cholerae*-DNA-containing clone that was isolated based on its ability to activate the *ctx-lacZ* fusion construct contained a second gene, designated *toxS*. Transcriptional activation of the *ctx-lacZ* fusion in *E. coli* required the ToxS protein. The postulated function of ToxS, which is active in the periplasm, is to help ToxR monomers form dimers. When researchers engineered a *toxR* gene to express a product that spontaneously formed dimers, the resulting ToxR protein no longer required ToxS to carry out its function; it could activate transcription in *E. coli* on its own. The formation of dimers may be important for the tight binding of ToxR to DNA, as is true for most other regulatory proteins in bacteria.

To determine what genes ToxR regulates, researchers fused the *toxR* gene to a constitutive promoter, and introduced this fusion construct into a collection of *V. cholerae* strains in which copies of the *lacZ* gene had randomly inserted around the chromosome. Those colonies expressing β-galactosidase (as shown by the blue color resulting from the splitting of the X-Gal substrate) contained *lacZ* genes adjacent to a promoter region regulated by *toxR*. In *V. cholerae*, these *lacZ* fusion genes must have been regulated by ToxR (at least indirectly) because all bacteria in the study contained the *toxR* gene fused to a constitutive promoter and thus were constitutive synthesizers of ToxR. However, when transferred into *E. coli*, this collection of genes was not regulated directly by ToxR. The lack of direct regulation by ToxR in *E. coli* triggered a search that culminated in the identification of an intermediate regulatory gene named *toxT*. The ToxT protein is a transcriptional activator that carries out its function by binding to the promoters of many genes, including *ctx* and the other virulence genes. While either ToxR/S or ToxT can activate the *ctx* genes that produce toxin, ToxT alone activates the additional virulence genes, which encode pili and other proteins that enable the bacteria to colonize the small intestine. Transcription of the *toxT* gene is regulated by ToxR/ ToxS.

A Model of Virulence Regulation Leaves Some Unanswered Questions

On the basis of these studies, researchers proposed the following model (Fig. 15.22): The membrane location of the ToxR protein enables it to respond to environmental cues that signal *V. cholerae*'s arrival in the small intestine; in response to these cues, the cytoplasmic domain of ToxR binds to the *ctx* promoter. In addition, ToxR binds to the promoter of *toxT*, activating the transcription of ToxT, which then activates the expression of *ctx* and the other virulence genes.

Despite this plausible hypothesis, several intriguing questions remain about the regulatory system that controls the expression of the virulence genes in cholera bacteria. Why is there a cascade (ToxR, ToxT) of regulatory factors? Since the promoter regions in the two genes recognized by ToxR do not have a common sequence of rotational symmetry (as has been found in other regulatory systems), what DNA sequence in the promoters does ToxR recognize? What is the signal that makes the cholera bacteria stop swimming and start to colonize (that is, adhere to the cells of) the small intestine? What molecular events differentiate swimming and colonization? How does the ToxR regulatory protein, located in the membrane, find binding sites on the chromosome?

Answers to these questions will help scientists complete the picture of how *V. cholerae* generate disease. And with a better understanding of pathogenesis, they will be able to devise more effective treatments for cholera as well as measures to prevent it.

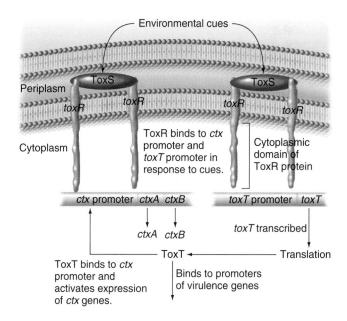

Figure 15.22 Model for how *V. cholerae* regulates genes for virulence. Environmental cues are sensed by ToxR, a transmembrane protein. ToxR is held in its active dimeric form by ToxS. In the cytoplasm, ToxR interacts with promoter of the *ctxA, ctxB, toxT* gene. ToxT in turn regulates the expression of many other virulence genes.

CONNECTIONS

Regulation in prokaryotes depends on the binding of regulatory proteins to specific DNA segments in the vicinity of a gene or group of genes. The existence of these regulatory elements adds another notch to the concept of the gene. Most geneticists would say that a gene consists of the nucleotides that specify amino acids in the gene's protein product or the ribonucleotides in the gene's RNA product, as well as the regulatory elements that influence the gene's transcription.

Some of the ways in which bacteria regulate their genes are available to eukaryotes as well. For example, both types of organisms can use diffusible regulatory proteins to start or stop transcription. By contrast, eukaryotes cannot regulate transcription by the attenuation mechanism described for the *trp* operon because their nuclear membrane prevents access to the growing transcript by the translational machinery. However, eukaryotic cells, with their larger genomes, have evolved many other mechanisms of gene regulation that go beyond those found in prokaryotic systems. In Chapter 16 we examine the regulation of eukaryotic genes, describing the special regulatory needs of eukaryotes and some of the solutions they have evolved.

ESSENTIAL CONCEPTS

1. Most mechanisms of gene regulation in prokaryotes block or enhance the initiation of transcription. In addition, later steps in gene expression are potential targets for fine-tuning the amount of gene products that accumulate in cells.

2. In the *lac operon* model proposed by Jacob and Monod, the binding of a *repressor* protein (encoded by the *lacI* gene) to the DNA *operator* prevents transcription of the structural genes *lacZ*, *lacY*, and *lacA* in the absence of the inducer lactose. When lactose is present, its binding to the repressor induces expression of the structural genes by causing the repressor to change its shape and lose its ability to bind to the operator.

3. A critical, general principle emerges from the *lac* operon studies: Regulatory genes usually encode *trans*-acting regulatory proteins that interact with *cis*-acting regulatory DNA elements located near the promoter (such as the *operator*). Negative regulatory proteins prevent or diminish the rate of transcription, while positive regulatory proteins enhance transcription.

4. Many types of coordinate gene regulation result from the clustering of genes into *operons* that are transcribed into a single polycistronic mRNA from a single promoter.

5. The binding of *repressor* proteins to *operators* can be influenced by either inducers (as for the *lac* repressor) or corepressors (as for the *trp* operon).

6. *Catabolite repression* regulates certain catabolite operons by preventing the CAP protein, a positive regulator, from binding to the operons' promoter region, in the presence of high concentrations of glucose.

7. Many regulatory proteins, both positive and negative, contain a helix-turn-helix motif, function as oligomers that bind to more than one DNA site, and interact with RNA polymerase to prevent or assist its function.

8. *Attenuation,* a form of fine-tuning for operons involved in the biosynthesis of amino acids, is based on the amount of premature termination of mRNA transcription. The termination, in turn, is determined by the intracellular concentration of tRNAs charged with the amino acid produced by the enzyme products of the structural genes in the operon.

9. Cells can express different sets of genes at different times or under different conditions by using alternate sigma factors or by producing novel RNA polymerases that recognize different classes of promoters.

SOCIAL AND ETHICAL ISSUES

1. The bacterium *Bacillus thuringiensis* (Bt) produces a toxin capable of killing many insects that attack crop plants. Bt has been used extensively as a biological pesticide for more than 20 years by spraying crops with the organism. Constant exposure to spraying has led to the development of some resistant strains already in the field. A cloned version of Bt has been constructed consisting of the gene fused to a regulatory region so that the toxin is expressed only when the crops are under stress—as in an insect infestation. Should farmers be forced to plant the transgenic crops rather than spray their fields so that this biological pesticide will still be effective? Are there ways to enforce this policy in the United States or worldwide?

2. Many scientists who are employed as professors at universities have become involved in biotechnology ventures outside the university. Companies often set up their laboratories near universities to take advantage of the brain resources there. Industries that develop near universities employ nonscientists in the community as well as scientists and are generally looked on as an asset in the community. The president of a major state university is encouraging faculty members to be active in development of associated industries, viewing their mission as academics to include community related activities. But legislators and the public whose tax dollars fund the university feel that the professors should

concentrate on teaching. Should faculty be encouraged to participate in outside ventures because it is a service to the community and potentially to humankind through development of new biotechnology or should they be held more strictly to a solely academic, teaching mission?

3. In studying virulence in bacteria, a group of researchers stumbled on ways of modifying a particular bacterial species in a way that increased the virulence. While their finding helped them understand the mechanism of virulence better, the group of researchers were split on whether to publish their findings or not. Some fear that this new information could be misused in developing biological warfare. Do scientists have a responsibility to withhold some information if it could be used in a harmful way?

S O L V E D   P R O B L E M S

I. In the galactose operon in *E. coli,* a repressor, encoded by the *galR* gene, binds to an operator site, *o,* to regulate expression of three structural genes, *galE, galT,* and *galK.* Expression is induced by the presence of galactose in the media. For each of the merodiploids listed below, would the cell show constitutive, inducible, or no expression of each of the structural genes? (Assume that $galR^-$ is a loss of function mutation.)

  a. $galR^-\ o^+\ E^+\ T^+\ K^+$
  b. $galR^+\ o^c\ E^+\ T^+\ K^+$
  c. $galR^-\ o^+\ galE^+\ galT^+\ galK^-\ /galR^+\ o^+\ galE^-\ galT^+\ galK^+$
  d. $galR^+\ o^c\ galE^+\ galT^+\ galK^-\ /galR^+\ o^+\ galE^-\ galT^+\ galK^+$

Answer

This problem requires an understanding of how regulatory sites and proteins that bind to regulatory sites behave. To predict expression in merodiploid strains, look at each copy of the operon individually, then assess what effect alleles present in the other copy of the operon could have on the expression. After doing that for each copy of the operon, combine the results.

  a. The *galR* gene encodes a repressor, so the lack of a GalR gene product would lead to *constitutive expression of the* galE, T, *and* K.
  b. The o^c mutation is an operator site mutation. By analogy with the *lac* operon, the designation o^c indicates that repressor cannot bind and there is *constitutive expression of* galE, T, *and* K.
  c. The first copy of the operon listed has a $galR^-$ mutation. Alone, this would lead to constitutive synthesis *galE* and *galT.* (*galK* is mutant, so there will not be constitutive expression of this gene.) The other copy is wildtype for the *galR* gene so produces a repressor that can act in *trans* on both copies of the operon, overriding the effect of the $galR^-$ mutation. Overall, there will be *inducible expression of the three* gal *genes.*
  d. The first copy of the operon contains an o^c mutation, leading to constitutive synthesis of *galE* and *galT.* The other copy has a wildtype operator so is inducible, but neither operator has effects on the other copy of the operon. The net result *is constitutive* galE *and* galT *and inducible* galK *expression.*

II. The *araI* site is required for induction of *araBAD.* I^- mutants do not express *araBAD.* Starting with an I^- mutant, a second mutation arose that resulted in constitutive arabinose synthesis. A Southern blot using a probe from the regulatory region and early part of the *araB* gene showed a very different set of restriction fragments than were seen in the starting strain. Based on the altered restriction pattern and constitutive expression, propose a hypothesis about the nature of the mutation.

Answer

To answer this question, you need to think how changes in restriction patterns could arise, what effects they could have, and what is necessary to get expression. The fact that the experiment began with a strain that lacked the inducing site, *I,* and that there is constitutive synthesis means that the normal regulation is lacking. Constitutive synthesis could result from a *deletion that fused the* araBAD *genes to another promoter* (one that is on under the growth conditions used). A deletion would lead to a different pattern of restriction fragments that could be observed in the Southern hybridization analysis.

III. Bacteriophage λ, after infecting *E. coli,* will either produce many progeny that are released by lysis of the cell (lytic growth) or the phage DNA integrates into the chromosome because transcription from the major promoters of the phage has been shut down. The repressor protein *cI,* encoded by phage λ, binds to two operator regions to shut down expression and therefore no phages are produced. Mutations in the *cI* gene that destroy the binding ability of the repressor lead to the lytic type of life cycle exclusively. That is, all cells infected by the phage will burst and release progeny phages. Another type of mutation gives the same phenotype-lytic growth only. Such mutations, called λ*vir,* arise at a much lower frequency than the *cI* mutations (about 1 in 10^{12} compared to 1 in 10^6 for *cI* mutants). What do you think these mutations are and why are they less frequent than *cI* mutations?

Answer

This problem requires understanding the types of regulatory mutations that can affect negative regulation. The lack of negative regulation (by *cI*) in the λ life cycle leads to the lytic cycle of growth only. Such mutations could be either in the gene encoding the negative regulator or the site to which the repressor binds. You were told that the *cI* mutations are defects in the gene encoding the repressor. The λ*vir* mutations could be mutations in the site to which the repressor binds, but since the repressor has to bind to two sites, *there must be two mutations in a* λvir *mutant. Therefore these would arise less frequently* (at a frequency predicted for two independent mutational events combined: 1 in 10^6 × 1 in 10^6 or 1 in 10^{12}).

P R O B L E M S

15-1   For each of the terms in the left column, choose the best matching phrase in the right column.

a. induction	1. glucose prevents expression of catabolic operons
b. repressor	2. protein undergoes a reversible conformational change
c. operator	3. often fused to regulatory regions of genes whose expression is being monitored
d. allostery	4. stimulation of protein synthesis by a specific molecule
e. operon	5. site to which repressor binds
f. catabolite repression	6. gene regulation involving premature termination of transcription
g. reporter gene	7. group of genes transcribed into one mRNA
h. attenuation	8. negative regulator

15-2   All mutations that abolish function of the Rho (termination) protein in *E. coli* are conditional mutations. What does this tell you about the *rho* gene?

15-3   Bacteriophage λ, after infection of a cell, can integrate into the chromosome of the cell if the repressor protein, cI, binds to and shuts down phage transcription immediately. (A strain containing a bacteriophage integrated in the chromosome is called a lysogen.) The alternative fate is the production of many more viruses and lysis of the cell. In an Hfr cross, a donor strain that is a lysogen was crossed with a lysogenic F⁻ recipient cell and no phages were produced. However, when the Hfr lysogen donor strain transferred its DNA to a nonlysogenic F⁻ recipient cell, the recipient cell burst, releasing a new generation of phages. Why did infection of a nonlysogenic cell result in phage growth and release but infection of a lysogenic recipient did not?

15-4   The promoter of an operon is the site to which RNA polymerase binds to begin transcription. Some base changes in the promoter result in a mutant site to which RNA polymerase cannot bind. Would you expect mutations in the promoter that prevent binding of RNA polymerase to act in *trans* on another copy of the operon on a plasmid in the cell or only in *cis* on the copy immediately adjacent to the mutated site?

15-5   For each of the *E. coli* merodiploid strains containing the *lac* operon alleles listed, indicate whether the strain is inducible, constitutive, or unable to express β-galactosidase and permease.
a. $I^+ o^+ Z^- Y^+ / I^+ o^c Z^+ Y^+$
b. $I^+ o^+ Z^+ Y^+ / I^- o^c Z^+ Y^-$
c. $I^+ o^+ Z^- Y^+ / I^- o^+ Z^+ Y^-$
d. $I^- p^- o^+ Z^+ Y^- / I^+ p^+ o^c Z^- Y^+$
e. $I^s o^+ Z^+ Y^+ / I^- o^+ Z^+ Y^-$

15-6   Mutants were isolated in which the constitutive phenotype of a missense *lacI* mutation was suppressed. That is, the operon was now inducible. These mapped to the operon but were not in the *lacI* gene. What could these mutations be?

15-7   For each of the following mutant *E. coli* strains, plot a 30-minute time course of concentration of β-galactosidase, permease, and acetylase enzymes grown under the following conditions. For the first 10 minutes, no lactose is present; at 10′ lactose becomes the sole carbon source. Plot concentration on the *y* axis, time on the *x* axis. (Don't worry about the exact units for each protein on the *y* axis.)
a. $I^+ p^+ o^+ Z^+ Y^+ A^+ / I^- p^+ o^+ Z^- Y^+ A^+$
b. $I^- p^+ o^c Z^+ Y^+ A^- / I^+ p^+ o^+ Z^- Y^+ A^+$
c. $I^s p^+ o^+ Z^+ Y^+ A^+ / I^- p^+ o^+ Z^- Y^+ A^+$
d. $I^- p^- o^+ Z^+ Y^+ A^+ / I^- p^+ o^c Z^+ Y^- A^+$
e. $I^- p^+ o^+ Z^- Y^+ A^+ / I^- p^- o^c Z^+ Y^- A^+$

15-8   You have isolated two different mutants (*reg1* and *reg2*) causing constitutive expression of the *emu* operon (*emu1 emu2*). One mutant contains a defect in a DNA-binding site, the other has a defect in the gene encoding a protein that binds to the site (loss of function).
a. Is the DNA-binding protein a positive or negative regulator of gene expression?
b. To determine which mutant has a defect in the site and which one has a mutation in the binding protein, you decide to do a merodiploid analysis. Assuming you can assay levels of Emu1 and Emu2 proteins, what results do you predict for the two strains below (i and ii) if *reg2* encodes the regulatory protein and *reg1* is the regulatory site?

c. What results do you predict for the two strains below (i and ii) if *reg1* encodes the regulatory protein and *reg2* is the regulatory site?
   i. F′ *reg1⁻ reg2⁺ emu1⁻ emu2⁺/reg1⁺ reg2⁺ emu1⁺ emu2⁻*
   ii. F′ *reg1⁺ reg2⁻ emu1⁻ emu2⁺/reg1⁺ reg2⁺ emu1⁺ emu2⁻*

15-9 You have isolated a protein that binds to DNA in the region upstream of the promoter sequence of the *sys* gene. If this protein is a positive regulator, which would be true?
   a. Loss of function mutations in the gene encoding the DNA-binding protein would cause constitutive expression.
   b. Loss of function mutations in the gene encoding the DNA-binding protein would result in no expression.

15-10 Seven strains of *E. coli* (mutants 1–6) that had one of the following mutations affecting the *lac* operon were isolated.
   a. deletion of *lacY*
   b. *oᶜ* mutation
   c. missense mutation in *lacZ*
   d. inversion of the *lac* operon (but not an inversion of the *lacI* gene)
   e. superrepressor mutation
   f. inversion of *lacZ, Y,* and *A* but not *lacI, p, o*

The entire *lac* operon (including the *lacI* gene and its promoter) from each of the seven *E. coli* strains was cloned into a plasmid vector containing an ampicillin resistance gene. Each recombinant plasmid was transformed into each of the seven strains to create merodiploids, and plated on ampicillin media in which lactose is the only carbon source. (Ampicillin was included to assure maintenance of the plasmid.) Growth of the transformants is scored below (+ = growth, − = no growth). Synthesis of β-galactosidase and permease are required for growth on this medium. Results of this merodiploid analysis are shown below. Which mutant bacterial strain contained each of the alterations listed above? (Mutation 1 was found to be a deletion of the *lacY* gene.)

	1	2	3	4	5	6
1	−	+	−	+	−	+
2	+	−	−	+	−	+
3	−	−	−	+	−	+
4	+	+	+	+	−	+
5	−	−	−	−	−	+
6	+	+	+	+	+	+

15-11 Given the data below, explain which strains and growth conditions are important for reaching the following conclusions.

a. Arabinose induces coordinate expression of the *araBAD* genes (encoding kinase, isomerase, and epimerase).
b. The *araC* gene encodes a positive regulator of *araBAD* expression.

Genotype	Arabinose in Medium	Kinase	Isomerase	Epimerase
1. *C⁺ B⁺ A⁺ D⁺*	no	−	−	−
2. *C⁺ B⁺ A⁺ D⁺*	yes	+	+	+
3. *C⁻ B⁺ A⁺ D⁺*	no	−	−	−
4. *C⁻ B⁺ A⁺ D⁺*	yes	−	−	−

15-12 For each of the growth conditions listed below, what proteins would be bound to *lac* operon DNA? (Do not include RNA polymerase.)
   a. glucose
   b. glucose + lactose
   c. lactose

15-13 Seven *E. coli* mutants were isolated. The activity of the enzyme β-galactosidase produced by cells containing each mutation alone or in combination with other mutations was measured when the cells were grown in medium supplemented with different carbon sources.

	Glycerol	Lactose	Lactose + Glucose
Wildtype	0	1000	10
Mutant 1	0	10	10
Mutant 2	0	10	10
Mutant 3	0	0	0
Mutant 4	0	0	0
Mutant 5	1000	1000	10
Mutant 6	1000	1000	10
Mutant 7	0	1000	10
F′ *lac* from mutant 1/ mutant 3	0	1000	10
F′ *lac* from mutant 2/ mutant 3	0	10	10
Mutant 3 + 7	0	1000	10
Mutant 4 + 7	0	0	0
Mutant 5 + 7	0	1000	10
Mutant 6 + 7	1000	1000	10

Assume that each of the seven mutations is one and only one of the genetic lesions in the list below. Identify the type of alteration each mutation represents.
   a. superrepressor
   b. operator deletion
   c. amber suppressor tRNA gene (assume that the suppressor tRNA is 100% efficient in suppressing amber mutations)
   d. defective CAP–cAMP binding site
   e. nonsense (amber) mutation in the β-galactosidase gene
   f. nonsense (amber) mutation in the repressor gene
   g. defective *crp* gene (encoding the CAP protein)

15-14 For each of the *E. coli* strains that follow, indicate the effect of the genotype on expression of the *trpE* and *trpC* genes in the presence and absence of tryptophan. (In the wildtype [$r^+ p^+ o^+ att^+ trpE^+ trpC^+$], *trpC* and *trpE* are fully repressed in the presence of tryptophan and are fully induced in the absence of tryptophan.)

r = repressor gene; r^n product cannot bind tryptophan; r^- product cannot bind operator
o = operator; o^- cannot bind repressor
att = attenuator; att^- is a deletion of the attenuator
p = promoter; p^- is a deletion of the *trp* operon promoter
$trpE^-$ and $trpC^-$ are null (loss of function) mutations

a. $r^+ p^- o^+ att^+ trpE^+ trpC^+$
b. $r^- p^+ o^+ att^+ trpE^+ trpC^+$
c. $r^n p^+ o^+ att^+ trpE^+ trpC^+$
d. $r^- p^+ o^+ att^- trpE^+ trpC^+$
e. $r^+ p^+ o^- att^+ trpE^+ trpC^- / r^- p^+ o^+ att^+ trpE^- trpC^+$
f. $r^+ p^- o^+ att^+ trpE^+ trpC^- / r^- p^+ o^+ att^+ trpE^- trpC^+$
g. $r^+ p^+ o^- att^- trpE^+ trpC^- / r^- p^+ o^- att^+ trpE^- trpC^+$

15-15 Clones of three adjacent genes involved in arginine biosynthesis have been isolated from a bacterium. If these three genes together make up an operon, what result do you expect when you use the DNA from each of these genes as probes in a Northern analysis? What result do you expect if the three genes do not make up an operon?

15-16 A molecular geneticist is investigating an operon by measuring the amount of expression of the four structural genes (*A, B, C,* and *D*) produced in wildtype and mutant bacterial cells after the addition of compound Z to a minimal medium. An additional protein (E) is of very small size (less than 20 amino acids) and cannot be measured by the same analytical system employed for the other proteins. Several of the mutations are nonsense mutations that have an effect on the genes transcribed after them in the operon. In addition to stopping translation of the gene in which the mutations lies, these nonsense mutations (called polar mutations) cause termination of transcription of the next genes in the operon. This means that genes downstream of the mutation are not expressed. (For example, in the *lac* operon, some *lacZ* nonsense mutations can result in no expression of *lacY* and *lacA*). The investigator has also obtained mutations in two other sites, *F* and *G*, closely linked to *A–D*. The graphs shown are all semilogarithmic. The percentage

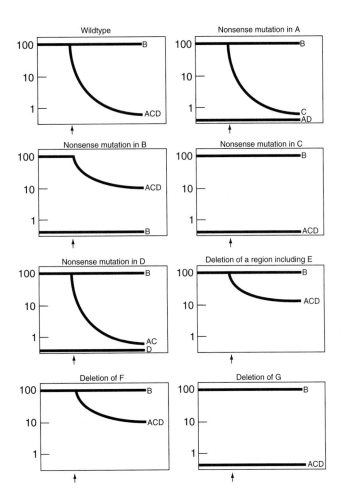

of maximal possible expression for a particular protein is plotted on the *Y* axis, while the *X* axis coordinate is time. Compound Z is added at the point specified by the arrow.

a. Is this operon likely to be involved in a pathway of biosynthesis or a pathway of degradation? Is the operon inducible or repressible?
b. Construct a map of this operon. Indicate the relative positions of genes *A, B, C, D,* and *E* as well as the sites *F* and *G*. List possible functions for all the genes and sites. (Possible functions [not all are necessarily represented] for the genes and sites *A–G:* promoter, operator, enzyme structural gene, CAP binding site, *crp* gene, attenuator region.)

15-17 Following is a sequence of the leader region of the *his* operon mRNA in *Salmonella typhimurium*. What bases in this sequence could cause a ribosome to pause when histidine is limiting in the media?

5′ AUGACACGCGUUCAAUUUAAACACCACCAUCAUCACCAUCAUCCUGACUAGUCUUUCAGGC 3′

15-18 Maltose utilization in *E. coli* requires the proteins encoded by genes in three different operons: *malE malF malG, malK lamB, malP malQ* operons. MalT is a positive regulator protein that regulates expression from each of these operons and the operon is catabolite sensitive. Bacteriophage λ uses one of the maltose transport proteins, LamB, that is found in the outer membrane. Synthesis of the λ receptor is induced by maltose in the media. *E. coli* cells that are resistant to infection by phage λ have been isolated. List the types of mutations in the maltose regulon that λ-resistant mutants could contain.

15-19 In an effort to determine the location of an operator site for a negatively regulated gene, you have made a series of deletions within the regulatory region. The extent of each deletion is shown by the line underneath the sequence and the resulting expression from the operon (*i* = inducible; *c* = constitutive; − = no expression).

```
. . . GGATCTTAGCCGGCTAACATGATAAATATAA...
. . . CCTAGAATCGGCCGATTGTACTATTTATATT...
1  i   ————
2  −   ————————————————————————
3  c   ——————————————————
4  −   ————————————————
5  c   ————————
```

a. What can you conclude from this data about the location of the operator site?
b. Why do you think deletions 2 and 4 show no expression?

15-20 The following data is from a DNase footprinting experiment in which either RNA polymerase or a repressor protein was added to a labeled DNA fragment, then the complex was digested with DNase. DNA sequencing reactions were also performed on the same DNA so the bases that were protected by proteins binding could be identified. (Notice that DNase does not cut at after each base in the DNA fragment.)
a. What is the sequence of the DNA in this fragment?
b. Mark on the sequence the region where the repressor binds.
c. Mark on the sequence the region where RNA polymerase binds.

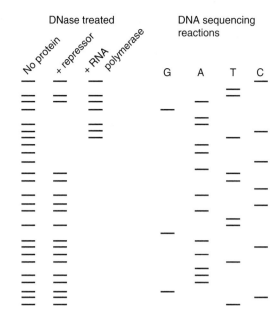

15-21 The original constitutive operator mutations in the *lac* operon were all base changes in O_1. Why do you think mutations in O_2 or O_3 were not isolated in these screens?

15-22 Cells containing mutations in the *crp* gene (encoding the positive regulator CAP) are Lac^-, Mal^-, Gal^-, etc. To find suppressors of the *crp* mutation, cells were screened to find those that were both Lac^+ and Mal^+.
a. What types of suppressors would you expect to get using this screen compared with a screen for Lac^+ only?
b. All suppressors isolated were mutant in the α-subunit gene of RNA polymerase. What hypothesis could you propose based on this analysis?

15-23 An operon fusion consists of a regulatory region cloned next to the coding region of the genes of an operon. A gene fusion consists of a regulatory region of a gene such as *lacZ* and the DNA encoding the first amino acids of the β-galactosidase protein cloned next to the coding region of another gene. What additional feature do you have to consider to create a functional gene fusion that is not necessary for an operon fusion?

15-24 Many genes whose expression is turned on by DNA damage have been isolated. Loss of function mutations in the *lexA* gene lead to expression of many of these genes even when there has been no DNA damage. Would you hypothesize that LexA protein is a positive or negative regulator? Why?

GENE REGULATION IN EUKARYOTES

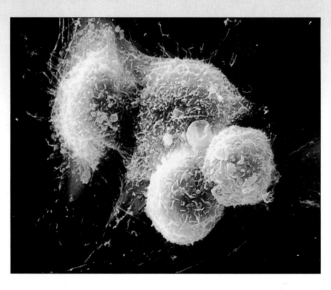

This false-color SEM shows dividing Hodgkin's cells magnified ×270. Hodgkin's disease is a cancer of the lymphreticular system—the mediator of nonspecific cell defense mechanisms and immune response.

When a *Drosophila* male courts a *Drosophila* female, he sings a species-specific song and dances an ancient dance, and, if successful, his instinctive behaviors culminate in mating. The male senses a female's presence by visual and tactile cues as well as by the pheromones she produces. After orienting himself at a precise angle with respect to his prospective mate, he taps his partner's abdomen with his foreleg and then performs his song by stretching out his wings and vibrating them at a set frequency; when the song is over, he begins to follow the female. If she is unreceptive (perhaps because she has recently mated with another male), she will run away, but if she is receptive, she will let him overtake her. When he does, he licks her genitals with his proboscis, curls his abdomen, mounts the female, and copulates with her for about twenty minutes.

Various mutations in a gene called *fruitless* produce behavioral changes that prevent the male from mating properly. Some mutant alleles alter the song to an unfamiliar air. Others diminish the male's ability to distinguish females from males; male flies with this mutation court each other. Still others reduce the male's ability to court either sex. Finally, some mutations create lethal null alleles that cause male flies to die just before they emerge from the pupal case, showing that the *fruitless* gene has other functions in addition to its effects on courtship and mating.

Cloned in 1996, *fruitless* is a large gene—roughly 150 kb in length—encoding a product that can inhibit transcription at several promoters. Alternative RNA splicing of the gene's several different transcripts can produce many variations of the fruitless protein. All of these gene products contain zinc-finger peptide motifs that facilitate binding to DNA, suggesting a role for the protein in the regulation of transcription (review Fig. 10.21a). Some forms of the fruitless protein are sex-specific, appearing only in males or only in females. By contrast, some fruitless proteins are expressed at low levels in many kinds of cells in both sexes.

The male version of the fruitless protein, although very similar to the female protein in amino-acid sequence, has an extra 101 amino acids at its N terminus, and this addition almost certainly determines the observed differences in male and female behavior. Remarkably, the male-specific fruitless mRNA is synthesized in only a few hundred of the tens of thousands of neurons that make up the male *Drosophila*'s nervous system. Most of these *fruitless*-expressing cells are located near motor neurons that control either wing movements (and thus possibly the song) or abdominal movements (and thus possibly the abdominal curling that immediately precedes mating). Work on the *Drosophila fruitless* gene has provided strong evidence that differences in gene expression can directly influence complex behaviors.

In this chapter, we see that **eukaryotic gene regulation**—the control of gene expression in the cells of eukaryotes—depends on an array of interacting regulatory elements that turn genes on and off in the right places at the right times. Some of these elements are specific DNA sequences in the vicinity of the gene to be regulated; others are DNA-binding proteins encoded by other genes; still others are proteins that bind to and modify the activity of DNA-binding proteins.

During the embryonic development of multicellular eukaryotic organisms like *Drosophila* or humans, gene regulation controls not only the elaboration of sex-related characteristics and behaviors, but also the differentiation of tissues and organs as well as the precise positioning of these tissues and organs within the multicellular animal. For example, because of the tightly controlled differential regulation of genes in different types of cells, red blood cell precursors selectively turn on the genes for making hemoglobin, while pancreatic cells turn on the genes for making insulin, even though both types of cells, like nearly all somatic cells, carry the same number of chromosomes with the same amount of DNA and the same complete set of genes.

In exploring the intricacies of eukaryotic gene regulation, it is helpful to bear in mind key similarities and differences between eukaryotes and prokaryotes (Table 16.1). In both types of cells, tran-scription is regulated through the attachment of DNA-binding proteins to specific DNA sequences that are adjacent to or near the transcription unit itself; and in both eukaryotes and prokaryotes, the same polypeptide motifs appear in many DNA-binding proteins. Indeed many, though not all, of the basic principles of gene regulation in prokaryotes apply to eukaryotes, but additional levels of complexity tailor regulation in eukaryotes to their requirements. First, because eukaryotes have up to 700 times more DNA and 20 times more genes than prokaryotes, they require additional biochemical mechanisms to achieve specificity; the mechanisms they have evolved include the higher-affinity binding of proteins to DNA. Second, because eukaryotic DNA is packaged in chromatin, the transcriptional machinery must be able to control chromatin compaction and decompaction. The control of chromatin structure over defined distances along the chromosomes provides eukaryotic cells with a level of gene regulation that is unavailable to prokaryotic cells. Third, eukaryotes accomplish a type of regulation, through the differential splicing of exons, unavailable to prokaryotes. Moreover, although the separation of transcription and translation into different cellular compartments in eukaryotes precludes certain forms of prokaryotic regulation, such as attenuation, it opens up the possibility of regulating gene function at different points in the expression process, from the initiation of transcription, through mRNA processing before transport from the nucleus, to protein synthesis and modification in the cytoplasm. Finally, in addition to these basic differences between eukaryotes and prokaryotes, multicellular eukaryotes must be able to use gene regulation to control cellular differentiation and the complex interactions of various types of differentiated cells within tissues and organs.

One multifaceted overarching theme emerges from our discussion of eukaryotic gene regulation. Like complex behaviors, most biological functions arise from the regulated interactions of

TABLE 16.1   Key Regulatory Differences between Eukaryotes and Prokaryotes

Characteristic	Prokaryote	Eukaryote
Control of transcription through specific DNA-binding proteins	Yes	Yes
Reutilization of same DNA-binding motifs by different DNA-binding proteins	Yes	Yes
Activator proteins	Yes	Yes
Repressor proteins	Yes	Yes
Specificity of binding to DNA by regulatory protein	Specific	Highly specific
Affinity of binding	Strong	Very strong
Role played by chromatin structure	No	Yes
Coordinate control achieved with operons	Yes	Rare
Differential splicing	No	Yes
Attenuation	Yes	No
RNA processing	No	Yes
Differential polyadenylation	No	Yes
Differential transport of RNA from nucleus to cytoplasm	No	Yes

large networks of genes. And because each gene in a network has multiple potential points of regulation, the possibilities for regulatory refinement are enormous. Thus, eukaryotic gene regulation is a modular system in which different combinations of elements interact with each other at specified times and places, as well as in response to changes in the cellular environment, to allow the synthesis of precisely modulated amounts of gene products.

As we examine the components and mechanisms of eukaryotic gene regulation, we describe

■ The use of genetics to study gene regulation, focusing on mutations that identify *cis*-acting control elements and *trans*-acting proteins.

■ Gene regulation at the initiation of transcription, including a discussion of how three different RNA polymerases

recognize three different classes of promoters; how *trans*-acting proteins help control class II promoters, causing the activation or repression of gene expression; how chromatin structure affects expression; how signal transduction systems work; and how DNA methylation, an epigenetic process, can regulate gene expression through silencing and genomic imprinting.

■ Regulation after transcription, including a discussion of the control of RNA splicing, RNA stability, mRNA editing, and translation, as well as the control of protein activity by posttranslational modification.

■ A comprehensive example of gene regulation during sex determination in *Drosophila*.

THE USE OF GENETICS TO STUDY GENE REGULATION

Genetic studies of various mutations have helped demystify the complexities of gene regulation. The techniques for analyzing these mutations in eukaryotes are similar to those researchers use in examining gene control in bacterial cells. These studies have established that for most, but not all, eukaryotic genes, the critical decisions controlling the amount of gene product synthesized are made during the initiation of transcription, when RNA polymerase starts to make a primary transcript, or RNA copy, of a gene's coding strand (review Fig. 7.13).

The Analysis of Regulatory Components Focuses on Mutations That Affect a Gene's Function but Do Not Affect the Amino Acids in the Gene's Product

Geneticists can study regulation by choosing a gene whose control they want to study (the "target gene") and then looking for mutations that affect the expression of that gene. To find such mutations, they observe changes in a phenotype that is extremely sensitive to the amount of gene product synthesized.

Regulatory Mutations That Map at or Near the Target Gene Help Define cis-Acting DNA Sequences That Influence Transcription

cis-acting elements at a gene are DNA sequences that serve as attachment sites for the DNA-binding proteins that regulate the initiation of transcription. The *promoter* is a *cis*-acting element that is very close to a gene's initiation site; its binding of RNA polymerase allows a basal level of transcription (see Fig. 16.1a). *Enhancers* are another class of *cis*-acting elements; their binding of proteins augments or represses basal levels of transcription. Enhancers are defined by their ability to retain original function even when moved far from the gene whose transcription they influence or when reversed in orientation.

(a) *cis*-acting elements

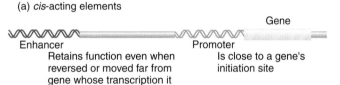

Gene

Enhancer
Retains function even when reversed or moved far from gene whose transcription it influences

Promoter
Is close to a gene's initiation site

(b) *trans*-acting gene products interact with *cis*-acting elements

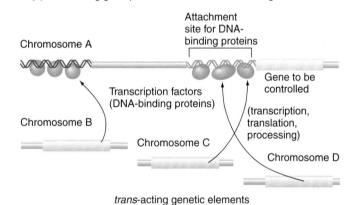

trans-acting genetic elements

Figure 16.1   How *cis*-acting and *trans*-acting elements influence transcription. (a) *cis*-acting regulatory elements are regions of DNA sequence that lie nearby to the gene they control. Promoter elements typically lie directly adjacent to the gene that they control. Enhancers that regulate expression can sometimes lie thousands of base pairs away from a gene. (b) *trans*-acting genetic elements encode protein products called transcription factors that interact with *cis*-acting elements.

Reporter constructs are synthesized in the laboratory to contain a gene's postulated regulatory region but not its coding region. They are one tool for studying gene regulation. Investigators can systematically identify promoters and enhancers by altering reporter constructs through *in vitro* mutagenesis across a presumed regulatory region and then reintroducing the reporter constructs into the genome by transformation. In assembling a reporter construct for this purpose, they replace the coding region of the gene whose regulation they are studying with the coding region of an easily identifiable product (the "reporter"), such as β-galactosidase or green fluorescent protein (GFP). (Recall that β-galactosidase turns blue in the presence of a substrate known as X-Gal [review Fig. 8.8]; similarly, GFP luminesces green when exposed to light of a particular wavelength.) Reporter constructs are particularly valuable for looking at mutations that affect gene expression rather than the amino-acid composition of the gene's polypeptide product. Mutations that alter the amount of reporter synthesized help define the elements necessary for a gene's regulation.

Regulatory Mutations That Map Far from a Target Gene Help Define trans-Acting Genetic Elements

These **trans-acting** elements are genes located elsewhere in the genome, that is, somewhere other than at the target gene; they encode proteins (or, in rare cases, RNAs) that interact directly or indirectly with the target gene's *cis*-acting elements to activate or repress expression of the target gene (Fig. 16.1b). *trans*-acting genes operate through their gene products, known as *trans-acting factors*. The *trans*-acting proteins that influence transcription are known generically as **transcription factors.**

Researchers can identify *trans*-acting elements in many ways. Mutations that alter phenotypes dependent on either the level of target gene expression or the level of expression of a reporter construct previously introduced into the genome, and that map far from the target gene or reporter construct, are likely to reside in *trans*-acting elements. Biochemical procedures are useful for isolating proteins that bind *in vitro* to *cis*-acting DNA sequences. Once researchers identify a *trans*-acting element, they can clone it for further study.

We now examine how *cis*-acting and *trans*-acting regulatory elements interact to initiate, increase, diminish, or prevent gene expression.

MOST GENE REGULATION OCCURS AT THE INITIATION OF TRANSCRIPTION

During transcription in eukaryotes, RNA polymerase makes a single strand RNA molecule—known as the primary transcript—that is complementary in base sequence to a gene's DNA template strand (see Chapter 7, Fig. 7.13). Many *cis*-acting DNA elements and *trans*-acting regulatory proteins help initiate or prevent this first step of gene expression.

In Eukaryotes, Three RNA Polymerases Transcribe Different Sets of Genes

One of the first indications that events at the initiation of transcription usually determine gene regulation was the finding that different classes of genes in eukaryotic cells are tran-

scribed by different RNA polymerases. The three types of RNA polymerases have some subunits in common, but contain some subunits that are distinct to each type. Each of the three RNA polymerases has a specific function (Fig. 16.2).

RNA Polymerase I is Responsible for the Transcription of rRNA

In most eukaryotic genomes, the genes for rRNA are arranged in tandem repeats at sites on one or more chromosomes. As described in Chapter 11, these sites loop out from their respective chromosomes and come together in the nucleolus. RNA polymerase I (pol I) transcribes rRNA from these clusters of genes in the nucleolus (review Fig. 11.18). Each tandem repeat has a promoter region just upstream of the transcription unit (Fig. 16.2a). The transcription of rRNA by pol I produces a large rRNA primary transcript. Subsequent RNA processing in the nucleolus breaks that transcript into a 28S rRNA and a 5.8S rRNA that become incorporated into the large ribosomal subunit and an 18S rRNA that becomes incorporated into the small subunit (Fig. 16.2a). These rRNAs associate with ribosomal proteins during assembly of the ribosome.

RNA Polymerase III Transcribes the tRNAs and Other Small RNAs

In addition to the tRNAs, which bring amino acids to the ribosome in response to codons in the mRNA, RNA polymerase III (pol III) transcribes a 5S rRNA that is yet another component of the large ribosomal subunit; it also transcribes some of the small nuclear RNAs (snRNAs) found in spliceosomes (Fig. 16.2b). As with pol I, the promoter elements in some of the genes recognized by pol III are just upstream of the gene's transcribed sequence. Surprisingly, however, studies of the transcriptional effects of small deletions in tRNA genes showed that the pol III promoter for these genes consists of sequences within the DNA transcribed to produce the tRNA (see Fig. 16.2b). These sequences thus serve two functions: They are promoter elements recognized by specific transcription factors; and, after the initiation of transcription, they serve as part of the tRNA gene that is transcribed into the primary transcript.

RNA Polymerase II Transcribes All Protein-Encoding Genes

Most of the primary transcripts produced by RNA polymerase II (pol II) undergo further processing to generate mRNAs (Fig. 16.2c). During mRNA formation, introns must be spliced out; in addition, ribonuclease cleaves pol II-transcribed primary transcripts to form a new 3′ end, to which the enzyme poly-A polymerase adds a poly-A tail; and the 5′ end of the transcript is chemically modified to produce a "5′GTP cap," which protects the molecule from degradation.

Measurements of mRNA and protein levels in eukaryotic cells revealed that, almost always, the more mRNA of a gene that accumulates in the cell, the greater the production of that gene's protein product. Further biochemical experiments showed that the control of mRNA production almost always

(a) Tandem repeats of rRNA genes are transcribed by RNA polymerase I

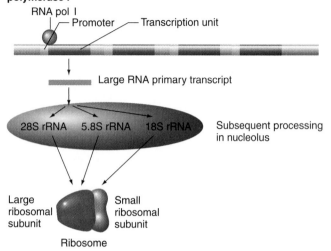

(b) Small RNA genes are transcribed by RNA pol III

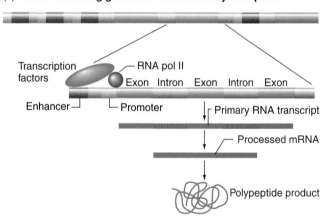

(c) Protein-encoding genes are transcribed by RNA pol II

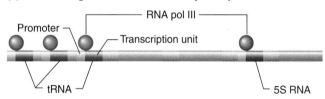

Figure 16.2 The three RNA polymerases of eukaryotic cells have different functions and recognize different promoters.
(a) RNA polymerase I recognizes promoters associated with ribosomal RNA (rRNA) genes, which always exist in tandem arrays. The primary rRNA transcripts are processed into smaller RNA molecules that form components of the ribosome. (b) RNA polymerase III recognizes promoters associated with a limited number of genes that encode small RNA molecules involved in the translation process. (c) RNA polymerase II recognizes promoters associated with all of the diverse protein-encoding genes in the genome.

reflects the rate at which transcription is initiated, rather than the rate of elongation or termination of transcription, or the stability of the mRNA.

cis-*Acting Regulatory Regions Recognized by pol II Consist of a Promoter and One or More Enhancers*

Although each of the regulatory regions of the thousands of pol II-transcribed genes in a eukaryotic genome is unique, they all contain two kinds of essential DNA sequences (Fig. 16.2c). The **promoter** is always very close to the gene's protein-coding region. It includes an initiation site, where transcription begins, and a "TATA" box, consisting of roughly seven nucleotides of the sequence T A T A (A or T) A (A or T), located 30 nucleotides upstream of the initiation site. **Enhancers** are regulatory sites that can be quite distant—up to tens of thousands of nucleotides away—from the promoter. In yeast, enhancer elements are often called *upstream activation sites,* or *UAS*s.

The enhancer regions of some class II genes are very large, containing multiple elements that make it possible to fine-tune regulation of the gene. This is particularly true of those genes in multicellular eukaryotes that must be expressed in many different tissues. The *string* gene in *Drosophila* is an example. The gene encodes a protein product that activates the 14th mitosis of embryonic development. This 14th mitosis begins just after membranes simultaneously form around the roughly 6000 nuclei of the giant syncytium that resulted from the first 13 mitoses (review the Fast Forward box in Chapter 3). What is interesting about the 14th mitosis is that cells in different areas, or domains, of the embryo enter it at different times in an intricate but reproducible temporal pattern (Fig. 16.3a). Thus, although the cells of each domain sustain the 14th nuclear division together, the time at which that mitosis takes place is different for different domains. Remarkably, the cells within each domain simultaneously express the *string* gene just before they enter mitosis; in fact, expression of the *string* gene induces their entry into the mitotic cycle. In *string* mutants, all embryonic cells arrest in the G2 stage of cycle 14 and never undergo mitosis. A roughly 35-kb region upstream of the *Drosophila string* gene contains binding sites for many transcription factors known to regulate formation of the *Drosophila* body pattern (Fig. 16.3b). The complex interaction of these factors ensures that the *string* gene is turned on in the cells of each embryonic domain at the correct time. Interestingly, the approximately 25 embryonic mitotic domains correspond, in part, to specific tissues or structures in the *Drosophila* larva.

trans-*Acting Proteins Control Transcription from Class II Promoters*

The binding of *trans*-acting proteins (transcription factors) to a gene's promoter and enhancer (or enhancers) controls the initiation of transcription from each class II gene. Different types of proteins bind to each of the *cis*-acting regulatory regions: **Basal factors** bind to the promoter; **activators** and **repressors** bind to the enhancers.

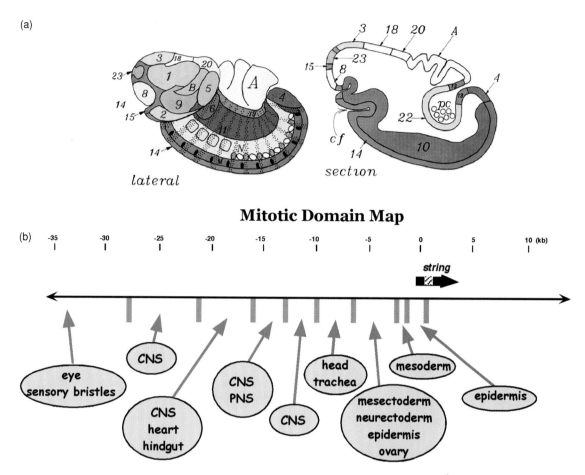

Figure 16.3 The large enhancer region of the *Drosophila string* gene helps create discrete mitotic domains during embryogenesis. (a) Colors indicate individual mitotic domains during the 14th cell cycle of the fruit fly embryo. The cells within each domain divide synchronously, but different domains initiate their divisions at different times. Both lateral and cross-sectional views of the embryo are shown. (b) Regulation of the *string* gene, a critical mitotic activator, determined at a slightly later time in development. A variety of enhancer regions ensure that *string* is turned on at the right time in each mitotic domain and tissue type (CNS: central nervous system; PNS: peripheral nervous system).

Basal Factors Are Required to Maintain a Basal Level of Transcription

Basal factors assist the binding of RNA polymerase II to the promoter and the initiation of low levels of transcription called basal transcription (from which the basal factors get their name). The key component of the basal factor complex that forms on most promoters is the TATA box–binding protein, or **TBP** (so named because it binds to the TATA box described above). The TBP is essential to the initiation of transcription from all class II genes that have a TATA box in their promoter. TBP associates with several other basal factors called TBP-associated factors, or **TAFs** (Fig. 16.4). The complex of basal factors bind to the proximal promoter in an ordered pathway of assembly. Once the complex has formed, basal transcription is initiated. Researchers have determined the structure of the TBP-TAF complex on the DNA at the TATA box (Fig. 16.4). The most striking feature of the structure is a sharp bend in the DNA at the TATA box, induced by TBP.

The primary sequence and three-dimensional structure of the basal factors are highly conserved in all eukaryotes, from yeast to humans, and this sequence conservation has facilitated identification of the factors. For example,

researchers isolated yeast TBP—the first basal factor to be purified—through its ability to substitute for mammalian TBP *in vitro*.

Activators Bind to Enhancer Sequences and Can Increase Transcription 100-Fold Above the Basal Level

Although similar sets of basal factors bind to all the promoters of the tens of thousands of genes in the eukaryotic genome, a cell can express different genes in widely varying amounts of mRNA. They accomplish this enormous range of gene regulation through the binding of different transcription factors to different enhancer elements associated with different genes. When activator proteins bind to enhancer elements, they can interact directly or indirectly with basal factors at the promoter to cause an increase in transcriptional initiation (Fig. 16.5a). Researchers have already identified hundreds of eukaryotic activator proteins, and it is likely that each eukaryotic genome encodes several thousand of them.

To carry out their function, activator proteins must bind to enhancer DNA in a sequence-specific way; and after binding, they must be able to interact with other proteins to activate transcription. Two structural domains within the activator

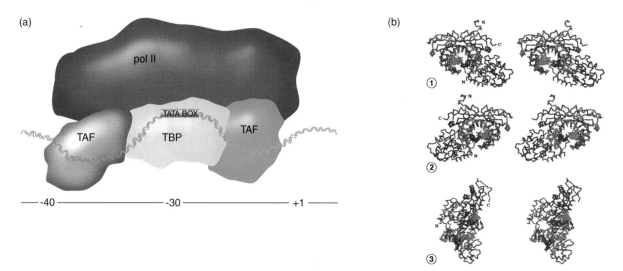

Figure 16.4 Basal factors bind to promoters of all protein-encoding genes. (a) Schematic representation of the binding of the various basal factors to the promoter DNA and the binding of RNA polymerase (pol II) to these basal factors. (b) Computer-generated image of the actual molecular structure of basal factors complexed to the promoter.

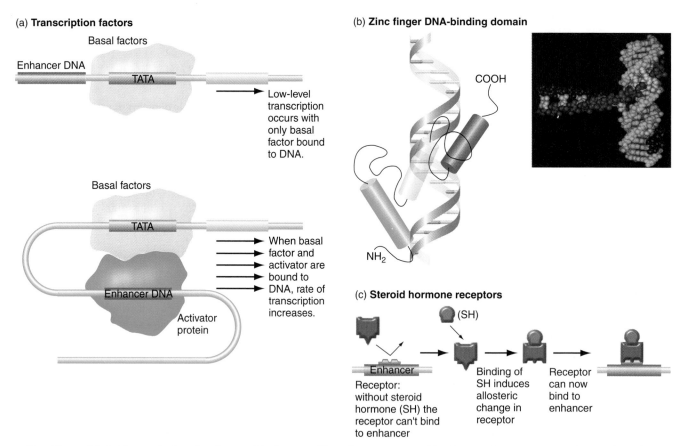

Figure 16.5 Transcriptional activators bind to specific enhancers at specific times to increase transcriptional levels. (a) In the presence of basal factors alone bound to the promoter, low levels of transcription occur. The binding of activator proteins to an enhancer element leads to an increase in transcription beyond the basal level. (b) Examples of peptide motifs that appear in the DNA-binding domains of activator proteins and their interaction with enhancer elements. A zinc-finger domain extends into the groove of the double helix. An asparagine-rich domain with positive charges interacts with the negative charges associated with phosphates in the DNA backbone. (c) Some activator domains are themselves activated into a DNA-binding conformation through allosteric changes caused by the binding of a steroid hormone molecule to another domain within the activator protein.

protein—the DNA-binding domain and the transcription-activator domain—mediate these two biochemical functions.

A small number of peptide motifs appear over and over again in the DNA-binding domains of many different transcription factors (Fig. 16.5b). The best characterized of these motifs are the zinc-finger, the helix-loop-helix, and the helix-turn-helix conformations. The general function of each of these motifs is to promote binding to the DNA double helix. Subtle differences in amino-acid sequence among activators can specify high-affinity binding to different DNA sequences associated with different enhancer elements. Extensive repetition of coding regions for the same peptide motif increases the genomic potential for evolving new activator proteins with new DNA-binding specificity; the duplication of an existing activator gene, followed by a slight divergence that by chance increases the binding affinity for a different DNA sequence, could generate a new activator protein that acts at a different target gene.

Many of the transcription-activation domains associated with different activators also have features in common. One subset is relatively acidic and rich in the amino acid asparagine; another is rich in prolines. Other transcription-activation domains are unique to the protein in which they occur.

Some activators have a third domain that is responsive to specific signals from the environment. An example of activators with this type of domain are the steroid hormone receptors (see Fig. 16.5c). Each receptor has a domain that is unique for a particular steroid. The binding of this steroid causes an allosteric change that greatly increases the affinity of the DNA-binding domain of the protein for its target enhancer sequence. Once bound, the hormone-receptor complex activates transcription of its target genes. In the absence of hormone, DNA binding does not occur, and target genes remain unactivated, that is, transcribed only at basal rates. There are no universal features of signal response domains.

Molecular biologists have identified and characterized the different domains associated with different activators through the use of recombinant DNA constructs that encode proteins with a well-defined domain from a well-characterized transcription factor fused to a series of polypeptide regions from the activator under analysis (Fig. 16.6). For example, the bacterial repressor lexA, which contains only a DNA-binding domain and no activator domain, binds to the same well-defined DNA sequence whether it is in its native prokaryotic cell or in an engineered eukaryotic cell. One can fuse different segments of an activator protein gene to *lexA* and produce this fusion protein in an engineered yeast cell in which the lexA-binding site has been placed upstream of a special reporter gene. Activation (or lack of activation) of the reporter gene can then signal that the segment from the activator under analysis does (or does not) have a complete activation domain.

Most Eukaryotic Activators Must Form Dimers to Function

Molecular analyses have shown that many eukaryotic transcription factors are **homomers** (that is, multimeric proteins composed of identical subunits) or **heteromers** (multimeric

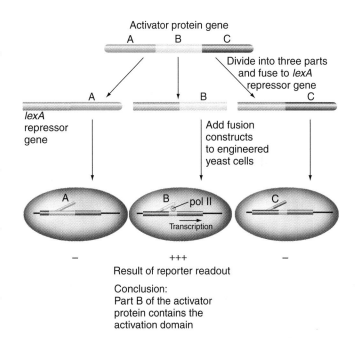

Figure 16.6 Localization of activator domains within activator proteins can be achieved with recombinant DNA constructs. The figure shows a strategy that can be used to determine which portion of an activator protein contains the actual domain that interacts with basal proteins to bring about activation. In this simplified example, the gene encoding the activator protein is divided into three parts, A, B, and C, and each part is fused to a portion of the bacterial *lexA* repressor gene that encodes a well-defined DNA-binding domain, and a yeast cell promoter. Each three-part fusion construct is placed into different yeast cells where the production of a fusion protein takes place. In this example, only part B of the original activator protein under analysis contains the activator domain and this allows the fusion protein to bind to a reporter gene enhancer element (through the *lexA* domain) and activate transcription of the reporter gene (through the part B element), which does not occur in the cells containing parts A or C of the activator protein.

proteins composed of nonidentical subunits; review Fig. 6.22). Among the best-characterized transcription factors of this type is Jun, which can form dimers (multimers composed of two subunits) with either itself or with another protein called Fos (Fig. 16.7a). The Jun-Jun dimers are *homodimers;* the Jun-Fos dimers are *heterodimers.*

Dimerization occurs through yet another transcription factor domain, the **dimerization domain,** which is specialized for specific polypeptide-to-polypeptide interactions. As with other transcription factor domains, certain motifs recur in dimerization domains. One of the most common is the leucine zipper motif (Fig. 16.7b), an amino-acid sequence that twirls into an α helix with leucine residues protruding at regular intervals. The motif received its name from the propensity of one leucine zipper to interlock like a zipper with a leucine zipper on another polypeptide. The ability of two leucine zippers to interlock depends on the specific amino acids that lie between the leucines.

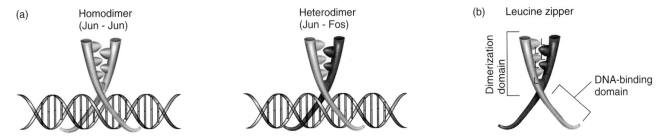

Figure 16.7 Most activator proteins function in the cell as dimers. (a) Homodimers contain two identical polypeptides, while heterodimers contain two different polypeptides. (b) The polypeptides within a dimer bind to each other through a dimerization domain. A common peptide motif present within dimerization domains is the leucine zipper, which is given this name because of the interlocking interaction between leucine residues present in the two polypeptides in a zipper-like fashion.

The Jun and Fos polypeptides both contain leucine zippers in their dimerization domains. A Jun leucine zipper can interact with another Jun leucine zipper or with a Fos leucine zipper. But the Fos leucine zipper *cannot* interact with its own kind to form a homodimer. Neither Jun nor Fos alone can bind DNA, so neither can act as a transcription factor as a monomer. Thus the Jun-Fos transcription factor system can produce only two types of transcription factors: Jun-Jun proteins or Jun-Fos proteins. Both bind to the same enhancer elements, but with different affinities.

The ability to form heterodimers greatly increases the number of potential transcription factors a cell can assemble from a set number of gene products. In theory, 100 polypeptides could combine in different ways to form 5000 different transcription factors; with 500 polypeptides, the number jumps to 125,000.

Repressors Diminish Transcriptional Activity

Some transcription factors suppress the activation of transcription caused by activator proteins. Any transcription factor that has this effect is considered a repressor. Different repressors act in different ways. Some compete with activator proteins for binding to the same enhancer (Fig. 16.8a). When a repressor binds to an enhancer, it blocks the activator's access to the same sequence. The Myc-Max system described in the following section provides an example of this type of activator-repressor competition.

Some repressors operate without binding DNA at all. Instead, in a mechanism called *quenching*, they bind directly to a specific activator (Fig. 16.8b). In one type of quenching, a repressor binds to and blocks the DNA-binding region of an activator, thereby preventing the activator from attaching to its enhancer. In another type of quenching, a repressor binds to and blocks the activation domain of an activator. These blocked activators still bind to their enhancers, but once bound, they are unable to carry out activation. An example of this type of repression is seen in the yeast GAL system, which we discuss later.

The repression resulting from both activator-repressor competition and quenching reduces activation, but has no effect on basal transcription. As in prokaryotes, however, some eukaryotic repressors act directly on the promoter to eliminate almost all transcriptional activity. They can do this by binding to DNA sequences very close to the promoter and thereby blocking RNA polymerase's access to the promoter. Or they

(a) Competition for binding between repressor and activator proteins

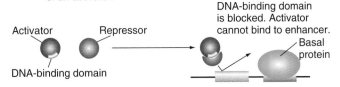

Binding of repressor to enhancer blocks binding of activator.

(b) Quenching

Type I: Repressor binds to and blocks the DNA-binding region of an activator.

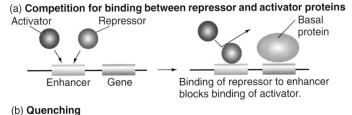

DNA-binding domain is blocked. Activator cannot bind to enhancer.

Type II: Repressor binds to and blocks the activation domain of an activator.

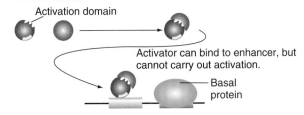

Activator can bind to enhancer, but cannot carry out activation.

Figure 16.8 Repressor proteins reduce transcriptional levels through competition or quenching. (a) Some repressor proteins act by *competing* for the same enhancer elements as activator proteins. But repressor proteins have no activation domain, so when they bind to enhancers, no activation of transcription can occur. (b) A second class of repressors act by binding directly to the activator proteins themselves to *quench* activation in one of two ways. Type I quenching is achieved when the repressor prevents the activator from reaching the enhancer. Type II quenching is achieved when the activator can bind to the enhancer but the repressor prevents the activation domain from binding to basal proteins.

can bind to DNA sequences farther from the promoter, then reach over and contact the basal factor complex at the promoter, causing the DNA between the enhancer and promoter to loop out and allow contact between the repressor and the basal factor complex. This second mechanism also denies RNA polymerase access to the promoter and reduces transcription below the basal level.

Whether a transcription factor acts as an activator or a repressor, or has no effect at all, depends not only on the cell type in which it is expressed, but also on the gene it is regulating. This is one reason why *cis*-acting elements that bind either activators or repressors are all referred to as enhancers, even though some may actually repress transcription when associated with the appropriate protein. As mentioned, all enhancers, whether they act to activate or repress transcription, are defined by their ability to function when placed at variable distances from a promoter and when reversed in orientation.

The specificity of transcription factors can be altered by other molecules in the cell. One example of this phenomenon is observed with the yeast α2 repressor, which helps determine the mating type of a cell. Yeast cells can be either haploid or diploid, and haploid cells come in two mating types: α and a. In α cells, the α2 repressor binds to enhancers that control the activity of a set of a-determining genes, whose expression would make the cell type a. The binding of the α2 repressor to these a-determining genes is one step in the generation of α cells (Fig. 16.9a). In diploid yeast cells, however, the same α2 repressor plays a different role. In such cells, expression of the polypeptide known as a1 occurs; the binding of a1 to the α2 repressor alters the repressor's DNA-binding specificity such that α2 now binds to enhancers associated with a set of haploid-specific genes, repressing the expression of those genes (Fig. 16.9b). To summarize, in diploid cells, the α2 repressor maintains the diploid state by repressing haploid-specific genes.

The Myc-Max System Is a Regulatory Mechanism for Switching between the Activation and Repression of Transcription

The ratio of heterodimer partners within a cell can have profound regulatory consequences. The Myc-Max transcription factor system, which evolved in multicellular eukaryotes, illustrates this point (Fig. 16.10). Through the identification of mutations affecting *myc* gene expression in one class of lymphocytes (found in a form of cancer called Burkitt's lymphoma), researchers showed that *myc* plays a critical role in the regulation of cell proliferation. However, although the genetic data suggested that the Myc protein is a transcription factor, biochemists could find no evidence for this function *in vitro*. Their experiments revealed that even though the Myc polypeptide contains both a helix-loop-helix (HLH) motif and a leucine zipper, it cannot bind to DNA or form homodimers. The apparent contradiction between genetic and biochemical results stymied the scientists who first associated mutations in the *myc* gene with Burkitt's lymphoma and other forms of cancer.

The discovery of the *max* gene product helped resolve this dilemma. Like Myc, Max contains a helix-loop-helix motif and a leucine zipper (Fig. 16.10a). Moreover, both Myc and Max contain another, more recently defined DNA-binding motif called a "basic motif" (because it contains mostly basic amino acids). Unlike Myc, however, the Max polypeptide can form homodimers. Further, when one mixes Max with Myc, heterodimers of the two polypeptides form.

In the Myc-Max system, the HLH and leucine zipper are both components of the dimerization domain (see Fig. 16.10a). The basic motif, by comparison, defines the DNA-binding domain, which functions only in dimeric proteins, never in monomers. This restriction of binding ability is a common feature of all transcription factor systems based on dimerization.

The Myc polypeptide contains an activation domain, but when the molecule is on its own, it cannot bind DNA and thus cannot serve as an activator. The Max polypeptide, on the other hand, can form homodimers and can bind DNA when present without Myc; but Max has no activation domain, so it cannot function as an activator even when it does bind to DNA. It is only when Myc and Max come together in a heterodimer that both DNA binding and Myc-directed activation become possible, and a transcriptional activator is born (Fig. 16.10b).

Myc-Max heterodimers and Max-Max homodimers both bind to the same enhancer sequences associated with genes that contribute to cell proliferation. The binding of the heterodimers results in transcriptional activation, whereas the binding of the homodimers results in transcriptional repression.

One final characteristic of this system is that Myc polypeptides have a much higher affinity for Max polypeptides than Max has for its own polypeptides. Thus, when Myc and Max are in solution together, the predominant dimer is a heterodimer. With this extensive background, we are ready to see how the cell uses the Myc-Max system to respond rapidly to signals that tell it to proliferate or stop proliferating.

The *max* gene is expressed in all cells at all times, but since its protein product does not carry an activation domain,

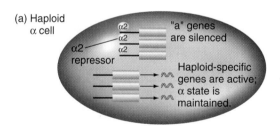

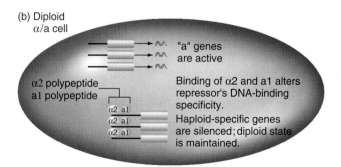

Figure 16.9 The same transcription factors can play different roles in different cells. (a) In haploid α yeast cells, the α2 factor acts to silence the set of "a" genes. (b) In α/a diploid yeast cells, the α2 factor dimerizes with the a1 factor and acts to silence the set of haploid-specific genes.

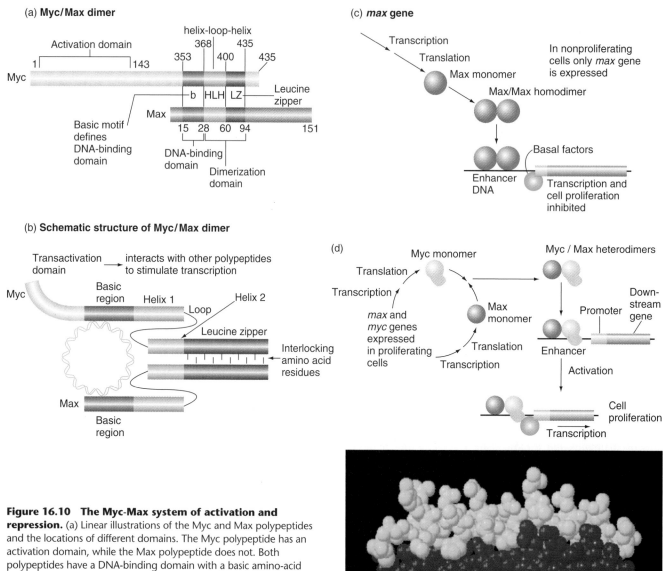

Figure 16.10 The Myc-Max system of activation and repression. (a) Linear illustrations of the Myc and Max polypeptides and the locations of different domains. The Myc polypeptide has an activation domain, while the Max polypeptide does not. Both polypeptides have a DNA-binding domain with a basic amino-acid motif, and a dimerization domain with adjacent helix-loop-helix and leucine zipper motifs. (b) Schematic diagram of the Myc-Max heterodimer binding to DNA. (c) Gene repression results when only the Max polypeptide is made in a cell. (d) Gene activation occurs when both Myc and Max are made in a cell.

Max-Max homodimers, when bound to enhancer DNA, inhibit transcription and therefore inhibit cell proliferation (Fig. 16.10c). By contrast, the *myc* gene is not universally expressed; rather the Myc polypeptide is normally synthesized in cells undergoing proliferation but not in cells at rest.

As soon as a cell expresses its *myc* gene, all its Max-Max homodimers convert to Myc-Max heterodimers that bind to the enhancers previously bound by Max-Max homodimers. Since the heterodimers include the Myc activation domain, the binding of Myc-Max complexes induces the expression of genes required for cell proliferation. Although researchers have not yet characterized all the genes activated by the Myc-Max dimer, they know that the genes guide the cell through its mitotic cycle. Thus, each cell in which *myc* is active divides to produce two daughter cells.

Interestingly, when cells reach a state of terminal differentiation, they stop expressing *myc,* and since the Myc protein is unstable and easily degraded, any Myc remaining from earlier *myc* expression quickly disappears. The loss of Myc allows the remaining Max polypeptides to form homodimers, whose binding to the enhancer once again represses the expression of genes required for cell proliferation. When this happens, terminally differentiated cells stop dividing.

In short, the Myc-Max system provides a rapid genetic switch for regulating cell division during the cell proliferation and terminal differentiation phases of development. The system functions by keeping the *max* gene turned on in all cells, carefully regulating the expression of *myc,* allowing homodimerization of Max but not Myc, allowing Myc-Max heterodimerization to outcompete Max-Max homodimerization, and allowing the rapid degradation of Myc. As a result, the expression of *myc* quickly turns on genes required for cellular proliferation and the cessation of *myc* expression just as quickly turns those same genes off. Mutations in the *myc* regulatory region can lead to constitutive expression of this critical gene, causing abnormal cellular proliferation and cancer.

The Yeast GAL System Is Another Complex Regulatory Mechanism

Like the Myc-Max system in multicellular eukaryotic organisms, the GAL system in unicellular yeast has been well characterized genetically and biochemically. The biochemical function of the GAL system is to modulate the expression of three linked genes—*GAL1*, *GAL7*, and *GAL10*—that encode the enzymes galactokinase, transferase, and epimerase (Fig. 16.11a). A cell must produce all three enzymes to use galactose

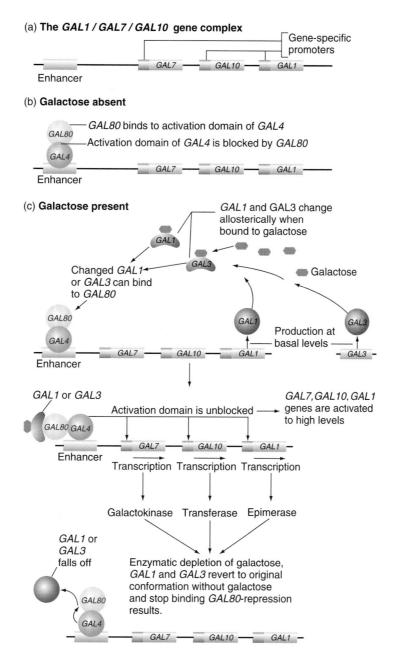

Figure 16.11 The *GAL1/GAL7/GAL10* gene system is a model of complex transcriptional regulation. (a) The three genes have independent promoters but are all controlled by the same enhancer element. (b) The activator protein GAL4 binds to this enhancer, but its activity is quenched by the repressor protein GAL80. (c) Galactose molecules in the cell bind to GAL1 and GAL3 polypeptides, which undergo allosteric changes that allow them to bind to GAL80 causing GAL80 to move into a position that is no longer quenching the activator GAL4. This allows activation of the *GAL1*, *GAL7*, and *GAL10* genes whose products bring about the enzymatic digestion of galactose. Once galactose is depleted, GAL1 and GAL3 can no longer bind to GAL80, and it moves back into position to quench GAL4.

as a source of energy. When galactose is present, the cell activates the genes encoding these enzymes. When galactose is absent, repression of the three genes allows the cell to avoid the unnecessary expenditure of energy and materials.

Since yeast cells always need the three enzyme at the same time, they economize by controlling the expression of all three structural genes from a single enhancer. Consequently, mutations in this enhancer can be readily distinguished by their effect on the regulation of all three gene products rather than just one.

Researchers identified the *GAL4* locus as a *trans*-acting genetic regulator of *GAL1, GAL7,* and *GAL10* through mutations that prevent galactose from activating expression of the three structural genes. Biochemical studies showed that the *GAL4* product has a DNA-binding domain containing multiple zinc-finger motifs and an acidic activation domain; it binds to an enhancer several hundreds of base pairs upstream of the *GAL1, GAL7, GAL10* genes and activates their transcription. Mutations in the *GAL4* coding region prevent its product from functioning as an activator.

Researchers identified another *trans*-acting genetic element—*GAL80*—through mutations that caused continuous expression of the three GAL enzymes, even in the absence of galactose. This genetic result suggested that the product of the *GAL80* gene is a repressor of *GAL1, GAL7,* and *GAL10*. When this repressor is absent, constitutive expression could occur.

On the basis of further genetic and biochemical studies, the researchers proposed that the GAL80 product binds to the activation domain of *GAL4* (Fig. 16.11b). GAL80 would thus be a repressor that does not bind directly to DNA; instead it quenches the activating ability of the GAL4 product by sequestering its activation domain. To test this model, they constructed double mutants in which they had knocked out both *GAL4* and *GAL80*. In these *GAL4⁻/GAL80⁻* cells, no induction of GAL1, GAL7, and GAL10 occurs. This means that, in genetic terms, the GAL4 phenotype is epistatic to the GAL80 phenotype. Such a result is consistent with the biochemical model because to function as a repressor, the GAL80 product needs to operate on the GAL4 product. In the absence of *GAL4*, mutations in *GAL80* will have no effect on gene regulation.

How does a normal yeast cell use this genetic regulatory system to control the expression of the *GAL1, GAL7,* and *GAL10* structural genes in response to the presence or absence of galactose? The answer is by controlling the physical placement of the GAL80 gene product. When galactose is present in the cell, it attaches to both the GAL1 product (a very small amount of which is produced through basal transcription), as well as to another polypeptide called GAL3. Association with galactose causes an allosteric change in GAL1 and GAL3 that gives both polypeptides a high-affinity binding site for the GAL80 product. When the two polypeptides bind to the GAL80 product, this binding somehow dislodges GAL80 from the GAL4 activation domain. With its activation domain now available, GAL4 is able to activate the *GAL1, GAL7, GAL10* gene complex (Fig. 16.11c). Later, when the cell has

depleted its supply of galactose, the GAL1 and GAL3 products again change shape, reverting to the conformation in which they do not have a high affinity for GAL80 and falling off. GAL80, when no longer bound to GAL1 or GAL3, moves back into position to block the GAL4 activation domain; and this binding of GAL80 to the activation domain of GAL4 once again reduces transcription of the three GAL structural genes to basal levels.

A Locus Control Region Is a cis-Acting Regulatory Sequence That Operates Sequentially on a Cluster of Related Genes

Promoters and enhancers are not the only kinds of *cis*-acting regulatory regions active in eukaryotic genomes. Clues that another kind of *cis*-acting element exists came from experiments in which the promoter and adjacent enhancer region of the β-*globin* gene were not sufficient to mediate its full transcriptional activation.

As we saw in Chapter 8 (introduction and comprehensive example), the β-*globin* gene is part of a cluster of related genes that function at different times during mammalian development to produce polypeptides that become incorporated into hemoglobin molecules with different kinetics of oxygen binding (Fig. 16.12a). The β-*globin* gene was among the first mammalian genes to be cloned, characterized, and used in the production of transgenic mice. Results from the earliest transgenic studies were disappointing, for even if researchers introduced the entire region of noncoding genomic DNA 5′ of the β-*globin* gene (and separating the gene from the flanking gene in the cluster) into transgenic mice, the level of β-*globin* transgene transcription in the mice was 10 times less than that observed with the endogenous gene.

Subsequent experiments showing that other transgenes with just the 5′-flanking sequences (containing the promoter and all of the enhancer elements) were fully expressed suggested that the β-*globin* gene was different; for full transcription, it required a *cis*-acting element that was missing from the DNA containing both the gene and the flanking sequence that separates it from its nearest neighbor (Fig. 16.12b).

Surprisingly, investigators found that the missing element, located 50 kb upstream of the β-*globin* gene, is separated from the β-*globin* gene by four other genes in the cluster. This regulatory element is called the **Locus Control Region (LCR)** because it exerts a higher level of control, influencing the transcription of all genes in the β-*globin* cluster. Like an enhancer, the LCR functions by binding to transcription factors with activation domains. Unlike most enhancers, however, the LCR-transcription factor complex interacts sequentially during development with other transcription factors at *cis*-regulatory regions that are directly adjacent to each gene within the β-*globin* gene cluster (Fig. 16.12c). An assembled LCR-transcription factor complex is known as an **enhancesome.** If any component of the enhancesome is missing, the program doesn't work, and no activation of any genes in the cluster take places. Cells other than the precursors of red

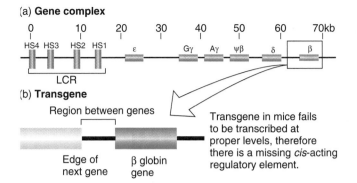

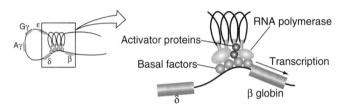

(c) Mechanism of transcriptional activation

Figure 16.12   A Locus Control Region (LCR) operates over a cluster of genes to activate transcription. (a) The human β-globin gene cluster contains five functional genes that can all undergo cis-regulation by a distant LCR. The LCR associated with this particular gene cluster contains four separate cis-acting DNA elements designated HS1 through HS4. (b) A transgene composed of an isolated β-globin gene together with the entire 5′ flanking region between it and the δ gene fails to undergo proper expression when it is inserted into the mouse genome. (c) The LCR at the 5′ end of the β-globin gene cluster is required for transcriptional activation of each of the genes in the cluster in hemoglobin-producing cells at different points in development. The four DNA elements of the LCR (HS1 through HS4) could effect activation of the distant β-globin gene through DNA looping that brings each of these elements and their associated transcription factors into contact with the β-globin promoter and its associated basal factors and RNA polymerase.

blood cells do not express the genes for enhancesome components; as a result, there is no functional enhancesome and no expression of any of the β-globin genes in these cells. The LCR for the β-globin cluster thus provides a single mechanism for ensuring that none of the genes in the cluster is expressed in cells outside the red blood cell lineage.

Complex Regulatory Regions Enable an Organism to Fine-Tune Gene Expression

In complex multicellular organisms, gene regulation is not just a matter of turning genes on and off. It also entails fine tuning the precise level of transcription—higher or lower in different cells, higher or lower in cells of the same tissue at different stages of development. It also includes mechanisms that allow each cell to modify its program of gene activity in response to constantly changing signals from its neighbors. Organisms accomplish the orchestration of transcription from each of tens of thousands of genes through cis-acting regulatory regions that are often far more complex than those we have so far described.

A regulatory region may contain a dozen or more enhancer elements, each with the ability to bind different activators and repressors, with varying affinities. At any moment, there may be dozens of transcription factors in the cell whose affinities for DNA or other polypeptides are being modulated by binding to hormones or other molecules. Different sets of these transcription factors compete for different enhancers within the regulatory region. The biochemical integration of all this information yields a precise level of transcriptional activation or repression.

As we saw earlier, a complex of regulatory factors together with the enhancer to which they are bound is an enhancesome. Slight changes in a cell's environment can dispatch signal molecules that cause changes in the balance of transcription factors, or in their relative affinities for DNA or for each other. These changes, in turn, lead to the assembly of an altered enhancesome, which recalibrates gene activity.

Molecular biologists and biochemists are not yet able to unravel the details of such complex regulatory networks, but their knowledge increases every day.

Chromatin Structure Plays a Role in Eukaryotic Gene Regulation

We know from Chapter 11 that the DNA of eukaryotic genomes does not float freely in the nucleus, but is normally packaged into chromatin (review Figs. 11.4 and 11.5 and Table 11.2). The basic repeating structural unit of chromatin is the nucleosome, which consists of a ball of histone proteins (two each of H2A, H2B, H3, and H4) around which is wrapped approximately 200 bp of DNA. Histone H1, which binds to short stretches of DNA that lie between nucleosomes, helps maintain chromatin structure. The binding of nonhistone proteins is the basis of higher order structures that further compact DNA.

The Normal Structure of Chromatin Provides a Brake on Runaway Basal Transcription

In vitro experiments show that basal factors and RNA polymerase readily bind to promoters on naked DNA and initiate high levels of transcription in the absence of activator proteins (Fig. 16.13a). One significant function of chromatin is the reduction of basal transcription from all genes to a very low level. In contrast to transcriptional modulation in prokaryotes, which requires active repression through the binding of repressors to cis-acting elements, in eukaryotes, the normal structure of chromatin is sufficient in and of itself to maintain transcriptional activity at the minimal, basal level (Fig. 16.13b). Thus, eukaryotic repressors play a different role than their prokaryotic counterparts. The main role of eukaryotic repressors is to modulate the activation of transcription caused by transcriptional activators. They do this through competition for DNA-binding sites, as described earlier for the Myc-Max system, or through quenching, as described for the yeast GAL system.

(a) Naked promoter binds RNA polymerase and basal factors.

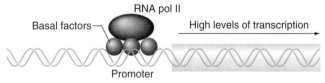

(b) Chromatin reduces binding to basal factors and RNA pol II to very low levels.

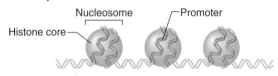

(c) Chromatin remodeling can expose promoter.

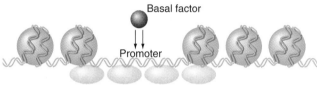

(d) DNase hypersensitive sites are at 5' ends of genes.

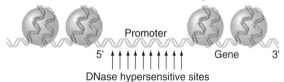

(e) The SWI-SNF multisubunit complex is an example of a remodeling complex.

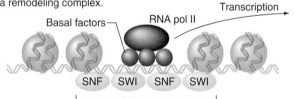

The SWI-SNF multisubunit destabilizes chromatin structure and gives transcription machinery access to the promoter.

Figure 16.13   Chromatin structure plays a critical role in eukaryotic gene regulation. (a) DNA molecules containing a promoter and an associated gene can be purified away from chromatin proteins *in vitro.* When basal factors and RNA polymerase are added to this purified DNA, high levels of transcription can be induced. (b) In its natural state within the eukaryotic nucleus, DNA is present within chromatin. The underlying structure of chromatin is based on repeating nucleosomes that each contain two loops of DNA wrapped around a histone core. Promoter regions are generally sequestered within the nucleosome and only rarely bind to basal factors and RNA polymerase. Thus, the chromatin structure maintains basal transcription at very low levels. (c) Chromatin remodeling can expose the promoter region. Remodeling proteins can cause specific nucleosomes to unravel in specific cells at specific times during differentiation or development. Exposed promoter regions can more readily bind basal factors. (d) Exposed regions of DNA sequence can be recognized by their hypersensitivity to DNase digestion. (e) The SWI-SNF protein complex is an example of a well-characterized remodeling apparatus that functions within yeast cells to expose promoter regions to basal factors, RNA polymerase, and transcriptional activation.

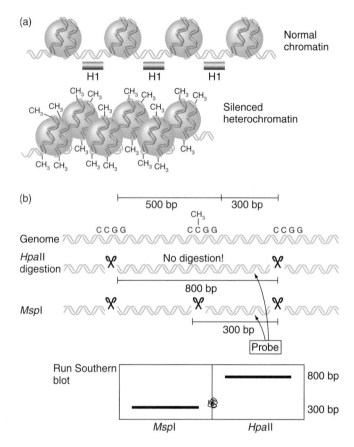

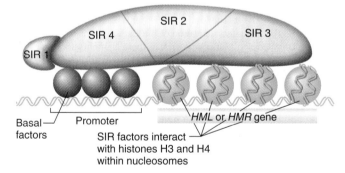

(c) SIR complex binds to basal factors and interacts with H3 and H4 components of histones

Figure 16.14   Heterochromatin formation can lead to transcriptional silencing. (a) Normal chromatin in standard nucleosome conformation can be converted into tightly packed heterochromatin with the addition of methyl groups to a series of cytosine bases within a local DNA region. (b) A determination of the methylation status of a DNA region can be made using a pair of restriction enzymes that both recognize the same base sequence, with one being able to digest methylated DNA while the other can't. In this example, the restriction site is C C G G, and the enzymes are *Msp*I, which can digest both methylated and unmethylated sites, and *Hpa*II, which cannot digest methylated sites. If a methylated site is present between two unmethylated sites, *Hpa*II digestion will leave a larger fragment than *Msp*I. After electrophoresis and Southern blot analysis with a probe that hybridizes to a sequence on one side of the methylated site, there will be a clearly observable difference in band size. (c) The SIR complex of polypeptides can bind to basal factors associated with the promoters of the *HML* and *HMR* genes. This binding, in turn, causes the SIR complex to interact with the histones H3 and H4 present in downstream nucleosomes associated with the gene itself. The result is complete transcriptional silencing.

The Remodeling of Chromatin Mediates the Activation of Transcription

An initial component of gene activation in eukaryotes is the remodeling of chromatin structure in the promoter region. Specialized proteins that unravel promoter DNA sequences away from the histone core are the agents of this remodeling. The freed DNA becomes much more accessible to transcription factors that then control transcription (Fig. 16.13c).

Chromosomal regions from which the nucleosomes have been eliminated are experimentally recognizable through their hypersensitivity to the enzyme DNase (as described in Chapter 11). When one scans a chromosome with the enzyme for the presence of DNase hypersensitive (DH) sites, the sites show up at the 5′ ends of genes that are either undergoing transcription or are being prepared for transcription in a later step of cellular differentiation (Fig. 16.13d). For example, DH sites appear at the 5′ end of the β-*globin* gene in human cells that are precursors to those in which the gene will be activated, but not in cells from other differentiative pathways.

Studies of chromatin structure show that nucleosomes sit atop the promoters of most inactive genes. Protein complexes that remove these promoter-blocking nucleosomes, or reposition them in relation to the gene, help put a gene in an activatable state. For example, the SWI and SNF proteins in yeast form a multisubunit complex that disrupts chromatin structure by removing or repositioning nucleosomes. The resulting chromatin decompaction gives basal factors much greater access to promoter regions, and consequently, transcription rapidly accelerates (Fig. 16.13e). Human cells contain related multisubunit protein complexes that also influence nucleosome position or structure, suggesting that this particular nucleosome-disrupting machinery has been conserved throughout evolution. The SWI-SNF protein complex represents just one of the many that help remodel chromatin at specific chromosomal locations in specific cells at particular points of development.

Hypercondensation over Chromatin Domains Causes Transcriptional Silencing

The heterochromatic regions of eukaryotic chromosomes, including parts of centromeres and telomeres and the whole of Barr bodies, are highly condensed. As a result, the genes contained in heterochromatin are completely silent without even basal transcription (review Figs. 11.16 through 11.20).

The **transcriptional silencing** of heterochromatin makes it impossible to activate genes within a heterochromatic region, no matter what transcription factors are active in the cell, as long as the silencing structure is in place. By contrast, as we have seen, substantial expression of genes in unsilenced regions depends on the outcome of competition between activators and repressors; the appropriate combination of signals can always activate these genes.

In mammals, the chromosomal regions that are silenced by heterochromatin vary with the cell type, but they are often associated with DNA methylation (Fig. 16.14a). Methylation—the addition of a methyl (CH$_3$) group—is specific for CG

dinucleotides in the DNA and occurs at the 5′ position of the cytosine. It is possible to determine the state of methylation of a DNA region by using two restriction enzymes that both cleave at a sequence containing a CG dinucleotide but have different sensitivities to the methylation of the DNA substrate. For example, *Hpa*II and *Msp*I both cleave at CCGG, but *Hpa*II will not cleave if the middle C of this site is methylated, while *Msp*I will (Fig. 16.14b). Thus, by digesting genomic DNA with *Hpa*II and *Msp*I and using a specific DNA probe on a Southern blot, you can determine whether a given CCGG sequence is methylated. Although methylation is associated with silencing, we know it cannot be the only mechanism because some organisms that show silencing, such as yeast, do not contain methylated DNA.

Deeper insight into the mechanism of silencing comes from studies of mutations that give rise to sterility in yeast. Recall that yeast cells come in two mating types—α and a—and the α2 gene product represses certain a-determining genes. The chromosomal locus of the α2 gene is known as MAT (for mating type). In normally mating yeast cells, there are two additional copies of the MAT locus called HML and HMR; located near the telomeres on each arm of chromosome III, these loci are transcriptionally silent. Mutations that reduce or destroy silencing at these loci cause sterility because they allow the simultaneous expression of α and a information, and the resulting cells, which behave as diploids, do not mate. Analysis of these mutations identified the family of *SIR* genes. The SIR polypeptide products associate to form a *trans*-acting complex that mediates silencing by acting at *cis*-acting sites near HML and HMR. Null mutations that eliminate the activity of any *SIR* gene or mutations that delete a *cis*-acting site abolish silencing. The SIR complex binds to other polypeptides and these larger complexes interact with histones H3 and H4 (Fig. 16.14c). These interactions with the histones establish a silenced chromosomal domain that remains hidden from the activators and repressors of transcription.

Genomic Imprinting Results from Chromosomal Events That Selectively Silence the Expression of Genes Inherited from One Parent or the Other

We saw in the Genetics and Society box in Chapter 11 that attempts to create viable mice in which the entire diploid content of the chromosomes came from a single parent did not succeed. In more precise genetic experiments, mice with both copies of a particular chromosome from a single parent often did not survive.

An understanding of the mechanism behind these failures came from studies of the transmission of a deletion in the chromosome 7 insulin-like growth factor gene (*Igf2*) in mice. Mice inheriting the deletion from the paternal side were small, while mice inheriting the same deletion from the maternal side were normal size (Fig. 16.15a). Recall that the phenomenon in which a gene's pattern of expression depends on the parent that transmits it is *genomic imprinting*.

The simplest explanation of these results is based on a model in which the *Igf2* gene copy inherited from the mother

A Promising Medical Tool: Synthetic Oligonucleotides That Selectively Reduce the Expression of Specific Gene Products

Oligonucleotides, often referred to as *oligos,* are short segments of DNA chemically synthesized to precise specifications. Most oligos are 10–30 nucleotides in length. Hybridization of an oligo to mRNA via complementary base pairing creates a DNA-mRNA duplex that can inhibit gene product formation by blocking translation (Fig. A). The DNA-mRNA duplex can also serve as a substrate for the cellular enzyme ribonuclease H, which destroys the integrity of the mRNA. In this duplex, the mRNA is often referred to as the "sense strand" because it contains information specifying the order of amino acids in a polypeptide; and the complementary sequence in the oligo is considered an "antisense sequence." The use of oligos to diminish the level of noxious gene products is known as "antisense therapy."

Clinical trials using antisense oligos as antiviral and anticancer agents have shown that antisense therapies may be able to help control both types of disease.

Oligos as Treatments for Serious Viral Infections

A key to the therapeutic use of oligos is their specificity. In theory, a well-designed antisense oligo complementary to one of a virus's

mRNAs would not bind to RNAs transcribed from the host cell genome. By binding to the viral RNA as it enters the host cell, such an antisense drug might help protect humans against infection by RNA viruses like the HIV retrovirus. Formation of the DNA-RNA duplex would block the reverse transcription on which the virus depends for creation of the proviral genome. Later in the viral cycle, after the viral genome has integrated into the host genome and then become activated for transcription, antisense oligos to viral pre-mRNAs could keep the mRNAs from leaving the nucleus for translation in the cytoplasm. Still later, antisense oligos to mature mRNAs could block translation in the cytoplasm. Preliminary clinical trials of antisense drugs against the AIDS virus are in the process of evaluating an oligo against the RNA product of the HIV *gag* gene, whose expression is required for viral replication.

Oligos as Cancer Therapies

The specificity and resultant selectivity of oligos also makes them good candidates for cancer therapy. In addition to their selectivity (which could enable them to generate fewer side effects than

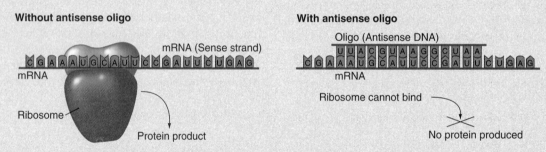

Figure A   Antisense oligonucleotides are designed to hybridize around the start codon (AUG) on the targeted mRNA.
Hybridization between mRNA and antisense forms a double helix that blocks association of ribosomes with the mRNA and thus prevents translation.

is normally silenced. Thus a deletion inherited from the mother produces no phenotypic effect because the maternal allele is not expressed anyway. If the deletion comes from the father, however, it produces a phenotypic effect because the animal is now unable to make any IGF2 products. As we noted in the box in Chapter 11, imprinting violates the tenet of Mendelian inheritance that says the parental origin of a given allele has no effect on the phenotype of the progeny.

Whatever genomic changes cause imprinting must be erased and reset during the process of gametogenesis (Fig. 16.15b). For example, an unexpressed *Igf2* gene inherited by a son from his mother must be reactivated in spermatogenesis, while an expressed gene inherited from the paternal side must be deactivated in oogenesis.

Insight into the mechanism of *Igf2* imprinting came from the surprising finding that a gene called *H19,* found just 70 kb downstream of *Igf2,* is also imprinted, but in the opposite way. With *H19,* it is the copy inherited from the father that is silenced and the copy inherited from the mother that is active in normal mice.

A model of how imprinting works at both *H19* and *Igf2* proposes that the two genes share an enhancer that is located between them. When the enhancer and its associated transcription factors are involved in the activation of one of the two genes, they are unavailable to activate the other gene (Fig. 16.15c). Because of proximity or affinity, the enhancer "prefers" interaction with the *H19* promoter over the *Igf2* promoter. But when the *H19* promoter is unavailable, the en-

chemotherapy), they are orders of magnitude smaller than the vesicles used to deliver some anticancer toxins to specific cellular addresses; as a result, they can penetrate capillary walls and the molecular meshes of intercellular space to reach tumor cells that may be otherwise inaccessible.

In theory, antisense oligos could fight cancer at several levels. They could suppress the expression of oncogenes—dominant mutant alleles that contribute to a cell's progression to malignancy (see Chapter 17). They could induce the redifferentiation of tumor cells by turning off expression of some of the transcription factors that allow proliferation, thus pushing the cell to terminal differentiation. They could suppress expression of protein factors that induce the growth of blood vessels into solid tumors (without these new vessels, the tumors cannot grow larger than 1 mm in diameter). And they could reverse the immunosuppression characteristic of many cancers by blocking production of the molecules that tumor cells secrete to shield themselves from destruction by the host's immune system.

Trials are underway in several laboratories to assess whether antisense oligos can prevent the progression of various cancers. In one study, clinicians are evaluating an antisense drug to see if it can eliminate cancer cells that have been removed from the bone marrow of leukemia patients. The target mRNA is from the c-*myb* gene, which promotes the proliferation of blood cells. Researchers think that abnormal regulation of the gene may contribute to several human leukemias.

Hurdles Remain in the Development of Oligo Drugs

While the preliminary laboratory tests on model animals and on small numbers of human patients have been very promising, antisense oligo developers are still grappling with crucial molecular and clinical considerations. In the molecular arena, some oligos work much better than others. Investigators trying to understand the basis for these differences are learning that oligos complementary to some points on an mRNA (usually those closer to the start-of-translation site) are more effective. The researchers are also trying to

determine the most appropriate length for an antisense oligo—short enough to retain its therapeutic properties, but long enough to avoid chance complementarity to points on untargeted mRNAs. In one study, for example, an oligo consisting of nine nucleotides whose target was a mutated *ras* gene mRNA bound to several hundred other completely complementary sites in a variety of mRNAs in the cell. Such an oligo would not be a useful therapeutic agent.

Once their molecular quirks are worked out, oligos must pass clinical muster, meeting the two basic criteria of safety and effectiveness. Preliminary trials have given them high marks in both areas, but some critical concerns remain. A few trials have shown inflammation of the injection site and toxicity of the liver, kidney, and blood in some animals at doses only slightly above therapeutic doses. For example, in one trial, monkeys suffered adverse cardiovascular effects after rapid infusion of 20-26-nucleotide oligos of varying sequences (the higher the dose, the worse the effects). Moreover, researchers do not yet know whether oligo breakdown products can produce adverse effects.

Thus, the clinically determined goals in the development of therapeutic oligos are identification of a safety threshold below which few side effects occur, and the production of oligos that are strong enough and stable enough to have a therapeutic effect at doses far below this threshold. A corollary is the identification of nonproblematic methods of delivery that make the drugs available to the organism at safe doses.

In summary, molecular and clinical studies suggest that antisense oligos able to attenuate the expression of specific gene products hold promise as short-term therapies for life-threatening conditions such as AIDS and cancer. If, with minor adjustments, they pass further clinical tests, they could become notable additions to the array of antiviral and anticancer drugs already available; doctors would most likely prescribe them in combination with one or more of these other drugs in an effort to halt the progression of serious disease. However, since oligos as currently synthesized are costly to produce and seem to have some cumulative side effects, they are probably not yet suitable for the treatment of less serious diseases or chronic conditions that require continuous, long-term therapy.

hancer interacts with the *Igf2* promoter to activate transcription of that gene.

According to the model, the molecular mechanism behind this either-or activation of *H19* and *Igf2* is as follows. Methylation at the paternally inherited *H19* promoter imprints this gene, leaving the paternally inherited copy of the enhancer free to activate transcription of the *Igf2* gene (see Fig. 16.15c). By contrast, there is no methylation of the maternal promoter of *H19*, which means the shared enhancer inherited from the mother activates the maternal *H19* gene and is unavailable to activate *Igf2*. Thus, the observed maternal imprinting of *Igf2* is an indirect consequence of competition for an enhancer shared with the adjacent *H19* gene.

Evidence supporting this model came from an analysis of gene expression in mouse embryos in which researchers have knocked out the gene for methylase, an enzyme that adds methyl groups to DNA. In these embryos, which die before birth, methylation of the paternally inherited *H19* gene cannot occur, and transcription of this copy of *H19* proceeds the same as transcription of the maternal copy; but neither copy of *Igf2* is transcribed. The model predicts these results since the shared *Igf2-H19* enhancer on both the maternal and the paternal chromosomes is engaged in activating *H19* and thus unable to activate *Igf2*.

The role of genomic imprinting is still under debate. Some scientists believe it is one of the many specialized tools

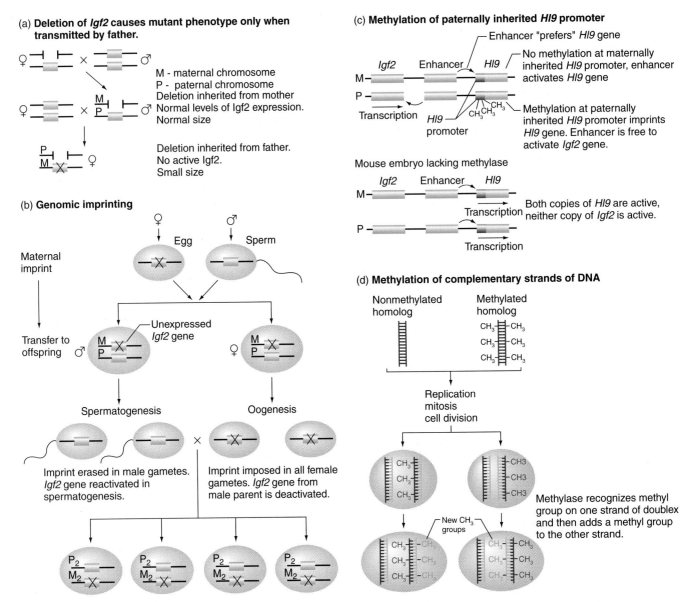

(a) Deletion of *Igf2* causes mutant phenotype only when transmitted by father.

M - maternal chromosome
P - paternal chromosome
Deletion inherited from mother
Normal levels of Igf2 expression.
Normal size

Deletion inherited from father.
No active Igf2.
Small size

(b) Genomic imprinting

Maternal imprint

Transfer to offspring

Egg   Sperm

—Unexpressed *Igf2* gene

Spermatogenesis   Oogenesis

Imprint erased in male gametes. *Igf2* gene reactivated in spermatogenesis.

Imprint imposed in all female gametes. *Igf2* gene from male parent is deactivated.

(c) Methylation of paternally inherited *HI9* promoter

Enhancer "prefers" *HI9* gene

Igf2   Enhancer   *HI9*

No methylation at maternally inherited *HI9* promoter, enhancer activates *HI9* gene

Transcription   *HI9* promoter

Methylation at paternally inherited *HI9* promoter imprints *HI9* gene. Enhancer is free to activate *Igf2* gene.

Mouse embryo lacking methylase

Igf2   Enhancer   *HI9*

Transcription   Both copies of *HI9* are active, neither copy of *Igf2* is active.

Transcription

(d) Methylation of complementary strands of DNA

Nonmethylated homolog   Methylated homolog

Replication mitosis cell division

New CH₃ groups

Methylase recognizes methyl group on one strand of doublex and then adds a methyl group to the other strand.

Figure 16.15 Genomic imprinting at the *Igf2* locus in the mouse. (a) The phenotypic effect of an *Igf2* deletion is determined by the parent transmitting the mutant locus. This parent-of-origin effect can be demonstrated in the two-generation cross illustrated here. In the first generation, a female transmits the deletion to her offspring. Although they carry only a single copy of the *Igf2* gene, all first-generation offspring express normal levels of the protein product and show no phenotypic abnormalities. But when a male of this generation transmits the deletion to second-generation offspring, they all fail to express any IGF2 product and they exhibit a small growth phenotype. (b) These results can be explained with females placing some kind of imprinting signal on all copies of their *Igf2* gene during oogenesis, and with males removing imprinting signals from all of their copies of the *Igf2* gene during spermatogenesis. Thus, all normal mice receive an imprinted and silenced copy of the gene from their mothers and a nonimprinted and functional copy of the gene from their fathers. (c) Reciprocal parent-of-origin expression occurs with the *Igf2–H19* gene pair. A single enhancer is shared by these two closely linked genes. When the *H19* promoter is available, the shared enhancer prefers to interact with it, and as a result, it is not available to interact with the *Igf2* promoter and no IGF2 product is produced. But, the *H19* promoter is methylated during spermatogenesis, and as a result, the paternal copy of *H19* is not available to the enhancer and is not expressed. On this chromosome, the enhancer is available to activate *Igf2*. In mouse embryos unable to produce methylase, the imprint of the paternal *H19* locus is lost, and the enhancer reverts back to activating this gene instead of *Igf2*. (d) An epigenetic state of DNA methylation can be maintained across cell generations. This is accomplished by the activity of DNA methylases that recognize methyl groups on one strand of a double helix and respond by methylating the opposite strand.

that cells use for gene regulation. Other scientists believe that genomic imprinting is not fundamentally a mechanism of gene regulation, but instead, an evolutionary response to competition between the sexes (as described in Chapter 11). In support of this evolutionary view, they point out that genomic imprinting is not universal: It is found only among mammals. Irrespective of the reason for its original emergence, genomic imprinting now plays a role in gene regulation, for a tiny number of genes in a small number of eukaryotic species.

Genomic imprinting is an example of an epigenetic phenomenon: a variant condition that does not involve a change in DNA sequence that is nevertheless inherited from one somatic cell generation to the next during the growth of the organism. Other examples of epigenetic phenomena described earlier include X chromosome inactivation and position effect variegation. All three of these epigenetic conditions arise from gene or chromosomal inactivation events that occur early in development or in each gamete before fertilization. Methylation seems to play an important role in maintaining these epigenetic conditions. Precisely methylated states can be passed from one cell generation to another by methylases, enzymes that recognize methyl groups on one strand of a DNA double helix and respond by methylating the newly synthesized complementary strand in the same dinucleotide position (Fig. 16.15d).

REGULATION AFTER TRANSCRIPTION INFLUENCES RNA PRODUCTION, PROTEIN SYNTHESIS, AND PROTEIN STABILITY

Gene regulation can take place at any point in the process of gene expression. While most regulatory mechanisms influence the initiation of transcription, some affect posttranscriptional events such as RNA splicing, RNA stability, mRNA editing, protein synthesis, and protein stability.

RNA Splicing Helps Regulate Gene Expression

One example of how RNA splicing contributes to the regulation of gene expression is the role it plays in maintaining sexual identity throughout the life of *Drosophila*. As we see in the comprehensive example at the end of this chapter, transcription factors in very early female (XX) embryos activate the expression of a key gene called *Sxl* through a promoter called the early promoter (P_e). The cellular machinery splices the resulting transcript to create an mRNA that is translated into the Sxl protein, which is essential to the female-specific developmental program. The *Sxl* gene is not transcribed in early male (XY) embryos, so these embryos do not make the Sxl protein (Fig. 16.16a).

Later in development, the transcription factors activating the *Sxl* early promoter in females disappear, yet, to develop as females, these animals still need the Sxl protein. How can they still make the Sxl protein they need? The answer is that later in embryogenesis, the *Sxl* gene in both males and females is transcribed from another promoter—the late promoter (P_L)

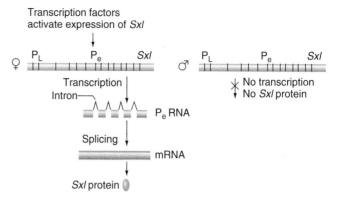

(a) Early embryo

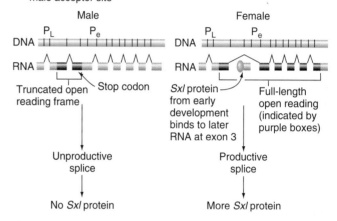

(b) *Sxl* protein regulates the splicing of its mRNA by blocking male acceptor site

Figure 16.16 Differential RNA splicing can regulate gene expression at the polypeptide level. (a) In the early female *Drosophila* embryo, transcriptional activators initiate transcription from the P_e promoter of *Sxl* to produce an mRNA that encodes the Sxl protein. Transcription of *Sxl* does not occur in the early male embryo. (b) Later in development, transcriptional activators that bind the P_L promoter are produced in both male and female animals. In the absence of any interference, the full-length RNA is spliced into a noncoding, and thus nonfunctional, RNA. When the Sxl protein is present, as it is in females, however, it binds to an RNA sequence at the 5' end of the third exon and causes the splicing apparatus to skip over this exon and splice exon 2 directly to exon 4. The resulting RNA molecule has an intact coding sequence and can be translated into more Sxl protein. This results in a feedback loop that maintains the presence of Sxl protein in females but not males.

(Fig. 16.16b). In males, splicing of the primary Sxl transcript generates an RNA including an exon (exon 3) that contains a stop codon in the reading frame of the protein. As a result, this RNA in males is not productive, that is, it does not generate any Sxl protein. In females, however, the Sxl protein previously produced by transcription from the early promoter influences the splicing of the primary transcript initiated at the late promoter. When the earlier made Sxl protein binds to the later transcribed RNA, this binding alters the splicing pathway such that exon 3 is no longer part of the final mRNA. Without exon 3, the mRNA is productive; that is, it can be translated to make more Sxl protein. Thus, a small amount of Sxl protein synthesized very early in development establishes a positive feedback loop that ensures more fabrication of Sxl protein later in development.

RNA Stability Provides a Mechanism for Controlling the Amount of Gene Product Synthesized

We saw in Chapter 7 and again in this chapter that most polypeptide-encoding RNAs transcribed in eukaryotes (by pol II) are modified at their 3′ ends by the addition of a poly-A tail (not encoded in the DNA) that can vary in length from fewer than 20 adenines to more than 200. Cellular enzymes in the cytoplasm slowly shorten the poly-A tails of all mRNAs after they leave the nucleus. Once the tail has disappeared, the core mRNA molecule is quickly degraded. This process provides the cell with a mechanism for fine-tuning the half-life of mRNAs produced from different genes. mRNAs that receive a longer poly-A tail last longer in the cytoplasm than mRNAs fitted with a shorter tail.

It might seem more efficient for a cell to keep mRNAs around for a long time to avoid using more energy in the continuous production of new transcripts. This is most likely true for the synthesis of many gene products in most cell types. During development and in fully differentiated cells, however, there are many instances in which a cell needs to implement a quick change in its genetic program by eliminating a gene product that it made and used earlier. Repression of transcriptional activation is part of the solution. But it is not enough if preexisting mRNAs and proteins are stable.

An extreme example of mRNA instability is observed in the class of histone transcripts, which are among the few protein-encoding RNAs that receive no poly-A tail at all. A cell requires high levels of histones during its S phase when it is adding these proteins to newly synthesized DNA. But as soon as S phase ends, the cell must rapidly eliminate all remaining histone messages. It achieves this goal by simply repressing transcription. Within minutes, all histone mRNAs, which are already without poly-A tails, are rapidly degraded.

In another more specialized example of regulation through changes in RNA stability, mammalian cells maintain the optimal concentration of iron for their activities by modifying the stability of mRNAs encoding a membrane-bound transferrin receptor that transports iron into the cell. The number of transferrin-receptor molecules that a cell makes depends on how much receptor mRNA is available to ribosomes for translation, and that, in turn, depends in part on how long a transferrin receptor mRNA lasts in the cell.

Figure 16.17 illustrates how cells regulate the stability of transferrin-receptor mRNA. The binding of extracellular iron to an extracellular protein called transferrin renders the iron soluble. The iron-transferrin complex gains entry into a cell via the transferrin receptor. The mRNA encoding the transferrin receptor contains sequences in the 3′ untranslated region known as iron response elements (IREs), and these IREs form stem-loop structures targeted by a cytoplasmic protein known as IRE-binding protein (IRE-BP). In the absence of iron, IRE-BP molecules bind to the IRE and stabi-

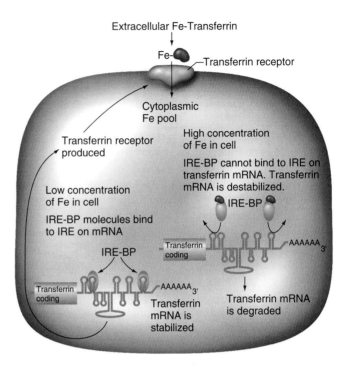

Figure 16.17   Transferrin receptor levels are regulated by intracellular iron concentration. In the absence of iron, IRE-BP binds to the transferrin receptor mRNA and prevents its degradation, which allows the production of higher levels of the product, which in turn allows the transport of more iron (bound to transferrin) from outside the cell to inside the cell. When iron levels get too high, expression of the transferring receptor is down-regulated. This occurs because iron binds to IRE-BP, inducing an allosteric change that no longer allows it to bind to and stabilize the mRNA.

lize the transferrin-receptor mRNA by protecting it from degradation. This makes biological sense because when little iron is available in the extracellular environment, a cell needs more transferrin receptors to transport as much as possible of the small amount of extracellular iron bound to transferrin. A high concentration of iron in the cytoplasm inactivates IRE-BP, preventing its binding to IRE, and thereby causing the degradation of the transferrin-receptor mRNA. This also makes biological sense, because when iron is abundant, it can gain access to the cell via other, lower affinity receptors.

mRNA Editing Can Affect the Biological Properties of a Gene's Product

Cells in the central nervous system of mammals carry AMPA membrane channel receptors that mediate excitatory postsynaptic currents. Comparisons of genomic DNA encoding the AMPA receptor with some cDNA copies of the mRNA show that an A G A (Arg) codon in the genomic DNA has become a G G A (Gly) codon in the mRNA (Fig. 16.18). This type of base substitution, occurring after transcription and RNA splicing, is known as **mRNA editing.** Researchers do not yet understand how the cellular machinery accomplishes such editing. They do know, however, that the biological properties of a Gly-containing membrane channel receptor differ from

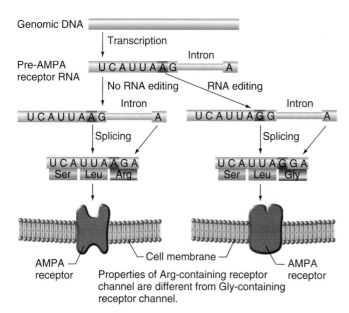

Figure 16.18   mRNA editing can regulate the function of the protein product that ensues. The AMPA receptor gene is transcribed into a sequence that contains an arginine codon at a particular point in the open reading frame. RNA editing replaces the first adenosine residue in this codon with a guanidine residue, and in so doing, conver4s this Arg codon into a Gly codon. The codon change has a significant effect on the activity of the protein product.

those of an Arg-containing channel, and that the extent of mRNA editing changes during brain development. The observation of multiple RNAs for the AMPA receptor raises the possibility that RNA editing may fine-tune the regulation of neuronal signaling by allowing cells to synthesize—from the same DNA blueprint—proteins with different properties, at different times in development.

Noncoding Sequences in mRNA Can Help Modulate Translation

Many mRNAs carry untranslated sequences that help modulate their translation. As mentioned earlier, the human body stores iron for future use in an iron-storage protein complex called ferritin. When there is no iron to store, the IRE-BP molecule just described binds to an IRE in the 5′ untranslated region of the ferritin mRNA, and this binding represses ferritin translation. When iron is abundant, the inactivated IRE-BP does not bind to the ferritin IRE, and translation proceeds. This on-off switch makes biological sense because the cell needs ferritin only when iron is present in the cytoplasm. Note that the interaction of IRE-BP with other IREs can influence mRNA stability as well as translation.

Protein Modifications After Translation Provide a Final Level of Control over Gene Function

The activity of a gene is reflected in the activity of its protein product, and various posttranslational modifications affect protein function. Many of these modifications occur ex-

tremely rapidly compared to the time it takes to activate gene transcription and accumulate sufficient protein product for a particular process, or to deactivate transcription and await the slow disappearance of a protein product. Thus, cells often rely on posttranslational modification in situations that require a rapid response to a stimulus.

Ubiquitination Targets Proteins for Degradation

Cells have many enzyme systems that destroy proteins. In one, ubiquitin—a small, highly conserved protein—functions as a marker. The covalent attachment of chains of ubiquitin to other proteins marks the ubiquitinized proteins for degradation by a large multienzyme complex known as the *proteosome.*

During the cell cycle, for example, the proper separation of sister chromatids at the beginning of anaphase depends on ubiquitination and its aftermath. As Fig. 16.19a shows, the separation of sister chromatids requires the elimination of specialized proteins that, like a glue, keep the chromosomes together. Elimination of these adhesive proteins requires the action of at least two large multienzyme complexes—the proteosome and the anaphase-promoting complex (APC). The APC brings together the gluelike proteins that must undergo ubiquitination with enzymes that catalyze the attachment of chains of ubiquitin to these targets. Inhibitors of sister chromatid separation are among the various targets of the APC. Ubiquitination of these inhibitors at the onset of anaphase marks them for degradation by proteosomes, which, in turn, allows the sister chromatids to separate. Researchers do not yet know exactly how the inhibitors carry out their function. One hypothesis is that some act as part of a proteinaceous glue that attaches sister chromatids at their centromeres; degradation of this glue as a result of APC and proteosome activity allows the sister chromatids to separate. Whatever the means, ubiquitination swiftly lifts the inhibition on sister chromatid separation, allowing the cell cycle to proceed.

Phosphorylation and Dephosphorylation Are Other Regulators of Protein Activity

In a process known as *sensitization,* many tissues exposed to hormones for a long time lose their ability to respond to the hormone. An example is the exposure of heart muscle to the stress hormone epinephrine. The binding of epinephrine to proteins called β-adrenergic receptors, which are located in the plasma membrane of heart muscle cells, normally increases the rate at which the heart contracts. But after several hours of continuous exposure to epinephrine, the heart muscle cells no longer respond in this way. Their sensitization is due to phosphorylation of the β-adrenergic receptors. The phosphorylation (addition of a phosphate) does not affect a receptor's ability to bind epinephrine, but it does prevent the receptor from transmitting the hormone signal into the heart muscle cells (see Chapter 17 for a detailed discussion of signal transduction). The phosphorylation itself depends in large part on the activity of kinase (phosphate-adding) enzymes that

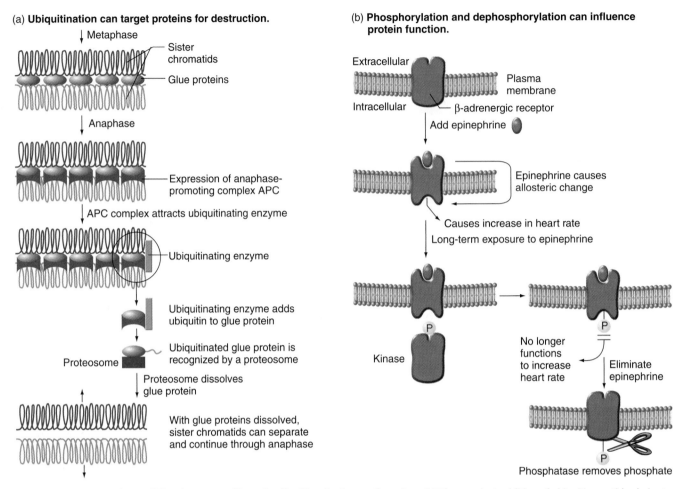

(a) Ubiquitination can target proteins for destruction.

Metaphase

Sister chromatids

Glue proteins

Anaphase

Expression of anaphase-promoting complex APC

APC complex attracts ubiquitinating enzyme

Ubiquitinating enzyme

Ubiquitinating enzyme adds ubiquitin to glue protein

Proteosome

Ubiquitinated glue protein is recognized by a proteosome

Proteosome dissolves glue protein

With glue proteins dissolved, sister chromatids can separate and continue through anaphase

(b) Phosphorylation and dephosphorylation can influence protein function.

Extracellular

Plasma membrane

Intracellular

β-adrenergic receptor

Add epinephrine

Epinephrine causes allosteric change

Causes increase in heart rate

Long-term exposure to epinephrine

Kinase

P

No longer functions to increase heart rate

P

Eliminate epinephrine

Phosphatase removes phosphate

P

Figure 16.19 Protein modifications can affect the final level of gene function. (a) The covalent addition of ubiquitin peptide chains to the gluelike proteins that hold together sister chromatids during the metaphase stage of mitosis marks them for digestion by proteosomes. This elimination must occur for mitosis to proceed from metaphase to anaphase. (b) Covalent phosphorylation of the β-adrenergic receptor has no effect on its binding to epinephrine, but blocks its downstream function of modulating heart rate.

phosphorylate the β-adrenergic receptor only when the receptor is bound to epinephrine (Fig. 16.19b). With the removal of epinephrine from the heart tissue, the kinases no longer act on the receptors, and phosphatase enzymes remove any phosphates already on the them. The removal of phosphate from the β-adrenergic receptors eventually restores the heart muscle's ability to respond to new doses of epinephrine.

SEX-DETERMINATION IN DROSOPHILA: A COMPREHENSIVE EXAMPLE OF GENE REGULATION

Male and female *Drosophila* exhibit many sex-specific differences in morphology, biochemistry, behavior, and function of the germ line (Fig. 16.20). By examining the phenotypes of flies with different chromosomal constitutions, researchers confirmed that ratio of X-to-A chromosomes helps determine sex, fertility, and viability (Table 16.2). They then carried out genetic experiments that showed that the X:A ratio influences sex through three independent pathways: One determines whether

the flies look and act like males or females; another determines whether germ cells develop as eggs or sperm; and a third produces dosage compensation through doubling the rate of transcription of X-linked genes in males. (Note that this strategy of dosage compensation is just the opposite of that seen in mammals, where the inactivation of one X chromosome in females equalizes the expression of X-linked genes with that in males.)

To simplify our discussion of sex determination in *Drosophila*, we focus on the first-mentioned pathway: the determination of somatic sexual characteristics. An understanding of this pathway emerged from analyses of mutations affecting particular sexual characteristics in one sex or the other. For example, as we saw at the beginning of the chapter, XY flies carrying mutations in the *fruitless* gene exhibit aberrant male courtship behavior, while XX flies with the same *fruitless* mutations appear to behave as normal females. Table 16.3 shows that mutations in other genes also affect the two sexes differently. Clarification of how these mutations influence somatic sex determination came from a combination of genetic experiments (studying, for example, whether one mutation in a double mutant is epistatic to the other) and molec-

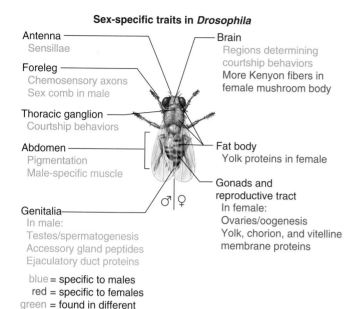

Sex-specific traits in *Drosophila*

Antenna
Sensillae

Foreleg
Chemosensory axons
Sex comb in male

Thoracic ganglion
Courtship behaviors

Abdomen
Pigmentation
Male-specific muscle

Genitalia
In male:
Testes/spermatogenesis
Accessory gland peptides
Ejaculatory duct proteins

Brain
Regions determining
courtship behaviors
More Kenyon fibers in
female mushroom body

Fat body
Yolk proteins in female

Gonads and
reproductive tract
In female:
Ovaries/oogenesis
Yolk, chorion, and vitelline
membrane proteins

blue = specific to males
red = specific to females
green = found in different
forms in the two sexes

Figure 16.20 Sex-specific traits in *Drosophila*. Objects or traits shown in blue are specific to males. Objects or traits shown in red are specific to females. Objects or traits shown in green are found in different forms in the two sexes.

TABLE 16.2 How Chromosomal Constitution Affects Phenotype in *Drosophila*

Sex Chromosomes	X:A	Sex Phenotype
Autosomal Diploids		
XO	0.5	Male (sterile)
XY	0.5	Male
XX	1.0	Female
XXY	1.0	Female
Autosomal Triploids		
XXX	1.0	Female
XYY	0.33	Male
XXY	0.66	Intersex

ular biology experiments (in which investigators cloned mutant and normal gene products for analysis). Through such experiments, *Drosophila* geneticists dissected various stages of sex determination to delineate the following complex regulatory network.

The X/A Ratio Regulates Expression of the Sex Lethal (Sxl) *Gene*

Recall from Chapter 3 that it is the ratio of X chromosomes to autosomes (A) that determines sex in *Drosophila*. Since in normal diploids, there are 2 copies of each autosome, the X:A

TABLE 16.3 *Drosophila* Mutations That Affect the Two Sexes Differently

Mutation	Phenotype of XY	Phenotype of XX
Sx^{fl}*	Male	Dead
Sxl^{ML}**	Dead	Female
transformer (tra)	Male	Male (sterile)
doublesex (dsx)	Intersex	Intersex
fruitless	Male with aberrant courtship behavior	Female

*Sx^{fl} is a recessive mutation of *Sex lethal*.
**Sxl^{ML} is a dominant mutation of *Sex lethal*.

ratio is 2/2 = 1.0 in a normal XX female and 1/2 = 0.5 in a normal XY male. In short, when the X:A ratio is 1.0, females develop; when the ratio is 0.5, males develop.

Key factors of sex determination are helix-loop-helix proteins encoded by genes on the X chromosome. Sisterless-A (Sis-A) and sisterless-B (Sis-B) are two such proteins. Referred to as *numerator elements,* these two proteins monitor the X:A ratio through the formation of homodimers containing two of the same kind of subunit or heterodimers containing two different subunits. The homodimers consists of two numerator elements, while the heterodimers are composed of one numerator element and one denominator element. *Denominator elements* are helix-loop-helix proteins that are encoded by genes on autosomes. Since the number of X chromosomes determines the ratio of numerator homodimers to numerator/denominator heterodimers, the homodimers of numerator elements provide a measure of the X:A ratio (Fig. 16.21).

The observation that in flies with a greater number of numerator homodimers, transcription of the *Sxl* gene occurs early in development suggests the following hypothesis for operation of the X:A ratio.

Numerator Subunit Homodimers May Function as Transcription Factors That Turn On Sxl

In this hypothesis, the association of denominator subunits with numerator subunits sequesters the numerator elements in inactive heterodimers that cannot activate transcription. Females produce enough numerator subunits, however, that some remain unbound by denominator elements. Homodimers formed from these free numerator elements act as transcriptional activators of *Sxl* at the P_e promoter early in development. Males, by contrast, carry only half as many X-encoded numerator subunits; thus, the abundant denominator proteins tie up all the numerator elements, and as a result, there are no free numerator elements in males to turn on the P_e promoter of the *Sxl* gene.

Although this model is likely to be an oversimplification, it suggests how different X:A ratios might activate and repress transcription of the *Sxl* gene.

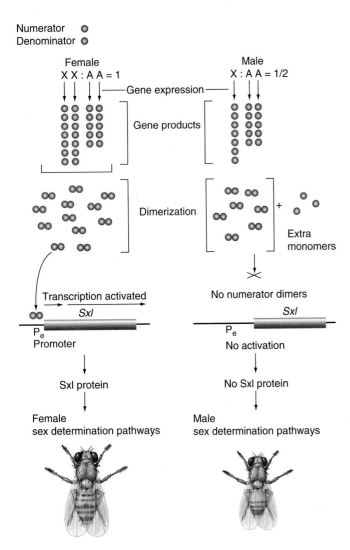

Numerator ⊙
Denominator ◉

Female
X X : A A = 1
← Gene expression →
Gene products

Male
X : A A = 1/2

Dimerization

Extra
monomers

Transcription activated
Sxl

P_e
Promoter

No numerator dimers

Sxl

P_e
No activation

↓
Sxl protein

Female
sex determination pathways

↓
No Sxl protein

Male
sex determination pathways

Figure 16.21   The X-to-autosome ratio determines the expression of *Sxl*, the first gene in the sex determination pathway. Numerator elements are produced by the X chromosome at a slightly higher level than denominator elements produced by autosomes. Denominator elements cannot form homodimers but they bind tightly to, and sequester, numerator elements. When the X:A ratio is one (in females), there are too many numerator elements to be occupied by denominators, and those not sequestered can form homodimers which act as activators of the *Sxl* gene. When the X:A ratio is 1/2 (in males), there are fewer numerators than denominators, and all the numerators get sequestered. Thus males are unable to produce numerator homodimers, and *Sxl* remains turned off.

The Sxl Protein Expressed in Early Development in Females Regulates Its Own Later Embryonic Expression through RNA Splicing

We have seen that the Sxl protein produced early in the development of female embryos participates later in an autoregulatory feedback loop (review Fig. 16.16). In this self-regulating system, the Sxl protein catalyzes the synthesis of more of itself through RNA splicing of the P_L-initiated transcript, which results in a productive mRNA. By contrast, in males where

there is no transcription of *Sxl* early in development, activation of the P_L promoter later in embryonic development results in an unproductive *Sxl* transcript containing a stop codon near the beginning of the message. Since no Sxl protein is present to splice out the problem stop codon, this unproductive transcript is not translated to protein; and males thus have no Sxl protein at any point in development.

The Effects of Sxl Mutations

Recessive *Sxl* mutations that produce nonfunctional gene products have no effect in XY males, but are lethal in XX females (see Table 16.3). This is because males, which do not normally express the *Sxl* gene, do not miss its functional product, but females, which depend on the Sxl protein for sex determination, do. The absence of the Sxl protein in females allows the expression of certain male-specific dosage-compensation genes that increase transcription of genes on the X chromosome; and the hypertranscription of these X-linked genes on two X chromosomes in mutant females proves lethal.

By comparison, rare dominant *Sxl* mutations that allow production of Sxl protein even in XY embryos are without effect in females, which normally produce the product, but lethal to males, which normally do not. In these mutants, the *Sxl* gene product indirectly represses transcription of genes that males need to express for dosage compensation. Without the products of these male-specific dosage-compensation genes, males cannot hypertranscribe X-linked genes and thus do not have enough X-linked gene products to survive.

Summary: The Sxl Gene Acts as an On/Off Switch That Is Sensitive to the X:A Ratio

Different proportions of numerator and denominator proteins reflecting the X:A ratio determine whether or not the *Sxl* P_e promoter is active in the early embryo. Later, when other transcription factors activate the upstream P_L promoter in both males and females, alternative RNA splicing after transcription generates more Sxl protein in females and none in males. Female development requires Sxl protein; male development requires its absence.

Sxl Triggers a Cascade of Splicing

In addition to splicing its own transcript, the Sxl protein influences the splicing of RNAs transcribed from other genes. Among these is the *transformer (tra)* gene. In the presence of the Sxl protein, the *tra* primary transcript undergoes productive splicing that produces an mRNA translatable to a functional protein. In the absence of Sxl protein, the splicing of the *tra* transcript results in a nonfunctional protein (Fig. 16.22).

The cascade continues. The functional Tra protein synthesized only in females, along with another protein encoded by the *tra-2* gene (which is transcribed in both males and females) influences the splicing of the *doublesex (dsx)* gene's primary transcript. This splicing pathway results in the production of a female-specific Dsx protein called Dsx-F. In males, where there is no Tra protein, the splicing of the *dsx* pri-

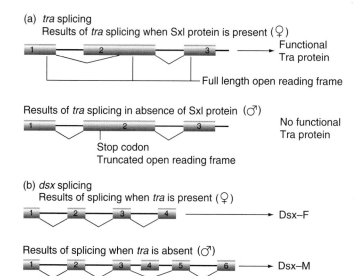

Figure 16.22   The expression or nonexpression of *Sxl* leads to downstream differences in expression of products that also play a role in the sex determination pathway. (a) The presence of *Sxl* alters the splicing of *tra* mRNA. Female transcripts produce functional Tra protein, while male transcripts have a truncated open reading frame and are unable to produce Tra. (b) Tra protein in turns plays a role in altering the splicing pattern of the *dsx* mRNA. A different Dsx product results in males (Dsx-M) than in females (Dsx-F).

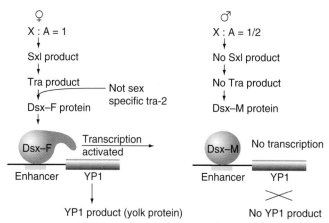

Figure 16.23   Alternative forms of the *dxs* gene product both bind to the same *YP1* enhancer but they have opposite effects on the expression of the *YP1* gene. Dsx-F acts as a transcriptional activator, while Dsx-M acts as a transcriptional repressor.

mary transcript produces the related, but different Dsx-M protein (Figure 16.22b). The amino terminal parts of the Dsx-F and Dsx-M proteins are the same, but the carboxy-terminal parts of the proteins are different.

Effects of Mutations in tra

Like mutations in *Sxl,* mutations in *tra* affect the two sexes differently. Since males do not normally express *tra,* null mutations of the gene have no effect on them. In females, however, the Tra protein helps implement female-specific patterns of RNA splicing. In the absence of Tra protein, *dsx* primary transcripts can only be spliced to make Dsx-M. XX Tra⁻ flies expressing Dsx-M are converted into flies having the phenotype of a male. The resulting XX males exhibit proper male sexual behavior, but without the Y chromosome, they are sterile because certain Y-linked genes are prerequisites of male fertility.

The Dsx-F and Dsx-M Proteins Are Transcription Factors That Determine Somatic Sexual Characteristics

Although both Dsx-F and Dsx-M function as transcription factors, they have opposite effects. In conjunction with the protein encoded by the *intersex* (*ix*) gene, Dsx-F functions mainly as a repressor that prevents the transcription of genes whose expression would generate the somatic sexual characteristics of males. Dsx-M, which works independently of the intersex protein, accomplishes the opposite: the activation of genes for the somatic sexual characteristics of males and the repression of genes that determine female somatic sexual characteristics. Interestingly, the two Dsx proteins can bind

to the same enhancer elements, but their binding produces opposite outcomes (Fig. 16.23). For example, both bind to an enhancer upstream of the promoter for the *YP1* gene, which encodes a yolk protein; females make this protein in their fat body organs and then transfer it to developing eggs. The binding of Dsx-F stimulates transcription of the *YP1* gene in females; the binding of Dsx-M to the same enhancer region inactivates (in conjunction with other transcription factors) transcription of *YP1* in males.

The Effect of dsx Mutations

Mutations in *dsx* affect both sexes because in both males and females, the production of Dsx proteins represses certain genes specific to development of the opposite sex. Null mutations in *dsx* that make it impossible to produce either functional Dsx-F or Dsx-M result in intersexes that cannot repress either certain male-specific or certain female-specific genes.

The Tra and Tra-2 Proteins Also Help Regulate Expression of the Fruitless Gene

We saw at the beginning of this chapter that the courting song and dance of male *Drosophila* are among the sexual behaviors under the control of the *fruitless* (*fru*) gene. As it turns out, the *fru* primary transcript is another regulatory target of the Tra and Tra-2 splicing factors (Fig. 16.24). In females, whose cells make both Tra and Tra-2 proteins, splicing of the *fru* transcript produces an mRNA that encodes a protein we refer to as Fru-F. In males, whose cells carry no Tra protein, alternative splicing of the *fru* transcript generates a related Fru-M protein with 101 additional amino acids at its N terminus. As we mentioned at the beginning of this chapter, these additions almost certainly determine some of the observed differences between male and female behavior. Since both Fru-F and Fru-M have the zinc-finger motifs characteristic of transcription factors, they probably activate and repress genes whose sex-specific products help generate courting behaviors.

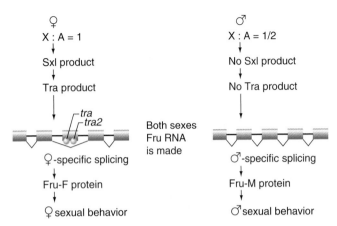

Figure 16.24 The primary *fru* RNA transcript is made in both sexes. Splicing occurs unhindered in males to produce an mRNA which is translated into the fru-M protein product. But tra protein (present only in females) causes alternative splicing of the *fru* transcript to produce an alternative mRNA which encodes an alternative protein product fru-F.

The sex-specific products of *fru*, we have seen, appear in only a few cells in the nervous system, and the location of these neurons is significant. Some are in regions known to help regulate the courtship song; others are in areas that process chemosensory information from the antennae (perhaps in neurons that receive pheromone signals); still others are in regions that control abdominal movements (suggesting how *fruitless* may influence the male's curling of the abdomen during mating). To understand precisely how changes in gene expression in these few cells control sexual behaviors, *Drosophila* researchers are now trying to discover which genes are the targets of transcriptional regulation by the fruitless protein.

Summary: A Complex Network of Molecular Interactions Regulates the Determination of Somatic Sexual Characteristics in *Drosophila*

Several kinds of gene regulation make the embryonic "decisions" that determine the sex of the fly. Controls of both transcriptional initiation and RNA splicing play critical roles in this biological process. First, X-encoded numerator and autosome-encoded denominator elements allow the early embryo to measure the X:A ratio. Presumably, numerator homodimers, present in sufficient quantity only in females, activate *Sxl* transcription in early embryos. Next, the Sxl protein acts as a splicing factor that perpetuates its own synthesis in females. There it initiates a cascade of splicing factors, each of which determines the splicing patterns of the primary transcript of the next gene in the pathway. These splicing cascades eventually produce female-specific versions of at least two transcription factors: Dsx-F and Fru-F. In males, default splicing of these genes produces the Dsx-M and Fru-M transcription factors. The female- and male-specific versions of the Dsx and Fru proteins help turn on and off genes whose products elaborate the sexual appearance and behavior of male and female *Drosophila*.

Genes other than those just described also contribute to sex-specific behaviors. For example, a gene called *dissatisfaction* affects courtship behavior in males and females. Mutant females resist copulating with males; mutant males are maladroit in copulation because they don't properly curl their abdomen. Genetic experiments suggest that the *dissatisfaction* gene is part of the same sex-determination pathway as *fru*, but its position in the pathway is not yet known.

CONNECTIONS

Multiple controls regulate gene function throughout gene expression, from the initiation of transcription, to the processing of RNA transcripts, to the chemical modification of final gene products. At the outset, the regulation of transcription occurs through the interaction of *cis*-acting DNA regions and a variety of transcription factors.

Accurate regulation of gene function is crucial for proper control of the cell cycle. Indeed, a critical network of *cis*-acting control regions, *trans*-acting factors, and protein modifications promotes cell growth, DNA replication, and cell division in response to certain environmental signals, and delays these events in response to other signals, such as DNA damage. Chapter 17 describes the regulatory network controlling the cell cycle in normal cells, and explains how mutations that disrupt one or more aspects of that network can result in cancer.

ESSENTIAL CONCEPTS

1. The activity of most genes in eukaryotic cells is regulated primarily at the point of transcriptional initiation. Analyses of mutations that affect a gene's function without changing the sequence of its product provided insight into this level of regulation. Through these mutations, researchers defined *cis*-acting DNA regulatory elements and *trans*-acting *transcription factors*.

2. The eukaryotic genome contains three classes of genes, each associated with a different type of promoter and transcribed by a different RNA polymerase. Class I genes encode the rRNAs. Class III genes encode the tRNAs as well as other small RNA molecules. Class II genes, by far the largest class, encode all proteins.

3. Two types of *cis*-acting regulatory regions—*promoters* and *enhancers*—are associated with class II genes. All promoters are located at the 5′ end of the gene they influence. The enhancers have a more variable location in relation to the genes they control.

4. The binding of nonspecific *basal factors* to promoters is a prerequisite for the transcription of a gene. Basal factors alone allow a low, nonspecific basal level of gene transcription.

5. The specific interactions of transcription factors with enhancer elements can increase transcriptional initiation above the basal level. Activation is mediated by transcription factors called *activators* that bind to enhancers and interact with basal factors at the promoter. Activation can be modulated by *repressors* that compete with activators for enhancer binding or quench the ability of activators to carry out their function. Many activators and repressors form homodimers and/or heterodimers. Formation of these dimers is a prerequisite for them to function as transcription factors.

6. Although promoters and enhancers are the most common and best characterized types of *cis*-acting regulatory elements, there are other elements that do not fit into either of these classes. One such element is the *Locus Control Region* associated with the β-*globin* gene cluster.

7. The regulation of transcription is a complex process dependent on the interactions of multiple enhancer elements with many competing transcription factors, and responsive to signals originating within and outside the cell. The integration of signals and regulatory components determines the precise level of transcription of a particular gene in a particular cell at a particular time.

8. The structure of chromatin around a gene contributes to the regulation of transcription. Normal chromatin structure prevents runaway transcription from genes that are not under activation. The unraveling of the DNA in chromatin is an initial step in activation. Hypercompaction of chromatin domains causes *transcriptional silencing* by blocking access to the promoter and enhancers of a gene and thereby preventing its activation even in the presence of activator proteins.

9. *Genomic imprinting* is an example of epigenetic control over gene expression. Imprinting operates on the copy of a gene received from one parent but not the other. *DNA methylation* plays a role in the maintenance of imprinting from one mammalian somatic cell generation to the next.

10. Although the regulation of most genes depends primarily on controls over transcription, in some cases, further regulation down the path to protein production also plays a role. Modulation of gene function can occur through changes in RNA splicing or RNA stability, mRNA editing, changes in the efficiency of translation, and chemical modification of the gene product.

S O C I A L   A N D   E T H I C A L   I S S U E S

1. Since the cloning of the human growth hormone gene *HGH* has become available as a drug and is prescribed for children with growth hormone defects, Fred and Susannah's twelve-year-old son Sam (who does not have a known growth hormone defect) is below average in height and his parents want the family doctor to prescribe growth hormone. They feel that Sam is suffering socially because of his small size and there will be a lasting psychological impact on him. The doctor is concerned that not enough is known about the effects of extra growth hormone on the balance of expression of many genes important in physiology and development. Do the parents have a right to demand that the doctor prescribe the drug? Should the prescription of such a drug be limited to specific cases?

2. One difficulty with transgenic technology for animals is the inability to direct where DNA integrates and how many copies of the DNA integrate. Yet germ cell treatments are proposed for improving livestock. Animal rights groups are protesting that there are too many unknowns involved in these experiments. Should the same criteria for experimentation be applied to animals as is applied to humans?

3. Experimentation has shown that fetal tissue can be used to treat conditions such as the neurological disorder Parkinson disease. Because this fetal tissue is at a very different stage in development, many proteins could be present that affect the expression of genes in the tissue into which it is transplanted. If this is an experimental technology, should the individuals who are taking part in the trials be obliged to undergo any testing that monitors changes in their physiology or should they be able to choose which follow-up studies they take part in?

SOLVED PROBLEMS

I. You are studying expression of a gene whose protein product is made after UV irradiation. You cloned the gene and made antibody to the protein.

a. If expression is regulated by turning on transcription after UV exposure, what results would you predict from hybridizing a DNA probe to RNA isolated from cells before and after UV irradiation (Northern analysis) and from incubating the antibody to proteins isolated from cells before and after UV treatment (Western analysis)?

b. If expression is regulated by preventing translation, what results would you predict from doing a similar Northern and Western analysis?

Answer

To answer this question, you need to consider the consequences of transcriptional and translational regulation on expression and think through what happens experimentally in Northern and Western analysis.

a. If a gene is transcriptionally regulated, the mRNA will not be present in cells that were not exposed to UV. *There will be no hybridizing band in the Northern analysis of mRNA from unexposed cells. The mRNA will be present in cells that have been treated with UV and there will be a hybridizing band.* Similarly, the protein will only be found in cells that were exposed to UV and *the antibody will bind to its protein target only in the protein preparation from exposed cells.*

b. If expression is regulated at the translation step, mRNA will be present in the cells whether they have been exposed to UV or not. *Hybridizing bands will be found in both RNA samples. The protein will be present only in those cells that were exposed to UV so signal will be seen only in the exposed preparation.*

II. The retinoic acid receptor (RAR) is a transcription factor that is similar to steroid hormone receptors. The substance (ligand) that binds to this receptor is retinoic acid. One of the genes whose transcription is activated by retinoic acid binding to the receptor is *myoD*. The diagram below shows a schematic of the RAR protein into which two different 12-base oligonucleotides were inserted at the sites indicated. For constructs a–e, oligonucleotide 1 (T T A A T T A A T T A A) was inserted. For constructs f–m, oligonucleotide 2 (C C G G C C G G C C G G) was inserted. Each mutant protein was tested for ability to: bind retinoic acid, bind to DNA, activate transcription of *myoD* gene. Results are presented below.

Mutant	Retinoic acid binding	DNA binding	Transcriptional activation
a	−	−	−
b	−	−	−
c	−	−	−
d	−	+	+
e	+	+	+
f	+	+	+
g	+	+	−
h	+	+	−
i	+	−	−
j	+	−	−
k	−	+	+
l	−	+	+
m	+	+	+

a. What is the effect of inserting oligonucleotide 1 anywhere in the protein?

b. What is the effect of inserting oligonucleotide 2 anywhere in the protein?

c. Indicate the three protein domains on a copy of the drawing above.

Answer

This question involves the concepts of domains within proteins and use of the genetic code to understand effects of oligonucleotide insertions.

a. Oligonucleotide 1 contains a stop codon in any of the three reading frames. This means it will *cause termination of translation of the protein wherever it is inserted.*

b. Oligonucleotide 2 does not contain any stop codons and so will *just add amino acids to the protein.* Because there are 12 bases in the oligonucleotide, it will not change the reading frame of the protein. *Insertion of the oligonucleotide can disrupt the function of a site in which it inserts.*

c. Looking at the data overall, notice that all mutants that are defective in DNA binding are also defective in transcriptional activation, as would be expected for a transcription factor that binds to DNA. The mutants that will be informative about the transcriptional activation domain are those that do not have a DNA-binding defect. Inserts a, b, and c using oligonucleotide 1, which truncates the protein at the site of insertion, are defective in all three activities. The protein must be made at least as far as point d before DNA binding or transcription activation are seen. These two activities must lie before d. Truncation at d is negative for retinoic acid binding but the truncation at e does bind to retinoic acid. The retinoic acid-binding activity must lie before e. Using the oligonucleotide 2 set of insertions, transcriptional

NH₂ ___f___ g ___h___ i ___j___ k ___l___ m ___ COOH
          a   b   c   d   e

activation was disrupted by insertions at sites g and h, indicating that this region is part of the transcriptional domain; i and j insertions disrupted the DNA binding; and k and l insertions disrupted the retinoic acid binding. The minimal endpoints of domains as determined from these data are summarized in the schematic below.

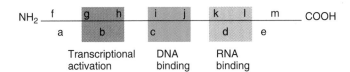

III. A cDNA clone that you isolated using pituitary gland mRNA from mice was used as a probe against a blot containing RNAs from embryonic heart (EH), adult heart (AH), embryonic pituitary (EP), adult pituitary (AP) and testis (T). The results of the hybridization are shown in next column.

a. What would you conclude about this gene based on the result with AH RNA?
b. How would you explain the result with testis RNA?

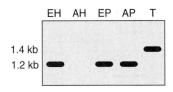

Answer

This problem requires an understanding of RNA and transcription.

a. No RNA from adult heart (AH) hybridized with the probe, indicating that the *gene is not transcribed in this tissue.*
b. *A different sized RNA is seen in the testis sample.* This could be due to *alternate splicing of the transcript or a different start site* in testis compared to other tissues.

P R O B L E M S

16-1 For each of the terms in the left column, choose the best matching phrase in the right column.

a. basal factors	1. influences transcription in a large region of DNA
b. transcriptional silencing	2. marks a protein for degradation
c. locus control region	3. pattern of expression is dependent on which parent transmitted the allele
d. activators	4. multimers of nonidentical subunits
e. imprinting	5. heterochromatin
f. ubiquitination	6. multimers of identical subunits
g. homomers	7. bind to enhancers
h. heteromers	8. bind to promoters

16-2 Does each of the following types of gene regulation occur in eukaryotes only? in prokaryotes only? in both prokaryotes and eukaryotes?
a. differential splicing
b. positive regulation
c. chromatin compaction
d. attenuation
e. negative regulation

16-3 What events occur during processing of a primary transcript?

16-4 Which fusion gene would you use for which purpose? (The slash indicates the fusion and the parts of each type of fusion are given in the order in which they would occur.)

Types of fusions:
a. DNA-binding domain of *lexA* gene/random mouse genome fragments
b. random mouse sequences/*lacZ* gene
c. mouse metallothionein promoter/a mouse gene

Uses:
i. try to identify genes turned on in specific cell types
ii. turn on expression of gene by including Zn in the diet
iii. search for transcription activation domains

16-5 Which eukaryotic RNA polymerase (RNA pol I, pol II, or pol III) transcribes which genes?
a. tRNAs
b. mRNAs
c. rRNAs

16-6 You isolated a gene expressed in differentiated neurons in mice. You made a fusion of the upstream DNA and beginning of the gene to *lacZ* (reporter gene) so you can monitor expression. Different fragments (shown as dark lines) were cloned next to the *lacZ* gene that lacked a promoter. The clones were introduced into neurons in tissue culture to monitor expression. From the results that follow, which region contains the promoter and which contains an enhancer?

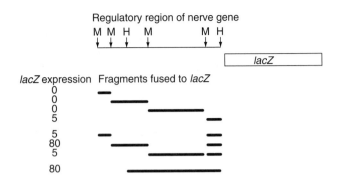

Regulatory region of nerve gene

16-7 Is each of the motifs listed below associated with DNA binding, transcription activation, or dimer formation?
a. zinc finger
b. helix-loop-helix
c. leucine zipper
d. acidic region
e. helix-turn-helix

16-8 From Northern analysis, you found that the *ADAG* gene is expressed only in brain. You examined expression in glial and neuronal cell lines (two types of cells in the brain) and found that only glial cells made *ADAG* mRNA. No one has characterized the *cis-* or *trans-*acting elements required for glial specific expression so you decided to do so. You made a set of deletions in the regulatory region and fused these to the *lacZ* gene so you can easily monitor the expression after introducing the clones into tissue culture cells. Deletions beginning at a site upstream of the gene and extending to base -85 (with the transcription start site considered $+1$ and bases prior to the start having negative designations) still retained full activity but a deletion to -75 left only 1% of the original activity. A DNaseI footprinting experiment reveals that a 19-bp sequence is protected. You also obtained evidence that the factor binds as a dimer although each monomer binds to a similar sequence. The sequence between -90 and -60 is shown below:

-90 -60
T A G G C T T C T G C G G A C C G C A A A C T G T G C T G G C

a. What is the most likely region that was protected from DNaseI treatment in the footprint experiment?
b. What is a likely consensus sequence for the monomer binding site?

16-9 A single enhancer site regulates expression of three adjacent genes *GAL1, GAL7,* and *GAL10* but the genes are not cotranscribed as one mRNA. How could you show experimentally that each gene is transcribed separately?

16-10 In yeast, the GAL4 protein binds to DNA to activate transcription of GAL7, or GAL10. GAL 80 represses expression by binding to GAL4 protein and preventing it from binding to DNA. In which genes should you be

able to isolate galactose constitutive mutations and what characteristics of the protein would the mutation disrupt?

16-11 *MyoD* is a transcriptional activator that turns on the expression of several muscle-specific genes in human cells. The *Id* gene product inhibits *MyoD* action. How could you determine if *Id* acts by quenching *MyoD* (as *GAL80* does to *GAL4* in yeast) or by blocking access to the enhancer? What differences would you expect to see experimentally?

16-12 How could you make a library of genes expressed during sporulation in yeast?

16-13 What experimental evidence indicates that chromatin structure acts to reduce basal levels of transcription?

16-14 You isolated nuclei from liver cells, treated them with increasing amounts of DNaseI, stopped the reactions, isolated the DNA, treated with restriction enzyme *Eag*I, electrophoresed the DNA, transferred to a blot, and hybridized with a probe from the gene you are studying. With no DNaseI treatment, there was a 20-kb *Eag*I fragment that hybridized with your probe. With trace amounts of DNaseI, two bands of 16 and 4 kb were present. The same DNaseI treatment of DNA from muscle cells produced only a 20-kb fragment. What does this result tell you about the region of DNA?

16-15 Which of the following would be suggested by a DNase hypersensitive site?
a. No transcription occurs in this region of the chromosome.
b. The chromatin is in a more open state than a region without the hypersensitive site.
c. Transcription terminates at this site.

16-16 Match the gene expression phenomenon with molecular components that modulate each.

a. transcriptional silencing   1. enhanceosome

b. imprinting   2. heterochromatin

c. LCR   3. TBF, TAFs

d. basal transcription   4. methylation

16-17 Follow the expression of a paternally imprinted gene through three generations. Indicate whether the copy of the gene from male in generation I is expressed in the germ cells and somatic cells of the individuals listed below.

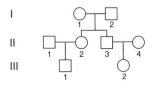

a. generation I male (I-2):   germ cells

b. generation II daughter (II-2): somatic cells

c. generation II daughter (II-2): germ cells

d. generation II son (II-3): somatic cells

e. generation II son (II-3): germ cells

f. generation III grandson (III-1): somatic cells

g. generation III grandson (III-1): germ cells

16-18 You are studying muscle cells and have found a protein that is only made in this tissue. The data below are from analysis of RNA (Northern blot) using a DNA probe from the gene and analysis of protein (Western blot) using antibody directed against the protein. Is this gene transcriptionally or translationally regulated?

16-19 a. Assume that two transcription factors are required to get expression of the blue pigmentation genes in pansies. (Without the pigment, the flowers are white). What phenotypic ratios would you expect from crossing strains heterozygous for each of the genes encoding these transcription factors?

b. Now assume that either transcription factor is sufficient to get blue color. What phenotypic ratios would you expect from crossing strains heterozygous for each of the genes encoding these transcription factors?

16-20 You isolated a cDNA from skin cells and when you hybridized that cDNA as a probe with a blot containing mRNAs from skin cells and nerve cells you saw a 1.2- and a 1.3-kb fragment, respectively. How could you explain the different sized cDNAs?

16-21 The *hunchback* gene, one of the genes necessary for setting up the dorsal-ventral axis of the *Drosophila* embryo, is translationally regulated. The location of the coding region within the transcript is known and there is additional DNA beyond the coding region at the 5′ and 3′ ends of the mRNA. How could you determine if the sequences at the 5′ or 3′ or both ends are necessary for proper regulation of translation?

16-22 From Northern and Western hybridization, the mRNA and protein from a tissue specific gene are present in brain, liver, and fat cells but you are detecting only protein activity in fat cells. Give an explanation for this phenomena.

CELL-CYCLE REGULATION
AND THE GENETICS OF CANCER

Killer cells surround a large cancer cell. Note the extended cellular
processes of the cancer cell.

The roughly 300 different types of cells that compose the brain, heart, lungs, liver, immune system, and other tissues of a healthy human adult proliferate only when necessary—to replace cells lost through normal attrition or injury, and in the case of the immune system, to attack infectious agents. Many cells, such as those that constitute the inner skin, blood, and intestinal lining, multiply daily to compensate for the senescence and death of their counterparts; others, such as those in the liver, rarely divide under normal circumstances but proliferate rapidly when toxins destroy some among them; still others, such as specific cells of the immune system, reproduce in response to particular bacteria and viruses; finally, some cells, such as the terminally differentiated cells of nerves and surface skin, never divide.

Most of us give little thought to this life-sustaining cycle of cell death and renewal, because despite the continuous comings and goings of our body's cells, we generally remain the same shape and size. But the appearance of an abnormal growth or a set of symptoms diagnosed as cancer startles us into the realization that we take for granted the intricate checks and balances that control cell division and behavior. These controls enable all cells of the body to function as part of a tightly organized, cooperative society. Cancer results when some cells divide out of control and eventually acquire the ability to spread beyond their prescribed boundaries. Cancer rarely occurs in children, but is a common disease of older adults, in whom it accounts for about one-fourth of all deaths. Although there are many types of cancer, they all result from excessive and inaccurate cellular proliferation (Fig. 17.1). Thus, the body's ability to regulate cell division is a foundation of health.

Scientists who study cancer seek information about the normal controls that govern cell proliferation as a basis for understanding what goes wrong in cancer. But they often work in reverse, using the abnormalities of cancer to retrieve information about the normal operation of the cell cycle. Observations of the differences in behavior between normal and cancerous cells, obtained from diseased individuals or from cultured cell lines, have

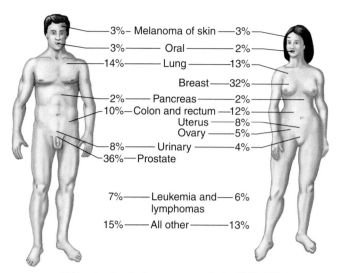

3%– Melanoma of skin —3%–
3%——— Oral ———2%–
14%——— Lung ———13%–
Breast——32%
2%——— Pancreas ———2%
10%—Colon and rectum —12%
Uterus ——8%
Ovary ——5%
8%——— Urinary ———4%
36%—Prostate

7%——— Leukemia and——6%
lymphomas
15%——— All other———13%

Table of estimated new cancers in the United States

Figure 17.1   The relative percentages of cancer that occur at different sites in the bodies of men and women.

In this chapter we examine the genes and gene products that control cell proliferation, including molecules that control the machinery of cell division, molecules that integrate the repair of DNA damage with progression through the cell cycle, and molecules that relay messages about whether conditions are right for cell division. We then describe how mutations in these genes and gene products lead to cancer.

Two unifying themes emerge from our discussion. First, cancer is ultimately a disease of the genes: The multiple phenotypes collectively referred to as cancer all result from mutations in genes that regulate a cell's passage through the cycle of growth and division. Chemicals in the environment that raise the rate of gene mutation increase the probability of cancer incidence. Second, cancer differs in two ways from cystic fibrosis, Huntington disease, and other genetic conditions caused by the inheritance of one or two copies of a single defective gene: (1) Most mutations that lead to cancer occur in the somatic cells of one tissue and (2) multiple mutations in an array of genes must accumulate over time in the clonal descendants of a single cell before the cancer phenotype appears. By contrast, the mutations that cause cystic fibrosis and Huntington disease are transmitted through the germ line; thus the mutant alleles of one particular gene appearing in all cells of all somatic tissues cause the disease in an affected individual.

Our presentation of the correlation between cell-cycle regulation and cancer covers

■   The normal control of cell division, including a description of the normal cell cycle and the molecular signals, machinery, and checkpoints that regulate passage through the cell cycle.

■   How cancer arises from malfunctions in controls over cell division, including a description of cancer phenotypes, an analysis of the clonal nature of tumors, and a description of the mutations in protooncogenes and tumor-suppressor genes that underlie all cancers.

■   A comprehensive example describing the progression of mutations leading from low-grade brain tumors to aggressive brain cancer.

made it clear that cancer cells can divide when normal cells cannot, and cancer cells can migrate from one tissue to another while normal cells cannot. Unfortunately, it is hard to study cellular behavior over time *in vivo* and *in vitro*. To study cells *in vivo*, it is necessary to see and manipulate them; but this requires cutting open an animal and carrying out extended observations, an approach that is rarely possible in humans and not easy even in model organisms like mice. It is also difficult to compare most types of tumor cells with normal laboratory-grown cells because most normal cell types do not grow in culture, and even those that do, have limited life spans. Genetic studies on the genomes of various eukaryotic organisms, from yeast to humans, have helped overcome these experimental impasses by identifying genes in which mutations lead to cancer; such studies continue to provide insights that are unavailable from other types of analyses.

THE NORMAL CONTROL OF CELL DIVISION

A variety of genes and proteins control the events of the cell cycle. These genes and proteins allow progression to the next stage of the cycle when all is well, but cause the cellular machinery to brake when damage to the genome or to the machinery itself requires repair. We now describe the molecules that control cell division.

Cyclin-Dependent Kinases Collaborate with Cyclins to Ensure the Proper Timing and Sequence of Cell-Cycle Events

Cell division, we saw in Chapter 3, requires the duplication of chromosomes and other cellular components as well as the precise partition of the duplicated elements to two daughter cells. During this complicated process, the cell coordinates the function of hundreds of different proteins. To see how the cell orchestrates the events of cell division, we first review the

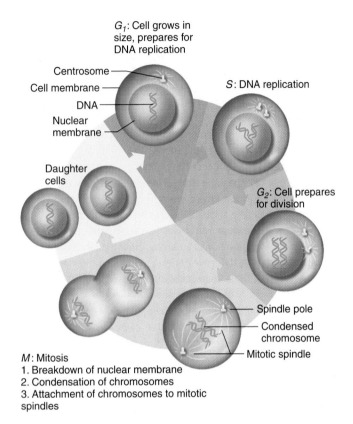

G_1: Cell grows in size, prepares for DNA replication

Centrosome
Cell membrane
DNA
Nuclear membrane

S: DNA replication

Daughter cells

G_2: Cell prepares for division

Spindle pole
Condensed chromosome
Mitotic spindle

M: Mitosis
1. Breakdown of nuclear membrane
2. Condensation of chromosomes
3. Attachment of chromosomes to mitotic spindles

Figure 17.2 The cell cycle is the series of events that transpire between one cell division and the next. After division, a cell begins in the G_1 phase, progresses into S phase, where chromosomes are replicated, to the G_2 phase, and to the M phase, where replicated sister chromosomes are segregated to daughter cells. In M phase, the nuclear membrane breaks down, centrosomes form the poles of the spindle, and microtubules construct a scaffold on which chromosomes migrate.

stages of the cell cycle and then look at some of the proteins that control progression through that cycle.

The Cell Cycle Has Four Phases: G_1, S, G_2, and M

G_1 is the period, or gap, between the end of a mitosis and the DNA synthesis that precedes the next mitosis (Fig. 17.2). During G_1, the cell grows in size, imports materials to the nucleus, and prepares in other ways for DNA replication. S is the period of DNA synthesis, or replication. It requires the timely, sequential activation of proteins that function at the replication fork, as well as the timely inactivation of these same proteins. G_2 is the gap between DNA synthesis and mitosis. During G_2, the cell prepares for division. M, the phase of mitosis, includes the breakdown of the nuclear membrane, the condensation of the chromosomes, their attachment to the mitotic spindle, and the segregation of chromosomes to the two poles; at the completion of mitosis, the cell divides. During M, the cell must coordinate the activities of a variety of proteins: those that cause chromosome condensation (see Chapters 3 and 11), tubulins that polymerize to form the mitotic spindle on which the chromosomes move, motor proteins in the centromeres that power chromosome movement, proteins that dissolve and re-form the nuclear membrane at the beginning and end of mitosis, and others.

Experiments with Yeast Helped Identify Genes That Control Cell Division

The budding yeast *Saccharomyces cerevisiae* and the fission yeast *Schizosaccharomyces pombe* have been instrumental in identifying the genes that control cell division. Several properties of both yeast species make them particularly useful for the genetic dissection of cellular processes. Both can grow as haploid or diploid organisms. As a result, it is possible to identify recessive mutations in the haploid cells, which carry no second, wildtype allele to obscure the mutant phenotype; and it is possible to construct diploid cells containing two mutations, allow them to proliferate, and test the resulting cell populations to determine the number of complementation groups defined by the mutations. The budding yeast *S. cerevisiae* has yet another property that facilitates cell cycle analysis. At the beginning of the cell cycle, toward the end of G_1, a new daughter cell arises as a bud on the surface of the mother cell. As the mother cell progresses through the division cycle, the bud grows in size; it is small during S phase and large during mitosis. Bud size thus serves as a marker of progress through the cell cycle. One can order cells in an asynchronous population according to position in the cell cycle by observing the relative sizes of their buds. A normal population of growing yeast cells contains nonbudding cells as well as cells with buds of all sizes.

Mutations that interfere with the cell cycle are lethal; and because cell proliferation depends on successive repeats of the cell cycle, a mutant unable to complete the cell cycle cannot grow into a population of cells. Researchers have obtained cell-cycle-defective mutants, however, by isolating cells with temperature-sensitive mutations (see the Fast Forward box in Chapter 6). In these mutants, a protein needed for cell division functions normally at a low *permissive temperature*, but loses function at a higher *restrictive temperature*. At the permissive temperature, the mutants grow nearly normally, producing a population of cells for study. A shift to the restrictive temperature causes the temperature-sensitive gene in the mutant population to become nonfunctional; researchers can then study the consequences of its loss.

To isolate temperature-sensitive mutations, investigators expose haploid cells to a mutagen and then plate them at the permissive temperature, allowing them to form colonies (Fig. 17.3). After the colonies grow up, each from a single mutant cell, the experimenters use replica plating to imprint them on two plates. They incubate one plate at the permissive temperature and the other plate at the restrictive temperature. Cells that have sustained a temperature-sensitive mutation grow at the permissive temperature but not at the restrictive temperature.

With this protocol, researchers have isolated thousands of temperature-sensitive mutations. These mutations could occur in any gene required for cell reproduction. Genes of particular interest for understanding the cell cycle are those whose protein product functions at only one stage in the cell cycle. It was possible to identify such mutants in *S. cerevisiae* by observing in the light microscope the shape and behavior of cells shifted from the permissive to the restrictive temperatures (Fig. 17.4). A population of cells growing at the permissive temperature

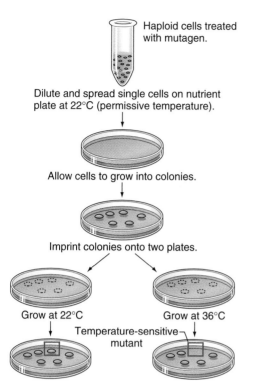

Haploid cells treated with mutagen.

Dilute and spread single cells on nutrient plate at 22°C (permissive temperature).

Allow cells to grow into colonies.

Imprint colonies onto two plates.

Grow at 22°C   Grow at 36°C

Temperature-sensitive mutant

Figure 17.3   The isolation of temperature-sensitive mutants of yeast. Mutations are induced in a culture of haploid cells by treating them with a chemical mutagen. The treated cells are distributed onto solid medium. Each cell reproduces a colony of cloned cells, passing on the mutation. Replicas of the colonies are imprinted onto solid medium. One is grown at the permissive temperature (22°C), one at the restrictive temperature (36°C). Colonies that grow on the former, but not the latter, carry a temperature-sensitive mutation.

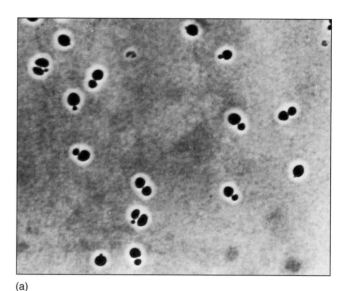

(a)

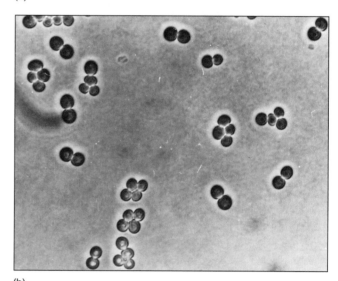

(b)

Figure 17.4   A cell cycle mutant of yeast. Cells of a temperature-sensitive mutant growing at the permissive temperature (a) display buds of all sizes. After incubation at the restrictive temperature (b), the same cells have arrested—all with a large bud. Cells that are early in the cell cycle at the time of the temperature shift (small buds in (a)) arrest in the first cell cycle and cells that are later in the cell cycle (large buds in (a)) finish the first cell cycle and arrest in the second.

includes unbudded cells as well as cells with the full range of bud sizes. After a cell-cycle mutant has grown at the restrictive temperature for about two cycles, however, the cells have a uniform appearance. In the mutant population shown in Fig. 17.4, for example, all cells have a single large bud. Moreover, the nuclei (although not visible in the figure) are also uniform, being located at a position between the mother cell and the daughter cell, as if beginning to divide. This uniformity identifies a particular cell-cycle mutant. Other cell-cycle mutants would arrest with different but also uniform morphologies, for example, with all unbudded cells. Thus, mutants that acquire a uniform bud-related morphology at the restrictive temperature are each defective at one stage of the cell cycle.

Further examination of cells transferred from permissive to restrictive temperatures illustrates another property of cell-cycle mutants—a requirement for the normal gene product at a particular stage of the cell cycle. Note that some of the cells in Fig. 17.4 formed one cell with a large bud at the restrictive temperature, while others formed two cells, each with a large bud. Note also that the former all had smaller buds than the latter at the time of the shift to the restrictive temperature. This observation indicates that cells early in the cell cycle at the moment of temperature shift arrested division in the first cell cycle, while those later in the cell cycle at the time of temperature shift finished the first cycle and became arrested only in the second

cell cycle. The point at which a cell acquires the ability to complete a cell cycle is the moment at which the temperature-sensitive protein has fulfilled its function in that cycle.

By analyzing the morphology of buds on cells shifted from permissive to restrictive temperatures, yeast geneticists have identified about 70 cell-cycle genes (see Table 17.1). Epistasis experiments comparing the phenotype of a double mutant to that of each of the two single mutants have enabled them to order some of these mutations with respect to one another (Fig. 17.5). One of the most important mutations, in the *CDC28* gene, identified the first step in the cell cycle, a step occurring prior to that catalyzed by the *CDC7* gene: After the *CDC28* step, the cell is committed to finishing the cycle it has begun before pursuing an alternative fate. Alternative fates

TABLE 17.1 Some of the Cell-Cycle Genes in Which Mutations Contribute to Cancer

Genes	Gene Products and Their Function
CDKs	Enzymes known as cyclin-dependent protein kinases that control the activity of other proteins by phosphorylating them
CDC28	A CDK discovered in the yeast *Saccharomyces cerevisiae* that controls several steps in the *S. cerevisiae* cell cycle
CDC2	A CDK discovered in the yeast *Schizosaccharomyces pombe* that controls several steps in the *S. pombe* cell cycle; also the designation for a particular CDK in mammalian cells
CDK4	A CDK of mammalian cells important for the G_1-to-S transition
CDK2	A CDK of mammalian cells important for the G_1-to-S transition
cyclins	Proteins that are necessary for and influence the activity of CDKs
cyclinD	A cyclin of mammalian cells important for the G_1-to-S transition
cyclinE	A cyclin of mammalian cells important for the G_1-to-S transition
cyclinA	A cyclin of mammalian cells important for S phase
cyclinB	A cyclin of mammalian cells important for the G_2-to-M transition
E2F	A transcription factor of mammalian cells important for the G_1-to-S transition
RB	A mammalian protein that inhibits E2F
p21	A protein of mammalian cells that inhibits CDK activity
p16	A protein of mammalian cells that inhibits CDK activity
p53	A transcription factor of mammalian cells that activates transcription of DNA repair genes as well as transcription of *p21*
RAD9	A protein that inhibits the G_2-to-M transition of *S. cerevisiae* in response to DNA damage
E6	A protein of the HPV virus that inhibits p53
E7	A protein of the HPV virus that inhibits Rb

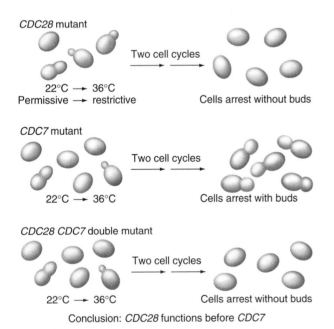

Figure 17.5 A double mutant reveals which gene is needed first. A temperature-sensitive *cdc28* mutant arrests as unbudded cells and a *cdc7* mutant as budded cells after a shift from permissive to restrictive temperature. The double mutant arrests with the phenotype of the *cdc28* mutant, revealing that this step occurs before the *cdc7* step.

include the arrest of cell division in response to nutrient starvation or fusion with a cell of opposite mating type to form a diploid cell (Fig. 17.6). The decision to pursue a particular fate occurs in the G_1 phase of the cell cycle at a step called "start." If cells have completed this *CDC28*-determined step, they cannot consider the alternatives until they have finished the cell cycle and arrived at the next G_1. The *CDC28* gene thus controls the commitment to proceed through the cell cycle.

The significance of the *CDC28* gene became apparent when geneticists identified related genes in other organisms. They found, for example, that the *CDC2* gene in fission yeast controls a step of commitment in that cell. They also learned that in extracts of *Xenopus* (African toad) embryos, the activity of a protein known as MPF (for maturation-promoting factor) controls the rapid early divisions. Sequences of the cloned budding and fission yeast genes revealed that they encode *protein kinases:* enzymes that add phosphate groups to their substrates. The *Xenopus* MPF also turned out to be a protein kinase. Moreover, genetic swapping experiments showed that the budding yeast *CDC28* gene and the fission yeast *CDC2* gene can replace one another in either organism, demonstrating that they encode proteins that carry out the same activity; the same is true of the *Xenopus* MPF-encoding gene.

Thus in three different organisms, genes that seem to be the central controlling element of the cell cycle encode functionally homologous protein kinases. Further work has shown that these kinases are **cyclin-dependent kinases** (**CDK**s), that is, they require another protein known as a **cyclin** for their activity.

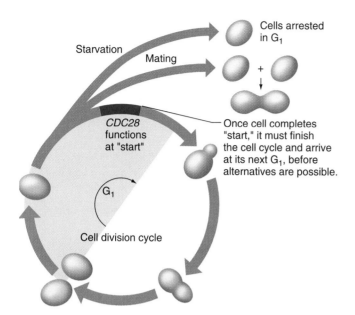

Figure 17.6 Yeast cells become committed to the cell cycle in G₁. A haploid yeast cell prior to "start" in the G₁ phase of the cell cycle is capable of arresting growth due to starvation, mating with a cell of opposite mating type, or dividing. Once the cell has passed "start" it is committed to completing the cell cycle.

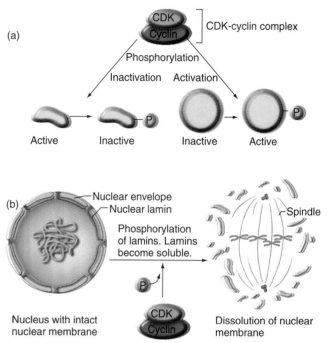

Figure 17.7 The cyclin-dependent kinases (CDK) control the cell cycle by phosphorylating other proteins. (a) A CDK combines with a cyclin and acquires the capacity to phosphorylate other proteins. Phosphorylation of a protein can either inactivate or activate it. (b) CDK phosphorylation of the nuclear structure proteins, lamins, is responsible for the dissolution of the nuclear membrane at mitosis.

Cyclin-Dependent Kinases and Cyclins Set the Times of Cell-Cycle Events

A protein kinase is an enzyme that adds a phosphate group to a target protein molecule; the addition either activates or inactivates the target protein. The CDKs are a family of kinases that regulate the transition from G₁ to S and from G₂ to M through phosphorylations that activate or inactivate target proteins. As mentioned and as their name implies, cyclin-dependent kinases function only after associating with another type of protein known as a cyclin. The cyclin portion of a CDK-cyclin complex specifies which set of proteins a particular CDK will phosphorylate; the CDK portion of the complex then performs the phosphorylation (Fig. 17.7a). One CDK-cyclin, for example, activates target proteins required for DNA replication at the onset of the S phase, whereas another CDK-cyclin activates target proteins necessary for chromosome condensation and segregation at the beginning of the M phase. The cyclins that guide the CDK phosphorylations appear on cue at each phase of the cell cycle, associate with the appropriate CDKs, point out the proper protein targets, and then disappear to make way for the succeeding set of cyclins. The cycle of precisely timed cyclin appearances and disappearances is the result of two mechanisms: gene regulation that turns on and off the synthesis of particular cyclins and regulated protein degradation that removes the cyclins.

The cyclin-dependent kinases are ideal enzymes for the coordination of complex events. Many copies of a particular CDK can simultaneously activate many different proteins throughout the cell, such as the proteins necessary for chromosome condensation, spindle assembly, and nuclear membrane breakdown; binding with the appropriate cyclin guides the CDK to these targets as the cell prepares for mitosis.

The target of an enzyme is its **substrate.** Although there is little direct information about most of the cellular proteins that serve as CDK substrates, scientists who study the cell cycle have confirmed one: the **nuclear lamins,** a group of proteins that underlie the inner surface of the nuclear membrane (Fig. 17.7b). The nuclear lamins probably provide structural support for the nucleus and possibly provide sites for the assembly of proteins that function in DNA replication, transcription, RNA transport, and chromosome structure. During most of the cell cycle, the lamins form an insoluble structural matrix. At mitosis, however, the lamins become soluble, and this solubility allows dissolution of the nuclear membrane into vesicles. Lamin solubility requires phosphorylation; mutant lamins that resist phosphorylation do not become soluble at mitosis. Thus, one critical mitotic event—dissolution of the nuclear membrane—is most likely triggered by CDK phosphorylation of nuclear lamins. It is highly likely that CDK phosphorylation of particular chromatin proteins activates other events of mitosis, such as chromosome condensation.

Genetic studies of yeast provided much of the evidence that CDK-cyclin complexes are key controlling agents in all eukaryotic cell cycles. In one series of studies, geneticists used yeast mutants that carry defective CDKs or cyclins to find the corresponding human genes (Fig. 17.8) and to show that the human CDKs and cyclins can function in yeast in place of the native proteins.

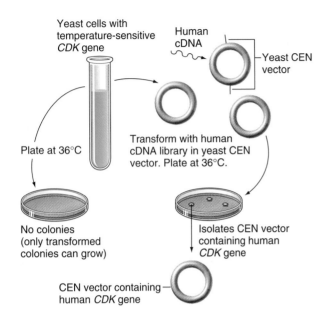

Figure 17.8 Mutant yeast permit the cloning of a human CDK gene. A culture of yeast cells containing a temperature-sensitive mutation in the yeast CDK gene were transformed with a library composed of human cDNA cloned into a yeast centromere-containing vector. The transformed yeast cells were spread on solid medium at the restrictive temperature. Only the rare transformants with a functional copy of the human CDK gene were able to grow.

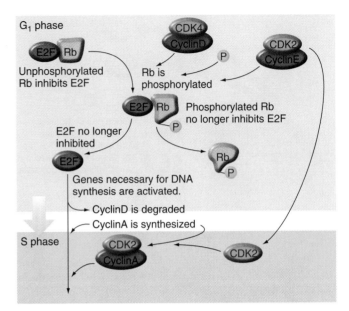

Figure 17.9 CDKs mediate the transition from the G_1 to the S phase of the cell cycle. In human cells CDK4 complexed to cyclinD and CDK2 complexed to cyclinE phosphorylate the Rb protein, causing it to dissociate from, and activate, the E2F transcription factor. E2F stimulates transcription of many genes needed for DNA replication. At the transition into S phase, cyclinD is destroyed, cyclinA is synthesized, and the CDK2–cyclinA complex activates DNA replication.

How Molecular Interactions Control the G_1-to-S Transition

Cell-cycle investigators have identified many of the molecular events controlling the transition from G_1 to S in human cells by analogy with similar events in the cell cycle of yeast. From their analyses, they have pieced together the following scenario.

The first CDK-cyclin complexes to appear during G_1 in humans are CDK4-cyclinD and CDK2-cyclinE (Fig. 17.9). These complexes initiate the transition to S by a programmed succession of specific phosphorylations, among which are phosphorylations of the protein product of the retinoblastoma (*RB*) gene. Unphosphorylated Rb protein inhibits a transcription factor, E2F. Phosphorylated Rb no longer inhibits E2F. Rb phosphorylation thus indirectly activates DNA synthesis by releasing the brakes on E2F and thereby allowing it to activate the transcription of genes necessary for DNA synthesis.

As cells initiate DNA synthesis, cyclinD and cyclinE are degraded, and a new cyclin—cyclinA—is synthesized. CyclinA serves as a kinase guide during S. As the cyclinA complexes with the appropriate CDK, the cell rapidly degrades the cyclinD it is replacing. The importance of cyclin degradation is evident from yeast cells in which mutations in genes controlling cyclin degradation prevent the destruction of G_1 cyclins; in these mutants, cells remain in G_1, unable to initiate DNA replication.

How Molecular Interactions Control the G_2-to-M Transition

Human cells appear to make the transition from G_2 to mitosis much as the well-studied cells of the yeast *S. pombe* accomplish the same transition. In the yeast, a CDK known as CDC2

(the second C replaces the K for historical reasons) forms a complex with cyclinB. Both the CDC2 kinase and cyclinB are present throughout G_2, but phosphorylation of a specific tyrosine residue on the cyclin-dependent kinase (by another protein kinase) keeps it inactive. When the time comes to initiate mitosis, a phosphatase enzyme removes the phosphate group from the CDC2 tyrosine, this removal activates the CDK, and the cell enters mitosis (Fig. 17.10).

Summary of Controls Intrinsic to the Cell Cycle

The activation and inactivation of CDK-cyclin complexes trigger transitions of the cell cycle. A variety of controls set the timing of these events. For example, phosphorylation and dephosphorylation at specific sites inactivate and activate CDKs. The transcriptional regulation of cyclin synthesis and the rapid degradation of cyclins at the completion of their specified tasks make these proteins appear and disappear at appropriate times.

Cell-Cycle Checkpoints Integrate Repair of Chromosomal Damage with Events of the Cell Cycle

Damage to a cell's genome, whether caused by environmental agents or random errors of the cellular machinery as it attempts to replicate and segregate the chromosomes, can cause serious problems for the cell. Damage to the cell-cycle machinery can also cause problems. It is therefore not surprising that elaborate mechanisms have evolved to arrest the cell cycle while repair takes place. These additional controls are called **checkpoints** because they check the integrity of the

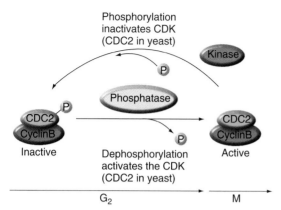

Figure 17.10   CDK activity in yeast is controlled by phosphorylation and dephosphorylation. The CDC2 protein complexed with cyclinB is inactivated prior to mitosis by phosphorylation of a specific kinase and then activated by dephosphorylation at the onset of mitosis by a specific phosphatase.

genome and cell-cycle machinery before allowing the cell to continue to the next phase of the cell cycle.

The G_1-to-S Checkpoint

When radiation or chemical mutagens damage DNA during G_1, DNA replication is postponed. This postponement allows time for DNA repair before the cell proceeds to DNA synthesis. Replication of the unrepaired DNA could exacerbate the damage; for example, replication over a single-strand nick or gap would produce a double-strand break. In mammals, cells exposed to ionizing radiation or UV light during G_1 delay entry into S phase by activating the p53 pathway (Fig. 17.11a). p53 is a transcription factor that induces expression of DNA repair genes as well as expression of the **CDK inhibitor** known as p21. Like other CDK inhibitors, p21 binds to CDK–cyclin complexes and inhibits

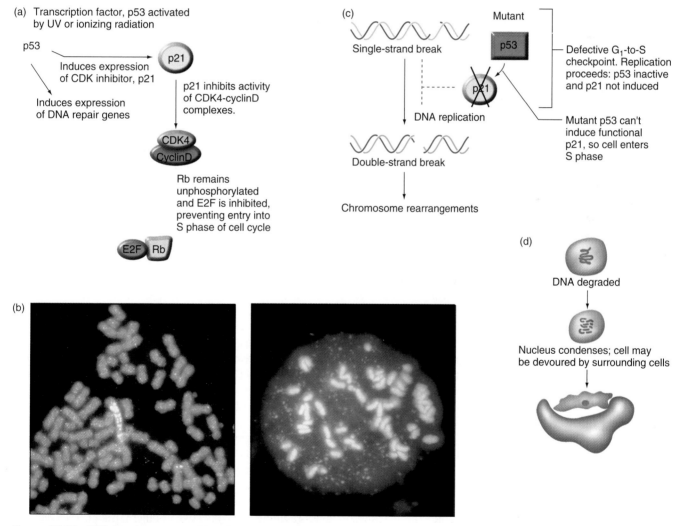

Figure 17.11   Cellular responses to DNA damage. (a) DNA damage activates the p53 transcription factor, which in turn, induces expression of the p21 gene. The p21 protein inhibits CDK activity, producing an arrest of the cell cycle in the G_1 phase. (b) Tumor cells exhibit amplified regions of DNA, unlike normal cells, that can appear as homogeneously staining regions (HSR) within a chromosome or as double minutes, small pieces of extrachromosomal DNA. (c) When the p53 gene is mutated in many cancers, p21 is not induced, cell cycle progress is not arrested, and cells replicate damaged DNA, producing DNA double-strand breaks from single-strand nicks or gaps. (d) DNA damage in normal cells often leads to cell death or apoptosis, in which cellular DNA is degraded and the cell is engulfed and digested by neighboring cells.

their activity; specifically, p21 prevents entry into S by inhibiting the activity of CDK4-cyclinD complexes.

Mutations in *p53* disrupt the G_1-to-S checkpoint. One sign of this disruption is a propensity for **gene amplification:** an increase from the normal two copies to hundreds of copies of a gene. This amplification is visible under the microscope, appearing as an enlarged area within a chromosome known as a homogeneously straining region (HSR) or as small chromosome-like bodies (called *minutes*) that lack centromeres and telomeres (Fig. 17.11b). Normal human cells do not generate gene amplification in culture, but *p53* mutants exhibit high rates of such amplification; *p53* mutants also exhibit many types of chromosome rearrangements. The explanation is as follows. Cells carrying mutations in the *p53* gene most likely have a defective G_1-to-S checkpoint that allows the replication of single-strand nicks; this replication produces double-strand breaks, which in turn, lead to chromosome rearrangements; some of the rearrangements generate gene amplification (Fig. 17.11c).

Wildtype cells able to produce functional p53 not only arrest in G_1 in the presence of DNA damage; if the damage is great enough, they commit suicide in a process known as **programmed cell death (PCD),** or **apoptosis.** During apoptosis, the cellular DNA is degraded and the nucleus condenses; the cell may then be devoured by neighboring cells or by phagocytes (Fig. 17.11d). Programmed cell death and the proteins that regulate it—including those that are part of the p53 pathway—appear in multicellular animals from round worms to humans. It makes sense for multicellular organisms to have a mechanism for eliminating cells that have sustained chromosomal damage. The survival and reproduction of such cells could generate cancer cells.

The G_2-to-M Checkpoint

Damage to DNA during G_2 delays mitosis, allowing time for repair before chromosome segregation (Fig. 17.12a). Researchers have isolated yeast mutants that, after sustaining radiation-induced DNA damage, do not arrest in G_2 and used these mutants to identify many genes that contribute to this checkpoint. One of these genes is *RAD9*. Whereas wildtype yeast cells can pause to repair as many as 100 double-strand breaks before entering mitosis, *RAD9* mutants fail to arrest in G_2 and die as a result of any double-strand breaks that were not repaired before mitosis.

A Checkpoint in M

During mitosis, one checkpoint oversees formation of the mitotic spindle and proper engagement of all pairs of sister chromatids (Fig. 17.12b). Observations of living cells reveal that as chromosomes condense and attach to the spindle, sometimes a single chromosome fails to attach at the expected time. When this happens, the cell does not initiate sister chromatid separation or anaphase chromosome movement until the lagging chromosome attaches to the spindle. Other studies show that in yeast cells exposed to an inhibitor that prevents assembly of a functional spindle, sister chromatids remain firmly

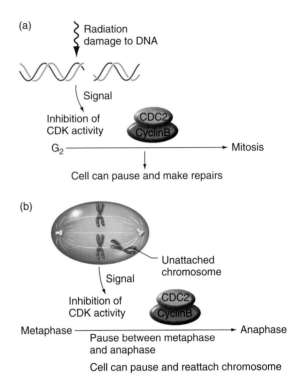

Figure 17.12 Two checkpoints act at the G_2-to-M cell cycle transition. (a) DNA damage, particularly double strand breaks, induce a signal that inhibits CDK activity preventing entry into mitosis. (b) Spindle damage resulting from failure of a chromosome to attach to the mitotic spindle results in a signal that inhibits CDK activity, preventing the metaphase to anaphase transition.

attached. These observations suggest the presence of a checkpoint that prevents chromosome segregation until all chromosomes are properly attached to the spindle. Mutations that eliminate the surveillance of chromosome behavior during mitosis have helped researchers identify several genes in yeast responsible for this checkpoint.

Checkpoints Ensure Genomic Stability

Checkpoints are not essential for cell division. In fact, experiments in mice and other animals demonstrate that mutant cells with one or more defective checkpoints are viable and divide at a normal rate. These mutant cells, however, are much more vulnerable to DNA damage than normal cells.

Knowledge of how checkpoints work hand in hand with repair processes to ensure the fidelity of DNA replication and chromosome segregation clarifies how checkpoints help prevent transmission of three types of genomic instability (described in Chapter 12): chromosome aberrations; aneuploidy (the loss or gain of one or more chromosomes); and changes in ploidy, for example, from *2n* to *4n* (Fig. 17.13). Single-strand nicks resulting from oxidative or other types of DNA damage are probably fairly common. A cell normally repairs such nicks to DNA in G_1 before it enters S phase. If the checkpoint that makes time for and coordinates this repair fails to function, however, the copying of single-strand breaks during DNA replication would produce double-strand breaks that could lead

(a)

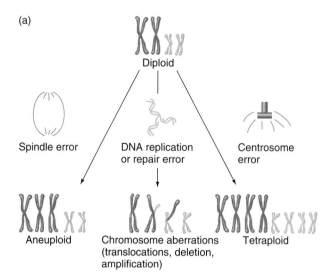

(b)

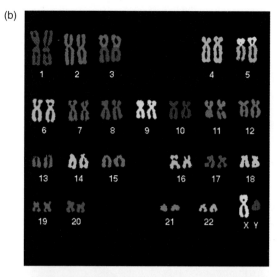

Figure 17.13   Three classes of error lead to aneuploidy in tumor cells. (a) Spindle errors can mis-segregate chromosomes resulting in whole chromosome aneuploidy; DNA replication and/or repair damage can lead to chromosome aberrations; centrosome errors can result in changes in cell ploidy. (b) The diagnosis of aneuploidy in tumors is improved by "chromosome painting" techniques that use a variety of fluorescent dyes attached to chromosomal DNA sequences. An appropriate choice of dyes and probes can cause each normal chromosome to appear relatively homogeneous with a unique color (top) while cancer cell chromosomes reveal many rearrangements and whole chromosome changes (bottom).

to chromosome rearrangements. Chromosome loss or gain can occur if a chromosome fails to attach to the spindle or if sister chromatids fail to disjoin. Normally, the G_2-to-M checkpoint recognizes such failures and prevents mitosis until the cell has fixed the problem. Cells without a functional checkpoint will produce daughter cells carrying one too few or one too many chromosomes. Changes in ploidy can occur if a cell begins S phase before completing mitosis or if a cell fails to replicate or to properly segregate its microtubule-organizing centers, or centrosomes. Checkpoints also recognize these errors, ensuring integration of the centrosome cycle with DNA replication and the formation and function of the mitotic spindle.

A Cascade of External and Internal Molecules Tells Cells Whether or Not to Initiate Division

How do cells know when to divide? To function according to the needs of the body as a whole, cells depend on signals sent from one tissue to another. These signals tell them whether to divide or metabolize (that is, make the product[s] they are programmed to make) or die. There are two basic types of signals.

Extracellular signals in the form of steroids, peptides, and proteins act over long or short distances and are collectively known as *hormones* (Fig. 17.14a). The thyroid-stimulating hormone (TSH) produced by the brain's pituitary gland, for example, travels through the bloodstream to the thyroid gland where it stimulates cells to produce another hormone, thyroxine, which increases metabolic rate. *Cell-bound signals,* such as the histocompatibility proteins that, like fingerprints, distinguish an individual's cells from all foreign cells and molecules, require direct contact between cells for transmission (Fig. 17.14b). The macrophages, helper T cells, and antibody-producing B cells of the immune system communicate via cell-bound signals about the presence of viral particles, bacteria, and toxins.

The Molecular Components of Each Signaling System: Growth Factors, Receptors, Intracellular Transducers, and Transcription Factors

Extracellular hormones and cell-bound signals that stimulate or inhibit cell proliferation are known as **growth factors.** Most growth factors deliver their message to specific **receptors**

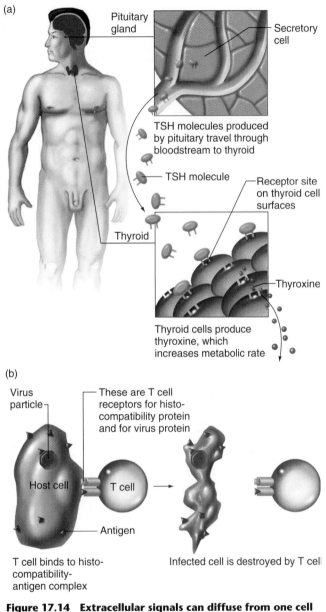

Figure 17.14 Extracellular signals can diffuse from one cell to another or be delivered by cell–cell contact. (a) The pituitary gland produces thyroid-stimulating hormone (TSH) that diffuses through the circulation to the thyroid gland, which produces another hormone, thyroxine, that acts on many cells throughout the body. (b) A killer T cell recognizes its target cell by direct cell–cell contact.

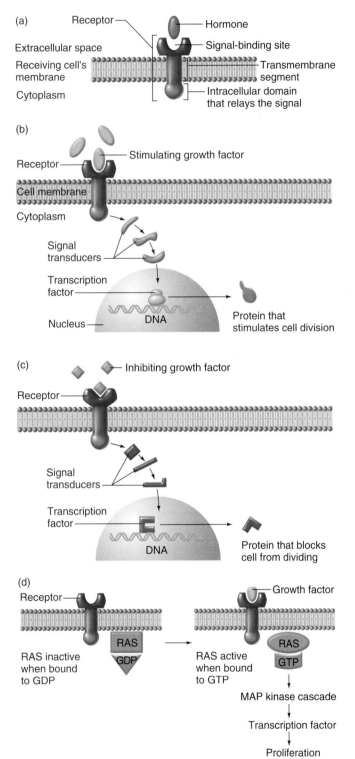

Figure 17.15 Many hormones transmit signals into cells through receptors that span the cellular membrane. (a) Hormones bind to a cell surface receptor that specifically recognizes their structure. The extracellular surface of the receptor transmits a signal to the intracellular domain of the receptor, which in turn, interacts with other signalling molecules in the cell either to stimulate growth (b) or to inhibit growth (c). The RAS protein is one intracellular signalling molecule that is induced to exchange a bound GDP (inactive) for a bound GTP (active) that binds to the receptor with which it interacts (d).

embedded in the membrane of the receiving cell (Fig. 17.15a). These receptors are proteins that have three parts: a signal-binding site outside the cell, a transmembrane segment that passes through the semipermeable cell membrane, and an intracellular domain that relays the signal (that is, the binding of growth factor) to proteins inside the cell's cytoplasm. These cytoplasmic proteins, which relay signals inside the cell, are known as **signal transducers.** The final link in this relay system is usually a **transcription factor** that activates the expression of specific genes in the nucleus, either to promote or to inhibit cell proliferation (Fig. 17.15b–c).

How the Molecules Interact to Relay a Signal

Binding of a growth factor to its specific receptor elicits a cascade of biochemical reactions inside the cell, often involving a large number of molecules. Each molecule in the cascade transmits the receptor's binding-of-messenger signal by activating or inhibiting another molecule. The activation and inhibition of intracellular targets after growth-factor binding is called **signal transduction.**

One example of a signal transduction system includes the product of the *RAS* gene (Fig. 17.15d). The RAS protein is a molecular switch that exists in two forms: an inactive form in which it is bound to guanosine diphosphate (RAS-GDP) and an active form in which it is bound to guanosine triphosphate (RAS-GTP). (GTP is also the nucleotide used to insert a G residue into an RNA molecule.) Once a growth factor activates a receptor, the receptor flicks the RAS switch to active by exchanging GDP for GTP. Next RAS-GTP activates a series of three protein kinases, and this trio, known as a MAP kinase cascade, activates a transcription factor.

The proteins in a signal transduction system are like the neurons in a nerve fiber: Each one serves as a link in a message-relay chain. In deciding whether or not to divide, the cell, like the brain, combines messages from many signal transduction systems and adjusts its behavior in response to the integrated information.

As we see in the next section, cells that have lost the ability to reproduce their genomes faithfully, due either to defects in the DNA repair systems described in Chapter 6 or to defects in checkpoints, are prone to mutations that cause cancer.

CANCER ARISES WHEN CONTROLS OVER CELL DIVISION NO LONGER FUNCTION PROPERLY

An understanding of the molecular basis of cell-cycle regulation sheds light on the life-threatening proliferative disease of cancer. The many genes contributing to the normal control of cell proliferation are all subject to mutations. Such mutations may produce inappropriate signals about the need for cell division, malfunctions in the CDK-cyclin complexes that control cell-cycle transitions, or checkpoint breakdowns that lead to genomic instability.

The Cancer Phenotype Results from the Accumulation of Multiple Mutations in the Clonal Progeny of a Cell

Cancer biologists now believe that most cancers result from the accumulation of many mutations during the proliferation of somatic cells. When enough mutations accumulate in genes controlling proliferation and other processes within a single clone of cells, that clone overgrows the normal cells that surround it, disseminates through the bloodstream to other parts of the body, and forms a life-threatening tumor, or cancer. (In this chapter, we use the term "tumor" to designate cancerous tissue; and the term "growth" to designate a benign mass.)

Epidemiological data, clinical studies, and experimental analyses of a range of cell types in a variety of species provide evidence for this gene-based view of cancer. Here we present some of that evidence as well as an interpretation of the evidence that explains how a cell progresses from normal to cancerous.

The General Cancer Phenotype Includes Many Types of Cellular Abnormalities

Theodor Boveri, one of the architects of the chromosome theory of inheritance, observed as early as 1914 that cells excised from malignant tumors have abnormal chromosomes. By the 1970s, when new staining techniques and improved equipment made it possible to distinguish each of the 23 different chromosome pairs in the human genome by their specific banding patterns, investigators noted that many different chromosomal abnormalities appear in tumor cells. Using tools developed in the 1980s, geneticists confirmed that most tumor cells exhibit karyotypic instability.

Figure 17.16 shows the main characteristics that distinguish tumor cells from normal cells. The cancer phenotype includes uncontrolled cell growth, genomic and karyotypic instability, the potential for immortality, and the ability to invade and disrupt local and distant tissues. Although no one cancer cell necessarily manifests all the phenotypic changes illustrated in Fig. 17.16, each cancer cell displays a number of them.

More than One Mutation Must Fuel the Progression from Normal to Cancerous

The large catalog of phenotypic changes seen in tumor cells suggests that many mutations—at least one for each phenotypic change—are necessary to convert a normal cell into a cancerous cell. In other words, a normal cell must change many aspects of its phenotype to become a life-threatening cancer cell, and this set of changes probably requires the alteration of many genes. Clinical analyses of human cells from colon tumors show that these cancer cells contain mutations in at least 5 to 10 genes. Since the intricate web of controls that organize and maintain the society of cells constituting our body consists of many gene-encoded proteins, the number of genes in which mutations can fuel the progression to cancer is probably quite large—at least 100 and possibly several hundred.

To study mutations associated with cancer, researchers initially identify and isolate a mutation of interest by linkage analysis of markers, traditional genetic mapping to a chromosome, and positional cloning (all techniques described in Chapter 10). It is possible to test in mice whether a mutation in a single gene associated with cancer is sufficient to induce a tumor. If the mutation acts in a dominant fashion, researchers insert a copy of the mutant allele into the mouse genome of a fertilized egg; if the mutation is recessive, they delete one copy of the homologous gene from the early embryonic mouse genome and then breed animals homozygous for the deletion. In gene-transfer experiments where a dominant cancer-causing mutation was inserted into a mouse genome under the control of a breast-cell-specific promoter, the transgenic mice

(a) MOST NORMAL CELLS   **MANY CANCER CELLS**

1. **Autocrine stimulation**

Absent   Present

2. **Contact inhibition**

Present   Absent

3. **Cell death**

Irradiation   Cell death   Irradiation

Present   Absent

4. **Gap junctions**

Present   Absent

a. Changes That Produce Uncontrolled Cell Growth

1. *Autocrine stimulation.* Most cells "decide" whether or not to divide only after receiving signals from neighboring cells, either positive signals that stimulate division or negative signals that prevent proliferation. Many tumor cells, by contrast, make their own stimulatory signals, in a process known as **autocrine stimulation,** or are insensitive to negative signals.

2. *Loss of contact inhibition.* Normal cells stop dividing when they come in contact with one another, as evidenced by the fact that the few normal cell types that grow in culture form sheets one cell thick. Tumor cells, which have lost the property of contact inhibition, climb all over each other to produce piles that are many cells thick. This change in behavior contributes to the disordered array of cells seen in tumors, which is a significant departure from the highly ordered patterns seen in normal tissues.

3. *Loss of cell death.* Normal cells die when starved of growth factors or when exposed to agents such as toxins or X rays that damage them. Their programmed cell death is activated by the expression of certain genes in the cell. Because it keeps cells from proliferating when they should not and eliminates badly damaged cells, apoptosis is probably a safeguard against the early stages of cancer. Most cancer cells are much more resistant than normal cells to programmed cell death.

4. *Loss of gap junctions.* Normal cells connect to their neighbors by small pores, or gap junctions, in their membranes. The gap junctions permit the transfer of small molecules that may be important in controlling cell growth. Most tumor cells have lost these channels of communication.

Feature Figure 17.16 Phenotypic changes that distinguish tumor cells from normal cells.

b. Changes That Produce Genomic and Karyotypic Instability

1. *Defects in the DNA replication machinery.* Cancer arises most often in cells that have lost the ability to reproduce their genomes faithfully. We saw in Chapters 5 and 6 that cells have elaborate systems for repairing DNA damage; these systems include the enzymatic machinery for mismatch repair and the repair of damage caused by radiation or ultraviolet light. Work on yeast and bacteria has shown that mutant organisms defective in DNA repair have enormously increased rates of mutation. And these increased mutation rates often lead to cancer. An example is the hereditary syndrome of xeroderma pigmentosum, which predisposes to skin cancer (see Fig. 5.6).

2. *Increased rate of chromosomal aberrations.* Tumor cell karyotypes often carry gross rearrangements, including broken chromosomes, with some of the pieces rejoined to other chromosomes; multiple copies of individual chromosomes, rather than the normal two; and deletions of large chromosomal segments and of whole chromosomes. Studies have confirmed that the fidelity of chromosome reproduction is greatly diminished in tumor cells. Normal fibroblast cells, for example, have an undetectable rate of gene amplification (an increase in the number of copies of a gene), whereas tumor cells have amplification rates as high as 1 in 100 cells. Some of the genes that control cell proliferation are inappropriately activated by such amplification.

Although research has confirmed that tumor cells have an increased rate of chromosomal aberration, probably only a small fraction of these chromosomal rearrangements lead to cancer; for example, tumors from solid tissues typically carry many chromosomal rearrangements, but most of these aberrations do not recur in all tumors. A few rearrangements, however, regularly appear in specific tumor types. Examples include the translocation between chromosomes 8 and 14 found in patients with certain kinds of lymphoma and the translocation between chromosomes 9 and 22 found in certain types of leukemias (see Fig. 12.12).

(b.1)

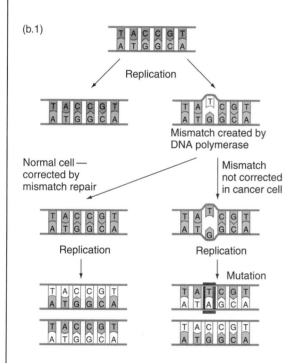

(b.2)

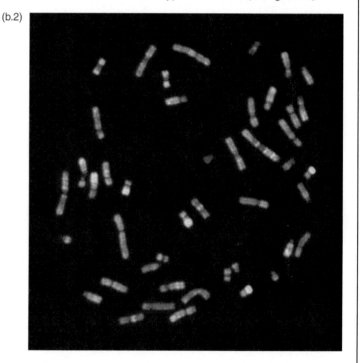

c. Changes That Produce a Potential for Immortality

1. *Loss of limitations on the number of cell divisions.* Most normal cells (except for the rare stem cells), die spontaneously after a specifiable number of cell divisions. The senescence and natural death of normal cells is evident both in culture and in the body. Tumor cells, by contrast, can divide indefinitely.
2. *Ability to grow in culture.* Cells derived from tumor cells usually grow readily in culture, making cancerous cell lines available for study. Normal cells do not grow well in culture. This difference between abnormal and normal is greatest in cells grown as isolated clones in soft agar, most likely because the normal cells must undergo more genetic changes to become competent to divide under the artificial conditions in culture than tumor cells.
3. *Restoration of telomerase activity* (not shown). Most normal human somatic cells do not express the enzyme telomerase, and this lack of telomerase expression prevents them from replicating the repeated sequences in the telomeres at the ends of their chromosomes (see Chapter 11). As a result, after a certain number of cell divisions, the telomeres shorten to the point where they contribute to cell senescence and death. Tumor cells have the ability to express telomerase, a feature that most likely contributes to their immortality.

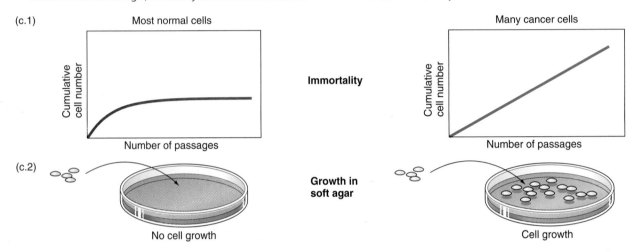

d. Changes That Enable a Tumor to Disrupt Local Tissue and Invade Distant Tissues

1. *The ability to metastasize.* Normal cells stay within rigidly defined boundaries. Tumor cells, by comparison, often acquire the capacity to invade surrounding tissues, and eventually to travel through the bloodstream to colonize distant tissues. Metastasis—the invasion of other tissues—is a complicated behavior requiring many genetic changes.
2. *Angiogenesis.* Once the adult human body has developed, new blood vessels do not normally form except to heal a wound. Tumor cells, however, secrete substances that cause blood vessels to grow toward them. The new vessels serve as supply lines through which the tumor can tap new sources of nutrients; they also serve as escape routes through which tumor cells can metastasize.
3. *Evasion of immune surveillance* (not shown). The human immune system may recognize cancer cells as foreign and attack them, thereby helping to eliminate tumors before they become large enough for clinical detection. As evidence, cancer patients often have antibodies and/or killer T cells directed against their cancer cells. Successful tumor cells, however, somehow develop the ability to evade detection by the immune system.

d.1 Metastasis

Tumor cells

Basement membrane

d.2 Angiogenesis

Blood vessel

Feature Figure 17.16 Phenotypic changes that distinguish tumor cells from normal cells (continued).

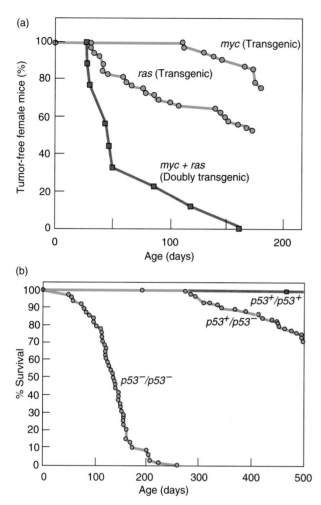

(a)

(b)

Figure 17.17   The percent of mice still alive as a function of age. (a) The activated *myc* oncogene produces tumors more slowly than the *ras* oncogene. Mice containing both oncogenes develop tumors even faster. (b) Homozygous *p53*⁺ mice rarely get life-threatening tumors, while those heterozygous for a *p53*⁻ mutation develop tumors late in life. Mice homozygous for the *p53*⁻ mutation develop tumors early in life.

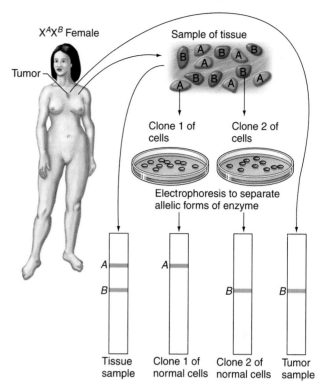

Figure 17.18   Polymorphic enzymes encoded by the X chromosome reveal the clonal origin of tumors. Each individual cell in a female expresses only one form of a polymorphic X-linked gene because of X chromosome inactivation. A patch of tissue will usually contain both types of cells. If single cells are grown into a clone they will exhibit one or the other enzyme form. A tumor also exhibits only one form, demonstrating that it arose from a single cell. The two forms are distinguished by electrophoresis.

produced a few breast tumors (Fig. 17.17a). Doubly transgenic mice made by breeding these transgenic mice carrying one mutated gene with transgenic mice carrying a different mutated gene implicated in cancer generated more tumors earlier. Even in these mice, however, only a small percentage of the transformed cells proliferated abnormally to form a tumor. These results support the idea that it takes several mutations in different genes to produce a cancer.

Studies of recessive mutations in the *p53* gene point to the same conclusion. Mice with both copies of the *p53* gene deleted from their genome develop relatively normally. The *p53* mutant mice, however, have shortened life spans and get a variety of tumors more frequently than wildtype mice (Fig. 17.17b). This experiment shows that the *p53* gene is not essential for development or for normal cell function, but it does play a role in preventing tumor formation. Consequently, deleting both copies of the wildtype gene from all of a mouse's cells does not convert every cell to a tumor cell, but it does increase

the probability that at least one cell will become cancerous. The conclusion is that mutations in *p53* are just one of the many genetic changes that may occur in a cell to produce cancer.

Cancers Are Clonal Descendants of One Cell

Karyotypes of cells from women heterozygous for X-linked alleles provide evidence that cancer originates in a single somatic cell (Fig. 17.18). Although the random inactivation of one of the two X chromosomes in each cell of a female means that individual cells express only one of the two X-linked alleles for any gene, in small samples of normal somatic tissues, one usually finds both alleles expressed. This is because most somatic tissues are constructed from many clones of cells. In contrast to normal tissue, tumors from females invariably express only one allele of an X-linked gene (review discussion of X-inactivation in Chapter 11). This finding suggests that the cells of each tumor are the clonal descendants of a single somatic cell that sustained a rare mutation.

Most Cancers Result from Exposure to Environmental Mutagens

Several epidemiological surveys support the hypothesis that most cancers arise by chance in somatic cells during their division and differentiation from fertilized egg to adult. The mutations that produce these cancers are not inherited through the germ line in a dominant or recessive pattern; rather they arise sporadically in a population as a result of chemicals or viruses in the environment. The evidence is as follows. First, the degree of concordance for cancers of the same type among first-degree relatives, such as sisters and brothers or even identical twins, is low for most forms of cancer in the population as a whole (we discuss specific exceptions later). If one sibling or twin gets a cancer, the other usually does not. Second, although rates for the incidence of specific cancers vary worldwide (Table 17.2), when populations migrate from one place to another, their profile of cancer incidence becomes more like that of the people indigenous to the new location. The change in cancer profile often takes decades, suggesting that the environment acts over a long period of time to induce the cancer. Third, epidemiological studies have established that numerous environmental agents increase the likelihood of cancer, and many of these agents are mutagens. These mutagens include cancer-causing viruses, some of which carry mutant forms of normal genes that control the cell cycle, as well as cigarette smoke. People who smoke for many years have a higher risk of lung cancer than people who do not smoke; and their risk increases with the number of cigarettes and the length of time they smoke.

Cancer Develops over Time

The data on lung cancer shows that decades elapse between the time a population begins smoking and the time that lung cancer begins increasing. In the United States, cancer inci-

dence in men rose dramatically after 1940, roughly two decades after men began frequent smoking; women did not begin frequent smoking until several decades after men, and lung cancer incidence in women did not begin its dramatic increase until after 1960 (Fig. 17.19a).

Epidemiological data also show that the incidence of cancer rises with age. The prevalence of cancer in older people supports the idea that cancer develops over time as well as the idea that the accumulation of many mutations in the clonal descendants of a somatic cell fuels the progression from normal to cancerous. If you assume that the rate of accumulation of cancer-causing mutations is constant over a lifetime, the slope of a curve plotting cancer incidence against age is a measure of the number of mutations required for cancer (Fig. 17.19b). Interestingly, the data for many types of tumors generate a similar curve in which the evolution of cancer requires 6–10 mutations. Thus, the correlation between cancer incidence and aging, as well as the time lag between exposure to carcinogens and the appearance of tumors, suggests that the mutations that produce cancer accumulate over time.

Although Most Cancers Are Sporadic, Some Cancers Run in Families

In some families, a specific type of cancer recurs in many members, indicating the inheritance of a predisposition through the germ line. We saw in Chapter 12 that retinoblastoma is an example of this type of cancer (review Fig. 12.6).

TABLE 17.2 **The Incidence of Some Common Cancers Varies between Countries**

Site of Origin of Cancer	High-Incidence Population		Low-Incidence Population	
	Location	*Incidence**	*Location*	*Incidence**
Lung	USA (New Orleans, blacks)	110	India (Madras)	5.8
Breast	Hawaii (Hawaiians)	94	Israel (non-Jews)	14.0
Prostate	USA (Atlanta, blacks)	91	China (Tianjin)	1.3
Cervix	Brazil (Recife)	83	Israel (non-Jews)	3.0
Stomach	Japan (Nagasaki)	82	Kuwait (Kuwaitis)	3.7
Liver	China (Shanghai)	34	Canada (Nova Scotia)	0.7
Colon	USA (Connecticut, whites)	34	India (Madras)	1.8
Melanoma	Australia (Queensland)	31	Japan (Osaka)	0.2
Nasopharynx	Hong Kong	30	UK (southwestern)	0.3
Esophagus	France (Calvados)	30	Romania (urban Cluj)	1.1
Bladder	Switzerland (Basel)	28	India (Nagpur)	1.7
Ovary	New Zealand (Polynesian Islanders)	26	Kuwait (Kuwaitis)	3.3
Pancreas	USA (Los Angeles, Koreans)	16	India (Poona)	1.5
Lip	Canada (Newfoundland)	15	Japan (Osaka)	0.1

*Incidence indicates number of new cases per year per 100,000 population, adjusted for a standardized population age distribution (so as to eliminate effects due merely to differences of population age distribution). Figures for cancers of breast, cervix, and ovary are for women; other figures are for men.
Adapted from V.T. DeVita, S. Hellman, and S.A. Rosenberg (eds.), *Cancer: Principles and Practice of Oncology*, 4th ed. Philadelphia: Lippincott, 1993; based on data from C. Muir et al., *Cancer Incidence in Five Continents*, Vol. 5. Lyon: International Agency for Research on Cancer, 1987.

(a) Lung cancer death rates, United States, 1930-91

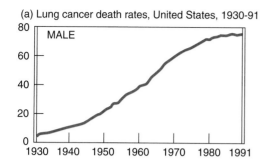

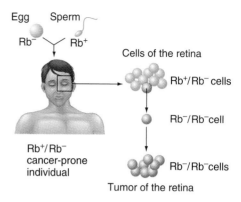

Figure 17.20   Individuals who inherit one copy of the Rb⁻ allele are prone to cancer of the retina. During the proliferation of retinal cells the Rb⁺ allele is lost or mutated and cancers grow out of the Rb⁻/Rb⁻ clone of cells.

Rates are per 100,000 and are age-adjusted to the 1970 U.S. census population.

(b)

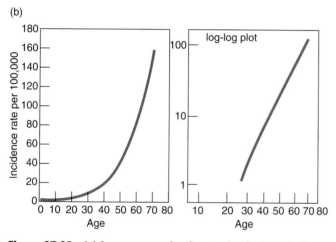

Figure 17.19   (a) Lung cancer death rates in the U.S. during the twentieth century began increasing rapidly for men in the 1940s and for women in the 1960s. This reflects the fact that smoking became prevalent among men about 20 years before it did in women. (b) Incidence of most cancers show a dramatic increase with age, a result thought to reflect the accumulation of mutations in somatic cells.

Half the individuals in families affected by retinoblastoma inherit a mutation in the *RB* gene from one parent. Since all their somatic cells carry one defective copy of the gene, a mutation in the single remaining wildtype copy of the *RB* gene in the cells that proliferate to produce the retina predisposes these cells to develop retinal cancer (Fig. 17.20). People who do not inherit a mutation in the *RB* gene need to experience a mutation in both copies of the gene in the same cell to develop cancer; this type of double hit is very rare. Interestingly, for nearly all common types of cancer that occur sporadically in a population, rare families can be found that exhibit an inherited predisposition to that cancer.

Interpretation of the Evidence: Cancer Develops When Cells in a Single Lineage Accumulate Enough Mutations to Overgrow Normal Cells

The epidemiological, clinical, and experimental evidence demonstrates that cancer cells have many phenotypic abnormalities. Underlying these phenotypic abnormalities are mutations. More than one mutation in the clonal descendants of a single cell is necessary to produce cancer. These mutations occur primarily in somatic cells and accumulate over time. Most mutations result from exposure to environmental mutagens and appear sporadically in a population, but the inheritance of certain mutations predisposes some families to cancers associated with those mutations.

A multistep model of cancer formation accounts for the evidence and provides a coherent interpretation of the disparate observations. According to this model, mutations accumulate over time in the cells of a clone, and this accumulation fuels the progression from normal to cancerous (Fig. 17.21).

The Mutations That Lead to Cancer Create Dominant Oncogenic Alleles or Recessive Tumor Suppressor Alleles

Research has not only revealed that cancer results from multiple genetic changes in the clonal descendants of one cell; it has also established that the mutations found in tumors are of two general types: those that improperly activate genes (for example, the genes responsible for promoting cell proliferation under clearly circumscribed conditions) and those that improperly inactivate genes (for example, the genes responsible for preventing excessive cell proliferation).

The mutant alleles that lead to cancer are referred to as **cancer genes;** but the term "genes" is a misnomer. *All cancer genes are, in fact, mutant alleles of normal genes.* When present in all or a subset of cells within an organism, these mutant alleles predispose the individual to develop cancer over a lifetime. Mutant alleles that act dominantly are known as

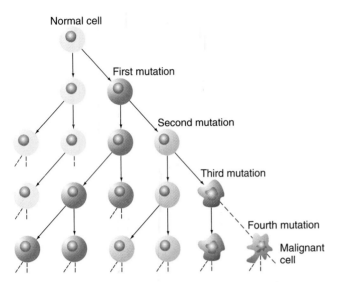

Figure 17.21 Cancer is thought to arise by successive mutations in a clone of proliferating cells.

oncogenes; in a diploid cell, only one mutant oncogenic allele alters the cell phenotype (Fig. 17.22a). Mutant alleles that are recessive are known as **mutant tumor-suppressor genes;** in a diploid cell, both copies of a tumor-suppressor gene must be mutant to make the cell abnormal (Fig. 17.22b).

Mutations That Create Oncogenes often Increase Cell Proliferation

Two approaches to identifying oncogenes are the study of tumor-causing viruses and the study of tumor DNA itself.

Tumor viruses are useful tools for studying cancer-causing genes because they carry very few genes and because they infect and change cultured cells to tumor cells, which makes it possible to study them *in vitro*. A large number of the viruses that generate tumors in animals are retroviruses whose RNA genome, upon infecting a cell, is copied to cDNA, which then integrates into the host chromosome (review the Genetics and Society box in Chapter 7). Later, during excision from the host chromosome, the virus can pick up copies of host genes. These normal genes change to abnormally activated onco-genes either through mutations that occur during viral propa-gation or through their placement near powerful promoters and enhancers in the viral genome (Fig. 17.23a). The oncogenes carried by tumor-producing viruses are thus mutated versions of normal host cell genes. The wildtype, or normal, genes that become oncogenes on mutation are known as **protoonco-genes.** When a virus carrying one or more oncogenes infects a cell, the oncogenes cause abnormal proliferation that can lead to the accumulation of more mutations and eventually to can-cer. The analysis of tumor-causing retroviruses led to the dis-covery of oncogenes in a variety of species (Table 17.3).

Some DNA viruses also carry oncogenes. An example is the human papilloma virus (HPV). HPV infection of a woman's cervical cells is probably the first step in the devel-opment of cervical cancer. The papilloma virus carries at least

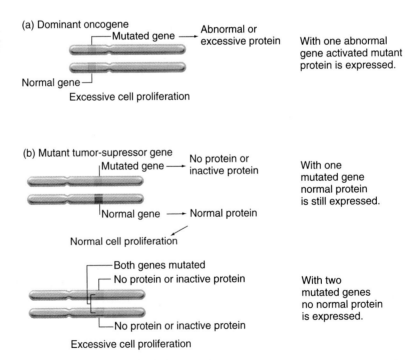

Figure 17.22 Cancer-producing mutations occur in two forms. (a) Dominant mutations produce oncogenes that exhibit an abnormal activity or excessive amount of protein. (b) Recessive mutations produce tumor suppressors that usually have little or no phenotype when they are heterozygous to a wildtype allele and that only affect cell proliferation when a second mutation inactivates the wildtype allele.

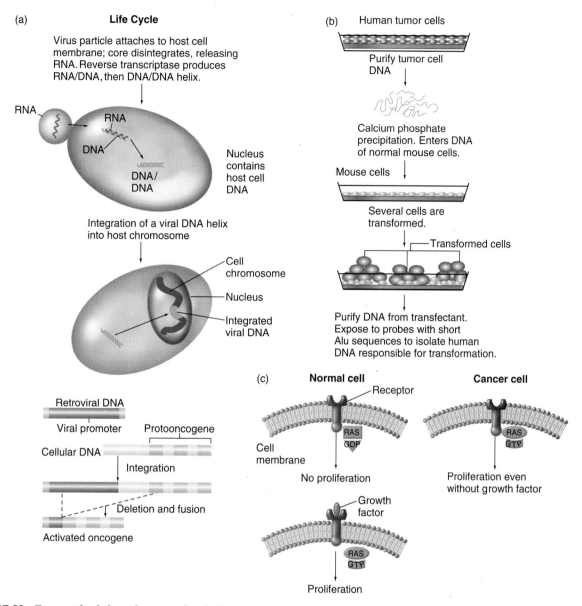

Figure 17.23 Two methods have been used to isolate oncogenes. (a) Retroviruses that cause cancer carry a mutant or overexpressed copy of a cellular growth-promoting gene. If a retrovirus integrates into the host chromosome near a protooncogene, the cellular gene may be packaged with the virus when it leaves the cell. (b) DNA isolated from some human cancers is able to transform mouse cells into cancer cells. These cells are found to contain a human oncogene. (c) The *RAS* oncogene is a mutant form of the *RAS* protooncogene in which the protein is locked into the GTP-activated form.

two oncogenes capable of transforming appropriate recipient cells in culture: *E6* and *E7*. The E6 and E7 proteins bind to and inactivate the normal products of the *p53* and *RB* genes. Only those HPV subtypes whose E6 and E7 proteins bind p53 and RB proteins are associated with cervical cancer in women. Progression of HPV-initiated cells to cancer requires additional mutations in genes not yet identified.

Scientists also identify oncogenes by isolating DNA from tumor cells and exposing noncancerous cells in culture to this tumor DNA. Some tumor DNA transforms cultured cells into cells capable of producing tumors (Fig. 17.23b). If one uses human tumor DNA in the transformation of mouse cells, for example, it is possible to identify the DNA responsible for the

transformation by reisolating the human DNA from the transformed mouse cells with probes for the short interspersed elements known as Alu sequences that appear only in the human genome (see Chapter 12). The oncogenes identified in this way, like those discovered in studies of tumor viruses, are oncogenic alleles of normal cellular genes that have mutated to abnormally active forms. Sometimes the two approaches have identified the same oncogene, for example, *RAS*. The oncogenic forms of the *RAS* gene generate proteins that are always (or constitutively) in the GTP activated form; hence, whether or not growth factor is present, a cell carrying a *RAS* oncogene receives signals to divide (Fig. 17.23c). Like the mutated *RAS,* many oncogenes continuously turn on one or

TABLE 17.3 Retroviruses and Their Associated Oncogenes*

Virus	Species	Tumor	Oncogene
Rous sarcoma	Chicken	Sarcoma	*src*
Harvey murine sarcoma	Rat	Sarcoma and erthyroleukemia	*H-ras*
Kristen murine sarcoma	Rat	Sarcoma and erthyroleukemia	*K-ras*
Moloney murine sarcoma	Mouse	Sarcoma	*mos*
FBJ murine osteosarcoma	Mouse	Chondrosarcoma	*fos*
Simian sarcoma	Monkey	Sarcoma	*sis*
Feline sarcoma	Cat	Sarcoma	*sis*
Avian sarcoma	Chicken	Fibrosarcoma	*jun*
Avian myelocytomatosis	Chicken	Carcinoma, sarcoma, and myeocytoma	*myc*
Ableson leukemia	Mouse	B cell lymphoma	*abl*

*Retroviruses identified as causative agents of tumors in animals contain oncogenes that were derived from a cellular gene.
Adapted from Lewin, *Genetics*, 1e, Oxford University Press, Inc. by permission.

TABLE 17.4 Oncogenes Are Members of Signal Transduction Systems*

Name of Oncogene	Tumor Associations	Mechanism of Activation	Properties of Gene Product
hst	Stomach carcinoma	Rearrangement	Growth factor
erb-B	Mammary carcinoma, glioblastoma	Amplification	Growth factor receptor
trk	Papillary thyroid carcinomas	Rearrangement	Growth factor receptor
Ha-ras	Bladder carcinoma	Point Mutation	GDP/GTP binding signaling protein
raf	Stomach carcinoma	Rearrangement	Cytoplasmic serine/threonine kinase
myc	Lymphomas, carcinomas	Amplification, chromosomal translocation	Nuclear transcription factor

*The roles of several of the oncogene products that are members of the signal transduction pathway and the ways in which they get activated in human cells are shown.

more of a cell's many signal transduction systems. They do this by encoding receptors, signal transmitters, and transcription factors that are active with or without growth factor (Table 17.4).

Excessive Proliferation Enhances the Potential for Mutation

Like the oncogenic *RAS* gene, many of the oncogenes so far identified affect cell-signaling pathways that tell a cell whether or not to proliferate. The importance of these genes in generating cancer is not just that they cause cells to proliferate, because an increase in proliferation alone, without other changes, generates benign growths that are not life threatening and can be removed by surgery. Rather, increased proliferation provides a large clone of cells within which further mutations can occur, and these further mutations may eventually lead to malignancy. The more cells that exist in a clone, the more likely that rare mutations will occur in the clone; and these additional mutations appear in a clone that already has the potential for rapidly propagating them.

Cancer researchers believe that oncogenes make up only a small percentage of the cancer genes identified to date; yet analyses of tumor viruses and of *in vitro* transformations by tumor DNA can identify only this subset. This is because both approaches depend on the addition of cancer-causing DNA to the wildtype genome of a normal cell. Under such circumstances, cancer can arise only if the cancer-causing allele is dominant to the normal counterpart in the host cell.

Mutations That Create Mutant Tumor-Suppressor Alleles Release a Brake on Cell Division and often Decrease the Accuracy of Cell Reproduction

Mutant tumor-suppressor genes are recessive alleles of genes whose normal alleles help put cell division on hold, whether in terminally differentiated cells or in cells with DNA damage. Targets for tumor-suppressor mutations include *RB, p53, p16,* and *p21.* One wildtype copy of these genes apparently produces enough protein to regulate cell division; the loss of both wildtype copies releases a brake on proliferation (see Fig. 17.22b). Researchers have identified dozens of tumor-

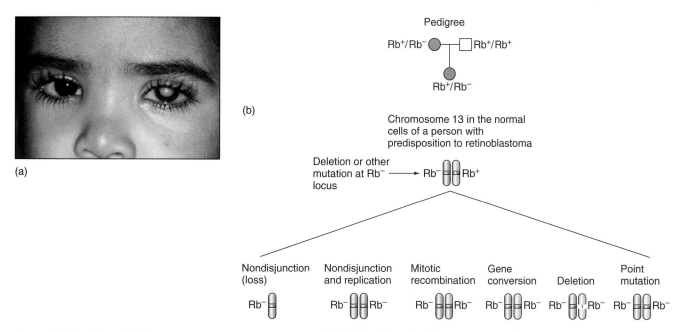

Figure 17.24 The retinoblastoma tumor suppressor gene. (a) A child with a retinoblastoma tumor in the left eye. (b) The Rb⁻ gene is inherited through the germ line as an autosomal recessive mutation. Subsequent changes to the Rb⁺ allele during somatic divisions generates a clone of cells homozygous or hemizygous for the Rb⁻ allele.

suppressor genes through the genomic analysis of families with an inherited predisposition to specific types of cancer or through the analysis of specific chromosomal regions that are reproducibly deleted in certain tumor types.

Retinoblastoma provides an example of this identification process. A cancer of the color-perceiving cone cells in the retina, retinoblastoma is one of several cancers inherited in a dominant fashion in human families (Fig. 17.24a). Roughly half of the children of a parent with retinoblastoma get the disease. Retinoblastoma tumors are easy to diagnose and remove before they become invasive. As we saw in Chapter 12, karyotypes of normal, noncancerous tissues from many people suffering from retinoblastoma reveal heterozygosity for deletions in the long arm of chromosome 13; that is, the patients carry one normal and one partially deleted copy of 13q (Fig. 17.24a). Karyotypes of the cancerous retinal cells from some of these same patients show homozygosity for the same chromosome 13 deletions that are heterozygous in the noncancerous cells (Fig. 17.24b). Although the deletions vary in size and position from patient to patient, they all remove band 13q14.

These observations indicate that band 13q14 includes a gene whose removal contributes to the development of retinoblastoma. *RB* is the symbol for this gene. The heterozygous cells in a patient's normal tissues carry one copy of the gene's wildtype allele (RB^+), and this one copy prevents the cells from becoming cancerous. Tumor cells homozygous for the deletion, however, do not carry any copies of RB^+, and without it, they begin to divide out of control.

Geneticists used their understanding of retinoblastoma inheritance to find the *RB* gene. They cloned DNA carrying the gene by looking for DNA sequences in band 13q14 that were lost in all of the deletions associated with the hereditary condition. They then identified the gene by characterizing a very small deletion that affected only one transcriptional unit—the *RB* gene itself. Analysis of the gene's function showed that it encodes a protein involved (along with many other proteins) in regulating the cell cycle. *RB* thus fits our definition of a tumor-suppressor gene; the protein it determines helps prevent cells from becoming cancerous. Cancer can arise when cells heterozygous for an *RB* deletion lose the remaining copy of the gene.

This picture of the genetics of retinoblastoma raises a perplexing question: How can the retinoblastoma trait be inherited in a dominant fashion if a deletion of the *RB gene* is recessive to the wildtype RB^+ allele? The answer depends on the specific phenotype under consideration: susceptibility to cancer versus the progression from a normal to a cancerous cell. At the level of the organism, *RB* deletions are dominant because there is a strong likelihood that in at least one of the hundreds of thousands of retinal cells heterozygous for the deletion, a subsequent genetic event, such as mitotic nondisjunction, mitotic recombination, or a new mutation, will disable the single remaining RB^+ allele, resulting in a mutant cell with no functional tumor-suppressor gene. This one cell lacking any RB^+ gene will then multiply out of control, eventually generating a clone of cancerous cells. Although the aberrant events that create mutant cells with no functional *RB* gene happen very rarely, there are enough cells in the retina that most patients who inherit one 13q14-deleted chromosome 13 develop multiple retinoblastoma tumors in both eyes. At the level of a single cell, however, *RB* deletions are recessive to RB^+ because a cell with only one copy of RB^+ produces enough functional protein to remain normal. Hence mutations in tumor-suppressor genes appear dominant when the phenotype under consideration is susceptibility of individuals to cancer, but they act in a recessive manner when the phenotype under examination is the progression from normal to cancerous of individual cells in a clone.

Geneticists first recognized the recessive *RB* mutation that leads to retinoblastoma through the genomic analysis of families inheriting a predisposition to the cancer. More recently, they noted that both copies of the *p16* gene on chromosome 9 are deleted in roughly 75% of all melanomas (a malignant skin cancer) and in approximately 85% of all gliomas (the most common form of brain cancer); the *p16* gene, we have seen, encodes a protein that binds to and inactivates CDK4. Finally, observations of deletions of both copies of a specific region of chromosome 18 in all colon cancers led to identification of the *DCC* (deleted in colorectal cancer) gene.

Many tumor-suppressor mutations occur in genes that control the cell cycle, and with it, the accuracy of genomic replication. It is important to distinguish those mutations that determine how the cell cycle is completed from mutations in genes that control proliferation. Alterations in genes that control proliferation result in an enlarged clone of cells, but aside from their increase in number, these cells—if they sustain no further mutations—are normal and thus form a benign growth. By contrast, mutations in genes that control the cell cycle can alter the accuracy with which a cell reproduces its genome. The resulting mutant cells can produce offspring with many

more mutations than occur in normal cells, and this increase in the frequency of mutation vastly increases the probability that the cascade of mutations necessary to produce the phenotypic changes of tumor cells will occur.

Because cancer arises in cells that have lost the ability to reproduce their genomes faithfully, it seems reasonable to conclude that a cell's primary safeguard against cancer lies in maintaining the integrity of its genome. Cells have extensive, elaborate systems for repairing damage to their DNA. Studies of yeast and bacteria have shown that mutant organisms defective in DNA repair often have enormously increased rates of mutation. One important system for repairing DNA damage is the mismatch repair system, described in Chapter 6. This system recognizes mismatches that result from errors in DNA replication and removes the mismatched bases, allowing substitution of the correct bases. Proper functioning of the mismatch repair system increases the accuracy of DNA replication 100- to 1000-fold over the accuracy obtained by DNA polymerase alone. Mutants with a defective system have mutation rates several orders of magnitude greater than wildtype cells. In the 1990s, cancer researchers discovered that some people with a hereditary predisposition to colon cancer are heterozygous for a mutation that inactivates a gene required for the normal functioning of the mismatch repair system; the cancers that develop in these individuals consist of cells that have lost the single remaining wildtype allele. This mismatch repair gene thus behaves like a classical tumor-suppressor gene. Presumably, the greatly increased mutation rate in a homozygous cell that has lost both wildtype alleles makes it easier for progeny cells to accumulate the large number of mutations necessary to produce a cancer cell. Why these cancers develop mainly in the colon rather than in other tissues is not clear. Table 17.5 describes several other tumor-suppressor genes that affect the accuracy of cell division.

TABLE 17.5 **Mutant Alleles of These Tumor-Suppressor Genes Decrease the Accuracy of Cell Reproduction***

Gene	Normal Function of Gene (if known), or Disease Syndrome Resulting From Mutation	Function of Normal Protein Product
p53	Controls G_1-to-S checkpoint	Transcription factor
RB	Controls G_1-to-S transition	Inhibits a transcription factor
p21	Controls G_1-to-S transition	Inhibits CDK
ATM	Controls G_1-to-S phase, and G_2-to-M checkpoint	DNA-dependent protein kinase
BS	Recombinational repair of DNA damage	DNA/RNA ligase
XP	Excision of DNA damage	Several enzymes
hMSH2, hmLH1	Correction of base pair matches	Several enzymes
FA	Fanconi anemia	Unknown
BRCA1	Repair of DNA breaks	Unknown
BRCA2	Repair of DNA breaks	Unknown

*Many tumor suppressor genes have been associated with a specific function in the cell cycle necessary for accuracy of cell division.

Genetic Testing Has Some Use in Predicting and Treating Cancer

Genetic tests for mutations in protooncogenes and tumor-suppressor genes can reveal whether a person has a higher probability of getting cancer at some point in his or her lifetime than a person without the mutations. But of those with an increased risk, some will and some will not get cancer. This is because, unlike many single-gene disorders where inheriting one dominant or two recessive mutant alleles means you will get some form of the condition, cancer is a complex disease resulting from mutations in many genes in the same cell. While a person who acquires one of these mutations through inheritance is pushed one step along the road to cancer, the other mutations must occur in the same cell by chance; and nongenetic factors such as exposure to radiation influence if, where, and when the mutations occur. Given this situation, what good is it to learn from a genetic test that you have an increased probability of getting cancer sometime in your life?

Predictive testing is useful if the means of medical surveillance make it possible to detect the cancer to which a mutation predisposes at an early stage. Thus testing for a genetic predisposition to skin, breast, or colon cancers would be useful because careful examination of the skin, breast, or colon with existing technology (microscopes, mammograms, or colonoscopies) can often detect cancers of these tissues in their earliest stages. A person whose genetic test shows a predisposition to colon cancer, for example, could undergo colonoscopy each year, and if one of these colon exams

disclosed a small cancer, medical practitioners could remove it by surgery or other means. Predictive genetic testing is not useful at this time for pancreatic cancer because there is as yet no way to detect tumors of the pancreas when they are small; by the time doctors identify this cancer, it has almost always reached the aggressive stage and metastasized to other tissues.

Once a cancer has been diagnosed, genetic testing of tumor cells may be useful for making a prognosis and determining a course of therapy. For example, if a breast tumor is very small (less than 1 mm in diameter) and has not spread to the lymph nodes, surgical removal and follow-up radiation usually give the patient a good chance of overcoming the cancer. But if the tumor, however small, carries mutations in the *p53* gene, the prognosis is poor. Breast tumors with absent or mutated p53 proteins tend to resist treatment with radiation and many anticancer drugs.

Genetic tests for oncogenic and mutated tumor-suppressor alleles thus have specific predictive and therapeutic uses. Understanding the limitations of such tests is important. The predictive tests can indicate only increased risk. If knowledge of such risks makes it possible to take preventive measures, the tests make sense. Tests to help determine the course of therapy are possible for only a few diseases in which specific mutations have been linked to specific prognoses.

Summary: The Accumulation of Oncogenic and Tumor-Suppressor Mutations Produces Cancer Cells with Grossly Altered Genomes

Cancer-causing mutations disrupt the normal controls that create a balance between activation and inhibition of cell division. Dominant mutations that change protooncogenes to oncogenes may overactivate expression of proteins that promote proliferation. Recessive mutations in tumor-suppressor genes may release the brakes that keep cells from proliferating. Both types of mutations may tip the balance toward excessive and inaccurate cell proliferation.

Mutations that disable one part of a cell's elaborate DNA repair system increase its mutation rate and thus its likelihood of becoming cancerous. Although no single mutation converts a normal cell to a cancer cell, if a cell has a mutation in one gene that predisposes to cancer, that cell has a higher than normal probability of becoming cancerous because it is already one step along the way. The early mutations in a cell's progression from normal to cancerous may lead to increased proliferation and affect the accuracy of cellular reproduction to allow the accumulation of several mutations. Other subsequent or simultaneous mutations may enable the abnormally and inaccurately proliferating cells of a single lineage to avoid

programmed cell death, evade the immune system, increase formation of blood vessels supplying the abnormal clone, alter the proteins that control tissue architecture, and invade nearby or distant tissues. Environmental factors such as radiation and mutagenic chemicals cause most of the mutations that result in cancer, but rare inherited defects can contribute the first step (see the Genetics and Society box "Genetic Testing Has Some Use in Predicting and Treating Cancer").

COMPREHENSIVE EXAMPLE: THE GENETICS OF BRAIN CANCER

Abnormal proliferation of cells at a critical site in the brain can threaten life even if cells in the resulting tumor do not acquire the ability to invade surrounding or distant tissues (that is, metastasize). Like tumorigenesis in other tissues, the development of a brain tumor depends on mutations that change protooncogenes to oncogenes or inactivate tumor-suppressor genes.

Cellular and Clinical Background

The most malignant form of brain cancer—glioblastoma multiforme (GBM)—is an aggressive cancer of the glial cells. The term "multiforme" signifies that GBM is a heterogeneous condition in which the same phenotype can result from mutations in different subsets of genes.

GBM Is the Highest Grade of Glial Cell Tumor

Glial cells, the most numerous cells in the human brain, come in three varieties: *astrocytes,* which provide support for the neurons; *oligodendrocytes,* which produce the myelin sheaths that vastly increase the speed of nerve impulse conduction; and *ependymal* cells, which line the brain cavities known as ventricles and help regulate production of cerebrospinal fluid (Fig. 17.25). Any of these glial cells can mutate to form a glial-cell tumor, or glioma. Gliomas containing only mutated astrocytes are called *astrocytomas,* tumors consisting of abnormal oligodendrocytes are *oligodendrogliomas,* and gliomas carrying both mutant astrocytes and mutant oligodendrocytes are *oligoastrocytomas.* Tumors composed of abnormal ependymal cells are extremely rare and will not be part of the following discussion.

Cancer specialists recognize different grades of gliomas—low grade (designated by a roman numeral II), anaplastic (III), and GBM (IV)—with each higher grade representing an increased likelihood of aggressiveness of growth. Unfortunately, low-grade gliomas have a tendency to undergo malignant transformations; most eventually develop into aggressive, fast-growing GBMs.

A person's prognosis depends on his or her age and the grade and type of tumor. For example, a 25-year-old with a low-grade oligodendroglioma can survive for as long as 10 years, whereas a 70-year-old with a GBM probably has less than a year to live.

A Large Number of Genes in Various Combinations Produce GBMs

A close look at the genetic basis of glioblastoma multiforme tumors clarifies how a cancer phenotype can arise from different combinations of changes in several genetic pathways (Fig. 17.26).

Some GBMs Evolve from Low-Grade Astrocytomas

Many early astrocytomas contain at least three types of mutations: those that inactivate the *p53* tumor-suppressor gene on chromosome 17; those that oncogenically activate expression of a growth factor known as PDGF (for platelet-derived growth factor) as well as its receptor; and those that delete an as yet unidentified tumor-suppressor gene on chromosome 22. We know that *p53* encodes a transcription factor active in cell-cycle arrest, response to DNA damage, and apoptosis. Most astrocytoma cells with no active *p53* gene have a mutation in

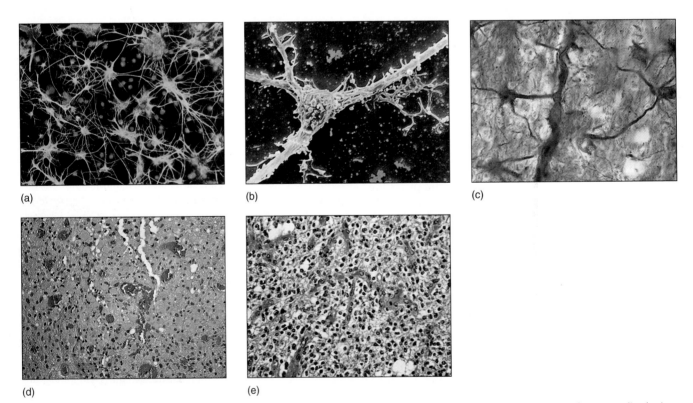

(a)

(b)

(c)

(d)

(e)

Figure 17.25 Mutations in three types of glial cells cause brain cancer. (a) Astrocyte brain cells from the cortex of a mammalian brain. (b) Oligodendrocyte cell (orange structure in the middle) attaching to an axon nerve cell at the start of myelination. (c) Ependymal cell. The astrocyte cells in (a) mutate to form astrocytomas (low-grade) shown in (d). The oligodendroglioma shown in (e) is the result of mutated oligodendrocyte cells (b). The tumors resulting from mutated ependymal cells are extremely rare and are not shown here.

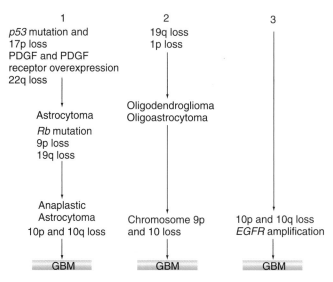

1
p53 mutation and
17p loss
PDGF and PDGF
receptor overexpression
22q loss

↓

Astrocytoma
Rb mutation
9p loss
19q loss

↓

Anaplastic
Astrocytoma
10p and 10q loss

↓

GBM

2
19q loss
1p loss

↓

Oligodendroglioma
Oligoastrocytoma

↓

Chromosome 9p
and 10 loss

↓

GBM

3

↓

10p and 10q loss
EGFR amplification

↓

GBM

Figure 17.26   Three routes for the evolution of glioblastoma multiforme have been identified. Different combinations of change along this genetic pathways give rise to a cancer phenotype.

one copy of the gene and have lost the other copy. These same low-grade astrocytoma cells also overexpress both the PDGF and its receptor, suggesting that the *p53* mutation may produce tumors only in the presence of PDGF overexpression. Finally, low-grade astrocytoma cells with no or inactive p53 and an overactive PDGF system have lost both copies of part of chromosome 22. Initial linkage studies suggest that the deleted region is near a telomere.

Astrocytomas that progress from low grade to anaplastic have three additional deletions—from 13q, 9p, and 19q. The deletion from the long arm of chromosome 13 carries the *RB* tumor-suppressor gene, which, as we have seen, encodes a

protein that contributes to cell-cycle arrest; this deletion thus removes a brake on abnormal proliferation. The region deleted from the short arm of chromosome 9 carries genes for two CDK inhibitors: p15 and p16; deletion of these inhibitors releases a brake on the activity of some of the cyclin-dependent kinases that help regulate the cell cycle. The loss of part of the long arm of chromosome 19 is a feature of all gliomas that does not appear in any other type of cancer. Linkage analysis has mapped the crucial tumor-suppressor region to 19q13.3. (Bear in mind throughout this discussion that if a cell loses a region containing a tumor-suppressor gene from one homologous chromosome, the remaining wildtype copy of the same gene must sustain a mutation to produce a cancer cell.)

Progression from aplastic astrocytoma to glioblastoma multiforme involves the loss of genes at particular points on both the long and short arms of chromosome 10. Deletion mapping shows that a primary tumor-suppressor gene resides at 10q25, while a secondary suppressor is located at an as yet unidentified point on 10p.

Thus, a high-grade GBM tumor evolves from a normal astrocyte through three levels of malignancy by the accumulation of at least 10 mutations: 2 generate oncogenes that overactivate the platelet-derived growth factor system and 8 inactivate tumor-suppressor genes on chromosomes 9, 10, 13, 17, 19, and 22, releasing a number of controls on cell division (Fig. 17.27).

Clinicians have compared mutations from early astrocytomas with mutations found in full-blown GBMs that appeared months or years later at the same site as the low-grade tumor. They found that the initial, low-grade tumors carried fewer mutations than the later more malignant GBMs; and the GBMs carried all the mutations found in the lower-grade cancers and then some. These data demonstrate that an accumulation of mutations in the clonal descendants of a single astrocyte cause the cell-cycle deregulation that produces this malignant cancer.

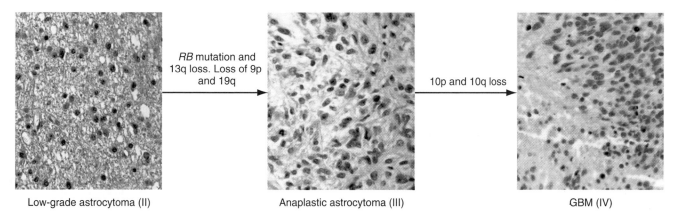

RB mutation and
13q loss. Loss of 9p
and 19q

→

10p and 10q loss

→

Low-grade astrocytoma (II)   Anaplastic astrocytoma (III)   GBM (IV)

Figure 17.27   Pathway from grade II astrocytoma to malignant GBM. These photos illustrate the development of a tumor from a low-grade (II) astrocytoma to high-grade (IV) GBM. The progress of the disease is traced through deletions from the tumor-suppressor regions of chromosomes 9, 13, and 19 and *RB* mutation in the step from grade II to grade III. The step from grade III to IV is marked by further deletions of the tumor-suppressor regions of chromosome 10. Note that necrosis has begun in the GBM photo.

Some GBMs Arise from Oligodendrogliomas or Oligoastrocytomas

Such GBMs carry a different combination of mutations from those found in the GBMs that develop from astrocytomas. Low-grade gliomas containing oligodendrocytes have lost part of the long arm of chromosome 19 (a deletion also seen in anaplastic astrocytomas). They have also lost a region on the short arm of chromosome 1 (see Fig. 17.26). The loss of one copy of both 9p and chromosome 10 converts these low-grade tumors to aggressive GBMs; the deletions are thought to carry at least three tumor-suppressor genes whose remaining, wildtype copies have no doubt sustained a mutation.

Some Rapidly Arising GBMs Have no Apparent Precursors

These glioblastoma multiforme tumors have an oncogenic amplification of the *epidermal-growth-factor-receptor* (*EGFR*) gene, and have also lost regions from both 10p and 10q (review Fig. 17.26). The tumors with these genetic alterations arise *de novo* or so rapidly that no precursors are detectable. Amplification of the *EGFR* gene almost never occurs in the astrocytoma-derived GBM tumors carrying *p53* mutations and 17p deletions. Moreover, GBMs with the *EGFR* gene amplification occur in significantly older adults than GBMs with mutant or deleted *p53* and chromosome 17 genes.

Summary

The GBM phenotype develops via different combinations of mutations in different genetic pathways: from lower-grade astrocytomas via *p53* and *RB* gene inactivations, from oligo- dendroglial tumors via deletions of chromosomes 1 and 19, or *de novo* via *EGFR* gene activation. This neat separation of mutational pathways is, however, an oversimplification. Not every GBM shows all the genetic changes described, and some GBMs derived from one type of cell exhibit mutations usually associated with another cell type. For example, rare astrocytoma-derived GBMs show a loss of chromosome 10 and amplification of the *EGFR* gene. The different combinations of mutations that subvert a cell's normal progression through the cell cycle can accumulate in the clonal descendants of a single cell either slowly over decades or rapidly in the course of a few months.

Knowledge of the Genetics of Brain Cancer May Lead to More Selective Therapies

Although cancer researchers have identified and isolated only a few of the genes that play a role in malignant gliomas, their insights have begun to point the way to more customized therapies for this type of brain cancer. Identification of the precise mutations in the cells of a patient's tumor can pinpoint the stage of disease, thus clarifying the benefits and disadvantages of certain therapeutic choices. An understanding of the genetic basis of cancer might also lead to the development of drugs that slow tumor growth by counteracting the effects of specific mutations in target cancer cells; such drugs would make it possible to zero in on a small subset of tumor cells and thereby produce fewer side effects than current chemotherapies, which indiscriminately kill normal and cancerous cells.

CONNECTIONS

The existence of numerous controls in each of several cell-cycle pathways suggests that evolution has erected many barriers in multicellular animals to the uncontrolled reproduction of "selfish" cells. At the same time, the hundreds of genes contributing to normal cell-cycle regulation provide hundreds of targets for cancer-producing mutations.

Variations on the theme of cell-cycle regulation play a key role in the development of eukaryotic organisms. During the development of multicellular organisms, cells must not only control their cell cycles, they must also adopt different fates and differentiate into different tissues. In *Drosophila,* for example, after fertilization, nuclear division occurs without cell division for the first 13 cycles; during these cycles, the nuclei go through many rapid S and M phases without any intervening G_1 or G_2 (Fig. 17.28). In cycles 10–13, the synthesis and degradation of cyclinB regulates mitosis. Sometime during cycles 14 to 16, a G_2 phase appears, and distinct patches of cells with different length cycles become evident within the embryo; the differences in cycle time between the different cell types is the result of variable G_2 phases. Late in G_2, *CDC25* activates cyclin-dependent kinases to control the timing of mitosis. Many tissues stop dividing at cycle 16, but a few continue. In the still-dividing cells, a G_1 phase appears. Some of these cells will arrest in G_1 during larval growth, only to start dividing again in response to signals relayed during metamorphosis.

In the next part of the text, we present details of developmental genetics in five model organisms.

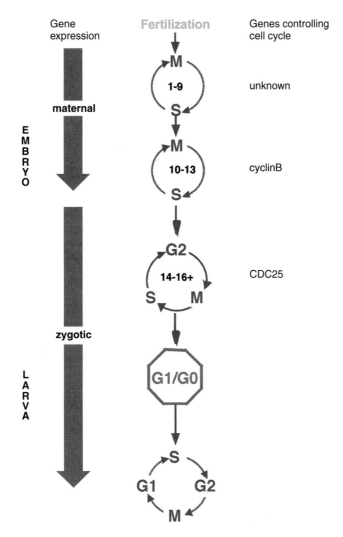

Figure 17.28   The regulation of the cell-cycle changes during *Drosophila* development. Each step of development has built-in regulators that act as barriers to uncontrolled reproduction of "selfish" cells. Some of these, such as cyclinB and *CDC25* are known, others are not.

ESSENTIAL CONCEPTS

1. Several genetic pathways help control cell division.
   a. The inhibition or activation of CDKs inhibits or activates G_1-to-S and G_2-to-M transitions.
   b. The measured synthesis and degradation of different cyclins guides CDKs to the appropriate targets at the appropriate times.
   c. Checkpoints that integrate repair of chromosomal damage with events of the cell cycle minimize the replication of DNA damage.
   d. The genes and proteins of various signal transduction systems relay signals about whether or not to enter the cell cycle.

2. Cancer is a genetic disease resulting from the growth of a clone of mutant cells.

3. A cell requires many mutations to become cancerous. Exposure to environmental mutagens probably generates most of these mutations.

4. Most mutations that lead to cancer jeopardize cell-cycle regulation.
   a. Mutations in growth factors, receptors, and other elements of signal transduction pathways can release cells from the signals normally required for proliferation.

b. Mutations in CDKs and the proteins that control them may also lead to inappropriate proliferation or genomic instability.

c. Mutations in DNA repair and checkpoint controls lead to genomic instability and, often, to loss of the surveillance system that kills aberrant cells by apoptosis.

d. Mutations that lead to genomic instability permit the rapid evolution of abnormal tumor cells.

5. The different symptoms and phenotypes of cancer depend on the cells and mutations involved.

S O C I A L   A N D   E T H I C A L   I S S U E S

1. Taxol is a chemical produced in the bark of Pacific yew trees. In experimental trials, taxol reduced ovarian tumors in up to 50% of terminally ill patients. When this effect was first discovered, ovarian cancer patients were extremely anxious to get the drug. Unfortunately, the Pacific yew tree is a slow-growing tree found in old growth forests of the Pacific Northwestern United States. For a series of treatments, each patient would require the bark of approximately three trees. Because of the scarcity of trees and the difficulty in obtaining taxol from the bark, the expense was high. In 1993, the cost of production by the National Cancer Institute was $1200 for the 2 grams needed for each patient. When a drug is scarce, who should get the drug? The clinical trials involved patients for whom no other treatment had worked. Is this the best use of limited resources of a potentially powerful anticancer drug? How would you decide who gets the drug? How should these decisions be made?

2. In 1976, cancer patient John Moore had his spleen removed at a University teaching hospital as a treatment for the hairy cell leukemia from which he suffered. Some of the spleen cells were cultured at that time and other cell types were cultured later, eventually leading to the establishment of a very useful T-cell line. The cell line produced the lymphokine GM-CSF, used in treating cancers and AIDS. A patent, granted to the University and the doctors for the cell line, was sold to two biotechnology companies. Moore filed a claim against the doctors, companies, and the University stating that he was entitled to some of the profits and that there had been a lack of informed consent. He had signed a consent form but was never informed of the potential for commercial exploitation. Several questions arise from the landmark legal case, Moore *v.* Regents of California. To what extent must a doctor disclose interests in a commercial venture to a patient? Does a person own rights to cells that have been removed from the body?

3. Suitable animal models for the effects of exposure to radiation on humans are not available. Only tests on humans can make it possible to evaluate the dangers of chromosome damage and cancer resulting from various types of exposure. Under these circumstances, is human experimentation acceptable? Who should participate in human trials? Volunteers who are well apprised of the risks? Prisoners on death row? Are there valid alternatives?

S O L V E D   P R O B L E M S

I. The addition of growth factors to tissue culture cells stimulates cell division. A number of candidate drugs can be tested for their ability to stop this stimulation of cell division. What do you think the target of these drugs could be?

Answer

This question concerns regulation of cell division. Growth factors are made by one cell and bind to receptors of another cell to stimulate the cell-division cycle. *A drug that binds to receptors* would block access for growth factor binding. Alternatively, the *drug could bind to the growth factor,* thereby preventing interaction with the receptor. (These are the most obvious targets. If you are familiar with the signal transduction pathway inside the cell, you might also propose that proteins in this pathway could be targets for drug development.)

II. The *p53* gene has been cloned and you are using it to analyze DNA in patients in which *p53* defects are involved in the development of their tumors. DNA samples were obtained from normal and tumor tissue of three different cancer patients, digested with *Bam*HI, electrophoresed on an agarose gel, and transferred to filter paper that was probed with a labeled *p53* fragment. Each of these patients inherited a *p53* mutation. DNA from an individual who did not inherit a *p53* mutation is shown in the lane labeled wildtype. Assuming the model of *p53* acting as a tumor-suppressor gene is correct and that *p53* defects are involved in each of these cancer patients, how would you describe the genetic makeup of the *p53* gene in the normal and tumor tissue of each of the three patients?

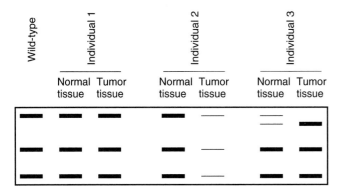

Answer

This question requires knowledge of tumor-suppressor genes. The wildtype *p53* region (as seen in the "wild-type" individual) has three hybridizing bands. Because *p53* is a tumor suppressor, it is recessive at the cellular level and both copies must be defective in the tumor cells. No observable changes are apparent in patient 1, so this patient *must have inherited a small mutation in* p53 *and in the tumor cells the second copy would also contain a small mutation* thereby inactivating both copies of *p53* in the tumor. *In patient 2, a small mutation must have been inherited. In the tumor, the whole region containing* p53 *was deleted (thereby removing the*

second copy of the gene), as seen by the loss of restriction fragments. In patient 3, a deletion mutation (smaller than that in the tumor cells of individual 2) , as seen in the altered restriction pattern, was inherited and in the tumor, the second copy of the gene was similarly deleted (probably by gene conversion since the pattern is the same for both copies).

III. The CDC28 protein of budding yeast *S. cerevisiae* and CDC2 protein of fission yeast *S. pombe* are protein kinases required at the "start" of the cell cycle. Both genes were identified by mutational analysis (temperature-sensitive mutations in each gene cause arrest in early G_1) and both genes have been cloned. How could you determine if one could substitute for the other functionally? (Be sure to mention sources of DNA and genotypes involved.)

Answer

The CDC28 *gene of* S. cerevisiae *could be cloned into a vector and transformed into a temperature-sensitive* cdc2 *mutant of* S. pombe. *If* CDC28 *has the same role (functionality), the transformed cell would now grow and divide at nonpermissive temperature. Conversely, a clone of the* CDC2 *gene of* S. pombe *could be cloned into a vector that could transform a temperature-sensitive* cdc2 *mutant of* S. cerevisiae. *If* cdc2 *of* S. pombe *can substitute for* CDC28 *of* S. cerevisiae, *the transformed cells would grow and divide at the nonpermissive temperature.*

P R O B L E M S

17-1 For each of the terms in the left column, choose the best matching phrase in the right column.

a. growth factor	1. mutations in these genes are dominant for cancer formation
b. tumor-suppressor genes	2. programmed cell death
c. cyclin-dependent protein kinases	3. series of steps by which a message is transmitted
d. apoptosis	4. proteins that are active cyclically during the cell cycle
e. oncogenes	5. control progress in the cell cycle in response to DNA damage
f. receptor	6. mutations in these genes are recessive at the cellular level for cancer formation
g. signal transduction	7. signals a cell to leave G_0 and enter G_1
h. checkpoints	8. cell cycle enzymes that phosphorylate proteins
i. cyclins	9. protein that binds a hormone

17-2 During which phase(s) of the cell cycle would the following enzymes or proteins be most active?
a. tubulins in the spindle fibers
b. centromere motor
c. DNA polymerase
d. CDC28 of *S. cerevisiae* or CDC2 of *S. pombe*

17-3 Conditional mutations are useful for genetic analysis of essential processes. For example, temperature-sensitive cell cycle mutations in yeast do not divide at 37°C (nonpermissive temperature) but will divide at 30°C. An alternative type of conditional mutation is a cold-sensitive mutation in which the nonpermissive temperature is low (23°C). List the steps you would go through to isolate cold-sensitive cell cycle mutants of yeast.

17-4 Many temperature-sensitive yeast mutants that showed defects in the cell cycle were isolated in the 1970s. The mutants that arrested at the unbudded stage were mated with each other to do a complementation analysis. A (+) on the chart on following page indicates that the resulting diploids grew at the high (nonpermissive) temperature. How many complementation groups (how many genes) are represented by these mutants?

	1	2	3	4	5	6	7	8	9
1	−	+	+	−	−	+	+	+	+
2	+	−	+	+	+	+	+	−	+
3	+	+	−	+	+	−	−	+	−
4	−	+	+	−	−	+	+	+	+
5	−	+	+	−	−	+	+	+	+
6	+	+	−	+	+	−	−	+	−
7	+	+	−	+	+	−	−	+	−
8	+	−	+	+	+	+	+	−	+
9	+	+	−	+	+	−	−	+	−

17-5 In 1951 a woman named Henrietta Lacks died of cervical cancer. Just before she died a piece of her tumor was taken and put into culture in a laboratory in an attempt to induce the cells to grow *in vitro*. The attempt succeeded, and the resulting cell line (known as HeLa cells) is still used today in laboratories around the world for studies of various aspects of cell biology. In the cell cycle of typical HeLa cells, G_1 lasts about 11 hours, S lasts about 8 hours, G_2 lasts 4 hours, and mitosis (M) takes about 1 hour.

   a. Cultured cells do not typically grow synchronously. That is, the individual cells in a culture are randomly distributed throughout the cell cycle. If you looked through the microscope at a sample of HeLa cells, in approximately what proportion of them would you expect the chromosomes to be visible? (Assume that cells do not split apart completely after cytokinesis so they can be counted as one cell.)

   b. Approximately what proportion would be in interphase?

17-6 Activity of key cell-cycle regulatory proteins is cyclical, appearing only when needed. What are three ways by which a cell can achieve this cyclical nature of protein activity?

17-7 True or false?

   a. CDKs phosphorylate proteins in the absence of cyclins.

   b. Degradation of cyclins is required for the cell cycle to proceed.

   c. CDKs are involved in checking for aberrant cell cycle events.

17-8 Checkpoints occur at several different times during the cell cycle to check that the DNA content of the cell has not been damaged or altered. Match the defect in a checkpoint with the consequences of that defective checkpoint.

Defective checkpoint	Consequences of defect in checkpoint
a. G_1 to S	1. aneuploidy
b. G_2 to M	2. single-strand nicks get replicated and amplification occurs
c. M	3. double-stranded breaks are unrepaired, resulting in broken chromosomes

17-9 Molecules outside and inside the cell regulate the cell cycle, making it start or stop.

   a. What is an example of an external molecule?

   b. What is an example of a molecule inside the cell that is involved in regulation?

17-10 a. Would you expect a cell to continue or stop dividing at a high temperature if you had isolated a temperature-sensitive *RAS* mutant that remained fixed in the GTP-bound form at nonpermissive temperature?

   b. What would you expect if you had a temperature-sensitive mutant in which RAS stayed in the GDP-bound form at high temperature?

17-11 Put the following steps in the correct ordered sequence.

   a. kinase cascade

   b. activation of a transcription factor

   c. hormone binds receptor

   d. expression of target genes in the nucleus

   e. RAS molecular switch

17-12 Mouse tissue culture cells infected with the SV40 virus lose normal growth control and become transformed. If transformed cells are transferred into mice, they grow into tumors. The SV40 protein responsible for this transformation is the T antigen. T antigen has been found to associate with the cellular protein p53. If the *p53* gene fused to a high-level expression promoter is transfected into tissue culture cells, the cells are no longer transformed by infection by SV40.

   a. Propose a hypothesis to explain how the high expression of p53 saves the cells from transformation by T antigen.

   b. You have decided to examine the functional domains of the p53 protein by mutagenizing the cDNA, fusing it to the high-level promoter, and transfecting into cells. Results are shown on the following page. How would you explain the effects of mutations 1 and 2 on p53 function?

   c. What is the effect of mutation 3 on p53 function?

p53 construct	Morphology	
	Noninfected cells	SV40-infected cells
None	Normal	Transformed
Wildtype	Normal	Normal
Mutation 1	Normal	Transformed
Mutation 2	Normal	Transformed
Mutation 3	Normal	Normal

17-13 What are four characteristics of the cancer phenotype?

17-14 Amplification of DNA sequences in *p53* mutants can be visualized using electron microscopy.
   a. Using a different technique, how could you detect amplification of a *specific sequence?*
   b. How could you detect gross rearrangements (> 10 Mb) of chromosomal DNA?

17-15 Some germ line mutations predispose to cancer yet often environmental factors (chemicals, exposure to radiation) are proposed to be major risks for developing cancer. Are these conflicting views of the cause of cancer or can they be reconciled?

17-16 The incidence of colon cancer in the United States is 30 times higher than it is in India. Differences in diet and/or genetic differences in the two populations may contribute to these statistics. How do you think you could assess the role of each of these factors?

17-17 Put the following steps in the order appropriate for positional cloning of a gene involved in predisposition to breast cancer, *BRCA1*.
   a. locate transcripts corresponding to the DNA
   b. use the physical map to get clones
   c. determine the tissues in which the transcripts are present
   d. look for homologous DNA in other organisms
   e. determine linkage to RFLPs and other molecular markers
   f. sequence the DNA from affected individuals

17-18 Because mutations occur in the development of cancer, researchers suspected that defects in DNA repair machinery might lead to a predisposition to cancer. Place the following steps in appropriate order for following a candidate gene approach to determine if defects in mismatch repair genes lead to cancer.
   a. Use molecular markers near the homologous gene to determine if the candidate gene is linked to a predisposition phenotype.
   b. Isolate a human homolog of a yeast mismatch repair gene.

   c. Compare DNA sequence of the mismatch repair gene of affected and unaffected persons in a family with predispositions.
   d. Determine the map location of the human homolog of the yeast mismatch gene.

17-19 Which of the following events is unlikely to be associated with cancer?
   a. mutations of a cellular proto-oncogene in a normal diploid cell
   b. chromosomal translocations with breakpoints near a cellular proto-oncogene
   c. deletion of a cellular proto-oncogene
   d. mitotic nondisjunction in a cell carrying a deletion of a tumor-suppressor gene
   e. incorporation of a cellular oncogene into a retrovirus chromosome

17-20 You have decided to study genetic factors associated with colon cancer. An extended family from Morocco in which the disease presents itself in a large percentage of family members at a very early age has come to your attention. (The pedigree is shown.) In this family, individuals either get colon cancer before the age of 16, or they don't get it at all.

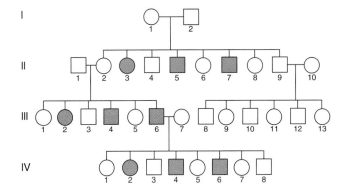

   a. Based on the information you have been given, what evidence, if any, suggests an inherited contribution to the development of this disease?
   b. You decide to take a medical history of all of the 36 people indicated in the pedigree and discover that a very large percentage drink a special coffee on a daily basis, while the others do not. The only ones who don't drink coffee are individuals numbered I-1, II-2, II-4, II-9, III-7, III-13, IV-1, and IV-3. Could the drinking of this special coffee possibly play a role in colon cancer? Explain your answer.

17-21 To further understand the basis for colon cancer, you find a family from the United States in which two members also get the disease before the age of 16. If there were a dominant inherited mutation segregating in this family, which of the individual(s) would you predict had the mutation in their colon cells but did not develop the disease?

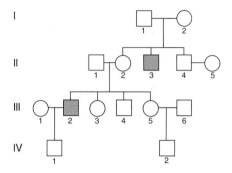

17-22 You suspect that a very specific point mutation in the *p53* gene is responsible for the majority of *p53* mutations found associated with tumors. Which combination of these techniques would you be most likely to use in developing a *simple assay* for predictive testing?
a. Polymerase chain reaction with oligos flanking the mutation
b. Restriction enzyme digestion followed by Southern blot
c. RNA isolation followed by Northern blot
d. Hybridization with allele-specific oligonucleotides

GENE REGULATION AND DEVELOPMENT: PORTRAITS OF MODEL EUKARYOTIC ORGANISMS

Using Genetics to Study Development: An Introduction to the Genetic Portraits

USING GENETICS TO STUDY DEVELOPMENT:

AN INTRODUCTION TO THE

GENETIC PORTRAITS

Evolutionary relatedness of all organisms permits extrapolation of information
from models to all living forms.

During human embryonic and fetal development, a single cell—the fertilized egg—divides and differentiates to form hundreds of intricate structures capable of complex behaviors (Fig. I.1a). The infant born at the end of this process has a nervous system, lungs, vocal cords, tongue, and gums; with these structures it can cry to communicate its need for nourishment and suckle to satisfy that need.

The development of a child from a fertilized egg is so miraculous that it is difficult to look at the process from a reductionist point of view, seeking to explain it on the basis of physical, chemical, and biological principles. Yet the more biologists study development at the microscopic level, the more clearly the miracle emerges as a precisely patterned set of cellular behaviors. Cells divide, change shape, specialize, and interact in highly reproducible ways. It is thus possible to look at development as a range of genetically specified cell behaviors, and to use genetics to study and compare these behaviors in model organisms.

Developmental biologists, for example, have used genetics to analyze the development of fruit fly embryos and found that in *Drosophila*, only a few dozen genes guide the formation of the early embryo's segmented body plan. Mutations in some of these genes eliminate specific body segments (Fig. I.1b). Once the embryo has divided into segments that will become the head, abdomen, and tail, the activation and inactivation of a different set of genes direct the development in each segment of specialized organs, such as wings and legs. The genes controlling the development of organs in *Drosophila* have counterparts in humans, and in many cases, specific human genes can substitute for the corresponding insect genes during fruit fly development.

Biologists who use genetics to study how the fertilized egg of a multicellular organism becomes an adult are called **developmental geneticists.** Their field of developmental genetics focuses on analyses of mutations that produce developmental abnormalities. An understanding of such mutations helps clarify

(a)

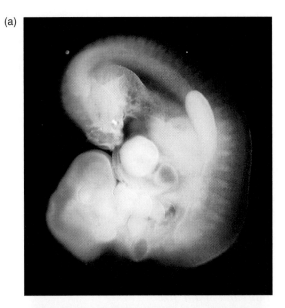

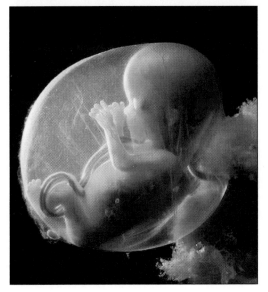

(b)

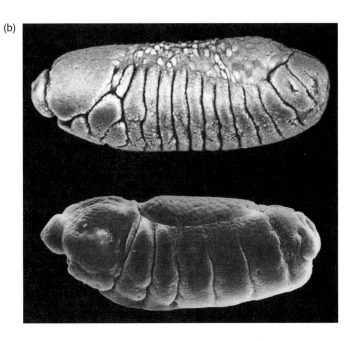

Figure I.1 Human and *Drosophila* development. (a) Human embryo at 35 days of development (top) and at the end of the first trimester (bottom). (b) Mutations in specific *Drosophila* genes can eliminate body segments; wildtype (top) and mutant (bottom) *Drosophila* bodies.

how normal genes control cell growth, cell communication, and the emergence of specialized cells, tissues, and organs. Many biologists extend the meaning of "developmental genetics" to include the study of single-celled organisms such as yeast or bacteria, and even the study of viruses. This is because these simpler organisms accomplish cell specialization and other basic tasks similar to those achieved by multicellular organisms.

Identification of the genes and proteins that control development has only just begun. Researchers seeking to understand the structure and function of these biological molecules must work with organisms that permit genetic analysis; that is, with organisms in which it is possible not only to produce and isolate mutations that disrupt development, but also to clone and sequence the defective genes and study the role of these genes' protein products. The deliberate production of mutants is, of course, not practicable in humans, but it is possible in model organisms,

which, as we saw in Chapter 10, are organisms that are amenable to experimental manipulation. Insights from model organisms can, in turn, illuminate general principles applicable to most other organisms, including humans.

In the remainder of this Introduction to the Genetic Portraits, we

- Present five model organisms that have become prototypes for research in developmental genetics.

- Explain how biologists learned that the evolutionary relatedness of all organisms enables extrapolation from the models to all living forms.

- Introduce the questions that motivate developmental geneticists and the genetic methods that enable them to answer these questions.

A SMALL NUMBER OF MODEL ORGANISMS HAVE BECOME PROTOTYPES FOR DEVELOPMENTAL GENETICS

Throughout the twentieth century, eukaryotic geneticists have concentrated their efforts on a small number of organisms that sample a range of species from many different phyla. The organisms that have contributed most to our understanding of development include the unicellular yeast *Saccharomyces cerevisiae* as well as several multicellular organisms: the plant *Arabidopsis thaliana*, the fruit fly *Drosophila melanogaster*, the nematode (or round worm) *Caenorhabditis elegans*, and the house mouse *Mus musculus*. Some developmental geneticists have studied other eukaryotic organisms, such as zebra fish and corn, and even viruses and prokaryotes, which have often provided paradigms for tackling developmental problems in the eukaryotic organisms.

Genome Size and Complexity Often Reflect Organismal Complexity

It seems reasonable to expect that a single-celled organism such as yeast would have a simpler structure, organization, and set of behaviors than a multicellular organisms such as a mouse. Perhaps the most meaningful measure of an organism's complexity is the number of genes in its genome. Table I.1 shows the DNA content of a haploid nucleus for each of the model organisms just mentioned. Indeed the DNA content seems to increase with the organisms' apparent complexity. DNA content, however, is not the whole story, since in many organisms, much of the DNA is noncoding. Researchers can estimate a genome's coding capacity by sequencing chromosomal regions to determine what fraction of the total DNA probably encodes RNA and proteins. They can also estimate DNA coding capacity by experiments that measure the complexity of messenger RNA. Table I.1 shows the estimated number of genes in the model organisms under consideration.

Biologists will truly comprehend the development of these organisms only when they understand the function and interactions of each of these many genes. At present, their knowledge is miniscule. In the best-understood eukaryote, *S. cerevisiae,* although they know the complete DNA sequence, they know something about the function of only approximately 2000—or 30% —of the genes; and they have far less knowledge of other eukaryotic organisms. It will thus be a long time before they approach a comprehensive understanding of the mouse, the nearest relative to humans on the list of models, with its roughly 100,000 genes.

Idiosyncrasies of Model Organisms Enhance Their Utility for Genetic Analysis and for Understanding Development

Each model organism has attracted a dedicated cadre of researchers for specific reasons.

The Model Organisms Used by Geneticists Share Three Useful Characteristics: Ease of Cultivation, Rapid Reproduction, and Small Size

These properties satisfy the geneticist's need to examine large numbers of individuals for the isolation of mutants, and to study the segregation of mutants through successive generations. Each model organism also has features specific to itself that make it valuable for particular types of genetic analyses. Yeast, for example, can grow vegetatively either as haploid or as diploid cells. This means that researchers can conveniently isolate mutants in haploids where each gene exists in only one copy and then use the mutants to construct diploid cells for complementation studies (Fig. I.2a). *D. melanogaster* contains polytene chromosomes in some tissues. Composed of up to 1000 identical chromatids bound together, polytene chromosomes make it possible to visualize the genome in great enough detail to locate genes and detect deletions or translocations (see Fig. 11.17). The mouse produces embryos that can be dispersed into undifferentiated cells (embryonic stem cells, ES cells) capable of growing in culture in unlimited numbers. Transformation with marked DNA and *in vitro* selection of cells with desired genotype (Fig. I.2b) can then derive cells with targeted mutations, and these cells can be reconstituted back into mouse embryos, and ultimately, mice with desired mutations can be produced.

TABLE I.1 DNA Content and Coding Capacity of the Haploid Genomes of Model Organisms

Model Organism	Haploid Genome Size (Mb)	Estimated Number of Genes
S. cerevisiae	13	6,022
D. melanogaster	170	15,000
C. elegans	100	14,000
M. musculus	3,000	100,000
A. thaliana	120 (estimated)	13,000–60,000
Homo sapien (not a model)	3,000	100,000

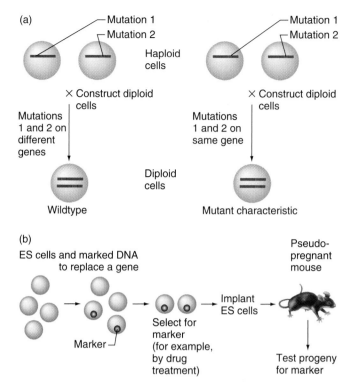

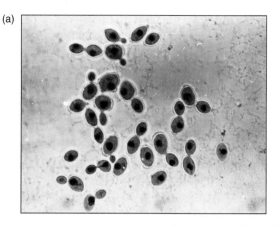

Figure I.2   How researchers make use of a model organism's characteristics to isolate mutants. (a) Yeast. (b) Mice.

Figure I.3   Model organism features that facilitate the study of development. (a) The budding of yeast cells. (b) The transparency of *C. elegans.*

Some Properties of Model Organisms Facilitate the Study of Development

Yeast cells, for example, produce progeny by budding. The size of the bud is a simple marker of the cell's position in the cell cycle (Fig. I.3a), and knowledge of this position is very useful in studies of cell cycle genetics. In another example, *C. elegans* is transparent (Fig. I.3b) and contains an invariant number of somatic cells as an adult—959 in the female/hermaphrodite and 1031 in the male. These properties make it possible to follow the lineage of each and every individual cell during development of the fertilized egg to the multicellular adult. And in female mice, the formation of a vaginal plug immediately after copulation enables researchers to determine the time of fertilization.

The Accumulation of Knowledge, Mutants, and Specialized Techniques Enhances the Utility of Model Organisms

Most of the model genetic organisms have a long tradition of study that has enabled researchers to gather a great deal of information about their normal development. This information is essential to an analysis of mutants that have developmental defects. Many mutants have been isolated that can serve as starting material in the isolation or study of additional mutants. Stock centers maintain these mutants so that they are available to the whole community of geneticists even decades after their original isolation. Recently, geneticists have applied

molecular techniques to the cloning and sequencing of each model organism's genome. They have already determined the complete sequence of the *S. cerevisiae* and *C. elegans* genomes and are proceeding rapidly with the sequencing of the genomes of *A. thaliana* and *D. melanogaster.*

As we will see, the availability of genetic maps detailing the locations of developmentally interesting genes combined with knowledge of the DNA sequence of the genome is furthering geneticists' ability to understand the functions of the genes influencing development.

THE EVOLUTIONARY RELATEDNESS OF ALL ORGANISMS ENABLES EXTRAPOLATION FROM THE MODELS TO ALL LIVING FORMS

In the last 150 years, biologists have come to realize that lifeforms are related on many levels. Even organisms that are evolutionarily quite distant share similarities in their anatomy and their basic strategies of development. For example, the cells of all eukaryotic organisms have many features in common that are recognizable in the light microscope—from a membrane-bounded nucleus containing linear chromosomes to mitochondria containing their own chromosome, to a network of membranous structures such as the endoplasmic reticulum and Golgi apparatus, which together facilitate the intracellular transport of molecules (Fig. I.4). Moreover, the metabolic pathways by which cells make or degrade organic molecules are virtually identical in all living organisms. For example, the enzymatic reactions that synthesize purines are the same in bacteria and humans, while nearly all cells use nearly the same genetic code to synthesize their proteins.

Studies Reveal the Conservation of Structure during Evolution

The sequencing of genes from a variety of organisms has made it possible to compare genes with the same function in different organisms. Biologists assume that similar genes are related to each other by evolutionary descent. Consider the histones, proteins that package DNA into higher-order chromosomal structures. Most eukaryotic cells have five different kinds of histone molecules: H1, H2A, H2B, H3, and H4. The amino-acid sequences of these proteins, determined by translating the DNA sequence of the gene using the genetic code dictionary, reveal a high degree of structural similarity between histones from widely divergent eukaryotic organisms (Fig. I.5).

Studies Also Reveal the Conservation of Function during Evolution

Recombinant DNA techniques enable researchers to go one step further in analyzing the degree of relatedness among the proteins and genes of different organisms. They can now remove a gene from one organism, replace it with the homologous gene from another organism, and ask whether the homologous gene can still function in the foreign host. Very often the answer is yes. Yeast geneticists, for example, have replaced many yeast genes with human genes. The resulting yeast cells are usually able to function quite well, even though humans and yeast branched in evolution about a billion years ago. Thus, although the sequences of most human and yeast genes have diverged considerably, the relatedness of their sequences is still apparent, and the human genes still retain their ability to function in yeast cells. This fact is useful in the cloning of unidentified human genes. If one transforms a mutant yeast cell with a human cDNA library, one can often rescue the human gene that encodes the function defective in the mutant yeast cell because the human gene has the ability to restore growth to the yeast (Fig. I.6).

Evolutionary Relatedness Makes It Possible to Construct the Universal Multicellular Organism

Multicellular organisms are composed of groups of specialized cells that communicate and cooperate with each other during the development and life of the organism. Developmental geneticists are learning from yeast, the plant *Arabidopsis,* fruit flies, nematodes, and mice how cells divide, differentiate, communicate, and cooperate to form a multicellular organism. Their studies have shown that the genes identified in one organism may have similar roles in the other organisms. This suggests that many organisms have similar strategies for determining the developmental fates of particular cells.

It is noteworthy, however, that organisms sometimes use disparate molecular principles to accomplish the same goal. An example is the development of male and female sexual differences. The molecular basis of sexual differentiation in *D. melanogaster* is quite different from the molecular basis of sexual differentiation in *C. elegans.*

HOW BIOLOGISTS STUDYING DEVELOPMENT USE GENETICS TO ANSWER THEIR QUESTIONS

Most animals develop according to a common embryonic plan, building first a ball of cells, or blastula, that folds in on itself during gastrulation. In these early stages of embryonic development, three different cell types appear: the endoderm, mesoderm, and ectoderm. Later, these cells give rise to a set of more differentiated cell types that form muscles, nerves, adipocytes, blood cells, cartilage, bone, and so forth.

Given these common patterns of embryonic development, biologists studying embryogenesis seek to answer the following questions. In general, how does development derive from cell behaviors? How do genes act to elicit these behaviors? In particular, what mechanisms regulate cell proliferation? What mechanisms control cell lineage, that is, the fate of a cell? What mechanisms control cell migration? How do cells become different? How do cells influence one another? How do cells cooperate to form structures? How do cells cooperate in functional ensembles? How do cells die? Insights gained from the study of different model organisms will help answer these questions (Table I.2).

Genetics Simplifies the Study of Development

Because proteins are the basic functional elements of cells, developmental biologists can analyze development in terms of the roles played by individual proteins. To do this, they eliminate all copies of a single protein species from a cell or organism and determine the consequences. From these consequences, they can often infer the function of the normal protein in normal development.

Genetics makes this task possible. All one has to do is isolate a mutant cell or organism with a specific, inactivated gene. The mutant with the altered gene will lack the protein encoded by that gene. Careful analysis of the mutant phenotype can reveal the importance of the protein in development.

A description of how development occurs from a genetic perspective would consist of a chart describing all the genes necessary for development and the phenotype of all mutants. This genetic description of development could serve as a blueprint for biochemical studies that seek to explain development in chemical terms.

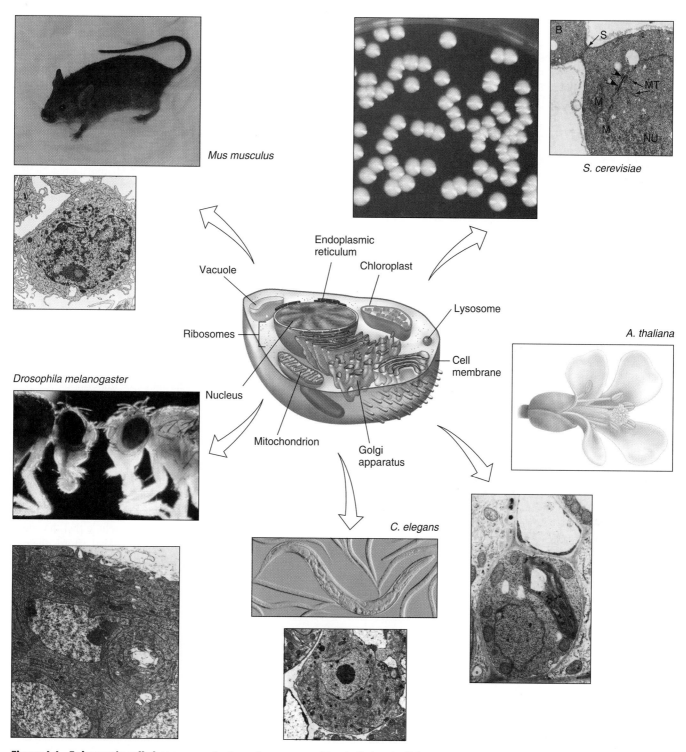

Figure I.4 Eukaryotic cells have many features in common. The similarity of cellular structure in our five model organisms is visible in the micrograph accompanying the photo of each of the entire organisms above. Each cell contains a variety of complex membranous organelles in the cytoplasmic matrix and has the majority of its genetic material contained within membrane-enclosed nuclei. Each model has characteristic variations, but the basic cellular plan is the same.

Alignment of Histone H4 from humans and four model systems.

H. sapiens	MSGRGKGGKGLGKGGAKRHRKVLRDNIQGITKPAIRRLARRGGVKRISGLIYEETRGVLKVFLENVIRDAVTYTEHAKRKT
M. musculus	MSGRGKGGKGLGKGGAKRHRKVLRDNIQGITKPAIRRLARRGGVKRISGLIYEETRGVLKVFLENVIRDAVTYTEHAKRKT
C. elegans	MSGRGKGGKGLGKGGAKRHRKVLRDNIQGITKPAIRRLARRGGVKRISGLIYEETRGVLKVFLENVIRDAVTYCEHAKRKT
D. melanogaster	MTGRGKGGKGLGKGGAKRHRKVLRDNIQGITKPAIRRLARRGGVKRISGLIYEETRGVLKVFLENVIRDAVTYTEHAKRKT
A. thaliana	MSGRGKGGKGLGKGGAKRHRKVLRDNIQGITKPAIRRLARRGGVKRISGLIYEETRGVLK IFLENVIRDAVTYTEHARRKT

* indicates identical amino acid
• indicates similar amino acid

Figure I.5   Comparison of amino-acid sequences in one region of a histone of four model organisms and *H. sapiens*. The most striking thing about this comparison is the similarity, not the difference between the sequences. Alignment of the amino acids from the histone H4 protein of five organisms shows the remarkable conservation of sequence. Note in the shaded area of the five organisms the only variation occurs in the second position of *D. melanogaster*. Researchers rely on this similarity when they generalize results from research on model organisms.

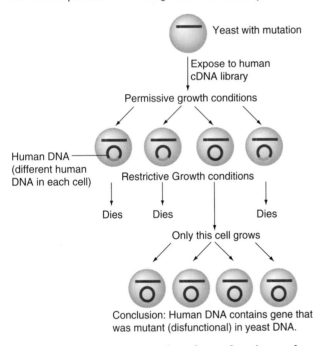

Figure I.6   Using the conservation of gene function to clone unidentified human genes in yeast. If a library of human cDNA clones is introduced into yeast cells containing a conditional mutation in an essential gene, the cell receiving the homologous human gene will be able to grow at the nonpermissive temperature.

The Genetic Dissection of Development Depends on a Comprehensive Set of Mutants

Developmental geneticists try to isolate many different mutations in a gene. If each mutation affects the function of the corresponding protein in a different way, studies of the phenotypes associated with the various mutations may shed light on the various roles the protein plays in the organism.

Loss-of-Function Mutants

To draw legitimate conclusions about the importance of a protein to development, it is important to study an organism that lacks the function provided by that protein, for example, an organism that carries a null allele of the gene. In an analogy, if you tried to ride a bicycle with no chain, you might conclude that the chain is required to move the wheels. If, however, you tried to ride a bicycle whose chain had a missing link (the equivalent of an individual who makes a partially defective protein) such that the bicycle would move but respond only erratically to your peddling, you might conclude that the chain is not critical to wheel movement and instead affords the cyclist some control over wheel movement. In Chapter 6 we saw that mutations that remove all function produce *null* alleles that are usually recessive to wildtype alleles (Fig. I.7a).

TABLE I.2   Insights from Different Model Organisms Help Answer Questions about Development

General Questions

How does development derive from cell behaviors?

How do genes act to elicit these behaviors?

Particular Questions	Organism Providing Major Insights
What mechanisms regulate cell proliferation?	Yeast, *Drosophila*
What mechanisms control cell lineage?	Yeast, *C. elegans*
What mechanisms control cell migration?	*C. elegans*, *Drosophila*
How do cells become different?	Yeast, *C. elegans*, *Drosophila*
How do cells influence one another?	Yeast, *C. elegans*, *Drosophila*, mice
How do cells cooperate to form structures?	*C. elegans*, *Drosophila*, *Arabidopsis*
How do cells cooperate in functional ensembles?	*C. elegans*, *Drosophila*, mice
How do cells die?	*C. elegans*, *Drosophila*, mice

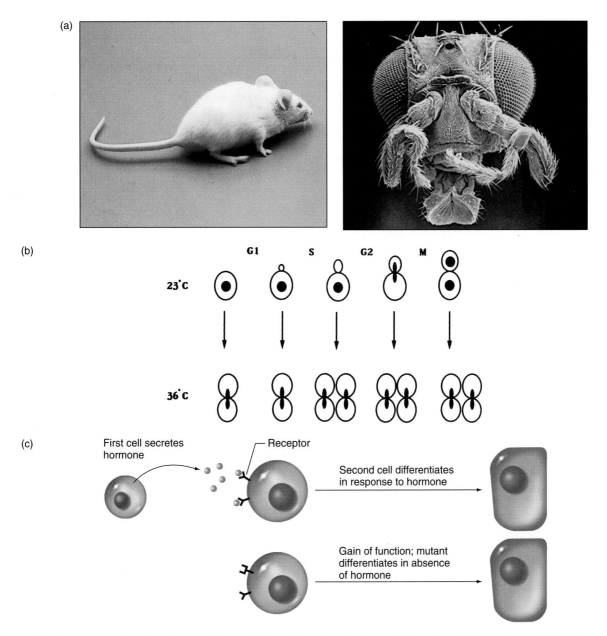

Figure I.7   Mutants used by developmental geneticists. (a) Functional mutants. (b) Conditional lethal mutants. (c) Gain-of-function mutants.

Mutants Carrying Conditional Lethal Mutations

As desirable as null alleles are for the analysis of development, it is not always possible to rely on them entirely. Some processes and the proteins that mediate them are essential to life; if you eliminate them, you have no organism to study. For example, if you want to study the role of the mitotic spindle in generating cell shape and polarity during gestation, you cannot conveniently study null alleles that eliminate essential components of the spindle. Without these essential components, there will be no cell division, and without cell division, there is no embryonic development. One solution is to isolate **conditional lethal mutations** that are lethal only under special circumstances (Fig. I.7b). The most commonly studied type of conditional lethal mutation, the temperature-sensitive

mutation, produces a protein that is functional at a lower, *permissive* temperature but is defective at a higher, *restrictive* temperature. In contrast, the protein product of the wildtype allele functions at both permissive and restrictive temperatures. It is best if the conditional lethal mutation produces completely nonfunctional proteins at the restrictive condition, but it is sometimes difficult to determine whether the resulting protein is completely nonfunctional or only partly functional.

Gain-of-Function Mutants

Mutations that produce proteins with a function not present in the wildtype protein are **gain-of-function mutations.** The alleles resulting from gain-of-function mutations are often distinguishable from recessive null alleles by the fact that they are frequently dominant to the wildtype allele. It is hard to understand unambiguously the role of a protein in development from a gain-of-function allele, however, because the mutation, rather than taking something away, adds something unusual to the organism, which might behave in an unpredictable way. Nevertheless, gain-of-function mutants can be useful. Consider a situation in which one cell secretes a hormone that, by interacting with a receptor on a second cell, causes the second cell to differentiate. It might be possible to identify the receptor gene through gain-of-function mutations that cause the second cell to differentiate in the absence of the hormone (Fig. I.7c).

Mutants That Express the Wildtype Gene at an Abnormal Place or Time

Recombinant DNA techniques that enable the isolation, alteration, and reintroduction of genes in model organisms enable researchers to construct loss- or gain-of-function mutations. The same technology makes it possible to alter genes in other ways. For example, changing a gene's promoter might cause expression of the gene in different tissues or at different times or in different amounts than are normal.

Using Mutants to Identify the Components of Development

To understand development, biologists seek to discover all the components that participate in the process as well as the roles they play.

Saturation Mutagenesis and Complementation Studies Are the First Steps

Saturation mutagenesis is the production of as many different mutants with related phenotypes as possible in an attempt to identify all the genes relevant to a process. Many mutations, of course, will identify the same gene, so it is necessary to group the mutations into genes by complementation analysis. When bacterial geneticists looked at the motility of *Escherichia coli* cells through saturation mutagenesis and complementation analysis, they identified 20 genes that contribute to the construction of the bacterial flagellum (see the comprehensive example in Chapter 13). Such studies, based on the purely genetic techniques of mutant isolation and complementation testing, provide a good estimate of the complexity of different biological processes.

In these studies, how do you know you have found all the relevant genes? One way is to isolate so many mutations that every complementation group contains several. At this point it is unlikely that you have missed any genes purely by chance. This approach, however, does not necessarily identify all the genes in the pathway, because some functions are performed by redundant proteins, that is, by either of two proteins encoded by different genes (Fig. I.8a). With this type of redundancy, a single mutation in either gene is without phenotype. Only when both genes sustain mutations in the same cell does

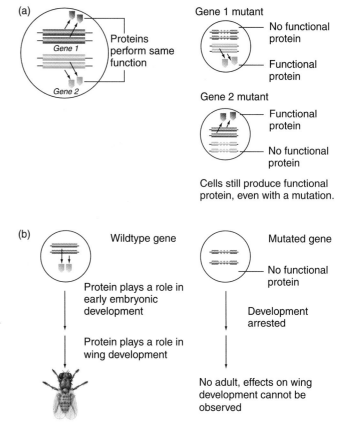

Figure I.8 Reason why saturation mutagenesis and complementation studies may miss some genes. (a) Some genes may be missed if two genes encode different proteins that perform the same function. (b) If one gene encodes a protein with two functions where one function is essential and one is not essential, some genes could be missed.

the mutant phenotype manifest itself. Another reason for missing important genes in a mutant hunt is that some proteins carry out more than one function (Fig. I.8b). For example, in *Drosophila*, some proteins play a role both early in embryonic development and later during the development of the adult. If you were looking for mutations that affected the structure of the adult wing, you might miss important genes if mutations in those genes arrested development earlier during formation of the embryo.

Are there ways of finding genes missed for reasons of protein redundancy or the masking of phenotypes? The answer is yes; there are several techniques geneticists use to reveal additional genes in a process.

Overexpression

There are many instances where the excessive expression of a gene can be detrimental and produce a recognizable phenotype. Overexpression phenotypes are dominant, and the redundancy of protein function does not prevent their appearance (Fig. I.9a).

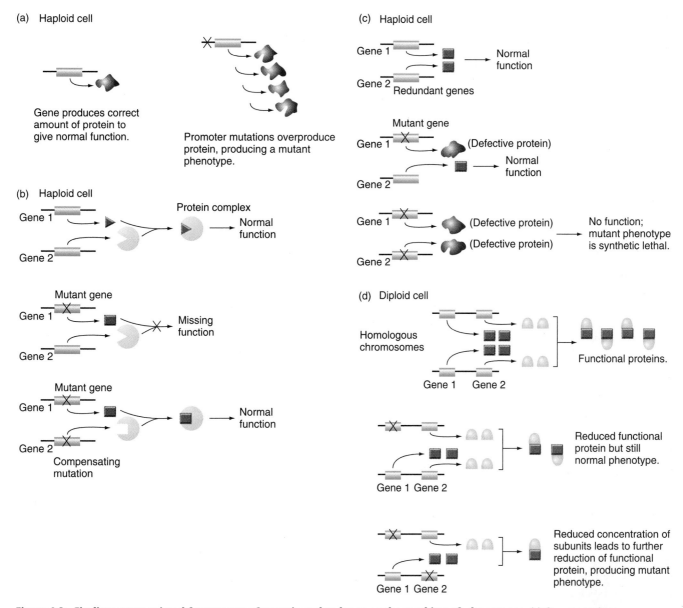

Figure I.9   Finding genes missed for reasons of protein redundancy or the masking of phenotypes. (a) Overexpression. (b) Suppressor mutations in a haploid cell. (c) Synthetic lethals in a haploid cell. (d) Nonallelic noncomplementation in a diploid cell.

Suppressor Mutations

If you begin with a mutation that disrupts a process, you can often isolate additional mutations that restore the missing function. That restoration sometimes results from a reversion at the original mutant locus, but at other times it results from a compensating mutation in some other important gene in the same process, which suppresses the effects of the first mutation (Fig. I.9b).

Synthetic Lethals

A strain containing a mutation in one gene may be sick but viable. In the presence of this mutation, additional mutations in genes whose protein products perform related functions can be lethal to the cell (Fig. I.9c). To analyze these related genes, you have to have a breeding scheme that enables you to recover the additional mutations in a viable form or to grow the organism under conditions where the genes in question are not needed.

Nonallelic Noncomplementation

Most mutations are recessive because an organism carrying one mutant and one wildtype allele has 50% of the normal gene product, which is usually enough to generate a wildtype phenotype. In some cases, however, two genes that are both heterozygous may generate a novel phenotype that does not appear when either gene is heterozygous and the other is homozygous. This phenomenon is known as **nonallelic**

noncomplementation. It usually implies that two proteins normally cooperate in some way. Decreasing to 50% the amount of two proteins that cooperate, for example, in building a structure, may reduce to 25% of normal the amount of structure that is built (Fig. I.9d).

Order-of-Function Analysis

Once you have isolated a comprehensive set of mutations and identified as many of the genes as possible that function in the biological process of interest, the next step is to establish the order of the functions performed by these genes.

Phenotypic Characterization

The most straightforward order-of-function analysis is in terms of the mutant phenotypes. In the study of a developmental process, for example, you may find that some mutants arrest development at early stages and some at later stages (Fig. I.10a). The differences suggest the order in which the genes function.

Double-Mutant Analysis

You can verify your proposed order of function by a double-mutant analysis. If two mutations define successive steps in a process, the double mutant should arrest with the phenotype characteristic of the earliest block in the process (Fig. I.10b). Many pathways of function are much more complex than a simple linear pathway. Gene products can function in parallel pathways, in branching pathways, and in redundant pathways. The phenotypes of double mutants can reveal some of this complexity.

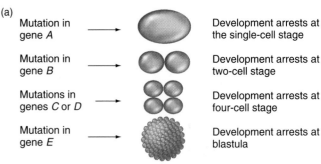

(a)

Mutation in gene A → Development arrests at the single-cell stage

Mutation in gene B → Development arrests at two-cell stage

Mutations in genes C or D → Development arrests at four-cell stage

Mutation in gene E → Development arrests at blastula

Suggests that gene A acts first, then gene B, then genes C and D, then gene E.

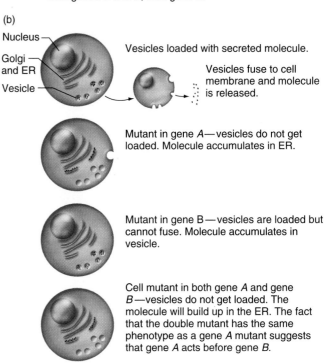

(b)

Nucleus
Golgi and ER
Vesicle

Vesicles loaded with secreted molecule.

Vesicles fuse to cell membrane and molecule is released.

Mutant in gene A—vesicles do not get loaded. Molecule accumulates in ER.

Mutant in gene B—vesicles are loaded but cannot fuse. Molecule accumulates in vesicle.

Cell mutant in both gene A and gene B—vesicles do not get loaded. The molecule will build up in the ER. The fact that the double mutant has the same phenotype as a gene A mutant suggests that gene A acts before gene B.

Figure I.10 Elements of an order-of-function analysis.
(a) Observing the stage of development at which mutations arrest.
(b) Double-mutant analysis.

CONNECTIONS

Mutational analysis is a powerful tool for identifying and dissecting the roles of genes that control development. With the combined insights from different organisms, developmental geneticists will be able to piece together a unified view of the development of multicellular organisms.

In the following chapters we discuss the yeast *S. cerevisiae,* the plant *A. thaliana,* the nematode *C. elegans,* the fruit fly *D. melanogaster,* and the mouse *M. musculus,* focusing on their contributions to an understanding of development. Our purpose is not to provide a comprehensive explanation of multicellular development. Rather, we seek to illustrate how genetic studies with these organisms are revealing the individual pieces of the development puzzle. For the model organism under consideration each chapter describes (1) the genome, including its size, structure, organization, and relevant peculiarities; (2) the life cycle, including which characteristics make the organism valuable as a model for genetic studies; and (3) a few topics in development that researchers have examined through comprehensive genetic analyses.

ESSENTIAL CONCEPTS

1. Development is the process by which cells selectively mobilize the genome to permit the generation of different cell types, as occurs during the emergence of a multicellular organism with complex structures and complex behaviors from a fertilized egg.

2. Developmental geneticists use model organisms as the basis for studying how a fertilized egg becomes a multicellular adult. Each model organism has special attributes that give it value as an experimental organism. The accumulation of knowledge, mutants, and techniques from long years of study enhances the utility of model organisms.

3. The evolutionary relatedness of all organisms, revealed by the conservation of structure and function, often enables extrapolation from the models to all living forms.

4. Genetics simplifies the study of development. A key to the genetic dissection of development is the isolation of a comprehensive set of mutations affecting a process. Such saturation mutagenesis enables the identification of many genes participating in a pathway. Multiple alleles of any one gene make it possible to analyze the many roles a protein might play in development.

5. Genetic analyses such as order-of-function tests can help examine the way in which proteins cooperate in developmental pathways. The results of such tests enable researchers to make predictions useful in guiding subsequent biochemical experiments.

CHAPTER

18

SACCHAROMYCES CEREVISIAE: A GENETIC PORTRAIT OF YEAST

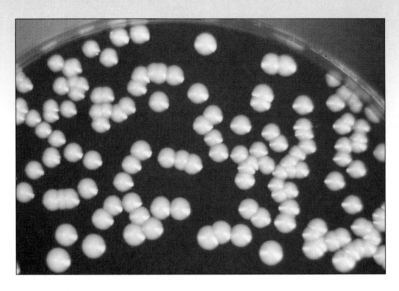

Yeast colonies, each containing about 10^7 cells, grow from single cells spread on solid media.

Saccharomyces cerevisiae, commonly known as baker's yeast, has been a preferred organism for genetic research since the mid-twentieth century (Fig. 18.1). The experimental value of this single-celled eukaryotic species lies in its simple life cycle, alternating haploid and diploid phases, short generation time, and easy-to-identify meiotic products. Recent technological advances enable yeast biologists to use genetic analysis to gain a deep understanding of the organization and regulation of eukaryotic cells.

Like all eukaryotic cells, yeast cells make decisions that determine their fate. They decide, for example, whether they are a or α cells, and whether to mate, enter the meiotic pathway, grow and divide by mitosis, or bud. Researchers can use genetic techniques to analyze the molecular events that influence these fate-determining decisions; and knowledge of these events, which include the activation and inactivation of transcription factors, can contribute to an understanding of how the cells of multicellular organisms differentiate.

Three main themes unify our genetic portrait of yeast. First, to identify the genes and proteins influencing development, re-

searchers isolate and study mutants defective for a specific process. Second, master regulatory genes in yeast control the expression of other genes, whose activities determine the phenotype of a cell. Third, cascades of proteins called *signal transduction systems* (described in Chapter 17) transmit the signals responsible for shifting a cell from one phenotype to another.

In examining yeast as a model organism for understanding development, we present

- An overview of yeast in the laboratory: current knowledge of the yeast genome, basic tools for looking at yeast, and significant details of the yeast life cycle.

- Cell differentiation: molecular mechanisms for determining cell type.

- Mating: how cell-to-cell communication through pheromones promotes the conversion of haploid cells to diploid cells.

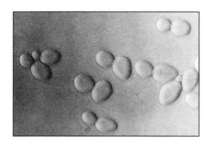

Figure 18.1 Yeast cells. Phase contrast micrograph of yeast cells.

AN OVERVIEW OF YEAST IN THE LABORATORY

Yeast geneticists have gathered enough information on *S. cerevisiae* to state with confidence that single-celled lower eukaryotes such as yeast express genes, organize themselves, perform biological functions, and differentiate using variations of the same processes found in multicellular, higher eukaryotes. Since cellular differentiation is a driving force behind the development of a multicellular organism from a fertilized egg, what we learn from yeast can very likely tell us a great deal about our own development, biochemistry, genetic makeup, and genetic history. This is the main reason why research on a seemingly irrelevant creature like yeast is funded by the National Institutes of Health under the umbrella of medical research. The experimental advantages of yeast include the fact that unlike the cells of most multicellular eukaryotes, yeast cells can grow as haploids or diploids. This makes it possible to identify recessive mutations in haploid cells and then combine mutations in diploid cells for complementation analysis. An equally important experimental consideration is that one can look at more cells in a shorter period of time with yeast than with any other eukaryotic organism; in fact, it is possible to obtain as many as 10^8 yeast cells in 1.0 ml of culture.

The Nuclear Genome of Yeast

Haploid *S. cerevisiae* has 16 linear chromosomes in the nucleus, all constructed according to the same blueprint. Each chromosome contains one centromere/kinetochore that binds to one microtubule, two termini consisting of longer subtelomeric repeats followed by short telomere repeats at the very ends, and multiple origins of replication spaced approximately 30–40 kb apart (of which only a small subset are used to initiate replication at any one time) (Fig. 18.2a). Yeast chromosomes are packaged into nucleosomes consisting of the core histones H2A, H2B, H3, and H4.

Physical Maps

Pulsed-field gel electrophoresis, which separates intact yeast chromosomes, produces molecular "karyotypes" of the yeast genome, such as the one shown in Fig. 18.2b. Geneticists use these karyotypes to confirm the number of yeast chromosomes and to observe major alterations in chromosome structure. As Table 18.1 indicates, chromosome I, about 230 kb in length, is the smallest yeast chromosome. Chromosome XII is the largest; its size varies between about 2060 and 3060 kb because of a variable number of tandem ribosomal RNA genes (rDNA)—usually in the range of 100–200 copies—in different strains. The total genome length is 12,000 kb plus the 2000–3000 kb of rDNA. DNA sequence data from the International Genome Project was available for 100% of the genome by April of 1996. Yeast DNA sequences can be found in the GenBank database. The yeast genome contains about 6000 open reading frames encoding polypeptides larger than 100 amino acids.

Genetic Maps

The classical genetic map of yeast currently shows the locations of over 1000 markers determined by tetrad analysis. The total genetic map length, a function of the frequency of meiotic recombination, is about 4400 cM. With an average of 3 kb/cM, genetic distance is roughly proportional to physical

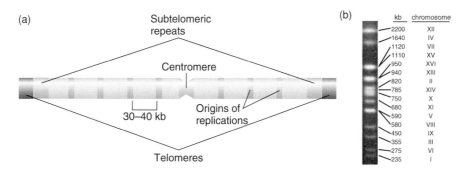

Figure 18.2 Yeast chromosomes. (a) Elements of a chromosome. Each linear chromosome contains one centromere, multiple origins of replication, subtelomeric repeats, and telomeres at the ends of the linear DNA. (b) Molecular karyotype of yeast chromosomes. Using pulsed field gel electrophoresis, the chromosomes can be separated by size.

TABLE 18.1	Yeast Chromosome Sizes in Kilobases
Chromosome	**Size (kb)**
I	230
II	813
III	315
IV	1522
V	575
VI	270
VII	1090
VIII	562
IX	439
X	745
XI	666
XII	2060–3060
XIII	924
XIV	784
XV	1091
XVI	948

Gene Arrangement and Genome Organization

The bulk of the yeast genome is composed of unique, nonreiterated DNA sequences. Approximately 80% of the genome is transcribed into RNAs; and most of these RNAs are nonoverlapping. These nonoverlapping RNAs represent about 6000 genes, about 7% of the estimated number of genes in humans. On average, there is one yeast gene for every 1.5–2.0 kb of DNA (Fig. 18.3). Genes related in function are often coregulated, but they are not usually clustered according to function or organized like bacterial operons. In the rare instances where functional clustering does occur, as with the *GAL* genes that encode enzymes active in the transport and degradation of galactose, the clustered genes are usually transcribed into mRNA from separate, divergent transcriptional promoters.

Although the yeast genome does contain some reiterated DNA sequences, the proportion of the genome devoted to reiterated DNA is much lower than in higher eukaryotes. The 100–200 copies of rDNA rank highest in abundance. Other repeated sequences of note include tRNA genes, which occur on average in five to seven dispersed copies per tRNA isoacceptor (that is, per group of different tRNAs that accept the same amino acid but possess different anticodons); the subtelomeric repeats, which are reiterated one to four times near the ends of the chromosomes; and a complex family of retroposons called Ty elements, of which there are about 50 dispersed copies per haploid genome.

The genetically mobile Ty elements resemble animal cell retroviruses except for the inability to undergo a lytic infection cycle. Ty elements are 5.4–6.3 kb in length and can pair through homologous sequences and recombine with each other even when copies of the elements are on different chromosomes. Such recombination between dispersed transposable elements may be a major source of reciprocal translocations, inversions, and deletions in yeast. Higher eukaryotes have a variety of transposable elements that may also be a significant source of these types of chromosomal aberrations.

distance (Fig. 18.3). Exceptions are the rDNA repeats, in which there is virtually no meiotic recombination, and the regions surrounding the centromeres, where recombination is modestly suppressed. Because *S. cerevisiae* has an unusually active recombination system, the total genetic map of yeast is extraordinarily long compared to the genetic map of other fungal organisms. *Neurospora crassa,* for example, has about the same DNA content as yeast but a genetic map that is only 15% as long as yeast's. The extremely active homologous recombination system in yeast might explain why plasmids in yeast integrate primarily by homologous recombination, whereas plasmids in other organisms integrate mainly by nonhomologous recombination.

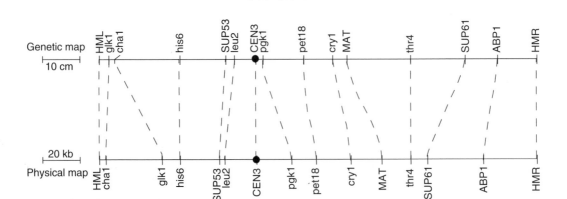

Figure 18.3 Comparison of genetic and physical maps. The location of genes on chromosome III as determined by recombination distances (genetic map) and sequencing DNA (physical map) are shown. While the order of genes is the same, some distances determined by these two measurements differ.

Gene Structure: Introns and Pseudogenes

Introns occur in only a few percent of all yeast genes. By comparison, most genes of higher eukaryotes are composed of several noncontiguous coding regions separated by introns. The low frequency of introns in yeast may be related to a similarly low frequency of "processed" pseudogenes, which like introns, are common in higher eukaryotes. Pseudogenes, we have seen, contain coding sequences that resemble those of real genes, but because the pseudogenes lack introns and transcriptional promoters, they are not expressed.

Basic Tools of the Yeast Geneticist

Several techniques of yeast genetics enable researchers to answer questions that are difficult, if not impossible, to resolve in other organisms. However, although many of the techniques described next were developed for the study of yeast, creative adaptations have turned them into similar tools for the study of other organisms.

Transformation: An Efficient System Relying on Homologous Recombination

Yeast cells are susceptible to efficient transformation by DNA molecules from external sources. To execute a transformation experiment, researchers construct a plasmid carrying a marker in the form of a functional yeast gene and expose a population of yeast cells to a solution of the plasmids. They then confirm the presence of the plasmid in transformed cells by ascertaining whether the marker gene has supplied a function missing in the transformation recipients, or host cells. For example, a plasmid carrying the yeast *LEU2* gene, coding for an enzyme in leucine biosynthesis, causes a host cell carrying a loss-of-function mutation in the *LEU2* gene to become Leu$^+$ in phenotype. The plasmids that transform yeast fall into two categories: those that cannot replicate on their own in yeast cells and those that can. YIp plasmids are circular plasmids that lack the ability to replicate on their own in yeast cells (Fig. 18.4a). YIps produce stable transformants by integrating into the chromosomal DNA through genetic recombination. Such integration in yeast almost always occurs through homologous recombination, that is, through alignment (pairing) and genetic recombination with a similar sequence. By contrast, autonomously replicating plasmids maintain transforming genes in an extrachromosomal state. Examples are yeast episomal plasmids (YEps) and yeast centromeric plasmids (YCps), both of which replicate in the yeast cell nucleus (Fig. 18.4b and c).

YEps replicate autonomously because they carry the replication origin from an endogenous yeast DNA plasmid known as the 2-μm circle. YEps are present in about 20 copies per cell. Because of this relatively high copy number, YEps are useful in experiments requiring the overproduction of a gene. YCps replicate autonomously because they carry DNA corresponding to a chromosomal origin of replication. They also carry the centromere of a yeast chromosome and, consequently, segregate like a chromosome. Because they contain a centromere, YCps have a copy number of one per cell, and the markers they transport segregate like chromosomes during

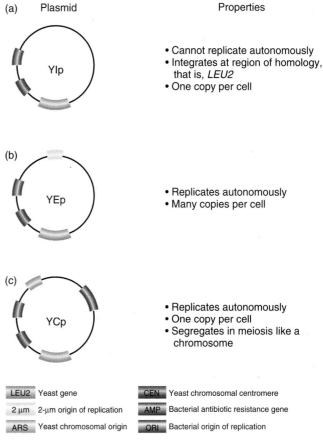

Figure 18.4 Plasmids in yeast. (a) YIp plasmids do not contain an origin of replication and therefore cannot replicate autonomously (separate from the chromosome). If the plasmid integrates into a chromosome by homology it is maintained and transmitted to subsequent generations of cells. (b) YEp plasmids contain an origin of replication from the 2-μm circle and therefore replicate autonomously. (c) YCp plasmids contain a centromere in addition to a replication origin, so are maintained autonomously and segregate during meiosis as a linear chromosome does.

meiosis. However, even though YCps behave like chromosomes, they are more prone than natural chromosomes to nondisjunction events that lead to duplications and deletions.

Moving Genes between Yeast and Bacteria

The autonomously replicating plasmids just described are designed to function as "shuttle vectors" able to ferry genes into both bacteria and yeast for replication. To this end, they all carry an origin of replication from a bacterial plasmid which allows replication as a multicopy plasmid in *Escherichia coli*. Most also carry genes conferring on *E. coli* resistance to ampicillin or tetracycline. Such resistance makes it possible to select for *E. coli* transformants that carry the plasmids. Completing the prerequisites necessary for shuttle vectors to move genes between yeast and bacteria are design features that enable the vectors to replicate and be selected for in yeast. With vectors carrying all these features, researchers can use recombinant DNA techniques on DNA prepared in *E. coli* and then place the DNA manipulated by recombinant technology back into yeast to test its function. For example, studies of the deletion of DNA sequences around the centromere have shown that a functional yeast centromere need contain only 120 bp of DNA.

Gene Replacement, or Genetic Microsurgery, Makes It Possible to Alter Genes without Resorting to a Formal Genetic Cross

Gene replacement is a very powerful technique that has a great advantage over classical methods of strain construction: Researchers can use it to alter at will a chromosomal gene in a living cell without changing the genome in any other way. This level of control over gene alteration makes it possible to compare the phenotypic effects of mutations in an otherwise isogenic (that is, genetically identical) background, and attribute any phenotypic differences to specific alleles.

For example, researchers can replace copies of genes under analysis with alleles constructed *in vitro*. To create such alleles, either they use recombinant DNA technology to delete or insert DNA in a cloned version of a gene, or they use a variety of *in vitro* methods to induce specific mutations. They then introduce the altered gene into yeast via a transformation that replaces the wildtype allele with the modified gene. Figure 18.5 illustrates gene replacement via transformation using a DNA fragment carrying a disrupted gene. The disrupted gene includes a functional *URA3* gene (used to select for transformants) inserted into the gene to be disrupted. Chromosomal integration of the disrupted gene occurs through the equivalent of a double crossover.

Since some genes are essential for growth and viability, researchers generally use disruption alleles to transform wildtype diploids. Such transformation produces −/+ heterozygotes containing one disruption allele and one wildtype allele. If the disruption allele confers recessive loss of viability, the diploid will be viable. In every tetrad generated by sporulation of the diploid, however, the two haploid spores that carry the disruption mutation will not survive, whereas the two spores that carry the wildtype allele will divide to form colonies (see Fig. 18.6).

The Yeast Life Cycle

Yeast cells alternate between haploid (1*n*) and diploid (2*n*) phases (Fig. 18.6). Haploid cells occur in two mating types, *a* and α. Both mating types can reproduce mitotically as stable haploid cells. Or they can engage in sexual reproduction, in which cells of opposite mating types communicate with one another by proteins known as pheromones that induce the dissimilar cells to undergo cell fusion followed by nuclear fusion. The zygote produced in this way has a single diploid nucleus and buds to produce diploid progeny. Diploid yeast cells can propagate as stable diploid cells by mitotic division. Starvation, however, induces them to undergo meiosis and sporulation. This allows the cells to "reshuffle" their genes when conditions are poor, perhaps enabling them to find a combination more suitable for survival in the environment at hand. Meiosis reduces the diploid nucleus to four haploid nuclei, which become encapsulated in four haploid spores. On

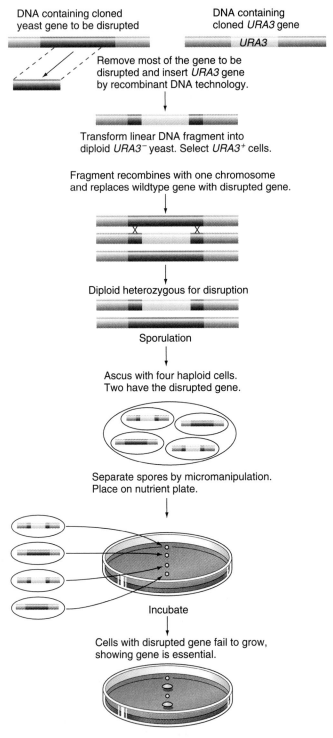

Figure 18.5 Gene disruption and replacement. A gene is disrupted by first constructing a linear fragment in which the *URA3* gene (to be used as a selectable marker) replaced the middle of the gene of interest. When the fragment is transformed into a diploid yeast cell, it can recombine with the gene in one of the chromosomes by homology, replacing the good copy with a mutant. The heterozygous diploid, when sporulated, produces two haploid spores containing the disrupted copy and two spores with the normal copy of the gene. If the gene is needed for growth, the spores with the disrupted gene copy will die.

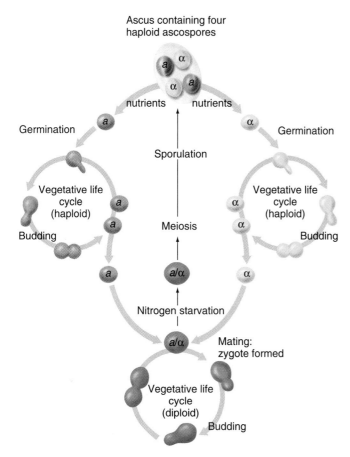

Figure 18.6   Life cycle of yeast. Yeast can grow vegetatively as either haploid or diploid cells. The transition from haploid to diploid occurs via mating, and the transition from diploid to haploid occurs via meiosis during sporulation.

the addition of nutrients, the haploid cells bud and can divide mitotically.

Sex in yeast is determined by the mating type locus (designated as MAT) on chromosome III. As already mentioned, there are two mating types: a and α. Mating ability segregates $2a:2\alpha$ in tetrads derived from MATa/MATα heterozygous diploids, indicating that the a and α mating types are specified by alleles of a single locus (MAT). MATa or MATα cells mate efficiently with cells of the opposite sex. Heterozygous MATa/MATα diploids are sterile, but it is possible to derive MATa/MATa or MATα/MATα diploid cells. These diploids will mate with cells of the opposite mating type, either haploids or diploids. Thus, the ability to mate is determined by the genetic configuration at the MAT locus and is not related to ploidy per se.

CELL DIFFERENTIATION IN YEAST: MECHANISMS FOR DETERMINING CELL TYPE

Like the specialized liver, muscle, and nerve cells of animals, the a, α, and a/α cells of yeast are differentiated cells that express different sets of genes. The three cell types in yeast dif-

fer in how they respond to signals from the environment and in their developmental potential. Three different genetic programs establish the differentiation pathways leading to each cell type. Because the establishment of cell identity is basic to the differentiation and development of multicellular organisms, our understanding of the process in yeast serves as a model for development in multicellular eukaryotes.

The MAT Locus Controls Expression of the Genes That Determine Cell Type

A cell with MATa information at the MAT locus has an a mating type. By contrast, a cell carrying MATα information at the MAT locus has an α mating type. What at the MAT locus determines these different mating types?

Clues to the role of the MAT locus first emerged from mutant analyses. When sterile mutants of α cells were isolated, one class of mutants had defects at the MAT locus. Surprisingly, the MATα mutations fell into two complementation groups, indicating that instead of a single gene at the MATα locus, there are two genes, $\alpha 1$ and $\alpha 2$. Mutations in the $\alpha 1$ and $\alpha 2$ complementation groups produce different molecular effects, which inform us of the roles of the genes. Strains carrying $\alpha 1$ mutations fail to synthesize α pheromone, a protein normally made by α cells. Strains carrying $\alpha 2$ mutations synthesize α pheromone but degrade it before it can be secreted; normally, only a cells express the phenotype of α pheromone degradation.

What hypothesis could explain the roles of $\alpha 1$ and $\alpha 2$ in determining cell type? According to the proposal illustrated in Fig. 18.7a, MATα is a complex locus consisting of two genes, $\alpha 1$ and $\alpha 2$. The $\alpha 1$ gene encodes an activator of the α-specific function required to elicit the α cell type. In the absence of a functional $\alpha 1$ product, α cells are sterile, in part because they do not produce α-specific proteins such as α pheromone. The $\alpha 2$ gene encodes a repressor of a-specific functions. Cells lacking a functional $\alpha 2$ gene product express both α-specific and a-specific genes, and the simultaneous expression of both sets of genes is incompatible with mating. $\alpha 1^{-}\alpha 2^{-}$ double mutants mate like a cells because they fail to turn off a-specific functions and fail to turn on α-specific functions. In effect, a simple network depending on two regulatory proteins—one an activator, the other a repressor—accounts for the a and α cell types. In this network, $\alpha 1$ and $\alpha 2$ act as master regulatory genes whose products control the expression of a large number of genes required for the appropriate behavior of a or α cells.

What is the role and makeup of the MAT locus in an a cell? Mutant analyses of the MATa locus indicate that it has a slightly different configuration from that of the MATα locus (Fig. 18.7b). No sterility-producing mutations have been isolated that map at the MATa locus. Haploid strains carrying MATa mutations mate normally and fall into a single complementation group known as $a1$. The lack of any detectable haploid phenotype for MATa mutations suggests that the a1 protein performs no function necessary for mating or differentiation into a cells. It is the lack of $\alpha 1$ and $\alpha 2$ functions that makes a cells what they are. a cells do not express α-specific

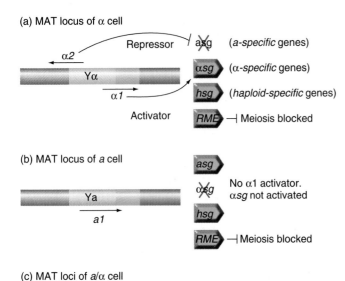

(a) MAT locus of α cell

(b) MAT locus of *a* cell

(c) MAT loci of *a*/α cell

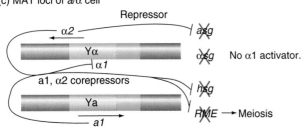

Figure 18.7   MAT a and MATα loci. Two different versions of the MAT locus, MAT*a* and MATα, share some common DNA sequences (shown in orange) but also contain unique sequences known as Ya and Yα. The presence of Ya, Yα, or Ya + Yα in a cell has different consequences on expression of the *a-specific*, α-specific, haploid-specific, and *meiosis* (*RME*) genes because activators and repressors are encoded by the different versions of the MAT locus. Arrows indicate the location of transcripts for the α*1*, α*2*, and *a1* genes at the MAT locus.

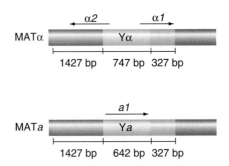

Figure 18.8   Structure of the MAT locus. The MAT locus consists of sequences that are found in both MAT*a* and MATα versions with the Ya or Yα block of DNA between them. The Y DNA contains promoters and early sequences for genes α*1* and α*2* in the Yα version, and *a1*, in the Ya version. (Part of each of these genes is encoded in the common sequences that surround Ya or Yα.)

blocks of similar sequences at each end totaling about 2500 bp, but they differ by the insertion of a unique 747-bp DNA sequence, known as Ya, in MATα and a unique 642-bp DNA sequence, known as Yα, in MAT*a*. The unique segments contain the promoters for transcription and parts of the coding regions of the α*1*, α*2*, and *a1* genes. Compared to other loci, the MAT locus is unique in that its two forms—MAT*a* and MATα—carry dramatically different genetic information.

Molecular analyses of RNA in mutant and wildtype cells have shown that the α1, α2, and a1 gene products act at the level of transcription. MATα1⁻ mutants (lacking a functional α1 gene) do not transcribe the α-specific genes known as *STE3* (which encode an a pheromone receptor) and *MFα1* and *MFα2* (which encode α pheromone); normal α cells do express these genes. Similarly, MATα2⁻ cells (lacking a functional α2 gene) transcribe and produce functional products of the *a*-specific genes that encode an α pheromone receptor and a pheromone; normal α cells repress expression of these genes. MAT*a*/MATα cells normally repress transcription of haploid-specific genes; however, loss of *a1* or α2 function in these cells results in the transcription of the haploid-specific genes, including genes that influence the pheromone response, cell fusion, and meiosis.

Studies of the binding to DNA of isolated α1 and α2 proteins have shown that the transcriptional activation of α-specific genes in an α cell requires the α1 gene product together with a protein known as Mcm1; in this same α cell, the DNA binding of the α2 gene product together with the Mcm1 protein represses the transcription of *a*-specific genes. By contrast, the transcriptional activation of *a*-specific genes depends only on Mcm1, requiring no product encoded by the MAT locus (Fig. 18.7b). The Mcm1 protein thus functions as a transcription factor in both *a* and α cells but actually has opposite effects in the two haploid cell types. This single transcription factor becomes an activator when combined with α1 protein and a repressor when combined with α2 protein. Furthermore, in *a*/α cells, a combination of the α1 and α2 proteins forms a new transcriptional repressor whose binding to the promoters of haploid-specific genes represses their transcription with no requirement for the Mcm1 protein. Thus the three cell types,

functions because they lack the α*1* activator, and they constitutively express a-specific functions because they lack the α*2* repressor. In short, the *a* cell type arises through a default pathway requiring no action on the part of the MAT-encoded regulatory proteins.

What happens in a MAT*a*/MATα heterozygous diploid? According to the model (Fig. 18.7c), the a1 and α2 regulators act in concert to perform a series of negative functions. The a1-α2 coregulators, acting together, repress expression of α1; as a result, the α*-specific* genes (or α*sg* genes) are not activated. The a1-α2 coregulators also repress expression of *haploid-specific* genes (*hsg* genes) that normally function in both *a* cells and α cells but not in *a*/α diploids, such as *HO* (see later). The coregulators further repress expression of Rme1, a protein whose presence inhibits the meiotic pathway and whose absence, therefore, confers competence for meiosis. Finally, the α2 regulator alone represses *a-specific genes* (*asg* genes). The net result is that *a*/α cells are sterile but competent to undergo meiosis.

Verification of the α1-α2 hypothesis, which was proposed on the basis of mutant analysis, has come from various molecular studies, including the cloning and sequencing of the MAT*a* and MATα alleles (Fig. 18.8). These alleles have

(a)

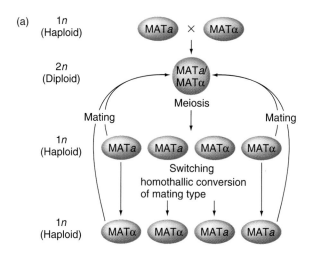

(b)

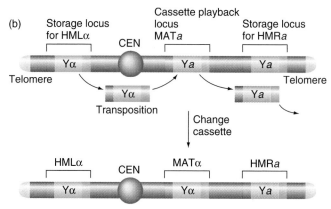

Figure 18.9   Mating type switching in homothallic strains. (a) Homothallic strains are able to switch mating types. After production of an *a/α* diploid through mating, the diploid can undergo meiosis during sporulation, producing two *a* and two α spores. Progeny of these spores can switch to the opposite mating types and mate back with a cell of opposite mating type. (b) Cassette model of mating type switching. Three copies of mating type information exist on chromosome III: HML, MAT, and HMR. Only the information at the MAT locus is expressed. HML and HMR contain the stored, silent information that can be copied into the MAT locus.

in which several cell-type-specific genes are expressed, arise from the activities of only three specific proteins that bind directly to the promoters of cell-type-specific genes to enhance or repress their transcription.

Mating Type Switching

Yeast strains differ from each other according to the stability of their mating type. Most laboratory strains have stable haploid mating types such that the progeny of successive mitotic divisions always have the same mating type as their parents. These strains with stable haploid mating types are known as **heterothallic** strains. By contrast, strains that can switch mating types are known as **homothallic** strains (Fig. 18.9a). In homothallic strains, switching occurs at a much higher frequency than one would expect if the switch were due to a point mutation. The homothallic phenotype arises from a dominant, functional *HO* allele at the *HO* locus on chromosome IV; by

comparison, heterothallism results from a recessive loss-of-function allele (*ho*) at this same locus.

The Roles of HML, HMR, and HO in Mating Type Switching

The key to understanding the mechanism by which homothallic (HO) strains switch mating type came from the discovery of two additional copies of the MAT locus called HML and HMR. Located near the telomeres on each arm of chromosome III, the HML and HMR loci are transcriptionally silent.

In wildtype strains, HML contains α information and HMR contains *a* information. The mechanism of mating type switching involves a transposition of mating type information from the transcriptionally silent HML or HMR locus into the transcriptionally active MAT locus. Switching occurs only in the *a* and α cell types. *a/*α cells do not switch even if they carry HO because the a1 and α2 regulators repress transcription of the *HO* gene.

A simple way to understand the mechanism of mating type switching in homothallic strains is to imagine an analogy with a cassette tape recorder. A cell can retrieve a cassette of mating type information located in an HML or HMR storage locus and insert it into the MAT playback locus. The cell retrieves information from the storage locus by a series of steps. The product of the *HO* gene, a site-specific endonuclease, initiates the process. The *HO*-encoded enzyme makes a double-stranded cut in chromosome III at a specific 18-bp recognition sequence just to the right of a segment known as the Y segment in MAT. Next, the double-stranded break is repaired through a recombination-like process in which homologous pairing occurs between sequences from one of the storage loci, such as HML, and identical sequences surrounding MAT (see Fig. 18.9b). In this example, HML DNA then serves as a template to resynthesize DNA at MAT. The result is the replacement of MAT DNA with DNA from the storage locus. Overall the process is conservative, with the duplication of storage-locus DNA balanced by an equal loss of DNA from MAT. The information in the storage locus remains unchanged.

The storage loci are silent in two respects. They are not transcribed, and even though they contain the 18-bp HO-protein-recognition sequence, they are not susceptible to double-stranded cutting by the HO enzyme. Both these properties result from the presence of "silencer" sequences at the boundaries of the storage loci, which are not present at the boundaries of MAT. In a process that may resemble the silencing of genes in higher eukaryotes, the silencer sequences in yeast contribute in a complex way to the configuration of the chromosome in the vicinity of each storage locus. RNA polymerases, transcription factors, and the HO endonuclease cannot gain access to these regions. As a result, the HO endonuclease cannot make double-stranded cuts at the storage loci. The silencing effect is general: Genes from other chromosomes that are normally active become transcriptionally silent when inserted into a storage locus.

Are there gene products that function to keep these storage loci silent? Mutational analyses show that the answer is yes. Loss-of-function mutations isolated in four *SIR* (*silent information regulator*) genes cancel repression and result in the transcriptional activation of the *α1*, *α2*, or *a1* genes located at HML or HMR. In haploid cells, transcription of the storage loci results in a phenotype resembling that of an *a/α* diploid because both MATa and MATα alleles are expressed. Moreover, in *sir⁻* strains, the HO enzyme cleaves the chromosome at both MAT and the storage loci. Depending on which site is cut in a given cell, HO-mediated transposition can occur from MAT to one of the storage loci or in the other direction. This observation demonstrates that in normal *SIR⁺* strains, the silencing of the storage loci ensures that MAT serves as recipient and the storage loci as donors in the switching process. As a consequence, in wildtype cells, mating type information flows in only one direction—from the storage loci to MAT.

The Role of HO in Mating Type Determination and Stability Serves as a Paradigm for Development

Pedigrees of cells undergoing HO-mediated mating type switching have revealed a few rules governing mitotic cell lineages during development (Fig. 18.10). In a typical pedigree analysis, investigators tease apart the mother and daughter cells in each cell division by delicate micromanipulation under a microscope so that they can examine each one individually. The first generation arises when a spore germinates to produce the spore (S) and the first daughter (D1) (Fig. 18.10a). In the second generation, the spore gives birth to the second daughter (D2), and the first daughter gives birth to its first daughter (D1-1).

Two general rules for HO-mediated switching have emerged from observations of many generations. First, cells always switch in pairs; if a mother cell switches, its immediate daughter also switches (Fig. 18.10b). Second, cells must gain competence to switch before actually switching; and they gain competence through the experience of undergoing at least one cell division. Thus, an "inexperienced" spore never switches at the first cell division. In addition, a daughter cell will not switch when it produces its first daughter cell. Once a cell has gained experience through cell division, however, it becomes competent to switch in subsequent generations (Fig. 18.10c).

What aspect of cell division triggers competence to switch? The answer, although not yet entirely clear, relates to the transcriptional activation of the *HO* gene. Occurring only during S phase, the transcription of *HO* is controlled in part by a transcription factor that regulates oscillations in the expression of other cell-cycle genes. The transcription factor regulates expression such that HO transcripts are made only in experienced mother cells and never in inexperienced daughters. Thus the inexperienced cells undergo a change of state in the process of cell division. There is an analogy here with the stem cells that give rise to differentiated cells in multicellular eukaryotes. The inexperienced daughter cells are like stem cells; they do not themselves express HO, but once they undergo cell division, they differentiate by acquiring the capacity to produce the HO protein.

In many cell lineages that arise during the development of an organism, programmed cell death, or apoptosis, destroys

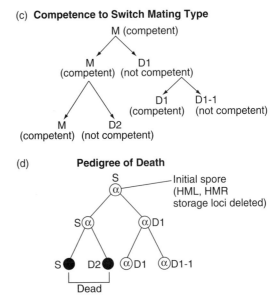

Figure 18.10 Pedigrees of mating type switching. (a) Pedigree of mating types in divisions of a homothallic (HO) spore cell. Yeast reproduce by budding. The inexperienced α spore cell (one that has not divided yet), produces a daughter cell (D1) in addition to the mother S and both are still α. With the next division of the spore cell, producing S and D2, mating type can change. (b) Symmetrical switching pattern. When a mother cell divides to produce a mother and daughter cell, mating type switching will occur in both resulting cells. When a daughter cell is produced, it always has the same mating type as the mother cell. (c) Competence to switch. The ability of cells to switch is dependent on the cell having undergone at least one cell division. Therefore, a newly arising daughter cell is not competent to switch, whereas the mother cell from which it arose is competent to switch. The competency to switch is therefore asymmetric. (d) Pedigree of death. In a strain containing a deletion of the storage cassettes (HML and HMR), a cell that tries to switch mating type will end up with a broken chromosome and therefore dies. Again the symmetry of switching pattern but asymmetry of competence to switch is seen in this pedigree of death.

some cells before the organism achieves adulthood. Studies of the nematode *Caenorhabditis elegans* have provided the most elegant demonstration of the role of cell death in development. But HO pedigrees in yeast provide clues to the genetic programming of cell death. The pedigree in Fig. 18.10d illustrates how this happens. This pedigree is identical to that shown in Fig. 18.10a, except that the initial spore carries complete deletions of the HML and HMR storage loci. Since there are no donor sequences available for mating type switching, the HO-mediated, double-stranded break at MAT cannot be repaired and becomes a lethal event. The resulting "pedigree of death" follows the rules described earlier; for example, experienced mothers who could switch now have a double-strand break and die. Some multicellular organisms use genetic mechanisms like the introduction of lethal breaks in DNA to eliminate cells that outlive their usefulness during development.

MATING: CELL-TO-CELL COMMUNICATION THROUGH PHEROMONES TRIGGERS THE CONVERSION OF HAPLOID a AND α CELLS TO a/α DIPLOIDS

Mating is the result of a genetic program that allows *a* cells and α cells to become *a*/α cells. During mating *a* and α cells communicate with one another through **pheromones:** hormones that play a role in eliciting sexual behavior. *a* cells secrete *a* pheromone and α cells secrete α pheromone. Each haploid cell type carries receptors for the pheromone of the opposite cell type; *a* cells carry receptors for α pheromone, and α cells carry receptors for *a* pheromone (Fig. 18.11). When bound to pheromone, the receptors activate a signal transduction system whose cascade of protein interactions ultimately promotes transcription of a specific set of genes. The yeast pheromones

provide an example of how small molecules produced in one cell type and secreted into the extracellular environment can influence the behavior of a neighboring cell. Through the isolation of mutants, it is possible to dissect genetically the processes by which cells produce and respond to pheromones.

Researchers discovered many of the genes responsible for the pheromone response system by the isolation of sterile mutants. Some mutants are sterile because they cannot signal to a mating partner; others are sterile because they cannot respond to a signal. A simple test enables one to distinguish between the two. Both *a* and α mating pheromones elicit an arrest of cell division in the receptive cell type. If one places a patch of cells of one mating type on a patch of cells of the opposite mating type, a zone of growth inhibition appears where the pheromone from one cell type has caused the arrest of division of the other cell type. Sterile mutants that produce no zone of growth inhibition either fail to secrete mating pheromone or they secrete mating pheromone but are unresponsive to pheromone from the opposite cell type (Fig. 18.12).

Pheromone Biosynthesis and Secretion

The biosynthesis of mating pheromone is one of the pathways regulated by the MAT-encoded transcription factors. Yeast mutants unable to make or secrete pheromones helped researchers define the steps of pheromone biosynthesis. We first describe the production of α pheromone, and then contrast its production with the biosynthesis of *a* pheromone.

α pheromone is one of a number of proteins that yeast cells deliver to the extracellular medium through the *secretion pathway*—a series of membranous organelles including the endoplasmic reticulum (ER), the Golgi apparatus, and cytoplasmic vesicles that fuse to the plasma membrane at the site of bud growth during the cell division cycle. Two unlinked α-specific genes, *MFα1* and *MFα2,* code for α pheromone. The two genes have a complex organization. Unlike most yeast

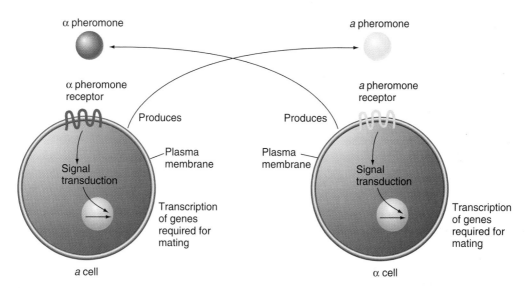

Figure 18.11 Pheromones bind to receptors and activate a signal transduction pathway. Chemical signals (pheromones) released by *a* and α cells bind to receptors on the cell surface of cells of the opposite mating type. Once bound to the receptor, a signal is sent to the nucleus via a signal transduction pathway and transcription of genes required for mating is turned on.

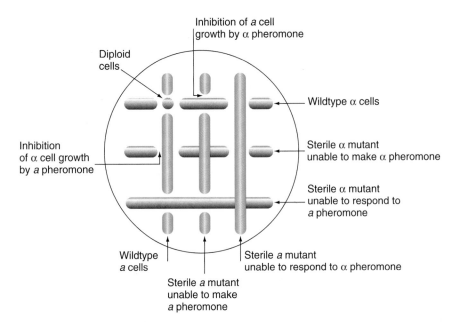

Figure 18.12 Mutants defective in producing or responding to mating signals. Strains can be cross-streaked on plates to test for defects in sending or receiving pheromone messages. Pheromones from a strain of one mating type will inhibit the growth of a strain of the opposite mating type. If a cell cannot produce or respond to one of the pheromones, there will be no inhibition where the streaks cross.

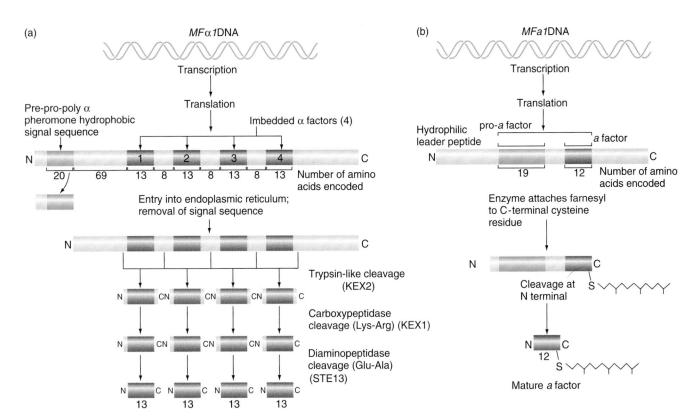

Figure 18.13 Pheromone biosynthesis. (a) α factor biosynthesis. Several copies of the coding region for α factor are present at the Mfα1 locus. After transcription and translation, the coding sequences are cleaved from the pre-pro-poly-α pheromone by three different enzymes. No active α factor is produced by cells containing mutations in genes encoding any of these enzymes. (b) a factor biosynthesis. The Mfa1 locus contains only one copy of the a pheromone coding sequence. After translation, the primary polypeptide is cleaved and processed by the addition of a farnesyl group.

genes, they contain more than one copy of the protein-coding sequence. *MFα1* has four adjacent copies of the sequence encoding α pheromone, and *MFα2* has two copies. The complex polypeptide synthesized from the adjacent copies of coding sequence is known as pre-pro-poly-α pheromone (Fig. 18.13a). Posttranslational processing transforms this polypeptide into four functional copies of mature α pheromone from the *MFα1* gene and two copies of mature α pheromone from the *MFα2* gene.

Mutant cells defective in the secretion of α pheromone defined three genes encoding proteases required for the processing of the pre-pro-α pheromone into a product 13 amino acids long. The extensive processing that occurs to produce α pheromone is similar to that which occurs during the production of many mammalian peptide hormones.

The biosynthesis of *a* pheromone differs slightly from that of α pheromone (Fig. 18.13b). Genes known as *MFA1* and *MFA2* each encode one copy of the *a* pheromone preceded by a leader peptide that is 19 or 21 amino acids long. Posttranslational processing by proteases modifies the longer pheromone-plus-leader polypeptide to the smaller, mature pheromone that is secreted. During this processing, an enzyme attaches palmitic acid to the C-terminal cysteine residue, which most likely directs the *a* pheromone to the plasma membrane; there it is secreted via a transport protein encoded by the *STE6* gene. Thus, *a* cells secrete *a* pheromone to the extracellular medium by a mechanism that does not involve the standard secretory system. There are two versions of mature *a* pheromone, one produced from the *MFA1* gene, the other from the *MFA2* gene; both mature *a* pheromones are 12 amino acids in length.

Cells Respond to the Binding of Pheromone through a Signal Transduction System

By analyzing mutants that are sterile but still synthesize pheromone, yeast researchers identified many components of the signal transduction systems through which cells respond to the binding of pheromone at their surface. Plasma membrane receptors at the cell surface mediate the pheromone response. *a* cells produce α-pheromone receptors, which are products of the a-specific *STE2* gene. Similarly, α cells produce *a*-pheromone receptors, which are products of the α-specific *STE3* gene. The receptors contain seven noncontiguous, hydrophobic domains long enough to span the lipid bilayer of the plasma membrane; as a result, each protein snakes back and forth through the plasma membrane, exposing some parts of the entire polypeptide to the cell's exterior and others to the cell's interior (Fig. 18.14). Although the receptors are cell-type specific, the remainder of the pheromone response pathway, depicted in Fig. 18.14, is controlled by the *haploid-specific* genes, or *hsg*, which are expressed in both *a* and α cells (review Fig. 18.7).

a cells and α cells communicate with each other for the purpose of mating through extracellular segments of the receptors that interact directly with the appropriate mating pheromone. Receptor segments exposed to the inside of the

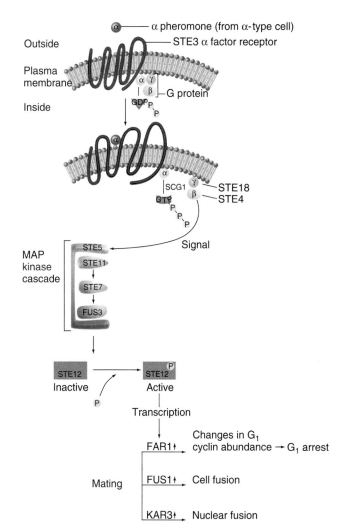

Figure 18.14   Signal transduction cascade in the mating response. Binding of pheromone to a receptor causes the conversion of GDP to GTP, which then activates the MAP kinase cascade, leading to activation of the STE12 transcriptional activator. Activated STE12 positively regulates genes involved in G₁ arrest of cells, cell fusion, and nuclear fusion. Mutations in genes in any of these pathways cause defects in mating.

cell are coupled to a membrane-bound G protein composed of α, β, and γ subunits, encoded respectively by the *SCG1*, *STE4*, and *STE18* genes. The mating pheromones, receptors, and G proteins act in concert as a molecular switch, with the binding of pheromone causing the conversion of Gα-GDP to Gα-GTP and this conversion, in turn, causing the dissociation of the Gβ/Gγ subunits from the heterotrimeric G-protein complex. The dissociated but still membrane-bound Gβ/Gγ dimeric complex initiates a signal that propagates through the series of proteins in the signal transduction pathway. The pathway ultimately leads to the transcriptional activation of genes whose products are required for G₁ cell cycle arrest, cell fusion, and nuclear fusion.

As Fig. 18.14 shows, Gβ/Gγ transmits the signal initiated by pheromone binding through a series of three protein kinases, known collectively as a MAP kinase cascade. The trio of kinases first phosphorylate each other and then promote phosphorylation of the *STE12* gene product, a transcription factor that is active only when phosphorylated. Since transcription occurs in the nucleus, the phosphorylation of Ste12 shows that the signal transduction pathway has transduced the pheromone-induced signal from the plasma membrane to the nucleus. Phosphorylated Ste12 activates the transcription of several genes whose products contribute to G_1 cell cycle arrest (for example, *FAR1*), cell fusion (for example, *FUS1*) and nuclear fusion (for example, *KAR3*). All these processes are necessary for mating. Components of the type of signal transduction pathway dissected in the yeast mating response by genetic and molecular analyses are remarkably conserved in all eukaryotes.

CONNECTIONS

Because it is relatively easy to find and analyze mutant yeast cells, as well as to replace yeast genes with engineered variants, yeast has become the model organism for studying eukaryotic cell biology. Three different genetic programs, each governing a particular pattern of gene expression, trigger the differentiation pathways that generate *a, α,* and *a/α* cell types in yeast. Mating between *a* and *α* cells is directed by another genetic program. Research has shown that the mechanisms for cell communication, the control of gene expression, and the differentiation of cell types are based on the same principles in yeast and multicellular, higher eukaryotes.

What we learn from yeast will very likely tell us a great deal about ourselves. Mounting evidence suggests that similar proteins in yeast and humans carry out most cellular processes. It also indicates that human genes closely related in sequence and function to the regulatory genes of yeast play similar regulatory roles in human cells. Geneticists have already sequenced the entire yeast genome, and by the year 2000 they will probably have produced a comprehensive catalog of the roughly 6000 gene functions required to make a eukaryotic cell.

In Chapter 19 we move from the differentiation of single-celled eukaryotes to the development of multicellular eukaryotic organisms as we present a genetic portrait of the simple, multicellular plant known as *Arabidopsis*.

ESSENTIAL CONCEPTS

1. Yeast geneticists use transformation studies, the ability to move genes between yeast and bacteria, and gene replacement to answer questions about eukaryotic cell functions that are difficult to answer with other organisms.

2. The nuclear genome of yeast contains 16 linear chromosomes carrying about 6000 genes; very few of these genes have introns. The yeast genome also contains mobile Ty elements, whose capacity for recombination may be a source of reciprocal translocations, inversions, and deletions.

3. During the yeast life cycle, cells alternate between haploid and diploid phases. Haploid cells occur in *a* and *α* mating types, which communicate with each other via pheromones; the fusion of cells of opposite mating types creates diploid zygotes.

4. Genes at the MAT locus act at the level of transcription to control the expression of other genes that determine cell type. MATα consists of two genes: *α1* and *α2*. *α1* encodes an activator of α-specific genes encoding proteins that elicit the α-type cell. *α2* encodes a repressor of *a*-specific functions. A simple network of the α1 activator and the α2 repressor accounts for both α and *a* cell types. By contrast, the a1 protein expressed in *a*-type cells performs no function necessary for mating or differentiation into *a* cells. The interaction between a1 and α2 coregulators makes MAT*a*/MATα heterozygous diploids sterile.

5. Heterothallic yeast strains are unable to switch mating type, while homothallic yeast strains can switch from *a* to α and back again. Mating type switching in homothallic strains is controlled by the activity of the MAT locus, the *HO* gene, the alleles at HML and HMR, and the four *SIR* genes. Two general rules govern HO-mediated switching: (1) Cells always switch in pairs and (2) cells must gain competence to switch by undergoing at least one cell division.

6. *a* and α cells communicate through pheromones determined by *a*-specific and α-specific genes. *a* cells secrete a pheromone, while α cells secrete α pheromone. Each cell type carries receptors for the opposite type pheromone. The binding of pheromone from the opposite cell type activates a signal transduction system that results in the transcription of genes whose products help arrest the cell cycle and contribute to nuclear and cellular fusion.

S O L V E D   P R O B L E M S

I. You want to study metabolism of purines in *S. cerevisiae* and begin your analysis by isolating auxotrophic mutants that cannot synthesize adenine. You mutagenized haploid *a* HO cells and screened for cells that require adenine for growth.
   a. Will you know if your mutations are dominant or recessive? If not, how could you find out?
   b. If you wanted to know if any of the mutations had an effect on sporulation, how could you test this? What strain would you need to construct?

Answer

This problem requires an understanding of the yeast life cycle, ploidy and the expression of dominant and recessive alleles, and segregation of alleles during meiosis.

   a. Because the newly isolated mutations are present in a haploid strains you will not know immediately if they are dominant or recessive. *To determine if a mutation is dominant or recessive, mate the mutant haploid strain with a wildtype haploid α strain.* If the diploid cannot grow without adenine in the medium, the mutation is dominant. If the diploid cells have the wildtype phenotype, the adenine mutation is recessive.
   b. *If the mutation is dominant, you can determine its effects on sporulation using the diploid strain constructed for part a. If the mutation is recessive, you would have to construct a homozygous recessive diploid.* You could do this by switching the mating type of the original strain to obtain an α haploid cell and mating this with the original *a* haploid cell.

II. You have cloned a gene encoding the histone subunit H2A.
   a. Now you want to do a gene replacement with a nonfunctional copy of the gene and observe the phenotype of the knockout mutation. Which of the steps below will you do and in what order?
      i.   transform haploid *a* cells
      ii. transform haploid α cells
      iii. select Ura$^+$ cells
      iv. transform diploid *a/a* cells
      v.   transform diploid *a/α* cells
      vi. sporulate the transformed cells
      vii. mate the transformed cells with cells of the opposite mating type
      viii. add the *URA3* gene to the plasmid vector
      ix. clone the *URA3* gene into the *H2A* gene on your plasmid
   b. You have obtained haploid cells that have the *H2A* gene knockout but there is no phenotype. The cells live and appear to grow normally. Offer two possible explanations for this result.
   c. How could you determine if there is more than one copy of the *H2A* gene in the yeast genome?

   d. You located a second copy of the *H2A* gene, determined the sequence, and found that the two copies had only a few base differences. Transcription of the first copy of H2A that you isolated is transcriptionally regulated during the cell cycle. How could you determine if the other copy is regulated in the same way?
   e. How could you determine if there are sites in the DNA 5′ to the *H2A* coding sequence that are responsible for the cell cycle transcriptional regulation?
   f. *HIR2* and *HIR3* are two genes that were identified as regulators of transcription of the histone genes based on the properties of mutations in these genes. Null mutations in either gene lead to constitutive expression of *H2A-H2B*. Do *HIR2* and *HIR3* encode positive or negative regulators?
   g. Histone gene transcription is also sensitive to the number of copies of the histone genes in the cell. Introduction of a high copy number plasmid containing H2A and H2B into the cell will repress transcription of H2A and H2B. How could you determine if the copy number control of transcription is dependent on *HIR2* and *HIR3*?

Answer

This problem involves molecular analysis of gene structure and expression and the use of mutants to determine the role of a gene.

   a. Because histones have such an essential role in packaging of DNA in the cell, you might predict that a knockout of function would be a lethal event. Therefore, the gene replacement should be done in a diploid cell, which can then be sporulated to determine phenotype of a knockout in a haploid cell. To do a gene replacement with a nonfunctional gene, the gene should be disrupted by a selectable marker like *URA3*. *The steps you would do, in the correct order are as follows:*
   clone *URA3* into the *H2A* gene on your plasmid (ix)
   transform diploid *a/α* cells (v)
   select Ura$^+$ cells (ii)
   sporulate the transformed cells (vi)
   b. *The H2A protein may not be an essential protein for cell viability* or *there may be another copy of the* H2A *gene in yeast.*
   c. To determine if there is another copy of the *H2A* gene, *the cloned* H2A *gene could be used as a probe in a Southern hybridization using total yeast genomic DNA.*
   d. The few base differences between the copies of the *H2A* gene could be used to distinguish transcription of the two genes. *Oligonucleotides that contain the specific bases differences between the two genes could*

be used as probes in two separate hybridizations to mRNAs that had been separated by electrophoresis and transferred to filter paper (Northern hybridization).

e. *If the DNA upstream (5′ to) the H2A gene was cloned next to a reporter gene such as* lacZ*, this construct could be used to analyze the role of the DNA in cell-cycle regulation. Different portions of the upstream region could be deleted, the resulting clones transformed into yeast, and expression (as measured by production of β-galactosidase) analyzed at different times in the cell cycle.*

f. Because the lack of functioning protein in the null mutations led to constitutive expression, HIR2 *and* HIR3 *must encode negative regulators.*

g. To determine if *HIR2* and *HIR3* are also involved in copy number regulated transcriptional control, the *high copy number histone gene plasmids should be transformed into* hir2 *or* hir3 *mutants.* If the level of expression is repressed to the same extent as in *HIR*$^+$ strains, then the *HIR2* and *HIR3* gene products are not required for copy number sensitive transcription.

P R O B L E M S

18-1 Choose the matching phrase in the right column for each of the terms in the left column.

a. pseudogene	1. silent copy of mating type information
b. shuttle vector	2. expressed copy of mating type information
c. homothallic	3. remnant of a gene
d. pheromone	4. plasmid that replicates in both *E. coli* and yeast
e. MAT	5. able to switch mating type
f. HML	6. unable to switch mating type
g. heterothallic	7. chemical signal

18-2 The yeast genome is approximately 12,000 kb in length and contains about 6000 open reading frames encoding an average of 500 amino acids. Using these numbers, calculate the proportion of the yeast genome that is protein coding DNA.

18-3 Chromosome XII in yeast is about 2060–3060 kb in length depending on the number of the rDNA repeats (100–200 copies of a 9-kb repeating unit). If you grow up 10 individual yeast cells as clones in liquid culture, how could you analyze each clone to see if there is variation in the size of the rDNA repeat in the clones?

18-4 Expression of genes involved in the same process (e.g., amino-acid biosynthesis) is often coregulated in yeast. For the following questions, assume that you have cloned the genes involved in histidine metabolism.
   a. How could you show if the *HIS* genes are transcriptionally coregulated?
   b. If there is transcriptional coregulation, what might you expect to see in the sequence in the regulatory region of these genes and why?

18-5 Several copies of subtelomeric repeats may be present at the ends of chromosomes, resulting in more than 20 kb of repeat DNA.
   a. How could you show which chromosomes contain repeated DNA sequences near the ends of each chromosome?
   b. If you had a genomic library of plasmid clones with an average insert size of 15 kb and were trying to make a physical map, how would you know which chromosome your cloned DNA came from?

18-6 In screening a population of diploid yeast for a defect in chromosome segregation, you found that the chromosome segregation defect segregated 2$^+$:2$^-$ in the spores of the ascus.
   a. Why is this evidence that the mutant phenotype is due to an alteration in one gene?
   b. Is the mutation dominant or recessive?

18-7 If a tetraploid *a/a/α/α* yeast strain was sporulated, what mating type phenotypes would you predict in the asci?

18-8 Which of the following statements apply to
   a. YIp vectors
   b. YEp vectors
   c. YCp vectors
      i. contains an origin of replication
      ii. high copy number
      iii. one copy per cell
      iv. contains a selectable marker
      v. integrates into chromosome via homologous recombination
      vi. exists autonomously

18-9 a. Describe the model for the roles of the Mcm1 protein in determining expression in *a*, α, and *a/α* diploids.
   b. What would be the mating phenotype of an α *mcm1*$^-$? a *mcm1*$^-$? *a/α mcm1*$^-$? Would they be sterile or fertile?

18-10 Seven haploid yeast strains auxotrophic for the compound sneezan were isolated. These strains were mated with each other, and the ability of the resultant *a/α* diploids to grow on the intermediate compounds dopan, sleepan, and happan are scored on the following page.

	1	2	3	4	5	6	7
1	+	+	+	+	+	+	+
2		+	+	+	+	+	+
3			+	+	+	+	+
4				+	+	+	+
5					+	+	+
6						+	+
7							+

Growth on dopan

	1	2	3	4	5	6	7
1	+	+	+	+	+	+	+
2		−	+	+	+	−	+
3			+	+	+	+	+
4				+	+	+	+
5					+	+	+
6						−	+
7							+

Growth on sleepan

	1	2	3	4	5	6	7
1	−	+	+	+	+	+	+
2		−	+	+	+	−	+
3			−	+	+	+	−
4				−	−	+	+
5					−	+	+
6						−	+
7							−

Growth on happan

	1	2	3	4	5	6	7
1	−	+	+	+	+	+	+
2		−	+	+	+	−	+
3			+	+	+	+	+
4				+	+	+	+
5					+	+	+
6						−	+
7							+

a. Are these results derived from a complementation or from a recombination experiment?

b. The a/α diploids from the experiment in part a are now induced to sporulate. Out of 400 spores examined from each a/α diploid, the numbers that were able to grow on minimal medium is scored below.

	1	2	3	4	5	6	7
1	0	100	5	100	100	100	5
2		0	100	3	3	0	100
3			0	100	100	100	0
4				0	0	3	100
5					0	3	100
6						0	100
7							0

How can you explain the two classes of results here (100 spores or ≤ 5)?

c. How many complementation groups involved in the biosynthesis of sneezan are represented in this collection of 7 mutations? Which genes are in which complementation groups?

d. The biochemical pathway involved in the biosynthesis of sneezan is linear. One of the steps of the pathway is catalyzed by a heterodimeric enzyme, with each subunit encoded by a different gene. What is the biochemical pathway? Identify the genes that encode the enzyme involved in each step. Indicate which enzyme is dimeric.

18-11 Six genes, each conferring the ability to utilize the sugar sucrose have been mapped to different loci in *Saccharomyces* strains.
   a. You have cloned the *SUC2* gene from a *S. cerevisiae* strain. How can you tell if the *SUC* loci in other strains contain DNA homologous to that at the *SUC2* locus?
   b. Other Suc$^+$ strains have DNA homologous to the *SUC2* gene but, in some cases, there is not enough information for full gene function. What would you call these partial genes?
   c. Starting with a *SUC2$^+$* strain, you have identified cells incapable of fermenting sucrose. Some of the mutants were defective in the gene at the *SUC2$^-$* locus (*suc2$^-$*). Other Suc$^-$ strains were pleiotropic. These genes, called *snf* (*sucrose nonfermenting*) genes, affect expression of galactose and maltose as well as sucrose. The common feature of the affected genes is that all are glucose repressible (not expressed when glucose is present). Expression of the *SUC2* gene was examined in the wildtype and mutant under repressing and nonrepressed conditions. How would you describe the transcription pattern?

18-12 To study DNA repair mechanisms, researchers isolated yeast mutants that were sensitive to various types of radiation. For example, they isolated mutants that were more sensitive to damage by UV light (and therefore died at a lower dose of UV radiation). Ten haploid mutants were isolated.

a. How would you do the complementation tests for these mutants in yeast?
b. Based on the complementation chart below, how many genetic loci are represented by these mutants?

	1	2	3	4	5	6
1	−	−	+	−	+	+
2		−	+	−	+	+
3			−	+	−	+
4				−	+	+
5					−	+
6						−

c. Double mutants were constructed containing some of the UV sensitive mutations. If the UV sensitivity of the double mutant is greater than that of the individual mutants, the mutations are in different repair pathways. How many different repair pathways do these mutations represent? What mutations are in each pathway?

Single mutant	Sensitivity to UV
1	+
3	+
6	+

Double mutant	
1,3	++
1,6	+
3,6	++

d. Assume you have cloned the genes and now want to ask whether genes identified in these mutant searches are expressed constitutively in cells or are expressed only in response to damage. How would you design your experiment?
e. To identify additional genes that might be responding to DNA damage, you have decided to construct a library of yeast DNA fused to the *lacZ* gene. How would you use this library to search for more genes that are induced by DNA damaging agents such as radiation?

CHAPTER 19

ARABIDOPSIS THALIANA:
GENETIC PORTRAIT OF A MODEL PLANT

The flower of *Arabidopsis thaliana.*

A. *thaliana,* a tiny weed whose common name is "mouse ear cress," grows low to the ground and produces clusters of small white flowers in meadows and laboratories around the globe (Fig. 19.1). Like more than 150,000 other species of flowering plants, including roses, daisies, tomatoes, peas, beans, and maple trees, *Arabidopsis* is a dicotyledonous angiosperm— **dicotyledonous** because the mature embryo carries two leaves, and **angiosperm** because its seeds are enclosed in an ovary within a flower. *Arabidopsis* shares three basic characteristics with other plants: It is an *autotroph,* that is, a nutritionally self-sufficient organism that produces its own food via photosynthesis; it is *nonmotile,* that is, rooted to one spot; and it produces new organ systems continuously throughout its life cycle (which consists of alternations between a diploid [$2n$] sporophyte generation, and a haploid [n] gametophyte generation).

The impetus for studying a small flowering weed with no commercial potential is straightforward: Plants are the chief source of food for life on our planet, and an understanding of how they grow and reproduce may make it possible to grow more nutritious, more copious crops in a wide range of environments. Classified in the same family as mustard and cabbage plants, *A. thaliana* has several features that make it an excellent experimental model. Its general strategies of growth, development, flowering, and seed production are the same as those of other higher plants, including many crop plants. However, it has a shorter generation time than most other plants, requiring only six weeks for seeds to germinate and develop into mature plants that produce more seeds. Moreover, like the peas that Mendel studied, it reproduces mainly by self-fertilization, yet cross-pollination—through the removal of stamen from one plant's flowers and the application of pollen from another plant's flowers onto the stigma of the stamen-deprived flowers—is possible. Whether self- or cross-pollinated, wildtype *Arabidopsis* plants produce a very large number of seeds—from 10,000 to 40,000 per plant—and these seeds have a high rate of germination. Such rapid, abundant reproduction makes it possible for geneticists to screen large populations of

Figure 19.1 *A. thaliana* **in the palm of the hand.**

(a)   (b)

Figure 19.2 The Pink Perfection camellia as a horticultural mutant. (a) Wildtype camellia. (b) Pink Perfection flower.

seedlings for specific phenotypes. Finally, *A. thaliana* grows well in the laboratory, requiring relatively little light (illumination from cool, white fluorescent bulbs is sufficient), and temperatures in the range of 22–26°C.

The goal of geneticists studying *A. thaliana* is to understand the physiology, biochemistry, growth, and development of a plant at the molecular level. With this knowledge, they may someday be able to produce plants containing more of particular parts prized for their nutritional content or floral display. Ornamental plant breeders have already accomplished this goal with some species. For example, the flowers of wild camellias contain stamens (male reproductive organs) and carpels (female reproductive organs) that enable them to produce seeds; but the flowers of a cultivated variety known as Pink Perfection lack both male and female reproductive organs and carry many extra petals in their place (Fig. 19.2).

In our examination of *A. thaliana* as a model plant, we present

■ The structure and organization of the genome.

■ The plant's anatomy and life cycle.

■ Techniques of mutational analysis, including chemical and radiation procedures, insertional mutagenesis, and the analysis of mutations to identify gene function.

■ Genetic analysis applied to various aspects of development, including embryogenesis, hormonal control systems, and responses to environmental signals.

■ The genetic analysis of flowering: a comprehensive example.

GENOME STRUCTURE AND ORGANIZATION

The genome of a model organism for genetic studies should be relatively small and representative of the genomes of the other organisms it represents. *A. thaliana* fills this bill to perfection. Its genome, composed of 100 Mb, is one of the smallest genomes known in the plant kingdom, notably smaller than the genomes of most other angiosperms. By contrast, the genome of the angiosperm maize (*Zea mays*) is 45 times larger. The *Arabidopsis* genome is about as large as the genome of *Caenorhabditis elegans* (see Chapter 20) and is 60% the size of the *Drosophila melanogaster* genome (see Chapter 21). The nuclear DNA of *Arabidopsis* is carried by five pairs of small chromosomes with well-defined banding patterns.

Little Repetitive DNA and a Tight Arrangement of Genes

The small size of the *A. thaliana* genome stems from the fact that it contains much less repetitive DNA than the genomes of other angiosperms. Molecular geneticists estimate that only 20% of the *Arabidopsis* genome consists of noncoding repeti-

tive DNA. In other angiosperms, repetitive DNA makes up the vast majority of the genome.

Single-copy DNA makes up the remaining 80% of the *Arabidopsis* genome. An estimated 15,000 to 20,000 genes are present in the five chromosomes, which means there is one gene for every 5 kb of genome. For a plant, this is a very tight arrangement of genes.

Comparing Genetic and Physical Maps

Geneticists study an array of *A. thaliana* **ecotypes:** plant varieties analogous to animal strains. The members of each ecotype share a common origin, shape, and other phenotypes. The uniformity of shape and characteristics among related individuals is a direct consequence of the plant's propensity to reproduce by self-pollination. The most commonly used *Arabidopsis* ecotypes—Columbia (Col) and Landsberg *erecta* (Ler)—are derived from the same heterogeneous population of seedlings, originally named Landsberg. Columbia arose by self-pollination from a plant isolated from the Landsberg population because of its ability for prolific reproduction. Landsberg *erecta*, by contrast, arose from a plant of the same population that, after gamma irradiation, was isolated for its small size and erect stature. A number of other ecotypes with well-defined phenotypes are natural isolates that come from and are adapted to specific ecosystems or regions around the world.

Arabidopsis ecotypes differ from each other not only in phenotypes but also in a significant number of DNA polymorphisms that are identifiable by RFLP or PCR analysis (as described in Chapter 9). Different ecotypes have different alleles at genetic markers defined at the molecular level. Researchers can use allelic variation at markers to follow the segregation of chromosomal regions carrying the markers in the progeny of self-pollinated plants derived from a cross between different ecotypes. They can also analyze the linkage between DNA markers as well as between DNA markers and morphological loci (defined by two or more alleles conferring alternative phenotypes on the different ecotypes). With the resulting data, they can create DNA marker and phenotypic maps, and ultimately integrate the two types of maps for all five chromosomes. At the writing of this chapter, a dense map of RFLP and PCR-based molecular markers is available for all five *A. thaliana* chromosomes. There are also some integrated maps correlating data from marker and phenotypic studies (Fig. 19.3).

Geneticists can use probes for molecular markers (spaced at an average distance of 1 cM or 185 kb) to identify large pieces of chromosomal DNA cloned into stable yeast or bacterial vectors. Two overlapping DNA clones representing neighboring regions of a chromosome constitute a *contig* of cloned chromosomal DNA sequences (as described in Chapter 10). By arranging the contigs on all five chromosomes with reference to their corresponding genetic maps, one produces a physical map of the organism's genome.

For example, the physical map of chromosome 4, one of the earliest to be completed, reveals that the chromosome is about 19 Mb in length and covers a genetic distance of about 83 cM (Fig. 19.3). A detailed comparison of the physical and genetic maps of chromosome 4 shows that the frequency of recombination varies greatly along the chromosome, with each centimorgan representing from 30 kb to 1600 kb (or 1.6 Mb). The small amount of repetitive DNA on chromosome 4 is found in the telomeric sequences, the ribosomal DNA sequences at the end of the short arm, and the 180-bp satellite sequences forming the heterochromatin in the vicinity of the chromosome's centromere. In addition, six families of repeated DNA sequences and the 5S-rDNA sequences exist as dispersed repetitive DNA in the vicinity of the centromere. The rest of the chromosome contains no sequences that hybridize to various repetitive DNA probes. This observation confirms earlier conclusions that interspersed repetitive DNA is found only infrequently in the *A. thaliana* genome and that the average uninterrupted, single-copy DNA sequence is about 125 kb long.

Researchers began constructing physical maps of *A. thaliana* chromosomes in the early 1990s. Their goal is to complete these physical maps and sequence the entire genome by the year 2000.

Summary

A. thaliana has a small genome carried by five chromosomes. Each chromosome contains a limited amount of repetitive DNA outside the centromeric and telomeric regions. Its genes, though similar in structure to the genes of other plants, are more tightly packed and contain only small introns. As a result, a much higher percentage of the genome is associated with coding and genetic function than is true of most plants. This higher signal-to-noise ratio is one of the features that make *A. thaliana* an excellent model for genetic analysis.

ANATOMY AND LIFE CYCLE

To understand a plant's growth and development at the molecular level, geneticists use mutational analyses to identify genes that influence the plant's anatomy and life cycle. A brief overview of the anatomy and life cycle of *A. thaliana* prepares the way for examining the techniques of genetic analysis in this model plant.

Anatomy: The Basic Body Plan

Like most higher plants, the adult *A. thaliana* is composed of only a few relatively simple organs and tissues (Fig. 19.4). The aboveground shoot connected to the below-ground root system represents the plant's main axis. Both root and shoot contain several cell types organized into the outer epidermal layer; the middle ground, or cortical, layer; and the inner vascular cylinder (which contains the water-conducting xylem vessels and the food-conducting phloem elements). Attached to the shoot are mature leaves also composed of three layers: an epidermal layer, containing both hair cells, or trichomes, and small pores, or stomata (singular, stoma), that regulate gas exchange; a mesophyll layer responsible for photosynthesis; and vascular tissues that conduct water and nutrients between the leaf and the stem.

The Apical Meristem Promotes Continuous Growth

Cells at the growing points of both shoot and root are organized into a tissue called the **apical meristem,** a group of undifferentiated cells that divide continuously. The root apical meristem, which consists of a central zone of meristem cells and a peripheral zone of differentiated cells, produces the cylindrical root through cell division, cell expansion, and cell differentiation. The shoot apical meristem, which consists of a central zone where cell divisions also maintain the meristem and a peripheral zone where cell divisions give rise to mature organs such as leaves and lateral shoots, generates the entire aboveground plant.

Vegetative versus Reproductive Development

During **vegetative development**—the period of growth before flowering—the shoot apical meristem initiates a spiral of leaves in precise sequence. When the plant switches to **reproductive development,** the shoot apical meristem becomes an inflorescence meristem, which, in turn, produces a series of lateral floral meristems. Each floral meristem produces a floral primordium that gives rise to a flower. The **inflorescence**

Classical Map of Mutant Genes

■ Uncloned ⎫ Mapped with mutant genes
■ Cloned ⎬ as reference points

■ Uncloned ⎫ Mapped relative to markers
■ Cloned ⎬ on the RI map

RI Map of Molecular Markers*

■ Location of representative markers indicated on the physical map

Length adjusted to match classical map

Physical Map

■ Regions covered by contigs
■ Regions not covered by contigs
■ Nucleolar organizer regions

Clones Used to Construct the Physical Map

■ YAC
■ BAC
■ PAC
■ TAC

BAC, bacterial artificial chromosome;
EST, expressed sequence tag;
PAC, P1 artificial chromosome;
RI, recombinant inbred;
T-DNA, DNA from *Agrobacterium* used to disrupt genes;
TAC, transformation-competent artificial chromosome;
YAC, yeast artificial chromosome;

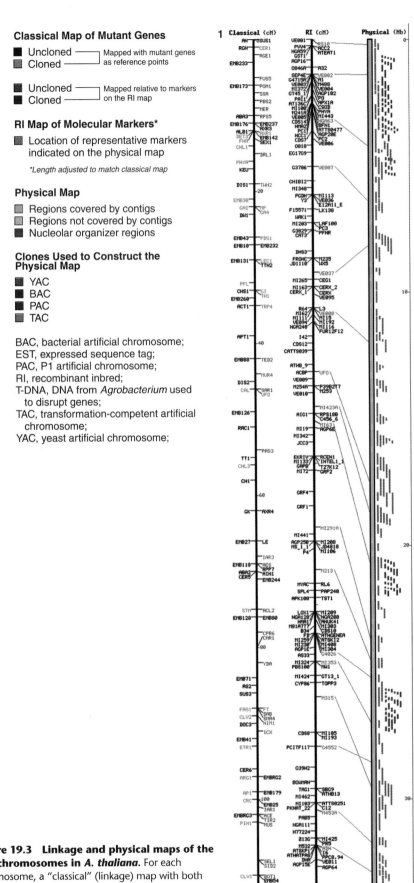
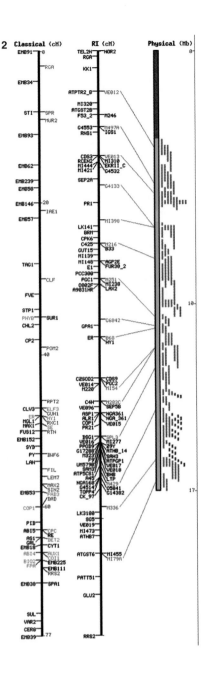

Figure 19.3 Linkage and physical maps of the five chromosomes in *A. thaliana*. For each chromosome, a "classical" (linkage) map with both phenotypic and molecular markers is shown alongside corresponding maps produced with a specialized mapping panel (containing RI or recombinant inbred strains) and physical mapping techniques. The physical map was constructed using the contigs containing four types of clones based on YAC, BAC, PAC, or TAC vector systems.

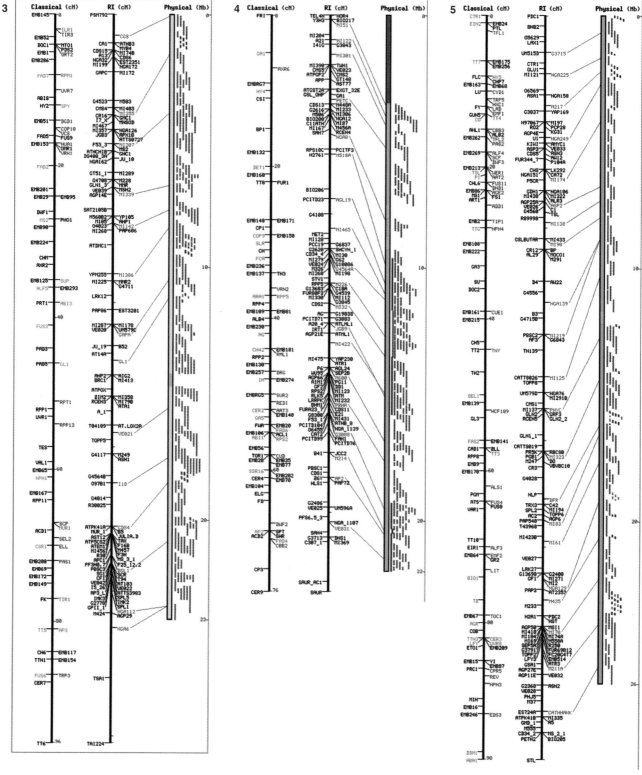

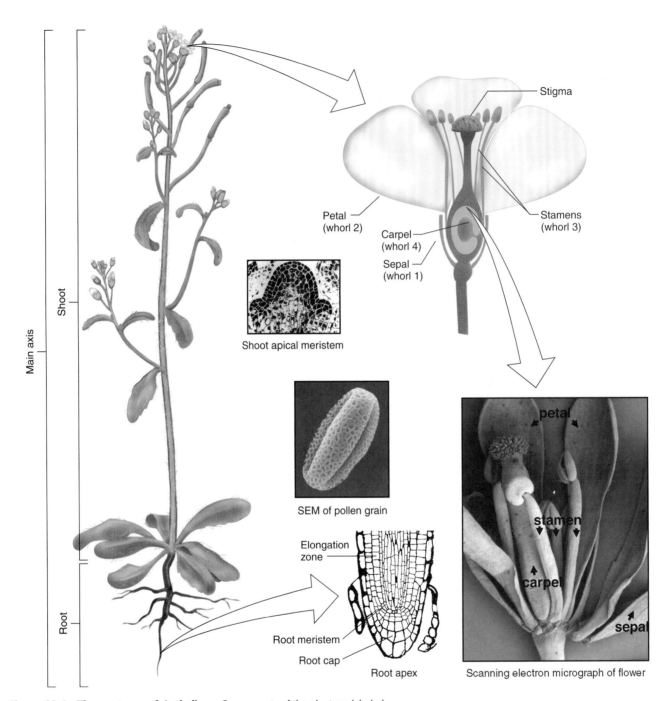

Figure 19.4 **The anatomy of *A. thaliana*.** Components of the plant are labeled.

is the flower or group of flowers at the tip of a branch. In *Arabidopsis,* the flowers are arranged in a spiral around the inflorescence stalk, as depicted in Fig. 19.4.

The Arabidopsis *Flower Is the Most Complex Set of Organs in the Plant*

The flower consists of a modified stem with four concentric regions, or **whorls,** of modified leaves, as shown in Fig. 19.4. The first whorl consists of four green leaf-like **sepals** (whorl 1); the second whorl is composed of four white

petals (whorl 2) that are leaf-like in shape but contain no photosynthetic cells; the third whorl is made of six **stamens** (whorl 3) (four long and two short) bearing the male gametes in the form of pollen; and the forth whorl is a cylinder composed of two fused **carpels** (whorl 4) housing the female gametes in the form of ovules. The fused carpels are part of a cylinder known as the **pistil** that consists of pollen-receptive **stigma** at the top, and a short neck, or **style,** leading to the **ovary,** which houses roughly 50 gamete-bearing ovules. After fertilization, the plant sheds the organs of the outer three whorls, and the pistil develops into the seed-bearing fruit.

Life Cycle: From Fertilization to Flowering to Senescence

As we have seen, *Arabidopsis* is a hermaphrodite capable of self-fertilization, although cross-fertilization by artificial means is easy to accomplish. Fertilization is the first step in a life cycle that also includes embryonic development, seed germination and vegetative growth, reproductive development, and senescence.

Arabidopsis *Undergoes Double Fertilization—A Process Unique to Higher Plants*

In *Arabidopsis,* each mature pollen grain contains two coupled sperm cells, while each ovule includes six mononucleate cells, each with one nucleus and a central cell containing two nuclei, all encased within a common cell wall (Fig. 19.5a). The landing of a pollen grain on the receptive tissue of the stigma initiates fertilization by triggering germination of the pollen grain (Fig. 19.5b). The emerging pollen tube migrates through the short neck of the pistil, also known as the *transmitting tract,* to an ovule in the ovary. There, one 1*n* sperm nucleus

from the pollen fuses with a 1*n* egg nucleus from the ovule to form the 2*n* nucleus of the zygote; the second 1*n* sperm nucleus fuses with the 2 1*n* nuclei within the ovule to form a 3*n* endosperm nucleus (Fig. 19.5c). After fertilization, the zygote divides mitotically to form the embryo, and the endosperm nucleus divides mitotically to form endosperm tissue that, like an animal yolk sac, will nourish the developing embryo. Meanwhile, the remaining 1*n* cells of the ovule degenerate, and the outer layer of the ovule sac hardens to form a seed coat.

The Embryo Develops within the Protective Maternal Seed Coat

A sequence of well-defined cell divisions leads the embryo through a series of stages (Fig. 19.6a). The first zygotic division gives rise to two cells, one small and one larger. The small cell constitutes the embryo proper, while the larger cell will divide only a few more times to produce the **suspensor,** a structure analogous to the umbilical cord in mammals. The embryo divides again, following a plane parallel to the first plane of division, after which two successive vertical divisions at a right angle to each other give rise to the **octant stage** embryo. Additional divisions parallel to the surface of the embryo result in the formation of a discrete outer cell

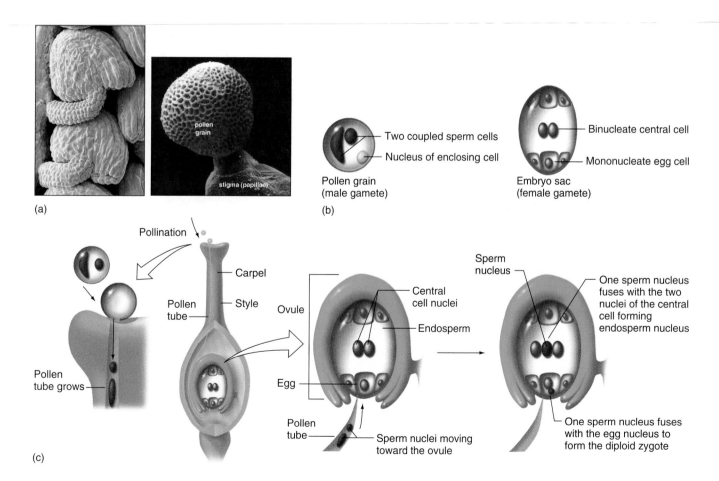

Figure 19.5   Double fertilization in *Arabidopsis*. (a) Micrographs of pollen tube growth (left) and of a pollen grain landing on a stigma. (b) Diagram of a pollen grain and an embryo sac. (c) Fertilization and postfertilization events.

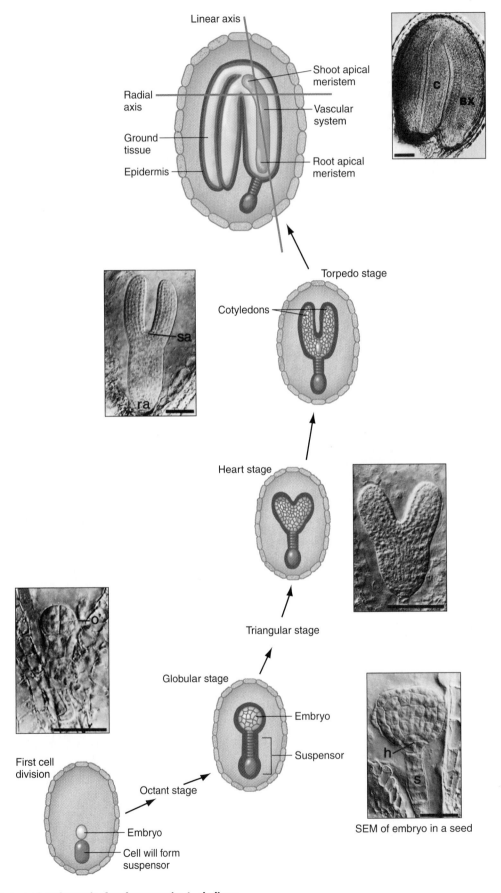

Figure 19.6 Stages of embryonic development in *A. thaliana*.

layer that will give rise to the epidermis of the mature embryo, and an innermost group of cells that acquire an elongated shape, the first evidence of differentiation toward the vasculature found in the mature embryo. The embryo has now reached the **globular stage.** The globular embryo rapidly loses its globular shape and acquires a more triangular form that defines the **transient** or **triangular stage** embryo, at which point the embryo becomes self-sufficient for growth, having used up most of its maternally deposited reserves. Next, the embryo moves through the **heart stage** to the **torpedo stage** as two protuberances expand and differentiate into two well-defined, discrete cotyledons.

By the end of this progression from octant to torpedo stage, cellular differentiation has established all the major tissues required for the future vegetative development of the plant along two main axes: the linear axis that makes up the shoot/root continuum and a radial axis moving from the center out to the edge of that continuum (Fig. 19.6b). The shoot/root axis includes two cotyledons, which store nutrients in the mature seed and will differentiate into the first rudimentary leaves after germination. The axis also includes a *hypocotyl,* which is a short section of embryonic stem between the two cotyledons and the embryonic root, and the root and shoot meristems. The radial axis carries the central vascular system, a middle layer of ground tissues, and the outer epidermis.

As the embryo reaches maturity, growth and development cease and desiccation (water loss) occurs. When the loss of water becomes extreme enough, the embryo enters a dormant state in which virtually all metabolic activity ceases. An embryo in the dormant state can withstand adverse environmental conditions such as winter in Madison, Wisconsin, or Buffalo, New York.

Favorable Environmental Conditions Trigger Seed Germination and Vegetative Growth

Under appropriate conditions, the *Arabidopsis* embryo absorbs water and rehydrates within the seed, which enables growth and development to resume. The enlarging embryo bursts out of the seed coat and begins to grow along its vertical axis. At this point, environmental cues influence the form the seedling will take. Both the root and the shoot can sense gravity; the root responds by bending toward it while the shoot responds by bending away from it. Light provides another developmental cue. If germination occurs in the dark, for example, under the surface of the soil, the hypocotyl, or embryonic stem, elongates rapidly; but development of the leaf-like cotyledons and activity of the shoot apical meristem are suppressed. The elongating hypocotyl is hooked near its apex, an adaptation that enables the seedling to push its way up through the soil while protecting the delicate meristem. When the elongating hypocotyl senses light, elongation ceases, the apical hook opens, and the cotyledons expand and turn green as chloroplasts begin to develop and establish the potential for self-nourishment through photosynthesis (Fig.

19.7a). This light-regulated developmental program is called **photomorphogenesis.**

With the establishment of the self-nourishing autotrophic lifestyle, the shoot apical meristem produces a compact spiral, or rosette, of leaves. Early flowering varieties of *Arabidopsis* may only produce 6–10 leaves before the shoot apical meristem converts to reproductive development. Later flowering

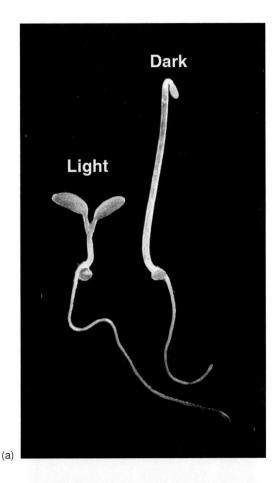

(a)

(b)

Figure 19.7 Emerging shoot in *A. thaliana.* (a) The embryonic stem reaches for the surface of the soil (right seedling) where the apical hook will open, allowing the greening of the cotyledons (left seedling). (b) In early flowering varieties, the shoot apical meristem may produce only a few leaves before developing flowers (left). Late flowering varieties may produce dozens of leaves before flowering (right).

varieties may produce a rosette containing dozens of leaves before flowering begins (Figure 19.7b).

Reproductive Development Begins When the Leaf-Producing Apical Meristem Switches to a Flower-Producing Meristem

Under the right conditions, the shoot meristem enlarges and switches from producing only cells that differentiate into leaves to producing floral meristems on its flanks. At this point, the shoot apical meristem becomes an **inflorescence meristem.** The floral meristems arise in a spiral around the inflorescence meristem and, in turn, give rise to flower primordia that differentiate into the four whorls of organs that make up a flower. Initiated one at a time, the flowers develop sequentially. Thus a view from the apex of the inflorescence meristem reveals the progression of floral developmental stages spiraling away from the shoot apex. Elongation of the main stem between each flower ultimately produces the inflorescence branch containing about 35 flowers (Fig. 19.8a).

Both environmental cues and endogenous developmental signals (resulting from the genetic program) control the timing of the switch from vegetative to reproductive development in *Arabidopsis*. Day length, or **photoperiod,** is a particularly important regulator of flowering. Laboratory strains, such as Columbia and Landsberg *erecta,* when grown under constant light or a long, 16-hour day, will flower after producing less than 10 rosette leaves. When grown in an 8-hour photoperiod, these same strains will produce over 30 leaves before flowering. Another significant environmental signal is temperature. If exposed to extended periods of low temperature during early vegetative development, normally late-flowering varieties of *Arabidopsis* will flower early. This phenomenon of exposing plants to the cold early in vegetative development is known as **vernalization.**

Since neither photoperiod stimulation nor vernalization is strictly necessary for flowering, researchers have inferred that an endogenous developmental clock is active in this process. However, they have not yet identified the genes responsible for this endogenous program.

Senescence: The Vegetative Plant Ages and Dies

Because plants can produce new organ systems continuously, they do not necessarily display the finite life spans characteristic of animal species. However, some plants, including *Arabidopsis,* do senesce and die after a single reproductive effort. In *Arabidopsis,* somatic tissues such as leaves and stems senesce a few weeks after they are initiated at the meristem. In leaves, senescence is a genetically programmed process in which yellowing occurs as salvage systems convert cellular constituents into mobile nutrients for transport out of the aging leaf (Fig. 19.8). The limited longevity of individual leaves does not limit the life span of the plant in the vegetative phase because the shoot apical meristem continuously produces new leaves. Rather, it is the conversion of the shoot apical meristem to reproductive development that delimits the production of new photosynthetic organs. In addition, the inflorescence meristem ceases proliferative activity after a set number of seeds have been produced. The parent plant fails to survive the massive reproductive effort that results in thousands of progeny-carrying seeds.

TECHNIQUES OF MUTATIONAL ANALYSIS

Researchers have used various techniques to identify mutations affecting many aspects of *A. thaliana* anatomy and development. Some of the mutations characterized in *A. thaliana* occur spontaneously in natural populations, but as described in Chapter 6, these spontaneous mutations appear at a low rate. Geneticists use mutagens to increase their frequency. The means of inducing mutations in *Arabidopsis* include exposure to mutagenic chemicals, exposure to radiation, and insertional mutagenesis.

Mutagenesis by Chemical and Irradiation Procedures Produces Different Ratios of Various Mutations

A mature seed carries a dormant embryo in which one to three cells from the apical meristem are destined to form the germ line (gametes) of the mature plant. If a mutation is to be transmitted to the progeny, it has to affect one of the two alleles carried by any of these cells. The phenotype derived from a recessive mutation induced by seed mutagenesis in *A. thaliana* will segregate in different ratios depending on how many gametophyte (gamete-producing) progenitor cells are present in the embryo at the time of mutagenesis. With a parental plant grown from a mutagenized seed in which only one gameto-

Figure 19.8 Reproductive development and senescene in *A. thaliana.* Young flowering plants next to older plants undergoing senescence.

phyte progenitor cell was present during mutagenesis, the phenotype associated with the recessive mutation will segregate into 1/4 of the progeny resulting from self-pollination. If two gametophyte progenitor cells were present in the parental seed during mutagenesis, the mutant phenotype will segregate into 1/16 of the selfed progeny. And if three progenitor cells were present, 1/36 of the progeny resulting from self-pollination will show the phenotype.

As in microorganisms, fungi, and animals, mutations resulting from mutagenesis vary with the nature of the mutagen. For instance, chemical mutagenesis using ethyl methane sulfonate (EMS) usually generates simple G to A transitions in the DNA, whereas radiation often generates deletions.

Arabidopsis *Researchers Use Two Types of Insertional Mutagenesis: Transformation by T-DNA and Transposon Tagging*

Strategies for insertional mutagenesis depend on the ability of certain classes of DNA molecules to integrate into the *Arabidopsis* genome. If an integration disrupts a gene, it represents a mutation.

In the T-DNA Approach, the Bacterium Agrobacterium tumefaciens, *Which Causes Crown Gall Disease in Plants, Is the Agent of Transformation*

A. tumefaciens can transfer a piece of plasmid DNA known as the T-DNA into the genome of wounded plant cells (Fig. 19.9a).

Experimenters can modify the T-DNA to eliminate its oncogenes and insert transformational marker genes that make the plant resistant to antibiotics (such as kanamycin and hygromycin) or herbicides (such as chlorsulfuron). In one protocol using modified T-DNA, *A. thaliana* is transformed by soaking the shoots of mature three-week-old seedlings or germinating seedlings in an *Agrobacterium* culture to which a strong vacuum has been applied. The vacuum enables the bacteria to infiltrate the shoots. As the bacterium-infiltrated plants grow and eventually self-pollinate, a few gametes or zygotes are transformed by *Agrobacterium,* and these produce a transformed seed that is resistant to antibiotics or herbicides. This resistance provides a means of selecting for transformed plants that have inserted the transgenic DNA into their genomes. Because the *A. tumefaciens* T-DNA integrates more or less randomly into the plant genome, it has the potential to serve as a mutagenic agent for any gene.

Researchers have used insertional mutagenesis to create collections of independent transgenic *A. thaliana* lines, which can be screened for specific morphological, developmental, or biochemical phenotypes. They have also modified the T-DNA to carry a *GUS* reporter gene with no promoter of its own. *GUS* encodes the enzyme β-glucuronidase, which converts a colorless soluble compound (X-Glu) into a blue insoluble product. Cells expressing *GUS* stain blue. Because the *GUS* gene in the modified T-DNA has no promoter, its expression depends on the insertion of the T-DNA into a plant gene; when that happens, the regulatory region of the disrupted gene controls the expression of *GUS*. The expression of *GUS* thus reports the

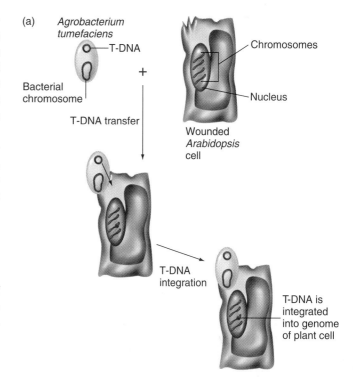

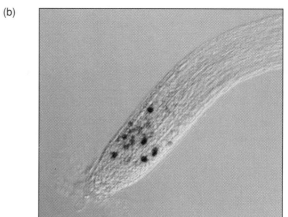

Figure 19.9   The natural process of *A. tumefaciens* transformation can be exploited for use as a method of gene transfer into *A. thaliana.* (a) Natural integration of T-DNA from *Agrobacterium tumefaciens* into *A. thaliana.* (b) Blue-staining cells expressing *GUS* gene.

expression pattern of the disrupted gene (Fig. 19.9b). The use of the GUS reporter allows not only the mutation of genes but also the determination of the disrupted genes' expression pattern (see the discussion of "enhancer-trapping" in *Drosophila* in Chapter 21).

*Transposon-Tagging Strategies Rely on the Transposable Elements from Corn (*Zea mays*) and* Arabidopsis

Transposable elements derived originally from corn can transpose to a new position in the *A. thaliana* genome when introduced by the *Agrobacterium*-mediated DNA transformation just described. For example, corn transposable elements of the families known as *Ac/Ds* and *Spm/dSpm* transpose when introduced into *Arabidopsis*. Similarly, transposable elements endogenous to *Arabidopsis* transpose in the plant's own genome.

Insertional Mutagenesis Not Only Allows the Generation of Mutations But Also Facilitates Their Molecular Characterization

One advantage of insertional mutagenesis over chemical and radiation procedures is that insertional mutagenesis produces mutated alleles carrying a specific DNA insert (the T-DNA or the transposable element). A copy of the specific T-DNA or transposable element that became the insert can therefore serve as a probe to identify and clone the gene the insert disrupts.

From Gene to Phenotype: Analyzing Mutations to Identify Gene Function

The genome sequencing project has identified genes with specific sequences, but the information in these sequences is often not sufficient to attribute a specific function to each gene. One way to solve this problem is to mutate the gene under study and analyze the phenotype of plants homozygous for the mutation.

Using PCR Protocols to Find Plants Carrying a Mutated Gene of Interest

Collections of insertional mutant lines contain many independent inserts, but it is now possible to identify in these collections particular lines carrying a DNA insertion (T-DNA or transposable element) in any gene of interest whose sequence is not known. The first step is to transform a large number of plant genomes with *A. tumefaciens* (Fig. 19.10a) The T-DNA integrates at a different random location in each genome. DNA purified from 100 different plant genomes is pooled. From 1000 genomes there would be 10 pools. Pooling reduces the amount of work required to screen a very large number of genomes for a rare insertion in the gene of interest. The next step is to attempt a PCR amplification in which one oligonucleotide hybridizes with a sequence in the T-DNA or transposon used to generate the collection, while a second oligonucleotide hybridizes with a sequence in the gene of interest. If the pool of individuals under study contains just one

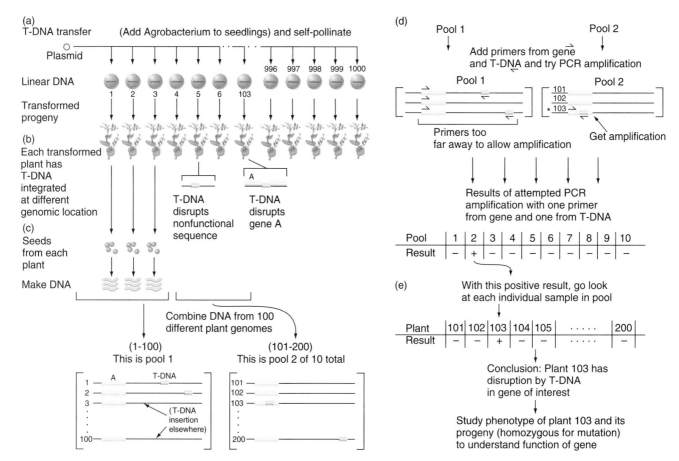

Figure 19.10   A PCR screening protocol enables rapid screening of a large collection of transformed plant genomes for a rare one containing disruption of the gene of interest. (a) Transformation of 1000 plant genomes with T-DNA. (b) Each genome has T-DNA integrated at a different location. (c) DNA is made from each genome. Pools containing 100 samples are made. With 1000 genomes, there are 10 pools. (d) PCR amplification is attempted on each pool with one primer from the gene of interest and one primer from the T-DNA. Only pools that contain a genome with T-DNA inserted inside the gene yield a positive result. (e) Each individual within a positive pool is subjected separately to PCR to identify the one plant of interest. This one plant is bred for genetic analysis of the gene of interest.

plant with an insertion in the gene of interest, the PCR protocol will amplify one DNA fragment carrying part of the disrupted gene and part of the DNA insertion. With the finding of such a band in one pool of lines, the next step is to repeat the strategy on smaller and smaller subsets to identify the one plant that carries the disrupted allele of the gene under study. With the plant in hand, it becomes possible to produce homozygotes for the mutation by allowing the heterozygous plant to self-pollinate. Study of these homozygous mutants can shed light on gene function.

A Less Random Strategy Relies on Gene Knockouts Generated by Homologous Recombination

Knockouts in *Arabidopsis* are similar to those described for yeast and mice (Chapters 18 and 22). While the transposon-mediated insertional mutagenesis strategy is random, homologous recombination protocols can target genes directly. In these types of protocols, the transforming DNA carries a specific version of the gene under study in which an internal fragment of DNA has been replaced by a selectable marker gene (for example *NPTII*). The selectable marker confers resistance to an antibiotic (for example, kanamycin) on any transgenic plants. The transgenic plants can be identified by PCR amplification using the following primers: an oligonucleotide that hybridizes with a region of the gene not found in the transforming DNA, and an oligonucleotide that hybridizes with the selectable transformation marker used to disrupt the gene. With these primers, PCR will amplify a DNA fragment only if the transforming DNA has integrated into the locus targeted by homologous recombination (see Fig. 19.10). This protocol enables the recovery of resistant transgenic plants and rare transformants carrying a copy of the transforming DNA integrated by homologous recombination into the gene of interest. Plants whose genomes carry these rare transformants can then be propagated for study. The advantage of the homologous recombination approach is that the precise mutant allele that is desired can be designed into the transforming DNA. The disadvantage is a very low efficiency in *A. thaliana*.

The use of insertional mutagenesis and homologous recombination strategies for moving from gene to phenotype will become more and more extensive as the *Arabidopsis* genome sequencing project progresses. When the sequence of the entire genome is in hand, the main role of the geneticist will be not only to identify the genes involved in specific functions but also to define what specific genes do, through knowledge of their sequences and the phenotypes induced by mutations.

THE GENETIC ANALYSIS OF DEVELOPMENT IN <u>ARABIDOPSIS</u>

While geneticists only partially understand the mechanisms of development in plants and animals, they do know that instructions for the formation of a body plan and the elaboration of specific functions are ultimately encoded in the DNA. Mutations can thus change the position, structure, or function of an organ or alter an individual's orderly passage through the life cycle. We now look at the use of genetic analysis in the study of embryogenesis, hormone control systems, and responses to environmental signals.

The Genetic Analysis of Embryogenesis

Mutations that kill the embryo or arrest its development have provided a basis for understanding embryonic development in *A. thaliana*. Screenings for mutations resulting in developmental arrest at a specific stage of embryogenesis have led to the identification of more than 500 embryo-lethal and embryo-defective mutations. A significant portion of the embryo-lethal mutants are arrested at the transient stage. Many of them carry mutations in housekeeping genes.

Researchers Have Found Very Few Maternal-Effect Mutations in Arabidopsis

Such mutations occur in maternal genes whose protein products are deposited in the egg during oogenesis. Maternal-effect mutations disrupt embryogenesis in all progeny of a plant carrying a particular mutant genotype.

The small number of maternal-effect mutations constitutes a major difference between plants and certain kinds of animals, such as *Drosophila,* that develop from large eggs. (In *Drosophila,* numerous maternal-effect mutations alter the early stages of embryogenesis, as described in Chapter 21.) A major reason for the near absence of maternal effects in *Arabidopsis* is that the plant's embryonic development does not depend on a large cytoplasm filled with maternal products (as happens in *Drosophila*). Instead, it depends from a very early stage on a cytoplasm newly produced from the zygote's own genome (as in mammals).

Also unlike what happens in *Drosophila,* in *Arabidopsis,* very few zygotic mutations (that is, mutations in the zygote's own genes) affect embryogenesis before the globular stage. At least part of this difference between *Arabidopsis* and *Drosophila* may derive from the fact that in plants many zygotic genes are expressed very early in embryogenesis (as they are in mammals), and these zygotic genes may encode protein products that are functionally redundant with those of maternal origin. In addition, many plant genes are members of multigene families in which each member encodes a protein whose function duplicates that of some of the other members. Hence, as in mammals, the scarcity of maternal-effect and early zygotic mutations in plants may simply reflect a large amount of functional redundancy between maternally deposited and zygotically derived products.

Screenings for Mutations That Arrest Development at a Specific Stage of Embryogenesis Enabled the Identification of Important Regulatory Processes in Plant Development

For example, a type of mutation known as *leafy cotyledon* (*lec*) results in embryos with cotyledons resembling mature leaves rather than wildtype cotyledons (Fig. 19.11a). As described earlier, cotyledons are embryonic structures that first

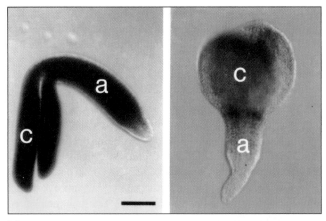

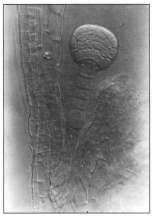

Figure 19.11 Examples of two mutations that arrest development. (a) Micrograph of *lec* mutant (right) versus a plant carrying *LEC* wildtype (left). The cotyledons in a *lec* mutant are abnormally rounded. (b) Micrograph of *twin* (right) versus *TWIN* embryos (left). a = axis; c = cotyledons.

appear late in the globular stage; leaves, by contrast, are products of the apical meristem that appear only after germination. Unlike cotyledons, leaves have trichomes—small, hairlike projections from the epidermis—and a leaf-specific pattern of vascularization. In addition to causing the conversion of cotyledons into leaves, *lec* mutations produce seeds unable to germinate because they cannot withstand the desiccation that accompanies seed maturation. These observations suggest that the wildtype *LEC* gene encodes a molecule that helps regulate the process of seed maturation in *A. thaliana*. Moreover, the fact that *lec* mutations are recessive suggests that expression of the wildtype *LEC* gene represses leaf development during embryogenesis. In fulfilling such a function, the *LEC* gene acts as a negative regulator. The *twin* mutation revealed another example of negative regulation during embryogenesis. In embryos carrying that mutation, some cells in the suspensor modify their developmental fate, resulting in the formation of a second embryo within the seed (Fig. 19.11b). From this observation, *Arabidopsis* geneticists inferred that the *TWIN*

gene probably encodes a molecule that represses the embryo development program in suspensor cells. To confirm these predictions, they need to clone the genes in question and analyze their products.

The Genetic Analysis of Hormonal Control Systems

The control of plant development requires complex communications within cells, between neighboring cells, and over long distances between different organ systems. **Plant hormones** are the main internal chemical signals that influence plant development. One group of hormones, the *auxins* (from the Greek *auxein,* to increase or augment), promotes shoot growth through cell elongation and lateral root formation and helps prevent senescence. The auxins also inhibit growth in lateral buds in favor of growth in the apical meristem (apical dominance), and inhibit cell elongation in roots. A second set of hormones, the *cytokinins,* promotes growth by stimulating cell division, including cytokinesis (hence their name). The cytokinins also play a role in the prevention of senescence but, unlike the auxins, promote the growth of lateral buds. Together, the auxins and cytokinins help regulate a range of processes including meristem induction, apical dominance, and root-shoot communication. A third type of hormone, *abscisic acid,* controls aspects of seed development and dormancy; it also mediates responses to water stress (too much or too little water). Yet another group of plant hormones, the *gibberellins,* regulates seed germination and stem elongation. Finally, *ethylene* promotes maturation of the seedling and influences the timing of leaf senescence.

Mutations That Disrupt Hormone Activity Can Clarify a Hormone's Biological Significance and the Biosynthetic Pathways Involved in Its Production

For example, researchers originally identified the *ga1-1* mutation, an X-ray-induced deletion, through a dwarf phenotype that could be rescued by gibberellin application (Fig. 19.12). Later, analysis of the cloned gene indicated that it encodes an enzyme that operates early in the gibberellin biosynthetic pathway. A second mutation, the *gai* mutation, produces a phenotype similar to that produced by *ga1-1,* but *gai* mutants are not rescued by gibberellin application. The inference from this observation is that *gai* most likely affects a plant's response to gibberellin rather than the hormone's biosynthesis. Researchers are currently using similar mutational analyses to identify the structural and regulatory genes contributing to the biosynthesis of auxins, abscisic acid, and ethylene.

Mutations That Render Arabidopsis Insensitive to a Hormone Help Reveal the Mechanisms by Which Plants Perceive and Transduce Hormone Signals

In the most advanced studies of this type, investigators are looking at ethylene signaling, using ethylene's effects on early seedling development as the basis for mutant screens. The exposure to ethylene of seedlings grown in the dark produces an inhibition of shoot and root elongation and an accentuation of the apical hook. Mutant seedlings that are insensitive to ethylene grow tall in the presence of ethylene.

Several loci that affect responsiveness to ethylene have been identified in *Arabidopsis* (Fig. 19.13). For example, dominant mutations at the *ETR1* locus cause insensitivity to ethylene, as do mutations at *EIN2* and *EIN3*. By contrast, mutations at the *CTR1* locus result in the constitutive activation of ethylene responses. All these mutations affect a wide range of ethylene responses in various tissues, suggesting that the genes influence the primary steps of ethylene signal processing. Analysis of various combinations of double mutants reveals that *CTR1* is epistatic to *ETR1*, while *EIN2* and *EIN3* are epistatic to *CTR1*.

Arabidopsis geneticists cloned the *ETR1* gene by chromosome walking and isolated the *CTR1* gene as a T-DNA-tagged allele. They then sequenced both genes. By aligning the sequences with sequences in the database, they found that *ETR1* is related to receptor protein kinases typical of bacterial sensory systems, while *CTR1* is related to eukaryotic protein kinases that initiate MAP kinase cascades in yeast and mammals (see Chapter 17). By expressing the *ETR1* gene in yeast and determining that this expression generated binding sites for ethylene, they also demonstrated that *ETR1* is the ethylene receptor. *CTR1* operates downstream from *ETR1* in the cytoplasm.

Arabidopsis also generates mutants that are insensitive to auxin, gibberellins, and abscisic acid. The isolation of genes determining the mutant phenotypes implicates the involvement of protein phosphatases, farnesyl transferases, and the ubiquitin protein degradation pathway in plant hormone signaling. The fact that many of the genes identified are homologous to genes in yeast and mammals is evidence that the basic information processing systems arose very early in the evolution of eukaryotes.

Figure 19.12 The *gal-1* mutation. The wildtype plant on the left compared to a dwarf plant with the *gal-1* mutation (right).

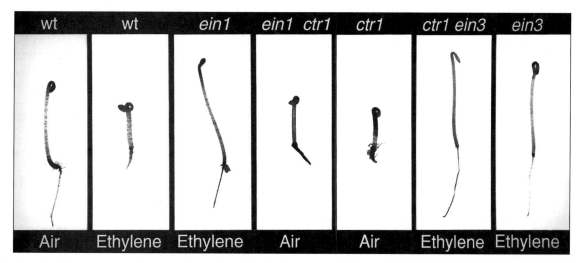

Figure 19.13 The effects of ethylene exposure on wildtype and mutant plants. Ethylene inhibits shoot and root elongation in wildtype seedlings. Single and double mutant analyses allows the ordering of genes in the signal pathway. Ein3 protein acts before Ctr1, which acts before Ein1.

The Genetic Analysis of Photomorphogenesis— The Regulation of Growth and Development in Response to Lighting Cues

During growth and development, plants continuously alter their responses to changing environmental signals. One of the most dramatic series of developmental changes takes place when the seedling grows up through the soil and encounters light. Collectively known as *photomorphogenesis,* these changes include the opening of the apical hook and the generation of chloroplasts. Two classes of photoreceptors— phytochromes and cryptochromes—mediate the changes. Receptors in the phytochrome family sense light primarily in the red and far-red wavelengths. Receptors in the cryptochrome family perceive blue wavelengths of light.

Mutational Analyses Have Helped Characterize the Photoreceptor Molecules by Which Plants Receive Light Signals

White light inhibits the elongation of the hypocotyl, or seedling stem, in *Arabidopsis.* A class of photomorphogenic mutants referred to as *hy* (for long hypocotyl) show reduced sensitivity to white light (Fig. 19.14a). Linkage map locations for two of the *hy* mutations, *hyx* and *hyy,* coincide with the locations of *phyA* and *phyB* members of the phytochrome gene family. The *hyx* mutant has an altered response to red wavelengths of light, while the *hyy* mutant has an altered response to far-red wavelengths of light, indicating that these two genes have different roles in mediating photomorphogenesis. Two other mutations, *hy1* and *hy2,* block the biosynthesis of the tetrapyrolle prosthetic group associated with all phytochromes.

The *hy4* mutant in *Arabidopsis* shows a decreased sensitivity to blue light, suggesting that the normal *HY4* gene product plays a role in the reception of blue light. Cloned using a T-DNA-tagged allele and then sequenced, the *HY4* locus exhibits homology with a bacterial enzyme that utilizes absorbed light energy to repair thymidine dimers in DNA. This conservation of a light-absorbing function is compelling evidence that the *HY4* gene product is a blue light receptor active in photomorphogenic responses.

Mutational Studies Have Also Clarified How Plants Process Light Signals

While *Arabidopsis* researchers have analyzed *hy* mutants for information on light perception, they have studied other sets of mutants known as *cop* (for constitutive photomorphogenesis) and *det* (for de-etiolated, that is, released from development without chlorophyll) to understand how plants process light signals into the multitude of changes that characterize morphogenesis. Mutant *cop* and *det* seedlings grown in the dark behave as though they have received the light signal—their apical hook opens, the cotyledons expand, light regulated genes are derepressed, and chloroplasts develop, all in the absence of light (Fig. 19.14b). The *det* and *cop* mutations are recessive. From characterizing the phenotypes generated by these recessive mutations, researchers have inferred that the normal *COP*

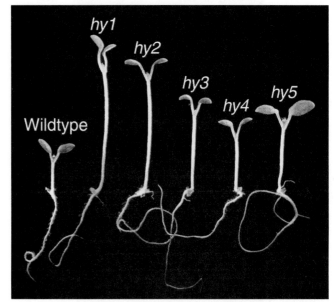

(a)

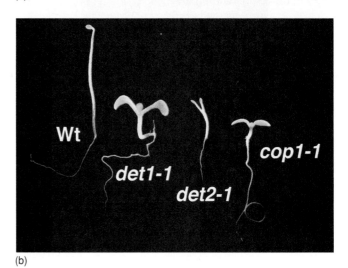

(b)

Figure 19.14   Mutations that affect the response to light. (a) Comparison of the response of a wildtype seedling to light, versus the response of seedlings with the *hy* mutation in one of five different "long hypocotyl" loci. (b) Comparison of the response of wildtype and mutant seedlings to dark. Three different mutant seedlings are shown that behave as if they've been exposed to light when they haven't.

and *DET* gene products somehow suppress all or some photomorphogenic responses until a light signal is received.

The *COP* genes affect all morphogenic responses. Several *COP* loci have been cloned, and biochemical analyses of the gene products reveal that some COP polypeptides associate to form a large protein complex in the nucleus. Interestingly, the *COP1* gene product, which contains zinc fingers indicative of a DNA-binding protein, is localized in the nucleus in the dark, but appears to migrate to the cytoplasm in response to light. This migration is probably mediated by other factors.

The *DET* genes affect only some photomorphogenic responses. The *det2* mutant, for example, shows derepression of some light-regulated genes, but does not develop chloroplasts in the dark. This mutant develops into a dwarf adult, indicating that the wildtype gene influences more than morphogenesis. The cloning of *DET2* led to the discovery that the gene is related to mammalian genes that operate in steroid biosynthetic pathways. Later, researchers found that the application of brassinosteroids, a class of steroid-like compounds found in plants, can rescue *det2* mutants. While botanists had long suspected that brassinosteroids contribute to the regulation of plant growth and development, the genetic analysis clarified that they play a role in the regulation of gene expression by light.

THE GENETIC ANALYSIS OF FLOWERING: A COMPREHENSIVE EXAMPLE

One of the major life events in the development of an angiosperm is the switch to reproductive growth that culminates in flowering. During the switch, reprogramming of the apical meristem from a leaf-producing to a flower-producing tissue gives rise to an astonishing series of events: The meristem no longer generates mature leaves in a specific spiral pattern. Instead, it becomes an inflorescence meristem (IM) that produces smaller (cauline) leaves, an elongated stem, and many side shoots. The meristems of the side shoots develop into floral meristems (FMs) that produce a single flower of specific design. The IM is *indeterminate* because it produces FMs indefinitely. By contrast, the FMs are *determinate* because they normally produce a fixed number of floral organs in a precise pattern of concentric rings, and in the process, they lose the potential for meristematic activity as cells of the meristem become incorporated into the maturing pistil (Fig. 19.15).

Arabidopsis accomplishes the reprogramming of the meristem through regulation of a complex network of intricately overlapping genetic pathways in response to environmental cues. Some determine the body plan of the flower; others specify the identity of the floral meristem; still others control the timing of floral meristem formation.

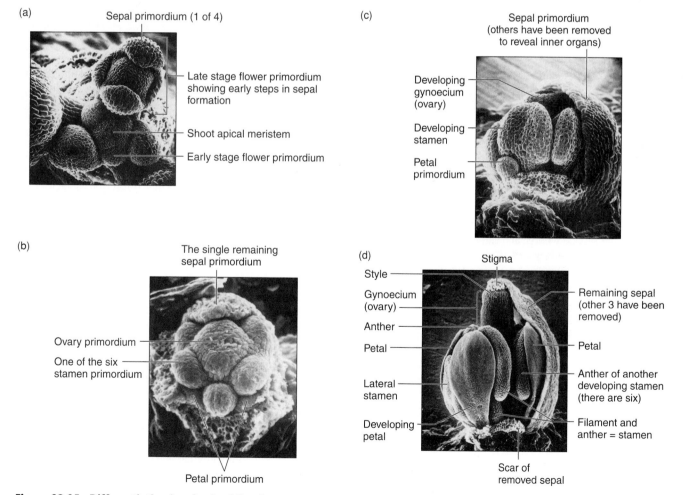

Figure 19.15 Differentiation in whorls of floral organs. This series of SEMs shows the earliest stage of development of the *A. thaliana* flower. In the earliest stage (a) sepals are just beginning to form. 3 of the 4 developing sepals have been removed (b) to allow the other organ primordia to be seen. Petal and stamen formation are noticeable in (c). The structure of the stamen and anthers become more defined in (d). Again, 3 outer sepals have been removed to show the developing flower parts; mature flower carrying 4 sepals, 6 stamen, and one gynoceum.

How Genes Determine the Body Plan of a Flower

The floral primordium is an outgrowth of undifferentiated cells near the tip of an inflorescence meristem. As cells in the primordium divide, differentiate, and elongate according to genetic instructions, a flower emerges. A wildtype *Arabidopsis* flower, we have seen, carries four green sepals in the outside whorl, four white petals just inside the sepals, six stamens in the next-to-center whorl, and a cylinder of two fused carpels in the center (review Fig. 19.4). Cascades of genetic signals alert cells to their position within the developing flower and thereby direct the proper growth and differentiation of the organs that make up a flower. Mutants can have an abnormal arrangement of organs.

Analyses of Homeotic Mutations Reveal That in Arabidopsis, *Three Types of Single-Gene Products Influence Floral Pattern Formation*

A **homeotic gene** is one that plays a role in determining a tissue's identity during development. A **homeotic mutation** causes cells to misinterpret their position in the blueprint and become normal organs in inappropriate positions. Homeotic mutations thus alter the overall body plan. In one study, researchers looked at mutants with an abnormal order and selection of floral organs. They obtained some of the mutants for their study from natural and laboratory populations, and used the mutagen ethyl methane sulfonate to generate more. Phenotypic screens of a large number of mutants showed they fell into three classes (Fig. 19.16a). Class A displayed carpels instead of sepals in the first whorl, and stamens instead of petals in the second whorl, for an overall radial pattern of carpels, stamens, stamens, carpels. In controlled crosses, the ratio of normal to abnormal phenotypes indicated that one gene was responsible for this abnormal radial pattern. Researchers named the gene carrying the mutations that caused this aberrant pattern, *APETALA2* (*AP2*) for the lack of petals arising as a result of the mutation. Class B mutants carried sepals in the first and second whorls, and carpels in the third and forth whorls. Mutations in either *APETALA3* (*AP3*) or *PISTILLATA* (*PI*) caused this deviation from the normal pattern (in botanical terminology, pistillate flowers lack stamen). Class C mutants had an abnormal radial pattern of sepals, petals, petals, sepals. Mutations in the *AGAMOUS* (*AG*) gene, named for the lack of gamete-forming tissues resulting from the mutations, determined these mutant phenotypes. The wildtype *AGAMOUS* gene also stops flower development after establishment of the fourth whorl.

The "Radial Pattern Model" Suggests How Four Genes of Three Classes Could Determine the Identity of Floral Organs

For the sake of simplicity, consider that class A mutants are missing class A gene activity, B mutants are missing B gene activity, and C mutants are missing C gene activity. The model makes three sets of assumptions about these three classes of genes (Fig. 19.16b):

1. A genes are active early in the emergence of the first and second whorls; B genes are active early in the definition of the second and third whorls; and the C gene is active early in the differentiation of the third and fourth whorls.

2. The gene products of these three types of genes, alone and in combination, determine the identity of the floral organs as they develop from the flower primordium. The A product determines sepals, A and B together determine petals, the combination of B plus C determines stamens, and the C product alone determines carpels.

3. The activities of A and C are mutually exclusive; that is, in areas where A is active, C is repressed, and in areas where C is active, A is repressed. Moreover, if a mutation incapacitates A, C is abnormally active and vice versa.

The model provides an elegant explanation of the mutational evidence. To confirm its predictions, the researchers carried out further genetic investigations. In one line of experiments, they bred double and triple mutants and predicted what they would look like. For example, they bred plants with mutations in both *AP2* (the A gene) and *AP3* (a B gene) and predicted that with only normal C activity in the flower primordium, the emerging flower would consist of all carpels. The resulting flowers confirmed the prediction (Fig. 19.16c). Similarly, they eliminated both B and C activity by breeding plants with mutations in the *AP3* and *AG* genes, and as predicted, the resulting mutants bore flowers consisting of only sepals (Fig. 19.16c). In triple mutants lacking A, B, and C gene activity, the flower organs differentiated as leaves (Fig. 19.16d).

The cloning and sequencing of the homeotic genes for floral identity have further validated the radial pattern model. All of the genes produce DNA-binding transcription factors. *In situ* hybridizations using DNA fragments complementary to A-, B-, and C-type mRNAs have monitored the activity of the normal genes and shown that the B and C genes (*AP3, PI,* and *AG*) are expressed specifically in the whorls predicted by the model (Fig. 19.16e). The A gene, *AP2*, however, is expressed in all four whorls of the flower although its product affects organ identity only in the first and second whorls. In *ap2* mutants, *AG* RNA appears in all four whorls of the flower, as predicted by the model. In these *ap2* mutants, however, only the first and second whorls have an altered phenotype. Researchers assume that another, not-yet-identified gene that may be redundant is expressed in the third and fourth whorls to maintain their identity.

Studies of the radial pattern model in snapdragons have shown that what happens in *Arabidopsis* also happens in other higher plants.

The A, B, C model explains one part of a complicated genetic process—the interpretation of cellular position within the flower primordium—but leaves unanswered the question of what establishes the positions. As we see next, the differentiation of floral organs, in addition to the appropriate expression of the A, B, and C genes, requires the activity of many other interacting genes at the proper time and place in an immense hierarchy of genetic control.

Earlier Acting Genes Specify the Identity of the Floral Meristem

The floral meristem produces the flower primordium that differentiates into four whorls of organs, two of which contain the gametes. But what tells cells in the side shoots of an inflorescence meristem to become a floral meristem? To answer this question, researchers analyzed loss-of-function single and double mutants (Fig. 19.17a and b) and gain-of-function transgenic plants. They constructed the transgenic plants by first fusing a nonspecific promoter called 35S with the DNA of the cloned gene under study and then inserting the promoter/gene fusion product into otherwise wildtype *Arabidopsis* plants. Table 19.1 describes six of the floral meristem development genes uncovered by these loss- and gain-of-function strategies. By comparing the effects of each gene's activity with that of the others, the researchers were able to suggest how they interact.

Model of the Progressive Specification of the Floral Fate

The first sign of floral development at the molecular level is expression of the *LEAFY* (*LFY*) gene. Activation of *LFY* marks the position of a future floral meristem and takes place before a visible FM appears. With inadequate *LFY* function, the meristem remains an inflorescence meristem. Next comes the expression *APETALA1* (*AP1*) and *CAULIFLOWER* (*CAL*), whose activities overlap in newly formed meristems. The orderly expression of *LFY, AP1,* and *CAL* promotes emergence of an unbranched flowering structure with four whorls of organs. Loss-of-function mutations in any of these genes prevents the shoot from flowering. Interestingly, mutations in the *AP2* gene enhance the effects of *lfy* and *ap1* mutations.

Expression of *AG* occurs at the center of the floral meristem where its product represses *AP1* expression and further commits the meristem to the determinate production of flowers (Fig. 19.17). A lack of *AG* activity can cause the developing FM to revert to an IM. With the expression of *LFY, AP1,* and *AG,* cells in the lateral, floral meristems become increasingly committed to reproductive development.

Wildtype *Arabidopsis* plants produce only lateral flowers because the *TFL1* gene prevents the expression of FM-producing genes in the center of the IM. As a result the indeterminate inflorescence meristem continues to grow and produce new side shoots even as older lateral shoots are activating the genetic program for flowering.

Some Genes Control the Timing of FM Formation and Flowering

Laboratory-grown, wildtype *Arabidopsis* plants adjust their flowering time in response to the amount of light they receive. Those that receive 16 hours of light a day begin to flower three weeks after germination, while those that receive only 8–10 hours of light do not initiate flowering until at least six weeks after germination. Moreover, the shorter days increase the duration of all phases of development, causing the production of more leaves in the rosette and more flowers on the inflorescence. Exposure to low temperatures for three to six weeks after germination accelerates the flowering of wildtype plants.

TABLE 19.1   Six of the *Arabidopsis* Genes That Control Meristem Identity

Arabidopsis Gene	Loss of Function (Mutant Phenotypes in Meristem Identity)	Gain of Function (Phenotypes of Transgenic Plants)
LEAFY (*LFY*)	Conversion of flower to shoot or shoot-like structures; floral reversion	Conversion of lateral shoots to flower; terminal flower; early flowering
APETALA1 (*AP1*)	Conversion of flower to shoot; branched flowers	Conversion of lateral shoots to flower; terminal flower; early flowering
CAULIFLOWER (*CAL*)	Enhances ap1 phenotypes	Not tested
APETALA2 (*AP2*)	Enhances ap1 and lfy phenotypes	Not tested
AGAMOUS (*AG*)	Indeterminate flowers; axillary buds; floral reversion	Conversion of lateral shoots to flower; terminal flower; early flowering; suppression of late flowering mutants
TERMINAL FLOWER (*TFL1*)	Terminal flower; early flowering	Not tested

Reference: TIG January 1998 Vol. 14 No. 1, "To Be, Or Not To Be, A Flower—Control of Floral Meristem Identity" by M. Hong.

(a)

Class A, B, and C mutants

Wild type

Class A mutants

Lack of activity of *APETELA2 (AP2)*

Class B mutants

Lack of activity of *APETELA3 (AP3)*
or *PISTILLATA (PI)*

Class C mutants

Lack of activity of *AGAMOUS (AG)*

Schematic key:

Sepal)

Petal (

Carpel

Stamen

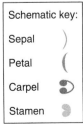

(b)

Assumption about the activity of Class A, B, and C mutants

A gene product ⟶ sepals

A+B gene product → petals

B+C gene product → stamens

C gene product ⟶ carpels

If A is active, C is repressed	result: sepals
If C is active, A is repressed	result: carpels
If a mutation incapacitates A, C is abnormally active	result: carpels
If a mutation incapacitates C, A is abnormally active	result: sepals

(c)

(1) *APETALA2 APETALA3* double mutants lack both A and B activities. The C activity expands into all the whorls, so all the organs are carpels.

(2) *APETALA2 AGAMOUS* double mutants have leaves in whorls 1 and 4 and organs intermediate between petals and stamens in whorls 2 and 3.

(3) *APETALA3 AGAMOUS* double mutants lack both B and C activities. The A activity expands into all the whorls, so all the organs are sepals.

(d)
TRIPLE MUTANTS that lack all three categories of genetic activities have flowers that consist entirely of leaves arranged in whorls

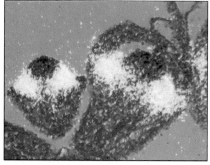

(e)
Probes for *APETALA3* RNA (yellow) bind to the regions that will differentiate as petals and stamens.

Figure 19.16   Homeotic mutants and flower formation in *Arabidopsis*. (a) Class A, B, and C mutants. (b) Diagram showing how three types of genes determine the identity of the floral organs. (c) *ap2/ap3, ap2/ag,* and *ap3/ag* double mutants. (d) *ap2/ap3/ag* triple mutant. (e) Hybridization probe visualizing *AP3* activity in the flower primordium.

(a)

(b)

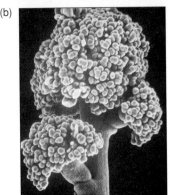

(c)

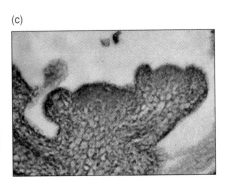

Figure 19.17 Identifying genes that influence development of the floral meristem. (a) Single mutants *lfy-6* (left), *ap1-1* (middle), and *tfl-2* (right). (b) Double mutant *ap1-1/cal-1*. (c) Hybridization probes visualize the activity of *AG* RNA in the center of the flower primordium.

As Table 19.2 shows, there are two types of mutations that influence flowering time: those that delay it, and those that precipitate it. Researchers identified many of the mutations that disrupt normal flowering by screening laboratory populations for their flowering responses. They also found that some mutations identified on the basis of other phenotypes affect flowering. For example, *phyB* mutants, which have an altered response to far-red wavelengths of light, and *cop1* mutants, which grow in the dark as if they have received a light signal, both flower early.

To understand the function of genes that affect flowering time, researchers have analyzed single mutants, double mutants, and transgenic plants containing extra copies of a particular gene. Among their findings is the observation that *cop1, phyB* double mutants show the Cop1 phenotype of short hypocotyls (rather than the phyB phenotype of expanded hypocotyls). This suggests that *COP1* acts downstream of *PHYB*. Since both mutations accelerate flowering, the normal gene products may be part of a pathway required to repress flowering and several overlapping pathways most likely lead

to flowering. Another finding is that environmental influences on flowering time affect the interactions between floral meristem identity genes. The *lfy* mutants, for example, show a stronger phenotype under short days than under long days; and there is more *COP1* RNA in young seedlings grown under long days than under short days, suggesting that the *COP1* gene helps promote flowering.

Hundreds of Mutational Analyses Have Generated a Preliminary Model to Guide Future Research on How Light Affects Flowering

Since there are probably hundreds of genes that contribute to the control of floral development, the genetic analysis of flowering is a work in progress. Nevertheless, certain observations enable us to build a preliminary genetic model of the process: First, three groups of mutations generate late-flowering plants: those that cause delays under long days, those that cause delays under short days, and those that cause delays under both long and short days. Second, most early flowering mutations are epistatic to late flowering ones, which suggests that the

TABLE 19.2   Some of the *Arabidopsis* Mutations That Affect Flowering Time

Full Name	Abbreviated Name	Probable Function of Gene Product
Mutations causing late flowering		
constans	*co*	Transcription factor
luminidependens	*ld*	Transcription factor
phytochrome A	*phyA*	Light receptor
gibberellic acid 1	*ga1*	Enzyme required for gibberellic acid biosynthesis
gibberellic acid insensitive	*gai*	Unknown
gigantea, frigida, fha, fy, fpa, fve, fca, fe, fwa, fd, ft	*gi, fri, fha, fy, fpa, fve, fca, fe, fwa, fd, ft*	Unknown
Mutations causing early flowering		
phytochrome B	*phyB*	Light receptor
constitutive photomorphogenic 1	*cop1*	Transcriptional repressor
early flowering 3	*elf3*	Unknown
early short day 4	*esd4*	Unknown
embryonic flower 1	*emf1*	Unknown
embryonic flower 2	*emf2*	
terminal flower	*tfl*	Unknown

Reference: TIG, October 1995 Vol. 11 No. 10 p. 395 "Genetic and environmental control of flowering time in *Arabidopsis*" by George Coupland.

genes promoting early flowering are downstream of those fostering late flowering; this, in turn, raises the possibility that the early flowering gene products may derepress genes turned off by the products of late flowering genes. Third, blue light and far-red light promote flowering.

To Flesh Out the Model, It Will Be Necessary to Identify More of the Genes That Influence Flowering

There are several reasons why some of these genes have not yet been identified: The genome is not yet saturated with mutations; some genes have redundant functions; and mutations in regulatory genes controlling earlier stages of floral development do not necessarily cause unique floral phenotypes. To find these hard-to-identify genes that influence flowering, researchers can look for new mutants, try to isolate suppressors and enhancers of existing mutations, and devise protocols that test for the late function of early acting genes.

In addition to identifying more of the genes active in the pathways that promote flowering, researchers will try to understand the mechanisms by which these genes work. To this end they will seek to demonstrate the activation of transcription by encoded proteins and to discover the target genes of the transcription factors. Because the food consumed by people around the globe consists largely of grains, fruits, and other flower products, an understanding of flowering at the molecular level may increase the chances of feeding an ever-expanding human population.

CONNECTIONS

Plants are fundamentally different from animals at the level of the whole organism. In contrast to animals, plants continue to produce new organ systems throughout their life cycles, are autotrophic, and do not move about. Despite these differences, the cellular machinery and genetic mechanisms of plants and animals are surprisingly similar. In fact, many of the genes involved in basic metabolic functions and information processing show a striking degree of homology between plants and animals. For example, the *AGAMOUS* gene specifies a DNA-binding protein that most likely regulates downstream genes. A human relative of the *AGAMOUS* gene makes a DNA-binding protein, known as serum response factor, that regulates the c-*fos* oncogene.

In the remaining genetic portraits, we look at genetic mechanisms that propel development in three model animals: the nematode *C. elegans,* the fruit fly *Drosophila,* and the house mouse *Mus musculus.*

ESSENTIAL CONCEPTS

1. *A. thaliana* makes an excellent model organism for genetic analysis because of its capacity for prolific reproduction and the relatively small size of its genome. The genome is carried in five chromosomes that contain little repetitive DNA and a tight arrangement of genes with a few, small introns.

2. *Arabidopsis* switches from vegetative to reproductive development when the leaf-producing apical meristem becomes a flower-producing meristem. After flowering, it undergoes double fertilization to produce an embryo that develops within a protective maternal seed coat. In response to proper conditions, including favorable light and temperature, the seed germinates and vegetative growth begins again. The vegetative plant eventually ages and dies.

3. Researchers use several techniques of genetic analysis in *Arabidopsis* to generate mutations and characterize them at the molecular level. Mutagenesis by treatment with chemicals or X-rays produces a variety of mutations for study. Insertional mutagenesis by *A. tumefaciens* and transposons produces mutated alleles carrying a specific DNA insert that can then serve as a probe for identifying and cloning the gene of interest. The ability to generate mutations makes it possible to move from identification of a gene to the phenotype it produces.

4. The genetic analysis of mutations that affect embryogenesis has revealed that *Arabidopsis* is more similar to mammals than to arthropods such as flies. Like mammals and unlike flies, plants have very few maternal-effects mutations and few zygotic mutations before the globular stage. Mutational analyses have also helped identify some of the regulatory processes involved in plant development. Future research will try to pin down the molecular mechanisms that control embryogenesis.

5. The genetic analysis of hormonal control systems has clarified the biological significance of hormones and the biosynthetic pathways that produce them. It has also shown how *Arabidopsis* perceives and transduces hormone signals. Many of the genes involved in hormone production and hormone responses are homologous to those found in yeast and mammals.

6. The genetic analysis of growth and development in response to light conditions, known as photomorphogenesis, has clarified how plants receive and process light signals. Some of the genes that contribute to photomorphogenesis are related to mammalian genes that operate in steroid biosynthetic pathways.

7. From the genetic analysis of flowering, researchers have proposed that four genes of three types determine the identity of floral organs. These four genes operate in the middle of a regulatory cascade in which earlier acting genes specify the identity of the floral meristem, while other genes control the timing of both floral meristem formation and flowering.

SOLVED PROBLEMS

I. The St ecotype of *A. thaliana* switches from vegetative to reproductive (flower) development late in the season compared to ecotype Limburg-5 that flowers early. How could you determine if flowering time is based on alleles of a single or multiple genes? Describe the results you would expect for the two different cases.

Answer

This problem requires an understanding of single and multigene inheritance. The two different ecotypes could be crossed and the hybrid selfed to produce an F_2 generation. *If the phenotype is due to a single gene, you would expect discrete phenotypes in the F_2 generation—* either two or three phenotypes, depending on whether there was complete or incomplete dominance. *If the trait is determined by several genes, there would be a spectrum of phenotypes in the F_2 generation* ranging from early to late flowering.

II. *HY5* is a gene that encodes a positive regulator of gene expression in response to light in *Arabidopsis*.
   a. What do you predict would happen to expression of light regulated genes in a loss of function *hy5* mutant?
   b. To examine the action of the Hy5 protein at the molecular level, the expression of two promoter-GUS reporter gene fusions was monitored. One of the fusions contained a light response element (G box) and the other lacked the light response element. Results are on the following page. (A + indicates an increase in expression.)

Promoter elements	Light-activated GUS expression			
	$HY5^+$		$hy5^-$	
	No light	+ Light	No light	+ Light
G box	–	+	–	–
No G box	–	–	–	–

What would you conclude about the action of Hy5?

c. If you wanted to determine if *HY5* regulates expression in nonphotosynthetic as well as photosynthetic tissues of the plant, how would you carry out the experiment?

Answer

This question requires an understanding of gene regulation.

a. Loss of function of a positive regulator would result in *no expression of genes in response to light.*

b. Both the *G box in the promoter and the* HY5 *gene product are needed to get increased gene expression in response to light.* (It is reasonable to hypothesize that *HY5* interacts directly or indirectly with the G box sequence.)

c. There are several ways in which you could examine the tissue specificity of light-regulated, *HY5*-dependent transcription. In all cases, the wildtype ($HY5^+$) and $HY5^-$ mutant should be compared with and without the light stimulation. *RNA can be isolated from photosynthetic and nonphotosynthetic tissues in the two strains, with and without light and hybridized using a probe from a light-regulated,* HY5-*dependent gene; or DNA from a* HY5-*dependent gene can be hybridized to fixed tissues in the different plants (in situ hybridization) or a reporter gene linked to a promoter of an* HY5-*regulated gene could be used to monitor expression in tissues.*

P R O B L E M S

19-1 For each of the terms in the left column, choose the best matching phrase in the right column.

a.	apical meristem	1.	light-regulated developmental program
b	photomorphogenesis	2.	variety of plant
c.	whorls	3.	growing point of a plant containing undifferentiated cells
d.	inflorescence	4.	concentric portions of a flower
e.	ecotype	5.	group of flowers at the tip of a branch

19-2 Why are the following considered to be advantages for the geneticist or molecular biologist using *Arabidopsis* as a model plant system compared to corn or peas?
a. generation time of six weeks
b. 10,000–40,000 seeds per plant
c. genome size similar to *C. elegans*

19-3 *Arabidopsis* has a small genome size yet carries out the same functions as plants such as tobacco and pea that have genomes 50–100 times larger. How can *Arabidopsis* accomplish similar physiology with so little DNA?

19-4 Different ecotypes have different seed set sizes (numbers of seeds produced).

a. If you wanted to determine whether this trait was due to a single gene or multiple genes, how would you test this?
b. If the trait was due to multiple genes, what characteristic of the different ecotypes would you use to locate genes involved?

19-5 What is the ploidy of each of the following tissues, what nuclei combined to form them, and how are they related genetically to the plant on which they are found?
a. endosperm
b. zygote
c. embryo sac cells

19-6 After germinating mutagenized seed, there may be some mutant phenotypes in that first generation. However, these are not usually the plants that a geneticist will choose to study. Why not?

19-7 Match the type of mutagenic agent with the best use (assume there is only one match for each agent).

a.	promoterless GUS-containing vector	1.	generate a collection of plants with different loss of function mutations
b.	EMS	2.	study expression of genes
c.	kan^r Ti	3.	generate temperature-sensitive mutants

19-8 a. Order the steps below for a reverse genetics approach to explore the function of a gene
   i. PCR amplify using primers from the gene and from the T-DNA sequence
   ii. homozygose plant with insertion
   iii. obtain a library of mutant lines containing T-DNA insertions
   iv. find a potential gene based on DNA sequence information
   b. An alternative to insertion mutagenesis with T-DNA is to construct a knockout mutant by homologous recombination. Describe the general characteristics of the DNA that you would transform into the plant to make a knockout mutation.

19-9 A sterile mutant was identified in which there was a random instead of directed growth of pollen tubes toward the ovary. The mutant was discovered after mutagenesis of seeds and self-fertilization of the progeny plants. It was not known therefore whether the sterility defect was a male or female defect. How could you test that?

19-10 How could you find out if an *APETALA* homolog exists in petunia, snapdragons, or potatoes?

19-11 The gravitropic response in plants includes upward curve of the hypocotyl and flowering stalk and downward growth of the root.
   a. It has been postulated that this response is auxin mediated. What is a genetic experiment to test this hypothesis?
   b. Auxin-regulated mRNAs have been identified in *Arabidopsis*. How would you determine the distribution of these auxin-regulated mRNAs in the part of the plant responding to gravity?

19-12 Ethylene affects plant growth and development in many ways including inhibition of stem and root elongation, stimulation of seed germination, and ripening of fruits. Ethylene biosynthesis occurs in the plant at several different developmental times. Mutations that are defective in the ethylene stimulation might be defective in the biosynthesis of ethylene by the plant or in the signaling pathway in cells that receives and processes the signal.
   a. The $ctr1^-$ mutant has a defect in the signaling pathway resulting in a constitutive ethylene response. If you treat this mutant with an inhibitor of ethylene biosynthesis, what effect do you expect?
   b. Ethylene overproduction mutants $eto1^-$, $eto2^-$, and $eto3^-$ are sensitive to inhibitors of ethylene biosynthesis. Would you conclude that the *eto* genes are biosynthetic or signaling mutants?
   c. No loss of function mutations have been isolated in the *ETR1* gene. This could indicate that the gene is required for growth or that there is redundancy of this gene. How could you look for redundancy of this gene in the genome?
   d. A series of bases in the promoter region of several ethylene responsive genes have been identified, as have a set of proteins, ethylene-responsive element-binding proteins (EREBPs), that bind to that DNA. The EREBPs have a similar region of the protein to that found in *APETALA2*. Why would these genes share a common region?
   e. Genetic engineering of *ETR1* and/or *CTR1* genes may prevent or modify the ethylene response and could be very helpful in regulating ripening of fruits and vegetables. If you need to control expression of the genes, what would you want to be part of the DNA construct that you injected into the plant?

CHAPTER

20

CAENORHABDITIS ELEGANS:
GENETIC PORTRAIT
OF A SIMPLE MULTICELLULAR ANIMAL

An adult *C. elegans* hermaphrodite surrounded by larvae of various stages.

The nematode *Caenorhabditis elegans*, one of the simplest multicellular organisms, lives in soils worldwide and feeds on soil bacteria. Adults are about 1 mm in length and contain an invariant number of somatic cells (Fig. 20.1). The mature "female," which is actually a hermaphrodite able to produce both eggs and sperm, has precisely 959 somatic cells that arose from progenitor cells by a reproducible pattern of cell division. The mature male, which produces sperm and has genitalia that enable it to mate with the hermaphrodite, includes precisely 1031 somatic cells that also arose by a reproducible pattern of cell division. *C. elegans* has a short life cycle and an enormous reproductive capacity, progressing in just three days from the fertilized egg of one generation to between 250 and 1000 fertilized eggs of the next generation. It is transparent at all stages of this development, and as a result, investigators can use the light microscope to track its development at the cellular level throughout its life cycle. Its small

size, precisely reproducible and viewable cellular composition, short life cycle, and capacity for prolific reproduction make *C. elegans* an ideal subject for the genetic analysis of development.

Although *C. elegans* and most other free-living species of nematodes are generally beneficial, they are related to nematodes that parasitize animals and plants, causing human disease and agricultural damage. Knowledge gained from the study of *C. elegans* will help combat these problems.

Three unifying themes surface in our discussion of *C. elegans*. First, the invariance of cell number and fates forms the basis of many of the experimental protocols used to study nematode development. Second, the reproducibility of cellular divisions and fates depends on a varied palette of developmental strategies. These include the distribution of particular molecules to particular daughter cells, inductive signals sent from one cell to influence the development of an adjacent cell, signal transduction pathways

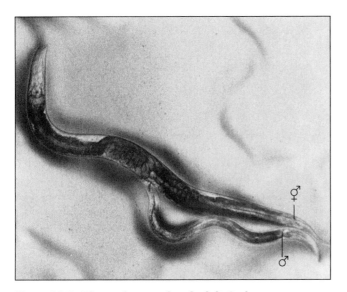

Figure 20.1 Photomicrographs of adult C. *elegans* **hermaphrodite (top) and male (bottom) caught** *in flagrante delicto.* These animals are oriented in opposite directions during their mating: the symbol denoting the male is placed near the head, while that for the hermaphrodite is near the tail. Note that the body is transparent, enabling visualization of internal structures. As an example, the eggs in the hermaphrodite are easily seen.

within each cell that respond to the arrival of an inductive signal, and a genetically determined program that causes the death of specific cells. Third, genetic studies on the development of *C. ele-*

gans reveal the simultaneous conservation and innovation of evolution. Because the nematode exhibits many features of development, physiology, and behavior found in other complex animals such as *Drosophila* and humans, studies of *C. elegans* can uncover developmental pathways and genes conserved throughout animal evolution. But because other features of *C. elegans* development, such as the precise reproducibility of cellular fates, are quite different from those found in other complex animals, studies of *C. elegans* provide a comparative counterpoint that deepens our understanding of the full range of genetic controls over development in multicellular eukaryotes.

Our genetic portrait of *C. elegans* presents

■ An overview of *C. elegans* as an experimental organism, including descriptions of its genome, its life cycle and anatomy, the precise patterns of its cell lineages throughout development, and techniques of genetic and molecular analysis that biologists use to study the development of *C. elegans.*

■ The genetic dissection of several developmental processes in *C. elegans,* including the specification of early embryonic blastomeres, the role of programmed cell death, and the timing of decision-making during larval development.

■ A comprehensive example on the use of genetics to probe a signaling pathway that helps control development of the hermaphrodite vulva.

AN OVERVIEW OF C. ELEGANS AS AN EXPERIMENTAL ORGANISM

As one of the model organisms selected for genomic analysis under the Human Genome Project, *C. elegans* is under intensive investigation.

The Nuclear Genome of C. elegans

The *C. elegans* genome is small for the genome of a multicellular animal, only 100 Mb in length. While this is about 10 times the length of the yeast genome, it is only two-thirds the size of the *Drosophila* genome. *C. elegans'* 100 Mb of DNA is packaged into six small chromosomes called I, II, III, IV, V, and X. These chromosomes are small and of approximately the same size, which makes it difficult to obtain much genetic information from cytogenetic and karyotypic analyses. Figure 20.2 shows simplified genetic maps of all six chromosomes.

Note that the maps lack a centromere, an important feature found on all other eukaryotic chromosomes described in this book. The *C. elegans* maps lack this feature because the nematode's chromosomes do not have a defined centromere. Instead, like the chromosomes of some insects and ciliates, *C. elegans* chromosomes are **holocentric;** that is, they have a dif-

fuse centromere. The distribution of centromeric activity along the length of the chromosome allows many points of spindle attachment during both meiosis and mitosis. This unusual centromere structure does not in general affect genetic analysis because the end result of chromosome segregation and recombination in *C. elegans* is similar to that of chromosomes in organisms with defined centromeres. As we see later, however, the behavior of holocentric chromosomes provides the basis of a technique important in the construction of *C. elegans* genetic mosaics.

The Construction of Physical Maps

Researchers have constructed a detailed physical map of the *C. elegans* genome by assembling overlapping contigs of cosmid and YAC clones covering each of the six chromosomes (see Chapter 10 for a description of contig assembly). The physical map has been useful not only for genomic sequencing, but also for the positional cloning of genes important in development.

Recent work in the mapping and cloning of many genes is providing an increasingly accurate comparison between the genetic and physical maps. Surprisingly, the relationship between recombination frequency and physical distance varies considerably in different parts of the *C. elegans* genome. In central regions of some chromosomes, for example, one recombination-determined map unit (or 1 cM) is the equivalent of 1500 kb of DNA, while in the more distal regions of these same chromosomes, one map unit equals 50 kb of DNA. The reasons for these discrepancies are not clear. One possibility is that the holocentric organization of the chromosome somehow impedes the resolution of recombination events occurring near the center of the centromere. Another possibility is that recombination-promoting sequences are found more frequently near the tips than near the centers of the chromosomes. Whatever the cause for the depression of recombination frequencies near the center of the chromosomes, the genetic map is clearly not an accurate reflection of the physical distribution of *C. elegans* genes. One symptom of this anomaly is that genetic maps based on linkage analysis show a marked clustering of genes near the center of the chromosomes (Fig. 20.2).

Genome Sequence and Organization

As of August 1998, researchers had completed a DNA sequence analysis of approximately 75% of the *C. elegans* genome. Initial data on the density of recognizable genes in the sequenced regions suggest that *C. elegans* has a total of about 14,000 genes. These genes are relatively closely packed for a genome in a multicellular animal, in large part because their introns are smaller on average than those of most other animals.

Two aspects of *C. elegans* gene expression are unique among multicellular animals. First, more than 70% of mRNAs have had one of several splice leader sequences trans-spliced onto the 5′ end of the message (Fig. 20.3). Pre-mRNAs do not contain these splice-leader sequences, but during processing

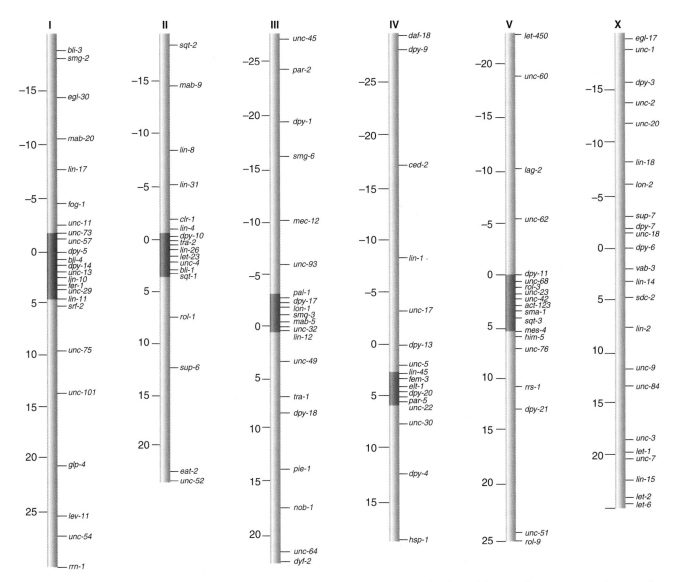

Figure 20.2 Skeleton genetic maps of the six *C. elegans* chromosomes. Only a small subset of the more than 1600 currently mapped genes is shown. Distances on these maps represent recombination frequencies in centimorgans. The zero position on each chromosome was chosen arbitrarily, as these chromosomes do not have defined centromeres. Each of the five autosomes has a central cluster (darker blue) in which the gene density is unusually high, reflecting a lower rate of recombination per kilobase of DNA.

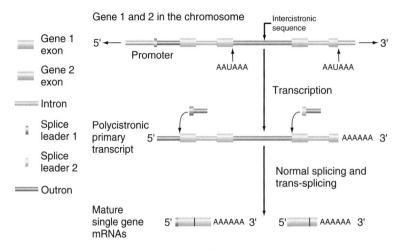

Figure 20.3 Polycistronic transcription and trans-splicing in C. elegans. Two or more genes close together in the genome can be transcribed together from a single promoter to make a multigenic primary transcript. By trans-splicing, splice leaders (encoded elsewhere in the genome) are added to the 5′ ends of what will become the individual mature mRNAs for each gene. Cleavage and poly-A addition signaled by AAUAAA also help divide the multigenic primary transcript into single-gene mRNAs. The sequence found at the 5′ end of the primary transcript that is removed by trans-splicing is technically not an intron (because it is not flanked on both sides by exons); it is sometimes called an *outron*.

in the nucleus, they are trans-spliced to acceptor sites near the 5′ ends of the corresponding primary transcripts. (Recall from Chapter 7 that in trans-splicing, the spliceosome joins parts of the primary transcripts of two genes; review Fig. 7.16b.) Second, some groups of adjacent *C. elegans* genes are transcribed as operons producing a polycistronic primary transcript (see Fig. 20.3). During the subsequent processing of this transcript, trans-splicing with specific splice-leader sequences produces constituent single-gene mRNAs. Before the discovery of these unique aspects of *C. elegans* gene expression, polycistronic transcripts had been reported only in prokaryotic cells and trans-splicing had been reported only in trypanosomes.

In addition to the roughly 14,000 protein-coding genes, there are several kinds of repetitive DNA sequences dispersed throughout the *C. elegans* genome. The best characterized of these repetitive sequences are six kinds of transposable elements, named Tc1 through Tc6. Different strains of *C. elegans* carry different numbers of these elements. In most strains, the elements are stable, but in some strains, they actively transpose, allowing transposon mutagenesis.

Life Cycle, Development, and Anatomy

We saw in the introduction to this chapter that *C. elegans* females are actually hermaphrodites that produce both oocytes and sperm. In the absence of males, they can reproduce by self-fertilization. Mating with males, however, produces cross-fertilized progeny. In the wild, mating promotes mixing of the gene pool; in the laboratory, it makes experimental genetics possible. Since a typical hermaphrodite produces 200–300 sperm and a larger number of oocytes, it can generate about 250 progeny by self-fertilization and over 1000 progeny when fertilized by a male. One external feature that distinguishes the sexes is the tail, which is simple and tapered

in the hermaphrodite but fan-shaped and specialized for mating in the male (see Fig. 20.1).

Gamete Production and Fertilization

The *C. elegans* hermaphrodite manufactures both oocytes and sperm in its bilobed gonad (Fig. 20.4a). The far (distal) end of each lobe contains germ line nuclei, not separated by cell membranes, that divide mitotically in a common core of cytoplasm (forming a syncytium). The nuclei move toward the other (proximal) end of the lobe, which terminates in the uterus. As the nuclei approach the bend between the distal and proximal parts of the lobe, they progress into meiosis, with individual nuclei pinching off as cells to form sperm or oocytes. The oocytes then move through the oviducts toward the uterus.

At the end of each oviduct lies a spermatheca, the storage site for sperm (from either the hermaphrodite or the male). These sperm are unusual in that they are ameboid rather than flagellated like the cells in most animals (Fig. 20.4b). Each spermatheca of a young, unmated adult hermaphrodite contains about 160 sperm. Fertilization occurs as oocytes pass through the spermatheca into the uterus where the fertilized eggs begin to develop. Eggs exit the uterus through the vulva, which forms a characteristic protrusion on the ventral side of the hermaphrodite.

To effect cross-fertilization, the male lies next to the hermaphrodite and slides its fan-shaped tail along the hermaphrodite's body surface until it contacts the vulva. The male then inserts specialized structures at the base of its tail into the hermaphrodite's vulva (Fig. 20.1). Sperm move from the male's tail through the vulva into the hermaphrodite's uterus. From the uterus, the sperm migrate to the spermathecae. There the male sperm somehow gain an advantage over the resident hermaphrodite sperm so that for several hours after mating, the hermaphrodite produces almost exclusively outcross progeny.

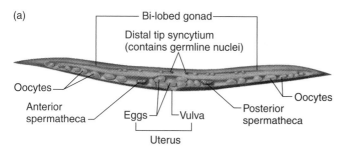

(a)

Bi-lobed gonad
Distal tip syncytium
(contains germline nuclei)
Oocytes
Anterior spermatheca
Eggs — Vulva
Uterus
Posterior spermatheca
Oocytes

(b)

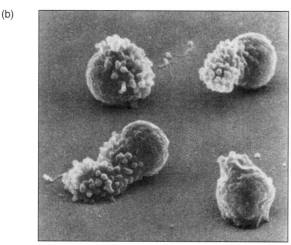

Figure 20.4 Reproduction in *C. elegans*. (a) The hermaphrodite reproductive system. Mitosis at the distal tip of each arm of the gonad produces a syncytium of germline nuclei. The nuclei undergo meiosis, become cellularized and then develop into eggs or sperm as they move around the bend of each U-shaped arm toward the proximal end at the uterus. During the life of a hermaphrodite, the first 40 or so germ cells that enter meiosis in each arm develop into approximately 160 sperm that are deposited into the spermatheca. Thereafter, the sexual fate switches so that all subsequent germ cells differentiate as oocytes. (b) Scanning electron micrograph of *C. elegans* sperm. These ameboid cells do not swim, but instead crawl along surfaces by extending and retracting pseudopods.

Development from Zygote to Adult

At room temperature, *C. elegans* progresses from fertilized egg to the fertilized eggs of the next generation in about three days.

Embryonic development begins in the hermaphrodite's uterus with secretion of a tough chitinous shell around the fertilized egg and continues after the hermaphrodite lays her eggs about 2 hours later. Figure 20.5a shows the progression of embryonic development. Many of the first rounds of cell cleavage are unequal and produce daughter cells of different sizes. Until the 28-cell stage, all cells contact the surface of the embryo. At the 28-cell stage, some of the cells begin to move to the embryo's interior, initiating gastrulation. Continued cell proliferation and cell movements during gastrulation produce the layers of cells that characterize the basic body plan of the larval and adult stages. Later in embryogenesis, cell divisions cease while the various tissues differentiate and become organized. Contractile proteins arranged into filaments around the circumference of the animal literally squeeze the embryo into its final wormlike shape. Roughly 14 hours after fertilization,

the first-stage juvenile, known as the L1 larva, hatches from the egg shell. It is 250 μm long and consists of 558 cells.

Over the next 50 hours, development proceeds through three additional larval stages—L2, L3, and L4—separated from each other by molts (Fig. 20.5b). During each molt, *C. elegans* synthesizes a new cuticle under the old cuticle, and then sheds the old. The structure of these cuticles differs from one stage to the next. In parasitic nematodes, the cuticles of different stages help the animal adapt to existence in a particular plant or animal host.

During larval development, most of the cells present at the hatching of the L1 larva do not undergo further cell divisions, although they may increase in size as the animal grows to adulthood. A few blast cells, however, do divide further during larval development to produce additional neurons, muscles, and structures involved in mating and reproduction. Sperm production begins in both hermaphrodites and males during the L4 stage. After the final L4-to-adult molt, the hermaphrodite germ line switches to oocyte production only. At this same molt in the male, specialized mating structures in the tail reach completion, and the animal becomes competent to mate.

In the laboratory, investigators usually grow *C. elegans* on agar plates covered with a lawn of *Escherichia coli* bacteria, which serve as food. They transfer individual animals from one plate to another with a flattened platinum wire. Under laboratory conditions, adult nematodes continue to reproduce for three to four days. However, they live up to two more weeks. The short life span makes *C. elegans* an excellent model system for the study of aging and its genetic control. One curious aspect of the life cycle adds interest to the question of how life span is determined. If food is scarce during early larval development, the L2 larva can molt to an alternative L3 form known as the **dauer larva** (see Fig. 20.5b). Dauer larvae do not feed and, in addition, have a specialized cuticle that resists dessication. They move around rapidly at first and then become more dormant, although they are still able to respond to stimuli. Dauer larvae that do not experience excessive dehydration can survive for at least six months. Whenever food becomes available within that time, they molt to L4 larvae and resume normal development. Apparently the *C. elegans* aging clock can stop and then restart in response to environmental signals.

Anatomy of the Adult

At 1 mm in length, *C. elegans* adults are just visible to the naked eye. As we saw earlier, hermaphrodites have 959 somatic cells, while males have 1031. The *C. elegans* body plan, typical of nematodes in general, is in essence two concentric tubes (Fig. 20.6a), with the outer tube composed of a cuticle, *hypodermis* (the "skin" of the worm), neurons, and muscles, and the inner tube consisting of an intestine that runs from the anterior *pharynx* to the posterior *anus*. The space between the two tubes is the *pseudocoelom;* in adults, it contains the gonad.

Note that although *C. elegans* is not segmented, its body plan is partially *metameric,* that is, comprised of repeating patterns of cell groups with similar functions. For example, along the ventral nerve cord, a set of 12 precursor cells located along

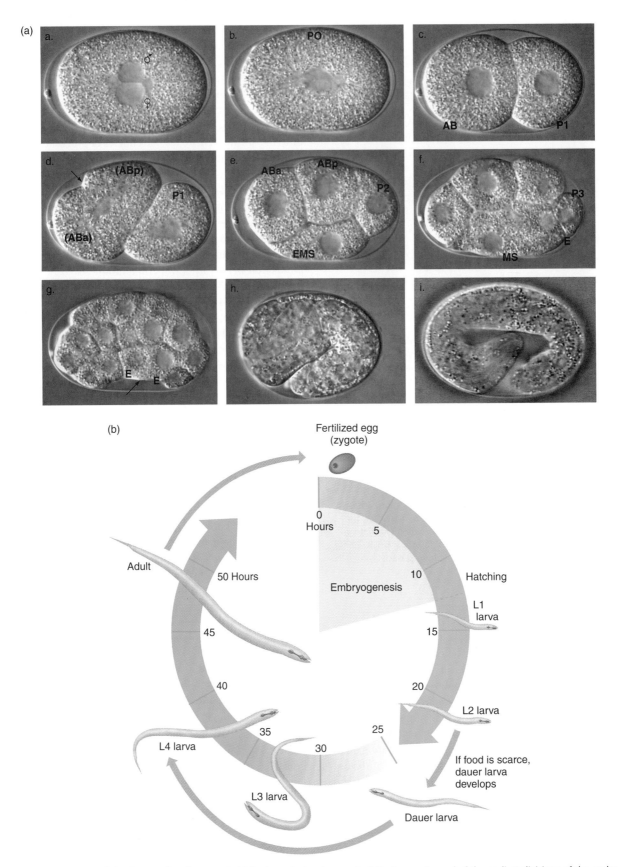

Figure 20.5 Stages of C. *elegans* development. (a) Embryonic development of *C. elegans.* Several of the earliest divisions of the embryo are asymmetric; for example, the first mitotic division produces a larger AB daughter cell and a smaller P_1 daughter cell (panel c). Gastrulation begins at the 28-cell stage as the two cells marked with a small E move into the embryo's interior (panel g). The wormlike shape of the animal becomes apparent later during embryogenesis (panel i). (b) The *C. elegans* life cycle. Embryogenesis, the time period illustrated in part (a), is highlighted in yellow. Under unfavorable environmental conditions, L2 larvae can molt into dormant, long-lasting dauer larvae.

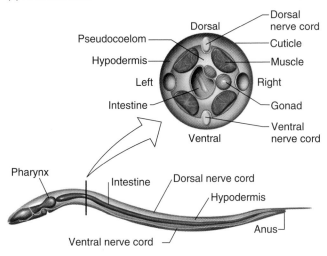

(a) **Cross section**

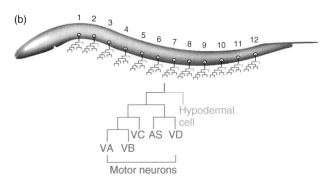

(b)

Hypodermal cell

VC AS VD

VA VB

Motor neurons

Figure 20.6 The body plan of *C. elegans*. (a) The basic features of *C. elegans* adult anatomy. (b) The body plan includes repeating sublineages. In the L1 larva, 12 precursor cells in the ventral nerve cord each initiate a similar series of divisions that will create a group of cells related by ancestry (a sublineage). Within each of these sublineages, one of the cells usually becomes a hypodermal cell (green lines), while the remaining cells become motor neurons (purple lines). The ultimate fate of corresponding cells in the 12 sublineages is not as invariant as suggested by this diagram. For example, the hypodermal cell in certain lineages may undergo additional rounds of cell division (not shown).

the body of the L1 larva undergo similar patterns of division to generate similar groups of cells known as *sublineages* (Fig. 20.6b). The cells in each of these 12 sublineages generally include five types of motor neurons and a hypodermal cell. There are, however, variations in cell fate specific to particular sublineages that can be likened to the differences in segment identity that arise during the embryogenesis of truly segmented animals such as *Drosophila* (described in Chapter 21).

Lineage Diagrams Show the Life History of Every Somatic Cell

Because *C. elegans* is transparent at all stages of its life cycle, it is possible to use light microscopy to observe and analyze

development at the cellular level, following living animals from fertilization to adulthood. A microscope equipped with differential interference contrast (Nomarski) optics provides a shallow depth of field with clear imaging of any plane in the specimen and little disturbance from out-of-focus information. With such a microscope, nematode investigators have observed and recorded the positions, movements, and divisions of every cell (review Fig. 20.5a). In this way, they learned that both embryonic and larval development proceed by an invariant temporal and spatial pattern of cell positions and divisions. From their observations, they have pieced together the entire cell lineage of *C. elegans,* pinpointing the ancestry of every cell in the adult animal all the way back to the fertilized zygote.

Cell-lineage diagrams, such as the one shown in Fig. 20.7, provide an unusual view of the life cycle. The vertical axis represents time, each horizontal line represents a cell division, and the end of each branch represents a cell in the adult. Because the history and fate of each of these cells is precisely determined, every somatic cell present at every stage of development has its own unique name. It should be cautioned that these lineage diagrams do not include the germline divisions that occur during larval development and into adulthood. This is because the number of germline divisions in either sex is not preordained, so different animals produce slightly different numbers of gametes.

The lineage diagram in Fig. 20.7a shows the complete history of all of the 959 somatic cells in the hermaphrodite. To study particular developmental processes in more detail, *C. elegans* researchers usually focus on a smaller time frame or region, as in the lineage diagram in Fig. 20.7b. This higher resolution view depicts the earliest stages of development, recording only the time elapsed between fertilization and the 28-cell stage when gastrulation begins (corresponding to the stages seen earlier in panels a–g of Fig. 20.5a). At the end of this time, the embryo consists of 28 **blastomeres** (early embryonic cells).

The first 100 minutes of development in *C. elegans* give rise to six **founder cells:** progenitors of the major embryonic lineages (Fig. 20.7b). Although progeny of the E (gut), D (muscle), and P_4 (germ line) founder cells populate only single tissues, the other three founder cells—AB, MS, and C— contribute to more than one of the three embryonic germ layers (ectoderm, mesoderm, and endoderm) and thus to more than one tissue. Conversely, several tissues include cells from more than one founder cell lineage. For example, in different parts of the body, the body-wall muscle derives from progeny of the MS, C, D, and AB founder cells.

Embryonic cell divisions are asynchronous (see Fig. 20.7b). For example, though EMS and P_2 are both the daughters of P_1, EMS divides a few minutes before P_2. As we see later, the timing of cell division in *C. elegans* is under genetic control, and considerable progress has been made in identifying the genes responsible for this kind of developmental regulation.

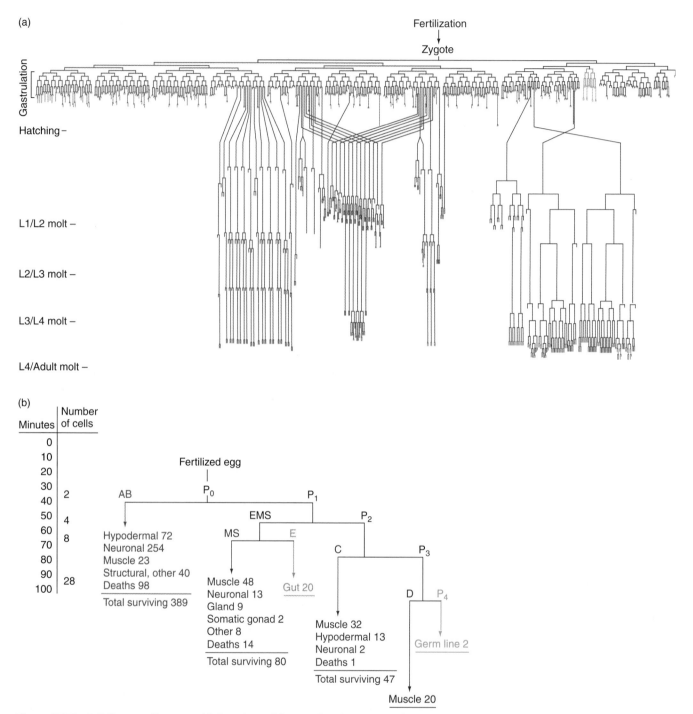

Figure 20.7 Cell lineage diagrams. (a) Complete cell lineage chart for *C. elegans* hermaphrodites. Progression from the top to the bottom of the chart corresponds to increasingly later times in development. Colors highlight the lineages established by the six founder cells. (b) Lineage pattern of most of the early cleavages in the *C. elegans* embryo, illustrating the origins of the six founder cells. (A few cleavages in the AB and E lineages providing some of the blastomeres at the 28-cell stage are not shown for simplicity.) Cells labeled in black—P_0 (the original zygote), P_1, P_2, P_3, and EMS—divide to give rise to one or two of the six founder cells. Below each founder cell is a list of the kinds and numbers of cells in its lineage at the time of hatching, color-coded as in part (a).

Laser Ablation: An Experimental Tool for Looking at Development and Cell Lineage

Researchers can deploy a laser beam to kill, or *ablate,* specific cells in *C. elegans,* and then determine the function of the ablated cells in development or behavior. Used in conjunction with an understanding of the genetics and molecular biology of the animal, and based on the invariance of most cell lineages, this laser technology is a powerful tool. It works as follows. A beam of laser light is fired into a microscope. The beam travels the same light path as the image from the microscope stage (there is no danger to the viewer) and is focused on the same point and focal plane that the viewer sees. The laser beam can fatally damage a single cell without hurting the cells above or below it. Depending on the experiment, investigators can kill as many as 20 or 30 cells in a single animal. Much of the knowledge of development and neurobiology presented in this chapter depends on laser ablation technology.

The central finding of cell ablation experiments in which researchers destroy a single cell is that every cell in the early embryo is essential for normal development, and the organism cannot replace destroyed cells. Worms thus provide an example of what is called **mosaic development:** The embryo is a collection of self-differentiating cells that at their formation receive a specific set of molecular instructions governing their unique fates.

The results of cell ablation experiments in *C. elegans* are very different from the results obtained in other animals, such as humans. For example, in Chapter 9, where we describe how to determine the genotype of an early human embryo by removing and examining a blastomere from the 6- to 10-cell stage, we mention that the remaining 5–9 blastomeres are still able to develop into an adult. Identical twins—formed when early blastomeres separate from each other—are another case in point. Early embryogenesis in humans depends on **regulative development,** in which blastomeres can regulate and adjust their development to make up for missing cells.

Laser ablation experiments in *C. elegans* illustrate another point about the development of this nematode. When laser irradiation destroys a cell, the effects on development are not always restricted only to the lineage of that cell. If the destroyed cell provides inductive signals needed for the proper development of adjacent cells, the treatment can also alter the fates of these other cells. The comprehensive example at the end of this chapter describes this type of cell-to-cell signaling in some detail.

The Use of Genetic Analysis and Recombinant DNA Technology to Study Development

C. elegans is a diploid organism, except for the sex chromosome in males. Males have only a single X and are described as XO; hermaphrodites are XX. Males arise spontaneously as the result of a low frequency of X-chromosome nondisjunction during hermaphrodite gametogenesis, which produces 0.1% to 0.2% XO embryos. Once males are present in a population, mating generates more of them in the following manner. When male sperm fertilize a hermaphrodite, half the progeny result from sperm that carry no X chromosome; these progeny are, of course, males.

Mutant Isolation

Chemical mutagens such as ethylmethane sulfonate (EMS) induce a high forward mutation rate in *C. elegans.* Because a mated hermaphrodite can produce self-progeny as well as outcross progeny, hermaphrodite genetics takes some getting used to, but it has several advantages. It is easy to recover recessive alleles following mutagenesis because a hermaphrodite heterozygous for a new mutation will produce 1/4 homozygous mutant progeny by self-fertilization. A further convenience is that it is possible to recover and propagate mutations causing very severe defects, for example, in the animal's movement, because the hermaphrodites do not have to mate to reproduce. Any hermaphrodite that can feed well enough to survive and has a functional reproductive system can produce progeny.

By mutagenesis, screening, and genetic analysis, geneticists have identified and placed on the genetic map well over 1000 genes. Easily recognized mutant phenotypes include animals with abnormal body shape, such as Dumpy (Dpy) or Long (Lon), and animals that move or respond to touch abnormally, designated as Unc for "Uncoordinated." By convention, *C. elegans* phenotypes are capitalized and not italicized, while genes and alleles are in lowercase, italic type. An example is *dpy-10(e128),* which designates the *e128* mutant allele of the *dpy-10* gene; this allele determines the Dpy phenotype.

Genetic Mosaic Analysis

Nematodes composed of wildtype and mutant cells are genetic mosaics. The analysis of such mosaics is one way to determine the **focus of action** of a gene of interest, that is, the cells in which the gene must be active to allow the animal to develop and function normally. It is possible to use chromosomal fragments generated by X or γ irradiation to construct genetic mosaics in *C. elegans.* Because *C. elegans* chromosomes are holocentric, even small chromosomal fragments have sufficient centromeric activity to segregate fairly normally during meiosis and mitosis. These fragments can thus be maintained as extrachromosomal, **free duplications** in a genetic stock. Even so, cells lose them at relatively high frequencies, ranging from one loss per 200 cell divisions for small fragments to much lower frequencies for larger fragments.

To carry out a mosaic analysis, a researcher constructs a *C. elegans* strain that is chromosomally homozygous for a recessive mutation in a gene of interest, and introduces into that strain a free duplication including the wildtype allele of the same gene. If the free duplication is lost during mitosis at some point during somatic development, a clone of mutant cells will arise (Fig. 20.8a). The inclusion of appropriate marker mutations in the experimental strain makes it possible to identify in adult mosaic animals those cells that have lost the free duplication. And since the entire cell lineage is known, it is usually possible to establish the precise division at which the fragment

(a)

Lin-12⁺ Mrk⁺ zygote carrying a free duplication

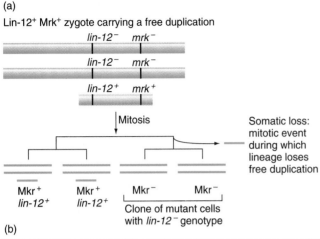

(b)

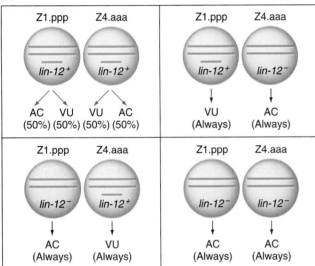

Figure 20.8 Mosaic analysis in *C. elegans.* (a) The experiment is set up so that mitotic loss of a free duplication carrying a wildtype allele of a gene of interest (*lin-12⁺* in this example) creates a clone of cells without any source of function for that gene. By examining many mosaic animals, researchers can determine which cells must have the wildtype allele for the animal to express a normal phenotype. The presence or absence of a marker gene (*mrk⁺*) phenotype expressed in each cell (such as the shape of the nucleus) indicates whether any particular cell is in the clone. (b) Results of mosaic analysis showing that a Z1.ppp or Z4.aaa cell must have *lin-12⁺* activity to assume a VU fate. The 50/50 distribution of fates when both cells are *lin-12⁺* occurs as a consequence of cross talk between these cells that randomly turns up expression of the gene in one of the cells (which will become VU) and down in the other (the presumptive AC).

was lost and thereby ascertain the genotype of every cell in the animal. Phenotypic analysis of a large number of mosaic animals can then reveal which cells must express the gene for an animal to develop and function normally.

Figure 20.8b illustrates a genetic mosaic experiment. In the developing somatic gonad, the cells Z1.ppp and Z4.aaa make cell-fate decisions requiring the function of the *lin-12* gene. In half of all wildtype animals, Z1.ppp becomes the anchor cell (AC) and Z4.aaa becomes the ventral uterine precursor cell (VU). The other half of the time, Z1.ppp becomes the

VU and Z4.aaa becomes the AC. In *lin-12* loss-of-function (null) mutants, both cell types become ACs. In *lin-12* gain-of-function (constitutively activated) mutants, both cell types become VUs. Using mosaic analysis of *lin-12,* researchers showed that Z1.ppp and Z4.aaa cells must express the *lin-12* gene product to become a VU. If one of these two cells lacked *lin-12* function, it always became an AC. If neither cell could manufacture Lin-12 protein, then both became AC. This is consistent with the anticipated role of the *lin-12* gene product, which is similar in amino-acid sequence to the membrane-bound Notch growth-factor receptor in *Drosophila.* The Lin-12 protein presumably receives the signal that induces a Z1.ppp or Z4.aaa cell to become a VU.

Gene Cloning and DNA Transformation

The ability to reintroduce cloned DNA sequences into the germ line facilitates the cloning of genes in *C. elegans* based on the position of these genes in the genome. One can clone genes identified through mutational studies by the following steps: (1) precise genetic mapping, (2) extrapolation to an approximate physical map position, and (3) phenotypic rescue by DNA transformation with one of an appropriate set of overlapping cosmid clones provided by the Worm Genome Project. Other cloning methods include the use of transposons, either for direct transposon tagging or as RFLP markers for pinpointing the location of a mutationally identified gene on the physical map.

To reintroduce cloned DNA sequences in *C. elegans,* experimenters microinject DNA through the cuticle into the syncytial distal gonad of the hermaphrodite (Fig. 20.9a). The injected DNA sequences recombine with each other to form large extrachromosomal arrays that are incorporated into oocytes and retained in most cells during embryogenesis. It is possible to demonstrate gene expression from such DNA by the phenotypic rescue of a mutant phenotype, if the corresponding wildtype gene is present in the injected DNA (as in positional cloning). It is also possible to demonstrate expression through the detection of a suitable reporter gene in the DNA, such as the *E. coli lacZ* gene for β-galactosidase, whose expression becomes apparent through X-gal staining of fixed preparations (Fig. 20.9b). Researchers can use reporter genes to study the regulation of gene expression in *C. elegans* as they do in other organisms (see Chapter 16).

If an embryo from an injected hermaphrodite retains the array of transforming DNA in its germ line, it can become the basis of a transmitting line in which the DNA passes from one generation to the next (although some animals in each generation will lose it). The irradiation of transmitting lines with X or γ rays often induces integration of the injected DNA at random into one of the chromosomes. Such integrated DNA is inherited in a completely stable fashion like any other chromosomal marker. However, integration of an injected transgene by homologous recombination into the precise region of the genome where the identical endogene resides occurs only at extremely low frequencies. (An *endogene* is a copy of a gene in its normal place in the genome.)

(a)

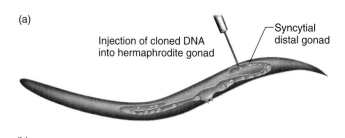

Injection of cloned DNA into hermaphrodite gonad — Syncytial distal gonad

(b)

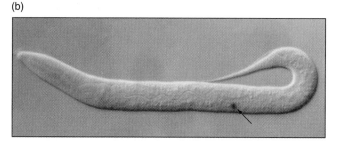

Figure 20.9 Transformation in C. elegans. (a) Researchers inject DNA into the distal syncytial gonad of hermaphrodites, where it can be taken up by any of the germ cell nuclei in the common syncytial cytoplasm. Although the transgenes originally form large extrachromosomal arrays that can be lost during cell division, irradiation promotes the integration of transgenes into random positions in the genome, stabilizing their transmission. (b) Reporter constructs show the expression of transgenes. Here, the *lin-3* gene (including its regulatory region and part of its coding sequence) was fused in frame to β-galactosidase-encoding sequences. Expression of this reporter gene in transformed animals is specific to the anchor cell.

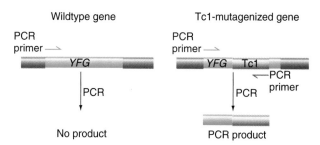

Figure 20.10 Identification of mutations caused by insertion of a transposable element. Researchers first construct many independent lines from "high-hopper" strains in which the Tc1 transposon is highly mobile. They isolate genomic DNA from a small sample of each line, and freeze the remaining worms (*C. elegans* can survive indefinitely in the frozen state). Using one primer specific to Tc1 and a second primer specific to their gene of interest (*YFG* for "your favorite gene"), a PCR product will be formed only if Tc1 has inserted into the gene. The worms in lines identified in this way can be thawed and grown for study.

Defining the Biological Function of a Gene Identified only by Its DNA Sequence

Many *C. elegans* genes of potential interest in development have been identified only by molecular methods. For example, they might be a DNA sequence from the genome project known to encode a protein homologous to a factor needed for development in *Drosophila*. To determine whether a gene identified in this way is essential for normal development or physiological function in *C. elegans*, it is necessary to reverse the normal genetic approach. Instead of starting with a mutation causing a specific phenotype and then cloning the corresponding gene, you start with the cloned gene. The challenge is to obtain mutations of the gene in the animal and then to analyze the phenotypic consequences of the mutations. As researchers complete the sequencing of the genome and find more and more potentially interesting *C. elegans* genes, a convenient method for isolating mutations in these genes becomes a priority. Though yeast and mouse geneticists have procedures for obtaining "knockout" mutations through the homologous recombination of suitably modified, injected cloned genes with their *in vivo* chromosomal counterparts, a reliable method for targeted homologous recombination in *C. elegans* has not yet been devised. It is possible, however, to obtain mutations in a cloned gene of known sequence, using a variation of the transposon-tagging approach. Figure 20.10 shows that the basis of this technique is a search for novel PCR amplification products from the genomes of animals with transposable element insertions into the gene of interest.

In the late 1990s, unexpected observations that copies of RNA transcripts interfere with gene function made it easier to define the biological function of many genes in the nematode. As it turns out, double-stranded RNA copies of an mRNA often interfere with the function of the gene transcribed into that RNA. When nematode researchers synthesize such a double-stranded RNA (dsRNA) in the test tube and then inject it into animals, they often see phenotypic effects that resemble those associated with a null mutation in the gene. The reason for this **RNA-mediated interference** is unknown, but the strategy of RNA injections provides a rapid means of assessing the possible phenotypic effects of disrupting the expression of particular genes, particularly those that influence embryonic development.

The availability of procedures for defining the biological function of genes known only by their DNA sequence substantially increases the usefulness of *C. elegans* as an experimental system. For example, it makes it possible to mutate *C. elegans* genes that are homologous to human genes involved in genetic diseases and analyze the nematode genes genetically and phenotypically. It may be possible to develop an animal model in *C. elegans* for some human genetic diseases far more quickly and at much less expense than in a mammalian species such as the mouse.

THE GENETIC DISSECTION OF DEVELOPMENTAL PROCESSES IN C. ELEGANS

Through a variety of clever mutant screens, nematode geneticists have identified genes that control a range of developmental and physiological processes in *C. elegans*. To understand how the protein products of these genes work together at the molecular level, the scientists carry out two types of studies: They perform more refined genetic analyses (through, for example, epistasis tests and the construction of genetic mosaics); and they clone the genes to obtain information about the biochemical reactions in which the gene products participate.

Studies utilizing these experimental strategies have furthered our knowledge of the principles governing *C. elegans* development. In addition to confirming that many developmental processes in *C. elegans* have been conserved in more complex animals, the studies have revealed that genes controlling aspects of development that seem to be unique in the nematode often have counterparts in other animals. This finding demonstrates that closely related proteins can play substantially different roles in different organisms.

We now describe genetic studies of several aspects of development in *C. elegans.* These discussions illustrate the power of genetic analysis in the dissection of development.

Specification of Early Embryonic Blastomeres

The lineage diagrams in Fig. 20.7 show that successive cleavages of embryonic cells produce blastomeres that give rise to different sets of tissues and structures during embryogenesis. Genetic screens for mutations that cause abnormal early cleavage patterns have identified several maternal genes that control early cell fates. Analyses of the functions of these genes suggest that the determination of blastomere fates depends on two factors: the segregation of cell-autonomous determinants received from the mother, which act independently in single cells; and the transmission of inductive signals between cells.

Early Cleavages Depend on Maternally Supplied Components

As in most animals, the transcription of genes in the embryonic nuclei of *C. elegans* begins only after several rounds of cell division have occurred. Thus, the fate of the earliest blastomeres must be guided by mRNAs or proteins present in the egg before fertilization. These molecules—the *maternal components* that the mother supplied to the egg during oogenesis—are the products of *maternal-effect genes.* Mutations in maternal-effect genes (known as **maternal-effect mutations**) influence the early events of embryogenesis in *C. elegans.*

A genetic screen devised to identify the maternal-effect genes transcribed during oogenesis is based on the fact that fertilization and early embryonic development occur in the hermaphrodite's uterus. If, for example, a hermaphrodite is unable to lay her fertilized eggs because of mutations that block development of the vulva, the eggs will grow into larvae within the uterus. These larvae will eventually eat the mother up from the inside and cause her to burst open. However, if the mother is homozygous for a mutation in a gene encoding a maternal component needed for early development (a maternal-effect lethal mutation), the embryos in her uterus will be unable to grow into larvae, and she will survive. Investigators can then look in the microscope at the improperly developing embryos to determine what aspect of their development the mutation has disrupted (Fig. 20.11).

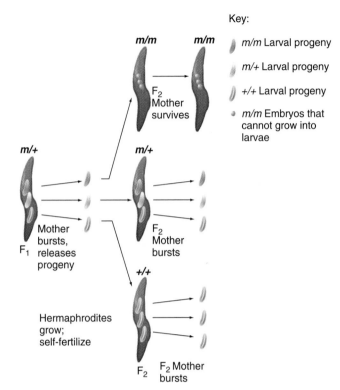

Figure 20.11 Identification of maternal effect lethal mutations in *C. elegans.* Mutagenized parental animals (not shown) produce F$_1$ hermaphrodites heterozygous for a maternal-effect lethal mutation (*m*). One-quarter of the self-fertilized progeny of such an *m/+* F$_1$ mother will be *m/m* F$_2$ hermaphrodites. The embryos made within these *m/m* mothers cannot hatch into larvae because they lack a maternally supplied component needed for their development. All hermaphrodites shown have no vulva, and will be eaten from the inside if their self-fertilized progeny hatch into larvae. In a screen for maternal-effect lethal mutations, researchers look for cultures derived from individual F$_1$ hermaphrodites in which some F$_2$ hermaphrodites (those that are *m/m*) survive because their progeny arrest development before hatching. The mutation can be maintained in *m/+* heterozygotes from the same population.

Cell-Autonomous Determinants Must be Distributed to the Cells in Which They Act

The unfertilized *C. elegans* oocyte has no predetermined anterior or posterior end. Instead, the polarity of the embryo depends on the site of sperm entry, which becomes the posterior end. The sperm entry point somehow directs flowing movements of the cytoplasm within the fertilized egg, and these cytoplasmic movements cause some molecules and molecular complexes to localize in an asymmetric pattern. An example of this phenomenon are the so-called P granules: ribonucleoprotein particles thought to be necessary for the development of the germline. Before fertilization, P granules are uniformly distributed in the egg. After sperm entry, the granules become localized to the posterior part of the one-cell zygote (Fig. 20.12a). After the first cleavage, they are found in the posterior P$_1$ daughter cell (from which the germline descends), but not in the anterior AB daughter cell. Another example of the polarity of the first division is the fact that the P$_1$ and AB cells are not the same size (see panel c in Fig. 20.5). How does an

(a)

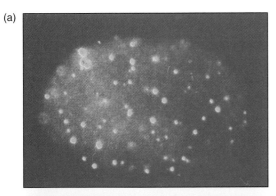

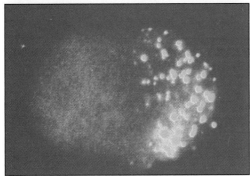

(b)

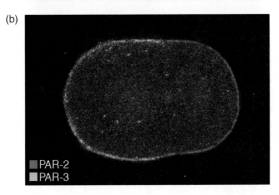

PAR-2
PAR-3

Figure 20.12 The PAR proteins help direct embryonic polarity. (a) In wildtype, P granules (green) are initially uniformly distributed (top), but then become localized to the posterior part of the zygote (bottom). The P granules are later sequestered to only one of the two daughter cells produced by the first mitotic division. In embryos of mothers homozygous for certain *par* mutations, P granules never accumulate at the zygote's posterior, and are thus distributed to both daughter cells (not shown). (b) Shortly after fertilization, PAR-2 localizes to the posterior cortex and PAR-3 to the anterior cortex.

embryo that starts its life without polarity develop an anterior and posterior end by the first cleavage?

To answer this question, researchers looked through collections of hermaphrodites for maternal-effect lethal mutants that produce 2-cell embryos with no evidence of polarity. The investigators identified mutations in six *par* (for "partitioning") genes. Mothers bearing these mutations produce embryos that had P granules in both the P_1 and AB cells and/or P_1 and AB cells that are the same size. Because this lack of polarity disrupts the patterning of the embryo, the embryos from mutant mothers eventually arrest their growth as amorphous masses of partially differentiated cells.

How the proteins encoded by the *par* genes respond to the cue given by entry of the sperm is not yet known, but molecular analysis of these genes is providing clues. Many of the PAR proteins become localized to the embryonic cortex (the part of the cytoplasm just under the cell membrane) at one or the other end of the embryo. For example, PAR-2 accumulates in the cortex at the posterior end, while PAR-3 is restricted to the cortex at the anterior end (Fig. 20.12b). Localization of PAR proteins occurs after fertilization but before the P granules move to the posterior part of the zygote, as would be expected if the PAR proteins regulate the positioning of the P granules. Further experiments show that interactions between PAR proteins are required for the establishment of polarity. For example, in eggs produced by a *par-2* mutant mother, the PAR-3 protein is distributed uniformly around the cortex; the same is true of the PAR-2 protein in eggs from a *par-3* mutant hermaphrodite. Thus, the PAR-2 protein excludes PAR-3 from its half of the embryo's cortex, and vice versa. Current research is focusing on the possibility that the PAR proteins influence microfilaments in the cortex of the one-cell embryo that are responsible for cytoplasmic flows.

Interestingly, even animals such as *Drosophila* and humans, whose early embryonic development contrasts sharply with that of *C. elegans,* have genes that are closely related to and probably homologous with many of the *par* genes. The *Drosophila* and human PAR-like proteins no doubt play some role in development, but not exactly the same role as that performed by the products of the *C. elegans par* genes.

Inductive Signals Control Cell Fates in Early Embryogenesis

Many aspects of *C. elegans* development depend on the asymmetric partitioning of determinants such as P granules during cell division, but other facets of nematode development involve communication between cells. Because cleavages are invariant and most cells do not move within the early embryo, embryonic cells remain in their proper places. As a result, the pattern of cell-to-cell interactions is reproducible from animal to animal.

A simple example of intercellular signaling is the different fates of the ABa and ABp daughter cells produced by the cleavage of the AB founder cell. The two daughter cells are developmentally equivalent when they arise at the second cleavage. Subsequent interactions of ABp with the P_2 cell makes them different. Normally P_2 contacts ABp but not ABa (Fig. 20.13a). If P_2 is removed by laser ablation, ABp adopts an ABa pattern of differentiation. Alternatively, when the embryo is manipulated so that ABa contacts P_2, the ABa lineage develops in an ABp-like fashion. These observations suggest that the P_2 cell provides an inductive signal receivable by cells that touch P_2.

To identify molecules involved in this cell-to-cell signaling, researchers examined collections of maternal-effect lethal mutations for those that disrupt the fate of the ABp lineage. They found that the receptor for the inductive signal is encoded by the *glp-1* gene. Like *lin-12, glp-1* is

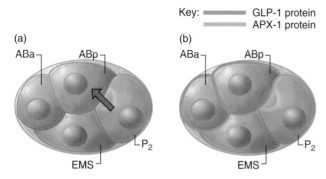

Key: ▬▬▬ GLP-1 protein
▬▬▬ APX-1 protein

Figure 20.13 Intercellular signaling determines the fates of ABa and ABp cells. (a) P_2 contacts ABp but not ABa. This contact enables P_2 to send a signal to ABp that is required for the proper fate of the ABp lineage. (b) The molecular basis of this intercellular signal. The P_2 cell expresses APX-1 protein on its surface, while ABa and ABp both have GLP-1 protein (the receptor for the APX-1 signal) on their surfaces. Binding of APX-1 to GLP-1 in the membrane junction between P_2 and ABp initiates the proper developmental program within ABp. Any condition that blocks the signal or its receipt (laser ablation of P_2, or mutations in the *apx-1* or *glp-1* genes) leads to a default program in which the ABp lineage develops like that of ABa. It is not clear why APX-1 localizes only to the surface of P_2 in contact with ABp; perhaps it is stabilized there by interaction with GLP-1.

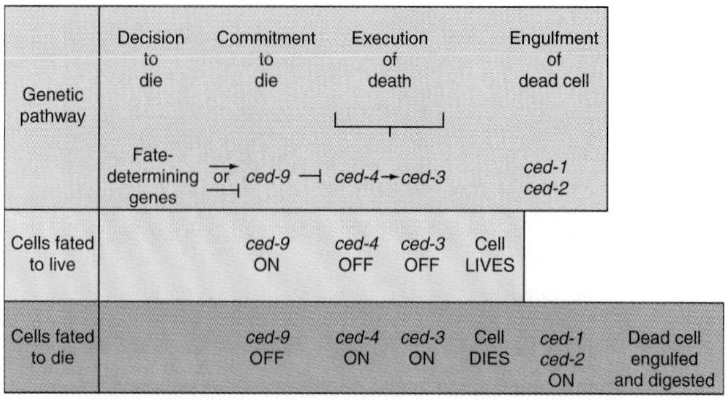

	Decision to die	Commitment to die	Execution of death		Engulfment of dead cell	
Genetic pathway	Fate-determining genes → or ─┤ *ced-9* ─┤		*ced-4*→*ced-3*		*ced-1* *ced-2*	
Cells fated to live	*ced-9* ON	*ced-4* OFF	*ced-3* OFF	Cell LIVES		
Cells fated to die	*ced-9* OFF	*ced-4* ON	*ced-3* ON	Cell DIES	*ced-1* *ced-2* ON	Dead cell engulfed and digested

Figure 20.14 Control of programmed cell death in C. elegans. The blue-shaded area describes four steps in the cell death program and some of the genes involved in each of these steps. In the genetic pathway, → indicates positive regulation (the gene to the left activates the gene to the right), while ─| indicates negative regulation (the gene on the left represses the gene to the right). It is not yet known whether the *ced-1* and *ced-2* genes act in the dying cell or in the cell engulfing it; all other genes shown must act within the cell whose fate is to be determined. Green- and red-shaded areas detail gene activity and subsequent events within cells that adopt alternative fates.

closely related to the *Drosophila Notch* gene, which encodes a growth factor receptor located in the membrane of cells that receive certain developmental signals. *glp-1* mRNA is present in all cells at the four-cell stage, but control of translation allows its expression only in AB cells and their descendants. The signaling molecule for ABp determination is the APX-1 protein. The protein, encoded by the *apx-1* gene, is made in P_2 cells where it localizes to the membrane junction between P_2 and ABp. There it is in a position to contact the GLP-1 protein (Fig. 20.13b). *apx-1* is a homolog of the *Drosophila Delta* gene, which encodes a molecule that binds to the Notch receptor. Thus, very similar signal molecules and receptors for these signal molecules foster cell-to-cell communication in both *C. elegans* and *Drosophila.*

Programmed Cell Death

About one-sixth of the cells that arise during *C. elegans* development are eliminated by programmed cell death (**apoptosis).** This cell death is a form of cellular suicide: During apoptosis, the affected cell shrinks and dies, after which it is engulfed and digested by a neighboring cell. In *C. elegans,* apoptosis, like most other aspects of development, is highly reproducible. The identity of the cells that die and the time at which they die is the same from animal to animal. The purpose of programmed cell death seems to be the elimination of unneeded cells that could be detrimental to the organism.

Mutations that cause the failure of all programmed cell deaths during development have helped identify the *ced* (for cell death) genes that control apoptosis. For example, loss-of-function mutations in the *ced-3* and *ced-4* genes and gain-of-function mutations in the *ced-9* gene cause the failure of programmed cell death. However, although animals bearing

such mutations have 12% more somatic cells than normal, they show only very subtle phenotypic defects in their growth rate and fertility. By contrast, loss-of-function *ced-9* mutations cause the death of many cells that would normally have lived, and this extensive cell death is lethal to the embryo.

Epistasis tests and phenotypic analysis of *ced* mutations led researchers to formulate the developmental pathway shown in Fig. 20.14. In that pathway, cells make a decision to die based on the action of upstream genes that control cell fates and function differently in different types of cells. These upstream genes activate or repress *ced-9*. Since most, if not all, cells require the function of the *ced-9* gene to prevent activation of the cell-death process, repression of *ced-9* allows activation of the events leading to apoptosis. Next, the *ced-3* and *ced-4* genes function (in an unknown manner) to carry out the cell-death program. One clue to the role of the products of these genes is that the CED-3 protein has proteolytic activity, although it is not yet clear what polypeptides CED-3 cleaves. Other genes, such as *ced-1* and *ced-2,* act in the last stages of the cell-death process. These genes do not directly contribute to cell death itself. Instead, they are required for the engulfment and digestion of dead cells by neighboring cells. Mutations in *ced-1* and *ced-2* cause a "persistent corpse" phenotype in which cells die as usual but remain in the worm for several days.

Programmed cell death in *C. elegans* appears very similar to apoptosis in mammals. In both types of organisms, the process helps control the size of cell populations by eliminating excess cells. As we saw in Chapter 17, the investigation of genes that control apoptosis is an important aspect of current cancer research. Although tumor formation often involves the release of normal controls on cell proliferation, it can also result from the failure of apoptosis. Sequence similarity between the membrane-associated protein product of the *C. elegans* *ced-9* gene and the protein product of the *bcl-2* oncogene, which helps cause tumors by blocking apoptosis in humans, underscores the close relation between the control of programmed cell death in *C. elegans* and mammals. In fact, although nematodes and humans are separated by almost a billion years of evolution, their apoptosis-controlling genes are still functionally interchangeable: The human *bcl-2* oncogene can restore the normal process of cell death when injected into a *C. elegans* *ced-9* mutant.

Control of Timing during Larval Development

Throughout development, control over when a cellular event takes place is as important as control over where it occurs. Researchers have used **heterochronic mutations,** that is, mutations resulting in the inappropriate timing of cell divisions and cell-fate decisions during larval development, to identify genes involved in temporal controls in *C. elegans.*

The genes identified by heterochronic mutations often behave as **switch genes** in which loss-of-function and gain-of-function mutations have opposite temporal effects. For example, loss-of-function mutations in the *lin-14* gene cause precocious development: Cell lineages that normally emerge late in development appear too early (Fig. 20.15a). Conversely, gain-of-function mutation in the *lin-14* gene causes retarded development in which lineage changes normally appearing only in early larval stages continue to appear at later larval stages (Fig. 20.15a). Mutations in *lin-14* not only change the timing of certain developmental programs, but can also alter the cell cycle. Cells that normally divide at a specific time in larval development will divide earlier than normal in *lin-14* null mutants and later than normal in *lin-14* gain-of-function mutants (Fig. 20.15b). In addition, heterochronic mutations can affect the formation of dauer larvae in starved animals. Instead of occurring normally at the L2 molt, the dauer larvae can form at the L1 or L3 molts. Investigators have identified mutations in several other genes that produce effects similar to those associated with *lin-14.* The interaction of all these genes thus sets the timing mechanisms that specify many aspects of larval development.

The pathway defined by the heterochronic genes appears to act ultimately through regulation of the *lin-14* gene. Epistasis tests have shown that one of the pathway's temporal control genes, *lin-4,* negatively regulates expression of *lin-14,* and as it turns out, the regulatory mechanism is a novel one. The LIN-14 protein, which shows no similarity to other known proteins, behaves as a central clock component whose concentration descends from a high in L1 animals to a low in L4 animals. During larval development, cells somehow use the relative abundance of LIN-14 to "tell time." The level of *lin-14* transcript, however, remains constant during development, which suggests that control of LIN-14 protein concentration is at the level of translation or protein stability. The *lin-4* gene encodes a small, untranslated mRNA whose sequence is complementary to sequences in the 3′ untranslated region of the

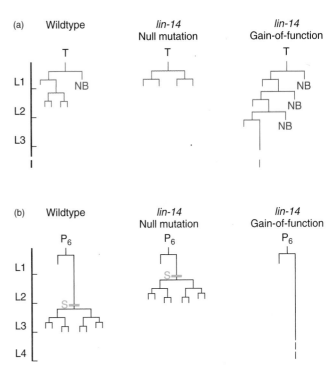

Figure 20.15 Effects of mutations in the heterochronic gene *lin-14.* (a) Transformations of cell fates in the T lineage. In wildtype, the T cell undergoes a pattern of division specific for the L1 larval stage (red; NB signifies a neuroblast), and one of the granddaughters of T undergoes an L2-specific cleavage pattern (blue). In *lin-14* null mutants, the T cell precociously initiates the L2 pattern in the L1 stage. In *lin-14* gain-of-function mutants, the lineage inappropriately reiterates the L1 pattern at later larval stages. (b) Effects on the timing of cell divisions in the P_6 lineage. Loss of *lin-14* function causes one of the daughters of P_6 to start cell division too early (S indicates the DNA synthesis phase of this division). Gain of *lin-14* function can delay this division indefinitely.

lin-14 mRNA. One proposal is that the *lin-4* mRNA regulates LIN-14 protein synthesis by forming a translationally inactive complex with the *lin-14* mRNA. This, together with a constant rate of LIN-14 protein degradation, would result in decreasing levels of LIN-14 during development.

USING GENETICS TO PROBE THE DEVELOPMENT OF THE HERMAPHRODITE VULVA: A COMPREHENSIVE EXAMPLE

A major contribution of developmental genetic studies in *C. elegans* has been the demonstration that a signaling pathway for cell-fate determination includes a gene similar to the *ras* oncogene in humans, which is responsible for a substantial fraction of malignant tumors arising from connective tissue. The *C. elegans* signal transduction pathway operates during larval development in the formation of the hermaphrodite vulva, the passageway through which sperm from the male enter and fertilized eggs exit the gonad.

Mutant Screens Identify Genes Involved in Vulva Formation

Within the developing somatic gonad, which lies above the hypodermis on the ventral side of the L3 larvae, there are six cells capable of giving rise to the vulva. These cells are numbered P3p through P8p and are known as vulval precursor cells, or VPCs. Three of the VPCs (P5p, P6p, and P7p) normally form the vulva in response to an inductive signal from the anchor cell (AC), as shown in Fig. 20.16a. Researchers inferred the existence of this signal from experiments showing that even though the AC does not become part of the vulva, elimination of this cell by laser ablation prevents vulva formation. The remaining three VPCs (P3p, P4p, and P8p) normally produce additional hypodermal cells, but they can adopt vulval fates under certain experimental conditions.

Worm geneticists have devised several powerful genetic screens to identify the genes involved in the AC induction of P5p, P6p, and P7p to produce vulval cell lineages. The screens all take advantage of the fact that a hermaphrodite can survive without a functional vulva.

In one procedure, investigators search for Vulvaless mutants. As we have seen, hermaphrodites without a vulva can self-fertilize their eggs. The resulting zygotes, which mature within the ovary, grow by devouring the mother's organs, and become a "bag of worms" that is easy to spot in a mutant hunt (Fig. 20.16b). Loss-of-function mutations that prevent formation of a vulva define genes required for vulval induction.

In another type of genetic screen, researchers look for mutants with a Multivulva phenotype. These animals have additional, ectopic vulvas that form from descendants of the P3p, P4p, and P8p cells, which do not normally participate in the process. Loss-of-function mutations that cause ectopic vulva formation define genes whose products act as negative regulators of induction.

From these screens, researchers identified various genes that participate in the inductive signaling pathway that specifies the fate of vulval precursor cells. Interestingly, many of these genes are essential for viability. The discovery of their role in vulval determination was possible only because the screens were so powerful that investigators could recognize animals with rare hypomorphic (partial loss-of-function) or hypermorphic (gain-of-function) mutations.

Organizing the Genes Contributing to Vulva Formation into a Signaling Pathway

By combining epistasis tests, mosaic analysis, and phenotypic characterization, researchers constructed a genetic pathway that clarifies the relationship between many of the genes involved in vulva formation (Fig. 20.16c). Subsequent cloning of the genes elucidated many of the pathway's protein components and led to an understanding of the pathway at the molecular level.

(a)

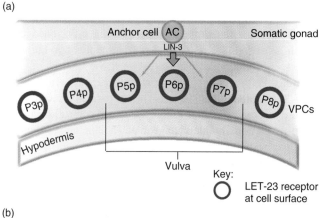

(b)

(c) Extracellular

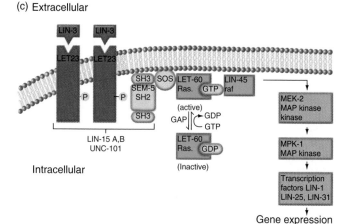

Figure 20.16 Cell signaling and signal transduction cascades in vulval development. (a) The anchor cell (AC) induces vulval fates in the vulval precursor cells (VPCs) P5p, P6p, and P7p. The signaling molecule provided by the AC is the LIN-3 protein; the receptor of this signal on the VPCs is the LET-23 protein. As shown here, differences between the fate of VPCs reflects their proximity to the AC, but this is only part of the story: The VPCs communicate with each other and with the underlying hypodermis as well. (b) A "bag of worms" in which a Vulvaless mother is consumed by her progeny. (c) The signal transduction pathway activated by the binding of the LIN-3 signal to the LET-23 receptor. Earliest events are to the left, subsequent events to the right. The signal-bound receptor activates the adaptor proteins SEM-5 and SOS, which in turn help catalyze the conversion of LET-60 into an active GTP-bound form. Activated LET-60 initiates a cascade of protein kinases; the phosphorylated form of the last of these kinases (MPK-1) turns on the activity of particular transcription factors in the nucleus. These transcription factors regulate the expression of the ultimate targets of the pathway: those genes whose products make vulval structures.

The Binding of a Signal Molecule to a Cell-Surface Receptor Initiates Events Leading to Vulva Formation

In this first step of the signaling pathway, the anchor cell emits a signal that must be received by the P5p, P6p, and P7p vulva precursor cells (Fig. 20.16). This AC signal is a small molecule encoded by the *lin-3* gene. That molecule shows some similarity to mammalian epidermal growth factor (EGF). The signal's probable receptor on the VPC cells is encoded by the *let-23* gene. The protein product predicted from analysis of the *let-23* DNA sequence is similar to mammalian EGF receptors. As expected for genes whose products are needed for vulval specification, partial loss-of-function mutations in both *lin-3* and *let-23* confer Vulvaless phenotypes. By contrast, overproduction of the LIN-3 protein causes a Multivulva phenotype. As required by the pathway model, LIN-3 (the signal) is expressed in the AC cell, while LET-23 (the receptor) is synthesized in the VPCs.

A Signal Transduction System Conveys the Signal Received at the Cell Surface to the Cell's Nucleus

After the LIN-3 signal binds to the LET-23 receptor, subsequent steps in the vulva-induction pathway occur within the P5p, P6p, and P7p cells (Fig. 20.16c). These intracellular steps together constitute a **signal transduction system** that converts the signal received at the cell surface into changes in the transcription of particular genes in the cell nucleus. It is these changes in gene activity that cause the cells to adopt vulval fates.

A key player in the signal transduction system for vulva induction is the *let-60* gene, which encodes a *C. elegans* RAS protein. Gain-of-function mutations in *let-60* produce alterations in the LET-60 protein identical to those that convert mammalian *ras* products to their oncogenic form. The mutant LET-60 becomes active even without a signal, and the ensuing inappropriate induction of vulval cell fates in P3p, P4p, and P8p leads to a Multivulva phenotype. By contrast, partial loss-of-function mutations in *let-60* cause a Vulvaless phenotype.

Like other RAS proteins, LET-60 is a guanine nucleotide-binding protein whose activity is indirectly controlled by an EGF receptor. Binding of the ligand signal activates the receptor, which in turn activates certain adaptor proteins. These adaptor proteins help transform the inactive GDP-bound form of LET-60 into an active GTP-bound form.

Also genetically identified in *C. elegans* are the downstream genes through which *ras* activity eventually influences events in the nucleus. These genes include *lin-45*, which encodes a protein kinase similar to one encoded by the mammalian protooncogene *raf*, as well as *mek-2* and *mpk-1*, which encode a MAP-kinase kinase and a MAP kinase, respectively. These proteins form a protein kinase cascade in which each member catalyzes the phosphorylation of the next member. Researchers do not yet know for sure which genes encode the transcription factors whose activity in the nucleus is eventually modified by the protein kinase cascade. Based on phenotypes and epistasis tests, however, they believe that *lin-1*, *lin-25*, and *lin-31* are promising candidate genes.

This Type of Signaling Pathway Has Been Conserved in Evolution

Signaling pathways including a signal molecule, a receptor for that molecule, and a signal transduction system activated by the binding of signal and receptor are a feature of intercellular signaling in many organisms. In *Drosophila*, for example, a system similar to the one we have described here operates in the development of the eyes; in humans, defects in signaling systems contribute to cancer formation.

Geneticists are thus very interested in using the genetic dissection of vulval development to identify additional genes that influence ubiquitous signaling pathway. They have recently begun two new types of searches. First, they are looking for mutations that suppress the Multivulva phenotype caused by *let-60* gain-of-function mutations. Second, they are screening for mutations that enhance the Vulvaless phenotype caused by *let-60* partial loss-of-function mutations. Through these mutant screens, they have already identified the kinase-encoding *ksr-1* gene, a component of the signal transduction system that eluded detection by other means (such as biochemical analysis). The results of these investigations may be of value in understanding and eventually controlling the production of tumors caused by the *ras* and *raf* oncogenes in humans.

CONNECTIONS

The special features of *C. elegans,* including its invariant cell structure and transparency at all stages of the life cycle, make it an excellent model for the genetic analysis of development in multicellular animals. With knowledge of the progenitors and position of every cell in the embryo, larvae, and adult, researchers can begin to correlate gene activity with cell differentiation to understand how genes influence development.

In the next chapter we present a genetic portrait of another multicellular animal: *Drosophila melanogaster.* In *Drosophila,* the developmental fate of individual cells is determined by a much more flexible mechanism than the invariant cell lineage seen in *C. elegans.* Through the analysis of these two model systems for animal development, biologists can compare the ways in which evolution has generated various structures and functions. Sometimes the strategies different organisms use to perform the same basic task are similar, but as the fundamental dichotomy in the mechanisms of cell determination suggests, sometimes they are very different.

ESSENTIAL CONCEPTS

1. *C. elegans* is an ideal subject for genetic analysis because of its transparency, invariant cellular composition, short life cycle, and capacity for prolific reproduction.

2. The small genome of 100 Mb is packaged into six small chromosomes. Its roughly 14,000 genes contain small introns. *Trans-splicing* and the production of *polycistronic transcripts* are two unusual aspects of *C. elegans* gene expression.

3. The life cycle from the fertilized egg of one generation to the fertilized eggs of the next takes about three days. Mating produces cross-fertilization. Hermaphrodites can self-fertilize.

4. Because the cell lineage of *C. elegans* is almost completely invariant, developmental biologists have been able to construct complete *cell lineage diagrams* that show the routes by which all the somatic cells in the adult arise from the fertilized zygote. Six *founder cells* give rise to the major embryonic lineages. The study of mutants that disrupt cell lineage patterns is a major focus of *C. elegans* research.

5. Determination of blastomere fate depends on inductive signaling between cells as well as on the segregation of signals that act independently in single cells.

6. Mutations that cause the failure of all programmed cell deaths during development helped researchers identify the genes that control *apoptosis*. Although nematodes and humans are separated by almost a billion years of evolution, many of their apoptosis-controlling genes are functionally interchangeable.

7. The interactions of various *heterochronic "switch" genes* set the timing of different stages of larval development.

8. Induction of the hermaphrodite vulva depends on a *signal transduction pathway* that includes a gene similar to the *ras* oncogene in humans. Identification of the genes active in this pathway in *C. elegans* may help researchers understand tumor formation in humans.

SOLVED PROBLEMS

I. You need to construct a strain of *C. elegans* that is homozygous for the *dpy-10* and *unc-54* mutations. The *dpy-10⁻* mutation is a defect in the collagen gene that results in a short, stubby-shaped worm (Dumpy). The *unc-54⁻* mutation, a defect in a myosin heavy-chain gene, affects muscle control so severely that the worm is paralyzed (Uncoordinated). *unc-54* and *dpy-10* are unlinked, autosomal genes and both mutations are recessive. Stock hermaphrodite worms that are homozygous for *dpy-10⁻* and another stock that is homozygous for *unc-54⁻* are available for your use. The first step is to cross wildtype males to homozygous *dpy-10⁻* hermaphrodites.

   a. What genotypes and phenotypes (including sex phenotype) do you predict in the progeny?

   b. The wildtype male progeny from the cross-fertilization are crossed to virgin hermaphrodites that are homozygous for the *unc-54⁻* mutation. (To isolate virgins, L4 larval stage worms are transferred individually to petri dishes, ensuring that mating has not occurred yet with any available males in the population.) What are the genotypes and phenotypes that result from this cross?

   c. What worms will you use from the second cross to get your double recessive *dpy-10⁻ unc-54⁻* strain? Will you be able to identify worms with the specific genotype you need for the next step?

   d. What is the phenotype of worms with the desired double recessive genotype? What proportion of the worms emerging from the last step of the crossing scheme will have the double recessive genotype?

Answer

This problem requires an understanding of basic Mendelian ratios and of *C. elegans* matings.

   a. The cross can be represented:

$$\frac{dpy\text{-}10^+}{dpy\text{-}10^+} \text{ males} \times \frac{dpy\text{-}10^-}{dpy\text{-}10^-} \text{ hermaphrodites}$$

Either the hermaphrodite will self-fertilize or there will be cross-fertilization between males and hermaphrodites. *When the hermaphrodite self-fertilizes, all progeny will be Dumpy hermaphrodites with the homozygous* dpy-10⁻/dpy-10⁻ *genotype.* (Males [XO] will also arise from self-fertilization due to nondisjunction of sex chromosomes during formation of gametes, but the frequency is low enough that we will generally ignore this group of males. These male worms would also have a Dumpy phenotype.) In a mating between the male and hermaphrodite, the male can produce only *dpy-10⁺* gametes. Half of these gametes contain an X chromosome and the other half contain no sex chromosome. *The outcross with males results in heterozygous* dpy-10⁺/dpy-10⁻ *hermaphrodites and males that are normal in phenotype and heterozygous at the* dpy *locus.*

   b. The second cross is:

$$\frac{dpy\text{-}10^+ \quad unc\text{-}54^+}{dpy\text{-}10^- \quad unc\text{-}54^+} \text{ males } \times$$

$$\frac{dpy\text{-}10^+ \quad unc\text{-}54^-}{dpy\text{-}10^+ \quad unc\text{-}54^-} \text{ hermaphrodites}$$

The hermaphrodite will produce, by self-fertilization, hermaphrodites that are normal shaped (nondumpy) but paralyzed and have the genotype dpy-10$^+$/dpy-10$^+$ unc-54$^-$/unc-54$^-$. From the outcross to males, the male can contribute either the *dpy-10$^+$* or *dpy-10$^-$* allele with the *unc-54$^+$* allele; whereas the hermaphrodite can produce only *dpy-10$^+$ unc-54$^-$* alleles. Half of the sperm from males have an X chromosome and half do not have a sex chromosome. *Results of cross-fertilization:*

XX *dpy-10$^+$/dpy-10$^+$ unc-54$^+$/unc-54$^-$*   wildtype hermaphrodites

XO *dpy-10$^+$/dpy-10$^+$ unc-54$^+$/unc-54$^-$*   wildtype males

XX *dpy-10$^+$/dpy-10$^-$ unc-54$^+$/unc-54$^-$*   wildtype hermaphrodites

XO *dpy-10$^+$/dpy-10$^-$ unc-54$^+$/unc-54$^-$*   wildtype males

c. *To obtain a doubly homozygous recessive, a double heterozygote must be selfed. Notice that half of the wildtype hermaphrodites will have the genotype desired for this cross, but you can't distinguish between the two different genotypes.* Pick several wildtype hermaphrodites and allow them to self-fertilize.

d. *The phenotype of the doubly recessive worm desired is a short, stubby (Dpy) paralyzed (Unc) worm. The proportion of worms with a double recessive phenotype from selfing heterozygotes is 1/4 for each gene. But only half of the worms picked were heterozygous. (1/4)(1/4)(1/2) = 1/32 worms are expected to have the double homozygous recessive genotype.*

II. In *C. elegans,* the cells at the tip of the gonad go through several mitotic divisions before switching to meiotic divisions and differentiating into germ cells. As a result of the mitotic and meiotic divisions, up to 1500 germ cells are produced. When the somatic cell at the very tip of the developing gonad (distal tip cell) is killed by laser ablation, the few germ cell precursors surrounding it will immediately go into meiosis instead of mitosis.

a. What effect would this ablation have on the number of germ cells produced?

b. The mutation *glp-1* mimics the laser ablation results. In the *glp-1* mutant, cells immediately go into meiosis, without mitotic divisions, and only a few germ cells are produced. The hypothesis is that the distal tip cell produces a molecule (inducer) that binds to a receptor to control mitotic and meiotic divisions. In what cells would you expect to see *glp-1* expression if it encodes a receptor?

c. The Glp-1 protein functions in other processes in addition to germ cell development. For example, Glp-1 is required in the earliest cell divisions in the embryo to establish the fate of cells that develop into the pharynx. In addition, some mutations in the *glp-1* gene affect only a subset of the *glp-1* activities. Speculate on how the Glp-1 can act as a receptor in processes that have a very different outcome.

Answer

The question involves thinking through cell-to-cell interactions and their consequences.

a. Without mitotic divisions, only a few cells are available to go through meiosis and produce gametes. *Many fewer gametes are produced.*

b. If *glp-1* encodes a receptor, it should be expressed in cells that receive and interpret the signal. Those cells are the *germ cell precursors.*

c. Glp-1 protein, because of its role in various cell-to-cell interactions, could be considered a general responding protein. This general protein could be *interacting with different cell-type-specific proteins to give an appropriate, specific response.* Alternatively, or perhaps in addition, the Glp-1 receptor *may have different regions that respond to different inducing signals* and send specific messages into the cell based on which region of the receptor has been contacted.

P R O B L E M S

20-1 Choose the best matching phrase in the right column for each of the terms in the left column.

a. ablation

1. having centromere activity along the length of the chromosome

b. apoptosis

2. cell death caused by outside forces such as lasers

c. heterochronic

3. extrachromosomal fragments of DNA

d. polycistronic

4. change in the time at which an event occurs

e. trans-splicing

5. programmed cell death caused by signals inside the cell

f. holocentric

6. sections of different RNAs are joined together

g. free duplication

7. containing several genes

20-2 To construct a library for which there is a 99% probability that any one genomic sequence will be represented, the amount of DNA represented in the independent clones should be approximately five times the genome size. Assuming an average cosmid size of 50 kb and a genome size of 100 Mb, how many cosmids would you need to have in your library to ensure that each sequence in *C. elegans* was represented at least once?

20-3 What is (are) the advantage(s) in using a hermaphrodite when screening for mutants?

20-4 Vulval development in the *C. elegans* hermaphrodite has provided an excellent model system for studying the molecular basis of cell interaction and signaling during development. Genetic analysis and cell ablation studies have been critical in identifying genes and determining the role of their gene products in vulval development. For which of the following purpose(s) would you want to
a. combine different vulval mutations into one strain
b. isolate suppressors of a vulval mutant phenotype
c. ablate a cell
    i. to identify mutations that bypass a normal mechanism
    ii. to identify genes that work in the same pathway
    iii. to determine the function of a cell during development
    iv. to determine if mutations are in the same gene

20-5 You isolated a cDNA for an actin gene. Which of the following results would suggest that trans-splicing occurs during processing?
a. the 5′ end of the cDNA hybridized with many DNA clones from the genome
b. the 5′ end of the cDNA hybridized with many RNAs in the worm
c. the 3′ end of the cDNA hybridized with many DNA clones from the genome
d. the 3′ end of the cDNA hybridized with many RNAs in the worm
e. the 5′ end of the cDNA hybridized with one DNA clone from the genome
f. the 5′ end of the cDNA hybridized with one RNA in the worm
g. the 3′ end of the cDNA hybridized with one DNA clone from the genome
h. the 3′ end of the cDNA hybridized with one RNA in the worm

20-6 Laser ablation of cells can be done during embryonic development, during larval stages, or in the adult. To determine if the anchor cell is involved in establishing the identity of surrounding vulval precursor cells, describe (in general) the time at which you would ablate the anchor cell.

20-7 The concentration of LIN-14 protein, high during the L1 larval stages and decreasing in later stages, regulates the timing of cell divisions in the T-cell lineage in *C. elegans.* According to the hypothesis presented in this chapter, *lin-4* RNA regulates the level of LIN-14 protein by binding to the *lin-14* mRNA and preventing translation. If you could introduce extra copies of *lin-4* into the cells and more *lin-4* RNA was produced, what effect do you think the extra copies of *lin-4* would have on translation of LIN-14? What effect could this have on cell lineages during development?

20-8 The *C. elegans* strain Bergerac has a mutation rate significantly higher than that seen in other strains. For example, spontaneous mutations in the *unc-22* gene arise at a frequency 100× higher than the frequency of mutation in the Bristol strain. You think the mutations may be caused by insertion of transposons. How could you test whether the transposition of the Tc1 element caused some or all of the mutations in the *unc-22* gene? (Assume that the *unc-22* gene and Tc1 DNA have been cloned.)

20-9 How would you construct a hermaphrodite with the genotype *dpy-10⁻/dpy-10⁻ daf-1⁻/daf-1⁻*? You are given stocks of wildtype males, *dpy-10⁻/dpy-10⁻* hermaphrodites and *daf-1⁻/daf-1⁻* hermaphrodites. (*dpy-10* and *daf-1* are unlinked autosomal genes.)
a. What is the first cross you would do?
b. Which worms would you use in the second cross?
c. What is the final cross you do to get the doubly homozygous worms? What proportion of the hermaphrodites do you expect to have the doubly homozygous genotype?

20-10 The *unc-54* (uncoordinated) and *daf-8* (dauer constitutive) genes are 18 m.u. apart on chromosome I of *C. elegans.* You want to construct a worm strain homozygous for mutations in both genes. Stocks of wildtype males and *unc-54⁻/unc-54⁻* hermaphrodites and *daf-8⁻/daf-8⁻* hermaphrodites are available for your use. The first cross is wildtype males × *unc-54⁻/unc-54⁻* hermaphrodites.
a. What progeny of this first cross will you use for the second cross?
b. What progeny arise from your second cross?
c. What worms will you use for the final cross to get the desired double homozygote?
d. What event has to happen to get a double homozygote?
e. What proportion of the progeny will have the desired genotype?

20-11 Muscles in the worm can be classified in four major groups: body muscles involved in locomotion, enteric muscles involved in defecation, pharynx muscles involved in feeding, and egg-laying muscles in the hermaphrodite. Uncoordinated (Unc) worms have been isolated that are defective in neuromuscular control, resulting in a variety of phenotypes from slow-moving to completely paralyzed.

a. *unc-103⁻* is a mutation that causes defects in all muscle groups except pharyngeal muscles and has the phenotype of complete paralysis. The approximate location of the gene was determined by genetic mapping. One gene already identified in the region encodes a potassium (K^+) channel protein. Because ion channels and membrane depolarization are involved in muscle response, it was hypothesized that the *unc-103* mutation might be in the channel gene. Taking a candidate gene approach, how would you proceed to determine if the *unc-103* mutation was in fact a mutation in the K^+ channel gene?

b. Some of the *unc-103* alleles were constructed by insertion of the Tc1 transposon. How could you use this type of mutation and the physical map information on the location of the K^+ channel gene to determine if *unc-103* is in fact the K^+ channel gene?

c. Depolarization of the membrane (increase in the positive charge inside the cell) is necessary for muscle contraction. Opening of a K^+ channel is involved in the repolarization event, allowing K^+ to flow out of the cell and reestablish the negative charge on the inside of the cell. *unc-103* gain-of-function alleles lead to an inappropriately open K^+ channel so potassium continues to flow out of the cell. Do you think the gain-of-function *unc-103* allele is dominant or recessive and why?

d. Another gene identified by mutations that cause defects in locomotion, defecation, and egg-laying is the *egl-2* gene. This gene also encodes a K^+ channel. When a fusion was made between the 5′ end of the *egl-2* gene and a reporter gene, the expression of the fusion gene could be traced in specific cells. It was found to be expressed in muscles and nerves required for egg-laying and defecation but also in a subset of the chemosensory neurons—the AWC but not the AWA cells. Each of these cell types is involved in sensing different chemoattractants. The AWC neurons sense volatile substances such as isoamyl alcohol and AWA neurons sense soluble substances such as diacetyl alcohol. How could you determine if the *egl-2* mutation disrupts chemosensory function in these cells?

20-12 Guidance nerve cell axons growing toward their ultimate targets are models for cellular movement during development. Because the location of all neurons in the *C. elegans* adult and the ancestry of all cells is known, much can be learned about the process of axonal growth using mutants defective in the final function of the nerve cells. Molecules involved in axon guidance have been identified by isolation of *unc* (Uncoordinated) mutants that are defective in locomotion. Proteins called netrins that act as guidance molecules along the path of growth have been identified in flies, worms, and humans. The *unc-6* gene product in *C. elegans* is a netrin and the Unc-5 protein is a receptor protein on some nerve cells that interacts with Unc-6 protein in the basal lamina (extracellular matrix) and receives guidance for growth.

a. Mouse Netrin-1 has similar properties to *C. elegans* Unc-6. How could you determine if mouse Netrin-1 could substitute for Unc-6 in *C. elegans?*

b. Other cells migrate during development in *C. elegans.* How would you know if these migrations require *unc-5* or *unc-6* function?

c. Unc-5 protein is present on the surface of the growing neuron and receives guidance from the Unc-6 protein for axon growth. Starting with an *unc-5* defective mutation (but not one completely lacking the protein), how could you isolate genes encoding proteins that interact with (and perhaps relay messages from) the Unc-5 receptor to the inside of the cell?

d. The appropriate growth of axons is hypothesized to involve a series of attractions and repulsions that may be dependent on the type of receptor expressed at a particular time by a nerve cell. If the time of expression of receptor/sensing genes such as *unc-5* was altered, there may be an effect on guidance. Describe a general approach you could take to express *unc-5* constitutively in the cells of *C. elegans.*

CHAPTER

21

DROSOPHILA MELANOGASTER: GENETIC PORTRAIT OF THE FRUIT FLY

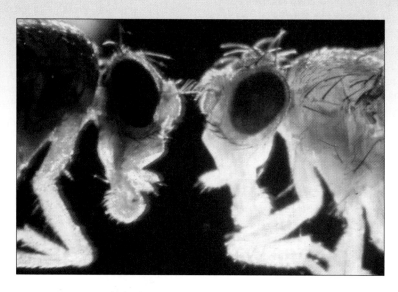

Adult fruit flies expressing β-galactosidase reporter genes.

A wildtype *Drosophila melanogaster,* or fruit fly, has multifaceted brick red eyes, a tan thorax studded with arched black bristles, a striped abdomen, and a pair of translucent wings (Fig. 21.1). The fly's rapid life cycle, low chromosome number, small genome size, and giant salivary gland chromosomes (review Fig. 11.21) are among the experimental advantages that made *D. melanogaster* the central organism in the study of transmission genetics in the first half of the twentieth century. These same features also made *Drosophila* a major model organism in genetic studies of animal development and behavior in the last quarter of the century.

Modern *Drosophila* genetics originated with Thomas Hunt Morgan's discovery of the sex-linked *white* eye mutation in 1910. In the next few decades, *Drosophila* studies by Morgan and his students at Columbia University in New York City laid the experimental foundation for the field of genetics (Fig. 21.2). Calvin Bridges, for his Ph.D. with Morgan, proved the chromosome theory of inheritance by showing that nondisjunction of the sex-linked *white* eye gene correlates perfectly with the nondisjunction

of the X chromosome (see Chapter 3). Alfred H. Sturtevant, as an undergraduate in Morgan's lab, generated the first chromosome map by measuring recombination frequencies among the known sex-linked genes (see Chapter 4). Herman J. Muller demonstrated the mutagenicity of X rays (see Chapter 6). These and other members of Morgan's "fly room" went on to establish many other basic features of transmission genetics, including the structure and genetic consequences of each major type of chromosome rearrangement (described in Chapter 12).

From the 1940s through the 1950s, bacteria and their viruses replaced *Drosophila* as the dominant organisms in genetic studies. The simplicity of these microbes and the ease with which they can be grown in vast numbers made them ideally suited for genetic and biochemical work on the basic structure and function of genes. During this time, *Drosophila* research went into almost total eclipse. Then, in the early 1970s, when many geneticists began to investigate how genes control the development and behavior of higher organisms, interest in *Drosophila* increased dramatically as

Figure 21.1   A colorized scanning electron micrograph of a *Drosophila* adult's head and wings.

researchers realized the great potential of combining the sophisticated genetic techniques of *Drosophila* studies with the newly developed methods of recombinant DNA technology. The successful marriage of these approaches depended in part on the fortunate fact that a handful of *Drosophila* geneticists had continued to work and maintain stock collections through the "dark" years.

In this chapter, we see that by combining modern molecular methods with the rich legacy of genetic techniques originally developed for *Drosophila,* researchers have made stunning progress in understanding how genes control the development and behavior of higher organisms. The impact of this work has been immense, and the accelerating pace of *Drosophila* genetics suggests that the fruit fly will remain a key model organism for the foreseeable future.

A major theme that emerges from our discussion of the genetic analysis of *Drosophila* is that many of the developmental genes and mechanisms operating in the fly have been conserved in a wide variety of animals. For example, most of the developmentally important genes first found in *Drosophila* have turned out to play similar roles in other animals, including humans. In fact, researchers identified most of the genes known to influence the development of humans as homologs of genes already discovered in *Drosophila*.

As we examine *D. melanogaster*'s use as a model organism, we present

- The structure and organization of the genome.

- The life cycle.

- Techniques of genetic analysis, including the construction and use of balancer chromosomes, the uses of P element transposons, the production of mosaics, ectopic gene expression, and the *Drosophila* Genome Project.

- The genetic analysis of body plan development in *Drosophila,* a comprehensive example.

Figure 21.2   The fountainhead of *Drosophila* genetics. Thomas Hunt Morgan and his students in the fly room at Columbia University, at a party in 1919 celebrating the return of Alfred Sturtevant from military service in World War I. Individuals whose work is discussed in this book include Morgan (back row, far left), Sturtevant (front row, third from the right), Calvin Bridges (back row, third from the right), and Herman J. Muller (back row, second from the left). The "honored guest" between Muller and Bridges has clearly seen better days.

STRUCTURE AND ORGANIZATION OF THE <u>DROSOPHILA</u> GENOME

The *Drosophila* genome is about 5% the size of the human genome and is packaged in far fewer chromosomes. It contains roughly 12,000 to 15,000 genes, about 20% of the number found in the human genome, which means that the density of genes on *Drosophila* chromosomes is higher than that of genes in the human genome. Even so, one-third of the fly genome consists of repetitive sequences that do not encode proteins or that act as transposable elements.

The Chromosomes of Drosophila

The haploid genome of *D. melanogaster* contains four chromosomes, designated numerically as 1–4. Chromosome 1 is the X chromosome; chromosomes 2–4 are autosomes. Sex determination, we saw in Chapter 3, is of the XY type, with females being XX and males XY. Unlike the situation in mammals, however, the actual determination of sex depends solely on the number of X chromosomes as it relates to the number of copies of each autosome (the X:A ratio). XXY zygotes with an X:A ratio of 1 (two copies of the X:two copies of each autosome) develop as normal females. By contrast, XO zygotes (carrying 1 X and no Y chromosome) have an X:A ratio of 0.5 (one copy of the X:two copies of each autosome) and develop as morphologically normal, but sterile, males. (Review Chapters 3 and 16 for more details about sex determination in *Drosophila*.) The Y chromosome, which by convention is not given a number, is required only to confer male fertility. The low chromosome number is one of *Drosophila*'s key advantages for genetic studies, as it simplifies most genetic manipulations.

Figure 21.3 shows the structure of *Drosophila* chromosomes prior to S phase DNA replication. The X chromosome is large and acrocentric; chromosomes 2 and 3 are large and metacentric, and chromosome 4 is a tiny acrocentric "dot" chromosome that is only about 2% the size of the major autosomes. Each chromosome contains a large block of heterochromatin near its centromere. Although a few genes are present in these heterochromatic blocks, for the most part, these regions contain highly repetitive DNA sequences and are transcriptionally silent. Centromeric heterochromatin comprises about a quarter of the length of each of the major (larger) chromosomes, as well as the majority of the small chromosome 4. The large submetacentric Y chromosome is almost entirely heterochromatic, and carries only a few (probably six) genes, which function in the male germ line. Nucleolus organizer regions are located within the short arm of the Y and the centromeric heterochromatin of the X. Each nucleolus organizer region carries 200–250 copies of the 18S and 28S ribosomal RNA genes in tandem array.

The Giant Polytene Chromosomes of the Larval Salivary Gland Are Key Tools for Drosophila *Genetics*

As we saw in Chapter 11, chromosomes in the fruit fly's salivary glands replicate 10–11 times without cell division. The

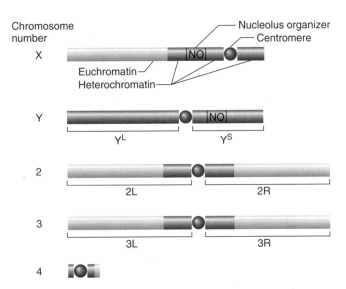

Figure 21.3 The *Drosophila* chromosome complement. Major blocks of all chromosomes, particularly in regions surrounding the centromeres, are heterochromatic. Key: darker shades of color, heterochromatin; lighter shades of color, euchromatin; circles, centromeres; NO, nucleolus organizer; Y^L and Y^S, long and short arms of the Y chromosome; and 2L, 2R, 3L, and 3R, left (L) and right (R) arms of chromosomes 2 and 3. See the Fast Forward box in Chapter 3 (pg. 80) for a picture of these same chromosomes in mitotic cells.

1024 or 2048 sister chromatids generated during these rounds of replication stay intimately associated with one another in perfect lateral register, producing a many-stranded polytene chromosome. These chromosomes are truly gigantic: Normal metaphase chromosomes are about as long as the polytene chromosomes are wide.

Regions of local condensation along a chromosome become amplified laterally in polytene chromosomes and are visible as distinct crossbands (separated by less condensed interbands) when viewed through a light microscope. A striking feature seen in cells carrying polytene chromosomes is that homologous chromosomes are tightly synapsed with each other (review Fig. 11.17). This is unusual because in most organisms extensive synapsis of homologs occurs only in prophase I of meiosis. In contrast to the euchromatin, regions of centromeric heterochromatin are not replicated during polytenization, and coalesce into a common region called the *chromocenter* (also shown in Fig. 11.17). As a result, in squash preparations of salivary gland nuclei, five major arms (the left and right arms of chromosomes 2 and 3, and the X chromosome) radiate out from the chromocenter. The much smaller chromosome 4 and the almost completely heterochromatic (and therefore underreplicated) Y chromosome (in males) are also present, but they are often hard to see.

The importance of polytene chromosomes in *Drosophila* genetics cannot be overemphasized. As we saw in Chapters 11 and 12, they have provided *Drosophila* geneticists with a ready-made detailed physical map of the fly genome. Since the finer visible bands contain only a few kilobases of DNA per chromatid, rearrangement breakpoints can often be mapped with very high precision simply by microscopic examination. Cloned DNA sequences can also be mapped very rapidly and precisely by *in situ* hybridization to polytene chromosomes. This has been of great utility in the positional cloning of many *Drosophila* genes, since it is possible to monitor the orientation and progress of chromosome walks when different probes along the walks are sufficiently far apart that they hybridize to different polytene chromosome bands.

The Drosophila *Genome*

The haploid genome contains about 170,000 kb of DNA. Roughly 21% of these kilobase pairs consists of highly repetitive "satellite" DNA that is concentrated in the centromeric heterochromatin and the Y chromosome. Another 3% consists of repeated genes that encode rRNA, 5S RNA, and the histones. Approximately 9% is composed of around 50 families of transposable elements, which are roughly 2–9 kb in length; the number of copies of each type of element varies from strain to strain, but is usually somewhere between 10 and 100. In addition to residing at scattered sites within the euchromatin, many transposable elements are found in the centromeric heterochromatin, while some are concentrated in telomeres (review Fig. 12.15). An unusual feature of *Drosophila* chromosomes is that the telomeres do not consist of simple repeats, such as the TTGGGG or TTAGGG repeats found at the ends of chromosomes in other well-characterized organisms. Instead, it appears that the rapid jumping into the telomeric region of certain kinds of transposable elements maintains the telomeres. The insertion into the telomeric region of the transposable elements counters the chromosomal shortening that inevitably occurs during the replication of the ends of linear DNA molecules.

The remaining 67% of the genome consists of unique DNA sequences in the euchromatic chromosome arms. We describe the nature and organization of the 12,000–15,000 genes present within these unique DNA sequences later in the chapter, in our discussion of the *Drosophila* Genome Project.

LIFE CYCLE

Fruit flies can be raised on a wide variety of media, but the standard formulation is a mixture of corn meal, brewer's yeast, sugar, agar, and a small amount of mold inhibitor. Flies bred on such a medium at 25°C complete their life cycle from zygote formation to emergence of the adult fly in 10 days (Fig. 21.4).

Drosophila has a very high reproductive capacity. A vigorous female can produce 3000 progeny in her lifetime; and a single male can sire well over 10,000 offspring. The female, with her special sperm storage organs (the spermatheca and seminal receptacle), can produce several hundred progeny after a single mating. Each *Drosophila* egg is about half a millimeter long and well supplied with yolk that can support eventual embryonic development. Fertilization normally occurs as the egg is being laid.

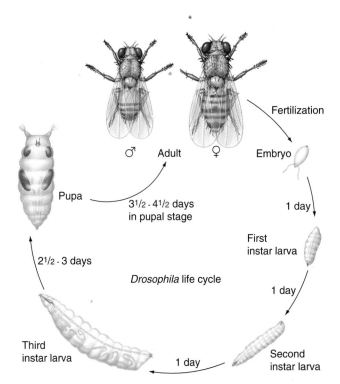

Figure 21.4 The *Drosophila* life cycle. Arrows are labeled with the number of days during which the animal remains in the stage starting at the beginning of the arrow. The transition from an embryo to a first instar larva is called *hatching*. The transitions between larval instars are *molts*. The process that converts a third instar larva to a pupa is *pupariation*. Emergence of the adult from the pupal case is called *eclosion*.

The embryo completes its development in just under 24 hours, and some of the early events happen so rapidly they can be watched in real time. For example, in the early stages of embryonic development, the zygotic nuclei divide every 10 minutes, and the morphogenic movements of gastrulation that produce the endoderm, mesoderm, and ectoderm tissue layers take only 20 minutes. Embryogenesis ends with the hatching of a wormlike first instar (or first-stage) larva that is specialized for feeding and grows dramatically. To allow for an increase in body size, the larva molts 24 and 48 hours after hatching to produce second and third instar larvae.

About 3 days after the second molt, the third instar larvae complete their growth, crawl out of the food, and pupate. Once inside the protective pupal case, the larvae undergo **metamorphosis:** a dramatic reorganization of the fly body plan. This reorganization takes about 4 days and consists of the disintegration of most larval tissues and their replacement through the proliferation and differentiation of cells that produce adult structures. Most structures specific to adults, such as the wings, legs, eyes, and genitalia, are generated from **imaginal discs:** flattened epithelial sacs that develop from small groups of cells set aside in the early embryo (see Fig. 21.5). Imaginal discs undergo extensive growth and development during the larval and pupal stages. Other adult-forming cells, sequestered in pockets within the larval

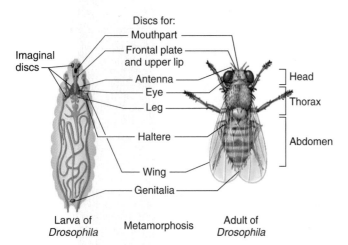

Figure 21.5 Many adult structures develop from imaginal discs in larvae and pupae. Lines connect particular imaginal discs in the larva with the structures they generate in the adult.

abdomen, eventually replace much of the larval epidermis and gut. At the end of metamorphosis, when the adult emerges from the pupal case, its wings expand and harden, and pigmentation proceeds to completion.

TECHNIQUES OF GENETIC ANALYSIS

The practice of genetics in *Drosophila* has a unique character that in part reflects the tools created by *Drosophila* researchers over many years and in part reflects certain distinctive features of fruit fly inheritance. (For a review of the symbols used by *Drosophila* geneticists, see the Chapter 3 discussion of eye-color inheritance in the fruit fly on page 93.)

In Drosophila, *Crossing-Over Occurs Only in Females*

Although the reason for the restriction to females of crossing-over is not clear, in the few other known species in which recombination is restricted to one sex, it is always the heterogametic sex that lacks crossing-over. This suggests that the prevention of crossing-over between different sex chromosomes has been of selective importance.

The Absence of Crossing-Over in the Male Has Considerable Technical Significance

It makes it possible to maintain linkage relationships simply by ensuring that a chromosome is inherited strictly through the male parent. This may seem trivial, but it greatly simplifies many genetic manipulations and allows a variety of selective genetic screens.

In Females, Only a Moderate Amount of Crossing-Over Occurs

Typically, only one or two crossovers take place in each major chromosome arm per meiosis, and essentially no crossing-

over occurs in the centromeric heterochromatin or anywhere in the fourth chromosome. As in other organisms, however, crossing-over is required for the accurate disjunction of the major chromosomes during meiosis I in females; in its absence, bivalents fall apart at the end of prophase I. A special mechanism, currently under active study, ensures a slightly less accurate segregation of the fourth chromosomes and the occasional large chromosome bivalents in which crossing-over does not occur. How the segregation of homologs is directed in males is not well understood, but it appears to depend on special sites along the chromosomes that mediate the association of homologs.

Balancer Chromosomes Help Preserve Linkage

Drosophila geneticists have used X rays as a mutagen to create many special chromosome types useful for genetic manipulations. By far the most important of these is the **balancer chromosome,** which carries multiple, overlapping inversions (Fig. 21.6a). Most balancers also contain a dominant marker that enables researchers to follow the balancer through crosses, and a recessive lethal mutation that prevents the survival of homozygotes. The utility of balancer chromosomes is that, like all inversion-containing chromosomes, they prevent the recovery of crossovers between normal sequence chromosomes and themselves. As a result, both balancer chromosomes and their homologs are inherited through crosses as intact units, and no recombinant chromosomes are passed on to the next generation.

Balancer Chromosomes Are Useful in Stock Keeping

The term "balancer" derives from the extensive use of these chromosomes in stock keeping. Experimenters establish a typical *Drosophila* stock by crossing males and females that carry a desired mutation heterozygous with a balancer chromosome homolog. If the mutation is a recessive lethal, as is often the case, and the balancer chromosome carries another recessive lethal, the only surviving progeny will, like their parents, be heterozygous for the chromosome bearing the desired mutation and the balancer (Fig. 21.6b). The only requirement for maintaining the mutation is to transfer progeny to fresh medium periodically. Stocks of this type are said to be "balanced lethals," hence the name "balancer."

Balancer Chromosomes Are also Useful in Genetic Manipulations and Mutant Screens

As an example, Fig. 21.6c shows a cross to determine by complementation analysis whether two independent recessive lethal mutations [*lethal₁* (l_1) and *lethal₂* (l_2)] on the third chromosome are allelic. The cross involves a second chromosome balancer, designated SM1, which carries the dominant wing mutation *Curly* (*Cy*). Males of genotype l_1/SM1 are crossed to l_2/SM1 females. If the mutations are allelic, only progeny with the Curly wing phenotype will emerge, whereas if the mutations complement each other, non-Curly (l_1/l_2) progeny will also appear.

(a) SM1 balancer chromosome

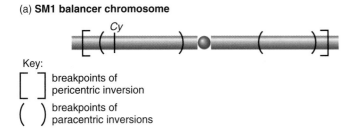

Key:

[ ] breakpoints of pericentric inversion

( ) breakpoints of paracentric inversions

(b) A balanced lethal stock

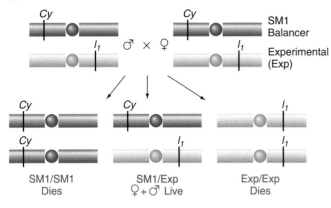

Cy   *Cy*   SM1
                                              Balancer

I_1 ♂ × ♀ I_1 Experimental
                                              (Exp)

Cy   *Cy* I_1

Cy I_1 I_1

SM1/SM1   SM1/Exp   Exp/Exp
Dies   ♀+♂ Live   Dies

(c) Complementation analysis

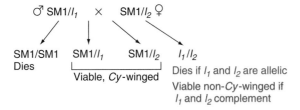

♂ SM1/I_1 ×   SM1/I_2 ♀

SM1/SM1   SM1/I_1 SM1/I_2 I_1/I_2
Dies   ‾‾‾‾‾‾‾‾‾‾‾‾‾‾‾‾‾‾‾‾   Dies if I_1 and I_2 are allelic
            Viable, *Cy*-winged   Viable non-*Cy*-winged if
                                     I_1 and I_2 complement

Figure 21.6 Balancer chromosomes in *Drosophila*. (a) SM1 (Second Multiple One), one of several balancers for chromosome 2, contains several overlapping inversions to suppress recombination, and carries the dominant marker *Cy* (Curly wings). *Cy* is also a recessive lethal, so that SM1/SM1 homozygotes do not survive. (b) A balanced lethal stock. When flies heterozygous for a lethal mutation lethal₁ (I_1) and a balancer chromosome such as SM1 are intercrossed, the only survivors are I_1/SM1 heterozygotes genetically identical to their parents. (c) Complementation analysis: Are two independently isolated lethal mutations (I_1 and I_2) allelic? Progeny of the diagrammed cross without Curly wings (of genotype I_1/I_2) will appear only if the mutations are in different genes.

P Element Transposons Have Played a Key Role in Drosophila *Molecular Genetics*

The discovery of P elements arose from research on a phenomenon called **hybrid dysgenesis.** P elements (Fig. 21.7a) are the causative agent of hybrid dysgenesis, which occurs when males from *Drosophila* strains carrying P elements (referred to as P strain males) are crossed with females that lack P elements (known as M strain females), and the P elements become highly mobile in the germ line of the F₁ hybrids (Fig. 21.7b and c). Chromosome breakage resulting from this high

(a)

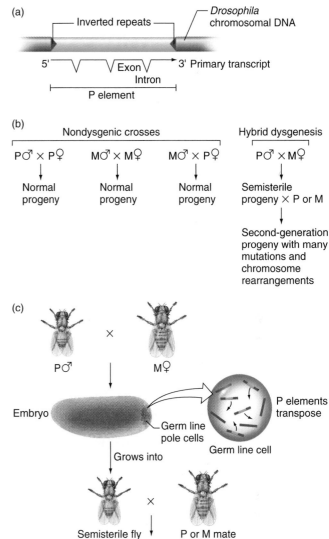

Inverted repeats ⌐ ⌐ *Drosophila* chromosomal DNA

5' ⌐_/‾_/‾ Exon _/‾→ 3' Primary transcript
              Intron
    ⌐‾‾‾‾‾‾ P element ‾‾‾‾‾‾⌐

(b)

Nondysgenic crosses   Hybrid dysgenesis

P♂ × P♀   M♂ × M♀   M♂ × P♀   P♂ × M♀
  ↓   ↓   ↓   ↓
Normal   Normal   Normal   Semisterile
progeny   progeny   progeny   progeny × P or M
                                             ↓
                                          Second-generation
                                          progeny with many
                                          mutations and
                                          chromosome
                                          rearrangements

(c)

P♂ × M♀

Embryo   P elements
Germ line   transpose
pole cells
                                Germ line cell
Grows into

Semisterile fly ♀ × P or M mate

Progeny with many mutations
and chromosome rearrangements
due to P-element transpositions

Figure 21.7 P elements and hybrid dysgenesis. (a) Structure of a P element. The P-element primary transcript encodes a transposase enzyme that acts on the 31-bp inverted repeats at P-element ends to catalyze transposition. (b) Hybrid dysgenesis occurs only when P males (with P elements) mate with M females (no P elements). The progeny of such a mating are semisterile because of active P-element transposition in their germ lines. Hybrid dysgenesis is not seen in M × M matings without P elements, nor in crosses involving P females, whose eggs contain high amounts of a repressor protein (encoded by an alternatively spliced P-element mRNA) that prevents transcription of the transposase gene. (c) Events during a dysgenic cross. Eggs produced by M females have no repressor protein, so P elements brought into the embryo in sperm from P fathers will transpose rapidly. Transposition only occurs in germ line cells because a splicing event needed to make transposase mRNA does not take place in somatic cells. The progeny of a dysgenic mating thus appear normal, but they are semisterile because their gametes carry many chromosomal aberrations caused by P-element movement.

transposon mobility causes reduced fertility in the hybrids; the few progeny that do survive carry many new mutations induced by the insertion of P elements at new sites in the genome. Hybrid dysgenesis does not occur in any other combination of M or P males and females, such as in crosses of P males and P females (Fig. 21.7b). The dramatic increase in transposition seen in crosses of P strain males to M strain females probably occurs because eggs from M strain mothers lack a cytoplasmic repressor of P-element transposition that is normally present in eggs from P strain mothers (Fig. 21.7b and c).

The high mobility of P elements during hybrid dysgenesis makes it possible to exploit these elements for two purposes: as transformation vectors and as genetic tags.

Transformation: The Introduction of Cloned DNA into Fruit Flies

Drosophila geneticists realized from the following experiment that P elements can serve as vectors to introduce cloned DNA into the organism. The researchers first constructed a recombinant plasmid that contained a fragment of *Drosophila* genomic DNA including an intact P element. When they injected this plasmid into early syncytial embryos from M strain mothers, the P element began to transpose and insert itself into the chromosomes of host germ line cells at a high frequency. This artificial situation thus mimics the phenomenon of hybrid dysgenesis. Additional experiments demonstrated that this transposition is catalyzed by the P-element-encoded transposase and requires the presence of the inverted terminal repeats normally found at P-element ends.

For germ line transformation, *Drosophila* geneticists use vectors that contain the P-element ends but not the gene for transposase. They clone novel DNA transgene sequences between the P-element ends and inject the resulting DNA construct into eggs from M strain mothers. A helper P element is coinjected to provide a source of the P-element transposase; this helper cannot itself integrate into a chromosome because it lacks both of the P-element ends. Figure 21.8 illustrates one protocol for using P elements to introduce cloned genes into fruit flies. Note that pCaSpeR, one of the most frequently used transformation vectors, contains a *white*$^+$ marker gene that makes it easy to trace the presence of the vector in the genome by the red eye color it confers.

P-Element Tagging Has Enabled the Cloning of Many Genes of Developmental Interest

The basic strategy is to isolate a mutant allele of the desired gene that arose from insertion of a P element, and then to recover sequences from the gene of interest that are in the DNA flanking the P element. A convenient way to accomplish this is first to mate two strains. One carries a chromosomal copy of a defective P element that provides a large amount of transposase but is itself not mobile. The other strain carries a specially engineered P element that contains both a marker such as the *w*$^+$ eye color gene that makes it possible to keep track of this P element, as well as bacterial plasmid sequences with an

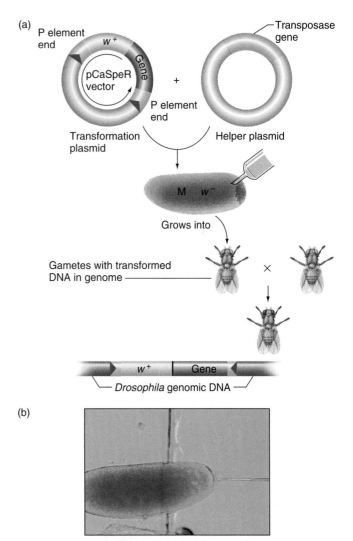

Figure 21.8   P-element-mediated transformation of *Drosophila*. (a) A transgene (Gene) is cloned into a transformation vector (pCaSpeR) containing the *white*$^+$ (*w*$^+$) marker gene between the P-element inverted ends, as shown. Researchers coinject the resulting transformation plasmid, along with a helper P element that can encode transposase but has no P-element ends (so it cannot move), into M-type *w*$^-$ fertilized embryos. The adult flies that grow from the injected embryos have white eyes because the *w*$^+$ gene cannot transpose into the genome in somatic cells. However, transposition of a segment containing the *w*$^+$ gene and the gene of interest occurs in some germ line cells. If these adults are mated with *w*$^-$ partners, some progeny in the next generation will have red (*w*$^+$) eyes and an integrated transgene. (b) Photomicrograph of the injection of DNA into a *Drosophila* embryo.

antibiotic resistance gene, whose purpose will become apparent in a moment. After the two strains mate, transposase in the germ line cells of the F$_1$ flies catalyzes transposition of the specially marked P element. Researchers can identify movements of the marked P element to new sites on other chromosomes by following the *w*$^+$ marker in the progeny of the F$_1$ flies.

If these new insertions cause mutations in interesting genes, the researchers can rapidly clone flanking DNA containing the genes of interest by a technique called *plasmid res-*

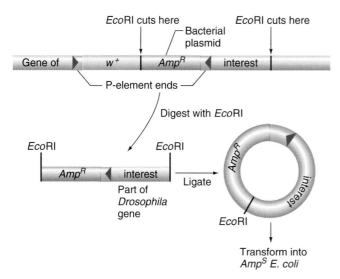

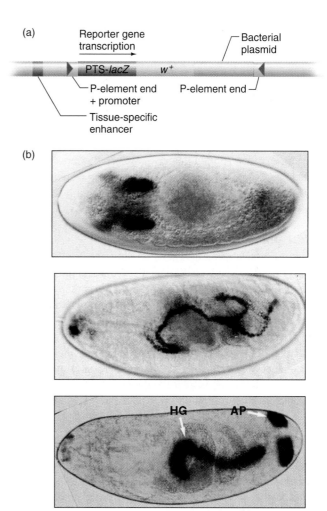

Figure 21.9 Plasmid rescue of a transposon-tagged gene. A specially engineered P element contains bacterial plasmid sequences including a gene for ampicillin resistance (Amp^R). This P element has jumped into a *Drosophila* gene, causing an interesting phenotype. To identify this gene, researchers digest genomic DNA from the mutant strain with an enzyme like *Eco*RI that cuts on one side of the bacterial plasmid but nowhere else in the P element. The next recognition site for this enzyme must be in the *Drosophila* genomic DNA flanking the P element. When the restriction-cut genomic DNA is circularized by treatment with ligase and transformed into Amp^S bacterial cells, the only ampicillin-resistant bacterial colonies will harbor a plasmid that includes the *Drosophila* DNA flanking the P-element insertion.

cue, illustrated in Fig. 21.9. In this technique, they simply isolate genomic DNA from the mutant insertion strain, cut the DNA with a restriction enzyme, circularize the restriction fragments with DNA ligase, and use the circularized DNA to transform *Escherichia coli.* The only bacterial colonies able to grow on antibiotic-containing medium will contain the bacterial plasmid sequences plus the gene-containing *Drosophila* DNA flanking the P-element insertion.

Enhancer Trapping: Using P Elements to Identify Genes by Their Expression Patterns

Drosophila developmental geneticists have constructed P elements in which the *E. coli lacZ* gene was fused in frame to the P transposase gene. Expression of *lacZ* in these elements is thus under the control of the P-element promoter (Fig. 21.10a). Remarkably, when an element is mobilized, over 65% of the new insertions isolated show patterned expression of the *lacZ* reporter during development, with the *lacZ* gene on in some tissues but off in others. It appears that the patterned expression results from the fact that the P-element promoter frequently comes under the control of nearby regulatory sequences (enhancers) of genes near its new insertion site. (See Chapter 16 for a definition and discussion of enhancers.) In support of this idea, many insertions into known genes have been recovered, and in most of these insertions, the expression pattern of β-galactosidase (the protein product of *lacZ*) accurately reflects the normal expression of the gene into which the element has

Figure 21.10 Enhancer trapping. (a) This "enhancer trap" contains a reporter gene in which coding sequences for the N-terminal region of P-element transposase (PTS) were fused in frame with *lacZ* sequences encoding β-galactosidase. The enhancer trap was introduced into the genome by transformation, then transposed to novel locations by mating with a strain carrying a source of transposase. The P-element promoter is by itself very weak, but transcription from the promoter will increased if the P element lands near an enhancer in the *Drosophila* genome. This will turn on the expression of the fusion protein reporter, which can be detected by staining for β-galactosidase activity. (b) Detection of tissue-specific enhancers. Late embryos from three lines of flies with the P-element "enhancer trap" inserted into different genomic locations were stained for β-galactosidase (blue). Expression of the reporter is seen in the salivary glands (top); in the Malpighian tubules (middle), and in the hindgut (HG) and anal pads (AP) in the bottom panel because of the activity of tissue-specific enhancers adjacent to the enhancer trap insertion sites.

been inserted (Fig. 21.10b). The identification of P-element insertion lines with particular β-galactosidase patterns is known as **enhancer trapping.** It is a powerful way to detect genes that are turned on in specific tissues. Since the most frequently used enhancer trap elements carry *E. coli* plasmid sequences, genes showing β-galactosidase expression patterns of particular interest can be rapidly isolated by plasmid rescue.

A particularly good example of the utility of P-element tagging and enhancer trapping is seen in the work of Thomas

Kornberg and his colleagues who have studied the function of a gene called *engrailed* (*en*), which plays a key role in establishing and patterning body segments. As described in a later section, *en* encodes a transcription factor that is expressed in a stripe in the posterior portion of each body segment. To identify target genes regulated by *en,* Kornberg's group screened for enhancer trap insertions that are expressed in the same or a complementary pattern. In this way, they identified several key targets of *en.* Among them is the crucially important gene called *hedgehog* (*hh*), which, as explained later, encodes a secreted protein responsible for patterning not only body segments, but many other structures as well.

The Production of Genetic Mosaics

Genetic mosaics are individuals composed of cells of more than one genotype. They enable scientists interested in *Drosophila* developmental genetics to answer many types of questions about the developmental fate of various cells and the functions of particular genes.

The most important method for generating mosaics is mitotic recombination, previously introduced in the Genetics and Society box in Chapter 4. Although crossing-over is normally restricted in *Drosophila* to the prophase of meiosis I in females, after irradiation of embryos or larvae, crossing-over between homologs occurs quite frequently in somatic mitotic cells. It is thought that X-ray-induced chromosome breakage initiates mitotic recombination. If mitotic recombination occurs in the G_2 portion of the cell cycle, and is followed by the appropriate segregation of chromatids in the subsequent mitotic division, the resulting daughter cells are each homozygous for all loci distal to the crossover (that is, for all genes

farther away from the centromere than the site of recombination). The mitotic descendants of these cells then produce genetically distinct patches in the emerging adults (review Fig. A on page 115 in Chapter 4).

We discuss here three of the many important applications of genetic mosaics in the analysis of *Drosophila* development.

Testing the Potential Ability of a Cell and Its Mitotic Descendants to Contribute to Particular Adult Structures

A cell produced by mitotic recombination and its clonal descendants may have a phenotype that marks them as different from their neighbors and thus enables them to be tracked through the organism's development. If the fate of the initial cell produced by mitotic recombination is already restricted to a tissue or region, all of the cells in the clone will be found only in that one part of the body. By contrast, if the initial cell produced by mitotic recombination has a less restricted developmental potential, the cells of the clone will end up in two or more different tissues or regions of the body.

In various studies of cell fate, developmental geneticists have observed that clones are in fact often limited to specific regions known as **lineage compartments.** In one experiment, for example, investigators exposed *Minute⁻/Minute⁺ Drosophila* embryos to X irradiation during an early stage of embryonic development to induce the formation of *Minute⁺/Minute⁺* (+/+) mitotic clones (Fig. 21.11). *Minute⁻* dominant mutations disrupt ribosomal protein genes, causing slow development and reduced bristle size. Thus, in a *Minute⁻/Minute⁺* background, the +/+ cells produced by mitotic recombination have a strong growth advantage over their neighbors and produce very large clones. These large *Minute⁺* clones can fill

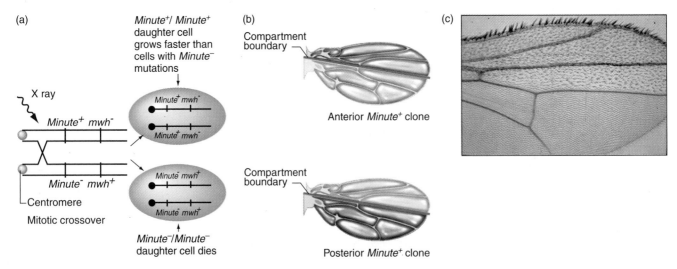

Figure 21.11 Mosaic analysis demonstrates the existence of developmental compartments. (a) Mitotic crossovers induced by X rays in the cells of embryos with the genotype shown at the left can generate daughter cells homozygous for *Minute⁺* and *mwh⁻*. Because these cells and their mitotic descendants are *Minute⁺*, they will outgrow the cells in which no mitotic recombination took place. The *Minute⁺* clone is marked by *mwh⁻*, a recessive mutation that causes each wing cell to produce several tiny hairs instead of the normal one. Homozygous *Minute⁻* daughter cells die and are not seen in the wing. (b) Large *Minute⁺* clones (dark gray) can fill the anterior or the posterior half of the wing, but will not cross the compartment boundary. (c) Photograph of an actual anterior *Minute⁺* clone; note that the wing cells in the clone are large and have several small hairs as they are homozygous for *mwh⁻*.

either the anterior half of the wing (one compartment) or the posterior half (another compartment), but they can never cross a boundary line separating the two. Thus, the *Minute⁺/Minute⁺* (+/+) cell formed at the time of mitotic recombination already has a restricted developmental potential: Its descendants can populate either the posterior wing compartment or the anterior wing compartment, but not both. By inducing mitotic recombination later in embryogenesis, researchers observed that the developmental potential of clones of cells becomes increasingly restricted to smaller and smaller subcompartments (for example, the anterior ventral or anterior dorsal regions of the wing) as development proceeds.

Using Mitotic Recombination to Determine When and Where a Gene Must be Expressed to Generate a Normal Phenotype

Ommatidia are the facets of the fly's compound eye (Fig. 21.12a). *Drosophila* geneticists have used mosaic analysis to study the normal development of the ommatidia. During the de-

velopment of the eye, a group of eight undifferentiated cells is sequentially recruited into a clump of cells that differentiate to eight photoreceptors, designated R1 to R8. These eight differentiated cells eventually become the light-sensing cells of a single facet. The last photoreceptor cell to differentiate is R7 (Fig. 21.12b). Geneticists have examined some aspects of the recruitment of R7 by isolating mutant flies whose ommatidia lack R7 but contain the other seven photoreceptor cells. One gene defined by such mutations is called *sevenless* (*sev*); another is *bride of sevenless* (*boss*). In what cells must these two genes normally be expressed if R7 is to become part of the eye facet?

To answer this question for the *sevenless* gene, biologists generated mosaics with marked cells that were mutant for both *white* and *sevenless* in an eye where other cells carried the dominant wildtype alleles for the two genes. Some facets contained a mix of red and white photoreceptor cells (Fig. 21.12c). But in every facet with an R7 cell, the R7 cell was red (that is, had both the *w⁺* and *sev⁺* genes), even if all the other photoreceptor cells in the same facet were homozygous for *w⁻* and *sev⁻*. There

(a)

(d)

(b)

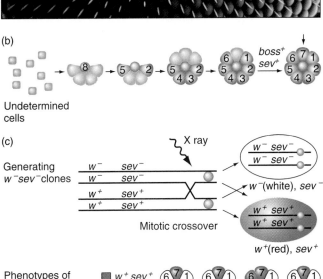

(c)

Figure 21.12 Using mosaics to determine if a gene product is cell autonomous or nonautonomous. (a) Scanning electron micrograph of a compound adult *Drosophila* eye, showing the individual facets (ommatidia). (b) During the development of the eye, eight photoreceptor cells are sequentially recruited into each ommatidium. Blue color indicates the commitment of cells to a photoreceptor fate. The products of the wildtype genes *sev⁺* and *boss⁺* are needed for the recruitment and specification of the last cell, which becomes photoreceptor 7 (R7). (c) Mosaic analysis shows that the requirement for the *sev⁺* gene product is cell autonomous: If the presumptive R7 cell is *sev⁺*, it develops into an R7 photoreceptor regardless of whether any or all of the other cells in the ommatidium are *sev⁺* or *sev⁻*. If the presumptive R7 cell is *sev⁻*, it never develops into an R7 cell and instead develops into a different kind of cell called a cone cell (not shown). (d) A model for the action of the Sev⁺ and Boss⁺ proteins in specification of the R7 cell. Blue color again indicates commitment of cells to a photoreceptor fate. The Boss⁺ protein is a ligand expressed on the surface of the R8 cell. Binding of the Boss⁺ ligand to the Sev⁺ receptor on the surface of the R7 precursor cell activates a signal cascade that regulates the expression of genes determining an R7 fate.

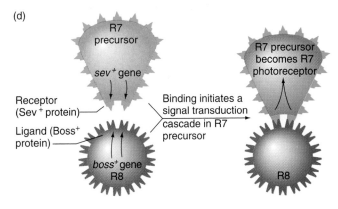

were no facets in which the R7 cell was white. These results suggest that the requirement for the *sev*$^+$ gene product is **cell autonomous:** To develop correctly, a presumptive R7 cell must make the Sev$^+$ protein, and it does not matter whether the protein is made in any other photoreceptor cell in the same facet (Fig. 21.12c). The Sev$^+$ protein affects only the cell in which it is made.

By contrast, when researchers performed the same type of analysis for *boss,* the results were very different. An R7 cell develops properly only if an R8 cell makes the Boss$^+$ protein; this is true even if the R7 cell itself is *boss*$^-$. If the R8 cell is *boss*$^-$, however, the facet does not have an R7 cell. These results show that the Boss$^+$ protein is not cell autonomous: It is produced in one cell (R8), but affects the differentiation of a different cell (R7).

Thus, if a trait is cell autonomous, all cells with a mutant genotype, and only those cells, will express the mutant phenotype, indicating that the gene product acts within the cell in which it is made. With a **nonautonomous trait,** genotypically mutant cells cause other cells to have a mutant phenotype regardless of the other cells' genotype, indicating that the gene product influences cells other than those in which it is made (for example, because the protein diffuses to or otherwise comes in contact with the other cells).

The cloning of the *sev* and *boss* genes made it possible to understand the different behavior of the two genes in the mosaic analysis in terms of the protein functions predicted by each gene's DNA sequence. The *sev* gene encodes a receptor protein embedded in the membrane of the developing R7 cell, while the *boss* gene encodes a protein located on the surface of the R8 cell. The binding of the Boss$^+$ protein on R8 (the ligand) to the Sev$^+$ receptor on the presumptive R7 cell initiates a cascade of gene activity (a signal transduction pathway; see Chapter 17) in the potential R7 cell that triggers its differentiation into a mature R7 cell (Fig. 21.12d).

Using Mosaic Analysis to Determine Whether Genes Needed for Viability of the Embryo Also Play a Role in the Formation of Structures Later in Development

Suppose you are studying a gene you know is needed for early embryogenesis; animals homozygous for mutations of this gene die before they hatch into larvae. You want to know whether the product of the gene does or does not play a role in the production of structures that appear much later in development, for example, the ovaries of adult females. You cannot answer this question by simply looking at the homozygous mutant, because it dies before the ovaries form. The use of mosaics circumvents this problem. It allows the growth of an adult heterozygous for the early lethal mutation, in which you can induce homozygous mutant clones in the structure of interest by mitotic recombination. Figure 21.13 illustrates an example of this kind of strategy: You make a female whose genetic constitution is

centromere – *ovo*$^+$ – *let*$^-$/centromere – *ovo*^{D} – *let*$^+$,

where *let*$^-$ is the early lethal mutation in question, and *ovo*^{D} is a dominant mutation that prevents any development of the ovaries. Mitotic recombination between the centromere and the

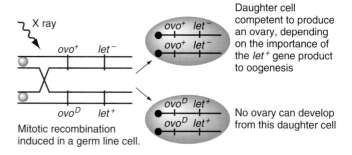

Figure 21.13 Using mosaics made by mitotic recombination to determine the importance of a gene product for oogenesis. A female of the genotype on the left cannot produce any ovaries because of the dominant *ovo*^{D} mutation, unless mitotic recombination creates an *ovo*$^+$/*ovo*$^+$ ovarian precursor cell. Such a cell will also be homozygous for the lethal mutation (*let*$^-$) in question. If the *let*$^+$ gene product is not needed for oogenesis, an ovary will be created, and a mature egg will be produced. If the gene product is needed for oogenesis, the only ovaries or eggs produced by this female will be defective. Phenotypic analysis of the defective ovaries or eggs reveals the importance of the *let* gene product for oogenesis.

ovo gene produces a germ line cell that is simultaneously homozygous for *ovo*$^+$ and *let*$^-$. This cell allows the generation of an ovary (because it is *ovo*$^+$), but if the missing *let*$^+$ gene product is needed in the female germ line for proper ovarian differentiation, these females will have defective ovaries or produce defective eggs.

Ectopic Expression

Although the effects of eliminating function (as in a null mutant) are of paramount importance in assessing the function of a gene, it is often helpful to determine the effects of **ectopic expression,** that is, expression outside the cells or tissue where the gene is normally expressed. Analysis of the *Drosophila eyeless* (*ey*) gene provides a good example of the usefulness of studying ectopic expression.

As indicated by the name of the gene, homozygotes for loss-of-function alleles of *ey* have no eyes at all, or at best, very small eyes (Fig. 21.14a). Surprisingly, mutations in the evolutionarily closely related genes *Pax-6* in mice (Fig. 21.14b) and *Aniridia* in humans also reduce or totally abolish eye formation. These observations suggest that *ey* and its mouse and human homologs play a significant role in eye development, but the precise nature of this role was for many years unclear. To deal with this problem, scientists sought to determine whether expressing the *ey* gene in tissues other than those that ordinarily produce eyes could cause development of an additional eye or any other effects.

In their studies, they placed a *Drosophila* cDNA sequence encoding the Ey$^+$ protein downstream and therefore under the control of a promoter for a "heat-shock" gene that is highly transcribed when the temperature rises. When they introduced this artificial heat shock promoter–*ey*$^+$ gene into the genome by P-element-mediated transformation, they could turn on expression of the Ey$^+$ protein in all cells by subjecting the flies to elevated temperatures. As Fig. 21.14c illustrates, the results were dramatic: Ectopic eye tissue could develop virtually any-

(a)

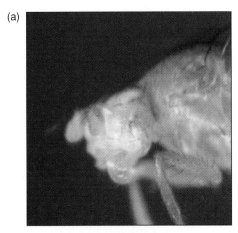

(b)

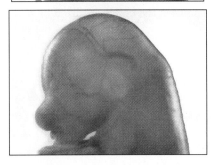

(c)   Ectopic red
       eye tissue

Ectopic red
eye tissue

Figure 21.14 The *eyeless* gene is a master control switch for eye development. (a) Null or hypomorphic mutations of the *eyeless* gene reduce the size of eyes or completely abolish them in adult flies. (b) Null or hypomorphic mutations in the homologous mouse gene *Pax-6* also reduce or abolish eye development. Top: wildtype mouse fetus. Bottom: *Pax-6* mutant fetus. (c) Ectopic expression of the *eyeless* gene causes the appearance of ectopic eye tissue in other parts of the fly's body.

where on the fly's body—such as in the antennae, legs, or wings! It thus seems that synthesis of the Ey$^+$ protein can activate a program of eye development. This fits with the fact that this protein appears to be a transcription factor able to turn on a number of target genes.

Surprisingly, the *Pax-6*$^+$ gene from the mouse has the same effect as *ey*$^+$ when it is expressed ectopically in *Drosophila*. This is remarkable because the eyes of vertebrates and insects are so dissimilar. Apart from sharing a similar light-sensing pigment (rhodopsin, described in Chapter 6), the eyes of insects and vertebrates have almost no features in common. For example, their structures and developmental origins are completely different: Insect eyes are composed of many ommatidia; the vertebrate eye is a single camera-like organ. Biologists had long assumed that the two types of eyes evolved independently, but the homology of *Pax-6* and *ey* suggests that the eyes of vertebrates and insects evolved from a single prototypical light-sensing organ, and that *ey/Pax-6* is a universal master control gene that can turn on programs for eye development in many animal phyla.

The Drosophila *Genome Project*

Drosophila, like other model organisms, is the subject of a very active genome project. The primary goal of the *Drosophila* project is to map and sequence the euchromatic portion of the genome. In addition, researchers on the *Drosophila* project seek to exploit the many genetic advantages of the fly to obtain functional as well as structural information about genes.

For example, a large-scale effort is under way to isolate P-element-insertion alleles of as many genes as possible. Studies of the phenotypes associated with these mutations should provide important information about the biological functions of the disrupted genes. The P elements used are enhancer traps that enable the rapid determination of the expression pattern of the disrupted gene, and also carry bacterial plasmid sequences that allow the easy cloning of flanking sequences by plasmid rescue. So far, investigators have isolated P-element disruptions for about 25% of the estimated 3600 essential *Drosophila* genes. Mapping of the P elements by *in situ* hybridization to polytene chromosomes reveals that the insertions are uniformly distributed throughout the genome. Since P elements tend to transpose only short distances, mobilization of existing insertions should make possible the rapid recovery of insertions in many of the remaining genes.

Based on the density of transcription units and measurements of transcript complexity, *Drosophila* genomists estimate that there are about 12,000–15,000 protein-coding genes in *Drosophila,* about twice the number of genes in yeast and 1/4 to 1/8 the number estimated for humans. Since genetic studies show that only about 3600 of these genes are essential, about 2/3 of *Drosophila* genes may be dispensable for the organism's survival, at least in the laboratory. A major challenge for the future will be to determine the functions of the nonessential genes. Isolation of P-element disruptions in these genes will be the first step toward identifying their functions. Some of these genes may appear to be inessential because other genes with identical or overlapping functions exist in the genome. The construction of strains carrying mutations in a group of related genes may reveal these overlapping functions. Another possibility is that mutations in nonessential genes could affect specific processes influencing such variables as fertility, life span, or a visible phenotype. The generation of ectopic expression by the methods described earlier also promises to reveal information about the functions of many genes identified by the *Drosophila* Genome Project.

THE GENETIC ANALYSIS OF BODY PLAN DEVELOPMENT IN DROSOPHILA: A COMPREHENSIVE EXAMPLE

The immensely powerful combination of genetic and molecular techniques available in *Drosophila* has revolutionized our understanding of development. And because most of the developmental regulators identified in *Drosophila* are highly conserved in other animals, the *Drosophila* model is having a profound influence on other fields, including medicine (in fact many developmentally important genes in humans were first identified as homologs of genes found in *Drosophila*). A vivid illustration of the impact of *Drosophila* is the work on genes controlling the fly's body plan.

Contributions from many laboratories have revealed how the body of *Drosophila* becomes specialized along both its anterior-posterior (AP) and dorsal-ventral (DV) axes. Together, these studies constitute one of the most successful genetic dissections ever performed. Here we consider only development of the AP body axis, that is, how the body differentiates along the line running from the front to the rear of the animal.

Very early in development, the action of a large group of genes known as the **segmentation genes** subdivides the body into an array of essentially identical body segments. Expression of a second set of genes called **homeotic genes** then assigns a unique identity to each body segment. In studies of the genes that specify and regulate development of the AP body axis, researchers have tried to answer two basic questions: How does the developing animal establish the proper number of body segments? and How does each body segment "know" what kinds of structures it should form and what role it should play in the animal's biology?

Early Development of the Basic Body Plan

To understand how the segmentation and homeotic genes function, it is helpful to consider some of the basic events that take place in the first few hours of *Drosophila* development (Fig. 21.15a and b). The egg is fertilized in the uterus as it is being laid, and the meiotic divisions of the oocyte nucleus, which had previously arrested in the metaphase of meiosis I,

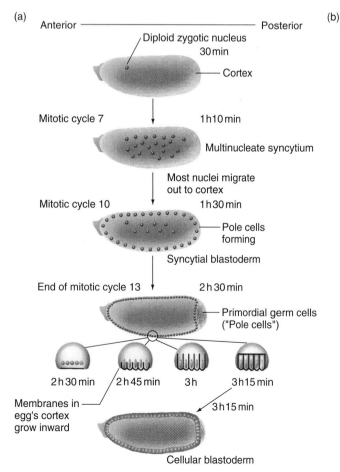

(a)

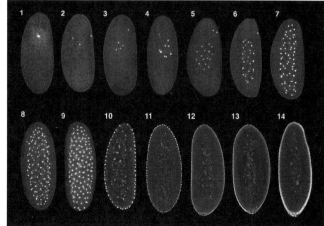

(b)

Figure 21.15 Early *Drosophila* development: From fertilization to formation of the cellular blastoderm. (a) A diagram of the first 3 hours postfertilization. The original zygotic nucleus undergoes 13 very rapid mitotic divisions in a single syncytium. By the ninth round of division, most nuclei migrate outward toward the egg surface (cortex); a few of these at the posterior end of the embryo are encased in membranes, forming the germ line *pole cells*. At the *syncytial blastoderm* stage, the egg surface is covered by a monolayer of nuclei (except for the pole cells at the posterior end). At the end of the thirteenth division cycle, cell membranes extend downward from the embryo surface, enclosing the nuclei into separate cells and thus producing a *cellular blastoderm*—a sheet of cells encircling the embryo's periphery. (b) Photomicrographs showing early embryonic stages in cross section. Nuclei are stained with a fluorescent dye for DNA. A different view of the syncytial blastoderm can be found in Fig. 3.10 on page 79.

resume at this time. After fusion of the haploid male and female pronuclei, the diploid zygotic nucleus of the embryo undergoes 13 rounds of nuclear division at an extraordinarily rapid rate, with the average time of mitotic cycles 2 through 9 being only 8.4–8.8 minutes. Nuclear division in early *Drosophila* embryos, unlike most mitoses, is not accompanied by cell division, so that the early embryo is a multinucleate syncytium. During the first 8 division cycles, the multiple nuclei are centrally located in the egg; during the ninth division, most of the nuclei migrate out to the cortex—just under the surface of the embryo—to produce the **syncytial blastoderm.** During the tenth division, nuclei at the posterior pole of the egg are enclosed in membranes that invaginate from the egg cell membrane to form the first embryonic cells; these "pole cells" are the primordial germ cells. At the end of the thirteenth division cycle, about 6000 nuclei are present at the egg cortex. During the interphase of the fourteenth cycle, membranes in the egg's cortex grow inward between these nuclei, creating an epithelial layer called the **cellular blastoderm** that is one cell deep (Figs. 21.15 and 21.16a). The embryo completes formation of the cellular blastoderm about 3 hours after fertilization.

At the cellular blastoderm stage, no regional differences in cell shape or size are apparent (with the exception of the pole cells at the posterior end). Experiments in which blastoderm cells have been transplanted from one location to another, however, show that despite this morphological uniformity, the segmental identity of the cells has already been determined. Consistent with this finding, molecular studies reveal that most segmentation and homeotic genes function during or before the cellular blastoderm stage.

Immediately after cellularization, **gastrulation** and establishment of the embryonic germ layers begin. The *mesoderm* forms by invagination of a band of midventral cells that extends most of the length of the embryo. This infolding (the ventral furrow; see Fig. 21.16b) produces an internal tube whose cells soon divide and migrate to produce a mesodermal layer. The *endoderm* forms by distinct invaginations anterior and posterior to the ventral furrow, one of these invaginations is the cephalic furrow seen in Fig. 21.16b. The cells of the endodermal infoldings migrate over the yolk to produce the gut. Finally, the nervous system arises from neuroblasts that segregate from bilateral zones of the ventral *ectoderm*. The first visible

(a)

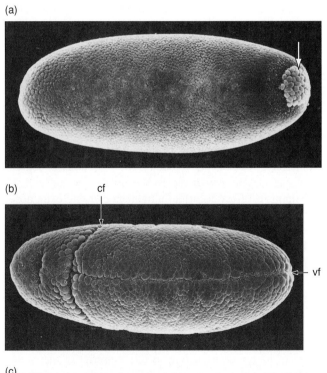

(b) cf

(c)

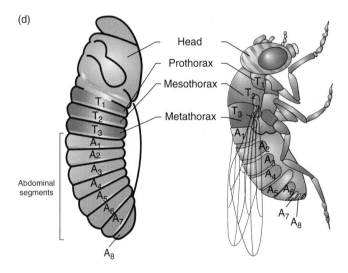

(d)

Figure 21.16 *Drosophila* development after formation of the cellular blastoderm. (a) Scanning electron micrograph of a cellular blastoderm; individual cells are visible at the periphery of the embryo, and the pole cells at the posterior end can be distinguished (arrow). (b) A ventral view of some of the furrows that form during gastrulation, roughly 4 hours postfertilization: vf, ventral furrow; cf, cephalic furrow. (c) By 10 hours postfertilization, it is clear that the embryo is subdivided into segments; Ma, Mx, Lb are the three head segments (mandibular, maxillary, and labial, respectively). CL, PC, and O refer to nonsegmented regions of the head. The three thoracic segments (labeled T_1, T_2, and T_3) are the prothorax, the mesothorax, and the metathorax, respectively, while the abdominal segments are labeled A_1 to A_8. (d) The identity of embryonic segments (left) is preserved through the larval stages and is also retained through metamorphosis into the adult (right). Head segments are not distinguished here for simplicity.

signs of segmentation are periodic bulges in the mesoderm, which appear about 40 minutes after gastrulation begins. Within a few hours of gastrulation, the embryo is divided into clear-cut body segments that will become the three head segments, three thoracic segments, and eight major abdominal segments of the larva that eventually hatches from the eggshell (Fig. 21.16c). Even though the animal eventually undergoes metamorphosis to become an adult fly, the same basic body plan is conserved in the adult stage (Fig. 21.16d).

The genes responsible for the formation of segments fall into four classes, and function in a regulatory hierarchy that progressively subdivides the embryo into successively smaller units. In the order of their expression, these classes are: (1) maternal genes; (2) gap genes; (3) pair-rule genes; and (4) segment-polarity genes.

Specification of Segment Number: Maternal Genes Interact to Produce Gradients of Morphogens

There is very little transcription of genes in the embryonic nuclei between fertilization and the end of the 13 rapid syncytial divisions that immediately follow fertilization. Because of this near (but by no means total) lack of transcription, developmental biologists hypothesized that formation of the basic body plan initially requires **maternally supplied components** deposited by the mother into the egg during oogenesis. How could they identify the genes encoding these maternally supplied components? Christiane Nüsslein-Volhard and Eric Wieschaus realized that the embryonic phenotype determined by such genes does not depend on the embryo's own genotype; rather it is determined by the genotype of the mother. By asking whether mothers homozygous for candidate mutations would produce defective embryos, they devised genetic screens to identify recessive mutations in maternal genes that influence embryonic development. To do this, they established individual balanced stocks for thousands of mutagen-treated chromosomes, and then examined the phenotypes of embryos obtained from homozygous mutant mothers. Through these large-scale screens, they identified a large number of maternal genes that are required for the normal patterning of the body. (For their work, they shared the Nobel Prize for physiology or medicine with Edward B. Lewis—whose work we describe later.) We focus here on two groups of genes: one group is required for normal patterning of the embryo's anterior; the other group is required mainly for normal posterior patterning. The genes in these two groups are the first genes activated in the process that determines segment number.

The finding that separate groups of maternal genes control anterior and posterior patterning in the embryo is consistent with the conclusions of classical embryological experiments. Studies in which polar cytoplasm was transplanted or preblastoderm embryos were separated into two halves by constriction of the embryo with a fine thread suggested that the insect body axis is patterned during cleavage by the interaction of two signaling centers located at the anterior and posterior poles of the egg. In a specific model, Klaus Sander proposed that each pole of the egg produces a different substance, and that these substances form opposing gradients by diffusion. He suggested that the concentrations of these substances then determine the types of structures produced at each position along the body axis. Molecular characterization of the maternal genes of the anterior and posterior groups indicates that the Sander model for body axis patterning is essentially correct. Substances that define different cell fates in a concentration-dependent manner are known as **morphogens.**

In Drosophila, the bicoid (bcd) Gene Encodes the Anterior Morphogen

Embryos from mothers homozygous for null alleles of *bcd* lack all head and thoracic structures. The protein product of *bcd* is a DNA-binding transcription factor whose transcript is localized near the anterior pole of the egg cytoplasm (Fig. 21.17a). Translation of the *bcd* transcripts takes place after fertilization, and the Bcd protein diffuses to produce a high-to-low, anterior-to-posterior concentration gradient that extends over the anterior two-thirds of the embryo by the ninth division cycle. This gradient determines most aspects of head and thorax development (Fig. 21.17b).

One of the first lines of evidence that the Bcd protein functions as a morphogen came from experiments in which the maternal dosage of the *bcd* gene varied (Fig. 21.17c). Mothers that carried only one dose of the *bcd* gene, instead of the normal diploid dose, incorporated about half the normal amount of *bcd* RNA into their eggs. As a result, translation yielded less Bcd protein, and the Bcd gradient was shallower and shifted to the anterior. In these Bcd-deficient embryos, the thoracic segments developed from more anterior regions than normal, and less of the body was devoted to the head. The opposite effect occurred in mothers carrying extra doses of the *bcd* gene. These and other observations suggested that the level of Bcd protein is a key to the determination of head and thoracic fates in embryo. Three other genes work with *bcd* in the anterior group of maternal genes; the function of the protein products of these three genes is to localize *bcd* transcripts to the egg's anterior pole.

The Bcd protein works in two ways: as a transcription factor that helps control the transcription of genes farther down the regulatory pathway, and as a translational repressor. The target of its repressor activity is the transcript of the *caudal* (*cad*) gene, which also encodes a DNA-binding transcription factor. The *cad* transcripts are uniformly distributed in the egg before fertilization, but because of translational repression by the Bcd protein, translation of these transcripts produces a gradient of Cad protein that is complementary to the Bcd gradient, with a higher concentration of Cad protein at the posterior end of the embryo and lower concentrations toward the anterior (Fig. 21.18). The Cad protein plays an important role in activating genes expressed later in the segmentation pathway to generate posterior structures.

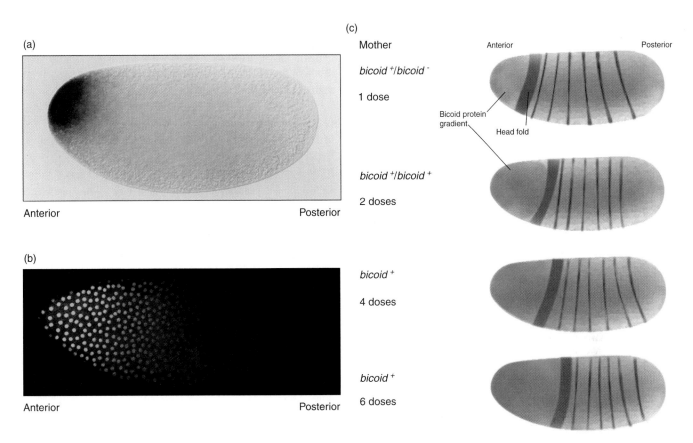

Figure 21.17   The Bicoid protein is the anterior morphogen. (a) The *bicoid* (*bcd*) mRNA (visualized by *in situ* hybridization of a *bicoid* probe to RNA) is concentrated at the anterior (left) tip of the embryo. (b) The Bicoid (Bcd) protein (seen by green antibody staining) is distributed in a gradient: high at the anterior end, and trailing off toward the posterior end. It can be seen that the Bcd protein (a transcription factor) accumulates in the nuclei of this syncytial blastoderm embryo. (c) The greater the maternal dosage of *bcd*+, the higher the concentration of Bcd protein in the embryo, and the more of the embryo that is devoted to anterior structures. This is because the concentration of Bcd protein that signals the position of the head fold invagination is found farther toward the embryo's posterior the greater the amount of *bcd* mRNA localized at the anterior tip. Head structures will develop to the left of the head fold; thoracic and abdominal structures to its right.

The nanos (nos) *Gene Encodes the Primary Posterior Morphogen*

The *nos* RNA is localized to the posterior egg cytoplasm by proteins encoded by other posterior group maternal genes. Like *bcd* RNAs, *nos* transcripts are translated during the cleavage stages. After translation, diffusion produces a posterior-to-anterior Nos protein concentration gradient. The Nos protein, unlike the Bcd protein, is not a transcription factor; rather, the Nos protein functions only as a translational repressor. Its major targets are the maternally supplied transcripts of the *hunchback* (*hb*) gene, which are deposited in the egg during oogenesis and are uniformly distributed before fertilization. For development to occur properly, the Hb protein (another transcription factor) must be present in a gradient with high concentrations at the embryo's anterior and low concentrations at the posterior. The Nos protein, which represses the translation of *hb* maternal mRNA and is present in a posterior-to-anterior concentration gradient, helps construct the anterior-to-posterior Hb gradient by lowering the concentration of the Hb protein toward the embryo's posterior pole (Fig. 21.18). The embryo also has a second mechanism for es-

tablishing the Hb protein gradient that functions somewhat later: It only transcribes the *hb* gene from zygotic nuclei in the anterior region (see below).

Figure 21.18 summarizes how the coordinated activity of the maternal genes establishes polarities in the embryo that eventually result in proper embryonic segmentation. In addition to distributing the Bcd and Cad transcription factors in gradients within the embryo, the maternal genes ensure that the maternally supplied *hb* RNA is not translated in the posterior part of the embryo.

Specification of Segment Number through the Activation of Zygotic Genes in Successively More Sharply Defined Regions of the Embryo

The maternally determined Bcd, Hb, and Cad protein gradients control the spatial expression of zygotic segmentation genes. Unlike the products of maternal genes, which are placed in the egg during oogenesis, the products of zygotic genes are transcribed and translated from DNA in the nuclei of embryonic cells descended from the original zygotic nucleus. The expression of zygotic segmentation genes begins in the

mRNAs in oocyte

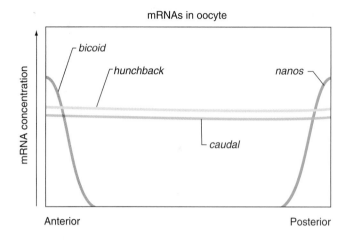

Proteins in early cleavage embryos

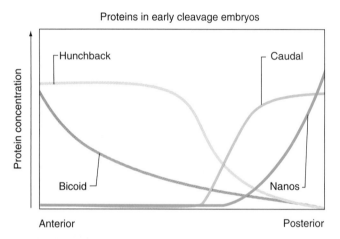

Figure 21.18 Distribution of the mRNAs and protein products of maternal-effect genes within the early embryo. (Top) In the oocyte prior to fertilization, *bicoid* (*bcd*) mRNA is concentrated near the anterior tip and *nanos* (*nos*) mRNA at the posterior tip, while maternally supplied *hunchback* (*hb*) and *caudal* (*cad*) mRNAs are uniformly distributed. (Bottom) By the early cleavage stages, the Bicoid (Bcd) and Nanos (Nos) proteins are translated and diffuse out from their sources at the tips. This creates a gradient of the Bcd protein high at the anterior and lower toward the posterior (A-to-P), and an opposite P-to-A gradient of Nos protein. The Bcd protein represses the translation of *cad* mRNA, producing a complementary P-to-A gradient of Caudal (Cad) protein. The Hunchback (Hb) protein is distributed in an A-to-P gradient for two reasons: Nos protein represses translation of the *hb* maternally supplied mRNA, and the Bcd protein activates transcription of the *hb* gene in anterior zygotic nuclei.

syncytial blastoderm stage, a few division cycles before cellularization (roughly cycle 10).

Most of the zygotic segmentation genes were identified in a second mutant screen also carried out in the late 1970s by Christiane Nüsslein-Volhard and Eric Wieschaus. In this screen, the two *Drosophila* geneticists placed individual ethylmethane-sulfonate (EMS)-mutagenized chromosomes into balanced stocks, and then examined homozygous mutant embryos from these stocks for defects in the segmentation pattern of the embryo. After screening several thousand such stocks

for each of the *Drosophila* chromosomes, they identified three classes of zygotic segmentation genes: gap genes (9 different genes); pair-rule genes (8 genes); and segment-polarity genes (about 17 genes). These three classes of zygotic genes fit into a hierarchy of gene expression.

Gap Genes, the First Zygotic Segmentation Genes to be Transcribed, Are Expressed in Broad Zones Along the Anterior-Posterior Axis

Embryos homozygous for mutations in the gap genes show a gap in the segmentation pattern (caused by an absence of particular segments) that corresponds to the position at which each gene is transcribed (Fig. 21.19a–c). How do the maternal transcription factor gradients ensure that the various gap genes are expressed in their broad stripes at the proper position in the embryo? Part of the answer is that the binding sites in the promoter regions of the gap genes have different affinities for the maternal transcription factors. For example, many gap genes are activated by the Bcd protein (the anterior morphogen). Gap genes such as *hb* with low-affinity Bcd protein binding sites will be activated only in the most anterior regions, where the concentration of Bcd is at its highest; by contrast, genes with high-affinity sites will be activated farther toward the posterior pole. Another part of the answer is that the gap genes themselves encode transcription factors that can influence the expression of other gap genes. The *Krüppel* (*Kr*) gap gene, for example, appears to be turned off by high amounts of Hb protein at the anterior end of its band of expression; activated within its expression band by Bcd protein in conjunction with lower levels of Hb protein; and turned off at the posterior end of its expression zone by the products of the *knirps* gap gene (Fig. 20.19c). (Note that the *hb* gene is usually classified as a gap gene, despite the maternal supply of some *hb* RNA, because the protein translated from the transcripts of zygotic nuclei actually plays the more important role.)

Pair-Rule Genes Subdivide the Body into Units That Are Two Segments in Length

After the gap genes have divided the body axis into rough, generalized regions, activation of the pair-rule genes generates more sharply defined sections (Fig. 21.20). These genes encode transcription factors that are expressed in seven stripes in preblastoderm and blastoderm embryos (Fig. 21.20a). The stripes have a two-segment periodicity; that is, there is one stripe for every two segments. Mutations in pair-rule genes cause the deletion of similar pattern elements from every alternate segment. For example, larvae mutant for *fushi tarazu* ("few segments" in Japanese) lack parts of abdominal segments A1, A3, and A7. Mutations in *even-skipped* cause loss of even-numbered abdominal segments.

There are two classes of pair-rule genes: primary and secondary. The striped expression pattern of the three primary pair-rule genes depends on the transcription factors encoded by the maternal genes and the zygotic gap genes. Specific elements within the upstream regulatory region of each pair-rule gene drive the expression of that pair-rule gene within a

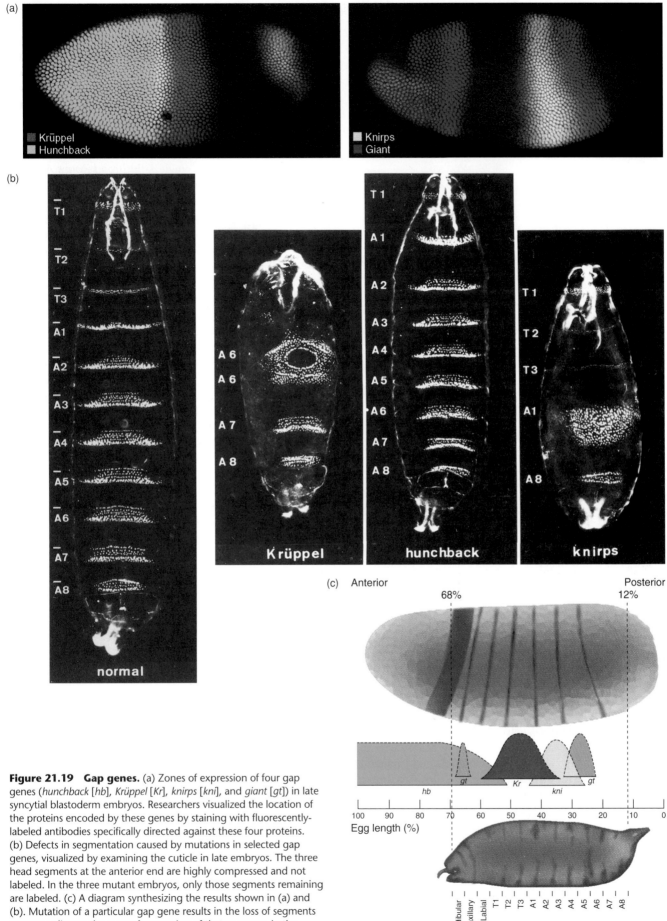

Figure 21.19 Gap genes. (a) Zones of expression of four gap genes (*hunchback* [hb], *Krüppel* [Kr], *knirps* [kni], and *giant* [gt]) in late syncytial blastoderm embryos. Researchers visualized the location of the proteins encoded by these genes by staining with fluorescently-labeled antibodies specifically directed against these four proteins. (b) Defects in segmentation caused by mutations in selected gap genes, visualized by examining the cuticle in late embryos. The three head segments at the anterior end are highly compressed and not labeled. In the three mutant embryos, only those segments remaining are labeled. (c) A diagram synthesizing the results shown in (a) and (b). Mutation of a particular gap gene results in the loss of segments corresponding to the zone of expression of that gap gene in the embryo.

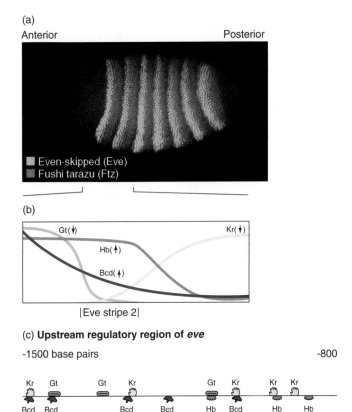

(a)

Anterior   Posterior

■ Even-skipped (Eve)
■ Fushi tarazu (Ftz)

(b)

Gt(↓)   Kr(↓)

Hb(↑)

Bcd(↑)

|Eve stripe 2|

(c) Upstream regulatory region of *eve*

-1500 base pairs   -800

Kr Gt   Gt Kr   Gt Kr   Kr Kr

Bcd Bcd   Bcd   Bcd   Hb Bcd   Hb   Hb

Figure 21.20 Pair rule genes. (a) Zones of expression of two pair-rule genes—*Fushi tarazu* (*Ftz*; shown in red), and *Even-skipped* (*Eve*; shown in green)—at the beginning of the cellular blastoderm stage. Each gene is expressed in seven stripes. Eve stripe 2 is the second green stripe from the left; the area detailed in part (b) is bracketed. (b) Eve stripe 2 is determined by the relative amounts of the Bcd protein and the proteins encoded by the gap genes *Kr, gt,* and *hb*. The formation of Eve stripe 2 requires activation of *eve* transcription by the Bcd and Hb proteins, and repression at its left and right ends by Gt and Kr proteins respectively. (c) The 700-bp upstream regulatory region of the *eve* gene that determines the expression of the Eve second stripe contains multiple binding sites for the four proteins shown in part (b).

particular stripe. For example, as Fig. 21.20b and c shows, the DNA element responsible for driving the expression of *even-skipped* (*eve*) in the second stripe contains multiple binding sites for the Bcd protein and the proteins encoded by the gap genes *Krüppel, giant* (*gt*), and *hb*. The transcription of *eve* in this stripe of the embryo is activated by Bcd and Hb, and repressed by Gt and Kr; and only in the stripe 2 region are Gt and Kr levels low enough and Bcd and Hb levels high enough to allow activation of the element driving *eve* expression. In contrast with the primary pair-rule genes, the five pair-rule genes of the secondary class are controlled by interactions with transcription factors encoded by other pair-rule genes.

Segment Polarity Genes Occupy the Lowest Level of the Segmentation Hierarchy

Many segment polarity genes are expressed in stripes that are repeated with a single segment periodicity, that is, there is one stripe per segment (Fig. 21.21a). Mutations in segment polarity genes cause deletion of part of each segment and its re-

placement by a mirror image of a different part of the next segment. For example, in *engrailed* mutants, the posterior compartment of each segment is replaced by a reversed duplication of the anterior compartment of the subsequent segment. The segment polarity genes thus function to determine certain patterns that are repeated in each segment.

The regulatory system that directs the expression of segment polarity genes in a single stripe per segment is quite complex. In general, the transcription factors encoded by pair-rule genes initiate the pattern by directly regulating certain segment polarity genes. Interactions between various cell polarity genes then maintain this periodicity later in development. Significantly, activation of segment polarity genes occurs after cellularization of the embryo is complete. Hence, the diffusion of transcription factors within the syncytium, of central importance in segmentation up to this point, ceases to play a role. Instead, intrasegmental patterning is determined mostly by the diffusion of proteins secreted between cells.

Two of the segment polarity genes, *hedgehog* (*hh*) and *wingless* (*wg*), encode secreted proteins. These proteins, together with the transcription factor encoded by the *engrailed* (*en*) segment polarity gene, are responsible for many aspects of segmental patterning (Fig. 21.21b). A key component of this control is that a one-cell-wide stripe of cells secreting the Wg protein is adjacent to a one-cell-wide stripe of cells expressing the En protein and secreting the Hh protein. The interface of these two types of cells is a self-reinforcing, reciprocal loop. The Wg protein secreted by the more anterior of the two adjacent stripes of cells is required for the continued expression of *hh* and *en* in the adjacent posterior stripe. The Hh protein secreted by the more distal stripe of cells maintains expression of *wg* in the anterior stripe. Gradients of Wg and Hh proteins made from these adjacent stripes of cells control many aspects of patterning in the remainder of the segment. This is because the products of both *wg* and *hh* appear to function as morphogens; that is, responding cells appear to adopt different fates depending on the concentration of Wg or Hh protein to which they are exposed.

Other segment polarity genes encode proteins involved in signal transduction pathways initiated by the binding of Wg and Hh proteins to receptors on cell surfaces. (Recall that signal transduction pathways enable a signal received from a receptor on the cell's surface to be converted through a series of intermediate steps to a final intracellular regulatory response, usually the activation or repression of particular target genes.) These signal transduction pathways determine the ability of cells in portions of each segment to differentiate into the particular cell types characteristic of that segment portion.

Homologs of the segment polarity genes are key players in many important patterning events in vertebrates. For example, the chicken *sonic hedgehog* gene (related to the fly *hh*) is critical for the initiation of the left-right asymmetry in the early chicken embryo as well as for the processes that determine the number and polarity of digits produced by the limb buds. The mammalian homolog of this gene has the same conserved function.

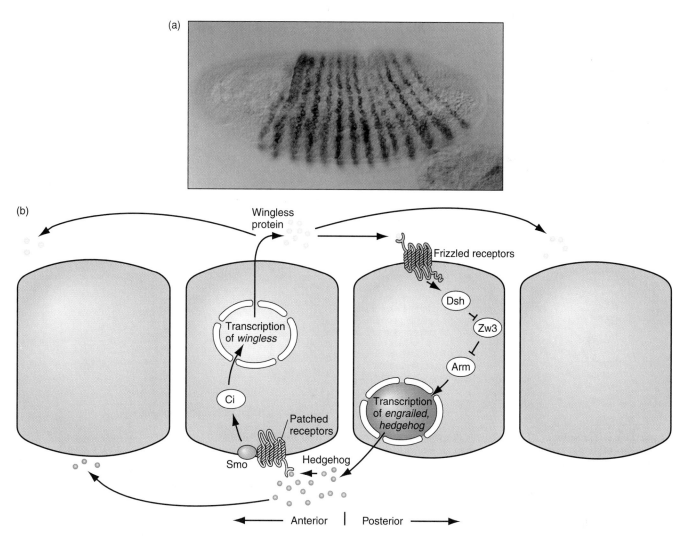

Figure 21.21 Segment polarity genes. (a) The embryo contains 14 stripes in which the segment polarity gene *engrailed* is expressed. (b) The border between posterior and anterior compartments of a segment is governed by the *engrailed* (*en*), *wingless* (*wg*), and *hedgehog* (*hh*) segment polarity genes. Regulation by pair-rule genes ensures that cells in the posterior compartment of each segment express *en*. The En protein activates the transcription of the *hh* gene, which encodes a secreted protein ligand. Binding of this Hh protein to the Patched receptor in the adjacent anterior cell initiates a signal transduction pathway (through the Smo and Ci proteins) leading to the transcription of the *wg* gene. Wg is also a secreted protein that binds to a different receptor in the posterior cell, which is encoded by *frizzled*. Binding of the Wg protein to this receptor initiates a different signal transduction pathway (some components of which are the Dsh, Zw3, and Arm proteins) that stimulates the transcription of *en* and of *hh*. The result is a reciprocal loop stabilizing the alternate fates of adjacent cells and the border between anterior and posterior compartments of a segment.

Summary of Segment Number Specification

For each class of segmentation gene, the pattern of expression of its members is controlled either by classes of genes higher in the hierarchy or by members of the same class, never by lower class genes (see Fig. 21.22). In this regulatory cascade, the maternal genes control the gap genes, the gap genes control themselves and the pair-rule genes, and the pair-rule genes control themselves and the segment polarity genes.

The expression of genes in successively lower parts of the hierarchy is restricted to more sharply defined embryonic regions. During the embryonic cycles of nuclear and cell divisions, the spatial restriction of gene expression is a response to maternally established transcription-factor gradients. In the syncytial blastoderm embryo, refinement of the spatial restriction of gene expression depends on the hierarchical action of the transcription factors encoded by the zygotic genes. Once cellularization is complete, intracellular communication mediated by secreted proteins encoded by certain segment polarity genes becomes the major mode of pattern formation. Although the cellular blastoderm, when viewed from the outside, looks like a uniform layer of cells (review Fig. 21.16a), the coordinated action of the segmentation genes has already divided the embryo into segment primordia. A few hours after gastrulation, these primordia become distinguishable morphologically as clear-cut segments, as in Fig. 21.16c.

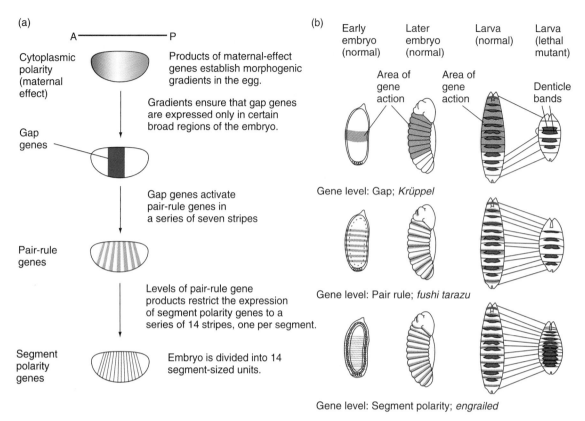

Figure 21.22   A summary of the genetic hierarchy leading to segmentation in *Drosophila*. (a) Genes in successively lower parts of the hierarchy are expressed in narrower bands within the embryos. (b) Mutations in segmentation genes cause the loss of segments (shown in yellow; right) that correspond to regions where the gene is expressed (also in yellow; left).

Each Segment Establishes Its Own Unique Identity through the Activation of Homeotic Genes

After the segmentation genes have subdivided the body into a precise number of segments, the homeotic genes help assign a unique identity to each segment. They do this by functioning as master regulators that control the transcription of batteries of genes responsible for the development of segment-specific structures. The homeotic genes themselves are regulated by the gap, pair-rule, and segment polarity genes so that at the cellular blastoderm stage, or shortly thereafter, each homeotic gene becomes expressed within a specific subset of body segments. Most homeotic genes then remain active through the rest of development, functioning continuously to direct proper segmental specialization.

Mutations in homeotic genes, referred to as **homeotic mutations,** cause particular segments, or parts of them, to develop as if they were located elsewhere in the body. Because some of the mutant homeotic phenotypes are quite spectacular, researchers noticed them very early in *Drosophila* research. In 1915, for example, Calvin Bridges found a mutant he called *bithorax* (*bx*). In homozygotes for this mutation, the anterior portion of the third thoracic segment (T3) develops like the anterior second thoracic segment (T2); in other words, this mutation transforms part of T3 into the corresponding part

of T2, as illustrated in Fig. 21.23a. This mutant phenotype is very dramatic, as T3 normally produces only small club-shaped balancer organs called *halteres,* whereas T2 produces the wings. Another homeotic mutation is *postbithorax* (*pbx*), which affects only posterior T3, causing its transformation into posterior T2. (Note that in this context, *Drosophila* geneticists use the term "transformation" to mean a change of body form). In the *bx pbx* double mutant, all of T3 develops as T2 to produce the now famous four-winged fly (Figure 21.23b).

In the last half of this century, researchers have isolated many other homeotic mutations. Most of these mutations map within one or two gene clusters. Mutations affecting segments in the abdomen and posterior thorax lie within a cluster known as the **bithorax complex (BX-C);** mutations affecting segments in the head and anterior thorax lie within the **Antennapedia complex (ANT-C)** (Fig. 21.24).

The Bithorax Complex

Edward B. Lewis shared the 1995 Nobel Prize for physiology or medicine with Christiane Nüsslein-Volhard and Eric Wieschaus for his extensive genetic studies of the BX-C. In his work, Lewis isolated BX-C mutations that like *bx* and *pbx* affected the posterior thorax, and additionally found novel BX-C mutations that caused anteriorly-directed transformations of each of the 10 ab-

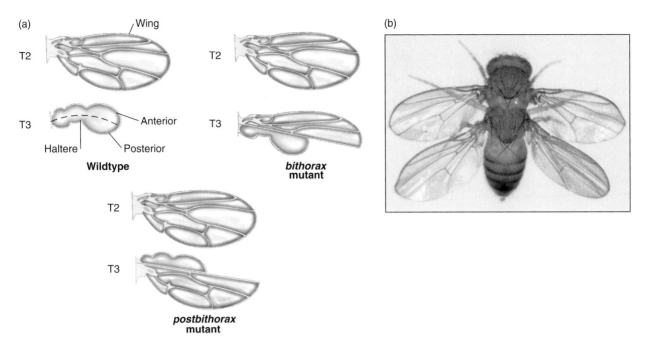

Figure 21.23   Homeotic transformations. (a) In animals homozygous for the mutation *bithorax (bx)*, the anterior compartment of T3 (the third thoracic segment that makes the haltere) is transformed into the anterior compartment of T2 (the second thoracic segment that makes the wing). The mutation *postbithorax (pbx)* transforms the posterior compartment of T3 into the posterior compartment of T2. (b) In a *bx pbx* double mutant, T3 is changed entirely into T2. The result is a four-winged fly.

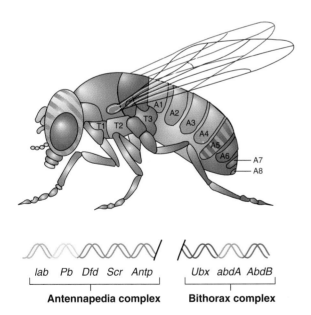

Figure 21.24   Homeotic selector genes. Two clusters of genes on chromosome 3—the Antennapedia complex and the bithorax complex—are responsible for determining most aspects of segment identity. Interestingly, the order of genes in these complexes is the same as the order of the segments each gene controls.

dominal segments. Lewis named mutations affecting abdominal segments *infra-abdominal (iab)* mutations, and he numbered these according to the primary segment they affect. Thus, *iab-2* mutations cause transformations of A2 toward A1, *iab-3* mutations cause transformations of A3 toward A2, and so forth.

Researchers initiated molecular studies of the BX-C in the early 1980s, and in fifteen years they had not only extensively characterized all of the genes and mutations in the BX-C at the molecular level, but also completed sequencing of the entire 315-kb region. Fig. 21.25 summarizes the structure of the complex. A remarkable feature of the complex is that mutations map in the same order on the chromosome as the anterior-posterior order of the segments each mutation affects. Thus, *bx* mutations, which affect anterior T3, lie near the left end of the complex, while *pbx* mutations, which affect posterior T3, lie immediately to their right. In turn, *iab-2*, which affects A2, is to the right of *pbx* but to the left of the A3-determining *iab-3*.

Because the *bx, pbx,* and *iab* elements are independently mutable, Lewis thought that each was a separate gene. However, the molecular characterization of the region revealed that the BX-C contains only three protein-coding genes: *Ultrabithorax (Ubx)*, which controls the identity of T3; *abdominal-A (abd-A)*, which controls the identities of A1–A5, and *Abdominal-B (Abd-B)*, which controls the identities of A5–A8 (Fig. 21.25). The expression patterns of these genes are consistent with their roles. *Ubx* is expressed in segments T3–A8 (but most strongly in T3); *abd-A* is expressed in A1–A8 (most strongly in A1–A4); and *Abd-B* in A5–A8. The *bx, pbx,* and *iab* mutations studied by Lewis affect large *cis*-regulatory regions that control the complex spatial and temporal expression of these genes within specific segments. The average size of each of these regulatory regions is about 15 kb. Researchers have not yet completely worked out the mechanisms by which single BX-C genes control multiple segment identities. However, it is likely that segment identity is determined by the

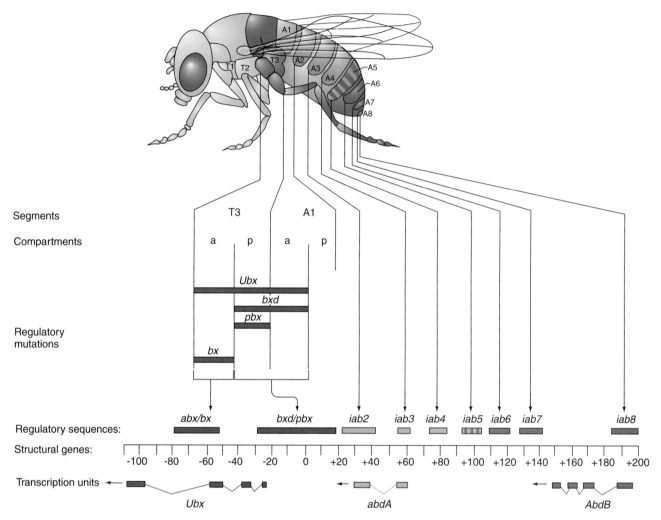

Figure 21.25   A close-up view of the 300 kb of the bithorax complex. There are only three homeotic genes in this complex: *Ubx, abdA,* and *abdB.* Many homeotic mutations such as *bx* and *pbx* affect regulatory regions that influence the transcription of one of these three genes in particular segments. For example, *bx* mutations (at the left end of the complex) prevent the transcription of *Ubx* in the anterior compartment of the third thoracic segment, while *iab-8* mutations (far right) affect the transcription of *AbdB* in segment A8. Note that the order of these regulatory regions corresponds to the anterior-to-posterior order of segments in the animal. *iab-5* mutations disrupt a regulatory sequence needed for the expression of both *abdA* and *abdB* in segment A5 (see also Fig. 21.24).

level or timing of BX-C gene expression, by the production of segment-specific protein isoforms through alternate splicing, or by the differential expression of cofactors that influence BX-C protein function.

The Antennapedia Complex and the Homeobox
Genetic studies in the early 1980s showed that a second homeotic gene cluster, the Antennapedia complex (ANT-C), specifies the identities of segments in the head and anterior thorax of *Drosophila* (see Fig. 21.24). The five homeotic genes of the ANT-C are *labial* (*lab*), which is expressed in the intercalary region; *proboscipedia* (*pb*), expressed in the maxillary and labial segments; *Deformed* (*Dfd*), expressed in the mandibular and maxillary segments; *Sex combs reduced* (*Scr*), expressed in the labial and T1 segments; and *Antennapedia* (*Antp*), expressed mainly in T2, although it is also active at lower levels

in all three thoracic and most abdominal segments. As with the BX-C, the order of genes in the ANT-C is the same (with the exception of *pb*) as the order of segments each controls. In addition to homeotic genes, the ANT-C contains several other important genes, including *zerknullt* (*zen*), which specifies dorsal embryonic structures; *fushi tarazu* (*ftz*), a segmentation gene of the pair-rule class; and *bicoid* (*bcd*), which, as we have seen, encodes the maternally supplied anterior determinant.

In the course of characterizing the ANT-C at the molecular level, researchers detected a region of homology between *Antp* and *ftz.* They subsequently found closely related sequences in all of the homeotic genes of the ANT-C and the BX-C, as well as in *zen* and *bcd* and many genes located outside the homeotic gene complexes. In each gene, the region of homology, called the **homeobox,** is 180 bp in length and is located within the coding sequence of the gene. The 60 amino acids encoded by the homeobox constitute a DNA-binding domain called the **homeodomain** that is structurally related to the helix-turn-helix motif of many bacterial regulatory proteins (see Fig. 15.15). We now know that the homeotic proteins of the ANT-C and BX-C are transcription factors in which the homeobox-encoded homeodomain is responsible for the sequence-specific binding of the proteins to their target genes. Surprisingly, however, DNA-binding studies have shown that most of these homeodomains have very similar binding specificities. The homeodomains of the Antp and Ubx proteins, for example, bind essentially the same DNA sequences. Since different homeotic proteins are thought to regulate specific target genes, this lack of DNA-binding specificity seems paradoxical. Much current research is directed toward understanding how the homeotic proteins actually target specific genes.

CONNECTIONS

The discovery of the homeobox was one of the most important advances in developmental biology in the last two decades of the twentieth century. Some scientists compare the homeobox to the Rosetta stone because it has enabled the isolation by homology of many other developmentally important genes in both *Drosophila* and other organisms. In the late 1980s and early 1990s, for example, the biological community was astonished to learn that the mouse contains clustered homeobox genes with clear homology to the ANT-C and BX-C genes of *Drosophila*. In the mouse, the homeobox genes are called *Hox* genes. The finding of *Hox* genes in mice is yet another demonstration of the evolutionary conservation of gene function. Chapter 22 presents a detailed discussion of homeobox genes in the mouse.

Some geneticists have voiced the opinion that the future of *Drosophila* is limited because the four- to eightfold higher gene number in mice, humans, and other mammals means that mammals have numerous genes with no homologs in *Drosophila*. However, more recent studies indicate that this difference in gene number results from the fact that mice and humans often have several closely related genes that all show homology to a single gene in flies. Although many of the multiple homologs in mice have undergone some functional divergence, in most cases, there is probably strong redundancy of function. This is certainly true of the mouse *Hox* genes. Since redundant genes are very difficult to detect by mutation, it is likely that *Drosophila*'s low gene number (and hence low level of genetic redundancy) will prove to be a major experimental advantage, rather than a disadvantage. The approach of identifying genes by mutation in *Drosophila*, followed by isolation and study of homologs in mice and humans, is likely to remain a central research paradigm. Thus, the evolutionary conservation of gene function combined with the many experimental advantages of the fruit fly ensure that *Drosophila* will remain at the forefront of biological research for the foreseeable future.

ESSENTIAL CONCEPTS

1. A small genome size and low chromosome number are among the features that make *D. melanogaster* an excellent model organism. The polytene chromosomes of the larval salivary glands, key tools in *Drosophila* genetics, provide a detailed physical map of the fly genome.

2. Several characteristics of *Drosophila* are useful for genetic analysis. A moderate amount of crossing-over occurs in females and no crossing-over occurs in males. *Balancer chromosomes* help preserve linkage. *P-element transposons* are a tool of molecular manipulation, including transformation, tagging, and enhancer trapping. *Genetic mosaics* resulting from mitotic recombination help track the roles of individual genes in the development of adult structures. *Ectopic expression* can help pin down the function of a gene.

3. The goal of the *Drosophila* Genome Project is to map and sequence the euchromatic portion of the genome, and to obtain as much information as possible about the structure and function of the 12,000 to 15,000 genes.

4. The development of the anterior-posterior (AP) axis in the fruit fly depends on the coordinated action of the *segmentation genes,* which divide the body into 3 head, 3 thoracic, and 8 abdominal segments; and the *homeotic genes,* which assign a unique identity to each segment.

5. The four classes of segmentation genes are expressed in the order maternal, gap, pair-rule, and segment polarity. The *maternal genes* produce gradients of *morphogens.* The *gap genes,* the first zygotic segmentation genes, are expressed in broad zones along the anterior-posterior axis. The *pair-rule genes* subdivide those broad areas into units that span two of the ultimate body segments. The *segment polarity genes* subdivide the two-segment units into individual segments. In the hierarchy of segmentation gene expression, each gene class is controlled by classes of genes higher in the hierarchy or by members of the same class.

6. The homeotic genes are master regulators of other genes that control the development of segment-specific

structures; the *homeobox* in each homeotic gene encodes a *homeodomain* that allows the gene products to bind to specific target genes and control their expression. The homeotic genes are, in turn, regulated by the earlier acting gap, pair-rule, and segment polarity genes. Homeotic mutations cause particular segments, or parts of them, to develop as if they were located elsewhere in the body.

Most homeotic mutations map to the bithorax and the Antennapedia complexes.

7. The finding of homeobox genes in other organisms such as the mouse has made it possible to identify developmentally important genes shared by all animals, and demonstrates the evolutionary conservation of gene function.

S O L V E D   P R O B L E M S

I. *Drosophila* geneticists try to minimize the work needed to maintain the many lines of flies they investigate. Ideally, they establish genetic stocks in which the flies in a culture bottle mate with each other so they produce the same kinds of progeny generation after generation. Because females homozygous for a maternal effect lethal (*mel*) mutation are sterile (their offspring die during embryogenesis), such mutations can be propagated only when females are heterozygous. A particular *mel* mutation on the second chromosome (an autosome) is kept as a heterozygote with a balancer chromosome carrying the *Cy* mutation, which is a recessive lethal but also produces a dominant Curly wing phenotype.

a. Suppose you start with equal numbers of male and female flies, both of genotype *mel/Cy*. What kinds of progeny would be produced in the next generation, and in what proportion? Would the culture bottles containing these progeny constitute a genetic stock?

b. What kind of flies would you select from the culture bottle to study the phenotypic effect of the maternal effect mutation in detail?

Answer

This problem involves an understanding of balancer chromosomes and maternal effect lethal mutations.

a. First determine all the ways gametes from these flies could combine and what the phenotype (include sex) would be. From a mating of *mel/Cy* females with *mel/Cy* males, the following kinds of zygotes would be formed: 1/8 *mel/mel* females (which survive but are sterile), 1/4 *mel/Cy* females (which survive and are fertile), 1/8 *Cy/Cy* females (which die before emerging as adults), 1/8 *mel/mel* males (which survive and are fertile because the mutation affects female but not male fertility), 1/4 *mel/Cy* males (which survive and are fertile), 1/8 *Cy/Cy* males (which die before emerging as adults).

Of the adults emerging from these zygotes, you would find: *1/6* mel/mel *females (sterile and* Cy$^+$*), 1/3* mel/Cy *females (fertile with Curly wings), 1/6* mel/mel *males (fertile and* Cy$^+$*), 1/3* mel/Cy *females (fertile with Curly wings).* If all the flies in this culture bottle were allowed to mate at random, the same types of adult flies would emerge in the subsequent generation

(you should work this out for yourself, remembering that only the heterozygous females are fertile). As a result, *this bottle constitutes a genetic stock: the same flies will be present in every generation.*

b. *You want* mel/mel *females, which can be recognized as the non-Curly (wildtype) females.* These females could be mated to any type of fertile male and the developing progeny examined to study the maternal effect lethal phenotype.

II. A particular gene in *Drosophila* called *no eclosion* (*nec*) encodes an enzyme needed for the fly to escape from the pupal case. The wildtype allele (*nec$^+$*) is dominant to a null mutant allele (*nec^1*). Animals homozygous for *nec^1* die because they cannot get out of the pupal case. The *nec^1* mutation (on an otherwise normal chromosome) is maintained in a stock over the autosomal balancer chromosome TM6C, which carries the *nec$^+$* allele and two dominant markers: *Stubble,* which affects bristle shape, and *Tubby,* which causes larvae and pupae to be short and squat. *Stubble* also behaves as a recessive lethal mutation, causing death during embryonic development (i.e., before the larva hatches). Several deletions (A–D) in the vicinity of the *nec* gene were generated by X rays. Each of these deletions is maintained in a stock over the same TM6C balancer chromosome. Crosses are made between the *nec^1*/TM6C stock and each of the deletion/TM6C stocks. Crosses with deletions A and D yield only Stubble adult flies, while crosses with deletions B and C yield both wildtype (non-Stubble) as well as Stubble flies.

a. Which deletions remove the *nec$^+$* gene?

b. If 1000 zygotes were generated from the crosses with deletions A and D that yielded only Stubble flies, how many of these zygotes should eventually produce adult flies?

c. Of the 1000 zygotes produced in part b, how many would be expected to develop into animals that die in the pupal case? Would these pupal cases be short and squat (Tubby) or long and thin (wildtype)?

d. From the crosses above with deletions B and C that yielded both wildtype (non-Stubble) as well as Stubble flies, what percentage of the adult flies that emerge should be non-Stubble?

e. *Stubble* and *Tubby* are located 10 m.u. apart. In a cross between *nec^1*/TM6C females and males true-breeding wildtype for all genes under consideration, how many of the 1000 adult progeny should have emerged from long, thin pupal cases and had Stubble bristles?

Answer

This problem involves an understanding of balancer chromosomes, the *Drosophila* life cycle, and complementation analysis.

a. In the cross between nec^1/TM6C and the deletion/TM6C strains, the following zygotes would be produced in the proportions listed. For simplicity, these are labeled as classes 1–4

Class 1	Class 2	Class 3	Class 4
1/4 nec^1/deletion	1/4 nec^1/TM6C	1/4 deletion/TM6C	1/4 TM6C/TM6C

Regardless of the relationship between nec^1 and a particular deletion, classes 2 and 3 should always survive, with animals developing as short, squat larvae and pupae (Tubby) that emerge as Stubble adults. In addition, class 4 should always die during embryonic development. Class 1 will form wildtype larvae, pupae, and adults if nec^1 and the deletion in question complement each other. If nec^1 and the deletion in question fail to complement each other, class 1 will not emerge from the pupal case; the pupal case will be long and thin.

Deletions A and D remove the nec *gene.* Because there are no wildtype adults in the matings with

deletions A and D, we know that class 1 dies (these animals have no functional nec^+ product). No complementation occurred, so the deletions include the *nec* gene. With deletions B and C, there was complementation (class 1 survived) so the *nec* gene is not within the deleted area.

b. We are looking here at crosses in which class 1 dies. In these crosses, zygotes of classes 1 and 4 would not develop into adults, but those of classes 2 and 3 would. Thus, *500 zygotes (1/2 of the total) would produce adult flies.*

c. Looking at the same kinds of crosses as in part b, only class 1 would die in the pupal case. Thus, *1/4 of the zygotes (250 animals) would die as pupae.* These animals do not have the balancer chromosome, so *their pupal cases would be long and thin.*

d. Here, we are looking at crosses in which class 1 survives and produces wildtype animals. In these crosses, classes 1, 2, and 3 would produce adults. Of these, only class 1 should be wildtype (non-Stubble), while classes 2 and 3 would have the *Stubble*-bearing TM6C balancer. Thus, of the emerged adults, *1/3 should be non-Stubble.*

e. *None.* Because of crossover suppression between the balancer chromosome and its homolog, viable recombinants that separate the *Tubby* and *Stubble* dominant markers on the balancer chromosome cannot be obtained.

PROBLEMS

21-1 For each of the terms in the left column, choose the best matching phrase in the right column.

a. morphogens	1. contains multiple overlapping inversions
b. chromocenter	2. reorganization of the body plan
c. balancer chromosome	3. making gene products in tissues where they are not normally made
d. enhancer trapping	4. define cell fate in a concentration dependent manner
e. ectopic expression	5. heterochromatin in polytene chromosomes
f. metamorphosis	6. locating regulatory regions based on patterns of expression of a reporter gene

21-2 A balancer chromosome that is often used in *Drosophila* genetics is called TM3. This balancer chromosome is a variant of the third chromosome (an autosome) containing multiple inversions, and the dominant mutation *Stubble,* which also has a recessive lethal phenotype. Ten different third chromosomes

carrying various recessive lethal point mutations were isolated and kept in stock as TM3 heterozygotes (i.e., lethal mutation/TM3). These stocks were then crossed to each other, and the number of wildtype progeny per 150 total adult progeny emerging from each of these crosses is recorded below.

Mutation number	1	2	3	4	5	6	7	8	9	10
1	0									
2	47	0								
3	51	49	0							
4	45	54	50	0						
5	53	44	51	48	0					
6	62	47	39	59	39	0				
7	55	50	43	0	48	52	0			
8	53	44	42	0	60	54	0	0		
9	56	42	38	57	0	50	51	48	0	
10	0	58	59	38	46	49	51	49	55	0

a. Does this experiment measure complementation or recombination between recessive lethal mutations?

b. In the cross between mutant 1 and mutant 2, there are 103 adult progeny that are not wildtype. What is their phenotype?

c. In the cross between mutant 1 and mutant 2, there are an additional 50 zygotes that never became adults (and therefore were not counted). What is the genotype of these zygotes?

d. How many genes are represented in this collection of 10 mutations? Which mutations are in the same genes?

e. Five different deletions in the third chromosome were isolated and balanced by TM3. These deletion stocks were crossed to the TM3-balanced point mutation stocks described above. The emergence of wildtype progeny is indicated in the following chart as "+"; the absence of wildtype progeny is indicated by "−". Does this second experiment measure complementation or recombination between the recessive lethal mutations and the deletions?

Point Mutation Number

Deletion	1	2	3	4	5	6	7	8	9	10
A	+	−	+	−	+	−	−	−	+	+
B	−	−	−	−	+	−	−	−	+	−
C	−	+	−	+	−	+	+	+	−	−
D	+	+	−	−	+	−	−	−	+	+
E	−	+	−	−	+	+	−	−	+	−

f. Using this new information, show the best possible map for the 10 point mutations and 5 deletions in the region.

21-3 For a particular P-element mutagenesis experiment, you want to obtain new insertions of a P element marked with the w^+ gene into any of the autosomes. Starting with a stock of flies containing a w^+ marked P element on the X chromosome, design an experiment that would enable you to obtain flies containing P-element insertions in autosomes.

21-4 Which of the following animals would show mutant phenotypes in somatic cells as a result of a *Drosophila* cross involving hybrid dysgenesis?
a. progeny of a P male × M female cross
b. progeny of a M male × P female cross
c. progeny of the progeny of a P male × M female cross
d. progeny of the progeny of a M male × P female cross
e. progeny of the progeny of a P male × P female cross

21-5 Which of the following is not a property of the *Drosophila hunchback* gene?
a. The *hunchback* mRNA is uniformly distributed in the egg by the mother.
b. Transcription of *hunchback* is enhanced by Bicoid (the anterior morphogen).

c. Translation of the *hunchback* mRNA is inhibited by Nanos (the posterior morphogen).
d. The Hunchback protein eventually is distributed in a gradient (anterior high; posterior low).
e. Hunchback protein directs the distribution of *bicoid* mRNA.

21-6 The *hunchback* gene contains a promoter region, the structural region (the amino-acid coding sequence) and a 3′ untranslated region (DNA that will be transcribed into sequences appearing at the 3′ end of the mRNA that are not translated into amino acids).
a. What important sequences required to control *hunchback* gene expression are found in the promoter region of *hunchback?*
b. What sequence elements that encode specific protein domains are found in the structural region of *hunchback?*
c. There is another important kind of sequence that turns out to be located in the part of the gene transcribed as the 3′ UTR (untranslated region) of the *hunchback* mRNA. What might this sequence do?

21-7 How do the segment polarity genes differ in their mode of action from the gap and pair-rule genes?

21-8 One important demonstration that Bicoid is an anterior determinant came from injection experiments analogous to those done by early embryologists. Injection experiments involve introduction of components such as cytoplasm from an egg or mRNA that is synthesized *in vitro* into the egg by direct injection. Describe injection experiments that would demonstrate that Bicoid is the anterior determinant.

21-9 In flies developing from eggs laid by a *nanos⁻* mother, development of the abdomen is inhibited. Flies developing from eggs that have no maternally supplied *hunchback* mRNA are normal. Flies developing from eggs laid by a *nanos⁻* mother that also have no maternally supplied *hunchback* mRNA are normal. If there is too much Hunchback protein in the posterior of the egg, abdominal development is prevented. What do these findings say about the function of the Nanos protein and of the *hunchback* maternally supplied mRNA?

21-10 Mutant embryos lacking the gap gene *knirps (kni)* are stained at the syncytial blastoderm stage to examine the distributions of the Hunchback and Krüppel proteins. The results of the *knirps⁻* and wildtype embryos stained for the Hunchback and Krüppel proteins is shown schematically below.

Hunchback protein   **Krüppel protein**

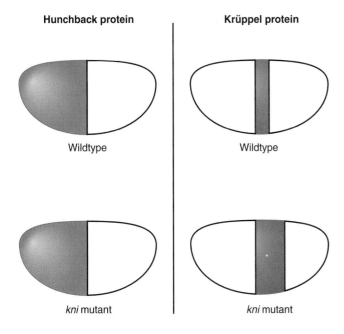

Wildtype   Wildtype

kni mutant   *kni* mutant

a. Based on these results, what can you conclude about the relationships among these three genes?
b. Would the pattern of Hunchback protein in embryos from a *nanos⁻* mutant mother differ from that shown above? If yes, describe the difference and explain why. If not, explain why not.

21-11 In the analysis of eye development, mosaic *Drosophila* were constructed in which some cells were $w^+ sev^+$ and others were $w^- sev^-$. Some eye facets (made up of eight cells) contained both white and red cells. When the results from many different mosaic facets were tallied, all possible combinations of red and white cells were seen. (a) What does this indicate about the origin of cells recruited for any one facet? (b) What does this indicate about the expression of the *white* gene?

21-12 Nüsslein-Vollhard and Wieschaus used 10 different mutant stocks of the form mutant/CyO where CyO is a *Drosophila* second chromosome balancer. Seven of these mutations were point mutations (mutations 1–7) and three of the mutations were cytologically detectable deletions (Del A, B, C), each removing a large number of genes. They crossed these stocks to each other and noted whether they obtained progeny without the *Cy* dominant marker on the CyO balancer chromosomes, and if so, whether such progeny

females were fertile when crossed with wildtype males. The results of this analysis are tabulated below.

Key: F, non-*Cy* progeny are obtained, the females are fertile
S, non-*Cy* progeny are obtained, but females are sterile
N, no non-*Cy* progeny are obtained

	Del A	Del B	Del C	1	2	3	4	5	6	7
Del A	N	N	N	N	F	S	S	F	F	F
Del B		N	F	F	N	F	S	F	F	N
Del C			N	N	F	F	F	S	S	F
1				N	F	F	F	F	F	F
2					N	F	F	F	F	N
3						S	F	F	F	F
4							S	F	F	F
5								S	S	F
6									S	F
7										N

a. Give the best genetic map of this region of the *Drosophila* second chromosome, showing the extent of the deletions with bars and positions of point mutations with an "X."
b. How many different genes are indicated by this data set? Which point mutations are in which genes?
c. How many genes needed for female fertility are indicated by this data set? Which point mutations are in which of these genes? How could you distinguish genes needed for oogenesis from maternal effect lethal genes that make otherwise normal eggs that cannot develop into adults after fertilization?
d. How many zygotic lethal genes are indicated by this data set? Which point mutations are in which of these genes?
e. In one sentence, explain why you could not determine the map distance between any two of the the above point mutations that both lie within the same gene, for any of the genes revealed by this data set, regardless of how many flies you analyzed or whatever experiment you used to try to determine this distance.

21-13 A particular variant of a P element was constructed containing a selectable marker (the *rosy⁺* gene, where *rosy⁻* homozygotes have rose-colored eyes) and the *lacZ* gene without a promoter region. When the P element inserts into a gene, the gene is inactivated, producing a mutant phenotype. The insertion of these P elements was used to produce and isolate a series of female sterile mutants and study cell activities during oogenesis. Egg development occurs in an egg chamber containing a single oocyte and 15 nurse cells all derived from a germ cell precursor. The chamber is surrounded by somatic follicle cells. Follicle cells and oocyte and nurse cells interact in establishment of polarity in the egg chamber.

a. What other information could you get about a developmental gene if the P element carrying the *lacZ* gene without a promoter inserted into it?

b. How would you maintain stocks containing recessive female sterile mutations? (What would be the genotype of the stock flies and how could you study the mutants?)

c. One of the sterile mutants isolated contained occasional mispositioned oocytes within the egg chamber and some defective egg chambers. The defective gene, called *homeless,* encodes a protein that shows similarity to RNA dependent ATPases and helicases, leading to the hypothesis that the protein may interact with, and help in the positioning of, particular RNAs. *bicoid* mRNA is a major anterior determinant for which localization is important. What experiment could you do to determine if *bicoid* mRNA is appropriately localized within the developing egg chambers in *homeless⁻* mutants? (Be sure to mention your control.)

d. What experiment could you perform to see whether *homeless* gene function is required either in the germline-derived cells (the nurse cells and/or oocyte), or instead in the somatic follicle cells? This experiment is made possible by the fact that various stocks are available in which the dominant *ovoᴰ* mutation, whose expression in germline-derived

cells blocks ovarian development, has been placed on a different chromosome arm. Describe how you would set up this experiment and the expected results.

e. Another P-element-induced mutation isolated in the same screen was a semilethal female sterile. By *in situ* hybridization, it mapped to a chromosomal region where the *ras1* gene had been mapped. Complementation analysis established that this mutation is an allele of *ras1*. Other alleles of *ras1* affect viability, eggshell development, and eye morphology. Moderate alleles affect eggshell appearance and viability but not eye morphology. Different alleles also show different levels of Ras1 activity. What does this information suggest about Ras1 protein requirements in different signaling pathways?

21-14 A protein called Bruno was found to bind to the 3′ UTR of several developmental mRNAs, including *oskar* mRNA (necessary for proper formation of the posterior part of the embryo). When the *bruno* gene was cloned, the predicted protein sequence contained an RNA binding domain consistent with the biochemical analysis of binding. The *bruno* gene mapped to a chromosomal region where the *arret* gene had previously been located. The *arret* gene, which was identified genetically in a screen for sterile mutants, is required for fertility in both sexes. How could you determine if *bruno* and *arret* are in fact the same gene that had been identified in different ways? What particular phenotypes would you examine?

MUS MUSCULUS: A GENETIC PORTRAIT
OF THE HOUSE MOUSE

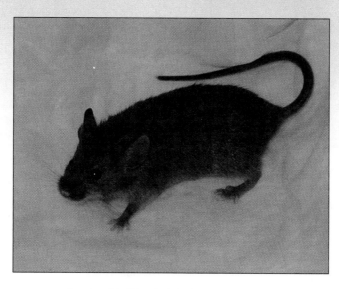

A member of the 129 strain of inbred mice commonly used in
targeted mutagenesis studies.

The common house mouse, *Mus musculus,* has played a prominent role in the study of genetics ever since C. Correns, H. DeVries, and E. Tschermak independently rediscovered Mendel's laws at the beginning of the twentieth century. Because these three scientists, as well as Mendel himself, performed their research entirely on plants, many in the scientific community questioned whether Mendel's laws could explain the basis for inheritance in animals, especially humans. The reason for this skepticism is easy to see. People, for example, differ in the expression of many commonly inherited traits—such as skin color, eye color, curliness of hair, and height—that show no evidence of transmission according to Mendel's laws. We now know that these traits result from the interaction of many genes with multiple alleles that segregate according to Mendel's first law even though the traits themselves do not. At the beginning of the twentieth century, however, a demonstration of the applicability of Mendel's laws to animal inheritance required the analysis of simple traits controlled by single genes.

M. musculus has many features that enhance its value as a model organism for genetic analysis, and foremost among these is the availability of hundreds of single-gene mutations. These mutations arose during the mouse's long history of domestication as a pet. Over the centuries, dealers in what became known as the "fancy mouse" trade selected and bred mice with numerous coat colors and other visible mutations, first in China and Japan, later in Europe (Fig. 22.1a). In contrast to the variation that occurs naturally in wild populations, new traits that appear suddenly in captive-bred mice are almost always the result of single-gene mutations. Early animal geneticists made note of this fact and used fancy mice to demonstrate that Mendel's laws apply to mammals, and by extrapolation, to humans.

In addition to providing a ready source of single-gene mutations, the house mouse has several other features that make it the mammal of choice for genetic analysis. Mice have a very short generation time of just eight to nine weeks. They are small enough so that thousands can live in relatively small rooms. They have large

(a)

(b)

Figure 22.1 The mouse is a model system for human biology. (a) Examples of visible phenotypes caused by single-gene mutations. (b) Mother mouse with her pups.

(a)

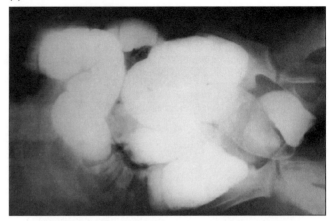

(b)

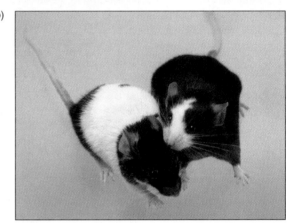

Figure 22.2 Hirschsprung disease: A human developmental disease that causes deformities of the colon (a) and (b) a mouse model of the disease with a classical mutation (piebald coat).

litters of eight or more pups. They breed readily in captivity. Fathers do not harm their young. And after centuries of artificial selection, domesticated mice are docile and easy to handle (Fig. 22.1b).

But why study a mammal at all when animals like fruit flies and nematodes are even smaller and more amenable to genetic analysis? The answer is that a major goal of current biological research is the understanding of human beings. And although many features of human biology, especially at the cellular and molecular levels, are common to a broad spectrum of life-forms, the most advanced organism-level human characteristics appear in a limited subset of animals. In fact, many aspects of human development and disease are common only to placenta-bearing mammals such as the mouse. Thus, the mouse provides a powerful model system for investigating the genetic basis of simple and complex human traits, especially those related to development and disease (Fig. 22.2).

Two general themes emerge from our presentation of *M. musculus*. First, because of the many similarities between mouse and human genomes, researchers can use *homology analysis* to identify and locate the same genes in both species. In this context, **homologs** are genes or regulatory DNA sequences that are similar in different species because of descent from a common ances-

tral sequence. Second, genetic analysis in mice exemplifies the combined use of molecular (that is, recombinant DNA) and classical breeding techniques to identify and understand the function of complex genetic systems.

Our genetic portrait of the house mouse describes:

■ An overview of *M. musculus* in the laboratory, including a look at the mouse genome, the mouse life cycle, and two powerful transgenic protocols: the addition of specific genes to the mouse genome by nuclear injection and the removal of specific genes from the mouse genome by targeted mutagenesis.

■ The uses of transgenic technology in determining the function of gene products, characterizing regulatory regions, establishing links between mutant phenotypes and particular transcription units, and creating a mouse model for a human disease.

■ The *Hox* genes: A comprehensive example.

AN OVERVIEW OF <u>MUS MUSCULUS</u> IN THE LABORATORY

The Mouse Genome

The most important feature of the mouse genome for contemporary geneticists is its close resemblance to the human genome (Table 22.1). The haploid genomes of humans and mice (and other placental mammals as well) contain approximately 3 billion base pairs of DNA. Nearly every human gene has a homolog in the mouse genome.

This does not mean that the two genomes are equivalent in content. Most differences, however, appear to result from species-specific additions to gene families that already existed in the common ancestor of mice and humans. Moreover, because mice and humans have been evolving apart for some 75 million years, other differences have arisen occasionally when gene homologs acquired different functions in the two species.

The human genome is distributed among 22 autosomes and 2 sex chromosomes, while the mouse genome is contained within 19 autosomes and 2 sex chromosomes. As we saw at the beginning of Chapter 12, examinations of mouse and human karyotypes under the microscope reveal no evidence of chromosome banding similarities between the two species. With the mapping of thousands of homologous genes in both species, however, a remarkable pattern has emerged. Genes that are closely linked in one species are usually closely linked in the other. When two or more loci are found to be linked in one species, they are said to be **syntenic** (meaning "on the same thread," or chromosome). When the same set of loci are also found to be linked in a second species, they are said to exist in a state of **conserved synteny.** A comparison of genetic maps that comprise the whole mouse genome with genetic maps that comprise the whole human genome shows that regions of conserved synteny extend across nearly the complete lengths of both. The average size of each conserved syntenic region is roughly 17.6 Mb. The implication of this finding is that during the 75 million years that mice and humans have been evolving apart from a common ancestor, their genomes have broken apart and rearranged some 170 times (17.6 Mb ×

170 = about 3000 Mb = the size of the mammalian genome). Conversely, if the proper genome-scale scissors and glue were available, one could break the mouse genome into about 170 pieces and reassemble those pieces—like a puzzle—in the form of the human genome.

In addition to its powerful evolutionary implications, conserved synteny is a useful tool for practicing geneticists. Once a researcher has mapped a locus in one species, he or she can look at a homology map and immediately identify its likely map position in the other species. Of course, for genes that have already been cloned, it is possible to use DNA-DNA hybridization to pick out gene homologs from the other species. But for loci characterized only by their phenotypic expression, conserved synteny enables geneticists to move back and forth between the analysis of a trait in humans and the analysis of a model for that trait in mice.

The discovery of a locus that predisposes female mice to excessive consumption of 10% ethanol (the concentration of alcohol found in most wines) provides an example of the use of conserved synteny for locating human homologs of mouse genes. Mouse geneticists used DNA markers (as described in Chapter 10) to map the *Alcohol-preference-2* (*Alcp2*) locus to the middle of mouse chromosome 11; even though they have not yet identified and cloned the gene, scrutiny of a conserved synteny homology map shows that the most likely location for the human homolog of *Alcp2* is on the long arm of human chromosome 17, close to the centromere (Fig. 22.3). With this information about the likely location of an alcohol-preference locus, human geneticists can now look for genetic linkage between DNA markers on human chromosome 17 and a gene with certain alleles that predispose people to some form of alcohol abuse.

The Mammalian Life Cycle

The mouse's life cycle is similar to that of humans and all other placental mammals, although the timing of events is unique to each species (see Table 22.1). For example, for mice, the average life span is just 2 years; for humans it is about 78 years. Gestation—the period from fertilization to

TABLE 22.1 Comparison of Mice and Humans

Trait	Mice	Humans
Average weight	30 g	77,000 g (170 lb)
Average length	10 cm (without tail)	175 cm
Genome size	~3,000,000,000 bp	~3,000,000,000 bp
Haploid gene number	~100,000	~100,000
Number of chromosomes	19 autosomes + X and Y	22 autosomes + X and Y
Gestation period	3 weeks	38 weeks (8.9 months)
Age at puberty	5–6 weeks	624–728 weeks (12–14 years)
Estrus cycle	4 days	28 days
Life span	2 years	78 years

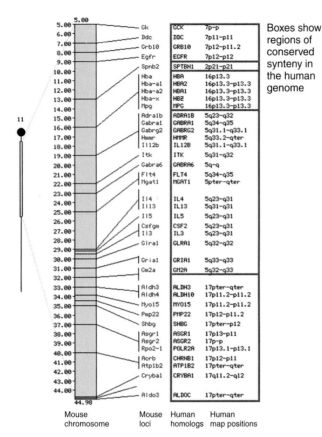

Figure 22.3 Conserved synteny: An example with mouse chromosome 11. The mouse chromosome is shown on the left with positions of loci with mapped homologs in the human genome as well. The map locations of human homologs are shown on the right.

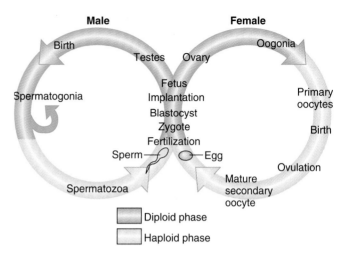

Figure 22.4 The mammalian life cycle. The life cycles of males and females is shown separately in the traditional format. Notice that primary germ cells are formed in the female before birth, but in the male after birth. Notice also that male germ cells are self-renewing, while in females, no new germ cells are formed after birth.

birth—lasts only 21 days in mice, but 8.9 months in humans. After birth, mice reach puberty in just 5–6 weeks, 100 times sooner than humans, who mature for roughly 12 years before they attain puberty and the ability to conceive children of their own (although in most societies, they wait 5–10 years longer). It is thus possible to go from the birth of one mouse to the birth of its offspring in just 8 weeks, whereas with humans, completion of the same cycle would take 13 to 25 years.

Even with significant differences in timing, however, the details of each stage of development, both before and after birth, are remarkably similar in all mammalian species. As a result, the overview presented here applies equally to mice and humans.

The mammalian life cycle, like that of all sexually reproducing species, can be visualized as a continuous circle (Fig. 22.4), with any point along the circumference marking the start. For ease of presentation, we begin with the haploid phase in both males and females.

Male Germ Cell Development

As we saw in Chapter 3, once a male mammal has reached puberty, he continuously produces a large number of haploid germ cells for the rest of his life. The mature haploid cell is a

sperm cell, or *spermatozoa,* and the process by which it arises is *spermatogenesis* (review Fig. 3.21). Mature spermatozoa released into the lumen at the center of the seminiferous tubule join millions of other sperm cells, which together pass through countless passageways to reach the epididymis; there they mature further and then continue on to the vas deferens where they await ejaculation during copulation.

Female Germ Cell Development

Unlike males, females are born with all the haploid cells they will ever have (~50,000 in the mouse; 1,000,000 in women). When mature, these haploid cells are known as *eggs,* or *oocytes,* and the process by which they arise is *oogenesis* (review Fig. 3.19). As we saw in Chapter 3, oogenesis begins inside the newly formed ovaries of the developing fetus. Long before birth, primordial germ cells differentiate into oogonia and enter meiosis, but stop at the diplotene stage of the first meiotic prophase. These primary oocytes remain arrested in suspended animation—for weeks in mice and many years in humans—until after puberty.

From this time on, the female progresses through an *estrus cycle* that lasts about 4 days in mice and about 28 days in humans. During each cycle, *primary oocytes* (8–10 in mice, usually only 1 in women) are stimulated to complete the first meiotic division and extrude the first polar body at the end of this division; the resulting *secondary oocyte* begins the second meiotic division but stops at metaphase and is released from the ovary in a process known as *ovulation.* Following ovulation, the secondary oocyte passes into an *oviduct* (called a *fallopian tube* in humans), where for a brief time, known as *estrus,* it remains alive and receptive to fertilization.

In nature, most mammals die while they still have the ability to reproduce. Many human females, however, live long enough to pass through *menopause,* during which they stop cycling through estrus, no longer ovulate, and thus lose the ability to reproduce.

Fertilization

Just before and during the estrus phase of the estrus cycle, female mammals in the wild release species-specific chemical signals, or *pheromones,* that stimulate male interest in sex. In response to these pheromones, a male will copulate with a female and ejaculate semen containing millions of sperm into her reproductive tract. The sperm swim from the vagina into the uterus and thence up the oviducts. Only 100 or fewer sperm survive this journey to the waiting eggs.

Fertilization is a multistep process. First, the surviving sperm bind to the zona pellucida—the thick solid shell composed of glycoproteins that surrounds the egg proper. The act of binding induces each sperm to release special proteases that enable it to "burn" its way through the zona pellucida into the space that surrounds the egg membrane. Although multiple sperm can make it into this space, normally only one fuses with the egg. This fusion causes rapid electrochemical changes in the egg membrane that prevent the entry of additional sperm. After fusion, the fertilized egg, or zygote, contains two haploid *pronuclei.* The two pronuclei never merge; instead, replication occurs within both of the pronuclei. The one-cell embryo carries two replicated pronuclei right up to the moment of the first mitosis, at which time the membranes of the two pronuclei break down, and the two sets of chromosomes, one from the paternal pronucleus, the other from the maternal pronucleus, align along the midplane of the fertilized egg and thence segregate chromatids into the two daughter cells.

The fusion of sperm and egg activates the newly fertilized egg to enter the pathway of animal development. For the purposes of analysis, scientists divide mouse development into two distinct stages of unequal length, separated by the moment of embryonic implantation into the uterus: a preimplantation stage that lasts 4 to 5 days, and a postimplantation stage that lasts about 16.5 days. During the preimplantation phase, the embryo is a free-floating object within the female's body. It is easy to remove this naturally free-floating preimplantation embryo from the animal, culture it in a petri dish, and expose it to genetic manipulation before placing it back in the reproductive tract of an adult female for development to a newborn animal. After implantation, however, such manipulation is no longer possible because the embryo, if removed from the adult's body, does not remain viable. The accessibility of the preimplantation embryo provides the basis for many of the genetic manipulations researchers use to study mammalian development.

For Most of the Preimplantation Stage, the Embryonic Cells Remain Undifferentiated

The preimplantation stage starts with the zygote at conception (Fig. 22.5). Development proceeds slowly in the beginning, with the first 22 hours devoted to the expansion of the highly compacted sperm head into a paternal pronucleus that matches the size of the egg's maternal pronucleus. After the paternal pronucleus has completed its expansion and replicated its chromosomes and the maternal pronucleus has replicated its chromosomes, the embryo undergoes the first of four equal divisions, or cleavages, that increase the number of cells from 1 to 16 over 60 hours.

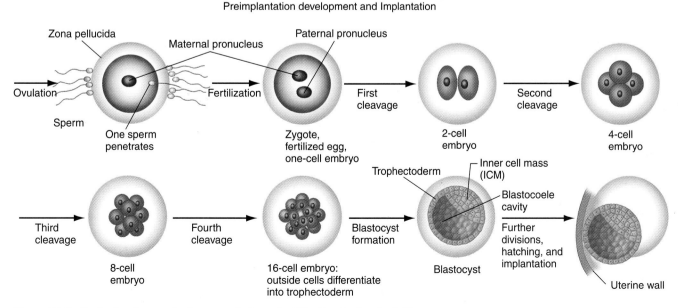

Preimplantation development and Implantation

Figure 22.5   Early development of mammals from fertilization to implantation.

The period of these four equal divisions is called the **cleavage stage.** During this stage, all the cells in the developing embryo are equivalent and **totipotent,** that is, they have not yet differentiated and each one retains the ability, or *potency,* to produce *every* type of cell found in the developing embryo and adult animal. This is very different from the developmental patterns found in many other animal species, such as *Caenorhabditis elegans,* where totipotency disappears as early as the two-cell stage. As a consequence of the mouse cells' totipotency, cleavage stage embryos can be separated into smaller groups of cells that each have the potential to develop into a normal individual. Identical human twins, or more rarely, identical triplets, are examples of the outcome of this process. (Twinning is impossible in *C. elegans* or *Drosophila melanogaster.*) In the laboratory, scientists have obtained completely normal mice from individual cells that they dissected out of the four-cell-stage mouse embryo and placed back into the female reproductive tract (Fig. 22.6a). This experimental feat demonstrates the theoretical possibility of obtaining four identical clones from a single embryo of any mammalian species.

Another more bizarre consequence of the equivalency of cleavage stage cells is the formation of **chimeras,** which are the opposite of clones (Fig. 22.6b). The term "chimera" comes from the Greek word for a mythological beast that is part lion, part goat, and part serpent. Geneticists use the term to designate an embryo or animal composed of cells from two or more different origins. The Polish embryologist A. K. Tarkowski reported the first mouse chimeras in 1961. To construct them, he removed the zona pellucida from two cleavage-stage mouse embryos, obtaining denuded cell masses that are naturally sticky; he then pushed the sticky denuded embryos up against each other. Denuded embryos pressed together in this way form a single chimeric cell mass that is capable of undergoing normal development within the female reproductive tract. If the two embryos of a chimera come from different females mated to different males, the resulting individual is *tetraparental,* that is, has four parents. It is also possible to produce *hexaparental* animals derived from a combination of three embryos. As we see later, the production of chimeric mice has been an essential component of the targeted mutagenesis technology that has revolutionized the use of the mouse as a model organism for studying human diseases.

A comparison of the early developmental program of placental mammals with that of other animals, including *C. elegans* and *D. melanogaster,* shows how different these programs can be. In nematodes, embryonic cells are highly restricted in their developmental potential, or fate, beginning at the two-cell stage; and in fruit flies, polarization of the egg before fertilization generates distinct cytoplasmic regions dedicated to supporting different developmental programs within the nuclei that end up in these locations. Consequently, half a nematode embryo or half a fly embryo can never give rise to a whole animal.

Events Restricting the Developmental Potency of Individual Cells Occur Near the End of the Preimplantation Stage

The first differentiation events of mouse embryogenesis occur in the 16-cell embryo (see Fig. 22.5). The cells on the outside of the embryo turn into a trophectoderm layer that will eventually take part in the formation of the placenta. Soon thereafter, the cells on the inside of the embryo compact into a small clump called the *inner cell mass,* or *ICM,* that remains attached to one spot along the inside of the hollow trophectoderm sphere. The embryo will develop entirely from the ICM. As Fig. 22.5 shows, compaction of the ICM causes the appearance of a fluid-filled space that is devoid of cellular material and surrounded by trophectoderm. This space is called the *blastocoel cavity;* the embryo develops within it. The embryo is now called a **blastocyst.** Two more rounds of cell division occur during the blastocyst stage, producing the 64-cell embryo that implants.

Throughout normal preimplantation development, the embryo remains protected within the inert zona pellucida. As a result, there is no difference in size between the 1-cell zygote and the 64-cell embryo. To accomplish implantation, the embryo must first "hatch" from the zona pellucida so that it can make direct membrane-to-membrane contact with cells in the uterine wall. Embryonic and fetal development within a uterus

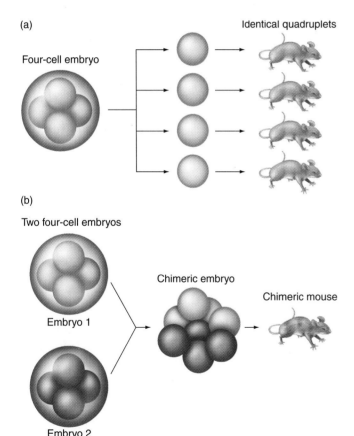

Figure 22.6 Early mammalian embryos are highly malleable in their development. There is no requirement for a one-to-one correspondence between embryo and adult. (a) Creating *identical* quadruplets from a single fertilized *egg.* (b) Creating a single chimeric animal from the fusion of two embryos.

inside the body of a female is a characteristic unique to all mammals except the primitive egg-laying platypus.

After Implantation, the Placenta Develops, the Embryo Grows, and the Tissues and Organs Emerge

Implantation initiates development of the placenta, a mix of embryonic and maternal tissues that mediates the flow of nutrients entering the embryo from the maternal blood supply and the flow of waste products exiting the embryo to the maternal circulation. The placenta maintains this intimate connection between mother and embryo, and later between mother and fetus, until the time of birth. Development of the placenta enables a period of rapid embryonic growth. Cells from the ICM differentiate into the three germ layers of endoderm, ectoderm, and mesoderm during a stage known as gastrulation. The foundation of the spinal cord is put into place, and the development of the various adult tissues and organs begins. With the appearance of organs, the embryo becomes a fetus, which continues to grow rapidly. Birth occurs at about 21 days after conception. Newborn animals remain dependent on their mothers during a suckling period that lasts 18–25 days. By five to six weeks after birth, mice reach adulthood and are ready to begin the next reproductive cycle.

Two Powerful Transgenic Techniques for Analyzing the Mouse Genome

Geneticists have capitalized on certain features of the mouse genome and the mouse life cycle to develop protocols that make it possible to add and remove specific genes from embryonic or germ cells.

The Addition of Genes to the Mouse Genome by Nuclear Injection

As we described in Chapter 10, the 1981 development of a method for inserting foreign DNA into the germ line of mice thrust a primarily observational discipline into the realm of genetic engineering with all its implications. Yet the incredibly powerful transgenic technology is based on a very simple process. Recall that a *transgene* is any piece of foreign DNA that researchers have inserted into the genome of a complex organism, such as a mouse or a pea plant, through experimental manipulation of early stage embryos or germ cells; any individual carrying a transgene is known as a *transgenic* animal or plant.

To create a transgenic mouse carrying a foreign DNA sequence integrated into one of its chromosomes, a researcher simply injects foreign DNA into a pronucleus of a fertilized egg and then places the injected one-cell embryo back into a female oviduct where it can continue its development. Roughly 25% to 50% of the time, depending on the skill of the investigator, the injected DNA will integrate at random into a chromosomal location. Integration can occur while the embryo is still in the one-cell stage, in which case the transgene will appear in every cell of the adult body. Or integration may occur somewhat later, after the embryo has completed one or two cell divisions; in this case, the mouse will be a mosaic of cells, some with the transgene and some without it. The relatively high rate of integration appears to be a consequence of

naturally occurring DNA repair enzymes present in all eukaryotic cells; during evolution, these enzymes acquired the ability to seek out and ligate together open-ended DNA molecules, which can result from mutagenesis. Figure 22.7 illustrates the details of the transgenic procedure.

Up to 50% of the mice born from injected embryos have the foreign DNA stably integrated into their genomes. They will thus transmit this DNA to their offspring. There are no limits to the type of DNA that can be injected. It can come from any natural source—animal, plant, or microbial—or directly from a DNA synthesizer. It is very common for investigators to construct DNA molecules (called "DNA constructs") composed of elements from different sources. For example, a DNA construct might have a coding region that is a composite of human and *Escherichia coli* sequences flanked by an upstream regulatory region that is a composite of mouse and synthetic sequences.

(a) Several embryos recovered from sacrificed female

Embryos transferred to a depression slide containing culture medium

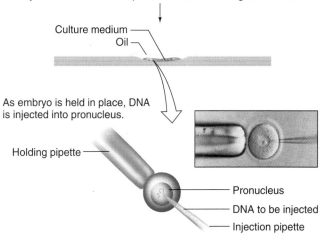

Several injected embryos are placed into oviduct of receptive female.

Figure 22.7 How transgenic mice are created. About 12 hours after conception, the female mouse is sacrificed and the embryos recovered. They are transferred to a depression on a specialized microscope slide containing culture medium. The slide is placed on the stage of an inverted microscope (as the name implies, the objective lens is beneath the stage rather than above it, as in a typical microscope). This arrangement enables the researcher to have space to manipulate the embryos from above. Suction holds the embryo in place on the tip of a special "holding pipette." A second type of pipette with a very narrow bore (the injection pipette) is used to inject the DNA solution through the plasma membrane and into the pronucleus where the DNA is released. The altered embryos are then placed into the oviduct of a receptive female.

Although embryonic nuclear injection is a powerful transgenic tool, it has two significant limitations. First, it can only add—not subtract—genetic material. Second, experimenters cannot target the insertion of foreign DNA to specific genomic locations. Consequently, transgenic mice produced by embryonic nuclear injection are useful only for the analysis of dominant phenotypes. By 1989, geneticists had developed a way to circumvent these limitations.

Targeted Mutagenesis, a Second Transgenic Technology, Makes It Possible to Remove, or Knock Out, Specific Sequences from the Mouse Genome

In addition to knocking out the function of a sequence altogether, targeted mutagenesis can produce an allele with an altered function. The emergence of targeted mutagenesis, which is technically more demanding and more complex than the nuclear injection technology described above, depended on two advances in cell culture techniques that occurred during the 1980s. The first was the establishment of *in vitro* conditions that enable researchers to place mouse embryos at the blastocyst stage into culture such that the embryonic cells continue to divide without differentiating. Cultured cells that behave in this way are called **embryonic stem cells,** or **ES cells** for short. ES cells appear to be similar in their state of differentiation to cells from the ICM. It is possible to grow cultures containing many millions of ES cells from a single embryo, and then recover a handful of cells from this culture for injection back into the blastocoel cavity of a normal embryo. Once inside the cavity, the ES cells can attach to the ICM, divide, and contribute to all of the tissues in the mouse that develops from the embryo. Most importantly for geneticists, the ES cells even contribute to the germ lines of these chimeric mice so that reproducing adults can transmit mutated genes present in the ES cells to future generations.

The second critical advance that provided a foundation for targeted mutagenesis was development of a protocol for homologous recombination in ES cells. The transformation of mammalian cells is known as **transfection.** During the transfection of mouse cells with mouse-derived DNA, the foreign mouse DNA almost always integrates at random into a chromosome at a site other than its point of origin. Occasionally, however, the added DNA will "find" and replace its homolog by homologous recombination. The frequency of homologous recombination events as a fraction of the total number of integrations is on the order of 10^{-3} to 10^{-5}.

If researchers transfect mouse ES cells with unaltered, cloned fragments of mouse DNA, homologous recombination events do not cause genomic changes. But with the recombinant DNA technology described in Chapter 8, investigators can modify cloned genes so that they no longer function; cloned genes modified in this way are known as **knockout constructs.** When homologous recombination occurs with a knockout construct, the nonfunctional knockout allele replaces the endogenous wildtype allele (Fig. 22.8). To construct mice in which homologous recombination has knocked out specific genes, researchers developed protocols for identifying and re- covering the very rare ES cells in which homologous recombination occurs.

HOW BIOLOGISTS USE TRANSGENIC TOOLS TO STUDY MICE AND CREATE A MOUSE MODEL FOR HUMAN DISEASE

Both add-on and knockout transgenic technologies have tremendous value in genetic research. The protocols enable researchers to determine the function of gene products, characterize genetic regulatory regions, establish links between mutant phenotypes and particular transcriptional units as an aid to verifying the identification of a cloned gene, and create mouse models of human genetic diseases.

Using Transgenic Technology to Determine Gene Function

In Chapter 10 we saw how geneticists used transgenic technology to demonstrate that the cloned *SRY* (for sex-determining region on the Y chromosome) gene could confer maleness on an animal without a Y chromosome. Incorporation of the *SRY* gene's coding region and regulatory sequences into a mouse embryo with two X chromosomes produced an animal with male genitalia and testes (Fig. 22.9). This result demonstrated that the product of a single gene on the Y chromosome is all that is needed to switch the developmental pathway of the fetus from female to male.

Biologists have used this same transgenic technology to examine the functions of many other genes. By combining a mouse gene of interest with regulatory regions from other mouse genes, they can cause transgenic mice to express the natural transgene product in an unnatural manner: at a higher-than-normal level, in an alternative tissue, or at an alternative developmental stage. They can then use the aberrant level, time, or place of expression to elucidate the normal function of the wildtype gene.

Experiments analyzing the *myc* gene demonstrate the power of transgenic technology for uncovering the function of genes in ectopic expression (that is, expression at an abnormal place, time, or level). Investigators originally discovered the *myc* gene in the genome of the myelocytomatosis chicken retrovirus; exposure to this virus caused cultured chicken cells to become tumorigenic. Hybridization studies demonstrated the existence of a *myc* gene homolog in the genomes of mice and other vertebrates, including humans (where it resides on the long arm of chromosome 8), but not in the genomes of nonvertebrate model organisms, such as *Drosophila*. Moreover, in human cancer cells obtained from patients with Burkitts lymphoma (a cancer of the immune system's B cells), the *myc* gene often appears close to one of the breakpoints of a reciprocal translocation characteristic of these cancer cells between the long arms of chromosomes 8 and 14; in this translocated position, the gene is usually expressed at a higher than normal level. Noncancerous animal cells have a very low level of *myc* gene expression. These findings constitute circumstantial evi-

(a) Construction of a knockout allele in ES cells

Early blastocyst
(× 10,000)

Culture into millions
of embryonic-like
(ES) cells.

Clone containing gene of interest

5' ———————————— 3'

Build knockout construct by
adding in selectable marker.

5' ———————————— 3'

Marker disrupts transcription
unit.

← Add cloned DNA to
culture of cells.

Homologous recombination

Added DNA construct

5' ———————————— 3'

ES cell chromosome with wildtype allele

↓

ES cell chromosome with knockout allele

Finding the cell with the knockout allele.
Subject culture to drug that kills all cells that do not contain
selectable marker.

Survivor cells have knockout allele (1% or less).
Begin new culture with survivor cells.

(b)

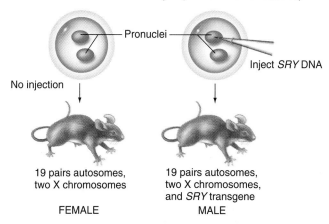

Figure 22.8 Knocking out a mouse gene in ES cells. (a) An early mouse blastocyst can be grown in culture under certain conditions that allow the cells of the ICM (inner cell mass) to remain undifferentiated as ES cells. A DNA clone, containing the gene of interest, can be modified in the laboratory into a disrupted allele with a selectable marker. This "DNA construct" is added to the ES cell culture where homologous recombination will occur. (b) A chimeric mouse composed of cells derived from a normal embryo (albino) and ones derived from the mutated ES cell (dark agouti).

Two one-cell female mouse embryos (with two X chromosomes)

← Pronuclei →

Inject *SRY* DNA

No injection

19 pairs autosomes,
two X chromosomes

FEMALE

19 pairs autosomes,
two X chromosomes,
and *SRY* transgene

MALE

Figure 22.9 The transgenic mouse protocol proves that *SRY* is the "testis-determining" locus responsible for the production of maleness during embryogenesis. DNA fragments containing the mouse *SRY* gene and its regulatory sequence were injected into a series of embryos that were allowed to develop into live animals. Normal XX animals without a transgene develop as females. But the presence of the SRY transgene in an XX embryo induces development as a male.

dence that abnormally high levels of *myc* expression might help transform a cell to a cancerous state. (The biochemical mechanism by which the myc gene product functions is described in Chapter 16).

The results of experiments using cells grown in culture support this hypothesis. In almost every case studied, however, cultured mammalian cells display programs of gene expression that do not correspond with those of the cells they are supposed to model. Thus, the results of cell culture studies do not necessarily reflect how cells *in vivo* (that is, within the body) behave in response to a change in gene activity. In addition to differences in gene activity, there are differences in chromatin structure and patterns of DNA methylation between cells growing *in vitro* and *in vivo*. These discrepancies are not surprising since cells in the body, unlike those in culture, exist in a complex environment that includes constant exposure to molecular signals released by other body cells. The living organism also has a pervasive immune system that is impossible to imitate *in vitro*. It is therefore possible that a phenotype observed in responses to the abnormal expression of a gene in cultured cells might be a consequence, not of one gene's abnormal expression, but of interactions with other genes that are expressed differently in cultured cells than in cells *in vivo*. For these reasons, it is not possible to rely on cell-culture results for an explanation of the true function of a gene; rather it is necessary to examine the effects of aberrant gene expression *in vivo*.

To learn whether increased expression of the *myc* gene affected tumor formation in various tissues of the mouse, researchers used transgenic technology (Fig. 22.10). In one experiment, they attached the immunoglobulin gene promoter to the *myc* coding sequence to produce a transgenic mouse line that expressed *myc* at high levels in the precursors to immunoglobulin-producing B cells. In another experiment, they attached tissue-specific promoters, including one for mammary gland expression, to the *myc* coding region. And in yet another experiment, they attached to the *myc* coding region a promoter that is stimulated in all tissue but only after the embryo's or animal's exposure to dexamethasone, a glucocorticoid hormone.

The results of all these experiments showed that overexpression of the *myc* gene does not have any effect on normal developmental processes; even when the gene was expressed at high levels in many developing tissues, normal animals were born. Moreover, even in the adult, most cells that overexpress the *myc* gene never display an aberrant phenotype. However, the rate of tumor formation increased significantly in most, but not all, types of tissue. The conclusion was that aberrant expression of the *myc* gene alone does not cause cells to become cancerous; but it can operate with other somatic mutational events to transform a cell to the cancerous state. These observations support the hypothesis that the cancer phenotype results from the accumulation of multiple mutations in the clonal progeny of one cell (see Chapter 17).

Geneticists still do not completely understand the normal function of the wildtype *myc* gene. But biochemical and genetic studies indicate that the gene product is a transcription factor that helps control the growth of cell populations in almost every tissue of the body (as described in Chapter 16).

Using Transgenic Technology to Characterize Regulatory Regions

To understand embryonic development, it is essential to learn not only how the proteins critical to this complex process function, but also how the genes that encode these proteins are regulated. The pattern of gene expression in a normally differentiating cell is finely tuned, with each gene expressed at exactly the right level and each gene turned on and off at exactly the right times. Geneticists know that cis-*acting regulatory regions* (that is, regulatory regions on the same chromosome) that are closely linked to each gene control the timing and intensity of each gene's expression; but determining the location and sequence of such regions is greatly aided with transgenic technology.

Researchers can characterize the *cis*-acting regulatory regions of single-cell organisms, such as bacteria and yeast, by random mutagenesis experiments. In these experiments, they select from cultures of millions of mutagenized cells those cells with single base changes or small deletions in genetic regions that influence the *cis*-regulation of a gene of interest. They can then use the regulatory mutations to identify the base pairs necessary for proper gene function. Because of the large number of organisms required, it is not possible to use a similar approach to study gene regulation in mice. Instead, mouse geneticists must turn to transgenics.

Once they have obtained a genomic clone of a mouse gene of interest, researchers can subclone flanking sequences that are likely to contain its regulatory region. In most cases studied to date, the regulatory regions have been confined to the 5–10 kb of DNA just upstream (to the 5′ side) of the coding sequence; thus, this is the first region an investigator examines at the start of a project. Although it might seem counterintuitive at first, the researcher has no use for the actual coding region of a gene when the aim is to study the gene's regulatory region. In fact, it is best to replace the true coding region with that of a "reporter gene" whose protein product is easy to detect. The most commonly used reporter gene is the β-*galactosidase* (*lacZ*) gene from *E. coli* bacteria (see Fig. 22.11). As we saw in Chapter 8, the β-galactosidase enzyme encoded by this gene will convert a special substrate known as X-Gal into a colored product that can be visualized under the microscope.

To characterize the regulatory region associated with a gene, you first create a series of different transgene constructs, each formed by splicing together different fragments or mutated forms of the putative regulatory region with the β-galactosidase reporter gene. You next establish a series of transgenic lines by injecting the transgene constructs into different mouse embryos, and then, after setting up timed matings within each of these lines, recover embryos at the developmental stage or stages during which the mouse gene normally undergoes expression. By examining the distribution

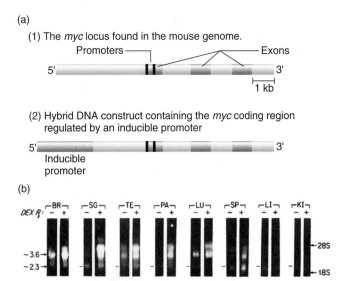

Figure 22.10 Transgenic expression of *myc* gene provides information on the gene's role in tumor formation.
(a) Construction of a transgene containing the *myc* gene under the control of an inducible promoter: (1) structure of the endogenous *myc* gene and (2) transgene construct with the MTV (dexamethasone-inducible) promoter attached to a portion of the *myc* gene that contains the coding region. (b) Northern blot showing induction of transgene expression in a range of adult tissues.

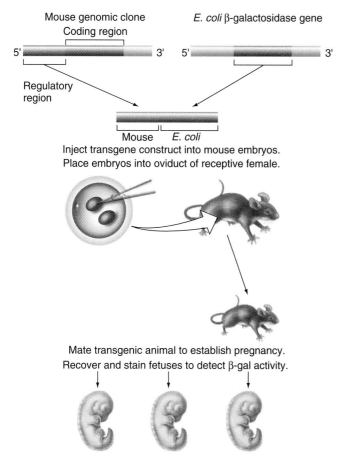

Figure 22.11 Transgenic technology can be used to analyze *cis*-acting regulatory regions. A DNA construct containing the mouse regulatory region of interest is attached to the *E. coli* reporter gene. The function of the regulatory region can be ascertained by observing β-gal expression in transgene fetuses.

of the colored β-galactosidase product at each embryonic stage and comparing it to the distribution of the natural gene product, it is possible to map the extent of the *cis*-acting regulatory region. The distribution of the natural gene product can be tracked with specific antibodies labeled for examination by immunohistochemistry. If the antibody label is a different color than the β-galactosidase label, the natural gene product and the β-galactosidase enzyme can be simultaneously observed and distinguished under the microscope. Further studies of the effects of individual base-pair substitutions or small deletions on details of the expression pattern can lead to a highly sophisticated understanding of the various *cis*-acting regulatory elements that together may determine the complex patterns of spatial and temporal expression of genes that play a role in development.

An example of the use of transgenic mice to study gene regulation is the analysis of a gene called *T complex protein-10bt* (*Tcp10b*^t^). Only differentiating male germ cells in the process of maturing into spermatozoa express *Tcp10b*^t^. Cells in the seminiferous tubules, called sertoli cells, mediate sperm differentiation. When investigators disrupt the testes and place sperm cells in culture, they cannot simulate natural conditions

completely; as a result, germ cell differentiation continues *in vitro* for only a brief time. But differentiation from stem cell to mature spermatozoa takes six weeks in mice.

Sperm differentiation is of interest to cell biologists because the mechanisms that regulate it may differ significantly from those that control the differentiation of other cells. This is, in part, because the transformations of differentiating sperm cells are much more dramatic than those experienced by other types of cells. Not only do the spermatogenic cells change in shape and size from large round stem cells to tiny, sleek spermatozoa almost without cytoplasm; they also drastically change their genetic program: Stem cells have a normal program of gene expression as well as chromosomes with a normal chromatin structure; in contrast, the chromosomes of differentiated sperm cells have a unique chromatin structure with no histones attached, and they exhibit no gene activity. To understand these differences, mouse geneticists have tried to characterize the regulatory regions associated with gene activity in spermatogenic cells.

To analyze the regulatory region associated with the *Tcp10b*^t^ gene, they first made a series of transgene constructs carrying different lengths of DNA from the 5′ flanking region of the *Tcp10b*^t^ gene, ligated to the *lacZ* coding sequence. The result was six DNA constructs containing from 0.6 to 1.6 kb of DNA from the putative *Tcp10b*^t^ regulatory region. The researchers injected copies of each DNA construct into multiple mouse embryos, obtaining at least four independent transgenic lines of mice for each construct. (It is important to use multiple transgenic lines in an experiment of this type to verify that sequences in the transgene construct itself, rather than sequences that coincidentally flank the transgene insertion site, are responsible for a particular phenotype.)

Figure 22.12 depicts how it was possible to map the regulatory region associated with *Tcp10b*^t^ by simply testing for the presence of *lacZ* transcripts in Northern blots of testicular RNA obtained from each transgenic line of mice. As the figure shows, there was no detectable transcription of the *lacZ* gene in testes from transgenic mice that carried 0.75 kb or less of flanking sequence 5′ to the *Tcp10b*^t^ gene; but with 0.97 kb or more of the flanking sequence, high levels of transcription occurred in the testes, but in no other tissue, of all mouse lines.

These observations located a critical testes-specific, *cis*-regulatory sequence within a 227-bp region between 746 and 973 bases upstream of the *Tcp10b*^t^ gene. With this information, it became possible to design additional experiments to examine the regulatory region in more detail and identify transacting proteins that bind to this region. Through these additional experiments, researchers identified specific testicular proteins that activate the *Tcp10b*^t^ gene.

Using Transgenic Technology to Link Mutant Phenotypes to Specific Transcription Units

A third significant use of transgenic tools is in establishing whether a cloned gene corresponds to a locus previously defined by a mutant phenotype. Consider, for example, the mouse *Brachyury* locus, symbolized by a *T*. This locus is

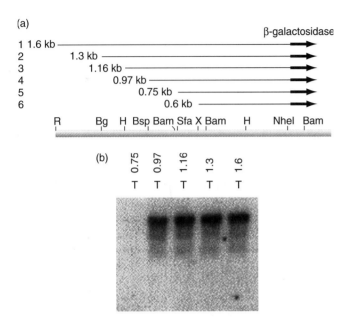

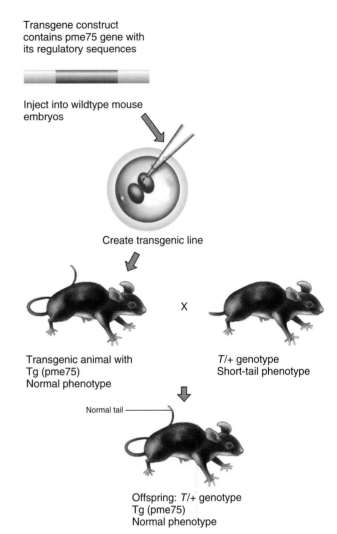

Figure 22.12 An example of transgenic technology used to map the *cis*-acting regulatory region associated with the *Tcp10b^t* gene. (a) Different hybrid DNA constructs were made with varying lengths of the 5' flanking sequence adjacent to the *Tcp10b^t* gene fused to the *E. coli* β-galactosidase gene, which acts as a reporter. (b) Testicular RNA was obtained from transgenic mice containing the various lengths of 5' flanking region (shown in kilobases). With 0.75 kb of flanking region, no transcription of the reporter gene was observed, but with all larger flanking regions, transcription did occur.

Figure 22.13 Transgenic technology can be used to identify the locus responsible for a mutant phenotype. A dominant deletion mutation at the *T* locus causes a short tail. A transgenic animal containing the pme75 transgene is mated with a mutant animal to create animals containing both the deletion and the transgene. A normal phenotype demonstrates that the deletion of the pme75 gene is responsible for the short-tail phenotype.

defined genetically by a dominant, X-ray-induced mutation that causes heterozygous *T/+* animals to develop short tails. Through high-resolution linkage analysis and positional cloning (described in Chapter 10), researchers showed that the *T* locus mutation is associated with a 200-kb deletion on chromosome 17. They also identified a transcription unit called pme75 that is located in this same 200-kb deletion; normally, pme75 is expressed in the embryonic tissue that develops into the tail. While highly suggestive, these data do not prove that the absence of the pme75 gene causes the short-tail phenotype. It is very likely that the 200-kb deletion also removes other genes, and on the basis of the genetic data alone, you cannot rule out the possibility that the absence of one of these undiscovered genes produces the mutant phenotype.

You can, however, resolve this impasse with the transgenic protocol illustrated in Fig. 22.13. The first step is to make a transgene construct containing the complete pme75 coding region together with its regulatory sequences. You next inject the construct into wildtype mouse embryos to create a transgenic line. By breeding this transgenic line to animals with the *T* locus mutation, you can obtain offspring that have the mutant *T/+* genotype as well as a second functional pme75 gene on a different chromosome (at the locus where the transgene construct integrated at random). These animals sport a tail of normal length. By demonstrating that the transgene can correct the mutant phenotype, this result proves that the *T* locus is the equivalent of the pme75 gene.

Using Targeted Mutagenesis to Create a Mouse Model for Human Disease

Researchers can use targeted mutagenesis, in combination with other protocols, to create a mouse model for human diseases that result from a loss of gene function. As an example, Fig. 22.14 illustrates the step-by-step creation of a mouse model for cystic fibrosis. With the identification and cloning of the human cystic fibrosis gene (designated *CFTR*), the first step toward a mouse model was the cloning of the mouse homolog. Development of the mouse model also required fabrication of a *CFTR* knockout construct, derivation of an ES cell culture from a mouse blastocyst, transfection of the ES cells with the *CFTR* knockout construct, selection of cells in which homologous recombination had replaced the wildtype *CFTR* gene with the mutant knockout allele, and finally, the production and analysis of chimeric mice and their offspring. Animals homozygous for the *CFTR* knockout allele display a mutant phenotype that is very similar to that expressed by humans suffering from cystic fibrosis. Thus, in developing drugs to alleviate CF symptoms in humans, pharmaceutical researchers can first test new products in mice to determine their efficacy.

THE Hox GENES: A COMPREHENSIVE EXAMPLE

A particularly striking example of how mouse biologists used both nuclear injection and targeted mutagenesis technologies to decipher gene function comes from an analysis of the mouse *Hox* gene family. The *Hox* genes are **homeotic selector genes,** that is, genes that control the development of body segments. Members of the *Hox* family are distributed among four unlinked gene clusters that each contain 9–11 genes (Fig. 22.15).

Researchers discovered the mouse *Hox* gene family through cross-hybridization studies using the homeotic selector genes of *D. melanogaster* as probes. In fact, homeotic selector genes were first identified on the basis of mutations that produced flies with four wings instead of two, flies whose mouthparts developed incorrectly as legs, and flies with legs instead of antenna growing out of their heads. The proteins encoded by these genes turned out to be transcription factors that act as on-off switches, instructing segments of the fly to develop into one type of tissue or another. William Bateson, the same man who coined the term "genetics," chose the designation of "homeotic selector" from the Greek word *homoios,* which describes a type of variation in which "something has been changed into something else." Homeotic genes appear to control the development of each *Drosophila* body segment (as discussed in Chapter 21). The bizarre phenotypes just described result when expression of a particular homeotic gene does not occur at the proper time and place. Lack of appropriate expression flicks the binary switch, transforming the recipient body segment into a different type of tissue.

Drosophila homeotic genes were first cloned in the early 1980s. By the end of that decade, it had become clear from cross-hybridization and cloning studies that homologs of these genes are likely to exist in every species of multicellular animal, from *C. elegans* to *Homo sapiens.*

In segmented animals such as flies, homeotic genes are active in the discrete segments that define the body plan, where they determine the proper differentiation of tissues. But what do they do in mice and humans, organisms that do not have obvious body segments? To answer this question, researchers had to overcome a serious drawback: the lack of known mutations at any of the *Hox* loci. This problem was not unique to understanding the functions of mammalian *Hox* genes. Rather, since the discovery of homeotic gene homologs in the mouse genome, it has become routine for developmental geneticists to use cross-hybridization protocols to look for mouse homologs of every *Drosophila* gene found to have a role in development. This strategy has led to the discovery of

dozens of new mouse genes, most of which were not associated with any known mutant phenotypes.

How Scientists Determine the Function of a Gene in the Absence of Previously Characterized Mutations

Analyses of Expression Patterns in Developing Embryos Can Provide a Clue to the Time and Location of Gene Action

A clone of a gene can be a tool for analyzing the gene's expression. One way to convert a clone into this type of tool is to label it, denature it, and use the resulting DNA strands as probes in *in situ* hybridization. To study development, investigators can perform *in situ* hybridization on the RNA present in fixed tissue sections obtained from embryos at different stages of development. When developmental geneticists examined the expression patterns of the *Hox* genes, they discovered that each one is transcribed along a portion of the developing embryonic axis that extends from the same most posterior point to a specific anterior boundary (Fig. 22.16). Analysis of the pooled data on *Hox* gene expression showed that the extent of spatial expression corresponds with the position of each gene in its cluster. Genes at the 5′ end of a cluster (for example, *D13*) have the least extensive expression patterns, while genes at the 3′ end of the cluster (for example, *A1* and *B1*) have the most extensive expression patterns. These data suggest that different *Hox* genes might be involved in controlling the development of different sections of the embryonic axis. Expression data alone, however, cannot provide conclusive evidence of function.

Ultimately, Only Genetic Tests Can Determine Gene Function

To understand what role a particular *Hox* gene plays in development, a scientist must be able to examine embryos that do not express that gene at all, or express it outside its normal time or place. Examining the changes in development that arise as a result of these genetic changes makes it possible to decipher the normal role of the gene and, in the case of the *Hox* family, test general hypotheses concerning the functional interactions of different *Hox* genes. We now present two examples of this approach.

Validating the Hypothesis That Expression of the 5′ Gene in a Hox Cluster Is Epistatic to Expression of the More 3′ Genes

The expression data show that some *Hox* genes are expressed across many segments of the embryo, even as those segments develop differently from each other. To account for this difference in the simplest way, researchers proposed that the only gene that counts in any particular embryonic segment is the most 5′ gene in a *Hox* cluster. In other words, expression of the most 5′ gene is epistatic to expression of the other, more 3′ *Hox* genes. By examining Figs. 22.15 and 22.16, you can see that this hypothesis could explain how each spinal segment develops in a different manner.

a. Researchers identified and cloned the mouse homolog of the human cystic fibrosis gene (known as the *CFTR* gene because its product is the cystic fibrosis transmembrane regulator protein). They accomplished this by using a cloned fragment from the human gene as a probe to screen a mouse genomic library in a phage lambda vector. At 250 kb, the *CFTR* gene is quite large, and the cloning capacity of lambda vectors is only 20 kb (see Chapter 8). This means that only a small portion of the gene can be recovered in any lambda clone. But, as we will see, this doesn't matter for the success of the homologous recombination protocol.

b. After recovery of a lambda clone carrying the first several exons of the mouse *CFTR* gene, the researchers digested the clone with restriction enzymes. They then combined the restriction fragments of the clone with a fragment including the bacterial gene known as *neo*, which provides resistance to the drug neomycin, a downstream region of *CFTR* genomic DNA, and a fragment including the thymidine kinase (*TK*) gene, which inhibits cell division. The final multipart construct contained portions of an original *CFTR* exon on either side of the *neo* fragment such that the coding sequence was disrupted. The researchers cloned this multipart construct within a plasmid vector and isolated purified insert DNA.

c. Meanwhile, they derived an ES cell culture from a mouse blastocyst recovered from a mating between two agouti parents of the 129/SvJ strain.

d. They then added the *CFTR* knockout construct to the cell culture and prevented any nontransfected cells from growing by exposing the culture to neomycin. The only cells remaining after this selection were those with an integrated copy of the *CFTR* construct. In most of those cells, the whole construct has integrated into a random

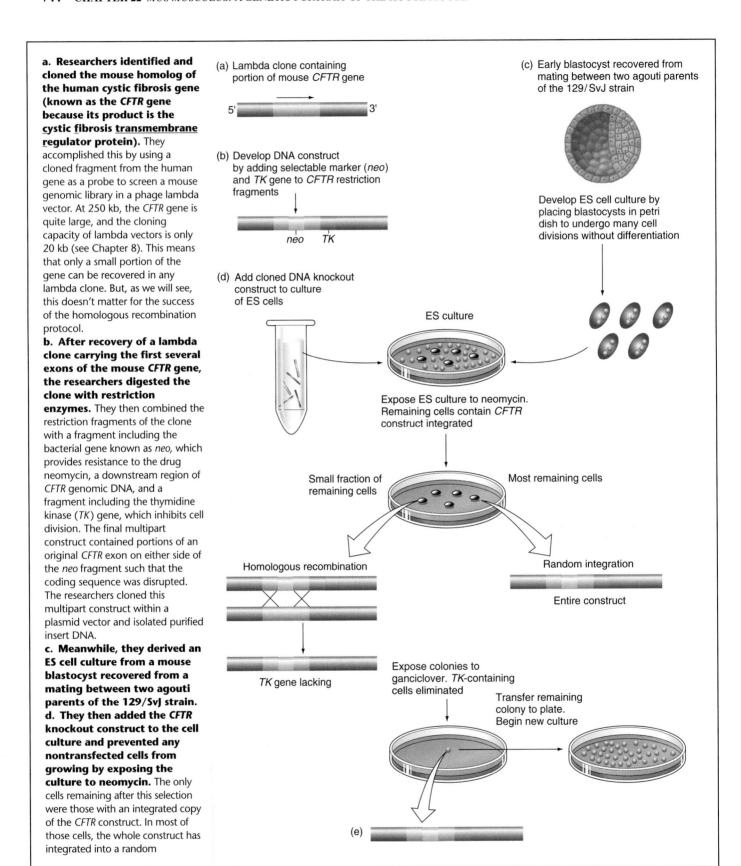

(a) Lambda clone containing portion of mouse *CFTR* gene

5'   3'

(b) Develop DNA construct by adding selectable marker (*neo*) and *TK* gene to *CFTR* restriction fragments

neo   *TK*

(c) Early blastocyst recovered from mating between two agouti parents of the 129/SvJ strain

Develop ES cell culture by placing blastocysts in petri dish to undergo many cell divisions without differentiation

(d) Add cloned DNA knockout construct to culture of ES cells

ES culture

Expose ES culture to neomycin. Remaining cells contain *CFTR* construct integrated

Small fraction of remaining cells

Most remaining cells

Homologous recombination

Random integration

Entire construct

TK gene lacking

Expose colonies to ganciclover. *TK*-containing cells eliminated

Transfer remaining colony to plate. Begin new culture

(e)

Feature Figure 22.14   Creating a mouse model for cystic fibrosis.

chromosomal location; but in a small number of the transfected cells, homologous recombination has resulted in the integration of only the *CFTR* sequences together with the *neo* gene (and not the *TK* gene). To distinguish those few cells in which homologous recombination has integrated only part of the construct, experimenters expose the neomycin-resistant colonies to a drug called ganciclover, which eliminates all cells containing the *TK* gene. The result of this double selection—for the presence of *neo* and the absence of *TK*—is a culture containing only those cells that have integrated the mutant *CFTR* gene and have lost the *TK* gene.

e. Molecular analysis of each colony of surviving cells identified at least one colony in which the mutant knockout allele had placed the wildtype *CFTR* gene. This colony was transferred to a separate plate to grow additional cells.

f. At this stage of the study, investigators set up a mating between two black mice of the B6 strain, and recovered embryos at the blastocyst stage. They then injected individual blastocysts with about 10 ES cells from the *CFTR* knockout line, and place these injected blastocysts back into the uterus of a foster mother. There the blastocysts continued their development into live-born mice. In these offspring, fur containing black and agouti stripes signaled chimeric mice sexual maturity, these mice could produce three different types of gametes: one derived entirely from the B6 line, a second derived entirely from ES cells and carrying the wildtype *CFTR* allele, and a third derived entirely from ES cells but carrying the knockout allele. These chimeric animals were now mated to B6 mice.

g. Since the agouti coat color is dominant to black, ES-derived offspring could be recognized by their completely agouti coat. By genotyping these offspring with the molecular tools described in Chapter 9, researchers identified those that carried the *CFTR* knockout allele.

h. They then set up matings between male and female carriers of the mutant allele to obtain homozygous mutant offspring that could serve as a model for the cystic fibrosis disease state.

(f) Mate B6 black mice. Embryos recovered from pregnant B6 female.

(+/+) × (+/+)

10 ES cells are placed in embryos which are returned to uterus of B6 foster mother.

(+/+)

ES cells with *CFTR* knockout (+/−)

Embryos develop into live-born mice

Chimera (+/−<−>+/+)   (+/+)

Mate chimera with B6 black mouse

(+/−<−>+/+) × (+/+)

(g) Obtain offspring, some of which are heterozygous for mutant allele. Breed heterozygotes.

(h) (+/+)   (+/−) × (+/−)   (+/+)

Offspring homozygous for mutant allele serve as models for CF disease state.

(+/−)   (−/−)   (+/+)

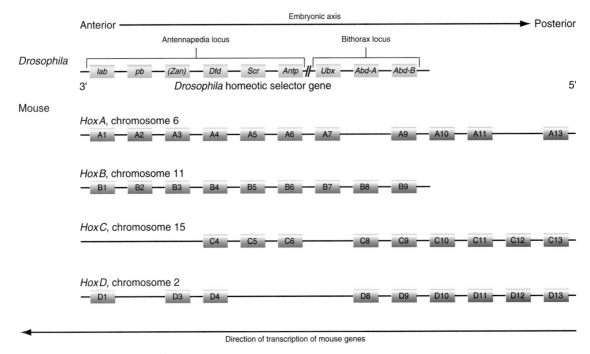

Figure 22.15 The mouse *Hox* gene superfamily contains multiple homologs of each member of the *Drosophila* homeotic selector gene family. Genes within the four mouse *Hox* clusters are lined up according to their homology with each other and specific *Drosophila* genes. Not all clusters have homologs of each *Drosophila* gene. From gene 9 and higher, there are multiple homologs of the *Drosophila* Abd-B gene in *Hox* clusters *A, C,* and *D.*

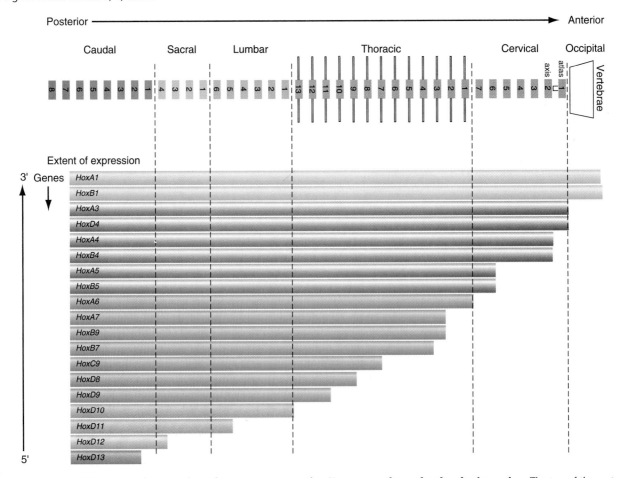

Figure 22.16 Spatial extent of expression of some representative *Hox* genes along the developing spine. The top of the mature spine is shown to the right, bottom to the left, with vertebrae numbered and grouped by name.

A Transgene Test Confirmed the Prediction of a Homeotic Transformation

As one test of this hypothesis, investigators made a transgene construct with a *HoxA1* regulatory region attached to a *HoxD4* coding sequence (Fig. 22.17a). According to the hypothesis, *HoxD4* normally controls the development of the C1 and C2 vertebrae in the cervical region of the spinal column, while *HoxA1* normally controls the development of the occipital bone at the base of the skull. In a transgenic fetus, however, the presence of the transgene construct causes expression of the *HoxD4* gene in the occipital region along with *HoxA1;* and as predicted by the hypothesis, a homeotic transformation converts the occipital bone into cervical vertebrae (Fig. 22.17b and c).

Knockout Studies Confirmed Predictions of Aberrant Phenotypes

The hypothesis of 5′ epistasis also leads to the prediction that aberrant phenotypes will arise when different *Hox* genes are knocked out by homologous recombination. With each knockout, one would expect a particular embryonic segment to become transformed to more anterior-like structures. The data obtained from knocking out various *Hox* genes support the hypothesis of 5′ epistasis. For example, a knockout of *HoxB4* produces a partial homeotic transformation of the second cervical vertebra from axis to atlas (see Fig. 22.16).

Transgenic Studies Lead to an Understanding of the Developmental Role of Hox and Other Homeotic Genes

With the accumulation of transgene and knockout data for many of the mouse *Hox* genes, a general understanding of the developmental role played by this gene family, not only in mice, but in other animals as well, has emerged. What the *Hox* genes and their homologs in other species apparently do is establish signals identifying the position of each region along the embryo's anterior-posterior axis. The genes do not determine the differentiation of any particular cell type or tissue. Rather, they provide positional information that other genes act on to promote the differentiation of particular tissues. The positional information as well as the genes that act on it vary from species to species.

It is likely that the emergence of the *Hox*-like gene family in our most recent single-cell ancestor played a fundamental role in the evolution of developmental complexity. Thus, the analysis of the *Hox* gene family is an example of how detailed genetic studies in one model species can provide a general understanding of gene function across large segments of the animal kingdom as well as clues to the evolution of developmental complexity.

(a)

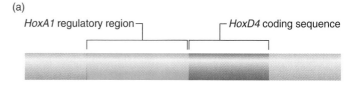

(b)

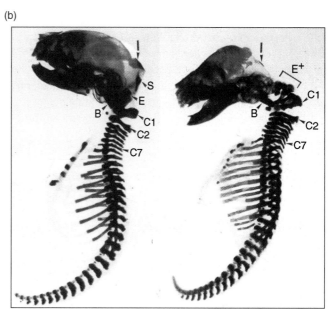

(c)

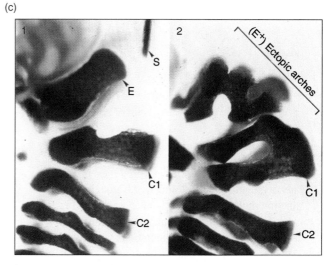

Figure 22.17 Transgenic technology provides support for the mode of action of the *Hox* gene family. (a) A transgenic construct is produced to missexpress *HoxD4* in a more anterior region where *HoxA1* is normally expressed (refer to Fig. 22.16 for the normal extents of expression of each of these genes). In panel (b), the complete skeleton of a wildtype animal is shown on the left, and that of an animal expressing the transgene construct is shown on the right. A blow-up of the cervical regions from both skeletons is shown in panel (c), again the wildtype is on the left and the transgenic construct is on the right. In transgenic newborns, what would have been occipital bone (region E in wildtype animal) has been transformed into ectopic arches (E+) that look like cervical vertebrae.

C O N N E C T I O N S

Geneticists can now produce transgenic animals by combining the classic tools of mutagenesis with an understanding of molecular biology and embryology. Transgenic technology has become so sophisticated that, in theory, it is possible to make any genetic change imaginable to the mouse genome and determine its effect on the individual that emerges.

With the ability to produce mice carrying add-ons and knockouts of *Hox* and other genes that play a role in development, mouse geneticists have begun to dissect the process by which the mouse embryo develops from the one-cell zygote, through gastrulation, and into the organ-building stage. Remarkably, the development of all mammals is similar, especially in these early stages. For example, all the genes important to mouse development are conserved in the human genome. Thus, much of what we learn about the genetic basis for mouse development will apply to normal and abnormal human development.

In Chapters 23 and 24, we change our focus from the analysis of gene activity in individuals to an analysis of gene transmission in whole populations, and an examination of how a population's genes evolve over time. Then, in the Epilogue, we explore how results obtained from mouse experiments can be combined with other types of molecular and experimental data to provide an increasingly penetrating understanding of human biology.

E S S E N T I A L   C O N C E P T S

1. The availability of hundreds of single-gene mutations and a short life cycle contribute to *M. musculus*'s value as a model organism.

2. The mouse genome closely resembles the human genome in size, gene content, and syntenic loci. Researchers can thus use *homology* and *conserved synteny* analysis to identify, locate, and determine the function of genes in both species.

3. The mouse life cycle is representative of the mammalian life cycle, although the timing of events is unique to each species. The *totipotency* of preimplantation cells makes it possible to create *chimeras.*

4. Researchers can use the *transgenic technology* of adding genes to the mouse genome by nuclear injection to determine gene function, characterize regulatory regions, and correlate mutant phenotypes with specific transcription units.

5. *Targeted mutagenesis* is the basis for creating a mouse model of human diseases caused by a loss of gene function.

6. Studies using transgenic technology have revealed that the *Hox* genes generate signals that identify the position of each region along a mouse embryo's anterior-posterior axis. Normal development of the embryo depends on this information.

7. Knowledge of how the *Hox* genes function in mice has elucidated the general role played by homeotic selector genes in the evolution of all metazoan organisms, including humans.

S O L V E D   P R O B L E M S

I. Gain-of-function mutations can produce a novel phenotype and act in a dominant fashion. In loss-of-function mutations no functional gene product is made; most, but not all, loss-of-function mutations are recessive.
   a. Which of these types of mutations would you study using add-on transgenic technology?
   b. Which type of mutation would you have to study with the implementation of homologous recombination in ES cells? Why?

Answer

This problem requires an understanding of add-on transgenic and homologous recombination techniques and their applications.

a. Dominant alleles caused by gain-of-function mutations are expressed phenotypically when a single copy of the dominant allele is present. *The effect of a dominant gain of function mutation can be characterized in transgenic mice,* where a transgene containing the dominant allele is inserted into the genome and the original gene copy is still present.

b. *Loss of function mutations would have to be studied using homologous recombination,* where a normal copy of the gene is replaced with a defective version. To observe a phenotype of a recessive loss-of-function mutation, there should not be any normal version of the gene present. After replacing one gene copy with a mutant version, heterozygous mice would be mated to obtain a mouse homozygous for the recessive allele and to observe the resulting phenotype.

II. You have isolated muscle cell RNA and made a muscle cDNA library. You want to study the regulatory region of one of the genes you identified as being expressed in muscle cells.

a. You need to isolate a genomic clone of this gene. Why is the cDNA clone not sufficient for your studies?

b. Outline the steps to isolate the genomic clone of this gene.

c. You isolated a genomic clone containing the entire coding region and 1.2 kb upstream of the coding region. A map of the genomic clone is shown below. To make a fusion construct of the presumed regulatory region attached to the reporter gene *lacZ,* you cloned the 1.3-kb *Bam*HI fragment into a vector containing the *lacZ* gene. The fusion construct, when injected into cultured muscle cells, expressed β-galactosidase protein. What experiment could you do next to determine if the fusion construct is developmentally regulated?

d. Using an antibody against the β-galactosidase portion of the fusion protein you analyzed the tissues of a transgenic mouse and found that the protein encoded by this gene is present also in bone cells. How could you determine if the same transcript was expressed in both cell types?

e. You decided to analyze the promoter region of the gene to identify sequences responsible for the expression in the two different cell types. The results obtained with constructs containing different fragments from the 5′ region flanking the gene to the *lacZ* coding region are shown below. What region(s) are important for expression in muscle? in bone?

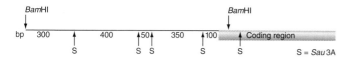

DNA in subclone	Expression in bone	Expression in muscle
1.	−	−
2.	−	−
3.	−	−
4.	+	−
5.	+	+
6.	+	+
7.	−	+

Answer

This problem requires familiarity with several molecular techniques including analysis of cDNAs and regulatory sequences in the promoter.

a. *The cDNA clone was derived from the mRNA and therefore it will probably not contain the regulatory region usually found upstream of (5′ to) the transcription start site.*

b. The cDNA clone is a good source of a probe for finding the genomic clone in a genomic library.
   1. *Isolate a DNA fragment from the cDNA clone and label for use as a probe.*
   2. *Grow clones (bacteriophages or cosmids) from a genomic library on plates.*
   3. *Transfer clones from plates to filter paper.*
   4. *Incubate the labeled probe with the filters to find a genomic clone containing the gene by hybridization.*

c. *To study developmental expression of this gene, the fusion construct would be transferred into the mouse genome, using transgenic techniques.* At different stages in development the tissues of the mouse can be tested for the presence of the mRNA or protein using the *lacZ* DNA or antibody that recognizes the β-galactosidase protein respectively.

d. To examine the transcripts in these two cell types, you would *prepare mRNA from isolated muscle and bone cells and use the cDNA clone as a probe against separated RNAs that have been transferred to a filter* (Northern hybridization analysis). Transcripts may be different sizes because of different transcription start sites or different processing (splicing) events.

e. The presence or absence of bone or muscle transcripts in mice that carry different transgene constructs provides a way to identify DNA fragments containing tissue-specific regulatory sequences. There are three constructs (4, 5, and 6) that express the transgene in bone tissue. The only portion of the regulatory region that they all have in common is a *50-bp Sau3A fragment. The bone-specific regulatory element must be within this fragment.* For muscle expression, there are *three clones (5, 6, and 7) that express the transgene in muscle tissue and these share the adjacent 350-bp Sau3A fragment.* This analysis shows that expression of the same gene is subject to different mechanisms of regulation in two different tissue types that use different regulatory sites in the 5′ region flanking the gene.

PROBLEMS

22-1 Choose the matching phrase in the right column for each of the terms in the left column.

a. stem cell	1. cells destined to become the fetus
b. conserved synteny	2. animals derived from two or more genetically different embryonic cells
c. inner cell mass	3. undifferentiated cells that serve as a source of two or more types of differentiated cells
d. trophectoderm	4. experimental elimination of gene function
e. embryonic stem cells	5. the same genes are genetically linked in two different species
f. chimeras	6. cells destined to become extraembryonic tissue such as the placenta
g. knockout	7. undifferentiated cells isolated from the blastocyst that are able to be reconstituted with normal cells in an embryo and differentiate normally into every tissue type

22-2 The mouse genome (3000 Mb) is 30 times larger than the *C. elegans* genome (100 Mb).

a. What challenges are there for a geneticist studying an organism that has a large genome size?

b. Despite the large genome size, many geneticists choose to use the mouse as a model system? Why?

22-3 The *CFTR* gene, which is defective in humans with cystic fibrosis, encodes a membrane protein that acts as a channel for the passage of Cl^- ions. Although one particular mutation (a deletion called $\Delta 508$) is predominant in Caucasians, more than 200 different mutations in the gene have been identified that cause the disease. Many Cl^- channel genes, including *CFTR* homologs, have been identified in other organisms such as *C. elegans* and yeast. For each of the research questions below, indicate whether you would be more likely to pursue an answer using yeast or mouse and why you would choose that organism.

a. Do mutations in different regions of the gene have the same effect on Cl^- channel function?

b. Do mutations in different regions of the gene affect channel function in all organs normally involved in cystic fibrosis?

c. What portion of the protein receives a signal to open the channel?

d. Are there drugs that can cause an opening of the channel?

22-4 A mouse gene was localized to a region of chromosome 4. From the synteny map, it appears this gene would be localized to chromosome 8 in humans.

a. How could you determine if the gene is in this region of human chromosome 8? (Do not use the approach of cloning the gene here).

b. Briefly outline how you could obtain the human clone containing the homolog of this mouse gene.

22-5 Why do identical twins occur in mice (and other mammals) but not in *Drosophila?*

22-6 How are different coat color alleles used in the protocol for making a gene knockout in mice?

22-7 You have discovered a gene, called *CTF*, that is expressed for several days during development of the embryonic heart and pituitary in mice. Expression continues in the pituitary in the adult and also is seen in the germ cells of the adult testis.

a. You bred a male mouse carrying one normal *CTF* allele ($+$) and one disrupted *CTF* allele ($-$) with a female heterozygous for the same alleles and examined 30 of their offspring. (The disrupted allele contains an insertion within the gene.) Twenty-one of the offspring were $+/-$ and 9 were $+/+$. What would you conclude about the *CTF* gene in this case?

b. Suppose instead that you obtained 30 offspring and seven of the mice that had the ($-/-$) genotype were half the normal size. Of these, all four males were sterile, whereas the three females were fertile. What would you conclude about the *CTF* gene from these results?

c. In the preceding experiments, what technique(s) were used to determine if the offspring mice were ($+/+$), ($+/-$), or ($-/-$)?

22-8 Retinoic acid, which acts on cells via a protein receptor, is thought to be important for limb development in mammals.

a. Several different strains of mice that have recessive defects in limb development are available. You have tested each of these strains for the presence of the retinoic acid receptor (RAR protein), RAR mRNA, and the RAR gene using Western (protein), Northern (RNA), and Southern (DNA) blots, respectively. Based on the data below, give a reasonable hypothesis regarding the defect in each mutant.

Strain	RAR protein	RAR mRNA	RAR gene
A	Normal*	Normal	Normal
B	Absent	Normal	Normal
C	Absent	Short	Normal
D	Absent	Short	Altered**
E	Absent	Absent	Altered**
F	Absent	Absent	Absent

*Normal means that the protein, RNA, and DNA bands were normal in size and abundance; normal does not refer to function of the protein.

**One of the bands hybridizing with the RAR cDNA clone migrated to a different position in strains D and E.

b. The limb deformity in strain A could be due to a defect in some gene other than *RAR*. Suggest a simple genetic experiment that would enable you to test this possibility.

c. If you wanted to test the developmental consequences of *RAR* gene expression in neurons, which do not normally make RAR, how would you do this?

22-9 When DNA is injected into the fertilized mouse egg, the DNA can insert at random in any of the chromosomes. Subsequent matings produce animals homozygous for the transgene insertion. Sometimes an interesting mutant phenotype is generated by the insertion event. In one case, after injection of DNA containing the mouse mammary tumor virus (MMTV) promoter fused to the c-*myc* gene, investigators identified a recessive mutation that causes limb deformity. In this mouse, the distal bones were reduced and fused together; the mutation also caused kidney malfunction.

a. The mutant phenotype could be due to insertion of the transgene in a particular region of the chromosome or a chance point mutation that arose in the mouse. How could you distinguish between these two possibilities?

b. The mutation in this example was in fact caused by insertion of the transgene. How could you use this transgene insertion as a tag for cloning?

c. The insertion mutation was mapped to chromosome 2 of mice in a region where a mutation called limb deformity (*ld*) had previously been identified. Mice carrying this mutation are available from a major mouse research laboratory. How could you tell if the *ld* mutation was in the same gene as the transgenic insertion mutation?

d. Analysis of transcripts from the *ld* gene showed that many different transcripts (formed by alternate splicing) were present in both the embryo and adult. Is this consistent with a role of the *ld* gene product in limb development in the embryo?

22-10 Several different mouse mutants have been identified that have an obese phenotype. The *Ob* gene, defective in one class of these mutants, was the first gene involved in obesity to be cloned. The *Ob* gene product is made in fat cells and is transported to the brain where it informs the brain that the animal is satiated (full).

a. You made an antibody to the Ob protein and used this to test predictions of the hypothesis that it is a satiety factor. You isolated protein from mice that had been eating normally, from mice that had been starved, and from mice that had been force-fed a high-calorie diet. You did a Western analysis using your antibody as a probe against proteins from these animals. Results are in the next column. Are these results consistent with the hypothesis for the role of the Ob protein? Why or why not?

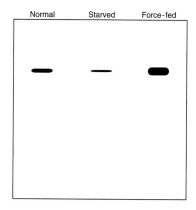

b. How could you determine if the *Ob* gene is transcriptionally regulated?

c. If the amount of mRNA was the same for all three types of mice, what type of regulation is involved?

d. You isolated RNA from several different *Ob* mutants and analyzed the RNA using Northern analysis. What conclusions could you reach about each mutation based on the results shown below?

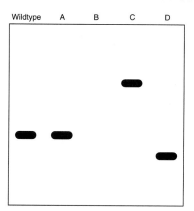

e. You have now decided to clone the receptor for the Ob protein and express the gene in mammalian cells. Which is the correct alternative for each of the following steps?
   1. A. Isolate brain mRNA.
      B. Isolate fat mRNA.
   2. A. Clone cDNA into a plasmid vector next to a mouse metallothionein promoter.
      B. Clone cDNA into a plasmid vector next to the yeast LEU2 promoter.
   3. A. Treat cells with zinc.
      B. Deprive cells of leucine.

f. You identified the *Ob* receptor clone using the screen outlined in part e and decided to make a knockout of the gene in mice. What phenotype would you predict for mice that lack the *Ob* receptor? (obese, slim, normal)

g. If you inject Ob protein into *Ob* mutant mice, what effect do you predict there will be on the phenotype?

HOW GENES CHANGE

Drosophila melanogaster are raised in sterilized milk bottles. The researcher at the light
microscope sorts the mutant from the normal flies. The number of mutants tells him
how the genes of the parent flies have interacted. Because of their short gestational
time, many generations may be studied in a relatively brief period of time.

THE GENETIC ANALYSIS OF POPULATIONS
AND HOW THEY EVOLVE

The enormous range of genetic diversity within our own species is easy to see.

Tuberculosis (TB) is an ancient and persistent human disease. Bone deformities typical of those produced by the infection are found in Egyptian mummies dating to 2000–4000 B.C.; and as recently as the mid–nineteenth century, TB was the leading cause of death in Europe and the urban United States. The microbe that causes TB is the bacterium *Mycobacterium tuberculosis*. In humans, populations of *M. tuberculosis* most often infect the lungs and lymph nodes, but sometimes they colonize the bones and skin of a patient (Fig. 23.1a). Severe infection by the tuberculosis bacterium damages the lungs, resulting in weakness, a persistent cough, bloody sputum, and ultimately death. *M. tuberculosis* bacteria can spread from person to person through the air when an infected individual exhales bacteria from his or her lungs during coughing.

Beginning in the late nineteenth century, improved sanitation and the quarantine of TB patients in sanitaria led to a steady decline in the death rate from TB in Europe and the United States (Fig. 23.1b). The introduction of antibiotics during the 1940s and

1950s further reduced TB mortality, and in the 1960s, when the death rate from tuberculosis had decreased to less than 2 deaths per 100,000, many people believed that the disease had been eradicated as a threat to public health, at least in the United States where the drugs were available. But 25 years later, the incidence of TB began to rise in urban areas around the globe. By 1994, in the Harlem area of New York City, there were 700 cases per 100,000 people. By 1996, TB annually claimed more than 2 million lives worldwide, and epidemiologists estimated that *M. tuberculosis* had infected close to one-third of the global population. Three factors contributed to this rapid increase in TB incidence: the emergence of AIDS, which among other effects, weakens the human immune system; protein deficiencies among the malnourished in many areas of the globe, which also weaken the immune system; and the widespread occurrence of *M. tuberculosis* strains that are resistant to one or more antibiotics.

Thus, despite the cultural, technological, and medical advances of the twentieth century, infectious microorganisms such as

(a)

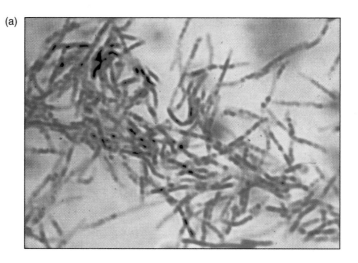

(b)

Death Rate from Tuberculosis in the United States

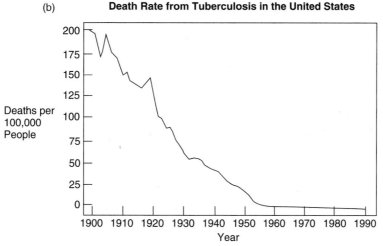

Prior to 1933 data are for only areas with death-registration;after 1933 data are for entire U.S.

Figure 23.1   Tuberculosis in human populations. (a) Photograph of *M. tuberculosis* bacteria colonizing the lungs and bones. (b) Death rate for tuberculosis in the United States from 1900 to 1990.

M. tuberculosis are still among the leading causes of death in many human populations. And in a related arena, populations of plant pests (such as the mites that prey on strawberry plants and almond trees) destroy a substantial fraction of human food supplies. How do new diseases emerge in human populations? Why do diseases persist in all living organisms? What causes diseases and pests long under control to resurge in frequency and intensity?

One way to answer these questions is to examine genetic variation and its expression as phenotypes within populations of organisms. The scientific discipline that studies what happens in whole populations at the genetic level is known as **population genetics.** It encompasses the evolutionary ideas of Darwin, the laws of Mendel, and the insights of molecular biology. To population geneticists, a **population** is a group of interbreeding individuals of the same species that inhabit the same space at the same time. An example would be all the white-tailed deer on Angel Island in San Francisco Bay in 1990 or all the rock cod at the mouth of the bay. The sum total of all alleles carried in all members of a population is that population's **gene pool.** In nature, the genetic makeup of a population changes over time as new alleles arise by mutation or are introduced by immigration, and as rare, preexisting alleles disappear when all individuals carrying them leave the population or die. Changes in the frequency of alleles within a population are the basis of **microevolution:** alterations of a population's gene pool.

In this chapter, we explore the nature of genotypic and phenotypic variation within populations and the role of this variation in evolution. We know from Chapter 1 that variation exists in all populations. To begin our examination of this variation, we analyze the incidence of genetic diseases, such as cystic fibrosis and retinoblastoma, that are determined by a single gene. Variations in the nature and number of wildtype and disease alleles determines the frequency of disease and how this frequency may change. Through our relatively simple analysis of single-gene disorders, we develop a framework for understanding how the frequency of a disease-causing allele determines the frequency of diseased individuals in a population.

We next examine how genetic variation within populations of pathogens and pests arises and how human efforts to control these destructive organisms forces changes in allele frequencies that determine the rate at which the pathogenic and pest populations evolve.

Finally, we consider variation in multifactorial traits, that is, in traits determined by two or more genes and their interaction with the environment (see Chapter 2). In fact, most aspects of disease susceptibility are multifactorial. To analyze the inheritance of multifactorial traits, population geneticists use quantitative tools that enable them to assess the relative contributions of genetic and environmental factors to individual differences in phenotype.

One general theme emerges from our discussions: Population geneticists rely on mathematical models in predicting a population's potential for evolution because most of the scientifically useful questions they ask are statistical. These questions include What is the frequency of genetic disease in a population? What fraction of the phenotypic variation in a trait is the result of genetic variation? Given certain quantifiable variables, how rapidly can a disease gain a foothold in a population and how long is it likely to persist? Simple mathematical models not only clarify the questions, they also serve as tools for analyzing data obtained from the field, and for making predictions about future populations.

As we present ways to examine the gene pools of stable and evolving populations over time, we describe

- The Hardy-Weinberg law: a model for understanding allele, genotype, and phenotype frequencies for single-gene traits in a genetically stable population.

- Calculations beyond Hardy-Weinberg: measuring how selection and mutations cause changes in the allele frequencies of a gene pool over time, including a comprehensive example of the evolution of drug and pesticide resistance.

- How to use the tools of quantitative genetics to analyze the inheritance of multifactorial traits

THE HARDY-WEINBERG LAW: A MODEL FOR UNDERSTANDING ALLELE, GENOTYPE, AND PHENOTYPE FREQUENCIES FOR A SINGLE-GENE TRAIT IN A GENETICALLY STABLE POPULATION

Suppose you were to look at a human population of 16 in which six people suffer from the recessive disease of cystic fibrosis caused by the r variant of the wildtype R allele. To predict how the number of diseased individuals in the population will change over time, you need to determine the frequencies of each genotype (homozygous RR, heterozygous Rr, and homozygous rr), each phenotype (normal and diseased), and each allele (R and r). Population geneticists define **phenotype frequency** as the proportion of individuals in a population that are of a particular phenotype. For our hypothetical population, the phenotype frequencies are 6/16 = 3/8 diseased (the number of homozygous rr individuals affected by the recessive condition) and 10/16 = 5/8 normal (the remaining fraction with either RR or Rr genotypes).

Genotype frequency is the proportion of total individuals in a population that are of a particular genotype. To determine genotype frequencies, you simply count the number of individuals of each genotype and divide by the total number of individuals in the population. For recessive traits like cystic fibrosis, it is not possible to distinguish between wildtype and heterozygous genotypes: Both give rise to nondiseased individuals. Thus, the only way to determine genotype frequencies directly at loci like the cystic fibrosis locus is to use molecular probes that distinguish between different alleles. For our hypothetical population, molecular analyses showed that 8 individuals (of 16) are of genotype RR, 2 are of genotype Rr,

and 6 are rr. This means that the RR genotype frequency is 8/16 = 4/8 = 1/2; the Rr genotype frequency is 2/16 = 1/8; and the rr genotype frequency is 6/16 = 3/8. Note that these three frequencies (4/8 + 1/8 + 3/8) sum to 1, the totality of genotypes in the whole population.

Finally, the definition of **allele frequency** is the proportion of all copies of a gene in a population that are of a given allele type. Because each individual in a population has two copies of each gene, the total number of gene copies is two times the number of individuals in the population. Thus, for our hypothetical population of 16 people, there would be 32 copies of the cystic fibrosis gene. Of course, both homozygotes and heterozygotes contribute to the frequency of an allele. But homozygotes contribute twice to the frequency of a particular allele, while heterozygotes contribute only once. To find the frequencies of R and r, you first use the number of people with each genotype to compute the number of R and r alleles:

$8RR \rightarrow$ 16 copies of R,
$2Rr \rightarrow$ 2 copies of R, $6rr \rightarrow$ 0 copies of R.
Together, 16 + 2 + 0 = 18 copies of the R allele;

$8RR \rightarrow$ 0 copies of r;
$2Rr \rightarrow$ 2 copies of r; $6rr \rightarrow$ 12 copies of r.
Together, 0 + 2 + 12 = 14 copies of the r allele.

Next, you add the 18 R alleles to the 14 r alleles to find the total number of gene copies:

18 + 14 = 32 copies of the gene, which is twice the number of people in the population.

Finally, you divide the number of each allele by the total number of gene copies to find the proportion, or frequency, for each allele:

For the *R* allele, it is 18/32 = 9/16 = 0.56.

For the *r* allele, it is 14/32 = 7/16 = 0.44.

Note that here again, the frequencies sum to a 1 representing all the alleles in the gene pool.

We have spent time defining and deriving the genotype, phenotype, and allele frequencies of a hypothetical population because population geneticists use these frequencies as the basis for calculating the genetic and phenotypic variations of a gene in a population and how they may change over time.

The Hardy-Weinberg Law Correlates Allele and Genotype Frequencies through a Binomial Equation

Many inherited diseases result from disease-causing alleles that are recessive to wildtype alleles. For scientists seeking to predict the potential incidence of such recessive conditions, an important question is How common are the heterozygous carriers of the disease-causing allele in a population? At the start, the scientists know only the phenotypic frequencies of healthy and diseased individuals. Can they use this information to predict the frequency of heterozygous carriers?

The key to answering this question lies in establishing a quantitative relationship between phenotype, genotype, and allele frequencies within and between generations. The **Hardy-Weinberg law,** named for the two men—G. H. Hardy and W. Weinberg—who independently developed it in 1908, clarifies the relationships between genotype and allele frequencies within a generation and from one generation to the next. To derive a general law describing the inheritance of alleles in a population, it is necessary to make certain simplifying assumptions about the nature of the population, the individuals within it, and the genes these individuals carry. Derivation of the Hardy-Weinberg law depends on five such assumptions:

1. The population includes a very large, virtually infinite, number of individuals.

2. The individuals mate at random in the sense that each individual's genotype at the locus of interest does not influence his or her choice of a mate.

3. No new mutations appear in the gene pool.

4. There is no migration into or out of the population.

5. There are no genotype-dependent differences in the ability to survive to reproductive age and transmit genes to the next generation.

Populations that satisfy all five of these assumptions are said to be at Hardy-Weinberg equilibrium.

The assumptions behind the Hardy-Weinberg equilibrium enable the mathematical derivation of an equation for calculating genotype (and thence phenotype) frequencies, even though no actual population is at Hardy-Weinberg equilibrium. All populations are finite; mutations occur constantly; migration into and out of a population is common; and many genotypes of interest, such as those that cause lethal childhood diseases,

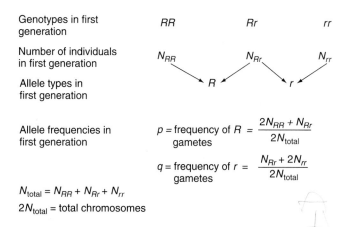

Figure 23.2 Computing allele frequencies from genotype frequencies.

$$N_{total} = N_{RR} + N_{Rr} + N_{rr}$$
$$2N_{total} = \text{total chromosomes}$$

affect the ability to survive and reproduce. Nevertheless, even when many of the assumptions of the Hardy-Weinberg equilibrium do not apply, the equation derived on the basis of these assumptions is remarkably robust at providing estimates of genotype and phenotype frequencies in real populations. Indeed, the equation has always been the most powerful mathematical tool available to population geneticists.

For a population of sexually reproducing diploid organisms, there are two steps in translating the genotype frequencies of one generation into the genotype frequencies of the next generation (Fig. 23.2).

First, if the likelihood that an individual will grow into an adult does not depend on the genotype (that is, if there is no difference in fitness among individuals), then the allele frequencies in the adults should be the same as in their gametes. For example, if *p* is the frequency of allele *R,* and *q* is the frequency of allele *r* in the adults, *p* and *q* will also be the frequencies of the two alleles in the gametes produced by the whole population of those adults.

Next, you can use the allele frequencies in the gametes to calculate genotype frequencies in the zygotes of the next generation—the Punnett square, which provides a systematic means of considering all possible combinations of uniting gametes, is the tool of choice (Fig. 23.3). For example, if the probability that an egg and a sperm unite does not depend on the genotype of the egg or sperm (that is, if fertilization is random among genotypes) and if the population of gametes is very large, then the following pattern emerges. Recall that *R R* zygotes result from fertilization of *R*-carrying eggs by *R*-carrying sperm. If *p* is the frequency of *R* gametes (eggs and sperm), then, applying the law of the product, the frequency of *R R* zygotes is *p* (frequency of *R* eggs) × *p* (frequency of *R* sperm) = p^2. Similarly, *r r* zygotes result from fertilization of *r*-carrying eggs by *r*-carrying sperm. If *q* is the frequency of *r* gametes (eggs and sperm), the frequency of *r r* zygotes will be *q* (the frequency of *r* eggs) × *q* (the frequency of *r* sperm) = q^2. Finally, *R r* zygotes result either from fertilization of *R* eggs

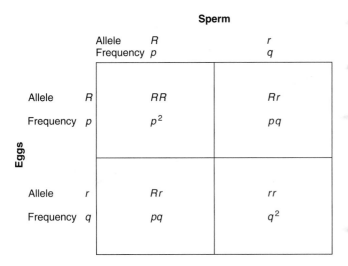

Figure 23.3 **Derivation of the Hardy-Weinberg equation.**
After mating, the genotype frequencies of the next generation will be p^2 for *RR*, $2pq$ for *Rr*, q^2 for *rr*.

by *r* sperm, with a frequency of $p \times q = pq$, or from the fertilization of *r* eggs by *R* sperm, occurring at a frequency of $q \times p = pq$. The total frequency of *Rr* zygotes is thus $pq + pq = 2pq$.

To summarize: The genotype frequencies of zygotes arising in a large population of sexually reproducing diploid organisms are p^2 for *RR*, $2pq$ for *Rr*, and q^2 for *rr* (Fig. 23.3). These genotype frequencies are known as the Hardy-Weinberg proportions; they exist in populations that satisfy the Hardy-Weinberg assumptions of a large number of individuals, mating at random, with no new mutations, no migration, and no genotype-dependent differences in fitness. Since these genotype frequencies represent the totality of genotypes in the population, they must sum to 1. Thus the binomial equation representing the Hardy-Weinberg proportions is

$$p^2 + 2pq + q^2 = 1 \qquad (23.1)$$

Because we have assumed no differences in fitness, the genotype frequencies of the zygotes will be the genotype frequencies of the adult generation that develops from those zygotes.

This equation thus enables us to use information on genotype and allele frequencies to predict the genotype frequencies of the next generation. Suppose, for example, that in a population of 100,000 people carrying the recessive allele *a* for albinism, there are 100 *aa* albinos and 1800 *Aa* heterozygous carriers. To find what the frequency of heterozygous carriers will be in the next generation, you compute the allele frequencies in the parent population:

> 98,100 *AA* individuals; 1800 *Aa* individuals,
> and 100 *aa* individuals → 196,200 *A* alleles +
> 1800 *A* alleles; 1800 *a* alleles + 200 *a* alleles.

Out of 200,000 total alleles the frequency of the *A* allele is

> 198,000/200,000 = 99/100 = 0.99; thus *p* = 0.99,

and the frequency of the *a* allele is

> 2,000/200,000 = 1/100 = 0.01; thus *q* = 0.01.

The Hardy-Weinberg equation for the albino gene in this population is

$$p^2 + 2pq + q^2 = 1$$
$$(0.99)^2 + 2(0.99 \times 0.01) + (0.01)2 = 1,$$
$$0.9801 + 0.0198 + 0.0001 = 1.$$

It thus predicts that in the next generation of 100,000 individuals, there will be

> 100,000 × 0.9801 = 98,010 *AA* individuals,
> 100,000 × 0.0198 = 1980 *Aa* individuals,
> 100,000 × 0.0001 = 10 *aa* individuals.

This example shows that in one generation, the genotype frequencies have changed slightly. A natural question is Have the allele frequencies also changed? Recall that the initial frequencies of the *R* and *r* alleles are *p* and *q*, respectively, and that $p + q = 1$. You can use the rules for computing allele frequencies from genotype frequencies (see Fig. 23.2 and 23.3) to compute the allele frequencies of the next generation. From the Hardy-Weinberg equation, you know that p^2 of the individuals are *RR*, whose alleles are all *R*, and $2pq$ of the individuals are *Rr*, 1/2 of whose alleles are *R*. Similarly, q^2 of the individuals are *rr*, whose alleles are all *r*, and $2pq$ of the individuals are *Rr*, 1/2 of whose alleles are *r*. If $p + q = 1$, then $q = 1 - p$, and the frequency of the *R* allele in the next-generation population is

$$p^2 + 1/2[2p(1 - p)] = p^2 + p(1 - p)$$
$$= p^2 + p - p^2 = p \qquad (23.2)$$

Similarly, $p = 1 - q$, and the frequency of the *r* allele in the next-generation population is

$$q^2 + 1/2[2q(1 - q)] = q^2 + q(1 - q)$$
$$= q^2 + q - q^2 = q \qquad (23.3)$$

Using these equations to calculate the allele frequencies of *A* and *a* in the second generation of 100,000 individuals, some of whom are albinos, we find

> for *p* 0.98 + 0.99 − 0.98 = 0.99 = the frequency of the
> *A* allele,
> for *q* 0.0001 + 0.01 − 0.0001 = 0.01 = the frequency of
> the *a* allele.

These frequencies are the same as those in the previous generation. Thus, even though the genotype frequencies have changed slightly from the first generation to the next, the allele frequencies have not. Note that this is true of both the dominant and the recessive alleles.

Consider, for example, the dominant human condition of *brachydactyly*, which causes people to have short stubby fingers. If a group of *B B* persons homozygous for the allele causing short, stubby digits were all to conceive children with *b b* mates homozygous for the allele determining normal finger length, all the progeny will be *B b* heterozygotes showing the brachydactyly phenotype. But even though the phenotype and genotype frequencies have changed drastically, the allele frequencies have not. In both generations,

$$1/2B + 1/2b = 1.$$

If no new alleles appear in the population by mutation or immigration and the heterozygotes of the second generation mate with each other at random, the third generation produced by *B b* × *B b* matings will contain the 3:1 ratio of brachydactyly to normal phenotypes, determined by the Mendelian genotype ratios of 1*B B*:2*B b*:1*b b*. Within this third generation, the allele frequency is 4/8*B* and 4/8*b*, the same as in the first generation in which *B B* mated with *b b*.

The Hardy-Weinberg law thus generates two significant conclusions.

Allele Frequencies Do Not Change from Generation to Generation in a Population at Hardy-Weinberg Equilibrium

This is one of the most important insights of the Hardy-Weinberg law: As long as a population is free of all influences that introduce or remove alleles, that is, as long as it is at Hardy-Weinberg equilibrium, its allele frequencies remain unchanged. The Hardy-Weinberg law thus provides a model for looking at genotype, allele, and phenotype frequencies in a genetically stable population.

A Hardy-Weinberg Population Achieves the Genotype Frequencies of p^2, 2pq, and q^2 in Just One Generation and Maintains These Frequencies in Subsequent Generations

This second significant implication of the Hardy-Weinberg law, which applies regardless of the genotype frequencies in the initial generation, enables us to answer the question posed at the beginning of this discussion: If you know only the phenotypic frequencies of healthy and diseased individuals in a population, can you use this information to predict the frequency of heterozygous carriers? For an autosomal recessive condition, all affected individuals are homozygotes for the recessive disease-causing allele. If you assume that the genotype and allele frequencies for the condition follow the Hardy-Weinberg law and that the frequencies of the normal and disease-causing alleles are *p* and *q*, respectively, then the genotype frequency of affected homozygotes is simply q^2, and the square root of this genotype frequency, *q*, is the allele frequency. Because the frequency of heterozygotes according to Hardy-Weinberg is 2*pq* and *p* = 1 − *q*, you can compute the frequency of heterozygous carriers of the disease.

Consider the inherited condition of phenylketonuria (PKU), which is caused by an autosomal recessive mutation that eliminates activity of the enzyme that converts the amino

acid phenylalanine to the amino acid tyrosine. As we saw in Chapter 2, without the enzyme, phenylalanine builds up, and this buildup interferes with normal brain development. Medical studies show that about 1 in 3600 Caucasians in the United States have PKU. Because PKU is an autosomal recessive condition, all the affected individuals are homozygotes for the recessive allele. Thus the genotypic frequency of PKU is 1/3600. Since according to the Hardy-Weinberg equation, the genotypic frequency of homozygotes for the recessive allele is q^2, you can say that q^2 = 1/3600, and then compute the frequency of the recessive PKU allele in the whole population by finding the square root of this number:

$$q = \sqrt{q^2} = \sqrt{1/3600} = 1/60 = 0.0167.$$

Next, because *p* = 1 − *q*, you can compute *p*:

$$p = 1 - 1/60 = 59/60 = 0.9833.$$

Knowing the frequencies of both *p* and *q*, we can compute the frequency of heterozygotes according to Hardy-Weinberg:

$$2pq = 2 \times (0.0167) \times (0.9833) = 0.0328.$$

Thus, if one assumes that the U.S. Caucasian population can be approximated by the conditions that define a Hardy-Weinberg equilibrium, it becomes possible to determine simply from observations of phenotypes that within this population, the frequency of heterozygous carriers is 0.0328 or roughly 1 in 30 people.

BEYOND HARDY-WEINBERG: MEASURING HOW MUTATION AND SELECTION CAUSE CHANGES IN ALLELE FREQUENCIES

One definition of **evolution** is changes in allele frequency in a population over multiple generations. Evolutionary scientists refer to this level of evolution as *microevolution* in comparison with **macroevolution**, which occurs over geologic time. Macroevolution refers to the process by which new species emerge from existing species, while microevolution refers to genetic changes that occur within a species. Many evolutionary biologists believe that macroevolution is simply a consequence of extended periods of microevolution. Population geneticists can use the Hardy-Weinberg law to look at microevolution. Because the law models populations in which genetic changes do not occur, it provides a theoretical standard against which to measure changes that do occur. In natural populations, conditions almost always deviate from the Hardy-Weinberg assumptions: Some populations are not very large, individuals do not always mate at random, new mutations do arise, there is migration into and out of the population, and different genotypes do generate differences in fitness. We now examine how simple modifications of the Hardy-Weinberg law make it possible to analyze the effects of these deviations, particularly those resulting from genotype-dependent differences in fitness, and new mutations.

Natural Selection Acts on Differences in Fitness to Alter Allele Frequencies

Contrary to the Hardy-Weinberg assumption that the probability of survival does not vary with the genotype, for many phenotypic traits, including inherited diseases, genotype does influence survival and the ability to reproduce. Thus in real populations, as opposed to model populations constructed according to Hardy-Weinberg assumptions, not all individuals survive to adulthood, and there is always some probability that an individual will not live long enough to reproduce. As a result, the genotype frequencies of real populations change as their individual members mature from zygotes to adults.

To population geneticists, an individual's relative ability to survive and transmit its genes to the next generation is its **fitness.** But although fitness is an attribute associated with each genotype, it cannot be measured within the individuals of a population because each animal with a particular genotype survives and reproduces in an unpredictable way. However, by considering all the individuals of a particular genotype together as a group, it becomes possible to measure the relative fitness for that genotype. Thus, for population geneticists, fitness is a statistical measurement only. Nevertheless, differences in fitness can have a profound effect on the allele frequencies of a population.

Fitness has two basic components: viability and reproductive success. The fitness of individuals possessing variations that help them survive and reproduce in a changing environment is relatively high; the fitness of individuals without those adaptive variations is relatively low. In nature, the process that progressively eliminates individuals whose fitness is low and chooses individuals of high fitness to survive and become the parents of the next generation is known as **natural selection.** The mechanisms of selection act independently of any individual; often, interactions between genetically determined phenotypes and environmental conditions are the agents of natural selection. For example, in a hypothetical population of giraffes browsing on the leaves of an ancient savanna, suppose that some of the animals had long necks and some short necks, and that each of these phenotypes resulted from variations in the genes contributing to neck length. If during a long drought, there were fewer low-hanging leaves than usual on the flora of the savanna, those long-necked individuals able to reach the higher morsels would have been able to harvest more food and survive to produce more offspring than individuals with shorter necks. Natural conditions would thus have selected the better adapted, long-necked giraffes—those with the higher fitness—to become the parents of the next generation (Fig. 23.4). Similarly, in the laboratory, scientists can establish experimental conditions, such as the absence of a nutrient or the presence of an antibiotic, that become the agents of **artificial selection.** In both natural and artificial selection, environmental and other external conditions interact with preexisting variations to "select" the parents of the next generation.

Field studies show that natural selection occurs for many phenotypic traits in a variety of natural populations. Since in human populations, genetic diseases frequently affect a per-

Figure 23.4 Giraffes on the savanna. A visible example of one outcome of natural selection for increased fitness.

son's survival and reproduction, the potential for natural selection in people affected by such diseases is considerable. How does selection alter our conclusions from Hardy-Weinberg?

We begin to answer this question with an analysis of the *R* gene in a population of zygotes in Hardy-Weinberg equilibrium. In this population, the genotype frequencies *R R, R r,* and *r r* are p^2, $2pq$, and q^2 respectively. Now suppose that the viability, that is the probability of surviving from zygote to adult, depends on genotype, while the second component of fitness—success at productive mating—is independent of genotype. If we define the relative fitness of the three genotypes as w_{RR}, w_{Rr}, and w_{rr}, respectively, the relative frequencies of the three genotypes at adulthood is $p^2 w_{RR}$, $2pq w_{Rr}$, and $q^2 w_{rr}$ (Fig. 23.5). These frequencies do not sum to 1 because they are relative frequencies for the three genotypes. To make the fre-

Gametes	R		r
Allele frequencies	p		q
Zygotes' genotypic frequencies	$G_{RR} = p^2$	$G_{Rr} = 2pq$	$G_{rr} = q^2$
Relative fitness: basis of selection	w_{RR}	w_{Rr}	w_{rr}
Adults' genotype frequencies	$\dfrac{p^2 w_{RR}}{\overline{w}}$	$\dfrac{2pq w_{Rr}}{\overline{w}}$	$\dfrac{q^2 w_{rr}}{\overline{w}}$
Gametes' (next generation) allele frequencies	$p' = \dfrac{p^2 w_{RR} + pq w_{Rr}}{\overline{w}}$		$q' = \dfrac{q^2 w_{rr} + pq w_{Rr}}{\overline{w}}$

$\overline{w} = p^2 w_{RR} + 2pq w_{Rr} + q^2 w_{rr}$

Figure 23.5 Changes in allele and genotype frequencies when selection acts on genotype-dependent differences in fitness.

quencies sum to 1, even with the inclusion of modifying terms that change the value of each genotypic frequency, it is necessary to "recalibrate" the Hardy-Weinberg equation.

This recalibration is accomplished by calculating the value of a numerical factor that each Hardy-Weinberg term can be divided into so that the terms all add up to 1. The calculation consists simply in setting the sum of the terms in the modified equation to a new variable, designated $\bar{w}$:

$$p^2 w_{RR} + 2pq w_{Rr} + q^2 w_{rr} = \bar{w} \qquad (23.4a)$$

Since $\bar{w}$ represents the sum of the individual fitness values, each multiplied by their relative occurrence in the population, $\bar{w}$ in fact, it represents the average fitness of the population.

In populations that satisfy the conditions of the original Hardy-Weinberg equilibrium, when the fitness for each genotype is 1, the value of $\bar{w}$ is also 1. However, when fitness varies from one genotype to another, $\bar{w}$ can also vary, but in a way that can be calculated as long as the value of each of the variables in the equation is known. The modified Hardy-Weinberg equation is thus recalibrated by dividing each term by $\bar{w}$ such that the new equation becomes

$$\frac{p^2 w_{RR}}{\bar{w}} + \frac{2pq w_{Rr}}{\bar{w}} + \frac{q^2 w_{rr}}{\bar{w}} = 1 \qquad (23.4b)$$

Each term in this recalibrated equation represents the actual frequency that each genotype will assume in the generation following the one used for the original calculation. Table 23.1 summarizes this process of calculation.

As an example, let us use the variables p' and q' to represent the frequencies of the R and r alleles in this next generation. Among the gametes produced by the original population, the frequency of allele r will be the result of contributions of r alleles from both Rr and rr adults relative to the number of individuals in the entire adult population. If q' represents the frequency of the r allele in the next generation adults, then

$$q' = \frac{q^2 w_{rr} + 1/2(2pq w_{Rr})}{\bar{w}} = \frac{q(q w_{rr} + p w_{Rr})}{\bar{w}} \qquad (23.5)$$

Thus, in one generation of selection, the allele frequency of r has changed from q to q'. It is often useful to know the change in allele frequency over one generation of selection. We can represent this change as

$$\Delta q = \frac{pq[q(w_{rr} - w_{Rr}) + p(w_{Rr} - w_{RR})]}{\bar{w}}. \qquad (23.6)$$

As these equations show, selection can cause the frequency of an allele to change from one generation to the next. Equation (23.6) shows that the change in allele frequency resulting from one generation of selection depends on both the allele frequencies and the relative fitnesses (in this case, the viability component of fitness) of the three genotypes. Note that if the fitnesses of all genotypes are the same, as in populations at Hardy-Weinberg equilibrium, then $\Delta q = 0$; in other words, if there are no genotype-related differences in fitness, there is no possibility of selection and if there is no possibility of selection there can be no evolution (defined as changes in allele frequency).

As an example, let us use Eq. (23.6) to look at how a recessive genetic condition, such as thalassemia, influences the allele frequencies of a population. If the disease, which results from an rr genotype for the R gene, decreases fitness by decreasing the probability of surviving to adulthood, then the fitness of RR and Rr individuals is the same, while the fitness of rr individuals is reduced. Because only the relative values of the fitnesses are important, it is useful to set the values of $w_{RR} = 1$, $w_{Rr} = 1$, and $w_{rr} = 1 - s$, where s is the selection coefficient against the rr genotype. This selection coefficient can vary from 0 (no selection against rr) to 1 (rr is lethal and no rr individuals survive to adulthood). For this example, we can rewrite Eq. (23.6) as

$$\Delta q = \frac{pq[q(1 - s - 1) + pq(1 - 1)]}{\bar{w}} = \frac{-spq^2}{\bar{w}}. \qquad (23.7)$$

Equation (23.7) has three interesting features. First, unless there is no selection and $s = 0$, Δq is always negative and the frequency of the r allele decreases with time.

TABLE 23.1 Calculating Natural Selection's Effect on Genotype Frequency

Genotypes	RR	Rr	rr	
Initial Genotypic Frequency	p^2	$2pq$	q^2	
Fitness	w_{RR}	w_{Rr}	w_{rr}	
Frequency after Selection	$p^2 w_{RR}$	$2pq w_{Rr}$	$q^2 w_{rr}$	$\bar{w} = p^2 w_{RR} + 2pq w_{Rr} + q^2 w_{rr}$
Relative Genotypic Frequency after Selection	$\dfrac{p^2 w_{RR}}{\bar{w}}$	$\dfrac{2pq w_{Rr}}{\bar{w}}$	$\dfrac{q^2 w_{rr}}{\bar{w}}$	

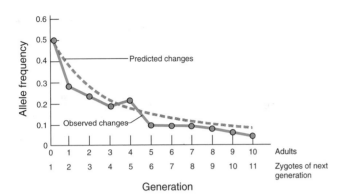

Figure 23.6 Decrease in the frequency of a recessive allele over time. The dotted line represents the mathematical prediction. The blue line represents the actual data obtained with an autosomal recessive allele.

Second, the rate at which q decreases over time depends on the allele frequencies; in particular, because Δq varies with q^2, the rate at which q decreases diminishes as q becomes smaller. (Recall that because q is always less than 1, $q^2 < q$.) To understand the effect of this correlation between the allele frequency and the rate at which q decreases over time, consider the special case of a lethal recessive disease for which $s = 1$. The dotted line in Fig. 23.6 shows the decrease in allele frequency predicted by Eq. (23.7), starting from an initial allele frequency of 0.5. The decrease in allele frequency is rapid at first, then slows. After 10 generations, the predicted frequency of the recessive disease allele is nearly 10%, even though the homozygous recessive genotype is lethal. The solid line in Fig. 23.6 plots actual data for the decrease in frequency of an autosomal lethal allele in an experimental population of *Drosophila melanogaster*. In Fig. 23.6, the predicted and observed changes in allele frequency match quite closely.

Why is selection unable to reduce the frequency of a recessive lethal allele to zero? The answer goes back to our consideration of the frequency of heterozygous carriers of a recessive disease allele. When q is small, individuals homozygous for the disease allele are very rare because most copies of the r allele occur in Rr heterozygotes who do not experience negative selection. By contrast, a lethal dominant allele will disappear from a population in a single generation of selection.

The third feature of Eq. (23.7) is that it predicts that the allele frequency q will continue to decline, albeit more and more slowly over time as q moves closer and closer to a value of zero. However, as we noted earlier, the reality is that many recessive genetic diseases persist in human populations at low but stable frequencies. What maintains these diseases despite continuing selection against them? One answer is that sometimes heterozygotes have a higher fitness than either homozygote, a situation referred to as **heterozygous advantage.**

It Is Possible to Calculate the Effect of Heterozygous Advantage on Allele Frequencies

We have seen that sickle-cell syndrome, which includes episodes of severe pain, serious anemia, and a probability of early death, is a recessive condition resulting from two copies of the sickle-cell allele at the β-*globin* locus. Yet the disease allele has not disappeared from several African populations, where it seems to have existed for a very long time. One clue to the maintenance of the sickle-cell allele in human populations lies in the observation that heterozygotes for the normal and sickle-cell alleles are resistant to malaria. This resistance is due, in part, to the fact that red blood cells infected by the malaria parasite, if they also contain a sickle-cell allele, break open, destroying the parasite as well as the red blood cell itself; by contrast, in cells with two normal hemoglobin alleles, the malaria parasite thrives. To set up a model of heterozygous advantage, let B_1 represent the normal β-*globin* allele and B_2 stand for the abnormal recessive sickle-cell allele; and for simplicity, assume that $B_1 B_2$ heterozygotes have the maximum relative fitness of 1, while the selection coefficient (representing the selective disadvantage) for $B_1 B_1$ homozygotes is $1 - s_1$ and the selection coefficient for $B_2 B_2$ homozygotes is $1 - s_2$. We can then represent the changes in allelic frequency resulting from selection as

$$\Delta q = \frac{pq(s_1 p - s_2 q)}{\overline{w}}. \qquad (23.8)$$

To maintain both alleles in the population using this equation, Δq must be 0 for some value of q between 0 and 1. The q value at which $\Delta q = 0$ is known as the **equilibrium frequency** of allele B_2. The value of q when $\Delta q = 0$ and both alleles are present occurs when the term inside the parentheses of Eq. (23.8) is 0, that is, when

$$s_1 p - s_2 q = 0.$$

Substituting $1 - q$ for p ($p = 1 - q$) and solving this equation for q reveals that the equilibrium frequency of B_2 (represented by q_e) is reached when

$$q_e = \frac{s_1}{s_1 + s_2}. \qquad (23.9)$$

Note that to find the equilibrium frequency, that is, the value of q at which $\Delta q = 0$ such that both alleles B_1 and B_2 persist in the population, you need know only the selection coefficients for the two homozygotes.

To understand the relationship between q, the change in q, and the equilibrium frequency q_e, you can formulate Δq using q_e:

$$\Delta q = \frac{-pq(s_1 + s_2)(q - q_e)}{\overline{w}}. \qquad (23.10)$$

From this formulation, you can see that when q is greater than q_e, Δq is negative; under these circumstances, q, or the frequency of allele B_2, will decrease toward the equilibrium frequency. By contrast, when q is less than q_e, Δq is positive and the frequency of B_2 will increase toward the equilibrium. Thus, the equilibrium frequency stabilizes, because a change away from it is always followed by a change toward it.

Now, if you assume that the African populations in which sickle-cell disease is prevalent are currently at equilibrium relative to their alleles at the β-*globin* locus, you can use the observed frequency of the sickle-cell allele in these populations to calculate the relative values of the selection coefficients. Field studies show that the actual value of q_e lies between 0.15 and 0.2 for an average value of 0.17. If you plug this number into Eq. (23.9), you get

$$0.17 = s_1/(s_1 + s_2).$$

This equation makes it possible to express either selection coefficient in terms of the other. For example, $s_1 = 0.2s_2$.

If you assume that $s_2 = 1$ (that is, those with sickle-cell trait never reproduce), as was most probably true before medical ad-

vances enabled the survival of children expressing the sickle-cell trait, you will find that $s_1 = 0.2$, which in turn means that the relative fitness of the wildtype genotype is 0.8. Recall, however, that we set the fitness of the heterozygote at 1.0. By dividing 1.0 by 0.8, you get 1.25, which represents the relative advantage in fitness that heterozygotes for the sickle-cell allele have over people who do not carry this allele in African populations exposed to malaria. The use of simple statistical methods to calculate this heterozygous advantage demonstrates how medical geneticists can use the tools of population genetics (Fig. 23.7).

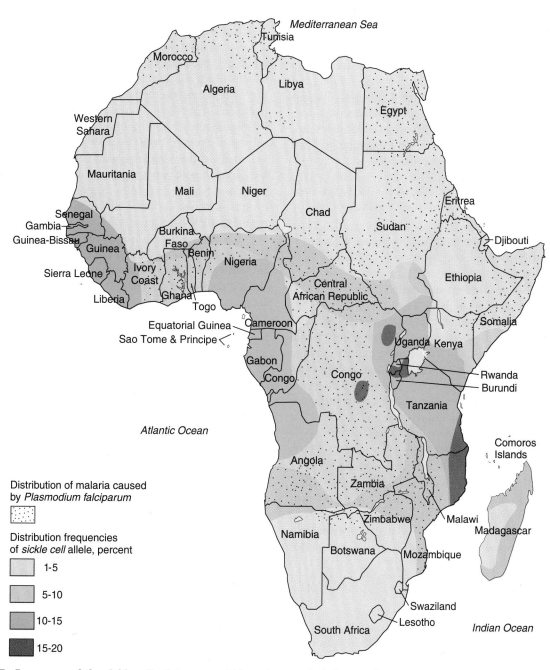

Figure 23.7   Frequency of the sickle-cell allele across Africa where malaria is prevalent.

Heterozygous advantage is one phenomenon that maintains recessive disease alleles in human populations. We now look at other phenomena that can counteract the effects of selection and maintain recessive alleles within a population.

Evolutionary Equilibrium: A Balance between Mutation to a New Allele and Selection against That Allele

Mutations, we have seen, are the ultimate source of genetic variation in a population. They arise from DNA damage due to environmental agents and from errors during DNA replication and the transmission of genetic information during cell division. Various studies show that in most animals, the mutation rate is roughly 10^{-4} to 10^{-6} mutations per gene per generation. Since we know that the human genome carries about 100,000 genes, we can infer that 0.1 to 10 new mutations arise somewhere in the genes of each human gamete per generation.

Many mutations that alter the amino-acid sequence of a gene product have deleterious consequences on survival and reproduction, that is, on fitness. Indeed, many genetic diseases result from mutations that give rise to a nonfunctional gene product; individuals homozygous for such disease alleles produce no functional copies of an enzyme or other polypeptide required for normal development or metabolism.

How Selection against a Disease Interacts with Mutation to the Disease Allele to Influence the Frequency of the Disease Allele in a Population

To address this question, let us again consider the example of a recessive genetic disease that occurs at low frequency in a population. Suppose that mutations from allele R (the dominant, normal allele) to allele r (the recessive, disease allele) occur at a rate of μ per generation. As a first approximation, we can ignore the very low level of reverse mutations from r to R because the frequency of r is so small. With this background, we can calculate the change in q (Δq) using Eq. (23.7) as the basis, but accounting for the fact that a fraction μ of the R alleles has mutated to r in each generation. In this equation, we multiply μ by $(1 - q)$ because only R alleles can mutate to r, and the frequency of R alleles is $(1 - q)$.

$$\Delta q = \frac{-spq^2 + \mu(1 - q)}{\overline{w}}. \tag{23.11}$$

Note that the effects of selection (s) and mutation (μ) on allele frequency are in opposite directions. Allele frequencies will thus evolve until $\Delta q = 0$. The allele frequency, labeled $\hat{q}$, at which evolution no longer continues is called the **evolutionary equilibrium.** By rearranging Eq. (23.11), you can see that this equilibrium occurs when

$$\hat{q}/\sqrt{(1 - \hat{q})} = \sqrt{(\mu/sp)} \tag{23.12}$$

In this equation, $\hat{q}$ represents the evolutionary equilibrium.

This simple equation shows how a balance between selection and mutation can maintain recessive genetic diseases in a population. Note that the equilibrium frequency of the disease allele can be quite high. For example, consider a lethal recessive disease (for which $s = 1$) in which all affected individuals die before reproducing and let the mutation rate = 1×10^{-6}. Substituting these values into Eq. (23.12) reveals that the equilibrium frequency of the disease allele is 1×10^{-3}, or 1 in 1000:

1. Let $\mu = 10^{-6}$ and $s = 1$.

2. Plug into Eq. (23.12).

3. Assume that $\hat{q}$ is small, which enables you to approximate $p = 1$ and $1 - \hat{q} = 1$.

4. Following these steps, you can conclude that $\hat{q} = 10^{-3}$.

Since the frequency of heterozygotes among zygotes is $2pq$, as predicted by the Hardy-Weinberg law, the frequency of heterozygous carriers of this deadly recessive condition will be approximately 2 per 1000.

For the many genetic diseases that have less severe effects on fitness and therefore sustain less negative selection, the allele frequency at evolutionary equilibrium will be even higher. For example, an inherited recessive disease that decreases the likelihood of surviving to reproductive age by only 10% ($s = 0.1$) would result in an equilibrium frequency of heterozygous carriers of about 63 per 1000 individuals.

In short, this simple, mathematical model suggests that, like heterozygous advantage, mutation is quite effective in maintaining recessive alleles in a population. As a result, at evolutionary equilibrium, substantial numbers of deleterious recessive alleles can remain in the heterozygous carriers of a population. However, as we see later, the mathematical model does not accurately predict what actually happens in human and other populations, because a critical assumption—namely, infinite size—does not reflect reality. Thus, the human population, with its millions of generations of evolution, is still nowhere near equilibrium.

The Time of Onset of a Disease Can also Influence the Frequency of Disease Alleles

The effects of many inherited human diseases become manifest only during middle age or later in life; the effects of others may appear earlier than middle age but not cause death until much later. From an evolutionary viewpoint, the important issue is how a disease reduces survival to reproductive age, for it is its impact on reproduction that determines a disease's potential effect on the strength of selection. Diseases that cause death after the completion of reproduction will sustain little or no negative selection; and with lower selection against them, their equilibrium frequency will be higher. One significant consequence of the relationship between the time of onset of disease and the strength of selection is that genetic diseases whose phenotypic effects appear late in life will experience little selection and may increase in frequency simply as a result of mutation. Aging, the increased rate of mortality with age that occurs after the completion of reproduction, may be the natural consequence of mutations whose effects appear late in life and accumulate in a population unopposed by natural selection.

Genetic Drift Has an Unpredictable Effect on the Evolutionary Equilibrium

Equation (23.12) models the balance between selection and mutation in a Hardy-Weinberg world where populations are very, very large—virtually infinite. But in addition to clarifying how mutations help maintain recessive diseases in a population, Eq. (23.12) leads to the untenable conclusion that in the absence of selection ($s = \phi$), $\hat{q}$ will become 1; that is, new alleles with no effect on fitness will invariably spread through the entire population. This is not what actually happens. In nature, where no population is infinite in size, new alleles with a neutral effect on fitness arise continuously, but most of these alleles drift to extinction in the course of several generations. Loss of an allele occurs when the individuals carrying that allele, by chance, fail to reproduce. By contrast, very rarely, a new recessive allele may increase by chance to high frequency, but even when it does, it will be replaced later by another new allele.

Population geneticists use the expression **genetic drift** to refer to unpredictable, chance fluctuations in allele frequency that have a neutral effect on fitness (Fig. 23.8). Genetic drift occurs in populations of any size, but the smaller the population, the more clearly visible are the effects of drift. To understand why this is so, consider the perpetuation of allele r, which occurs in 10% of the individuals in a population of sunflowers. If a chance occurrence such as a severe summer hailstorm strikes a population of 1,000,000 plants and half die, it is likely that among the 500,000 survivors, roughly 50,000 (10%) will still carry the r allele. If the same storm hits a field of just 10 plants, only 1 of which carries the r allele, there is a 50% chance that the single plant bearing the r allele will not be among the survivors. If it isn't, the total loss of this one allele from the population will reduce the allele frequency in an unpredictable way to 0. If, however, the lone individual carrying r does by chance survive, the frequency of allele r will increase unpredictably to 20%.

One example of genetic drift in human populations is the **founder effect,** which occurs when a few individuals separate from a larger population and establish a new one that is iso-

lated from the original population. The small number of founders in the new population carry only a fraction of the alleles from the original population, and this discrepancy may change the allele frequencies. The 14,000 people living in the Amish community in eastern Pennsylvania are the descendants of about 200 individuals who emigrated from Germany in the early eighteenth century. Among the Amish, there is a much higher incidence of manic-depressive illness than among the larger original Swiss population, most likely because several founders carried alleles producing this disease.

Summary of the Balance between Mutation and Selection

New alleles (some of which are disease alleles) arise in a population by mutation. When a new allele has an effect on fitness that differs from the effect of the wildtype allele, selection will drive the frequency of the new allele toward an equilibrium with the wildtype allele. The equilibrium value will be determined by the relative selection coefficients of individuals heterozygous and homozygous for the new allele compared to individuals who are wildtype. In contrast, when an allele has a neutral effect on fitness, its frequency will depend on unpredictable genetic drift. Most neutral alleles become extinct, but occasionally one "drifts up" to become the new wildtype allele in the population.

Comprehensive Example: How Human Activity Affects the Evolution of Human Pathogens and Crop Pests

Infectious diseases have been a major killer throughout human history, and as the AIDS epidemic illustrates, previously unknown diseases continue to emerge. In the twentieth century, discovery of a variety of vaccines, antibiotics, and other drugs has made it possible to combat infectious diseases such as small pox (vaccines) and tuberculosis (antibiotics) with great success. In the last 25 years, however, many formerly surefire drugs have lost their effectiveness because populations of pathogens have evolved resistance to them. Similarly, populations of agricultural pests have evolved resistance to the pesticides used to control or eradicate them.

At the beginning of this chapter we posed three questions: How do new diseases emerge in human populations? Why do diseases persist in all living organisms? What causes diseases and pests long under control to resurge in frequency and intensity? We have answered the first two. New diseases emerge in human populations as a consequence of new mutations. Diseases persist because changes in allele frequency tend toward an evolutionary equilibrium in which mutation balances selection. To answer the third question, we turn to an examination of how pathogens and pests interact with their hosts.

The Evolution of Drug Resistance in Pathogens

Many of the bacterial agents of tuberculosis, we have seen, are resistant to several antibiotics. We now know that a major factor contributing to this evolution of multidrug resistant TB strains is the failure of patients to complete the lengthy drug

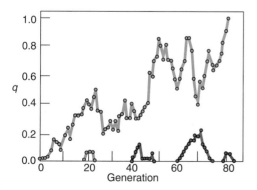

Figure 23.8 The effect of genetic drift. New alleles with no effect on fitness are usually lost rapidly from a population as indicated with the red and green lines. Rarely, a new allele will increase in frequency as shown with the blue line.

regimens required for a cure. The two most widely used drugs for TB, isoniazid and rifampicin, require ingestion for six months and have side effects that include nausea and loss of appetite. However, the symptoms of TB can begin to disappear after only two to four weeks of treatment.

Imagine a TB patient with a persistent cough, shortness of breath, and general weakness. This individual harbors a large, actively growing and dividing population of TB bacteria in his lungs. At first, these bacteria are susceptible to antibiotics, but occasional mutations conferring partial resistance appear at random (Fig. 23.9). The patient's physician prescribes a six-month course of treatment with the antibiotic isoniazid. After a few weeks of treatment, the bacterial population in the patient's lungs has decreased considerably, and the patient's symptoms have abated, although the negative side effects of the drug continue to cause discomfort. However, the composi-

tion of the bacterial population has now changed so that the remaining bacteria are likely to include a high proportion of mutant bacteria possessing partial resistance to the antibiotic. If the patient continues his course of treatment, the persistent dose of antibiotic will eventually kill all of the bacteria, even those with partial resistance, eliminating the infection. By contrast, if the patient stops treatment prematurely, the remaining (partially resistant) bacteria will proliferate, and within three to four weeks reestablish a large population. Subsequent treatment of the same individual upon relapse or of a new patient to whom the partially resistant bacteria had spread would permit a second cycle of selection. New mutations could then convert the partially resistant bacteria to fully resistant microbes.

It is easy to see how repeated cycles of antibiotic treatment with multiple drugs, coupled with premature cessation of treatment, can promote the evolution of fully resistant bacterial populations and even bacterial strains resistant to more than one drug within a single patient. Initially individual mutant cells that express genes conferring partial resistance increase in frequency; subsequent mutations in some of the mutant bacteria will increase resistance, and incomplete drug dosages will select for the resistant strains. Surveys of resistance illustrate the effects of this process on the frequency of resistance. A 1982–1986 U.S. Centers for Disease Control (CDC) survey showed that 9% of TB strains isolated from patients who had not previously received treatment were resistant to antibiotics, compared with 25% of strains isolated from previously treated, relapsed patients. More recently, a 1991 survey in New York City showed that 23% of strains isolated from previously untreated patients were resistant to one or more antibiotic drugs and 7% were resistant to both isoniazid and rifampicin; moreover, 44% of isolates from relapsed patients were resistant to one or more drugs and 30% were resistant to both isoniazid and rifampicin. These data support the idea that patient noncompliance with drug treatments is a major factor in the evolution of antibiotic resistance.

Several factors contribute to the rapid evolution of resistance in bacterial pathogens. The short generation times—often only a few hours—and rapid rate of reproduction under optimal conditions allow evolution to proceed quickly relative to a human life span. The large population densities typical of bacteria, which may exceed $10^9/cm^3$, ensure that rare resistance-conferring mutations will appear in the population. The strong selection imposed by antibiotics increases the rate of evolution in each generation, unless the bacterial population is entirely eliminated.

The large variety of ways by which bacteria can acquire genes also contributes to the rapid evolution of resistance. Many genes for resistance are found on plasmids, and as we saw in Chapters 12 and 13, the capacity of plasmids to replicate and be transmitted among bacteria allows the amplified expression of resistance genes in bacterial populations. Plasmids also provide a means for the genetic exchange of resistance genes among bacterial populations and species through transformation, conjugation, and transduction. Laboratory

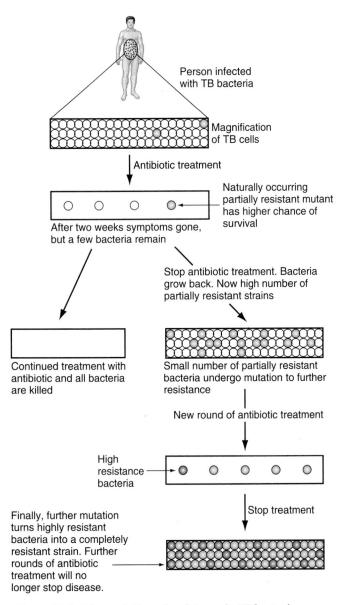

Figure 23.9 The evolution of resistance in TB bacteria.

studies have demonstrated the ready transfer of plasmid-borne resistance genes to new bacterial species.

The Evolution of Pesticide Resistance

Like infectious bacteria, many agricultural pests spawn large populations because of their short generation times and rapid rates of reproduction. These large, rapidly reproducing populations evolve resistance to the chemical pesticides used to control them via selection for resistance-conferring mutations. Our understanding of this familiar pattern of selection and rapid evolution is most complete for certain insecticides, pesticides that target insects.

The large-scale, commercial use of DDT and other synthetic organic insecticides, begun in the 1940s, was initially highly successful at reducing crop destruction by agricultural pests such as the boll weevil and medical pests such as the mosquitoes that transmit malaria and yellow fever. Within a few years, however, resistance to these insecticides was detectable in the targeted insect populations. Since the 1950s, resistance to every known insecticide has evolved within 10 years of its commercial introduction. By 1984 there were reports of more than 450 resistant species of insects and mites (Fig. 23.10). Because different populations within a species can become resistant independently of other populations, the number of times insecticide resistance has evolved probably exceeds 1000.

Genetic studies show that insecticide resistance often results from changes in a single gene, and that several significant mechanisms of resistance are similar to those seen in infectious bacteria. DDT, for example, is a nerve toxin in insects. House flies and some mosquitoes develop resistance to DDT from dominant mutations in a single, enzyme-encoding allele. The mutant enzyme detoxifies DDT, rendering it harmless to the insect. As we saw earlier, even at low frequencies, dominant alleles can experience strong selection, because of heterozygous advantage, that is, selection for the heterozygotes. Consider, for example, dominant mutation R (for insecticide resistance), which occurs initially at low frequency in a population. Soon after the mutation appears, most of the R alleles are in SR heterozygotes (in which S is the wildtype susceptibility allele). With the application of insecticide, strong selection favoring SR heterozygotes rapidly increases the frequency of the resistance allele in the population. A field study of the use of DDT in Bangkok, Thailand, to control *Aedes aegypti* mosquitoes, the carriers of yellow fever, illustrates the rapid evolution of resistance. Spraying of the insecticide began in 1964 (Fig. 23.11). Within a year, DDT-resistant genotypes emerged and rapidly increased in frequency. By mid-1967, the frequency of resistant RR homozygotes was nearly 100%.

The Biological Costs of Resistance

Since DDT no longer controlled mosquito populations in the region, the insecticide program was stopped. The response of the mosquito population to the cessation of spraying was intriguing: The frequency of the R allele decreased rapidly, and by 1969, RR genotypes had virtually disappeared. The precipitous decline of the R allele suggests that in the absence of

(a)

(b)

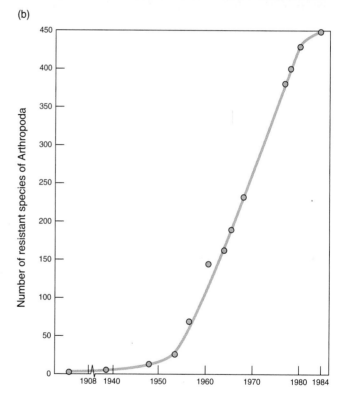

Figure 23.10 Increase of insecticide resistance from 1908 to 1984. (a) Insecticide resistance evolved with the practice of aerial spraying of DDT begun in the 1940s. Within 10 years, resistance to an insecticide becomes widespread as the insects evolve defenses to the insecticides just as they would evolve defenses against infectious bacteria. (b) The graph depicts the evolution of resistance among Arthropoda.

DDT, the RR genotype produces a lower fitness than the SS genotype. In other words, the homozygous resistance genotype imposes a **fitness cost** on individuals such that in the absence of insecticide, resistance is subject to a negative selection that decreases the frequency of R in the population.

To understand the biological basis of fitness costs, consider how rats evolve resistance to warfarin, a pesticide introduced in the 1940s and 1950s to control small mammals, among other pests. Warfarin interferes with blood clotting by blocking the recycling of vitamin K (a cofactor in the clotting

(a)

(b)

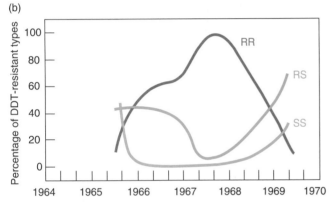

Figure 23.11 How genotype frequencies among populations of *A. aegypti* mosquito larvae change in response to insecticide. (a) Mosquitoes and larvae. (b) Changing proportions of resistance genotypes of *A. aegypti* (larvae) under selection with DDT, and after selection was relaxed, in a suburb of Bangkok, Thailand.

cascade). When a rat ingests warfarin, the inability to form a clot leads to a fatal loss of blood following any internal or external injury. In Europe in the 1960s, the extensive use of warfarin for rat control fueled the evolution of a single-gene resistance allele in many rat populations. The frequency of resistance, however, did not increase to 100%; instead, in most populations, it leveled off at 30% to 60%. Apparently, some mechanism was maintaining both the R (resistance) and the S (susceptibility) alleles in the presence of warfarin.

Further study showed that in the presence of warfarin, the relative fitnesses were 0.37 for the SS genotype, 1.0 for the SR genotype, and 0.68 for the RR genotype. These figures suggest that heterozygous advantage, the higher fitness of heterozygotes, was maintaining both alleles in the populations. Further investigation revealed that two factors caused the observed differences in fitness. Both the SR and the RR genotypes were relatively resistant to the effects of warfarin, and thus provided higher fitnesses than the susceptible SS genotypes. However, RR homozygotes suffered from a vitamin K deficiency because of the less efficient vitamin K recycling during blood

clotting, and this deficiency reduced the rate of survival when the diet did not contain a large amount of vitamin K. In the absence of warfarin, therefore, RR homozygotes had a lower fitness than SS homozygotes because of a nutritional deficiency. The biological costs of fitness, which are widespread and occur by various mechanisms, are very likely a major reason why resistance alleles occurred in very low or undetectable frequencies before the routine use of pesticides.

Ecological Considerations

Selection for resistance in the presence of pesticides is not the only factor influencing the rate of pesticide resistance evolution among pest populations. The ecological context also plays a significant role. One illustration of this phenomenon comes from studies of insect predators and parasites, and insect pests. Most predators and parasites on agricultural insect pests are also insects. It would be logical to expect that applications of insecticides to agricultural fields would expose pests and predators alike, fostering the evolution of resistance in both classes of insects. Observations, however, do not support this logic. Of the more than 450 species of insects, mites, and ticks known to be resistant to insecticides, less than 10% are predators or parasites. What causes this pattern of resistance and what are its consequences?

Simple ecological and evolutionary models provide an explanation. Consider a pest species and a predator species that coexist in an agricultural field. In the absence of insecticide, the population densities of pest and predator change over time, but both persist (Fig. 23.12a). Suppose that in each species, a single allele R at one locus determines resistance to a particular insecticide, and that initially the frequency of the S susceptibility allele is nearly 100% in both populations. The grower then applies heavy doses of insecticide to the field every two weeks (Fig. 23.12b). Because most individuals in both the pest and the predator populations are SS, high mortality from the insecticide decreases the numbers of pests and predators. For the few surviving pests, conditions are now ideal: Plant food is abundant, while competitors as well as predators are few. These pest survivors thus grow and reproduce at a rapid rate. By contrast, the predators remaining after exposure to pesticide at first face low densities of prey, which retards their growth and reproduction. This initial impairment may be so great that the predator population dies out within a few generations, before it has had a chance to evolve pesticide resistance. Now, the pest population, which has no predators to contend with, goes on to evolve a high frequency of resistance as a result of strong selection promoted by the insecticide. In these circumstances, the application of large doses of insecticide yields a worst-case scenario of a resistant pest population and a suppressed or extinct predator population. In sum, the different ecological consequences of insecticide application for prey and predators have created differences in the rate and likelihood that each population will evolve resistance. As Fig. 23.12c shows, changes in the intensity and timing of insecticide application may diminish these differences by allowing predators more time to recover before the next expo-

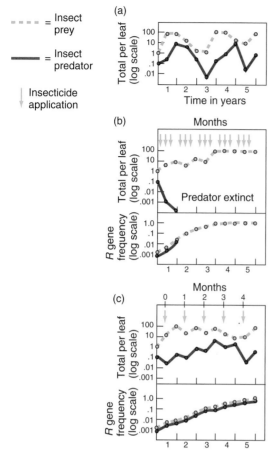

Figure 23.12   Changes in population density (insects per leaf) of prey and pest over time, with and without insecticide spraying. Fluctuations (a) in the absence of insecticide, (b) with biweekly applications of insecticide, and (c) with monthly applications of insecticide.

sure. These models suggest that both genetics and ecology have an impact on the use of predators to control pests.

ANALYZING THE QUANTITATIVE VARIATION OF MULTIFACTORIAL TRAITS

Unlike cystic fibrosis, albinism, PKU, and thalassemia, the majority of inherited human disorders are influenced by more than one gene. For example, at least 12 separate regions of the human genome are associated with the development of insulin-dependent diabetes. Indeed, in addition to the many inherited abnormalities, most aspects of normal size, shape, physiology, and behavior in humans and other organisms are complex traits determined by multiple genes that interact with each other as well as with the environment.

Complex traits of this type are called **multifactorial** because more than one factor, either genetic or environmental, contribute to the phenotype. We saw in Chapter 2 that many mutlifactorial traits are continuous, or **quantitative,** that is, they vary over a continuous range of measurement—such as

the length of a tobacco flower in millimeters, the amount of milk produced by a cow per day in liters, or the height of a person in meters. Figure 2.21 shows how quantitative variation can arise when the alleles of two or more genes affect the same trait. From a review of the Chapter 2 figure, you can see that two alleles—one dominant and one recessive—at one locus produce two discrete (discontinuous) phenotypes (AA and Aa versus aa); two alleles at two genes generate 5 discrete phenotypes; two alleles at six genes can produce 13 phenotypes that appear to vary over a continuous range. As the number of phenotypic categories for a trait increases, the trait's variation appears more and more continuous.

Quantitative traits controlled solely by the alleles of two or more genes are **polygenic traits** that are, by definition, multifactorial. But not all multifactorial traits are polygenic (some may be determined by interactions of the alleles of one gene with each other and with the environment). Moreover, not all quantitative traits are polygenic (continuous variation may result from environmental effects on the expression of one gene), and not all polygenic traits are quantitative (some traits determined by the interaction of several genes may have discrete, discontinuous phenotypes).

With these subtle distinctions in mind, we now examine how population geneticists study quantitative traits. The continuous variation of such traits depends on the number of genes that generate the trait as well as the genetic and environmental factors that affect the penetrance and expressivity of these genes (see Chapters 2 and 10). One of the goals of quantitative analysis is to discover how much of the variation in a particular trait is the result of genotypic differences among individuals in a population and how much arises from differences in the environment.

Genes Versus the Environment

To sort out the genetic and environmental determinants of phenotypic variation in a population, consider a series of experiments on a population of dandelions, a common weedy plant in lawns and other disturbed areas throughout North America (Fig. 23.13a). Dandelions have a long tubular stem and a large, yellow composite flowering structure composed of many small individual flowers; each of these flowers can produce a single, tufted, diploid seed, dispersible by the wind. Most dandelion seeds arise from mitotic, rather than meiotic, divisions such that all the seeds from a single plant are genetically identical. Your goal is to compare the influence of genes and environment on the length of the stem at flowering.

To distinguish environmental from genetic effects on phenotypic variation, you need to quantify one variable, say the environment, while controlling for the other one; that is, while holding the genetic contribution steady. You could begin by planting half of the genetically identical seeds from a single plant on a grassy hillside and allowing them to grow undisturbed until they flower. You then measure the length of the stem of each flowering plant and determine the mean and variance of the distribution of values for this trait in this dandelion population. As Fig. 23.13b shows, you find the **mean**

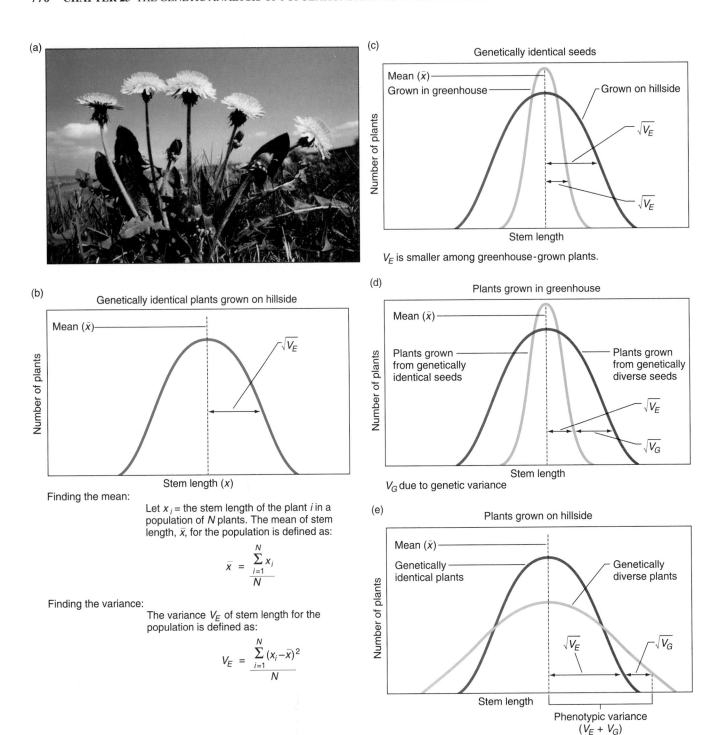

Figure 23.13   Studies of dandelions can help sort out the effects of genes versus the environment. (a) The familiar dandelion (*Taxaracum* sp.) is a useful model when studying population variations. (b) Finding the mean and variance of stem length. (c) Variance of genetically identical plants grown in a greenhouse and grown on a hillside. (d) Genetic variance of plants grown in a greenhouse. (e) Phenotypic variance of plants grown on a hillside.

by summing the values of all stem lengths and dividing by the number of stems; you then find the **variance** by expressing the stem lengths as plus or minus deviations from the mean, squaring those deviations, and again dividing by the number of stems.) Because all members of this population are genetically identical, any observed variation in stem length among individuals should be mainly a consequence of environmental

variations, such as different amounts of water and sunlight at different locations on the hillside. When represented as a variance from the mean, these observed environmentally determined differences in stem length are called the **environmental variance**, or V_E (Fig. 23.13b).

To refine your estimate of environmental variance, you plant the second half of the genetically identical seeds from the single test plant in a controlled greenhouse in which growth conditions are everywhere the same (Fig. 23.13c). Because environmental conditions are very similar for all these genetically identical plants, the amount of environmental variance (V_E) among greenhouse plants is much smaller than among hillside plants. In theory, in a perfectly controlled greenhouse, growth conditions would be the same for all plants, the V_E would be zero, and all plants would have identical stem lengths (within measurement error). In reality, there is no such thing as perfect control, and the greenhouse V_E will have some value greater than zero. Nonetheless, the difference between the V_E of the dandelions grown on the hillside and the V_E of dandelions grown in the greenhouse is a measure of the impact of the more diversified hillside environment on the phenotypic variation of stem length.

Even though the greenhouse V_E will have some value greater than zero, for the sake of simplicity in the following discussion, we assume that it is small.

To examine the impact of genetic differences on stem length, you take seeds from many different dandelion plants growing in many different locations, and plant them in a controlled greenhouse (Fig. 23.13d). Because you are raising genetically diverse plants in a relatively uniform environment, observed variation in stem length (beyond that found in the genetically identical population) is the result of genetic differences promoting **genetic variance, or V_G.**

Now, to determine the total impact on phenotype of genes and environment, you take the seeds of many different plants from many different locations and grow them on the same hillside (Fig. 23.13e). For the population of dandelions that grow up from these seeds, the **total phenotype variance (V_P)** in stem length will be the sum of the genetic variance (V_G) and the environmental variance (V_E). The environmental variance is determined directly from the phenotypic variance found in the initial population of genetically identical plants grown on the hillside. It thus becomes possible for you to calculate the genetic variance in the second, mixed population as the difference between the phenotypic variance found in this genetically mixed population and the phenotypic variance found in the genetically identical population. For natural populations of dandelions, both genetic variation among individuals and variation in the environmental conditions experienced by each plant contribute to the total phenotypic variation.

Heritability Is the Proportion of Total Phenotype Variance Attributable to Genetic Variance

With the ability to determine the relative contributions to phenotypic variation of genes and environment, geneticists have developed a mathematical definition of the **heritability (h^2)** of a trait: it is the proportion of total phenotypic variance ascribable to genetic variance:

$$h^2 = \frac{V_G}{V_G + V_E} = \frac{V_G}{V_P} \qquad (23.13)$$

Because the amounts of genetic, environmental, and phenotypic variation may differ among traits, among populations,

and among different environments, the heritability of a trait is always defined for a specific population and specific set of environmental conditions. If you know any two of the three variables of total phenotype variance, genetic variance, and environmental variance, you can find the remaining unknown variance.

How to Measure Heritability

In analyzing the contributions of genes and environment to dandelion stem length, you measured the phenotypic variation among genetically identical individuals in a range of specified environments and compared it to the phenotypic variation among all individuals in the population. Of course, most organisms are not as easy to clone as dandelions. The key to generalizing from the dandelion example, however, is to recognize that genetic clones are simply a special case of the broader notion of genetically related individuals, or genetic relatives. Various kinds of genetic relatives share certain alleles because they have one or more common ancestors. To quantify this idea, we can define the **genetic relatedness** of two individuals as the average fraction of common alleles at all gene loci that the individuals share because they inherited them from a common ancestor.

To determine the genetic relatedness between two siblings, for example, all you have to do is calculate the probability they received the same allele at any locus from the same parent. If you assume that one sibling received allele *A1* from an *A1/A2* heterozygous parent, the probability that the second sibling received the same allele is 0.5. Because this simple calculation holds for every locus transmitted by both parents, the total genetic relatedness of two siblings is 0.5.

If genetic similarity contributes to phenotypic similarity for some trait, it is logical to expect that a pair of close genetic relatives will be more phenotypically similar than a pair of individuals chosen at random from the population at large. Thus, by comparing the phenotypic variation among genetic relatives with the phenotypic variation of the entire population over some range of environments, it is possible to estimate the heritability of a trait.

The finches observed by Darwin in the Galapagos Islands (often referred to as "Darwin's finches") provide an example of a population for which geneticists have measured the heritability of a trait under natural conditions in the field. Scientists studied the medium ground finch, *Geospiza fortis*, on the island of Daphne Major by banding many of the individual birds in the population (Fig. 23.14a). They then measured the depth of the bill for the mother, father, and offspring in each nest on the island and calculated how the bill depth of the offspring correlated with the average bill depth of the mother and father (called the *midparent value*). The results, depicted in Fig. 23.14b show a clear correlation between parents and offspring; parents with deeper bills had offspring with deeper bills, while parents with smaller bill depth had offspring with smaller bill depth. In the figure, the heritability of bill depth, as represented by the slope of the line correlating midparent bill depth to offspring bill depth, is 0.82. This slope means that roughly 82% of the variation in bill depth in this population of

(a)

(b) Correlation between parents and offspring

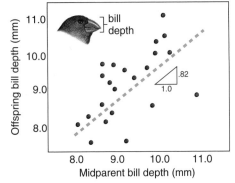

(c) If the heritability were 1.0

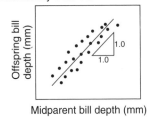

Midparent bill depth (mm)

(d) If the heritability were 0.0

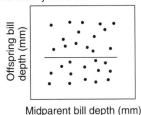

Midparent bill depth (mm)

Figure 23.14 Measuring the heritability of bill depth in populations of Darwin's finches. (a) *G. fortis* with bands placed by scientists. (b) Results of the field study with a heritability of 0.82. (c) Results if heritability were 1.0. (d) Results if heritability were 0.0.

Darwin's finches is attributable to genetic variation among individuals in the population. If there were no correlation between bill depth in parents and bill depth in offspring, the data points would be randomly distributed and the line would be horizontal (Fig 23.14d). And if there were a population in

which the environment had no influence at all on the trait, the slope of the line representing the heritability of bill depth, that is, correlating bill depth in parents with bill depth in offspring, would be 1.0 (Fig. 23.14c).

Now consider instead a population in which the bill depth for parents and their offspring is, on average, no more or less similar than the bill depths for any pair of individuals chosen from the population at random. In this case, a plot of midparent and offspring bill depths produces a circular "cloud" of points with a slope of zero (Fig. 23.14d), indicating that the heritability of this trait in this population is zero. From these examples, you can conclude that phenotypic similarity among genetically related individuals may provide evidence for the heritability of a trait.

Conversion of the phenotypic similarity among genetic relatives to a measure of heritability depends on a crucial assumption: that there is a random distribution of genetic relatives with respect to environmental conditions experienced by the population. In the finch example, we assumed that parents and their offspring do not experience environments that are any more similar than the environments of unrelated individuals. In nature, however, there may be reasons why genetic relatives violate this assumption by inhabiting similar environments. With finches, for example, all offspring produced by a mother and father during a breeding season normally hatch and grow in a single nest where they receive food from their parents. Because bill depth affects a finch's capacity to forage for food, the amount of feeding in a nest correlates with parental bill depth, for reasons quite distinct from genetic similarities. One way to circumvent the confounding variable of environmental similarity is to remove eggs from the nest of one pair of parents and randomly place them in nests built by other parents in the population; this random relocation of eggs is called **cross-fostering.** In heritability studies of animals that receive parental care, cross-fostering helps randomize environmental conditions. Controlling for both environmental conditions and breeding crosses is a fundamental part of the experimental design of heritability studies carried out on wild and domesticated organisms.

Measuring the Heritability of Polygenic Traits in Humans
Mating does not occur at random with respect to phenotypes in human populations, yet researchers cannot apply techniques for controlling environmental conditions and breeding crosses to studies of such populations. Nonetheless, in most human societies, family members share similar family and cultural environments. Thus phenotypic similarity between genetic relatives may result either from genetic similarities or similar environments (or both). How can you distinguish the effects of genetic similarity from the effects of a shared environment?

One way is to study monozygotic, "identical" twins given up for adoption shortly after birth and raised in different families. In such a pair of identical twins, any phenotypic similarity should be the result of genetic similarity. At first glance, then, the study of adopted identical twins eliminates the confounding effects of a similar family environment. Further scrutiny shows that this is usually not true. Many pairs of twins are adopted by different genetic relatives; the adoptions often occur in the same geographic region (usually in the same state and even the same city); and families wishing to adopt must satisfy many criteria, including job and financial stability and a certain family size. As a result, the two families adopting a pair of twins are likely to be much more similar than a pair of families chosen at random, and this similarity can reduce the phenotypic differences between the twins.

A better approach is to compare the phenotypic differences between different sets of genetic relatives, particularly different types of twins (Fig. 23.15a). For example, monozygotic (MZ) twins, which are the result of a split in the zygote after fertilization, are genetically identical because they come from a single sperm and a single egg; they share all alleles at all loci and thus have a genetic relatedness of 1.0. By contrast, dizygotic (DZ) twins, which are the result of sperm from a single father fertilizing two different maternal eggs, are like any pair of siblings born at separate times; they have a genetic relatedness of 0.5. Comparing the phenotypic differences between a pair of MZ twins with the phenotypic differences between a pair of DZ twins can help distinguish between the effects of genes and family environment.

Consider a trait in which the differences in phenotype among individuals in the population arise entirely from differences in the environment experienced by each individual, that is, a trait for which the heritability is 0 (Fig. 23.15b). For this trait, you would expect the phenotypic differences among many pairs of MZ twins to be as great as the differences among many pairs of DZ twins. The fact that the MZ twins are more closely related genetically has no effect.

Now consider a trait for which differences in phenotypes among individuals in a population arise entirely from genetic differences, that is, a trait for which the heritability is 1.0 (Fig. 23.15c). Because DZ twins have a genetic relatedness of only 0.5 while MZ twins have a genetic relatedness of 1.0, you would expect the phenotypic differences among many pairs of DZ twins to be twice as great as the differences among many pairs of MZ twins. Thus, by comparing measures of differences among MZ and DZ twins, it is possible to assess the relative effects of genetic relatedness and family environment on phenotypic variation. Recent studies comparing MZ and DZ twins to measure the heritability of some aspects of personality and cognition have shown that there is a genetic component to memory, extroversion, and verbal reasoning, among other behavioral traits. It is important to remember, however, that although heritability is a useful measure of the relative contributions of genetics and environment, as with any measurement, it must be properly interpreted. The heritability of a trait applies only for a specific set of environments and specific population; the heritability of the same trait may be different in a different environment or in a different population.

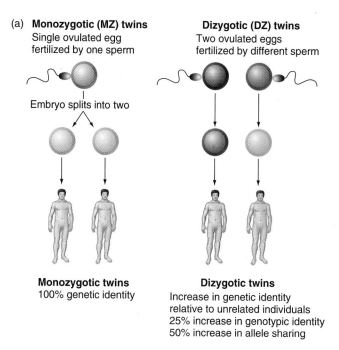

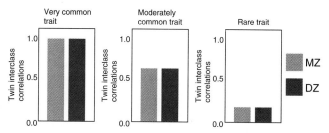

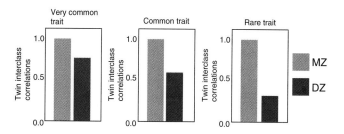

Figure 23.15 Comparing phenotypic differences between sets of MZ and DZ twins to show the heritability of a trait. (a) The difference between MZ and DZ twins. (b) Bar graphs of a trait with 0.0 heritability (MZ versus DZ twins). (c) Bar graphs of a trait with 1.0 heritability (MZ versus DZ twins).

The Fact That a Trait Has a Large Heritability Does Not Mean That the Trait is Genetically Determined and Therefore Unaffected by the Environment

Changes in the environment can, in fact, cause large changes in a phenotypic trait, even if that trait has a high heritability. For example, the trait of adult height in humans has a heritability greater than 0.8 in many populations; but European adult men in the twentieth century are much taller, on average, than European adult men were in the Middle Ages. This increase in average height is largely the result of vast improvements in diet. Similar changes in height associated with dietary changes have also occurred in other human populations. Because heritability is defined as the ratio of genetic variation to total phenotypic variation, the heritability of a trait decreases as the proportion of environmental variation increases. Thus, changes in the environment can alter the heritability of a trait. Unfortunately, many reports of supposed genetic determinism in the popular and scientific literature fail to take this fact into account.

The Heritability of a Trait Determines Its Potential for Evolution

We saw earlier how the selection of preexisting mutations generates evolutionary change. Since the heritability of a multifactorial trait is a measure of the genetic component of its variation, heritability quantifies the potential for selection and thus the potential for evolution from one generation to the next. A trait with high heritability has a large potential for evolution via selection, because the amount of evolution (change in allele frequency) generated by selection depends on the degree of heritability. To grasp the relationship between heritability, selection, and evolution, consider the trait of number of bristles on the abdomens of fruit flies in a laboratory population of *D. melanogaster*. This fruit fly population exhibits some phenotypic variation in bristle number. If the trait has a high heritability in the population, the offspring of this original population will closely resemble their parents in bristle number (Fig. 23.16). If, however, you select as parents of the

next generation only those flies among the top 15% in bristle number, the average bristle number among these breeders of the next generation will be greater than the average bristle number in the population as a whole. This artificial selection in conjunction with the high heritability of the trait will produce an F_1 generation in which the average bristle number is greater than the average bristle number in the previous generation. In other words, the artificial selection imposed by the experimenter will induce an evolutionary change whose magnitude is related to the heritability of the trait. If the heritability of bristle number were zero, there would be no evolutionary change.

A Mathematical Model of the Relation between Heritability and Evolution

If you let S represent the average trait value (in this case, the average bristle number) of breeding individuals in the parental population, measured as the difference between the value of this trait for parents and the value of the trait in the entire parental population of both breeding and nonbreeding individuals, then S is a measure of the strength of selection on the trait; as such, it is called the **selection differential.** Now let R represent the average trait value in the offspring of these breeding parents, again measured as the difference between the trait's value for parents and its value in the entire parental population of breeding and nonbreeding individuals. Used in this way R signifies the **response to selection,** that is, the amount of evolution, or change in mean trait value, resulting from selection. The heritability of the trait (arbitrarily designated as h^2), as seen in the slope of the line relating parental to offspring trait values, determines the relationship between S and R:

$$R = h^2S. \qquad (23.14)$$

In other words, the strength of selection (represented by S) and the heritability of a trait (h^2) directly determine the trait's amount or rate of evolution in each generation. This is the primary reason population and evolutionary geneticists consider the ability to measure heritability so important.

Variations in Polygenic Traits Arise at a Rapid Rate Because a Change at Any One of Many Loci Can Cause a Change in Phenotype

Geneticists have long used bristle number in *Drosophila* as a model for understanding the variation, selection, and evolution of quantitative traits. Early laboratory studies of selection acting on bristle number showed that the trait has substantial heritability in *Drosophila,* and evolves rapidly in response to selection for either high or low bristle number (Fig. 23.17). Two results from these early studies were particularly striking. First, selection can rapidly lead to phenotypes not seen in the original population. After 35 generations of artificial selection for high or low bristle number, there was no overlap in the unselected and the selected populations. Some of the change in bristle number probably arose from reassortment and changes in frequency of existing alleles at multiple genes affecting the

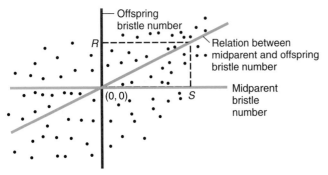

Figure 23.16   Relationship between midparent number of abdominal bristles and bristle number in offspring for a hypothetical laboratory population of *Drosophila*.

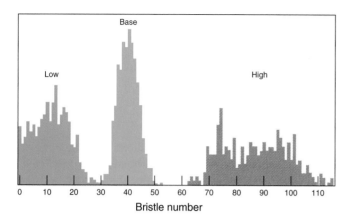

Figure 23.17   Evolution of abdominal bristle number in response to artificial selection in *Drosophila*. Effects of bristle number distributions with different populations under selection for low or high values, compared to population not subjected to selection (Base).

only in the presence of new mutations affecting bristle number. Quantitative analyses revealed a significant selection-driven evolution of bristle number, which means that new mutations affecting bristle number arise in a population at a substantial rate (Fig. 23.18). These results highlight a key characteristic of polygenic traits: If many genes contribute to a trait, new variation in the trait may arise rapidly even if the mutation rate per gene is low, because a change at any one of many loci can cause a phenotypic difference.

Populations Eventually Reach a Selective Plateau

The bristle-number experiments with *Drosophila* showed that after many generations, populations eventually reach a selective plateau at which, even with continued selection, the average bristle number does not change for many more generations. The existence of such evolutionary plateaus suggests that selection can, for a time, eliminate all genetic variation in a trait and that the potential for new mutations allowing further extremes in phenotypes has been exhausted.

How Many Loci Determine a Quantitative, Polygenic Trait

In theory, if a quantitative trait were determined by thousands or millions of loci, selection would never exhaust the trait's genetic variation. This observation of evolutionary plateaus thus begs the question of how many loci determine a trait. While it is hard to come up with an exact answer, recent studies on bristle number provide some tantalizing clues. According to genetic and molecular characterization of QTLs (quantitative trait

trait, without the appearance of new alleles. However, traits such as bristle number continue to evolve in response to selection for many generations. This observation suggests that new mutation is an additional source of variation in the population.

Experimenters have examined the contribution of mutation to genetic variation in bristle number (and by extension, other quantitative traits) through studies of highly inbred lines of *Drosophila* that at first had low or no genetic variation in bristle number. With these inbred lines, selection could occur

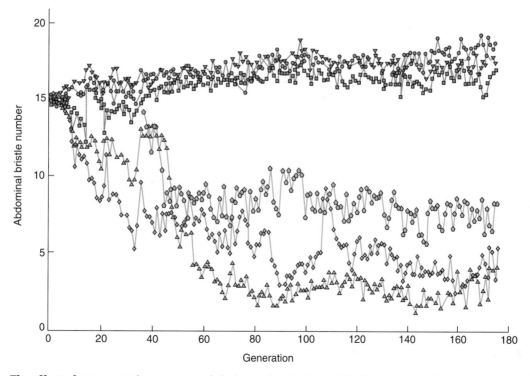

Figure 23.18   The effect of new mutations on mean bristle number in *Drosophila*. The average bristle number in population under artificial selection for a reduced number over many generations is indicated with the diamond, pentagon and up-triangle data points. Populations not under selection are indicated with the down-triangle, circle, and square data points.

loci, see Chapter 10) affecting bristle number in *Drosophila*, 10–20 loci contribute the majority of genetic variation in bristle number, and substantial variation is attributable to alleles with large phenotypic effects at a small number of loci. It is thus likely that many polygenic traits are determined by tens, rather than hundreds or thousands, of loci, and that alleles of large effect at relatively few loci have considerable influence on such quantitative traits. If this turns out to be a general pattern, it will greatly simplify genetic analysis of normal and disease-causing variation in human and other populations.

CONNECTIONS

We have seen that populations at Hardy-Weinberg equilibrium have unchanging allele frequencies, and in one generation achieve genotype frequencies of p^2, $2pq$, and q^2 which are subsequently maintained. In nature, where populations are rarely at Hardy-Weinberg equilibrium, however, natural selection acts on differences in fitness to alter allele frequencies. New mutations, genetic drift, and heterozygous advantage can also alter the allele frequencies of a population. For quantitative traits influenced by the alleles of two or more genes as well as by the interaction of those alleles with the environment, the heritability of a trait, that is, the proportion of its phenotypic variation attributable to genetic variation, determines its potential for evolution by mutation, selection, and genetic drift.

In Chapter 24, we look at evolution from the point of view of the molecular mechanisms that propel it. In that chapter, we examine the various ways in which changes at the genomic level continually reshuffle the genetic deck to create the ever-changing abundance of life-forms that inhabit the earth.

ESSENTIAL CONCEPTS

1. The Hardy-Weinberg law uses the binomial equation of $p^2 + 2pq + q^2 = 1$ to correlate allele, genotype, and phenotype frequencies in a very large ideal population; in this ideal population, individuals mate at random, no new mutations appear, no individuals enter or leave, and there are no genotype-dependent differences in fitness. In the binomial equation, p represents the frequency of one allele and q represents the frequency of the other, and p^2, $2pq$, and q^2 represent the frequencies of the two homozygous and one heterozygous genotypes.

2. A population satisfying the Hardy-Weinberg assumptions is said to be at Hardy-Weinberg equilibrium. In such a population, allele frequencies remain constant from one generation to the next, and the genotype frequencies of p^2, $2pq$, and q^2 appear in one generation, after which they are maintained.

3. Evolution consists of changes in allele frequency over time. Selection acting on genotype-dependent differences in fitness can drive evolution. Selection does not entirely eliminate deleterious recessive alleles from a population. One reason for this is heterozygous advantage.

4. The existence of an evolutionary equilibrium is another reason deleterious recessive alleles persist in populations. The evolutionary equilibrium is a balance between mutation to a new allele and selection against that allele. Late onset of a disease undermines selection against the disease allele, while genetic drift has an unpredictable effect on the evolutionary equilibrium.

5. For quantitative traits, the environmental variance is a measure of the influence of environment on phenotypic variation. Similarly, genetic variance measures the contribution of genes to phenotypic variation. Total phenotype variance is the sum of genetic variance and environmental variance.

6. Measures of environmental, genetic, and total phenotype variance make it possible to define the heritability of a trait as the proportion of total phenotypic variance attributable to genetic variance. With traits for which the number and identity of contributing genes remain unknown and there is no way to obtain genetic clones, it is possible to correlate phenotypic variation with the genetic relatedness of individuals—that is, the average fraction of common alleles at all gene loci that the individuals share because they inherited them from a common ancestor—to measure the heritability of a trait.

7. To ascertain the heritability of a human trait, population geneticists often turn to studies of twins. The most useful approach is to compare the phenotypic differences between pairs of monozygotic and dizygotic twins. Environmental changes can always influence heritability.

8. The heritability of a trait quantifies its potential for sustaining selection and thus its potential for evolution from one generation to the next.

9. Variations in polygenic traits arise at a rapid rate through selection because changes at many loci contribute to changes in phenotype. Nevertheless, it is possible that many polygenic traits are determined by tens, rather than thousands, of loci.

SOCIAL AND ETHICAL ISSUES

1. A new strain of cotton was developed in which a gene conferring resistance to a major cotton pest was introduced. The U.S. company that developed the strain is testing the strain in Bolivia. They chose this site because it is an area in which native cotton grows well. However, because native cotton grows here, there is concern that the genes from the genetically engineered plants may be transferred to the wild strains. Should there be international laws or regulations established to ensure that testing is being done in an environmentally responsible way? What responsibility does the company have to the country in which they are doing the testing? What type of remuneration should there be for the testing? Who is responsible if spread to native populations becomes a problem long after the testing is done?

2. Zoos often house endangered species and breed them to increase the population size and reintroduce into the wild. A problem with some endangered species is that, in the wild as well as in the zoos, there is very little genetic diversity remaining. Should we try to save a species by mating with other related species to generate more genetic diversity with the hope of increasing the possibility of survival? Or should we breed existing animals to try to increase their numbers even though the inbreeding may lead to the expression of additional recessive diseases? Are these activities that you would want your city zoo to spend money on? What is the human responsibility in preserving other species?

3. Periodically in the last 30 years there have been efforts to determine the heritability of IQ. Results on the heritability of IQ in one particular population are not applicable to other populations, but unfortunately such results have been applied sometimes to serve a political agenda. Is it ethical to even do studies on heritability of IQ or other politically or socially charged topics when the data are easily misinterpreted or misused? Is there a way to prevent these studies from being done or to prevent inappropriate application?

SOLVED PROBLEMS

I. A population called the "founder generation," consisting of 2000 *A A* individuals, 2000 *A a* individuals, and 6000 *a a* individuals is established on a remote island. Mating within this population occurs at random, the three genotypes are selectively neutral, and mutations occur at a negligible rate.
   a. What are the frequencies of alleles *A* and *a* in the founder generation?
   b. Is the founder generation at Hardy-Weinberg equilibrium?
   c. What is the frequency of the *A* allele in the second generation (i.e., the generation subsequent to the founder generation)?
   d. What are the frequencies for the *A A, A a,* and *a a* genotypes in the second generation?
   e. Is the second generation at Hardy-Weinberg equilibrium?
   f. What are the frequencies for the *A A, A a,* and *a a* genotypes in the third generation?

Answer

This question requires calculation of allele, genotype frequencies, and an understanding of the Hardy-Weinberg equilibrium principle.

   a. To calculate allele frequencies, count the total alleles represented in individuals with each genotype and divide by the total number of alleles.

Number of individuals	Number of A alleles	Number of a alleles
2000 *A A*	4000	0
2000 *A a*	2000	2000
6000 *a a*	0	12,000
Total	6000	14,000

The frequency of the A *allele* (p) = 6000/20,000 = 0.3.
The frequency of the a *allele* (q) = 14,000/20,000 = 0.7.
   b. If a population is in Hardy-Weinberg equilibrium, the genotype frequencies are p^2, $2pq$, and q^2. We calculated in part a that $p = 0.3$ and $q = 0.7$ in this population. Therefore,

$$p^2 = 0.09,$$
$$2pq = 2\,(0.3)(0.7) = 0.42,$$
$$q^2 = 0.49.$$

For the population of 10,000 individuals, the number of individuals with each genotype frequencies for a population in equilibrium with the allele frequencies of $p = 0.3$ and $p = 0.7$ would be: *A A,* 900; *A a,* 4200; and *a a* 4900. *The founder population described therefore is not in equilibrium.*
   c. Given the conditions of random mating, selectively neutral alleles, and no new mutations, allele frequencies do not change from one generation to the next; p = *0.3 and* q = *0.7.*

d. *The genotype frequencies for the second generation would be those calculated for part b because in one generation the population will go to equilibrium. A A = p² = 0.09; A a = 2 pq = 0.42; and a a = q² = 0.49.*

e. *Yes, in one generation a population not at equilibrium will go to equilibrium if mating is random and there is no selection or significant mutation.*

f. *The genotype frequencies will be the same in the third generation as in the second generation.*

II. Two alleles have been found at the X-linked phosphoglucomutase gene in *Drosophila persimilis* populations in California. The frequency of the Pgm^A allele is 0.25, while the frequency of the Pgm^B allele is 0.75. Assuming the population is at Hardy-Weinberg equilibrium, what are the expected genotype frequencies in males and females?

Answer

This problem requires application of the concept of allele and genotype frequencies to X-linked genes. For X-linked genes, males (XY) have only one copy of the X chromosome so the genotype frequency is equal to the allele frequency. Therefore, $p = 0.25$ and $q = 0.75$. *The frequency of male flies with genotype $X^{Pgm^A}Y$ is 0.25; the frequency of males with genotype $X^{Pgm^B}Y$ is 0.75. Three genotypes exist for females: $X^{Pgm^A}X^{Pgm^A}$, $X^{Pgm^A}X^{Pgm^B}$, X^{Pgm^B} and X^{Pgm^B} corresponding to p^2, 2pq, and q^2. The frequencies of female flies with these three genotypes are $(0.25)^2$, $2(0.25)(0.75)$, and $(0.75)^2$ or 0.0625, 0.375, and 0.5625.*

III. Two hypothetical lizard populations found on opposite sides of a mountain in the Arizonan desert have two alleles (A^F, A^S) of a single gene A with the following three genotype frequencies:

	$A^F A^F$	$A^F A^S$	$A^S A^S$
Population 1	38	44	18
Population 2	0	80	20

a. What is the allele frequency of A^F in the two populations?

b. Do either of the two populations appear to be in Hardy-Weinberg equilibrium?

c. A huge flood opened a canyon in the mountain range separating populations 1 and 2. They were then able to migrate such that the two populations, which were of equal size, mixed completely and mated at random. What are the frequencies of the three genotypes in the next generation of the single new population of lizards? ($A^F A^F$, $A^F A^S$, and $A^S A^S$).

Answer

This question requires calculation of allele frequencies and genotype frequencies in existing and in a newly created population.

a. The frequency of allele A is calculated in the following way:

38 $A^F A^F$	76 A^F alleles
44 $A^F A^S$	44 A^F alleles
	120 A^F alleles

120 A^F alleles/200 total alleles = *0.6*.

b. For population 1, the allele frequencies are $p = 0.6$ and $q = 0.4$. Genotype frequencies when the population is in equilibrium are

$$p^2 = (0.6)^2 = 0.36,$$

$$2pq = 2(0.6)(0.4) = 0.48,$$

$$q^2 = (0.4)^2 = 0.16.$$

For population 1 which consists of 100 individuals, the equilibrium would be 36 $A^F A^F$, 48 $A^F A^S$, and 16 $A^S A^S$ lizards. *Population 1 does seem to be in equilibrium.* (Sampling error and small population size could lead to slight variations from the expected frequency.) For population 2, the allele frequency (p) is based on the number of A^F alleles from the 80 $A^F A^S$ individuals. The total number of alleles = 200, so the frequency of A^F alleles is 80/200 or 0.4. The genotype frequencies for a population at equilibrium would be

$$p^2 = (0.4)^2 = 0.16,$$

$$2pq = 2(0.4)(0.6) = 0.48,$$

$$q^2 = (0.6)^2 = 0.36.$$

Population 2 does not seem to be in equilibrium.

c. The combination of the two populations of lizards results in one population with the following allele frequencies:

A^F alleles:		
$A^F A^F$	38 × 2	76
$A^F A^S$	44	44
$A^F A^S$	80	80
Total:		200

A^S alleles:		
$A^F A^S$	44	44
$A^F A^S$	80	80
$A^S A^S$	18 × 2	36
$A^S A^S$	20 × 2	40
Total:		200

The allele frequencies are 200/400 or 0.5 for both p and q. *The genotype frequencies in the next generation will therefore be*

$$p^2 = (0.5)^2 = 0.25,$$

$$2pq = 2(0.5)(0.5) = 0.5,$$

$$q^2 = (0.5)^2 = 0.25.$$

PROBLEMS

23-1 Choose the best matching phrase in the right column for each of the terms in the left column.

a. fitness

b. gene pool

c. fitness cost

d. allele frequency representation

e. genotype frequency representation

f. heterozygote advantage

g. equilibrium frequency

h. genetic drift

i. heritability

1. the genotype with the highest fitness is the heterozygote

2. chance fluctuations in allele frequency

3. ability to survive and reproduce

4. proportion of total phenotypic variance attributed to genetic variance

5. collection of alleles carried by all members of a population

6. p^2 and q^2

7. p and q

8. the advantage of a particular genotype in one situation is a disadvantage in another situation

9. frequency of an allele at which $\Delta q = 0$

23-2 In a certain population of frogs, 120 are green, 60 are brownish green, and 20 are brown. The allele for brown is denoted G^B, while that for green is G^G and these two alleles show incomplete dominance relative to each other.
a. What are the genotype frequencies in the population?
b. What are the allele frequencies of G^B and G^G in this population?
c. What are the expected frequencies of the genotypes if the population is at Hardy-Weinberg equilibrium?

23-3 A large, random mating population is started with the following proportion of individuals for the indicated blood types:

$$0.5 \ \text{MM}$$

$$0.2 \ \text{MN}$$

$$0.3 \ \text{NN}$$

This blood type gene is autosomal and M,N alleles are codominant.
a. What will be the allele and genotype frequencies after one generation under the conditions assumed for the Hardy-Weinberg equilibrium?
b. What factors cause a departure from a Hardy-Weinberg equilibrium?

23-4 A dominant mutation in *Drosophila* called *Delta* causes changes in wing morphology in *Delta/+* heterozygotes. Homozygosity for this mutation

(*Delta/Delta*) is lethal. In a population of 150 flies, it was determined that 60 had normal wings and 90 had abnormal wings.
a. What are the allele frequencies in this population?
b. Given that there is random mating, no migration, no mutation, and ignoring the effects of genetic drift, what are the expected numbers of the different genotypes in the next generation if 160 viable offspring of the population in part a are counted?

23-5 Which of the following populations are in Hardy-Weinberg equilibrium?

Population	AA	Aa	aa
a	0.25	0.50	0.25
b	0.10	0.74	0.16
c	0.64	0.27	0.09
d	0.46	0.50	0.04
e	0.81	0.18	0.01

23-6 A gene called Q has two alleles Q^F and Q^G that encode alternative forms of a red blood cell protein that allows blood group typing. A different, independently segregating gene called R has two alleles R^C and R^D, permitting a different kind of blood group typing. A random, representative population of football fans was examined, and on the basis of their blood typing, the following distribution of genotypes could be inferred (all genotypes were equally distributed between males and females):

$Q^F Q^F R^C R^C$	202
$Q^F Q^G R^C R^C$	101
$Q^G Q^G R^C R^C$	101
$Q^F Q^F R^C R^D$	372
$Q^F Q^G R^C R^D$	186
$Q^G Q^G R^C R^D$	186
$Q^F Q^F R^D R^D$	166
$Q^F Q^G R^D R^D$	83
$Q^G Q^G R^D R^D$	83

This sample contains 1480 fans.
a. Is the population at Hardy-Weinberg equilibrium with respect to either or both of the Q and R genes?
b. After one generation of random mating within this group, what fraction of the *next* generation of football fans will be $Q^F Q^F$ (independent of their R genotype)?
c. After one generation of random mating, what fraction of the *next* generation of football fans will be $R^C R^C$ (independent of their Q genotype)?

d. What is the chance that the first child of a $Q^F Q^G R^C R^D$ female and a $Q^F Q^F R^C R^D$ male will be a $Q^F Q^G R^D R^D$ male?

23-7 A population with an allele frequency (p) of 0.5 and a genotype frequency (p^2) of 0.25 is in equilibrium. How can you explain the fact that a population with an allele frequency (p) of 0.1 and a genotype frequency (p^2) of 0.01 is also in equilibrium?

23-8 When an allele is dominant, why does it not always increase to produce the phenotype proportion of 3/4 : 1/4 (3:1) dominant : recessive individuals in a population?

23-9 It's the year 1998, and the men and women sailors (in equal numbers) of the American ship the *Medischol Bounty* have mutinied in the South Pacific and settled on the island of Bali Hai where they have come into contact with the local Polynesian population. Of the 400 sailors that come aboard the island, 324 have MM blood type, 4 have the NN blood type, and 72 have the MN blood type. Already on the island are 600 Polynesians between the ages of 19 and 23. In the Polynesian population, the allele frequency of the *M* allele is 0.06, and the allele frequency of the *N* allele is 0.94. No other people come aboard the island over the next 10 years.
a. What is the allele frequency of the *N* allele in the sailor population that mutinied?
b. It's the year 2008, and 1000 children have been born on the island of Bali Hai. If the mixed population of 1000 young people on the island in 1998 mated randomly and the different blood group phenotypes had no effect on viability, how many of the 1000 children would you expect to have MN blood type?
c. In fact, 50 children have MM blood type, 850 have MN blood type, and 100 have NN blood type. What is the observed allele frequency of the *N* allele among the children?

23-10 A new university on a Caribbean island has recruited its 700 faculty members directly after they received their Ph.D.s from colleges in France and Kenya. Five hundred came from France and 200 came from Kenya with equal numbers of men and women in both cases. Upon arrival, you notice that 90 of the French and 75 of the Kenyans express a peculiar trait of rolling their eyes up into their sockets when asked a stupid question. Upon studying this trait, you discover that it is always due to the expression of allele 1 at a single gene called *Ugh*. This unusual phenotypic expression of allele 1 is dominant to the non-eye-rolling expression of the only other known allele (2) at the *Ugh* locus. Field trips taken to both Kenya and France indicate that the two alleles at the *Ugh* locus are in Hardy-Weinberg equilibrium in both of these *separate*

populations. You also determine that the trait is only expressed with 50% penetrance under all circumstances. All of the faculty members arrived in Granada single, but after teaching for a few years, they all married, and in all cases, they married other faculty members. If the faculty members marry each other in a completely random manner to produce 1000 children in total, how many of these children do you expect will express the Ugh phenotype?

23-11 You have identified a new human immune system gene that encodes a protein you refer to as antibody X. In the general *adult* population, you find only two codominant alleles of this gene. Allele 1 produces $\sim 100 \pm 50$ times as much of the same product as allele 2. Your semiquantitative survey of the general population for antibody X production shows that 50% have high levels of production (spread over a broad range) and 50% have a *very low* production. You then perform a detailed analysis of the population and discover that high levels of antibody production provide a five-year-old child with complete resistance to a newly evolved disease that attacks *only children of this age group*. Ten percent of all five-year-old children are attacked by the virus that causes this disease and 80% of those with low levels of antibody X die, whereas none of those with high levels of antibody X die. Assume that these infection rates remain constant as far as you can see into the future.
a. What is the frequency of allele 1 in the current adult population?
b. In the babies born in the following year, what will the frequency of allele 1 be?
c. In individuals who were born in the following year (referred to in part b), what will the allele 1 frequency be after these babies grow up?

23-12 A mouse mutation with incomplete dominance in mice ($t = tailless$) causes short tails as a heterozygote (t^+/t). Homozygotes (t/t) are lethal in utero. In a population consisting of 150 mice, 60 are t^+/t^+ and 90 are heterozygotes.
a. What are the allele frequencies in this population?
b. Given that there is random mating among mice, no migration, no mutation, and ignoring the effects of random genetic drift, what are the expected numbers of the different genotypes in this next generation if 200 offspring are born?
c. Two populations (called Dom 1 and Dom 2) of mice come into contact and interbreed randomly. These populations initially are composed of the following numbers of wildtype (t^+/t^+) and tailless (t^+/t) heterozygotes.

	Dom 1	Dom 2
Wildtype	16	48
Tailless	48	36

What are the frequencies of the two genotypes in the next generation?

23-13 The phenotypic frequency of colorblind males in the Caucasian population is .08. If the population is in equilibrium, what would be the frequency of color-blind females?

23-14 For the autosomal recessive condition cystic fibrosis, affected individuals in the African-American population are found at a frequency of 1/17,000. Assuming Hardy-Weinberg equilibrium conditions and no advantage or disadvantage of heterozygotes, what is the frequency with which carriers are present in the African-American population?

23-15 In *Drosophila*, the vestigial wings recessive allele, *vg*, causes the wings to be rudimentary. A geneticist crossed some true-breeding wildtype males to some vestigial virgin females. The male and female F_1 flies were wildtype. He then allowed the F_1 flies to mate and found that 1/4 of the male and female F_2 flies had vestigial wings. He dumped the vestigial F_2 flies into a morgue and allowed the wildtype F_2 flies to mate and produce and an F_3 generation.
   a. What will be the frequencies of wildtype and vestigial flies in the F_3?
   b. Assuming he repeated the selection against the vestigial F_3 flies (i.e., he dumped them in a morgue and allowed the wildtype F_3 flies to mate at random), what will be the frequency of the wildtype allele and mutant alleles in the F_4 generation?
   c. Now he lets all of the F_4 flies mate at random (i.e., both wildtype and vestigial flies mate). What will be the frequencies of wildtype and vestigial F_5 flies?

23-16 Would you expect to see a greater Δq from one generation to the next in a population with an allele frequency (q) of 0.2 or in a population with an allele frequency of 0.02? Assume relative fitness is the same in both populations and that the equilibrium frequency for q is 0.01.

23-17 Why is the elimination of a fully recessive deleterious allele by natural selection difficult?

23-18 An allele of the *G6PD* gene acts in a recessive manner to cause sensitivity to fava beans, resulting in a hemolytic reaction (lysis of red blood cells) after ingestion of the beans. The same allele also confers dominant resistance to malaria. The heterozygote has an advantage in a region where malaria is prevalent. Will the equilibrium frequency (q_e) be the same for an African and a North American country? What factors affect q_e?

23-19 Human geneticists have found the Finnish population to be very useful for studies of a variety of conditions. The Finnish population is small; Finns have extensive church records documenting lineages, and few people have migrated into the population. The frequency of some recessive disorders is higher in the Finnish population than elsewhere in the world. Diseases such as PKU and cystic fibrosis that are common elsewhere do not occur in the Finnish population.
   a. How would a population geneticist explain these variations in disease occurrence?
   b. The Finnish population is also being used as a source of information for the study of quantitative traits. The genetic basis for schizophrenia is one question that can be explored in this population. What advantage(s) and disadvantage(s) can you imagine for studying complex traits based on the Finnish population structure.

23-20 How can each of the following be used in determining the role of genetic and/or environmental factors in phenotypic variation in different organisms?
   a. genetic clones
   b. human monozygotic versus dizygotic twins
   c. cross-fostering

23-21 Which of the following statements would be true of a human trait that has high heritability in a population of one country?
   a. The phenotypic difference within monozygotic twin pairs would be about the same as the phenotypic differences among members of dizygotic twin pairs.
   b. There is very little phenotypic variation between monozygotic twins but high variability between dizygotic twins.
   c. The trait would have the same heritability in a population of another country.

23-22 Two alleles at one locus produces 3 distinct phenotypes. Two alleles of two genes lead to 5 distinct phenotypes; two alleles of six genes leads to 13 distinct phenotypes.
   a. Derive a formula to express this relationship. (Let n equal the number of genes.)
   b. Each of the most extreme phenotypes for a trait determined by two alleles at one locus are found in a proportion of 1/4 in the F_2 generation. If there are two alleles of two genes that determine a trait, each extreme phenotype will be present in the F_2 as 1/16 of the population. In common wheat (*Triticum aestivum*) kernel color varies from red to white and the genes controlling the color act additively, that is, alleles for each gene are semidominant and each gene contributes equally to the color. A true-breeding red variety is crossed to a true-breeding white variety, and 1/256 of the F_2 have red kernels and 1/256 have white kernels. How many genes control kernel color in this cross?

23-23 In a certain plant, leaf size is determined by four independently assorting genes acting additively. Thus, alleles *A, B, C,* and *D* each adds 4 cm to leaf length and alleles *A', B', C',* and *D'* each add 2 cm to leaf length. Therefore, an *A/A; B/B; C/C; D/D* plant has leaves 32 cm long and an *A'/A'; B'/B'; C'/C'; D'/D'* plant has leaves 16 cm long.

  a. If true-breeding plants with leaves 32 cm long are crossed to true-breeding plants with leaves 16 cm long, the F$_1$ will have leaves 24 cm long, i.e. *A/A'; B/B'; C/C'; D/D'*. In the F$_2$, what proportion of the plants will have leaves that are 32 cm long?

  b. Now assume in a randomly mating population that the following gene frequencies are found:

frequency of *A* = 0.9,

frequency of *A'* = 0.1,

frequency of *B* = 0.9,

frequency of *B'* = 0.1,

frequency of *C* = 0.1,

frequency of *C'* = 0.9,

frequency of *D* = 0.5,

frequency of *D'* = 0.5.

In the F$_2$, what proportion of the plants will have leaves that are 32 cm long?

EVOLUTION AT THE MOLECULAR LEVEL

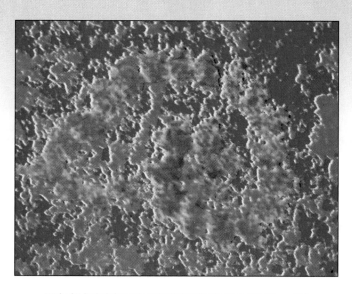

Beads of individual ribosomes connected by slender strands of messenger RNA
(mRNA) from a human brain cell (magnification ×240,000).

From December 1831 to October 1836, Charles Darwin circled the globe as naturalist for the HMS *Beagle* (Fig. 24.1). He was 22 years of age when he set sail. With indefatigable energy and insatiable curiosity, he dredged the oceans of the world for samples of the myriad organisms they concealed. He scoured the pampas of Uruguay for fossils and unusual contemporary species of lizards, birds, and mammals; scouted the no-man's-land of Tierra del Fuego for signs of unexpected life-forms; and climbed the highest peaks of the Andes, where he found deposits of seashells thousands of feet above sea level. Darwin also collected sundry specimens of the various species inhabiting different islands of the Galapagos. Whenever possible, he preserved and shipped his discoveries back to England. Some 23 years later, in 1859, he published *The Origin of Species,* a comprehensive distillation of his thinking on what he had observed. The first printing of 1250 copies sold out in one day.

In *The Origin of Species,* Darwin uses an extensive comparative analysis of thousands of specimens and fossils as the basis for proposing that "The similar framework of bones in the hand of a

man, wing of a bat, fin of the porpoise, and leg of the horse—the same number of vertebrae forming the neck of the giraffe and of the elephant—and innumerable other such facts, at once explain themselves on the theory of descent with slow and slight successive modifications." This observation leads him to the stunning conclusion that "all organic beings which have ever lived on this earth may be descended from some one primordial form."

While Darwin was not the first to suggest that species could undergo evolution, he was the first to suggest a mechanism by which evolution could occur. He based his theory of evolution on three principles—each obvious in and of itself—whose combination had revolutionary implications. First, within any species, there is variation among the individuals of a population in the expression of numerous traits. Second, variant forms of traits can be passed down through inheritance from one generation to the next. Third, some variant traits give the individuals that express them a greater chance of surviving and reproducing. (This is the so-called "survival of the fittest" principle.)

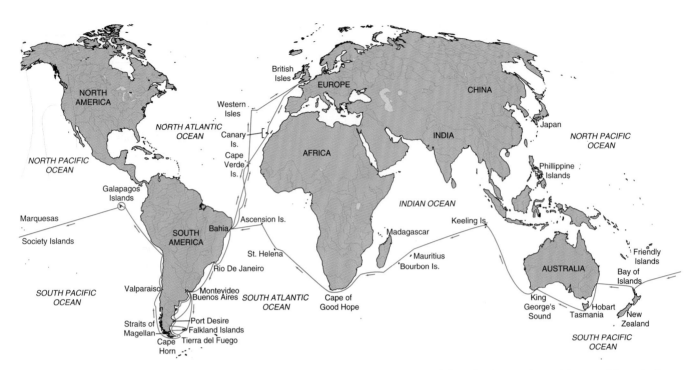

Figure 24.1 The voyage of the HMS Beagle. Charles Darwin spent 5 years circling the globe collecting specimens. The material acquired on this grand tour provided the basis for his comparative analysis *The Origin of the Species,* published almost 24 years later.

Darwin recognized that an advantageous trait that at first appears in only one or a few individuals could allow those individuals and their descendants to out-compete individuals who do not express the trait. This process of "natural selection," that is, of natural conditions selecting for an advantageous trait, would result in the transmission of the trait to a greater proportion of the population at each successive generation. Ultimately, many generations of natural selection would produce a population whose members all expressed a particular advantageous trait. Meanwhile, variant forms of other traits would arise in different individuals in each generation, and natural selection would, for each trait, determine which variants survive over time. Darwin's understanding of the process of continuous evolution by natural selection became the cornerstone of his theory of evolution. In *The Origin of Species,* he nevertheless cautions that "the natural selection of numerous successive, slight, favorable variations . . . has been the most important, but not the exclusive, means of modification."

Despite the enormous range of his revolutionary insight, Darwin was at a loss to explain the source of the visible variation on which natural selection acts. He knew inheritance plays a role: "Everyone must have heard of cases of albinism, prickly skin, hairy bodies, etc., appearing in several members of the same family. If strange and rare deviations of structure are really inherited, less strange and commoner deviations may be freely admitted to be inheritable." But he did not understand how "The laws governing inheritance are for the most part unknown." Mendel published "Experiments on Plant Hybrids" in 1866 (just seven years after Darwin's *The Origin of Species* appeared), but as we saw in Chapter 1, although Darwin received a copy of Mendel's paper, he most likely never read it.

Today, 140 some years after the publication of *The Origin of Species,* biologists accept Darwin's theory as a foundation of modern biology. Moreover, thanks to Mendel (and many other geneticists), they now understand the basic principles of heredity. And thanks to Watson and Crick (and many other molecular biologists), they know that ultimately, evolution is a process that begins at the molecular level, inside the double helix of DNA.

An example of short-term competitive evolution at the molecular level is seen in the battle between cells of the human immune system and the AIDS virus. Both evolve by the diversification of progenitor populations followed by the selective amplification of some divergent forms. When an invading virus activates the immune system, a virus-specific immune response ensues. As we saw in the Fast Forward box in Chapter 12, specific human immune responses depend on the cellular diversity of the immune system, which includes a trillion (10^{12}) circulating white blood cells, or lymphocytes, divided into two types—T cells and B cells. Populations of lymphocytes have the genetic capacity to synthesize a large and varied group of cell-surface immune receptors collectively capable of recognizing foreign (that is, nonself) macromolecular structures on virtually any virus, bacterium, or other pathogenic microorganism (Fig. 24.2). Gene rearrangements in the cells of the developing immune system enable this enormous diversification of receptors from a relatively small repertoire of genes.

However, although populations of lymphocytes carry a great diversity of immune receptors, each individual lymphocyte synthesizes only one type of immune receptor. This attribute of one type of immune receptor per cell is the key to the specificity of an immune response. When the multiple copies of that one type of

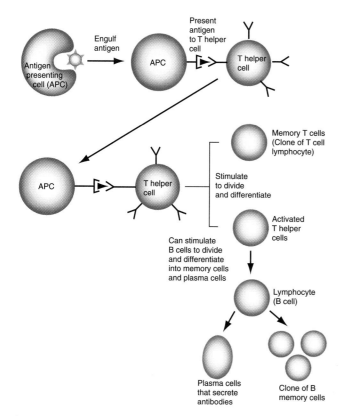

Figure 24.2 The immune response. Diverse receptors (on lymphocytes) recognize different molecular structures from invading pathogens. T cells employ T-cell receptors. B cells employ antibodies and differentiated B cells, plasma and memory cells, and actually secrete antibodies that destroy or neutralize antigen by a variety of mechanisms. Expanded numbers of memory T cells and B cells represent a memory response that can rapidly respond to an antigen seen a second time.

receptor in a cell's membrane encounter complementary foreign structures (called *antigenic determinants*), interaction between the immune receptors and the antigenic determinants triggers the lymphocyte to divide and differentiate.

The rapid proliferation of just a few selected lymphocytes to a few clones of genetically identical cells is the first step of an immune response. Among the cells in each expanded clone are *memory cells,* which may live as long as 40 years, and *effector cells,* which actually carry out (or effect) the immune reactions that help dispose of the microorganisms. These effector cells include T effector cells that bind to the antigenic determinants of a bacterium or virus and B effector cells that secrete antibodies, which in turn bind to antigenic determinants. As the immune response progresses, some of the B effector cells mutate their membrane receptors (generating further diversification). The altered receptors with the closest fit to the antigenic determinants bind the determinants more tightly, and this high-affinity binding drives the amplification of the lymphocytes that carry the altered receptors (selective amplification). The differentiated effector cells become involved in a multitude of effector mechanisms that ultimately destroy the pathogenic organism carrying the antigenic determinants targeted by the im-

mune response. Thus, the immune system's generation of specific immune responses is a marvelous example of molecular evolution: the diversification of lymphocytes into many variants and the selection by antigenic determinants of just one or a few of these variants for amplification over a period of weeks.

The AIDS virus (HIV) is a worthy adversary for the human immune system because it is able to diversify and amplify via selection far faster than the immune response itself. HIV, we have seen, is a retrovirus with an RNA genome (review Fig. A in the Genetics and Society box in Chapter 7). The virus makes its own reverse transcriptase available to the cells it infects (which include the T lymphocytes), forcing the cellular machinery first to copy the viral RNA genome into a DNA copy and then to generate a complementary DNA strand to form a proviral double helix. The infected cells integrate the double-stranded viral DNA into their own genomes (review Fig. B in the Genetics and Society box in Chapter 7). We have seen that the HIV reverse transcriptase has a very high mutation rate of roughly 1 in 5000 nucleotides incorporated into the viral cDNA. Since the HIV genome is about 10 kb in length, each replicated virus carries an average of two mutations (the basis of rapid diversification).

When HIV replicates explosively on infection, it generates billions of variant viruses before the host's immune response has a chance to take hold. Many of the variant viral particles are beyond the reach of the immune response because the antigenic determinants targeted by the initial immune response have mutated beyond recognition by the time the originally activated T cells and B cells have proliferated (selection by lack of immune cell recognition). Although some of the new antigenic determinants may be recognized by other cells among the diverse population of lymphocytes, subsequent viral mutations may alter these viral targets in good time. Eventually, the speed of viral evolution outstrips the ability of the immune response to keep pace not only because the viral genome mutates at a high frequency but also because a viral generation is so short.

A new therapeutic approach to AIDS, which tries to tip the evolutionary balance in favor of the human immune system, has been remarkably effective in prolonging the initial symptom-free phase of the disease. Called triple drug therapy, it entails the simultaneous delivery of three different anti-HIV drugs; two block the function of the HIV reverse transcriptase through different mechanisms, and the third blocks the functioning of a viral protease critical to HIV reproduction. By reducing the rate of viral replication, the triple drug therapy significantly decreases the virus's ability to develop variants that are simultaneously resistant to the human immune response and all three drugs. As a result, the HIV population in patients diminishes dramatically with the initial treatment. Thus, the triple drug therapy shifts the ability to diversify and amplify by selection in favor of the human immune response.

In this chapter, we examine in detail the basic components of evolution at the molecular level: **diversification** into many variants, followed by **selection** of one or a few variants **for amplification** in a population over many generations. The same process is at work in nearly all organisms, from tiny, subcellular viruses to large, multicellular plants and animals. Unlike the evolution of the human immune response and the AIDS virus, however, most molecular evolution in higher, more complex organisms occurs over millions of years.

As we describe how diversification and selection for amplification operate at the molecular level, we discuss

■ The origin of life on earth, including a possible scenario for descent from a single ancestor; the idea that RNA may have been the original replicator; and a probable timetable for the evolution of complex life-forms from a sea of molecules.

■ The evolution of genomes, including a review of the various types and effects of mutations; a discussion of how an increase in genome size correlates with the evolution of

complexity; a description of how larger genomes evolve through duplication and divergence; and an overview of molecular archaeology based on an understanding of gene duplication, diversification, and selection.

■ The organization of genomes, including a description of genes, gene families, and gene superfamilies; repetitive nonfunctional DNA families; simple sequence repeats that dot the genome at random; and repeat sequences in centromeres and telomeres.

■ The immunoglobulin gene superfamily: A comprehensive example of molecular evolution.

THE ORIGIN OF LIFE ON EARTH

To many biologists, the similarity of all living things on earth is more striking than the differences between them. Not only are all organisms composed of cells, but these cells work in essentially the same way, using the same complex molecules and the same type of genetic material read according to virtually the same genetic code. As we have seen, the flow of information in most living organisms from DNA via RNA to protein follows a well-defined pathway. DNA contains a digital code with a four-letter alphabet encoding the instructions for the construction and development of organisms. The units of information in the DNA code are genes that can be differentially expressed as RNA. Some RNA molecules (for example, mRNA) contain codes for proteins; other RNA molecules can fold into three-dimensional molecular machines and function as enzymes. The proteins synthesized from the information carried in RNA are three-dimensional objects that catalyze the chemistry of life and give organisms their shape and form. In complex organisms, the proteins, together with other macromolecules, assemble to form organelles, cells, tissues, and organ systems.

Descent from a Single Ancestor: Speculations on How the First Cell Arose

There is no fundamental law of biochemistry that says all living cells have to be constructed in the way that they are; indeed, an imaginative biochemist could think of an almost infinite number of ways to build a functioning cell that, like the organic life-forms around us, is based on the laws of chemistry. The observation that the cells of all plants, animals, fungi, and microorganisms analyzed so far are extremely similar in subcellular organelles, biochemistry, and genetic processes suggests that the abundant variety of life-forms alive today descend from the same original cell that happened to begin life with the particular genetic code scientists now consider universal (even with the minor differences seen in some organelles and microorganisms).

The First Step on the Pathway to Life Must Have Been a Replicator Molecule That Could Copy Itself

While the molecular evidence for the evolution of viral genomes and immune responses is accessible and available for examination, no evidence remains of the earliest steps of molecular evolution, during which self-replicating macromolecules became enclosed in membranes and ultimately formed cells, the basic units of life as we know it today. Scientists have therefore inferred and imagined these earliest steps, building their inferences and imaginings on a foundation of macro- and microbiological observations and analyses.

A key step in molecular evolution was the emergence of a molecule that could replicate itself. Although one can only speculate on the nature of the original replicator, it must have been simple enough to form spontaneously after a half billion years of atoms bumping into each other on the surface of the primordial earth. To give rise to the living cells we see today, it would have had to fulfill three requirements: (1) encode information by the variation of letters in strings of a simple digital alphabet; (2) fold in three dimensions to create molecules capable of self-replication and ultimately other functions; (3) expand the population of successful molecules through selective self-replication. What was the original self-replicating molecule and where did it originate?

The Original Replicator May Have Been RNA

In the 1980s, Thomas Cech and colleagues discovered that RNA, in addition to its ability to carry genetic information like DNA, can catalyze chemical reactions. Before this discovery, molecular biologists had assumed that only proteins have the chemical flexibility to fold into enzymes. RNA molecules that can act as enzymes to catalyze specific chemical reactions are called **ribozymes.** Although the substrates of most naturally occurring ribozymes are other RNA molecules, their potential range of enzymatic activity may be much broader.

The discovery of ribozymes gave molecular biologists an ideal candidate for the original replicator: a hypothetical RNA molecule that on its own could have (1) encoded information as linear strings in a four-letter alphabet; (2) folded into three-dimensional molecular machines able to execute critical functions of life (such as those currently carried out by polymerases, nucleases, and ligases); and (3) reproduced itself. The hypothetical primordial world in which this RNA became the first replicator has been termed the **RNA world.** The fact that all living organisms share the four letters of the RNA alphabet (A, U, G, and C) is testimony to the idea that successful, self-replicating RNA strings probably had a single origin, and this, in turn, could have provided the common evolutionary origin of the information in all living organisms.

The earliest RNA strings most likely had informational motifs, or coding regions (for example, encoding a simple polymerase), separated by noncoding regions of "background noise." The ribozymes of contemporary organisms that can cleave, join, copy, and even modify informational strings may reflect these early RNA activities. The pattern of coding regions separated by noncoding introns could have been established by the early random assembly of RNA strings. The evolutionary challenge was how to use the information and separate it from the noninformational regions. Accordingly, there were evolutionary pressures to evolve RNA coding regions that could cut (like an endonuclease) and splice (like a ligase), as well as synthesize (like a polymerase) RNA molecules.

However, multipurpose RNA strings that both stored information and folded into molecular machines would have had several intrinsic disadvantages. First, RNA is a relatively unstable molecule, readily susceptible to chemical and enzymatic hydrolysis. Second, RNA has only four letters in its alphabet, which are chemically similar, and is thus less capable of folding into complex three-dimensional structures than other polymers with more complex alphabets. This is so because each letter in a molecular alphabet represents a particular shape and charge; in fact, proteins with their 20 chemically divergent amino-acid letters can fold into more complex structures than RNA with only four. Accordingly, there were evolutionary pressures to evolve more stable storage molecules (DNAs) and more flexible molecular machines (proteins).

No Record Exists of the Intermediates between the RNA World and the Organized Complexity of the Cell

This lack of evidence makes it difficult to provide solid answers to questions about evolutionary events at the precellular stage. If RNA was indeed the original replicator, was it available for the first cells to use as an informational molecule, with the DNA and protein alphabets evolving subsequently? How did membrane compartments that concentrate the molecules of life and facilitate their sorting arise? What metabolic pathways gave some of the early cells an evolutionary edge? Many

biologists speculate that life began around the volcanic thermal vents of the ancient seas more than 3.5 billion years ago. In this high-temperature environment, informational subunits could have emerged and then polymerized into strings of informational molecules. It is likely that with each of these informational intermediates was a proto-life-form that reproduced itself into a large number of slightly different copies (diversification) until by chance one gained a better mode of survival and outcompeted all its cousins in the next round of reproduction (selection). This two-step process of diversification and selection repeated itself over and over again until the first cell appeared. From that point on, the history of life's evolution is easier to discern.

The Evolution of Living Organisms: Inferences from the Fossil Record

It is an enormous jump from a microscopically small single-cell organism to, for example, a human being composed of 100 trillion cells. However, like Darwin, no current biologist doubts that the one evolved into the other. This confidence arises not just from what Darwin's theory says *should* happen, but from the possibility of seeing the critical intermediate stages of biological complexity, all the way up the ladder from single cells to sponges, to worms, to fish, to reptiles, to mammals, and to primates and humans.

The Intermediate Stages of Evolution from a Single Cell to Complex Multicellular Organisms Exist in the Fossil Record and in the Living Fossils That Surround Us

While the fossil record enables us to date the initial appearance of each intermediate stage, the living representatives of each stage, in conjunction with our ability to study them with the tools of molecular biology, give us a glimpse into how evolution occurs step by step. It is fortunate for the scientists who study evolution that so many critical intermediate forms of life kept on reproducing in their less complex state, even as their cousins went on to evolve to the next stage of complexity.

Single-Celled Organisms without a Membrane-Bound Nucleus Emerged First

Scientists agree that planet earth coalesced some 4.5 billion years ago. By 4.2 billion years ago, enormous oceans covered the planet, and the first informational RNA molecules may have emerged around the high-temperature volcanic ocean vents. The first living organisms, consisting of a membrane surrounding information-replicating and information-executing machinery, evolved about 3.7 billion years ago; these were the precursors of present-day cells.

Fossils laid down 3.5 billion years ago cells near North Pole, Australia (a small town in an arid, rocky region), are the earliest evidence so far uncovered of distinct cells (Fig. 24.3a). Once life in cellular form emerged, living organisms evolved into three distinct domains: archaea, bacteria, and eukarya (Fig. 24.3b). Contemporary representatives of archaea and bacteria include only single-celled organisms whose genomes

(a)

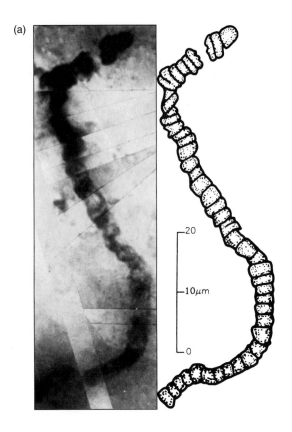

(b)

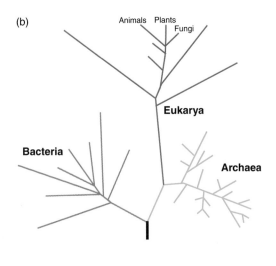

Figure 24.3 The earliest cells evolved into three kingdoms of living organisms. (a) The oldest fossilized cells. (b) The distinct branches represent different organisms in each kingdom. The length of the branches is proportional to the times of species divergence.

carry tightly packed genes. Because their genes do not contain introns, some biologists have concluded that introns were a late evolutionary elaboration, appearing with the emergence of eukaryotes. An alternative explanation is that archaea and bacteria, requiring great efficiency in their use of energy, relinquished, in the course of evolution, the flexibility of discontinuous coding segments for the more energy-efficient synthesis of genes without introns.

More Complex Cells and Multicellular Organisms Appeared After More Than 2 Billion Years of Cellular Evolution

Eukarya emerged about 1.4 billion years ago with the incorporation of single-cell organisms that evolved to become intracellular organelles, and the complex compartmentalization of the cell's interior, including the segregation of DNA molecules into the nucleus. The evolution of these relatively complex eukaryotes from the earliest cells thus took almost 2.3 billion years. About 1 billion years ago, the single-celled ancestors of contemporary plants and animals diverged. Then 0.7 billion years ago, one of the most remarkable events in the evolution of life occurred: the explosive appearance of a multitude of multicellular organisms, both plants and animals. The multicellular animals are often referred to as **metazoans.**

The Burgess Shale of Southeastern British Columbia in Canada Captured the Drama of Metazoan Evolution

The shale arose from a mud slide that trapped a wide variety of different organisms in a shallow Cambrian sea. Amazingly,

events and conditions during and after the slide conspired to achieve the nearly perfect preservation of the three-dimensional structure of the entrapped specimens' soft body parts. Three aspects of the Burgess organisms are remarkable. First they represent a myriad of very different body plans (Fig. 24.4). For example, paleobiologists have distinguished 20–30 classes of arthropods in the Burgess sea shale, a striking contrast to the three contemporary classes of arthropods. Second, this emergence of metazoan organisms occurred over a remarkably short (in evolutionary terms) period of time, perhaps just 10–20 million years. This rapid evolution is an example of **punctuated equilibrium:** the tendency of evolution to proceed through long periods of stasis (lack of change) followed by short periods of explosive change. It was as if the evolution of life represented a supersaturated salt solution where the addition of a final salt crystal catalyzed a remarkable solutionwide cooperative crystallization of many different body plans. Third, it seems that all the basic body plans of contemporary organisms initially established themselves in the metazoan explosion. For example, the ancestor of contemporary vertebrates depicted in Fig. 24.4c probably emerged at about the same time as the ancestor of all contemporary invertebrates.

Many Species Resulting from the Metazoan Explosion Have Disappeared

The enormous diversity of metazoan body plans that materialized 0.7 billion years ago has by now become tremendously reduced, in part through four to six abrupt extinction events

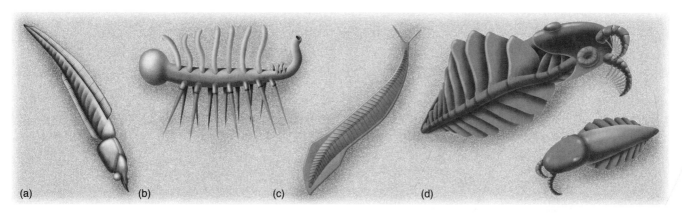

Figure 24.4 Some of the body plans found in Burgess shale. Although all are now extinct, we have seen in fossils their diverse body plans: (a) the *Nectocaris,* (b) the *Hallucigenia,* (c) the *Pikaia,* and (d) the *Anomalocaris.* The *Pikaia* is a vertebrate with a notochord, which makes it an ancestor to modern-day vertebrates.

that each destroyed 70%–95% of the existing organisms. The most recent example was the global decimation 65 million years ago that led to the extinction of the dinosaurs. Many scientists believe this extinction was a consequence of a large meteorite impact in the Yucatan region of present-day Mexico that dramatically changed earth's climate by propelling enormous amounts of dust into the higher atmosphere. Scientists hypothesize that this thick cloud of dust dispersed and shrouded the globe for several years, preventing solar rays from reaching earth's surface; the lack of solar heat led to a nuclear-winter-like scenario in which the demise of all green life caused the demise of all large animals, such as dinosaurs, that depended for survival either directly on plants or indirectly on animals that ate plants.

Some smaller animals presumably survived this long sunless winter because their lesser size allowed them to get by on seeds alone. When the sun returned, the seeds lying dormant on the ground sprang to life and the world became an abundantly fertile place. In the absence of competition from dinosaurs, mammals became the dominant large animal group, diverging into numerous species that could take advantage of all the newly unoccupied ecological niches. Some eventually evolved into our own species.

The Evolution of Humans

Humans arose from an ancestor common to most contemporary primates that existed 35 million years ago. They diverged from the ancestors of their closest primate relatives, the chimpanzees, about 6 million years ago (Fig. 24.5). While paleobiologists have not yet sorted out the immediate evolutionary ancestors of *Homo sapiens,* the recent typing of fossil DNA suggests that one previous candidate, the Neanderthal lineage, is not on the direct human evolutionary line (see the Fast Forward box in Chapter 14). Perhaps the single most fascinating aspect of human evolution is the striking changes in the evolution of the brain since the human lineage diverged from the

chimpanzee line. These changes include an increase in size, folding complexity, and nerve cell density, which together provide people with the capacity for a level of intelligence sufficient to appreciate the universe and the history of life.

Remarkably, on average, the chimpanzee and human genomes are approximately 99% similar. Moreover, as we saw in Chapter 11, the chimpanzee and human karyotypes are nearly the same. And in every comparison to date of chimpanzee and human DNA sequences, the observed differences between the two have been insignificant in terms of gene function. What these data suggest is that the evolution from a common primate ancestor to the modern human species can probably be accounted for by a very small number of isolated genetic changes yet to be uncovered. While these changes may have occurred in protein-coding sequences, many evolutionary biologists think it more likely that the changes occurred in regulatory sequences. Such changes would alter the regulatory sequences controlling when and how master regulatory genes produce transcription factors, and when and how ordinary structural genes or batteries of genes respond to these regulatory molecules. This idea is supported by the amazing ability of genes from one species to substitute for the absence of homologous genes in other species, in some cases even when the species are as different as yeast and humans. If homologous coding sequences from very different species are functionally indistinguishable, it is reasonable to speculate that species-specific differences in phenotype may arise from species-specific differences in gene expression. Thus, what makes humans and chimpanzees different may not be the proteins that their genes encode, but rather when, where, and at what level those proteins are expressed during development. It is possible, for example, that the difference between chimpanzee and human consciousness can be accounted for simply by changes in the regulatory sequences of master regulatory genes and other genes encoding proteins that play a role in the development of the cerebral cortex.

(a) (b) (c) (d)

Figure 24.5 Humans diverged from chimpanzees about 6 million years ago. The evolutionary relationship of common primates including humans. (a) orangutan; (b) gorilla; (c) chimpanzee; and (d) human being.

THE EVOLUTION OF GENOMES

Although Darwin developed his theory of evolution—based on the selection of preexisting variations—without much knowledge of the molecules that make up living systems, evolution is very much a molecular process that operates on genetic information. In particular, the variation that initiates each step in the evolutionary process occurs within the genetic material itself in the form of new mutations.

DNA Alterations Form the Basis of Genomic Evolution

We have seen that mutations arise in several different ways (review Figs. 6.6–6.10). One is the replacement of individual nucleotides by other nucleotides. Replacements occurring in a coding region are silent, or synonymous, if because of the degeneracy of the genetic code, they have no effect on the amino acid encoded; by contrast, they are nonsynonymous if the change in nucleotide determines a change in amino acid or a premature termination codon, leading to a truncated gene product. Molecular biologists further distinguish between nonsynonymous changes that cause conservative amino-acid changes (for example, from one acidic amino acid to another) and those that cause nonconservative changes (for example, from a charged amino acid to a noncharged amino acid). Other gene mutations, arising from errors in replication or recombination, consist in the deletion from or insertion into genes of a DNA sequence of any length.

Different mutations can be *deleterious, neutral,* or *favorable* to the organisms that inherit them. In multicellular organisms with large genomes, such as corn or humans, genes and their regulatory sequences make up only a small fraction (~5%) of the total genetic material. As a result, random mutations occur most often in DNA that plays no role in the de-

velopment or function of an organism. Such mutations are presumably neutral. It might seem that synonymous mutations within coding regions would also be neutral, but there is some evidence that even changes in the codon used to produce a particular amino acid can provide a minute advantage or disadvantage to the organism based on the availability of different tRNA molecules and their associated synthetases. And while conservative amino-acid replacements were once considered neutral, current evidence suggests that they can have an impact on the growth and survival of an organism. Nonconservative amino-acid changes and changes such as deletions and insertions involving larger portions of a gene almost certainly have an impact on gene function.

Genetic Changes That Are Truly Neutral Are Unaffected by the Agents of Selection

Such mutations have a neutral effect on fitness and thus are not susceptible to selection. They survive or disappear from a population through *genetic drift,* which is the result of chance reproductive events. The smaller the population, the more rapidly genetic drift exerts its effect.

Mutations That Have Only Deleterious Effects Will Disappear from a Population by Negative Selection, That Is, Selection Against an Allele

However, as we saw in Chapter 23, some mutations (such as those that cause sickle-cell anemia) are deleterious to homozygotes for the mutation, but advantageous to heterozygotes. Because of this *heterozygous advantage,* selection retains these mutations in a population at a low equilibrium level (review Fig. 23.7). However, even deleterious mutations for which there is no heterozygous advantage are eliminated very slowly if they are fully recessive.

Some Extremely Rare Mutations Give an Organism a Significant Advantage

Because of this advantage, individuals carrying the mutation are more likely to reproduce, and in each succeeding generation, the frequency of the mutation increases (review Fig. 23.4). This is positive selection (selection for an allele). Ultimately, the allele that began as a mutation is present in nearly every member of the population on both chromosomes. At this stage, the allele has become *fixed* in the population.

An Increase in Genome Size Correlates with the Evolution of Complexity

Consider that although both bacteria and mammals evolved from a common cellular ancestor, the contemporary *Escherichia coli* genome is about 5 Mb in length, while in humans, the genome is about 3000 Mb long. How has evolution fashioned such different genomes from the same original material? The answer lies in the evolutionary potential for increasing the size of the genome through the duplication and diversification of genomic regions. In these processes, new DNA is born out of old.

Genomes Grow in Size through Repeated Duplications

Duplication events can occur at random throughout the genome, and the size of the duplication unit can vary from a few nucleotides to duplication of the entire genome. When a duplication segment contains one or more genes, either the original or the duplicated copy of each gene is free to accumulate function-destroying mutations (diversify) without harm to the organism, since the other "good" copy with original function is still present. With duplications acting as such an important force in evolution, it is critical to understand the two main ways in which they arise.

Some Duplications Result from Transpositions

As we saw in Chapter 12, transposition, the transfer of one copy of a chromosomal sequence from one chromosomal site to another, can occur in various ways: through the direct movement of a DNA sequence from one chromosomal site to another, through an RNA intermediate that is copied into a DNA intermediate, leaving the original DNA site intact, or through a DNA intermediate (Fig. 24.6; review Figs. 12.17 and 12.18). When the genomic DNA (rather than its RNA or DNA proxy) moves to a new site, the duplication of genetic material occurs only after the altered chromosome receiving the DNA segregates together with the unaltered homolog of the chromosome containing the original locus, into an egg or sperm. When the gamete with the duplication unites with a normal gamete, the resulting zygote has three copies of the original locus (Fig. 24.7). In subsequent generations, the new transposition element may become fixed in the population.

Other Duplications Result from Unequal Crossing-over

Normal crossing-over, or recombination, occurs between equivalent loci on the homologous chromatids present in a

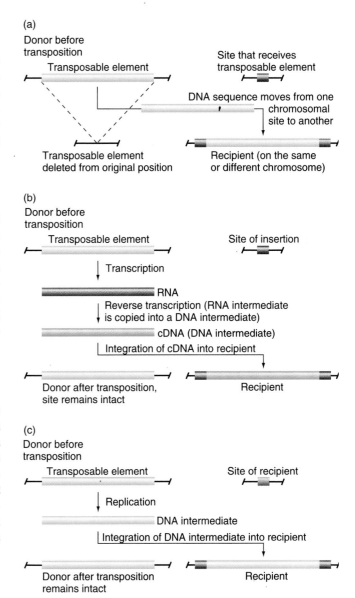

Figure 24.6 Duplication by transposition: Three possibilities. Transposition may occur by (a) excising and reinserting the DNA segment; (b) by making an RNA copy, which is then converted to a DNA copy for insertion; or (c) by making a DNA copy for integration. In the latter two examples, the transposon is duplicated.

synaptonemal complex that forms during the pachytene stage of meiosis. Unequal crossing-over, also referred to as illegitimate recombination, occurs between nonequivalent loci (review Fig. 6.9a). Unequal crossing-over is most often initiated by related sequences located close to each other in the genome. Although the event is unequal in the exchange of nonequivalent segments of DNA, it is still mediated by the sequence similarities at the two separate loci.

So-called **nonhomologous unequal crossing-over** also occurs, although much less often than homologous crossing-over. These nonhomologous crossovers may in fact be mediated by at least a short stretch of sequence homology, coding or noncoding, at the two sites at which the event is initiated.

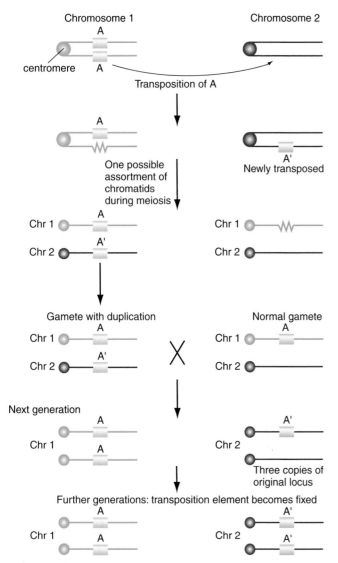

Figure 24.7 Transposition through direct movement of a DNA sequence. A DNA sequence may be transposed from one chromosome to a second chromosome in a sex cell. In subsequent generations, this transposed element may be fixed.

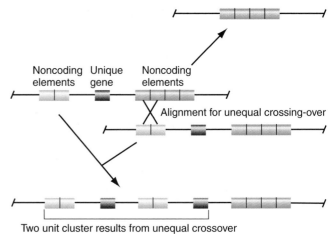

Figure 24.8 Duplications arise from unequal crossing-over. Crossover involving dispersed repeat elements (blue boxes). The pink boxes denote a unique gene. In the crossover event, this gene is deleted in one chromosome and duplicated in the other.

An initial duplication by unequal crossing-over that produces a two-unit cluster may be either homologous or nonhomologous, but as Fig. 24.8 illustrates, once two units of related sequence are present in tandem, further rounds of homologous unequal crossing-over between nonequivalent members of the pair readily occur. Thus small clusters can easily expand to contain three, four, and many more copies of an original DNA sequence.

The result of unequal crossing-over between homologous chromosomes is always two reciprocal chromosomal products: one carrying a duplication of the region located between the two crossover sites and the other carrying a deletion that covers the exact same region (review Fig. 6.9). Unlike retrotransposition, unequal crossing-over operates on genomic regions without regard to functional boundaries. Regions duplicated by unequal crossing-over can vary from a few base pairs to hundreds of kilobases in length, and may contain no genes, a portion of a gene, a few genes, or many genes.

Random Genetic Drift and Mutations Can Turn Duplications into Pseudogenes That Eventually Dissolve into Random DNA Sequences

Duplicated regions, like all other genetic novelties, must originate in the genome of a single individual. At first, the survival of such duplications in at least some animals in each subsequent generation of a population is, most often, a matter of chance. This is because the addition of one extra copy of a chromosomal region, including most genes, to the two already present in the diploid genome usually causes no significant harm to the individual animal. In the terminology of population genetics, the duplicated units are neutral with regard to genetic selection. They are thus subject to genetic drift, inherited at random by some offspring but not others. By chance again, most neutral genetic elements disappear from a population in several generations.

When a duplicated region that includes a functional gene survives for hundreds or thousands of generations, random mutations in the gene may turn it into a related gene with a different function, or into a nonfunctional *pseudogene*. Some of the mutations generating a pseudogene lead to a loss of regulatory function; others change one or more critical amino acids in the gene product; still others cause premature termination of the growing polypeptide chain, or change the translational reading frame of the gene, or alter the RNA splicing patterns. Pseudogenes, because they serve no function, are subject to mutation without selection and thus accumulate mutations at a far faster pace than the coding or regulatory regions of a functional gene. Eventually, nearly all pseudogene sequences mutate past a boundary beyond which it is no longer possible to identify the functional genes from which they derived. Thus, continuous mutation can turn a once functional sequence into an essentially random sequence of DNA.

Diversification of a Duplicated Gene Followed by Selection Can Produce a New Gene

Every so often, the accumulation of a set of random mutations in a spare copy of a gene leads to the emergence of a new functional gene that provides benefit and, consequently, selective advantage to the organism in which it resides. Because it provides a selective advantage, the new gene persists in the population. Although its function is usually related to that of the original gene, it almost always has a novel pattern of expression—in time, in space, or both—which most likely results from alterations in *cis*-regulatory sequences that occur along with codon changes. For example, a duplicated copy of the original human β-*globin* gene evolved into the *myoglobin* gene, whose protein product has a higher affinity for oxygen than the hemoglobin molecules composed in part of β-globin polypeptides. The *myoglobin* gene is active only in muscle cells, while the β-*globin* gene is expressed only in red blood cell precursors. Thus, duplication, divergence, and selection generated a new gene function from a previously functional gene.

Molecular Archaeology Based on an Understanding of Gene Diversification and Selection

Before the advent of molecular biology, researchers determined the genealogies of organisms by calculating the rates of evolution in phenotypic traits such as teeth and vertebrae. Then, in the 1950s, Linus Pauling and Emile Zuckerkandl analyzed hemoglobin and cytochrome C protein sequences from different species and noted that the rates of amino-acid substitution in each type of protein are similar for various mammalian lineages. On the basis of this observation, they postulated that for a given protein, the rate of evolution is constant across all lineages. They called this idea of a constant rate of change for each type of molecule a **molecular clock.** Its existence enables biologists to determine from molecular data the approximate times when species diverged and then use these dates to reconstruct genealogical phylogenetic trees. Although the molecular clock hypothesis is not perfect, in some instances where it has been tested, it has produced a reasonably good estimate of the time of divergence between two types of organisms. Today molecular biologists compare the nucleotide sequences of genes as well as the amino-acid sequences of proteins to determine phylogenetic divergences.

Scientists use molecular data in various ways to construct **phylogenetic trees** that illustrate the relatedness of homologous genes or proteins. A phylogenetic tree consists of **nodes** and **branches** (Fig. 24.9). The nodes represent the taxonomic units, which may be species, populations, individuals, or genes, while the branches define the relationship of these units. The branch length suggests the amount of time that has elapsed based on the number of molecular changes that have

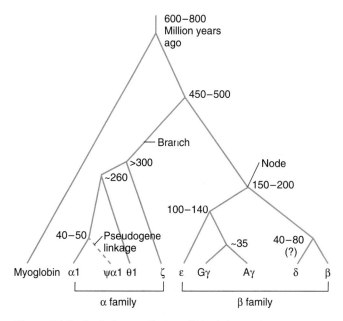

Figure 24.9   A phylogenetic tree. This phylogenetic tree demonstrates the evolutionary history of human hemoglobin genes. The broken line denotes a pseudogene linkage. Only one of the two α *hemoglobin* genes is shown in the figure because their date of divergence is uncertain.

occurred. Because different genes accumulate changes at different rates, different types of genes are best suited to the construction of different types of phylogenetic trees. For example, the genes for fibrinopeptides, which are not under strong selective pressure, evolve quickly. In contrast, the ribosomal genes, which are highly conserved, evolve slowly. As a result, the genes encoding fibrinopeptides are useful for looking at recent evolutionary events among very closely related species, while the ribosomal genes are useful for looking at ancient evolutionary events such as the relationships of phyla to each other. Indeed, phylogenetic trees based on ribosomal genes helped rewrite the most fundamental domain classifications of multicellular organisms (Fig. 24.3).

THE ORGANIZATION OF GENOMES

With this understanding of the basic mechanisms by which genomes evolve, we turn our attention to the results of genome evolution as seen in the organization of contemporary genomes. Our focus is on the various organizational features of the enormous mammalian genome, which evolved from the much simpler bacteria-like genomes through eons of duplication, diversification, and selection.

Molecular geneticists have accumulated data on genome organization from hybridization studies and genomic sequence analyses of chromosomal DNAs from humans and mice, from the completely sequenced genomes of *E. coli,*

At lowest level, exons duplicate or shuffle

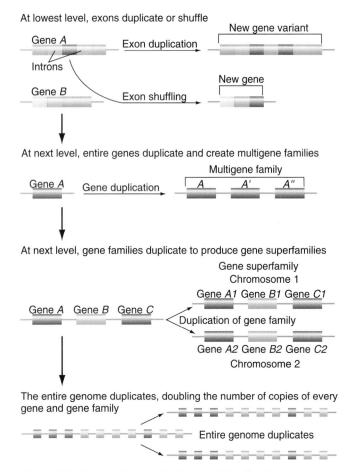

Figure 24.10   Four hierarchical levels of duplications increase genome size. Genome size increases through duplication of exons, or genes, of gene families, and finally of entire genomes.

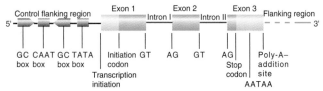

Figure 24.11   The basic structure of a gene. Schematic diagram of typical eukaryotic protein-coding gene. The large dark boxes represent the 5′ and 3′ untranslated mRNA sequences. The small open region represents regulatory sequences where gene-controlling transcription factors bind. The eukaryotic gene may range in size from a few hundred base pairs to more than 2 million base pairs.

yeast, and more than 25 other microbial organisms, and from the completed genomic sequence of the nematode *Caenorhabditis elegans* (which was the first animal to have its genome completely sequenced). The single-celled organisms exhibit genomes with densely packed genes and few if any introns. The mammalian genomes, with their far less densely packed genes, have several distinct features dominating their landscape: genes and families of genes; dispersed repetitive elements constituting more than a third of the genome; simple sequence repeats composed of single nucleotides or di-, tri-, tetra-, pentamers, and so forth; simple repetitive elements serving as a core for centromeres and telomeres; and unique nongene sequences. We now describe how these genomic features could have evolved.

Mechanisms Behind the Expansion from Genes to Multigene Families to Gene Superfamilies

Four levels of duplication (followed by diversification and selection) have fueled the evolution of complex genomes. At the lowest level, exons duplicate or shuffle to change the size or function of genes. At the next three increasingly complex lev-

els, entire genes duplicate to create multigene families; multigene families duplicate to produce gene superfamilies; and the entire genome duplicates to double the number of copies of every gene and gene family (Fig. 24.10). At each of these successively higher levels of organization, the duplication of larger and larger units leads to the hierarchical generation of greater and greater amounts of new information.

Basic Gene Structure: All Genes Have Several Components
We have seen that genes consist of many different components: exons containing coding regions and 5′ and 3′ untranslated regions; spliced out introns; and associated control regions (Fig. 24.11). Many of the control regions lie just 5′ to the transcribed region; but some, such as the locus control region of the β-*globin* gene family (described in Chapter 16), lie far outside the gene and appear to play a role in opening up the chromatin of the gene family locus so that gene expression can proceed at the appropriate time and level. Until all associated regulatory elements have been defined, the boundaries of a genetic locus remain uncertain.

The Duplication and Shuffling of Exons Can Elongate Genes and Enable Them to Encode Proteins with More Than One Functional Domain
Many proteins carry discrete compact domains, some of which perform a specific function, while others sustain molecular structure. In many genes, the discrete exons encode the structural and functional domains of a protein. Genes may elongate by the duplication of these exons to generate tandem exons that determine tandem functional domains such as those seen in antibody molecules (Fig. 24.12a). The functions of tandem domains may eventually diverge.

Entirely new genes may arise from **exon shuffling:** the exchange of exons among different genes. Exon shuffling produces mosaic proteins such as tissue plasminogen activator (TPA), a molecule with four domains of three distinct types: kringle (K), growth factor (G), and finger (F). The gene for TPA captured exons governing the synthesis of four domains from the genes for three other proteins: K from the gene for plasminogen, G from the gene for epidermal growth factor, and F from the gene for fibronecten (Fig. 24.12b). The mechanism by which exon shuffling occurs is unclear; it can, however, create proteins with two or more distinct functions.

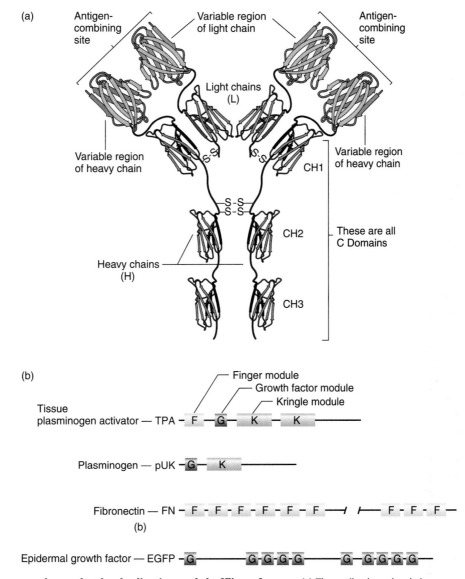

Figure 24.12 Genes can change by the duplication and shuffling of exons. (a) The antibody molecule is composed of two identical pairs of light (shaded) and heavy (white) chains. Each domain in a chain carries out separate functions. (b) The tissue plasminogen activator has evolved by exon shuffling from plasminogen, fibronectin and epidermal growth factor.

One or More Duplications of an Entire Gene Can Create Multigene Families with Homologous Members

A **multigene family** is a set of genes descended by duplication and diversification from one ancestral gene. The members of a multigene family may be either arrayed in tandem (that is, clustered on the same chromosome) or distributed on different chromosomes (Fig. 24.13a). Unequal crossing-over can expand and contract the number of members in a multigene family cluster (Figure 24.13b).

Genetic Exchange between Related DNA Elements by Gene Conversion Can Increase the Variation among Members of a Multigene Family

There are many places in the genome where there appears to have been a flow of genetic information from one DNA element to other related, but nonallelic elements located nearby or on different chromosomes. Such information flow between related DNA sequences occurs through an alternative outcome of the process responsible for unequal crossing-over. This

(a)

Tandem gene family: Members of the multigene family are clustered on the same chromosome

Dispersed gene family: Members of the multigene family are on different chromosomes

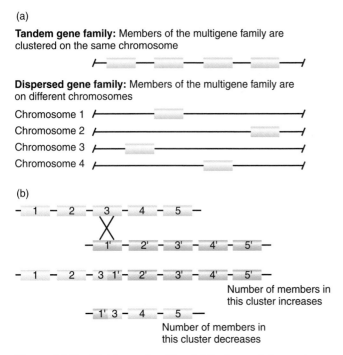

(b)

Figure 24.13 **Multigene families.** (a) Tandem and dispersed multigene families on segments of the indicated chromosomes. (b) A schematic illustration of the expansion and contraction of gene number in a multigene family by crossover.

alternative is known as **intergenic gene conversion** (Fig. 24.14a). Recall from Chapter 5 that a recombination intermediate can resolve itself in two ways: one leads to crossing-over and the other allows gene conversion to occur without crossing-over. The same alternative outcomes can occur with unequal recombination intermediates. The gene conversion outcome of unequal crossing-over allows the transfer of information from one gene to another. In special cases, the flow of information from such intergenic gene conversion has been so extreme that it has caused all members of a gene family to co-evolve with near identity. And in at least one case—that of the class I genes of the major histocompatibility complex (MHC)—selection has acted on information flow in only one direction, causing information transfer from a series of nonfunctional pseudogenes to a small subset of just 2 to 3 functional genes (Fig. 24.14b). In this unusual case, the pseudogene family members serve as a reservoir of genetic information that produces a dramatic increase in the amount of polymorphism (that is, the number of alleles) in the small number of functional gene members.

Concerted Evolution Can Make the Members of a Multigene Family Virtually Identical

A few multigene families have evolved under a special form of selective pressure that requires all family members to maintain essentially the same sequence. In these families, the high number of gene copies does not result in variations on a theme;

(a) Gene family

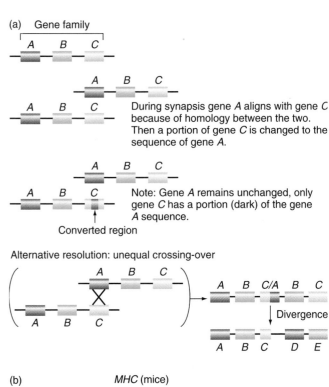

During synapsis gene *A* aligns with gene *C* because of homology between the two. Then a portion of gene *C* is changed to the sequence of gene *A*.

Note: Gene *A* remains unchanged, only gene *C* has a portion (dark) of the gene *A* sequence.

Converted region

Alternative resolution: unequal crossing-over

(b) *MHC* (mice)

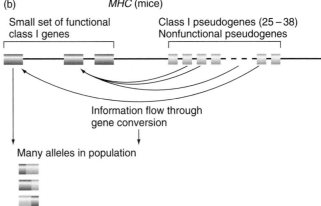

Figure 24.14 **Intergenic gene conversion.** (a) An illustration of the process of gene conversion. One gene is changed, the other is not. (b) Polymorphism can be increased in functional *MHC* class I genes in mice through gene conversion events from pseudo-class I genes.

rather it supplies a cell with a large amount of product within a short period of time. Among the gene families with identical elements is the one that produces RNA components of the cell's ribosomes, the one that produces tRNAs, and the one that produces the histones (which must rapidly generate enough protein to coat the new copy of the whole genome replicated during the S phase of every cell cycle).

Each of these gene families consists of one or more clusters of tandem repeats of identical elements. There is strong selective pressure to maintain the same sequence across all members of each family because all must produce the same product. Optimal functioning of the cell requires that the products of any individual gene be interchangeable in structure and function with the products of all other members of the same family. The problem is that the natural tendency of duplicated sequences is to drift apart over time. How does the genome counteract this natural tendency?

When researchers first compared ribosomal RNA and other gene families in this class, both between and within species, a remarkable picture emerged: Between species, there was clear evidence of genetic drift, but within species, all sequences were essentially equivalent. Thus it is not simply that some mechanism suppresses mutational changes in these gene families. Rather, there appears to be an on-going process of **concerted evolution,** which allows changes in single genetic elements to spread across a complete set of genes in a particular family. With this knowledge, we can narrow the previously posed question about counteracting the natural tendency toward drifting apart to How does concerted evolution occur?

Concerted evolution appears to occur through two different processes. The first is based on the expansion and contraction of gene family size through sequential rounds of unequal crossing-over between homologous sequences (Fig. 24.15a). Selection acts to maintain the absolute size of the gene family within a small range around an optimal mean. As the gene family becomes too large, the shorter of the unequal crossover products becomes selected; as the family becomes too small, the longer of the products become selected. This cyclic process causes a continuous oscillation around a size mean. However, each contraction results in the loss of divergent genes, whereas each expansion results in the indirect "replacement" of those lost genes with identical copies of other genes in the family. With unequal crossovers occurring at random positions throughout the cluster and with selection acting in favor of the least divergence among family members, this process can act to slow down dramatically the natural tendency of genetic drift between family members.

The second process responsible for concerted evolution is intergenic gene conversion between nonallelic family members. Although in each case, the direction of information transfer from one gene copy to the next is random, selection will act on this molecular process to ensure an increase in homogeneity among different gene family members (Fig. 24.15b). Information transfer (presumably by means of intergenic gene conversion) can also occur across gene clusters that belong to the same family but are distributed to different chromosomes.

Thus, with unequal crossing-over and intergenic gene conversion (which are actually two alternative outcomes of the same initial process) as well as selection for homogeneity, all members of a gene family can retain nearly the same DNA sequence. However, although concerted evolution maintains the close similarity of the members of multigene families within a species, divergences between the whole gene families present in different species nevertheless occur. This is because different sequences serve as the prototype for gene family homogenization in different species.

(a)

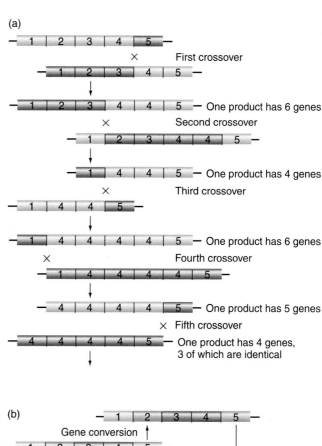

(b)

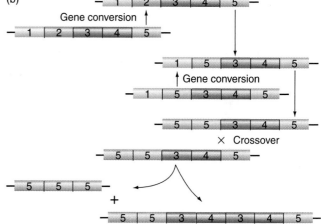

Figure 24.15   Concerted evolution can lead to gene homogeneity. Repeat cycles of unequal crossover events (a) or gene conversion (b) cause the duplicated genes on each chromosome to become progressively more homogenized.

The Evolution of Gene Superfamilies

Molecular geneticists use the phrase "gene superfamily" to describe a large set of related genes that is divisible into smaller sets, or families, with the genes in each family being more closely related to each other than to other members of the larger superfamily. The multigene (or single-gene) families that compose a gene superfamily reside at different chromosomal locations. A prototypical small-size gene superfamily is the very well-studied globin genes illustrated in Fig. 24.16.

The superfamily has three branches: the multigene family of β-like genes, the multigene family of α-like genes, and the single myoglobin gene. The duplications and divergences that produced the three superfamily branches occurred early in the evolution of vertebrates; as a result, the three branches of the superfamily are found in all mammals. All functional members of this superfamily play a role in oxygen transport (see Chapter 8: introduction and comprehensive example). As we have seen, the products of the α and β globin are active in red blood cells, while the product of the myoglobin gene transports oxygen in muscle tissue.

The primordial globin gene gave rise to the myoglobin and α/β precursor genes by gene duplication and transposition. The primordial α- and β-globin genes probably arose by a complete genome duplication (tetraploidization).

The β-like branch of this gene superfamily arose by duplication via multiple unequal crossovers. The duplication products diverged into five functional genes and two β-like pseudogenes. In the mouse, all the β-like genes are present in a single cluster on chromosome 7 (see Fig. 24.16). The β-like chains encode similar polypeptides, each of which has been selected for optimal function at a specific stage of mouse development: one functions during early embryogenesis, one during a later stage of embryogenesis, and two function in the adult.

The α-like branch also evolved by unequal crossovers and divergences that generated a cluster of three genes on chromosome 11: one functions during embryogenesis, and two function in the adult (see Fig. 24.16). The two adult α-*globin* genes are virtually identical at the level of DNA sequence, which suggests that the duplication producing them occurred very recently (on the evolutionary time scale). In addition to the primary α-like cluster, there are two nonfunctional α-like genes—pseudogenes—that have dispersed via transposition to locations on chromosomes 15 and 17. Pseudogenes existing in isolation from their parental families are called *orphons*. Interestingly, the α-globin orphon on chromosome 15 (named *Hba-ps3*) has no introns and thus appears to have arisen through a retrotransposition event involving mRNA copied back to DNA; whereas the a-globin orphon on chromosome 17 (*Hba-ps4*) contains introns and may have arisen by a direct DNA-mediated transposition.

The single myoglobin gene on chromosome 15 has no close relatives either nearby or far away. The *globin* gene superfamily provides a view of the many different mechanisms that the genome can employ to evolve structural and functional complexity.

The mouse *Hox* gene superfamily (discussed in Chapter 22) provides an alternative prototype for the evolution of a gene superfamily (Fig. 24.17). At first, there was one *Hox* gene that had evolved to produce a protein product that could bind to DNA enhancer regions and thereby regulate the expression of other genes. Unequal crossover events predating the divergence of insects and vertebrates some 600 million years ago produced a cluster of five related genes encoding DNA-binding proteins that regulated the expression of other genes encoding

spatial information (that is, instructions for the spatial positioning of tissues and organs) in the developing embryo. The original *Hox* gene family then duplicated en masse and dispersed to four locations—on chromosomes in a common ancestor to all vertebrates. Because of the order of duplication events leading to the superfamily—unequal crossing-over to expand the original cluster size, followed by transposition en masse of the expanded cluster to create the superfamily—an evolutionary tree would show that a single gene family within the superfamily has actually splayed out physically across the four gene clusters, as shown in Fig. 24.17. After the en masse duplication that generated the gene superfamily, smaller duplications by unequal crossing-over added genes to some of the dispersed clusters and subtracted genes from others, thereby generating differences in gene number and type within a basic framework of homology among the different clusters. Each of the four *Hox* clusters in the mouse superfamily currently contains from 9 to 12 homologous genes.

Repetitive "Nonfunctional" DNA Families Constitute More Than One-Third of the Genome

Many repetitive nonfunctional DNA families consist of retroviral elements. Retroviruses, we have seen, are RNA-containing viruses that can convert their RNA genome into circular DNA molecules through the viral-associated RNA-dependent DNA polymerase known as reverse transcriptase, which becomes activated on cell infection (see Figs. A and B in the Genetics and Society box in Chapter 7). The resulting DNA can integrate itself at random into the host genome, where it becomes a provirus that retains the genetic information of the retroviral genome. Under certain conditions, the provirus can become activated to produce new viral RNA genomes and associated proteins, including reverse transcriptase, which can come together to form new virus particles that are ultimately released from the cell surface by exocytosis. By contrast, many stably integrated retroviral elements appear to be inactive.

Once integrated into a host chromosome, the provirus replicates with every round of host DNA replication, irrespective of whether the provirus itself is expressed or silent. Moreover, proviruses that integrate into the germ line, through the sperm or egg genome, segregate along with their host chromosome into the progeny of the host animal and into subsequent generations of animals as well. The genomes of all species of mammals contain inactive integrated proviral elements.

It is of evolutionary interest to ask where retroviruses originated. Since retroviruses cannot propagate in the absence of cells, but cells can propagate in the absence of retroviruses, it seems likely that retroviruses derived from nucleotide sequences that were originally present in the cell genome. The first retrovirus might have been able to free itself from the confines of the cell nucleus through an association with a small number of proteins that allowed it to coat and thus protect it-

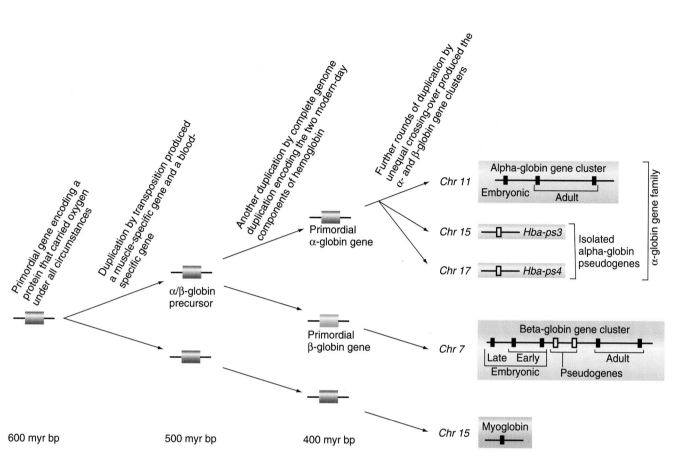

Figure 24.16   Evolution of the mouse globin super family. Repeated gene duplication by various mechanisms has given rise to the globin supergene family with two multigene families (α and β) and one single gene (myoglobin). The β family has both tandemly arrayed and dispersed gene members.

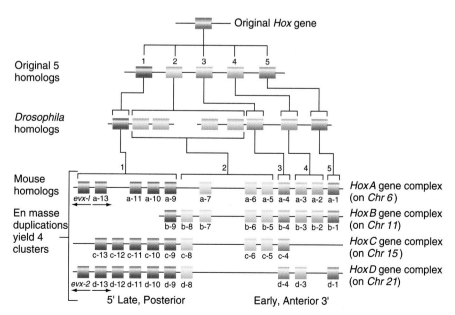

Figure 24.17   Evolution of the *Hox* gene superfamily of mouse and *Drosophila*. This super gene family arose by a series of gene duplications. Four multigene families are present in mouse and one in *Drosophila*.

self from the harsh extracellular environment. Of course the protein most critical to the propagation of the retrovirus would have been the reverse transcriptase that allows the retrovirus to reproduce. But where did this enzyme come from? Reverse transcriptase catalyzes the production of single-stranded complementary DNA molecules from an RNA template, but this enzymatic activity is not required for any normal cellular process known in mammals. So, how could an activity without any apparent benefit to the host organism arise *de novo* in a normal cell? One possible answer is that reverse transcriptase did not evolve because it benefitted the organism itself; rather it evolved because it benefitted selfish DNA elements within the genome that utilize it to propagate themselves within the confines of the genome. We now describe one such selfish element.

The LINE Family of DNA Elements Are Retroposons That Appear to Derive from a Selfish DNA Sequence Encoding Reverse Transcriptase

The LINE family of DNA elements is very old. Homologous families of repetitive elements exist in a wide variety of organisms, including protists and plants. Thus, LINE-related elements, or other DNA elements of a similar nature, are likely to have been the source material that gave rise to retroviruses.

Dispersion to new positions in the germ line genome presumably begins with the transcription of LINE elements in spermatogenic or oogenic cells. The reverse-transcriptase-encoding region on the transcript is translated into an enzyme that preferentially associates with and utilizes the transcript it came from as a template to produce LINE cDNA sequences, but the reverse transcriptase often stops before it has made a full-length DNA copy of the RNA transcript. The resulting incomplete cDNA molecules can nevertheless form a second strand to produce truncated double-stranded LINE elements that integrate into the genome but remain forever dormant (Fig. 24.18a).

The SINE Family Consists of a Second Type of Repetitive, Nonfunctional DNA Element That Did Not Originally Encode Reverse Transcriptase

The Alu family in the human genome is an example of a highly repetitive, widely dispersed SINE family. Over 500,000 Alu elements are dispersed throughout the human genome. At 300 bp in length, the Alu element is far too short to encode a reverse transcriptase. Nonetheless, like LINE elements, these and other SINE elements are able to disperse themselves through the genome by means of an RNA intermediate that undergoes reverse transcription. Clearly, SINES depend on the availability of reverse transcriptase produced elsewhere, perhaps from LINE transcripts or the proviral elements of retroviruses.

All SINE elements in the human genome, as well as in other mammalian genomes, appear to have evolved out of small cellular RNA species, most often tRNAs, but also the 7S

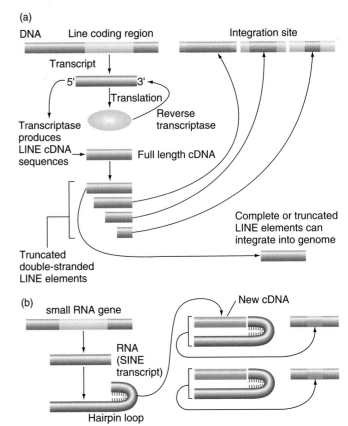

Figure 24.18 The creation of LINE and SINE gene families.
(a) A complete LINE sequence can be copied into RNA. It encodes a reverse transcriptase that can make cDNA copies from the RNA. These copies may be complete or truncated. Both may then integrate into other sites on chromosomes. (b) SINE elements can be transcribed and, because they form 3' hairpin loops, can be copied into cDNAs by reverse transcriptase. These cDNA copies then integrate into the genome.

cytoplasmic RNA that is a component of the signal recognition particle (SRP) essential for protein translocation across the endoplasmic reticulum. The defining event in the evolution of a functional cellular RNA into an altered-function, self-replicating SINE element is the accumulation of nucleotide changes in the 3' region that lead to self-complementarity with the propensity to form hairpin loops. Reverse transcriptase can recognize the open end of the hairpin loop as a primer for strand elongation. Since it is likely that the hairpin loops form only rarely among normal cellular RNAs, a cell will preferentially use its SINE transcripts as templates for the production of cDNA molecules that are somehow able to integrate into the genome at random sites (Fig. 24.18b).

Comparing LINE and SINE Families

Like LINE families, the Alu SINE family appears to be evolving by episodic amplification followed by sequence degradation. Unlike LINE families, however, SINE families present in the genomes of different organisms appear, for the most part, to have independent origins.

Some Selfish Elements May Confer a Selective Advantage

While SINE and LINE elements may have amplified themselves for selfish purposes, they have had a profound impact on whole-genome evolution. In particular, homologous SINE or LINE elements located near each other can, and will, catalyze unequal but homologous crossovers that result in the duplication of single-copy genes located between the homologous elements; such duplications initiate the formation of multigene families. In addition, some selfish elements appear to have evolved a regulatory role—acting as enhancers or promoters—through chance insertions next to open reading frames. Thus, selfish elements may confer a selective advantage by facilitating duplication through unequal crossing-over or by becoming regulatory elements.

Genomic Stutters—the Simple Sequence Repeats Known as Microsatellites, Minisatellites, and Macrosatellites— Dot the Mammalian Genome

Through large-scale sequencing and hybridization analyses of mammalian genomes, researchers have found tandem repeats of DNA sequences with no apparent function scattered throughout the genome. The size of the repeating units ranges from two nucleotides (C A C A C A C A . . .) to 20 kb, and the number of tandem repeats varies from two to several hundred.

The mechanism by which these tandem repeats originate may be different for loci having very short repeat units as compared with loci having longer repeat units. Tandem repeats of short di- or trinucleotides can originate through random changes in nonfunctional sequences. By contrast, the initial duplication of larger repeat units is likely to be a consequence of unequal crossing-over. Once two or more copies of a repeat unit exist in tandem, an increase in the number of repeat units in subsequent generations can occur through unequal crossovers or errors in replication (see Fig. 24.8). It is not yet clear whether random mechanisms alone can account for the rich variety of tandem repeat loci in mammalian genomes or whether other selective forces are at play. In any case, tandem repeat loci continue to be highly susceptible to unequal crossing-over and, as a result, tend to be highly polymorphic in overall size.

Repeat Sequences in Centromeres and Telomeres

Highly repetitive, noncoding sequences shorter than 200 bp are found in and around centromeres. For example, in the human genome, *alphoid*, a noncoding sequence 171 bp in length, is present in tandem arrays extending over a megabase in the centromeric region of each chromosome (review Fig. 11.10). In addition, several similar size repetitive sequences unrelated to alphoid are found in some centromeres. These regions are sites of interaction with the spindle fibers that segregate chromosomes during meiosis and mitosis. Selection may have acted to retain the thousands of copies of centromeric repeat elements in each centromeric region because they increase the efficiency and/or accuracy of chromosome segregation.

A second type of repeated genomic element with a special location is the hexamer T T A G G G found in the telomeres of all human chromosomes. The six-base unit is repeated in tandem arrays 5–10 kb in length at the ends of human and all other mammalian chromosomes (review Fig. 11.12). Selection may have conserved this repeat element because it plays an essential role in maintaining chromosomal length.

Unique Nongene Sequences

The mutation and information degradation of extra copies of duplicated genes, which produces pseudogenes, can ultimately result in unique nongene sequences. The mutation and divergence of genomic repeat elements can also lead to unique nongene sequences.

THE IMMUNOGLOBULIN GENE SUPERFAMILY: A COMPREHENSIVE EXAMPLE OF MOLECULAR EVOLUTION

More than 200 different types of genes belong to the immunoglobulin gene superfamily (Table 24.1). Most encode cell-membrane proteins that recognize either soluble molecules or molecules on the surface of other cells. All members of the immunoglobulin gene superfamily share a common feature: the **immunoglobulin homology unit,** so-called because it was first identified in the genes for immunoglobulin (antibody) molecules (Fig. 24.19). The homology unit encodes about 100 amino acids, which fold into a characteristic three-dimensional structure termed the *immunoglobulin fold* (Fig. 24.19). The unique design

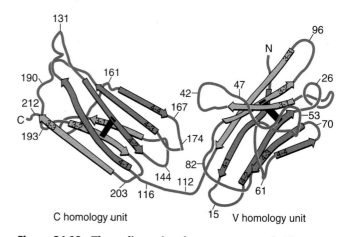

Figure 24.19 Three-dimensional structures encoded by immunoglobulin homology units. Tertiary organization of V and C homology units. This illustrates the tertiary structure of an immunoglobulin light chain. The β strands and their orientation are shown as *flat arrows.* Opposite faces of the sandwich structure are indicated by either blank or hatched β strands. The disulfide bond is indicated with the *solid bar.* The V and C homology units are remarkably similar in structure. The variable loops indicated by *96, 26,* and *53* fold to form the walls of the antigen-binding site.

TABLE 24.1 **Examples of Immunoglobulin Gene Superfamily Members**

Protein	Molecular wt. (kDa)	Ig domains C1	C2	V	Human chromosome	Function
Ig heavy chain	55	3–4	—	1	14q32.33	Recognition of Ag
Ig κ light chain	24	1	—	1	2p12	Recognition of Ag
Ig λ light chain	24	1	—	1	32q11.12	Recognition of Ag
B29 (Ig-β/Ig-γ)	39	(1 Ig domain)			(Mouse)	Component of surface Ig complex
MB-1 (IgM-α)	34	(1 Ig domain)			(Mouse)	Component of surface Ig complex
TCR α chain	45–60	1	—	1	14q11.12	Recognition of peptide-MHC
TCR β chain	40–50	1	—	1	7q32	Recognition of peptide-MHC
TCR γ chain	45–60	1	—	1	7p15	Recognition of peptide-MHC
TCR δ chain	40–60	1	—	1	14q11.2	Recognition of peptide-MHC
CD3 ε chain	25–28	—	1	—	11q23	TCR signal transduction
CD3 δ chain	20	—	1	—	11q23	TCR signal transduction
CD3 ε chain	20	—	1	—	11q23	TCR signal transduction
MHC Class I or chain	44	1	—	—	6p21.3	Peptide binding
β2 microglobulin	12	1	—	—	15q21-q22	Stabilize MHC Class I α
MHC Class II or chain	32–34	1	—	—	6p21.3	Peptide binding
MHC Class II β chain	29–32	1	—	—	6p21.3	Peptide binding
Qa/TL heavy chain	48	1	—	—	(Mouse)	Restriction element for γ/δ TCR?
CD1	43–49	1	—	—	1q22-q23	Restriction element for γ/δ TCR?
CD2	50	—	2	—	1p13	Ligand for LFA-S (CD58)
LFA-3 (CDUB)	40–65	—	2	—	1	Ligand for CD2
ICAM-1 (CD64)	80–114	—	5	—	19	Ligand for LFA-1
CD4	55	—	1	2	12pter-q12	Binds MHC Class II
CD8 α chain	34	—	—	1	2p12	Binds MHC Class I
CD8 β chain	34	—	—	1	2p12	Binds MHC Class I
CD7	40	—	—	1	17q25	T cell activation
CD22	130/140	—	5	—	Unassigned	B cell adhesion
CD28	44	—	—	1	2q33-q34	T cell activation
CD33	67	(2 Ig domains)			19q13	Myeloid cell adhesion?
CD48 (blast-2)	41–45	—	1	1	1q21-q23	B cell adhesion
Thy-1	17.5–18.7	—	—	1	11q23	T cell activation and adhesion
MRC OX2	41–47	—	1	1	3	Unknown
PolyIgR	100	—	1	4	1q31-q42	Polymeric Ig receptor
FcγRI (CD6)	70	—	3	—	1q23-q24	High-affinity IgG receptor
FcγRII (CD32)	40	—	2	—	1q23-q24	Low-affinity IgG receptor
FcγRIII (CD16)	50–70	—	2	—	1q23-q24	Medium-affinity IgG receptor
FcγRI or chain	25	—	2	—	1q23-q24	High-affinity IgE receptor
FcαR	60	—	2	—	Unassigned	IgA receptor
IL-1 receptor (type 1)	82	(8 Ig domains)			2q12	Cytokine receptor
IL-6 receptor	80	—	1	—	Unassigned	Cytokine receptor
M-CSF receptor (c-fma)	165	—	4	1	5q33.2-q33.3	Colony stimulating factor (CSF-1) receptor
G-CSF receptor	150	—	1	—	ND	Colony stimulating factor (G-CSF) receptor
SLF-HGF receptor (c-kit)	145	—	4	1	4q31-3	Hematopoietic growth factor receptor
FDGF receptor	180	—	4	1	5q31-q32	Growth factor receptor
NCAM (CD58)	97–220	—	5	—	11q23	Adhesion of neural cells
MAG	67/72	—	5	—	(Rat)	Axon-glial interactions of myelination
P₀ myelin protein	28–30	—	—	1	(Rat)	Myelin compaction
Neuroglian (NGL)	167	—	6	—	(Drosophila)	Neural cell adhesion
Falciculum II	96	—	5	—	(Grasshopper)	Axonal fasciculation
Almagam	38–50	—	2	1	(Drosophila)	Neural cell adhesion
Contactin	180	—	6	—	(Aves)	Neurite fasciculation
F11	180	—	6	—	(Aves)	Neurite fasciculation and elongation
L1	200	—	6	—	(Mouse)	Neurite fasciculation and elongation, neuronel migration
Ng-CAM (G4)	200	—	6	—	(Aves)	Homolog of L1
NILE	200	—	6	—	(Rat)	Homolog of L1
F3	155	—	6	—	(Mouse)	Unknown
TAG-1	135	—	6	—	(Rat)	Growth and guidance of axons
Sk. muscle C-protein	140	—	6	—	(Aves)	Binds myosin H chain and F-actin
Carcinoembryonic Ag	175–200	—	2–6	1	19q13.1-q13.2	Family of molecules; adhesion?
Pregnancy-specific Ag	54–72	—	3	1	19q13.2-q13.3	Family of molecules; adhesion?
α1B glycoprotein	83	—	5	—	Unassigned	Unknown
B, Memb. link, protein	44.5–48.8	—	—	1	Unassigned	Locks proteoglycan to haluronate chain
Twitchin	600	—	5	—	(Nematoda)	Regulation of myosin activity
CaVPT	30	—	2	—	(Amphioxus)	Unknown
Basigin	43–66	(1 Ig domain)			(Mouse)	Unknown
LAR	215–250	—	3	—	Unassigned	Adhesion?

of the immunoglobulin fold enables it to (1) interact with other immunoglobulin folds to form dimeric molecules, (2) display an enormous diversity of recognition sites while still maintaining a highly conserved three-dimensional framework, and (3) resist proteolysis in the blood. The folds encoded by different homology units may vary by as much as 90% of their sequences while still preserving their folded framework because of a small but critical number of highly conserved structural features. The evolutionary success of the immunoglobulin homology unit is reflected in its deployment in more than 200 types of genes.

Evolution of the Immunoglobulin Homology Unit

The immunoglobulin homology unit probably emerged early in the evolution of metazoans. Scientists believe this to be true for three reasons. First, both vertebrates and invertebrates employ cell-surface recognition molecules containing the immunoglobulin homology unit, but such molecules are not found in prokaryotes or single-cell eukaryotes. Since the common ancestor of present-day vertebrates and invertebrates was present around the time when metazoan organisms emerged, it is likely that the immunoglobulin gene superfamily also came into existence at that time. Second, the need for molecular recognition at the cell surface increased exponentially with the evolution of multicellular organisms, and it appears that the immunoglobulin homology unit evolved coincidentally, allowing metazoans to fill their need for diverse cell-surface recognition units. Third, the early primordial homology unit quickly duplicated to make three distinct types of homology units: variable (V), constant (C), and homology (H) (Fig. 24.20a). The V, C, and H units are equally related to one another and are deployed in the diverse cell surface receptors needed by all metazoans (Fig. 24.20b). The inescapable conclusion from these observations is that the highly diverse immunoglobulin gene superfamily evolved relatively rapidly, and its rapid creation may have been a driving force in generating the cell-surface recognition required by emerging metazoan organisms.

Evolution of the Immunoglobulin Gene Superfamily

Molecular evolution, we have seen, can occur very rapidly because of the hierarchical nature of changes in structural information: Nucleotide substitutions can change exons; the duplication, divergence, and shuffling of exons can create more complex genes; the duplication of genes can create multigene families; and the duplication of multigene families (and single-gene members) can create gene superfamilies.

The immunoglobulin gene superfamily evolved by exon shuffling to create mosaic proteins (Fig. 24.20c), by exon duplication to create very long genes, by gene duplication to create multigene families, and by the duplication of whole gene families to create the gene superfamily depicted in Fig. 24.20 and Table 24.1.

The First Members of the Superfamily to Evolve Encoded Specific Antigen Receptors

These receptors included antibodies, T-cell receptors, and MHC molecules. Antibodies recognize the three-dimensional shapes of their cognate antigens. T-cell receptors recognize peptide fragments of antigen presented by MHC molecules; the T-cell receptors also react with coreceptors, including the CD4 and CD8 molecules (which are also members of the immunoglobulin gene superfamily), to orchestrate immune responses. Elaborate recognition mechanisms operating at the cell surface are the most common functions of the immunoglobulin superfamily members. Other members of the immunoglobulin gene superfamily evolved to recognize immunoglobulin molecules at the cell surface and/or to transport immunoglobulins across membrane compartments.

Surprisingly, Many Members of the Immunoglobulin Superfamily Carry Out Functions Unrelated to Vertebrate Immune Recognition

Some components, such as link protein and HSP-PBPG7, help recognize the cellular matrix. Other components make up the carcinoembryonic antigen (CEA) gene family, which has 10 or more members all involved in specific cell adhesions; the recognition of specific cells by other cells is critical for the development and structural integrity of tissues and organs. Several cell-adhesion molecules guide the growth of neurons in vertebrates (CAM LI and contactin) and invertebrates (falciculum I and almagam). Two other members of the superfamily (MAG and P_0) appear to play a role in the wrapping and compaction of the myelin sheath. Interestingly, several other superfamily molecules, including Thy I and MRC OX2, are part of the vertebrate immune and nervous systems, raising the possibility that these molecules play a key role in communication between these two complex systems. Other members can encode growth factors or their receptors (PDGF receptor). The existence of so many immunoglobulin superfamily members that do not play a role in the immune response demonstrates how powerful the forces of duplication, divergence, and selection are at making new functional units from older units with a different function.

Immune-Cell Receptors Provide the Backbone of the Immune Response

We saw at the beginning of this chapter that the vertebrate immune response develops the capacity to recognize virtually any foreign macromolecular pattern (antigenic determinant) through the maturation of its two major functional branches. The B cell, or humoral system, manufactures antibodies that bind with and clear unprocessed antigen in viral particles and microbial macromolecules. The T cell, or cellular, immune system uses T-cell receptors and two classes of major histocompatibility complex (MHC) receptors to recognize the processed peptide products of protein antigens.

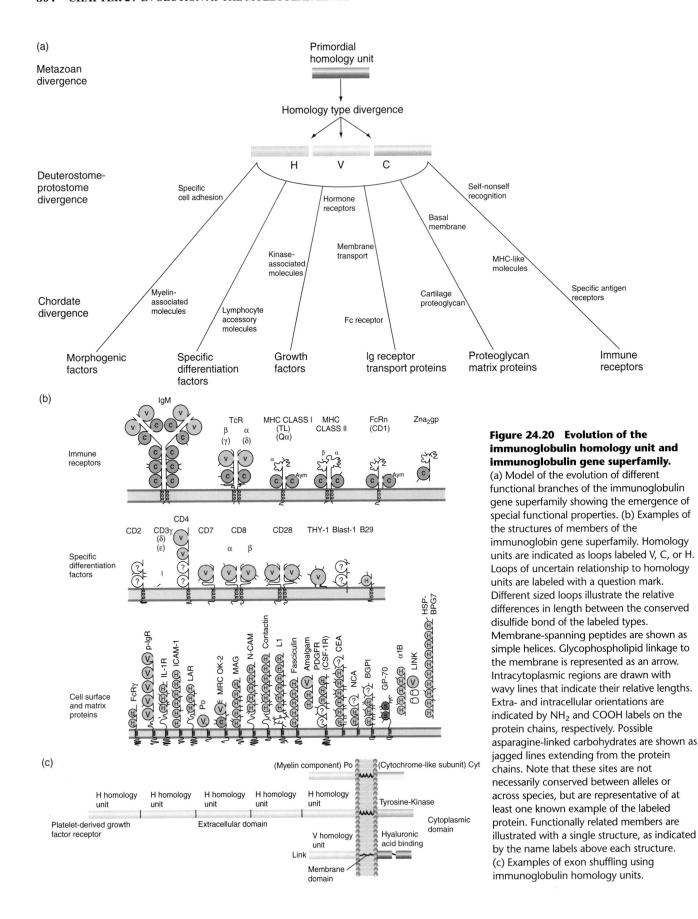

Figure 24.20 Evolution of the immunoglobulin homology unit and immunoglobulin gene superfamily.
(a) Model of the evolution of different functional branches of the immunoglobulin gene superfamily showing the emergence of special functional properties. (b) Examples of the structures of members of the immunoglobin gene superfamily. Homology units are indicated as loops labeled V, C, or H. Loops of uncertain relationship to homology units are labeled with a question mark. Different sized loops illustrate the relative differences in length between the conserved disulfide bond of the labeled types. Membrane-spanning peptides are shown as simple helices. Glycophospholipid linkage to the membrane is represented as an arrow. Intracytoplasmic regions are drawn with wavy lines that indicate their relative lengths. Extra- and intracellular orientations are indicated by NH_2 and COOH labels on the protein chains, respectively. Possible asparagine-linked carbohydrates are shown as jagged lines extending from the protein chains. Note that these sites are not necessarily conserved between alleles or across species, but are representative of at least one known example of the labeled protein. Functionally related members are illustrated with a single structure, as indicated by the name labels above each structure. (c) Examples of exon shuffling using immunoglobulin homology units.

There are two general types of T cells: T cytotoxic cells, which destroy foreign cells as well as intracellular infections and neoplastically transformed cancer cells, and T helper cells, which facilitate the differentiation of B-cell and T-effector-cell responses.

A Close Look at Immune-Cell Receptors

The immune receptors on both B cells and T cells are heterodimers. The antibodies on B cells consist of light and heavy chains; T-cell receptors are composed of α and β chains; class II MHC molecules carry α and β chains, while class I MHC molecules consist of a heavy chain plus a β microglobulin chain (which is an immunoglobulin superfamily member containing only a single immunoglobulin fold) (Fig. 24.21).

The T-cell and B-cell systems employ very different strategies for the recognition of antigen (see Figs. 24.12a and 24.21). The antibodies of B cells view the native three-dimensional surface of their corresponding antigen by a direct molecular complementarity. T-cell receptors recognize the peptide fragment of their corresponding antigens as presented in a molecular groove on the surface of the MHC molecules. Class I MHC molecules on antigen presenting cells generally present peptide fragments to T cytotoxic cells, whereas class II MHC molecules generally present peptides to T helper cells. Antibodies and T cell receptors both employ two types of homology units: variable (V) and constant (C).

The primary structure of the antibody molecule provided the first evidence of the homologous repeat units that are the functional building blocks of the immunoglobulin gene superfamily. In three dimensions, the homology units form a sandwich of β pleated sheets connected by loops that can be highly variable. It is in the variation of these loops that much of the immune cells' ability to recognize specific antigen resides (review Fig. 24.19).

Immune-Cell Receptors Define the Basic Features of an Immune Response

An immune response has three characteristics: it is highly **specific,** it is **adaptive,** and it has a **memory.** Immunologic *specificity* arises because each type of T cell or B cell expresses a single type of receptor. The binding of receptor to complementary, or cognate, antigen drives the clonal expansion of B cells and T cells carrying suitable receptors, and thus provides the basis for the specificity of immune responses (see Fig. 24.2).

The fact that each lymphocyte expresses only a single receptor is interesting in two regards. It signifies that in a large multigene family, only one gene undergoes expression; and of the two copies of this gene—on the paternal and maternal chromosomes—only one copy is expressed. (This second phenomenon—the expression of only one allele of the gene—is known as *allelic exclusion.*) The DNA rearrangement process for creating the highly variable and thus highly specific antigen-recognition regions is probably responsible for the expression of just one immune receptor gene and just one allele of this gene (review the Fast Forward box in Chapter 12).

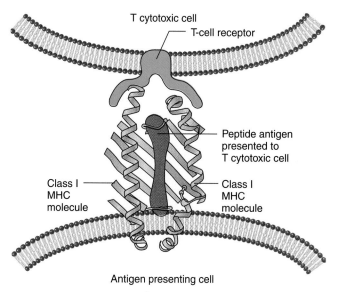

Figure 24.21   Immune cell receptors. Schematic model of the T-cell receptor recognizing a peptide antigen presented by an MHC molecule. The T cell recognizes both the peptide fragment and portions of the MHC molecule.

The immune response is *adaptive* because it has a large repertoire of different lymphocytes (approximately 10^{12}) that circulate in the blood and lymph. With this constant circulation, foreign organisms receive continuous exposure to many different T-cell and B-cell receptors. This enables specific lymphocyte receptors complementary to the pathogen's antigenic determinants to recognize and interact with them. The large repertoire of lymphocytes collectively has sufficient diversity to respond to all antigens, even the many natural and synthetic ones that did not exist when the immune system emerged. Moreover, in the B cell, or antibody response, there is a process called *affinity maturation* in which as the immune response progresses, the fit of the antibody becomes better and better. This affinity maturation is the consequence of *somatic hypermutation,* which we discuss later.

Immunologic *memory* arises when individual lymphocytes are stimulated by antigen to undergo clonal expansion and two types of cells result: effector cells that lead to the destruction of the antigen and memory cells that can remain in the organism for many decades, after which time the same antigen that caused the production of the memory cells can restimulate them to proliferate and produce more effector cells.

Gene Rearrangements and Other Recombinatorial Mechanisms Generate the Immense Diversity of Immune Receptors

Immunoglobulin (or antibody) molecules are encoded by three families of genes: one heavy-chain gene and two types of light-chain genes, kappa (κ) and lambda (λ) (Fig. 24.22). The genes for the variable regions of each heavy chain contain three

Human
Heavy chain

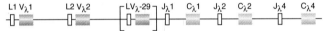

Figure 24.22 There are three families of antibody genes. The three families of antibody chains include two light—λ and κ—and one heavy. The letters L, V, D, J, and C indicate leader, variable, diversity, joining, and constant regions, respectively.

distinct types of segments: variable (V), diversity (D), and joining (J). The genes for the light-chain variable regions encode V and J gene segments. During the differentiation of lymphocytes that occurs as an individual's immune system develops, the gene segments that will make up the gene to be expressed in each lymphocyte are brought together by DNA rearrangements (Fig. 24.23). Specific signals at the boundaries of the gene segments to be joined and site-specific recombination enzymes mediate these rearrangements. The heavy gene segments undergo rearrangement early in development—first joining a specific D segment to a specific J segment, then joining a particular V to the DJ combination. After the heavy gene rearrangements have occurred, one V to J rearrangement occurs in either the κ or the λ light-chain gene. Allelic exclusion ensures that each B cell expresses only one type of light and one type of heavy chain, which prevents the formation of unproductive antibodies. Each light-chain and each heavy-chain rearrangement is ultimately selected to produce one useful antibody; if a cell made more than one light chain and more than one heavy chain, unproductive dimers would form. The assembled V gene for both heavy and light chains is joined with an assembled constant gene during RNA splicing to generate a complete mRNA for the immunoglobulin chain.

Five distinct mechanisms of diversification generate the enormous repertoire of specific antibody receptors:

1. *Germ line diversity.* A multiplicity of V, D, and J gene segments exist for each of the gene families. These are the raw materials for the rearrangements that occur during development.

2. *Combinatorial joining.* With heavy chains, any D can join to any J and any V and join to any DJ. Thus, there are a large number of possible V, D, and J combinations.

3. *Junctional diversity.* Two mechanisms operate to create diversity at the junctions of joined gene segments. Exonucleases can trim back one or both of the gene segments to be joined; and the enzyme terminal deoxynucleotide transferase can add random nucleotides not encoded in the chromosome to the junctional region. These two mechanisms together generate enormous diversity in the joined regions, which in the resulting folded protein happen to represent a major portion of the antigen-binding site.

4. *Somatic hypermutation.* As the immune response proceeds with the proliferation of specific cells into large clones of millions of identical cells, the joined V genes undergo random *somatic hypermutation*, that is, somatic mutation at a rate that is orders of magnitude higher than the typical mutation rate. A single V gene may have 10–15 changes. These somatic hypermutations are often rather extensive. Hypermutation produces many different forms of the gene; those mutations encoding V regions with a better fit for the antigen will be selectively amplified through tighter binding to create "affinity maturation" during the immune response.

5. *Joining light and heavy chains.* Any light chain can associate with any heavy chain, so once again, there is a combinatorial amplification of diversity through the manipulation of information at the protein level. For example, if two cells happen to produce the same light chain, it is highly unlikely they will also produce the same heavy chain; and the joining of a different heavy chain with the same light chain produces two distinct immunoglobulin molecules with distinct antigen-binding sites.

Most of these mechanisms of diversification also apply to T-cell receptors. The two types of T-cell receptors, α/β and γ/δ, are encoded by three distinct gene families: α/δ, β, and γ. The T-cell receptors employ all the diversification mechanisms discussed for the immunoglobulin receptors, except for somatic hypermutation. The MHC receptors do not employ any special mechanisms for generating diversity; they have achieved their great range of variation via evolutionary mutation and divergence.

Evolution of the Vertebrate Immune Response: A Possible Scenario

The evolution of the vertebrate immune response demonstrates the enormous flexibility and extendibility of the immunoglobulin gene superfamily. The first member of the superfamily may have encoded nothing more than a cellular glue to adjoin various cells of the earliest metazoan organisms. With gene duplication and divergence, various members of the superfamily became tissue-specific cell adhesives. From the DNA evidence, we know that the three types of homology units emerged in this early period of superfamily evolution, and their appearance initiated the

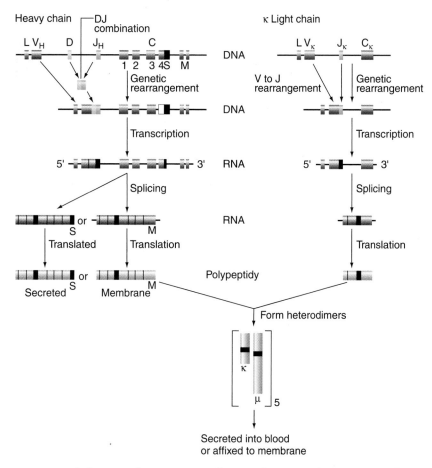

Figure 24.23   DNA rearrangements bring together segments of a gene for expression. Expression of rearranged heavy- and light-chain genes in Bμ cells. The genetic rearrangements occur during development of the Bμ. For simplicity only, a single gene segment of each kind is represented. The letters L, V, D, J, and C are leader, variable, diversity, joining, and constant regions, respectively, and S and M indicate the secreted and membrane tails of the μ chain, which are generated by alternative patterns of RNA splicing. Thus the IgM molecule may be attached to the membrane (M) serving as a receptor to trigger B-cell differentiation or secreted to serve as a blood factor molecule of B-cell immunity.

rapid acquisition of a variety of different functions (review Figure 24.20).

The Key Event in the Evolution of the T-Cell Receptor Gene Family Was the Acquisition of the Ability to Rearrange Gene Segments

This probably arose from the insertion of a transposable element in the 3′ end of the primordial V gene segment. Once the DNA rearrangement machinery of this transposon came under precise developmental control, the first T-cell receptor gene family acquired the ability to rearrange gene segments. Since all of the rearranging gene families employ similar recognition sequences for their DNA rearrangements, it is safe to assume that the rearranging gene family for immune recognition emerged once—in the T cells of early vertebrates; then duplicated to make successive T-cell receptor gene families; and subsequently, duplicated to generate the antibody gene families. Presumably the primordial T-cell receptor gene family was able to make homodimers and then duplicate to make the early heterodimeric T-cell receptor gene families. Subsequent duplication of one or more of these gene families could have

led at a later time to the acquisition of rearranging gene families for the evolution of antibody molecules.

Comparisons of the Sequences of T-Cell Receptor Gene Families Provide a Snapshot of Evolution

The Human Genome Project has created the resources to determine the DNA sequence of entire family branches of the immunoglobulin gene superfamily. As a result, immunogeneticists can study evolution by comparing, for example, the individual genes of the β T-cell receptor family in one species to one another and then comparing β T-cell receptor families of different species (Fig. 24.24). The availability of large-scale DNA sequence analyses that enable one to compare gene families is transforming our understanding of molecular evolution.

The Human β T-Cell Receptor Family

The β T-cell receptor family of humans spans 700,000 bp of DNA on chromosome 7. It contains 45 functional V β gene segments and 19 pseudogene segments, 2 clusters consisting

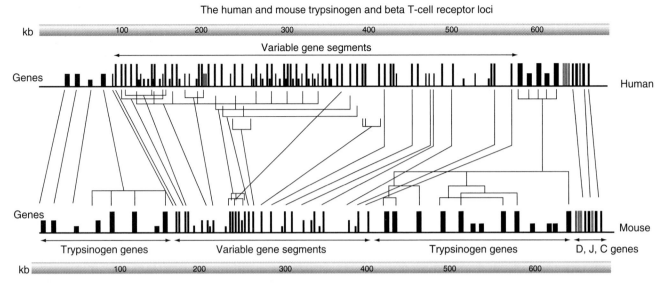

Figure 24.24 Diagrams of the human and mouse β T-cell receptor loci. The black bars indicate T-cell receptor elements and the orange bars trypsinogen genes. The full length bars represent functional gene element, the half bars pseudogenes and the quarter bars are badly decoded genes. The vertical lines between the mouse and human gene families indicate orthologs. The V, D, J and C gene segments are indicated.

of 1 D segment and 5 or 6 functional J gene segments, with each DJ region associated with a different C β gene (Fig. 24.24). The β locus also has trypsinogen genes or pseudogenes on either side of the V β cluster of genes. This gene family has several interesting features.

Approximately 36% of the locus consists of repeat units such as LINE and *Alu* sequences. Since the age of many of these repeat sequences can be approximated by noting the extent of their divergence from the ancestral repeat sequence (which itself can be determined because so many repeat element sequences are available), the repeat sequences have been very useful in reconstructing the complicated molecular archaeology of this family.

Within the human β T-cell receptor family, there are seven different genomic modules (duplicated chromosomal segments) ranging in size from 0.7 to 32 kb. One contains the two D, J, C clusters; a second the five trypsinogen genes that are intercalated between the V and D gene segments. The remaining five modules are associated with the amplification of two or more V gene segments. Indeed, more than 47% of the locus is composed of these modular units.

A portion of the 3′ end of the β T-cell receptor locus has been duplicated and translocated from human chromosome 7 to human chromosome 9. The translocated segment on chromosome 9 contains seven V gene segments and one functional trypsinogen gene. The V gene segments are not functional because the corresponding D, J, and C elements were not translocated. Thus, translocation is one mechanism for the birth of a new multigene family.

Analysis of the promoter regions of the 45 functional V β gene segments reveals that a conserved motif is present adjacent to most of the genes. Presumably this motif represents a site at which transcriptional factors bind to regulate the expression of these genes.

Rearrangement signals from the V, D, and J gene segments are strikingly similar to their counterparts in antibody genes, which reaffirms the idea that these rearranging gene families had a common origin.

Comparing Human and Mouse β T-Cell Receptor Gene Families Enables One to Ask and Tentatively Answer Questions about Evolution

One can search the sequences of the two species for conserved regions where recognizable sequence homology has been retained through 65 million years of mammalian species divergence. The conserved regions might represent coding regions, regulatory regions, or regions that are associated with various chromosomal functions, such as DNA replication or compaction.

Detailed comparisons of the V, D, and J genes in the two species can help reveal orthologs (equivalent homologs in the V gene families of different species) and make it possible to determine how these genes and their organization have changed in 65 million years of divergence. Preliminary comparisons have shown that very different repetitive sequences have been sprinkled through mouse and human loci (about 45% of the mouse locus consists of repetitive sequences), and the nature of the homology units that have been expanded in each species is quite different. Nevertheless, identification of a number of different orthologs suggests that the general organization of these genes has been conserved in the two species.

In both mouse and human genomes, trypsinogen genes are associated with the V β genes. However, there has been an enormous expansion of trypsinogen genes in mouse as compared to in humans. Conversely, there has been an enormous expansion of V genes in the human genome as compared to in the mouse. The net result of these two expansions is that the loci in both species are roughly the same 700,000 bp in length (Fig. 24.24). The trypsinogen genes are also associated with the T-cell receptor β genes in the chicken, a species that diverged from mammals more than 350 million years ago.

One as yet unanswered question raised by the conserved association of T-cell receptor genes and trypsinogen genes is whether genes that live together work together; that is, do the trypsinogen and T-cell receptor genes share common functional and/or regulatory strategies? This question must be answered separately for each genomic region. Its answer is crucial for understanding the regulation of the genes in a particular region.

CONNECTIONS

A retrospective bird's-eye view of key events that led to an understanding of evolution at the molecular level goes something like this. In 1859, Charles Darwin published *The Origin of Species* in which he inferred from the visible evidence of descent through modification that the diverse organisms alive today evolved from a single primordial form, in large part, by a process of natural selection. Several years later, Mendel published "Experiments on Plant Hybrids" in which he applied the laws of probability to the visible evidence of heredity, inferring the existence of hereditary units that segregate during gamete formation and assort independently of each other. In the early twentieth century, Thomas Hunt Morgan and coworkers gave Mendel's units of heredity a physical location in the cell,

establishing the chromosomal basis of heredity and showing not only that genes have chromosomal addresses but also that recombination can separate otherwise linked genes. In the 1940s several people, including Oswald Avery, Martha Chase, and Alfred Hershey, showed that the molecule of heredity is DNA. Then in 1953, James Watson and Francis Crick deciphered the structure of DNA and proposed a mechanism by which the molecule replicates. By the end of the twentieth century, extensive genomic analyses had made it possible to explain how DNA mutates, duplicates, diverges, and is acted on by selection to generate the diversity of life we see around us.

The Epilogue gives our view of the direction and impact of genetics in the twenty-first century.

ESSENTIAL CONCEPTS

1. All forms of life on earth are descendants of a single cell—a common ancestor that existed approximately 3.7 billion years ago.

2. Charles Darwin explained how biological evolution occurs through a process of natural selection.

3. Natural selection operates on variant forms of inherited traits. The particular variant of a trait that provides the highest degree of reproductive fitness is selected over many generations to become the predominant form in the entire population.

4. New mutations provide a continuous source of variation.

5. Mutations with no effect on fitness are considered neutral. Neutral mutations are not acted on by selection and are subject instead to genetic drift. Mutations with a deleterious effect on fitness are selected against, while extremely rare mutations with a positive effect on fitness are selected for.

6. RNA has a unique combination of properties: It can carry genetic information as well as catalyze chemical reactions. These two properties led scientists to speculate that RNA may have predated the cell as the original independent replicator.

7. The fossil record as well as living organisms of all levels of complexity provide scientists with a detailed picture of the evolution of complex life from the first cell to human beings.

8. The evolution of organismal complexity depends on an increase in genome size, which occurs through repeated duplications. Some duplications result from transpositions, while others arise from unequal crossing-over.

9. Mutations rendering genes nonfunctional turn many duplicated genes into pseudogenes that over time diverge into random DNA sequences. However, rare advantageous mutations can turn a second copy of a gene into a new functional unit able to survive and spread through positive selection.

10. Sequence comparisons make it possible to construct phylogenetic trees illustrating the relatedness of species, populations, individuals, or molecules.

11. The mammalian genome contains genes, multigene families, gene superfamilies, genomewide repetitive elements, and simple sequence repeats; repetitive elements in centromeres and telomeres; and unique nongene sequences.

12. Complex genomes arose from four levels of duplication followed by diversification and selection: exon duplication to create larger, more complex genes; gene duplication to create multigene families; multigene family duplication to create gene superfamilies; and the duplication of entire genomes.

13. Genetic exchange between related DNA elements by intergenic gene conversion most often increases the variation among members of a multigene family. Sometimes, however, it can contribute to concerted evolution, which creates a family of nearly identical genes.

14. The enormously diverse immunoglobulin gene superfamily encodes cell receptors that carry out different recognition functions at the cell surface in different types of cells. In the cells of the vertebrate immune system, these receptors enable both the specificity and diversity of immune responses. Gene rearrangements and other recombinatorial mechanisms generate the immense diversity of immune receptors during the development of each individual.

SOCIAL AND ETHICAL ISSUES

1. Jill' husband's boss Mary always seems to be going to the doctor for some ailment and has built up quite a store of antibiotics. Whenever she gets a cold, she starts taking antibiotics. She continues them only until she begins to feel a little better. She is now starting to give her children medications in a similar way even though many of the illnesses can be caused by viruses that do not respond to antibiotics. Jill knows from her training as a physician's assistant that this practice can also lead to an increase in bacteria that are resistant to antibiotics as the bacterial population evolves by selection. However, she hesitates to interfere with her husband's boss. Should she intervene? What should she do?

2. In the novel *Jurassic Park,* extinct dinosaurs were recreated starting from bits of DNA found in fossilized insect blood. While there were several unrealistic steps in the recreation of the dinosaurs, it raised the issue whether there should be controls or laws that would prevent individuals from pursuing this type of recreation. Who could or would enforce them? What are the problems associated with introducing an extinct organism into today's environment?

3. Evolution is a challenging area of study because many of the hypotheses are not directly testable but require drawing inferences of what happened in the past by what is present today. When new data or new analyses seem to contradict or refute an existing major hypothesis, who makes the decision that a particular hypothesis is no longer valid? Who makes the decision when one hypothesis has been strongly enough disproven that it is no longer a viable hypothesis?

S O L V E D   P R O B L E M S

I. The sequence of two different forms of a gene starting with the ATG is shown below. Which of the base differences in the second sequence are synonymous changes and which are nonsynonymous changes?

form 1   ATGTCTCATGGACCCCTTCGTTTG
form 2   ATGTCTCAAAGACCACATCGTCTG

Answer

The key to answering this question is understanding the difference between synonymous and nonsynonymous changes in DNA sequence.

Synonymous changes are nucleotide substitutions that do not change the amino acid specified by the DNA sequence. Nonsynonymous changes are nucleotide changes that result in a different amino acid in the protein.

Looking at the amino acids specified by the base sequence,

           Met Ser His Gly Pro Leu Arg Leu
form 1   ATGTCTCATGGACCCCTTCGTTTG

           Met Ser Gln Arg Pro His Arg Leu
form 2   ATGTCTCAAAGACCACATCGTCTG

(The base substitutions in form 2 are underlined.) *The first, second, and fourth A substitutions are nonsynonymous changes; the third A substitution and the C substitution are synonymous changes.*

II. What difficulties would there be if you tried to derive a molecular clock rate using a noncoding sequence in some species and a coding sequence in other species?

Answer

To answer this problem you need to think about how molecular clocks are derived and the constraints on base changes in coding and noncoding sequences.

The evolution of coding sequences is restricted by the fact that functionality needs to be maintained for the gene product to function. The sequence of noncoding regions generally can tolerate many base substitutions without selection on these sequences. *Therefore, you would expect more substitutions in the noncoding sequence. The result would be an inconsistency in your clock rate if you are using coding region in some species and noncoding DNA in other species.*

III. If these chromosomes misalign by pairing of repeated sequences (shown as solid blocks) and crossing-over occurs, what will be the products?

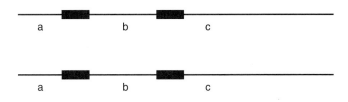

Answer

This question requires an understanding of crossing-over via homology. When you align the homologous repeated sequences out of register, one of the resulting products will have a duplication of the region between the repeats and three copies of the repeated sequence and the other product will be deleted for the DNA between the repeats and contain only one repeated sequence.

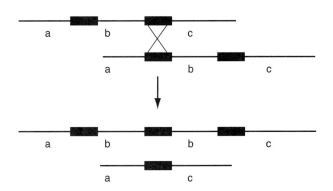

P R O B L E M S

24-1 For each of the terms in the left column, choose the best matching phrase in the right column.

a. ribozymes	1. constant rate of change in amino-acid sequence
b. retrotransposition	2. sudden explosive evolutionary changes
c. SINES	3. exchange of pieces of genes
d. punctuated equilibrium	4. RNA molecules that catalyze specific chemical reactions
e. molecular clock	5. short repeated sequences in human genomes
f. phylogenetic trees	6. RNA intermediate in duplication of genetic material
g. exon shuffling	7. representation of evolutionary relationships

24-2 What observations support the unity of life concept that life-forms evolved from a common ancestor?

24-3 Which of the following statements is support for RNA being the first replicator molecule?
a. RNA molecules can function as enzymes.
b. DNA is more stable than RNA.
c. Information can be encoded in RNA.

24-4 a. In what ways is RNA not a good information storage molecule?
b. In what ways is RNA not as good as protein as a molecular machine?

24-5 a. A particular chemical reaction that occurs between two proteins is sensitive to RNase treatment *in vitro* (i.e., it will not occur if the reaction mix is pretreated with RNase) but is not sensitive to protease treatment. What would you propose about the nature of the enzyme that carries out the reaction?
b. Another chemical reaction is sensitive to both RNase and protease treatment. What would you propose about the nature of the enzyme that carries out this reaction?

24-6 Humans and chimps have a 1% difference in their genomic sequence, while two humans have a 0.5% sequence difference. How could something as small as a 1% difference in DNA sequence lead to dramatic differences seen in chimps versus humans? Speculate on the types of differences the 1% variation may represent.

24-7 Rates of synonymous and nonsynonymous amino-acid substitutions for three genes that were compared in humans and mice or rats are in next column. (Rates are expressed as the average number of substitutions per

amino acid per billion years with the standard deviation given.)
a. Why do the rates of nonsynonymous substitutions vary among these genes?
b. Why are the rates of synonymous substitutions similar?

Protein	Nonsynonymous substitutions	Synonymous substitutions
Histone 3	$0.0 \pm 0.0 \times 10^9$	$6.38 \pm 1.10 \times 10^9$
Growth hormone	$1.23 \pm 0.15 \times 10^9$	$4.95 \pm 0.77 \times 10^9$
β globin	$0.8 \pm 0.13 \times 10^9$	$3.05 \pm 0.56 \times 10^9$

24-8 Synonymous mutations are more prevalent than nonsynonymous mutations in most genes. The immunoglobulin genes (encoding antibody subunits) are an exception where nonsynonymous base changes outnumber synonymous changes. Based on the function of the Ig genes, why do you think this might be true?

24-9 Mutations in the *CF* gene that cause the disease cystic fibrosis are carried by 1/20 of individuals of Caucasian ancestry. The disease is clearly detrimental, yet the allele is maintained at a relatively high level in population. What does this paradox suggest about the effect of the mutation?

24-10 Unequal crossing-over between two copies of a gene can lead to duplication and deletion on the two homologs involved. How could a single gene become duplicated?

24-11 Human beings have three-color vision, while most other species of animals have two-color vision or one-color vision (i.e., black and white). Three-color vision is produced by the products of a three-member, cross-hybridizing, gene family that encodes light-sensitive pigments active in different ranges of the color spectrum (red, green, and blue). What is the most likely molecular explanation for the evolution of the three-color vision in the ancestor to human beings?

24-12 How do transposition and unequal crossing-over differ based on the location of final copies of the duplicated sequence?

24-13 You have identified an interesting new gene that appears to be involved in brain development in humans. You have discovered three cross-hybridizing copies of this gene within the human genome (*A*, *B*, and *C*). In the mouse genome, there is only a single copy of this gene (named *M*) and in the frog *Xenopus*, there is also only a single copy (named *X*). You have sequenced the same 10,000 bp of open reading frame from each of these genes and calculated the number of base-pair differences that exist between different pairs with the following results:

Comparison	Number of base-pair differences
A versus B	300
B versus C	10
A versus C	300
A versus M	600
A versus X	3000

Assume that a constant rate of evolution has occurred with all members of this gene family, and assume that mice and humans evolved apart 60 million years ago.

a. How long ago did frogs evolve apart from the line leading to mice?

b. How many gene duplication events are observable within the data that are given? At what time in the past did each of these events occur?

c. Mapping of the *A, B,* and *C* genes show that *A* and *C* are very closely linked, while *B* shows independent assortment with either of these genes. From this linkage information, what can you say about the molecular nature of the duplication events that have occurred along the evolutionary line leading to human beings?

24-14 Organisms have a characteristic percentage of G C bases in their genomes. In a species of halophilic (salt-loving bacteria) containing two essential plasmids and one chromosome, the plasmids have a different G C content than the chromosome. What could you hypothesize about the origin of the plasmids?

24-15 When a phylogenetic tree was constructed using comparison of glucose-6-phosphate isomerase amino acid sequences from a wide variety of species including animals, plants, and bacteria, the bacterium *E. coli* was placed on a branch of the phylogenetic tree with a flower. What explanation could there be for this unusual association based on this one protein-coding sequence? (Recall that bacteria and plants are in very different locations on the evolutionary trees of organisms derived from analyses of several gene sequences or morphology and physiology.)

24-16 Phylogenetic trees of primates constructed using chromosome mutations (deletions, insertions, inversions, etc.) show the same relationships of species as those constructed using base substitutions within a gene. Which type of genetic alteration would you expect to have the greater impact on evolution of chimp and man from a common ancestor. Why?

24-17 What is the unit that can get duplicated and modified to form
a. a new gene?
b. a multigene family?

24-18 In the evolution of vertebrates, it has been hypothesized that there were two successive doublings of the genome (tetraploidization) to produce the vertebrate genome.
a. How does the *Hox* super family fit with this hypothesis?
b. Within the *Hox* gene family clusters, there is variation among vertebrates in the numbers of genes present. How could these arise?

24-19 LINES and SINES are considered to be selfish DNA, yet they can, in some instances, confer a selective advantage for the organism. What are two ways in which a LINE or SINE can change the genome?

24-20 How is the size polymorphism of dinucleotide repeated sequences thought to occur?

24-21 a. What is an example of repetitive noncoding DNA for which we know no physiological function?
b. What is an example of repetitive noncoding DNA for which we know a cellular function?

24-22 Match the observation on the right with the event that it suggests occurred during evolution.

Event	Evidence in the genome
a. concerted evolution	1. genes in different individuals of the same species have 6, 8, or 10 homology units in an immunoglobulin gene family member
b. exons and introns of a gene	2. several copies of the same gene that are identical in the genome
c. unequal crossing-over	3. poly-A sequences at the end of coding regions in genomic DNA
d. retrotransposition	4. blocks of DNA sequence conserved between species separated by nonconserved blocks

HUMAN GENETICS AND THE FUTURE OF BIOLOGY

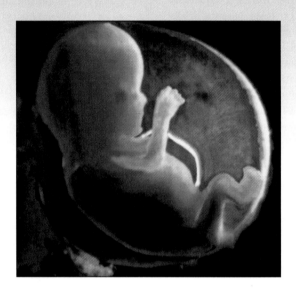

HUMAN GENETICS IS EMERGING AS THE OVERARCHING FIELD OF BIOMEDICAL RESEARCH

As we enter the second Mendelian century, humans are moving to the center stage of genomic study because researchers have overcome some of the species' experimental drawbacks and turned others to advantage. For the first 80 years after the rediscovery of Mendel's laws in 1900, humans were considered a poor choice of species for genetic analysis. Among their disadvantages were a long breeding cycle, a small litter size, and a large amount of genetic heterogeneity and polymorphism. In addition, researchers could not control the choice of breeding partners or the environment in which individuals lived; nor could they subject humans to mutagenesis studies or *in vivo* experimental manipulation. As a result, until the explosion of recombinant DNA technology in the 1980s, most scientists who wanted to use genetic tools to understand hu-

man biological systems chose to study the genes of nonhuman organisms—bacteria, yeast, *Drosophila*, and mice—whose experimental virtues we have already described.

It is still not (and never will be) possible to breed people according to plan, but the development of new statistical methods amenable to computer implementation has made it possible to extract linkage data from existing families; with the help of church and civil records, these data can provide information on genetic relationships extending back many generations. It is still not (and never will be) acceptable to mutate humans purposefully for experimental study, but subtle mutations that occur spontaneously will catch the eyes of medical professionals; and with a world population of roughly 6 billion and typical per-locus mutation rates of 10^{-5} to 10^{-6}, one would expect to find thousands of individuals carrying mutations that are not prenatally lethal at every locus in the genome. Finally, the numerous polymorphisms that once caused frustration today provide a rich source of genetic markers for both linkage analysis and disease association studies. Because of these changes, the human organism can now claim equal status with the model organisms profiled earlier in the book, in terms of its power to approach genetic problems.

The driving force behind the rapid emergence of human genetics has been the development of ever more sophisticated and automated tools for deciphering genetic information. And the analytic scope of the new tools has produced a paradigm shift in thinking about biology.

IN THE NEW PARADIGM, BIOLOGY IS A SCIENCE BASED ON THREE LEVELS OF MOLECULAR INFORMATION

The science of biology is as old as human civilization itself. Traditionally, biologists divided their field into a series of separate disciplines, each with its own jargon and methods of analysis. One type of division was based on systematic classifications: zoologists studied animals, botanists studied plants, and microbiologists studied everything too small to see with the naked eye. Another type of division was based on functionality: immunologists studied the immune system, embryologists studied the embryo, cardiologists studied the heart, and so forth.

Beginning with the discovery of DNA structure and the genetic code, and helped by ideas derived from computer science and physics, a dramatic shift in biologists' concept of the nature of living things took place. In the new paradigm, all living organisms (as well as the organs, tissues, and other functional systems within them) develop and operate according to the same basic principles of information processing; and life is considered an emergent property associated with networks of molecules interacting with each other. Through the organism's expenditure of energy, these molecular networks gain the ability to maintain structure and information against the constant pull of information-destroying entropy. When sufficient energy is not available, information degenerates and life ceases to exist.

Thus, in the new paradigm, biology is a science of information, and biologists study how that information is stored, how it is used, and how it evolves. In this pursuit, many different types of biologists can take advantage of the same molecular tools to analyze the variously defined biological systems, comparing and contrasting the details of information storage and processing that define them.

The information associated with living organisms exists at three levels of complexity.

The First Level Is Genetic Information Encoded in DNA

Although DNA is a three-dimensional molecule, the information within this molecule is one-dimensional and digital. It is one-dimensional because it is encoded along the length of a molecule. It is digital because the basic informational unit—the base pair—can exist in only one of four discrete states.

Because genetic information is digital, it can be stored as readily in computer memories as in DNA molecules. Indeed, the combined power of DNA sequencers, computers, and DNA synthesizers makes it possible to store, replicate, and transmit genetic information electronically from one place to another anywhere on this or any other planet. This electronic wizardry works something like this: A DNA sequencer reads the base sequence of a molecule. The sequence information is stored in a computer. The computer transmits the information via satellite from New York to a receiver in Hong Kong or from Paris to a receiver on Mars. There, the information is fed into a DNA synthesizer, which makes an exact replica of the originally sequenced DNA molecule.

Just as the limited number of letters in a written alphabet places no restrictions on the stories one can tell, so too the limited number of letters in the genetic code alphabet places no restrictions on the kinds of organisms genetic information can define. As we have seen, these organisms range from viruses and bacteria to complex metazoans such as fruit flies, mice, and humans. The basic genetic language is virtually the same for all; the differences are in the content and amount of information. Although it may seem incredible, it takes only three billion base pairs of genetic information to develop a human being, from its basic body plan to the initiation of consciousness. To appreciate the long journey from a finite amount of genetic information (easily storable on a computer disk, once the Human Genome Project makes it known) to human consciousness, it is necessary to follow the flow of information from DNA to the second and third levels of biological information.

The Second Level of Biological Information Is Embodied in Proteins

As we have seen, there is a direct correlation between the DNA within a gene and the three-dimensional shape of the protein it encodes. Proteins, synthesized from the 20-letter alphabet of amino acids, are strings of 5 to 3000 letters that fold into a precise three-dimensional shape dictated by the order of letters in the string. Thus, the one-dimensional information of a gene becomes the three-dimensional information within a protein. In addition to containing three-dimensional information, proteins differ from DNA in the nature of their activity. While genetic information is inert—that is, it can be read by the cellular machinery but it doesn't actually do anything by itself—proteins operate as molecular machines that carry out the functions of life by virtue of their shape and chemical properties.

The Broad Third Level of Biological Information Encompasses Dynamic Interactions among Molecules as well as Interactions among Cells and Tissues

These complex interactive networks represent **biological systems** that function both within individual cells and among groups of cells within an organism. Here we use "biological system" to mean any complex network of interacting molecules or groups of cells that function in a coordinated manner through dynamic signaling. There are several layers of biological systems. The human pancreas, for example, is an isolated biological system that operates within the larger biological system of the human body and mind. A whole community of animals, such as a colony of ants that functions in a highly coordinated manner, is also a biological system.

The information that defines any biological system is four-dimensional because it is constantly changing over the three dimensions of space and the one dimension of time.

One of the most complex examples of the third level of biological information (other than an entire human being) is the human brain with its 10^{11} neurons connected through perhaps 10^{18} synapses. From this enormous biological network arise systems and emergent properties such as memory, consciousness, and the ability to learn.

If Biology Is an Informational Science, Then the Focus of Biological Research Must Be on Deciphering Biological Information

The term "decipher" means both "to define" and "to understand." Defining the DNA sequences of the 23 human chromosomes (the first level of biological information) is the goal of the Human Genome Project, scheduled for completion by the year 2005. Understanding how this information—accumulated over 3.7 billion years of evolution—gives rise to a living system (the second and third levels) will be the goal of biomedical scientists for many years to come. At both these higher levels, it will be necessary first to define information (in terms of *structures* and *connections* among network components) and then to understand how this information *functions*. For example, it is one thing to define the three-dimensional structure of a protein and quite another to understand how that structure carries out function. The same is true for a system: One can define its elements and connections and still have the difficult task of explaining how the interactions of multiple units create emergent properties that do not exist in each of the system's components.

USING THE TOOLS OF GENOMIC AND PROTEIN ANALYSIS TO DECIPHER BIOLOGICAL NETWORKS AND SYSTEMS

The simplest living entity—a small prokaryotic cell—is more structurally complex than the most sophisticated computer so far invented; and multicellular biological systems, like the human brain, have a structural complexity that is orders of magnitude greater. Even the simplest biological system is too complex to study as a whole. To understand how it works, scientists must first break it down through successive subsystems that still contain some properties of the whole system, but are accessible to molecular analysis. Once they understand the

molecular interactions of smaller subsystems they can build back up through successively more complex layers to an understanding of the whole system.

Consider how you would use such an approach to study the brain. First, you would divide the whole brain into subsystems represented by particular regions like the cerebellum or hippocampus. Next, you would repeatedly subdivide the subsystems down to the level of individual neurons and their accessory glial cells. A single neuron can have synaptic connections to thousands of other neurons. Synapses along the neuron's many dendrites provide a pathway for incoming signals. Each neuron processes the array of signals it receives and "decides" whether to send out a signal of its own to other neurons via synapses along its lone axon. As a further response to certain levels of incoming signals, a neuron can undergo *long-term potentiation*—a form of neuronal memory—which alters the way the neuron responds to the same set of signals in the future.

To understand how a neuron receives, processes, and transmits signals, as well as how it accomplishes long-term potentiation, you would break the individual cells down into subcellular components, such as the plasma membrane, the cytoskeleton, and the nucleus, and then reduce these microscopic systems into their individual molecules, such as neurotransmitter receptors on the cell surface, microtubule units along the axon, phosphorylating enzymes that carry signals in the cytoplasm, and transcription factors that alter gene activity in response to signals received from the cytoplasm.

With information gained from these reductionist studies, you could begin to build back up through increasingly complex subsystems to reconstruct the whole system. With an understanding of the role that individual molecules play in signal processing, you could decode the connections among them to begin to understand how the whole neuron functions. Next you could decipher the interactions among neurons and how they function to define a particular brain region. Then you could analyze the interactions among different regions.

This is just one example of determining the connections within and among subsystems to define the pathways along which information is processed in a larger system. A second, complementary approach, which is just as powerful, is to take advantage of the large amount of genetic variation within the human population to understand how changes (both subtle and not so subtle) in particular genes lead to changes at the highest informational level of the whole human system. Analyses of this type, which depend on traditional pedigree studies as well as genotype-phenotype association studies, provide insight into both the genetic and the nongenetic components of biological systems.

Moving from informational pathways to systems properties will require the enormous modeling skills of computer scientists and applied mathematicians. Ultimately, it will be necessary to model whole systems on powerful future computers to understand how the systems operate.

Automated High-Throughput Tools of Genomic and Protein Analysis Provide the Keys to Deciphering Biological Systems

The global tools of genomics—such as high-throughput DNA sequencers, genotypers, and large-scale DNA arrays (also called DNA chips)—and the tools of protein analysis—such as two-dimensional gels and mass spectrometry—have the capacity to analyze thousands of genes or gene products rapidly and accurately. These global tools are not specific to a particular system; rather, they can be used to study the genes, proteins, and higher level systems of all living things. For example, large-scale DNA sequencing has delivered more than 1,000,000 human expressed sequence tags (ESTs) representing roughly 40,000 genes and 90 Mb of human chromosomal sequence to the database. (ESTs are not full coding regions, but "tags" for particular coding regions made from the ends of cDNAs obtained from cDNA libraries. Researchers can use these tags to map the genes from which they come and to compare the genes' transcripts to others in the database.) Large-scale DNA sequencing has also enabled researchers to determine the complete genomic sequences of *Escherichia coli* (4.5 Mb), yeast (12 Mb), and *Caenorhabditis elegans* (97 Mb). Knowledge of these sequences has provided powerful new approaches to systems biology.

The DNA chip is the most powerful example to date of a global genomic tool. Produced by several techniques, individual chips are subdivided into arrays of microscopic blocks that each contain a unique oligonucleotide. On exposure of a chip to a complex mixture of fluorescently labeled nucleic acid—such as DNA or RNA from any cell type or sample—each block acts as a hybridizing detector for a specific complementary sequence. A computer-driven microscope can then analyze hybridization to each of the hundreds of thousands of blocks on the chip, and special software can enter this information into a database.

As mentioned in Chapter 9, the potential power of DNA chips is enormous for both research and clinical purposes. Already chips with over 400,000 oligonucleotide detectors can provide simultaneous information on the presence or absence of 400,000 discrete DNA or RNA sequences in a complex sample. And they can do it within hours. Examples of potential uses of DNA chips include the following:

■ Unique oligonucleotides representing each of the 100,000 human genes could be placed on a chip (once the sequence of all human genes is known) and used to determine the complete set of expressed genes, as well as their level of expression, in any human cell type at any stage of development or differentiation. By comparing cells at different points in a differentiation pathway, it would be possible to correlate precise changes in gene expression with particular changes in cell function. Such computer-driven comparisons could provide a powerful tool for the analysis of whole systems. Similarly, by comparing normal and cancerous cells, it could become possible to pinpoint the particular genes involved in establishing the cancerous phenotype.

■ As discussed in Chapter 9, a single nucleotide change in a complementary sequence can eliminate hybridization to a short nucleotide. Researchers could use this property to develop DNA chips that are loaded with hundreds of thousands of oligonucleotides, each one containing a human genomic sequence that includes a central nucleotide known to be highly polymorphic among people. A dense, randomly scattered array of single nucleotide polymorphisms (SNPs) from across the entire genome could facilitate genetic mapping in humans as well as in other organisms. Presently, a single genotyping matching can determine up to 2000 genotypes a day. The DNA chip could provide several orders of magnitude more, at a fraction of the cost and in a fraction of the time.

■ The DNA chip may make it possible to synthesize whole genomes from scratch. For example, the entire double-stranded chromosome of a microorganism could be synthesized as overlapping 30-mers that are cleaved from the glass chip, allowed to self-assemble by hybridization, and then ligated. In this way, it would be possible to engineer an entire microbial genome in perhaps 2 hours. In another 2 hours, the same genome could be engineered with several gene substitutions to facilitate a systems analysis of interesting biological properties. Genome engineering of this type could be an enormously useful tool for deciphering informational pathways in microbial organisms and for designing new types of organisms of particular use in the agricultural, pharmaceutical, and energy industries.

■ A modification of the technology used to create DNA chips could probably be used to create analogous peptide chips. Peptide chips with thousands of unique amino-acid domains could be used as global probes of protein-protein interactions within a complex mixture of proteins taken from cells of different types and at different stages of development. Ultimately such peptide chips might also be used as sensitive detectors of changes in individual protein conformations and activities as the cell undergoes differentiation.

■ Similarly, double-stranded DNA could be synthesized on chips and used to examine interactions between proteins (such as transcription factors) and DNA. Analyses of this type could provide insight into the regulation of all the genes in a biological system.

In Summary: The Tools of Global Analysis Serve at Least Three Important Functions

First, they can decipher biological information. Second, they can create large databases of biological information. Finally, they can measure how biological systems change their gene expression and protein translation patterns when they are perturbated biologically (for example, the induction of DNA synthesis in yeast) or genetically (for example, the engineering of gene knockouts).

STUDIES USING THE TOOLS OF GLOBAL ANALYSIS REVEAL THE CONSERVATION AND UNITY OF ALL THREE LEVELS OF BIOLOGICAL INFORMATION

Although the powerful tools of genomics and protein analysis will help decipher the biological systems that define us as human beings, there is a limit to how far the direct study of human systems can go. Even with the most powerful tools, the experimental manipulation of whole human systems, in any but the most superficial way, is ethically unacceptable. Yet, the perturbation of a biological system has always been one of the most powerful tools for understanding its biological function. Findings that demonstrate the conservation and unity of biological information have offered a way around this dilemma.

In the 1980s and early 1990s, when the tools of molecular biology were first applied in earnest to various model organisms, a remarkable picture of the unity of life on earth began to emerge. Not only do most living things use the same molecule (and code) to carry (and decipher) genetic information, but at the second level of biological information—the structure and function of proteins—there is remarkable conservation and unity as well. As we saw in Chapter 24, for example, many of the homologous proteins from yeast and humans are interchangeable. Even more surprising is the conservation and unity observed at the third level of biological information—in the development of whole systems.

One of the most striking examples of conservation at this level was uncovered in studies of eye development. Both insects and vertebrates (including humans) have eyes; but they are of very different types. Biologists had long assumed that the evolution of eyes occurred independently in the lineages leading to present-day insects and present-day vertebrates. Indeed, in many evolution textbooks, eyes are used as an example of convergent evolution, that is, evolution in which structurally unrelated but functionally analogous organs emerge in different species as a result of natural selection. Studies of the *Pax6* gene have turned this view upside down.

Mutations in the *Pax6* gene lead to a failure of eye development in both people (with a condition known as aniridia) and mice, and molecular studies have suggested that *Pax6* might play a central role in the initiation of eye development in all vertebrates. Remarkably, when the human *Pax6* gene is expressed in cells along the surface of the *Drosophila* body, it induces numerous little eyes to develop there. This result demonstrates that there was a single origin of the eye in an ancestor common to flies and people and that, after 600 million years of divergent evolution, both vertebrates and insects still share the same master control switch for initiating eye development. Studies of many other genes have shown that the entire developmental program of flies and people uses many of the same genes. These genes have duplicated and evolved divergent functions, but they still retain their ancestral markings.

Because of the Unity of Biological Information, Model Organisms Are Rosetta Stones for Deciphering Human Informational Pathways

The utility of the finding of conservation and unity at all three levels of biological information cannot be overstated. It means that in many cases, the experimental manipulation of model organisms can shed light on complex networks in humans. If homologs of human genes function in simple model organisms such as *Drosophila, C. elegans,* yeast, and *E. coli,* one can determine gene function and regulation in these experimentally manipulable organisms and bring these insights to the human organism. The same is true of the shared informational pathways such as DNA replication, protein synthesis, and many metabolic and signal transduction cascades. The mouse, a more complex model organism than those just mentioned, shares many of the sophisticated informational networks found in humans.

An example of the model organism approach to fathoming human informational pathways is recent work in *Drosophila* and mice that led to an understanding of the human disease known as Alagille syndrome. Alagille syndrome is a dominant disorder characterized by abnormal development of the liver, heart, skeleton, eyes, face, and less frequently, the kidney; the expression of phenotypes within each of these organs is highly variable. The great variety of symptoms made it hard to pinpoint the underlying cause of the disease.

The solution to the origin of Alagille syndrome came with its genetic mapping to the short arm of chromosome 20 (band 20p12). Within the region where the disease gene mapped, researchers uncovered a gene with homology to a *Drosophila* gene called *Delta.* In *Drosophila,* the product encoded by the *Delta* gene binds to a cell surface receptor encoded by the *Notch* gene. Subsequent studies have shown that the *Delta/Notch* signaling system, analyzed in both flies and mice, plays a crucial role in the control of early development. Medical researchers and clinicians can now apply this information, obtained from the study of model organisms, to the study of Alagille syndrome. Although knowledge of the genetic origin does not provide a cure for the disorder, it offers the possibility of a molecular understanding of the disease. And that understanding offers hope of a therapy or cure.

A Systems Approach to Human Genetic Disease Has Led to the Concept of Disease Stratification and, with It, New Approaches to Drug Therapy

We now know that common diseases such as prostate cancer, obesity, and rheumatoid arthritis are generally not a single disease with a single, well-defined cause; rather, they are several diseases, each caused by different aberrations in informational pathways, that ultimately result in similar pathologic phenotypes. The process of dividing a single general condition into different diseases on the basis of the underlying molecular defects is known as **disease stratification.**

For example, severe obesity can result from a breakdown in the informational system that normally tells our brain there is no longer a need to eat more food. In normal individuals, fat cells produce a hormone called leptin that travels to the brain, where it binds to a specific receptor; this binding ultimately changes the behavioral patterns of the person. The more fat storage cells present in a person's body, the more leptin he or she produces, and the greater the signal to stop eating. A defect in the leptin gene can keep the "stop eating" signal from being produced, leading to obesity. A defect in the brain receptor can stop the signal from being received in the brain, also leading to obesity. Other defects in the informational pathway, downstream of the leptin receptor (and yet to be uncovered), can keep the "stop eating" signal from being processed appropriately, again resulting in obesity. Finally, cultural and environmental influences (including the societal view toward eating and the type of food consumed) can also play a role in the final phenotype.

The global tools of genomics and protein analysis can help identify diagnostic markers characteristic of each division, or type, of a stratified disease. For a genetic disease, the implication is that different abnormal genes or combinations of genes cause each subtype of the disease.

An understanding of the stratification of disease is important in developing therapeutic or preventive drugs, because each layer of a disease may require a different therapy. Implicit in the systems view of disease is the idea that to identify effective therapeutic or preventive drugs, it is necessary not just to identify the predisposing genes, but also to understand the informational pathways in which they operate. Indeed, the most effective means of therapy may entail using systems knowledge to circumvent the limitations of the genes.

Thus, the analytic tools that led to a paradigm shift in scientific thinking about biology have provided insights and new approaches to the most challenging questions concerning the genetics of complex human systems.

HUMAN GENETICS IS LEADING US TOWARD PREVENTIVE MEDICINE

Over the next 25 years, geneticists will identify hundreds of genes with variations that predispose people to many types of disease: cardiovascular, cancerous, immunological, mental,

metabolic. As we know, some mutations will always cause disease; others will only predispose to disease. For example, a glutamine-to-valine substitution at codon 6 in the β-*globin* gene will nearly always cause sickle-cell syndrome. By contrast, a mutation in the *breast cancer 1* (*BRCA1*) gene has only a 70% chance of causing breast cancer in a 60-year old woman carrying one copy of the mutation; this conditional state arises either because the *BRCA1* gene interacts with environmental factors that affect the probability of activating the cancerous condition or because various forms of other genes modify expression of the *BRCA1* gene. Defining and analyzing the multiple factors contributing to genetic predispositions will be essential for understanding and designing therapies for each disease.

As scientists come to understand the complex systems in which disease genes operate, they will be able to design therapeutic drugs designed to block and/or reverse the effects of mutant genes. Then, DNA diagnostics—a collection of techniques for detecting specific mutant genes at any stage in the lifetime of an individual—will make it possible to determine a predictive health history revealing the probability that an individual will acquire any of these gene-based diseases and at what period of his or her life. With the ability to predict individual health histories and an understanding of the informational systems within which the defective genes are embedded will come the potential for designing preventive drugs. If taken before the onset of disease, such drugs could prevent occurrence of the gene-based disease.

The development of preventive drugs will signal the advent of preventive medicine with a focus on keeping individuals healthy rather than on trying to heal the sick. In this new era of preventive medicine, many individuals could live into their 80s and 90s, physically fit and mentally alert. At this point, society would have to rethink how it treats the elderly, as they would represent an enormous reservoir of able, talented people capable of making significant contributions to society. Society would also have to redesign the education of physicians, other health professionals, and the public to include a perspective on preventive medicine.

THE NEW PARADIGM, THE NEW SCOPE OF HUMAN GENETICS, AND THE NEW POTENTIAL OF PREVENTIVE MEDICINE INTENSIFY THE NEED TO CONFRONT MANY SOCIAL ISSUES

Genetics began as a separate biological discipline dedicated to determining the rules governing the frequency of appearance of alternative traits in siblings and other related individuals. At the threshold of the twenty-first century, it has become the central focus and tool in the study of complex biological systems created by the interactions of molecular entities. Although biological information is similar to other types of information from a strictly technical point of view, it is as different as can be in its meaning and impact on individual human beings and human society as a whole. The difference lies in the personal nature of the unique genetic profile carried by each person from birth. Within this first level of biological information are complex life codes that provide greater or lower susceptibility or resistance to every kind of disease, as well as greater or lesser potential for the expression of every physiologic, physical, and neurological attribute that distinguishes people from each other. Until now, almost all this information has remained hidden away. But if research continues at its present pace, in less than a decade it will become possible to read a large part of a person's genetic profile; and with this information will come the power to make predictions about future possibilities and risks.

As we have seen in many of the Genetics and Society boxes throughout this book, society can use genetic information not only to help people, but also to restrict their lives (for example, by denying insurance or employment). We believe that just as our society respects individuals' right to privacy in other realms, it should also respect the privacy of an individual's genetic profile and work against all types of discrimination.

Another issue raised by the potential for detailed genetic profiles is the interpretation or misinterpretation of that information. Without accurate interpretation, the information becomes useless at best and harmful at worst. Proper interpretation of genetic information requires some understanding of statistical concepts such as risk and probability. To help people understand these concepts, widespread education in this area will be essential. Since the media plays an enormous role in the lives of most people, public education could begin with media reports on new genetic findings that are well reasoned and accurate. It will also be essential to bring kindergarten through high school education up to date so that children can learn the concepts and implications of modern human biology as a science of information.

Yet another pressing issue concerns the regulation and control of the new technology. While the sequencing of the entire human genome with the use of governmental funds was hotly debated in the early 1990s, in 1998 the genome sequencing project was taken up by a business consortium able to carry it forward with private funds. As a result, government funds appropriated for the sequencing project can now be directed toward analyzing genetic variation among humans as well as the second and third levels of biological information. The question of whether the government should establish guidelines for the use of biological information, reflecting society's social and ethical values, remains in open debate.

To many people, the most frightening potential of the new genetics is the development of technology that can alter or add to the genes present within the germ line of single-cell human embryos. As mentioned in Chapter 10, this technology (referred to as "transgenic technology" in scientific discourse and "genetic engineering" in public discussions) has become routine in hundreds of laboratories working with various animals; and for the first time, it appears that the technology could be made safe and efficient for use in humans.

Some people caution that developing the power to alter our own genomes is a step we should not even think of taking, arguing that if genetic information and technology are misused (as they certainly have been in the past), the consequences could be horrific. Others warn that we should be careful before rushing to such conclusions, because if a technology could help children and adults lead healthier, happier lives, we need to think very carefully about whether the reasons for objecting to its use for any purpose are valid.

Ultimately, the biological revolution we are living through will have a greater impact on human society than any technological revolution of the past. Education and public debate are the key to preparing for the consequences of this revolution.

There are inconsistencies within the various branches of genetics on some nomenclature—because it is a relatively new area of scientific investigation, the consistency present in more basic sciences has not been established. The authors debated whether they should try to impose a consistency on the entire topic area and decided against that path. As the study of genetics matures, the process itself will create a more consistent nomenclature. The following guidelines can be applied to all chapters in this book.

General Rules

■ Names of genes are in italics (*lacZ, CDC28*)

■ Names of proteins are in regular (roman) type with an initial cap (LacZ, Cdc28)

■ Chromosomes: sex chromosomes are represented by a capital letter in roman type (X, Y); autosomes are designated by a cardinal number (1, 2, 21, 22)

■ Names of transposons are in roman type (Tn10)

Specific Rules for Different Organisms

■ **Bacteria:** lowercase italics for genes (*lac, ara*), with the addition of a capital letter to designate a specific gene in a pathway or operon (*lacZ, lacA, araB*); numbers (not superscript) for alleles (*trpC2; hisB2*); superscript "+" (plus) for wildtype alleles; superscript "−" (minus) for mutant alleles (*lacZ$^+$, lacZ$^-$*)

■ **Yeast:** all caps for wild-type alleles (*CDC28*); all lowercase for mutant alleles, with a cardinal number indicating the specific mutation (*cdc28-1*)

■ **Arabidopsis:** all caps for wild-type alleles (*LEAFY*, abbreviated *LFY*); all lowercase for mutant alleles (*lfy*)

■ **C. elegans:** lowercase italics for genes and wild-type alleles (*dpy-10*); mutant alleles in parentheses following the gene [*(dpy-10(e128)*]

■ **Drosophila:** Many genes are named for the mutant phenotypes that revealed them. If the mutation causing the phenotype is dominant, the allele designation has an initial cap (*Deformed*, abbreviated *Dfd*); if the mutation that revealed the gene is recessive, the allele designation is all lowercase (*white*, abbreviated *w; wingless*, or *wg*)

■ **Mice:** Genes are now named for the protein they encode and designated by an initial cap followed by any mix of letters or numbers (for example, *Tcp1* designates the gene for T complex protein 1); alleles are designated by superscripts. For multiple wild-type alleles, the superscripts may be *a, b, c,* or *1, 2, 3 (Tcp1^a, Tcp1^b)*; some wild-type alleles are indicated by a superscript plus "+" (*Kit$^+$*). For mutant alleles, the superscripts often describe the resulting phenotype (the *Kit^w* mutant allele causes white spotting).

■ **Humans:** All caps for genes (*HD, CF*); wild-type alleles designated by a superscript plus, "+" (*HD$^+$, CF$^+$*); mutant alleles designated by a superscript minus "−" (*HD$^-$, CF$^-$*).

Answers are provided here to selected odd-numbered problems.

Chapter 1

1. a. 4; b. 3; c. 6; d. 7; e. 11; f. 13; g. 10; h. 2; i. 14; j. 9; k. 12; l. 8; m. 5; n. 1.

3. The dominant trait (short tail) is easier to eliminate from the population by selective breeding. You can recognize every animal that has inherited the allele, because only one dominant allele is needed to see the phenotype. Those mice that have inherited the dominant allele can be prevented from mating.

5. a. *aaBbCcDDEe;* b. *aBCDE* or *aBcDE* or *aBCDe* or *aBcDe* or *abCDE* or *abCDe abcDE* or *abcDe.*

7. Short hair is dominant to long hair.

9. a) Dry is recessive; sticky is dominant. b) The 3:1 and 1:1 ratios are obscured because the results of crosses that give these phenotypes are combined with results of other crosses.

11. a. 1/64; b. 1/8.

13. The genotype can be determined by performing a test cross; that is, crossing your fly with the dominant phenotype (but unknown genotype) to a fly with the recessive (short wing) phenotype. If the fly with the dominant phenotype has the homozygous dominant genotype the progeny in this case would be Ww and would have the dominant phenotype. If the fly being tested had a heterozygous genotype, 1/2 of the progeny would be Ww and have the dominant phenotype, and 1/2 of the progeny would be ww and have the recessive phenotype.

15. No. If you mate several red horses to each other and also mate several black horses to each other, the crosses that always yield offspring with the parental phenotype must have been between homozygous recessives.

17. P = purple, p = white; S = spiny, s = smooth.
a. $PpSs \times PpSs$; b. $PPSS \times P{-}ss$ or $P{-}Ss \times PPss$;
c. $PpSS \times ppSS$ or $Pp{-} \times ppSS$ or $PpSs \times ppS{-}$;
d. $ppSs \times PpSs$; e. $Ppss \times Ppss$; f. $ppSs \times ppSs.$

19. a) Rough and black are the dominant alleles (R = rough, r = smooth; B = black, b = white). b) a 1:1:1:1 ratio of rough, black; rough, white; smooth, black; smooth, white.

21. a. 3/16; b. 1/16.

23. a. Recessive. Two unaffected individuals have an affected child and it was a consanguineous marriage that produced the affected child. Affected = *aa;* carrier = Aa b. Dominant.

The trait is seen in each generation; affected person has an affected parent; III-3 is unaffected even though both his parents are (this would not be possible for a recessive trait). Affected = *AA;* carrier, not applicable. c. Recessive. Two unaffected parents have an affected child. Affected = *aa;* carrier = *Aa.*

25. a. 2/3; b. 1/9; c. 3/9.

27. Recessive; common.

Chapter 2

1. a. 2; b. 6; c. 11; d. 8; e. 7; f. 10; g. 10; h. 3; i. 5; j. 4; k. 1.

3. Incomplete dominance; red = *YY,* yellow = *Yy,* white = *yy.*

5. L (long) allele is completely dominant to l (short) allele. Flower color trait shows incomplete dominance of two alleles.

7. a. *ii* (phenotype O) or *iI^A* (phenotype A) or *iI^B* (phenotype B); b. $I^B I^B$, $I^B i$ or $I^B I^A$.

9. a. There will be three different phenotypes: spotted and dotted, marbled, and spotted in a 1:2:1 ratio; b. marbled and dotted.

11. 2/6 yellow, 3/6 albino, and 1/6 agouti progeny.

13. a. $A^Y a\, Bb\, Cc \times Aa\, bb\, Cc$; b. six phenotypes: albino, yellow, brown agouti, black agouti, brown, black.

15. Two genes are involved. The black mare was *AA bb* and the chestnut stallion was *aa BB,* the liver horses were *aa bb,* the bay horses were *Aa Bb.*

17. *Aa Bb, aa Bb, Aa bb* and *aa bb* in equal proportions: 1/4 green and 3/4 yellow fruit.

19. Dominance relationships are between alleles of the same gene. Only one gene is involved. Epistasis involves two genes. The alleles at one gene affect the expression of a second gene.

21. 1/4 would appear to have O type blood, 3/8 have A, 3/8 have AB.

23.

	I-1	I-2	I-3	I-4	II-1	II-2	II-3	III-1	III-2
Phenotypes	AB	A	B	AB	O	O	AB	A	O
Genotypes	$I^A I^B$	$I^A?$	$I^B?$	$I^A I^B$	*ii*	$I^A?$	$I^A I^B$	$I^A?$	??
	Hh	*Hh*				*hh*	*Hh*		*hh*

25. 27/64 wildtype; 37/64 mutant.

27. 44/56.

Chapter 3

1. a. 5; b.7; c.11; d.10; e.12; f. 8; g. 9; h.1; i. 6; j. 4; k.3; l. 2.

3. a. 23; b. 4 autosomes and 2 sex chromosomes; c. 23; d. one.

5. a. brown females and ivory-eyed males; b. female with brown eyes and males with ivory or brown eyes.

7. a. iii; b. i; c. iv; d. ii; e. v.

9. a. Mitosis, meiosis I, II; b. mitosis, meiosis I; c. mitosis; d. meiosis I; e. meiosis I; f. none; g. meiosis I; h. meiosis II, mitosis; i. mitosis, meiosis I.

11. 2/4.

13. a. 400 spermatozoa; b. 100; c. 100; d. none.

15. a. Only females; b. males; c. males; d. 1/5 ZZ males and 4/5 ZW females.

17. Vestigial wings is autosomal; body color is X-linked recessive.

19. a. Nonbarred females and barred males; b. barred and nonbarred females and barred and nonbarred males.

21. a. Recessive; b. autosomal; c.aa; d.Aa; e. Aa; f. Aa; g. Aa; h. Aa.

23.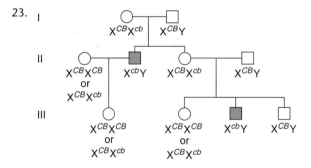

25. a. 3; b. 1 or 3.

27. Mate potential mutant females (with genotype X^y/X^y) with X^y/y^+· Y males. Nondisjunction mutants would have elevated numbers of brown-bodied female and yellow-bodied male progeny.

Chapter 4

1. a. 8; b. 4; c. 1; d. 11; e. 2; f. 5; g. 6; h. 3; i. 10; j. 7; k. 9.

3. a. $Oo\,Bb$; b. 9:3:3:1; c. between 0.5 and 0.1.

5. a. 20% Ab and aB, 30% Ab and ab; b. 30% Ab and aB, 20% AB and ab.

7. 10%

9. a. A = normal pigmentation, a = albino allele, Hb^A = normal globin, Hb^S = sickle allele 49.5% aHb^A, 49.5% AHb^S, 0.5% aHb^S, 0.5% aHb^A; b. 49.5% aHb^S, 49.5% AHb^A, 0.5% aHb^A, 0.5% AHb^S; c. .002.

11. a. 1/4 black, 1/2 albino, 1/4 brown; b. 34 m.u. apart.

13. a. 0; b. 1%; c. 5%; d.10%; e. 50%.

15. 360 $a^+\ b^+\ c^+$; 360 $a\ b\ c$; 90 $a^+\ b\ c$; 90 $a\ b^+\ c^+$; 40 $a^+\ b^+\ c$; 40 $a\ b\ c^+$; 10 $a^+\ b\ c^+$; 10 $a\ b^+\ c$

17. a. $sc\ ec\ cv/ + + +$ and $b/+$; b. $sc\ ec\ cv/+ + +$ sc-cv = 9 m.u., ec-cv = 10.5 m.u.; c. predicted DCO = 0.009, observed DCO = 0.001, interference.

19. $$\frac{dwp/pld^+ \quad\quad rv^+ \quad\quad rmp}{dwp^+\,pld \quad\quad rv \quad\quad rmp^+}$$ rv-rmp 10 m.u.; pld/dwp-rv 5 m.u.

21. ◯ $\underline{c \quad 8.2 \qquad a}$
    ◯ $\underline{0.45\ b}$

23. a. First group: $met^+\,lys^+$ and $met^-\,lys^-$; second group: $met^+\,lys^+$, $met^+\,lys^-$, $met^-\,lys^+$, $met^-\,lys^-$. b. 5.5 m.u. c. $met^-\,lys^+$ and $met^+\,lys^-$.

25. $ade2^-/ade2^-$. mitotic recombination.

Chapter 5

1. a. 6; b.11; c. 9; d. 2; e. 4; f. 8; g.10; h.12; i. 3; j.13; k. 5; l. 1; m. 7.

3. c

5. Tube 1, nucleotides; tube 2, base pairs (without the sugar and phosphate) and sugar phosphate chains without the bases; tube 3, single strands of DNA.

7. a. 20% C; b. 30% T; c. 20% G.

9. Determine the percentage of each base.

11. 3′ G G G A A C C T T G A T G T T T C G G C T C T A A T T 5′.

13. 5′ U A U A C G A A U U 3′.

15. After one additional generation, 1/4 HL; after two additional generations, 1/8 H/L

17.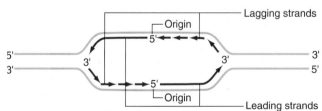

19. a. Relieves the stress of the overwound DNA ahead of the replication fork; b. unwinds the DNA; c. synthesizes a short RNA oligonucleotide; d. joins the sugar phosphate backbones.

21. Regardless of which strands are cut during resolution (to result in crossing-over or no crossing-over) mismatches within the heteroduplex region can be corrected to the same allele, resulting in gene conversion.

23. Would not undergo recombination.

Chapter 6

1. 1. a, b, h; 2. b, c; 3. f, j; 4. g, j; 5. a, b, h; 6. f, i, k; 7. g, k; 8. d, i; 9. e, f, i.

3. 9.5×10^{-5}; higher than normal rate.

5. If phages induce resistance, several appear in random positions on each of the replica plates. If the mutations preexist, the resistant colonies would appear at the same locations on each of the three replica plates.

7. Female A has a white-eyed mutation on the X. Female B has a recessive mutation on the X. Female C is mosaic with a lethal mutation in one strand and wildtype sequence in the other strand of one X chromosome.

9. Yes; Liver converts substance X into a mutagen.

11. If in another gene, suppressor and original mutation are separable during meiosis.

13. a. Three groups: 1,8; 2; 3,4,7; b. 3:1 ratio of Lys^-:Lys^+;
    c. 2? | 8   1 | 4   3   7 | 2?
            | deletion 6 |
                | deletion 5 |

15. a. 3, 6, and 7 are deletions (nonreverting);
    b. (2,5)   1 | 8   4   9
             |deletion 6 |
      |deletion7| |deletion 3 |

    c. Use other deletions in crosses with mutants 2 and 5.

17. a. Parental ditype: All spores Arg⁻. Nonparental ditypes: 2 Arg⁻ and 2 Arg⁺; b. Two of the PD spores grow on either ornithine, citrulline, arginosuccinate, or arginine; the other two grow with arginine only. NPD Arg⁻ grow with arginine only.

19. 45 purple:16 green:3 blue.

21. a. Mutant   18   14   9   10   21
    Pathway   ⟶D ⟶B ⟶A ⟶C ⟶thymidine

    b. 9 and 10 accumulates B; 10 and 14 accumulates D.

23. Outside

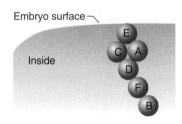

25. One chromosome with b d/b d; another with b/d only.

27. a. b^+/b^o and $b^+ b^{20}$; b. b^o:null, b^{20} 20%, b^+ 50%; c. 1/4; d. 1/4.

Chapter 7

1. a. 5; b. 10; c. 8; d. 12; e. 6; f. 2; g. 9; h. 14; i. 3; j. 13; k. 1; l. 7; m. 15; n. 11; o. 4.

3. a. GU GU GU GU GU or UG UG UG UG; b. GU UG GU UG GU UG GU UG GU UG GU; c. GUG UGU GUG U etc.; d. GUG UGU GUG UGU GUG UGU GUG UGU GU (depends on where you start); e. GUGU GUGU or UGUG UGUG (depends on where you start).

5. Hb^C therefore precedes Hb^S in the map of β-globin gene.

7. In vitro translation system.

9. Gene F: bottom strand; gene G: top strand.

11. Three.

13. a) Mutant 1: transversion Pro; mutant 2: deletion; mutant 3: transition Thr; mutant 4: insertion; mutant 5: transition to stop; mutant 6: inversion. b) EMS: 1, 3, 5. Proflavin: 2, 4.

15. Transcription: adding appropriate ribonucleotide complementary to template; translation: codon in mRNA and anticodon in tRNA are complementary.

17. a. UGG changed to UGA, so the DNA change was G to A. b. If the Arg codon is CGU, single base insertion leads to UAA. Mutation of A to T in the first base of the Lys codon leads to UAA. If the Gly codon is GGA, mutation of G to T creates stop codon. If Val codon is GUG, a deletion of the first base creates a UGA stop.

19. a. 3′ AUC 5′; b. 5′ CAG 3′; c. minimum two genes.

21. Missense mutations change identity of a particular amino acid inserted in many normal proteins but nonsense suppressors only make proteins longer.

23. a. Very severe; b. mild; c. very severe; d. mild; e. no effect; f. mild to no effect; g. severe; h. severe or mild.

Chapter 8

1. a. 10; b. 1; c. 9; d. 7; e. 6; f. 2; g. 8; h. 3; i. 5; j. 4.

3. Means to select cells in the transformation mix that received the vector.

5. a. 100%; b. 25%; c. none; d. 1/2 chance; e. 3/4.

7. a. Five; b. divide the number of base pairs in the genome by the average insert size, then multiply by 5.

9. d; g; e; h; c; f; b; a.

11. Shorter molecules slip through pores more easily; large molecules get caught.

13. EcoRI: 49 and 2400-bp fragments; MboI: 705-, 944-, 500-, and 300-bp or 905-, 744-, 500-, and 300-bp fragments.

15. a. A; b. 10 kb; c. 10 kb.
    d.

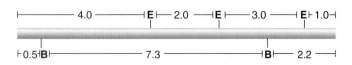

17.

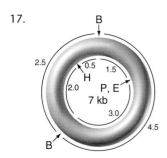

19. a. 1.83 kb.

b.

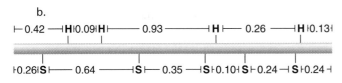

c.

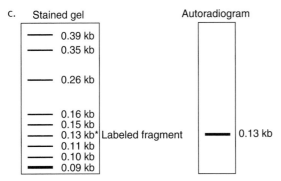

21. a. Use a probe of the same gene from another organism or if the amino-acid sequence is known, reverse translate to DNA sequence and make an oligonucleotide to use as a probe; b. 7 kb.

c.

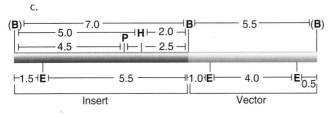

23. UAGAUAAGGAAUGUAAGAUAU<u>A</u>ACUGAGAUUUAAC

25. b.

Chapter 9

1. a. 5; b. 3; c. 8; d. 6; e. 2; f. 7; g. 1; h. 4.

3. a. Different numbers of the simple sequence repeats; b. slippage of DNA polymerase during replication; c. unequal crossing-over.

5. a. 3. b. Dad is heterozygous for 3 and 4 + 2 alleles. Mom is heterozygous for the 6- and 3-kb alleles. Sy is heterozygous for the 6 and 4 + 2 alleles.

7. a. 13-kb form; b. 7-kb form; c. 7.6-kb form; d. 0; e. 1/9.

9. Individuals should be heterozygous for a polymorphism.

11. a. Y-linked trait. II-3: XY^s; II-2 and II-4: XX. b. X-linked recessive. II-1: X^BY; II-3: X^bY; II-4: X^BX^b. c. Autosomal recessive. II-1: *Ss;* II-2: *ss;* II-4: *Ss.*

13. *In vitro* fertilization: man donates sperm, eggs collected from female. After fertilization, a single cell from eight-cell embryo is removed, DNA prepared and genotype analyzed using PCR. Appropriate embryos are implanted.

15. c; e; a; b; d.

17. An 18-base oligonucleotide including the mutant base.

19. Use a closely linked RFLP for which parents are heterozygous or look for polymorphism using SSCP.

21. a. Man could be the father; b. man could not be the father.

Chapter 10

1. a. 9; b. 6; c. 1; d. 5; e. 7; f. 3; g. 4; h. 2; i. 8.

3. a. Cut the clone into subfragments to use as probes and look for single bands in hybridization; b. oligos used to detect STSs surrounding a microsatellite could be used to amplify DNA from a large number of individuals in the population.

5. a. Beta; b. 16%.

7.

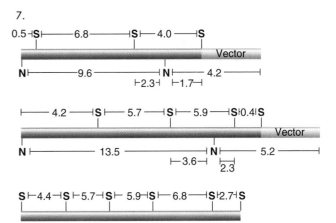

9. Two cosmids probably contain repeated DNA.

11. One: search for transcribed regions; two: search for open reading frames and splice sites; three: hybridize to look for cross-hybridizing DNA in other species.

13. 4.1-kb mRNA: from DNA between A and G with introns covering regions B, C, E; 3.4-kb mRNA: from DNA between B and E with intron covering region D; 1.8-kb mRNA: from DNA D–G with at least one intron in this region.

15. Different splicing of the same primary RNA.

17. Pseudogenes from cDNAs lack introns found in functional copy of the gene.

19. a. Protein motifs; b. tissue- and development-specific expression; knocking out the gene.

21. mRNAs present in all tissues could represent alternately spliced messages leading to different proteins that are not recognized by the antibody. Or, the two other messages are not translated.

23. a. Cloning a cDNA; b. clone into an expression vector.

25. Identify subcategories of the same disease and look for a different genetic basis for the subcategories.

Chapter 11

1. a. 4; b.10; c. 5; d. 8; e. 9; f. 2; g. 3; h. 6; i. 1; j. 7.

3. Interphase: 40-fold compaction; metaphase: 10,000-fold compaction.

5. Proteins involved in packaging of chromosomes, replication, segregation, transcriptional regulation and processing, recombination.

7. a. Unique DNA found next to the repeated T T A G G G sequences; b. heterogeneous telomere length

9. a. Integrates; b. maintained separately but mitotically unstable; c. maintained separately but mitotically unstable.

11. 0.6-kb *Sau*3A.

13. More genes adjacent to boundary would be expressed.

15. a. Constitutive heterochromatin: centromere DNA, facultative heterochromatin: *white* gene; b. constitutive heterochromatin: centromere DNA, facultative heterochromatin: inactive X.

17. a. 1; b. 0; c.1; d. 1; e. 3; f. 0.

19. Some patches of cone cells in the eye in which the X chromosome carrying the *CB* allele was inactivated would be defective in color vision. If several patches occurred next to each other, there might be partial color vision.

21. Mother.

23. a. *Aa*; b. *Aa*; c. *Aa*.

Chapter 12

1. a. 4; b. 8; c. 6; d. 5; e. 7; f. 3; g. 2; h. 1.

3. In a duplication, there would be a repeated set of bands; in a deletion, bands normally found would be missing.

5. a. and b.

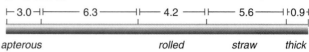

apterous ⊢3.0⊢⊢── 6.3 ──⊢⊢── 4.2 ──⊢⊢── 5.6 ──⊢⊢0.9⊢ rolled straw thick

7. a. 2 *URA3⁺ ARG9⁺* spores and 2 *ura3⁻ arg9⁻* spores; b. 2 spores die, 1 *URA3⁺ ARG9⁺* and one *ura3⁻ arg9⁻*; c. 4 viable spores, 2 *URA3⁺ ARG9⁺* spores and 2 *ura3⁻ arg9⁻* spores.

9. a. 2, 4; b. 2, 4; c. 2; d. 1, 3.

11. a. 1/4 will be fertile and green, 1/4 fertile and yellow-green, 1/4 semisterile and green, 1/4 semisterile and yellow-green; b. 1/2 fertile and yellow-green and 1/2 semisterile and green; c. from crossing-over between the translocation chromosome and homologous region on the normal chromosome.

13. a. 1, 3, 5, 6; b. 2, 4; c. 1, 3; d. 5, 6.

15.

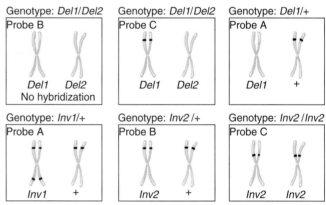

17. *Lyra* males and wildtype (*Lyra⁺*) females

19. Ds is a defective transposable element and Ac is a complete, autonomous copy.

21. Use probe of DNA from the sequence preceding the 200 A residues to hybridize to genomic DNA.

23. a. 7; b. sand oats: diploid, slender wild oats: tetraploid, cultivated oats: hexaploid; c. sand oats: 7, slender wild oats: 14, cultivated wild oats: 21; d. same answer as c.

25. A: meiosis II in father; B: meiosis I in mother; C: meiosis I in father; D: meiosis II in mother.

27. Treat with colchicine.

29. In autopolyploids, banding patterns of homologs should be the same; in allopolyploids, different banding patterns will be seen for chromosomes from different species.

Chapter 13

1. a. 4; b. 5; c. 2; d. 7; e. 6; f. 3; g. 1.

3. a. iv; b. iii; c. ii.

5. Southern hybridization using an IS1 DNA as a probe.

7. Transform the plasmid into a nontoxin-producing recipient strain.

9. *pyrE xyl mal arg met tyr his.*

11. *lac arg102 arg101 arg103.*

13. a. F plasmid must have integrated into the chromosome near the *mal* genes.

    b.

    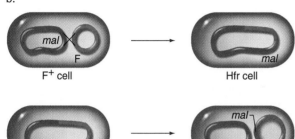

15. a)

    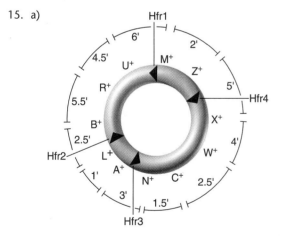

    b) Use an Hfr strain that transfers the gene late and screen for a derivative that transfers the gene early.

17. DNA being transferred by transformation is degraded; DNase has no effect on DNA being transduced.

19. Generalized transduction: transducing phage particles contain chromosomal DNA only; specialized transduction: particles contain phage and chromosomal DNA.

21. a. Cross Hfr (Tn10) strains with F⁻ strain containing the mutation. Tet^r is the selectable marker in the donor Hfr and the F⁻ has the Str^r counterselectable marker. To cover

genome, you need 20 different Hfrs. Replica mate and plate on selective media, then onto a plate at 30°C and 42°C to determine which Hfr transfer the gene early. b. 255 plasmids.

Chapter 14

1. a. 6; b. 8; c. 7; d. 2; e. 1; f. 3; g. 5; h. 9; i. 4.

3. a. both; b. both; c. neither; d. both.

5. a)

UGG	CAU CAC	AUA	AUG
Trp	His	Ile	Met

b)

UGG	CAU	AUC	AUG
UGA	CAC	AUU	AUA
Trp	His	Ile	Met

7. a, c, e.

9. Demonstrates it is not a nuclear gene that is getting passed on from the fertile male.

11. a. 3; b. 1; c. 2.

13. Resemble mother because sperm does not contribute cytoplasmic material to the zygote.

15. a. Resistant to erythromycin (genotype *ery^r*); b. genotype *ery^r/ery^s* and the phenotype reflects the dominant allele; c. sporulate the diploid. If the gene is nuclear, two spores would be *ery^s* and two would be *ery^r*. If the gene were mitochondrial, all four spores would be *ery^r*.

17. a. 2; b. 1; c. 4; d. 3.

19. a. Mother could have a small proportion of mutant genomes and the proportion in her daughter is higher or the mutation occurred in the germ line of I-2. b. Examine mother's somatic cells.

21. Differences in the time during development when mutation occurred or the proportion of mutant genomes in the cells in each tissue.

Chapter 15

1. a. 4; b. 8; c. 5; d. 2; e. 7; f. 1; g. 3; h. 6.

3. Nonlysogenic recipient cell did not have the cI (repressor) protein, so incoming infecting phage could go into the lytic cycle.

5.
β-galactosidase	Permease
a. constitutive	constitutive
b. constitutive	inducible
c. inducible	inducible
d. no expression	constitutive
e. no expression	no expression

7.

9. b.

11. a. 1 and 2; b. 2 and 4.

13. a. 4; b. 6; c. 7; d. 2; e. 3; f. 5; g. 1.

15. If the three genes make up an operon, they are cotranscribed as one mRNA and only one band should appear on a hybridization analysis using any of the three genes as a probe versus mRNA. If the genes are not part of an operon, there would be three hybridizing bands.

17. His codons (CAC or CAU).

19. a. Operator begins on the left site after the endpoint of deletion 1 and before the endpoint of deletion 5. The right endpoint cannot be determined by this set. b. removed bases within the promoter.

21. Mutations in O_2 or O_3 alone have only small effects in synthesis levels and would therefore be difficult to detect in the screens.

23. Protein coding region of your gene must be in the same frame as the *lacZ* gene.

Chapter 16

1. a. 8; b. 5; c. 1; d. 7; e. 3; f. 2; g. 6; h. 4.

3. Introns are spliced out, ribonuclease cleaves the primary transcript near the 3′ end, poly-A tail is added.

5. a. pol III; b. pol II; c. pol I.

7. a. DNA binding; b. DNA binding; c. dimerization; d. transcription activation; e. DNA binding.

9. Using probes from each of the three genes you would expect to see different signals corresponding to three different transcripts.

11. If Id acts by quenching, it interacts with MyoD, whereas if it blocks access to an enhancer, it binds to DNA. Experimentally, you could look for binding to the regulatory DNA of a gene regulated by MyoD.

13. RNA polymerase transcribes naked eukaryotic DNA at high levels *in vitro*.

15. b.

17. a. No expression; b. no expression; c. expression; d. no expression; e. no expression; f. expression; g. no expression.

19. a. 9:7 ratio of blue to white flowered plants; b. 15:1 ratio of blue to white.

21. The 5′ and 3′ regions could be cloned at the 5′ or 3′ ends of a reporter gene that is transformed back into *Drosophila* early embryos to see if either sequences affect the translatability of the reporter protein.

Chapter 17

1. a. 7; b. 6; c. 8; d. 2; e. 1; f. 9; g. 3; h. 5; i. 4.

3. Replica plate colonies onto two sets of plates—one incubated at 30°C and the other at 23°C. Cold-sensitive mutants defective in cell cycle genes should die at the nonpermissive temperature (23°C).

5. a. 4%; b. 96%.

7. a. F; b. T; c. F.

9. a. growth factors; b. cyclin, cyclin-dependent protein kinase, any molecule in the signal transduction pathway, receptor or molecule that transmits a second signal.

11. c; e; a; b; d.

13. Uncontrolled growth, genomic instability, potential for immortality, ability to invade and disrupt local and distant tissues.

15. The commonality is an accumulation of mutations necessary to cause cancer.

17. e; b; a or d or c; f.

19. c.

21. II-2 and I-1 or I-2.

Chapter 18

1. a. 3; b. 4; c. 5; d. 7; e. 2; f. 1; g. 6.

3. Isolate DNA from culture of each of the 10 clones, use pulse-field gel electrophoresis to compare sizes of chromosome XII in the samples.

5. a. Pulse-field gel electrophoresis, transfer to filter and probe with a subtelomeric repeat piece of DNA; b. use cloned piece of DNA as a probe to hybridize to a blot of chromosomes separated by gel electrophoresis.

7. *a*-maters, α-maters, and nonmaters.

9. a. α cell: *α1* + Mcm1 protein act as a positive regulator of α-specific gene transcription; *α2* + Mcm1 protein act to repress a-specific genes. *a* cell: Mcm1 protein transcribes *a*-specific genes. *a*/α cell: Mcm1 has no effect. b. α *mcm1⁻*: *a*-mater; *a mcm1⁻*: sterile; *a*/α *mcm1⁻*: sterile

11. a. Southern hybridization using *SUC2* DNA from *Saccharomyces cerevisiae* as a probe against genomic DNA from other *Saccharomyces* species; b. pseudogene; c. larger transcript requires SNF1 gene product, smaller one does not.

Chapter 19

1. a. 3; b. 1; c. 4; d. 5; e. 2.

3. Extra DNA is repetitive DNA that is not necessary for the physiology of the plant.

5. a. 3*N* formed by fertilization of one sperm nucleus and two 1*N* nuclei in the ovule, genotype is like that of embryo (from double fertilization event); b. 2*N* formed by fertilization of one sperm nucleus with the 1*N* egg, genotype is like endosperm; c. 2*N* cells from fertilization that lead to the plant—not the same genotype as the embryo.

7. a. 2; b. 3; c. 1.

9. Do manipulated crosses in which pollen from the sterile plant is used to fertilize a wildtype plant and the reciprocal cross (pollen from fertile plant fertilizes ovule of plant with sterile genotype).

11. a. Examine the gravitropic response in an auxin mutant; b. hybridize using an auxin-regulated gene as a probe versus ground up tissue from different parts of the plant or fixed tissue from the plant *or* make a fusion between an auxin regulated gene and a reporter gene such as *GUS* and look at the distribution of expression of the reporter gene in the intact plant.

Chapter 20

1. a. 2; b. 5; c. 4; d. 7; e. 6; f. 1; g. 3.

3. Self-fertilization makes it easier to recover recessive alleles.

5. b.

7. Lin14 protein level would be reduced further and cells influenced by Lin14 would differentiate at inappropriate times.

9. a. Cross males to the homozygous *dpy-10⁻/dpy-10⁻*; b. wildtype progeny from first cross are crossed to *daf-1⁻/daf-1⁻*; c. self-fertilization of the wildtype hermaphrodites from cross 2, 1/32 will have genotype desired.

11. a. Sequence of DNA from the mutant determined to look for a mutation in that particular K⁺ channel gene or introduce a wildtype copy of the candidate gene into the mutant; b. Southern hybridization or PCR; c. dominant; d. compare the phenotypic response to isoamyl alcohol and diacetyl alcohol in the wildtype and mutant strains.

Chapter 21

1. a. 4; b. 5; c. 1; d. 6; e. 3; f. 2.

3. Mate males having insertion(s) of the *w⁺* marked *P* elements on the X chromosome with females homozygous for the *w⁻* mutation (white eyes). Male flies in the next generation that have pigmented eyes should now have the *P* element on one of the autosomes.

5. e.

7. Segment polarity genes encode signal transduction proteins and act after cellularization has occurred. Gap and pair-rule genes encode transcription factors and act before cellularization.

9. Maternally supplied *hunchback* mRNA is completely dispensable and the function of the Nanos protein is only needed to restrict the translation of the maternally supplied *hunchback* mRNA in this posterior of the egg.

11. a. More than one progenitor cell forms any one facet. b. White gene is cell autonomous.

13. a. Time of expression during development and the cells in which the gene is expressed; b. heterozygous with a balancer chromosome; c. compare the pattern of *bicoid* mRNA in *homeless* vs wildtype embryos by hybridization using a *bicoid* DNA probe; d. Make females with the genotype centromere–*ovoᴰ–ho⁺*/centromere–*ovo⁺–ho⁻*, and induce mitotic recombination. If sterile, *homeless* is needed in the germ line. If fertile, *homeless* is needed in non-germline cells. e. Ras is needed for several different pathways, but the level of Ras protein required is different in the various signaling pathways.

Chapter 22

1. a. 3; b. 5; c. 1; d. 6; e. 7; f. 2; g. 4.

3. a. yeast; b. mouse; c. yeast; d. yeast and mouse.

5. In *Drosophila*, the zygote is a syncytium with polarity and division of the embryo is two results in loss of important information.

7. a. *CTF* is an essential gene; b. *CTF* gene affects growth in both males and females and is necessary for fertility in males; c. different restriction patterns.

9. a. MMTV c-*myc* gene should segregate with the phenotype; b. hybridization of MMTV sequences vs a library of genomic clones from the mutant; c. in the *ld* mutant, sequence the gene into which the transgene inserted in the other mutant; d. consistent.

Chapter 23

1. a. 3; b. 5; c. 8; d. 7; e. 6; f. 1; g. 9; h. 2; i. 4.

3. a. *M,* 0.6; *N,* 0.4; *MM* 0.36; *MN* 0.48; *NN,* 0.16; b. Nonrandom mating, small population size, migration into the population, mutation, or selection.

5. a, e.

7. Each allele frequency has a different set of genotype frequencies at equilibrium.

9. a. 0.1; b. 478; c. 0.1.

11. a. 0.29; b. 0.29; c. 0.30

13. 0.0064.

15. a. Wildtype, 0.888; vestigial, 0.111; b. *Vg,* 0.75; *vg,* 0.25; c. wildtype, 0.9375; vestigial, 0.0625.

17. A fully recessive allele is not expressed in a heterozygous organism, so there is no selection against the heterozygotes.

Sometimes there is a heterozygote advantage and mutation to the recessive allele will contribute to more mutant alleles.

19. a. founder effect; b. advantages: genetic homogeneity and fewer genes that may affect a polygenic trait, disadvantages: some mutations are not found in the population that are in the general population.

21. b

23. a. 1/256 or 0.0039; b. 0.0016.

Chapter 24

1. a. 4; b. 6; c. 5; d. 2; e. 1; f. 7; g. 3.

3. a, c.

5. a. The enzyme consists of an RNA molecule; b. the enzyme has both an RNA and a protein component.

7. a. Different constraints on the functions of each of the proteins; b. Rates are more constant because these base changes do not affect function of the gene product.

9. Suggests there is some benefit to the *CF* allele in the heterozygous state.

11. Duplication followed by evolutionary divergence.

13. a. 240 million years; b. two; *C* allele arose 30 million years ago; *B* allele arose 1 million years ago; c. duplication of *B:* transposition; duplication of *C:* misalignment and crossing-over.

15. This gene was introduced from a different species.

17. a. Exons; b. genes.

19. Mediate genome rearrangements or contribute regulatory elements adjacent to a gene.

21. a. SINES or LINES; b. centromere satellite DNA.

Note: An italicized word in a definition indicates that word is defined elsewhere in the glossary.

A

acentric fragment a chromatid fragment lacking a centromere; usually the result of a crossover in an inversion loop.

activator *trans*-acting proteins (transcription factors) that bind to enhancers. Compare *basal factors* and *repressors*.

adaptation the ability to stop responding when the stimulus is present. Refer *biased random walk*.

adenine (A) a nitrogenous base, one member of the *base pair* A–T (adenine-thymine).

adjacent-1 segregation pattern one of two patterns of segregation resulting from the normal disjunction of homologs during metaphase. Homologous centromeres disjoin so that one translocation chromosome and one normal chromosome go to each pole, resulting in each gamete containing a large duplication region and a correspondingly large deletion, making them genetically unbalanced. Contrast *alternate segregation pattern*.

adjacent-2 segregation pattern a pattern of segregation resulting from a nondisjunction in which homologous centromeres go to the same pole. The resulting genetic imbalances are lethal after fertilization.

allele frequency the proportion of all copies of a gene in a population that are of a given allele type.

allele specific oligonucleotides (ASOs) short oligonucleotides that hybridize with alleles distinguished by a single base.

alleles alternative forms of a single *gene*.

allopolyploids hybrids in which the chromosome sets come from two or more distinct, though related, species.

allosteric proteins proteins that undergo reversible changes in conformation when bound to another molecule.

alternate segregation pattern one of two patterns of segregation resulting from the normal disjunction of homologs during metaphase. Two translocation chromosomes go to one pole, while two normal chromosomes move to the opposite pole resulting in gametes with correct haploid number of genes. Contrast *adjacent-1 segregation pattern* or *adjacent-2 segregation pattern*.

amino acids the building blocks of proteins.

amphidiploids organisms produced by two diploid parental species; they contain two diploid genomes, each one derived from a different parent.

anaphase the stage of mitosis in which the connection of sister chromatids is severed, allowing each chromatid to be pulled toward the spindle pole to which it is connected by its kinetochore microtuble.

aneuploid an individual whose chromosome number is not an exact multiple of the haploid number for the species.

angiosperm plant descriptive meaning the seeds of the plant are enclosed in an ovary within the flower.

angstrom (Å) a unit of wavelength light equal to one ten-billionth of a meter.

anonymous locus a designated position on a chromosome with no known function; see *locus*.

antenna-pedia complex (ANT-C) in *Drosophila;* a homeotic mutation affecting segments in the head and anterior thorax.

apical meristem a group of undifferentiated cells that divides continuously, located at the growing points of both shoots and roots in some plants.

apoptosis programmed cell death (PCD); a process in which cellular DNA is degraded and the nucleus condenses; the cell may then be devoured by neighboring cells or phagocytes.

archaea one of the three major evolutionary lineages of living organisms known as domains. Analyses of rRNA indicates that the archaea represent a primary biological line of evolution related to both bacteria and eukaryotes.

artificial selection the purposeful control of mating by choice of parents for the next generation. Contrast *natural selection*.

artificial transformation a process by which bacteria transfer genes from one strain to another; process in which laboratory procedures weaken cell walls and make membranes permeable to DNA; this step is required before most bacterial species are able to take up DNA. Contrast *natural transformation*.

ascospores in some fungi, the haploid cells that result from meiosis. Also known as *haplospores*.

ascus sac-like structure in some varieties of fungi that houses all four haploid products of meiosis.

autopolyploids a kind of polypoid that derives all of its chromosome sets from the same species.

autosome a chromosome not involved in sex determination. The diploid human genome consists of 46 chromosomes, 22 pairs of autosomes, and 1 pair of sex chromosomes (the X and Y chromosomes).

autotroph a plant that is nutritionally self-sufficient; it produces its own food via photosynthesis.

auxotroph a mutant microorganism that can grow on minimal medium only if it has been supplemented with one or more growth factors not required by wild-type strains.

B

bacterial chemotaxis bacteria movement up and down gradients of chemical attractants to avoid or seek chemical diffusing through their growth medium. Interaction between genes and proteins are the mechanisms of movement.

bacterial chromosomes essential component of a typical bacterial genome; a single circular molecule of double-helical DNA.

bacteriophage See *phage*.

balancer chromosome special chromosome often created by use of X-rays for purposes of genetic manipulation; these chromosomes carry multiple, overlapping inversions that enable researchers to follow them through crosses, and a recessive lethal mutation that prevents the survival of homozygotes.

Barr bodies inactive X chromosomes observable at interphase as darkly stained heterochromatin masses.

basal factors *trans*-acting protein (transcription factors) that binds to promoters. Compare *activators* and *repressors*.

base analogs   mutagens that are so similar in chemical structure to the normal nitrogenous bases in DNA that the replication mechanism can incorporate them into DNA in place of the normal bases. Because the base analog may have pairing properties different from those of the base it replaces, it can cause base substitutions on the complementary strand synthesized in the next round of DNA replication.

base pair (bp)   two nitrogenous bases held together by weak bonds. The two strands of DNA are held together in the shape of a double helix by the bonds between base pairs. Adenine pairs with thymine and guanine pairs with cytosine.

base sequence   the order of nucleotide bases in a DNA molecule.

base sequence analysis   a method for determining a *base sequence.*

basic chromosome number   the number of different chromosomes that make up a single complete set.

b-form DNA   the most common form of DNA in which molecular configuration spirals to the right. Compare *Z-form DNA.*

biased random walk   bacterial movement resulting from the addition of an attractant or repellant. The time spent in a straight run is longer immediately after the addition, the movement over time is biased toward or away from the chemical gradient, though each direction change continues to be random.

biochemical pathway   the orderly series of reactions that allows an organism to obtain simple molecules from the environment and convert them step-by-step into successively more complicated molecules.

biological systems   complex interactive networks that function both within individual cells and among groups of cells within an organism.

biotechnology   a set of biological techniques developed through basic research and now applied to research and product development. In particular, the use by industry of recombinant DNA, cell fusion, and new bioprocessing techniques.

bithorax complex (BX-C)   in *Drosophila;* homeotic mutation that affects the segments in the abdomen and posterior thorax.

bivalent   containing pairs of synapsed homologous chromosomes.

blastocysts   describes the embryo at the 16-cell stage of development through the 64-cell stage when the embryo implants.

blastomeres   early embryonic cells.

C

cancer genes   mutant alleles that lead to cancer; mutant alleles of normal genes.

carpels   structure in a plant comprising the fourth whorl, usually two are fused together; houses the female gametes in the form of ovules. The fused carpels are part of a cylinder known as the *pistil.*

cDK inhibitor   See *cyclin-dependent kinases.*

cDNA libraries   libraries which store sequences copied into DNA from RNA transcripts; these sequences carry only the exon information for making proteins.

cell cycle   repeating pattern of cell growth, replication of genetic material, and mitosis.

cellular blastoderm   in *Drosophila* embryo; one-cell-deep epithelial layer resulting from the egg cortex growth inward during the fourteenth cycle of nuclear division.

centimorgan (cM)   a unit of measure of recombination frequency. One cM is equal to a 1% chance that a marker at one genetic *locus* will be separated from a marker at a second locus due to *crossing-over* in a single generation.

centromere   a specialized chromosome region to which spindle fibers attach during cell division.

checkpoints   mechanisms that have evolved to arrest the cell cycle while repair takes place; called checkpoints because they check the integrity of the genome and cell-cycle machinery before allowing the cell to continue to the next phase of the cell cycle.

chiasmata   observable regions in which nonsister chromatids of homologous chromosomes cross-over each other.

chimeras   an embryo or animal composed of cells from two or more different organisms; the opposite of *clones.*

chi square (χ^2) test   a statistical test to determine the probability that an observed deviation from the expected event or outcome occurs solely by chance.

chromatin   the generic term for any complex of DNA and protein found in a cell's nucleus.

chromocenter   the dense heterochromatic mass formed by the fusing of the centromeres of polytene chromosomes.

chromosomal interference   the phenomenon of crossovers not occurring independently. Refer to *crossing-over.*

chromosomal puff   the region of a polytene chromosome that swells to from a large, diffuse structure when high rates of transcription cause the band containing that gene to decondense.

chromosomal rearrangement   mutations that produce rearrangements of normal DNA sequences.

chromosome theory of inheritance   the idea that chromosomes are the carriers of genes.

chromosomes   the self-replicating genetic structures of cells containing the cellular DNA that carries in its nucleotide sequence the linear array of genes.

cis-acting elements   DNA sequences that serve as attachment sites for the DNA-binding proteins that regulate the initiation of transcription. Categories of *cis*-acting elements include promoters, enhancers, and reporter constructs.

cleavage stage   in early embryonic development, the stage of the first four equal cell divisions.

clones   a group of cells derived from a single ancestor.

cloning   the process of asexually producing a group of cells (clones), all genetically identical, from a single ancestor. In recombinant DNA technology, the use of DNA manipulation procedures to produce multiple copies of a single gene or segment of DNA is referred to as cloning DNA.

cloning vector   DNA molecule originating from a cell of a higher organism into which another DNA fragment of appropriate size can be integrated without loss of the vectors' capacity for self-replication. Vectors introduce foreign DNA into host cells, where it can be reproduced in large quantities. See *vector.*

codominant   expression of heterozygous phenotype resulting in offspring that resemble both parents. Parental traits show up equally and fully in the offspring.

coefficient of coincidence   the ratio between the actual frequency of double crossovers observed in an experiment and the number of double crossovers expected on the basis of independent probabilities.

colinearity   the parallel between the sequence of nucleotides in a gene and the order of amino acids in a polypeptide.

colony   a mound of genetically identical cells.

competent   description of state of cells able to take up DNA from the medium.

complement   process in which the alleles on each of the two homologs make up for the defect in the other chromosome, generating enough of both gene products to yield a normal phenotype.

complementary base pairing   during DNA replication, base paring in which a complementary strand aligns opposite the exposed bases on the parent strand to create the nucleotide sequence of the new strand of DNA. Refer to *base pair, complementary sequences.*

complementary gene action   genes working in tandem to produce a particular trait.

complementary sequences   nucleic acid base sequences that can form a double-stranded structure by matching base pairs; the complementary sequence to G T A C is C A T G.

complementation group   a collection of mutations that do not complement each other. Often used synonymously for gene.

complementation table   a method of collating data that helps visualize the relationship among a large group of mutants. See figure 6.13c.

complementation test   method of discovering whether a particular phenotype arises from mutations in the same or separate genes.

complete coverage   in mapping of DNA, when the number of markers on a *linkage map* is sufficient so that the locus controlling the phenotype in question can be linked to at least one of those markers.

complete digest   a method of cutting a DNA fragment in which the fragment is exposed to an enzyme for enough time to allow the restriction enzyme to cut every one of the recognition sites on the DNA fragment. Compare *partial digest.*

complete genomic library   a hypothetical collection of DNA clones that includes one copy of every sequence in the entire genome.

complex   refers to the multiple types of variation that can exist at alternative alleles, including more than one nucleotide substitution, a substitution in combination with a small deletion, duplication, or other insertion.

complex haplotype   a set of linked DNA variations along a chromosome, with the possibility of many differences in alternative alleles. See *haplotype.*

concerted evolution   process that allows changes in single genetic elements to spread across a complete set of genes in a particular gene family.

conditional lethal   an allele that is lethal only under certain conditions.

conjugation   one of the mechanisms by which bacteria transfer genes from one strain to another; in this case, the donor carries a special type of plasmid that allows it to transfer DNA directly when it comes in contact with the recipient. The recipient gene is known as an exconjugant. Contrast *transformation.*

conjugative plasmids   plasmids that initiate conjugation because they carry the genes that allow them to transfer themselves to the recipient.

consanguine   related by a common ancestor.

consanguineous mating   mating between blood relatives.

conserved synteny   state in which the same two or more loci are found to be linked in several species. Compare *syntenic.*

constitutive heterochromatin   chromosomal regions that remain condensed in heterochromatin at most times in all cells.

continuous trait   inherited trait that exhibits many intermediate forms; determined by segregating alleles of many different genes whose interaction with each other and the environment produces the phenotype. Also called quantitative trait. Compare *discontinuous trait.*

core histones   proteins that form the core of the nucleosome; identified as H2A, H2B, H3 and H4.

corepressor   an effector molecule whose binding to the actual *repressor* protein allows the negative regulator to bind to DNA and inhibit transcription of the genes in the *operon.*

cosmids   hybrid plasmid-phage *vectors* that make use of a virus capsule to infect bacteria; constructed with plasmid-derived selectable markers and two specialized segments of phage λ DNA known as *cos* (for *co*hesive end) sites.

cotransformation   simultaneous transformation of two or more genes. See *transformation.*

criss-cross inheritance   inheritance pattern in which males inherit a trait from their mothers, while daughters inherit the trait from their fathers.

cross   the deliberate mating of selected parents based on particular genetic traits desired in the offspring.

cross-fertilization   to brush the pollen from one plant onto the female organ of another plant, thereby creating offspring with the particular traits of the selected parent plants.

cross-fostering   random relocation of offspring to the care of other parents, typically done with animal studies to randomize the effects of environment on outcome.

crossing-over   during *meiosis,* the breaking of one maternal and one paternal chromosome, resulting in the exchange of corresponding sections of DNA, and the rejoining of the chromosome. This process can result in the exchange of alleles between chromosomes. Compare *recombination.*

crossover suppressors   result of inversions, in which no recombinants among the viable progeny of an inversion heterozygote are possible.

c-terminus   the end of the polypeptide chain that contains a free carboxylic acid group.

cyclin-dependent kinases (CDKs)   a protein kinase is an enzyme that adds a phosphate group to a target protein molecule; in this case, the kinases are dependent on a protein known as cyclin for their activity. CDKs regulate the transition from G_1 to S and from G_2 to M through phosphorylations that activate or inactivate target proteins.

cytokinesis   the final stage of cell division, which begins during anaphase but is not completed until after telophase. In this stage the daughter nuclei emerging at the end of telophase are packaged into two separate daughter cells.

cytosine(C)   a nitrogenous base, one member of the base pair G-C (guanine and cytosine).

D

dauer larva   in nematodes, an alternate L3 form that does not feed and has a specialized cuticle that resists desiccation; survival mechanism when food is scarce allows larva to survive at least six months; when food becomes more available, they return to the normal development.

deamination   the removal of an amino (NH_2) group from normal DNA.

degrees of freedom (df)   the measure of the number of independently varying parameters in an experiment.

deletion   occurs when a block of one or more nucleotide pairs is lost from a DNA molecule. Compare *insertion.*

deletion loop   an unpaired bulge of the normal chromosome that corresponds to an area deleted from another homolog. Contrast *duplication loop.*

denatured   unfolding of proteins, usually due to exposure to urea, heat, or changes in pH that disrupt the interactions between amino acids that normally stabilize the *secondary* and *tertiary* structures.

deoxyribonucleic acid (DNA)   See *DNA.*

deoxyribonucleotide   See *nucleotide.*

depurination   DNA alteration in which the hydrolysis of a purine base, either A or G, from the deoxyribose-phosphate backbone occurs.

dicentric chromatid   a chromatid with two centromeres.

dicotyledonous   plant descriptive meaning that the mature embryo carries two leaves.

dihybrid   an individual that is heterozygous for two genes at the same time.

dimerization domain transcription factor domain, specialized for specific polypeptide-to-polypeptide interactions. Certain motifs recur, such as the leucine zipper motif.

dipeptide a pair of amino acids connected by a peptide bond.

diploid containing a full set of genetic material; zygotes and other cells carrying two matching sets of chromosomes are described as diploid (compare *haploid*).

discontinuous trait phenotypes that are expressed in clear-cut variations. Compare *continuous trait.*

discrete trait inherited trait that exhibits clearly an either/or status (that is, purple versus white flowers).

disease stratification the process of dividing a single general condition into different diseases on the basis of the underlying molecular defects.

DNA deoxyribonucleic *acid*; the molecule of heredity that encodes genetic information.

DNA clones replicated DNA fragments with identical genomic DNA.

DNA fingerprint the multilocus pattern produced by the detection of genotype at a group of unlinked, highly polymorphic loci.

DNA markers See *marker.*

DNA probes short, single-stranded stretches of DNA of known composition, from 10 to several thousand nucleotides in length. Usually labeled with either radioactive nucleotides or fluorescent dyes for use in identifying clones that contain complementary DNA sequences and diagnosis of DNA alterations that underlie genetic diseases.

DNA sequence the relative order of *base pairs,* whether in a fragment of DNA, a *gene,* a *chromosome,* or an entire *genome.*

DNA topoisomerases a group of enzymes that help relax *supercoiling* of the DNA helix by nicking one or both strands to allow the strands to rotate relative to each other and thereby restore normal coiling density.

DNase hypersensitive (DH) sites sites on the DNA strand that contain few, if any, nucleosomes; these sites are susceptible to DNase enzymes, allowing the DNA strand to be fragmented when treated with very small amounts of DNase.

domain a discrete portion of a protein with its own function. The combination of domains in a single protein determines its overall function.

dominance series dominance relations of all possible pairs of alleles are arranged in order from most dominant to most recessive for a particular genotype.

dominant allele an allele that expresses its phenotype even when heterozygous with a recessive allele.

dominant epistasis the allele causing the epistasis is dominant. Compare *recessive epistasis.*

dominant negative mechanism of dominance in which some alleles of genes encode subunits of multimers that block the activity of the subunits produced by normal alleles.

donor in gene transfer in bacteria, the cell that provides the genetic material. See *recipient, transformation, conjugation, transduction.*

double helix the shape that two linear strands of DNA assume when bonded together.

downstream the direction traveled by RNA polymerase as it moves from the 5′ end of a gene from point A to point B. Compare *upstream.*

duplication events that result in the increase in the number of copies of a particular chromosomal region. See *tandem duplications, nontandem duplications.*

duplication loop a bulge in the *Dp*-bearing chromosome that has no similar region with which to pair in the unduplicated normal homologous chromosome. Contrast *deletion loop.*

E

ecotypes plant varieties analogous to animal strains.

ectopic expressions gene expressions that occur outside the cell or tissue where the gene is normally expressed.

electrophoresis a method of separating large molecules (such as DNA fragments or proteins) from a mixture of similar molecules. An electric current is passed through a medium containing the mixture and each kind of molecule travels through the medium at a different rate, depending on its size and electrical charge. Separation is based on these differences.

elongation the second step in DNA replication, in which proteins connect the correct sequence of nucleotides on to a continuous new strand of DNA. Compare *initiation.*

embryonic stem cell (ES cells) cultured embryonic cells that continue to divide without differentiating.

endosymbiont theory proposes that chloroplasts and mitochondria originated when free-living bacteria were engulfed by primitive nucleated cells. Host- and guest-formed cellular communities in which each member adapted to the group arrangement and derived benefit.

enhancer trapping in *Drosophila,* the identification of P-element insertion lines with particular β-galactosidase patterns.

enhancers a *cis*-acting regulatory site, which can be quite distant from the promoter. In yeast, enhance elements are often called upstream activation sites, or UASs.

enhancesome an assembled LCR-transcription factor complex. Refer to *Locus Control Region.*

environmental variance deviation from the mean attributed to the influence of external, noninheritable factors. Compare *genetic variance.*

epigenetic literally, outside the genes.

episomes plasmids, like the F plasmid, that can integrate into the host chromosomes.

epistasis a gene interaction in which the effects of an allele at one gene hide the effects of alleles at another gene.

equilibrium frequency the *q* value at which $\Delta q = 0$; the allele frequency required to maintain an allele in the population.

euchromatin chromosomal region of cells that appears much lighter and less condensed when viewed under a light microscope. Contrast *heterochromatin.*

eukaryotes one of the three major evolutionary lineages of living organisms known as domains; differentiated as organisms whose cells have a membrane-bound nucleus. Contrast *prokaryotes.*

eukaryotic gene regulation the control of gene expression in the cells of *eukaryotes.*

euploid cells containing only complete sets of chromosomes.

evolution in the study of genetics, changes in allele frequency in a population over multiple generations; the basic components of evolution at the molecular level include diversification, selection, and amplification.

evolutionary equilibrium the allele frequency at which evolution no longer continues; represented by p̂.

excision repair DNA repair mechanism in which specialized proteins recognize damaged base pairing, and remove the damaged section so that DNA *polymerase* can fill in the gap by complementary base pairing with the information from the undamaged strand of DNA.

exons the protein-coding DNA sequences of a gene. See *introns.*

exon shuffling the exchange of exons among different genes, producing mosaic proteins with two or more distinct functions.

expression vectors cloning vehicles, which promote the expression of a polypeptide product from the gene inserts they carry.

expressivity the degree or intensity with which a particular genotype is expressed in a phenotype.

F

f plasmid the first conjugative plasmid to be identified; carries many genes required for the transfer of DNA. Cells carrying F plasmid are called F^+ cells. Cells without the plasmid are called F^- cells.

f′ plasmid plasmid variation that carries most of the genes of F plasmid plus some bacterial DNA; this variant is particularly useful in genetic complementation studies.

facultative heterochromatin regions of chromosomes (or even whole chromosomes) that are heterochromatic in some cells and euchromatic in other cells of the same organisms.

filial generations the successive offspring in a controlled sequence of crossbreeding, starting with two specific parents (the P generation) and selfing or intercrossing the offspring of each subsequent generation . The first generation is referred to as the first filial (F_1) generation, the second as the second filial (F_2) generation, and so on.

FISH (fluorescence in situ hybridization) a *physical mapping* approach that uses fluorescein tags to detect hybridization of probes with metaphase chromosomes and with somatic interphase chromatin.

fitness cost negative impact of the development of a homozygous resistance genotype.

fluctuation test basis of the Luria-Delbrück experiment to determine the origin of bacterial resistance. Variations in the generation of mutant colonies showed up as fluctuations in the numbers of resistant colonies growing in different petri plates.

focus of action in a gene of interest; the cells in which the gene must be active to allow the animal to develop and function normally.

forward mutation a mutation that changes a *wild-type allele* of a gene to a different *allele*.

founder cells in nematode development, progenitors of the major embryonic lineages.

founder effect variation of genetic drift, occurring when a few individuals separate from a larger population and establish a new one that is isolated from the original population, resulting in altered allele frequencies in the new population.

G

frameshift mutation changes that alter the grouping of nucleotides into codons. Refer to *reading frame*.

free duplications small chromosomal fragments maintained as extrachromosomal in a genetic stock.

G

gametogenesis the formation of gametes.

gastrulation folding of the cell sheet early in embryo formation; usually occurs immediately after the blastula stage of development.

gene basic unit of biological information; specific segments of DNA composed of distinctive sets of nucleotide pairs in a discrete region of a chromosome that encodes a particular protein; a coding *locus*.

gene amplification an increase from the normal two copies to hundreds of copies of a gene; often due to mutations in *p53*, which disrupt the G_1-to-S checkpoint.

gene conversion any deviation from the expected 2:2 segregation of parental alleles.

gene dosage the number of times a given gene is present in the cell nucleus.

gene expression the process by which a gene's coded information is converted into the structures present and operating in a cell. Expressed genes include those that are transcribed into mRNA and then translated into protein and those that are transcribed into RNA but not translated into protein.

gene family set of closely related genes with slightly different functions that most likely arose from a succession of gene duplication events.

gene function generally, to govern the synthesis of a polypeptide; in Mendelian terms, a gene's specific contribution to *phenotype*.

gene pool the sum total of all alleles carried in all members of a population.

generalized transduction enzymatic process, which can result in the transfer of any bacterial gene between related strains of bacteria.

genetic code the sequence of *nucleotides*, coded in triplets (*codons*) along the *mRNA* that determines the sequence of amino acids in protein synthesis.

genetic drift unpredictable, chance fluctuations in allele frequency that have a neutral effect on fitness of a population.

genetic linkage See *linkage*.

genetic markers identifiable genes that can serve as points of reference in determining whether particular progeny are the result of recombination. Compare *physical markers*.

genetic relatedness the average fraction of common alleles at all gene loci that

individuals share because they inherited them from a common ancestor.

genetic screen an examination of each colony in a population for its phenotype.

genetic variance (V_G) deviation from the mean attributable to inheritable factors. Compare *environmental variance*.

genetics the science of heredity.

genome the entire collection of chromosomes in each cell of an organism.

genomic equivalent represents the number of clones needed in a perfect library; to calculate size for any genome, simple divide the length of the genome by the average size of the inserts carried by the library's vector.

genomic imprinting the phenomenon in which a gene's expression depends on the parent that transmits it.

genomics a branch of biology dedicated to the development and application of more effective mapping, sequencing, and computational tools.

genotype the actual alleles present in an individual.

genotype frequency proportion of total individuals in a population that are of a particular genotype.

germ cells specialized cells that incorporate into the reproductive organs where they ultimately undergo meiosis, thereby producing gametes that contain half the number of chromosomes as other body cells. Germ cells are responsible for transmitting genes to the next generation of an organism. Compare *somatic cells*.

germ line gene therapy a genetic engineering technique that modifies the *germ cells* that are passed on to progeny.

α-globin locus chromosomal region carrying all of the α-globin-like genes.

β-globin locus chromosomal region carrying all the β-globin-like genes.

globular stage embryo stage of development in a plant in which cell divisions give rise to the first evidence of differentiation found in the mature embryo. The innermost group of cells acquire an elongated shape and a discrete outer cell layer is present.

growth factors extracellular hormones and cell-bound signals that stimulate or inhibit cell proliferation.

guanine (G) a nitrogenous base, one member of the base pair G–C (guanine and cytosine).

H

haploid a single set of chromosomes present in the egg and sperm cells of animals and in the egg and pollen cells of plants (compare *diploid*).

haplospores See *ascospores*.

haplotype specific combination of linked alleles in a cluster of related genes. A contraction of the phrase "haploid genotype."

Hardy-Weinberg law defines the relationships between genotype and allele frequencies within a generation and from one generation to the next; named for two scientists, G.H. Hardy and W. Weinberg, who independently developed it early in the 20th century.

heredity the way genes transmit biochemical, physical, and behavioral traits from parents to offspring.

heritability (h^2) the proportion of total phenotypic variance ascribable to *genetic variance.*

heterochromatin highly condensed chromosomal regions of cells that appear much darker when viewed under a light microscope. Contrast *euchromatin.*

heterochronic mutations mutations resulting in the inappropriate timing of cell division and cell-fate decisions during development.

heteroduplex region site that marks the spots of crossing-over. The segment of the DNA molecule located between two breakpoints have reannealed; beyond this region both strands of one DNA molecule have switched places with both strands of its homolog, or when the initiating and resolving cuts are on the same DNA strand, the one short segment of one strand may trade places with one short segment of a homologous nonsister strand. In either case, mismatch repair may alter one allele to another.

heterogeneous trait a mutation at any one of a number of genes can give rise to the same phenotype.

heterogenotes partial diploids carrying different alleles of the same gene. Contrast *merozygotes.*

heteromers multimeric proteins composed of nonidentical subunits. Compare *homomers.*

heteroplasmic genomic makeup of a cell's organelles characterized by a mixture of organelle genomes. Contrast *homoplasmic.*

heterothallic strains of organisms with stable haploid mating types; progeny of successive mitotic divisions always have the same mating type as their parents. Contrast *homothallic.*

heterozygous a genotype in which the two copies of the gene that determine a particular trait are different. See *hybrid.*

heterozygous advantage the situation where heterozygotes have a higher fitness than either homozygote; sited as the explanation for the survival of recessive genetic diseases.

heterozygous carrier unaffected parents who bear a dominant normal allele that masks the effects of an abnormal recessive one.

hfr bacteria that produce a *h*igh *f*requency of *r*ecombinants for chromosomal genes in mating experiments.

high-density linkage map a *linkage map* that shows one gene or marker for each centimorgan of a genome.

histocompatibility antigens three major genes that encode a class of cell surface molecules that play a critical role in stimulating a proper immune response that destroys invaders while leaving the body's own tissue intact; known as *HLA-A, HLA-B,* and *HLA-C.*

histones small proteins with a preponderance of the basic, positively charged amino acids lysine and arginine. The histones' strong positive charge enables them to bind to and neutralize the negatively charged DNA throughout the chromatin.

homeobox in *homeotic genes,* the region of homology, usually 180 bp in length, located within the coding sequence of the gene.

homeotic gene a gene that plays a role in determining a tissue's identity during development.

homeotic mutation mutation that causes cells to misinterpret their position in the blueprint and become normal organs in inappropriate positions; such a mutation can alter the overall body plan.

homeotic selector genes genes that control the development of body segments.

homologous chromosomes (homologs) chromosomes that match in size, shape, and banding. A pair of chromosomes containing the same linear gene sequence, each derived from one parent.

homologs genes or regulatory DNA sequences that are similar in different species because of descent from a common ancestral sequence.

homomers multimeric proteins composed of identical subunits. Compare *heteromers.*

homoplasmic genomic makeup of a cell's organelles characterized by a single type of organelle DNA. Contrast *heteroplasmic.*

homothallic strains of organisms that can switch mating types; progeny of successive mitotic divisions may not have the same mating type as their parents. Compare *heterothallic.*

homozygous a genotype in which the two copies of the gene that determine a particular trait are the same.

hot spots sites within a gene that spontaneously mutate more frequently than others.

Human Genome Program collective name for several projects begun in 1986 by DOE to (1) create an ordered set of DNA segments from known chromosomal locations, (2) to develop new computational methods for analyzing genetic map and DNA sequence data, and (3) develop new techniques and instruments for detecting and analyzing DNA. The national effort led by DOE and NIH is known as the Human Genome Project.

hybrid dysgenesis mutation in which chromosome breakage resulting from high transposon mobility causes reduced fertility in hybrid progeny; the result of crossing *Drosophila* males carrying the P element with females that lack the P element.

hybridization the process of joining two complementary strands of DNA or one each of DNA and RNA to form a double-stranded molecule.

hybrids offspring of genetically dissimilar parents.

hydrogen bonds weak electrostatic bonds that result in a partial sharing of hydrogen atoms between reacting groups.

hypermorphic mutation produces an allele generating either more protein than the wild-type allele or the same amount of a more efficient protein. If excess protein activity alters phenotype, the hypermorphic allele is dominant. Compare *hypomorphic mutation.*

hypomorphic mutation produces either much less of a protein or a protein with a very weak but detectable function. Compare *hypermorphic mutation.*

I

imaginal discs flattened epithelian sacs that develop from small groups of cells set aside in the early embryo from which adult specific structures like wings, legs, eyes and genitalia develop; in *Drosophila,* they undergo extensive growth and development during the larval and pupal stages.

immunoglobulin homoly unit common feature of the immunoglobulin gene superfamily; encodes about 100 amino acids, which fold into a characteristic three-dimensional structure termed the immunoglobulin fold.

incomplete dominance expression of heterozygous phenotype resulting in offspring that do not resemble either parent. Offspring phenotype is intermediate between those of the parents.

independent assortment the random distribution of genes during gamete formation. See *Mendel's second law.*

inflorescence   in a plant, the flower or group of flowers at the tip of a branch.

initiation   first stage in DNA replication, in which the proteins open up the double helix and prepare it for complementary base pairing. Compare *elongation*.

initiation codon   nucleotide triplet that marks the precise spot in the nucleotide sequence of an mRNA where the code for a particular polypeptide begins. Compare *nonsense codon*.

insertion   the addition to a DNA molecule of one or more nucleotide pairs.

insertion sequences (ISs)   small transposable elements that dot the chromosomes of many types of bacteria; they are transposons that do not contain selectable markers.

intercalators   class of chemical mutagenic compounds composed of flat, planar molecules that can sandwich themselves between successive base pairs and disrupt the machinery of replication, recombination, or repair.

intergenic gene conversion   information flow between related DNA sequences that occurs through an alternative outcome of the process responsible for unequal crossing-over.

interphase   the period in the cell cycle between divisions.

intragenic suppression   the restoration of gene function by one mutation canceling another mutation in the same gene.

introns   the DNA base sequences interrupting the protein-coding sequences of a gene; these sequences are transcribed into RNA but are cut out of the message before it is translated into protein. See *exons*.

inversion heterozygotes   in cells heterozygous for inversion ($In/+$), the chromosome carrying an inversion pairs with its homolog at meiosis, creating an inversion loop.

inversion loop   formed when one chromosomal region rotates to conform to the similar region in the other homolog. This allows the tightest possible alignment of homologous regions on a DNA segment to occur.

K

karyotype   the visual description of a complete set of chromosomes in one cell of an organism; usually presented as a photomicrograph of an individual cell with the chromosomes arranged in a standard format showing the number, size, and shape of each chromosome type; used in low-resolution physical mapping to correlate gross chromosomal abnormalities the characteristics of specific traits or diseases. See *physical map*.

kinetochore   a specialized structure composed of DNA and proteins that is the site at which chromosomes attach to the spindle fibers.

knockout constructs   cloned genes modified so that they no longer function.

L

lac **operon**   a single DNA unit, composed of the *lacZ*, *lacY*, and *lacA* genes together with the promoter (p) and operator (o), that enables the simultaneous regulation of the three structural genes in response to environmental changes.

late-onset   a genetic condition in which symptoms are not present at birth, but manifest themselves later in life.

law of independent assortment   See *Mendel's second law*.

law of the product   states that the probability of two or more independent events occurring together is the product of the probabilities that each event will occur by itself.

law of segregation   See *Mendel's first law*.

law of the sum   states that the probability of either of two mutually exclusive events occurring is the sum of their individual probabilities.

lawn   bacteria immobilized in a nutrient agar, used as a field on which to test for the presence of viral particles.

lethal gene   a gene whose expression results in death.

library   a collection of DNA clones that contains multiple copies of nearly every fragment in the whole genome inserted into a suitable vector and placed into storage.

LINE   long *in*terspersed *e*lements; one of the two major classes of transposable elements.

lineage compartments   specific regions of cells to which clones are often limited.

linkage   the proximity of two or more markers on a chromosome; the closer together the markers are, the lower the probability that they will be separated during DNA repair or replication processes, thereby increasing the probability that they will be inherited together.

linkage group   a group of genes chained together by linkage relationships. See *linkage*.

linkage maps   depict the distances between *loci* as well as the order in which they occur on the organism.

linker DNA   a strand of 40 base pairs of DNA that connect one nucleosome with the next.

locus   a designated location on a chromosome. See *α-globin locus* and *β-globin locus*.

locus control region (LCR)   a *cis*-acting regulatory element that influences the transcription of all genes in the β-globin cluster; it functions by binding to transcription factors with active domains.

losses or gains   event in which the genome is reshaped by the losses or gains of entire chromosomes or sets of chromosomes. Compare *rearrangements*.

lysate   population of phage particles released from the host bacteria at the end of the lytic cycle.

lysogenic bacterium   bacterial cell that carries a prophage; lysogen.

lysogeny   the integration of the viral DNA into the host chromosome.

lytic cycle   bacterial cycle resulting in cell lysis and release of progeny phage.

M

macroevolution   in the study of genetics, the process by which new species emerge from existing species. Contrast *microevolution*.

mapping   the process of determining the locus of a gene on a particular chromosome.

mapping function   mathematical equation that compensates for the inaccuracies inherent in relating recombination frequencies to physical distance.

mapping panels   a set of DNA samples used in multiple linkage analysis tests.

marker   an identifiable physical location on a chromosome, whose inheritance can be monitored. Markers can be expressed regions of DNA (genes) or some segment of DNA with no known coding function but whose pattern of inheritance can be determined.

maternal-effect mutations   mutations caused by maternal components (those supplied by the mother) of inheritance.

mean   statistical average; the middle point.

meiosis   the process of two consecutive cell divisions in the diploid progenitors of sex cells. Meiosis results in four rather than two daughter cells, each with a haploid set of chromosomes.

Mendel's first law   the law of segregation states that the two alleles for each trait separate (segregate) during gamete formation, then unite at random, one from each parent, at fertilization.

Mendel's law of segregation   See *Mendel's first law*.

Mendel's second law   the law of independent assortment states that during gamete formation, different pairs of alleles segregate independently of each other.

merozygotes partial diploids in which the two gene copies are identical. Contrast *heterogenotes*.

messenger RNA (mRNA) RNA that serves as a template for protein synthesis. See *genetic code*.

metabolism the chemical and physical reactions that convert sources of energy and matter to fuel growth and repair within a cell.

metamorphosis a dramatic reorganization of an organism's body plan; e.g., transition from larval stage to adult insect stage.

metaphase a stage in mitosis or meiosis during which the chromosomes are aligned along the equatorial plane of the cell.

metazoans the first multicellular animals, which appeared about 0.7 billion years ago.

methylated cap formed by the action of the enzyme methyl transferase at the 5′ end of eukaryotic mRNA, critical for efficient translation of the mRNA into protein.

methyl-directed mismatch repair DNA repair mechanism that corrects mistakes in replication through enzyme action of the methyl groups on the parental strands of DNA.

microevolution alterations of a population's *gene pool*. Contrast *macroevolution*.

microsatellite DNA element composed of 15–100 tandem repeats of one-, two-, or three-base sequences.

minisatellite DNA element composed of longer (10–40 bp) tandem repeating units of identical sequence lined up like boxcars in a freight train.

mitosis the process of nuclear division in cells that produces daughter cells that are genetically identical to each other and to the parent cell.

mitotic nondisjunction the failure of two sister chromatids to separate during mitotic anaphase generates reciprocal trisomic and monosomic daughter cells.

model organisms used in genomic analysis because they have many genetic mechanisms and cellular pathways in common with each other and with humans. These organisms lend themselves well to classical breeding experiments and direct manipulation of the genome.

modification the phenomenon in which growth on a restricting host changes a virus so that succeeding generations grow more efficiently on that same host.

modification enzymes enzymes that add methyl groups to specific DNA sequences during restriction.

modifier genes produce a subtle, secondary effect on phenotype by altering the alleles of other genes. There is no formal distinction between major and modifier genes, rather it is a continuum of degrees of influence.

molecular clock hypothesis stating that one can evaluate a constant rate of change for each type of molecule in proteins as a means of determining the genealogies of organisms.

molecular cloning the multistep process that takes a collection of restriction enzymes and uses living cells to purify and make many exact replicas of individual genomic fragments.

molecular marker a segment of DNA found at a specific site in the genome of an organism, that has inherent properties that make it easily recognizable when subjected to technology tools such as gel electrophoresis. See *physical marker, genetic marker.*

monohybrid crosses crosses between parents that differ in only one trait.

monohybrids individuals having two different alleles for a single trait.

monomorphic a gene with only one wild-type allele.

monosomic individual lacking one chromosome from the diploid number for the species.

morphogens substances that define different cell fates in a concentration-dependent manner.

mosaic mistakes in mitotic division resulting in an organism containing tissues of different genotypes.

mosaic development the embryo is a collection of self-differentiating cells that at their formation receive a specific set of molecular instructions governing their unique fates. Contrast *regulative development.*

mRNA a type of base substitution occurring after transcription and RNA splicing.

multifactorial traits determined by two or more factors, including multiple genes interacting with each other or one or more genes interacting with the environment.

multihybrid crosses crosses between parents that differ in three or more traits.

mutagen any physical or chemical agent that raises the frequency of mutations above the spontaneous rate.

mutation alterations of the genetic material occurring in gamete-producing cells that are heritable. Researchers define as changes in the base sequence that affect phenotype.

N

n-terminus the end of a polypeptide chain that contains a free amino group that is not connected to any other amino acid.

natural selection in nature, the process that progressively eliminates individuals whose fitness is low and chooses individuals of high fitness to survive and become the parent of the next generation. Contrast *artificial selection.*

natural transformation a process by which bacteria transfer genes from one strain to another; in a few species of bacteria DNA fragments are spontaneously picked up from their surroundings. Contrast *artificial transformation.*

nondisjunction failures in chromosome segregation during meiosis; responsible for defects such as trisomy.

nonhomologous unequal crossing-over recombination resulting from unequal crossing-over, believed to be mediated by a short stretch of sequence homology, coding or noncoding, at the two sites at which the event is initiated.

nonparental ditype (NPD) a tetrad containing four recombinant spores; the two parental classes have recombined to form reciprocal, nonparental combinations of alleles.

nonsense codons the three stop codons that terminate *translation.* Compare *initiation codon.*

nontandem duplications sometimes called dispersed duplications; two or more copies of a region that are not adjacent to each other and may lie far apart on the same chromosome or on different chromosomes. Contrast *tandem duplications.*

Northern blot analysis protocol for determining whether a fragment of DNA is transcribed in a particular tissue. In this protocol, RNA transcripts in the cells of a particular tissue are separated by gel electrophoresis. Compare *Southern blot analysis.*

nuclear lamins type of cellular proteins that serve as CDK substrates, in particular, a group of proteins that underlie the inner surface of the nuclear membrane.

nucleolar organizer clusters of rRNA genes on long loops of DNA from several chromosomes within a nucleolus.

nucleolus large sphere-shaped organelle visible in interphase eukaryotic cells with the use of a light microscope. The most obvious nuclear structure of a eukaryotic cell.

nucleosome rudimentary DNA packaging unit, composed of *core histones.*

nucleotide a subunit of DNA or RNA consisting of a nitrogenous base (adenine, guanine, thymine, or cytosine in DNA; adenine, guanine, uracil, or cytosine in RNA), a phosphate molecule, and a sugar molecule. Thousands of nucleotides are linked to form a DNA or RNA molecule.

null hypothesis   a statistical hypothesis to be tested and accepted or rejected in favor of an alternative; the prediction that an observed difference is due to chance alone and not due to a systematic cause.

null mutations   mutations that abolish the function of a protein encoded by the wild-type allele. Such mutations either prevent synthesis of the protein or promote synthesis of a protein incapable of carrying out any function.

O

octant stage embryo   early stage of embryo formation during cell divisions of a plant.

oligonucleotides   DNA probes composed of short, synthesized chains of nucleotides.

oligopeptide   several amino acids linked by peptide bonds.

oncogene   a gene, one or more forms of which is associated with cancer. Many oncogenes are involved, directly or indirectly, in controlling the rate of cell growth.

oogonia   diploid germ cells in the ovary.

open reading frame (ORFs)   amino-acid sequences identified as genes that encode proteins with functions that have not yet been assigned; characterized by long stretches of codons in the same reading frame uninterrupted by stop codons.

operator site   site created by the binding of repressor proteins to a DNA sequence near the promoter of the lactose utilization genes; functions to block the promoter, and occurs only when lactose is not present in the medium.

operon   a unit of DNA composed of specific genes, plus a promoter and/or operator, which act in unison to regulate the response of the structural genes to environmental changes.

oxidative phosphorylation   a set of reactions requiring oxygen that creates portable packets of energy in the form of ATP.

P

p-**value**   numerical probability that a particular set of observed experimental results represents a chance deviation from the values predicted by a particular hypothesis.

parental ditype (PD)   a tetrad that contains four parental class haploid cells.

parental (P) generation   pure-breeding plants selected as the mating individuals whose progeny will be studied for specific traits. Refer to *filial generations*.

partial digest   a method of cutting a DNA fragment by controlling the amount of enzyme or the amount of time the DNA is exposed to the restriction enzyme, thereby producing fragments averaging 20 kb in length (cuts occurring at approximately one in every five target sites). Compare *complete digest*.

pedigree   an orderly diagram of a family's relevant genetic features, extending back to at least both sets of grandparents and preferably through as many more generations as possible.

peptide bond   a covalent bond that joins amino acids during protein synthesis.

peptide motifs   stretches of amino acids conserved in many otherwise unrelated polypeptides.

petals   the structure composing the second *whorl* of a flower, leaf-shaped but containing no photosynthetic cells, therefore not green.

phage   short for *bacteriophage;* a *virus* for which the natural host is a bacteria cell; literally "bacteria eaters."

phage induction   process of phage DNA excision from bacterial chromosome and entry into the lytic cycle.

phage lambda (λ)   a naturally occurring double-stranded DNA virus that infects *E.coli*. A common plasmid *vector* used to study DNA structure of other organisms.

phenocopy   a change in phenotype arising from environmental agents that mimic the effects of a mutation in a gene. Phenocopies are not heritable because they do not result from a change in the gene.

phenotype   an observable characteristic.

phenotype frequency   the proportion of individuals in a population that are of a particular phenotype.

pheromones   hormones that play a role in eliciting sexual behavior by releasing chemical signals.

phosphodiester bonds   covalent bonds joining one *nucleotide* to another. Phosphodiester bonds between nucleotides form the backbone of DNA.

photomorphogenesis   light-regulated developmental program of a plant.

photoperiod   day length; period of light exposure for plants.

phylogenetic tree   diagram composed of nodes, which represent the taxonomic units, and branches, which represent the relationship of these units.

physical map   a map of locations of identifiable landmarks on DNA (e.g., restriction enzyme cutting sites, genes)

regardless of inheritance. Distance is measured in base pairs. For the human genome, the lowest-resolution physical map is the banding patterns on the 24 different chromosomes; the highest resolution map would be the complete nucleotide sequence of the chromosomes. See *karyotype*.

physical markers   cytologically visible abnormalities that make it possible to keep track of specific chromosome parts from one generation to the next. Compare *genetic markers*.

pistil   floral structure that consists of fused carpels, with pollen receptive stigma at the top, and a short neck or style, leading to the ovary.

plaque   during a test for bacteriophage activity, a resulting clear area of the petri dish, devoid of living bacterial cells, indicating the presence of viral particles.

plasmids   minute circles of double-stranded DNA; the simplest vectors that can gain admission to, and replicate in, the cytoplasm of many kinds of bacterial cells, independently of the nuclear chromosomes.

pleiotropy   phenomenon in which a single gene determines a number of distinct and seemingly unrelated characteristics.

point mutation   a mutation of one nucleotide.

polar body   the smaller, diploid sister cell resulting from meiosis I.

polarity   an overall direction.

poly-A tail   the 3′ end of eukaryotic mRNA consisting of 100–200 As, believed to stabilize the mRNAs and increase the efficiency of the initial steps of translation.

polygenic   traits controlled by multiple genes. See *continuous traits*.

polymer   a linked chain of repeating subunits that form a larger molecule; DNA is a type of polymer.

polymerase chain reaction (PCR)   a faster, less expensive method of replicating a DNA sequence once the sequence has been identified; a reiterative loop in which an operation is repeatedly applied to the products of earlier rounds of the same operation.

polymerase, DNA or RNA   enzymes that catalyze the synthesis of nucleic acids on preexisitng nucleic acid templates, assembling RNA from ribonucleotides or DNA from deoxyribonucleotides.

polymorphic   genes that have more than one wild-type allele.

polypeptides   amino-acid chains made up of proteins containing hundreds to thousands of amino acids joined by *peptide bonds*.

polyploids euploid species that carry three or more complete sets of chromosomes.

polytene chromosome giant chromosomes consisting of many identical chromatids lying in parallel register.

population a group of interbreeding individuals of the same species that inhabit the same space at the same time.

population genetics scientific discipline that studies what happens in whole populations at the genetic level.

positional cloning the process that enables researchers to obtain the clone of a gene without any prior knowledge of its protein product or function. It utilizes large-scale maps to make the molecular connection between phenotype and genotype.

primary structure the linear sequence of amino acids within a polypeptide.

primary transcript the single strand of RNA resulting from *transcription.*

primer short, pre-existing polynucleotide chain to which new DNA can be added by DNA polymerase.

primer walking a common approach to directed sequencing of DNA that begins with the application of two vector-based primers flanking the two ends of an insert. The primers make it possible to sequence into the unknown insert for 600–800 nucleotides.

programmed cell death (PCD) See *apoptosis.*

prokaryotes one of the three major evolutionary lineages of living organisms known as domains; characterized by the lack of a nuclear membrane. Contrast *eukaryotes.*

prokaryotic cells or organisms lacking a discrete, membrane-bound nucleus. Compare *eukaryote.* See *chromosome.*

prokaryotic gene regulation coordinated control of gene expression in a bacterial cell via mechanisms such as activating, increasing, diminishing, and preventing the *transcription* and *translation* of specific genes or groups of genes.

promoters DNA sequences near the beginning of genes that signal RNA polymerase where to begin transcription. Compare *terminators.*

prophage the integrated phage genome.

prophase the phase of the cell cycle marked by the emergence of the individual chromosomes from the undifferentiated mass of chromatin, indicating the beginning of mitosis.

proteins large polymers composed of hundreds to thousands of amino-acid subunits strung together in a specific order into long chains. The order is determined by the base sequence of nucleotides in the gene coding for the specific protein. Proteins are required

for the structure, function, and regulation of the body's cells, tissues, and organs.

protooncogene a *gene* that can mutate into an *oncogene*—an *allele* that causes a cell to become cancerous.

pseudogenes DNA sequences that look like, but do not function as, a gene. They occur in all higher eukaryotic genomes.

pulse-field gel electrophoresis a special type of electrophoretic protocol that allows an extreme extension of the normal separating capacity of the gels. In this protocol, the DNA sample is subjected to pulses of electrical current that alternate between two directions. The range of sizes separated in this manner is a function of pulse length. Compare *electrophoresis.*

punctuated equilibrium the tendency of evolution to proceed through long periods of stasis (lack of change) followed by short periods of explosive change.

pure-breeding lines families of organisms that produce offspring with specific parental traits that remain constant from generation to generation.

Q

quantitative trait See *continuous traits.*

quantitative trait loci (QTLs) loci that control the expression of *continuous traits.*

quaternary structure structure in a complex protein made up of the three-dimensional configuration of subunits in the multimer.

R

radial loop-scaffold model the model of looping and gathering of nonhistone proteins that results in extreme compaction of chromosomes at mitosis.

random amplification of polymorphic DNA (RAPD) a protocol designed to detect single-base changes at polymorphic loci throughout a genome.

random walk description of movement of bacteria to achieve chemotaxis reflecting nonpredictable changes in the direction of movement. Contrast *biased random walk.*

reading frame the partitioning of groups of three nucleotides such that the sequential interpretation of each succeeding triplet generates the correct order of amino acids in the resulting polypeptide chain. This assumes that each codon is a trio of nucleotides, and for each gene there is a starting point.

rearrangements event in which the genome is reshaped by the reorganization of DNA sequences

within one or more chromosomes. Compare *losses or gains.*

receptors three-part proteins composed of a signal-binding site outside the cell, a transmembrane segment, and an intracellular domain that relays the signal to proteins inside the cell's cytoplasm; they receive the signals sent by *growth factors.*

recessive allele an allele whose phenotype is not expressed in a heterozygote.

recessive epistasis special case of epistasis, in which the allele causing the epistasis is recessive. Compare *dominant epistasis.*

recipient during gene transfer in bacteria, the cell that receives the genetic material. See *donor, transformation, conjugation,* and *transduction.*

reciprocal crosses crosses in which the male and female traits are reversed, thereby controlling whether a particular trait is transmitted by the egg or the pollen.

reciprocal translocation results when two breaks, one in each of two chromosomes, yield DNA fragments that do not religate to their chromosome of origin; rather they switch places and become attached to the other chromosome. Compare *translocation.*

recombinant DNA molecules a combination of DNA molecules of different origin that are joined using recombinant DNA technologies. A cutting and splicing together of vector and inserted fragment from different origins.

recombinant DNA technology tools of genetic analysis that allow researchers to search through enormously long strings of information for genes that may be only several thousand *base pairs* in length.

recombinant type a phenotype reflecting a new combination of genes occurring during gamete formation.

recombinants chromosomes that carry a mix of alleles derived from different homologs.

recombination the process by which offspring derive a combination of genes different from that of either parent; the generation of new allelic combinations. In higher organisms, this can occur by *crossing-over.*

recombination frequency (RF) the percentage of recombination; can be used as an indication of the physical distance separating any two genes on a chromosome. See *centimorgan.*

regulative development developmental process in which blastomeres can regulate and adjust their development to make up for missing cells. Contrast *mosaic development.*

replicon (replication unit) the DNA running both ways from one origin of replication; the endpoints where it merges with DNA from adjoining replication forks.

repressor a negative regulator. Refer to *prokaryotic gene regulation.*

λ repressor protein proteins that bind to the infecting phage DNA, preventing the new phage from transcribing its genes, thus making the host cell immune to infection by incoming λ particles.

reproductive development period in plant development that produces flower and seed. Contrast *vegetative development.*

response to selection (R) the amount of evolution, or change in mean trait value, resulting from selection. Compare *selection differential.*

restriction the bacterial capacity for limiting viral growth.

restriction enzymes proteins that recognize specific, short nucleotide sequences and cut DNA at those sites. Bacteria contain over 400 such enzymes that recognize and cut over 100 different DNA sequences.

restriction fragment length polymorphism (RFLP) variation between individuals in DNA fragment size cut by specific *restriction enzymes; polymorphic* sequences that result in RFLPs are used as markers on both *physical maps* and *genetic linkage* maps. RFLPs are usually caused by *mutation* at a cutting site.

restriction map a linear diagram that illustrates the order of different size fragments within a clone.

retroposons genetic elements that transpose via reverse transcription of an RNA intermediate. One class of *transposable elements.* Contrast *transposons.*

retroviruses viruses that transmit their genetic information in molecules of RNA and carry as part of their gene transmission kit the unusual enzyme *reverse transcriptase.*

reverse mutation a mutation that causes a novel mutant to revert back to wild-type.

reverse transcriptase RNA-dependent DNA *polymerase.*

reverse transcription variation of transcription in which a double strand of DNA is constructed from an RNA template. Compare *transcription.*

reversion See *reverse mutation.*

ribonucleic acid (RNA) a chemical found in the nucleus and cytoplasm of cells; it plays an important role in protein synthesis. It has a structure similar to DNA. There are several classes of RNA molecules, including messenger RNA (mRNA), *transfer RNA*

(tRNA), ribosomal RNA (rRNA), and other small RNAs, each serving a different purpose.

ribosomes small cellular components composed of specialized ribosomal RNA (rRNA) and protein; the site of protein synthesis.

ribozymes RNA molecules that can act as enzymes to catalyze specific chemical reactions.

RNA See *ribonucleic acid.*

RNA editing the process that converts pre-mRNAs to mature mRNAs. Essential step for the expression of mitochondrial genes, because without RNA editing, the pre-mRNAs do not encode polypeptide.

RNA-mediated interference mutations that result from the activity of double-stranded RNA that interferes with the function of the gene transcribed onto that RNA.

RNA splicing a process that deletes *introns* and joins together adjacent *exons* to form a mature mRNA consisting only of exons.

RNA world hypothetical primordial world in which RNA became the first replicator.

Robertsonian translocation translocation arising from breaks at or near the centromeres of two acrocentric chromosomes. The reciprocal exchange of broken parts generates one large metacentric chromosome and one very small chromosome. Named for W.R.B. Robertson who first suggested their mechanism of translocation.

S

satellite DNAs the blocks of repetitive, simple noncoding sequences containing centromeres; these blocks have a different chromatin structure and different higher-order packaging than other chromosomal regions.

scaffold-associated regions (SARs) special, irregularly spaced repetitive base sequences of DNA that associate with nonhistone proteins to define chromatin loops. SARs are most likely the sites at which DNA is anchored to the condensation scaffold.

secondary structure characteristic geometric localized region of a polypeptide chain generated by the *primary structure.*

segmentation gene in *Drosophila,* a large group of genes responsible for subdividing the body into an array of identical body segments.

segregation equal separation of alleles for each trait during gamete formation, in which one allele of each gene goes to each gamete.

selectable markers vector genes that make it possible to pick out cells harboring a particular DNA molecule.

selection a processes that establishes conditions in which only the desired mutant will grow.

selection differential (S) measure of the strength of selection on a trait. Compare *response to selection.*

self-fertilization (selfing) fertilization in which both egg and pollen come from the same plant, resulting in offspring with the same genetic traits as the single parent.

semi-conservative replication a pattern of double helix duplication in which complementary base pairing followed by the linkage of successive nucleotides yields two daughter double helices that each contain one of the original DNA strands intact (conserved) and one completely new strand.

semi-sterility a condition in which the capacity of generating viable offspring is diminished by at least 50%.

sepals green leaf-like structure composing the first *whorl* of a flower.

sequence tagged sites (STSs) one-of-a-kind markers that tag positions along the DNA molecule.

sequencing determining the order of *nucleotides* (base sequences) in a DNA or RNA molecule or the order of amino acids in a *protein.*

sex chromosomes the X and Y chromosomes in human beings that determine the sex of an individual. Females have two X chromosomes in diploid cells; males have an X and a Y chromosome. The sex chromosomes comprise the 23rd chromosome pair in a karyotype. Compare *autosome.*

sex-influenced traits traits which show up in both sexes but are expressed differently in each sex due to hormonal differences.

sex-limited traits traits that affect a structure or process that is found in one sex but not the other.

signal transducers cytoplasmic proteins that relay signals inside the cell.

signal transduction the activation and inhibition of intracellular targets after growth-factor binding.

signal transduction system in mutants in the *che* (chemotaxis) gene; the mechanism by which messages from the cell surface, where nutrients are detected, are relayed to the motor, where the messages control the frequency with which the direction of rotation of the flagella changes. Refer to *signal transduction* and *signal transducers.*

SINE short *interspersed elements;* one of the two major classes of transposable elements. Contrast *LINE.*

somatic cells any cell in an organism except gametes and their precursors. Compare *germ cells*.

somatic gene therapy remedial measures in which a replacement gene is inserted into affected tissue to compensate for a faulty gene.

Southern blot analysis protocol developed by British scientist, Edward Southern, to provide information on the location of a gene within a DNA fragment. DNA fragments are separated by gel electrophoresis in order to identify specific base sequences. Compare *Northern blot analysis*.

specialized transducing phages *bacteriophage* carrying mainly phage DNA but also one or a few of the bacterial genes that lie near the site of *prophage* insertion. They can transfer genes from one bacterium to another in the process known as specialized transduction.

spermatogenesis the production of sperm.

spliceosome a complicated intranuclear machine that ensures that all of the splicing reactions take place in concert.

stamen structure comprising the third *whorl* of a flower that bears the male gametes in the form of pollen.

stigma pollen receptive structure at the top of the pistil of a flower.

subcloning cloning smaller pieces of DNA that are enriched for sequences of interest.

substitution occurs when a base at a certain position in one strand of the DNA molecule is replaced by one of the other three bases.

substrate the target of an enzyme, usually cellular proteins.

supercoiling additional twisting of the DNA molecule caused by movement of the replication fork during unwinding. Refer to *DNA topoisomerases*.

suspensor in plants, a structure analogous to the umbilical cord in mammals.

switch genes genes identified by heterochronic mutations that effect loss-of-function and gain-of-function mutations causing them to having opposite temporal effects.

syncytial blastoderm in *Drosophila* embryos, result of the ninth division, when most of the nuclei migrate out to the cortex just under the surface of they embryo.

syncytium an animal cell with two or more nuclei.

syntenic relationship of two or more loci found to be linked in one species; literally "on the same thread" or chromosome. Compare *conserved synteny*.

T

TAFs *TBP-associated factors*; one form of basal factors.

TBP *TATA box-binding protein*; key component of basal factors that assist the binding of RNA polymerase II to the promoter and the initiation of low levels of transcription called basal transcription; forms on most promoters.

tandem duplications repeats of a chromosomal region lie adjacent to each other, either in the same order or in reverse order. Contrast *nontandem duplications*.

telomerase an enzyme critical to the successful *replication* of DNA.

telomeres specialized terminal structures on eukaryotic chromosomes that ensure the maintenance and accurate replication of the two ends of each linear chromosome.

telophase the final stage of mitosis in which the daughter chromosomes reach the opposite poles of the cell and re-form nuclei.

temperate marked by moderation; in bacteriophages; after infecting the host, they can enter either the lytic cycle or the alternative lysogenic cycle, during which their DNA integrates into the host genome and multiplies with it, doing little harm to the host. Compare *virulent*.

template strand the one strand of the double helix that serves as the pattern for the *mRNA* and is complementary to both the RNA-like DNA and the mRNA.

teratype (T) a tetrad carrying four kinds of haploid cells; two different parental class spores and two different recombinants. Teratypes result from a *crossover* between one of the two genes and the centromere of the chromosome on which it is located.

terminalization shifting of the chiasmata from their original position at the centromere toward the chromosome end or telomere. Refer to *chiasmata*.

terminators sequences in the RNA products that tell RNA polymerase where to stop transcription. Compare *promoters*.

tertiary structure ultimate three-dimensional shape of a protein.

tetrad the assemblage of four ascospores (resulting from meiosis) in a single *ascus*.

tetrad analysis the use of *tetrads* to study gene *linkage* and *recombination* during *meiosis*.

tetrasomic organism with four copies of a particular chromosome.

thymine a nitrogenous base, one member of the base pair A–T (adenine–thymine).

torpedo stage embryo stage of embryonic development in plants where two protuberances expand and differentiate into two well-defined, discrete cotyledons.

total phenotype variance (VP) population deviation calculated as the sum of the *genetic variance* and the *environmental variance*.

totipotent description of cell state during early embryonic development in which the cells have not yet differentiated and retain the ability to produce every type of cell found in the developing embryo and adult animal.

transcript the product of *transcription*.

transcription the conversion of DNA-encoded information to its RNA-encoded equivalent.

transcription bubble the region of DNA unwound by RNA *polymerase*.

transcription factors generic name for the *trans*-acting proteins that influence transcription by activating the expression of specific genes in the nucleus, either to promote or inhibit cell proliferation.

transcriptional silencing hypercondensation over chromatin domains makes it impossible to activate genes within a heterochromatin region, no matter what transcription factors are active in the cell.

transduction one of the mechanisms by which bacteria transfer genes from one strain to another; donor DNA is packaged within the protein coat of a bacteriophage and transferred to the recipient when the phage particle infects it. Recipient gene is known as a transductant.

transfection transformation of mammalian cells via the process of targeted mutagenesis.

transfer RNA (tRNA) small RNA adaptor molecules that place specific amino acids at the correct position in a growing polypeptide chain.

transformation one of the mechanisms by which bacteria transfer genes from one strain to another; occurs when DNA from a donor is added to the bacterial growth medium and is then taken up from the medium by the recipient. The recipient gene is known as a transformant. Contrast *conjugation*.

transgene any piece of foreign DNA that researchers have inserted into the genome of a complex organism, through experimental manipulation of early stage embryos or germ cells.

transgenic any individual carrying a *transgene*.

transient (or triangular) stage embryo stage following *globular stage* in plant development; embryo becomes self-sufficient for growth, having used up most of its maternally deposited reserves.

translation   the process in which the genetic code carried by mRNA directs the synthesis of proteins from amino acids. Compare *transcription.*

translocation   a rearrangement of a DNA sequence that occurs when parts of two nonhomologous chromosomes change places. Compare *reciprocal translocation.*

transposable elements   all DNA segments that move about in the genome, regardless of mechanism.

transposition   the movement of small fragments of DNA from one position on the genome to another.

transposons   whole units of DNA that move from place to place within the genome without an RNA intermediate; sometimes causing a change in the DNA function when they insert themselves in a new chromosomal location.

trisomic   individual having one chromosome in addition to the normal diploid set of the species.

U

unequal crossing-over   change in DNA caused by erroneous recombination in which one homologous chromosome ends up with a duplication while the other homolog sustains a deletion.

uniparental inheritance   transmission of genes via one parent. Most species transmit mtDNA and cpDNA through the mother.

upstream   movement opposite the direction RNA follows when moving along a gene. Compare *downstream.*

uracil   a nitrogenous base normally found in RNA but not in DNA; uracil is capable of forming a *base pair* with adenine.

V

variance   statistical measurement of deviation from the mean (middle); typically expressed in plus or minus terms referring to the relationship to the mean.

vector   a specialized DNA sequence that can enter a living cell, signal its presence to an investigator by conferring a detectable property on the host cell, and provide a means of replication for itself and the foreign DNA inserted into it. A vector must also possess distinguishing physical traits by which it can be purified away from the host cell's genome. See *cloning vector.*

vegetative development   in plants, the period of growth before flowering. Contrast *reproductive development.*

vernalization   the process of exposing plants to cold early in the vegetative development to promote earlier flowering.

virulent   marked by a rapid, severe, and malignant course. In bacteriophages, after infecting the host, they always enter the lytic cycle, multiply rapidly and kill the host. Compare *temperate.*

virus   a noncelluler biological entity that can reproduce only within a host cell. Viruses consist of nucleic acid covered by protein. Inside an infected cell, the virus uses the synthetic capability of the host to produce progeny virus.

W

whorls   in the structure of a flower, the concentric regions of modified leaves including *sepals, petals, stamens,* and *carpels.*

wild-type allele   an allele whose frequency is greater than 1%, designated by a superscript plus sign ($^+$). See *allele frequency.*

Y

YAC   See *yeast artificial chromosome.*

yeast artificial chromosome (YAC)   a vector used to clone DNA fragments (up to 400 kb); it is constructed from telomeric, centromeric, and replication origin sequences needed for replication in yeast cells. Compare *cloning vector, cosmid.*

Z

Z-form DNA   DNA in which the nucleotide sequences cause the structure to assume a zig-zag shape due to the helices spiraling to the left. Compare *B-form DNA.* The significance of this variation on DNA structure in unknown at this time.

zygote   the diploid cell formed by the fertilization of the egg by the sperm during sexual reproduction.

Text and Line Art

Chapter 1

Box 1 Figure B: From E. J. Mange & A. P. Mange, *Basic Human Genetics*, 1994. Sunderland, MA: Sinauer Associates.

Chapter 2

Box 2.1*a*: Provided by Institut National d'Etudes Demographiques. Reprinted with permission from SCIENCE, "Fragment of a Family Tree," April 19, 1991, p. 369 © American Association for the Advancement of Science.

Chapter 5

Figure 5.10: From *Molecular Cell Biology* by H. Lodish, et al. © 1986, 1990, 1995 by Scientific American Books. Used with permission by W. H. Freeman and Company.

Chapter 8

Figure 8.1: From Weatherall, DJ, and Clegg, JB. *The Thalassemia Syndrome*, 3rd Ed. Oxford: Blackwell Scientific Publications, 1981.

Chapter 9

Figure 9.19: From Cell Press, Huntington's Disease Gene, p. 977.

Chapter 10

Figure 10.2*a,b*: Provided by Dr. Mark Guyer. Reprinted with permission from SCIENCE, vol. 258, p. 79. © 1992 American Association for the Advancement of Science. **Figure 10.13*a*:** Reprinted with permission from Gusella, JF, et al. "DNA markers for nervous system diseases," SCIENCE 225: 1320. © 1984 American Association for the Advancement of Science. **Figure 10.13*b*:** From *American Journal of Human Genetics*, 39:382–391 (1980) by Magenis, et al. (James Gusella Lab). Published by The University of Chicago Press. **Figure 10.15:** From *Nature Genetics*, vol. 7, 1994, p. 475. New York: Nature Publishing Co. **Figure 10.19*b*:** Edwin H. McConkey, *Human Genetics: The Molecular Revolution.* © 1993 Sudbury, MA: Jones and Bartlett Publishers, *www.jbpub.com.* Reprinted with permission.

Chapter 11

Figure 11.2*b(2)*: Provided by Daniel Starr. **Figure 11.12*b*:** Courtesy of B. Alberts, et al., *Essential Cell Biology.* New York: Garland Publishing, 1998.

Chapter 12

Figure 12.1: Derived in part from MF Seldin, et al., *Mammalian Genome,* 4, S10 (1993); Ibid., S47 (1993).

Chapter 13

Figure 13.26: Illustration by Bunji Tagawa.

Chapter 14

Figure 14.15*b*: Shoffner, et al., "Myoclonic Epilepsy and Ragged-Red Fiber Disease (MERRF) Is Associated with a Mitochondrial DNA tRNA (Lys) Mutation," *Cell* 61:931–37 (1990), p. 932.

Chapter 15

Figure 15.6*c*: Provided by Mitchell Lewis. Reprinted with permission from SCIENCE, "View of the *lac* repressor-DNA Complex Monomer," vol. 271, p. 250. © 1996 American Association for the Advancement of Science. **Figure 15.12*a*:** Courtesy of B. Alberts, et al., *Molecular Biology of the Cell,* 3rd edition. New York: Garland Publishing, 1994.

Chapter 17

Figure 17.2: Adapted from V. T. DeVita, et al., *Cancer: Principles and Practices of Oncology,* 4th edition. Philadelphia: Lippincott. Data from C. Muir, et al., *Cancer Incidence in Five Continents,* Vol. 5, IARC. **Figure 17.17*a*:** Reprinted from *Trends In Genetics,* "Oncogene Synergism *in vivo.*" April 1993, vol. 9, no. 4, with permission from Elsevier Science. **Figure 17.19:** Reprinted by the permission of the American Cancer Society, Inc. **Figure 17.19*b*:** Data from C. Muir, et al., *Cancer Incidence in Five Continents,* Vol. 5. Lyon: IARC. **Figure 17.21:** Jared Schneidman Design. **Figure 17.24:** From *Introduction to Genetic Analysis* by Anthony J. F. Griffiths, Jeffrey H. Miller, David Suzuki, Richard Lewontin, William M. Gelbart © 1996, 1993, 1989, 1986, 1981, 1976 by W. H. Freeman and Company. Used with permission.

Chapter 19

Table 19.1: Reprinted from *Trends In Genetics,* January 1998, vol. 14, no. 1, with permission from Elsevier Science. **Figure 19.5:** Reprinted from *Trends In Genetics,* April 1995, p. 148–149, vol. 11, no. 4, with permission from Elsevier Science. **Table 19.2:** Reprinted from *Trends In Genetics,* "*Arabidopsis* and *Antirrhinum* Genes Controlling Meristem Identity," October 1995, vol. 11, no. 10, p. 395, with permission from Elsevier Science. **Figure 19.16:** Provided by George Retseck.

Chapter 20

Figure 20.5: From *In Development: The Molecular Genetic Approach,* ch. 19, p. 273, V. Russo, S. Brody, D. Cove, S. Ottolenghi. 1992. New York: Springer-Verlag. **Figure 20.6*c*:** Cold Spring Harbor Symposium on Quantitative Biology, vol. 58 (1983). Cold Spring Harbor, NY: Cold Spring Harbor Laboratory Press. **Figure 20.7:** Adapted from W. B. Wood, 1988 [Ref: Wood, W. B., "Embryology," in *The Nematode Caeuorhabditis elegans,* Wood, W. B. (ed)]. Cold Spring Harbor, NY: Cold Spring Harbor Laboratory Press, p. 215–241 (1988). **Figure 20.11:** Provided by Dr. Kenneth Kemphues, et al. From *Cell,* "Identification of Genes Required for Cytoplasmic Localiation in Early *C. elegans* Embryos," Vol. 52, p. 318. Feb. 12, 1988. Copyright 1988 Cell Press. **Figure 20.16:** From *Current Opinion in Genetics & Development,* 5:38–43, p. 39.

Chapter 21

Figure 21.12*b,c*: From *An Introduction to Genetic Analysis* by Anthony J. F. Griffiths, Jeffrey H. Miller, David Suzuki, Richard Lewontin, William M. Gelbart © 1996, 1993, 1989, 1986, 1981, 1976 by W. H. Freeman and Company. Used with permission. **Figures 21.17*c*, 21.19*c* & 21.23*a*:** From "Making of a Fly," by P. A. Lawrence, 1941, p. 29. Oxford: Blackwell Scientific Publications. **Figures 21.18, 21.22*a,b*, 21.24 & 21.25:** From Gilbert, S., *Developmental Biology,* 4th edition. Sunderland, MA: Sinauer Associates. **Figure 21.20*b,c*:** From Smold, et al., *Genes and Development,* vol. 5, 827–839 (1991). Cold Springs Harbor, NY: Cold Spring Harbor Laboratory Press. **Figure 21.20*d*:** From St. Johnston and Nösslein-Volherd, *Cell,* 68:201–19, (1992). **Figure 21.21:** From "Making of a Fly," by P. A. Lawrence, 1941, p. 73. Oxford: Blackwell Scientific Publications. **Figure 21.21*a,c*:** From Smold, et al., *Genes and Development,* vol. 5, 827–839 (1991). Cold Spring Harbor, NY: Cold Spring Harbor Laboratory Press.

Chapter 22

Figure 22.3: This figure was adapted with permission (J. T. Eppig, Mouse Genome Database (MGD, The Jackson Laboratory, Bar Harbor, ME)). Data depicted are from MGD as of 4/98. Current information can be assessed at *http://www.informatics. jax.org.* **Figure 22.10*a*:** From

"Consequences of Widespread Deregulation of the *c-myc* Gene in Transgenic Mice: Multiple Neoplasms and Normal Development," by A. Leder, et al., *Cell,* vol. 45, p. 485–495; May 23, 1986, Cell Press. **Figure 22.12***b:* Reprinted by permission of Wiley-Liss, Inc., a subsidiary of John Wiley & Sons, Inc. **Figures 22.15 & 22.16:** We thank R. Krumlauf and W. McGinnis for permission.

Chapter 23

Table 23.1: From *Genetics* by Peter J. Russell, Fifth Edition. Copyright © 1998 by Peter J. Russell. Reprinted by permission of Benjamin/Cummings Publishing Company. **Figure 23.6:** Data from Wallace (*Amer. Natur.* 97:65–66, 1963). The University of Chicago Press. **Figure 23.8:** Philip Hedrick, *Population Biology.* © 1984 Sudbury, MA: Jones and Bartlett Publishers. *www.jbpub. com.* Reprinted with permission. **Figure 23.10***b:* Reproduced from Georgiou, G. P. (1986), p. 14–43 in *Pesticide Resistance and Tactics for Management,* National Academy Press, Washington, DC. **Figure 23.12:** From B. Tabashuik. 1986. *Bulletin of the Entomological Society of America,* 32: 156–161. **Figure 23.16:** Reprinted with permission from SCIENCE, "Heritability of Personality Trait," Vol. 264, p. 1733–39. © 1994 American Association for the Advancement of Science.

Chapter 24

Figures 24.1, 24.10, 24.20*a,b,c* **and 24.22:** From S. Owasa & T. Honjo. *Evolution of Life: Fossils, Molecules, & Culture.* New York: Springer-Verlag, 1991. **Figure 24.2:** From W. Klug & M. Cummings, *Concepts of Genetics,* 5th edition, p. 617 (1997) Prentice Hall. **Figure 24.3:** Provided by V. Morrell. Reprinted with permission from SCIENCE, vol. 276, May 2, 1997, p. 701. American Association for the Advancement of Science. **Figure 24.6:** From *Fundamentals of Molecular Evolution* by W. H. Li, D. Graur. Sunderland, MA: Sinauer Associates, Inc., 1991, p. 174. **Figure 24.8:** From *Fundamentals of Molecular Evolution* by W. H. Li, D. Graur. Sunderland, MA: Sinauer Associates, Inc., 1991, p. 167. **Figure 24.9:** From *Fundamentals of Molecular Evolution* by W. H. Li, D. Graur. Sunderland, MA: Sinauer Associates, Inc., 1991, p. 152. **Figure 24.11:** From *Fundamentals of Molecular Evolution* by W. H. Li, D. Graur. Sunderland, MA: Sinauer Associates, Inc., 1991, p. 7. **Figures 24.12***a* **& 24.23:** From *IMMUNOLOGY,* Second Edition, by Leroy Hood, Irving Weissman, John H. Wilson & William Wood; Copyright 1984, by BENJAMIN/CUMMINGS PUBLISHING COMPANY, reprinted by permission. **Figure 24.12***b:* From *Fundamentals of Molecular Evolution* by W. H. Li, D. Graur. Sunderland, MA: Sinauer Associates, Inc., 1991, p. 154. **Figure 24.17:** Reprinted from *Trends In Genetics,* © 1994, vol. 10, no. 10, with permission from Elsevier Science.

Photos

Part Openers

Two: © Keith V. Wood/Visuals Unlimited; **3:** © Joseph Hacia, National Human Genome Research Institute/ National Institute of Health; **4:** © Mark Harmel/Tony Stone Images; **5:** © Chad Ehlers/Tony Stone Images; **7:** © Dr. Jeremy Burgess/Science Photo Library/Photo Researchers, Inc.

Prologue

Page 1: © James Strachan/Tony Stone Images; **p. 5:** © J. William Schopf, UCLA. Reprinted with permission from *Science* 260: 640–646,1993. "Microfossils of the Early Archean Apex Chert: New Evidence of the Antiquity of Life." © 1993 American Association for the Advancement of Science.

Chapter 1

Page 8: © Doug Sokell/Visuals Unlimited; **p. 10:** © Brian Stablyk/Tony Stone Images; **1.1:** © Bruce Ayres/Tony Stone Images; **1.2:** © /Mendelianum Institute, Moravian Museum; **1.3:** © Saudjie Cross Siino/ Weathertop Labradors; **1.5a,b:** © Mendelianum Institute, Moravian Museum; **1.4:** © The Metropolitan Museum of Art, Gift of John D. Rockefeller, Jr., 1932 (32.143.2), Photograph ©1996 The Metropolitan Museum of Art; **1.6:** © Klaus Gulbrandsen/ Science Photo Library/Photo Researchers, Inc.; **1.7a:** © Dwight Kuhn Photography; **1.18a:** © Science Photo Library/Photo Researchers, Inc.; **1.18b–d:** © Mendelianum Institute, Moravian Museum.

Chapter 2

Page 38: © Tom & Pat Leeson/Photo Researchers, Inc.; **2.1:** © The Bergman Collection; **2.3a:** © John D. Cunningham/ Visuals Unlimited; **2.7 (top left):** © McGraw-Hill Higher Education Group, Inc./Jill Birschbach, Photographer; Arranged by Alexandra Dove, McArdle Laboratory, University of Wisconsin-Madison; **2.7 (top right,bottom):** © Charles River Laboratories; **2.10a:** © Stanley Flegler/ Visuals Unlimited; **2.12a:** © William H. Allen Jr./Allen Stock Photography; **2.18a:** © Lynn McLaren/Photo Researchers, Inc.; **2.19a:** © Renee Lynn/Photo Researchers, Inc.; **2.20a:** © Rudi Von Briel/PhotoEdit.

Chapter 3

Page 70: © Adrian T. Sumner/Tony Stone Images; **3.1:** © Richard Hutchings/Photo Researchers, Inc.; **3.2:** © L. West/Photo Researchers, Inc.; **3.3a:** © Biophoto Association/Photo Researchers, Inc.; **3.6:** © Scott Camazine/Photo Researchers, Inc.; **3.8a-f:** © Dr. Conly L. Reider; **3.9a:** © David M. Phillips/Visuals Unlimited; **3.9b:** © R. Calentine/Visuals Unlimited; **3.10:** © Dr. Byron Williams, Cornell University; **Figures A, B:** © Dr. Michael Goldberg, Cornell University; **Figure C:** © Dr. Roger

Karess; **Figure E:** © Christine Hunter/The Bergman Collection; **3.16:** © Dr. Leona Chemnick, Dr. Oliver Ryder/San Diego Zoo,Center for Reproduction of Endangered Species; **3.22:** © Reproduced with permission from *Ishihara's Tests for Colour Blindness,* published by Kanehara & Co., Ltd., Tokyo, Japan, but tests for color blindness cannot be conducted with this material. For accurate testing, the original plates should be used.

Chapter 4

Page 105: © Rudi Von Briel/PhotoEdit; **4.16a:** © J. Forsdyke/Gene Cox/Science Photo Library/Photo Researchers, Inc.; **4.16b:** © James W. Richardson/Visuals Unlimited; **4.21:** © Dr. Eric Alani, Cornell University.

Chapter 5

Page 144: © Tony Stone Imaging/Tony Stone Images; **5.1a:** © George Bernard/ Animals Animals; **5.1b:** © William Hauswirth; **5.3:** © Evanston Hospital Microbiology; **5.4b:** © The Bergman Collection; **5.6:** © Science Source/Photo Researchers, Inc.; **5.9a:** © A. Barrington Brown/Science Source/Photo Researchers, Inc.; **5.11a:** © Biophoto Associates/Science Source/Photo Researchers, Inc.; **5.11b:** © Microworks/Dan/Phototake; **5.11c:** © Ross Inman & Maria Schnös, University of Wisconsin, Madison, WI; **5.11d:** © Jack D. Griffith/University of North Carolina Lineberger Comprehensive Cancer Center.

Chapter 6

Page 179: © Milkie Studio, Inc./Tony Stone Images; **6.1:** © Dr. Don Fawcett/J. R. Paulson & U. K. Laemmli/Photo Researchers, Inc.; **6.3 (both):** © Charles River Laboratories; **6.7b:** © The Bergman Collection; **Figure A:** © Science VU/Visuals Unlimited; **6.13 (all):** © Carolina Biological/Photo Researchers, Inc.; **p. 198 (top):** © The Bergman Collection; **p. 198 (bottom):** © Seymour Benzer, Professor Emeritus, California Institute of Technology; **6.25a:** © Courtesy of Dr. Patricia Olds-Clarke and Dr. Stephen Pilder, Temple University School of Medicine, Philadelphia, PA; **6.25c:** © Tom Vasicek; **6.25d:** © Science VU/F. R. Turner/Visuals Unlimited.

Chapter 7

Figure 7.1a: © Sinclair Stammers/Science Photo Library/Photo Researchers, Inc.; **p. 234:** © Professor Oscar Miller/Science Photo Library/Photo Researchers, Inc.; **7.12:** © Dr. Thomas Maniatis, Thomas H. Lee Professor of Molecular and Cellular Biology, Harvard University.

Chapter 8

Page 262: © Cytographics/Visuals Unlimited; **8.1a:** © Stanley Flegler/Visuals Unlimited; **8.14:** © Overseas/Phototake;

p. 283 (left, right): © Stratagene;
p. 283 (bottom): © Amersham Pharmacia Biotech; **p. 287:** © James King-Holmes/ Science Photo Library/Photo Researchers, Inc.; **8.22:** © Jean Claude Revy/Phototake; **8.26a:** © Omikron/Science Source/Photo Researchers, Inc.; **8.26b:** © Professor Sir David Weatherall, University of Oxford.

Chapter 9

Page 308: Courtesy of Ronald Carson, The Reproductive Science Center of Boston; **9.18:** © Meckes/Ottawa/Photo Researchers, Inc.; **9.21c:** © Larry Mulvehill/Photo Researchers, Inc.

Chapter 10

Page 341: © Photograph by Eric Lessing/ Art Resource, NY; **10.2 (both), 10.3a:** © Lee Silver, Princeton University; **10.4b:** © Dr. Gopal Murti/Science Photo Library/ Photo Researchers, Inc.; **10.7:** © Dr.Simon Foote, The Walter and Eliza Hall Institute, Royal Melbourne Hospital. Reprinted with permission from *Science: 258,62–63.* © 1992 American Association for the Advancement of Science; **10.16b, 10.17b:** Reprinted with permission from *Nature* 1990 Jul 19;346(6281):216–7, Sinclair et al. © 1990 Macmillan Magazines Limited; **p. 364:** © Brigid Hogan, Vanderbilt University; **p. 364:** © Dr. Robin Lovell-Badge, MRC National Institute for Medical Research; **10.19c:** © Johanna Rommens, Hospital For Sick Children, Toronto; **10.21a:** © Diane Schepis/The Bergman Collection; **10.23a:** © Calgene, Inc.

Chapter 11

Page 390: © David M. Phillips/Visuals Unlimited; **11.1:** © Dr. Robert Moyzis/ University of California-Irvine, Department of Biochemistry; **11.2 a:** © Dr. Don Fawcett/ U.K. Laemmli/Photo Researchers, Inc.; **11.2b:** © Courtesy Daniel A. Starr, University of Colorado; **11.3a:** © Ada L. Olins, University of Tennessee/Biological Photo Service; **11.3b:** © Dr.Gerard J. Bunick, Oak Ridge National Laboratory; **p. 396c:** © Dr. Barbara Hamkalo/University of California-Irvine, Department of Biochemistry; **p. 396 (right):** © Dr. Don Fawcett/H. Ris & A. Olins/Photo Researchers, Inc.; **11.5:** © Dr. Don Fawcett/J. R. Paulson, U. K. Laemmli/Photo Researchers, Inc.; **11.6a:** © Jacques Giltay, Center for Medical Genetics, Utrecht; **11.7a:** © H. Kreigstein and D. Hogness "Mechanism of DNA Replication in Drosophila Chromosomes: Structure of Replication Forks and Evidence of Bidirectionality", Proceedings of the National Academy of Sciences USA, 71(1974): 135–39; **11.11a:** © Biophoto Associates/Photo Researchers, Inc.; **11.11b:** © Jeremy David Pickett-Heaps, School of Botany, University of Melbourne; **11.14a:** © Stan Fakan, University of Lausanne; **11.14b:** © Doug Chapman, University of Washington Medical Center, Cytogenetics Laboratory; **11.15 (top,bottom):** © Dr. Clinton Bishop, Department of Biology, West Virginia University; **11.16a:** © George

Wilder/Visuals Unlimited; **11.17a (left):** © Jack M. Bostrack/Visuals Unlimited; **11.17a (right):** © David M. Phillips/Visuals Unlimited; **11.17b:** © David M. Phillips/ Visuals Unlimited; **11.17c:** © Dr. Michael Goldberg,Cornell University; **11.18a:** © David Phillips/Science Source/Photo Researchers, Inc.; **11.18c:** © O. L. Miller, B. R. Beatty, D. W. Fawcett/Visuals Unlimited.

Chapter 12

Page 419: © Dr. Michael Goldberg, Cornell University; **12.4b:** © Dr. Ross MacIntyre, Cornell University; **p. 426:** © Courtesy of the Centers for Disease Control; **12.5b:** © Photo courtesy of Dr. Sally Camper, University of Michigan. Reprinted with permission from *Science* 280:29 May 1998, p. 1444. "Correction of deafness in shaker-2 mice by an unconventional myosin in a bac transgene", Probst, et al. © 1998 American Association for the Advancement of Science; **12.8 (both):** © Cabisco/Visuals Unlimited; **12.10a:** © Provided by M. J. Moses, Duke University, from Poorman, Moses, Davisson and Roderick: *Chromosoma* (Berl.) 83:419–429 (1981); **12.11b:** © Lisa G. Shaffer, Ph.D, Baylor College of Medicine; **12.12:** © Dr. E. Walker/Science Photo Library/Photo Researchers, Inc.; **12.12 (right):** © Photo Researchers, Inc.; **12.13d:** © M. G. Neuffer, University of Missouri; **12.15a:** © Dr. Michael Goldberg, Cornell University; **12.15b:** © Dr. Michael Goldberg, Cornell University; **12.16:** © Dr. Nina Fedoroff, Pennsylvania State University; **12.24:** © Leonard Lessin/Peter Arnold, Inc.; **12.27:** © Davis Barber/ Developed by Adam Lukaszewski, University of California- Riverside/PhotoEdit.

Chapter 13

Page 461: © Pat O'Hara/Tony Stone Images; **13.2a:** © Dr. Jeremy Burgess/ Science Photo Library/Photo Researchers, Inc.; **13.2b:** © Stephen Frisch Photography; **13.3:** © David M. Phillips/Visuals Unlimited; **13.4a:** © Dr. Gopal Murti/Science Photo Library/Photo Researchers, Inc.; **13.8a:** © S. P. L./Science Source/Photo Researchers, Inc.; **13.22a:** © Jack D. Griffith/University of North Carolina Lineberger Comprehensive Cancer Center; **13.28:** © Julius Adler, University of Wisconsin; **Figure A:** © Veronika Burmeister/Visuals Unlimited; **Figure B:** © Oliver Meckes/Photo Researchers, Inc.

Chapter 14

Page 501: © Newcomb & Wergin/Tony Stone Images; **14.1:** © Eric L. Heyer/Grant Heilman Photography; **14.4:** © Dr. Steve Hajduk, University of Alabama at Birmingham; **p. 513:** © John Reader/ Science Photo Library/Photo Researchers, Inc.; **14.11c:** © Carolina Biological Supply/ Phototake; **14.1 (both):** © Jim Strawser/ Grant Heilman Photography; **14.14a:** © Dr. John Sanford/Sanford Scientific, Inc.; **14.14b:** © Ronald A. Butow, Ph.D, Southwestern Medical Center, University of Texas; **14.15a:** © Reprinted from *Trends In*

Genetics, January 1989, Vol. 5, No. 1, Dr. Douglas C. Wallace, p. 11. © 1989, with permission from Elsevier Science.

Chapter 15

Page 530: © Illustration courtesy of Mitchell Lewis, University of Pennsylvania. Reprinted with permission from *Science* 271:1247, from M. Lewis, G. Chang, N. C. Horton, M. A. Kercher, H. C. Pace, M. A. Schumacher, R. G. Brennan and P. Lu, © 1996 American Association for the Advancement of Science; **15.1:** © London School of Hygiene & Tropical Medicine/ Science Photo Library/Photo Researchers, Inc.; **p. 534:** © Bettmann/UPI/Corbis; **Figure A:** © Dr. Ann Hirsch, University of California-Los Angeles.

Chapter 16

Page 558: © Dr. Andrejs Liepins/Science Photo Library/Photo Researchers, Inc.; **16.3:** © Bruce Edgar, Fred Hutchinson Cancer Research Center; **16.4b:** © Reprinted with permission from *Nature,* Vol. 377, 14 Sept. 1995, fig. 2, p. 124. © 1995 Macmillan Magazines Limited; **16.5:** © S. L. McKnight; **16.10:** © Courtesy Andreas Bauxevanis, National Human Genome Research Institute, National Institute of Health.

Chapter 17

Page 590: © photo Lennart Nilsson/Albert Bonniers Forlag/Boehringer Ingelheim International GmbH; **17.4a,b:** © Lee Hartwell, Fred Hutchinson Cancer Research Center; **17.1 (both):** © Thea Tlsty, University of California-San Francisco Medical Center-Pathology; **17.13 (top, bottom):** © Michael R. Speicher and David C. Ward, "The Coloring of Cytogenetics." *Nature Genetics,* 2:1046–1048, 1996, figs. 2 and 3. Photos courtesy of David C. Ward; **p. 603:** © Dr. Joe Gray, University of California-San Francisco; **17.24a:** © Custom Medical Stock Photo; **17.25a:** © Nancy Kedersha/Science Photo Library/Photo Researchers, Inc.; **17.25b:** © Dr. John Zajicek/Science Photo Library/Photo Researchers, Inc.; **17.25c:** © Martin M. Rotker/Photo Researchers, Inc.; **17.25d,e:** © Frederick C. Skvara, M.D./The Bergman Collection; **17.27 (all):** © David N. Louis, M.D./Massachusetts General Hospital; **17.28:** © Bruce Edgar, Fred Hutchinson Cancer Research Center.

Introduction to Portraits

Page 626: © Jewish Museum, Zeldis, Malcah/Art Resource, NY; **I.1a (top, bottom):** © Lennart Nilsson/Albert Bonniers Forlag, from *A Child Is Born,* Dell Publishing Company; **I.1b:** © Photomicrographs by F. Rudolph Turner/The FlyBase Consortium, 1999; **I.3a:** © Lee Hartwell, Fred Hutchinson Cancer Research Center; **I.3b:** © Sinclair Stammers/Science Photo Library/Photo Researchers, Inc.; **I.4 (Clockwise from top right pair):** © E. Koneman/Visuals

Unlimited; © Breck Byers and Loretta Goetsch, 1974 Reprinted with permission from Cold Spring Harbor Laboratory Press; © Illustration-Katherine Sutliff/Reprinted with permission from *Science,* Oct. 23, 1998, Genetic and Physical Maps-Genome Maps 9. © 1998 American Association for the Advancement of Science; © Sara Patterson, University of Wisconsin, Department of Botany; © Sinclair Stammers/Science Photo Library/Photo Researchers, Inc.; © David Hall, Department of Neuroscience/Albert Einstein College of Medicine; © Dr. Volker Harentstein; © Dr.Michael Goldberg,Cornell University; © Elizabeth Hirst, Neurobiology, National Institute for Medical Research, London, UK; © Myung Shin/The Bergman Collection; **I.7a (left):** © Charles River Laboratories; **I.7a (right):** © Science VU/ F.R. Turner/Visuals Unlimited.

Chapter 18

Page 638: © E. Koneman/Visuals Unlimited; **18.2b:** © Dan Gottschling, Fred Hutchinson Cancer Research Center.

Chapter 19

Page 655: © Illustration-Katherine Sutliff/ Reprinted with permission from *Science,* Oct. 23, 1998, Genetic and Physical Maps-Genome Maps 9. © 1998 American Association for the Advancement of Science; **19.1, 19.2a,b:** © Dr. Elliot Meyerowitz, California Institute of Technology; **19.3:** © Reprinted with permission from *Science,* Oct. 23, 1998, Genetic and Physical Maps-Genome Maps 9. © 1998 American Association for the Advancement of Science; **19.4 (top,center,right):** © John Bowman; **19.5 (left):** © Courtesy of Bernard A. Hauser and Charles S. Gasser, University of California-Davis; **19.5b:** © Sara Patterson, Tony Bleecker, University of Wisconsin-Madison; **19.6b-f:** © John Harada, University of California-Davis. Copyrighted by the American Society of Plant Physiologists and is reprinted with permission. *Plant Cell* 5:1361–1369.; **19.7a:** © Christian Fankhauser/Salk Institute; **19.7b:** © Amasino; **19.8:** © Sarah Patterson, Tony Bleecker, University of Wisconsin-Madison; **19.9b:** © John Celenza, Boston University; **19.11a:** © John Harada, University of California-Davis. Copyrighted by the American Society of Plant Physiologists and is reprinted with permission. *Plant Cell* 6:1731–1745; **19.11b:** © Photographs courtesy of Daniel M. Vernon, Whitman College, Department of Biology; **19.12:** © John Bowman; **19.13:** © Dr. Joe Kieber, University of Illinois, Reprinted with permission from *Cell* 427–421, Kieber et al.

1993.; **19.14a:** © Bridey Maxwell/Salk Institute; **19.14 b:** © Michael Neff/Salk Institute; **19.15a-d:** © John Bowman; **p. 674 (all), p. 675 (all):** © Dr. Elliot Meyerowitz, California Institute of Technology; **19.17 (all):** © Detlef Weigel, Salk Institute, La Jolla, CA; **19.17b,c:** © Dr. Elliot Meyerowitz, California Institute of Technology.

Chapter 20

Page 681: © Sinclair Stammers/Science Photo Library/Photo Researchers, Inc.; **20.1:** © Paul W. Sternberg, Cal Tech Biology Division. Neuron, Jan. 1995; **20.4b:** © Courtesy of Samuel Ward, University of Arizona. From Nelson, Roberts and Ward 1982, *Journal of Cell Biology* 92:121–131, fig. 1.; **20.5a:** © Dr. Fabio Piano; **20.9:** © Courtesy of Dr. Paul Sternberg, Cal Tech Biology Division. Reprinted with permission from *Nature* (Vol. 358, page 475, 6 Aug. 1992), © 1992 Macmillan Magazines Limited.; **20.12a:** © Dr. Susan Strome, Department of Biology, Indiana University; **20.12b:** © Dr. Ken Kemphues and Bijan Etemad-Moghadam, Cornell University; **20.16b:** © Paul W. Sternberg/Cal Tech Biology.

Chapter 21

Page 702: © Dr. Michael Goldberg, Cornell University; **21.1:** © David Scharf/Peter Arnold, Inc.; **21.2:** © California Institute of Technology-Archives; **21.8b:** © John Lis; **21.10:** © Hugo J. Bellen, Baylor College of Medicine. Reprinted with permission from *Genes & Development 3,* 1288–1300, fig. 2 page 1291, Cold Spring Harbor Laboratory Press; **21.11c:** © Dr. Gines Morata; **21.12a:** © Manfred Kage/Peter Arnold, Inc.; **21.14a:** © Dr. Walter Gehring, Biozentrum, University of Basel; **21.14b:** © Helen Pearson, Western General Hospital/MRC Human Genetics Unit; **21.14c:** © Dr. Walter Gehring, Biozentrum, University of Basel; **21.15b:** © Bill Sullivan, University of California-Santa Cruz; **21.16a-c:** © Dr. Rudi Turner and Dr. Tom Kaufman, Indiana University; **21.17a:** © Steve Small, New York University; **21.17b, 21.19a:** © David Kosman and John Reinitz, Mount Sinai School of Medicine; **21.19b(all):** © Dr. Eric Wieschaus, Princeton University. Reprinted with permission from *Nature* 287:795–801. "Mutations affecting segment number and polarity in Drosophila", Nusslein-Volhard, C. and E. Wieschaus, 1980. © 1980 Macmillan Magazines Limited.; **21.20:** © David Kosman and John Reinitz, Mount Sinai School of Medicine; **21.21a:** © Steve Small, New York University; **21.23b:**

© Edward Lewis, California Institute of Technology.

Chapter 22

Page 731 © Myung Shin/The Bergman Collection; **22.1a:** © Carolyn A. McKeone/ Photo Researchers, Inc.; **22.1b:** © Tom McHugh/Photo Researchers, Inc.; **22.2a:** © The Bergman Collection; **22.2b:** © William Pavan/National Institute of Health; **22.7b:** © Brigid Hogan, Howard Hughes Medical Institute, Vanderbilt University; **22.8b:** © GenomeSystems Inc./ Photo provided by Kearns Communication Group; **22.10b:** © Aya Leader, Harvard Medical School; **22.12b:** © John Schimenti, The Jackson Laboratory. Reproduced from "Promoter Mapping of the Mouse Tcp-10bt Gene I Transgenic Mice Identifies Essential Male Germ Cell Regulatory Sequences", Ewulonu et al., *Molecular Reproduction and Development* 43:290–297, 1996.; **22.17 (all):** © Photographs courtesy of Dr. Thomas Lufkin, Reprinted with permission from *Nature* 29 Oct; 359 (6398):835–41, Lufkin, Mark, Hart, Dolle, LeMeur, Chambon. © Macmillan Magazines Limited.

Chapter 23

Page 754 © Jim Pickerell/Tony Stone Images; **23.1a:** © A. M. Siegelman/Visuals Unlimited; **23.4:** © Schafer & Hill/Peter Arnold, Inc.; **23.10a:** © PhotoDisc; **23.11:** © Robert Noonan/Photo Researchers, Inc.; **23.13a:** © Dr. Eckart Pott/OKAPIA/Photo Researchers, Inc.; **23.14a:** © Frans Lanting/ Photo Researchers, Inc.

Chapter 24

Page 783 © Secchi-Lecaque-Roussel-UCLAF/CNRI/Science Photo Library/Photo Researchers, Inc.; **24.3a:** © J. William Schopf, UCLA. Reprinted with permission from *Science* 260: 640–646, 1993. "Microfossils of the Early Archean Apex Chert: New Evidence of the Antiquity of Life." © 1993 American Association for the Advancement of Science.; **24.5 (top left):** © Gerard Lacz/Animals Animals; **24.5 (top center):** © Daniel J. Cox/Tony Stone Images; **24.5 (top right):** © Joe McDonald/Animals Animals; **24.5(bottom left):** © Renee Lynn/Tony Stone Images; **24.5 (bottom right):** © Roger De La Harpe/Animals Animals.

Epilogue

Page E-1 © B.S.I.P./Custom Medical Stock Photo.

Note: Page numbers followed by *f* refer to illustrations; page numbers followed by *t* refer to tables.

wg gene
    of *Drosophila melanogaster,* 720, 721*f*
white gene, of *Drosophila,* 93–96, 93*f,* 96*f,*
    107, 107*f,* 405–407, 406*f*
Wilson, E. B., 97

X

X chromosome, 72–73, 73*f,* 75, 75*t*
    aneuploidy for, 443
    in colorblindness, 96–97, 97*f,* 105, 106*f,*
       134, 134*f*
    fragile, 188–189, 189*f*
    gene linkage on, 107–109, 107*f,* 108*f*
    in hemophilia, 96–97, 97*f,* 105, 106*f,*
       134, 134*f*
    inactivation of, 407, 407*f,* 410, 443

    map of, 134, 134*f*
    nondisjunction of, 95–96, 96*f*
    reactivation of, 443
    in vitamin D—resistant rickets, 97, 98*f*
Xenopas, mtNDA inheritance in, 511, 511*f*
Xeroderma pigmentosum, DNA damage
    repair defect in, 184, 186*f*
Xist gene, 410
X-rays, in mutation, 184, 185*f,* 190, 190*f*

Y

Y chromosome, 72–73, 73*f,* 75, 75*t*
    physical map of, 352*f*–353*f*
    sex-determining region of, 174, 174*f*
Yanofsky, Charles, 225, 231
Yeast. *See also Saccharomyces cerevisiae*

    gene nomenclature for, 821
Yeast artificial chromosomes, 269,
    269*t,* 272*f*
    in chromosome function research,
      403–404, 403*f*
    construction of, 403, 403*f*
    subcloning for, 275, 275*f*

Z

Zebrafish, non-tail gene of, 367–369, 368*f*
Zinc finger motif, in protein analysis,
    367, 368*f*
Zygote, 17, 71, 73–74, 73*f*
Zygotene, of meiosis I, 83, 84*f,* 86*f*